RAPID EXCAVATION
and
TUNNELING CONFERENCE
2 0 0 7   P R O C E E D I N G S

Edited by
Michael T. Traylor
John W. Townsend

Sponsored by
Society for Mining, Metallurgy, and Exploration, Inc.

Cosponsored by
Construction Institute of ASCE

Published by
Society for Mining, Metallurgy, and Exploration, Inc.

Society for Mining, Metallurgy, and Exploration, Inc. (SME)
8307 Shaffer Parkway
Littleton, Colorado, USA 80127
(303) 973-9550 / (800) 763-3132
www.smenet.org

SME advances the worldwide mining and minerals community through information exchange and professional development. SME is the world's largest professional association of mining and minerals professionals.

ISBN: 978-0-87335-256-7

Printed in the United States on acid free, elementally chlorine free, recycled paper by Sheridan Books, a member of the Green Press Initiative.

Printed on recycled paper

CONTENTS

Preface   xi

Executive Committee   xii

Session Chairs   xiii

International Committee   xiv

1 Design and Planning of Underground Projects I

Challenges and Solutions to Designing Deep Excavations for Urban Transit Stations • *Andrew H. Liu, Ching-Liu Wu, John Hawley, James Chai* 2

Cracking the Code—Assessing Implementation in the United States of the Codes of Practice for Risk Management of Tunnel Works • *Robert J.F. Goodfellow, Terry W. Mellors* 12

NATM Design for Stanford LINAC Coherent Light Source Tunnels • *Irwan Halim, Frederick (Rick) Vincent, Jonathan Taylor* 21

New York Subway Stations and Crossover Caverns—Update on Initial Support Design • *Dru Desai, Hannes Lagger, Charles Stone* 32

Risk Management in Action—Controlling Difficult Ground by Innovation • *A. Moergeli* 44

Tunneling Alternatives for Subway Connection in Downtown Chicago • *V. Nasrl, A. Ayoubian* 56

2 Design and Planning of Underground Projects II

Designing to Protect Adjacent Structures During Tunneling in an Urban Environment • *Marco D. Boscardin, Paul A. Roy, Andy J. Miller, Kevin J. DiRocco* 70

Groundwater Control for Two-Pass Soft Ground Tunneling • *Cary Hirner, David Stacey, William A. Rochford* 80

New Crystal Springs Bypass (Polhemus) Tunnel—Design Challenges • *Roy C. Fedotoff, Gregg Sherry, Jon Hurt, Gilbert Tang* 91

New Test Methodology for Estimating the Abrasiveness of Soils for TBM Tunneling • *B. Nilsen, F. Dahl, J. Holzhäuser, P. Raleigh* 104

Outline Design and Construction of the 10 km (6.25 mile) Croydon Cable (High Voltage) Tunnel Through Groundwater Source Protection Zones, United Kingdom • *Colin Rawlings, David Keeble, John Mathews, Christopher Darton, Peter Townsend, Maria Karakitsiou, Russell Bowler, Scott Sadler, Simon Pepper* 117

Tunnel Interaction: Hydraulic, Mechanical, and Hydromechanical
Approach • *Sunghoon Choi, Kyle R. Ott, Arne Fareth, John Kinneen,
Edward S. Barboe* **137**

3 Difficult Ground Conditions I

Excavation and Support of a Water Tunnel Through the Hayward
Fault Zone • *Sarah Holtz Wilson, David F. Tsztoo, Carl R. Handford,
Kenneth Rossi* **148**

Forensics of Difficult Ground Conditions Leading to Advanced Solutions
for Recovery of an Abandoned Tunnel Project • *Faruk Oksuz,
Phillip L. Covell, Martin Doll* **160**

Gas Mitigation in the Mill Creek Tunnel • *M. Schafer, R. Pintabona,
B. Lukajic, M. Kritzer, S. Janosko, R. Switalski* **168**

Inner City Slurry Tunneling—Project ASDAM, Antwerp, Belgium •
Klaus Rieker **176**

The Sir Adam Beck II Intake Tunnels—Homage to the Builders •
David Heath, Clair Murdock **185**

Support of the Leipzig City—Tunnel Extension Using a Tie-Back
Supported Freeze Wall • *Helmut Hass, Michael B. Gilbert* **197**

4 Difficult Ground Conditions II

Design and Construction Service Tunnels on Palm Jumeirah Island •
Michael Ryjevski, Ben Hayes **208**

EPB Tunnel Boring Machine Design for Boulder Conditions •
Michael A. DiPonio, David Chapman, Craig Bournes **215**

Piercing the Mountain and Overcoming Difficult Ground and Water
Conditions with Two Hybrid Hard Rock TBMs • *Brian Fulcher,
John Bednarski, Michael Bell, Shimi Tzobery, Werner Burger* **229**

Practical Consequences of the Time-Dependency of Ground Behavior
for Tunneling • *George Anagnostou* **255**

Tunneling Through Faults in Volcanic Rocks for the Construction of
the La Joya Project (Costa Rica) • *R. Grandori, L. Pellegrini* **266**

Tunneling Through Mountain Faults • *Don W. Deere* **280**

5 Focus on Canada

Design/Build Agreement for the Niagara Tunnel Project • *John Tait,
Russel Delmar, Harry Charalambu, Rick Everdell* **288**

Environmental Assessment Study for Extension of Spadina Subway
in Toronto • *Verya Nasri, Scott Thorburn* **300**

Niagara Tunnel Project • *Doug Harding* **312**

Permitting New Tunnels in Southern Ontario, Canada—Panacea, Pragmatism, and Professionalism • Adrian Coombs, Paul Magowan, Derek Zoldy 322

Tunneling Under Glaciers Galore Creek Mine Access Tunnel, BC, Canada • Dean Brox, Jason Morrison, Peter Procter 332

6 Geotechnical and Ground Improvements

Beacon Hill Station Dewatering Wells and Jet Grouting Program • Zephaniah Varley, Richard Martin, Red Robinson, Paul Schmall, Dominic Parmantier 346

Ground Freezing Under the Most Adverse Conditions: Moving Groundwater • Paul C. Schmall, Arthur B. Corwin, Lucian P. Spiteri 360

Innovative Retaining Wall with Jet Grout for New Metro Tunnel Underneath Amsterdam Central Station • B.J. Schat, A.W.O. Bots, O.S. Langhorst 369

Monitoring Used as an Alarm System in Tunneling • G. Borgonovo, A. Contini, L. Locatelli, M. Perolo, E. Ramelli 381

Pre-Excavation Grouting Through Water-Bearing Zone Under High Pressure Under Extreme Flow Conditions • Robert Fu, Adam Shang, Alex Naudts 396

Program-Wide Geo-Instrumentation Monitoring for the MTA East Side Access Project • Daniel A. Louis, W. Allen Mawr, Jesus Schabib, Dilip I. Patel 407

7 Innovations in Underground Construction

Alternative Contracting Methods • John Reilly 418

Evaluating Ground Conditions Ahead of a TBM Using Probe Drill Data • Steven K. Duke, Jay Arabshahi 430

Evaluation of Geological Conditions Ahead of TBM Using Seismic Reflector Tracing and TBM Driving Data • Takuji Yamamoto, Suguru Shirasagi, Koji Murakami, Jozef Descour 442

Guidance for Partial Face Excavation Machines • Nod Clarke-Hackston, Jochen Belz, Allan Henneker 457

Innovations in Underground Communication Infrastructures • Darrell Gillis 466

Los Angeles Metro Gold Line Eastside Extension—Tunnel Construction Case History • Brett Robinson, Christophe Bragard 472

8 Mega Projects

Designing the Alaskan Way Tunnel to Withstand a 2,500-Year Earthquake • Celal Kirandag, James R. Struthers, Gordon T. Clark 496

Kárahnjúkar Hydroelectric Project, Iceland Extreme Underground
Construction • *Johann Kröyer, Bernhard Leist, Holger Evers,
William D. Leech* **507**

The Kensico-City Tunnel Will Add Reliability to New York City's Water
Supply System • *Kevin Clarke, Eric Cole, Heather M. Ivory, Verya Nasri*.... **520**

Tunneling for the Eastlink Project, Melbourne, Australia •
*Harry R. Asche, Edmund G. Taylor, P. Scott Smith, Peter Campi,
Ian P. Callow* **534**

Water Tunnel in Liaoning Province, Northeast China •
I. Michael Gowring, Zhong Gang, Chen Yong Zhang, Li Jiuping **551**

9 New and Future Projects I

The Dulles Corridor Metrorail Project—Extension to Dulles International
Airport and Its Tunneling Aspects • *J. Rudolf, V. Gall* **564**

Durban Harbour Tunnel—First Use of a Slurry Tunnel Boring Machine
in South Africa • *J. Andrew Hindmarch, A.L. Griffiths, Andrew K. Officer,
Gert Wittneben* **579**

Joint Water Pollution Control Plant Tunnel and Ocean Outfall Project •
Steve Dubnewych, Jon Kaneshiro, Calvin Jin **594**

New York City's Harbor Siphons Water Tunnel • *Michael S. Schultz,
Colin Lawrence* **605**

Preliminary Design of the Caldecott Fourth Bore • *Bhaskar B. Thapa,
Michael T. McRae, Johannes V. Greunen* **616**

Upper Rouge Tunnel CSO Control • *Wern-ping Chen, Mirza M. Rabbaig,
Robert Barbour, Jean Habimana, Jinyuan Liu* **627**

10 New and Future Projects II

The DC WASA Anacostia River CSO Control Tunnel Project •
*Larry Williamson, Leonard R. Benson, David C. Girard,
Ronald E. Bizzarri* **638**

The Lake Mead Intake 3 Water Tunnel and Pumping Station, Las Vegas,
Nevada, USA • *Michael Feroz, Marcus Jensen, James E. Lindell* **647**

Moving Forward on New York's Second Avenue Subway • *Anil Parikh,
Geoffrey Fosbrook, Donald Phillips, David Caiden, Jon Hurt,
Jaidev Sankar* **663**

Planning and Design of the A3 Hindhead Tunnel, Surrey,
United Kingdom • *Tom Ireland, Tony Rock* **673**

Planning and Design of the Bay Tunnel • *R. John Caulfield,
Victor S. Romero, Johanna I. Wong* **685**

Rock Engineering Aspects of Designing Large-Span Caverns at
Depth in the Deep Underground Science and Engineering Laboratory •
C. Laughton 694

11 Pressurized Face Tunneling I

The Big Walnut—A Planned Approach • Steve Skelhorn, Tom Szaraz 704

Construction of the Big Walnut Augmentation/Rickenbacker
Interceptor Tunnel (BWARI, Part 1) —Columbus, Ohio •
Jeremy P. Theys, Mina M. Shinouda, Gary W. Gilbert, Glen D. Frank 712

Construction of the Sacramento River Tunnels on the Lower Northwest
Interceptor Sewer, Sacramento, California • Tolga Togan, Dan Martz,
Wally Chen, Matthew Crow, David Young, Bill Moler, Steve Norris 741

Factors in the Variability of Measured Surface Settlements over
EPB Driven Tunnels in Soft Clay • J.N. Shirlaw, S. Boone 757

The Madrid Renewal Inner Ring Calle 30 with the Largest EPB
Machines—Planning and Results • Enrique Fernandez 769

12 Pressurized Face Tunneling II

Design Principles for Soft Ground Cutterheads • Werner Burger 784

Driving the Twin Bore Running Tunnels by Earth Pressure Balance
Machine on the Sound Transit C-710 Beacon Hill Station and Tunnels
Project • Steve Redmond, Chris Tattersall, Nestor Garavelli,
Tomyuki Kudo, Peter Raleigh, Michael J. Lehnen 793

Planning and Construction of the Metro Gold Line Eastside Extension
Tunnels, Los Angeles, California • Eli Choueiry, Amanda Elioff,
Jim Richards, Brett Robinson 808

Slurry Shield Tunneling in Portland—West Side CSO Tunnel Project •
James A. McDonald, James J. Kabat 830

Soft Ground Tunneling Issues Handled by Partnering—A Win-Win
Project in Rancho Cordova, California • Tom Martin, Lysa Voight,
John Forero, Jack Magtoto 843

Soil Deformation Analysis During TBM Excavation: Comparing Three
Different TBM Technologies Used During the Construction of the
Toulouse Metro Line B • Jacques Robert, Emilie Vanoudheusden,
Fabrice Emeriault, Richard Kastner 852

13 Rock Tunneling Project Case Histories

Construction of the Powerhouse Cavern for the Pumped Storage
Scheme Kops II in Austria • Helmut Westermayr, Michael Tergl 870

Design-Build of the Lake Hodges to Olivenhain Pipeline Tunnel and
Shaft • M. Luis Piek, Jon Y. Kaneshiro, Sean A. Menge, Brian Barker 882

Elm Road Generating Station Water Intake Tunnel • *Jon Isaacson,
Brendan Reilly, Paul McDermott* **889**

An Innovative Approach to Tunneling in the Swelling Queenston
Formation of Southern Ontario • *Michael Hughes, Paul Bonapace,
Stephen Rigbey, Harry Charalambu* **901**

Urban Blasting Essential to Schedule-Driven West Area CSO Tunnel
and Pumping Station Project • *John MacGregor, James McNally,
Adam Stremcha* **913**

14 Sequential Excavation Methods

Construction of a Mixed Face Reach Through Granitic Rocks and
Conglomerate • *Michael A. Krulc, John J. Murray, Michael T. McRae,
Kathy L. Schuler* **928**

Construction of the C710 Beacon Hill Station Using SEM in Seattle—
"Every Chapter in the Book…" • *Satoshi Akai, Mike Murray,
Steve Redmond, Richard Sage, Rohit Shetty, Gerald Skalla,
Zephaniah Varley* **943**

Evaluation of Large Tunnels in Poor Ground—Alternative Tunnel
Concepts for the Transbay Downtown Rail Extension Project •
*Steve Klein, David Hopkins, Bradford Townsend, Derek Penrice,
Ed Sum* **964**

Monitoring Successful NATM in Singapore • *K. Zeidler, T. Schwind* **976**

NATM Through Clean Sands—The Michigan Street Experience •
Paul H. Madsen, Mohamed A. Younis, Vojtech Gall, Paul J. Headland **988**

Overcoming Challenges in the Construction of the Dublin Port Tunnel •
Peter Jewell, Tim Brick, Stephen Thompson, Simon Morgan **998**

15 Shafts and Outfalls—Design and Construction

Construction of Nine Shafts for the Manhattan Section of the New
York City Water Tunnel No. 3 • *Shemek Oginski, Patrick Finn,
Yifan Ding* **1012**

Design Risk Mitigations for Contract CQ028 of the East Side Access
Project • *Jeffrey L. Rice, Avelino Alonso, Wai-shing Lee, Samer Sadek,
David I. Smith* **1023**

Development of Slurry Wall Technique and Equipment and CSM Cutter
Soil Mixing for Open-Cut Tunnels • *W.G. Brunner* **1031**

Drop Shafts for Narragansett Bay Commission CSO Abatement Program,
Providence, Rhode Island • *Rafael C. Castro, Geoffrey Hughes,
Fredrick (Rick) Vincent, Philip H. Albert* **1047**

Shaft Construction in Toronto Using Slurry Walls • *Vince Luongo* **1058**

16 TBM Case Studies

Construction of the West Area CSO Tunnels and Pumping Station •
Ray Hutton, Darrell Liebno, Taro Nonaka 1064

Guadarrama Tunnel Construction with Double Shield TBMs •
F. Mendaña 1079

The Relationship Between Tunnel Convergence and Machine
Operational Parameters and Chip Size for Double Shield TBMs—
A Case History of Ghomroud Water Conveyance Tunnel •
Ebrahim Farrokh, Jamal Rostami.. 1094

Successful Tunneling for Water Tunnel No. 3, New York City,
New York • Thomas P. Kwiatkowski, Joginder S. Bhore, Jose Velez,
Christopher D. Orlandi... 1109

TBM Excavability: Prediction and Machine–Rock Interaction •
Z.T. Bieniawski, Benjamín Celada, José Miguel Galera 1118

Use of Tunnel Boring Machines at Depth: Extending the Limits •
Bruce Downing, Trevor Carter, Richard Beddoes, Allan Moss,
Peter Dowden... 1131

17 Tunnel and Shaft Rehabilitation

Emergency Repairs to Beach Interceptor Tunnel • David Jurich,
Joseph N. McDivitt 1144

Heartland Corridor Tunnels • Frank P. Frandina, Michael J. Loehr,
Paul E. Gabryszak... 1152

Inspection of a Brick-Lined Aqueduct • Julie Frietas, Fidel L. Mallonga,
Arne Fareth, Francis Lo, Paul S. Fisk 1162

Middle Rouge Parkway Interceptor Extension—A Case History •
Harry R. Price, Charles J. Roarty, Jr., Behrooz Tahmasbi........................... 1172

Rehabilitation of a Brick-Lined Aqueduct • Nikolas Sokol,
Taehong Kim, Carl Pannuti, Walter Herrick 1181

18 Tunnel Lining Technology

Design, Testing, and Production of Steel-Fiber Reinforced Concrete
Segmental Tunnel Lining for the East Side Combined Sewer Overflow
Tunnel • R.F. Cook, K. Wongaew, J.E. Carlson, C.R. Smith, T. Cleys 1192

Development of Steel-Fiber Reinforced High Fluidity Concrete Segment
and Application to Construction • Hiroshi Dobashi, Mitsuro Matsuda,
Yoshinari Kondo, Aki Fujii 1205

Seismic Response of Precast Tunnel Linings • Gary J.E. Kramer,
H. Sedarat, A. Kozak, A. Liu, J. Chai 1225

Steel-Fiber Reinforced Self-Compacting Concrete on the San Vicente
Aqueduct Tunnel • M.R. King, C.D. Hebert... 1243

Waterproofing of a Subsea Tunnel with a Unique Sprayable Membrane—
The Nordöy Road Tunnel, Faroe Islands • *Sigurd L. Lamhauge,
Karl Gunnar Holter, Svein E. Kristiansen* **1252**

19 Underground Project Case Studies

Breakthrough Outage at the Second Manapouri Tailrace Tunnel
Project, New Zealand • *Ken Smales, Ron Fleming, Charlie Watts* **1262**

Case History of Trenchless Construction at the Lower Northwest
Interceptor Program, Sacramento, California • *William Moler, Debi Lewis,
Matthew Crow, Patrick Doig, Michael Cole, Craig Pyle, Steven Norris*........ **1274**

Design and Construction of Dulles Airport Tunnels • *David P. Field,
Allan Sylvester, Diane R. Hirsch, Paul E. Gabryszak*................................... **1286**

Portland, Oregon's Alternative Contract Approach—A Final Summary •
Paul Gribbon, Greg Colzani, Julius Strid, Jim McDonald............................ **1297**

Singapore's Deep Tunnel Sewerage System—Experiences and
Challenges • *Robert H. Marshall, Richard F. Flanagan* **1308**

Index   1321

PREFACE

The program chairman, the Executive Committee, and the SME staff warmly welcome attendees of the 2007 RETC in Toronto, Canada. This conference will present technical papers from around the world on many of the most challenging and interesting underground construction projects currently being designed or under construction. The success of this conference is due to the efforts of authors who have taken time from their busy schedules to share with their industry colleagues both successes and failures. Your participation will benefit the industry as a whole, and for that we thank you. RETC would not be complete without the participation of enterprise owners, engineers, contractors, and suppliers, and their willingness to engage and debate challenging issues. This is an exciting time for the tunneling industry, with many complex and demanding underground projects being designed and constructed to meet global infrastructure needs. By providing a forum for the exchange of ideas, RETC serves to advance the tunneling industry throughout the world. This year's exhibit hall will be one of the largest and represents a key element of the conference. Suppliers will showcase the latest advances in technology and their newest products.

The chairs wish to extend a warm welcome to students who are attending through the RETC scholarship program. This conference offers an opportunity to encourage young people embarking on a career in the tunneling industry. We hope that all attendees will use this occasion to introduce themselves to these students.

The program chairs also offer their gratitude to the session chairs, co-chairs, and other members of the RETC executive committee for volunteering their valuable time. We especially thank the dedicated staff of SME in Littleton, Colorado, for providing the behind-the-scenes management that makes RETC one of the most successful tunneling conferences in the world. We thank all of you for your dedication, encouragement, and enthusiastic support.

EXECUTIVE COMMITTEE

SESSION CHAIRS

D. Adams
Jacobs Associates

S. Boone
Golders Associates

B. Fulcher
JF Shea Co., Inc.

T. Gregor
Hatch Mott MacDonald

B. Hansmire
Parsons Brinckerhoff

C. Hebert
Traylor Shea Ghazi Precast JV

S. Hunt
CH2M HILL

G. Kramer
Hatch Mott MacDonald

J. Laubbichler
Dr. Sauer Group

J. MacDonald
Consultant

P. Madsen
Kiewit Construction Co.

M. Malott
McNally International, Inc.

P.S. McDermott
Kenny Construction Co.

M. McRae
Jacobs Associates

G. Raines
MWH Americas, Inc.

P. Rice
Parsons Brinckerhoff

M. Roach
Traylor Bros., Inc.

J. Rostami
CDM, Inc.

S. Yanagisawa
Kiewit Construction Co.

INTERNATIONAL COMMITTEE

Australia	**Tony Peach** Terratec Asia-Pacific Pty. Ltd.
Austria	**Manfred Jaeger** Jaeger Baugesellschaft MHB
Canada	**Rick P. Lovat** Lovat Tunnel Equipment Inc.
England	**Alan P. Finch** Mott MacDonald
Germany	**Otto Braach** Consultant
Italy	**Remo Grandori** SELI Societa Esecuzione Lavori
Japan	**Yoshihisa Obayashi** Obayashi Corp.
Korea	**Nam-Seo Park** Daeduk Consulting & Construction Co.
Mexico	**Roberto Gonzalaz Izquierdo** Moldequipo Internacional, SA
Spain	**Enrique Fernandez** Dragados Y Construcciones SA
Switzerland	**Fredric Chavan** Prader AG Tunnelbau

1

Design and Planning of Underground Projects I

Chair

D. Adams

Jacobs Associates

CHALLENGES AND SOLUTIONS TO DESIGNING DEEP EXCAVATIONS FOR URBAN TRANSIT STATIONS

Andrew H. Liu

Hatch Mott MacDonald

Ching-Liu Wu

Bechtel Civil

John Hawley

Hatch Mott MacDonald

James Chai

Santa Clara Valley Transportation Authority

ABSTRACT

The Hatch Mott MacDonald/Bechtel Joint Venture team is designer and construction manager for the 5.25-mile Tunnel Segment of the overall 16-mile Silicon Valley Rapid Transit Project, which extends the existing Bay Area Rapid Transit (BART) system southward from Fremont to San Jose. The Tunnel Segment includes twin bored tunnels, mined cross passages, and three deep station excavations situated in highly variable ground, high ground water table, and a dense urban setting with high-rise buildings adjacent to the largest excavation. This paper will discuss design challenges that have arisen with the excavations and the proposed solutions.

INTRODUCTION

The design and construction of underground structures is dependent upon uncertainties associated with ground behavior in relation to the engineered works. When these uncertainties are combined with highly variable ground conditions in an urban setting, they become magnified. This paper discusses some topics unique to underground projects that engineers may need to consider during the design and planning stages of a project. The Hatch Mott MacDonald and Bechtel (HMM/Bechtel) Joint Venture Design Team's experiences from the Tunnel Segment of the Silicon Valley Rapid Transit Project are used as an illustrative example for discussion. Though this project contains elements of bored tunneling, mined cross passages, and deep station excavations, the emphasis of the paper will be on specific challenges experienced with the design of the deep excavations and general geotechnical considerations.

PROJECT DESCRIPTION

The Santa Clara Valley Transportation Authority (VTA) intends to construct the Silicon Valley Rapid Transit (SVRT) Project in San Jose, California. This will be a 16.3-mile long extension of the Bay Area Rapid Transit (BART) system from its planned terminus at the end of the Warm Springs Extension in Fremont, to San Jose. The alignment includes six stations (three above- and three below-grade), a proposed future station, and vehicle storage and maintenance facilities.

2

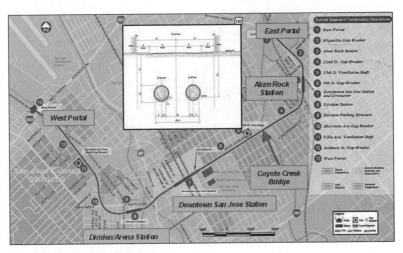

Figure 1. Overview of SVRT project tunnel segment

The alignment is comprised of three major segments: a Line Segment that will be approximately 11-miles of at-grade, elevated and cut-and-cover track from Warm Springs to San Jose, a 5.1 mile long Tunnel Segment, predominantly consisting of twin-bored tunnels and cut-and-cover structures through downtown San Jose, and a Yard & Shops Segment with a surface station that will serve as a connection to the San Jose International Airport. The Tunnel Segment, shown in Figure 1, includes at-grade and open cut track, three cut-and-cover stations and a cut-and-cover track crossover structure that together total approximately 4,970 feet. The remaining 4.3 miles of the alignment will be twin circular tunnels with mined cross passages connecting the twin tunnels for emergency egress. The tunnel alignment follows Santa Clara Street between 24th Street and the Downtown San Jose Station.

TUNNEL SEGMENT SITE CONDITIONS

Regional Geology

The Tunnel Segment of the SVRT Project is located in the Santa Clara Valley, which is bounded by San Francisco Bay to the north, the Diablo Range to the northeast and the Santa Cruz Mountains to the southwest. The valley is covered by alluvial fan, levee, and active stream channel deposits with marine estuary deposits along the Bay margins. The Tunnel Segment is located on alluvial deposits that are underlain, at depths greater than 1,000 ft, by Tertiary-age (65 to 1.8 million years old), Upper Cretaceous-age (78 to 65 million years old) marine sedimentary rocks, and Cretaceous-age (144 to 65 million years old) Franciscan Complex bedrock.

Local Soil Conditions

The alluvial deposits within the depth of the station excavations are variable across the alignment but generally consist of a slightly overconsolidated stiff silty clay with some

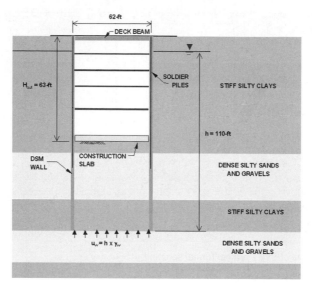

Figure 2. Typical soil stratigraphy with respect to typical station deep excavation

thin sand layers interbedded within. This clay layer is commonly separated into an upper and lower clay by the highly permeable upper aquifer which can be described as a granular layer with either dense silty sand or silty gravel. Below the upper aquifer is a highly variable interlayering of the clay and granular sand and gravel strata that corresponds to the depths of the cutoff wall toe. The clay layers within this interlayered strata serve as the major aquitard to the lower aquifer which is well below the bottom of the cutoff wall. One of the major challenges in the investigation and design, elaborated in further sections, is defining a consistent clay layer within this interlayered strata to serve as an embedment layer for the cutoff walls. Figure 2 shows a simplified view of the stratigraphy with respect to a deep excavation for a typical underground station.

Ground Water Conditions

In addition to the highly variable geotechnical conditions, the ground water conditions along the Tunnel Segment add additional difficulties and risk to the construction of the cut-and-cover excavations, bored tunnels, and mined cross passages. For the majority of the alignment, the ground water table is predicted to be within 10-ft to 15-ft below ground surface during construction. However, at particular locations, ground water monitoring during the 35% Preliminary Design phase from 2004–2005 indicated ground water table levels up to the ground surface and even possible artesian conditions during the wet season (January through May). When combined with the highly permeable sands and gravels in the upper aquifer, the ground water conditions can pose significant challenges for tunnel stability and base heave of deep excavations during construction. Thus, a comprehensive historical study and field investigation and monitoring program of the ground water regime were needed. Subsequent sections describe the field investigation program and how the results are integrated within the design process.

TUNNEL SEGMENT STRUCTURES AND DESIGN OVERVIEW

Structures

As previously described, the Tunnel Segment features multiple cut-and-cover excavations for underground stations, auxiliary structures, and tunnel portals that are interconnected by twin bored tunnels. Due to the nature of the ground conditions along the alignment, closed face Earth Pressure Balance Machines (EPBM) are most appropriate for excavating the interconnecting tunnels. Specifically, the particle size distribution of soils, typical permeability, soil density, and hydrostatic head were evaluated to determine the preference of an EPBM over a Slurry Tunnel Machine. In the soil zone where the tunnel will break-in and break-out, ground treatment such as jet grouting or soil mixing will be used to reduce the risks of problematic behavior near the tunnel eyes. In addition to the TBM tunnels, the project will include mined cross passages for emergency egress between the twin-bored tunnels that will require ground treatment or improvement prior to mining. However, since the deep cut-and-cover excavations are the main focus of this paper, the remaining sections will focus on their design aspects and some general geotechnical aspects that are important and unique to the challenges associated with this type of underground construction.

The cut-and-cover station excavations are typically 60-ft wide by 60-ft deep in section and are temporarily supported by internal steel bracing system in the majority of the station excavations. The longest excavation to support construction of the continuous Downtown San Jose Station and Crossover Structure (shown in Figure 3), is approximately 1,500-ft in length and is situated in a densely populated business district with many adjacent high-rise buildings that are as close as 16-ft to the excavation support walls. Because of the sensitivity of these surrounding buildings, controlling ground movements and protecting the existing buildings and foundations will be critical. In addition, because of the heavy traffic in the Downtown San Jose business district, the need to minimize disruption to normal business will require the use of temporary precast concrete street decking to maintain traffic flow during excavation of the deeper portions of the station.

Multiple construction methodologies have been considered for supporting the excavation and for constructing the cutoff walls, but after careful consideration of the ground conditions, sensitivity of adjacent structures, and overall cost, the project has determined the following systems to be the recommended approach. For ground water control during construction, Deep Soil Mix (DSM) cutoff walls will be constructed to a depth of approximately 120-ft prior to excavation. To support the lateral soil pressures and minimize surrounding ground deformation, two of the three stations (Downtown San Jose and Diridon/Arena) will utilize an internal steel bracing system and will have soldier piles inserted within the DSM walls to span vertically between the walers of the internal bracing system. The current design contains four levels of struts within the depth of excavation that will need to support the open excavation for a period of up to two years while tunneling is completed before construction of the permanent structure begins. The third station (Alum Rock) will utilize soil tiebacks to support lateral soil pressures and control deformation. The use of soil tiebacks was chosen to maximize working space for construction of the permanent structure because the TBMs will be launching from this station. The permanent structures of the stations, ventilation structures, and portals are heavy reinforced concrete structures that will be cast in place after the tunneling operation is completed. It is worthwhile to note that careful coordination procedures between the design teams need to be planned for and implemented on issues such as design of strut locations and strut removal sequences. The next section discusses the design process in more detail.

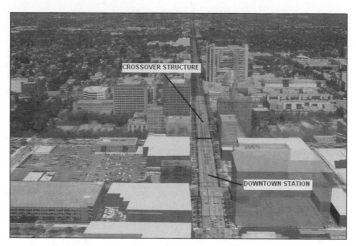

Figure 3. Cross section of downtown San Jose station excavation

Design Overview—Temporary Structures

The design and analysis process for temporary excavations consists of three major steps: simplified analyses and studies, detailed numerical analyses, and integration of the first two analyses into the design through reviews which consider historical case studies, including local experience.

Step 1: Simplified Analyses and Studies. Simplified analyses, which consist of analytical and empirical analyses, were completed during the 35% Preliminary Design phase, which ended in December 2005. These analyses were carried out simultaneously with the geotechnical field investigation program and were based accordingly on preliminary soil parameters. In addition, miscellaneous studies were carried out to evaluate different design options that formed the basis for the 35% Preliminary Design submittal. The different analyses and studies that were performed are summarized as follows:

1. *Stability analyses of retained cuts and related ground deformations:* Analyses of the stability of retained open cuts were performed for the design of each station. The purpose of the calculations was to verify the overall stability of the planned excavation against ground pressures and hydrostatic uplift, and to provide input for the structural design of the ground supporting system. These analyses also predicted anticipated ground displacements resulting from the excavations. Well-established empirical methods such as Clough et al. (1989) and Clough and O'Rourke (1990) were used to obtain the first order approximation of required supporting system stiffness to satisfy the project minimum factor of safety and maximum tolerable ground displacement criteria.

2. *Calculation of static lateral earth pressures:* Lateral at-rest (for non-yielding walls) and active (for yielding walls) earth pressures were calculated based on the Rankine theory (Rankine, 1857) for vertical walls with horizontal backfill. The results were used in preliminary sizing of the excavation support system.

3. *Seismic analysis of the temporary excavation support system:* Seismic loads were based on a design earthquake with 10% of probability of exceedance in 10 years, but not less than 0.2g. Dynamic loads on the retaining structure were then determined from dynamic soil pressure increments using the Mononobe-Okabe method (Okabe, 1926; Mononobe and Matsuo, 1929).

4. *Simplified liquefaction potential evaluation:* Simplified liquefaction potential analyses, based on the NCEER Liquefaction Proceedings (NCEER, 1997), were performed for the soils surrounding each of the excavation locations.

5. *Potential impacts of the excavations:* Calculations were performed to evaluate factors of safety against bearing capacity failure and base heave due to hydrostatic uplift. The high shear strength of the soils and the ground water conditions indicated that bearing capacity failure would not be of major concern but that base heave would govern the design. Thus, the design approach was to increase resistance against hydrostatic uplift by designing the cutoff walls to an adequate penetration depth of 120-ft to 130-ft to counterbalance the hydrostatic uplift forces as shown in Figure 2.

6. *Evaluation of temporary excavation support and bracing systems:* To support the baseline design of DSM walls and shoring systems described in the previous section, alternative systems were studied and evaluated. Production rates, costs, noise impacts, and advantages and disadvantages for each system were analyzed.

7. *Evaluation of excavation base stabilization systems:* Alternative approaches to stabilizing the bottom of excavation against base heave were evaluated and studied. The studies resulted in different approaches at different station locations due to the highly variable ground conditions. For instance, the station identified as having the highest probability of base heave was Alum Rock Station because of the thick gravel layer beneath the bottom of excavation and the lack of a continuous low permeability cutoff layer at the depth of the cutoff wall toe. As a result, studies were conducted to analyze different approaches to mitigating the problem including permeation grouting the blanket of soil between the bottom of excavation and cutoff wall toe, dewatering for deep pressure relief, or extending the cutoff walls to a deeper elevation where a continuous cutoff layer can be found. Field investigations in the next phase of design, which include deeper boreholes, pumping tests, and grouting trials, will help assess the preferred solution by taking into account the cost and risk of each approach. Decisions made as a result from these studies will then be integrated within the next step of the design and analysis process; detailed numerical analyses.

Step 2: Detailed Numerical Analyses. In advancing the design to the 65% level, detailed numerical analyses will be performed for the static and dynamic performance of the excavations. The static numerical analysis will be carried out using the computer program FLAC to predict the soil and DSM wall displacements and forces and moments that develop in the structural members as the excavation sequence progresses and as temporary struts are removed to accommodate construction of the permanent structures. The modeling will simulate the site-specific soil profiles and properties, undrained versus drained soil behavior with respect to time, and the dewatering process in order to capture as much detail as is practical for engineering purposes. In addition, the numerical modeling will also be used to aide in developing factors to account for variations of the baseline design conditions. Some examples of variations include, but are not limited to, the following:

1. *Asymmetric geology:* As experience has shown at Islais Creek, San Francisco (Clough and Reed, 1984) and Nicoll Highway in Singapore (Committee of Inquiry, 2004), not accounting for asymmetric soil profiles along the cross section of the excavation can contribute to the failure or near-failure of a deep excavation. A careful study of potential asymmetric geology based on field data needs to be developed and accounted for in the numerical analyses.

2. *Over-excavation:* Since construction is not always carried out precisely as defined in specifications, it is important to analyze the effects of over-excavating prior to installing a level of struts and adopt a reasonable safety factor.

3. *Load combinations and accidental loading:* Similar to the over-excavation problem, it is important to analyze the affects of accidental loading on the excavation.

4. *Thermal effects:* As temperatures change throughout the seasons, it is important to account for thermal effects on structural members.

The dynamic analysis for seismic loads will be based on equivalent-linear dynamic soil structure interaction analysis. This analysis will include a soil site response analysis using the computer program SHAKE and a dynamic soil-structure interaction analysis using a computer program like SASSI. Alternatively, a nonlinear soil-structure analysis for a selected structure may be considered for comparison purposes.

Step 3: Integration of Simplified and Numerical Analyses. The third and final step in the design and analyses process is to integrate the simplified and numerical analyses to support the final design of the excavations. More specifically, the results of the numerical analyses will be used to check the validity of the 35% Preliminary Design baseline design and to supplement the simplified analyses to bring the design to a 65% level. In doing so, it is important to know the strengths and weaknesses of both the simplified and numerical approaches. For instance, the simplified empirical predictions of ground displacement profiles behind the excavation wall are validated by numerous case histories but are not the most robust way for analyzing site-specific soil and excavation geometry-specific conditions. Numerical analyses, on the other hand, are excellent for capturing site-specific soil conditions and detailed geometry, but are known to have shortcomings in predicting distribution of ground movement behind the excavation wall due to overestimating the settlement zone of influence, which can lead to under-predictions of angular distortion, and hence, impacts on existing structures. However, by utilizing both methods, the design team can predict settlements at the excavation wall more accurately using numerical analyses and distribute the settlement behind the wall using empirical analyses for the most robust answer.

After the HMM/Bechtel team has the opportunity to integrate all the aspects of the analyses into the design, internal peer reviewers and an external board of specialist consultants with vast design and analysis experience will review the results. Historical case studies, particularly those for transit facilities in urban U.S. cities, will be investigated and compared to the SVRT Tunnel Segment design. Design adjustments, if necessary, will then be made accordingly.

Design Overview—Permanent Structures

The general design and analysis process of the permanent structures, which include portals, ventilation structures, and station structures, are similar to that for the temporary structures. However, for seismic design, the seismic demand will be based on the larger of an earthquake with 10% probability of exceedance in 50 years or a deterministic median value of three nearby earthquake faults. According to the project criteria, the transit system should be designed to return to revenue operation within 72 hours after a major earthquake. Thus, only minor damage such as small cracks and spalling that can be repaired during off revenue hours is allowed. Due to this performance requirement, the displacement ductility ratio of the reinforced concrete structure is limited to 1.5. Also, the underground permanent box structures need to be designed to resist floatation from the long-term high ground water table and the artesian water pressures that exists in certain areas in the Santa Clara Valley.

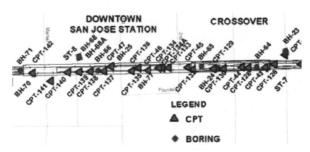

Figure 4. Summary of 35% preliminary design geotechnical borehole and CPT locations

CHALLENGES WITH SITE INVESTIGATION AND CHARACTERIZING GROUND CONDITIONS

Investigation Challenges and Solutions

Prior to advancing too far in the design process outlined in the previous section, a thorough geotechnical investigation program needed to be performed to sufficiently capture all of the soil and ground water data at the most cost effective price. Since the investigation program and design often progress simultaneously, it is important to integrate the field investigation into the overall project schedule. The spectrum of topics relating to geotechnical investigations is beyond the scope of this paper but a few general considerations relating specifically to underground construction are discussed in the following four sections. The first two sections discuss topics relating to placement of investigation locations and sampling. The third section discusses maximizing efficiency through use of field and laboratory correlations and the fourth section discusses characterization of the ground water regime.

Longitudinal and Transverse Spacing. One of the challenges in developing a site investigation program for a project with a large longitudinal extent in comparison to the width is to establish a tight, but practical, range of spacing between investigation points. For the SVRT Tunnel Segment, typical spacing between investigation locations was approximately 200-ft between investigation locations along the bored tunnel and 75-ft between investigation locations at the cut-and-cover excavation locations. For the 5.1-mile (roughly 27,000-ft) long alignment, a total of 76 boreholes and 146 Cone Penetration Tests (CPTs) were drilled during the 35% Preliminary Design geotechnical investigation program. For a typical station excavation length of approximately 900-ft, approximately 6 boreholes and 8 to 10 CPTs were conducted. Figure 4 gives a view of the geotechnical investigation coverage at the Downtown San Jose Station and Crossover Structure excavation. Additional boreholes and CPTs will be drilled during the next phase of design to fill in data gaps that have developed from changes in the tunnel alignment.

In addition to the longitudinal spacing, it is important to investigate the soil variability in the transverse direction. To do so, the HMM/Bechtel investigation plan was to position borings and CPTs at locations that bestride the footprint, particularly at the deep excavation locations. Though this proved difficult while working in a dense urban setting that has inherent drilling limitations, the value added by obtaining data on both sides of the excavations helped guide the design team in assessing the potential asymmetries that may exist in the geology over a short width of 60-ft across the excavations.

Sampling Depth and Intervals. Though continuous sampling is oftentimes recommended for any type of project, its use is essential for underground construction so that the ground continuum as a whole, including any minor layers and lenses that can

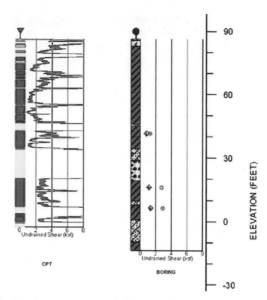

Figure 5.   Example calibration between a geotechnical borehole and CPT

be missed by regular sampling intervals, is captured. It is also worthwhile to note that the geotechnical borings with continuous sampling serve as an excellent choice for calibrating CPT interpretations of soil type since both give an uninterrupted view of the soil stratigraphy.

Maximizing Investigation Efficiency. There are many areas where a well thought out geotechnical investigation program can help the client save money without sacrificing quality. One example is maximizing the usage of CPTs to supplement the borehole program. Generally, CPTs are quicker and thus more cost-effective to complete, have minimal public disruption, and can offer a continuous view of the soil stratigraphy whereas typical boreholes are limited to discrete sample locations. They are also capable of incorporating specialized in-situ testing such as dissipation tests and seismic velocity tests without significantly interrupting the drilling process. However, the main drawback to the CPT is that it is not typically used to retrieve soil samples for laboratory testing that help in assessing soil parameters for design. To overcome this drawback, HMM/Bechtel performed a number of calibration CPTs that were drilled within close proximity of a geotechnical boring with soil samples that had exhaustive laboratory tests performed on them. These calibration pairs showed an excellent correlation between the CPTs and the geotechnical borings (as seen in Figure 5) so the CPT logs were capable of being used to derive not only the soil stratigraphy, but also soil parameters such as undrained shear strength, friction angles, and shear-wave velocity. Using these and similar correlations, the overall cost of a geotechnical investigation program can be controlled by reducing the number of geotechnical boreholes, in-situ field tests, and laboratory tests needed.

Investigations for Ground Water. During the 35% Preliminary Design geotechnical investigation program, a monitoring well and piezometer system was installed and a monthly monitoring program was implemented to observe the ground water table levels and pressures over time. In addition, a series of slug tests were performed along the alignment to measure in-situ localized permeability values for the soils.

Because of the ground water issues that were identified in the last round of investigations, the next phase of design will include a series of field pumping tests at the locations of the deep excavations. These pumping tests will assist in assessing potential hydraulic connections between different granular layers, macroscopic permeability values with respect to each excavation area, and effects construction dewatering would have on relieving pressure at the base of excavation. The results from the pumping tests will be incorporated into the numerical analyses and also be considered in selecting the methodology to stabilize the excavation base.

CONCLUSIONS

The HMM/Bechtel Design Team is taking on a major task in designing the underground structures for the Tunnel Segment of the SVRT Project. The involvement of multiple underground structures, engineering disciplines, design teams, and engineering tasks requires careful coordination and integration between the designers and the geotechnical engineering team. In addition, because of the complexities related to underground construction, a comprehensive investigation program needs to be carried out. Though this program is critical to the success of the design and construction bid packages, it needs to be well thought out and carefully planned to minimize the cost to the client and disruption to the community.

ACKNOWLEDGMENTS

The authors would like to thank our client, Santa Clara Valley Transportation Authority, for allowing us to present the information contained herein. We would also like to acknowledge the HMM/Bechtel Design Team, our peer reviewers and our board of consultants for their contributions along the way. We would also like to thank Dr. Ignacio Arango for review of this manuscript.

REFERENCES

Clough, G.W., and Reed, M.W. (1984). "Measured behavior of brace wall in very soft clay," *ASCE J. of Geotech. Engrg.*, Vol. 110(1), 1–19.

Clough, G.W., Smith, E.W., and Sweeney, B.P. (1989). "Movement control of excavation support systems by iterative design," *Proceedings, ASCE Foundation Engineering, Current Principles and Practices,* Vol. 2, pp 869–884.

Clough, G.W., and O'Rourke, T.D. (1990). "Construction induced movements of insitu walls," *Design and Performance of Earth Retaining Structures,* Geotechnical Special Publication No. 25, ASCE, 439–470.

Committee of Inquiry (2004). "Report of the Committee of Inquiry into the incident at the MRT Circle Line worksite that led to the collapse of the Nicoll Highway," Vol. I and II, Singapore.

Okabe, S. (1926). "General theory of earth pressures," *Journal of the Japan society of Civil Engineering,* Vol. 12, No. 1.

Mononobe, N., and Matsuo, H. (1929). "On the determination of earth pressures during earthquakes," *Proceedings, World Engineering Congress*, 9p.

NCEER (1997). "Workshop on evaluation of liquefaction resistance of soils," *Proceedings, NCEER-97-0022,* edited by T.L. Youd and I.M. Idriss, Utah.

Rankine, W.J.M. (1857). "On the stability of loose earth," *Phil. Trans. Roy. Soc., London*, 147, Part 1, 9–27.

CRACKING THE CODE—ASSESSING IMPLEMENTATION IN THE UNITED STATES OF THE CODES OF PRACTICE FOR RISK MANAGEMENT OF TUNNEL WORKS

Robert J.F. Goodfellow

Black & Veatch

Terry W. Mellors

Mellors and Associates

ABSTRACT

The underground industry has been forced to reconsider risk management since the introduction and implementation of the "Joint Code of Practice for Risk Management of Underground Works in the UK" published in 2003 and the subsequent international version published by the 'International Tunneling Insurance Group' (ITIG) in January 2006.

What does this mean for major tunnel projects in the US and elsewhere?

The discussion will highlight the implementation of the Codes to date, risk management principles embodied in the Codes and the use of risk registers as a project management tool and the management of risk.

INTRODUCTION

The Codes were published in 2003 and 2006 respectively and are being applied to over 20 large projects involving tunneling worldwide.

The international version of the Code (published in 2006) was based on the UK Code (published in 2003) but modified following discussion with the International Tunneling Association (ITA) to recognize that some of the provisions may not necessarily be appropriate or legal in all countries. The ITA eventually supported and agreed with the principles of the international Code. At the time, the Codes were already being applied internationally because of the number of projects that require insurance underwriting from the London Insurance Market, where the principles of the Codes are applied to all underground projects.

The magnitude of recent insurance losses led the insurance market to reassess the practices of their customers and their own underwriting process of accepting these practices.

The paper addresses the need for and generation of the Codes of Practice and how the Codes provide the means to ensure that best practices are applied in the Industry by addressing the issues of the requirements of the Codes and how they can be adopted into the design and construction process for a tunnel project.

THE NEEDS FOR A CODE OF PRACTICE

Insurers' Needs

By 2001, the Association of British Insurers (ABI) representing insurers and reinsurers on the London–based Insurance Market had fully recognized that insurance losses in underground construction in the previous decade or so far outstripped the premiums received. Statistics showed the loss ratio (the difference between sums paid out in relation to sums received via premiums etc.) had been over 500%. This situation was clearly unsustainable and, with the events of September 11 2001, a major reevaluation of profitable and nonprofitable insurance sectors took place.

Insurers were faced with choices of either (a) withdrawing from the tunneling market (as some had done already), or (b) increasing terms (premiums, excesses and/or restricting cover) or (c) seeking better risk management on tunneling projects by working with the tunneling industry to produce a code of 'best practice' for tunneling works. This latter approach was adopted in the light of the perceived effectiveness by the insurance industry of the 'Joint Code of Practice on the Protection from Fire of Construction Sites and Buildings Undergoing Renovation' introduced in 1992 which, it was considered, had made a significant contribution to the reduction in the number of fires on construction and building sites and also in the number of claims made against insurance policies.

With this experience in mind, an approach was made to the British Tunneling Society to develop guidelines on best practices, initially for application in the UK and then, once accepted by the insurance market, for use internationally by appropriate and suitable amendment and revision.

Industry Needs

The tunneling industry has long considered how to mitigate uncertainties and ensuing impacts associated with construction in the ground—a medium that is *"supplied by nature and seldom to specifications that engineers would choose"(CIRIA report 79)*, not fully exposed nor the properties reliably known before excavation is carried out.

Public and political perceptions of tunneling being a high cost and high risk activity have been addressed over the years both in the US and worldwide by the introduction of improved contract practices. These have included the application of 'Differing Site Conditions' (DSC) or their equivalent, Geotechnical Baseline Reports (GBR) and the use of Dispute Resolution Boards (DRB). However, benefits obtained from the use of these mechanisms are still frequently overshadowed when high profile underground projects suffer from chronic cost and schedule overruns as well as safety concerns, apparently from the lack of management of identifiable risks.

The tunneling industry needs to be able to demonstrate for political as well as social and environmental reasons that it is capable of managing and can and does manage risks in a transparent manner.

Confluence of Needs

There is an obvious synergy between Insurers' needs and those of the tunneling industry—i.e., a transparent risk based approach to projects and demonstration that identified risks are being managed. The Codes of Practice provide this through stipulated 'project management procedures' and the systematic application of risk assessments and the preparation of risk registers.

The authors are of the opinion and contend that the use of risk registers and adherence to the principles of the Codes of Practice have benefits that go beyond those of any perceived 'monitoring' actions of the insurance industry. Risk management is (or should be) a fundamental requirement of any project management system, either imposed by the Owner, or at least enforced by the Designer and adopted subsequently by the Constructor during the works. With appropriate use and consideration, risk registers should be considered as a powerful project management tool.

REQUIREMENTS OF THE CODES

A review of the principles of the Codes was made by Mellors and Southcott (2003). The objective of the Codes is to promote and secure best practice for the minimisation and management of risks associated with the design and construction of tunnel works including the renovation of existing underground structures. Risk management is only part of best project management practice and in this regard, the Codes merely set out a framework for risk management through 'best practice' for project management. Risk management *per se* is encompassed in the Codes through the requirement for formalized risk assessments and the preparation of risk registers. Compliance with the Codes should "minimize the risk of physical loss or damage and associated delays." (ITIG, 2006)

The Codes cover all project activities from project feasibility studies, through geotechnical investigations, design studies (preliminary through to final), selection of a Contractor to construction and include recognizing the qualifications and experience of participants and the management of changes during the project.

The Codes identify Owners' roles and responsibilities. These are seen to be especially important given that the wish of some Owners is to offload risks on to others. However, it should be borne in mind that it is the Owner who sets or should set the tone for risk management throughout the project through, for example, a Risk Management Policy for a project (Eskesen et al., ITA 2002). Consequently, Owners should be constantly aware of and understand their obligations on each and every Contract. The Contract terms, including for example, allowable methods of procurement, bidding regulations, differing site condition clause language and dispute resolution systems are all under the direct control of the Owner.

The Codes recognize that Owners may not have the appropriate background and experience to project manage an underground construction and consequently an Owner may appoint a representative (most often the Design Consultant Engineer). The implication of this is that the appointed Engineer has significant responsibility for Owner 'education' and informing the Owner on issues arising which are likely to include the development and design stages of a project and contractor procurement.

The Codes identify that a comprehensive risk management program should be instigated and implemented at the beginning of a project. Project alternatives can be assessed partly on the basis of the risk register associated with each alternative relating to their technical feasibility, cost and schedule.

Geotechnical reference conditions or use of a GBR and their inclusion in the contract documents for procurement of a Contractor is a requirement of the Codes. This should be readily appreciated in the US as this has become the standard industry practice over the past 10 years.

Risk assessments and the development of risk registers are required by the Codes to clarify ownership of risks and how they are to be allocated, controlled, mitigated and managed. Risk registers are intended to be 'live' documents that should be continuously reviewed and revised as appropriate. The Codes are not prescriptive in respect of the forms and formats of risk assessments and risk registers. However, they do stipulate that risk registers should identify hazards, consequent risks, mitigation and contingency measures, proposed actions, responsibilities, critical dates for completion of actions and when actions have been closed out. These are to be perceived as valuable project management actions.

Example formats for a risk assessment form and a register are given in Tables 1 and 2 on pages 18 and 19 respectively.

The risk register should be transferred during the bid process to the Contractor for completion and submission with his bid for modification to reflect his particular means and methods as appropriate and discussion during bid review and prior to award of contract.

Continual review of the risk register during construction should be led by the Owner or Owner's representative collectively with the Designer and the Contractor. Whilst an Owner may perceive that he has transferred risks to the Contractor, it is in his best interest to ensure that those transferred risks are being managed adequately and appropriately in respect of impacts associated with reputation and indirect liability (third party damage, health and safety, etc).

The Codes make specific provision for the Contractor to have sufficient time to put together a viable bid. This recognizes that a too short bid period is a major hazard to successful bidding and therefore potentially to successful completion of the project.

The management of change to the Contract is another potential hazard that is recognized and specifically addressed in the Codes. Any change of design or method of construction including Value Engineering changes that increase risk to the Project are required to be recognized and acknowledged.

The Codes provide (in Appendix B) a schedule of 'deliverables' for all stages of a project from project development through to Construction and Commissioning. This schedule can be considered as a 'checklist,' but which is by no means comprehensive, for project management considerations as well as consideration by Insurers.

PROJECT APPLICATION OF THE CODES

As the purpose of the Codes is to promote the use of best industry practices to assess, mitigate and manage project risks, all tunnel projects should abide by the principles of the Codes. This is not to say, however, that every project would be subject to inquiry or regular audit by insurers. While attention to risk management is essential on every project, common sense must be applied in the application of risk management as with all design tools. Using scheduling as a comparative example, a small project may be scheduled with a simple spreadsheet, while a large project will have a complex Work Breakdown Structure (WBS) schedule with critical path analysis and a scheduler dedicated to the Project Team. A similar level of effort would be exhibited on risk management for small and large projects.

The cost impact of using the Codes should not be onerous in any way. As the Codes represent 'best practice,' their implementation should be a matter of course and hence should not have any real affect in terms of cost. The development of a risk register may be intensive in its initial setup and assessment stages—but no more so, in our view, than initially populating a detailed WBS schedule (continuing the comparison with project scheduling). The demand for the use of best practices, by definition, means that most if not all of the activities demanded by the Codes should already be carried out during the course of a project.

Risk registers can be used for many purposes and their spreadsheet format gives them great power to sort and arrange information in different ways. Risk registers can be used, for example, as a permitting activity checklist for mitigating actions for the hazard of not obtaining a required permit, a project "To Do" list that can be sorted by responsible party name to identify outstanding action, items of sub-consultants on the team and other project management functions. They are most beneficial when developed on a project activity basis according to the major activities shown on a schedule.

Noncompliance with the principles of the Codes is potentially catastrophic. The lack of management of the project risks in underground projects will inevitably lead to increased cost, schedule delay and potential consequences for the health and safety of both workers and the public. A lack of proper regard for project risk may also threaten the viability of the Contractor or design firm organization. In short, proper application of the Codes and a clear presentation of the project risk profile using a risk register help all parties to gain an objective view of how risks are to be mitigated during design and construction.

The use or presence of a risk register however does not necessarily mean that risks are being managed. The management of risks should be seen as a separate exercise which requires concerted effort and action to demonstrate adequate and appropriate mitigation of risks in a structured way that can then be monitored. Continuing to use scheduling as our example, it is clear that simply having a printed schedule on the office wall does not mean that the project is on schedule nor does it mean that the schedule is being managed effectively.

APPLICATION OF THE CODES IN THE UNITED STATES

Adherence to the principles of the Codes is currently being requested directly by insurance and reinsurance companies for more than half a dozen major projects in North America. The principles of the Codes are being applied to many more projects worldwide. The application and endorsement of the Codes has met with some institutional resistance in the US, however, with several unsubstantiated, unfounded and bizarre observations and criticisms being leveled against its implementation into the US marketplace which have proved to be of limited value since the Codes are being applied. This section of the paper briefly discusses application of the Codes to US projects and is followed by a case history example of risk management principles and tools applied to the Bi-County Water Main in accordance with the principles of the Codes.

The US Insurance Market

One phrase that is heard frequently is that the US insurance market is unique and therefore the Codes do not apply in this marketplace. Unfortunately this statement is essentially false for two major reasons. Firstly, different insurance terminology does not necessarily mean different insurance—Contractor's all-risk referenced in the Codes is similar to Property and Builder's Risk Insurance available for heavy construction in the US. Secondly, the majority of major tunnel project insurance and almost all the reinsurance on these projects are placed internationally and most if not all on the London insurance market. The major insurers and reinsurers on the London market apply the requirements of the Codes in their assessment of tunnel projects.

Far from being unique, the US insurance market is, as would be expected, only part of an international web of insurance and reinsurance underwriters and brokers that have common objectives and requirements. This community of underwriters for both insurance and reinsurance of large heavy civil projects is quite small and most of the insurance companies involved in underwriting tunnel projects played an active part in the generation of the Codes and fully support, if not require, the application of their principles.

When Do Insurers Require Adherence to the Codes?

At some point, insurers ask for adherence to the Codes and this generally occurs when the cost of project insurance is large enough to require insurance companies to join together to spread the risk between several companies. At this stage of the insurance assessment process underground insurance specialists become involved and these personnel are most often based in London and use the Codes as the basis of their project review. The dollar value threshold for this activity may be larger than the $2 million quoted in the Codes and could be in the $50–$80 million project range and above.

Insurers recognize that, for various reasons, strict compliance with the Codes may not be possible. The degree of noncompliance and its impact in relation to project management practices is then taken into consideration in the underwriting exercise in relation to premiums, excesses and limitations of cover.

BI-COUNTY WATER MAIN TUNNEL—RISK MANAGEMENT CASE HISTORY

The Bi-County water supply main project in the Maryland Suburbs of Washington, DC involves constructing a new 84-inch diameter water main for the Washington Suburban Sanitary Commission (WSSC). This project is approximately 5.5 miles long and is between 120 and 300 feet deep with hard rock geology anticipated to consist of the Piedmont province metamorphic schist and gneiss intruded by the Georgetown suite of mafic igneous rocks.

Based on hydraulic modeling by WSSC, the design flow in the Bi-County Water Supply Main is approximately 100 mgd. Due to various elevation and flow requirement factors, the design working pressure is 175 psi with a maximum surge pressure for design of 240 psi.

Risk Management Measures

Risk management has been the primary focus on this project for the owner and the design team from the beginning of planning and alignment selection. The Owner has previous experiences with a tunnel project that did not meet its functional purpose and has demanded strict control of risk on this project as a consequence. The risk management discussion in this paper is focused on final design and three major categories of project risk:

1. Geotechnical conditions and the use of baseline statements,
2. Payment conditions linked to the baseline statements,
3. Procurement tools to mitigate risk.

Geotechnical hazards have been considered using a Differing Site Conditions (DSC) clause in the general conditions of contract. This is coupled with the use of a Geotechnical Baseline Report (GBR) and linked with appropriate payment conditions. This approach is reasonably conventional US practice. Linking geotechnical baselines to unit price payment conditions is somewhat less conventional in the US but is not unknown and this approach is particularly effective when considering activities where quantities are difficult to predict, such as preexcavation grouting of fractured rock masses.

Procurement hazards to the project include receiving bids higher than that expected and can be caused by many factors, including unqualified bidders or a misunderstanding of the bid documents. These hazards can be mitigated in part by measures such as prequalification of contractors and allowing these prequalified contractors to review the design documents at, say a 60% level of completeness, giving comments to the Project Team before the bid advertisement.

Another hazard in procurement is the inclusion of unbalanced liability and dispute resolution clauses in the contractual general conditions. Owners often feel that these clauses provide protection in case of claims. However, unbalanced clauses are a primary cause of Contractors choosing to not pursue projects or to include large contingencies and should therefore be avoided in underground work. One recent project in Washington DC, saw insurance and liability clauses in the contract that caused severe objections and conditions imposed by insurance and surety companies on the Contractors. This resulted in the eventual shelving of the project due to a lack of bidders.

Table 1.   Risk register example: hazard identification and initial assessment

| Hazard | Cause of Hazard | Potential Consequence | Risk Likelihood | Risk consequence | | | | | | Risk Score |
				Financial	Project Schedule	Corporate Reputation	Regulatory/Legal	Health and Safety	Environment	
Bid Price from Contractor higher than budget allowance of Owner	Insufficient bid Competition	Schedule delay and cost increase	4	4	3					16

For the Bi-County Water Main, and after much discussion, WSSC accepted that their standard conditions were not well adapted for bidding major underground construction projects. The Owner's team decided to use a set of standard industry conditions of contract. Special Conditions of Contract addressing key concerns of WSSC and specific to this project were appended to the standard contract documents.

The Codes demand that engineering best practice is used in prosecution of underground projects to reduce the overall project risk profile. For the Bi-County water main, as we have discussed briefly here, we are using all the tools at our disposal to mitigate the geotechnical and procurement risks associated with the project. A risk register is used to present and manage these risks.

The format of our risk register is such that not only does the register present the project hazards and initial assessment of those hazards (Table 1) it allows clear presentation of project action items, mitigation measures and the responsible party for each action item (Table 2). In short, our register is both a presentation tool and a powerful risk management tool. The Design team uses the register daily and reviews it weekly at progress meetings as the action items and mitigation measures are reviewed with consequent reassessment when appropriate. The Owner is briefed on the risk profile of the project at monthly project meetings. The Owner then uses the Risk Register internally for presentation to upper management and Commissioners as needed to convey how the project risk profile is being managed.

CONCLUSION

The magnitude of recent insurance losses led the insurance market to reassess the practices of their customers and their own underwriting process of accepting these practices. The process has resulted in publishing Codes of practice for risk management of tunnel projects. The following conclusions have been drawn from the Author's experience of using the Codes over the past few years:

- Use of the Codes has been endorsed by the ITA.
- The Codes require that the Project Team demonstrates the use of best industry practices to mitigate and manage project risk.
- There are over a half a dozen US projects where the principles of the Codes have been requested by insurers or have been applied by the design team.
- While issues have been raised against the Codes, its implementation has proceeded and is expanding.

Table 2. Risk register example: risk management and continuing assessment

Control Measures Implemented (Actually in place at time of assessment)	Indicators or Metrics (Measuring the effect of Control Measures)	Residual Likelihood—After Mitigation	Residual Consequence—Once Controls in Place						Residual Risk Score—After Mitigation Action	Action Item for Risk Mitigation	Action Item Completion Date (Target Date)	Risk Owner (Name of Individual)
			Financial	Project Schedule	Corporate Reputation	Regulatory/Legal	Health and Safety	Environment				
Personal contact from Design team and advertising of project ahead of bid date	Letters of interest/response from Contractor Community	2	4	3					8	Prequalification of Contractors	February 15	J. Smith

- The Codes require the use of reference ground conditions—or baseline conditions—widely used already in US practice.
- The Codes require the use of a risk register to present the project hazards and risk profile clearly to all project team participants and observers.
- Risk registers can be used for various project functions, including "to do" lists, permit completion checklists, health and safety hazard mitigation plans etc as well as demonstrating risk mitigation measures.
- The risk register is a powerful project management tool to clearly present hazards in nontechnical language so that political decision-makers and other nontechnical observers can see that risks are being managed and mitigated.
- Having a risk register does not mean that project risk is being effectively managed. Systematic attention to hazards is required for mitigation and management of project risk.

The Codes and by extension, the risk register is a powerful tool to add to the existing array of tools that help project teams in underground work manage risks to the benefit of all parties.

ACKNOWLEDGMENTS

The authors acknowledge the personal communication provided by members of the insurance industry as well as the owners and engineers on the various unnamed projects from which the experience was drawn to make this paper possible.

REFERENCES

"Tunneling—Improved Contract Practices"—CIRIA report 79, 1978. Construction and Research Information Association (CIRIA), London, UK.
Eskesen, S.D., Tengborg, P., Kampmann, J., and Olsen, T.H. 2002. Guidelines for Tunneling Risk management (Final Draft). International Tunneling Association, Working Group No. 2.
The Joint Code of Practice for Risk Management of Tunnel Works in the UK. Mellors, T.W., and Southcott, D., eds. British Tunneling Society and Association of British Insurers, 2003.
Mellors, T.W., and Southcott, D. 2004. A Code of Practice for Tunneling. Proceedings of the 30th ITA-AITES World Tunnel Congress, Elsevier, London.
A Code of Practice for Risk Management of Tunnel Works, ITIG, 2006.

NATM DESIGN FOR STANFORD LINAC COHERENT LIGHT SOURCE TUNNELS

Irwan Halim

Jacobs Engineering Group, Inc.

Frederick (Rick) Vincent

Jacobs Engineering Group, Inc.

Jonathan Taylor

Jacobs Engineering Group, Inc.

ABSTRACT

The new underground facilities for the LINAC Coherent Light Source project at the Stanford Linear Accelerator Center in Menlo Park, CA, are constructed in very weak sedimentary rock interspersed with uncemented zones. In addition to several 21-foot wide (excavated) tunnels, the NATM excavations include a large, 49-foot wide (excavated) cavern, an unprecedented size in this geology. The critical design issues include characterization of the weak rock, some shallow rock cover areas, and very stringent criteria for final invert slab movements within the tunnel. The design also incorporated severe seismic criteria, since the site is in close proximity to the San Andreas Fault.

INTRODUCTION

The Stanford Linear Accelerator Center (SLAC), operated by Stanford University in Menlo Park, CA, just outside of San Francisco, is embarking on construction of its LINAC Coherent Light Source (LCLS) Project, funded by the United States Department of Energy. The project consists of ½ mile of underground construction, including 1,700 ft of tunneling in rock 10 to 100 ft deep, to be excavated using the New Austrian Tunneling Method (NATM), also called herein as the Sequential Excavation Method (SEM). Three 19.5 ft (finished) tunnels and one 46 ft (finished) cavern make up this length, along with eighteen 6" to 18" ID shafts. All linings are synthetic fiber reinforced and plain shotcrete, with welded wire fabric and lattice girders and serve as both initial and final linings. In the cavern, 16 or 20 ft galvanized cement grouted rock dowels are also specified.

The Linac Coherent Light Source (LCLS) will be the world's first X-Ray free electron laser. The x-rays are emitted in the form of a laser beam, with a brightness that is 10 billion times greater and several orders of magnitude shorter than that of any existing x-ray source on earth. These characteristics will enable new science, including discovering and probing new states of matter, understanding and following chemical reactions and biological processes in real time, imaging chemical and structural properties of materials on the nanoscale, and imaging non-crystalline biological materials at atomic resolution. The project site and beam alignment are depicted in Figure 1.

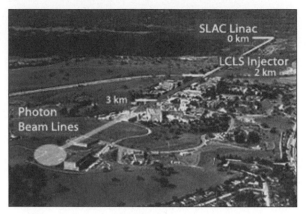

Figure 1.   Project site aerial view with superimposed beam alignment

NATM DESIGN FOR THE TUNNELS AND CAVERN EXCAVATIONS

Geologic Considerations

As shown on the geologic profile of Figure 2, the tunnels and cavern are to be located within the Ladera Sandstone formation, which is a poorly graded, fine to medium grained sandstone. The rock is generally weathered and moderately cemented, with occasional zones or pockets of poorly cemented and uncemented sands, which would tend to ravel when excavated. The average compressive strength of the rock is only approximately 160 psi based on laboratory tests, which is indicative of a soft or very weak rock, and sometimes can be classified as Intermediate Geo Material (IGM—materials with strength in-between those typically of rocks and soils). Therefore, one of the most difficult tasks during the design was to find the most appropriate manner to characterize the rock mass conditions that would best reflect its behavior during the tunnel construction. Traditional rock mechanics approaches need to be supplemented by engineering judgments to consider the very weak and sometimes uncemented nature of the rock. And the NATM, was deemed early on to be the most appropriate excavation and initial support design approach under these conditions.

Design rock mass properties are normally derived by calibrating or reducing the intact rock properties obtained from laboratory analyses, using the widely accepted rock mass classification indices such as RQD, Q, and RMR systems (Kirkaldie 1988). However, by definition, RQD cannot be applied to rock cores with either low hardness or high degree of weathering. Therefore, the Q and RMR indices, of which RQD is a component, had to be used with caution in this case, and other geologic considerations including observations of actual ground behavior during previous tunneling experience in the area became of the utmost importance. For example, based on the Q system, the ground was rated as a very poor rock with practically no stand-up time. It was believed in this case that the Q values have been over-penalized by the low or non-existent RQD component and thus grossly underestimate the rock mass quality. On the other hand, the rock RMR rating indicated generally fair to poor or better quality rock than indicated by the Q system.

Based on the experience during previous tunnel excavations at the site, the typical ground is anticipated to behave more like fair to good rock with limited zones of poor to very poor rock. Furthermore, to bridge the apparent gap between the Q and RMR systems, an alternate system recommended by Hoek (Hoek and Marinos

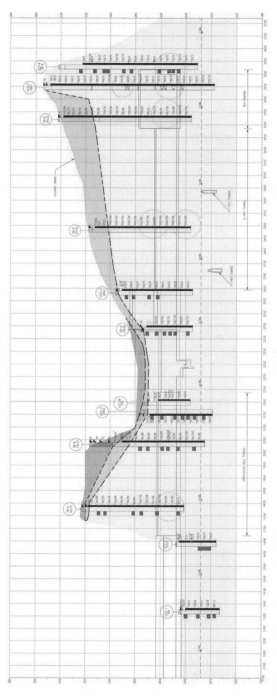

Figure 2.   Subsurface profile along tunnel alignment

Table 1.   Summary of Ladera Sandstone design properties

Classification Ratings:	
RMR	30 to 50
Q	0.04 to 0.5
GSI	15 to 50
Strength Criteria:	
Typical/Average Rock:	
Hoek-Brown	mb = 1.65, sb = 0.0026, ab = 0.509
Equivalent Mohr-Coulomb	c = 7.5 psi, f = 31°
Low/Poor Rock:	
Hoek-Brown	mb = 0.61, sb = 0.0001, ab = 0.553
Equivalent Mohr-Coulomb	c = 4 psi, f = 23°
Modulus of Deformation:	
Typical/Average Rock	43.5 ksi
Low/Poor Rock	20.3 ksi

2000), the Geological Strength Index (GSI), was used for the final rock mass properties estimation. This index can be directly related to the RMR for better quality rock
masses, and has been extended for weak or poor quality rock masses where the
RMR correlation does not normally work such as in this case. For the typical or
"standard" design condition, the ground condition was estimated to be blocky and
seamy, with interlayers of sandstone and/or siltstone with fair to good surface conditions of discontinuities. For the worst ground condition or contingency design, the
ground condition was estimated to be crushed or tectonically deformed, heavily
folded/faulted siltstones and sandstones forming an almost chaotic structure with
poor to very poor surface conditions of discontinuities. Table 1 presents the final
design rock mass properties based on these rock conditions. The rock mass strength
properties are given for the generalized Hoek-Brown failure criterion (Hoek et al.
2002), and their equivalent Mohr-Coulomb parameters. The rock modulus of deformation is based on laboratory measurements of intact rock samples, field pressuremeter test results, and empirical correlations with other rock mass properties.
The in-situ stress condition is estimated based on the regional geologic condition
and the laboratory Poisson's Ratio measurement.

Undulator Hall, X-Ray and Access Tunnels Excavation and Initial Support

Due to the weak nature of the on-site rock, the tunnel excavations would likely be
performed by mechanical means such as using roadheader, with relatively minor disturbance to the rock surrounding the tunnel. The Undulator Hall, X-Ray, and FEH Access
tunnels are all expected to be excavated in a single heading and bench excavation
sequence. The design relies as much as possible on the ability of the surrounding rock
to support itself during tunneling, and employs relatively flexible, passive support systems, that will allow controlled ground deformations and prevent the tunnel failure or
collapse at the same time. Therefore, the sequence of excavation and the timing of support installation are of paramount importance and specified for construction. Figure 3
presents the contract requirements for the Undulator Hall Tunnel excavation. In areas
with minimum or shallow covers such as near the tunnel portals, the amount of rock
cover would not be adequate to maintain stability, and therefore pre-support systems by

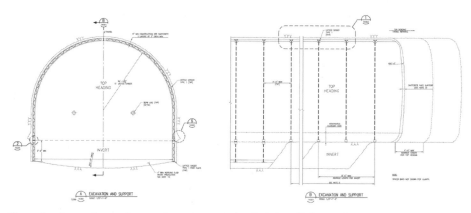

Figure 3.   Excavation and support requirements for Undulator Hall tunnel

forepoling or spiling are used. In addition, a set of contingency measures or tool box is incorporated into the design to account for the varying ground conditions. The application of these contingencies would be based on the actual ground behavior observed during construction.

In general, the typical or average tunneling ground condition is anticipated to be firm to slowly raveling, with pockets of weakly cemented or uncemented sandstone which could lead to local instability in the tunnel perimeter or heading. For tunnel excavations in this typical ground condition, initial tunnel support system consisting of a combination of Synthetic Fiber Reinforced shotcrete (SFR) and regularly spaced lattice steel girders with spot rock dowels (as needed) are employed. Due to the high susceptibility of the rock to slaking when subjected to a wetting and drying cycle, the initial shotcrete layer needs to be immediately applied at or close to the tunnel heading to protect the rock surface against deterioration. The very strict movement tolerance/criterion of the final invert slab in the Undulator Hall tunnel requires the slab subgrade to be stable and undisturbed at all time during construction. For this purpose and to minimize ground heave during construction, excavation for the Undulator Hall tunnel is specified with a slightly curved bottom (not required for the other tunnels). Furthermore, a minimum 3-foot thick invert bench and immediate installation of mud slab (not required for the other tunnels) are also specified to stabilize and protect the final tunnel invert against construction traffic. Limits are placed on excavation equipment to be used for the bench, and the amount of bench excavation that can be performed before installing the mud slab.

Based on the available subsurface information, the stand-up time in the typical tunnel opening is expected to range from several hours to several days, depending on the size of openings and ground condition. Therefore, the maximum length of unsupported excavation round and timing of support installation are specified as part of the support requirements. For the Undulator Hall, X-Ray, and Access tunnels, the main heading excavations are to be conducted in a 4-foot maximum round length, the initial shotcrete layer to be applied within 4 hours of excavation, and the initial support system to be completed at a maximum 10-foot distance behind the excavation face or within 24 hours of the excavation opening. Using this sequence, the average advance rate of the tunneling is anticipated to be in the order of about 6 to 8 LF of tunnel length per typical working shift/day.

In areas with shallow rock covers, pre-support measure with forepoling by grouted pipe arches or canopies are used. Contingencies for the worse than typical ground conditions and the anticipated zones or pockets of unstable or "running" sands include several pre-support and face stability measures such as self-drilled and grouted pipe spiling, flashcrete and fiberglass bolting at the tunnel face. The needs for all the special measures and contingencies (non-standard support system), including the tunnel ground behavior that warrant the application of these special measures are defined in the Geotechnical Baseline Report (GBR). The extent and timing of applications of these various measures are defined in the contract documents. Geotechnical instrumentation is normally an integral part of NATM tunneling. However, due to the limited extent of the tunnels at SLAC, only limited in-tunnel instrumentation, consisting of optical convergence/survey monitoring, is being used. This is complemented by instrumentation and monitoring of the ground above and around the tunnel using deformation points and multi-point probe extensometers (for the Far Experimental Hall cavern described below).

Far Experimental Hall Excavation and Initial Support

The design approach for the Far Experimental Hall (FEH) cavern excavation and support is similar to the one for the tunnels described above. However, due to its relatively large size, the FEH would need to be excavated in multiple headings and benches, the size and sequence of which are specified to maintain the excavated openings stability, and also to facilitate construction and the initial support installation. The FEH excavation would likely start with a top center heading followed by two side drifts. A single level of benches (i.e., a center and two side benches) would be needed to complete the cavern excavation. The initial support for the cavern consists of fully grouted dowels, synthetic fiber reinforced shotcrete, and lattice girders, which are installed immediately in conjunction with the staged excavations. The initial support system would be integrated into the final support system for the tunnels and cavern, as described in the following section.

Because of the very weak or soft nature of the sandstone rock, the failure mode in the tunnel excavations are controlled by the plastic failure or yielding of the rock mass surrounding the excavation opening, instead of rock discontinuities. Figure 4 shows a typical result of the final stress distribution around the cavern, and also the stresses and forces in the initial support system, using a continuum mechanics finite element method. The Phase2, 2-dimensional program, by RocScience (2002) is used to perform the analyses in this project. The primary function of the dowels in this case is to "knit" the rock together and improve the rock mass quality as a whole, and therefore limit the plastic zone formation around the excavation opening.

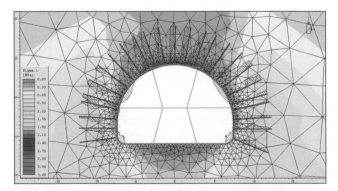

Figure 4.   Typical Phase2 results of final stress distribution around FEH cavern

In some areas, such as the tunnel portals and intersections, and where the rock is relatively good and its discontinuities are well defined, spiling bolts and spot bolting are specified as needed to stabilize potentially unstable blocks or wedges, and to prevent those rock blocks from falling out. Additional dowels are also specified for the areas where high stress concentrations are expected. Due to the nature of the rock and lack of joints definition, no formal wedge stability analysis could be performed in this project.

FINAL LINING DESIGN FOR TUNNELS AND CAVERN

Final Lining Concept

The concept for the final lining for the three tunnels and the FEH cavern is a composite system including both the SFR shotcrete initial support layers and lattice girders, and the additional layers of plain shotcrete placed after excavation and initial support is installed. The plain shotcrete is reinforced with welded wire fabric (WWF) for long-term static and seismic loadings. The 6" SFR shotcrete layers are combined with the plain shotcrete layers to achieve an overall minimum 12" thick final lining in the tunnels and a 15" thick final lining in the FEH cavern. The final support system in the FEH cavern also relies upon 1" diameter cement grouted dowels for long-term support, which are installed as part of the initial support system. All tunnels and cavern also have a reinforced cast-in-place invert slabs that are tied through the linings to the rock, to be structurally integrated with the tunnel liners. While most slabs are flat, in the case of the Undulator Hall, where strict final slab movement criteria was required, the slab bottom is constructed with a minimum radius to divert upward ground movements to the undisturbed ground at the tunnel sidewalls, promoting the arching effect. Where the final tunnel linings meet with cut-and-cover cast-in-place concrete construction, special seismic connection details were developed to allow for the large predicted relative movements at this joint.

Lining Materials

The minimum 28-day compressive strength of all shotcrete is 5,000 psi, and due to the weak nature of the rock, the average toughness absorption of the SFR shotcrete at 28 days, per ASTM C1550, is 500 Joules using 40 mm central deflection. In addition to the 500 Joule requirement, the macro-synthetic fibers were also specified with a minimum dosage of 18 lbs/CY. However, as a value engineering item, using newly established methods of quantitative correlation between design values and toughness performance level developed by Chen (2006), the dosage was allowed to be reduced to an amount that would satisfy only 400 Joules of energy absorption. The lattice girders are composed of smooth bars with minimum yield strength of 70 ksi, while the invert slabs contain grade 60 reinforcement and the WWF is in accordance with ASTM A185. Final cavern rock dowels are ASTM A615 grade 75 minimum, hot-dip galvanized.

Loading Criteria

The design loading criteria for the final lining are based on the static (i.e., gravity) load of rock above the tunnels, the tunnel lining dead loads, and the seismic structural requirements governed by the project. Since the tunnels and cavern will be located above known groundwater table at the site, the hydrostatic load acting on the tunnel structure is expected to be insignificant, and therefore not considered during the design, although saturated unit weights for rock were used in all analyses. The long-term gravity rock load is based on the anticipated ground behavior during excavation (as predicted by the analytical models), and widely accepted empirical formulas using the rock mass characteristics, but in no case greater than one tunnel width or diameter of rock load above the tunnel crown (from prior experience in similar tunneling ground conditions). As previously mentioned for the FEH cavern, the main function of the dowels is to improve the rock mass condition to reduce the long-term loading into an amenable amount.

The seismic performance criteria for the tunnel and cavern structures are based on the widely accepted methods of seismic design and analysis of underground structures, as well as all applicable codes and SLAC project specific seismic criteria. SLAC LCLS criteria included the Geotechnical Data and Investigation Report (Rutherford & Chekene 2005), as well as the Specification for Seismic Design of Buildings, Structures, Equipment and Systems at the Stanford Linear Accelerator Center (2000). In addition, the designer referred to: Seismic Design and Analysis of Underground Structures (Hashash et al. 2001); Seismic Considerations for Design of SLC Tunnels, Appendix C of SLAC Architecture/Engineering Design Guidelines (Earth Sciences Associates 2003); CALTRANS Seismic Design Criteria, Version 1.3 (2004); and other structures and building codes as appropriate.

The seismic analyses produce normal, longitudinal and shear incremental strains in the lining elements from design seismic events. The resulting incremental strains are combined with the strains calculated from the static loadings described above and are then checked against the inelastic lining capacities to determine the final seismic reinforcement requirements. Also, a similar analysis was performed by combining the strains produced by a racking analysis with the static loading strains. The controlling seismic case proved to be the shear strains (ovaling), which is predicted to produce minor local yielding of the WWF steel but significant amounts of ductility exists within the yield plateau and conservative assumptions were used in the analysis, such as the project-criteria to sum absolute strains which in reality cannot exist simultaneously and ignoring the contribution of the lattice girders and synthetic fibers.

Lining Tolerances and Serviceability Concerns

Since the shotcrete lining will be the final surface in the manned tunnels and cavern walls, for serviceability purpose, strict surface tolerances are specified. The tolerances were set by SLAC to satisfy equipment operational tolerances and included both minimum and maximum final lining targets. The final tunnel and cavern lining must fall within an annulus centered on the tunnel/cavern centerline that is 4" in total thickness, with the actual inside surface falling no greater than 1" inside or 3" outside the theoretical inside tunnel profile. This tolerance is specified not only to permit proper placement of the high-level physics equipment to be installed, but to facilitate the ventilation design which establishes the controlled environment (temperature and air velocity) that the equipment and experiments require.

As mentioned previously, the invert of the tunnel is above the water table. However, as the tunnels and cavern will permit manned access and, in the case of the cavern, will be occupied during some experimentation periods, the use of low-permeability shotcrete is specified. Also, if some perched water does seep into the areas, a drainage system consisting of sumps embedded within the final invert slabs is provided to carry water out of the facilities via pumps. In the final condition, the shotcrete surfaces will be painted with an epoxy to minimize dust and improve aesthetics and lighting. Aside from its greater post-crack energy absorption capacity, the use of synthetic fibers over steel fibers in the shotcrete lining was also chosen for serviceability considerations. Its use was advantageous in that it would prevent any possibility of magnetic effects on the extremely sensitive nanoscale experiments to be performed.

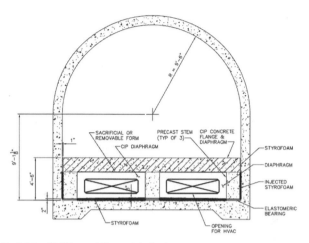

Figure 5.   Early Undulator Hall tunnel floor slab design concept

Another major serviceability consideration that the designer was required to follow was the strict final movement tolerance criteria established for a facility whose laser will function only under the tightest operational tolerances, complicated by the fact that the site is only 2.5 miles from the San Andreas Fault. This criterion was primarily focused upon very strict differential settlement limits for the invert slab of the Undulator Hall Tunnel. As the most sensitive underground area, the Undulator Hall (UH) will house a series of undulator magnets that a laser will pass through, and microscopic movements in the magnets will misalign the laser, so the criteria was required to prevent frequent adjustments to the magnets during experimentation. Based on the operational requirements and experience with the other similar tunnel facilities in the area (PEP and LINAC tunnels), it was determined that a vertical differential movement criterion of 2–3 microns per week over points at 10-m apart was both reasonable and feasible (Welch 2004).

During design development, several methods were considered to achieve this criterion. The early design concept involved a precast "Tri-T" beam resting on an elastic foundation (Figure 5). The major intent of the design concept was to allow for a separation between the slab and the surrounding ground, thus significantly reducing or minimizing the slab differential movements from those occurring in the ground itself. In order to do so, known foundation material properties with high degree of analytical certainty would be needed, in order for the predicted stiffness and deflection of the "floating" system to accurately reflect the as-built conditions over the long term. However, this Tri-T beam on elastomeric bearing concept and associated construction sequence was ultimately abandoned for the final design. The reasons included the perceived uncertainty in the proposed elastic foundation material properties and behavior, construction complexity (labor cost); materials cost (due to reported limited precast concrete availability); and the perceived risk associated with an unproven "state of the art" system.

Based upon those factors, it was decided that a similar approach to the existing PEP and LINAC tunnels is preferred, and that the instruments supported on the slab, rather than the slab itself, would be required to make up any differential movement that is experienced in the tunnel. To achieve a similar to or better movement performance in the proposed UH than the existing facilities, the cast-in-place (CIP) concrete invert slab is designed to obtain an equal or greater overall structural stiffness than the existing tunnel structures. Other structural details are incorporated that should allow for

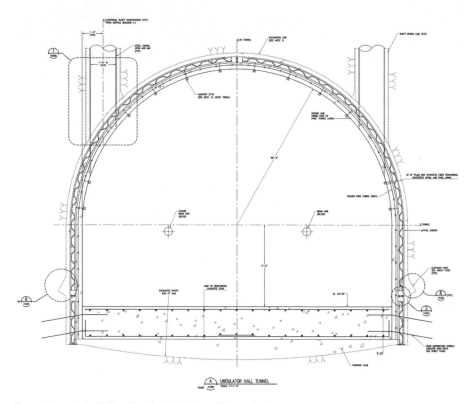

Figure 6.   19.5-ft ID SLAC Undulator Hall (UH) tunnel

improved performance. For example, in the PEP tunnel, the floor slab was simply poured on a flat invert after completion of the tunnel lining, and was only nominally connected to the tunnel sidewalls via a shear connection dowels. In the UH tunnel design two grouted rock dowels are provided on each side of the reinforced floor slab, which will provide moment capacity to the floor slab in the transverse direction and make it integral with the remainder of the tunnel. The concave shape of the tunnel invert will also potentially reduce the upward heaving of the final subgrade and bending movements in the floor slab. Finally, the concrete mix for the invert floor slab has been specified as a low-shrink mix to assure that, together with the grouted dowels, the invert slab will be in full contact with the shotcrete walls and thus provide the full shear transfer across that vertical joint (Figure 6).

CONCLUSIONS

The NATM design approach for Stanford LINAC Coherent Light Source Tunnel Project has been described in this paper. The important geologic consideration includes the very weak nature of the rock (strength in-between those of soils and rocks), that makes traditional rock mass characterization more difficult. To alleviate this problem, several complementary rock classification systems were used, in conjunction with assimilation of actual tunneling experience learned from previous projects in the area. To address the very strict movement tolerance requirements (on the order of microns), some unique structural details were employed for the final lining, as well as to satisfy the seismic design criteria in what is considered a very active seismic region. The initial support system is being integrated into the final lining design of the tunnels and cavern to optimize or reduce the project cost. Furthermore, a specific inner surface profile criterion is used for the finishing shotcrete in the tunnel walls, especially for the Undulator Hall tunnel, which is required for the installation of some precision equipment inside the tunnels.

ACKNOWLEDGMENTS

The authors wish to acknowledge SLAC of Stanford University as the Owner of the project described in this paper. This paper represents the opinions and conclusions of the authors and not necessarily those of SLAC. This paper shall not be used as evidence of design intent, design parameters or other conclusions that are contrary to the expressed provisions in contract documents for the project.

REFERENCES

Kirkaldie, L. (ed.), 1988, Rock Classification Systems for Engineering Purposes, ASTM STP 984, Philadelphia: American Society for Testing and Materials.

Hoek, E., Carranza-Torres, C., and Corkum, B., 2002, Hoek-Brown Failure Criterion—2002 Edition, Proceedings of the 5th North American Rock Mechanics Symposium and 17th Tunnel Association of Canada, NARMS-TAC, Toronto, Canada, Vol. 1, pp. 267–273.

Hoek, E. and Marinos, P., 2000, GSI—A geologically friendly tool for rock mass strength estimation, Proceedings of the GeoEng Conference, Melbourne, Australia.

Phase2 v5.0, 2002, Two-dimensional finite element program, Rocscience Inc.

Chen, W., 2006, Design and specification of fiber reinforced shotcrete for underground supports, Proceedings of the North American Tunneling, Chicago, Illinois.

Rutherford & Chekene Consulting Engineers, 2005, Geotechnical Data and Investigation Report, Linac Coherent Light Source (LCLS) Project, Stanford Linear Accelerator Center, Menlo Park, California, 21 January.

Specification for Seismic Design of Buildings, Structures, Equipment and Systems at the Stanford Linear Accelerator Center, December 4, 2000.

Hashash, Y., Hook, J., Schmidt, B., and Yao, J., 2001, Seismic design and analysis of underground structures, Tunneling and Underground Space Technology 16.

Earth Sciences Associates, 2003, Seismic considerations for design of SLC tunnels, Appendix C of SLAC Architecture/Engineering Design Guidelines, October.

CALTRANS Seismic Design Criteria, Version 1.3, February 2004.

Welch, J., 2004, Ground motion expectations for the LCLS Undulator Hall, LCLS-TN-04-14, November.

NEW YORK SUBWAY STATIONS AND CROSSOVER CAVERNS—UPDATE ON INITIAL SUPPORT DESIGN

Dru Desai

DMJM+HARRIS•Arup JV

Hannes Lagger

DMJM+HARRIS•Arup JV

Charles Stone

DMJM+HARRIS•Arup JV

ABSTRACT

The design methodology of the planned Second Avenue Subway Project (SAS) led to development of the initial support design to estimate cost and quantities in the Preliminary Engineering (PE) stage. Careful selection of the cavern design rules that recognized the impact of rock quality, rock cover, excavation geometry, and excavation sequence helped to overcome the design challenges of large, shallow caverns for the initial ground support.

In the course of the design verification process prescribed by the design methodology, the design team was continuously reviewing and updating the support quantities. The design procedure includes methods using different approaches of empirical, continuum, and discontinuum analyses, which are currently advanced in the Final Engineering (FE) stage. The paper is a continuation of the RETC 2005 "Trials and Tribulations" and focuses on the procedures that led to the changes in the initial support as a result of the ongoing design development.

INTRODUCTION

The Second Avenue Subway (SAS) project is being planned for construction on the east side of Manhattan Island for the New York City Transit Authority by the DMJM+HARRIS•Arup Joint Venture design team. The design of several major mined caverns was initiated during the concept engineering phase of the project and has continued into the final engineering (FE) phase that is currently underway. A design methodology was developed that employed a careful selection of cavern design rules that recognized the impact of rock quality, rock cover, excavation geometry, and excavation sequences to overcome the design challenges of large, shallow caverns for the initial ground support. Preliminary results were described at the RETC 2005 in "Trials and Tribulations." The design team has continued to develop the initial support for the cavern design. Initial ground support requirements have been updated based upon the results of numerical modeling. This paper is a continuation of the earlier paper and focuses on the procedures that led to the changes in the initial support as a result of the ongoing design development.

METHODOLOGY

Large excavation spans, low rock cover, and variable geotechnical conditions, all located within a complex urban environment characterize the design challenges of the caverns of the SAS project. The methodology adopted by the SAS design team accommodates these mined cavern characteristics by applying several design concepts as follows:

- Verification of initial support design by four inherently different design methods.
- Development of an initial ground support requirements (IGSR) chart including cavern design rules and design ranges of five support classes, with the associated support quantities.
- Rock load as one of the parameters to categorize the support type over broad ranges in the IGSR chart, and provide a measure of applicability of the verification methods.

The four differing design approaches that were used to verify and optimize the IGSR chart are as follows:

- Empirical methods (Q, [Barton & Grimstad 2004], RMR, [Bieniawski 1989]).
- Stress-strain method of discontinuum (UDEC, [Itasca 2003]).
- Force-equilibrium method (UNWEDGE, [Rocscience 2006]).
- Plane stress models, for shotcrete stresses (STAAD, [REI 2005]).

Each method was used independently and given equal importance during the design process although they differ in the range of their applicability. None of the methods is considered an inferior or superior analytical tool, but the methods were applied in appropriate circumstances. In this paper the software names of UNWEDGE and STAAD are used to represent force-equilibrium methods and structural design methods, respectively. Other verified software of force-equilibrium methods or structural design methods could also be used.

Initial Ground Support Requirements (IGSR) Chart

The IGSR Chart (Table 1) is used to select support for the tunnels and caverns and to ensure that a consistent design approach is applied for all the mined structures along the project alignment. The ranges of the support types encompass lengths of similar ground conditions, "design slices," to minimize frequent changes in construction methods, while concurrently matching the ground support to the expected rock conditions.

The functions of the design chart are as follows:

- Define the design ranges of five support types by cavern design rules that are selected to estimate ranges of similar ground behavior.
- Estimate quantities of materials associated with the five support types.
- Provide guidance for excavation sequence design.
- Determine design rock loads.

CAVERN DESIGN RULES

In order to standardize the cavern designs across areas of the project and to standardize the design with respect to other projects as well, a set of cavern design rules was established. During the concept, preliminary, and final design stages of the project these rules were employed to serve as a basis of the empirical design with confirmation from numerical analysis. In general, these rules arose from empirical design methodologies and were subject to numerical analysis during the design phase of the project. By the end of PE there were five cavern design rules, of which four were considered as input in the selection of the cavern design, and one of which

Table 1. Initial ground support requirements chart

was considered as both input and output. The first four rules are the Q, RMR, shallow cavern, and rock mass class rules. The fifth rule is rock load which is considered as an input as far as establishing lining rock loadings is concerned, but is also considered as an output, since the rock loading estimated is based upon the four other cavern rules serving as an input to the design. The five design rules were incorporated into the IGSR chart, which incorporated five classes of ground support and several cavern widths.

The five ground support classes were developed to cope with the anticipated ranges of ground conditions and boundary conditions in the vicinity of a cross section (overburden, tunnel intersections, building loading, etc.). The ground behavior rules, which recognized impacts of both the rock quality and the excavation geometry, arranged the five support types in a hierarchy from bolt-reinforced rock to structural shotcrete arch in recognition of a boundary between "good" and "bad" ground behaviors. There are two other cavern design rules that do not serve as a basis for the design; proximity to adjacent structures and peak friction angle. These are taken into account in the design via the UDEC and 3-DEC analysis, but are not discussed in this paper.

#1 Norwegian Geotechnical Institute's Rock Tunneling Quality Index—Q

The Q-system developed by Barton is an empirical method of predicting probable ground behavior considering discontinuous geotechnical and stress-strength relationships. As required for the design of support systems, the Q-system has been developed with a view to determining the mechanism and mode of failure in the rock mass based roughly on the block size, inter-block shear strength, and the active stress regime, with the aim of evaluating stability as one of the first steps in designing an underground excavation.

Originally developed in 1974, the system was updated in 1993 to include steel fiber reinforced concrete. This update included an additional sample of 1050 sections of highway tunnels and hydropower tunnels where support was selected by experienced engineers. A subsequent update in 2003 was primarily concerned with identifying the requirements for reinforced ribs of shotcrete where the toughness and energy adsorption of sprayed concrete has been taken into consideration in bad ground conditions where deformation may be expected.

For the SAS project the Q system provides an evaluation in terms of both rock quality and cavern width. The Q rating has been employed to provide a first indication of initial ground support, and it is employed as a cavern rule of thumb to put the design on the same page as other tunneling projects worldwide.

#2 Geomechanics Classification, Rock Mass Rating—RMR

The RMR or Geomechanics classification was developed by Bieniawski in 1972 and was updated in 1976. The system is based on 351 case histories in various applications in hard rock mining. The RMR classification is an empirical method of rating relative rock mass qualities for mining and construction activities. Initially, it was intended to represent a structural region of a discontinuous rock mass by a uniform index value. The system was updated in 1976 to clarify the significance of some of the input parameters. It has been widely used and modified in 500 to 1000 mining and tunneling case histories.

Various correlations between the RMR and the rock mass modulus have been investigated. The RMR value provides an estimate of the degree of ground support required and expected failure modes. Applying the RMR values for SAS, prudent ground stand-up times have been estimated, which then have been used to determine appropriate limitations on excavation sequences. For the SAS project various correlations between the RMR and the rock mass modulus value have been investigated.

#3 Rock Cover to Span Ratio

Empirical data shows that there is a breakdown of the natural arching concept below some minimum cavern rock cover to span ratio. The ⅓ third span rule has long been used as a rule of thumb in the mining industry based on empirical evaluation of mining conditions. It can be expressed as follows; leave intact rock cover over the cavern equal to at least ⅓ of the cavern span. This design rule was investigated by numerical analysis, via UDEC, employing representative geotechnical parameters, and found to be acceptable for the anticipated ground conditions. During PE the shallow cavern rule could be stated as follows: Normal construction conditions in rock apply down to a rock thickness/span ratio of 0.33. Rock cover is considered to be adverse, type IV, and worst case, type V, when the rock cover-to-span ratio was less than 0.33 and 0.25 respectively. During the PE and FE design phases most of the cavern sections were kept above the 0.25 limit to avoid heavy support requirements. During FE detailed crown pillar analyses were undertaken in particular where the ratios were found below 0.25 and where poor geotechnical, geological, and geometrical conditions (penetrations) have been encountered. Numerical analysis shows that due to the breakdown of the natural arching concept, shallow caverns lead to higher crown displacements, higher arch bolt forces, and higher shotcrete thicknesses. These results are reflected in the IGSR charts.

#4 Rock Mass Class

For the SAS the rock mass class was considered adverse when the rock was blocky to very blocky with shear zones present and worst case when at least three shear zones were striking along the cavern axis based upon a classification scheme presented by Cording. This paper reflected the experiences gained during subway construction in the Washington D.C. area. Cavern geometry was similar to planned SAS excavations with cavern width 60–70 ft, height 44–55 ft, length >600 ft, and rock cover 25–40 ft. The following rock mass classes have been employed by the SAS project which are equivalent to Cording's Classification:

- Type I, II, and III–Rock Mass Class 1—Moderately jointed to blocky. Shears and shear zones not extensive. Rock is usually more gneissic than schistose. Foliation is not pronounced.
- Type IV–Rock Mass Class 2—Blocky to very blocky and seamy. Shears and shear zones are present but are not as frequent or as wide as in Class 3.
- Type V–Rock Mass Class 3—Very blocky and seamy. Heavily sheared, with 3–10 major shear zones striking along the tunnel axis in each of the chambers. Rock is usually schistose, with pronounced foliation.

#5 Rock Load

Empirical data from previously constructed mined station caverns in Washington, D.C. and New York showed that rock loading should be considered applicable to a determined ground support class rather than as an input to the determination of the class. Rock loads based upon measured rock loading values from these existing caverns were correlated with geological data and cavern geometry using the Q value and scaled crown span. An example of these results is shown in Figure 1.

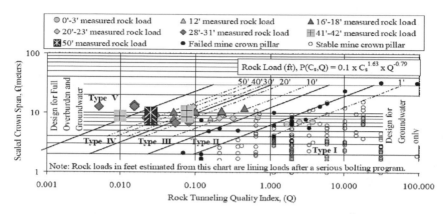

Figure 1. Estimation of rock loads for shallow caverns

Rock load is required as an input parameter in the design of the shotcrete initial linings. It summarizes the impacts of all of the other ground conditions and represents a convenient measure of ground behavior. The rock load is first estimated by conventional rules and then verified by empirical, force-equilibrium, and discontinuum methods. The magnitude of the rock load carried by the support is determined by the support stiffness and the rock arching capacity to support the hoop stresses generated by the excavation. In shallow excavations that require stiff support, the rock load can be expected to approach 80% to 100% of the in place vertical stress for type IV and type V conditions respectively.

BASIC CRITERIA

The initial ground support is designed as a temporary structure to be stable during the period of time between the initial cuts in the excavation sequence through the completion of the final lining. For support classes I to III the design is based upon basic criteria for rock bolting (Table 2) and thin shotcrete layers (Table 3) including:

- Non-factored loads.
- Strength reduction factors in accordance with:
  - ACI 318-05 (Shotcrete structural arch)
  - PTI (Bolts)
  - AISC (Steel sets)
  - Moy & Bandis 2004 (Bolt grout-rock bond strength)
- Safety factors (Force equilibrium wedge stability generally 1.4–1.5).

For ground support classes IV and V the structural shotcrete arch is supported by footings and by the passive resistance of surrounding rock which generate axial thrust and enhances the flexural strength of the structural arch.

The strength envelope of the moment-thrust diagram determines the shotcrete arch strength in flexure and compression (ACI-318-05). The lattice girder's contribution to the shotcrete compressive and shear strength is not considered. The shear strength of the shotcrete arch is verified according to ACI-318-05. Fiber contents of 0.6, 0.9, and 1.2 percent by volume may account for 20%, 35%, and 40% shear strength increases, respectively, (ACI 544.4R-88).

Table 2.   General rock bolting criteria

Bolts	
Maximum bolt force (kips)	Minimum ultimate strength $\times$ 0.6
Grout-rock bond strength (psi)	Peak shear strength $\times$ 0.5
Maximum yield strain	4%
Minimum bolt length (ft)	10 (DHA, 2006a)
Minimum bolt diameter (in)	1 (NYCT 1998)

Table 3.   General shotcrete criteria

Shotcrete plain, with wire mesh, fiber, or lattice girder	
Unit weight, γ (pcf)	150
Compressive strength, f'_c (psi)	6,000
Young's modulus, E (ksi)*	2,870
Poisson's ratio, ν	0.17
Adhesion to rock, c_a (psi)	75 (Clements 2003)
Friction angle on rock contact (deg)	25
Shear strength, f_v (psi)	0.05 f'c (Cording 1984)

* Elastic modulus reduced by 35% to account for plastic deformation, and microcracking (Bickel et al. 1996).

FINAL DESIGN PROCEDURE

Based on the development of the IGSR chart a design procedure was developed for the two shallow SAS caverns; the 72nd Street Station and associated crossover caverns, as well as the 86th Street Station Cavern. It was established by the design team in conjunction with a detail design criteria manual, to assure a constant design quality and to ease the Quality Assurance/Quality Control procedure. The goals are to verify the chosen IGSR, which are based on the empirical rules described above, and compare and verify them with numerical and structural calculations along the cavern. The design guideline is based on eight steps to ensure a sophisticated and robust cavern design solution. These steps are described in the following pages.

#1 Employ the IGSR Chart to Create Design Slices along the Cavern

The first step in the design procedure is to divide the cavern into design slices that serve as a basis for modeling the cavern based upon the five cavern design rules as presented in the IGSR chart: Q-values, RMR values, rock cover thickness, rock mass classes, and expected rock load. The cavern is divided into zones of similar rock support classes based upon the cavern design rules. These rules apply to certain areas of the cavern with distinct geotechnical conditions. These so called "design slices" are evaluated by numerical analyses that use geotechnical parameters based upon the geotechnical conditions encountered by the geological investigation within the design slice. Every design slice should include the geological information of a minimum of two boreholes. Each cavern penetration (entrances, auxiliary space etc.) is a separate design slice, including 15-ft on either side of the penetration as shown in Figure 2. A first guess initial ground support requirement is developed for each design slice.

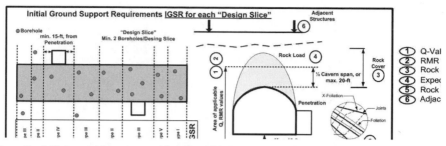

Figure 2.   Initial ground support requirement (IGSR)

#2 Evaluate the Intact Rock and Rock Mass, Rock Joint, and Soil Properties per Design Slice

For each design slice the available in-situ data for the intact rock, rock mass, rock joint and soil properties are interpreted and best estimate and lower bound values are given. For the intact rock mass the Hoek-Brown Strength Criterion Edition 2002 is used to determine the appropriate strength properties for the intact rock mass. Using the Rocklab 2005 Software package, by Rocscience, the input parameters c' rock and ϕ' rock, as well as the rockmass modulus E_m rock, are calculated from lab data for each design slice. For the rock joint model, a Mohr-Coulomb joint model is adopted, using the existing shear box data where available to calibrate the joint behavior. Where geometrically possible, the correct geometry of the joint is modeled, e.g., instead of a plane linear joint the waviness of the joint may be incorporated.

#3 Evaluate Structural Properties of Materials

The initial ground support elements for the cavern consists of resin grouted, tensioned rock bolts, Swellex rock bolts, and un-reinforced and fiber reinforced shotcrete. Lattic girders are not included in the structural capacity calculations for the lining. Shotcrete arches reinforced with steel sets, where used are incorporated with increased stiffness properties.

#4 Development of Geometry, Boundary Conditions and Adjacent Structure Loads

The mesh size for UDEC analysis is set to 12*D (D = Cavern width) width and the lowest point of the excavation is 2*D away from the boundaries of the mesh to avoid boundary effects and to achieve reasonable computation times as represented in Figure 3. Where the excavation areas occur the FEM model is represented as a discontinuum. The rest of the model is defined with continuum elements. For rock, an elasto-plastic material model of the Mohr-Coulomb type is assumed, where the soil is modeled as a stiff elastic media on top of the rock. Other boundary conditions include the following;

- Building loads are applied on top of the soil according to the estimated loads from a building survey.
- Horizontal stress ratio K_0 is applied according to the Geotechnical Interpretive Report.
- No water pressures are applied because the caverns are drained.

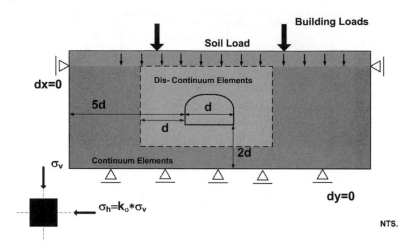

Figure 3.   Finite element mesh geometry

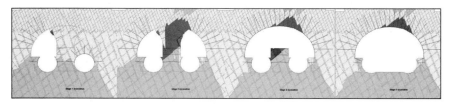

Figure 4.   Vertical displacements, UDEC 2D-Model

#5 Develop a 2D UDEC Numerical Model

The Universal Distinct Element Code (UDEC) 2D is used to evaluate the global stability of the main cavern sections due to the excavation, i.e., to evaluate the response of
the discontinuous media (such as a jointed rock mass) subjected to static loading. For
each different IGSR class, different excavation sequences have been developed during
FE. For Type I to IV either side drift excavation (multiple drifts) or center cuts can be chosen by the contractor. For Type V ground (adverse ground conditions) only side drifts will
be allowed. To account for the three dimensional effects of the rock mass, a ground
relaxation function is introduced in the 2D model (Panet 1984). A relaxation of 30% of
the insitu-stress is applied prior to installing the initial ground support. Using the UDEC
code these excavation sequences are modeled to ensure the stability of the rock mass
and adjacent structures and to limit displacements to the specified levels. Examples of
some typical UDEC results are presented in Figure 4.

#6 Kinematic Block Calculation via UNWEDGE

One of the major concerns in rock tunnels and caverns is the kinematic failure of
discrete blocks in the excavation area, i.e., rock blocks become loose and fall out of the
rock matrix, destabilizing the rock mass. Therefore the local stability of the rock mass
between the rock support necessary for the global stability is checked using
UNWEDGE. Based on engineering geology input data (dip and dip direction) and
excavation direction, the maximum geometrically formable blocks between the rock

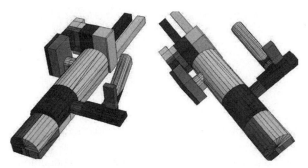

Figure 5.   Three dimensional model, 3-DEC, 86th Street Station, geometry

bolts are calculated. The rock wedges are scaled in terms of excavation area, no larger than the bolt spacing of the surrounding rock support. Using gravity loading and hydrostatic water pressure, the factor of safety (FS_{min}=1.5) is then calculated for the local stability of the rock blocks between the rock bolts.

#7 Shotcrete Stress Evaluation via STAAD

After the global and local stability of the rock mass is achieved, the internal forces of the shotcrete shell are checked using a beam spring model typically using the STAAD program. Based on the IGSR chart only Type IV and Type V of the initial ground support requires a structural fiber reinforced shotcrete shell to support the surrounding rock arch.

#8 Evaluate the Stability of Cavern Penetrations via a 3DEC—3D Model

During PE the global stability of the cavern section was checked using empirical models and plain strain discontinum 2D UDEC models. The global stability in areas where the cavern cross sections were penetrated by entrances, ventilation structures, and auxiliary buildings have been calculated using a reduced Q-value. Barton 1974 and 2003, recommend that due to changes in the in-situ stress state (transfer of membrane stresses) around a tunnel junction the calculated Q-values should be divided by a factor of 3 to account for this stress transfer. Using this method, initial ground support in the junction and penetration areas have been designed using Q-values only during PE to evaluate an indicative initial ground support for cost estimate reasons.

During FE a more sophisticated analysis was chosen to determine a more accurate ground support system in the areas of the penetration. The main cavern cross section is designed using the seven previous steps with a 2D UDEC model. This model incorporates various excavation sequences, overburden and building loads, etc. to evaluate the global stability of the main cavern section.

For the areas of penetrations and junctions a three dimensional numerical model 3-DEC has been applied to evaluate the effect of the stress change in the initial ground support of the main cavern station and to determine a detailed updated ground support in the penetration areas.

The 3-DEC calculation is based on the same discontiuum algorithm as 2D-UDEC code, but in all three dimensions (Figure 5). It is possible with the computer technology today that the entire station can be modeled in three dimensional space, including detailed modeling of joint systems, geometries and excavation sequences, and ground support. To reduce the calculation time and ease the interpretation of the results the following simplifications have been made to the 3-DEC model:

- Adjacent cut and cover boxes and shafts will be excavated instantaneously and the necessary ground support will be installed to achieve stability.
- Individual stand alone 3D-penetration models used rather than modeling of the entire station.
- Type IV support for the penetrations is assumed for the calculations.
- The jointing system in the entire model is modeled as joint clusters.
- In rock pillars and in close proximity to the penetration the joint clusters are replaced with discrete joint sets and shear zones if encountered.
- The excavation sequence of the main cavern section is not simulated. The top heading is modeled as excavated first in an instantaneous excavation. The same initial ground support as designed with the 2D plain strain models is applied prior to the excavation of the penetration.
- After excavation of the top heading in the main cavern, the penetration top headings are excavated sequentially and initial ground support is applied. Bolt and shotcrete forces as well as displacements in the initial ground support of the main cavern are monitored, and if necessary additional ground support is installed in the main cavern section to ensure global stability of the station cavern.
- When the top headings of the penetrations are excavated the main cavern bench and invert is excavated instantaneously, and the ground support is installed.
- The penetration bench and inverts are modeled in sequence, and the necessary ground support is applied.

CONCLUSION

The approach to cavern design for the large shallow caverns of the SAS has developed during the design phase into a robust procedure consisting of the best-available design tools. This procedure has been employed to verify the designed ground support system, incorporating cost savings opportunities into the IGSR chart where possible. Numerical modeling via 3-DEC has been added during FE to verify ground support systems around penetrations. In this manner, the high risk arising from unknown conditions that could have severe impacts have been met with a comprehensive analysis.

Several advantages are offered by this approach. The design can be implemented step by step through out the design stages of the project. Designed support can be modified and support quantities easily recalculated throughout the project. Multiple approaches are employed to cover the bases where geology may differ from what is anticipated. Finally, it provides for some consistency with and between large scale projects.

REFERENCES

Caiden, D., Sankar, J., Parikh, A., and Redmond, R. 2006. Rock Tunnels for the Second Avenue Subway. North American Tunneling Conference (NAT), Chicago, IL.

Cording, Edward J. 1984. State of the art: rock tunneling. Tunneling in Soil and Rock.

Desai, D., Naik, M., Rossler K. and Stone, C. 2005. New York Subway Caverns and Crossovers—A tale of trials and tribulations. RETC, Seattle, WA.

Grimstad, Eystein, et al. 2003. Q-System advance for sprayed lining. Tunnels and Tunneling International. 1/44.

Parikh, A., Fosbrook, G. and Phillips, D. 2005. Second Avenue Subway—Tunneling Beneath Manhattan. RETC, Seattle, WA.

ACKNOWLEDGMENTS

The authors wish to thank Jeff Fosbrook, Project Manager of DMJM+HARRIS•Arup Joint Venture, and Anil Parikh, Program Manager, of NYCT for allowing the publishing of project information and data. They would also like to thank Jessica Moeller for proof-reading and Dave Caiden for reviewing the paper.

RISK MANAGEMENT IN ACTION—
CONTROLLING DIFFICULT GROUND BY INNOVATION

A. Moergeli

Moergeli + Moergeli Consulting Engineering

ABSTRACT

The Swiss are building the world's longest tunnel, the 57-Kilometer (35-mile) long Gotthard Base Tunnel. Lot 252, Tunnel Amsteg: Two Hard Rock Tunnel Boring Machines (TBM) crossed hydro-thermally decomposed granite. While the East TBM advanced slowly but steadily, the West one got stuck for five months and had to be recovered by special means and methods. Both TBMs have finished their more than 11-Kilometer (7 miles) long drives successfully. Lot 360, Tunnel Sedrun: For the first time in tunneling, roadway support machines are used through most challenging, squeezing geology. They constantly advance in full face mode for about one Tunnelmeter/24 hours.

WHAT IS DIFFICULT GROUND?

There are many known definitions about what difficult ground is or may be. For the purpose of this paper (and identical in our contribution to the 2005 RETC) difficult ground is defined as ground—known or unknown in advance—that requires a major alteration of the general driving method.

GOTTHARD BASE TUNNEL

General Project Overview

The Gotthard Base Tunnel (GBT) has been subject of numerous publications. For updated information please feel free to check the owner's website http://www.alptransit.ch/pages/e/index.php or have a closer look at the author's publications http://www.moergeli.com/dldocuebersichte.htm—thank you.

For readers not (yet) familiar with project details and for an easier reading please find a brief overview in Table 1.

Lot 360, Tunnel Sedrun

Sedrun Overview. For readers not (yet) familiar with project details and for an easier reading please find a brief overview in Table 2.

How Do You Cross Difficult Ground Productively Without Compromising Safety? The Tavetscher Sub-Massif represents the geologically most challenging section of the GBT. Typically alternating sequences of cataclastic phyllite, schist and gneiss are sandwiched between more competent to intact schists and gneisses. The combination of weak ground with the estimated stress level results in two primary risks: large (homogeneous or anisotropic) deformations and face instabilities, in addition to high pore water pressure. This results—at an overburden of about 800 m—in squeezing rock.

Table 1.   Selected project data of Gotthard Base Tunnel (GBT, 12/2006)

Owner	AlpTransit Gotthard AG (ATG, http://www.alptransit.ch)
Purpose	New highspeed railway
Location	In the heart of Switzerland, Cantons of Uri (UR) & Tessin (TI)
Total tunnel length	56.8 km (ca. 35.3 statute miles)
5 Main tunneling lots	Lot 151, Tunnel Erstfeld Length 7.4 km Award pending
	Lot 252, Tunnel Amsteg Length 11.4 km Under construction
	Lot 360, Tunnel Sedrun Length 6.8 km Under construction
	Lot 452, Tunnel Faido Length 14.6 km Under construction
	Lot 554, Tunnel Bodio Length 16.6 km Under construction
Construction method	Mainly TBM (exception: Drill & Blast @ Lot 360, Tunnel Sedrun)
Forecast final costs	Ca. CHF 7.7 billion* (ca. $6.4 billion[†])
Forecast construction time	1996—ca. 2016[‡]
Project status (12/2006)	Ca. 70% of excavation complete 90% of he inner lining complete in the Bodio section
Author's contracts	Contractor's Safety & Health Engineer for all tunneling lots under construction so far

* Excl. Swiss VAT (Value Added Tax).
† US$1 ≈ CHF 1.20 (Swiss Franc, 12/2006).
‡ According to best current knowledge.

Table 2.   Selected additional project data of Lot 360, Tunnel Sedrun

Location	Canton of Grisons (GR)
Total tunnel system	Ca. 20 km (ca. 12.4 statute miles)
Cross section	Ca. 69 m^2—135 m^2 (in squeezing ground)
Construction method	Drill & Blast at up to six faces simultaneously
Contractor	Joint Venture TRANSCO Sedrun (TRANSCO, http://www.transco-sedrun.ch)
Contract value	CHF 1.165 billion* ($0.971 billion[†])
Geotechnical boring	5 Bore holes @ total of 5,602 m for Tavetscher Sub-Massif (Keller 2004)
Forecast contract time	1996—ca. 2011[‡]
Author's contract	Safety & Health Engineer (2nd Level support for Site Safety Officer)

* Excl. Swiss VAT (Value Added Tax).
† US$1 ≈ CHF 1.20 (Swiss Franc, 12/2006).
‡ According to best current knowledge.

Early on, the owner took some strategic decisions with regard to a pro-active Risk Management

- Full-face excavation and full circle profile
- Combine the yielding principle with defined rigidity according to the ground's characteristic pressure-support-behavior

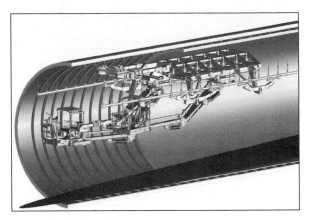

Figure 1.   RSM, front part (Courtesy of Rowa)

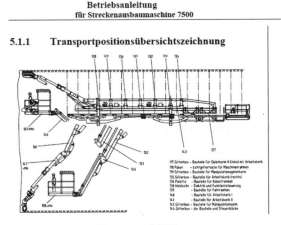

Figure 2.   RSM, excerpt from manual (Courtesy of GTA)

- Use steel ring arch lining immediately behind the face (up to 16 elements, set up to a minimum spacing of only 0.33 m, TH profile, yielding from the largest possible diameter of about 13 m to 11 m in the event of tunnel convergence)
- Use shotcrete as well as radial and long face anchors
- Systematically monitor all deformations in real-time.

By Using a Roadway Support Machine. Due to the large cross-section and varying, but huge quantities of excavation support—steel arches, anchors and wet shotcrete—a mechanization system is key for structural and process safety and contractor's performance. The basic concept, proposed by the owner and its consulting engineers, engineered by Rowa Tunneling Logistics AG (http://www.rowa-ag.ch), is a world's first (Figure 1).

Although the roadway support machine (RSM) concept has been known in mining for a long time, never before it has been used in a scale up to the Sedrun dimensions.

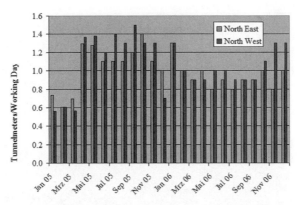

Figure 3.   Performance of special drives over 24 months, data courtesy of TRANSCO

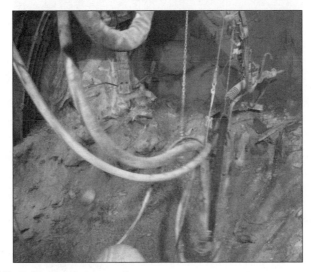

Figure 4.   RSM North East, pocket excavation

In cooperation with competent mining engineers from GTA Maschinensysteme GmbH (Figure 2; http://www.gta-maschinensysteme.de), Rowa has further developed the RMS for the Sedrun tunnel heading for

- Working on two levels (more or less) independently of each other,
- Steel arch installation in a safe and efficient manner,
- Increasing performance, productivity and safety in the same process.

Performance. Depending on geology (as always in a tunnel) and a lot of other factors average daily performance of both special drives approximates around one Tunnelmeter/working day (TM/d). (Figure 3)

Lessons Learned. It is always too early to debrief a tunnel that has not yet been broken through (Figures 4 and 5).

Figure 5.   RSM North West, two working levels

However, after two years into the drives it may be time for a—of course only preliminary—look back.

- The predicted ground was met and behaved as expected, no big surprises so far.
- Radial deformations of up to 0.7 meters occurred, faster than expected and asymmetric.
- The ground was less squeezing and the face less stable than expected.
- The overall system works perfectly well in a unique combination of large tunnel dimensions and squeezing ground that might not have been considered as feasible for tunneling just a few years ago.
- Adapting the original standard operating procedure to shotcreting the steel arches in the invert immediately at the face, helped in speeding up cycle times.
- The learning curve was there, and it may have taken longer than expected. However, it will always take extra time to familiarize a new crew while going through the commissioning of a world's true first innovation and taking off into difficult ground immediately. And today, it looks like everything that could be optimized has been done so far.
- Both special drives are working constantly in a very stable, controlled process mode now. A remarkable success story so far!

Current Status and Outlook. At the end of 2006 (the time of submittal of this paper) the drives were 391 TM in the East Tunnel and 352 TM in the West Tunnel short of breakthrough to the section boundary of Lot 252, Tunnel Amsteg.

Lot 252, Tunnel Amsteg

Amsteg Overview. For readers not (yet) familiar with project details and for an easier reading please find a brief overview (Table 3).

Table 3.   Selected additional project data of Lot 252, Tunnel Amsteg

Location	Canton of Uri (UR)
Total tunnel system	Ca. 26 km (ca. 16.2 statute miles)
Cross section	Ca. 71.6 m^2 (TBM excavation diameter of 9.55 m)
Construction method	2 Tunnel Boring Machines (TBMs) with Back-up trailers
Contractor	Joint Venture Gotthard Nord (AGN, http://www.agn-amsteg.ch)
Contract value	CHF 0.627 billion* ($0.523 billion[†])
Geotechnical boring	2 Bore holes @ 340 m + data from a nearby head race tunnel
Forecast contract time	2002—ca. 2009[‡]
Author's contract	Safety & Health Engineer (2nd Level support for Site Safety Officer)

* Excl. Swiss VAT (Value Added Tax).
† US$1 ≈ CHF 1.20 (Swiss Franc, 12/2006).
‡ According to best current knowledge.

Table 4.   Chain of events in the East Tunnel—brief summary

2005	The TBM ("Gabi I") in the East Tunnel…
June 5	reaches the predicted 11 m long fault zone A13. Advance rate slows down.
June 6–27	crosses hydro-thermally decomposed granite adjacent to the A13 fault zone. Low performance, challenging rock support work immediately behind the cutter head. Complete stop for one day (June 7). Water was no factor.
June 28	find its way back to "normal" performance (12.10 TM/day)

The East TBM Crosses Difficult Ground. As presented by the author at the 2005 RETC, both TBMs have already successfully crossed the expected Intschi fault zone and were well on their way, until… (Table 4 and Figure 6).

The West TBM Crosses Difficult Ground. (Table 5 and Figure 7)

Recovery of the West TBM (Gabi II). Expect more in-depth information by our presentation at the 2007 RETC. (Table 6 and Figures 8 and 9)

Lessons Learned. Minor cause—huge impact!

Periodic seismic investigation was a SOP from the leading East Tunnel.

It showed the A13 fault zone.

However, no (warning) signs about the adjacent hydro-thermally decomposed granite. Therefore, there was no reason for additional probing ahead at this time. With hindsight, probing ahead would have had a fair chance of detecting the water in the West Tunnel. However, even more than 40 borings for injections, probing and drainage could not drain the water adequately. Hard to tolerate, but the residual risk of an open hard rock TBM to become blocked by inflowing loose material cannot be controlled by efficient and cost-effective measures well in advance up to this day. Tunneling is and remains a risky business—and therefore definitely not everyone's business. Fortunately, no one was seriously injured this time.

Current Status and Outlook. At the end of 2006 the West TBM is being dismantled, while injections are applied in the East Tunnel.

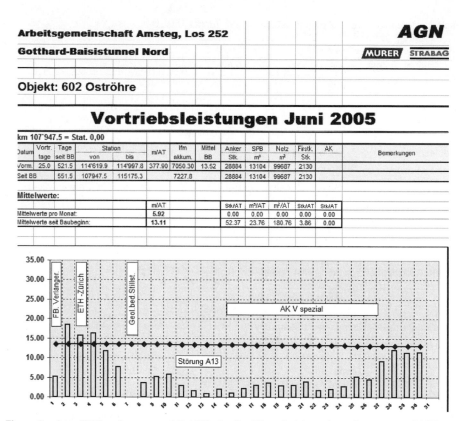

Figure 6. June 2005 performance of Gabi I in the East Tunnel, data and graph courtesy of AGN

Table 5. Chain of events in the West Tunnel—brief summary

2005	The TBM ("Gabi II") in the West Tunnel…
June 13	reaches the predicted fault zone A13. No measurable impact.
June 14–17	crosses hydro-thermally decomposed granite adjacent to the A13 fault zone. More compact than in the East Tunnel, a better TBM advance rate results. Water was no factor so far.
June 18	is hit by a water inrush (of about 2–3 lt/sec) that flushes loose material into the cutter head during standard maintenance procedures. As a result the TBM is blocked and cannot be moved again, in no direction

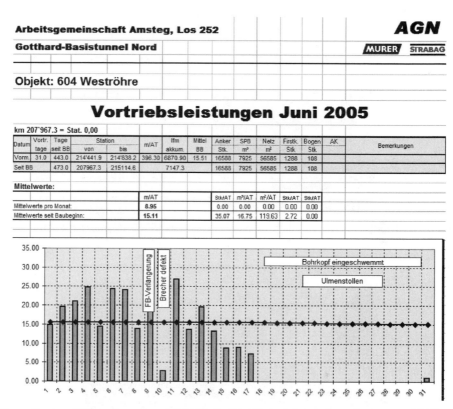

Figure 7.   June 2005 performance of Gabi II in the West Tunnel, data and graph courtesy of AGN

Table 6.   Efforts to recover Gabi II (the West TBM)—brief summary

Project Management	Task Force by the owner, engineer, construction management and the contractor
Probing ahead	More than 30 Self Bore Anchors from above the blocked TBM 11 cored @ 26–45 m long borings from peripheral niches
Injection tunnel	Circular profile of 18 m^2, to a distance of about 15 m to the blocked TBM
Injections	More than 120 borings, more than 2,800 boring meters More than 50 to of HydroBloc Polygel 530 injections to seal the cutter head More than 110 to of Duroflow R cement compensation grouting
Access tunnel	35 m^2, in solid rock from the East Tunnel 2 barrel vaults to cross the loose ground and reach the blocked cutter head
Time	June 18–November 23, 2005: ca. 5 months
Costs	Ca. CHF 10 Mio
Fatalities	Fortunately none

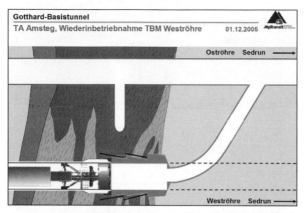

Figure 8. Overview of means and methods to recover the blocked TBM (Courtesy of ATG)

Figure 9. And how this looks in situ

ACKNOWLEDGMENTS

The author thanks and acknowledges the very competent support and kind provision of plans, schemes and pictures.

Last but not least, my thanks go to Mrs. S. Tschupp for her always very competent support in improving my use of the English language.

Without their big help this document would not have been possible.

The biggest thanks go to all crews on site, safely coping with the unforeseeable as their daily routine. Every day they move into places where no human being has ever been before. Always just one small step for a man, but a giant leap for mankind (Figures 16 and 17).

The author's apologies go to all readers for any inconvenience dealing with small print, compressed tables and pictures.

An update of our paper will be offered through our presentation at the Rapid Excavation and Tunneling 2007 Conference (RETC), June 10–13, Toronto, Ontario, Canada.

Figure 10.   AlpTransit Gotthard AG
(http://www.alptransit.ch), the Owner

FIGURE 12. Joint Venture AMSTEG,
Los 252, Gotthard-Basistunnel Nord
(http://www.agn-amsteg.ch)

Figure 14.   Rowa Tunneling Logistics AG,
Back-up Manufacturer (http://www.rowa-ag.ch)

Figure 11.   Joint Venture TRANSCO Sedrun
(http://www.transco-sedrun.ch)

FIGURE 13. Herrenknecht AG, TBM Manufacturer
(http://www.herrenknecht.de)

Figure 15.   GTA Maschinensysteme GmbH,
Back-up Manufacturer
(http://www.gta-maschinensysteme.de)

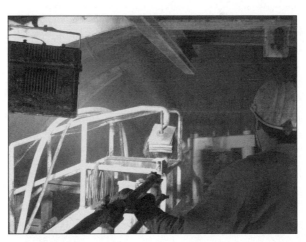

Figure 16.   Amsteg West Tunnel

Some interesting videos are expecting you. Hope to see you there!

Both our paper and presentation will be available for download at http://www.moergeli.com/dldocuebersichte.htm shortly after the conference.

All abbreviations/terms can be used in singular or plural, with or without capital letters and with or without a period when abbreviated. All abbreviations/terms are explained in the text upon their first use.

Figure 17.   Sedrun North West

REFERENCES

Gruber, L.R. 2006, Lehner, J. 2006, Scherrer, R. 2006, ARGE TRANSCO Sedrun (TRANSCO), private & unpublished.

Kapeller, E. 2006, Sempelmann, F. 2006, Arbeitsgemeinschaft Gotthard Nord (AGN), private & unpublished.

Theiler, P., 2005, Geologie und Geotechnik der Basistunnels am Gotthard und am Lötschberg, Symposium Geologie Alptransit, 26.—28.09.2005, vdf, Publisher Simon Löw.

Wildbolz, A. 2005, AlpTransit Gotthard AG (ATG), sia—FGU, Swiss Tunnel Congress 2005.

FOR FURTHER INFORMATION

AlpTransit Gotthard AG (ATG), Zentralstrasse 5, CH-6003 Luzern/Switzerland, http://www.alptransit.ch.

Herrenknecht AG Tunneling Systems, Schlehenweg 2, D-77963 Schwanau/Germany, http://www.herrenknecht.de.

Joint Venture Amsteg, Lot 252, Gotthard Base Tunnel North (AGN), Grund, CH-6474 Amsteg/Switzerland, http://www.agn-amsteg.ch.

Joint Venture TRANSCO Sedrun, P.O. Box 67, CH-7188 Sedrun/Switzerland, http://www.transco-Sedrun.ch.

Rowa Tunneling Logistics AG, Leuholz 15, CH-8855 Wangen SZ/Switzerland, http://www.rowa-ag.ch.

moergeli + moergeli consulting engineering, Rosengartenstrasse 28, CH-8716 Schmerikon/Switzerland, http://www.moergeli.com.

GLOSSARY

Abbreviation/ Acronym	Definition/Explanation
#	Number, No
$	US dollar, $, USD
ā	about, circa
(...)	option or issue date
"..."	Quotation (within quotation marks)
/	or
[...]	dimension and/or directions/instructions
+	and
=>	[goal/target]
abbr	abbreviation
ARGE	joint venture (German: Arbeitsgemeinschaft)
ca	circa, about
CH	Switzerland
CHF	Swiss Franc, always without VAT if no other mention
d	day
D & B	Drill & Blast
Doc(u)	document
e.g.	for example (Latin: example gratia)
engl	English
excl.	excluding (without)
FGU	German: Fach-Gruppe Untertagebau), section of sia
GBT	Gotthard Base Tunnel
incl.	including (with)
km	Kilometer (1 Statute Mile = 1.609 km)
lt	liter
m	meter
m(+)m	moergeli + moergeli consulting engineering
Mio	Million
sec	second
sia	Swiss society of engineers and architects (German: Schweizerischer Ingenieur- und Architekten-Verein)
SOP	Standard Operating Procedure
TBM	Tunnel Boring Machine
TM	Tunnelmeter
to	tonne
USD	US Dollar
VAT	Value Added Tax

TUNNELING ALTERNATIVES FOR SUBWAY CONNECTION IN DOWNTOWN CHICAGO

V. Nasri

STV Incorporated

A. Ayoubian

STV Incorporated

ABSTRACT

This Tunnel Connection project will include construction of two short tunnels to connect the existing subway tunnels (Red and Blue lines) in the northern loop area of downtown Chicago through a new station. The Southeast Tunnel will connect the Red Line to the new station below the intersection of State and Washington Streets and the Northwest Tunnel will connect the Blue Line to the new station below the intersection of Dearborn and Randolph Streets. The connection tunnels are located in a very dense urban area of downtown Chicago and under several important historical buildings. Both tunnels are curved at the intersections and located at a depth of about 7.5 m below ground surface in saturated soft clay with an average undrained shear strength of approximately 25 kPa. The design calls for minimizing surface disruption and ground displacement while maintaining normal operation of the existing subways during the construction of the connections tunnels. This paper presents mining and cut and cover tunneling alternatives for the project.

INTRODUCTION

To create an express train service linking O'Hare International and Midway Airports, the Chicago Transit Authority (CTA) decided to build two short tunnels to connect the Blue and Red transit lines in Chicago. The connection will be at Block 37, located between State and Dearborn Streets, and will link the two subways through a new station (Figure 1). The new Block 37 station structure will be constructed as part of the Block 37 development project using slurry wall construction and a top down technique. The new connection tunnels will be constructed using mining or cut and cover methods. Mining operations will originate from access shafts to be built within Block 37. The project will also include modifications to the continuous platforms at the existing State Street and Dearborn Street subway tunnels to accommodate the new connections. This project will be accomplished under a complex and tight schedule and must be completed with minimal disruptions to the non-stop subway service. The anticipated subsurface conditions and the stringent project settlement criteria create difficult conditions for the design and construction of the tunnels.

The soft deposits at the project site with their low shear strength can experience large deformations during tunneling. For the mining alternative, application of a pre-support interlocking pipe arch method and fiberglass face reinforcement technique ahead of the advancing tunnel face are proposed to limit the ground deformation.

Continuous and swift excavation and support installation combined with an appropriate staging and support system is necessary to minimize short and long-term deformations. The tunnels will be constructed using fiber reinforced shotcrete and

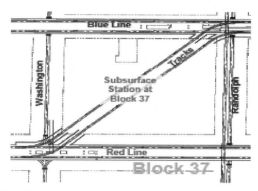

Figure 1. Plan view of project site

lattice girders as an initial support lining. The final lining will be installed using a composite system of steel fiber reinforced shotcrete and lattice girders.

For the connection of the new short tunnels to the existing subway tunnels, a new larger span tunnel will be constructed using the interlocking pipe arch technique, which will be parallel to the existing Tunnel. The new expanded tunnels will transfer the imposed loads to the column line of the existing tunnels. To facilitate the construction of the new expanded tunnels, two cross adits will be constructed perpendicular to the existing tunnels. These cross adits will originate from within Block 37 and extend below State and Dearborn Streets and above the existing Track B and Track C. This construction approach allows all tunneling work, including excavation, temporary and final lining work, to take place outside of the existing tunnel envelopes. Revenue operation can continue practically uninterrupted until the final breakthrough of the existing tunnels is required.

Structural modification and reinforcement of the existing tunnels must take place prior to the excavation of the new tunnels. These structural modifications and reinforcement consist of platform demolitions, erection of new tunnel roof structural support systems and installation of permanent tie down anchors. Removal of some of the existing columns and longitudinal beams will take place after construction of the new tunnels and breakthrough of the sidewall of the existing tunnels. Most of these construction activities will be performed using enclosures separating the work areas from the revenue operations.

EXISTING SUBWAY TUNNELS

The initial subway system of Chicago included two routes, State Street Route and Dearborn Street Route. The tunnel portion consisted of 7.7 miles of double tube and station sections. All the tunnels were entirely constructed in Chicago clay. The bottom of the tunnels is located at a depth of about 15 m below the street surface.

For constructing the tunnels two different methods were used: the liner plate method and the shield method. In the liner plate method the pressure acting on the roof of the tunnel is carried by steel ribs, which transfer the load onto the footings. In order to use this method the clay located beneath the bottom of the tunnel must be stiff enough to withstand the downward pressure exerted by the footings. This condition was satisfied north of the river, where liner plate method was used. On the other hand, the clay at the south of the river was too soft to support the heavily loaded footings.

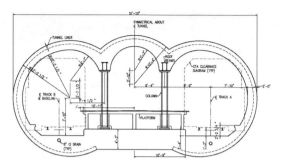

a) Station cross section (triple arch)

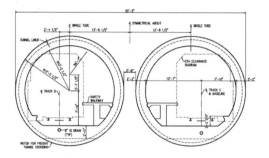

b) Tunnel cross section (circular arch)

c)   Existing subway station (triple arch)

Figure 2.   Existing tunnels

Inadequate support of the footings of the arch ribs in a liner plate tunnel could cause not only tunnel roof failure but also collapse of adjacent buildings. Therefore, in the loop district, shield method was employed in spite of the higher construction cost. Figure 2 shows cross sections of the existing station and tunnels built in 1940s. The outside diameter of the individual tubes is about 7.5 m and the minimum horizontal distance between them at the mid-height of the tunnels is about 2'-8" (Figure 2).

SUBSURFACE INVESTIGATIONS

The subsurface investigation for this project consisted of performing geotechnical borings with standard penetration tests and cone penetration tests with shear wave velocity measurements. Laboratory tests were also conducted on representative and undisturbed samples of soils, which consisted of index tests, unconsolidated-undrained triaxial tests, consolidated-undrained triaxial compression and extension tests in conjunction with laboratory shear wave velocity measurements.

The subsurface soils consist of an upper layer of miscellaneous sandy fill under-lain by layers of glacial deposits above the limestone bedrock. The glacial deposits are Blodgett, Deerfield, Park Ridge, Tinely and Valparaiso. The three layers of Blodgett, Deerfield, and Park Ridge are predominantly clay ranging in strength from soft to stiff. The top of the clay is located at a depth of about 4.5 m below the street surface, approximately at the level of the Lake Michigan. The existing tunnels and the proposed construction are predominantly in Blodgett layer.

Blodgett stratum is characterized as very soft to soft silty clay with trace of fine sand and gravel and shale and mostly shows a variable distribution of water content. Red and Blue line subways were constructed in this layer and the proposed connection at Block 37 is expected to be constructed entirely in this stratum. The upper part of the Blodgett layer to a depth of about 1.5 m is desiccated clay crust and much stiffer than the underlying soft clay. However, the clay crust is very variable in thickness and strength and there-fore, the entire stratum was considered as Blodgett layer with softer clay characteristics. The groundwater table is approximately at the top of the Blodgett stratum.

DESIGN CONSIDERATIONS FOR MINING ALTERNATIVE

Minimum surface settlement, minimum disturbance to the street traffic and mini-mum impact on the operation of the existing tracks were the main requirements for the design. The critical design criteria consisted of strict limitation of surface subsidence due to tunneling at the location of existing building foundations. Based on these criteria the angular distortion at these locations cannot exceed 1/1000, and no surface point shall subside more than 12 mm below its pre-construction location. Presence of important high-rise historical buildings supported on shallow foundations in the zone of mining influence dictated such strict settlement criteria.

The state of the art geotechnical testing and interpretation program has been carried out by Northwestern University to define the behavior of different Chicago clays at the project site. The type of ground in which the new tunnels will be constructed is classified as soft clay with very low strength characteristics. Different ground improvement techniques and various pre-support methods were considered in initial phases of design. Ground improvement approaches such as jet grouting and ground freezing were excluded due to their unsatisfactory performance in similar geological settings and also their incompatibility with the densely distributed utilities and other site constraints.

Use of forepoling and interlocking pipe arch techniques as pre-support methods were studied in detail. The interlocking pipe arch concept was selected for the final design to achieve the above described surface settlement criteria. An interlocking pipe installation technology was recommended to minimize installation induced soil distur-bance, excess pore water pressure generation and surface settlement. This technique combines drilling with continuous controlled jacking and auguring.

To limit the face extrusion and its resulting ground deformation and to increase the excavation stand up time, fiberglass face reinforcement bars were designed as an ele-ment of the pre-support system. The reinforced core in the excavation face reduces

ground deformations including extrusion, convergence and pre-convergence components (Lunardi, 2000). The design provides either interlocking pipe support or fiberglass reinforcement in all the exposed excavation surfaces including crown, sidewalls,
invert and face of all excavations.

The sequential excavation method has been selected as a means of progressively
excavating and supporting the ground. The longitudinal tunnels and cross adits are to
be excavated with two headings in accordance with the specified sequences designed
to stabilize the ground in limited excavation rounds. The excavation sequences of
specified round lengths consist of application of steel fiber-reinforced flashcrete to
newly exposed surfaces, installation of lattice girders at predetermined intervals, and
application of fiber reinforced shotcrete support with variable thicknesses depending
upon final opening dimensions.

Due to its particular geometry, the State Street tunnel BD is to be excavated full
face. The proposed steel fiber reinforced shotcrete of initial and final liners are combined and the lattice girder is not used as a support element. For shorter connections
in State and Dearborn streets a horizontal secant concrete cylinders concept is used
as pre-support system. The design of State Street tunnel BD and the longitudinal tunnel were adapted to accommodate the presence of the paid transfer tunnel and the
abandoned freight tunnel. The design of the longitudinal tunnel at Dearborn Street was
modified to account for the presence of the active and abandoned freight tunnels in
Randolph Street.

The final tunnel lining will be designed to carry the combination of full overburden
and hydrostatic groundwater loads as well as surcharge loads. Due to the varying
tunnel shapes and short tunnel lengths, it is proposed to use high quality shotcrete for
the permanent tunnel support lining.

Because of very low permeability of the ground and small water head, relatively
thick shotcrete layers of initial and final liners with appropriate mix design are
considered sufficient to provide acceptable watertightness for the new tunnels. Fuko
grouting hoses at construction joints and intersections can be used for repairing the
possible leaks.

Monitoring of surface movements will be conducted using surface settlement
points installed in a series of arrays perpendicular to tunnel centerline across the
expected settlement trough. In addition, inclinometers, extensometers and piezometers will be installed to detect horizontal and vertical ground movements and ground
water level. Total station technology will be used to determine buildings deformation in
the zone of influence of tunneling operation. Monitoring points and tiltmeters will be
installed to observe any building movement. Tunnel monitoring will primarily comprise
of convergence monitoring of the new and existing tunnels and deformation and stress
measurements of liner using strain gauges and concrete and soil pressure cells.

CONSTRUCTION ELEMENTS OF MINING ALTERNATIVE

The construction elements and the corresponding sequences are as follows:

- Building a stairway tunnel under the center arch of the existing tunnel (Figure 3),
- Reinforcement of existing tunnels by erecting new columns, beams and tempo-
  rary columns (shoring) and installing tie down anchors for both existing Red and
  Blue line tunnels,
- Installation of 0.6 m diameter interlocked steel pipe roof system and filling them
  with cement grout (Figure 4),

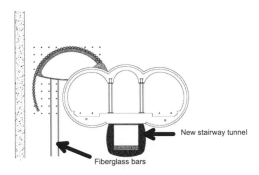

a)   Top heading excavation

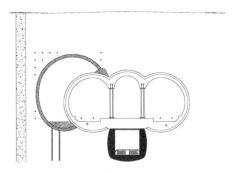

b) Completed initial liner

Figure 3.   New expanded tunnel (during construction)

- Construction of two cross adits under State and Dearborn Streets based on prescribed excavation sequence and support of face, side walls and invert by steel fiber reinforced shotcrete and fiberglass bars (Figure 4),
- Construction of the new expanded tunnel based on prescribed excavation sequence and installation of initial steel fiber reinforced shotcrete and lattice girder support (Figure 3),
- Completing final support system consisting of a combination of cast in place reinforced concrete framing and steel fiber reinforced shotcrete and lattice girder allowing for opening arches at breakthrough points,
- Construction of Tunnel BD at State Street side using interlocking steel pipe arch and fiberglass face reinforcement techniques based on prescribed excavation sequence together with steel fiber reinforced shotcrete as the final liner,
- Connection of track BD at Dearborn Street to the new expanded tunnel at the cross adit location using a short reinforced concrete box structure,
- Connection of Track AC at State Street and Dearborn Street with two new expanded tunnels at intersection points using horizontal secant concrete cylinders pre-support system and steel fiber reinforced shotcrete and lattice girder as the final liner,

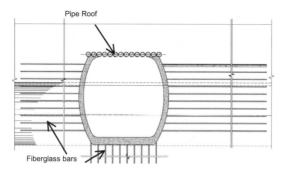

a)   Longitudinal section

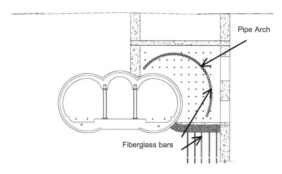

b)   Transverse section

Figure 4. Cross adit sections

- Backfilling both cross adits based on prescribed backfilling schedule,
- Breakthrough sidewalls of existing tunnel (Figure 5),
- Transferring the loads of removed beams and columns of existing tunnels to new framing systems by means of jacking.

The excavation of both expanded new tunnels will be performed in two stages of top heading and bench. Maximum longitudinal distance between faces of top heading and bench will be 9 m. The more desirable full face excavation cannot be applied due to constructability constraints, which would require hand mining or the use of a movable platform for mechanical excavation. The Tunnel BD at State Street side will be excavated full face in one stage.

NUMERICAL ANALYSES AND MODELING

The three layers of Blodgett, Deerfield, and Park Ridge along with the miscellaneous fill were modeled in the finite element analysis using PLAXIS software (Brinkgreve 2002). The Hardening Soil (H-S) model (Schanz, 1999) was used for modeling Blodgett, Deerfield and Park Ridge strata. H-S model is an advanced soil model with stress-dependant soil stiffness using the following three different input stiffnesses: the triaxial loading stiffness (E_{50}), the triaxial unloading-reloading stiffness (E_{ur}) and

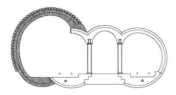

a)   Cross section at triple-arch tunnel

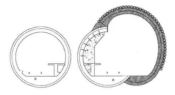

b) Cross section at circular-arch tunnel

Figure 5.   New expanded tunnel (after construction)

the oedometer loading stiffness (E_{oed}). The limiting states of stress are defined by the friction angle (ϕ), cohesion (c) and dilatancy angle (Ψ). For miscellaneous fill, Mohr-Coulomb soil model was used in the analysis. Material properties used for analysis of Block 37 tunnel connections were obtained from Geotechnical Baseline Report (Patrick Engineering, Inc., 2005) and hardening soil parameters developed for this project at Northwestern University (Finno, 2005).

The excavation and support of various components of the project such as the extended stairway tunnel, cross adits, expanded tunnels and tunnel BD have been modeled according to the construction stages of the existing and new tunnels using PLAXIS software. For example, the stages used to model the construction of expanded tunnels are as follows:

1. Initialization of stresses. In this stage the soil layers and ground water table are defined and vertical and horizontal stresses are generated before the construction of existing tunnel.

2. Construction of existing tunnels (initial liner). Construction of existing tunnels, which took place in 1940s, was modeled by creating their geometry as closely as possible and then modeling the excavation using undrained analysis. The initial liner consisting of a steel segmental liner was installed in this stage. This liner was modeled using a structural beam element. The beam element section properties were obtained from the existing drawings. To model the effect of compressed air used during the construction of these tunnels, the groundwater pressure inside the excavated tunnel was kept active, since the pressure of air was originally designed to compensate the existing groundwater pressure during the excavation of these tunnels. Moreover since the concrete final liner was installed at a considerable longitudinal distance from the face of excavation, 100% of excavation induced stresses are released at this stage. A uniform surcharge of 12.5 kPa was applied at ground surface at this stage representing the traffic live load. This surcharge was kept active for the rest of the analysis.

3. Construction of existing tunnels (final liner). In this stage, the concrete final liner for the existing tunnels was installed and the groundwater pressure inside the tunnels was deactivated to model the removal of compressed air.

4. Consolidation for 65 years. This stage was added to create the existing state of stress in soil layers, as closely as possible, before the new construction for Block 37.

5. Pre-construction support. Prior to new construction, the existing tunnels were supported by installation of tie-down rock anchors. The displacements from previous stage were set to zero at this stage, which marks the start of Block 37 construction. Undrained analysis was used for this stage and the following stages.

6. Construction of new arch (top heading 1). In this stage the top heading of the expanded tunnel was excavated after activating the longitudinal pipes of the pipe arch. It is assumed that 50% of the excavation induced initial stresses are released in this stage.

7. Construction of new arch (top heading 2). In this phase, top heading initial liner was activated (shotcrete and vertical fiberglass bars at invert) and 100% of the initial stresses at the excavation boundary were released.

8. Construction of new arch (bench 1). The bench of the expanded tunnel was excavated and 50% of the initial stresses were released.

9. Construction of new arch (bench 2). In this phase, bench initial liner was activated and 100% of the initial stresses at the excavation boundary were released.

10. Cutting the existing tunnel liner. At this stage final liner was installed and part of the existing tunnel liner was removed.

11. Consolidation for 100 years. A 100-year consolidation was performed to calculate long-term ground movements and stresses in surrounding soil and liner.

The tensile stresses in the existing tunnel liner were checked against the rupture modulus of concrete at every stage of construction. The analysis indicates that the surface settlement and angular distortion were within the acceptable limits.

CUT AND COVER DESIGN ALTERNATIVE

An alternative design concept for tunnel connections proposed by the contractor's team is being evaluated for construction and is based on cut-and-cover approach (Kiewit-Reyes, 2006). According to this design, two cut-and-cover excavations will be performed. One at the intersection of State Street and Washington Street for Blue line connection and another at the intersection of Dearborn Street and Randolph Street for Red line connection. The excavation at the intersection of State Street and Washington Street is about 52 m × 30 m in plan with a curved SPTC wall and the excavation at the intersection of Dearborn and Washington Street is approximately 48 m × 16 m in plan.

This proposed design uses a top-down construction with installation of retaining structures for ground support and placement of new roof and invert slabs for the new tunnels. Figure 6 presents cross sections of the existing triple arch and double tube tunnels in the proposed design. This design involves the following main features:

Management of Underground Utilities

The top 3 to 5 meters of the site is covered with a variety of utility lines. Test pitting will be performed to map the location of the utilities. When necessary, existing utility concrete vaults will be demolished and the utilities will be supported within temporary wood boxes and hung from the deck beams during the excavation. Once the excavation is backfilled, the boxed utilities will be restored to their original location.

Traffic Control and Lane Closure

The project site is located in a very active and busy part of Downtown Chicago with heavy pedestrian and vehicular traffic. Access to all businesses will remain available during the construction. Depending on the stage of construction, single lane, multi lane and full intersection closure will be used. Full intersection closure will be used at minimum and only on the weekends.

Installation of SPTC (Soldier Pile Tremie Concrete) Wall

The SPTC wall will be used to support the excavation on the side of the Block 37 slurry wall at locations where this slurry wall is not present.

Installation of Soldier Pile and Lagging Wall

Soldier pile and lagging wall will be constructed on the opposite side of the excavation across from SPTC wall or Block 37 slurry wall and also on both ends of the excavation. This wall will be on top of the existing tunnel along the length of the excavation. The wall will not carry any vertical load and its top will be just below the deck beams.

Installation of Braces

Lateral support of the excavation will be provided in three levels as follows:

- **Deck Beams.** Deck beams will be installed after a few feet of excavation at 1.8 m intervals. Excavation for deck beams and top level of walers requires full street closure and will be done during the weekend. Prior to installation of deck beams, a long strip footing will be placed parallel to the excavation and just next to the soldier pile and lagging wall. The strip footing will carry the loads transferred from the deck beams. Deck beams will rest on the strip footing on one side of the excavation and on SPTC wall or Block 37 slurry wall on the other side. Steel decking plates will be used to cover the excavation and transfer the traffic load to the deck beams. Steel decking plates are removable so workers can enter the excavation when necessary. In addition to carrying the traffic surcharge and transferring them to the strip footing and SPTC wall or Block 37 slurry wall, the prestressed decking beams serve as the first level of lateral support for the excavation.

- **Cross-lot Braces.** Cross-lot braces will be prestressed and installed every 3.6 meter as the second level of lateral support for the excavation.

- **Diaphragm Slab.** Diaphragm slab serves as the permanent roof for the new tunnel and also as the third level of lateral support for the excavation. It will be constructed when excavation reaches the top of the existing tunnels. The slab also will have upturned beams, which will span across the new roof of the tunnel to support the roof when some of the existing columns are removed for accommodating the new tracks. The diaphragm slab will carry the backfill overburden load above the slab. At the space between the double tube tunnels under Dearborn Street, mini-piles will be installed for vertical support of the diaphragm slab as shown in Figure 6.

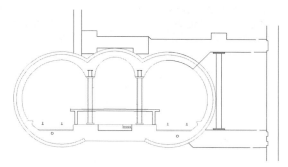

a) Cross-section of the exiting triple arch tunnels

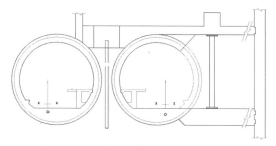

b) Cross-section of the exiting double tube tunnels

Figure 6.   Cut and cover tunneling alternative

Undermining the Diaphragm Slab

After diaphragm slab is installed across the entire excavation, it will be undermined to the invert slab elevation. Undermining of the diaphragm slab will be accomplished by leaving an opening in the diaphragm slab such that the equipment and workers can excavate under it and remove the soil.

Installation of invert slab and new structural elements. After the excavation reaches the proposed elevation for bottom of the invert slab, the invert slab will be poured. Once the invert slab is in place and cured, new structural elements such as columns, tie-down anchors and mini-piles will be installed to support the new tunnel roof slab.

Column Removal

Installation of the new columns will be followed by removal of some of the existing columns, which are in the way of the new tracks.

Connection to Existing Tunnels

Connection of the new tunnel to the existing tunnels will be completed by cutting the existing concrete liner at locations where the new tracks will merge with the existing tracks as shown in Figure 6.

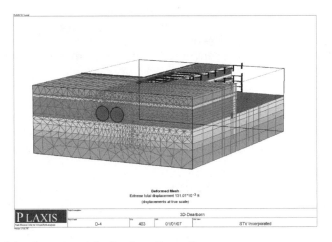

Figure 7.   3D finite element mesh for Dearborn Excavation

Backfilling

Light flowable concrete will be used to backfill above the diaphragm slab to the bottom of the lowest utilities. The remaining space will be backfilled with granular fill. During backfilling cross-lot braces and decking beams will be removed.

Instrumentation

Instrumentation will be one of the key components of this project. The major instruments that will be used for monitoring the response of the ground and existing tunnels and other structures before and during construction include:

- Inclinometers, for measuring lateral ground movements outside of the excavation
- Extensometers, for measuring vertical ground movements inside and outside of the excavation
- Piezometers, for measuring pore pressures within ground
- Total station devices, for measuring movements of ground surface and existing structures during construction
- Strain gauges, for measuring deformation of concrete and steel structures
- Tiltmeters, for monitoring tilt of retaining walls and buildings

Design of the new tunnel connections include geotechnical and structural analyses for excavation support system (soldier pile and lagging wall and SPTC wall), diaphragm slab, invert slab, removal and additions of structural elements, overall tunnel stability and excavation-induced stresses in the existing tunnel.

Geotechnical Analysis

As part of the efforts in analyzing the new tunnel connections, finite element analyses are carried out in 2D and 3D spaces using PLAXIS and PLAXIS TUNNEL 3D, respectively. The finite element analyses are undertaken to predict the response of the existing tunnels to the proposed construction and estimate the magnitude of the ground movements.

Given the complicated geometry of this project and its inherent three-dimensional nature, 3D finite element analyses are carried out in conjunction with 2D analyses for evaluating the stability and serviceability of the existing tunnels and ground deformations. A view of the 3D finite element model of the Dearborn Street connection is presented in Figure 7.

Structural Design

Structural design of the new tunnel connections are performed along with geotechnical design and involved design of diaphragm slab, invert slab, members of excavation support system, new columns, etc.

CONCLUSIONS

A brief history of tunnel construction in Chicago area in the past century including the complexities and challenges involved was presented followed by an overview of tunnel connection construction activities for development of Block 37 in downtown Chicago. A comprehensive geotechnical investigation has been carried out to identify soil parameters required for the design of the connection tunnels at Block 37.

The methodology applied for selection of optimum construction techniques and their complexities were discussed. To verify the efficiency of the applied construction technique, series of finite element analyses were performed to investigate the initial state of stress in soil due to construction of the existing tunnels, the soil-structure interaction during various stages of the proposed construction, the response of the existing tunnels during construction, and the construction-induced ground movement.

REFERENCES

Brinkgreve, R.B.J., 2002. PLAXIS 2D Version 8 Manual. Balkema Publishers, the Netherlands.

Finno, R.J., 2005. Hardening Soil Parameters for Block 37 Tunnel Project. Internal Report.

Keiwit-Reyes, 2006. Construction of Block 37 Tunnel and Track Connection. Internal Report.

Patrick Engineering Inc., 2005. Geotechnical Baseline Report for Block 37 Tunnel Connections.

Lunardi, P., 2000. The design and construction of tunnels using the approach based on the analysis of controlled deformation in rocks and soils. Tunnels & Tunneling International, May 2000, pp. 3–30.

Peck, R.B., and Reed, W.C., 1954. Engineering Properties of Chicago Subsoils. Engineering Experiment Station, University of Illinois, Vol. 47, No. 423, 63p.

Schanz, T., Vermeer, P.A., and Bonnier, P.G., 1999. The hardening soil model—formulation and verification. Proc. Plaxis Symposium Beyond 2000 in Computational Geotechnics, Belkema, Amsterdam, The Netherlands, pp. 281–296.

Terzaghi, K., 1942. Shield Tunnels of the Chicago Subway. Journal of the Boston Society of Civil Engineers, Vol. 29, No. 3, pp. 163–210.

2

Design and Planning of Underground Projects II

Chair

T. Gregor

Hatch Mott MacDonald

DESIGNING TO PROTECT ADJACENT STRUCTURES DURING TUNNELING IN AN URBAN ENVIRONMENT

Marco D. Boscardin
Boscardin Consulting Engineers, Inc.

Paul A. Roy
DMJM Harris, Inc.

Andy J. Miller
Faber Maunsell

Kevin J. DiRocco
GEI Consultants, Inc.

ABSTRACT

The North Shore Connector Project for the Port Authority of Allegheny County involves TBM mining a pair of subway tunnels through soil and rock under the Allegheny River and a narrow city street. Along the tunnel alignment are a historic building on shallow foundations and modern buildings on deep foundations, one constructed directly over the alignment during the latter phases of design. This paper provides an overview of the general site conditions, and technical basis for excavation support and tunneling decisions. Deformation limits, ground improvement, monitoring and mitigation plans, geotechnical baseline report development, and risk management will also be discussed.

INTRODUCTION

Downtown Pittsburgh is southwestern Pennsylvania's central business district, as well as its cultural and recreational capital. The compact, vibrant, urban center of Pittsburgh is known as the Golden Triangle. The new North Shore Connector (NSC) Light Rail Transit (LRT) benefits the community and economy of Pittsburgh tremendously by offering Pittsburgh residents a convenient connection to the city's central downtown business and recreational districts from the North Shore and by offering convenient access from downtown to the newly constructed PNC Park and Heinz Field sports complex, new business and residential developments, and cultural museums. The NSC project will promote future North Shore development and possible LRT service to Pittsburgh International Airport and local hospitals.

PROJECT DESCRIPTION

The NSC Project, constructed by The Port Authority of Alleghany County (PAAC), is a 2,250 m extension of the existing LRT system from Gateway Center in the Golden Triangle to the North Shore. The project consists of twin-bored tunnels from the launch pit, beneath the Allegheny River, to the reception pit using a tunnel boring machine (TBM); a cut-and-cover tunnel along a city street (Reedsdale Street), and a 245 M aerial viaduct structure to the North Shore terminus (see Figure 1).

70

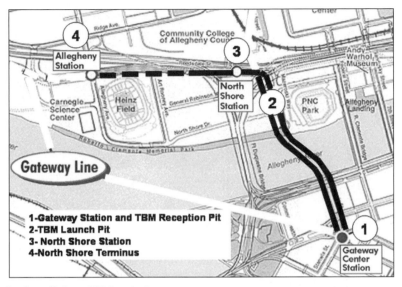

Figure 1. Overall view of NSC project

Figure 2. Looking north down Stanwix Street

The southern end of the NSC alignment is at the existing Gateway Center Station and TBM reception pit (Location 1—Figure 1). At the completion of the TBM drives, a new cut-and-cover tunnel will be constructed in the reception pit to connect the bored tunnels to a new station constructed at the location of the existing Gateway Station.

The TBM drive starts at the launch pit (Location 2—Figure 1) on the North Shore and proceeds south at a relatively steep ±7.0% downward slope. The tunnels pass beneath the newly constructed Equitable Resources Building, River Front Park, a river-side bulkhead wall, and the Allegheny River. Beneath the river the drive transitions from a downward slope to a relatively steep ±7.0% upward slope. At the south side of the river, the tunnels will pass beneath a riverside bulkhead wall, two major roadways (10th Street Bypass and Fort Duquesne Boulevard), and continue beneath a city street (Stanwix Street). Figures 2 and 3 show the relationship of the twin bores to adjacent structures along Stanwix Street.

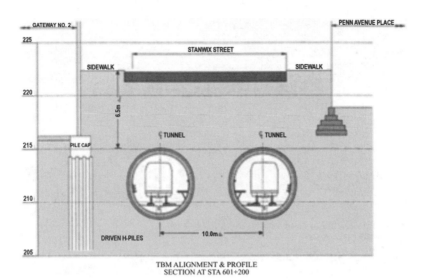

Figure 3.   Section through Stanwix Street

North from the Launch Pit, the tunnels will be constructed using cut-and-cover methods. The cut-and-cover tunnels will traverse westward underneath a highway aerial section (State Route (S.R.) 279) to a new underground station referred to as the North Side Station (Location 3, Figure 1). The station will be constructed using slurry walls, with the station platform about ±15 m below the ground level. Beyond this station, the alignment remains underground and crosses under an aerial approach ramp to the historic Fort Duquesne Bridge (S.R. 0065). The tunnels then emerge from the ground in the new Heinz Field parking lot where they make a transition into an aerial structure behind Heinz Field. The aerial structure connects to the North Shore Terminus platform station at Heinz Field.

REGIONAL GEOLOGY

The project is located in the Allegheny River valley at the confluence of the Allegheny and Monongahela Rivers where they merge to form the Ohio River. The project site is located in the Pittsburgh Low Plateau section—prototypical area of the Appalachian Plateaus Province.

The natural soils consist of unconsolidated deposits of the Quaternary Period The natural soil deposits are referred to as valley fill, and include moderately to poorly sorted clay, silt, sand, and gravel with some cobbles and boulders. Two distinct units of valley fill deposits are encountered. The basal unit, fluvioglacial deposits, was deposited during the Pleistocene, and the upper unit, alluvial deposits, was deposited during the Holocene. The fluvioglacial deposits are generally coarse-grained and consist of mostly sand and gravel from melting glaciers, and the alluvial deposits are more fine-grained and typically consist of clay, fine silt, and sand. The near-surface soils are anthropogenic fills have been placed and extensively modified due to residential, commercial, and industrial construction. Considerable fill has been placed along the river banks to extend developable land.

Bedrock in the project area belongs to the upper members of the Glenshaw Formation of the Conemaugh Group of the Pennsylvanian Period. The rock types along

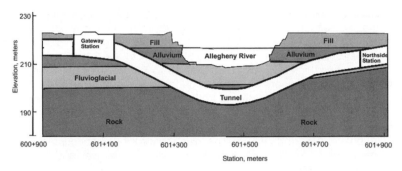

Figure 4. Tunnel profile and geology

the project alignment are the sandstones and shale of the Saltsburg member of the Glenshaw Formation. The stratigraphic horizon of the Pittsburgh Coal near the project area is well above the stratigraphic horizon of the proposed work. The sedimentary strata in the Pittsburgh Low Plateau Section—prototypical area are only mildly deformed, and folding is of very low amplitude. Bedrock is either horizontal or dipping less than 2 degrees toward the southeast. Some faults occur in this section, but are generally not readily apparent topographically and are not expected to impact the project. Joints, in contrast, are obvious in some areas.

PROJECT GEOTECHNICAL CONDITIONS

In the project area, bedrock is overlain by about 15 to 22 m of overburden soils on land and 6 to 9 m in the river. A generalized subsurface profile along the project alignment is shown in Figure 4.

Fill forms the uppermost 2.4 to 4.6 m of the soil for most of the alignment, but increases to as much as 11 m along Stanwix Street and is generally absent in the river. Typical of urban fill, the fill along the alignment is highly variable consisting of reworked alluvium, gravel, and miscellaneous debris, mixed with silty-clay and decomposed rock fragments. The maximum fill thickness is typically near the river banks.The fill varies significantly in composition and density, but is predominantly coarse-grained with pockets of fine-grained materials.

Alluvium is encountered beneath the fill along most of the alignment. The alluvium ranges in thickness from about zero to 4 m on the south shore and averages about 4.5 m in thickness on the north shore with increasing thickness to the north away from the river. The alluvium generally consists of slightly over-consolidated silty-clay or clayey sand, and exhibits a fining upward from more sandy soils at the bottom to more silty and clayey soils at the top. The alluvial soils are soft to stiff when fine-grained and very loose to loose when sandy, and generally exhibit low shear strength and low permeability. The alluvium is generally not within the tunnel bore or at the bearing elevation of cut-and-cover tunnel segments. Its principal impact will be on excavation support performance.

The fluvioglacial deposits form the lowermost part of the soil zone and are generally thicker along the south shore of the river than beneath the river or along the north shore. The fluvioglacial deposits are typically coarse-grained consisting of medium to coarse sands and gravels with cobbles encountered near the bedrock surface. These deposits are typically medium dense to very dense with high permeability. Most of the landside portion of the tunnel bore will be in the fluvioglacial soils, and cut-and-cover tunnels will be founded on these soils.

Bedrock elevation along the bored tunnel section rises gently from south to north. The stratigraphic sequence consists of an upper layer of silty shale with uniaxial compressive strengths (UCS) ranging from 37 to 81 Mpa overlying a 3- to 6-m-thick layer of soft calcareous claystone with UCS of 1.4 to 30 MPa. A thin, 0.3 to 0.6-m-thick, discontinuous layer of hard limestone with UCS of 16 to 62 MPa, overlies the claystone, except beneath the river, where an up to 6-inch-thick, discontinuous, bony coal seam is present over the claystone. Experience in the area indicates that the strength of this coal seam is likely to be around 7 MPa. The next lower rock unit is a 3- to 4-m-thick layer of thinly bedded, blocky to massive, siltstone with hard calcareous inclusions having UCS between 37 and 76 MPa. The lowermost rock unit encountered in the borings is a fine-grained thinly to medium bedded, blocky to massive sandstone with UCS ranging from 29 to 61 MPa.

Ground water in the general area of the project exists under phreatic conditions. The valley fill deposits provide a relatively permeable aquifer that is continuously recharged by the adjacent rivers, and the aquifer generally responds quickly to the river pool elevations.

THIRD PARTY INTERFACES

An important aspect of the project design has been identifying and managing interfaces with nearby properties and people using those properties. The third party interfaces include: city streets; public utilities; state highway ramps; existing buildings, including a historic building; new buildings adjacent and over the tunnel alignment that were under construction during tunnel design; and private high-capacity cooling water wells adjacent to the tunnel alignment. Selected major third party interfaces are described in more detail below.

Reception Pit and South Shore Tunnels

The reception pit is located in Stanwix Street near Gateway Center, and the tunnels follow Stanwix Street and pass beneath Fort Duquesne Boulevard, 10th Street Bypass, and River Front Park. The tunnels are adjacent to Penn Avenue Place (six to eight stories) to the east side of Stanwix Street, and Two Gateway Center and Allegheny Towers (both 24 stories) on the west side of Stanwix Street. Two Gateway Center is a steel frame structure constructed in 1952 and Allegheny Towers is a concrete frame building constructed in 1964. The buildings are supported by steel H-pile deep foundations bearing in bedrock well below the bottom of the tunnels. Historic Penn Avenue Place, formerly the Joseph Horne Building, was built in three stages between 1892 and 1922 and appears to be a concrete encased steel frame building, with a brick and stone masonry exterior, supported by stacked stone and concrete encased steel grillage spread footings bearing on fluvioglacial soils about 6 to 8 m below the ground surface. The tunnel invert near the buildings ranges from 11 to 18 m below the ground surface. The buildings along Stanwix Street have at least one level of basement with vaults that extend under the sidewalks. North of Stanwix Street the tunnels pass beneath Fort Duquesne Boulevard and the 10th Street Bypass, which are bordered by pile-supported retaining walls. In this area, the tunnel passes under the roadways and through the retaining wall foundation piles, near a sanitary sewer drop shaft and siphon chamber, and under a riverside bulkhead wall.

Tunnel Under the Allegheny River

Although there are no structures in the river near the tunnel alignment, concerns were raised regarding the impact of the construction on the river bottom habitat and species, in particular native oysters.

TBM Tunnels from the North Shore of the River to the Launching Pit

The TBM tunnel extends from the river under a sheet pile bulkhead on the north shore of the Allegheny River,past PNC Park, to the TBM launching pit. After passing below the tips of the bulkhead sheet piles, the tunnels pass under Riverfront Park and then an area presently occupied by parking lots and two city streets. To the north, the tunnel alignment passes directly beneath the Equitable Life Building, which was designed and constructed during final design of the tunnel project. The building is a six-story steel frame structure supported on H-piles driven to bedrock except that the building's basement floor and service entrance structures are constructed using slabs-on-grade and shallow foundations.

TBM Launching Pit, North Shore Cut and Cover Tunnels, and North Side Station

The TBM launching pit is about 17 m deep and located in a city street (General Robinson Street). The area around the launching pit is relatively open and used for parking. PNC Park, supported on pile foundations, is located about 45 m east of the tunnel alignment. North and west of the launching pit, the cut-and-cover alignment crosses under the I-279 viaduct to the Fort Duquesne Bridge passing between two twin level rectangular steel box girder viaduct bents constructed in the early 1970s and founded on steel piles to bedrock. There are headroom constraints below the I-279 viaduct, as well as potential interference issues with the battered foundation piles of the bents.

Beyond the I-279 viaducts, the North Side Station will be constructed along the south side of and parallel to a city street (Reedsdale Street). A concourse will be constructed south of the main station box to connect the new station to the adjacent existing SEA Option Area Garage. The station excavation will be approximately 15 m deep, and the concourse excavation will be approximately 9 m deep. Beyond North Side Station, the tunnel alignment continues parallel to Reedsdale Street and across Tony Dorsett Street where it passes directly beneath Pier bents of the Ohio River Boulevard (SR 0065) approach to the Fort Duquesne Bridge. The tunnel portal location is in the newly constructed Heinz Field parking.

MITIGATING THIRD PARTY IMPACTS

Mitigation of third party impacts started early in design. Early on, the project team met with owners and operators of properties that might be affected by the construction to obtain details of potentially affected facilities and to ascertain the sensitivity of each property to the construction. Agreements were made between the Project and property owners to permit the Project access to potentially affected properties for evaluation of existing conditions and assessment of construction impacts and to allow access for instrumentation installation and monitoring during construction. A first step in the development of construction impact mitigation plans is the delineation of the zone of tunnel-induced settlement in the absence of special measures.

Estimated Settlement Due to Tunneling

Tunneling induced settlement typically occurs due to ground loss into and around the TBM, commonly referred to as "face loss," and is measured as a percentage of the theoretical tunnel bore volume (percent face loss). Face loss occurs during tunneling due to stress relief of the surrounding ground and 'over-boring' causing ground loss in the vicinity of the tunnel, which results in settlement troughs over the tunnels. The settlement profile depends on the external tunnel diameter, tunnel depth, percentage face loss, and settlement trough width parameter. The percentage face loss is dependent on the type of ground, the tunneling method, the quality of workmanship, and the rate of tunnel advance. To minimize tunneling induced settlement, a pressurized, closed-face TBM system is required for this project. Using the pressurized, closed-face TBM system, 1.5% face loss was assumed for design in the fluvioglacial deposits and 1% face loss was assumed for design in bedrock. The settlement trough width parameter varied depending on the ground conditions. Evaluation of ground movements was generally performed using empirical methods, except at the Equitable Resources Building where a finite element analysis was performed for a more adetailed estimate of ground movement patterns due to the close proximity of this structure to the tunnels.

Calculated greenfield vertical ground movements along the tunnel alignment ranged from 20 to 70 mm above the tunnel centerline. Estimated settlement of the spread footing-type building foundations, which are located 6 m or more from the tunnel centerline, ranged from 12 to 20 mm. Settlements of buildings supported on deep foundations are expected to be negligible, as are effects of lateral ground movements on the deep foundations. Calculated ground settlement of slabs-on-grade and sidewalk vaults range from about 6 to 30 mm. Estimated maximum movements of the Penn Avenue Place sidewalk vault (30 mm vertical) and beneath the Equitable Resources Building base slab (40 mm vertical) assuming untreated ground, represent unacceptable levels of movements that are expected to result in damage to these structures.

The effects of tunneling induced settlement were evaluated for critical structures located along the bored tunnel alignment. On the south side of the Allegheny River, Penn Avenue Place is the primary structure at risk of damage due to tunneling induced settlement because it is the only structure constructed on shallow foundations. Other buildings on the south side of the river along the alignment are founded on deep foundations, and are generally at less risk of damage due to tunneling induced settlement. To control the movements on Penn Avenue Place ground improvement was selected in lieu of structural underpinning because of significant disruption to building operations expected during underpinning installation and because building settlements during underpinning were expected to be in the 10 to 15 mm range. The goal was to keep angular distortions in the range of 1/600 or less so that damage, if it developed, would be limited to the range of negligible to slight categories and only require cosmetic repairs.

As a result, ground improvement was incorporated in the design to reduce settlements to acceptable levels. In areas where ground improvement is provided, tunnel construction related ground movements were re-estimated assuming that the effective face loss will be reduced to 0.75%, and result in settlemente estimates about one-half of the greenfield settlements.

Ground Improvement

Ground improvement consisting of jet grouting and compensation grouting will be performed at several locations along the bored tunnel alignment. Ground improvement for the project, starting at the launching pit and proceeding south toward the receiving pit is described below. An approximately 32-m-wide by 14-m-high by 7-m-thick jet grout block is to be installed prior to the start of tunneling at the breakout area of the launching pit to assist with alignment of the TBM and to an electric manhole from tunneling-related ground movements.

Compensation grouting is to be performed beneath the Equitable Resources Building basement slab to reduce potential damage to the slab from tunneling-related ground movements. Compensation grouting is to be performed as the tunnel passes beneath the building to fill voids that develop beneath the slab as soil below the slab settles away from the slab.

Jet grouting is planned at three retaining walls along Fort Duquesne Boulevard and the 10th Street Bypass where foundation piles supporting the retaining walls will be removed to facilitate tunneling. Portions of the retaining walls will be removed to facilitate pile extraction, and jet grouting beneath the bottom of the retaining walls to form a foundation for reconstructed portions of the retaining walls. Prior to jet grouting, voids remaining from pile extraction are to be backfilled. The jet grout will provide vertical support for and minimize the tunneling-related ground movements of the reconstructed retaining walls. To limit ground movements, piles that do not interfere with tunnel boring operations are left in place, requiring that jet grout columns be located to minimize shadowing effects from the piles.

Ground modification design for northernmost 24 m of tunnel bore along Penn Avenue Place consists of low mobility compensation grouting as the tunnel advances. Compensation grouting was selected to minimize the potential for interfering with the high capacity (>45 L/sec) cooling water well located at the northwest corner of the building. The compensation grouting will consist of compaction grouting methods to consolidate the ground above the tunnel quarter arch and to offset the ground movements due to ground loss around the TBM.

Ground improvement, consisting of jet grouting, is designed for 144 m of Stanwix Street in front of Penn Avenue Place to reduce the building movements to tolerable levels. A jet grout zone will be constructed from the reception pit north along the tunnel adjacent to Penn Avenue Place prior to tunnel boring operations. The jet grouted zone will create a shroud around the tunnel and will extend about 1.5 m beyond the limits of the tunnel bore. In addition, the jet grouting performed in Stanwix Street will serve to protect the west tunnel (driven first). Due to the close proximity of the bored tunnels to each other in Stanwix Street, ground stabilization is required between the bored tunnels to protect the lining in the first tunnel from damage during the boring for the second tunnel.

Due to the shallow cover over the tunnels and close proximity of the tunnels to each other near the receiving pit, a block of jet grout will be created at the break in area at the north face of the receiving pit. The 2 m of jet grout closest to the receiving pit will extend to the top of rock, and the remaining jet grout will extend to a minimum depth of 2 m below the bottom of the tunnels (approximately 12 to 14 m below existing grade). The purpose of this grouting is to improve stability of the TBM portal wall and to reduce the need for bracing at the portal wall.

Support of Excavation

For cut-and-cover structures, stiff support of excavation systems (SOE) are necessary to protect streets, to safeguard adjacent buildings, foundations and utilities, to control groundwater infiltration, and to limit the area of excavation during construction. Except for the North Side Station SOE, the Contractor is responsible for final selection and design of the SOE to meet the ground movement and building movement criteria set for the project. At the North Side Station, the Contractor is required to use concrete diaphragm walls, because the diaphragm walls will be used as the permanent exterior structural walls of the station and because of the proximity of the existing SEA Garage.

I-279 Viaduct. Where the tunnel alignment for the cut-and-cover tunnel segment is constructed below the I-279 viaduct, there are head room restrictions affecting the placement of the SOE. The SOE system is adjacent to the battered piles of the I-279 viaduct bents, which are sensitive to movement. As a result, more restrictive movement criteria are specified for the SOE system in this area than in other areas. In addition, the battered piles for the viaduct foundations may physically interfere with the SOE system installation at three locations. The SOE is required to "bridge over" these battered piles.

S.R. 0065 Viaduct. The SOE system designed by the Contractor for the excavation beneath the underpinned SR 0065 bents is required to meet more stringent wall movement criteria than other SOE systems in less critical areas. The more restrictive wall movement criteria are specified for the protection of the underpinning system and protection of the pile foundations that support portions of bents, which are located outside of the excavation and not underpinned. The SR 0065 underpinning includes (8) 1.5-m-diameter drilled shafts (caissons) and one 2.1-m-diameter caisson into bedrock.

Instrumentation Monitoring

Construction-related movements of adjacent streets, structures, and utilities are anticipated along the tunnel alignment in response to: excavation support movements for cut-and-cover construction; tunnel boring and construction operations; ground improvement operations; dewatering operations and underpinning operations at the SR 0065 bents. An instrumentation monitoring program was developed to verify that construction-related movements meet the performance criteria for the project and to quickly identify construction-related movements that exceed performance criteria or that could adversely impact streets, structures, and utilities so that mitigating and corrective actions can be implemented. In addition, the program will provide data regarding the performance of the TBM so that the TBM operator can adjust the tunneling operations to minimize ground movements. The Contractor is responsible for installing and monitoring the instrumentation.

Instrumentation was included in the monitoring program to measure the following: vertical and horizontal ground movements related to TBM performance; vertical and horizontal ground movements related to tunnel construction; vertical and horizontal ground movement adjacent to structures and utilities; vertical and horizontal movements of structures; strains in the SR 0065 highway bridge bents; tilt of highway bridge bents; groundwater levels; and construction-related vibrations and noise.

The instrumentation monitoring and reporting program has been designed to alert the Contractor and the Engineer when construction activities could potentially impact critical structures and utilities and to allow the contractor to adjust construction procedures or methods before significantly impacting structures and utilities. The important features of the program are the action levels (response values) and the monitoring and reporting frequency. Two action levels, referred to as the response values, are used for each instrument. Response values were established for each critical structure and utility based on their sensitivity to construction-related movements, on the type of nearby construction activities, and on the expected movements due to construction. The threshold value is the first response value and is established to alert the contractor and engineer that construction-related movements are reaching potentially damaging levels. If a threshold limit is reached, the contractor is required to implement an in-place action plan, developed prior to reaching threshold limits,to reduce the potential for impacting structures and utilities. The limiting values is the second response value and has been established to alert the Contractor that construction related movements have reached or exceeded expected movements and that construction activities are expected to exceed tolerances of and cause damage to structures and utilities if movement continues. If a limiting value is reached, the Contractor is required to stop work in the area the limiting values is reached with the exception of any activities needed to prevent readings from exceeding the limiting value and to quickly meet with the Engineer to discuss and implement an action plan to mitigate and prevent further damage and to maintain movements and responses associated with the remaining construction within threshold limits.

The monitoring and reporting frequency is also an important feature of the instrumentation monitoring and reporting program. Required monitoring frequencies vary depending on the type and proximity of construction activities to structures and utilities, the sensitivity of structures and utilities, and the type of instrument. In general, the more likely a construction activity is to impact a structure or utility and the closer the activity is to the structure or utility the higher the monitoring frequency. Typical monitoring frequencies range from daily to monthly depending on the proximity of construction activities, but if response values are reached, monitoring frequencies increase. For normal reading frequencies, the Contractor is required to report the instrumentation data to the Engineer within 24 hours of data collection, but if a response value is reached, the contractor is required to notify the engineer immediately.

SUMMARY AND CONCLUSIONS

Underground construction in a crowded urban environment can be performed to manage and control underground construction impacts on adjacent properties, related users and traffic, but it does not eliminate them and may introduce new ones. This project illustrates that attention to third party issues starting at the early phases of design and continuing through construction allows the project team to design systems to control and manage those impacts in a cooperative working relationship with the various stakeholders.

ACKNOWLEDGMENTS

The authors acknowledge the following firms and individuals for the information upon which this paper is based. Firms: The PAAC—Project Owner; DMJM Harris, Inc.—Prime Design Consultant; GeoMechanics, Inc.—Geotechnical Consultant; and GEI Consultants, Inc.—Building Protection and Instrumentation Consultant. Individuals: G. Yates; J. Prizner; D. Haines; K. Chong; D. Veights; S. Woodrow; T. Boscardin and J. Alvi.

GROUNDWATER CONTROL FOR
TWO-PASS SOFT GROUND TUNNELING

Cary Hirner

Black & Veatch

David Stacey

Jay Dee Contractors

William A. Rochford

U.S. Army Corps of Engineers

ABSTRACT

Design considerations and construction techniques were implemented to address groundwater infiltration from a confined aquifer during the excavation and lining of the Cady Marsh Flood Relief Tunnel. This 3-m finished diameter, 2-km long tunnel was successfully completed cost-effectively in an urban setting near East Chicago, Indiana. The soft ground was mined using a conventional tunnel boring machine with hydraulically actuated flood doors and supported with steel ribs and timber lagging. The tunnel was lined with cast-in-place concrete. Dewatering was implemented in settlement prone, boulder-laden ground in residential neighborhoods during construction to depressurize and partially dewater the aquifer at the tunnel horizon. This paper describes the design decisions and construction methods utilized to mitigate and limit infiltration into the tunnel during construction and operation, as well as manage ground settlement.

INTRODUCTION

The Cady Marsh Flood Relief Tunnel is a significant component of the U.S Army Corps of Engineer's regional scheme to address flooding along the flat lying lowlands of northwest Indiana. The project involved the construction of a soft ground tunnel to divert flood water from the Cady Marsh Ditch to provide regional flood control benefits. The Cady Marsh Ditch is a minor tributary to the Little Calumet River System and drains an area of about 44 square km (10,900 acres).

The project alignment and profile are shown on Figure 1. The major components of the project consist of the following:

a. Approximately 30 linear meters of Cady Marsh Ditch improvements

b. 2-km long, 3-m diameter Cady Marsh Tunnel

c. Three 1.5-m diameter intermediate access shafts

d. Inlet structure consisting of a side overflow weir with trash rack

e. 6.1-m diameter inlet shaft with an anti-vortex wall

f. 6.1-m diameter outlet shaft with a discharge weir and trash rack

g. Outlet channel consisting of an energy dissipating structure and approximately 60 m of riprap lined channel

h. Tunnel dewatering sump and two submersible chopper pumps

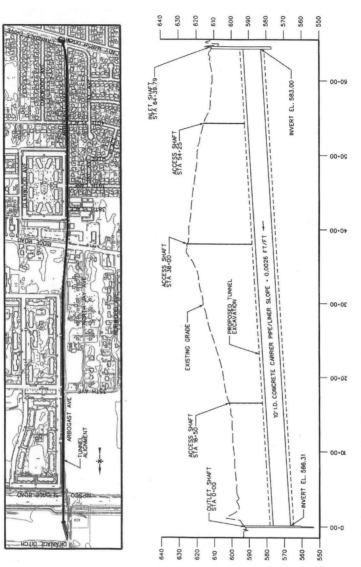

Figure 1.   Project plan and profile

Originally, Black & Veatch was retained to assist the Corps in delivering this project using a design-build contract (as presented at RETC 2003). Both open-cut and tunneling installations were to be permissible. However, during project planning and contract document preparation residential and community concerns associated with deep large diameter open-cut excavations in the established residential neighborhood led to the removal of the open-cut excavation option. Given this change in philosophy, the project design proceeded as a two-pass tunnel to limit community impacts and the contract delivery method was changed to traditional design-bid-build.

GEOLOGIC CONDITIONS

The Cady Marsh Tunnel and shafts traverse through three soil layers. These layers are designated from ground surface as follows:

- Calumet Aquifer (sand)
- Clay Aquitard
- Valparaiso Aquifer (sand)

The Calumet Aquifer consists of sand with traces of silt and gravel. This layer is continuous across the project area, and extends from the ground surface to a maximum depth of 8 m below ground surface (bgs). Groundwater is encountered at depths in this shallow aquifer ranging from near ground surface to 2 m bgs. The tunnel envelope is located below this sand unit. The clay aquitard underlying the Calumet Aquifer is soft to very stiff silty clay that contains traces of sand and gravel, and occasional cobbles and boulders. Silt and sand lenses are present within the clay as well as at the contact with the Calumet and Valparaiso aquifers, which it separates. The tunnel is located for the majority of the alignment in this clay. In the project area the Valparaiso Aquifer consists of sand with traces of gravel and silt. The aquifer is present below the clay aquitard along the southern end of the tunnel alignment. This aquifer is artesian, flowing in some instances, with depths to water in piezometers ranging from ground surface to 4 m bgs. The aquifer is present in the invert of the tunnel for 120 m and directly underlies approximately 335 m of the southern portion of the tunnel.

DESIGN CONSIDERATIONS AND CONTRACT REQUIREMENTS

The tunnel system was designed to convey water from the Cady Marsh Ditch to the Little Calumet River at a peak inflow rate of 914,000 liters per minute (lpm) (538 cfs), which is the 100-year flood event on the Cady Marsh Ditch. The Corps of Engineers established the tunnel diameter based on studies previously conducted and set the tunnel system inlet and outlet weir elevations to operate under the tunnel using gravity flow during storm events as an inverted siphon. In order to empty the tunnel following storm events the tunnel is dewatered using the submersible pumps housed in the outlet shaft. The design was focused on providing the most cost effective tunnel that was compatible with ground conditions and would meet the project criteria while limiting short and long term impacts and disruptions to the community.

In addition to Plans and Specifications, a Geotechnical Baseline Report (GBR) and Geotechnical Data Report (GDR) were part of the Contract Documents. Also, a Disputes Review Board consisting of three independent tunneling experts was established.

Tunnel Design

The tunnel was designed to be constructed as a two-pass tunnel using a fully shielded conventional tunnel boring machine (TBM). The initial tunnel support was to

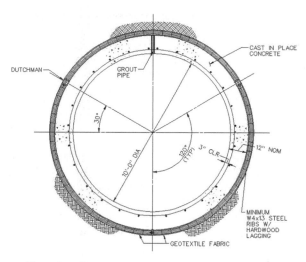

Figure 2.   Soft ground tunnel section

consist of conventional steel ring beams with wood lagging spanning between the steel beams and the final liner was to be cast-in-place concrete or reinforced concrete pipe at the Contractor's option (Figure 2). While the Contract Documents did provide a minimum initial support design, the design of the initial tunnel support was the responsibility of the Contractor to account not only for ground and surcharge loads but also loads associated with thrusting the TBM forwarded off of the initial support. The following design criteria were used for the tunnel:

- Tunnel system convey 914,000 lpm (538 cfs)
- Minimum soil cover of 1.25 times the diameter of the tunnel
- Tunnel sloped downward at 0.26% grade from south to north
- Minimum initial support of full circumference steel ribs every 1.2 m with hardwood lagging placed between the steel ribs
- Geotextile fabric placed full circle between the hardwood lagging and ground
- 30.5 cm (12 inch) thick cast-in-place reinforced concrete liner throughout tunnel.

Shaft Design

The inlet and outlet shafts were required to be supported during excavation. Dewatering of the Valparaiso Aquifer during inlet shaft construction was required during excavation and concrete lining if the initial ground support used by the Contractor was not water tight.

Managing Groundwater

Dewatering to depressurize the Valparaiso Aquifer was required to limit groundwater infiltration during excavation and concrete lining of the southernmost 335 m of tunnel and the inlet shaft. It was estimated during design up to 10,600 lpm (2,800 gpm) would need to be remove to drawdown and maintain the piezometric surface below the tunnel invert. With this dewatering, steady-state and transient heading inflow rates into the tunnel of 375 lpm (100 gpm) and 75 lpm (20 gpm), respectively, were established in the GBR.

Laboratory tests indicated that prolonged dewatering periods would cause consolidation settlement of the clay. This data showed dewatering the clay layer above the Valparaiso Aquifer for greater than 60 days would cause more than 2.5 cm (1 inch) of settlement, which was determined to be the maximum acceptable consolidation settlement that would not damage structures along the alignment. Historical experience with the glacial clay would indicate that a significantly longer period of time would be required before 2.5 cm of consolidation settlement would be experienced. In addition, the settlement would be expected to be uniform in nature and not differential. However, given the laboratory data results and the presence of soft silt lenses that could readily consolidate in the clay unit the contract documents required the Contractor to limit the tunnel dewatering period for 60 days or prove with monitoring data consolidation settlement was within the acceptable limits prior to extending the dewatering period.

Control of Settlement Damage Caused by Ground Loss

Settlement caused by ground loss is inevitable using a conventional soft ground TBM with limited ground cover above the tunnel crown. Therefore, the design accounted for settlement. It was estimated that approximately 3.8 cm of settlement at the tunnel centerline could be expected in areas of shallow ground cover and the settlement would gradually decrease away from the tunnel. It was calculated to be negligible 9 m off of the tunnel centerline.

The tunnel was located near the center of the street that it underlain to the extent practical to provide adequate distance between utilities paralleling the street and the tunnel, and to prevent damaging sidewalks and private property. The contract required the street to be resurfaced following construction to repair the settlement damage.

Accounting for Boulders

Boreholes advanced during the geotechnical investigations encountered boulders in the glacial clay aquitard and at the interfaces of this clay with the Calumet and Valparaiso aquifers. To account for these ground conditions it was required that the TBM be capable of passing cobbles and boulders up to 46 cm (1.5 feet) in size through the cutterhead and that the face was accessible to shoot larger boulders, if required. Baseline conditions for 300 boulders sized between 46 and 90 cm (1.5 and 3 feet), 30 boulders 90 to 150 cm (3 to 5 feet) and 2 greater than 150 cm in size were established in the GBR to account for larger boulders that could be encountered. When amassed together these boulders constituent 0.3 percent of the mined area of the tunnel. This estimated volume was based on correlations between the geotechnical borings and limited available data from other tunnels driven in similar ground conditions in the Chicago/Milwaukee area. Other estimating methods were used during design which provided a wide range of anticipated number of boulders. After much evaluation the baseline numbers were established at a workshop held with the Corps of Engineers where it was determined that this baseline would provide a conservative estimate at a reasonable project cost on bid day.

CONSTRUCTION METHODS

The construction contract was awarded to the Joint Venture formed between Jay Dee Contractors and Kenny Construction (Jay Dee/Kenny) for $18.4M in September 2004. Jay Dee/Kenny was selected via a Best Value contracting method, specifically

Figure 3. TBM cutterhead

"Technical Acceptable—Low Price" for this contract. The selection was based on the Contractor adequately demonstrating capability in the construction approach, dewatering approach and ability to effectively manage the schedule and project costs. Contractors that met the minimum criteria for these factors were then considered for award based on their proposed fee.

Tunnel

The tunnel was excavated starting at the outlet shaft and proceeding to the inlet shaft (north to south) using a re-furbished Lovat M-157 Series 4200 soft ground TBM. The diameter of this machine was 4 m with a cutting edge diameter 23 mm greater in diameter than the TBM (Figure 3). The TBM was fully shielded, full circle and equipped with hydraulically actuated control doors that allowed the tunnel face to be completely closed off if ground conditions warranted. The cutterhead was equipped with spade teeth. From the outlet shaft to Station 54+88 a steel hood was present in front of the crown of the TBM until it was damaged and subsequently removed. It is presumed a boulder encountered near Station 54+72 forced this hood back into the cutterhead binding it and suspending mining until the hood was removed. An oversized hole was dug out in front of the machine through the TBM to remove the hood. This void was subsequently filled with grout.

Clay and silt was encountered across the tunnel horizon throughout the tunnel with the exception of occasional localized sand lenses and boulders present in the clay and the last 46 m (150 feet) of the tunnel which contained up to 1.2 m of sand in the invert of the excavation.

Tunnel excavation proceeded in 1.2-m intervals to facilitate initial support installation, which was erected within the tail shield of the TBM (Figure 4). The TBM was equipped with a hydraulically actuated rib expander that was used to place 4×13 Grade 50 steel ribs around the tunnel excavation circumference at 1.2-m spacings. These ribs had an outside diameter of 4 m and were pushed up against the excavation and held in-place using Dutchmen to maintain the rib expansion. Nominal 7.6-cm (3-inch) thick hardwood lagging was placed between the steel ribs to fully support the tunnel excavation. Geotextile fabric was placed full circle between the ground and hardwood lagging to prevent the migration of fines with groundwater infiltration into the tunnel. Tunnel excavation progress varied but 30 meters per day (8 hour shift) was often achieved.

Figure 4. Tunnel excavation support

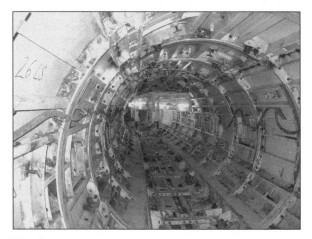

Figure 5. Tunnel lining forms

At the conclusion of tunnel excavation the TBM was removed through the inlet shaft and cast-in-place concrete lining operations took place. Concrete was generally placed in the tunnel at 30- to 46-m intervals. Prior to each placement, the tunnel was cleaned, rebar tied and hung, and concrete forms set using spuds to position the forms. Collapsible steel forms 3 m in diameter were used (Figure 5). Concrete was pumped behind the tunnel forms through guillotine values at the tunnel crown beginning at the upslope end of the forms and proceeding to the bulkhead (Figure 6). Contact grouting was conducted at the tunnel crown on 3-m centers to fill any void space between the liner and initial support developed during concrete lining operations.

Inlet and Outlet Shafts

The inlet and outlet shafts were excavated in stages by a backhoe and crane utilizing a clam shell bucket. To support the excavations 8.8-m diameter W6×25 steel ribs spaced at 1.2-m intervals with wood lagging between them were used (Figure 7).

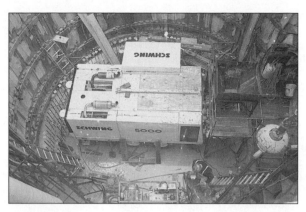

Figure 6. Concrete pump in shaft

Figure 7. Holing through inlet shaft braced excavation

Excavation preceded by digging 1.2 m vertically and installing the next set of steel ribs and lagging boards. Hand shovels and pneumatic chisels were used to shape the excavation for initial support installation. Void spaces behind the initial support were packed with straw or backfilled with sanded grouted. W14×109 solder beams and pieces of W6×25 steel ribs were used to frame the tunnel eye to launch and receive the TBM.

A 1.8-m thick concrete plug was placed at the base of each shaft to counteract buoyancy prior to placing the shaft walls. Following completion of the tunnel excavation, 6.1-m diameter shaft lining forms were used to construct the reinforced concrete walls.

Intermediate Shafts

Three access shafts along the tunnel alignment were constructed as drilled shafts using a Texoma 600 auger rig. Drilling fluid consisting of water and bentonite, and water was used to drill through the Calumet Aquifer and into the underlying clay unit, respectively. Corrugated steel casing was placed in the drilled shaft and the annular space between the ground and casing grouted prior to dewatering the shaft. Various diameters of steel casing ranging from 2.4 to 3.4 m were used to step-in the excavation during drilling. Inside the dry casing, 1.5-m diameter pre-cast reinforced concrete pipe manholes were installed and structurally connected to the crown of the tunnel. The annular space between the manhole and casing was filled with structural grout.

CONSTRUCTION IMPACTS AND CHALLENGES

The primary construction challenges associated with the ground conditions that were addressed included negotiating and managing the artesian Valparaiso Aquifer and limiting settlement (ground loss and consolidation) in the residential neighborhood. Boulders present in the tunnel horizon increased the potential for ground loss settlement during tunnel excavation.

Managing the Artesian Aquifer

Groundwater inflow was minimal through the portions of tunnel mined in the clay. However, dewatering of the Valparaiso Aquifer was required to prevent excessive groundwater and soil inflows into the tunnel in the southernmost portion of the tunnel. Using additional monitoring wells installed during construction Jay Dee/Kenny determined that dewatering of the aquifer was necessary over the last 335 m of the tunnel alignment. Nine dewatering wells were installed at approximately 45-m spacings between the sidewalk and street and four wells were installed near the inlet shaft. Discharge piping and electrical conduit to power the pumps were trenched in place along the street below the pavement during dewatering operations. The maximum withdrawal rate of the total dewatering system approached 4,600 lpm (1,200 gpm). Dewatering continued 119 days while excavating and concrete lining this portion of the tunnel. The dewatering timeframe exceeded the limitation of 60 days for dewatering during tunnel construction; however, the settlement data provided by Jay Dee/Kenny indicated that minimal consolidation settlement well less than 2.5 cm occurred over the dewatering period. Besides the installation and removal of the wells and discharge piping, no adverse short or long term impacts caused by the dewatering operations to the community and project were observed.

Groundwater inflow was minimal during tunnel mining and the tunnel was sloped so infiltration would flow to the outlet shaft were it was removed by a sump pump. Areas of water dripping into the tunnel through the wood lagging were observed infrequently. Following the dewatering period, which included concrete lining the tunnel past the Valparaiso Aquifer inflows into the entire tunnel were measured to be around 125 lpm (33 gpm). Of this amount, 75 lpm (20 gpm) were estimated to be infiltrating through the concrete lined portions of the tunnel in contact with the aquifer. Most of the infiltration through the tunnel lining was attributed to construction joints, seepage cracks, and spud holes. The nature of this infiltration was deemed to be unacceptable. Therefore, additional contact grouting was conducted at the tunnel crown and around the perimeter at four construction joints that yielded the most water. Follow-up inspections indicated that this contract grouting coupled with the spud holes and many of the seepage cracks seal-healing with calcite deposits was successful in minimizing the volume of groundwater inflow into the tunnel. The final inflow measurement collected prior to tunnel filling was 23 lpm (6 gpm).

Table 1.   Boulders encountered in tunnel

Tunnel Station	Number of Boulders	Size Category (cm)	Boulder Size (cm)
19 + 35	1	46–90	76 × 61 × 46
32 + 80	1	46–90	61 × 53 × 38
41 + 25	1	46–90	33 × 48 × 61
46 + 00	1	46–90	46 × 38 × 51
52 + 94	1	46–90	61 × 66 × 38
55 + 99	1	46–90	51 × 69 × 46
56 + 28	1	Unknown, required blasting	Unknown
61 + 92	1	Unknown, required blasting	Unknown
61 + 93	1	Unknown, required blasting	Unknown

Settlement Due to Ground Loss

Surface, utility and structure settlement monitoring points were measured prior to, during and following construction. Surface monitoring points were installed at 61-m intervals along the centerline of the tunnel alignment, in arrays perpendicular to the alignment in two locations and throughout the neighborhood in the dewatering area.

Settlement along the tunnel centerline occurred as predicted along most of the alignment. Generally settlement was limited to a maximum of 2.5 to 3.8 cm along the centerline of the tunnel. The higher settlement valves consistently were associated with areas of lower cover above the tunnel crown. Settlement measured 3 to 6 m from the tunnel centerline varied from nonexistent to as great as 20 mm (0.8 inches). Settlement was not observed at the monitoring points established 9 m and further from the tunnel centerline. In addition, no appreciable movement of the structure monitors or settlement monitors established in the neighborhoods to monitor dewatering effects was observed.

Excess settlement did occur on an oil line that was being monitored and at the location were the oversized hole was dug in front of the TBM to remove the hood from the TBM cutterhead. The excessive oil line settlement was caused by the TBM being left over the weekend directly under the oil line. The excavation cut of the machine is a larger diameter than the tail shield, which allowed the ground to settle around the tail shield causing settlement.

Boulders Encountered

Boulders were encountered during shaft and tunnel excavation. Table 1 summarizes the number and locations of boulders that were encountered during tunnel excavation that were above the baseline size of 46 cm (1.5 feet). Although hundreds of cobbles and boulders sized below 46 cm were encountered, the number of boulders actually encountered above the baseline size was significantly less than estimated in the GBR. This difference in values can be explained by the given uncertainties associated with predicting the number and size of boulders in the subsurface based on geotechnical borings and the team's desire to use conservative baseline numbers that added a minimal cost to the project. It should be noted, with the exception of the boulder that was encountered near Station 54+72 that bent the steel hood back into the TBM, none of the boulders were problematic to remove or caused a significant delay in mining. Three boulders required shooting, the remainder passed through the cutterhead during normal mining process.

SUCCESS STORY

The Cady Marsh Flood Relief Tunnel had design and construction challenges to overcome associated with managing groundwater inflows and limiting ground settlement in residential neighborhoods while keeping project costs to a minimum. Proven technologies with appropriate levels of control established during design were employed to construct this project cost effectively. The acquisition method that allowed the project owner an alternative to a low-bid contract created an environment for better sharing of project risks, which resulted in improved working relationships between the Government and Jay Dee/Kenny and helped achieved a successful project. While a Disputes Review Board was established at the onset of the project, no disputes were heard by the board and no third party claims filed. The minor settlement impacts that occurred along the tunnel alignment were anticipated and addressed within the parameters of the Contract Documents. Jay Dee/Kenny achieved substantial completion of the operational project on budget in September 2006.

NEW CRYSTAL SPRINGS BYPASS (POLHEMUS) TUNNEL— DESIGN CHALLENGES

Roy C. Fedotoff

San Francisco Public Utilities Commission

Gregg Sherry

Brierley Associates

Jon Hurt

Arup

Gilbert Tang

San Francisco Public Utilities Commission

INTRODUCTION

The 1,380 m (4,200 ft) long New Crystal Springs Bypass (Polhemus) Tunnel (NCSBT) will replace the existing Crystal Springs Bypass Pipeline. This tunnel project is part of the SFPUC's $4.3 billion capital improvement program to repair and seismically upgrade their aging pipelines, tunnels, reservoirs and dams that provide drinking water to more than 2.4 million customers in the San Francisco Bay Area.

This paper describes the development of the tunnel alignment and design, and the complex ground conditions of the Franciscan Mélange. Of particular interest is the challenge of tunneling with a possible downhill grade of 3%, a rigorous environmental process and incorporating the lessons learned from 1960s construction of the adjoining Crystal Springs Bypass Tunnel.

SFPUC WATER SUPPLY IMPROVEMENT PROGRAM

The San Francisco Public Utilities Commission's (SFPUC) massive rebuild of its landmark Hetch Hetchy water delivery system (Figure 1) includes multiple program elements to improve treatment, transmission and storage facilities along the 269 km (167 mile) gravity fed regional water system (Armistead, 2006). Each project is tied to levels of service for water quality, seismic and delivery reliability, and water supply reliability through 2030 for the nation's sixth largest public utility agency.

Built in the early- to mid-1900s, many parts of the regional water system are nearing the end of their useful service lives. In addition, crucial portions of the system cross over or near three major earthquake faults in the Bay Area. In 2002, the SFPUC teamed with the Bay Area Water Supply and Conservation Agency (BAWSCA) representing the 26 cities and water districts plus two private utilities that purchase water from San Francisco on a wholesale basis, to launch a major initiative to repair, replace, and seismically upgrade the system's aging pipelines, tunnels, reservoirs, pumping stations and dams. The mandate for this program included a legislative act, known as AB 1823 that provides a process, established by the State Legislature, to ensure that the San Francisco Bay Area regional water system is rebuilt as soon as possible.

In 2002, the program was known as the Capital Improvement Program (CIP). Presently, this $4.3 billion Water System Improvement Program (WSIP), the largest

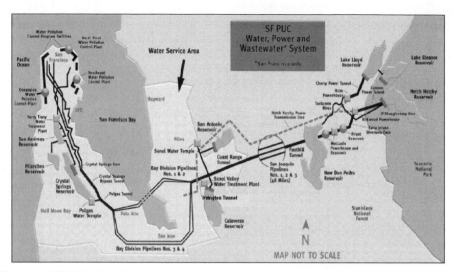

Figure 1.   SFPUC's Hetch Hetchy water delivery system

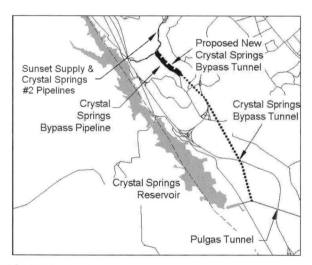

Figure 2.   Crystal Springs bypass system

public works project in San Francisco history, includes more than 75 San Francisco and regional projects, to be completed by the end of 2014. The current WSIP program evolved when the new General Manager requested an independent assessment of the revised capital improvement program in 2004. Parsons Water & Infrastructure, Inc. with CH2M Hill conducted this review. The revised program was formally adopted on November 29, 2005.

BACKGROUND

The Crystal Springs Bypass system (Figure 2) is a critical component of the SFPUC water transmission system that provides Hetch Hetchy and Sunol water to the north Peninsula and San Francisco, bypassing the Crystal Springs Reservoir. The system consists of the Crystal Springs Bypass Tunnel and the Crystal Springs Bypass Pipeline. Under normal operating conditions, water is transported from the Pulgas Tunnel to the Crystal Springs Pipeline No. 2 and Sunset Supply Pipeline via the tunnel and pipeline. Additionally, in an emergency where the Irvington Tunnel may be out of service, flow could be reversed and this system used to transport Peninsula reservoir water south from the Harry Tracy Water Treatment Plant to customers in the South Bay.

NEED FOR THE PROJECT

The CSB Pipeline is a 2.4 m (96 inch) diameter pipeline connecting the existing tunnel to the Crystal Springs No. 2 and the Sunset Supply pipelines. The 1,385 m (4,545 ft) long pipeline, runs nearly parallel to and below Polhemus Road in San Mateo County, and is comprised of prestressed concrete cylinder pipe. This pipeline is connected to the Polhemus chemical feed station (chlorine and fluoride) plant at the south end and crosses beneath San Mateo Creek with a siphon structure at the north end.

During the winter of 1996–1997, a landslide occurred along the northeast side of Polhemus Road threatening the integrity of the pipeline. The pipeline did not fail. However, during inspections in 1998 and 1999, some separation and reparable circumferential mortar cracking were observed along one joint of the pipeline within the landslide area. The potential for the occurrence of future landslides in other areas along Polhemus Road and the consequences of a potential pipeline failure raised concerns for the safety and reliability of the existing pipeline.

To address the above concerns, eight alternatives were studied including replacement with new parallel pipeline, hybrid pipeline/tunnel, or tunnel. The following were key objectives to be met by the chosen alternatives:

- Increase reliability by avoiding or reducing potential damage to the conveyance facility to the maximum extent possible from slope and seismic hazards;
- Provide a safe, reliable, and adequate water supply to customers in San Francisco and San Mateo County;
- Reduce physical, social, and economic impacts associated with the rupture of the existing pipeline (e.g., flooding, erosion, traffic interruption);
- Minimize environmental impacts associated with the construction of a new tunnel;
- Provide consistency with SFPUC policy of replacing of prestressed concrete cylinder pipelines within the entire regional water system;
- Minimize costs to SFPUC customers, and
- Increase redundancy in the Bypass System.

Based on an evaluation of the viable alternatives, using a weighted criteria system, the tunnel alternative was determined to be the most favorable and recommended for implementation.

PROJECT DESCRIPTION

The project calls for construction of a bypass tunnel about 1,280 meters long (4,200 feet, or 0.8 mile) with an inside diameter of 2.4 m (96 inches). The tunnel alignment is almost parallel to the Polhemus Road crossing and traverses under San Mateo

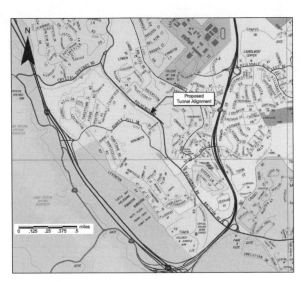

Figure 3. Site vicinity map

Creek, as shown in Figure 3. It is below ground and would not be noticeable at the surface, except at the two ends of the tunnel. The existing pipeline would be taken out of
service but would remain in-place. It would be used when the new tunnel needs to be
shutdown for maintenance or other causes. The proposed configuration of the new
tunnel is shown in Figure 4.

At each end of the new tunnel, 2.4 m (96 inch) diameter tunnel risers connect the
tunnel with the surface pipelines. At the south end, the new tunnel merges with a new
2.4 m (96 inch) diameter pipeline segment that connects to the existing facilities
(Figure 5). At the north end, the new tunnel will connect to the existing Sunset Supply
Line and Crystal Springs Pipeline #2 via a system of new pipes and valves (Figure 6).
The final pipeline will consist of welded steel pipes throughout the project.

During construction, the south shaft will be used as the main construction/mainte-
nance entry. It is anticipated that the construction shaft will be excavated at a diameter
of about nine meters (30 ft) to allow efficient access for tunneling and pipe placement.
At the north shaft, there is limited available space for the shaft, with existing SFPUC
pipelines, the San Mateo Creek and Crystal Springs Road all placing limits on the construction shaft diameter.

The south shaft will be constructed on land owned by the SFPUC. There is sufficient space to support the tunneling operation and to allow disposal of the excavated
material on-site. The existing hillside will be regraded, avoiding the need for costly
transport of the spoils to landfill sites.

GROUND CONDITIONS

The entire length of the proposed tunnel will be constructed in bedrock contained
within the Franciscan Complex. Franciscan Complex rocks are characterized by lack of
internal continuity of contacts or strata and by the inclusion of fragments and blocks of
all sizes embedded in a fragmented matrix of finer-grained material (Figure 7). Franciscan rocks along the proposed tunnel alignment are classified as mélange as viewed
at many scales: 1) At a regional scale the mélange encompasses the California Coast

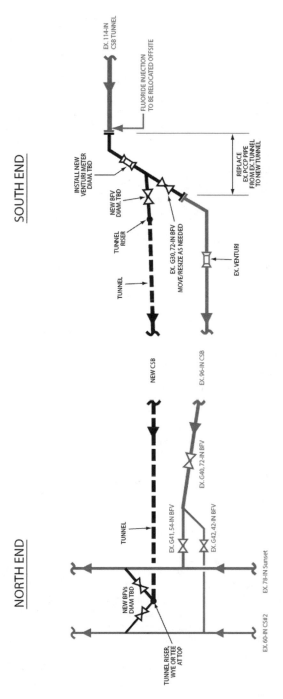

Figure 4.   Schematic configuration of New Crystal Springs bypass tunnel

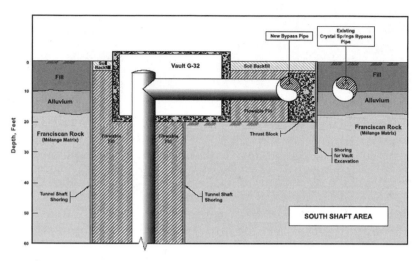

Figure 5.   Connection into existing pipeline at south end of the new tunnel

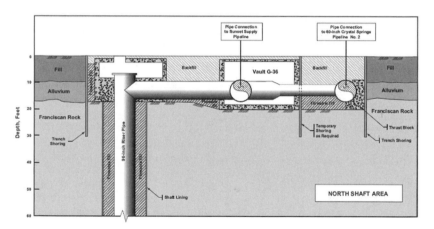

Figure 6.   Connection into existing pipelines at north end of the new tunnel

Range and includes blocks the size of mountains; 2) At the scale of an exploratory drill hole, a block might fit into one core box. Studies by Medley (2002) can be used to determine the block size ("characteristic engineering dimension") of significance for the proposed new tunnel with the smallest block size or block/matrix threshold being 125 to 150 mm (5 to 6 inches) and the largest block size 365 m (1,200 feet). Considered in this light, the entire tunnel will be constructed in Franciscan Complex mélange as depicted on smaller scale regional maps.

As part of the preliminary studies for this project, the SFPUC undertook a series of investigations in 2003, which included field explorations and testing as well as geologic investigations within the limits of the project. Further investigations were carried out in 2005 by the Arup/Brierley design team to supplement the available information and fill recognized data gaps. The major rock types identified during field mapping and from subsurface explorations along the proposed tunnel alignment were Mélange

Figure 7.   Franciscan Complex exposed on a nearby project

Matrix, Sandstone, Shale, Siltstone, Greenstone, Chert, and Serpentine. The Mélange Matrix (35–55%) and Sandstone (30–40%) are the dominant rock types.

The Mélange Matrix ranges in character from intensely fractured or "flaky" shale with low recovery, to cohesive clay with complete recovery. Core recovery in the mélange matrix materials ranged from 0 to 100 percent with an average core recovery of 89 percent. The fracture frequency observed in core samples of these materials ranged from 0 (intact core) to crushed. The matrix material consisting of cohesive clay with clasts of intact materials generally resulted in high core recovery. It is anticipated that the mélange matrix will experience swelling after excavation due to the relief of in situ stresses. However, the amount of deformation that will result from swelling should be expected to be small and to occur relatively quickly. Squeezing may also occur due to yielding of the mélange matrix and both the primary lining and the tunnel boring machine shield must be designed to accommodate the resulting loads.

The Sandstone along the tunnel alignment varies in character from completely crushed to massive. The RQD index was found to range between 0 to 100 percent, with an average of 23 percent. As part of the geotechnical investigation for the new tunnel, a large number of point load index tests were performed, from which the uniaxial compressive strengths of the sandstone can be estimated. In addition, 29 uniaxial compression tests were performed, including 18 tests as part of the 2003 (SFPUC) investigation and 11 tests as part of the 2005 (Arup/Brierley) investigation. Uniaxial compressive strengths measured ranged from less than 7 MPa (1,000 psi) to 193 MPa (28,000 psi).

Soil deposits overlie the Franciscan Complex. At the south shaft the soil is colluvium down to a depth of just over six meters (20 ft), comprising stiff to very stiff gravelly clay and clayey sand and gravel, indicating that some alluvial soils may be mixed with the colluvium. At the north shaft location, adjacent to the San Mateo Creek, there is approximately 3 meters (10 ft) of fill and a further 1.8 meters (6 ft) of alluvium.

The ground water table is approximately 10 feet below ground surface. As would be expected from the mixed nature of the Franciscan Complex, measured permeabilities varied. Even in the regions of higher permeability, the volume of water that can flow into the excavation is anticipated to be limited and with low recharge potential. The possible exception is under the San Mateo Creek. Two additional inclined boreholes were drilled at the creek to increase the available data on the elevation of top of rock and determine if any infilled erosional channels were present. Horizontal boreholes from the north shaft will also be drilled during construction to identify any areas of high permeability prior to tunneling.

TUNNEL ALIGNMENT AND DESIGN

Alignment. The proposed new tunnel alignment is shown on Figure 8 and was chosen using the following criteria:

- Develop an alignment roughly parallel to the existing Crystal Springs pipeline,
- Maintain the tunnel alignment outside and away from the previous landslide area, avoiding the new retaining wall structure,
- Maintain the tunnel alignment within land owned by the City&County of San Francisco and within the public right of way,
- Maintain two tunnel diameters (7.5m (25ft)) spacing between the tunnel and Crystal Springs pipeline,
- Locate the north shaft on the northern side of the San Mateo Creek on the gravel pull-in adjacent to Crystal Springs Road,
- Locate the south shaft to make connection and perform shaft construction on the six acres of property owned by SFPUC near the west portal of the existing tunnel.
- Locate the north shaft such that the TBM can be driven beyond the north shaft and abandoned or dismantled below Crystal Springs Road,
- Locate the tunnel entirely in rock,
- Provide sufficient depth to the tunnel crown below the San Mateo Creek so as to have a minimum depth of one tunnel diameter of good rock above the tunnel, and
- Provide a constructible cost effective vertical grade.

The horizontal alignment as shown on Figure 4 starts from the South Shaft (station 10+00) on the north side of Polhemus Road near the portal of the existing CSBT. The new tunnel would then follow closely along the north side of Polhemus Road, generally on a straight alignment with three 300 m (1,000 ft) radius curves. At the north end of the alignment, the tunnel would cross under Polhemus Road and then under San Mateo Creek, ending at the North Shaft (station 51+76).

The vertical tunnel alignment studied during the alternatives assessment had a downhill gradient of 3%, approximately following the natural topography along the alignment. On selection of the tunnel as the preferred option and with commencement of the design, it was recognized that imposing a downhill gradient is not desirable for tunnel construction. Working with a downhill gradient imposes additional costs and restrictions on muck and materials haulage and introduces safety concerns with water inflows and runaway trains. Cost estimates demonstrated that, within the range of accuracy of the estimates, there was little difference between tunneling with a 3% downgrade, and sinking the south shaft an extra 36 m (120 ft) deeper allowing level tunneling. For this reason, it has been decided that the final vertical alignment will be chosen by the contractor to allow for competitive bids. Two limiting vertical alignments are shown on Figure 4, one with a 3% grade downhill from the south shaft and the second with a 0.2% slight downhill grade. The contractor will be allowed to identity in his bid and construct a vertical alignment at or within these two limiting vertical alignments.

Tunnel Design. As previously noted, the final pipeline will consist of welded steel pipe. The initial support for the tunnel will be a continuous lining of precast concrete segments. Due to the nature of the ground with materials such as the mélange matrix, use of a gripper machine is not possible. Torque generated when cutting through the

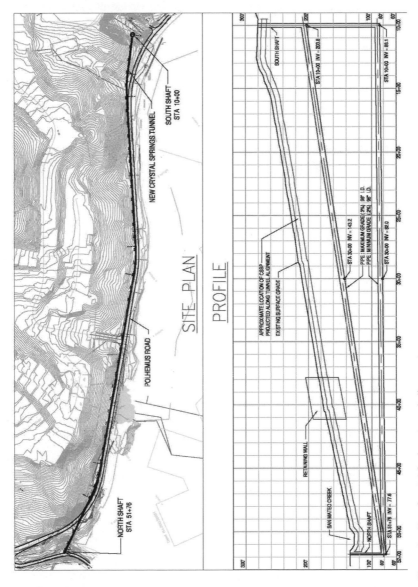

Figure 8. New Crystal Springs tunnel alignment

material at the face will need to be resisted by a combination of friction around the shield and the initial liner. Given the variability of the ground, the potential exists for the material at the front of the machine to generate a high torque but the lining resisting the torque to be in weaker ground (may want to rework last portion of sentence). In addition, the mélange matrix will deteriorate when exposed to water. For these reasons, the precast concrete segmental lining will be specified in order to provide a sufficient thrust reaction for the tunnel machine.

Environmental Issues. As previously described, the shaft sites are located in close proximity to residential properties. Particularly at the south shaft, these properties are exposed to noise and vibration generated at construction locations. This will require extensive noise control measures to allow for 24-hour workday operations while tunneling. These measures are likely to include noise walls, noise enclosures for equipment such as ventilation fans, and restrictions on the use of non-essential equipment. There will also be time limitations on blasting.

Dust control during shaft and tunnel excavation and spoil disposal will also require the Contractor's attention. It is anticipated that serpentine, which can contain asbestos, could account for up to 1% of the volume of excavated materials, and the works will be subject to the California Air Resources Board's "Asbestos Airborne Toxic Control Measure for Construction, Grading, Quarrying, and Surface Mining Operations" Act that became effective in July, 2002. This requires that an Asbestos Dust Mitigation Plan be prepared, submitted to and approved by the Regional Air Quality Board District before the start of construction activities.

CONSTRUCTION OF ADJOINING 1960s
CRYSTAL SPRINGS BYPASS TUNNEL

This existing tunnel was built in the late 1960s to the south of the planned alignment of the proposed new tunnel. The northern end of the existing tunnel is in the vicinity of the south shaft of the NCSBT, and the ground conditions are similar. The project was described in a couple of (sometimes contradictory) magazine articles *Western Construction 1967* and *California PB&E, 1968*. The location of the existing tunnel is shown in Figures 1 and 2.

The Crystal Springs Bypass system was built to allow water to be transmitted directly from the Hetch-Hetchy reservoir in Yosemite National Park to San Francisco Bay Area without passing through the intermediate Crystal Springs reservoir. During the rainy season, run-off into the Crystal Springs reservoir was causing turbidity in the water and as a result discoloration, which the bypass system would avoid. The bypass system as shown in Figure 2 consisted of a tunnel and a pipeline now being replaced with the proposed new tunnel.

The existing tunnel is three and a quarter miles long with a final internal diameter of 2.9 m (114 inches). The tunnel was constructed by a joint venture of Gates & Fox Inc, Gordon H. Ball Enterprises and Granite Construction between 1966 and 1969. The low bid price was $8.6 million.

The project is described in the *Western Construction,* February 1969 as the first project in the western United States to use a Loren Scott "mole," or wheel excavator. The 7.6m (25ft) long mole was held in position by eight jacks, four near the front and four near the rear. The machine was advanced by two 900 mm (36 inch) long thrust cylinders. The 3.8 m (12 foot 6 inch) cutter wheel itself was powered by two, 300 volt, 60 horsepower Westinghouse motors. While the cutting wheel is not described in detail in the available magazine articles, it is thought that the wheel would consist of a relatively open face with four spokes equipped with cutter teeth. A partial shield extended over an arc of 120 degrees for the first 12 feet behind the cutter wheel. Where the ground was too strong for the tunneling machine, or small blocks of stronger material were present, blasting was used to progress the tunnel drive.

The tunneling machine was selected as it allowed continuous mining. Limitations on blasting due to nearby residential areas would have limited conventional drill-and-blast tunneling to two shifts. In addition, since the required shape of the final tunnel was circular, use of the tunneling machine avoided over-excavation of the invert that would have occurred with drill-and-blast. The existing tunnel was mined from the same worksite proposed for the new tunnel, and the excavated material was also placed in the same location as is planned for the NCSBT. The tunneling machine performed well and the project was completed approximately a year ahead of schedule. As far as can be determined, the tunneling machine covered 4,300 m (14,100 ft) in about 15 months, which equates to a weekly progress rate of 70 m (225 ft).

Initial support to the tunnel consisted of 7.3 kg (16.2 lb), 100 mm (4 inch) steel H ribs. The spacing of the ribs varied from 600 mm (two feet) to 1,200 mm (four feet), the maximum spacing allowed in the specifications. Coeur d'Alene lagging was used between the ribs. In one 60 m (200 ft) long section, squeezing ground was encountered which required the ribs to be placed at the 600 mm spacing.

The final lining for the existing tunnel was cast-in-place concrete lining, with a minimum thickness of 300 mm (12 inches). This was generally placed as plain concrete, although reinforcement was used in the section of squeezing ground. The lining was cast using continuous pneumatic placement and telescopic forms.

The original geological logs from the construction of the existing tunnel are available and provide interesting details on the construction progress. The beginning of the drive (Figure 9) illustrates the varied nature of the Franciscan Complex. The tunnel started at station 172+30, and began mining through soft sandstone and shale. Daylighting occurred after 22 m (75 ft) between stations 171+55 and 171+36 due to the blocky nature of the ground. After a further 30 m (100 ft), squeezing ground required the removal of the shield and loaded the steel sets. Within another 30 m, the ground was hard sandstone that required blasting and the steel sets were not being loaded. While subsequent progress was somewhat less eventful, the variation continued between stronger rock requiring blasting, and weaker squeezing ground.

It would appear from the geological logs that water inflows were low. Some locations indicate an initial flush flow that then decreased to just drips within a few days. This would be consistent with the anticipated behavior of the Franciscan Complex, where discrete areas of discontinuities may be filled with water and cause an initial flow. However, with no continuous water path for recharge the flows quickly reduced.

PLANNED BID SCHEDULE

In May 2006, the proposed tunnel project environmental status was changed from a Mitigated Negative Declaration to a full environmental analysis. While the tunnel design is expected to be complete in the first Quarter of 2007, it awaits final environmental certification, now expected in early 2008. The environmental documents are currently under preparation in accordance with the California Environmental Quality Act (CEQA). The Bid and Award phase of the project is expected to commence in the fourth Quarter of 2007, with the prequalification of Bidders expected to happen in Fall 2007.

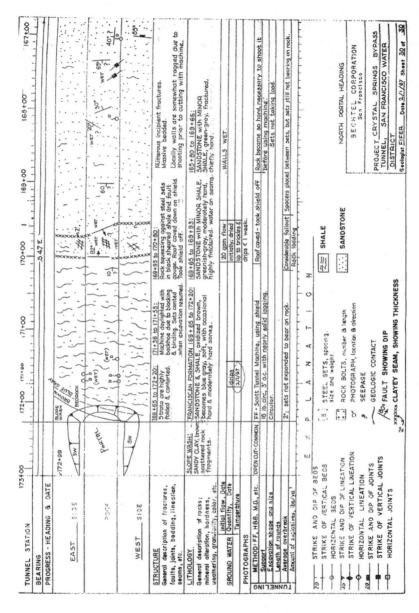

Figure 9.   Geological log from construction of Crystal Springs bypass tunnel

CONCLUSIONS

Following the landslide which threatened the integrity of the existing pipeline, construction of a new tunnel was determined to be more favorable than reinforcing the existing pipeline because it allowed the SFPUC to provide both greater water delivery reliability and redundancy for maintenance or in the case of an emergency. Given the demanding environmental setting, the tunnel alternative reduced physical, social and other impacts on the surrounding community.

The ground conditions in this area are highly variable. Tunneling in the Franciscan Complex will result in rapid changes of ground conditions from competent sandstone to potentially squeezing mélange matrix. The contractor will be given the flexibility to choose the vertical alignment, allowing the most cost-effective choice to be made between tunneling down grade and construction of a deep shaft. Control of noise, vibration and dust during construction will be important, and will require appropriate mitigation measures.

The construction of the existing adjoining tunnel in the late 1960s provides a good indication of the variable nature of the ground conditions, and demonstrates that tunnels can be excavated at good rates of progress with the right equipment. The use of a tunneling machine was beneficial and allowed project to complete ahead of schedule. The geological logs from the project are most beneficial in helping to understand the ground conditions and water inflow for the new tunnel.

ACKNOWLEDGMENTS

The authors acknowledge the permission of the San Francisco Public Utilities Commission to publish this paper.

Other individuals should be acknowledged for playing a major role in the development of this project. These include Tasso Mavroudis, SFPUC Project Manager during the conceptual engineering and early design phases; Holly Chan, SFPUC engineer who coordinated much of the design work; Gary Brierley of Brierley Associates who served as the design consultant's Project Manager; and Demetrious Koutsoftas of Arup who managed the comprehensive geotechnical investigation. Joe Sperry, advisor to the SFPUC, provided the historical information describing the construction of the existing tunnel.

REFERENCES

Medley, E. 2002. Estimating Block Size Distributions of Mélanges and Similar Block-in-Matrix Rocks (Bimrocks). *Proceedings, 5th North American Rock Mechanics Symposium*, University of Toronto, Toronto, Canada, pp. 599–606.

Western Construction, February 1967. Mole and conventional mining methods used on Northern California Water tunnel, pp 63–65.

California PB&E, April 5, 1968. Crystal Springs Tunnel is One Year Ahead of Schedule.

Armistead, Thomas F. July 17, 2006. Hetch Hetchy System Launches Upgrade. ENR *Engineering News Record,* pp. 10–12.

NEW TEST METHODOLOGY FOR ESTIMATING THE ABRASIVENESS OF SOILS FOR TBM TUNNELING

B. Nilsen

Norwegian University of Science and Technology

F. Dahl

SINTEF Rock and Soil Mechanics

J. Holzhäuser

Smoltczyk & Partner GmbH

P. Raleigh

Jacobs Associates

INTRODUCTION

Tunnel excavation using tunnel boring machines (TBMs), has become increasingly common in recent years, despite the fact that precise evaluation of certain risks have not kept pace with the use of these machines. One of the risks easily overlooked by Engineer and Contractor alike are the effects of abrasive ground on the costs and schedule of a given project. The impacts of worn and damaged TBM cutter heads have been observed on hundreds of tunnel projects around the world. It would appear that a reliable prognosis of the abrasiveness of soils on a project would be of great importance for designers, clients and contractors alike. Several well acknowledged test and prognosis methods already exist for rock, however there is only very limited knowledge available to describe the abrasiveness of soil and its impact on soft ground TBMs. This paper will examine approaches to this problem and suggest a new approach based on a current project undergoing design.

DEFINING WEAR

For the purposes of the following discussion it will be necessary to introduce the terms to be used, *primary wear* and *secondary wear*. By *primary wear* we refer to the expected wear on the excavation tools and surfaces such as drag bits, disc cutters, scrapers and buckets etc. which are designed for excavation and require "normal" replacement at appropriate intervals. *Secondary wear*, on the other hand, is an unplanned wear and occurs when the primary wear on the cutting tools described above is excessive leading to wear of the structures designed to hold or support the tools in place such as cutting head spokes or cutter mounting saddles and wear on other surfaces not anticipated by the designers and TBM manufacturers (Herrenknecht and Frenzel, 2005).

IMPACT OF ABRASIVE GROUND ON TBM TUNNELING

In abrasive ground, wear can occur on several parts of the TBM, including wear on the excavation tools, front, rear and periphery of the cutterhead structure, bulkhead and plunging wall structures, on outlet devices such as screw conveyors on EPB-TBMs or slurry pipes, valves and pumps on Slurry-TBMs. It is clear that during

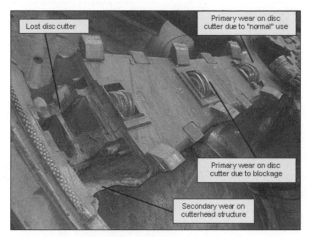

Figure 1.   Excessive wear on the cutterhead of a slurry—TBM Ø 11.7 m

the design phase, TBM manufacturers should have access to objective wear characteristics of the ground to be encountered in order that a rational approach to TBM component selection and wear protection may be adopted. Moreover, during the operational phase when the TBM components are exposed to the abrasive ground, an agreed plan for scheduled inspections and maintenance should be prepared by the Contractor. Daily cutter head inspections are common in hard rock TBM drives where cutter head access is relatively easy, however cutterhead inspections on soft ground TBM projects are typically executed where convenient or as indicated due to reduced TBM performance. Typically the presence of groundwater in soft ground tunnels makes cutter head interventions more complicated and time consuming compared to hard rock tunnels.

The examples in Figures 1 and 2 illustrate the extent of wear which can be observed on soft ground TBM tunnel projects. If primary wear remains undetected and the carbide inserts on drag bits or the disc cutter ring steel and hub body of these tools fitted to the face of the cutterhead become excessively worn, subsequent secondary wear on the cutterhead structure itself can develop rapidly as observed on the periphery of the cutterhead after breakthrough of the first tube on the Wesertunnel in Germany, shown in Figure 1. Sticky clay can block disc cutters from rotating, so that they remain in one position and are ground down on one side (flat-spotted).

Wear on the outside of the cutter head rim caused by inappropriate gauge cutter material that resulted in failure and loss of originally fitted chromium carbide wear plate on the cutterhead rim is shown in Figure 2, as observed on the ECIS project in Los Angeles. Here the cutterhead radius shows the loss of 2 cm of carbide plate in addition to 2 cm of structural wear. This loss increased the required thrust force applied and slowed the TBM progress rate. Extensive underground repair works were required, delaying the works for several months.

On Slurry-TBMs *secondary wear* can occur if the rear part of the cutterhead turns within the shield (Nilsen et al. 2006a)). The excavated material drops down into the bottom of the excavation chamber where the cutterhead must then plough through a volume of accumulated spoil (Babendererde et al., 2000). For example on the 4th Elbe tunnel project in Germany (bore diameter 14.2 m) severe wear occurred in this area of the TBM and had ground down the steel structure of the cutter head from 80 mm thickness to just 15 mm.

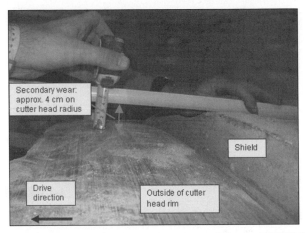

Figure 2.   Excessive secondary wear on the outside cutterhead rim of an EPB-TBM Ø 4.7 m

Figure 3.   A bucket tool before repair showing wear through fixing bolts (left) and after repair (right)

On EPB-TBM tunnel drives significant *secondary wear* can occur while the excavation chamber is filled with excavated material and pressurized. As the pressure within the excavation chamber increases, the *secondary wear* increases as a function of pressure, as has been observed on major projects in such places as the Porto Metro in Portugal and the MTA in Singapore.

Figure 3 illustrates the peripheral area of the cutterhead before and after repairs had been affected underground adjacent to the 24 do Agosto station on the Porto Metro. The wear was largely due to the abrasive Porto granite in its various states of weathering. The use of closed mode (EPB) operation where the TBM excavated mixed soil conditions also contributed to the observed wear requiring six weeks of around-the-clock working in order to complete the cutter head refurbishment.

Figure 4 summarizes the cutter consumption, ground conditioning used and the ground type actually encountered along the Porto Metro line S. The ground conditions

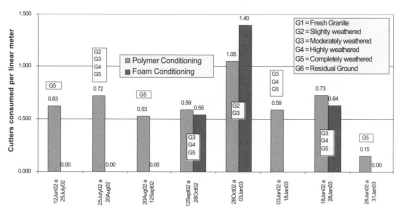

Figure 4. Disc cutter consumption for polymer and foam ground conditioning in the various weathered granites found along the Porto Metro line S.

range from G1–fresh granite to G6–residual ground. As can be noted there does not appear to be a great difference between the types of ground conditioning employed, however the degree of weathering seems to play a crucial role in determining wear. It was quite typical to change 4 to 5 cutters per day which required almost an entire shift to accomplish, thus permitting only 7.5 to 9m per day of advance.

The examples as previously described give a first indication of the variety of wear problems in soft ground TBM tunneling. Up to now there has been no generally acceptable method for estimating the amount of wear to be expected in relation to objectively tested soil properties other than recourse to the anecdotal references of adjacent projects. It is clear that the ground abrasiveness characteristic is only one of the factors which affect both the primary and secondary wear observed. TBM operational modes, the type of TBM be it EPB or Slurry, and the additives used for ground conditioning and timely maintenance are among other important factors. However the characterization of the abrasive properties of the ground plays the most important role in the development of effective strategies for dealing with the problem of wear.

EXISTING TEST METHODS TO DESCRIBE
THE ABRASIVENESS OF ROCK AND SOIL

All rocks and soils consist of minerals, which all have their distinctive scratch hardness. To define the hardness, the Moh's hardness scale is most commonly used. The scale is divided into ten increments, ranging from talc, with a hardness of 1, as the softest to diamond (hardness 10) as the hardest. The scale is linear from hardness of 1 to 9, with each mineral being able to scratch the one below it in the scale.

Among the most common minerals, mica and calcite are very soft (hardness 2.5 and 3, respectively), while feldspar, pyroxene and amphibole may be characterized as medium hard (hardness 6). Quartz and garnet are very hard (hardness 7 and 7–7.5, respectively), and to a great extent, determine the degree of cutter wear.

Cutter life can be estimated from the relative percentage of minerals of different Moh's hardness classes (>7, 6, 4–5 and <4). For coarse grained rock and soil this is most commonly determined by petrographic analysis in microscope. For fine grained rock and soil it is most commonly determined by X-ray diffraction (XRD), some times supplemented by differential thermal analysis (DTA). The higher the percentage of hard minerals found at the face, the more abrasive the soil or rock, and the shorter the cutter life.

In addition to mineral composition, many other textural features however also influence on TBM performance, such as: grain size, shape and elongation, grain orientation, degree of anisotrpy, grain suturing, interlocking, micro fractures and pores.

The use of Moh's hardness therefore is restricted mainly to preliminary estimates of cutter wear. As far as is known, Moh's hardness is not used directly as input in any TBM performance prediction model.

Test Methods for Rock

For rocks several methods for estimating abrasiveness exist already. The most commonly used are (Ozdemir & Nilsen, 1999 and Büchi et al. 1995):

1. The Vickers test, giving the Vickers Hardness Number (VHN)
2. The Cerchar test, giving the Cerchar Abrasivity Index (CAI)
3. The LCPC abrasimeter test, giving the LCPC abrasivity index (ABR)
4. The NTNU abrasion test, giving the Abrasion Value (AV/AVS)

These methods normally give a fairly reliable estimation of the abrasiveness. The greatest challenge in most cases is to collect representative samples. The first three methods are briefly discussed in the following part of this section, while the NTNU-test is discussed in more detail below.

Vickers hardness defines the micro-indentation hardness of a mineral, and provides a Vickers hardness number (VHN). The hardness number is defined as the ratio of the load applied to the indenter (gram or kilogram force) divided by the contact area of the impression (square millimeters). The Vickers indenter is a square based diamond pyramid with a 130° included angle between opposite faces, so that a perfect indentation is seen as a square with equal diagonals. A virtually linear relationship has been found between Moh's hardness and VHN (in log-scale). As with Moh's hardness, the use of VHN is primarily for the purpose of preliminary estimates of abrasivity and the expected cutter wear.

The Cerchar test is performed by scratching a freshly broken rock surface with a sharp pin of heat-treated alloy steel The Cerchar Abrasivity Index (CAI) is then calculated as the average diameter of the abraded tip of the steel pin in tenths of mm after 1 cm of travel across the rock surface. The advantage of this test is that it can be performed on irregular rock samples. The CAI value is related directly to cutter life in the field. The CAI values vary between less than 0.5 for soft rocks such as shale and limestone to more than 5.0 for hard rocks such as quartzite.

The LCPC abrasimeter test was developed in France to test the "abrasivity" and "breakablity" of granular material such as crushed rock or synthetically created materials (Büchi et al. 1995). Layout of the test apparatus and the procedure of the LCPC test are described in the French Code P18-579. As only the 4 mm to 6.3 mm fraction is used, coarse grained material, such as rock samples, have to be crushed and sieved after drying at typically 105 °C (limestone at <50 °C). Fine grained (<4 mm) and very coarse grained material (>6.3 mm) are not included in the original LCPC test.

A total amount of 500 g of the 4 mm to 6.3 mm fraction is filled into a steel cylinder at an internal diameter of 93 mm (approx. 4 inches). Within the cylinder a rectangle steel propeller is rotated at 4,500 rpm for 5 min (Figure 5). The propeller is made of relatively soft steel (Rockwell B 60-75), which can be easily scratched with a

Figure 5.   The LCPC abrasimeter test apparatus

knife. The abrasion coefficient ABR corresponds to the weight loss of the propeller per tonne of sample.

The NTNU Abrasion Test (AV/AVS)

A methodology for estimating the drillability of rocks by percussive drilling was developed at the Engineering Geology Laboratory of the Norwegian Institute of Technology (NTH) already in the early 1960s (Lien, 1961). Abrasion testing of crushed rock particles <1.0 mm, as illustrated in Figure 6, was then introduced together with the Brittleness test and the Sievers-J miniature drill test for estimating the drillability parameters DRI (Drilling Rate Index) and BWI (Bit Wear Index).

Since the early 1980s, the tests have been used mainly for predicting hard rock TBM wear performance according to the method developed by the NTH (since 1996 named NTNU) Department of Building and Construction Engineering, Bruland, Dahlø & Nilsen (1995). For TBM cutter wear prediction, a test piece of steel taken from a cutter ring is used instead of the tungsten carbide test piece used for percussive drilling estimation, and the parameter CLI (Cutter Life Index) is calculated instead of BWI. The NTNU prognosis model has been continuously revised and improved as new tunneling data has become available, and is now based on data from about 250 km of bored tunnels in Norway and many other countries around the world (NTNU, 1998).

The Abrasion Values AV/AVS represent time dependent abrasion of tungsten carbide/cutter steel caused by crushed rock powder. The same test equipment as for the AV is used to measure the AVS.

The two tests are defined as follows:

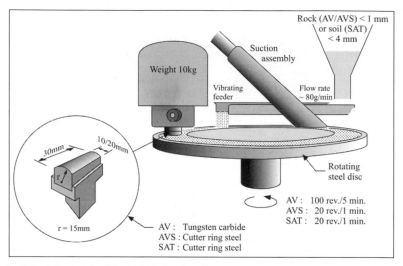

Figure 6. Principle sketch of the NTNU abrasion tests

AV The Abrasion Value is the mean value of the measured weight loss in milligrams of 2–4 tungsten carbide test bits after 5 minutes, i.e., 100 revolutions of testing, by using an abrasion apparatus and crushed rock powder.

AVS As described for AV, but with 1 minute, i.e., 20 revolutions of testing.

Test Methods for Soil

For soils the situation is quite different. There are only very few test methods to describe the abrasive characteristic of soils. Typically tests are limited to describe the hardness of minerals, such as the Vickers Hardness Number (VHN), Mohs hardness, quartz content and abrasive mineral content (AMC), but grain size of the soil is not taken into account.

Additionally there exist some abrasivity model tests for soils, such as the Los Angeles Abrasion Test, the Nordic Ball Mill Test (NBMT) and Dorry's Abrasion Test, which were developed to study the abrasion of aggregates to be used in road pavement works (Gudbjartsson and Iversen, 2003).

- The Los Angeles Abrasion Test rig consists of a rotating circular drum (∅ 0.7 m; L = 0.52 m) which is filled with cast iron spherical balls ∅ 48 mm along with the aggregates (5–10 kg). The cylinder is rotated at a speed of 30 to 33 rpm for 500 to 1,000 revolutions. Then the material is sieved through 1.7 mm sieve and the passed fraction is expressed as a percentage of the total weight of the sample. This value is called "Los Angeles abrasion value."

- The Nordic Ball Mill Test is common in Scandinavia and Iceland (similar to L.A. abrasion test).

- Dorry's abrasion test uses the resistance of aggregates to surface wear by abrasion induced by a rotating steel plate and is determined by measuring the volume loss of the aggregate specimen.

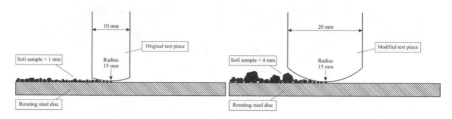

Figure 7.   The original (left) and modified SAT test pieces (right)

The former three tests are suitable to measure the abrasion of soil grains due to abrasion induced by steel or by contact to other soil grains but they are not valid to determine the abrasion of steel induced by soil, which is the case in TBM tunneling.

On a Slurry-TBM abrasion can have an adverse impact on the slurry discharge components, pipes and pumps. Particularly on long tunnel drives severe abrasion can occur due to the long period of exposure of the discharge components to flowing slurry mixed with excavated soil.

In the USA there is a standardized test, the so called Miller test (ASTM G75-01), which was originally developed in the oil industry for deep vertical borings, but deals with a similar abrasion problem as on Slurry-TBM drives. This test can be used to collect data from which the relative abrasivity of a slurry related to a standardized steel surface can be known, additionally the response of different materials to an abrasive slurry can also be investigated.

As described above, only few abrasion test methods are available for soils. They provide information on the abrasion characteristic of minerals within the soil and of slurry-soil mixtures, which is important information, but limited to specific aspects of the abrasion problem.

As will be described in the following, a new attempt has been made for an abrasion test for soils, the NTNU Soil Abrasion Test (SAT), describes the abrasiveness of soils in a more objective way. The initial testing has given quite promising results, and the test is believed to have a great potential for soft ground.

THE NEW NTNU SOIL ABRASION TEST (SAT)

Test Procedure

The new NTNU Soil Abrasion Test (SAT) is a further development of the existing abrasion tests for rock. Compared with the AVS test only one detail has been changed: instead of crushed rock powder <1 mm a sieved soil sample <4 mm is used in the SAT test. The initial SAT tests were performed with an upper grain size limit of 1 mm (Nilsen et al. 2006a to c), but has now by a modification of the original test pieces, as shown in Figure 7, been increased to 4 mm.

To enable comparison with previous test results and to take advantage of the extensive NTNU database it is considered important to follow the standardized NTNU abrasion test procedures as closely as possible. The following preparation of soil samples therefore is recommended, and has been followed for the soil testing described here.

In order to reduce or avoid changes of the original properties, soil samples should be dried gently in a ventilated oven at 30°C for 2–3 days. The following techniques should be used after drying in order to disintegrate and separate the particles for the abrasion powder:

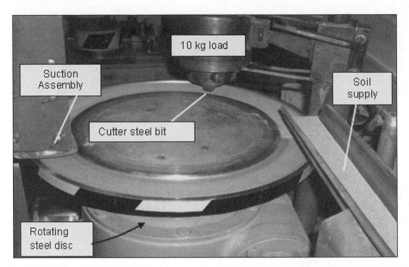

Figure 8.   SAT testing in the NTNU abrasion test rig. The test piece is clamped under the 10 kg load and is running on sand supplied on the rotating disc by the vibrating feeder.

1. Disintegration by use of a soft hammer.
2. Sieving with steel balls as gentle milling/disintegration aid.
3. Initial disintegration in a jaw crusher if the samples contain very hard lumps of cohesive material after drying. Crushing of intact grains should be avoided.

Action (1) had to be carried out for most of the samples described below, action (2) for all of the samples and action (3) for some of the samples.

SAT-testing of the sieved fraction was then carried out according to the same procedures as for AVS-testing, see Figure 6, and the SAT-value is calculated as the mean value of the measured weight loss in mg (to be accepted, the results of 2–4 parallel tests should not deviate by more than 5 units). SAT-testing in progress is illustrated in Figure 8 and examples of the appearance of test pieces after completed tests are shown in Figure 9.

Descriptions of Samples

The soil samples all originated from planned or completed TBM soft-ground tunnel projects in the Seattle area, USA, comprising the Brightwater Conveyance System, Henderson and Alki projects. They represent 4 typical soil types which may be described as follows:

- Clay: Low and high plasticity clays, low and high plasticity silts and scattered organic zones. Liquid limit average 46; plastic limit average 24. Average moisture content 24%. About 80% passes the No. 200 sieve.
- Silt: Non-plastic silt. Average moisture content 23%. About 75% passes the No. 200 sieve.
- Sand: Silty sand, poorly-graded/silty sand and poorly-graded sand. About 18% passes the No. 200 sieve.
- Gravel: Poorly-graded gravel, silty gravel, well-graded sand and well-graded gravel. About 10% passes the No. 200 sieve.

Figure 9. Abrasion of test pieces (L = 30 mm) after Soil Abrasion Test (SAT) (minimum 2 test runs per soil sample)

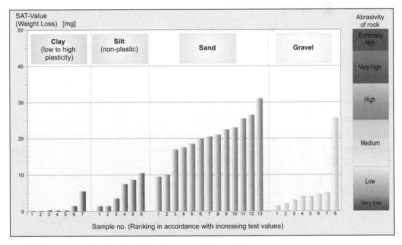

Figure 10. SAT test results for the Seattle area soil samples

Test Results

As can be seen in Figure 10, there is a considerable variation in abrasiveness, also within each soil type. This reflects the different mineralogical compositions of the samples and other features such as grain shape. Apart from a few exceptions the tested Sand has the highest SAT-values, and Clay the lowest.

EVALUATION OF SAT TEST RESULTS AND POTENTIAL OF METHODOLOGY

Compared to the standard AVS-classification based on testing of rock samples as shown in Table 1, the SAT-results based on sieved material <1.0 mm classify as "High" for 4 of the samples, and "Medium" for 8. The remaining 17 samples classify as "Low" to "Extremely low" concerning AVS.

Table 1.　AVS classification for rocks based on the NTNU/SINTEF database of 1590 rock samples

Category	% of Total	AVS
Extremely low	5	<1
Very low	10	2–3
Low	20	4–2
Medium	30	13–25
High	20	26–35
Very high	10	36–44
Extremely high	5	>44

Table 2.　AVS values for some sedimentary rocks and quartzite tested at NTNU/SINTEF

Rock Type	Number of Samples	AVS
Limestone	17	0.2–1.4
Shale	17	0.4–10
Siltstone	4	0.4–44
Sandstone	36	0.4–52
Quartzite	20	17–63

As already mentioned, the NTNU-abrasion test was developed originally for rock testing, and no previous experience exists for use of this test related to tunnel projects in soil. Tunneling in soil is quite different from TBM rock excavation, but there are also many similarities concerning cutter-tool abrasion.

These similarities can be evaluated and compared to the results in the extensive NTNU/SINTEF rock test database.

In order to get the maximum benefit from the SAT-testing, the test conditions and procedure have been kept as similar as possible to the conditions and procedure for rock testing. Although no database for SAT-testing exists, this testing is believed to give a fair indication of the abrasiveness of the various soils.

Based on rock testing, the content of quartz and other hard minerals like garnet and epidote have a major impact on the abrasion on the test pieces, but grain shape and grain binding may also contribute substantially. In Table 2, AVS-results for some sedimentary rocks tested at NTNU/SINTEF are shown, illustrating that there is a considerable difference in AVS-values between the softest (i.e., limestone) and hardest (i.e., quartzite) rocks. As also shown, the AVS-value may also differ significantly within one type of rock.

As illustrated by Table 2, the Seattle area soil samples have SAT-values similar to the AVS-values of rocks. Since a SAT-database for soil and a performance prediction model for soil tunneling based on SAT-values does not yet exist, the soil test results can not be used directly for estimation of wear of cutter tools. The test results do, however, provide a good basis for comparing the abrasiveness of the respective soils, and by comparing the results with those for rock, useful indications of relative abrasiveness may be obtained.

Table 3.   Comparison between SAT test and LCPC test

	SAT Test	LCPC Test
Type of material to be tested	Soil	Crushed rock
Grain size range	<4 mm Clay, Silt, Sand, Gravel (partly)	4 mm to 6.3 mm Gravel only (partly)
Material of test piece which is subject of abrasion	Cutter ring steel	Very soft steel
Rotation speed within contact surface which is subject of abrasion	20 rpm	4,500 rpm
Type of contact causing abrasion on test piece	Friction at **low** velocity within contact surface	Friction due to hitting impulse and due to **high** velocity within contact surface

COMPARISON BETWEEN SAT TEST AND LCPS TEST

As the SAT test and the LCPC test follow different approaches it seems useful to give a short comparison of both test methods concerning the key aspects. (Table 3)

Compared to the LCPC test it appears that the conditions of the SAT test are closer to the in-situ conditions during a TBM tunnel drive, where the cutter head rotates at a relatively low velocity of typically 1.5 to 2 rpm in soft ground. Due to the high rotation speed of 4,500 rpm, the abrasion produced within the LCPC test is caused predominantly by a hitting impulse and at high velocity within the contact surface of steel propeller and test material. The kinetic energy increases in square order at increasing velocity and increases linear with increasing mass and grain size respectively. This is probably the main reason for the LCPC test results published by Thuro et al. (2006), which indicate that "the LCPC abrasion values (ABR) increase more than linear with increasing grain size and that sand, silt and clay do not play a significant role on abrasion even at high quartz content." These results published by Thuro et al. (2006) do not correspond to the SAT test results, which indicate the highest SAT abrasion values for sand and lower values for gravel and silt/clay respectively.

Currently there is no database available which shows the actually encountered wear on TBM drives, abrasion values of lab test results (such as SAT or LCPC values) and type of soil or grain size distribution. We are currently on the way to develop such a database which will remain a key task over the next few years. Nevertheless, on several completed projects it appears that sand can be very or extremely abrasive, sometimes even more abrasive than gravel or sand/gravel mixtures.

CONCLUSION

The abrasiveness of soils:

- Results in schedule delays to many tunneling contracts
    - due to a failure to fully evaluate abrasion at the design stage
    - due to a failure of the contractors to properly inspect and maintain the TBM
- Is a soil property, which needs to be determined during design phase (site exploration)
- Should be taken into account by designers/owners in relation to

- number and duration of inspections and cutter changes (and additional measures if required)
- time
- cost
- development of contract documents
- Should be taken into account by contractors in relation to
  - pricing during bidding process
  - scheduling to accommodate reasonable TBM maintenance

REFERENCES

ASTM G75-01 Standard Test Method for determination of slurry abrasivity (Miller number) and slurry abrasion response of materials (SAR number). Sep 2001.

Babendererde, S., Babendererde, J. & Holzhäuser, J. (2000): "Difficulties with Operation of Slurry Tunnel Boring Machines." Proc. NAT Conference 2000, Boston, 317–326.

Bruland, A., Dahlø, T.S. & Nilsen, B. (1995): "Tunneling Performance Estimation Based on Drillability Testing." Proc. 8th ISRM Congress, Tokyo, 1995, 123–126.

Büchi, E., Mathier, J.F. & Wyss, Ch. (1995): "Rock abrasivity—a significant cost factor for mechanical tunneling in loose and hard rock." Tunnel 5/95, 38–44.

Gudbjartsson, J.T. & Iversen, K. (2003): "High-quality wear-resistant paving blocks in Iceland." Proc. 7th Int. Conf. in Concrete Block Paving, Sun City 12th–15th Oct. 2003.

Herrenknecht, M. & Frenzel, C.: "Long Tunnels in Hard Rock—A Preliminary Review." Bauingenieur 80 July/August 2005, 343–349

Lien, R. (1961): "An indirect test method for estimating the drillability of rocks." Dr. thesis, NTH Dept. of Geology, 90 p. (in Norwegian).

NTNU-Anleggsdrift (1998): "Hard Rock Tunnel Boring." Norwegian University of Science and Technology, Dept. of Building and Construction Engineering, Report 1B-98, 164 p.

Nilsen, B., Dahl, F., Holzhäuser, J., Raleigh, P. (2006a): Abrasivity of soils in TBM tunneling. Tunnels & Tunneling International, March 2006, 36–38.

Nilsen, B., Dahl, F., Holzhäuser, J., Raleigh, P. (2006b): Abrasivity testing for rock and soils. Tunnels & Tunneling International, April 2006, 47–49.

Nilsen, B., Dahl, F., Holzhäuser, J., Raleigh, P. (2006c): SAT: NTNU's new soil abrasion test. Tunnels & Tunneling International, May 2006, 43–45.

Nilsen, B. & Ozdemir, L. (1999): "Recent developments in site investigation and testing for hard rock TBM projects." Proc. RETC-Conference, Orlando 1999, 715–731.

Ozdemir, L. & Nilsen, B. (1999): "Recommended laboratory rock testing for TBM projects." AUA News 14:2, 21–35.

Thuro, K., Singer, J., Käsling, H. (2006): "Abrasivitätsuntersuchungen an Lockergesteinen im Hinblick auf die Gebirgslösung." Proc. Baugrundtagung 2006 in Bremen (Germany), VGE-Verlag, Essen, 283–290

Young, B.B. & Millmann, A.P. (1964): "Microhardness and deformation characteristics of ore minerals." Trans. Inst. Min. Metall., 73:437–466.

OUTLINE DESIGN AND CONSTRUCTION OF THE 10 KM (6.25 MILE) CROYDON CABLE (HIGH VOLTAGE) TUNNEL THROUGH GROUNDWATER SOURCE PROTECTION ZONES, UNITED KINGDOM

Colin Rawlings

Kellogg Brown & Root

David Keeble

Kellogg Brown & Root

John Mathews

Kellogg Brown & Root

Christopher Darton

Kellogg Brown & Root

Peter Townsend

Kellogg Brown & Root

Maria Karakitsiou

Kellogg Brown & Root

Russell Bowler

RBE Environmental

Scott Sadler

National Grid Company plc

Simon Pepper

National Grid Company plc

ABSTRACT

National Grid needs to maintain a connection between two existing electricity sub-stations (Beddington & Rowdown) in the Croydon (south London) area, and allow the replacement of two 275 kV cable circuits that are buried under roads. This is to be achieved using a single 400 kV cable circuit within a 10 km (6.25 mile) long, 3 m (9.8 ft) diameter high voltage cable tunnel which is being driven through water bearing Chalk, at depths ranging from 5 m (15 ft) to 80 m (262 ft), with four access shafts (12.5 m (41 ft) in diameter) ranging in depth from 10 m (31 ft) to 40 m (131 ft). The Chalk in this area acts as a source of public drinking water supply (PWS) and is protected by areas designated Groundwater Source Protection Zones. To minimise the groundwater risk the design and construction of the cable tunnel and shafts has paid particular attention to the construction materials used, including testing of these construction materials prior to use, developing construction methods (including the design of the Earth Pressure Balance TBM) and maintaining a groundwater monitoring & testing programme prior to and during construction to ensure the integrity of the Chalk aquifer is maintained as a drinking water source. This paper describes the tunnel & shaft design and construction in relation to these particular groundwater environmental issues which set new precedents for tunnel assessment, design and construction requirements in the United Kingdom.

INTRODUCTION

National Grid Company plc (National Grid) owns, operates and maintains the high voltage electricity transmission network (known as the "National Grid") in England & Wales and is a wholly owned subsidiary of National Grid Electricity Transmission plc. The National Grid transmission system operates mainly at 275,000 volts (275 kV) and 400,000 volts (400 kV). The primary role of this transmission system is to provide connection between the generating stations and distribution companies (Regional Electricity Companies) and a few large industrial users. As the holder of the Transmission Licence for England & Wales, National Grid is required by the Electricity Act 1989 to develop and maintain an efficient, co-ordinated and economical system of electricity transmission and to facilitate competition in the generation and supply of electricity. The steady growth in demand for electricity in London and South East England and the fact that many circuits installed in the highways in the 1960s are coming to the end of their life has made necessary a major refurbishment programme for primary distribution circuits. The replacement of cable circuits buried in trenches in highways has a very significant disruptive effect on the urban environment, particularly since there has been a rapid growth in the number of services installed in the highways and in vehicular traffic since the 1960s. The need to minimise disturbance to traffic and the local communities arising from long sections of open trenches has resulted in the construction of a number of deep level tunnels to accommodate 400 kV and 275 kV circuits.

Capital cities throughout the world are laying their primary high voltage transmission networks in tunnels, and this is becoming increasingly common in constrained urban environments. The advantages of cable tunnels are:

- Surface disruption is minimal and limited to point (shaft) locations
- Future maintenance and replacement of cables in the tunnel causes minimal surface disruption
- Upgrading and adding more cables is possible with minimal disruption
- Accidental damage from other utility excavation works within the highway is eliminated
- Improved security of system against third party damage is obtained.

In the United Kingdom National Grid uses powers granted as a transmission licence holder which are contained in Schedule 4 of the Electricity Act, 1989, to implement the development of new electricity transmission tunnels. In addition, the New Roads and Streetworks Act allows utility companies to tunnel under public highways without seeking specific planning permission. In addition, the New Roads and Streetworks Act, 1991 allows Utility Companies to tunnel under public highways without seeking specific planning permission. The 1989 Electricity Act allows for a necessary easement or wayleave (which is an agreement with a landowner for right of passage) provides additional safeguards to enable preservation of the right where the electric line passes through or below private property, but these provisions do not extend to dwellings. The 1989 Act also provides that a transmission licence holder can follow procedures for compulsory acquisition, which would be the only way to acquire a right, in the absence of a willing owner, if the cable tunnel were to pass under dwellings. National Grid's permitted development rights (Town & Country Planning (General Permitted Development) Order 1995, Schedule 2 Part 17G) grant planning permission for construction of cable tunnels and associated subsurface access/egress shafts.

The Croydon Cable Tunnel is designed to house 400,000 volt circuits that will maintain a connection between existing substations (Beddington & Rowdown), replacing two existing oil filled 275,000 volt circuits which were buried under the highway in 1964 and are reaching the end of their design life. These inter-connections are necessary to ensure that the electricity supply is maintained to statutory standards required by National Grid's licence. National Grid has a further existing 400 kV buried circuit between these substations that can be accommodated in the cable tunnel when it is replaced in the future.

It is noted that despite the scale of the engineering works, cable tunnel and shafts of this nature are not covered by Town & Country Planning (Environmental Impact Assessment) (England & Wales) Regulations 1999. For this reason and to address the requirements of the Electricity Act 1989 (Schedules 9 & 38), National Grid as a matter of policy prepared an extensive environmental report which assessed the proposals and was presented to interested parties and the general public. National Grid also has duties under Schedules 9 & 38 of the Electricity Act 1989 to preserve environments of special interest and to mitigate the impact of its activities on the environment.

Were it not for the need to route the tunnel through a major aquifer and across areas designated Groundwater Source Protection Zones 1 & 2, the environmental implications and mitigation measures necessary for the scheme would have been relatively straightforward. These groundwater sensitivities however, set new precedents for tunnel assessment, design and construction requirements in the United Kingdom.

CROYDON CABLE TUNNEL PROJECT

The Croydon Cable Tunnel (CCT) Project (see Figure 1) involves the construction of a tunnel linking National Grid's existing substations at Beddington (in the London Borough of Sutton) and Rowdown (in the London Borough of Bromley) by a 3 m (9.8 ft) internal diameter and 10 km (6.25 mile) long tunnel constructed at depths below ground level ranging from approximately 5 m (15 ft) to 80 m (262 ft). The tunnel has four shafts (12.5 m (41 ft) in diameter) connecting it with the surface, one at each end (within the existing substations) and two at intermediate points along the route to provide ventilation to the tunnel and a safe means of access and egress. The intermediate shafts are located at Lloyd Park, and adjacent to the A2022, at Kent Gate Way. The shaft at Kent Gate Way is the drive shaft site for the construction of the tunnel. All shafts require a headhouse (about 16 m (52.5 ft) square in plan and 7 m (23 ft) high) at the surface to accommodate and protect the associated mechanical and electrical equipment. The tunnel will house three 400 kV Cross Linked Polyethylene (XLPE) Cables with a cross-sectional area of about 2,500mm^2 copper conductor.

The current project status (January 2007) is that the detailed civil, mechanical & electrical and cable design and civil construction has started. Construction of the drive shaft at Kent Gate Way is complete. Tunnel boring (using one TBM) is due to commence in Spring 2007 from Kent Gate Way to Rowdown (and then from Kent Gate Way to Beddington) with completion of the tunnel and shaft construction by Autumn 2008. The headhouse construction and all mechanical & electrical fit out are due to be completed by late 2009. The 400,00 V XLPE cable is due to be manufactured in 2009 and installed in the tunnel, shafts and culvert between 2010 to 2011. The cable circuits are programmed to be commissioned by National Grid in late 2011.

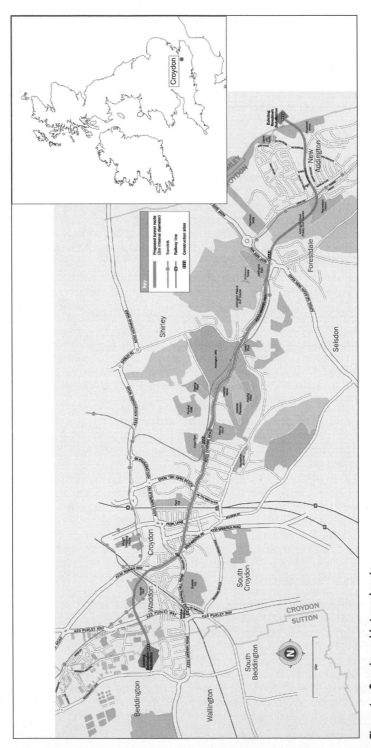

Figure 1. Croydon cable tunnel route

The following are the main organisations working on the Cable Tunnel Project: Kellogg, Brown & Root (KBR)—project manager on behalf of National Grid from feasibility study through outline design to detailed design and then construction management; Gardiner & Theobald—cost consultant; Morganest—main civil engineering contractor; and NKT—cable supplier and manufacturer with Balfour Beatty Power Networks Ltd carrying out the XLPE cable installation.

Tunnel and Shaft Lining Design

The tunnel and shaft bolted segments are manufactured by Morganest for a design life of 120 years. The tunnel and shaft linings consist of segmental precast plates (12 No per ring for the shafts; 6 No per ring for the tunnel) of high grade (C60) reinforced concrete designed to withstand the permanent dead and imposed loads of the soil and water. The design accords with the requirements of BS 8007, Code of Practice for design of concrete structures for retaining aqueous liquids and BS 8110, Structural use of Concrete. The shaft linings use conventional reinforcement whilst the tunnel linings are steel fibre reinforced. The concrete mix has been designed to provide a dense impermeable and fire resistant (use of polypropylene fibres) concrete. Phoenix gaskets provide watertight seals in the joints between segments. Due to the sensitive nature of the groundwater through which the tunnel is constructed a strict water tightness criterion was applied i.e., the completed tunnel, shafts and culverts are required to meet maximum water seepage as defined by Class 3 Capillary Dampness in the STUVA (Studiengesellschaft fur Unterirdische Verkerhsanlagen e.V.) recommendations for allowable water seepage into tunnels.: The twelve shaft smooth bore segments are 12.5 m (41 ft) internal diameter, 325 mm (13 inches) thick and 1.2 m (3.9 ft) long. The shaft segments have integral bentonite injection tubes, the facility for external introduction of radial joint bolts (ten ordinary segmental plates and two tapered plates). The six trapezoidal tunnel segments are 3 m (10 ft) internal diameter, 180 mm (7 inches) thick and 1.2 m (3.9 ft) long.

Tunnel and Shaft Construction

Due to the predominantly Chalk sub strata ground conditions at the tunnel horizon and the likely presence of groundwater at pressure over much of the tunnel length, tunneling will be carried out using an Earth Pressure Balanced full face tunnel boring machine (details provided later in this paper). Shaft construction through the overlying deposits/strata and the Chalk is being performed using caissons (dry or wet caissons). To ensure that shafts are sunk in a controlled manner, a computer control unit linked to the twelve hydraulic jacking rams is used to control the rate of sinking and provide an even advance across the shaft.

Tunnel Horizontal and Vertical Alignments

Horizontal Alignment. The 1989 Electricity Act provides the right for the electric line to pass through or below private property, but these provisions do not extend to dwellings. ["Dwellings" means normal residential property, whether occupied or not, and includes gardens, outbuildings, etc.] As a consequence of the need to avoid private property especially dwellings the horizontal alignment follows highways where possible. Extensive stakeholder dialogue necessitated a series of feasibility study reports to consider various horizontal alignment route options. This included a further cost benefit analysis route review which was prepared and presented to the Environment Agency (EA) and Thames Water Utilities Limited (TWUL). In general the alignment is designed for a minimum bend radius of 250 m (820 ft) to allow for future cable replacement possibly by Gas Insulated Line (GIL) as well as the minimum bend radius for a Tunnel Boring Machine.

Vertical Alignment. The vertical alignment was selected to avoid, as far as practicable, the "pay zones" of the Public Water Supply wells and the Private Water Abstractors as well as the zone of seasonal fluctuation of the groundwater table within the Chalk aquifer. In addition, shaft locations were influenced by the need to locate the shaft locations outside a Groundwater Source Protection Zone 1. Since the outline design was carried out by KBR in 2001/2002 there have been a number of exceptionally dry winters which have resulted in lowering of the groundwater table. This together with the use of a steeper vertical gradient has enabled the vertical alignment to be raised in the detail design carried out by Morganest in 2006.

Design of the alignment also needs to ensure that impact on other surface and underground structures is minimised. Approvals in Principle based on ground movement predictions were prepared, monitoring of ground movement will be carried out and condition surveys of structures will be carried out pre, during and post tunnel and shaft construction.

400 kV Cross Linked Polyethylene (XLPE) Cable Design

At 400 kV three single core cables (see Figure 2) are required to make up one three phase circuit. The 400 kV XLPE cables are manufactured by NKT for a design life of 40 years. This means that the tunnel can be used for two future cable renewals as well as an additional 400 kV circuit replacement and then renewals. The power cables are designed to meet with IEC 62067 "power cables with extruded insulation for rated voltages above 150 kV (U_m (maximum system voltage) = 170 kV) up to 500 kV (U_m = 550 kV). The design is a 420 kV XLPE cable with a compacted copper Milliken conductor. The conductor, which has a cross sectional area of 2,500mm^2 and of a Milliken configuration, is manufactured with powder within the individual wire strands, which swells on contact with water, to ensure longitudinal water tightness along the conductor. The insulation of the cable is obtained using cross-linked polyethylene (XLPE) with an extruded inner and outer semi-conducting layer. In addition, to the metallic screen of copper wires laid around the insulated core, the cable is provided with tape (which swells on contact with water) and coated aluminum foil laminate to ensure both longitudinal and radial water tightness. The outer polyethylene sheath will be extruded in two layers, the inner red and the outer black to enable a visual check that no damage to the sheath has occurred before the cable is jointed, commissioned or during operation. The overall weight of the completed cable is 42 kg/m (28lb/ft) length and will be delivered to site in lengths of approximately 1,100m length. The impact of short circuits and other electrical activity in the cable system have been calculated and taken into account in the design of the cable fixing cleats and has necessitated the need of "snaking" the cable during installation.

In addition to the power cable system, a distributed temperature monitoring system, a fire detection system, a Partial Discharge monitoring system and a sheath voltage monitoring system form part of the overall design of the cable system.

400 kV Cross Linked Polyethylene (XLPE) Cable Installation

The cable system will be installed in nine cross-bonded sections, each approximately 1,130 m (3,707 ft in length. The individual cable section lengths will arrive at site on cable drums with a combined weight approaching 50 tonnes. The cable will be drawn off from the drum at ground level and lowered down the shaft at Kent Gate Way via an adit constructed as part of the spoil removal system during the tunnel construction. A winch and bond/clamp arrangement will be used to guide the cable through the

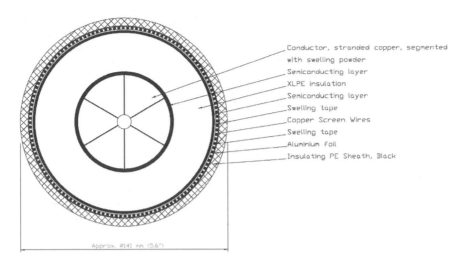

Figure 2.   Section showing a cross linked polyethylene (XLPE) high voltage (400kV) cable

adit to avoid straining the cable. Methods available to transport the cable along the tunnel include power rollers spaced along the tunnel invert and the use of the mechanised transport system provided for the operational use of the tunnel.

The cables will be transported to their relevant position in the tunnel using the monorail beam installed for the final operational vehicle system and then placed on cable cleats fixed to the support brackets. Joint bays are established along the tunnel where the cable section lengths end and subsequently jointed together. Transposition of the cable conductor and the cable screens occurs at each joint bay position to further improve the current carrying capacity of the cable and the cable screens at each joint bay position will be protected by sheath voltage limiters located close to the main cable joint. The cables will be supported vertically in the shafts and horizontally in the tunnel by means of cleats secured to vertical steelwork fastened to the concrete elements forming the tunnel and shaft lining. In the transitional bends from the tunnel to the shafts, independent steel structures, on which cleats may be fastened at suitable locations, will be constructed. The main supports and cleats in the tunnel will be located at approximately 7.20 m (24 ft) centres. At intermediate points between the main supports short circuit straps of a non-magnetic material will secure and maintain the spacing between cables. The cables will be allowed to sag by one cable diameter depth ("snaked") over the 7.2 m (24 ft) distance between two fixed clamps.

Prefabricated cross bonding joints will be provided with connectors to enable the screens to be cross-bonded and sheath voltage limiters to be installed in the immediate vicinity of the joint itself thus obviating the need of link boxes. The joints will be supported on isolated pedestals. The joints will be supplied and installed complete with sensors and terminals to enable remote monitoring of partial discharge activity within the joint. Sheath transient voltage limiters (SVL) will also be supplied with a remote system to monitor failure. The cable terminations will be provided with connectors to enable the system to be connected via a link box to earth. Commissioning of the cables will be in accordance with National Grid's testing and commissioning procedures.

Headhouses

At each shaft site there will be a headhouse about 16 m by 16 m in plan with a height of about 7 m. The headhouses are simple in design with steel frames, brick and block walls and suitable cladding to meet the Local Planning requirements. The roof will be pitched with a standing seam roof. The headhouses at Beddington and Rowdown are located within the existing substation. At Lloyd Park and Kent Gate Way, to take into account the local planning authority requirements and reduce visual impact, a sedun mat "green" curved sloping roof will be used. The headhouses accommodate the mechanical and electrical systems required to maintain a safe and secure environment for the cables to operate in and for personnel to access the tunnel for operational & maintenance requirements. The cable system can operate with a maximum conductor temperature of 90°C giving a potential maximum tunnel temperature of 50°C depending on the air inlet ambient temperature. A forced air ventilation system will be installed in the tunnel to remove the heat generated by the cables under normal use. The cable tunnel and associated M&E systems is designed to accommodate a second 400 kV cable when required.

Reversible, variable speed extract fans will be provided at the four shafts to meet the ventilation and personnel safety design criteria. The ventilation system will incorporate speed control of the fans to ensure that energy consumption is minimised. The use of fan speed control also has the advantageous effect of reducing the fan noise because periods of higher fan velocities will be minimised. Fan operation will be dependent on cable operating conditions and climatic factors consequently the fans will not operate all the time. During normal operations fresh air input will be provided at both Lloyd Park and Rowdown with extraction at Beddington and Kent Gate Way shafts. Air intake and exhaust will be through high-level louvers in the headhouses above the shafts. The intermediate headhouses, which are located within Metropolitan Green Belt, were the only buildings linked with the Croydon Cable Tunnel Project requiring planning permission from the local authority. For this reason the planning permissions drew considerable interest and a Section 106 (planning gain) agreement was developed with the local authority, generating separate project and programme risks.

GEOLOGY, HYDROGEOLOGY AND GROUNDWATER SOURCE PROTECTION ZONES

Geological Setting, Ground and Groundwater Conditions

Croydon lies on the southern part of the London Basin (Figures 3 and 4) which is a synclinal structure underlain by the Chalk aquifer, which outcrops to the north and the south (including the Croydon area of the North Downs). Above the Chalk lies a sequence of strata: Thanet Sands, overlain by the Lambeth Group, then the London Clay; and finally the Terrace Deposits. The Chalk is subdivided into Lower, Middle and Upper Chalk. The Croydon Cable Tunnel is located wholly within the Upper Chalk.

The Terrace Deposits (Kempton Park, Hackney & Lynch Hill Gravels) occupy the current and historical river channels and comprise medium dense to dense orange brown sandy fine to coarse gravel ranging in thickness from a few to several metres. The London Clay is present sporadically at the surface in the area to the north of Addington Hills, around Addiscombe. The deposit is up to 5 m (16 ft) thick and consists of dark grey pyritous silty clays with common courses of claystone. The Harwich Formation in the Croydon area consists of the Blackheath Beds. These deposits reach a maximum thickness of 15 m (49 ft) in the area and often locally areas of higher ground. The Blackheath Beds comprise cross-bedded dense green brown silty fine to medium

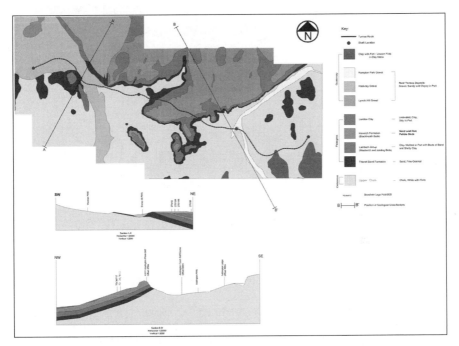

Figure 3.   Geological map and cross sections

sands with flint gravel and rounded pebbles of flint. The Blackheath Beds in-fill channels cut into the Lambeth Group. The Lambeth Group, comprising the Woolwich and Reading Beds, underlies the Harwich Formation. The Lambeth Group reaches a maximum thickness of 20 m (66 ft) in the area. It comprises interbedded stiff to very stiff grey—brown silty clays, dense to very dense light brown silty fine sands and sandy silts and green grey medium sands with glauconitic pebbles. The Woolwich Beds contain a basal conglomeritic layer containing rounded flint pebbles and nodular flints. The Thanet Sand comprises dense to very dense yellow brown silty fine sand which tends to be clayey and more silty in the lower part. At the base of the formation a bed of gravel and cobble sized glauconite-coated flint up to 0.5 m (1.6 ft) thick occurs, known locally as the Bullhead Bed. The formation caps hill tops in the study area and is found preserved locally in solution features in the Chalk. The Upper Chalk occurs below the Thanet Sand and is the stratum through which the tunnel construction will occur. The thickness of the Upper Chalk varies between about 65 and 100 m (213 to 328 ft) in the area. Structurally, the dip of the Chalk in the area of the tunnel route is similar to that seen elsewhere in the North Downs with the strata dipping gently at 3° to 8° to the north north east. Associated faults are predominantly sub vertical.

The ground conditions at the four shaft sites were studied from boreholes that were drilled in three separate Site Investigations which were supervised by KBR during 2003 and 2005 and Morgan Est during summer 2006. These are as follows:

- Shaft 1—Beddington Substation: Made Ground up to 1m below ground level, Hackney Gravel and Thanet Sand Formation up to 13m below ground level following by Upper Chalk

- Shaft 2—Lloyd Park: Made Ground up to 0.60m below ground level and then Upper Chalk

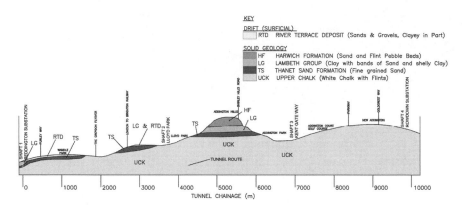

Figure 4. Geological cross section showing tunnel route (schematic vertical scale)

- Shaft 3—Kent Gate Way: Topsoil and Kempton Park Gravel up to 4.5m below ground level followed by Upper Chalk
- Shaft 4—Rowdown Substation: The shaft will be constructed through the Upper Chalk for the whole depth.

Thus in summary the whole tunnel construction will be in Upper Chalk as well as the shafts at Lloyd Park, Kent Gate Way and Rowdown. The shaft at Beddington will be constructed through the Hawich Formation/Lambeth Group/Thanet Sands and then Upper Chalk.

Details of the groundwater levels from the groundwater monitoring are given in the following sections.

Hydrogeology and Groundwater Source Protection Zones

The Chalk is Southern England's major aquifer which is both confined and unconfined with water production based largely upon fracture flow. The aquifer is in hydraulic continuity with the overlying layers and has complex hydraulic properties. Chalk has a porous matrix providing groundwater storage but, because of the pores and their interconnections being so minute, the matrix permeability is low. Chalk also has secondary porosity due to the presence of joints and fractures. These fractures are best developed at shallow depths and beneath valleys. These fractures provide rapid groundwater transport and contain the most mobile groundwater. In the area along the tunnel route the regional groundwater flow is to the north-northwest. Using the results of the ground/groundwater investigations, groundwater modelling and monitoring, stakeholder discussions and early contractor involvement the tunnel route was, as far as possible, selected to avoid the major zones of fracturing and high groundwater transmissivities.

The Chalk beneath the project area comprises part of the North Downs aquifer. This is heavily exploited for public and private water supply. There are four active public water supply (PWS) sources within the area of the tunnel, one in reserve, and several private abstractions operated by industrial users and golf courses. The PWSs are:

- Addington
- Stroud Green (held in reserve)
- Surrey Street
- Waddon
- West Wickham

The Environment Agency (EA) has a primary remit, under the Environment Act 1995 and the Water Resources Act 1991 to protect and monitor the integrity of aquifers and groundwater. The Groundwater Regulations 1998 (SI 2746) implement the EC Groundwater Directive (80/68/EEC) and provide the main significance criteria to assess groundwater quality and control the discharge of certain chemicals. The aquifer in the Croydon area (see Figure 5) has been subdivided by the Environment Agency (EA) into three classes (1, 2 & 3) of Groundwater Source Protection Zones (SPZ) as follows.

Groundwater Source Protection Zone 1 (Inner Source Protection Zone). This is defined by a 50-day groundwater travel time to the abstraction, based on the decay of biological contaminants, and must have a minimum radius of 50 m (164 ft). Zone 1 is designed to protect against activities that have an immediate effect on the source. Within Zone 1 operating procedures should be in place to minimize potential for pollution. This Zone is not usually defined where an aquifer is confined by a substantial and continuous covering strata of very low permeability since in such cases the cover will prevent infiltration. There must be no conduits, such as solution features, that may allow "short-circuiting" from the surface to the groundwater. The attenuating properties of a deep unsaturated zone or thick cover of Drift deposits may be sufficient to prevent contamination from some sources; this is not considered when developing Zone 1 SPZs.

Groundwater Source Protection Zone 2 (Outer Source Protection Zone). Zone 2 is defined by a 400-day groundwater travel time to the abstraction, and is designed to offer some protection against slowly degrading contaminants, through attenuation processes. This zone is generally not designated for confined aquifers.

Groundwater Source Protection Zone 3 (Source Catchment). Zone 3 encompasses the entire catchment for the groundwater source. It is defined as an area required to support an abstraction from long-term groundwater recharge. Where an aquifer is confined beneath impermeable strata, the source catchment may be a significant distance from the abstraction.

LIAISON WITH THE ENVIRONMENT AGENCY, THAMES WATER UTILITIES LIMITED AND OTHER THIRD PARTIES

Protection of the Groundwater—Compliance Concerns

The Groundwater Regulations 1998 require specific measures to be taken to prevent List 1 substances from entering the groundwater and to restrict the entry of List 2 substances so as to prevent pollution (see Table 1). The EA set as the compliance point, for the Croydon Cable Tunnel Project, the extrados of the tunnel and shaft linings. The EA and TWUL raised a number of concerns relating to groundwater impacts from the tunnel and shafts including:

- Inadvertent release of List 1 or List 2 substances into the groundwater because of leaching of substances used to construct the tunnel and shafts
- The need to carry out tests to demonstrate if leaching was likely
- Potential reduction in the water available in the PWS wells, and an associated increase in drawdown
- An increase in the turbidity at the PWS wells
- Risk of groundwater pollution over the life of the tunnel and shafts (this refers to possible pollution pathways created by the presence of the tunnel and shafts).

Risks potentially affecting the quantity and quality of groundwater are considered to be associated mainly with construction activities and include: tunnel grouting; issues of operating and maintaining a tunnel boring machine (TBM); shaft excavation; and

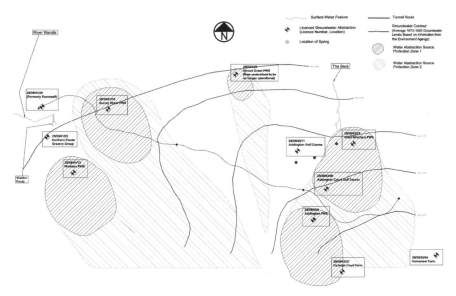

Figure 5.   Location of surface waters, licensed groundwater abstractions and source protection zones

dewatering associated with shaft excavation. Throughout the project a phased, risk based approach was used to address these concerns.

Stakeholder Dialogue

From feasibility study to detail design National Grid has worked closely with the EA, TWUL, the relevant local authorities and general public. Many meetings have been held with the EA and TWUL to address issues relating to the groundwater resource. The project has benefited from close liaison and the continuity of the EA project technical personnel throughout its development. Aspects which needed to be agreed with the EA included:

- Tunnel route—vertical and horizontal alignment
- Groundwater monitoring, sampling and testing including control and trigger levels as well as contingency measures
- Construction materials—leaching tests
- Methods of shaft construction
- Tunnel Boring Machine requirements
- Temperature effects of the cable
- Risk Assessments for the effects of tunnel & shaft construction of groundwater.

TWUL as the major abstraction licence holder has also played a role, influencing tunnel route and mitigation solutions, to ensure that their water quality and abstraction capability (i.e., quantity) is maintained. To implement these requirements work carried out by the project team includes:

- Groundwater monitoring, sampling and testing including control and trigger levels as well as contingency measures

Table 1.   Families and groups of substances listed under the Groundwater Directive (80/68/EEC)*

List 1	List 2
Organohalogen compounds (and substances which may form such compounds in the aquatic environment), i.e., any organic compound which contains one or more covalently bonded halogen atoms.	The following metalloids and metals and their compounds: zinc, copper, nickel, chrome, lead, selenium, arsenic, antimony, molybdenum, titanium, tin, barium, beryllium, boron, uranium, vanadium, cobalt, thallium, tellurium and silver.
Organophosphorus compounds: any organic compound which contains one or more covalently bonded phosphorus atoms.	Biocides and their derivatives not appearing in List 1.
Organotin compounds: any organic compound which contains one or more covalently bonded tin atoms.	Substances which have a deleterious effect on the taste and/or odour of groundwater, and compounds liable to cause the formation of such substances in such water and to render it unfit for human consumption.
Mercury and its compounds	Toxic or persistent organic compounds of silicon, and substances which may cause the formation of such compounds in water, excluding those which are biologically harmless or are rapidly converted in water into harmless substances.
Cadmium and its compounds	Inorganic compounds of phosphorus and elemental phosphorus
Cyanides	Fluorides
Substances which are carcinogenic, mutagenic or teratogenic in or via the aquatic environment	Ammonia and nitrites
Mineral oils and hydrocarbons	

*JAGDAG (the Joint Agency Groundwater Directive Advisory Group) determines whether specific dangerous substances are appropriate to List 1 and List 2 and includes these in a database.

- Risk assessments for the effects of tunnel and shaft construction of groundwater
- Support to the development of mitigation solutions.

In addition specific consultations including public exhibitions, presentations at community 'Neighbourhood Partnerships,' communications with private water abstractors and presentations to local authorities have been made by National Grid and the project team.

GROUNDWATER MONITORING, SAMPLING AND TESTING PROGRAM

Groundwater Monitoring

An extensive pre-construction groundwater monitoring, sampling and testing programme was developed and undertaken through consultation with the EA and TWUL. Hydrochemical data have been obtained from the EA from monitoring wells, some TWUL abstraction wells and from monitoring wells installed as part of the Croydon Cable Tunnel project. Rainfall and pressure data have been collected at weather stations installed at Beddington and Rowdown substations.

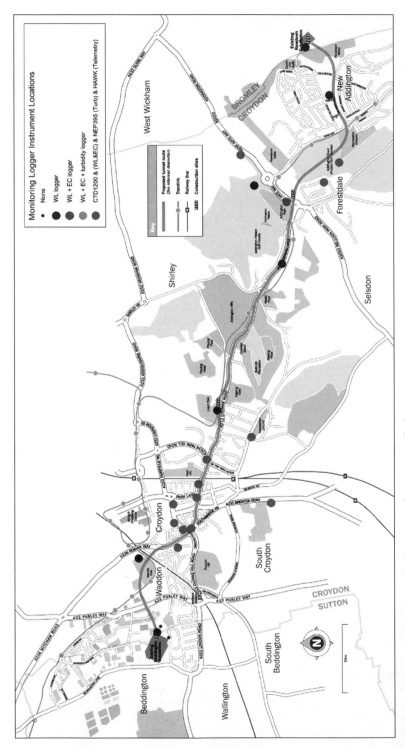

Figure 6. Project groundwater monitoring network (pre-construction)

A total of nineteen monitoring wells were installed and form part of the groundwater monitoring network for the project that operated for approximately 2 years prior to tunnel construction. A further ten clustered monitoring wells were installed (during June–August 2006) and will form part of the monitoring network during construction (see Figure 6).

At the time the outline design was carried out by KBR in 2001/2002 the depth of the groundwater (as measured by the monitoring boreholes) beneath the ground surface at the shaft locations were as follows: 1 m (3.3 ft) Beddington; 23 m (75 ft) Lloyd Park; 9 m (30 ft) Kent Gate Way; and 38 m (125 ft) Rowdown. When the detailed design was carried out by Morganest in 2006 the depth of the groundwater (as measured by the monitoring boreholes) beneath the ground surface at the shaft locations were as follows: 2 m (6.6 ft) Beddington; 28 m (92 ft) Lloyd Park; 24 m (79 ft) Kent Gate Way; and 52 m (171 ft) Rowdown. This, together with a further assessment of the tunnel vertical alignment (including the willingness of Morganest to drive the tunnel downhill at steeper gradients than indicated during the early contractor involvement exercise), has enabled the vertical tunnel alignment to be raised. The revised depths for the vertical alignment below ground level as follows: Lloyd Park—38 m (125 ft) (previously 60 m (197 ft)); Kent Gate Way—12 m (39 ft) (previously 61 m (200 ft)); Rowdown—10 m (31 ft) (previously 78 m (256 ft)). The original deep level alignment had been agreed by the EA and TWUL at tender stage, as a viable vertical alignment going through a zone 1 SPZ. The alignment was below the 'pay-horizons' of the extraction wells. The alternative shallower alignment in the eastern section proposed by the Contractor was later agreed by all parties as the vertical alignment was above historic measured water levels. This resulted in a steeper tunnel gradient and considerably shallower shafts at Kent Gateway and at Rowdown.

Groundwater levels and hydrochemistry are being monitored using pressure transducer, turbidity and electrical conductivity sensors with data loggers and via groundwater sampling for laboratory analysis. The logged data have been downloaded and samples taken on a monthly basis.

For the construction phase, thirteen of the project monitoring network sites have been fitted with telemetry. Data from these sites are acquired by polling the outstations from a remote office based server. The server is programmed to poll each outstation on a daily basis and to receive the data stored. The server is also programmed to send text messages to the mobile phones of key personnel should a pollution event occur.

Laboratory test data are transferred to electronic data deliverables (EDDs). These are spreadsheets with pre-defined column headings and file names allowing all data to be tracked and quality control to be applied. The EDDs are then loaded onto an electronic environmental data management system and a short summary report produced on a quarterly basis.

Prior to construction the monitoring programme indicated that groundwater quality is generally good across the network with sporadic elevated concentrations. The main points from the pre-construction monitoring period include:

- Turbidity was not recorded at the PWS boreholes analysed (only a very limited number of samples have been provided). On-line turbidity and conductivity loggers have recorded, however, elevated turbidity within monitoring boreholes

- Micro-biological contaminants are present in a number of the boreholes across the whole of the monitoring area.

- Heavy metals were recorded at background and occasionally elevated levels above the current regulatory limits for treated drinking water across the monitored area

- Pesticides were recorded at slightly elevated concentrations in a number of locations across the whole of the monitoring area.

Control and Trigger Levels During Construction Monitoring

A control and trigger levels for groundwater quality in the immediate area surrounding the tunnel will be used to identify adverse trends that are indicative of impacts from the development. The control and trigger levels assigned will allow for baseline variations (variations due to natural and background anthropogenic influences) in groundwater quality.

Trigger levels will serve as a threshold in such a manner that if groundwater quality trends appear to near, reach or exceed the trigger level for a particular determinand then action may be taken. Action taken may include the investigation of the causes of the trend and notification of the trend or exceedance to the EA, TWUL, the Private Water Abstractors and the project team. The results of the investigation will determine whether a course of remedial or preventative action is required by the project team or if the trends are due to other outside influences.

Control levels will be set at a lower level compared to trigger levels and provide an early warning that an adverse trend has developed, prior to the concentration reaching such a high value that an impact will occur. This may allow the project team to identify the cause of the trend prior to an impact occurring. Determinands considered relevant include:

- Turbidity
- Electrical conductivity
- Organics (possibly naphalene and toluene).

Turbidity is a key concern for the TWUL PWSs. Increased turbidity is known to be an issue in the Croydon area, with fine sand particles entering the Chalk fissure system following periods of rainfall. Trigger levels will be derived from logger data only as it is considered that laboratory data will not be representative of actual turbidity fluctuations which are likely to fluctuate widely over short timescales due to the 'flashy' nature of this fissured aquifer.

Electrical conductivity (EC) provides a generic assessment of the relative concentration of ionic species in groundwater. The tunnel grout is largely composed of such ionic species and it is considered that if migration of grout into the fissures was to occur it is likely to result in an increase in electrical conductivity of the groundwater. It is considered appropriate to assign control and trigger levels to individual or groups of monitoring wells as the wells monitored to date show very different ranges and fluctuations of EC. Control and trigger levels are based on sensor data from each individual well as these data are considered to be more representative of fluctuations in specific wells. A control level equal to the average plus two standard deviations and a trigger level equal to the average plus three standard deviations is being used. Control and trigger levels for these wells and the remaining wells which are to be installed with EC sensors and loggers (i.e., wells located within Source Protection Zone 1) will be assigned/reviewed, as necessary, following collection of additional time series data.

A trigger level for organics equal to the current limit of detection is to be set for naphthalene and toluene.

CONSTRUCTION MATERIALS—LEACHING TESTS

Concerns were raised about substances leaching from shaft and tunnel grouts and the TBM tail skin sealant into the groundwater. The Groundwater Regulations designate certain substances which must be controlled either as List 1 or List 2. In the case of List 1 substances, their direct or indirect release is prohibited, whilst for List 2 substances such a release should be minimised so as not to cause pollution. The EA set as the compliance point for the List 1 and 2 substances the extrados of the tunnel and shaft linings.

At the outline design stage a risk based approach was taken to investigate the chemical composition of construction substances and the potential risk of leaching from typical construction techniques. The objective of the assessment was to confirm that works could be undertaken in compliance with the requirements of the Ground-water Regulations. Construction materials suppliers and construction contractors co-operated by providing technical information. After consultations with the EA a series of leaching tests for grout and tail skin sealant were undertaken. The tests followed a pro-tocol developed specifically for the Croydon Cable Tunnel Project. The test procedure was based on a draft European protocol (CEN/TC292/WF2 DOC140, titled 'Character-isation of Waste-Leaching—Compliance Test for leaching of monolithic waste of regu-lar shape,' Draft prEN, 2000.06.02). The method was developed and agreed with the EA to be representative of material leaching behavior in Croydon groundwater condi-tions. The leachant used was raw groundwater obtained from Croydon PWS monitor-ing boreholes. The test was dynamic and involved the removal and renewal of leachant five times over a 48 hour period. The laboratory analysed samples for listed sub-stances, comparing the results to an analysis of the raw groundwater, UK drinking water standards (UK DWS) and laboratory detection limits.

During the outline design, initial tests considered grouts that included a mix of Portland Cement, Pulverised Fuel Ash (PFA), polymeric gelling additive (for saturated conditions) and an accelerator. The analysis of the grouts and tail skin sealant indi-cated that List 1 substances would not be released into the groundwater and that List 2 substances will not leach at levels that will give rise to pollution of the groundwater at the compliance point.

This initial assessment was then developed by the civil contractor (Morganest) to re-test alternative materials. To reduce pollution risk, a non PFA grout mix was used for the leaching tests. For the List 2 substances, a dilution factor was also used to predict down-hydraulic gradient concentrations for the leached substances, in order to compare them against the UK DWS. A sensitivity analysis was then introduced using a range of esti-mated hydraulic conductivities and a variation of grout thicknesses. The assessment concluded that there will be no discharge of List 1 substances from the shaft grout and following dilution with groundwater, there will be no pollution by List 2 substances.

DESIGN AND BUILD CONTRACT DOCUMENTATION

Conditions of Contract

The conditions of contract are IChemE Green Book—Reimbursable Contracts, Second Edition 1992 amended with a Target Cost from the Institution of Chemical Engineers for the civil engineering contract. During the design & build tender assess-ment the weighting for scoring was 33% Technical; 33% Safety, Health & Environment; and 33% Financial. The cable contract was let using the FIDIC Yellow Book—Condi-tions of Contract for Electrical and Mechanical Works, Third Edition 1987 (amended).

Construction Materials

The leaching test requirements for construction materials as detailed earlier were incorporated into the contract documentation as well as dialogue with the EA

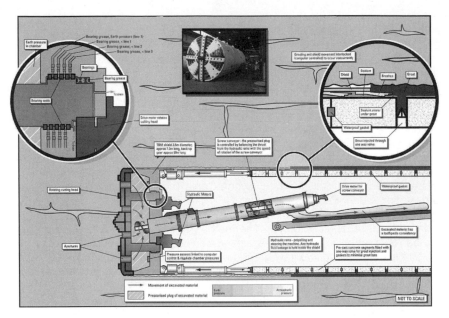

Figure 7.   Section through tunnel boring machine

on construction materials and the compliance point as the extrados of the tunnel and shaft linings.

Construction Methods

The requirements for the design & build contractor to allow the EA to comment on the method statements for shaft and tunnel construction were incorporated into the contract documentation.

Tunnel Boring Machine (TBM) Specification

The requirements for List 1 and 2 substances were applied to the TBM specification to ensure that a new bespoke TBM was manufactured for this project to cover the specific environmental requirements. The TBM (see Figure 7) is a LOVAT RME 140SE mixed face earth pressure balance (up to 8 bar) with a cut diameter of 3.584 m (11.8 ft). Particular features for this project include:

- Bearing grease is only in contact with the spoil in the plenum chamber
- Can cope with water pressures up to 8 bar
- Electric drive motors instead of hydraulic drive motors
- A non PFA cementitious grout using a limestone aggregate, with grouting through the tailskin.

Groundwater Monitoring, Sampling and Testing Programme

The detailed groundwater monitoring, sampling & testing programme requirements were included in the contract documentation as well as the establishment and agreement with the EA and TWUL on the control and trigger levels.

PROJECT MANAGEMENT

KBR acted as project manager on behalf of National Grid from feasibility study through outline design to detailed design and then construction management. Some of the key issues of project management have included:

- Environmental dialogue with stakeholders & third parties that is integrated at all phases with the design & construction process
- Obtaining planning consents for the intermediate headhouses in Metropolitan Green Belt
- Having two contractors (civil & cable) working on design and construction and the associated interfaces
- Managing interfaces with contractors working on separate contracts, but associated works to upgrade electrical substations at Beddington & Rowdown
- Risk management
- Risks associated with tunnel & shaft construction in an urban environment where the ground and groundwater conditions may vary and unforeseen obstructions may materialise.

CONCLUSIONS

Environmental and Planning constraints, requirements and regulations are becoming increasingly onerous for major civil engineering projects including underground structures. The Croydon Cable Tunnel is a project with particular groundwater environmental issues which have had a major influence on: the tunnel & shaft design; the horizontal & vertical alignment; the location of intermediate shafts; the construction materials used; the construction methods; the design of the TBM; and the monitoring, sampling & testing programme to monitor the environmental effects of tunnel & shaft construction. For this project these environmental issues have been successfully addressed by:

- Early and continued stakeholder and third party involvement from feasibility study to outline design to detailed design and finally construction
- Inclusion in the design and build contract documentation of the particular environmental and planning requirements
- Ensuring that environmental performance is a significant factor in tender evaluation for the design and build contract
- Bespoke TBM procurement
- An extensive monitoring, sampling and testing programme to ensure environmental compliance during construction
- Ensuring that the design & build contractor continues the stakeholder & third party involvement.

ACKNOWLEDGMENTS

The authors wish to gratefully acknowledge the assistance of National Grid in the preparation of this paper and for permission to publish. The authors wish to gratefully acknowledge the work of their present and former colleagues at KBR, the cost consultant Gardiner & Theobald, the main civil engineering contractor, Morganest, the cable contractor, NKT, NKT's sub-contractor—Balfour Beatty Power Networks Ltd, and the various sub consultants and subcontractors who have been involved with the design and construction of this project. In this paper the opinions expressed by the authors do not necessarily reflect the opinions of National Grid, Kellogg, Brown & Root and RBE Environmental.

REFERENCES

BS 8007 Code of Practice for design of concrete structures for retaining aqueous liquids.

BS 8110 Structural use of concrete.

CEN/TC292/WG2 DOC 140 Characterisation of Waste-Leaching—Compliance Test for leaching of monolithic waste of regular shape, Draft prEN, 2000.06.02.

Electricity Act 1989.

Environment Act 1995.

Environment Agency's 1998 Policy and Practice for Protection of Groundwater.

Environment Agency, 2003. Hydrogeological Risk Assessments for Landfills and the Derivation of Groundwater Control and Trigger Levels. Landfill Directive. Environment Agency, LFTGN01.

Federation Internationale Des Ingenieurs-Conseils (FIDIC) 1987 Conditions of Contract fro Electrical and Mechanical Works (including erection on site) with forms of tender and agreement.

Groundwater Regulations 1998 (SI 1998 No 276).

Groundwater Directive (86/68/EEC).

IChemE Green Book Conditions of Contract for Process Plant 1992 Second Edition Reimbursable Contracts Institution of Chemical Engineers.

IEC 62067 "power cables with extruded insulation for rated voltages above 150 kV (U_m = 170 kV) up to 500 kV (U_m = 550 kV).

Marsland, P.A. and Carey, M.A., 1999. Methodology for the Derivation of Remedial Targets for Soil and Groundwater to Protect Water Resources. R&D Publication 20, Environment Agency.

New Roads and Streetworks Act, 1991.

STUVA (Studiengesellschaft fur Unterirdische Verkerhsanlagen e.V.) recommendations for allowable water seepage into tunnels.

UK Drinking Water Standards.

TUNNEL INTERACTION: HYDRAULIC, MECHANICAL, AND HYDROMECHANICAL APPROACH

Sunghoon Choi

Parsons Brinckerhoff

Kyle R. Ott

Parsons Brinckerhoff

Arne Fareth

New York City Department of Environmental Protection

John Kinneen

Metcalf & Eddy

Edward S. Barboe

Hazen and Sawyer

ABSTRACT

A new water treatment plant is to be constructed for a portion of the water supply system of New York City. The project includes construction of new water conveyance tunnels crossing perpendicular and sub-parallel to the existing pressurized water tunnel at two locations. A study was conducted to evaluate the impact of new tunnel construction on stability of the existing tunnel and hydrojacking potential in the surrounding rock mass. This paper introduces a practical approach to evaluate tunnel interaction, expressed in terms of rock-liner interaction, stress-induced stability and hydrojacking failures. A simple method to evaluate hydraulic and mechanical properties of concrete liners and jointed rock mass is also presented.

INTRODUCTION AND ASSUMPTIONS

New York City has one of the most extensive water supply systems in the world. A new water treatment plant (WTP) will be constructed to provide filtration and disinfection of a portion of the water supply system. The plant is intended to meet the public water supply and public health needs of the City of New York and to comply with State and Federal drinking water standards and regulations. The plant will have a capacity of 1.1 million cubic meter per day (m^3/day). The project includes a new raw water tunnel connection to the plant, raw water pumping facilities, treatment processes, treated water pumping, and a treated water tunnel with two separate pipelines to the City of New York's High and Low Level water delivery services (Figure 1A). The new water tunnels will cross over the existing pressurized water tunnel in two locations; the Treated Water Tunnel (TWT) crossing sub-parallel with a vertical separation of 26.5 m and the Raw Water Tunnel (RWT) crossing perpendicular with a vertical separation of 21 m (Figure 1B).

It is not uncommon to construct new tunnels near existing underground structures in urban environments. As a new underground structure is added, interaction effects need to be considered with regard to stability and leakage and to mitigate construction

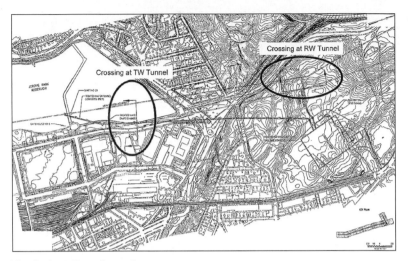

Figure 1A.   Project site and crossings

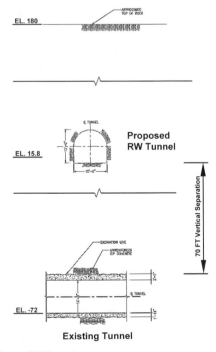

Figure 1B.   Typical crossing at RW Tunnel

impacts on the existing (and even the new) structures. In the past, a rule of thumb was used to maintain "safe distances" between new and existing tunnels such that in general, tunnels can be constructed side-by-side by maintaining a distance of two tunnel diameters of separation, measured from centerline to centerline (a pillar width of one diameter). Most transit tunnels use a pillar width of one tunnel diameter. For pressurized water supply tunnels, a 9.6 m diameter Mainstream tunnel was built for the City of Chicago, which crossed below the Chicago Avenue water supply tunnel with a vertical separation distance of 12.2 m crown to invert. The water supply tunnel was in operation with an internal water pressure of 0.5 MPa during the crossing. No rock movement during or after excavation was observed.

This paper introduces a practical approach to evaluate mechanical and hydraulic tunnel interactions, such as stress induced stability, rock-liner interaction, and hydrojacking. An interaction study provides a general assessment of the adequacy of the proposed tunnel construction. The proposed approach focused on the global behavior of the rock mass, treating the rock mass as a continuum rather than as discrete materials. To account for the presence of discontinuities, modulus and strength of the rock mass (mechanical properties) were reduced from those measured from intact rock samples. However, if the rock mass contains predominant weak planes and those are continuous and oriented unfavorably to the excavation, then the analysis should consider incorporating specific characteristics of these weak planes. The study also idealized the concrete liner and rock mass as isotropic, homogeneous media. The actual flow regime may be a fluid flow through the fracture networks; therefore, the absolute value of the hydraulic response may differ from the prediction based on the assumption of isotropic, homogeneous, porous media. However, the relative significance of the variables involved and the sensitivity of the rock-liner system to changes in these variables can be adequately evaluated with the current approach.

The primary objectives of this study are as follows:

- To evaluate the mechanical impact of the proposed tunnel construction on the global stability of the existing tunnel, taking into account various geostatic in-situ stresses. The mechanical impacts are examined in terms of changes in stress fields and deformations.

- To predict groundwater flow regime due to the hydraulic interaction between the proposed tunnel construction and the existing tunnel, while the existing tunnel is pressurized. The hydraulic interaction is investigated in terms of groundwater pressure distribution and water inflow into the proposed tunnels during the construction.

- To assess hydrojacking potential in the surrounding ground by comparing total minimum stresses in the rock mass and water pressures induced by internal water pressure of the existing tunnel.

SUBSURFACE CONDITIONS

The regional basement rock is Fordham Gneiss consisting of Proterozoic metasedimentary and metavolcanics rocks that were deformed during the Grenville orogeny and later intruded by the Yonkers Gneiss. The Fordham Gneiss (Middle Proterozoic) is the rock unit believed to underlie the entire project site. It is fine- to medium-grained, quartzo-feldspathic gneiss, with foliation well-defined by compositional layering. Although one member is reportedly distinctly schistose, the Fordham Gneiss's most characteristic feature is its folded and intensely distorted compositional foliation. The Fordham Gneiss has been subdivided into four informal members (A, B, C, and D) based largely on composition and the interbedded relationships of different rock types at their contacts. Three of these members, described below, are believed to underlie the project site. Other rock types, including Dolomite, Yonkers Gneiss, and Inwood Marble are also observed in the project site.

Overburden in the Bronx, northern Manhattan, and southern Westchester is predominantly dense glacial till consisting of a mixture of clay, silt, sand, gravel, and boulders. The till varies in thickness from less than 1.5 m to more than 30.5 m, averaging about 7.6 m in areas with few rock outcrops. In upland areas, rock outcrops are numerous. Kame and outwash deposits of sand and gravel are also present within the region.

EVALUATION OF MECHANICAL AND HYDRAULIC PROPERTIES

This section briefly introduces a method to evaluate hydraulic and mechanical properties of concrete liners and jointed rock mass using rock-liner interaction. When a concrete liner is subjected to internal water pressure in a pressure tunnel, the concrete liner will expand, but it will be restrained by the surrounding rock mass. During the initial loading, the radial displacement of the concrete liner and the rock mass are equal at the liner-rock mass boundary (continuity of displacement), and the magnitude of the displacement is determined by the mechanical interaction of the liner and the surrounding rock mass. Under this condition, part of the net internal water pressure, defined as the difference between internal water pressure and hydrostatic water pressure, is transferred to the surrounding rock mass and part is absorbed by the concrete liner. Consequently, the surrounding rock mass resists changes in liner displacement so that a passive reaction develops to the active loading (mechanical rock-liner interaction) where the concrete liner and the surrounding rock mass are in contact.

The hydraulic property of concrete liners is governed by crack width and spacing, depending on the tensile circumferential strain induced by internal water pressure. The circumferential strain in turn is controlled by mechanical interaction between the rock mass and concrete liner. The rate of infiltration therefore is determined not only by the permeability of the rock masses surrounding the tunnel but also by the permeability of the concrete liner.

Evaluation of Cracked Concrete Liner Permeability Subject to Internal Water Pressures

The outward displacement of the concrete liner is a function of not only stiffness of the concrete liner but also stiffness of the surrounding rock mass. The magnitude of outward displacement of the tunnel walls determines the level of the tensile strain of the concrete liner. In a concrete liner, the magnitude of tensile strain, in turn, controls the cracking and thus the permeability of the concrete liner (hydromechanical rock-liner interaction). The hydromechanical rock-liner interaction determines the degree of cracking of the liner due to the internal water pressure. The amount of leakage through cracks is dependent upon the number and size of longitudinal cracks, the internal water pressure, the hydraulic pressure at the liner-rock boundary, and the thickness of the liner. The amount of leakage into the surrounding rock mass is dependent upon the hydraulic pressure at the liner-rock boundary, the free-field groundwater pressure at tunnel elevation, and the equivalent permeability of the rock mass. The circumferential strain in the concrete liner, $\varepsilon_{\theta L}$, assuming an elastic response in both the concrete liner and surrounding rock mass, was proposed by Fernandez (1994) as follows:

$$\varepsilon_{\theta L} = \frac{\Delta P_w(1 + \nu_m)}{M}\frac{}{E_m}; \quad M = 1 + \frac{E_L}{E_m}\frac{t_L}{a}(1 + \nu_m) \tag{1}$$

where ΔP_w is net-driving water pressure $(P_i - P_o)$; E_m is average rock mass modulus; ν_m is Poisson's ratio of the rock mass; E_L is modulus of elasticity of the liner; t_L is average thickness of the liner; and a is radius to the liner center. When the concrete liner is cracked, equivalent permeability of the concrete liner can be estimated using the cubic law for viscous flow between parallel plates. For laminar flow along the longitudinal cracks with an aperture, w, evenly distributed around the perimeter of the liner, the equivalent permeability of the cracked concrete liner can be expressed as:

$$k_L = \frac{\gamma_w}{12\mu}\left(\frac{w^3}{S}\right); \quad w = \varepsilon_{\theta L}S \tag{2}$$

where γ_w is unit weight of water; w is average crack width; S is average crack spacing, and μ is dynamic viscosity of the water. For an unreinforced concrete liner, two cracks typically develop parallel to the plane of minimum in situ stress. Therefore, the spacing of the cracks can be estimated as $S = \pi a_1$, where a_1 is the inside radius of the tunnel.

In reinforced concrete, the crack spacing, S, can be obtained as $S = 0.1 d/\rho$, where d is the diameter of the reinforcing bar and ρ is the ratio of area of steel to area of concrete (Hendron et al., 1987). Hydraulic and mechanical rock-liner interaction is utilized to estimate permeability of the concrete liner in the existing tunnel when the liner is subjected to internal water pressure. The existing tunnel is 4.6 m in diameter with 0.38 m thick unreinforced concrete liner. Then, the liner permeability of the existing pressure tunnel is calculated, using equations (1) and (2) above, as 5.3×10^{-5} cm/sec nearby the TWT and 9.7×10^{-6} cm/sec nearby the RWT when the concrete liner is subjected to internal water pressures of 1.1 MPa.

Evaluation of Equivalent Rock Mass Permeability

Fluid flow through a jointed rock mass is a complex phenomena and it is extremely difficult (almost impossible) to measure hydraulic properties across a large area. Borehole packer tests conducted in a small borehole only represent local hydraulic properties of the area influenced by a single test. The hydraulic properties obtained from borehole packer tests, while informative, are somewhat different from the equivalent hydraulic properties representative of a large area, which may be influenced by a pressure tunnel. One method to predict equivalent permeability of the surrounding rock mass is to back-calculate it using measured water infiltration rates during tunnel construction. Fernandez (1994) developed an analytical solution to evaluate magnitude of the rate of flow into the rock mass when the tunnel is subject to internal pressure. The equation is modified for zero internal pressure, such as during tunnel excavation, and then can be modified as follows:

$$k_m = \frac{Q\ln(L/b)}{2\pi h_o} \tag{3}$$

where k_m is permeability of surrounding rock mass; Q is water infiltration rate during construction; h_o is external hydrostatic water head; L is distance between center of the tunnel and imaginary mirror source/sink $(= 2h_o)$; and b is excavated radius of the tunnel. When there is no available data, an alternative method may be used to predict the water infiltration rate during construction. For example, a semi-empirical method to predict the rate based on borehole packer test results was proposed and later extended by Heuer (1995, 2005). With water infiltration rates calculated from Heuer's method, the equivalent rock mass permeability is estimated, using equation (3) above, as 1.3×10^{-4} cm/sec nearby TWT and 2.4×10^{-5} cm/sec nearby RWT.

Table 1. Mechanical properties of rock mass used in the analyses

		Near TWT	Near RWT
Total unit weight		2,723 kg/m^3	2,723 kg/m^3
Modulus of deformation		33,095 MPa	35,853 MPa
Poisson's ratio		0.25	0.25
Hoek & Brown Strength	m	13.5	14.0
	s	0.07	0.08
	ucs	103 MPa	157 MPa

Evaluation of Mechanical Properties of Rock Mass

In evaluating the stress and deformation behavior of a rock mass, the strength and deformation modulus obtained from intact rock samples must be reduced to account for the presence of discontinuities within the rock mass. For example, the correlation between the Rock Quality Designation (RQD) and the modulus reduction ratio, E_M/E_L, has been widely used in practice to reduce the laboratory modulus determined from small intact rock samples to the mass modulus of deformation. An alternative method to obtain rock mass modulus is to estimate it using empirical relationships from rock mass classifications (Q, RMR, and GSI). Hoek and Brown's shear strength parameters can also be derived from the empirical equations based on rock mass classifications. The estimated mechanical properties of the rock mass in the project area are presented in Table 1.

TUNNEL INTERACTION

One of the most critical aspects in the design of pressure tunnels is the prevention of excessive leakage and excessive water pressure due to hydrojacking of the rock mass, which, in some cases have triggered landslides that have buried access roads and/or threatened surface power plants. It is known that when the effective stress in a single rock joint becomes zero, the joint opens and the flow rate increases sharply, defining the onset of hydrojacking. Therefore, when the water pressure in the joint equals the total normal stress, hydrojacking takes place. Numerous case reviews and numerical studies have been conducted to evaluate the initiation and evolution of hydrojacking in fractured rock masses surrounding unlined pressure tunnels. The key unknown in the evaluation of hydrojacking failure may be the minimum in-situ total normal stresses in the joints.

When a concrete liner, whether it is reinforced or not, is used, the hydraulic, mechanical, and its coupled rock-liner interaction in a pressure tunnel are of additional concern. Of particular interest are the head losses across cracked liners, crack openings in the concrete liner induced by radial displacement of the liner, flow rate into the rock masses, and hydrojacking in the rock mass.

Mechanical Interaction

In mechanical interaction, stress fields in the surrounding rock mass are computed using finite element analyses and the results before and after the proposed tunnel construction are compared. The overall stress-induced stability and deformation at the concrete liner of the existing tunnel are of concern. A parametric study for various in-situ initial conditions was conducted, using in-situ horizontal to vertical stress ratios (k_o) of 1, 3 and 5. Stability of surrounding rock mass is investigated using the strength/stress

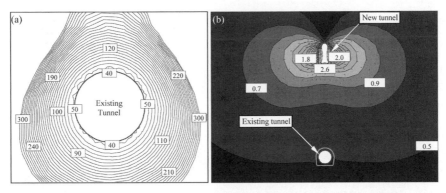

Figure 2.   Mechanical tunnel interaction, k_o = 1: (a) strength factor and (b) displacment in mm

ratio. The strength factor (SF) against shear failure is defined as $(\sigma'_{1f} - \sigma'_3)/(\sigma'_1 - \sigma'_3)$, where $(\sigma'_{1f} - \sigma'_3)$ is the strength of the rock mass from the Hoek-Brown strength parameters and $(\sigma'_1 - \sigma'_3)$ is the induced stress. A SF greater than 1.0 indicates that the rock mass strength is greater than the induced stress, which means there is no overstress in the rock mass, while a SF less than 1.0 indicates that the induced stresses are greater than the rock mass strength. In this case, the rock mass is overstressed and behaves in the plastic range.

Based on SF contour plots presented in Figure 2a, the minimum SF against shear failure near the existing tunnel is 40, which means the rock mass strength is 40 times the induced stresses, indicating that the entire domain is not over-stressed and no stress-induced stability problems are anticipated. As presented in displacement contours (Figure 2b), the maximum anticipated displacement is 2.6 mm at the crown and invert of the new tunnel when k_o = 1. Negligible displacements (0.5 mm at the liner when k_o = 1) are predicted at the existing tunnel for all cases considered.

The normalized stresses before and after new tunnel construction, $\sigma_{after}/\sigma_{before}$, are plotted along the section centerline between two tunnels (Figure 3). As shown in Figure 3a, the maximum normalized horizontal stresses (stress amplification) are observed at the proposed tunnel boundary and diminish before reaching the existing tunnel. The normalized stress is equal to one at the existing tunnel, representing no changes in stresses after the proposed tunnel construction. When k_o = 1, the vertical stresses are relieved at the vicinity of the proposed tunnel and the values do not exceed one (stress relief) as shown in Figure 3b. For higher in-situ stress ratios (k_o = 3 or 5), the vertical stresses are also relieved at the vicinity of the proposed tunnel. However, those become amplified, reaching a maximum at approximately one tunnel width (4.3m) below the invert of the proposed tunnel. Vertical stresses then diminish further, as distances increase from the proposed tunnel. Based on the results, there is no stress amplification or relief expected in the vicinity of the existing tunnel for the in-situ stress conditions considered.

In summary, the mechanical interaction analysis indicates that new tunnel construction at the proposed depths would not cause stress-induced stability problems and negligible incremental deformation is anticipated at the existing tunnel. Changes in stresses induced by the proposed tunnel construction are highest at the new tunnel boundary and diminish as distances increase from the new tunnels. For all in-situ stress conditions considered (k_o = 1, 3, and 5), negligible changes in stresses are indicated at the existing tunnel. Considering pressurization of the existing tunnel, the

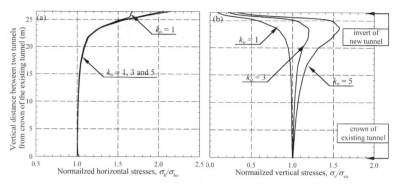

Figure 3. Normalized stresses along the centerline from crown of existing to invert of new tunnels

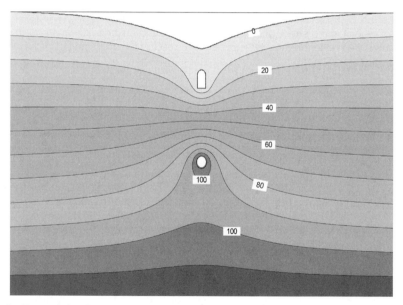

Figure 4. Pressure head contours (meters)

stresses generated in the rock mass between tunnels are generally low compared to the strength of the rock mass.

Hydraulic Interaction

In hydraulic interaction, groundwater flow regime and pressure distribution are predicted. The pressure head contours (Figure 4) indicate that the existing pressure tunnel is a flow source and the proposed tunnel acts as a sink, representing the new tunnel construction stage. Water infiltration rates at the construction stage are expected to be 0.02 m³/min/3-m of tunnel length in the TWT close to the crossing of the existing tunnel. The rate reduces as the TWT moves away from the existing tunnel. For the RWT, the maximum water infiltration rate is expected to be 0.05 m³/min/30-m of tunnel length at the crossing of the existing tunnel.

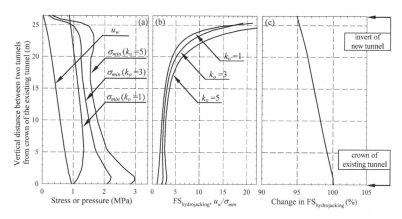

Figure 5. Hydrojacking potential (u_w = water pressure, σ_{min} = minimum total stress)

Coupled Hydromechanical Interaction: Hydrojacking Potential

One of the most critical hydromechanical impacts is hydrojacking of the rock mass. It is known that when the effective stress in a single joint becomes zero, the joint opens and the flow rate increases sharply, defining the onset of hydrojacking. In other words, when the water pressure at the joint equals the minimum total normal stress, hydrojacking takes place. Calculated water pressures and the minimum total normal stresses along the vertical plane between the new tunnel and the existing tunnel are shown in Figure 5a. Factors of Safety (FS), which is defined as the ratio of water pressure to minimum total stress at the point, are plotted in Figure 5b. The $FS = 1$ near the existing tunnel and increases significantly as distances increase from the tunnel. The changes in FS against hydrojacking before and after new tunnel construction, defined as FS_{after}/FS_{before}, are presented in Figure 5c, which shows that a maximum of 5% reduction in FS is observed at the invert of the new tunnel. It is concluded that hydrojacking potential due to pressurization of the existing tunnel and the proposed tunnel construction is assessed to be extremely low for the in-situ stress conditions considered.

CONCLUSIONS AND RECOMMENDATIONS

Construction of the new water conveyance tunnels in New York City at the proposed depths and alignments will not affect stress induced stability or the hydraulic function of the existing pressure tunnel based on the coupled hydromechanical interaction study. The stability of the surrounding rock mass is also expected to be preserved. The conclusions based on the presented interaction analyses are summarized as follows:

- Finite Element analyses indicate that construction of the TWT and RWT at the proposed depths and locations would not result in tunnel interaction causing stress-induced stability problems in either new tunnels or existing tunnel and negligible incremental deformation is anticipated in the existing tunnel as a result of construction of the proposed tunnels.

- Changes in stresses induced by the proposed tunnel construction are maximum at the excavation boundaries of the proposed tunnels and diminish with distance from the proposed tunnels. For all in-situ stress conditions considered

(k_o = 1, 3, and 5), no change in stress is observed at the existing tunnel due to the proposed tunnel construction.

- Considering the effects of pressurization of the existing tunnel, the stresses generated in the rock mass between the existing tunnel and the proposed tunnels are generally low compared to the strength of the rockmass. However, local stress concentrations should be expected in the vicinity of the proposed tunnel boundaries.

- Considering the effects of pressurization of the existing tunnel, the water inflow during construction stage is expected to be 20 liters per minutes per 3 m of tunnel length in the Treated Water Tunnel (at the point closest to the existing tunnel), which crosses sub-parallel to the existing tunnel. The water inflow rate becomes smaller as the TWT moves away from the existing tunnel.

- For the RWT (which crosses perpendicular to the existing tunnel), the water inflow during the construction stage is expected to be approximately 50 liters per minutes per 30 m of tunnel length at the point closest to the existing tunnel crossing.

- Water inflow rates toward the proposed tunnels during construction are considered to be manageable by suitably equipped contractors.

- Hydrojacking potential due to pressurization of the existing tunnel and the proposed tunnel construction is assessed to be low in all in-situ stress conditions.

- Additional measures to evaluate the interaction effects and to reduce potential impacts are recommended. These recommended measures include: (1) design documents may incorporate the use of steel or reinforced lined sections of the new tunnels in the vicinity of the crossings as an added precaution and (2) to mitigate vibration effects, special construction measures will be included. In the area of the crossings, the technical specifications for drill and blast excavation will require perimeter drilling around the new tunnel; reduced round lengths; reduced explosives per delay; and peak particle velocity limits at the existing tunnel, which should be monitored using deep vibration monitoring points, installed as geophones.

ACKNOWLEDGMENTS

The authors are grateful for the approval to publish details on this project provided by the New York City Department of Environmental Protection.

REFERENCES

Fernandez, G., 1994. Behavior of pressure tunnels and guidelines for liner design. J. Geotechnical Engrg. ASCE, Vol.120. No.10. pp.1768–1791.
Hendron, A.J., Fernandez, G., Lenzini, P., Hendron, M.A., 1987. Design of pressure tunnels. The Art and Science of Geotechnical Engineering at the Dawn of the Twenty First Century, A Volume honoring Ralph B. Peck. pp.161–192.
Heuer, R.E., 1995. Estimating rock tunnel water inflow. Proceedings of Rapid Excavation and Tunneling Conference. pp. 41–60.
Heuer, R.E., 2005. Estimating rock tunnel water inflow-II. Proceedings of Rapid Excavation and Tunneling Conference. pp. 394–407.

3

Difficult Ground Conditions I

Chair
G. Raines
MWH Americas, Inc.

EXCAVATION AND SUPPORT OF A WATER TUNNEL THROUGH THE HAYWARD FAULT ZONE

Sarah Holtz Wilson

Jacobs Associates

David F. Tsztoo

East Bay Municipal Utility District

Carl R. Handford

Atkinson Construction

Kenneth Rossi

EPC Consultants, Inc.

ABSTRACT

The East Bay Municipal Utility District's Claremont Tunnel is an 18,000-foot long water tunnel that crosses the Hayward Fault zone near Oakland, California. It was originally completed in 1929. This project involved the construction of a bypass tunnel through the fault zone, including an enlarged vault section designed to accommodate up to 8.5 feet of horizontal offset. Ground conditions encountered included sheared and crushed serpentinite and clayey fault gouge, and presented specific challenges for construction. A special excavation and support sequence used concrete-backfilled side drifts as foundations for the vault steel sets. It also allowed for a 20-foot-wide top heading and bench excavation in soft and squeezing ground conditions.

INTRODUCTION

The Claremont Tunnel is the main water transmission facility of the East Bay Municipal Utility District (EBMUD). The 3.4-mile tunnel is located beneath the Oakland-Berkeley Hills between EBMUD's Orinda Water Treatment Plant (WTP) and the Claremont Center in Berkeley, California. The tunnel conveys up to 175 million gallons per day of treated water from the Orinda WTP to over 800,000 customers in Richmond, Oakland, San Leandro, and surrounding communities.

Originally constructed between 1926 and 1929, the Claremont Tunnel has a 9-foot diameter horseshoe cross-section and a nominal one-foot thick concrete final lining. The majority of the tunnel lining was not reinforced, and only a small portion of the final lining was contact grouted to fill in voids between the lining and the surrounding ground.

The tunnel crosses the Hayward Fault approximately 130 feet below Tunnel Road in Berkeley, and is highly vulnerable to damage from major earthquakes along the fault. In a moment Magnitude (M_w) 7.0 earthquake, the tunnel must accommodate up to 7.5 feet of horizontal offset and approximately 0.5 foot of vertical displacement along the primary trace of the fault, and up to 2.25 feet of sympathetic movements within a 920-foot-wide secondary fault zone straddling the primary fault. In addition to discrete fault offset, 1 foot of active fault creep had to be accommodated within the design (Figure 1). Such displacements would cause collapse of the existing tunnel lining and likely block water flows for many months until repairs could be completed.

Figure 1.   Steel sets installed inside existing Claremont Tunnel. Arrow indicates deformation due to fault creep.

In 1994, EBMUD approved a $189 million Seismic Improvement Program (SIP) to proactively reinforce and protect its water system from the damaging effects of earthquakes, and avoid severe disruptions to its customers' water service. The Claremont Tunnel Seismic Upgrade Project became a key element of the program. Completion of the tunnel in June 2007 is the last major piece of the SIP. The program will have met its goal of ensuring that sufficient water is available to customers following a major earthquake.

During the design phase of the project, contact grouting and lining repairs were planned for the existing tunnel and construction of a 1,570-foot Bypass Tunnel was planned as a replacement for the most vulnerable part of the tunnel—the section that crosses the Hayward Fault Zone (HFZ). A 460-foot access adit was planned at Claremont Center to provide construction access to the Bypass Tunnel and future access for maintenance.

The Bypass Tunnel features a special vault section spanning the active strand of the Hayward Fault and designed to accommodate large earthquake movements along the fault. This paper generally discusses the design and construction of the Bypass Tunnel but focuses particularly on the construction of the vault section and the ground condition challenges encountered. For details about other portions of the project, which included contact grouting and repairs to the existing tunnel, and a thickened tunnel lining through the adjacent Subsidiary Fault Zones (SFZ), refer to Caulfield et al. 2005.

Project construction began in July 2004. The Bypass Tunnel and Access Adit were excavated between September 2004 and October 2005; vault excavation occurred between late April and mid-August 2005. Final tie-in connections to the existing Claremont tunnel were made while it was taken out of service in December 2006.

Project completion is scheduled for June 2007.

GEOLOGIC SETTING AND GROUND CONDITIONS

The anticipated HFZ location and width were based on regional USGS maps, surface mapping performed and borings drilled for this project, historic logs for the BART Oakland-Berkeley hills transit tunnels and the original Claremont tunnel, and fault

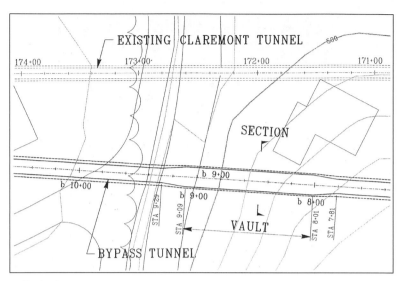

Figure 2.   Vault plan

creep documented during prior inspections of the existing tunnel. Within the antici-pated 920-foot HFZ, the maximum primary displacement zone, or active strand of the fault, was expected to span a length of 60 feet of the Bypass Tunnel alignment. The rest of the HFZ consisted of subsidiary fault zones (Geomatrix 2002; Jacobs Associ-ates 2004).

On the west side of the main active strand, the tunnel was excavated through alluvial and colluvial Quaternary sediments as well as Franciscan mélange and silica carbonate rock. East of the main active strand, serpentinite was encountered; at first, it was pervasively sheared and crushed, but it became more blocky and less deformed at the eastern end of the side drifts (Brown 2005; Geomatrix 2006). The serpentinite encountered during the construction of this project contained small amounts of naturally-occurring chrysotile asbestos and heavy metals, which required special handling and disposal of muck, as well as personal protective equipment for construction personnel.

Additionally, it was determined during the design that the tunnel would be classi-fied as "gassy" by the California State OSHA Mining and Tunneling Unit, based on doc-umented gas intrusion in the original Claremont Tunnel construction. This required that equipment used for the tunneling had to be rendered non-flammable, spark arrested, or acceptable to Cal/OSHA.

DESIGN OF THE VAULT TO ACCOMMODATE SEISMIC OFFSET

Vault Design

The Bypass Tunnel was designed to withstand offset due to seismic activity and remain in service to provide essential water for service immediately after an earth-quake. The size of the vault section was based on the anticipated fault zone width and anticipated offset (Figures 2 and 3). It is a 108-foot-long section of the tunnel alignment

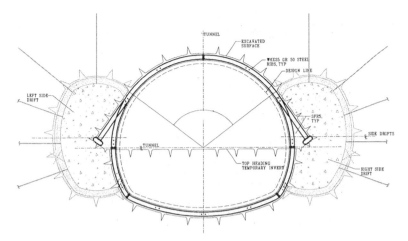

Figure 3.   Vault section

with finished inside width of 16.5 feet. This section can accommodate the entire 8.5 foot lateral design offset in one location. However, shear fuses were incorporated into the final concrete lining to provide weak zones, encouraging rupture every 11 feet along the vault, in order to provide some residual structural capacity that will help facilitate repair and removal of debris after the earthquake.

A steel carrier pipe was included in the design to keep rock and concrete debris out of the water flow. The pipe is 85 feet long, has an inside diameter of 6 feet, with a 3-inch wall thickness, and is lined and coated with cement mortar. If the vault completely collapses in a seismic event, this carrier pipe will support the full overburden load of the slope above it. It is designed to pass the essential lifeline flow of at least 130 million gallons of water per day.

Vault Sequencing

The initial support and excavation sequence was intended to allow an enlarged opening to be constructed in crushed and squeezing ground (as defined in the Geotechnical Baseline Report (2004) and based on Terzaghi's ground classifications), and to protect the surrounding rock from erosion following offset. Egg-shaped side drifts approximately 10 feet wide and 14 feet high were designed to be excavated for the entire length of the vault and backfilled with concrete prior to vault excavation. Following seismic offset, these side drifts are intended to protect the surrounding rock from potential erosion caused by water passing through the tunnel, and to mitigate intrusion of the surrounding rock material into the water flow.

Following backfill of the side drifts, the vault was anticipated to be excavated using top heading and bench construction sequencing. The steel sets for initial support were designed with a special shape so that they would transfer ground load to the concrete in the side drifts, rather than to the crushed and squeezing ground. They were designed to be installed on 4-foot centers, but a provision was made in the contract requiring that the contractor furnish jump sets, or additional sets enabling installation on 2-foot centers, for the entire length of the vault to protect the tunnel excavation from squeezing ground.

Figure 4.   Grouting spiling over inbound vault transition

Figre 5.   Mini excavators mining top heading of outbound vault transition (inset shows roadheader attachment)

CONSTRUCTION OF THE ENLARGED VAULT SECTION
IN CHALLENGING GROUND

Excavation

Excavation of the Bypass Tunnel and the vault was primarily completed by mechanical means with an Alpine AM 75 roadheader (and AM 50 roadheader for the tie-in connections). Where harder rock was encountered, excavation was occasionally supplemented by drill-and-blast. In some areas of softer ground, IHI mini-excavators fitted with bucket, hoe-ram, and Voest-Alpine mini-roadheader attachments were used in lieu of the roadheader. Cover over the tunnel ranged from 45 feet to 230 feet, and water inflow in the excavation heading was less than 30 gallons per minute during the entire excavation. The main active strand of the fault was essentially dry.

Figure 6.   Reinforcing of pillar at vault turnunder

Vault Transitions

The first activity for the construction of the vault was widening the bypass tunnel in a 20-foot long inbound transition from Sta. b9+49 to b9+29. Custom steel sets were fabricated by American Commercial on a slightly different spacing than was called for in the design. The transition required the contractor to modify the mining cycle, and it was challenging to use even the mini-excavators to mine a face wider than the previously installed steel set. Spiling was also required over the entire 20-foot long transition from the last Bypass Tunnel steel set, as well as over the outbound transition from Sta. b8+01 to b7+81 (Figures 4 and 5).

Ground Support at Vault Turnunder

The side drifts and vault were required to begin at approximately Sta. b9+29. The design called for a nominal 15-foot pillar to be maintained between the two side drifts. The contractor installed approximately 8 to 10 inches of steel-fiber-reinforced shotcrete on the entire tunnel face, and further reinforced the pillar with welded wire fabric pinned up with 4-foot split sets (Figure 6). Grouted R32S hollow, self-drilling spiles were also installed over the last transition steel set at Sta. b9+29, between approximately the 10 o'clock to 2 o'clock positions. Spiling was not required in the remainder of the main vault. Following the installation of initial and presupport, the left side drift excavation began.

Vault Side Drifts

The shape of the side drifts was modified slightly from that shown on the drawings to provide clearance for ventilation lines and allow continued use of Wagner ST 3.5 Scooptrams (LHDs) for muck removal. Two IHI mini-excavators were used to mine the side drifts. Their electrical and exhaust systems were modified by the contractor to conform to the Cal/OSHA gassy tunnel permit. The side drifts were excavated in a top heading and bench sequence, with the bench taken out every third 4-foot round, or every 12 feet (Figure 7). They were supported with 8-foot-long grouted fiberglass rock dowels and steel–fiber-reinforced shotcrete. The fiberglass rock dowels in the interior top heading area were subsequently removed during excavation of the top heading.

Figure 7.   Left and right side drift excavation

Instrumentation Readings Used to Modify Ground Support

The instrumentation and monitoring program for this portion of the project included convergence points read by tape extensometer on a daily basis. In early June 2005, readings taken from the convergence points installed at Sta. b8+55 and Sta. b8+30 in the left side drift showed that the side drift was converging at a slightly accelerating rate (Figures 8 and 9). It was assumed that this convergence extended no further than Sta. b8+80, as the convergence point at that station showed virtually no movement. The contractor quickly completed the excavation of the remainder of the drift to Sta. b8+01, and placed concrete backfill in approximately the bottom 3 feet of the drift. The convergence was arrested immediately, and the remainder of the drift was backfilled with concrete. Because this convergence likely indicated the presence of squeezing ground, the EBMUD engineer required installation of jump sets in the vault between Sta. b8+82 and Sta. b8+01.

Excavation Sequence Changes During Construction

The original contract documents specified that the left side drift would be excavated first. Then, while the left side drift was being backfilled, the right side drift would have been excavated. However, during construction, the combination of reasonably stable ground conditions and the desire to make up time in the schedule drove a change in the contract which allowed the drifts to be excavated simultaneously with a minimum lag distance of 30 feet between excavation faces. Once the drifts were completed, the 20-foot wide vault top heading was excavated for the entire vault length, and the bench was removed from the heading end (Sta. b8+01) backward to Sta. b9+29 (Figures 10 through 12).

Geologic Mapping Used to Verify Carrier Pipe Placement

Geologic mapping was an important part of the construction intended to verify the locations of the Hayward Fault primary strand and the vault, and the final location of the

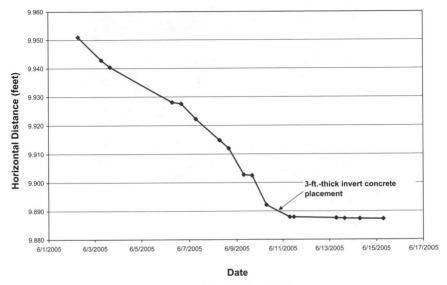

Date

Figure 8.   Convergence readings in the left side drift, Sta. b8+55

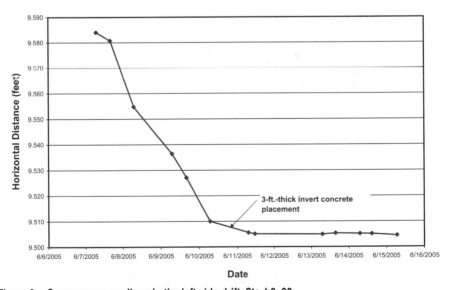

Date

Figure 9.   Convergence readings in the left side drift, Sta. b8+30

Figure 10.   Roadheader in vault top heading

Figure 11.   Vault top heading sets with custom shape to bear on side drift concrete

Figure 12.   Mining the vault bench

Figure 13.   Tunnel face at Sta. b8+80

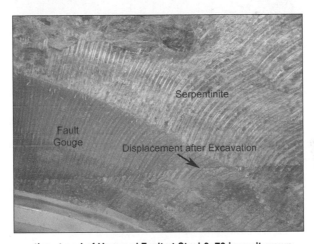

Figure 14.   Primary active strand of Hayward Fault at Sta. b8+78 in vault crown

carrier pipe which would later be installed. The side drifts and the vault were mapped by geologists from Geomatrix Consultants and EPC Consultants during the brief time between excavation and placement of initial support shotcrete (Geomatrix 2006). This mapping of the Hayward Fault primary strand verified the location of the vault and was used to determine that the optimal location for the carrier pipe was about 13 feet west of its design location (originally centered about Sta. b8+55; moved to b8+68)(Figures 13 and 14).

CONCLUSIONS

The Claremont Tunnel Seismic Upgrade Project has safeguarded the water lifeline of over 800,000 EBMUD customers. The project offers assurance that an essential, adequate water supply will be available immediately after a major earthquake.

Table 1.   Summary of construction duration of vault excavation and initial support

Structure	Stationing	Mining and Initial Support	Notes
Inbound Transition	b9+49–b9+29	29 shifts	
Left Side Drift	b9+29–b8+01	93 shifts	+ 7 shifts backfill concrete
Right Side Drift	b9+29–b8+01	77 shifts	+ 5 shifts backfill concrete
Vault Top Heading	b9+29–b8+01	88 shifts	Includes 4 shifts to complete initial support shotcrete
Vault Bench	b9+29–b8+01	42 shifts	
Outbound Transition	b8+01–b7+81	15 shifts	
Total		356 shifts	

Figure 15.   Cast-in-place concrete and reinforcing in vault. Inside geometry was slightly modified so the contractor could reuse the bypass tunnel steel forms

The project also revealed new information about the Hayward Fault. Much information about the fault has been developed over the last several decades through studies, geologic investigations, and construction projects. However, as this project demonstrated, there is still much to be learned, as evidenced by the need to shift the carrier pipe some 13 feet to match the approximate center location of the Hayward Fault primary strand. The ground conditions encountered during tunneling for this project may have added incrementally to what is known about constructing tunnels through active faults. More likely, the project exposed how much more there is to learn about in-depth fault ground conditions, and why innovative, practical means are necessary to overcome the challenges they pose to tunnel construction (Table 1; Figure 15).

ACKNOWLEDGMENTS

The authors would like to thank the East Bay Municipal Utility District, Jacobs Associates, EPC Consultants, Atkinson Construction, and Geomatrix Consultants for their support during this project.

REFERENCES

Brown, Justin P., 2005, *Excavation of the Claremont Tunnel Seismic Upgrade Project.*

Caulfield, R.J., Kieffer, D.S., Tsztoo, D.F., and Cain, B., 2005: *Seismic Design Measures for the Retrofit of the Claremont Tunnel,* RETC 2005.

Geomatrix Consultants, 2006, *Location and Characteristics of the Hayward Fault Zone Exposed in the Claremont Bypass Tunnel,* Berkeley California, dated January 2006.

Geomatrix Consultants, 2002, *Geologic Interpretive Report for the Claremont Tunnel Seismic Upgrade Project,* Berkeley, California, dated March 5, 2002.

Jacobs Associates, 2004, *Geotechnical Baseline Report for the Claremont Tunnel Seismic Upgrade Project,* dated January 2004.

FORENSICS OF DIFFICULT GROUND CONDITIONS LEADING TO ADVANCED SOLUTIONS FOR RECOVERY OF AN ABANDONED TUNNEL PROJECT

Faruk Oksuz
ARCADIS

Phillip L. Covell
ARCADIS

Martin Doll
ARCADIS

INTRODUCTION

A series of difficult ground conditions causing high groundwater inflows and hazardous gas intrusion abruptly halted the construction of a 6.4-meter (21-foot) finished diameter and 1,890 meters (6,200 feet) long tunnel planned to serve as a 4-million cubic meters (1-billion gallon) per day capacity river outfall for treated wastewater. Some 808-meters (2,650-feet) section of the constructed tunnel, TBM, and all other tunneling equipment were completely flooded with hydrogen sulfide laden artesian groundwater. The subsequent pumping and tunnel recovery efforts were unsuccessful and the tunnel was abandoned. In the end, the construction contract was terminated and options to redesign and reconstruct the project were re-evaluated to maximize the use of already constructed tunnel facilities such as the construction access and intermediate shafts and six outfall diffuser shafts in the river.

The ARCADIS team experts conducted a thorough examination of possible causes that led to the abandonment of the tunnel and developed alternatives to complete this $90 million project. This paper presents the results of a structured alternative screening and decision-analysis workshop that lead to a recommended solution for the project recovery and summarizes the efforts undertaken to find this solution within a 40-day response period. The discussions include the following:

- Information gathering and problem definition phase
- Function analysis, risk management strategy and risk registry
- Screening of project recovery and completion alternatives
- Completion of a "40-day Report" with recommendations and probable range of costs and schedule.

The study concluded that the tunnel failure was caused by a suspected hydrofracturing or blow out of a horizontal bedding plane that had been missed or insufficiently grouted in the pre-excavation grouting program, which ultimately led to the inflow of high quantities of hydrogen sulfide laden groundwater that forced evacuation of the tunnel. We concluded that a reconstruction of the tunnel was possible but required the use of a different approach in tunneling technology to deal with the difficult rock conditions.

It is noted that the ARCADIS team's purpose was not to determine responsibility for the events surrounding the project difficulties, but through evaluation of historical data, determine what conditions and/or practices may have contributed to the flood event, and identify actions for recovery of the project. Most importantly, the team has prepared information that will be useful in avoiding similar events on future tunneling projects in similar difficult ground conditions, and succeeded in identifying both a recovery plan and an option for designing and reconstructing the project.

BACKGROUND AND GEOLOGIC SETTING

The 6.4 meter (21-foot) finished diameter tunnel was being constructed using a shielded, open-face, main-beam TBM and one-pass precast concrete segmental liner in dolomitic limestone at approximately 90 meters (290 feet) below ground surface. Pre-excavation grouting ahead of TBM was specified and implemented to cutoff groundwater inflow during tunnel advancement. The tunnel advance rates were already significantly impaired by extensive pre-excavation grouting work required. Approximately 808 meters (2,650 feet) of tunnel was completed prior to project failure.

The subsurface profile at this site consists of overburden materials to a depth of 25 to 30 meters (80 to 100 feet) extending to an irregular interface with the underlying rock formations. The lower few feet of the overburden layer is described as a till hardpan. Most boring logs describe the hardpan as being composed of sand, gravel, clay, and silt. The hardpan is frequently overlain by sandy, silty clay. The underlying rock formations are dolomitic limestone. These formations are characterized by solution-enlarged joints and bedding planes. The boring logs and rock core samples confirm the presence of gypsum and anhydrite beds within the limestone formations. The cavities and open joints encountered in the borings and in the tunnel grouting were frequently partially filled with clay, silt, and sand.

The extent and geometry of the open joint and cavities within the rock mass are not known with certainty and this is typical of karstified limestone masses. Based on the inspection of the cores and review of relevant geotechnical information including core recoveries, the cavity size and opening widths are likely to be up to several inches in thickness, although isolated larger voids can reasonably be expected.

The dolomitic limestone formations constitute an artesian aquifer with piezometric pressure that can exceed 3 meters (10 feet) above ground surface. Groundwater pressures measured during tunnel grouting operations were frequently up to the full 9 bars (130 psi) at the invert. The water contained in the aquifer is believed to be confined primarily to channel flow in the solution-enlarged joints and bedding planes that control the location and development of cavities. There appears to be a network of connecting open joints and cavities that allows the aquifer to function generally as a unit. The hardpan at the contact between the overburden and the top of rock serve as an aquiclude/aquitard (confining layer) that provides confinement for the groundwater in the limestone.

Based on review of the permeability packer test results and the boring logs, the upper portion of the rock profile has the highest permeability and the greatest frequency of cavities and open joints. The tunneling records and tunnel grouting records, however, show that high-pressure groundwater flows and partially filled cavities and open joints presented a major tunneling problem also within the rock mass where the tunnel is located, principally in the last 183 meters (600 feet) and predominantly at or below the tunnel invert. The control of the atmospheric concentration of hydrogen sulfide gas in the tunnel was also a critical safety issue during the entire project.

TUNNEL ABANDONMENT

The tunnel abandonment occurred when the face was at 808 meters (2,650 feet) away from the construction access shaft. Excessive groundwater inflow and associated hazardous hydrogen sulfide gas release into tunnel atmosphere, forced immediate tunnel evacuation and ultimately shut down of tunneling operations.

The tunnel daily log indicates that the pumping rate from the tunnel was roughly steady at 2.27 cubic meters per minute (1,429 gallons per minute or gpm) one day before abandonment occurred. The logbook notes that pumping continued at a rate of 2.27 cubic meters per minute (1,429 gpm) until an overwhelming inflow of approximately 23.4 cubic meters per minute (6,200 gpm) occurred from the TBM face. During the last day-shift, emergency pumping capacity was added, thus increasing the total discharge capacity to an estimated 17 cubic meters per minute (4,500 gpm). However, due to continuous inflows, the tunnel's pumping system eventually failed after about 28 hours of non-stop pumping. The tunnel heading had already been evacuated due to the high concentration of toxic gas. Due to swift actions of the construction team, fortunately there were no injuries or casualties to personnel as a result of tunnel flooding or release of hazardous gases.

Probable Causes of Flooding and Abandonment

The tunnel was being excavated under artesian groundwater conditions and approximately 9 bars (130 psi) of hydrostatic head. The groundwater contained hydrogen sulfide in dissolved phase, at an average estimated concentration of 100 ppm (parts per million). Accordingly, the more groundwater entering the tunnel, the more gas is released. Hydrogen sulfide is a highly corrosive and odorous substance. It is denser than air, colorless, and toxic if inhaled, and can result in a variety of ailments at higher concentrations.

The tunnel water (and also hydrogen sulfide gas) control primarily relied on pre-excavation grouting and one-pass precast segmental lining installation. The pre-excavation grouting ahead of tunnel face was intended to control and cutoff groundwater inflow into the heading. The installation of precast, bolted and gasketed concrete segmental lining was expected to cutoff inflow behind the advancing TBM and tunnel face.

The pre-excavation grouting has been carried out with various application pressures, grout types, and grout mixes (cement to water ratios). While there were numerous reported difficulties and significant overrun of quantities in pre-excavation grouting, the tunnel construction and advance rates were also hindered by additional difficulties in placement of backfill or annulus grouting behind precast segments. These difficulties included high backpressure, longitudinal water flow developed in the annulus, and other limitations in the segment and TBM design to effectively apply and receive backfill annulus grout to tightly secure the segments in place and cutoff the water flow. Although the backfill grouting application was modified to use expandable grout bags, they were still difficult to place behind the segments and not able to withstand high pressure grouting to displace water and residual materials in the annulus and within the surrounding fractured rock mass. These difficulties in backfill grouting resulted in some squatting and misalignment of segments which further jeopardized water tightness of segmental lining. Other available means to stop water or backfill grout flowing into heading were spring plates on TBM but they were not designed for or expected to be effective against pressures up to 9 bars (130 psi).

Effects of Geologic Features

The cavities, open joints and other karstic features of fractured dolomite encountered in the borings and in the tunnel grouting were frequently filled with gouge, consisting of clay, silt, and sand material. This infilling or residual erodible material significantly reduces the effectiveness of grouting when it is not removed or replaced with stable grout mix. The resulting pre-excavation grout curtain or envelope established around tunnel perimeter remains highly susceptible and can be washed out under high hydro-static heads and large gradient flows. These conditions develop when hydrostatic pressure at advancing tunnel face is not counter-balanced and unsealed annulus space or other open-channel flow conditions remain behind segments that generate or contribute to a gradient and longitudinal flow.

As previously stated, the segment backfill grouting difficulties were compounded by water moving longitudinally along the annulus and washing out the backfill materials toward the cutter head. This water resulted from the residual permeability of the pre-grouted envelope or "curtain" around the tunnel, which increased with time due to washout/solution, driven by the ever-present artesian pressure. However, there is evidence that the construction team was being successful in reducing the normalized rate of flow during the last six months or so of tunneling, as a result of the routine implementation of revised backfilling plan with expandable grout bags. Therefore, there was no evidence that the flood developed anywhere other than within a few feet of the advancing face, while the construction team's observations suggest the source was at or below the tunnel invert.

The pre-excavation drilling and grouting records show that a significant recurrent (if not totally continuous) feature was encountered under and around the invert of the tunnel from approximately 670 meters (2,200 feet) onward. The feature contained soft material and/or clay and was under full artesian head. It was most clearly intersected during the last grout curtain installation, approximately 15 meters (48 feet) ahead of the face, and 1.5 to 2.0 meters (5 to 7 feet) below the invert. The feature accepted relatively large volumes of grout. However, there is no verification the feature was filled completely due to the impracticality of drilling additional "split spaced" holes in the inner arc, given the fixed locations of the guide ducts through the TBM. Flooding occurred when tunnel face was at 808 meters (2,650 feet).

It is reasonable to surmise that the relatively thin stratum (say 1.5 to 2 meters or 5 to 7 feet) of sound rock (and/or the clay infill material it contained) under the invert, overlying the feature, was simply unable to resist the artesian pressure (9 bars or 130 psi) and that it ruptured (by hydro fracture), allowing full hydraulic connection to the feature up into the face. As far as we understand the configuration of the TBM and the significance of opening caps on the cutter head (to relieve face water pressure), this water would have simply found its way into the tunnel regardless, through and in front of the rings which had been placed but not grouted. This inflow was initially estimated by the construction team as approximately 1.9 cubic meters per minute (500 gpm) and was noted to be muddy and emanating from the invert of the tunnel. It appears that the heading was initially evacuated not due to the rate of water inflow (which was potentially manageable, given the pumping capacity available in the tunnel), but due to the high concentration of hydrogen sulfide that the flow carried.

There is no evidence that an old brine well or an ungrouted site investigation hole may have caused the flood. However, the possibility that such a vertical conduit may have allowed the feature to be fully "charged" cannot be discounted, unlikely though it is. In any event, such an artificial conduit was not hydraulically necessary to charge the feature. Likewise, the existence of some vertical/sub vertical fault connecting the feature, and charging it in this way, cannot be dismissed given the limitations of the site investigation conducted.

ALTERNATIVE SCREENING AND EVALUATION

ARCADIS was contacted approximately 14 months after project failure for its expertise in innovative tunneling technologies including European tunnel projects. The scope of work included a synthesis of conditions that led to the abandonment of the tunnel and development of feasible recovery options for the project.

The first order of business was to gather all design and construction data and perform an independent analysis of ground conditions, tunneling methods, notably the pre-excavation and backfill grouting efforts. ARCADIS' first recommendation was to stop the artesian tunnel water flow from the existing construction shaft by either raising the steel shaft lining to a level above the static water level or to seal it off completely. This action eliminated the need for continuous treating of tunnel water for removal of hydrogen sulfide and total suspended solids before discharging to surface waters.

As part of the systematic project recovery and completion efforts, ARCADIS arranged an alternatives evaluation and screening workshop and brought the project stakeholders together. The evaluation team consisted of ARCADIS and independent tunnel experts, and also included representatives from the original design, construction, and construction management groups of the abandoned tunnel. The team was facilitated by a certified ARCADIS Value Engineering (VE) specialist and followed the six-phase VE Job Plan to guide its deliberations:

- Information gathering and problem definition
- Function identification
- Creative idea generation (solutions for recovery and completion)
- Evaluation of creative ideas
- Development of selected alternatives
- Presentation of results

The alternative evaluation was strictly focused on project recovery and completion while maximizing the benefits of already constructed facilities and minimizing risks inherent to construction and completion of the tunnel. A risk registry has also been developed as a result of final alternative selection to identify and address the risks.

Prior to and during workshop, eleven alternatives were identified and evaluated using a weighted-average criteria scoring system. These criteria included noneconomic function analysis such as the system operations and hydraulics, regulatory compliance, right-of-way and easements, risk of ground conditions, and safety. Four were immediately dropped out from further consideration because their total scores were significantly lower than others. Other remaining alternatives carried through for further analysis and included the following:

- A new rock tunnel within existing easement and starting from the existing construction shaft and using the existing outfall structures. The tunnel would be constructed using a sealed and pressurized-face slurry (SPFS) TBM and at a higher elevation than the abandoned tunnel.
- Four alternatives of new tunnel in soft ground within different alignment and outfall options. None of these alternatives were able to connect to existing outfall structure and most of them required new easements and permitting that could deem project infeasible.
- A new shorter soft ground tunnel to a cut-and-cover box structure for pretreatment, and utilization of a pump station and a smaller diameter force main for direct discharge to river.
- Two or three new replacement soft ground tunnels along the same alignment and constructed with smaller diameter Earth Pressure Balance Machines and connected to new outfall structures.

The short soft-ground tunnel and pump and treat option did not exactly meet the regulatory criteria and naturally had significantly higher long-term operation costs compared to a siphon or gravity operating deeper tunnel option. The evaluation team ranked remaining rock and soft ground tunneling alternatives fairly even, and it was evident that the cost and constructability risks would be the tie breakers. The cost estimating included inquiries with the TBM manufacturers and contractors. The project risks were identified using ARCADIS' comprehensive project risk management approach and using a project specific risk registry. The risk registry was very important in the quantification and ranking of project risks for design, construction, and operation, and ultimately guided the decision making.

The risk registry grouped the project risks under five categories:

- Contract documents and general conditions
- Design
- Construction activities that were separated for shafts, tunnel, and connections to existing facilities
- Project completion and operations
- Third party claims.

The risk ratings were developed using likelihood and consequence of each risk and represented probable incremental cost to the project. The risks were also prioritized based on these ratings. The risk registry was later expanded into the comprehensive risk registry for completion of the rock tunnel option. A risk management and mitigation strategy was mandated for each risk item and the responsible party or parties were identified. The ARCADIS project manager had the ultimate responsibility for tracking the risk registry and ensuring that all risk items were properly addressed in the selection and recommendation of design and reconstruction alternative.

SELECTED ALTERNATIVE AND RECOMMENDED SOLUTION

The ARCADIS experts concluded that the project can be completed by using advanced and innovative tunneling techniques and a sealed and pressurized-face slurry (SPFS) TBM. This determination also backs the original design concept for the rock tunnel except the tunneling method uses a completely sealed tunneling environment and the SPFS TBM nearly completely eliminates the risks associated with groundwater and gas inflows into the tunnel. Additionally, the hydrostatic head and high gradient flows become much more manageable by bringing the tunnel invert about 30 meters (100 feet) higher while still allowing connection to already constructed outfall structures.

This recommended reconstruction solution and TBM technology is an innovative and proven tunnel construction technique that is used in numerous projects in Europe and Asia, but still novel to the U.S. To safely advance the tunnel and with due consideration given to the possibility of another sudden water inflow and the resulting high concentration of hydrogen sulfide, we recommended the use of a SPFS TBM. Even if a fully charged, highly permeable feature is encountered in the rock mass, the SPFS TBM would not allow water and gas to enter the tunnel, thanks to the confined working chamber which is sealed off against the tunnel and counteracts and controls the pressure at the tunnel face. In order to provide a completely sealed tunnel, this tunneling system requires construction of a watertight lining behind the TBM itself. The rock mass has proven to be mechanically amenable to mining with the TBM method, and with a SPFS TBM, there will be no need for systematic pretreatment of the rock mass in advance of the face.

Competency of Recommended Solution for Geologic Conditions

In this type of adverse geologic environment, it is advantageous to maintain continuous support of the ground and water pressure at the tunnel face. To develop such continuous support, the tunnel face in front of the cutterhead must be pressurized either by air, water, or slurry. In order to maintain the tunnel face continuously pressurized, the cutterhead and the working chamber behind it are separated and sealed from the rest of the TBM assembly with a steel bulkhead, thus the designation of a "sealed" TBM. The presence of the hydrogen sulfide in the tunnel indicated that a sealed TBM using pressurized slurry as a medium to transfer the counter-pressure to the tunnel face would be needed. The slurry in the working chamber is kept pressurized by compressed air acting on it by means of hydrostatic communication on both sides of a submerged wall separating the working chamber in two compartments, which can communicate via an opening in the bottom. The pressure can thus be adjusted to counterbalance pressure from soil, rock, or groundwater at the tunnel face. In tunneling practice today, the maximum pressure used in sealed machines can be as high as 13 bars (190 psi), i.e., the Hallandsastunnel in Sweden, but most slurry machines operate under pressures ranging from 5 to 7 bars (75 to 100 psi). Given the ground conditions at the site, the mining depth would be limited to about 60 meters (200 feet) below ground surface.

Traditional disc cutters are generally used to break the rock, and the slurry must be continuously circulated and processed to remove cuttings. Repair and maintenance of the cutters are accomplished using a hyperbaric personnel lock access into the working chamber and to the cutterhead and tunnel face. Recent developments in the cutterhead arrangements have now significantly eliminated the hyperbaric access needs by allowing replacement of cutters behind the cutterhead wheel and under atmospheric conditions.

The sealed TBM not only prevents the water and hydrogen sulfide from entering the tunnel face, but in so doing, it eliminates the needs for systematic probing and permeation grouting in front of the heading. This feature is a significant advantage over the regular shielded TBM. A sealed TBM removes the dependence on the effectiveness of the perimeter grouting program necessary for a regular TBM to be successful. However, if needed, the sealed TBM also provides the ability to probe and grout ahead of the tunnel excavation.

Furthermore, with this recommended solution, the originally constructed construction shaft and river outfall shafts would be utilized to full extent. However, it is recommended that the existing tunnel is completely abandoned. This approach appeared to save cost, time and reduce risks associated with building a new shaft and connecting to the existing tunnel, plus avoided the risks of plugging, dewatering and repairing of the existing tunnel. Also, and of significant importance, the hydraulic deficiencies resulting from the exiting liner segment joint offsets could also be mitigated by new tunnel liner and a slightly larger diameter tunnel.

The SPFS TBM application would be the first application of a large diameter (i.e., greater than 6.0 meters or 20 feet) for a rock tunnel in the U.S. This method has been successfully used on a number of projects, including the SOCATOP tunnel near Paris, France, the EOLE tunnels for the Paris metro, and the Lefortovo Tunnel in Moscow where ARCADIS experts were consulted or involved. Mixed ground and soft ground applications of basically the same technique are commonly referred to as the Slurry TBMs or Mixshield TBMs depending on their technological specifications and mode of operation.

Tunnel Lining

Under protection of the shield behind the pressurized working chamber, the reinforced concrete lining segments are erected in the SPFS TBM tunnels. The tunnel lining has a very significant role in the success of the tunnel construction. The lining must withstand ground pressures, water pressures, chemical attack, grouting pressures, thrust forces, probable rock falls, dead loads, and plus a variety of other forces. For circular tunnels, the lining ring typically consists of five to seven segments. The number and geometry of the segments is optimized based on the project requirements and is influenced by tunnel diameter, transport method, the erector mechanism used by the TBM, and the thrust jack configuration.

Regarding the actual project, advancing the tunnel using a SPFS TBM, makes it doubtful that the existing tunnel ring segments can be used. With consideration of the higher jacking forces from the sealed TBM, the design of a more efficient sealing system, and the development of modified annulus and contact grouting methods, it was evident that new liner segments would be necessary for the project.

Experience shows the extreme importance of controlling the grouting pressure and filling of the annular space in order to securely lock the lining ring in position. Without controlled, continuous, effective back-grouting, the lining can be subject to significant deformation and ultimately joint leakage. With the proposed SPFS TBM, the annulus grouting can be conducted continuously and effectively with constant control as the machine advances.

Vertical and Horizontal Alignment

The ARCADIS evaluation team recommended raising the tunnel vertical alignment to keep the external groundwater pressure within practical and economically feasible operating limits of the SPFS TBM as discussed above. The new tunnel would be located in an area of rock that was characterized as "poorer quality" than the deeper rock, however, using the SPFS TBM mining can effectively address this concern. No pre-excavation or permeation grouting would be required and the water-tight lining would be installed immediately behind the shield. There were no significant modifications to horizontal alignment so that there would no additional easement needed and the reconstructed tunnel would connect to existing outfall diffuser shaft structures as originally planned.

ACKNOWLEDGMENTS

The authors acknowledge and appreciate the team effort involved in this project. The entire project team is recognized for their diligent efforts and valuable contributions to develop a sound project recovery and completion alternative within a 40-day period. The authors further acknowledge the tunneling, geotechnical, and water conveyance systems experts, including Jacques Robert, Dr. Donald Bruce, Dennis Kamber, Robin Dill, Pascal Guedon, Rohit Trivedi, Paul Booth, and Dr. John Richardson, and VE facilitator Howard Greenfield with LZA a subsidiary of ARCADIS.

GAS MITIGATION IN THE MILL CREEK TUNNEL

M. Schafer

MWH Americas, Inc.

R. Pintabona

MWH Americas, Inc.

B. Lukajic

MWH Americas, Inc.

M. Kritzer

Northeast Ohio Regional Sewer District

S. Janosko

Northeast Ohio Regional Sewer District

R. Switalski

Northeast Ohio Regional Sewer District

ABSTRACT

Methane gas caused an eight-month shutdown of mining operations at the Mill Creek, Phase 3, Tunnel. The gas-related shutdown occurred as the tunnel was advanced to an approximate distance of 2,700 feet, which constitutes approximately 18% of the total tunnel length. The construction of an emergency ventilation shaft, a comprehensive program of de-gassing wells and an expanded gas monitoring system were used to mitigate the gas condition. When complete, this tunnel will be utilized to convey and store combined storm and sanitary sewage collected from a portion of Cleveland, Ohio and ten suburbs.

This paper will provide an overview of gas related remedial measures and explain how these measures were integrated into the project to ensure safe tunneling conditions.

INTRODUCTION

The Mill Creek, Phase 3, Tunnel (MCT-3) is currently under construction with planned completion in the year 2008. It is one of the largest tunneling projects undertaken by the Northeast Ohio Regional Sewer District (NEORSD) to date. The tunnel horizon is situated within the Devonian Chagrin Shale rock formation at an average depth of 280 feet. The tunnel was excavated using a two-pass method. A full face, fully shielded, Robbins, 23.8 ft diameter tunnel boring machine (TBM) was used to excavate approximatly 15,000 feet of tunnel and facilitate installation of initial supports (first pass). The final lining (second pass) consists of 12-inch thick cast-in-place reinforced concrete and integral low flow channel. A total of seven (7) shafts were constructed on the project with the excavation of the tunnel commencing at Shaft 14, and proceeding down grade to Shaft 9, the terminus shaft.

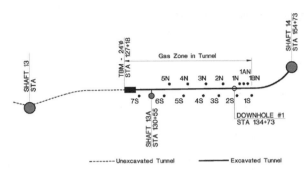

Figure 1. Location plan

Just prior to the eight-month gas related shutdown, several unusually large plumes of methane gas entered the tunnel while mining was proceeding. Each of these occurrences required that mining operations be suspended and all personnel be evacuated from the tunnel. The first incident of gas in the tunnel occurred when the tunnel intersected Down-Hole No.1 (Figure 1) on August 9, 2004. However, due to the frequency and volume of subsequent gas occurrences, the decision was made by the District (NEORSD) to suspend mining operations. Additional steps were then taken to mitigate the gas including drilling de-gassing wells, installing a comprehensive gas monitoring system and constructing a new 14-ft diameter ventilation shaft. Shortly before the suspension of mining operations, the Contractor Kassouf, Murray Hill, Mole & Kenny, a Joint Venture (KMM&K) attempted to mitigate the gas occurrences by installing an additional blower and bag line that evacuated air from the vicinity of Down-Hole No. 1 and discharged it at Shaft 14. This improvement proved insufficient in quickly dissipating gas occurrences.

NATURAL GAS IN CHAGRIN SHALE

Although not frequent, methane gas has been encountered in previously mined tunnels within the Chagrin Shale Formation. The subsurface investigations confirm that low levels of methane could be encountered in this rock formation and that the gas appears to be confined to openings along joints and bedding planes. It is also known that shallow gas wells in the project region produce some amount of gas for domestic consumptions.

POTENTIALLY GASSY TUNNEL

A tunnel is classified as potentially gassy if either of the following circumstances are anticipated or occur:

- When air monitoring shows 10% or more of the lower explosive limit (LEL) for methane or other flammable gases, measured at 12 inches (304.8 mm) from the roof, face floor or walls in any underground work are present for more than a 24-hour period.

- When the geological formation or history of the area indicates that 10% or more of LEL for methane or other flammable gases is likely to be encountered in such underground operations.

Table 1. Summary of gas emissions in tunnel

Date	LEL%	Location of Gas Detection	Remarks
April 2, 2004		At downhole Sta.134+73	Rapid gas discharged for 30 hours
August 9, 2004	118	At tunnel heading Sta. 134+73	Brief tunnel evacuation
August 19, 2004	146	At Station 134+73	Tunnel evacuated; gas clear in one hour
August 20, 2004	123	At Station 128+00	Tunnel evacuated; 3-hour shut down
August 20, 2004	195	Behind the TBM-Station 129+13	Tunnel evacuated-shift ended
August 23, 2004	65	Behind the TBM-Station 129+13	Ceased mining to increase ventilation
August 26, 2004	76	At Station 134+73, hand-held monitor	Tunnel evacuated; gas clear in one hour
August 30, 2004		TBM at Station 127+18	Start of eight-month shutdown
April, 2005		TBM at Station 127+18	Resumption of mining operation

The MCT-3 Contract Documents classified this tunnel as "potentially gassy." The Contract Geotechnical Baseline Report (GBR) states that quantities of natural combustible gases (primarily methane) under pressure shall be anticipated in the shafts and tunnels. The GBR further stipulates that encounters of gas requiring temporary suspension of construction operations should be expected.

It is important to note that the construction equipment utilized to mine the tunnel met the OSHA requirements for a potentially gassy tunnel.

GAS EMISSIONS CAUSING WORK SUSPENSION

On April 2, 2004, prior to the start of the mining operations, gas was encountered at a depth of 230 feet while the Contractor was drilling a down-hole along the tunnel alignment (Down-Hole No.1), Figure 1. It took approximately thirty hours for high-pressure gas to dissipate from the down-hole. Because gas was frequently encountered during previous subsurface investigation borings, there were no immediate concerns. The downhole was left open to vent any additional gas that developed. The first incident of gas inflow in the tunnel occurred on August 9, 2004, when the tunnel reached its intersection with Down-Hole No. 1. On that and subsequent dates, gas was released into the tunnel in a sudden cloud and under pressure, which quickly overwhelmed the ventilation system. The frequency of gas emissions into the tunnel continued to intensify between August 19 and 26, 2004, causing several tunnel evacuations. It is important to note that these gas emissions did not cause tunnel conditions to meet OSHA's requirements for a gassy tunnel. Evacuation of the tunnel is required when the lower explosive limit (LEL) reaches 10%. Based on the numerous readings and high LEL levels during this period, the decision was made to suspend mining operations on August 30, 2004. At the time of shutdown, the tunnel was advanced to Station 127+18, an approximate distance of 2,700 feet from Shaft 14. The location of the TBM at the time of shutdown is illustrated in Figure 1.

The following table demonstrates the frequency of emissions, which eventually lead to a decision to suspend mining operations (see Table 1).

GAS MITIGATION MEASURES

Initial Ventilation Capacity

Initially, the tunnel ventilation system in the MCT-3 Tunnel consisted of a series of fans and ducts. This arrangement pulled air from the tunnel face and exhausted it to the ground surface. The primary exhaust line was a 60-inch diameter bag line and the secondary exhaust was a 30-inch steel duct, which together provided a total of 50,000 cubic feet per minute (CFM) of tunnel ventilation. Prior to tunnel shutdown, the Contractor upgraded the tunnel ventilation by introducing another 60-inch diameter bag line on August 20, 2004. This provided an additional 30,000 CFM, totaling 80,000 CFM in tunnel ventilation capacity.

Required Additional Ventilation Capacity

Based on initial readings of LEL levels at various locations in the tunnel and knowing the existing ventilation rates, we were able to estimate a sudden peak methane inflow of 3,000 CFM. These estimates indicated that an additional ventilation capacity of 300,000 CFM was needed to safely dissipate additional occurrences of gas of similar volume.

Degassing Wells

In order to allow the release of gas from the bedrock and provide data on gas conditions along the tunnel, a total of fourteen (14) de-gassing wells were drilled in the area of the potential gas source, from Station 135+75 to 129+75. As illustrated in Figure 1, the relief wells were located on each side of the tunnel at 50-ft centers and were drilled to an approximate depth of 500 feet. The top portion of each well, between the ground surface and Chagrin Shale contact, was cased and grouted, while the remaining portion of each well within the Chagrin Shale was left open to allow gas and water to flow into the well, as illustrated in Figure 2. Each well was furnished with a pump to maintain the groundwater level at a depth of approximately 140 feet below the tunnel grade.

Figure 3 shows a typical record of gas inflow into the borehole during drilling operations. From these records, most wells commonly show gas emissions occurring below a depth of 300 feet, and the tunnel grade. Initially, methane was detected at or above 100% of LEL in each of the 14 wells. However, shortly after completion of drilling operations, gas emissions into the de-gassing wells subsided considerably, with the majority of the wells discharging methane gas generally below 5% of LEL. Examples of this can be found in Figure 4. All 14 wells remained in operation during subsequent tunnel mining and lining activities.

Measures to Enhance Ventilation Capacity

In order to meet the ventilation criteria outlined in the preceding paragraphs, plans were developed to increase the ventilation capacity of the tunnel. The measures taken would need to be capable of quick dissipation of a peak methane inflow of 3,000 CFM to a safe level (less than 10% LEL). Part of the solution was for the Contractor to retain a specialized gas mitigation Consultant. KMM&K retained Weir International Mining Consultants, Inc. (WIMC) to provide an independent evaluation of the gas issue. Based on the recommendations of WIMC, the Contractor proceeded to increase the ventilation capability from the existing 80,000 CFM to 300,000 CFM. This was accomplished by constructing a new 14-ft diameter ventilation shaft equipped with four 75-hp

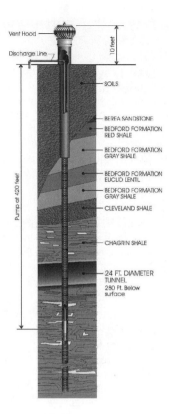

Figure 2. Profile of degassing well

blowers. New ventilation Shaft 13A was located about 400 feet downgrade from Down Hole No. 1 and about 300 feet behind the TBM cutter head (see Figure 1). It was determined that this additional ventilation shaft, with the help of baffles in the tunnel, would effectively dilute any methane inflows to below shutdown levels, and thereby satisfy the requirements for safe resumption of tunnel operation. Baffles were installed in the gas zone to create a turbulent air stream, thereby ensuring that a concentrated plume of gas entering the tunnel is quickly mixed with air. This additional ventilation feature reduces the possibility of gas layering in the crown of the tunnel in the event of new gas emissions. The shaft was constructed off line of the tunnel and was connected to the main tunnel through a 10-ft long adit, (see Figure 5). Around-the-clock (6 days/ week) operations were implemented to construct the shaft as quickly as possible.

Tunnel Monitoring System

As part of the gas mitigation program, the tunnel was equipped with an upgraded gas monitoring system to continuously read and record methane levels in the tunnel. New gas monitoring instrumentation, in addition to the existing monitors, was installed in three areas as indicated below.

1. Five (5) stationary gas monitors were installed in the area of the suspected gas source, about 18 feet above the tunnel inverts.

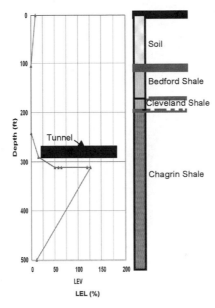

Figure 3. Typical drill log of degassing well

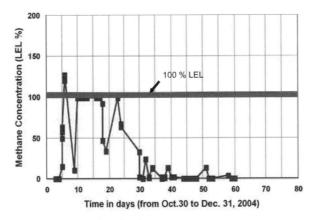

Figure 4. Typical gas discharge during shutdown

2. Two (2) new sensors, in addition to the existing three (3), were installed in the cutting and gripper area of the TBM.

3. Two (2) new sensors were installed at the rear of the TBM trailing gear.

The function of the new monitors was to provide enhanced gas detection and immediate shutdown of the TBM upon a 10% LEL reading. Data from the instruments was recorded at a central location on the Shaft 14 Site.

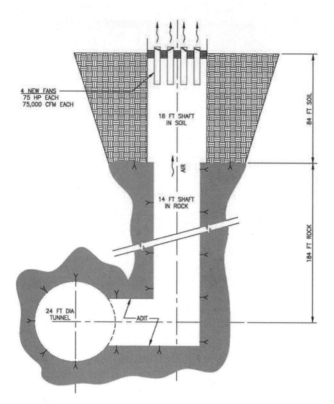

Figure 5.   Ventilation shaft 13A

Tunnel Maintenance During Shutdown

During the shutdown period maintenance crews entered the tunnel on a daily basis. To ensure their safety, the Contractor developed entry procedures on the presumption that the gas was still present in the tunnel. The maintenance staff exercised the mining equipment for short durations each day to ensure its readiness for resumption of mining operations. The TBM cutter head was advanced forward and backward and rotated to ensure that the bearings and hydraulic systems remained in good operating condition. The conveyor system was operated daily to keep the bearings lubricated and maintain belt flexibility.

Resumption of Mining Operation

Full scale mining operations resumed in April 2005. Tunnel excavation proceeded by advancing the tunnel heading at an average rate of 200 feet per week and was completed in January, 2006. While driving the remaining 12,300 feet of the tunnel, there were no further occurrences of methane gas. However, the readiness for action was required while mining was in progress. The ventilation system and de-gassing wells properly performed their intended function. The monitoring equipment and ventilation system remained in place throughout the mining operation.

Tunnel Retains Its Potentially Gassy Status

Although the gas occurrences were encountered while conducting mining operations, the tunnel continued to meet the OSHA criteria for potentially gassy, therefore, reclassifying the tunnel as gassy was not required.

A tunnel is classified as "gassy" if air monitoring discloses 10% or more of the lower explosive limit (LEL) for methane or other flammable gases, measured at 12 inches (304.8 mm) from the roof for three consecutive days. In such cases additional measures, such as elimination of the ignition sources, posting a sign at tunnel entrance indicating a gassy tunnel, prohibiting smoking, and maintaining a fire watch during hot work, would be required as a minimum.

CONCLUDING REMARKS

Excessive gas occurrences in MCT-3 tunnel resulted in work suspension and implementation of comprehensive remedial measures to mitigate the gas. These measures required degassing wells, a ventilation shaft and an expanded gas monitoring system in the tunnel. Even though the sudden gas occurrences temporarily exceeded 10% of LEL, the specified potentially gassy tunnel condition remained in effect throughout the tunnel construction. Much of the success for gas mitigation measures can be attributed to timely response and innovative planning by the project team members from NEORSD, MWH, PBQD, KMM&K and WIMC. This contributed to the safe completion of the mining operation.

ACKNOWLEDGMENTS

The authors would like to thank Northeast Ohio Regional Sewer District and Charles Vasulka, Director of Engineering for his review and approval to publish the paper. The authors wish to acknowledge Barry Doyle's technical contribution to the gas mitigation program and Kyle Ott of Parsons Brinckerhoff for his role in the initial exploratory program. Special thanks go to Carol Chavis for managing the paper and communicating with the RETC organizing committee.

REFERENCES

Hazardous Gases Underground, Application to Tunnel Engineering, Barry R. Doyle, Marcel Dakker Inc.

Schafer, M. et al. 2004, Rock Tunneling at the Mill Creek Project, North American Tunneling Conference, Atlanta, GA.

INNER CITY SLURRY TUNNELING—PROJECT ASDAM, ANTWERP, BELGIUM

Klaus Rieker

Wayss & Freytag Ingenieurbau AG

ABSTRACT

North of Antwerp central train station, two single-track tunnels for the high-speed train line between Brussels and Amsterdam were constructed. At the same time, the central train station was modified to a through station. Both 1.23 km tunnels are situated in the Antwerp sands and were constructed using a slurry shield machine.

Due to the inner city location, the tunnels run over their entire length underneath structures, in some cases with only minimal clearance. The paper deals with these challenging situations.

In addition, the tunnel boring machine, the tunnel lining and measures to limit settlements will be introduced.

INTRODUCTION

For decades, Antwerp central station had been a dead-end station accessible only from the south. In order to enter or leave the station, trains to and from The Netherlands therefore had to bypass the inner city via an orbital rail system. This meant an additional 15 minutes' traveling time on a railway line already used to full capacity for freight haulage as well.

This situation was no longer acceptable for the high-speed line from Brussels to Amsterdam (Figure 1). For years, even before the high-speed line was on the agenda, people had been thinking about converting the central station into a through station, so that the north of the city would be better linked to the centre. Furthermore, there was the wish to get the commuters living to the north of Antwerp away from the motorway, also already used to full capacity. A through station would make the city center more easily accessible, thereby enhancing its economic development. In addition, the listed station building (constructed in 1905) urgently needed to be restored.

These constraints led to the implementation of a construction project that was divided into three sections. The first section comprised the complete structural alteration of the existing accesses to the central station. Thus, additional accesses at levels −1 and −2 were added to the existing ones at level +1. In a further section, a tunnel was driven underneath the central station at level −2 under the protection of a pipe-roof system.

PROJECT DESCRIPTION

The third construction section is described below. It comprised the driving of two single-track tunnel tubes between Astrid Square in front of the central station and Dam Square in the north of Antwerp. The contract also included the construction of the ramp area located there to bring the tracks back to surface level. The project derived its name from the squares at each end of the tunnel, AStrid Square and DAM Square (Figure 2).

As can be seen in Figure 2, the project was located in the densely populated city center of Antwerp. The total length of the construction lot was 1.44 km. The central station can be seen in the lower right hand corner, the ramp area and the start shaft for

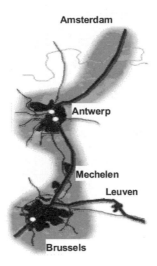

Figure 1.   High-speed railway line

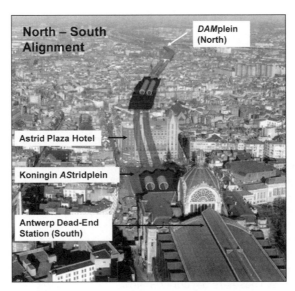

Figure 2.   Project overview

the two tunnel drives at the top. The bored tunnel section is 1.23 km long. The outside diameter of the tunnel tubes is 8.0 m.

GEOLOGY

The tunnels were mainly driven in the Antwerp Sands. The layers encountered near the ground surface consisted partly of fills of different soil types and partly of quaternary soil varying in thickness from 2.0 to 4.0 m. Some of the Tertiary sands

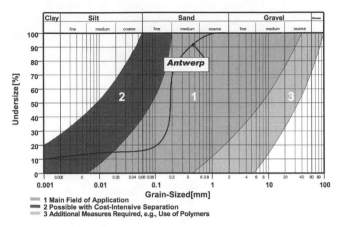

Figure 3.   Grain-size distribution curve

situated below these were interspersed with silt. The groundwater table lay 4 m above the tunnel tube.

Figure 3 shows the grain-size distribution curve of the sands encountered. It is in a range that is considered to be the main field of application for Hydroshields. They can be used in soils lying outside this range, but then extra measures or additional, cost-intensive separation work become necessary.

PROJECT CHARACTERISTICS

The project was commissioned by the Belgian Railway (NMBS). Construction supervision was carried out by Tucrail consulting engineers (a wholly-owned subsidiary of NMBS) and De Lijn (a Flemish public transportation company). The contract value amounted to approx. €70 million at tender submission stage, but increased to approx. €80 million due to additional work performed.

The contract was awarded on March 29, 2001. Driving work for the two tubes was carried out between October 2002 and September 2003. The project was handed over to the client on schedule on March 7, 2005.

SPECIAL FEATURES

Due to its inner city location the project presented numerous technical challenges and special features. These are described below.

The tunnel tubes were built from north to south using a shield machine. The east tube was driven first, followed by the west tube (Figure 4). From the launch shaft the whole of the tunnel alignment passes under buildings.

Close Proximity of Tunnel Tubes and Minimal Cover at the Launch Shaft

Due to the given circumstances, tunneling had to begin with a cover of only approx. half the tunnel diameter. In addition, the horizontal clearance between the tunnels, at only 1.26 m (Figure 5), was extremely small. The approx. 35 m long launch shaft itself was enclosed by diaphragm walls. In order to launch the tunneling machine, a short sealing block consisting of cement and bentonite was also built using diaphragm walls.

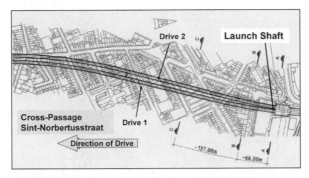

Figure 4.   Alignment, northern part

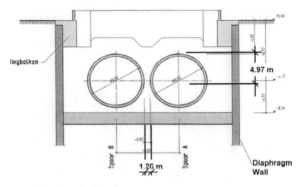

Figure 5.   Cross-section launch situation

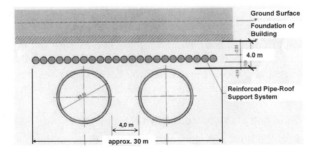

Figure 6.   Pipe-roof support system

Protection of Buildings Using a Pipe-Roof Support System

After some 50 m had been driven, the tunnel passed under buildings with a clearance of only 4 m between the foundation of the buildings and the top of the tunnel roof. To protect the buildings a pipe-roof support system consisting of 1.2-m-diameter tangent reinforced concrete pipes was installed ahead of the tunnel drive (Figure 6) using the pipe jacking technique. The reinforced concrete pipes were jacked forward from a shaft in Visestraat (Figure 7) and then filled with reinforced concrete.

Figure 7. Launch shaft and jacking shaft

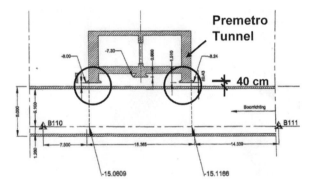

Figure 8. Tunneling under the metro tunnel

Tunneling Under the Existing Metro Tunnel

In the southern part of the drive, the tunnel had to pass under an existing double-track metro tunnel with a clearance of only 40 cm (Figure 8). Due to the extremely small cover it was impossible to carry out any protective measures in advance so that high-quality driving work alone had to avoid settlements.

Tunneling Under Astrid Plaza Hotel with a Small Clearance Between the Tunnels

At the end of the drives the tunnels had to be driven under the Astrid Plaza Hotel. The clearance between the base of the structure and the tunnel roof was 3.5 m (Figure 9) at that point, but the allowable settlement was limited to only 10 mm. The job was further complicated by the fact that the tunnels were again located close to each other with a horizontal clearance of 1.5 m. Since it was impossible for the shield machine to enter the target structure, the shield skins had to remain in the ground.

PARALLELS TO PREVIOUS CONSTRUCTION PROJECTS

Extensive experience in the construction of inner city tunnels was necessary to master the special requirements of the ASDAM project. Wayss & Freytag can look back on a long tradition of similar construction projects. The project Mecoma 1, 2 and

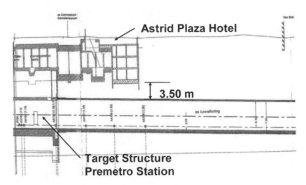

Figure 9.   Tunneling under Astrid Plaza Hotel

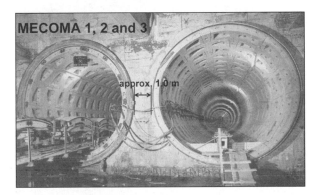

Figure 10.   Mecoma Tunnels, horizontal arrangement

3, which was successfully completed between 1977 and 1981, is described below as a representative example (Figure 10 and Figure 11).

In Antwerp, Mecoma was a forerunner of ASDAM. For one thing, the horizontal clearance between the tunnel tubes was similarly small; for another, the vertical clearance between the tunnel tubes was just less than 2 m. The machine then used to drive the tunnels was also a Hydroshield. The shield diameter was 6.56 m. The tunnel tubes were steel lined using coffered segments.

DRIVING METHOD

As already mentioned, in view of the geology encountered a Hydroshield was ideally suited for driving the tunnels. The custom-made tunneling machine had an outside diameter of 8.27 m (Figure 12). Including the three back-ups, the tunneling machine manufactured by Herrenknecht, Schwanau, Germany, had a total length of 60 m and a total weight of 860 tonnes. To allow safe entry into the working chamber the shield was provided with four face-supporting plates, which could be moved forward between the five spokes of the cutting wheel to the working face, thus permitting safe working.

The 9 m long shield was advanced by means of 28 driving jacks with a total power of 52,000 kN. A 350-mm-diameter feed and discharge line connected the machine to

Figure 11.   Mecoma Tunnels, vertical arrangement

Figure 12.   Hydroshield

the separation plant installed at the ground surface where the excavated soil was separated from the bentonite suspension in several steps and stored until removal.

A correct face support was indispensable at all times to ensure that the high requirements of the projects regarding low-settlement tunneling (max. 20 mm, 10 mm under the Astrid Plaza Hotel) were met. Three calculation models and conditions were used to determine the working face support required. These models were: full liquid support, partial air support and full air support. On this basis, the limits shown in Figure 13 were calculated. They represent both the maximum allowable and the

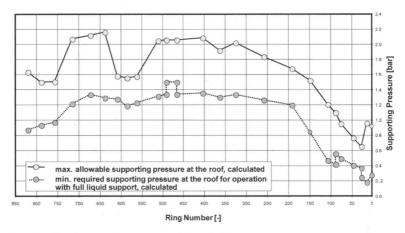

Figure 13. Results of supporting pressure determination

Figure 14. Connection of cross-passage

minimum required supporting pressure at the tunnel roof. The shield operators had to keep the supporting pressure between these two limits.

TUNNEL LINING

The single-pass tunnel lining has an inside diameter of 7.30 m and a wall thickness of 35 cm. It consists of seven plus one segments. The ring width is 1.50 m and the ring taper 50 mm. Waterproof concrete B45 with a reinforcement content of 100 kg/m^3 was used. The rings were coupled using the cam and pocket system developed by Wayss &

Figure 15.   Finished tunnel tube

Freytag. Two bolts per segment were installed in the longitudinal joint and four bolts per segment in the circumferential joint.

CROSS-PASSAGES AND EMERGENCY EXITS

The two tunnel tubes are connected by cross-passages every 300 m so that three of them had to be driven over the whole tunnel length. They were constructed using the pipe jacking technique with an open cutting sleeve under the protection of a low-ered groundwater table and were linked to the tunnel tubes by means of steel seg-ments (Figure 14). In addition, an emergency exit shaft with a staircase had to be built at the middle cross-passage.

SUMMARY

Despite the great technical challenges, such as close proximity of the tunnel tubes, tunneling under a number of structures with minimal cover and the adherence to small settlement tolerances, tunnel driving was carried out without any incidents and high-quality, dry tunnels were built (Figure 15) and handed over to the client on schedule.

THE SIR ADAM BECK II INTAKE TUNNELS—
HOMAGE TO THE BUILDERS

David Heath

Ontario Power Generation (OPG)

Clair Murdock

Consultant

ABSTRACT

The 1,280 MW Sir Adam Beck II Hydroelectric Project was constructed in the 1950s, at Niagara Falls, Ontario, Canada. The twin intake tunnels total over 17 kilometers in length and pass deep below the City of Niagara Falls. The tunnels were excavated using drill and blast methods, with a top heading and bench approach. The shot rock was loaded with electric shovels and transported to the five shafts with off-highway trucks. The tunnel muck was hoisted to the surface by skip and transported by truck to disposal areas. The volume of underground excavation for the 15.55 meter diameter tunnels and the shafts exceeds 3,500,000 cubic meters, then and now the largest underground civil works excavation by drill and blast in the Western Hemisphere. The lining required over 900,000 cubic meters of concrete. All of this work was completed in three years, remarkable progress, considering the technology of the day, comparing very well with the rapid excavation and tunneling methods of the twenty-first century. Big, fast, and deadly—at least 20 lives were lost in construction accidents.

SIR ADAM BECK

Beck's Accomplishments

Adam Beck was an early proponent of publicly owned electricity grids, opposing the privately owned companies which he felt did not adequately serve the needs of the people. With the slogan "Power at Cost," he was appointed in 1906 as the first chairman of the Hydro-Electric Power Commission of Ontario, better-known as Ontario Hydro, now called Ontario Power Generation. He created the Queenston-Chippawa power station, which was renamed after him. The subject of this paper, the Sir Adam Beck II Project (SAB II) was also dedicated to his memory, to recognize his enormous contributions to the progress of the people of Ontario.

Beck's Life

Adam Beck was born in 1857 in Baden, Ontario to German immigrants. Beck worked in his father's foundry as a teenager and later went into business with his brother William, manufacturing cigar boxes. Their business flourished and Adam Beck became a wealthy man. He was elected mayor of London, Ontario several times and also was elected to the Ontario Legislature, where his public power campaign took flight. Knighted by King George V in 1914, Beck died in 1925.

Beck's Words

Beck—"The gifts of nature are for the people." His statue still commands the intersection of University Avenue and Queen Street in Toronto. "Nothing is too big for us. Nothing is too visionary." The essence of Beck's leadership was in his vision.

THE NIAGARA RIVER

Drainage Area

At Niagara Falls, the Niagara River drains an area in Canada and the United States of 583,000 square kilometers. This is greater than all of France (547,030 square kilometers) or greater than the Czech Republic, Slovakia, Germany and Austria combined (568,602 square kilometers).

Flow

The 140 year average instantaneous flow of the Niagara River is 5,876 cubic meters a second. Because of the modulating effect of the Great Lakes upstream (Lakes Superior, Michigan, Huron and Lake Erie) the flow at Niagara is very steady, as compared to other major rivers of the world. The maximum flow is less than triple the minimum flow, whereas other major rivers' flow may vary by factors of 15 to 30, passing great torrents where there was previously only a trickle.

The Falls

Niagara Falls is, by volume, the world's greatest waterfall. The American Falls are 260 meters wide and carry 10% of the flow. The Canadian, or Horseshoe Falls are 670 meters wide, and carry 90% of the flow. The Falls were formed some 25,000 years ago, as the retreating glaciers exposed the Niagara Escarpment, permitting the waters of Lake Erie, which previously flowed south, to flow northward into Lake Ontario. The Falls have eroded the soft shales and limestones of the escarpment at an average rate of 1.2 meters a year and they now stand 11 kilometers from their origin at Queenston. The rate of erosion has been substantially reduced, due to hydroelectric diversions and riverbed remediation work above the Falls.

History

Known to the Iroquois people for thousands of years, the name Niagara means, from their language "thunder of the waters." The awesome spectacle was described by explorer Father Louis Hennepin in 1678, who called them "a vast and prodigious Cadence of Water." Charles Dickens wrote "I seemed to be lifted from the earth and to be looking into Heaven." The first exploitation of the power of the falls came in 1759, when Daniel Joncaire built a small canal above the falls to power his sawmill. The first generation of electricity at Niagara Falls, in 1881, was sufficient to light the Village of Niagara Falls, New York, power a few local mills and illuminate the Falls at night, using 18 carbon arc lights.

Tourists and Tourist Power

Over 14 million people visit Niagara Falls each year. The 1950 Niagara River Water Diversion Treaty between Canada and the United States provides for the most beneficial use of the waters, with varying minimum flow requirements over the Falls, depending on the time of day and the time of year. In the daytime, from April 1st to October 31st, the flow must be not less than 2,832 cubic meters per second. For the remaining five months of the year and at night during the other seven months, the flow over the Falls may be reduced to not less than 1,416 cubic meters per second. These figures represent roughly ½ and ¼ of the Niagara's average flow. Power production in Canada and the US is exactly tailored to fit water availability, with a very close compliance to the minimum flow requirements.

Sharing the Water

Canada and the United States share the available water equally, with a small adjustment of 141 cubic meters per second in Canada's favour, recognizing two Canadian diversions into the Great Lakes watershed. The Long Lac and Ogoki Rivers originally flowed into the Hudson's Bay watershed.

Installed Hydroelectric Capacity

On the US side of the Niagara River, the Robert Moses Generating Station and the associated Lewiston Pump GS (pump-turbine) have a total capacity exceeding 2,575 MW. On the Canadian side, the Sir Adam Beck I and Sir Adam Beck II Generating Stations, along with the SAB Pump GS (pump-turbine) and two associated smaller plants at St. Catharines, Ontario have a total capacity of 2,235 MW. In total then, with 64 generators (18 of these being reversible pump-turbines), Canada and the US have 4,810 MW of generating capacity at Niagara. Recent upgrades to the US and Canadian generating units are reflected in the figures above. Excavation of the third intake tunnel for the Sir Adam Beck complex is now underway, to augment OPG's diversion capacity. A fourth tunnel remains under consideration, along with additional OPG generating capacity.

THE MAJOR COMPONENTS OF THE SIR ADAM BECK II PROJECT

The Control Dam on the Niagara River

The Control Dam extends 600 meters into the Niagara River from the Canadian shore. It has 18 sluices, each with a clear span of 30.5 meters, separated by piers 4.25 meters wide and 28 meters long. The 18 sluice gates are of the 'fish belly' type, hinged at the upstream sill, affording no obstruction to the passage of ice when in their lowered position. The Control Dam ponds the water, called the Grass Island Pool. This is the primary control of intake water levels and diversion volumes for both the US and Canadian sides of the river.

Two Intakes of the Johnson-Wahlman Type

The Johnson-Wahlman design looks like a very long harmonica lying on its side. Ice on the Niagara River is of concern every winter. Following extensive model-testing, each intake was designed to be 152 meters long, with varying heights and depths for the intake slots. The sills are set 7.6 meters below river level, while the lintels vary from 5.8 to 2.4 meters below the water level. Behind the slots is a gathering tube of increasing cross section leading to the intake gates. The two 180 tonne gates (one for each tunnel) have a clear span of 13.7 meters and are 17.7 meters high.

Two Parallel Concrete-Lined Tunnels

The two tunnels are on 76 meter centres, excavated to a diameter of 15.55 meters and lined to a finished diameter of 13.72 meters. The nominal thickness of the unreinforced lining is 0.91 meters. Tunnel No. 1, which follows the longer outer course around the curving Niagara River, is 8,707 meters long, while Tunnel No. 2, following the shorter inner course, is 8,323 meters long. Each tunnel is designed to carry 565 cubic meters of water per second, although testing showed a possible capacity of 650 cubic meters per second. Water velocities at these two capacities are 3.8 meters per second and 4.4 meters per second respectively.

An Open Canal

As the tunnels approach the open canal, they rise at an angle of 30 degrees, to discharge their flow into an open canal leading to the forebay. The portals are at the southerly side of the St. David's Gorge, a prehistoric river channel filled to a great depth with glacial drift materials and debris, unsuitable for tunneling. (The much-deeper third tunnel, now under excavation, goes below the St. David's Gorge.) The canal is concrete lined through the Gorge section. The overall canal length is 3,622 meters, with a width of 61 meters.

A Forebay with Headworks

The forebay widens from the canal, reducing water velocity to about 0.9 meters per second. The headworks structure is 267 meters long, the entire width of the forebay. It is here that control of the water entering the penstocks is maintained. There are 32 openings on the upstream side of the headworks, two for each of the 16 penstocks.

Sixteen Concrete-Encased Steel Penstocks

Each of the 16 penstocks is 5.8 meters in diameter and 152 meters in length, of welded steel construction. Plate thickness at the upper end is 16mm, increasing to 36mm at the lower end. Each penstock was built of 68 prefabricated rings, brought to the site by road. They were installed on the 60 degree slope of the excavated cliff face and welded in place, followed by an X-ray inspection of each circumferential weld. The penstocks were encased in 60 centimeters of concrete to protect them from rapid and extreme temperature variations.

A 16-Unit Powerhouse

The powerhouse is located on the left river bank in the Niagara Gorge, just upstream of the SAB I powerhouse. The SAB II powerhouse is 350 meters in length, 19.2 meters in width and 15.2 meters in height. The 16 turbines are of the Francis type, operating at a head of 89 meters. The main transformers which increase the generator voltage from 13.8kV to 230kV are located on an outside deck behind the powerhouse superstructure. Three single-phase transformers are provided for each pair of generating units, for a total of 24 transformers. Excavation for the draft tubes was carried out behind a cofferdam in the Niagara River, to a depth of 13 meters below water level.

A 6-Unit Pumped-Storage Powerhouse

The pumped-storage generating station pumps water into a storage reservoir during low demand periods, for reuse in high demand periods. The reservoir area is

304 hectares, with a drawdown of 7.6 meters. Usable storage is 19 million cubic meters, equal to about 4½ hours of full flow from the two tunnels. The reservoir is bounded by a dike of about 7,400 meters in length, varying in height from 5 to 20 meters. The perimeter dike contains 2 million cubic meters of zoned fill, mostly built with tunnel muck. The reversible pump-turbines operate through head ranges of 18–27 meters in pump mode and 12–24 meters in generating mode.

Overall Cost and Schedule for SAB II

The final cost for the SAB II project was Can$343,742,000, giving a unit cost of Can$245.00 per kW of installed capacity, about ⅒ of today's hydroelectric costs. Construction started in April 1951 and the first unit was commissioned on April 30, 1954, just 37 months later. By mid-1955 12 units were on line, and all 16 units were producing power in early 1958. Ontario Hydro decided, in response to their mid-1952 forecast of sharply-increasing demand (and a year after construction had started on the first tunnel) to add the second intake, the second tunnel and six more units, in addition to the six units originally planned. Later, in 1953, Hydro decided to add four more, bringing the final total to 16 units, plus adding the pumped-storage project. The availability of the last four units delayed the overall project completion into 1958, but all the 'original' 12 units were on line within 52 months of breaking ground.

Employment, Work Regime and Contracting

Project employment peaked at 7,655 persons. The nominal work week was 5 days. Tunnel workers worked 9.6 hours a day, surface workers 8.8 hours, giving work weeks of 48 and 44 hours—the difference was for underground travel and cage time. Workers requiring accommodation were housed in three temporary camps—Chippawa Camp, near the intakes; Whirlpool Camp, near the downstream portals and Queenston Camp, near the powerhouse. Each camp had a cafeteria that would seat 600 men. All of the work (with the exception of the intake tunnels and the 5 shafts) was done by Ontario Hydro personnel. The tunnels were competitively bid, with each bid package comprising two shafts and about 3,600 meters of single tunnel. One contract was awarded to Rayner-Atlas, the other to Perini-Walsh. Ontario Hydro's mid-1952 decision to add the second tunnel required negotiations with both contractors, as work was then under way on the first tunnel. The remaining portion of 1,100 meters of tunnel at the intake end was subsequently awarded to Perini-Walsh, who were out-performing Rayner-Atlas.

SHAFTS

Locations

Shafts are located midway between the two tunnels, which are at 76 meter centres. The five shafts range in depth to tunnel invert level from 60 to 100 meters. The reasons for this range are three—the topography; the slope of the tunnel invert and, the depth to the underside of the Irondequoit limestone. This is the most competent rock strata and it was decided to make the underside of the Irondequoit limestone the crown of the tunnel. The Irondequoit slopes up from the inlet end of the tunnels at a gradient of approximately 0.5%, with the tunnel crown following that relatively strong and consistent 2.4 meter thick band.

Table 1. Thicknesses of the various strata encountered

Strata Description	Depth (m)	Total Depth (m)	Notes and Observations
Clay	7.3	7.3	From the surface
Sandy silt	16.8	24.1	Water table at 14m depth
Boulder clay, some slabs	9.1	33.2	To top of rock at 33.2m
Guelph Lockport dolomite	12.2	45.4	Water-bearing rock
Dolomite and Gasport limestone	8.5	53.9	Water-bearing rock
Gasport limestone	8.3	62.2	Water-bearing rock
Rochester shale	18.9	81.1	Tight—no water
Irondequoit limestone	2.4	83.5	Tunnel crown—the strongest rock
Reynales dolomite	4.3	87.8	Within the tunnel horizon
Neagha shale	1.8	89.6	
Thorold sandstone	2.4	92.0	
		99.1	Tunnel invert in Grimsby's stone
Grimsby sandstone	12.5	104.5	Muck pocket and sump
Power Glen shale	13.4	117.9	
Whirlpool sandstone			

Geology

Following are the depths of various strata encountered in the excavation of Shaft #1. The total depth is also shown in Table 1.

Dimensions

Shafts #1 and #2, done by Rayner-Atlas, were sunk to rock with 9.75 meter inside diameter concrete caissons, with a wall thickness of 0.76meters, giving an outside diameter of 11.27 meters. For Shafts #3, #4 and #5, Perini-Walsh chose to use sheet pile cofferdams, partly due to the lesser overburden depths at those locations. A tight international supply situation for steel then prevailed, another factor for Rayner-Atlas in choosing to use the concrete caissons. On reaching rock level the shafts were continued to tunnel invert grade at a finished diameter of 9.75 meters. Temporary timber sinking head-frames were used, these replaced with steelhead-frames of greater hoisting capacity, after the tunnel invert level was achieved. The muck loading pocket and the sump at each shaft were excavated a further 21 meters, the sump to a diameter of 5 meters.

Overburden Excavation Methods—Rayner-Atlas

The top few meters of soil inside the caissons was excavated by backhoe, followed by clamming. After the third and final lift of concrete was placed for the caisson, the total weight exceeded 1,100 tonnes but it just wouldn't go down. In addition, uncontrolled inflows of water under the cutting edge flooded the previously-dry excavation. Then ensued a series of ineffectual missteps, truly a comedy of errors, including: attempts to block the water inflow from the inside of the caisson; placing 25 kilogram blocks of dry ice in pits dug inside the cutting edge, trying to freeze the soil, discontinued after five weeks because the carbon dioxide gas released by the melting dry ice became hazardous to the crew; grouting around the perimeter, this discontinued after 3 days of effort; driving an external ring of sheet pile, with this attempt much obstructed by the earlier grouting program, all of these measures with little success. It was then decided to continue with wet clamming but the caisson still wouldn't drop. Soil losses from the exterior piped to the surface, requiring backfill. The next effort was top-ballasting the caisson. With a top load of 562 tonnes of sheet pile, the caisson finally moved, bottoming-out on the rock, and after 3 months of frequently-futile effort. By contrast, Perini-Walsh, at their three shafts, and using the sheet pile rings proceeded uneventfully and in much less time.

Rock Excavation Methods—Rayner-Atlas

Work in each shaft was carried out on 8 hour shifts, working around the clock. Each shift had a 22 man crew. The average sinking rate achieved was 0.9 meters per day. Hand-held pneumatic pluggers, with drill steels of 0.9, 1.8 and 2.4 meters were used to drill 65 holes on half the bench, with each round pulling about 1.9 meters. Drilling off half a bench took about 2.5 hours. Mucking half a bench took about 12 hours, all mucking done by hand. Crew access to the bench was by the muck buckets, rated at 1.4 cubic meters or 11 men. Mucking took over 70% of the total shaft-sinking time, exclusive of the timbering, which was done on Sundays by a special crew.

Shaft Bottom Complications and Hoisting Arrangements

Substantial structural steel arches and underpinning were required at the shaft bottom, to permit the safe excavation of the cross drifts to the main tunnels. The hourly-paid underground work force at Shaft #1 during this portion of the work totalled 240 men. During this period the permanent structural steel head frames and the electric hoists were installed. A typical shaft arrangement provided for 2 counter-balanced muck skips each of 10 cubic meter capacity and an 18 tonne capacity cage with counterweight. The cage was used for men, materials and the construction equipment, which included 15 tonne capacity Euclid Model 87FD off-highway trucks and 2 cubic meter Northwest Model 80D electric shovels. The Northwest shovel booms were shortened for tunnel clearance reasons.

TUNNEL EXCAVATION

Planning for the 15.55-Meter Diameter Face

Rayner-Atlas intended a modified full-face approach, using two multi-deck jumbos, one for the top portion and another for the lower, with mucking to be done with 0.7 cubic meter Eimco overhead loaders dumping onto a conveyor, to load 4 cubic meter muck cars, for locomotive haulage back to the shaft. This arrangement was shown to be inadequate, impractical, slow and unsafe. With Ontario Hydro's 'strong encouragement,' Rayner-Atlas hastily adopted the more productive top heading and bench approach that Perini-Walsh was using, even copying the Perini-Walsh plans for the drill jumbos. This approach divided the face into a 9 meter high top heading excavated the full length of the section, followed by a 6.5 meter high bench, with this work done as a second pass.

Drilling and Loading the Heading

For Rayner-Atlas' modified approach, the heading was drilled with a multi-deck rail-mounted jumbo, fitted with 23 Ingersoll-Rand pneumatic drifters, 16 of which were column and bar mounted. Only three drills on the top deck and four more near the rail bogeys were mounted on hydraulic booms, which worked only in the vertical plane—the horizontal movement was manual. Jacklegs supplemented the four hydraulic booms for drilling the bottom lifter holes. A round required 173 holes, drilled to 6 meters, using tungsten carbide bits. Electric blasting caps were used, with an overall powder factor of 1.6 kg/m^3. The toe of the muck pile was usually thrown about 40 meters from the face, exceptionally as far as 55 meters. Considerable oversize was encountered, and the grizzly at the loading pocket was reset, from 60 cm × 60 cm to 90 cm × 90 cm openings. Rocks sometimes had to be hoisted off the grizzly by air tuggers and loaded into small trucks, for cage transport to the surface, when the grizzly became clogged with oversize. Up to 10 electric-driven air compressors supplied as much as 212 m^3 of air per minute. In contrast, Perini-Walsh used fewer drills—17 Joy drifters, but all on hydraulic booms and pulled rounds averaging 4.8 meters. A Perini-Walsh heading round produced 900+ cubic meters of muck.

Scaling and Rock Support

Scaling after the blast was difficult and dangerous, as much scaling was done manually from the jumbo decks. Additional scaling was done using the Northwest shovels and from the muck pile. Steel ribs (8W × 20) 20 cm × 65 kg were set at 1.2 meter centres throughout the tunnel, a total of over 14,000 ribs. Blocking and timber and steel lagging were installed. The ribs were fabricated in 4 pieces, hoisted onto the jumbo decks by air tugger and then man-handled into place. A quarter section of one rib weighed in excess of 200 kilograms. Several rock falls occurred in Tunnel #1 during August of 1952, requiring that 80 meters of tunnel be re-timbered. Intermediate ribs were placed in some areas where the rock was suspect. On the break-throughs between headings, the alignments compared within a few centimeters—on the money for the surveyors, without lasers.

Crew Size and Cycle Times for Two-Heading Work

After abandoning their fatally-flawed full-face approach and after copying the Perini-Walsh jumbo, Rayner-Atlas still didn't get it right, going cheap with their choice of drill booms, with larger crews and higher labour costs as the results. Nonetheless, Rayner-Atlas excavated 173 meters of tunnel in 21 shifts, working two headings at once, during their best-ever 7-day production week. This advance required excavating 20,400 cubic meters, or almost 3,000 cubic meters a day. (Tables 2 and 3)

Mucking the Heading and the Bench

Both contractors used similar equipment for loading and hauling—Northwest Model 80D electric shovels and 10 cubic meter Euclid Model 87FD off-highway trucks. The shovel booms were shortened to give better clearance under the ribs and lagging. Perini-Walsh, always the innovator, found a novel way to eliminate long back-ups with the Euclids—they drove the shovel-bound Euclid onto a large steel plate in the invert, oiled the plate and spun the unloaded truck around with an air tugger. Rayner-Atlas wasn't too far behind. All the Euclids were scrubber-equipped. Muck pile cleanup and invert maintenance were by crawler tractor. The drilling crew worked on Heading A while the mucking crew worked on Heading B, alternating headings.

Table 2. The Rayner-Atlas underground crew

Crew	Men	Notes and Observations
Drilling	51	23 drills + jack-legs for lifter holes—generally a driller & chuck-tender
Mucking	35	Drilling on Heading A, Mucking on Heading B
Steel and timber	23	4+ steel sets required per round
Ventilation	4	Single shift only
Mechanics	19	
Total Men	128	Add for: electricians; hoisting; surface haulage, etc.

Table 3. Typical Rayner-Atlas cycle for a round

| Operation | Time Required | | Notes and Observations |
	Hours	Minutes	
Smoke out after blast	0	30	
Cleanup and scale	0	45	
Mucking	4	50	Perini-Walsh' time—4 hr-45 min.
Move in jumbo	0	30	Parked on rails about 60 meters back from the face
Drill, powder, out	4	35	Opportunity for dozer to do road maintenance—Euclid trucks
Total Cycle	11	10	Possibility of 2 rounds every 24 hours—three 8-hour shifts

Drilling the Bench

After cleaning the bench with compressed air and water jets, the drilling proceeded using a rail-mounted vertical-drilling jumbo equipped with 17 drills, mostly Gardner-Denver. This jumbo ran on rails attached to the bottom of the ribs and it was moved forward with air tuggers. Each vertical hole was drilled to a calculated depth, producing a nearly circular invert shape. The overall powder factor for the bench excavation was 1.45 kg/m^3, using packaged explosives and electric caps. A muck pile about 15 meters long was left in place before blasting, to reduce fly rock. Rounds were loaded as drilling progressed and were blasted at the end of each 8 hour shift—shoot what you drill. The average daily advance in Tunnel 2 was 30 meters, done over 3 shifts.

Ventilation

Steel ventilation ducts 1.2 meters in diameter were brought up to within 60 meters of the top heading face. Reversible electric fans at the surface and booster fans at 250 meters intervals supplied 1,840 cubic meters of fresh air per minute to the face. Reversing the fans expelled the blast fumes. The duct and fans were removed for the benching operations. Some supplemental unducted airplane-type fans were used to improve circulation while benching.

CONCRETE OPERATIONS

Arrangement

The nominal excavation size was 15.55 meters diameter, the finished size 13.72 meters. The 20 cm ribs and blocking are within the nominal 15.55 meters. With an average over-break of 25 cm, the concrete volume is 54 cubic meters per meter of tunnel. The lining was placed in 3 operations. First, two small curbs each 30 degrees above the invert, on which steel rails were laid for the concreting equipment; secondly, the invert, covering the bottom 60 degrees of the circular lining; and, finally, the upper 300 degrees.

Concrete Supply and Delivery Arrangements

All concrete was supplied by Ontario Hydro from a wet mix central batch plant at Chippawa. This plant supplied much of the SAB II project, including the Control Dam and the intakes, in addition to the tunnels. Rayner-Atlas used up to 17 three-axle dump trucks to haul the mix to delivery point drop hoppers, at distances as great as 5 kilometers, travelling on city streets. Along the entire 17 kilometer length of the two tunnels, at intervals of about 150 meters, 25 cm diameter steel drop pipes were installed, into vertical holes drilled in advance of the tunnel excavation. Hoppers in the tunnel caught the falling concrete. There were no re-mix facilities, but concrete segregation appears not to have been a problem. For the invert, concrete was transported within the tunnel to the placement location with a travelling and self-propelled pot gantry that rode the rails set on the top of the curbs—the workers called this gantry "The Queen Mary." For the arch concrete, another gantry was positioned under the drop pipe and the concrete was directed by chutes to twin Pumpcretes, from which 20 cm diameter steel pump lines transported the concrete to the arch forms, which were never more than 200 meters from the drop points. The form was fitted with numerous concreting access doors and another gantry supported the slick lines for the crown of the arch. Moving the gantry back withdrew the slick line as the concreting advanced, accompanied by removing sections of pump line on the invert.

Invert Concrete

Steel formwork panels were placed along the top sides of the invert section. These panels were inclined to reflect radial lines. The top of the side forms was the rolling surface and grade control for the invert drag screed and also supported the concrete finishers' curved access bridge. The drag screed was ballasted with rock, to prevent flotation and weighed about 11 tonnes. It was moved ahead using 3 tonne capacity air tuggers on 5 part lines, anchored to steel pins set in the rock invert ahead. For batch plant and transport considerations, concrete placement was done on a single shift per day, although clean-up and form-setting went on around the clock. The record length placed in one 8 hour shift was 86 meters (requiring about 800 cubic meters of concrete), while the average pour length was 40 meters, representing 350–400 cubic meters. The Queen Mary was also used to move the formwork and rails ahead. Stripping of the side forms was started after a minimum 8 hours of curing. Following the wood float and steel trowel finishing, the invert concrete was covered with Fibreen bitumastic paper, for curing and protection. Richmond screw anchor inserts were embedded near the top of the invert concrete, as anchors for buttoning-up the arch form later on.

Arch Concrete

Before placing the arch concrete, it was necessary to remove all the loose rock that had accumulated behind the steel and wood lagging, along with removing most of the lagging itself, to ensure complete filling of the excavated shape. This work was done from a re-timbering gantry which also ran on the curb rails. The arch forms were manufactured by Mayo, and were built as designed by Rayner-Atlas. Always ready to get it wrong, Rayner-Atlas specified forms that were too lightly built and that blocked supply traffic on the invert, because the flimsy formwork required numerous bottom struts. Proving to be too light, the formwork had to be substantially reinforced underground, with a doubling of the ribs and the replacement of inadequate structural steel support members. The skin plate couldn't be replaced underground and it became distorted, requiring continual repair and straightening. Even the cast iron wheels carrying the arch form had to be changed for solid steel wheels, due to failure. Two Mayo arch forms, of different lengths were used by Rayner-Atlas, never great at anticipation or standardization. One form was originally bought at 18.3 meters length, the other at 24.4 meters length. Both were later lengthened by 3 meters, giving final form lengths of 21.3 and 27.4 meters. The forms ran on rails supported by bevelled timbers resting on the curved invert concrete. The forms were built in sections 6.1 meters long and these were the lengths moved ahead on shifting the form to the next pour. The vertical movement for stripping purposes was 15 cm and each side panel could be swung in by 22 cm—a two-hinge form, with the hinges located just above the spring line. It was generally possible to pour the form every day, with concreting done on a single shift, with 12 hours allowed for curing and then four hours to strip and move the form ahead and prepare for the next pour. After moving the form into the next location and jacking it up to grade, the bottom edges of the two side flaps were bolted-in, using the Richmond inserts that were set in the top of the invert concrete. The average placing rates using the two Pumpcretes exceeded 100 cubic meters per hour. An arch pour of 24 meters length required approximately 1,100 cubic meters of concrete, overbreak-dependent.

Grouting

Following 60 days of concrete curing, the Cementation Company (Canada) Ltd., working as a sub-contractor to Rayner-Atlas, commenced their grouting work. Cementation equipped the job with a C.F. Johnson surface batch plant and a Colcrete mixer. Grout was pumped in 40mm steel lines to the underground injection plant. This consisted of three receiving agitator tanks for receiving and remixing grout from the surface and low and high pressure grout pumps, all equipped with pressure-limiting devices. All of this was mounted on a too-light gantry (done in the typical Rayner-Atlas style), composed of sections of modular scaffolding. The flimsy gantry gave ongoing and completely avoidable aggravation and delay. Grout hole drilling was done off the gantry using jacklegs and stoppers. For the invert, the drilling was done with wagon drills—wheeled undercarriages with mast-mounted drifters—the evolutionary ancestor of current crawler drills. Wagon drills were not self-propelled, and were moved either by muscle-power or by their on-board air tuggers, whilst steering the front caster wheel by hand. After a low-pressure grouting sequence with sand-cement grout, the high-pressure grouting using cement-only grout was undertaken. This was followed by a series of test holes—drilling anew, followed by grouting—very little grout was taken on the test sequences. The total grout take on the job was one third of expectations, mostly due to the excellent filling at the crown of the tunnel form with the air sluggers (a pneumatic placer) and the Pumpcretes. In Tunnel #1, Rayner-Atlas gave very little attention to timely clean-up of excess grout in the invert and did a lot of chipping of the pooled and hardened grout with pneumatic hammers. For Tunnel #2, timely grout clean-up was made routine, and with a lesser final cost.

ACKNOWLEDGMENTS

The authors wish to thank the following engineers for their reviews, comments and other generous assistance in the preparation of this paper.

John Boots, BC Hydro, Vancouver, Canada
William Earis, BC Hydro, Vancouver, Canada
Scott Sylte, EBA Engineering, Vancouver, Canada
James Thomson, ex-Perini, Oakville, Ontario, Canada

REFERENCES

Central Intelligence Agency World Fact Book.

Hale, John (Mr. Hale was a surveyor for Ontario Hydro on the SAB II tunnels). Conversations in 2006 with the co-author.

Hearn, Richard L. (In 1956, Dr. Hearn was Chairman of the Hydro-Electric Power Commission of Ontario) The New Sir Adam Beck Generating Station at Niagara. Proceedings of the Institution of Electrical Engineers, London, UK. Paper No. 2204-S—Pages 41–48.

Norton, Dean (Public Affairs Officer—Niagara Plant Group, OPG). Sir Adam Beck 2 G.S.—A Commemorative Retrospective—The First 50 years—1954–2004. A publication of Ontario Power Generation (OPG)–Toronto.

Ontario Hydro News—HEPC of Ontario—Various issues: Pattern of Progress, January 1954; Big Job, March 1954; Romance of the Niagara, July–August 1954; From Queenston's Cliffs, September 1954.

Rayner-Atlas Limited—Description of Operations, June 1954. Volume 1, Excavation, Pages 1–86; Plates 1–69; Volume 2, Lining and Services, Pages 87–160; Plates 70–111.

Thomson, G. James (Mr. Thomson was an engineer for Perini-Walsh on the SAB II tunnels. A Ring of Urgency: An Engineering Memoir, Pages 58–77, Abbeyfield Publishers, Scarborough, Conversations in 2006 with the co-author.

SUPPORT OF THE LEIPZIG CITY—TUNNEL EXTENSION USING A TIE-BACK SUPPORTED FREEZE WALL

Helmut Hass

CDM Consult GmbH

Michael B. Gilbert

CDM Inc.

INTRODUCTION

The extension of the commuter railway system (S-Bahn) in Leipzig, Germany will require the construction of an under-ground line crossing the downtown area of the city to tie the Bavarian Station in the South to the Main-Station in the North. The City-Tunnel-Leipzig is currently one of the largest infrastructure projects in Europe. This system extension will have a length of approximately five kilometers. 3.9 kilometers of that length are currently being constructed using either cut & cover construction or tunneling methods. The design and construction of this project currently is and will continue facing very significant challenging urban and demanding geotechnical conditions.

The project has been divided in two main contracts (Lot B and Lot C). Construction of the City-Tunnel-Leipzig Lot B contract started in 2005 and the Lot C contract started in 2006. As part of the projects, the following underground stations (HP) still have to be built:

- Bayrischer Bahnhof (Bavarian Station)
- Wilhelm Leuschner Platz (Wilhelm Leuschner Place)
- Marktplatz (Market Place)
- Hauptbahnhof (Main Station)

Figure 1 shows the routing of the City-Tunnel in downtown Leipzig with the four underground stations.

The track tunnels are being constructed as two parallel single track tubes with an inside diameter of approximately 7.7 meters using pressure balanced hydro-shield tunnel boring machines (TBM). Three of the underground stations are currently under construction using cut & cover techniques.

The alignment of the City-Tunnel subway requires the construction of a segment of track to pass beneath the Main Station. That construction segment includes the Hauptbahnhof or Main Station, which will be constructed underground. The work has been divided in two cut & cover tunnel sections approaching the existing station from the north and south and one 80 m long section directly underneath the station building. Construction of this 80-m section will use ground freezing with tie-backs as the lateral excavation support system. Ground freezing operation will start summer/fall of 2007 as part of the Lot C contract.

MAIN STATION BUILDING

The main station of Leipzig is one of the oldest and largest terminal stations worldwide. The original construction took 13 years. The main station started operation in 1915, included 26 platforms. During World War II several bombings in 1943 and 1944 destroyed large parts of the station (see Figure 2).

Figure 1.   Routing of City-Tunnel-Leipzig

Figure 2.   Main station building after bombing on July 7, 1944

After the War the station was rebuilt from 1954 through 1962, at an approximate cost of 40 Billion Mark. After Germany's reunification in 1990 the main station building was reconstructed in 1996 and 1997. Included in this reconstruction were building and subsurface extensions. Now the structure serves as both a main commuter rail station and for the long distance heavy railway traffic. It also has an attractive shopping mall with associated vehicle and pedestrian traffic.

TUNNEL SECTION UNDERNEATH THE MAIN STATION BUILDING

The lobby of the station is in a four-story building with a length of 298 meters. During that post-unification construction, drilled foundation piles were utilized to support the building extensions. At that time, consideration was given to future track modifications including the work currently being undertaken. The drilled piles, having a diameter of 1.5 meters, were installed in three rows to provide space for the construction of a train tunnel between the pile rows. The building slab was structurally designed to span the loads between the piles so that future excavation of soil underneath the building will be possible between the pile rows. In Figure 3 presents a typical cross-section underneath the existing main station building with the pile foundation of the slab.

Remnants of the ground support systems including tiebacks and soil modified with silicate grout and jet grout as part of the 1996 building modification and extension work are still in the underground. In addition, soldier piles and older, abandoned structures can be locally encountered. Figure 4 presents a 3-D model showing the existing

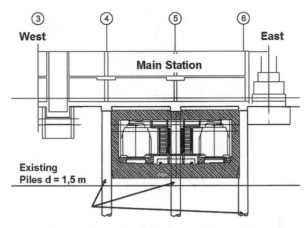

Figure 3. Tunnel cross-section underneath existing main station building

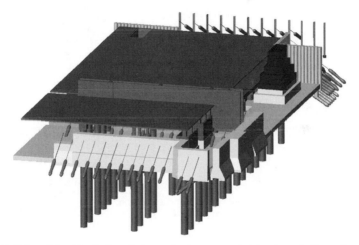

Figure 4. Building slab with pile foundation and other remnants/structures in the underground

foundation and possible remnants in the underground in the area where the City-Tunnel has to be built underneath the building.

In the southern part of the main terminal building, the foundation level is at approximately El 108 m (datum is MSL) over a length of 30 meters. In the northern part of the structure, the foundation level is at approximately El 103 m and has a length of 56 meters. In the basement pit foundation level is at El 101.1 m and includes areas for drainage collection that includes a grease separator and an elevator shaft. The average groundwater level has been observed at El 105.6 m. This is as much as three to four meters higher than the lower levels of the northern foundation. The bottom of the new tunnel is located at El 91.4 m. Because of these elevations the excavation zone for the tunnel will be continuously located below groundwater level.

To adequately support the excavation for the new subway tracks, maintain the integrity of the existing drilled pile foundation supporting the four-story City-Tunnel Terminal building and control groundwater movement watertight retaining walls parallel

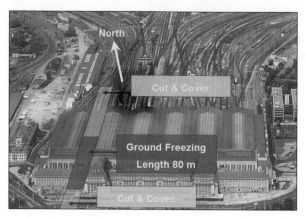

Figure 5. Tunnel alignment underneath Leipzig Main Station

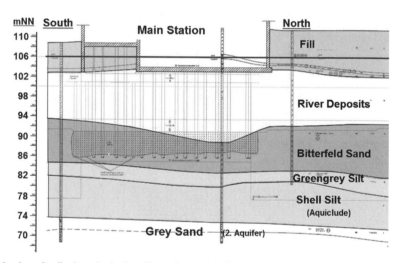

Figure 6. Longitudinal geological profile underneath Leipzig Main Station

to the outer pile rows are required. Other construction or tunneling methods were evaluated and were not appropriate considering the given boundary conditions and the required elevation of the tunnel alignment.

After investigation and evaluation of different methods the ground freezing method was deemed as the most appropriate and safest construction method for the retaining walls considering the present complex boundary conditions. The aerial photo of Figure 5 shows the main station building and the proposed 80 meters long tunnel area where ground freezing will be applied. The adjacent sections will be constructed using deep diaphragm walls for ground support (cut and cover).

SUBSURFACE AND GROUNDWATER CONDITIONS

Figure 6 presents a longitudinal section underneath the main station building with the geological profile.

The alluvial river deposits consist of sand and gravel with trace to no silt. This stratum's density ranges from medium dense to very dense. Based on the explorations it is anticipated that the lower portion of the stratum may contain cobbles and boulders. Underlying the alluvial river deposits is a sand stratum, identified as the Bitterfeld Sand. This layer consists of uniformly graded fine to medium sand with trace silt. This stratum is dense to very dense. Underlying the Bitterfeld Sand is a silt stratum identified as Greengrey silt. This stratum generally consists of silts with large amounts of fine sand and clay. The consistency is firm to stiff. Underlying the Greengrey silt is another silt stratum identified as Shell Silt consisting of fine sandy silts and clays. This lower Shell Silt layer, which has a low permeability, is considered as a natural aquiclude to an underlying grey sand stratum that is an aquifer.

The hydrostatic head in the upper aquifer ranges between El 101.7 m and 106.9 m and is in the governing aquifer. The long-term average elevation is El 105.6 m.

The proposed excavations for the tunnel construction are mainly located in the following two layers: the alluvium river deposits (geotechnical layer III) and the Bitterfeld sands (geotechnical layer VI). To attain a bottom cut-off it was determined that the freeze walls must penetrate at least two meters into the low permeable Shell Silt which is the natural aquiclude. This cut-off requirement means that the freeze wall will also extend through the Greengrey Silt (geotechnical layer VII) and into the Shell Silt (geotechnical layer IX).

Within the lower portion of the alluvium stratum with the cobbles and boulders locally high groundwater velocities with values of more than 10 m/d were also taken into account in the design.

DESIGN OF GROUND FREEZING

The City-Tunnel excavation will be laterally supported by two 80 m long freeze walls, which serve as retaining wall and for water cut-off. At the interfaces with the abutting cut & cover tunnels two deep excavations with structural concrete walls will have been built in advance.

The freeze walls serve as retaining walls and have to resist the lateral earth loads and full water pressure. The freeze walls will be reinforced with four to five rows of tieback anchors. The concentrated tieback loads will be evenly distributed to the freeze wall via a shotcrete lining. The anchors will be pre-stressed. The reason for the pre-stressing is to minimize the deformations. The contractor will need the ability to restress the installed anchors during excavation to accommodate the creep behavior of frozen soil. The short-term loading conditions during excavation considered for the freeze wall design included the full lateral loading of soil and hydrostatic load without support from the tiebacks and shotcrete. After strengthening of the shotcrete lining with tiebacks the retaining wall consists of a combination of freeze wall and tie-backed shotcrete lining. The retaining walls will serve both as a structural wall for lateral support and for water cut-off until the final tunnel structure will be finished and watertight connected to the adjacent sections.

After the final tunnel structure is completed below the Main Station Terminal building the freezing operation will be turned off. With the thawing of the frozen soil the watertight retaining wall will be gone and groundwater can flow again in the aquifer below the tunnel.

Implementing a ground freezing system will require freeze pipes to be installed in the ground to create the freeze walls. Two methods were possible—horizontal or vertical drilling of the freeze pipes. The first alternative would be to drill horizontal freeze pipes. For the given actual conditions and site constraints this would require that some of the horizontal drillings only could be accessed from one side of the excavation. This

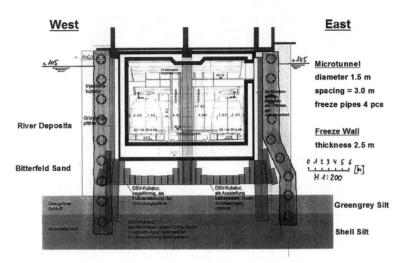

Figure 7.   Initial consideration with horizontal freeze pipes being installed in microtunnels

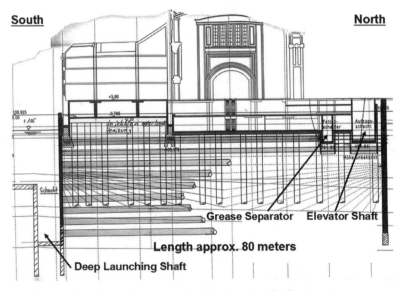

Figure 8.   Initial consideration with horizontal freeze pipes—longitudinal section

would result in drill lengths of 80 meters. The necessary accuracy for the freeze pipe locations would only be achievable using microtunneling techniques. The installation of freeze pipes inside microtunnels has been successfully performed for the Fahrlachtunnel, Mannheim, Germany, which is a 180 meters long road tunnel. This same method is also currently being used for the construction of the underground Brandenburg Gate station as part of the subway line U5 in Berlin. These freeze lines had lengths of 90 meters. Figures 7 and 8 present a possible microtunneling/freeze pipe layout along with a longitudinal section with the arrangement of microtunnels.

For the application of the microtunneling technique alternative two major risk aspects had to be considered. The ability to advance the microtunnel TBM and to control the line and grade was of concern regarding the existing remnants of previous ground modifications and abandoned structures within the microtunnel horizon. If a microtunnel drive were stuck, installation of a recovery shaft would not be possible because of the main station building preventing surface access. Also, there was the economical factor to be considered. To attain groundwater cutoff the freeze wall must penetrate into the Shell Silt stratum. To install freeze elements at the required locations to achieve this penetration into the Shell Silt would require very deep launching shafts. These deep shafts would have to resist significant loads including a surcharge loading from the massive four-story station building. In addition, the risk associated with the higher probability of these very deep excavations creating a cross connection between the upper aquifer and the second aquifer below the Shell Silt had to considered.

The associated risk and financial penalties with that alternative lead to the decision to go with the second alternative installation of slightly inclined vertical freeze pipes. To attain access at the required locations to install these freeze pipes two horizontal auxiliary tunnels will have to be advanced under the terminal building. Installing the freeze pipes from within these two auxiliary tunnels (2.4 meters inside diameter) will require drilling against the hydrostatic pressure of the groundwater. Figure 9 presents a typical cross section of the proposed auxiliary tunnel in the area of the deep slab below the building.

The preliminary design considered an auxiliary tunnel right underneath the building slab with slightly inclined freeze pipes underneath the auxiliary tunnel. The auxiliary tunnel would also serve as the watertight connection to the building slab. At that stage of the design it was planned to advance the auxiliary tunnel with the sequential excavation method (SEM) and shotcrete lining under pressurized air. For safety reasons the minimum thickness of the freeze wall was to be constructed as a minimum of 3.0 meters with an average freeze wall temperature not warmer than −10 °C across the wall width. As the design advanced the option to increase wall thickness as a function of the calculated structural wall thickness requirements was still a consideration. To achieve the 3.0 meter frozen ground wall thickness and potential for a greater thickness two rows of freeze pipes would be required. The layout of the pre-design is shown on the left of Figure 9.

The general contractor changed the tunneling method during the final design. The contractor will now advance the auxiliary tunnel using the pipe jacking method. The contractor adjusted the elevation of the auxiliary tunnel to avoid the same risk issues regarding construction debris and isolated zones of modified ground that were taken into account for microtunneling discussed for the first alternative. The tunnel elevation was lowered to a vertical horizon where no debris could reasonably be expected to extend. The right side of Figure 9 shows the adjusted layout for the freeze pipes and the auxiliary tunnel. Because of the lower elevation of this tunnel, freeze pipes will be installed both below and above the auxiliary tunnel to achieve a watertight connection between the freeze wall and the building slab.

Once the freeze walls are in place and functioning and excavation for the subway tunnel has advanced to a depth of about 3 to 4 meters, the resulting unsupported length of the middle row of piles will require additional support. This support is required to account for the increased bending moment induced by the increased unsupported length. This additional lateral ground support, as shown in Figure 9, will be achieved by grouting the ground below the elevation of the proposed excavation to provide additional lateral resistance to the middle row of piles. The outside row of piles will be embedded in the frozen soil with a significant higher resistance force that were sufficient so that grouting measures are not anticipated for the piles of the outer rows.

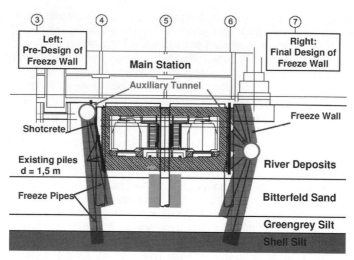

Figure 9.   Typical cross-section pre-design (left) and final design (right)

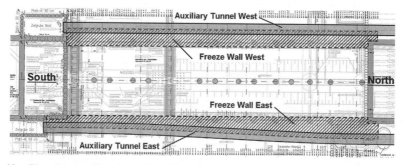

Figure 10.   Plan view auxiliary tunnels and freeze walls

Figure 10 presents a plan view of the tunnel section to be constructed. For the auxiliary tunnels additional small launching shafts will be constructed.

SPECIAL ITEMS

The complex character of the existing conditions causes numerous challenging individual tasks that were considered in the design process. Two of these more challenging tasks are the geometric layout to mitigate excessive eccentric loading on the foundation due to the freeze wall and the effects of groundwater flow velocity on the freeze system. These issues are described in the following text.

The layout requires the eastern auxiliary tunnel to be advanced underneath a heavily loaded column foundation. To avoid eccentric loading on the building foundation elements the eastern auxiliary tunnel will be skewed to the centerline of the building foundation. The tunnel alignment geometry is set so that the distance between the outer pile row and the auxiliary tunnel increases versus the tunnel length. The layout of the freeze pipes will also be adjusted accordingly to achieve the required freeze wall. In Figure 11 a section underneath the column foundation with the adjusted auxiliary

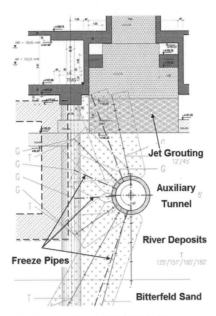

Figure 11. Section underneath high loaded column foundation

Figure 12. High loaded column

tunnel and freeze pipe layout is shown. Figure 12 gives a good impression of the massive column and the high loads.

Groundwater flow can have a major impact on the freezing operation. Based on borehole tracer dilution tests effective groundwater velocities of $v_{eff} = 0.2$ and 1.4 meters/day have been calculated for the groundwater and used as the design parameter value to estimate freeze times during the construction phase. With further evaluations based on strata permeability developed from grain size distribution curves higher theoretical values, in excess of $v_{eff} = 10$ m/d could not be excluded in the lower portion of the base of the River Deposits containing the cobbles and boulders. To address this theoretical worst case condition regarding parameter values the two

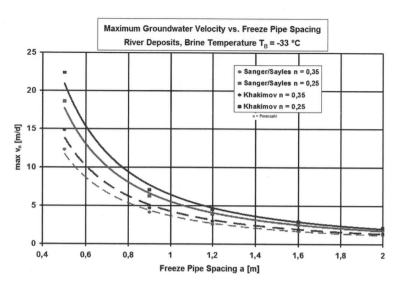

Figure 13. Critical groundwater velocity versus freeze pipe spacing

governing rows of freeze pipes will be staggered so that the spacing between the pipes is around a = 0.6 meters.

Using the analytical method of Sanger/Sayles and Khakimov the critical groundwater velocities can be determined related to the soil type, freeze pipe spacing and freezing temperature. Figure 13 shows the results of the calculations. Considering the given boundary conditions groundwater velocities of v_{eff} = 9 to 15 m/d can be handled with the planned system. In addition, during the drilling operation for the installation of the freeze pipes, grout will be injected to the ground. Based on the close spacing of only a = 0.6 meters the grouting will significantly reduce the high permeability in the critical very coarse areas.

SUMMARIZED DATA

- Freeze wall length: approx. 80 m
- Freeze wall thickness: $t \geq 3.0$ m
- Total frozen soil volume: approx. 15,500 m
- Total length of freeze pipes: approx. 7,500 m
- Total length of temperature pipes: approx. 1,500 m
- Number of temperature devices: approx. 650 pcs
- Freezing temperature (brine): $T \leq -35$ °C

REFERENCES

"Marktplatz Leipzig Hauptbahnhof," Helge-Heinz Heinker, 2. erw. Aufl.—Stoneart-Verlag., 1999
Unpublished Project Documents
Website: www.city-tunnel.de

4

Difficult Ground Conditions II

Chair
G. Kramer
Hatch Mott MacDonald

DESIGN AND CONSTRUCTION SERVICE TUNNELS ON PALM JUMEIRAH ISLAND

Michael Ryjevski

Al Naboodah Engineering Services

Ben Hayes

Al Naboodah Engineering Services

INTRODUCTION

The Palm Jumeirah is the smallest of the three Palm Islands (Palm Jumeirah, Palm Jebel Ali and Palm Deira). It is located on the Jumeirah coastal area of the emirate of Dubai, in the United Arab Emirates (UAE). The unique man made island is built in the shape of a date palm tree and consists of a trunk, a crown, 17 fronds and a surrounding crescent island that will form a breakwater. The Jumeirah Palm Island is primarily a retreat and residential area for living, relaxation, and leisure. It will contain themed boutique hotels, three types of villas, shoreline apartment buildings, beaches, marinas, restaurants, cafes and a variety of retail outlets.

As part of the infrastructure installation it was necessary to create permanent physical links between the main island and the surrounding crescent island to facilitate the transfer of services including electricity, water, sewage and gas. Specialist local contractor Al Naboodah Engineering Services (ANES) proposed to install the links using Microtunneling (also know as pipe jacking) and Horizontal Directional Drilling (H.D.D) techniques. Microtunneling was proposed to create two cable tunnel links for the transfer of high voltage electricity from the main island to the crescent. The remaining gas, water and sewage service links were to be installed using H.D.D.

Originally the project was designed as two concrete culverts, to be laid in a trench along the sea bed. Although more risky and challenging than the original design, ANES proposal offered considerable savings to the developer. Subsequently the project was awarded to ANES on a design and construction basis. ANES in house design team, technical department, produced the full design of the project and employed Hyder Consulting to carry out their third party verification. Site supervision was carried out by Parsons De Leuw Cather Overseas Ltd, who reported to Parsons Brinkerhoff designer to the developer Nakheel.

The proposed microtunneling works included two tunnels linking Fronds M and D to the crescent (Figure 1).

Both tunnels were designed with an internal diameter of 2.4 m and an external diameter of 3.0 m, the lengths were 760 m from Frond M and 650 m from Frond D. Each tunnel was designed with a vertical curve at the end of each drive to reduce the depth of the reception shaft.

The design includes two access shafts for each tunnel which were used for launching and recovering tunnel boring machine during construction period of time and to accommodate the required cables during tunnels operation. All shafts were designed with circular shape. Launching shafts were constructed on Fronds M and D and they were designed of 12 m diameter. Both receiving shafts were constructed on the Crescent Island with diameter of 9 m.

Soils investigation at the fronds found that to reduced level −11.0 m soil conditions were mainly loose heavy saturated sands and gravels with some shell fragments. This material was the reclaimed dredged material placed to form the island. From −11.0 m to

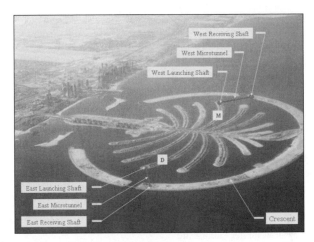

Figure 1. Service tunnels on Palm Jumeirah Island

–31.0 m conditions went from dense sand, to cemented sands and sandstone. Ground conditions at the tunnel level were mainly calcareous sandstone interbedded with cemented sand and some gypsum crystals. The ground water table was encountered at depths ranging from 2.02 m to 4.5 m i.e., at a reduced level +1.58 m to –0.30 m below the respective borehole level.

Based on the ground investigation, length of drive and drive alignment ANES opted to use a Herrenknecht AVND 2500 Mixshield TBM, which employs a slurry system for spoil removal. Supplied with the TBM was a VMT SLS-RV guidance system for curved pipe jacking. Separation of the spoil from the slurry was by a Schauenburg MAB 400 separation plant.

Shaft construction consisted of contiguous secant piles designed as a shoring system for excavation in unstable soil conditions. The drive shafts were sunk to depth of 28.0 m with a inner diameter of 12.0 m, the depth of secant piles at the reception shafts were 19.0 m with a reduced inner diameter of 9.0 m. A larger diameter launching shaft was required to accommodate the 6 cylinder telescopic jacking rig (Figure 2).

Prior to commencing the microtunnel the permanent lining of the shaft was constructed up to the top level of the tunnel. This lining consisted of a reinforced concrete wall with nominal thickness 450 mm, the first lift of was designed to incorporate a reinforced thrust wall and launch seal bulkhead. The jacking equipment was then installed within the finished shaft directly to the permanent concrete base slab. Special designed waterproofing system was installed between the secant pile shoring system and permanent concrete wall to protect the shaft against water infiltration to the shaft. Pre-cast concrete roof slabs were designed to cover shafts after tunnel excavation, recovering TBM and cable installation.

ANES engaged Gollwitzer Gmbh—the services of pipe designer to provide the pre-cast concrete jacking pipe design, and employed local firm Emirates Pre-Cast, part of the Al Naboodah Group of companies, to carry out the manufacturer of the jacking pipes. The pipes produce were of 3.0 m O.D and 2.4 m I.D. with a length of 3.0 m. These dimensions were opted for to increase the weight of the pipe to reduce the uplift force as the pipeline was to be installed below the water table. ANES were concerned that the uplift forces may cause alignment control problems, particularly on the vertical curves. The joint detail consisted of a spigot and socket with the socket being formed by a continuous steel band anchored into the concrete at one end of each pipe and the

Figure 2.   Launching shaft for microtunneling

spigot formed by the cast concrete. This design of pipe joint is fundamental to the pipe jacking procedure. The spigot fits the socket in such a way that the external surface of the joint is smooth, a smooth external joint is vital to keep friction at a minimum and consequently jacking pressures low.

Microtunneling usually employs a remote control tunnel boring machine (TBM) and uses concrete pipes as a thrust medium. Initially the TBM was lowered into the shaft where previously a jacking rig consisting of high capacity hydraulic cylinders as been installed. Once the TBM is positioned on the jacking rig the cylinders are extended and subsequently push the TBM, initially through the shaft wall and then into the ground, excavating as it goes. Once the cylinders have been fully extended the excavation sequence stops. The cylinders are then fully retracted and a concrete jacking pipe is lowered on the main jacking station. The concrete pipe is then connected to the TBM by a spigot and socket joint. The cylinders are then extended once again and push the pipe and the TBM further into the ground. Once the cylinders fully extended the operation again stops and another pipe is connected. This procedure is repeated until the TBM reaches the reception shaft. Intermittently special pipes known as intermediate jacking stations (IJS) are installed. IJS pipes are manufactured in such a way that they can hold small hydraulic cylinders. IJS pipes are used to advance the pipe string in sections should the force required by the main jacking rig to move the pipe string become too great. When this situation arises the cylinders in the IJS are extended and thus advance the pipe string ahead of the IJS. Once these cylinders are fully extended the following IJS or the main jacking station cylinders are extended to move the following section of the pipe string and close the open IJS. On long pipelines installed using microtunneling it may be necessary to install several IJS and move the pipe string in several sections in order to keep jacking forces at acceptable levels. During the jacking operation a lubricant is pumped into the over cut. The TBM excavates a hole in the ground larger than the pipes that are following it and this gap is referred to as the over cut. The lubricant used is normally bentonite, however in certain circumstances it may be necessary to introduce additive to the bentonite mix to improve its lubricating properties.

When considering the design of the tunnel alignment ANES opted to incorporate vertical curves at the end of each drive to reduce the depth of the reception shafts which offered a considerable saving on the shaft construction. The depth of the tunnel alignment was determined by the underlying bedrock. ANES opted to maintain a minimum cover of 6.0 m, which was twice the outside diameter of the TBM, of bedrock above the machine at all times. It was believed that maintaining such cover would eliminate any problems that may occur from the potential instability of the sea bed material.

The first drive, the longest at 760 m, commenced at the beginning of August 2005. Once the machine exited from the shaft it was immediately apparent that the ground conditions were not exactly as expected. The material being excavated was sandstone as expected. However water infiltration rates into the cutter head chamber indicated the ground to be far more permeable than was originally thought from the borehole results. As it was mainly broken sandstone being recovered from the separation plant and the tunnel boring machine was developing minimal cutter head torque during excavation it was concluded that the ground conditions were solid weak sandstone, but far more fractured than originally thought and the water ingress was through the fractures and fissures as opposed to the material itself. The TBM was equipped to operate in D-mode which utilizes an air bubble chamber to control water and material ingress in unstable ground conditions, however ANES opted to proceed in slurry mode as the face was not believed to be unstable.

Initially production rates of only 2 to 3 pipes per day were achieved, as the crew became more familiar with the new equipment production increased to a steady production of approximately 100 m/week. Production continued to go well, however from approximately chainage 400 m it was noticed that there was a steady increase in jacking pressure. Various changes to bentonite mix were made including the introduction of polymers at varying dosages in an attempt to reduce the jacking pressure. After consulting a lubrication specialist supplied by TBM supplier Herrenknecht the a bentonite mix of 3–5% bentonite to water with 0.2% Bariod PAC_R polymer, added to prevent water loss from the bentonite that would normally occur in salt water conditions, was adopted. However this mix did not reduce jacking pressure. It was then concluded that the high jacking pressures being experienced were not due to poor lubrication. The use of the 4 intermediate jacking stations installed showed that that there was exceptionally high jacking forces between the main jacking station in the drive shaft and interjack 3 installed at pipe number 116. Jacking pressure required to close interjack number 3 with the main jacks had reached 1,550 ton by pipe 167. During the jacking of pipe 167 cracks appeared on pipes 145 through 150. The jacking operation was immediately halted whilst investigations were carried out.

ANES gathered experts from within their own technical department, Herrenknecht, Ingenieur-Buro Dr.-Ing H.P. Uffmann and Gollwitzer pipe designer. Initially all were in agreement that an unknown object had somehow attached itself to the pipe string and was exerting an external point load on the outside of the pipe. The decision was made to proceed with caution in the hope that the object would detach itself or be broken up by the jacking process. Over the course of jacking the next 5 pipes the damage to pipes increased and the decision was made to stop and find another solution.

Initially the decision was made to give some internal support to the damaged pipes in the form of steel in-liners. The support was installed as a safety measure and offered protection to workers in the pipes in the event that the condition of the pipes deteriorated. The pipes were visibly heavily cracked internally, however it was impossible to know the external condition and consequently their ultimate strength.

From the nature of the cracking it was thought highly likely that the unknown object was attached to the underside of the pipe. There was visible evidence that the damage was localized to an area of 5 pipes, based on this evidence a decision was made to attempt to physically destroy the object.

A proposal was put forward to core 150 mm diameter holes in invert of the pipe and then using very high pressure water jetting, attempt to break up the unknown object and consequently free up the pipe string. This proposal was accepted and work commenced.

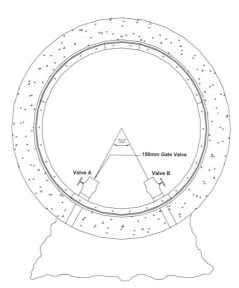

Figure 3.   Core holes for jetting arrangement

Each pipe had 2 No. holes drilled offset from the centre in the invert section of the pipe. Two holes were opted for to allow high pressure jetting through one and the removal of material from the other, each hole was drilled through a 200 mm diameter valve to stop the water ingress when not jetting. The arrangement is shown on the Figure 3.

Once all holes had been drilled each valve was opened to briefly inspect the discharge. Natural ground water pressure forces large quantities of a substance that appeared to be highly compressed clay like material and sand. Each hole was naturally flushed until a quantity of compressed clay like material and sand mixture was recovered from all of them. The team decided that due to the amount of material recovered they would try and push before doing any jetting to see whether the removal of this material had made a difference. The team was surprised and relieved to discover that the jacking pressure had reduced to 750 t on the restart. The decision was then made to proceed and monitor the situation. It was noticed after the next 5 to 6 pipes that the jacking pressure had steadily risen to over 1,200 t, the valves were then opened again and more material recovered and the jacking pressure then reduced. The team concluded that the unknown object was possibly not the entire problem but that the unknown object was collecting material as the pipe string advanced and it was this material that was building up, compressing and exerting the point load to the pipe.

It was decided that in order to continue it would be necessary to regularly flush the area in which the pipes were damaged. The team opted to use the TBM excavation slurry circuit to make this easier. The TBM slurry circuit consisted of 2 No. 200 mm diameter steel pipes that ran the length of the tunnel. One of the steel pipes was used to supply clean water to the cutter head chamber of the TBM, this pipe is known as the feed line. Once the clean water arrives in the cutter head chamber it is mixed with the excavated material to form slurry. This slurry is then pumped via the second steel pipe to the separation plant. The second steel pipe is known as the slurry line. A system was set up whereby half the cored holes were connected with pipe work to the feed line and the other half via pipe work to the slurry line. Through a system a valves the team was

Figure 4. Tunnel break though

able to flush and remove material from under the pipes via the slurry circuit whilst excavating, this kept the jacking forces at acceptable levels until the completion of the drive. The TBM successfully broke through on December 28th 2005 (Figure 4).

Upon breakthrough the water ingress to the shaft coming through the TBM over cut was far greater than ANES had expected. The AVN D 2500 has an over cut of approximately 35 mm. ANES estimated that approximately 800 m³/hour of water was flowing into the shaft via this over cut at approximately 2.5 bar water pressure. ANES brought in 8 high head submersible pumps to keep the water level at a level that enabled work on sealing the over cut to commence. ANES finally sealed the over cut by the injection of a combination of polyurethane foam and cement grout, however the time taken to complete this work was far greater than programmed. This delay meant that the grouting of the entire over cut was delayed. ANES realized a different procedure had to be adopted for the second drive.

The machine was recovered and all equipment moved across the island to frond D. Frond D drive commenced on February 4th 2006. Production averaged at 85 m per week, without any major incident, and the TBM broke through April 3rd 2006.

Upon break through the water ingress was at a similar level to the first drive so ANES implemented a new procedure to seal the over cut. Once the TBM had broken into the shaft the first TBM section was partly pushed onto steel beam reception rails. At this stage the shaft was allowed to flood thus balancing the water pressure and ensuring there was no flow through the over cut. ANES in house grouting specialists then moved their equipment along the 650 m pipe line to the back of the TBM. The grouting team then proceeded to inject polyurethane foam through the first 5 pipes of the pipe string. The shaft was then pumped out and ANES observed a significant reduction in water ingress, however 4 high head submersible pumps were still required to control the water. The TBM was then pushed fully onto the reception rails and the shaft allowed to flood once more. The grouting team once again injected polyurethane foam into the over cut through pipes 1 to 5. The shaft was once again pumped out this time ANES were pleased to see that only one pump was required to control the water ingress. The TBM was subsequently recovered and the operation to grout the entire over cut commenced.

Figure 5. Completed cable tunnel

On completion of the over cut grouting operation and sealing all leaks the tunnel was ready to receive the mechanical and electrical installation. ANES opted to design and manufacture a bracket system in house to support the high voltage cables to be installed in each tunnel. The bracket system was designed in such a way that drilling into the concrete pipes was not required initially. The design enabled ANES to install the brackets in a relatively short period of time and the project was handed over to the developer for the cable installation to commence (Figure 5).

EPB TUNNEL BORING MACHINE DESIGN
FOR BOULDER CONDITIONS

Michael A. DiPonio

Jay Dee-Michels-Traylor JV

David Chapman

Lachel, Felice & Associates

Craig Bournes

Lovat, Inc.

ABSTRACT

Historically, boulders are a frequent source of problems in soft ground tunneling. During tunnel construction, the need to manually break and remove boulders as obstructions causes delays to the project. Repairs to a tunnel boring machine (TBM) can also be a source of delays.

Managing these problems is difficult since normal soil investigation techniques do not accurately predict the presence or frequency of boulders. This has lead to considerable number of claims for extra costs and delays during the construction of soft ground tunneling projects. These issues are exacerbated in pressurized face tunneling systems where there is limited access to the TBM cutterhead for obstruction removal and/or cutterhead maintenance.

The project team that built the Big Walnut Augmentation/Rickenbacker Interceptor (BWARI) Project for the City of Columbus successfully managed the design and construction of a 4,267 mm (14 ft) diameter soft ground tunnel with high boulder concentrations in a complex geologic setting. The management of the boulder issue took place during several phases of the project.

BOULDER INVESTIGATION

Geologic research and reconnaissance early in the project indicated that boulders were a major issue for the project. The project setting is the Till Plains Physiographic Province, where thick sheets of glacial till cover deep buried valleys filled with glacial outwash from earlier glaciations. The till is weathered and oxidized at the ground surface, but at tunnel depth it is extremely compact (unit weight of 140–145 pcf) with a mixture of grain sizes encompassing the entire range of soil texture from clay sizes to gravel with cobbles and boulders. The glacial outwash also includes concentrations of cobbles and boulders, particularly at contact zones atop till sheets. Although not detected in very many of the standard split spoon borings, boulders excavated from fields and house cellar holes can be observed guarding driveways at numerous locations in the surrounding countryside.

It was recognized that characterizing the boulder fraction of the ground would be a critical element of the geotechnical investigation and development of the Geotechnical Baseline Report and other components of the contract document package, particularly related to risk allocation. In addition, recovery of continuous samples using rotasonic boring techniques showed the extreme nature of the stratification which is not as accurately depicted using standard split spoon borings with the normal 5-foot sampling interval and potential loss of samples in wet granular soils or if a gravel particle becomes lodged in the split spoon.

The geotechnical investigation for the project was documented in Frank and Chapman, (2001). One special measure taken for the characterization of the coarse fraction included the performance of large diameter borings 914 mm (36 inches) and 1,067 mm (42 inches) from which the large cobble fraction (rock retained on a 152 mm (6-inch) grid) was separated for volume determination within each 1.5 m (5-foot) vertical interval. Another was provided by the fortuitous presence of a gravel mining operation adjacent to a portion of the tunnel route, which afforded the opportunity to perform counting of boulders from 10 days production of gravel mined with a dragline and scalped over an 457 mm (18-inch) grizzly. These datasets were later used to develop a mathematical model for quantification of the boulder fraction for presentation in the Geotechnical Baseline Report (Frank and Chapman, 2005). In addition, rock type was determined for boulders in boulder stockpiles, and boulders were cored to obtain samples from which to determine the range of compressive strengths. As expected, many of the boulders were of igneous or metamorphic origin and were inferred to have been transported from the area of the Canadian Shield, that in itself demonstrating their resistance to destruction. Compressive strengths ranged up to nearly 45,000 psi.

PRESENTATION OF GROUND CONDITIONS IN THE GBR

The boulder quantification described by Frank and Chapman (2005) is shown in Figure 1. During development of the contract documents, the appropriateness of presenting the boulder baselines was debated, since with the planned tunneling methods, it would not be possible to count the boulders or to know if the baseline was exceeded. It was eventually decided to present the information as a means of communicating the potential magnitude of the boulder fraction to the contractor.

In addition, the layering detected in the rotasonic borings was interpreted and several important trends were identified and were also presented as "geotechnical factors," determined to be "useful in TBM selection and in ensuring application of means and methods which result in ground and face control by the TBM in order to meet settlement requirements." (GBR, May, 2003) Subsurface conditions along the tunnel route were interpreted and summed to estimate total expected footages and distribution by stationing along the tunnel to which the defined conditions applied. The factors were divided into "TBM Selection Factors" and "Tunnel Face, Ground, and Settlement Control Factors," and presented graphically as per the excerpt shown in Figure 2. It can be seen that these factors describe such characteristics as extremely coarse grained soils; glacial till; mixed face conditions such as flowable granular soils in the tunnel crown overlying hard, resistant materials lower in the tunnel face; and low ground cover areas. It was reasoned that these descriptions would help the contractor to visualize the range of challenging conditions and their extent to aid in selection of tunneling equipment as well as in planning of tunneling procedures and selection of ground conditioning agents.

SPECIFICATION OF TUNNEL BORING MACHINE

In conjunction with the provisions of the Geotechnical Baseline Report, development of the specification for the TBM was given careful attention relative to risk management for the project. The tunnel design engineers recommended purchase of a new tunnel boring machine to ensure the capabilities required to provide the best possible

BOULDERS AND COBBLES EXPECTED DURING EXCAVATION									
			Boulders						
	Cobbles	Small Boulders		Obstructions					
Cobble or boulder size in feet	0.5 - 1	1 - 1.5	1.5 - 2	2 - 3	3 - 4	4 - 5	5 - 6	> 6	
Number expected in shafts and open cut (10,000 cyds)	14,250	616	38	10	1	1	0	0	
Number expected in tunnel (10,000 lft of tunnel)	113,000	3,146	415	126	16	4	1	1	
Number expected in 500 lineal feet of microtunnel	460	21	5						
Number expected in 200 lineal feet of bore and jack	184	8	2						
* Maximum for shafts and open cut	285	18	2						
** Maximum for tunnel (per 50 lineal feet of tunnel)	11,300	300	50						

Total number of obstructions expected in 10,000 lineal feet of tunnel   563
Average number of lineal feet of tunnel between obstructions   18

* Occurrences are for volumes of 100 square feet in horizontal area and 5 feet in vertical thickness.
(Areas of highest concentration are expected to encompass 10,000 square feet.)

** No more than 5 obstructions will be encountered simultaneously

Figure 1.   Boulder and cobble study results

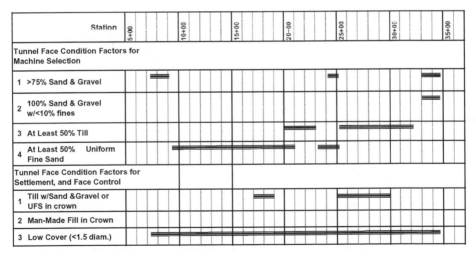

Figure 2.   Geotechnical factors by station along centerline

opportunity to successfully mine through all of the ground conditions. This was based on the challenges presented by the geologic environment and on the tunnel size, which meant that there were few if any existing machines in the required size range that had the specified range of capabilities. It was desired to level the playing field among bidders and ensure that the bidding would not be won by virtue of a marginally capable TBM that could not be rejected but that would potentially be ineffective under the difficult mining conditions. This recommendation was accepted by the City of Columbus and new TBMs were specified for both Part 1 and Part 2 of the project.

It was considered by most involved with the design that the specifications should allow either an earth pressure balance (EPB) or slurry machine. Consideration was given to the provision of differential contingency levels for EPB versus slurry machines. This concept was based on the views of some that this might help to balance the higher cost of slurry machines and potentially permit slurry machines to be more competitive. Limiting the specification to slurry machines was considered as well. According to some recognized European standards the geotechnical conditions pointed strongly to the Slurry TBM approach. These concepts were both ultimately rejected based on previous successful completion (in North America) of tunnels in difficult granular conditions with EPB machines, e.g., the South Bay Ocean Outfall Tunnel.

Highlights of the tunnel specifications included:

- Slurry or EPB TBM allowed

- Design TBM for 3 bar operating pressure

- Design, build and maintain the TBM to operate in all ground conditions indicated in the Contract Documents.

- Fit the TBM with a compressed air lock(s) and associated compressed air equipment designed of 3 bars (45 psi) of working air pressure

- Design the TBM to be capable of excavating through ground containing cobbles, small boulders, and obstructions (boulders exceeding 457 mm (18 inches) in dimension) of the number and size and in all conditions indicated in the Contract Documents without interruption of the excavation operations

- Design the TBM to accommodate both ripper teeth and disc cutters capable of cutting and removing boulders.

TBM DESIGN AND PROCUREMENT

During the bidding stage for the project, the contractor team of Jay Dee–Michels–Traylor Joint Venture considered both Slurry and EPB TBMs. There was a substantial cost advantage to the EPB for the procurement of the TBM and the support systems for each tunneling process. Additionally, we were unable to quantify any advantage in the operational cost of Slurry tunneling vs. EPB tunneling that would offset the higher initial outlay for the Slurry TBM. As such, EPB was deemed to be the most cost effective. The Slurry TBM alternative would require any soil particles larger than 152 mm (6 inches) to be reduced to this size. In addition to anticipated boulder quantities exceeding 7,000 for the project, the number of cobbles larger than 152 mm (6 inches) were projected in the hundreds of thousands. EPB was therefore considered to be technically superior because of the lesser degree of size reduction of boulders and cobbles required for muck ingestion by the EPB TBM as compared to the Slurry TBM.

The Contractor selected an EPB TBM designed and manufactured by Lovat, Inc. of Toronto, Canada. The main emphasis of the TBM design was building a machine with high torque, the ability to pass large sized particles and robust cutterhead and drive system. The diameter of the screw conveyor was maximized to gain the ability to pass large sized rocks. A rear discharge screw conveyor was adopted to minimize locations that a large particle could hung-up. A center stem auger was used instead of an open center or ribbon auger. Ribbon augers can pass larger sized particles for the same screw diameter. This design however was not considered to be sufficiently robust for this application. Further, there were concerns about this design's ability to achieve a soil plug that would withstand the hydrostatic head throughout the high permeability reaches along the alignment.

The Contractor requested and was allowed to deviate from the original Contract Document requirements for airlocks mounted to the TBM and for the location of the screw conveyor at bottom of the cutterhead chamber. Instead, airlocks mounted in the tunnel behind the TBM would be used if compressed air was necessary for cutterhead maintenance. If compressed air was needed for cutterhead entry, it was determined that there was substantial risk for sudden air loss through the highly permeable sand and gravel deposits. The small pressure reservoir afforded by the TBM mounted airlock would subject workers in the pressurized environment to injury from a sudden pressure drop during such an event in addition to the threat from water and soil inflows. The room afforded by removing the airlock from the TBM allowed for the screw diameter to be increased by at least 152 mm (6 inches), thereby reducing the number of rocks that would have to be broken before they could be ingested by the TBM.

Another deviation from the original Contract Specification involved raising the screw location off the bottom of the cutterhead chamber. This had several advantages. Locating the screw at the bottom of the cutterhead chamber requires it to be positioned outside of the cutterhead drive. With the screw conveyor positioned within the interior of the bearing; a larger, more robust bearing with more drive locations could be incorporated into the design. Further, it was identified during testing of soil conditioners for the project that the larger soil fraction (cobbles and boulder fragments) would sink to the bottom when left undisturbed for a few hours. With the screw conveyor positioned at the bottom of the cutterhead chamber, the screw would get buried under a pile of large rocks during a shutdown. This could create a condition with resumption of tunneling that would require considerable effort by the screw to re-mix the separated large fraction with the balance of the soil in the cutterhead chamber. Having the screw located up off the bottom allowed for the rotating cutterhead structure to pass beneath the screw and sweep the pile of rocks up off the bottom so that they could be redistributed throughout the conditioned soil paste.

The cutterhead drive was configured with eight 150 hp electric motors coupled through mechanical gearboxes to pinions engage the main gear at eight locations. The 1,200 hp drive was an increase from the 900 hp drive that was included in Lovat's original proposal. The 150 hp electric motors were controlled with variable frequency drives (VFD) such that the cutterhead could be rotated at speeds from 0 to 3.4 rpm. The VFD's also allowed for startup of the electric motors while directly connected to the cutterhead without a substantial rise in amperage in the electrical system. This drive system was selected over a more common hydraulic drive system due to the increased efficiency of the electric direct drive. During the course of tunneling a substantial benefit was identified from the contribution of the increased inertia of the rotating cutterhead with the direct coupling of the electric motors to the cutterhead. This will be discussed later in this paper.

The cutterhead was designed with 32 cutter mounts that could be interchanged in the field from ripper teeth to disc cutters. There were an additional 6 mounts for ripper teeth only. All cutter mounts were designed as rear loading for cutter replacement from within the cutterhead chamber. The 14 outer-most cutter positions were oriented to be redundant such that multiple cutters cut the same kerf.

There has been considerable debate amongst the Contractor's team and generally throughout the industry on the proper cutters for dealing with boulders in soft ground. There has been considerable use of disc cutters in soft ground applications on slurry TBMs. However their use on EPB TBM applications is in dispute. After considerable consideration, it was determined to configure the TBM with six disc cutters located at or near the gauge of the TBM. These disc cutters were redundant to rippers

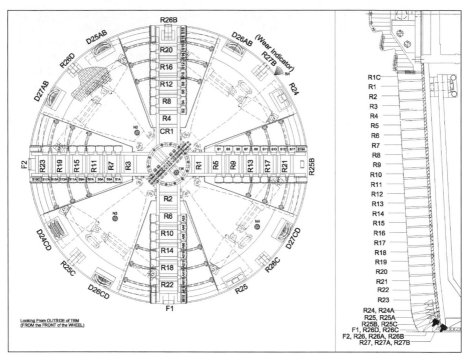

Figure 3.   TBM cutterhead configuration

that cut the same kerf and would operate as back-up cutters to ensure that the gauge was cut in the event the ripper cutters were damaged from a large boulder that may be only partially into the tunnel face. The remaining cutter mounts were outfitted with ripper teeth. A diagram of the cutterhead as configured in shown as Figure 3.

The openings through the cutterhead were equipped with closure doors that could be activated to fully breast the face during cutterhead maintenance. The cutterhead was armored with 19-mm (¾-inch) thick chromium carbide plate. The outer-most ripper tooth, one of three cutter positions that cut the gauge, was outfitted with a hydraulic wear detection system. A 6 mm (¼-inch) diameter hole was drilled to within 25 mm (1 inch) of the end of the ripper. It was then plumbed with a hydraulic line back to the TBM operator's controls. The TBM operator could identify when the ripper was worn or broke to expose the end of this hole by it inability to maintain hydraulic pressure.

The 915 mm (36 inch) diameter screw conveyor had center stem auger that was able to readily convey rocks up to 305 mm (12 inch) diameter. Three grizzly bars mounted across each of the cutterhead openings prevented passage of rocks larger than what could pass through the auger (Figure 4).

The TBM was christened 'Mary Margaret' and shipped to the Columbus for assembly and launch.

TBM PERFORMANCE

The Contractor made some minor configuration changes to the TBM during assembly on site. These included the addition of chromium carbide plating on the interior surface of the cutterhead chamber and surrounding the EPB sensors. Further,

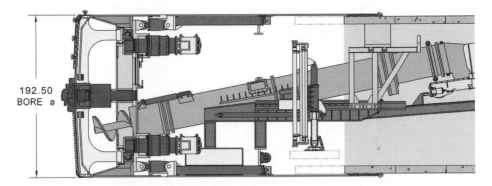

Figure 4. TBM profile showing screw conveyor position

Two of the three grizzly bars were removed to help prevent rocks from nesting in the cutterhead openings and causing a blockage. (This was later determined to be ill advised in that it led to the screw becoming plugged with an oversized rock on two occasions.)

The six disc cutters that were mounted on the cutterhead, proved to be useless. They were recessed from the engagement point of the ripper teeth by approximately 102 mm (4 inches). This meant that they didn't engage undisturbed soil unless the adjacent ripper teeth were gone. This occurred at one location on the project where the wear indicator ripper showed that the gauge rippers needed replacement when we were approximately 30 m (100 ft) from a manhole location. It was decided to wait until the manhole was reached before changing cutters. At that manhole the two discs that engaged the ground were worn flat. One of these disc cutters is shown in the photo of the front of the TBM at manhole #2 as Figure 5.

The ripper cutters wore at a rate that was proportional to the distance that each cutter traveled. This was indicated by the outside cutters wearing at a faster rate than the inside cutters. In order to predict the cutterhead maintenance intervals, the amount of cutter wear was recorded during cutterhead maintenance and this data was plotted against the distance that the cutter traveled since it was last replaced. A general relationship was identified by this plot. A representative sample of this data plot is shown in Figure 6. The relationship varied from one maintenance interval to the next. This was partly due to the varying soil conditions; however, there were several other factors that affected the cutter wear rate. A large proportion of the ripper cutters were breaking from impacts with boulders. After the ripper cutter was broken, the rippers that cut the adjacent kerfs would wear at an advanced rate.

The wear of the rippers could be predicted and cutter changes could be scheduled; whereas ripper breakage was unpredictable. During the first 1,829 m (6,000 ft) of tunneling one-third of all the ripper cutters were being broken by boulders.

Another factor affecting the cutter wear rate was the application of soil conditioners. Soil conditioning was applied at the TBM operator's discretion to optimize the TBM's performance. A major consideration in controlling the soil conditioning was the appearance of the excavated muck as it discharged from the screw conveyor. The operator was trained to condition the muck at rate such that the muck at the discharge point would closely resemble fresh concrete as is exits a transit mix truck. After reaching the first manhole location where ready access to the cutterhead was available, the soil conditioning swivel was found to be damaged and soil conditioner was being injected into the chamber behind the cutterhead instead of in front of the cutterhead. The rate of cutter

Figure 5.   Front of TBM at manhole #2

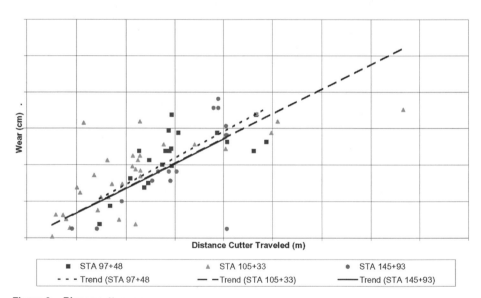

Figure 6.   Ripper cutter wear

wear through this reach of the tunnel was 3 times that of the remainder of the tunnel. This proves one of many major benefits of soil conditioning in EPB tunneling; the reduction in the cutter wear rate.

The screw conveyor generally performed well in handling large cobbles and boulder fragments. There were three instances where considerable downtime was caused by jamming of the screw conveyor. Two of these instances were caused by boulders that were too large to pass getting lodged within the screw conveyor and preventing the auger from rotating. A considerable amount of time was expended in trying to regain rotation of the auger such that the rock could either be pulled through the auger or ejected back into the cutterhead chamber. Eventually inspection doors to the screw were opened, EPB sensor cells removed and ultimately small holes were cut through the conveyor casing to locate the rock. Once the first rock was located, a corner of the rock was broken off with a rivet buster which allowed it to pass through the screw. The second rock was lodged in the same location within the screw but attempts to break it mechanically failed. It was then drilled and blasted with explosives in its position within the screw. The resulting gravel readily passed through the auger.

The third instance of downtime from screw jamming occurred after a prolonged cutterhead maintenance period in wet sand and gravel face conditions. Entry into the cutterhead was in areas of high permeability if the water head was not severe (less than 6 m (20 feet) above tunnel crown). In these locations, the face doors would be closed and the cutterhead chamber depressurized. Personnel would enter the cutterhead and perform cutter replacement on the upper part of the cutterhead as the screw conveyor was operated as an Archimedes pump to control the water level in the chamber. The water inflow would typically be 200 to 700 gpm. A considerable amount of fine soil particles would be washed in with this water, leaving only the coarse fraction remaining in front of the TBM. When tunneling was resumed after this event the lack of fines caused the screw conveyor to lock up. After a considerable effort to free the screw with various methods of soil conditioning injection, the cutterhead and screw were opened up and manually mucked out. The screw was loaded with cobbles—no sand or gravel at all. When the jammed screw was freed, 50lb bags of bentonite were added to the cutterhead chamber when tunneling resumed to create a soil paste that could flow through the muck system without separation.

ADJUSTMENTS MADE DURING TUNNELING

At the first manhole crossing (S/M #2) all of the grizzly bars were re-installed to prevent rocks larger than could be handled by the screw from passing through the cutterhead. The screw conveyor performed with-out any lock-ups after this change.

Once it was identified that ripper cutter breakage was occurring at a high rate, a remedy to this problem was researched. Several different ripper designs were tested utilizing modified shapes and various high strength/high wear resistant materials. None of these showed good results. It was decided to increase the size of the ripper cutters. The original cutters were made of high strength steel with a cross section of 64 mm × 140 mm (2-½" × 5-½"). These rippers were changed to larger rippers having a cross section of 76 mm × 152 mm (3" × 6"). These larger rippers required the installation of new adaptor boxes that fit into the mounting for the disc cutter. The outer-most 18 of the 32 cutter mounts that are interchangeable with disc cutters were replaced with adaptor boxes for the larger ripper cutters. The remaining 14 inner-most ripper cutters were shortened by 51 mm (2 inches). It was reasoned that the shortened rippers would have a lesser tendency to break and the extra wear length was not necessary since the travel distance for these teeth is considerably less than the gauge cutters. The remaining disk cutters were replaced with ripper teeth as well. This change-over could not be performed from within the cutterhead, so the modification was implemented at the second access manhole crossing (S/M #3) which was 2,007 m (6,585 ft) into the drive.

The ripper size increase was very successful. Ripper cutter breakage was reduced from one third of the rippers breaking to less than one out of twenty rippers breaking. The rate of wear for the outside cutters was reduced with the addition of the ripper cutters mounted in the locations that were previously occupied by the disc cutters.

The new adaptor boxes had thinner walls to accommodate the larger rippers. These new adaptor boxes began to crack with one of them failing entirely. Ultimately

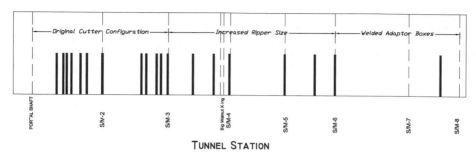

Figure 7.   TBM ripper cutter change locations

after 4,485 m (14,715 ft) into the drive at the fifth manhole crossing (S/M #6), the adaptor boxes were welded directly to the cutterhead structure. This eliminated the possibility of changing these rippers cutters to disc cutters while the TBM was in the ground; although this was not likely to happen on this project anyway.

In addition to the configuration changes made to the cutter head, the operation of the TBM was adjusted. The TBM operator increased the cutterhead rotation rate from 1.7 rpm to 2.3 rpm. This allowed the ripper cutters to take a smaller 'bite' with each pass. The rippers would now engage about 25 mm (1 inch) of undisturbed soil per revolution instead of 38 mm (1-½ inches). The higher rotation rate would increase the amount of cutter and component wear but would lend to the reduction in cutter breakage. It was not identified at the time, but further analysis indicates that the higher speed added to the momentum of the high inertia cutterhead drive. This increased momentum would increase the impact load on the engaged boulders thereby increase the TBM's capacity to break the boulders. This is discussed later in the paper.

The modifications made during the tunneling drive reduced the amount of downtime required for cutterhead maintenance. The average cutter change frequency was reduced from approximately one cutterhead maintenance per 152 m (500 ft) to less than one cutterhead maintenance per 914 m (3,000ft). A chart showing the cutterhead maintenance intervals is shown in Figure 7.

A picture of the front of the TBM after hole thru is shown in Figure 8. This picture shows the excellent condition of the cutterhead after only one cutterhead maintenance event while tunneling through more than 1,875 m (6,000 feet) sand and glacial till including cutting through more than 1.2 m (4 feet) of concrete slurry wall.

Of further note, the project locations that possessed the coarsest soils which were considered by many as being well beyond the capabilities of EPB and the basis for recommending the use of slurry methods, were areas that the TBM achieved its best performance. While tunneling through the areas with 100% sands and gravels and less than 10% fines the TBM was able to advance at its maximum rate with less than 75% of its full torque.

ANALYSIS OF PERFORMANCE

As mentioned earlier, there was a great deal of discussion regarding the appropriate dress of cutting tools that should be applied to the cutterhead. Disc cutters are the standard tool for cutting rock, but it is debatable if boulders can be cut effectively with disc cutters. Disc cutters break rock by loading the rock at the contact point sufficiently high to create shear stresses on either side of the contact point. Single disc cutters are known to perform better than multiple disc cutters for this reason. In order

Figure 8.   TBM front at Shaft 8

for disc cutters to perform effectively in soft ground with boulders the rock has to be held in place with sufficient security such that the point load from the cutter can mobilize the shear stresses in the rock for it to break. In order to stabilize the rock and keep it from shifting position, multiple disc cutters are frequently used. There less efficient cutters are also used to minimize the side loading on the cutter bearings that can result from either a glancing impact on a single disc cutter or from the rock shifting during the impact. When the disc cutter breaks the rock, it produces relatively small fragments or chips. These small fragments can be effectively digested in a slurry mucking system; however, an EPB mucking system does not require the rocks to be broken into chip size fragments. If the rock is not held securely it will either be pushed through the ground until it is either pushed aside or it will eventually be broken up by repeated impacts from the cutterhead or other rocks.

In soft ground tunneling, the ability of the soil matrix to hold the rock in place such that the rock can be cut may be possible in many instances but it is not common. Boulders are usually broken by impact. If you have a boulder on the ground in front of you, would you use knife to cut it or a hammer to break it? When a disc cutter encounters a boulder in a soft ground matrix, the disc will impact the rock as the cutterhead rotates. When a ripper cutter encounters a boulder in a soft ground matrix, it too will impact the rock as the cutterhead rotates. Both tools can act as a hammer to break the rock. The question is which style of hammer will readily survive the impact. The use of disc cutters in soft ground requires a reduced preload on the bearings in the cutter such that the cutter can be readily turned by hand. Otherwise, when engaged against a soft ground matrix there will not be sufficient frictional forces for the cutter to roll across that matrix. In EPB tunneling this issue is worse since the excavated paste that fills the chamber behind the cutter will provide additional rolling resistance than if the cutterhead chamber was empty (as in non-pressurized face TBM) or filled with liquid (as in a slurry TBM). The EPB muck also enhances the possibility that the disc will get soil particles lodges between the sides of the disc and cutterhead structure causing it to stop turning. If the disc doesn't turn in contact with the ground, it will wear flat and fail. On this TBM the disc cutters require at least five times the effort to replace over that required for replacement of ripper cutters.

Figure 9. Rock fragments

In order for a ripper cutter to survive an impact with a boulder it must have sufficient strength to handle the impact loading. The smaller rippers that were mounted on the TBM when it was originally configured did not have sufficient strength. When these cutters were replaced with the larger cross section ripper cutters the tools were able to survive the impact.

WHERE ARE THE BOULDERS?

During the course of tunneling we awaited a call from the heading that the TBM had encountered a rock that has stopped the cutterhead and is unable to advance. This call never occurred. There were several instances where the TBM operator was able to identify that there were substantial rocks being engaged by the cutterhead. We were unable to identify the number or size of these rocks. On one of these occasions we separated the muck from that push and spread the excavated material across the ground. An inspection of the muck identified many rock fragments that exhibited what appeared to be recent fracture surfaces. A sample of the rock fragments is shown in a photograph as Figure 9. However, we did not find several fragments of identical rock that could be pieced together to identify its original size.

We reviewed the data log of the TBM drive system on a regular basis to try to identify torque or amperage spikes that would be an indication of a boulder encounter. We were hoping to find a pattern in the data in the log that would occur during a boulder encounter that could be used as a 'boulder signature.' There were continual torque variations during each advance with generally higher torque during pushes that the operator reported that he was engaging boulders, but there was no identifiable boulder signature that could be used to identify a boulder encounter.

We were puzzled by the lack of boulder problems in terms of no shut-downs to clear obstructions. We did have a considerable amount of downtime for cutterhead repairs, but there seemed to be no stopping of this machine. Could there be enough stored energy in the rotating cutterhead to break the rocks that we were encountering without the need to draw additional power?

We asked the engineers from Lovat to analyze the inertial aspects of the drive. Here is their analysis.

INERTIA FACTOR IN BREAKING ROCK OR CUTTING TOOLS

The total torque (T_{tot}) delivered to the TBM Cutting Head (CH) during mining is distributed as follows:

$$T_{tot} = T_{inertia} + T_{resist} + T_{cut} \tag{1}$$

where:

T_{resist} = Torque required to overcome friction forces resisting CH rotation,

T_{cut} = Torque required to overcome CH cutting forces,

$T_{inertia}$ = Torque required to overcome inertia forces of the rotating components,

In case of the incidental rapid increase of tangential resistance force (F_{tan}) on one of cutting tools caused by hard boulders, the angular impulse on the tool ($I_{ang} = R_i F_{tan} dt$) is generated by angular momentum AM = $J_M \omega$ of all rotating components. where:

R_i = Equivalent radius of the tangential force,

J_M = rotating parts mass moment of inertia,

ω = angular velocity of rotating mass

The following equation can be used to find forces, moments and/or energy to overcome these resistances:

$$\int_{t_1}^{t_2} R_i F_{tan} dt = J_M d\omega \tag{2}$$

The right side in equation (2) can be elaborated for each TBM drive relative to the CH angular speed as shown in the following equations:

$$J_M d\omega = (J_{CH} + J_{MG} + Nr_1 J_{GB} + Nr_2 J_{EM}) d\omega_{CH} \tag{3}$$

where:

J_{CH} = CH mass moment of inertia
(for RME193SE J_{CH} = 77,023.4 kg·m^2)

ω_{CH} = angular velocity of CH (for RME193SE nominal ω_{CH} = 0.16 rad/s and maximum ω_{CH} = 0.36 rad/s)

J_{MG} = Main Gear mass moment of inertia (for RME193SE J_{MG} = 13,354 kg·m^2)

J_{GB} = Gearbox equivalent mass moment of inertia (for RME193SE J_{GB} =12 kg·m^2)

J_{EM} = Electric motor rotor mass moment of inertia (for RME193SE J_{EM} =5.8 kg·m^2)

N = Number of drives in Main Drive (for RME193SE N = 8)

r_1 = Main gear ratio (for RME193SE r_1 = 9.82)

r_2 = Total MD gear ratio (for RME193SE r_1 = 1,049.8)

$$J_M d\omega = (77,023.4 + 13,354 + 942.7 + 48,710.7)(kg·m^2)d\omega$$

$$= 140,030(kg·m^2)d\omega \tag{3a}$$

Therefore, equation (2) for RME193SE can be simplified to the following:

$$\int_{t_1}^{t_2} R_i F_{tan} dt = 140,030 \ (kg·m^2)d\omega \tag{4}$$

In the case of a hydraulically driven cutterhead, the electrical motor inertial component is not present. The resulting angular momentum would be reduced.

$$J_M d\omega = (77{,}023.4 + 13{,}354 + 942.7)(\text{kg·m}^2)d\omega$$

$$= 91{,}320.1(\text{kg·m}^2)d\omega \qquad (3b)$$

The direct connection of the electric drive increases the angular momentum of the cutterhead drive system by more than 53%.

Considering the operational change of increasing the rotation rate of the cutterhead from 1.7 rpm to 2.3 rpm will further increase the angular momentum.

CONCLUSION

Maximize the particle size that can be ingested by the TBM by maximizing the screw diameter. This needs to be considered in coordination with other project parameters, such as the soil permeability and the hydrostatic head in that a large diameter screw may need to be longer to effect a soil plug that will dissipate the head without a blow-in.

The best tool for breaking rocks is a big hammer. An electric direct drive cutterhead provides the high inertia that a big hammer features when breaking rock. The teeth should be sized to withstand the cutterhead drive torque plus the impact loading from the cutterhead momentum.

Unless there is bedrock expected in the tunnel face, do not use disc cutters for cutting boulders with an EPB TBM.

ACKNOWLEDGMENT

The support of the author's employers is gratefully acknowledged.

REFERENCES

Geotechnical Baseline Report—Big Walnut Augmentation.Rickenbacker Interceptor Sewer Project—City of Columbus, Ohio, CIP No. 491.1, May, 2003.

Frank, Glen and Chapman, David; Geotechnical Investigations for Tunneling in Glacial Soils, 2001 RETC Proceedings, Society for Mining, Metallurgy, and Exploration, Inc., Littleton, Colorado; pages 309–324.

Frank, Glen and Chapman, David; A New Model for Characterizing the Cobble and Boulder Fraction for Soft Ground Tunneling, 2005 RETC Proceedings, Society for Mining, Metallurgy, and Exploration, Inc., Littleton, Colorado; pages 780–791.

PIERCING THE MOUNTAIN AND OVERCOMING DIFFICULT GROUND AND WATER CONDITIONS WITH TWO HYBRID HARD ROCK TBMS

Brian Fulcher

Shea-Kenny JV and Shea Construction Co.

John Bednarski

Metropolitan Water District of Southern California

Michael Bell

Hatch Mott MacDonald

Shimi Tzobery

Hatch Mott MacDonald

Werner Burger

Herrenknecht, AG

ABSTRACT

The Arrowhead Tunnels Project represents the final portion of a 70 km (44-mile) long water conveyance facility that will bring up to 28 m³/sec (1,000 ft³/sec) of water into Southern California. The 13 km (8-mile) tunneling project is well underway and consists of two, 5.8 m (19-foot) diameter TBM bores through extremely variable geological conditions. The tunnels lie near the base of the San Bernardino Mountains and several significant faults including the San Andreas Fault run within one kilometer (0.6 miles) of the tunnel alignment. Numerous other significant faults cross the tunnel alignment and water pressures in the tunnel have been recorded as high as 20 bar (300 psi). This paper describes some of the problems and challenges encountered and the solutions implemented to advance the two hybrid TBMs under conditions that ranged from hard rock with no water inflows to full-face granular material under 10-bar pressure and in excess of 32 l/sec (500 gpm) inflows. Due to the range and severity of ground and water conditions actually encountered, the Owner-Contractor-Designer team eventually found that adjusting the means and methods to be more responsive— virtually on a day-by-day basis—was needed to suit the conditions.

INTRODUCTION

The Arrowhead East and Arrowhead West Tunnels, both of which have a finished inside diameter of 3.65 m (12 feet), traverse the mountainous terrain north of the City of San Bernardino. The work includes completing the East Tunnel, constructing the West Tunnel, finishing the connections between the tunnels with pipelines and site restoration. The remaining 6,840 m (22,443 LF) of the East Tunnel excavation must proceed in a down-grade drive from the Strawberry Creek Portal. Construction of the 6,062 m (19,890 LF) West Tunnel is being completed from the Waterman Canyon Portal. Figures 1 and 2 illustrate the site location east of Los Angeles.

Figure 1.   Local area map and project location approximately 130 km (80 miles) east of downtown Los Angeles, CA

Figure 2.   Inland Feeder system layout showing the Arrowhead Tunnels at top

Purpose and Design of the Tunnels

The Arrowhead Tunnels were designed to convey water from the California Aqueduct system to the recently completed Diamond Valley reservoir. They are also intended to provide a second source of supply in the event of an earthquake and reduce the risk of interruption of a sole water source feeding the new reservoir. The final tunnel alignment was situated to avoid existing local urban development and private land ownership restrictions as well as important forest reserves and significant topographical variances in the terrain.

Construction Challenges to Overcome

This project benefited from extraordinary cooperation and understanding from all parties and stakeholders involved and required a wide range of methods to be implemented to deal with the actual ground conditions encountered. The Arrowhead Tunnels Project was not by any measure a routine hard rock tunnel located adjacent to an urban development. It tested the resolve of the project team to meet and succeed in often difficult and challenging conditions to successfully pierce the mountain and complete the tunnel that at this stage is 60% complete overall.

Excavation of both tunnels was through variable ground conditions ranging from highly fractured to massive, moderately jointed igneous and metamorphic rocks. Numerous faults and shear zones containing clay gouge, crushed rock, brecciated and highly fractured rock have been encountered. The rock mass can be completely to moderately weathered at or near the existing ground surface and ranges from highly altered to fresh at tunnel depth. The East Tunnel is being excavated beneath ground cover ranging from 15 m (50 ft) to 630 m (2,065 ft) while the West Tunnel has ground cover that ranges from 15 m (50 ft) to 335 m (1,100 ft). High groundwater inflows with high hydrostatic pressures were anticipated through the highly fractured rock mass along both tunnel alignments. Much of the high volume, high pressure groundwater inflows have been controlled using an effective pre-excavation grouting program. However, the limitations of such a program in a variable and highly fractured rock mass were constantly reviewed and modified. Figures 3 and 4 illustrate the site conditions at the East and West Tunnel portals.

SITE CONDITIONS AND PERMITS

The Arrowhead Tunnels lie parallel to and within one kilometer (0.6 miles) of the well known San Andreas Fault. In fact the tunnel alignment crosses active major splays of this fault. Amongst the risks of working in the dry canyons of Southern California were fire and flood (landslides). Seasonal dry periods elevated the risk of fire to the point where frequent and unexpected forest fires delayed the job. In 2003 for example, a massive flood and slide in a nearby canyon inundated the West Tunnel portal resulting in a three-month delay to this tunnel alone (Figures 3 and 4).

The performance of the work was controlled by the requirements of US Forest Service Special Use Permit that stipulated a maximum of 37 l/sec (580 gpm) for the West Tunnel and 33 l/sec (520 gpm) for the East Tunnel. Additional permits restricted the discharge of tunnel water and the injection of specific grouting materials to control groundwater inflows. These conditions made it necessary to accurately measure and record the rate that groundwater inflows into the tunnels at all times. The Owner employed full time hydrologists and environmental technicians to monitor surface environmental features as well as biological resources, notwithstanding the depth of the tunnels.

GEOLOGY AND GROUND CONDITIONS

Both tunnel alignments lay beneath mountainous terrain that is part of the San Bernardino National Forest. The geology of the area is highly complex (Figures 5 and 6) and range from intact to highly weathered and altered gneiss, marble/calc-silica gneiss, and granitic rocks. These challenging conditions mandated the design of two, one-of-a-kind hard rock TBMs which were subsequently modified once experience with ground conditions and behavior while tunneling was obtained to address specific geologic and water inflow conditions described as follows.

Figure 3. Aerial view of the Strawberry portal with the assembled TBM equipment and storage areas

Figure 4. Aerial view of the Waterman tunnel portal with the assembled TBM, trailing gear and storage areas

Geological Conditions—Ground Water

A key construction requirement was to protect the water resources within the forest and neighboring private properties during tunnel construction. Effective ground water inflow control measures during and after tunnel excavation were mandatory and a key element to the success of the project. To achieve these limitations, the Contractor was required to implement specific groundwater control measures including the following.

- Use of a new tunnel boring machine (TBM) for each tunnel designed and fabricated to be compatible with the anticipated ground and groundwater conditions and to operate in an open-mode (not pressurized face or EPBM) up to 3.0 bar and a static pressure capability of up to 10 bar
- Implementation of a specified probe-hole and pre-excavation grouting program in open-mode or in a closed-mode TBM configuration
- Installation of a bolted, gasketed, precast concrete segment lining for primary ground support and as a reliable groundwater inflow barrier
- Construction of a watertight final lining assembled from reinforced concrete cylinder pipe (RCCP) (Figures 5 and 6)

Geological Conditions—Rock Mass

The rock mass soil deposits encountered during the excavation of the East and West Tunnels included the following.

- Granite rocks including mainly quartz monzonite, quartz diorite, and granodiorite
- Gneissic rocks including quartzo-feldspathic gneiss with amphibolite and biotite gneiss
- Marble and calc-silicate gneiss
- Fault gouge, crushed rock, and breccia

Geological Conditions—Faults and Shear Zones

Both tunnel alignments cross numerous faults that impact tunnel construction with as many as sixteen separate faults that have been mapped across the West Tunnel alignment in addition to five other faults across the East Tunnel. The fault locations and lengths of tunnel intersected by these faults were based on surface projections to tunnel depth and owing to the limitations on the geotechnical studies, locations and quantities may vary considerably.

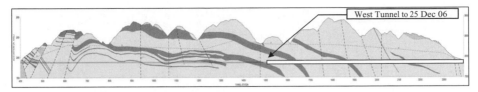

Figures 5.   Geological profile of the West Tunnel

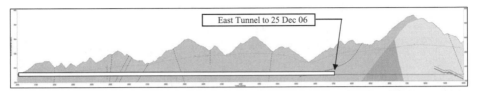

Figures 6.   Geological profile of the East Tunnel showing the variable conditions including faults and shear zones and progress to date. The extent of the ground cover restricted the number of geological bore holes throughout the tunnel length but nevertheless revealed considerable faulting, and highly altered rock zones in addition to groundwater pressures and rock mass permeability.

In addition to the mapped faults, unidentified faults consisting of zones of crushed rock, gouge and brecciated rock could occur randomly anywhere along the alignments. A baseline for the anticipated lengths of tunnel to be affected by unidentified faults was established for each reach of tunnel excavation. The materials encountered in these fault zones consisted of crushed and brecciated rock with sandy and clay gouge. The crushed and brecciated rocks resulted in raveling or running ground when groundwater was absent or flowing ground when groundwater was present (and the material was completely pulverized or crushed to the consistency of sand or gravel). Sandy gouge materials behaved as flowing ground below the groundwater table. Considering the cover above the tunnels, clay gouge was anticipated to exhibit the characteristics of squeezing ground. Tests of samples indicated a low potential for swelling of the rock mass.

Shear zones contained infillings consisting of breccia and/or clay material and other minerals. Rock core data indicated the pervasiveness of thin, curving shear planes at seemingly random orientations in or close to fault zones. In the highly weathered rock near both portals of the West Tunnel, thin shear planes were pervasive in the rock mass in a zone extending several hundred feet from the Arrowhead Springs Fault. Shears could act as sources of groundwater inflows or alternately as groundwater barriers in the tunnels.

PROJECT SCHEDULE AND QUANTITIES

The original Contract provided for 1,050 work-days (wd) plus the actual time for Owner-directed pre-excavation grouting operations to complete the project following Notice-to-Proceed for construction. An initial and separate Notice-to-Proceed for technical submittals allowed for an additional seven months of time for planning and procurement activities primarily related to the TBM design. The base construction time of 1,050 wd did not include time required to perform the pre-excavating grouting program based on the Bill-of-Quantity units of work. This time could be substantial and would be periodically added to the base Contract time by the Owner. Figure 7 provides some

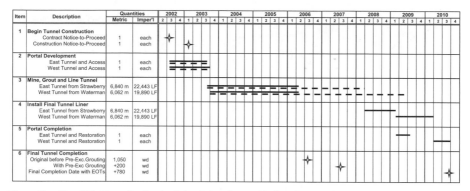

Figure 7. Simplified bar chart schedule of the planned (—) and actual (- - -) performance to date for the mining and initial lining operations in the East and West Tunnels. Time for pre-excavation drilling and grouting is also included.

descriptive data on the planned and actual schedule performance to date for the East and West Tunnels.

At this point in the project and after approximately 60% of the tunnel excavation has been complete (to 25 Dec 06), the pre-excavation drilling and grouting operations together with mining has consumed over three years with advance rates as listed in Table 1 for both headings. It is apparent, therefore, that a variety of conditions in addition to the pre-excavating drilling grouting program have contributed to a significant amount of extra time needed to complete the work to date. It is anticipated that an additional 780 wd in addition to the 200 wd awarded before to 25 Apr 06 will be required to finish the project as illustrated in Figure 7.

The time needed for the mining and the precast liner work alone historically and prior to 25 Apr 06 been approximately 30% whereas the remainder of the time was spent on grouting and delays. It should be noted though that the distribution of time between mining and pre-excavation grouting has varied greatly and especially in cases were the ground is dry and massive versus cases where the ground is unstable and water inflows are substantial.

TUNNELING PRODUCTION AND IMPACTS

This is a unique tunnel that required not only specified tunnel excavation equipment, but also a capability to perform repetitive pre-excavation drilling and grouting operations through various geological conditions. Accordingly, the TBMs may be generally described as fully-shielded hard rock machines (cutterhead) with soft ground muck handling and segment erection capabilities. The drilling and grouting equipment included four drills (initially) and several grout plant/pumps as described below. Figure 8 shows the "twins" in the shop. Figure 9 is the starter tunnel.

Tunneling Advance Rates

The tunneling advance rates since the start of the work have varied substantially and were dependent on the ground and water inflow conditions encountered. Generally, the more water that was encountered at the face, the slower the overall mining advance rates. This was due to the need to perform additional pre-excavation drilling and grouting, water handling as well as the presence of excess quantities of fine mate-

Figure 8. Twin Herrenknecht 5.8 m diameter fully-shielded hard rock TBMs in the shop after testing

Figure 9. Starter tunnel and bored tunnel interface showing various utility systems and ground support

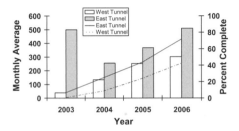

Figure 10. Average annual excavation quantities in the West and East Tunnel since start of the project

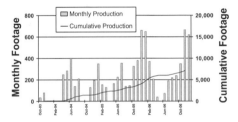

Figure 11. Monthly advances in the West Tunnel only, including time for pre-excavation grouting

rial often found with the muck. Fault, gouge and shear zones have also contributed to slower advance rates, in addition to encountering highly altered rock that behaved poorly in the presence of ground water pressure and a free face in front of the TBM. The tables and charts below will summarize the average productions per month in both tunnels and the distribution of the productions since the start of the work in the West Tunnel which was 42% complete (25 Dec 06) with an overall average monthly advance rate of 69 m (227 ft). Figure 10 is the total tunneling since the start of work for both tunnels while Figure 11 illustrates the advance per month for the West Tunnel—including time for pre-excavation grouting. As of 25 Dec 06, the East Tunnel was 72% complete with an overall average monthly advance rate of 116 m (381 ft).

Table 1 lists the actual best and average tunnel production rates to date for both the East and the West Tunnels whereas Table 2 lists examples of the influences and conditions affecting tunnel mining cycles and, therefore, overall advance rates. Table 2 should also be read in conjunction with Table 5 for impacts to mining operations.

Based on the above, it is apparent that when drilling conditions for probe and water inflow verification holes was considered satisfactory and without difficulty caused by collapsing probe holes, the corresponding mining phase was performed promptly and without delay or incident. Conversely, when the drilling and grouting program suffered difficulties, these frequently foretold the ground conditions and behavior during the upcoming mining cycle.

Table 1.   Tunnel lengths and overall advance rates—to 25 December 2006

Description	East Tunnel (AET)		West Tunnel (AWT)	
Tunnel total lengths	6,840 m	(22,443 LF)	6,062 m	(19,890 LF)
Tunnel (TBM drives)	6,765 m	(22,195 LF)	6,023 m	(19,760 LF)
Best month to date	361 m	(1,185 LF)	203 m	(665 LF)
Best week to date	102 m	(335 LF)	62 m	(205 LF)
Best day to date	29 m	(95 LF)	21 m	(70 LF)
Average month advance	116 m	(381 LF)	69 m	(227 LF)
Average week advance	29 m	(95 LF)	17 m	(56 LF)
Average day advance	5.8 m	(19 LF)	3.5 m	(11.3 LF)

Table 2.   Influences and conditions affecting tunnel advance rates

Advancing grouted plug ahead of TBM	6 to 12 m (20 to 40 ft) depending on conditions
Probe holes drilled ahead of TBM face	15 to 30 m (80 to 200 ft) depending on conditions
Probe holes drilled per mining cycle	2 to 6 holes depending on conditions
Water inflow verification holes per mining cycle	2 each depending on quantity and certainty of flow
Minimum grout holes per mining cycle	2 each at a 4° angle
Maximum grout holes per mining cycle	15 each at 1.5° angle through the face plus, 19 each at 4° angle, plus 11 each at 8° angle in shield

PRE-EXCAVATION DRILLING AND GROUTING OPERATIONS

Probing and grouting ahead of the TBM (probe hole drilling and pre-excavation grouting) was required for both tunnels to achieve the following objectives. Figure 12 provides a sectional view of the TBM showing the drills.

- Minimize and control the effects of tunnel excavation on the groundwater resources in the project area to satisfy the requirements of the USFS Special Use Permit
- Reduce groundwater inflows into the tunnel to facilitate tunnel excavation and the installation of the primary support system that would ultimately benefit the installation of the final tunnel liner
- Improve the ground conditions and, therefore, mining tunneling advance rates

The U.S. Forest Service Special Use Permit limited groundwater inflows into the tunnel heading (defined as the portion of the tunnel 180 m (600 ft) behind the advancing face back towards the portal) to no more than 37 l/sec (580 gpm) for the West Tunnels and 33 l/sec (520 gpm) for the East Tunnel inflow. Control of the heading groundwater inflows to this limit was accomplished by a combination of the following operations.

- Pre-excavation water control and ground improvement grouting in multiple stages as needed
- Filling of discontinuities in the rock mass as needed and as encountered to avoid running or raveling ground conditions at the face, around the shield and the primary segmental tunnel liner

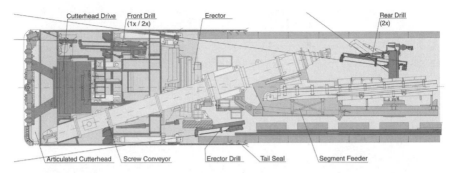

Figure 12. Section through the 5.8 m diameter Herrenknecht fully-shielded hard rock TBM showing the principle components, screw auger and drills for probing and grouting, segment feeder and cutterhead drive arrangement

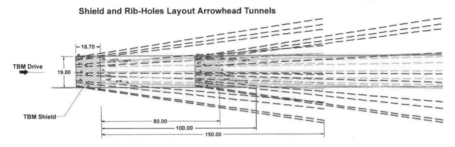

Figure 13. General arrangement of the drilling patterns for pre-excavation grouting. A 6 to 9 m (20 to 30 ft) grouted plug was constantly maintained in front of the TBM at all times and adjusted with ground conditions.

- Installation of a watertight primary support system that provided full ground support with backfill grouting
- Installation of inflatable segment collars, in combination with backfill grouting
- Installation of drain holes through the precast tunnel lining in specific locations as needed to relieve excess ground water pressure and to improve ground conditions for mining operations

Pre-excavation grouting was initially performed, as directed by the Owner with the intent of reducing groundwater inflows to typically 16 l/sec (250 gpm) in the 30 m (100 ft) interval from the tunnel face back towards the portal. However, the extent and potential for improvement in ground stability could not be directly quantified by performing additional grouting. Furthermore, it was always uncertain if probe drilling would consistently be successful in detecting groundwater inflows and geological features that would impact the mining operations. Figure 13 illustrates the general arrangement of the probing and grouting drill hole patterns and approximate lengths.

Pre-excavation grouting was initially planned to be performed along any length of tunnel depending on measured probe hole inflows from two verification holes. The Owner established an initial criteria value wherein a probe hole inflow which exceeded 0.06 l/sec/m (0.3 gpm/ft) of probe hole (or a concentrated flow of 30 gpm at any location in the probe hole) represented the threshold at which pre-excavation grouting was

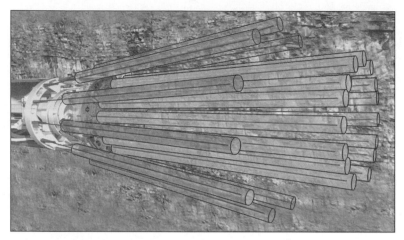

Figure 14.   Rendering of the general arrangement of the pre-excavation grouting scheme using 1.5° and 4° drill ports though the TBM face

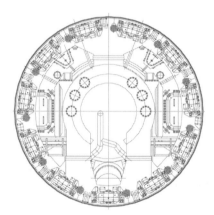

Figure 15.   Inside view of the forward bulkhead with face drill ports laid-out

initiated. Drilling and grouting equipment and methods were designed for stage grouting since some drill holes would not stay open due to unstable ground conditions; e.g., highly fractured rock, gouge, weathered and altered rock. Actual experience found that because of the uncertainty of groundwater inflows through a fractured rock mass, probe holes would not necessarily detect all potential groundwater inflows as well as areas where mining operations would be difficult. (Figures 14 and 15)

Probe hole drilling was performed ahead of the tunnel face through 15 drill ports at 1.5° in the TBM face and 19 locations at 4° also through the face. Two probe holes were normally drilled approximately 180-degrees apart along the longitudinal axis of the tunnel. A minimum overlap of 6.5 m (20 ft) was required at the face between successive set-ups of probe/grout holes in order to maintain a grouted plug. Probe holes were limited to a maximum of 60 m (200 ft) in effective length and grout holes limited to a maximum of 45 m (150 ft) due to the uncertainty in hole location for longer holes and the reduced effectiveness of grouting long holes away from the face.

Table 3.   Tunnel grouting materials and corresponding operations

Item	Description	Pre-Excavation Grouting			Annular Grouting	
		Water	Ground	Voids	Stage I	Stage II
A	**Cementitious Products**					
1	Portland Cement—Type II	●	●	●	●	●
2	Portland Cement—Type III	●	●	●		
3	Rapidset Cement	●	●		●	
4	Micro-Fine Cement	●	●	○		
5	Flyash		●		●	
B	**Chemical Products**					
1	Polyurethane grout	●	●		○	
2	Sodium silicate	◆			◆	
3	Colloidal silica	●			○	○
4	Minova Celbex 802				◆	

●   Primary grouting material.
◆   Secondary grouting material.
○   Trial and test grouting material.

To date, a substantial amount of time spent driving the tunnels was consumed with pre-excavation drilling and grouting operations before the TBM was able to advance. Notwithstanding, this operation was essential to successfully and safely mine the tunnels through highly variable strata and in sometimes unknown and uncertain geological and water bearing features that needed grouting for water control and/or ground improvement.

ADDITIONAL GROUTING MEASURES AND MATERIALS UTILIZED

In addition to the pre-excavation grouting program described above, the project required the implementation of several other grouting programs for the control of water inflows as the TBMs advanced. These included the following as briefly described below and as listed in Table 3.

- **Stage I Backfill Grouting.**   Injected as the TBM advanced and filled the annular void surrounding the segmental tunnel lining with a sand-cement (Type II) grout. Experiments have been performed using various anti-dispersion agents and grout accelerators with some success.

- **Stage II Contact Grouting.**   Injected periodically as the TBM advanced and completed the filling of the annular void surrounding the tunnel liner with neat cement (Type II) grout.

- **Drainage Holes.**   Installed at 20° through the segmental tunnel lining to relieve excess hydrostatic pressure and inflows into the cutterhead plenum and to assist with ground stabilization and pre-excavation grouting procedures.

Several grouting materials have been used for the control of inflows and ground improvement. Some are still in use depending on mining conditions in the heading. Table 3 provides a summary of the use of these materials. Additional materials are currently under test and may be used later for both water and ground control needs.

- **Micro-Fine Cement.** Used for most rock and rock-like conditions to curtail water inflows and improve structural properties of the rock mass for mining. Various mix designs and pressures were routinely used.

- **Types II and III Cement.** Used for most rock conditions to curtail water inflows and improve properties of the rock mass for mining following the completion the grout injection sequence with micro-fine cement.

- **Colloidal Silica Grout.** Used in strata that while producing water inflows, resisted penetration from micro-fine cement grouts—even in diluted concentrations. This material was successful in reducing water inflows.

- **Polyurethane Grout.** Used in the tunnels with limited success to control ground water inflows and filling voids. Based on our experience this material did not work well under high water pressures when injected in bore holes.

- **Chemical Grout.** Sodium silicate was used in the tunnels and portals to consolidate loose and brecciated rock and to control ground water too. This material had limited success in tight rock formations and where clay was present.

Segmental Tunnel Lining with EDPM Gaskets

A bolted and gasketed, precast concrete segmental lining was specified as the primary support system for the entire length of both tunnels. This construction has been substantially addressed in the referenced papers and will only be briefly described here. Table 4 will provide physical and dimensional details of the segmental tunnel lining.

In the design of the primary support system the maximum groundwater head during construction for the West Tunnel was 170 m (550 ft) and for the East Tunnel 275 m (900 ft). Accordingly, the primary segmental tunnel liner was designed to resist these loads as well as provide a reliable reaction block to the tremendous thrust forces ultimately applied by the TBM. The maximum thrust force actually applied was periodically almost four times the original jacking force to move through difficult ground causing apparent squeezing and packing on the TBM shield.

The annular void outside the precast segmental liner was filled with a sand-cement grout, commencing a two-stage backfill grouting process (Stage I and Stage II) though grout sleeves and one-way valves embedded in the segments. Admixtures required for successful backfill placement included accelerators, anti-washout agents, fluidizers, and other materials. Stage I backfill grouting was performed continuously as the segmental rings were pushed-out from the TBM tail shield. See Figure 16 for details.

Final Tunnel Lining—RCCP Pipe Inner-Lining

The final tunnel lining will be installed in the tunnels to provide a hydraulically efficient system for conveying water and to prevent infiltration and exfiltration of groundwater. In addition, the final tunnel lining must support long-term rock and hydrostatic loads, provide erosion protection, and facilitate dewatering and inspection of the tunnels. There are three options for the final tunnel lining—depending on location in the tunnels and outside at the portals.

- Reinforced Concrete Cylinder Pipe (RCCP)—to be used for most of the tunnel length

- Welded Steel Pipe (WSP) with stiffeners—to be wherever the tunnel crossed an identified fault zone

- Plain Welded Steel Pipe (without stiffeners)—to be used wherever the tunnel crossed an identified fault

Table 4.   Segmental tunnel liner dimensions, weights and features

Description	Details and Dimensions	Description	Details and Dimensions
General Arrangements:		**Annular Void and Theoretical Grout Volumes:**	
• Typical and key	5 pieces + key segment	• Annular gap	142 mm (5.5")
• Tapered rings	Left and right hand— 38 mm	• Annular area	2.52 m² (27.2 ft²)
• Grout ports	2 each through segment	• Annular volume	3.78 m³ (5 CY/ring)
• Handling means	Vacuum lifting only	**Precast Concrete:**	
Quantity and Dimensions:		• Concrete T-1	41 MPa (6,000 psi)
• Extrados	5.537 m (18'-2")	• Concrete T-2	55 MPa (8,000 psi)
• Intrados	4.877 m (16'-0")	**Reinforcing Steel:**	
• Thickness	330 mm (13")	• Welded steel mat	100 kg/m³ (167 lb/CY)
• Sectional area	5.4 m² (58 ft²)	• Rebar A496	480 MPa (70 ksi)
• Length per ring	1.524 m (5'-0")	**EDPM Gaskets per Segment:**	
Volumes and Weights:		• Gaskets	One continuous piece
• Volume per ring	8.2 m³ (10.7 CY)	• Gasket design	275 m (900 ft) pressure
• Linear weight	12.9 t/m (8,700 lb/ft)	• Gasket profile	Phoenix M 385 73
• Ring weight	19.37 t/ring (43,500 lb/ring)		
• Heaviest piece	3,850 kg (8,500 lbs)		

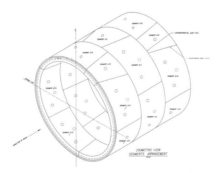

Figure 16.   Isometric view of the assembled precast segmental tunnel liner showing the joint configuration, grout ports and gasket grooves for the 5+1 piece ring.

GROUND BEHAVIOR WHILE TUNNELING

The Contract documents described the subsurface conditions and behavior anticipated during construction of the East and West Tunnels. These documents and in particular, the Geotechnical Baseline Report (GBR) presented the Owner's best judgment based on data from surface geologic mapping, exploratory boreholes placed along and in the general area of the tunnel alignments and assumptions regarding the Contractor's

Table 5.   Ground conditions, behavior and impacts to mining operations

Squeezing and Packing Ground Around the TBM:	
• Squeezing ground	Extremely high thrust pressures needed to advance
• Packing ground	Extremely high thrust pressures needed to advance
• TBM seized in ground	Thrust force up to 114 MN applied to the shield
• Unstable ground/squeezing	Hand-mining around the shield to free the TBM
Raveling Ground Conditions at the Face:	
• Tunnel face ravels	Delay to the tunnel advance rate due to loss of face
• Tunnel face loose and fall-out	Delay to allow for backfilling the face with sand-cement grout
• Tunnel face partial collapse	Breasting and special grouting operations to stabilize
Weak Rock Drilling Conditions at the Face:	
• Weak rock at face	Slow penetration rates due to hole collapse
• Raveling drill holes	Hole collapse and loss of steel needing casings to advance
• Seating grout packers	Ineffective and time consuming due to loss of ground
• Seating grout packers	Multi-stage drilling and grouting needed to be effective
Hard Rock Conditions at the Face:	
• Very hard rock	Slow advance rates, heat build-up in TBM, cutter wear
• Very hard massive rock	More cutter wear, machine repairs, more time to advance
Water Inflows and Pressures:	
• Excess water pressure	Delay to drain and secure the heading from excess water
• Excess water inflows	Additional grouting to seal and secure vital equipment
• Excess fines in muck	Mucking and slurry separation systems overwhelmed
Tunnel Plug Unstable, Weak or Leaking Grout:	
• Unstable plug at face	Delay to drilling & grouting until grout sets in multiple stages
• Weak plug at face	Grout leakage and pressure loss limiting effectiveness

means and methods for tunnel excavation and support. Additionally, the minimum requirements for tunneling equipment specified by the Owner was coordinated with the anticipated ground conditions and behavior.

Actual Ground Behavior Experienced to Date

While the anticipated ground conditions were as generally described in the GBR as variable, what was actually encountered was a wide spectrum of ground types and behaviors. The rate of change from one type to another was often frequent and even dramatic to the extent that extremely hard material could be encountered on the same day as soft and raveling material. Additionally the face could exhibit both very competent hard material (with and without water inflows) at the same time as loose ground in another portion of the face. Periodically, acute geological features would release significant flush inflows and even flowing ground resulting in the formation of a cavity that required an immediate halt to mining operations and then backfilling. Table 5 provides a brief summary of impacts to mining.

On occasions to date, decomposed rock acting like sand and silt, blocky ground with large pieces dislodged from the face, both white and gray marble as well as very competent granitic material have been encountered. High in-situ water pressures and

Table 6.   TBM basic specifications, weights and measures

Description	Size or Capacity		Description	Size or Capacity	
Design Parameters:			**Dimensions and Weights:**		
TBM design	Fully-shielded articulated		Cutterhead diameter	5.820 m	19'-1"
	Cutterhead & tail skin		Front shield diameter	5.790 m	18'-11.5"
Static pressure design	147 psi	10 bar	Center shield diameter	5.770 m	18'-11"
Operating pressure	44 psi	3.0 bar	Tail shield diameter	5.750 m	18'-10.5"
Power and Thrust:			TBM total length	10.35 m	34'-0"
Cutterhead power	2,000 kW	2,680 HP	TBM weight	392 t	433 tons
Cutterhead torque	2,000 kNm	1.5 M-ft-lb	TBM & trailing gear	142 m	465 ft
Maximum torque	2,346 kNm	1.7 M-ft-lb	TBM & trailing gear	820 t	906 tons
Cutterhead drives	7 hydraulic motors		**Cutting Tools and Muck Handling System:**		
Rotational speed	Variable 0 > 9.75 rpm		Disk cutters	39 ea, 432 mm	17 in Ø
Thrust cylinders	22 each, 2.300 m stroke		Maximum load/cutter	250 kN	56,000 lbs
Thrust force	29,270 kN	3,300 tons	Cutterhead buckets	3 face + 3 periphery	
Thrust speed (max)	80 mm/min	3.1 in/min	Screw auger size	700 mm Ø	2'-4" Ø
Total installed power	3,600 kW	4,825 HP	Screw auger power	160 kW	215 HP
			Screw auger capacity	175 m³/hr	230 CY/hr
			Maximum particle	200 mm	8 inch

flows have from time to time been very adverse and have daunted the pre-excavation grouting program to the extent that this program was often adjusted before each mining cycle and constantly assessed during the actual grouting program to be responsive to the actual conditions encountered. From time-to-time unusual ground conditions have been encountered that have resulted in the following situations and delays to the work.

The Contractor's tunneling crews together with the Owner's geological and grouting experts developed a toolbox of flexible and innovative means and methods to cope with a wide range of ground and water conditions with the goal of achieving effective pre-excavation grouting and to optimize the TBM advance safely while constantly controlling water and ground conditions.

TBM SPECIFICATIONS, MODIFICATIONS, AND PERFORMANCE

Two specially-designed Herrenknecht TBMs were purchased for the project and included a full complement of supporting systems including power, segment handling, ventilation, waste water management, grouting material storage and mixing equipment mounted on the trailing gear. The TBMs had the specifications as listed in Table 6 that were subsequently adjusted as described below as part of a program of improvements to improve performance while coping with widely variable ground and water conditions. Figure 17 illustrates the general arrangement of the principle components of the TBM and the first deck of the trailing gear—as supplied to the project.

Unique Characteristics of the TBMs

The TBM were hybrids in that they needed to deal with rock having less than 2 MPa to more than 310 MPa (300 to 45,000 psi), in addition to static ground water pressures up to 10 bar (145 psi). The TBMs were designed to continuously operate

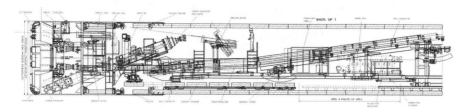

Figure 17.   Section through the 5.8 m diameter Herrenknecht fully-shielded hard rock TBM showing the principle components, screw auger, drills, segmental tunnel lining installation and forward deck of the trailing gear

with 3.0 bar (44 psi) ground water pressure (cutterhead bearing seal capacity in dynamic mode)—provided hard rock conditions were present. Whereas the cutterhead was equipped with considerable rock cutting tools and the available thrust was 250 kN (56,000 lbs) per disk cutter it was not designed to excavate extensive lengths in "soft ground" conditions and mining in a closed-mode (EPBM) operation.

The screw auger was needed to handle water-laden muck but not to develop a plug and dissipate groundwater pressures in the muck handling system. The cutterhead was not originally designed to act as a pressurized plenum in the same manner as an EPBM system and was only later equipped with pressure cells, mixing bars, air cannons, and injection systems for foam, bentonite and other fluidizing admixtures. Other important features of these highly specialized TBMs include the following items and as illustrated in Figures 18 and 19.

- Inflatable main bearing seal—good for up to 10 bar static pressure in case of an emergency
- Vacuum lifting systems for precast tunnel segments
- Hydraulic linear feeder for precast segment supply to the erector
- Bulk storage of cementitious materials in the trailing gear
- Twin Hany cement grout mixing and pump plants—good for 65 bar (950 psi) and expandable to 100 bar
- VMT computer guidance and ring-building optimization software

Friction Forces Analysis on the Shield

As part of the TBM performance evaluation, a TBM friction trial test was conducted during the initial stages of the project to assess the machine's ability to overcome the anticipated ground pressures. The study was performed on the East Tunnel through 15 m (45 ft) of excavation where the TBM parameters were recorded along with observations of the radial clearance between the excavated rock surface and the extrados of the shield to determine the area of the shield actually in contact with the excavated ground surface. Based on a fully-equipped shield weight including the cutterhead, the coefficient of friction was derived using the following simplified equation.

$$\text{Coefficient of friction} = \frac{\text{Friction forces}}{\text{Ground forces around the shield} + \text{Shield weight}}$$

In order to calculate the friction forces induced on the shield, the cutterhead thrust; i.e., contact force and some minor frictional effects, were removed from the total TBM thrust applied for advance. For the analysis, a minor frictional force was computed

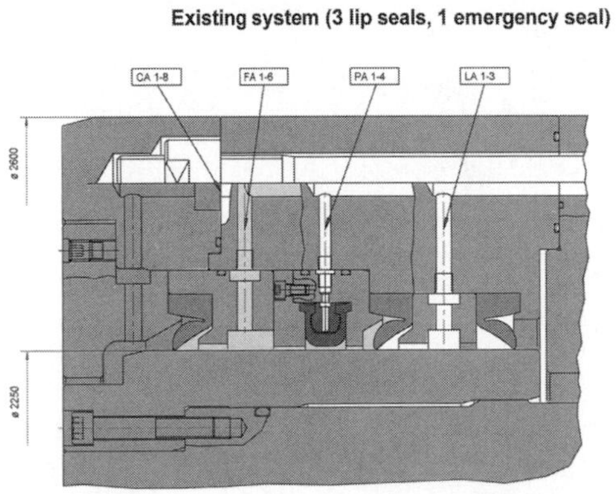

Existing system (3 lip seals, 1 emergency seal)

Figure 18.   Section through the inflatable main bearing seal showing the four stages of protection

Figure 19.   Hany batch/pumping plants for supply of cement grouts for the pre-excavation grouting program

based on the recorded trailing gear drag and stabilizer contact. Throughout most of the trial, the radial gap around the shield was free of ground contact as confirmed by the amount of Stage I annular grout injected at those locations. The ground pressures were deemed minor and irrelevant for the friction trial results. The derived average coefficient of friction ranged between 0.37 and 0.45, with the higher values occurring at the start of the TBM shoves. The results were consistent with published friction coefficients of metal on rock. As an exercise, the coefficient of friction was projected using maximum expected ground pressures equivalent to a rock height of three tunnel diameters vertically and one tunnel diameter horizontally. Using a maximum anticipated ground-unit weight of 27 kPa (170 lb/CF) for a gneissic rock, the originally provided TBM thrust of 29,270 kN (3,300 tons) was confirmed to have sufficient capacity to overcome anticipated squeezing ground conditions.

In light of the experience gained during the tunnel drive, it was well recognized that the actual mechanisms of ground pressure, TBM attitude, actual ground adhesiveness, and friction factors were too variable to accurately predict TBM behavior. As a general rule the actual performance of a TBM during operation, in a range of ground

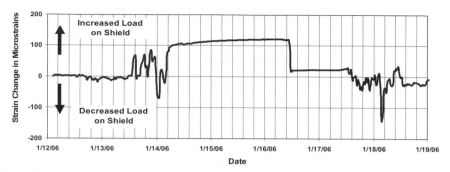

Figure 20. Strain gauge recording of loads on the shield presenting the fluctuating loading conditions during a typical mining cycle. The sustained positive loading for 2.5 days represents loading on the shield before release.

conditions encountered while mining, is the only ultimate test. The effects of highly to extremely weathered rock with traces of sandy and clayey gouges, combinations of adhesive fines packing behind the shield, TBM in "iron-bound" situations, plowing tendency of shield, and squeezing or raveling ground conditions have a tremendous impact on the mining production. High-friction forces develop on the shield, leading to entrapment of the machine, which in turn results in downtime and work stoppages. In addition to the installation of several significant modifications to improve TBM performance as described, some other measures were used to maintain movement of the TBM. These included bentonite injection, ground consolidation, adjustable gauge cutter over-cut, retracted movements of the articulated tail shield and cutterhead, and high-pressure water-jet blasting through shield ports.

A total of 11 strain gauges were installed in two rows on the internal skin of the TBM shield at angles of 0, 45, 90, 270 and 315 degrees. As shown in Figure 20, a higher change in the strain recorded in micro-strains was interpreted as gauge compression due to a potentially increased load on the shield. Following a short learning curve, this system was recognized as a useful tool to assist in detecting localized tight spots where increased thrust had to be used.

Alignment Steering Difficulties in Tunnel Curves

As experienced by mining through two of the project's five curves, it was determined that the packed fine material around the shield was associated with difficulties in negotiating the 365 m (1,200 ft) radius horizontal curves. Although assertive actions were taken, including intensive bentonite injections around the shield, differential pressure applied to the thrust jacks, articulated cutterhead tendency adjustments, and tail-shield free movement, the TBM deviated from the planned alignments. A deviation of approximately 200 mm (8 in) was successfully restored within approximately 60 m (200 ft) of a right-hand curve. Conversely, during the excavation of a left-hand curve, the TBM could not maintain the required tendency and horizontally drifted away from the planned alignment by up to 2.2 m (7 ft). As a result, a corrective action adjustment of 4 mm (0.16 in) per 1.5 m (5 ft) of TBM advance had to be used throughout hundreds of shoves to redirect the TBM to the planned alignment. The difference in performance between left and right-hand curves is attributed to the accumulated build-up of fine material (fines packing) on the left side of the shield as a result of the cutterhead's clockwise rotation—looking towards face.

Table 7. Summary of changes and improvements to the TBMs

Description	Original TBM Design		Modified TBM Design	
Cutterhead operational torque	2,000 kNm	1.5 million ft-lbs	3,520 kNm	2.6 million ft-lbs
Cutterhead exceptional torque	2,323 kNm	1.7 million ft-lbs	4,400 kNm	3.3 million ft-lbs
Cutterhead speed (max)	9.75 rpm		7.5 rpm	
Cutterhead rear annular gap	100 mm	4 inches	75 mm	3 inches
Shield thrust	29,270 kN	3,300 tons	58,540 kN	6,600 tons
Probe drilling holes	26 each (1.5° and 8° lookout angle)		45 each (1.5°, 4° and 8° look-out)	
Probe drills	4 each; 1 front, 1 erector, 2 rear		5 each; 2 front, 1 erector, 2 rear	
Screw auger	13.35 m (44 ft) single piece unit		4-piece (3.35 m each piece)	
EPBM capabilities	None		Sensors, mixing arms, foam mixer	
Slurry handling/separation system	None		54 l/sec (850 gpm) maximum	

Modifications to the TBM and Trailing Gear Systems and Equipment

Based on experience and performance in variable ground and water conditions, several significant modifications were made to the TBMs in 2005 to cope with ground control and water inflows. The following provides a brief summary of the scope and benefits of the TBM modifications that were installed and refined over a 12-month period. Table 7 summarizes the original and modified capacities based on the various modifications made to date.

- **Cutterhead Shroud Ring**—Installed on the trailing edge of the articulated cutterhead to close-up the gap between the leading edge of the forward shield. This modification was needed to control raveling ground inflows and particle intrusions into the plenum that periodically slowed or stopped cutterhead rotation and production.

- **Cutterhead Bucket Closures**—This modification was temporarily needed to improve breasting of the face in poor and raveling ground conditions. It provided better control in blocky ground conditions, but resulted in increased wear, maintenance and stress on the cutterhead even though the need for improved face control was apparent. This modification may have also contributed to stress cracks and additional maintenance on the cutterhead.

- **Cutterhead Torque Increase**—The original cutterhead torque was 2,000 kNm, (1.5 million ft-lbs) but after several months excavating in variable geological conditions including soft ground conditions, it was determined that an increase in torque together with a decrease in cutterhead speed was desirable to improve performance, reduce torque-limited performance and to provide for an "EPBM mode" in soft ground conditions. Accordingly, the main drive gearboxes were changed resulting in a torque increase of 75% to 3,520 kNm (2.6 million ft-lbs) and the maximum cutterhead rotational speed was reduced to 7.5 rpm from 9.75 rpm.

- **Thrust Pressure Increase**—Overall TBM thrust was increased to provide improved ability to move the shield in squeezing and packing ground conditions. Increased thrust was not, however, delivered to the disk cutters since the cutterhead articulation cylinder capacity remained as per the original design and was, therefore, limited to 250 kN (56,000 lb/cutter). The maximum cutter load effectively buried the cutters in the face and did not contribute to improved excavation rates. In order to achieve a two-fold increase in thrust, the propel cylinders were changed to operate at 700 bar (10,000 psi) hydraulic pressure with a corresponding change to the hydraulic pumps and power units. The total theoretical thrust on the shield was increased by two times—from 29,270 kN (3,300 tons) to 58,540 kN (6,600 tons). Notwithstanding this increase on several occasions in both the East and West tunnels, additional auxiliary hydraulic jacks had to be installed to free the shield and maintain forward progress. In these cases, thrust was increased by an additional 55,000 kN (6,200 tons) using 11 portable 565-ton hydraulic jacks. Total thrust applied to the TBM in this case was 114 MN (12,800 tons) or four times the original TBM design thrust.

- **Four-Degree Grout Holes**—An additional 19 grout holes at 4° were installed through the forward bulkhead and the cutterhead to provide access for additional and potentially more effective grouting operations. Whereas the TBMs were originally designed to have 11 grout holes mounted at 8° and 15 grout holes at 1.5°, additional grout holes were deemed necessary to improve the overall effectiveness of the pre-excavation grouting operations.

- **Screw Auger Modifications**—The screw auger has undergone several modifications—principally to address wear and maintenance issues and improve the operational service life. Modifications to date have included the following.

  - Installation (welded) of Herrenknecht hardened tool steel wear plates on the flights above the bolt-on plates
  - Installation of 25 Cronatron high-wear resistant "chalky" bars full length on the inside of the auger barrel
  - Development of a split-sided barrel with bolted connections 13.5 m (44 ft) long for access and maintenance
  - Development of a sectional auger for ease of replacement. This included changing the single-piece 13.5 m (44 ft) auger with four-piece auger having 3.35 m (11 ft) sections with bolted and welded connections
  - Installation of 560 Cronatron "chalky" bars on the auger flights (periphery and flats) to reduce wear rates

- **Additional Probe Drill—Total of Five Installed**—Four probe drills were originally supplied and installed per the TBM design specifications—one front and two in the rear and another on the segment erector arm. Owing to the ground conditions encountered and the pre-excavation grouting program implemented in response to these conditions, one additional front probe drill was installed. This modification provided flexibility while reducing drilling times, particularly with emphasis on the 4° face holes as shown in Figures 21 and 22.

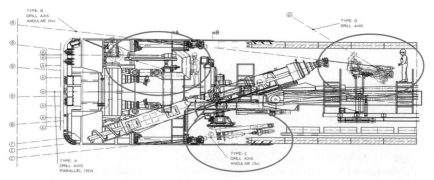

Figure 21. Section through the Herrenknecht TBM showing the location of the five drills—two in front and two in the rear plus an erector-mounted drill

Figure 22.   View of the forward TBM bulkhead and 4° drill ports

- **TBM Bathtub Modifications**—Added solids separation screens as part of the additional slurry handling and separation system. This was needed to limit the particle size handled by the primary Habermann slurry discharge pump. A later modification included air and water flow mixing jets to improve the suspension of solids in the bathtub before discharge into the slurry system running the length of the tunnel.

- **Slurry Handling and Separation System**—A unique slurry handling and sep-aration system was developed on-site and integrated into the existing TBM and trailing gears water supply and discharge system. This additional muck handling system proved to be a major asset to improved mining production in ground conditions that resulted in the production of excess fine materials that was diffi-cult to handle. This system will be described in detail below.

- **Earth-Pressure Balanced Operations**—The TBM was modified to address a "semi-EPBM" capability due to encountering periodic soft ground conditions. Pressure sensors and mixing arms were installed in the plenum and a foam mixing and injection system was installed to provide foam to the plenum and the screw auger.

Changing Worn Screw Augers and Revisions to the Original Design

The TBM screw augers are an essential component in the mucking system but due to the nature and abrasiveness of the fragmented rock muck, considerable wear was experienced. On average, an auger was replaced every 1,200 m (3,900 LF)—each requiring approximately 15 shifts to complete. Figure 23 illustrates a section through an auger.

The auger replacement work was a significant task and involved the removal of various internal decks, the guillotine gate and planetary drive motor assembly. Whereas the original auger design was a single piece—13.5 m (44 ft) long and a chal-lenge to install at 710 kg/m (480 lb/ft)—replacement augers were supplied in 3.35 m (11 ft) sections with a special alignment coupling that was ultimately bolted and welded together to complete the installation.

The auger design was the subject of progressive modifications to improve the design and material selection with the goal to increase its service life and minimize the costs and time needed to install a replacement. These design refinements included

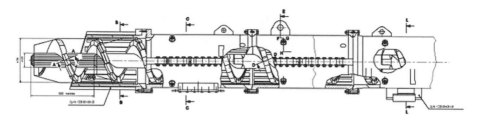

Figure 23. Revised screw auger design showing the bolt-on wear plates in the lowermost 2 m (7 ft) (left side of illustration) and the split barrel configuration with access inspection doors for maintenance and inspection

various replaceable bolt-on wear plates on the lowermost 2 m (7 ft) and tungsten carbide impregnated wear plates welded on the flights throughout the remaining length. The auger generally wore down fastest in the lowermost 2 m (7 ft) on both the flights' thickness and the periphery. Flights were originally made of cast steel and did not have any special wear resistant hard-facing applied except on the lowermost 2 m (7 ft). Lesser wear rates were experienced on the upper half of the auger—possibly due to less regrinding and particle impacts from the muck. Please note that the auger was used solely as a material conveyor and not intended to dissipate hydrostatic pressures as in the design of an EPBM TBM. There were few clay and silt-sized materials ever present in the auger to form a plug sufficient to withstand pressures from the plenum and to control inflows from the face.

Impacts from worn augers increased the muck-out times substantially since it had difficulty in clearing the plenum of muck and especially "fines" (sand-sized) particles. A secondary problem would occur in the plenum as the added cutterhead rotations would further regrind and reduce the average muck particle size—further challenging the worn auger to convey muck upwards towards the guillotine gate. It was also noticed that when the plenum was slow to be emptied, additional fines would build-up under the forward shield and contribute to steering difficulties.

SLURRY HANDLING AND SEPARATION SYSTEM

The Herrenknecht TBMs were not originally equipped with a slurry handling and separation system. The TBMs did, however, have a powerful dewatering system in accordance with the specifications and were designed to pump a high volume of water from the heading out of the tunnel. These TBMs were truly hybrid rock cutting machines designed to operate with a screw auger for muck handling but not with a pressurized plenum chamber. They were not designed to handle high volumes of fines (sand-sized particles) that in practice from time to time could be far in excess of 50% of the material excavated from the tunnel face. Additionally, and to compound the problem, excess water conditions periodically developed in the heading that overwhelmed the muck handling system and, therefore, impacted the mining production to the point where an alternate or supplemental means of handling a significant and problematic fraction of the muck was needed. Figure 24 illustrates the principle slurry handling and separation system components within the TBM that were added to cope with muck conditions.

In order to cope with the periodic excess fines and water inflows, a water slurry handling and separation system was designed by Shea-Kenny in conjunction with Herrenknecht engineers. The system had the characteristics listed in Table 8 in response the muck and water conditions encountered to date and included a feed water return circuit to

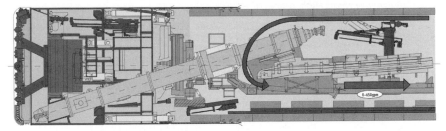

Figure 24.   Cross-section through the TBM showing the principle muck handling components and slurry separation scheme through the bathtub. Slurry system capacity is variable to 54 l/sec (850 gpm) depending on flow character.

Table 8.   Summary of slurry separation system design and operating characteristics—original design

Description	System or Capacity	Primary Equipment
Peak flow rate	54 l/sec (850 gpm)	Habermann slurry pump
Nominal flow rate	41 l/sec (650 gpm)	Habermann slurry pump
Solids separation	Screens, clarifier and centrifuge	Brandt Equipment located at the portal
Fines suspension	<6% nominal to a 15% maximum (by volume) and dependent on flow character	
Slurry medium	Water in a closed re-circulation circuit with additives injected at the plant	

provide additional water to the heading to flush fines that could not be otherwise handled by the existing mucking equipment—designed principally for rock chips and fragments.

This innovative system was integrated into the existing TBM and trailing operating systems and equipment wherever possible so as to become a seamless function and controlled with the overall mining operations. The primary features and components of the integrated slurry handling and separation system included the following.

- Installation of a modified "bathtub" below the screw auger to conveyor transfer point to capture water and fine material being washed-in and to channel flows to the Habermann slurry pump
- Installation of a Habermann slurry pump to convey water and suspended fine material to the slurry discharge system mounted in the tailing gear
- Use of the existing Warman centrifugal pump (formerly part of the tunnel water discharge system) to expel the slurry into the existing tunnel water discharge system and to the portal area
- Conversion of the existing tunnel water discharge system (piping, pumps, holding tanks, valves and controls) located in the trailing gear to handle both incoming feed water and discharge slurry flows
- Installation of a solids separation plant at the portal including;
  - Screening plant for particles > 9 mm nominal size
  - Vertical clarifier to remove sand-sized particles
  - Horizontal centrifuge to remove silt
  - Flocculant injection system to accelerate separation of sand and silt particles from the slurry

- Only Yellow Mine Pipe made of PVC was used to connect all facilities and to convey the water slurry
- No bentonite was used in the slurry system since water was used as the suspension media
  - Water velocities were maintained at generally above 2.1 m/sec (7 ft/sec)
  - Slurry flow rates varied widely and in accordance with the quantities of fines and water inflows in the face—generally from 13 to 32 l/sec (200 to 500 gpm)
  - Make-up water was automatically regulated at a holding tank mounted in the trailing gear

The water slurry and separation system generally worked very well whenever excess water and fines were encountered at the face. This system provided for a more continuous and self-regulated slurry and fines handling system that was integrated into the TBM mining control systems. It also provided an alternate materials handling system whenever the belt conveyor proved to be unsuitable to handle the fines fraction of the muck.

The slurry system flow was regulated to suit the fines and/or water inflows generally to a maximum of 15% fines (by volume). But as the density of the slurry flow increased, the resulting discharge flow decreased substantially due to a maximum horsepower available in the system—principally at the Habermann slurry pump mounted in the TBM. Several upgrades, modifications and improvements have been made to the slurry and water handling system to increase capacity and to handle flows on occurrence up to 69 l/sec (1,100 gpm) depending on slurry density.

- Installation of a larger hydraulic power unit for the Habermann pump to increase the peak output flow
- Adjustments to the flow control logic system on both the feed water and discharge sides of the circuit
- Additional discharge line was added to relieve reliance on the single 150 mm (6") line in the case of excess flows and potential blockages the East Tunnel—being excavated on a downhill grade
- Water jets mounted in the bathtub to improve flows and suspension of fine material drawn into the system

Surprisingly, the Yellow Mine Pipe has shown very little wear even after 4,900 m (16,000 LF) of tunnel has been mined. No fluidizers or flow enhancing agents have been introduced into the slurry flow to reduce friction and wear.

CONCLUSIONS AND RECOMMENDATIONS

The Arrowhead Tunnels Project is probably one of the most challenging tunnel projects presently underway in the United States and it has taken considerably longer to excavate than planned. The ground and ground water conditions are extremely variable and could not be well predicted largely due to the extent of the ground cover as well as the degree, frequency and severity of the alterations in the rock mass including the influence from nearby faults. The pre-excavation grouting program was intended to address and arrest ground water inflows into the tunnel as a measure to comply with the US Forest Service Special Use Permit. Ultimately, this essential drilling and grouting program was adjusted to be more flexible and used to accomplish the following goals.

- Install probe holes to measure water inflows and to assess the rock mass relative hardness and stability
- Inject cement grouts and/or colloidal silica grout to curtail water inflows and to improve the quality of the rock mass for mining. Alternate grout materials have been tested and used periodically
- Improve the effectiveness of Stage I and II backfill grouts at enveloping the segmental tunnel liner without the adverse influence of excess water inflows surrounding the tunnel and the TBM
- Help minimize the stress relaxation of the rock mass as a result of the modest over-excavated volume needed to advance the TBMs in all ground conditions

The TBMs as originally designed, modified and adjusted throughout the tunnel drives should be considered as a significant achievement to improving tunneling technology and mechanical excavation capability in very adverse ground conditions and an example of the combined use of new and innovative systems.

The stout design and high-quality manufacture of the precast segmental tunnel liner was also a hidden benefit and especially when it was tested in-situ with as much as 114 MN (25.6 million pounds) thrust or four times the original TBM thrust load in order to move the TBMs forward in very difficult squeezing and packing ground conditions.

An effective pre-excavation drilling and grouting program was absolutely essential to successfully mine the tunnel in the presence of high ground water pressures and flows. Optimization of the program was a continuous task and just as the ground conditions were variable, the pre-excavation drilling and grouting program was equally flexible and adjusted to suit. A prescriptive and inflexible program would have proven to be unworkable and not been responsive to the ground conditions and behavior encountered. The Owner, Designer and Contractor constantly worked together to discuss and evaluate various means to drill and grout to control water (first) then to improve the ground (second) so as to be able to mine the tunnel faster and reduce risks. Ground water pressures and flows and their effects on the ground behavior became the "enemy" and were eventually treated in a manner that not only provided control of inflows but also improved the safety and production of the overall mining operation.

ACKNOWLEDGMENTS

This paper includes key contributions from a number of persons in addition to the co-authors. We would, therefore, like to acknowledge Stuart Warren of Hatch Mott MacDonald, the Construction Manager, John Wagonner of GeoPentech Geotechnical and Geoscience Consultants, the geotechnical consultant and Tom Hibner of the Metropolitan Water District of Southern California, the project Owner. In addition, the authors would like to acknowledge Ed Marcus, Robert Gordon and Stuart Lipofsky of Shea-Kenny, the Joint Venture Contractor for their valuable and timely assistance in the preparation of this paper.

There have been many recent papers written on the design and development of the Arrowhead Tunnels Project as part of the Inland Feeder Program for the Metropolitan Water District of Southern California. While they pre-date the events of the site work described in this paper, they should be considered as useful pre-construction references.

REFERENCES

Hibner, T., Shamma, J., Tempelis, D., 2003, Managing the Groundwater Impacts of Tunnel Construction, *Proceedings of the Rapid Excavation and Tunneling*

Conference 2003, Robinson and Marquardt (eds.), p. 592–599, 2003. Littleton: Society for Mining, Metallurgy, and Exploration, Inc.

King, M., Ghazi, M., Hebert, C., 2003, The Design and Manufacture of the Arrowhead Tunnels Project Precast Segmental Lining, *Proceedings of the Rapid Excavation and Tunneling Conference 2003, Robinson and Marquardt (eds.), p. 688–696, 2003.* Littleton: Society for Mining, Metallurgy, and Exploration, Inc.

Lum, H., Crouthamel, D., Hopkins, D., Cording, E., Shamma, J., 2002, Behavior of a Single Gasket System for a Precast Concrete Segmental Liner Subject to High External Hydrostatic. *Proceedings of the North American Tunneling 2002, Ozdemir, L. (ed.), p. 141–150, 2002.* Lisse: Swets & Zeitlinger.

Shamma, J., Tempelis, D., Duke, S., Fordham, E., Freeman, T., 2003, Arrowhead Tunnels: Assessing Groundwater Control Measures in a Fractured Hard Rock Medium, *Proceedings of the Rapid Excavation and Tunneling Conference 2003, Robinson and Marquardt (eds.), p. 296–305, 2003.* Littleton: Society for Mining, Metallurgy, and Exploration, Inc.

Swartz, S., Lum, H., McRae, M., Curtis, D., Shamma, J., 2002, Structural Design and Testing of a Bolted and Gasketed Precast Concrete Segmental Lining for High External Hydrostatic Pressure. *Proceedings of the North American Tunneling 2002, Ozdemir, L. (ed.), p. 151–160, 2002.* Lisse: Swets & Zeitlinger.

Swartz, S., Tzobery, S., Hemphill, G., Shamma, J., 2005, Trapezoidal Tapered Ring— Key Position Selection in Curved Tunnels, *Proceedings of the Rapid Excavation and Tunneling Conference 2005, Hutton and Rogstad (eds.), p. 957–969, 2005.* Littleton: Society for Mining, Metallurgy, and Exploration, Inc..

SELECTED ADDITIONAL READINGS

Crouthamel, D., Farmer, D., Klein, S., Phillips, S., 2005, Pre-Excavation Grouting Guidelines in Hard Rock Excavations, *Proceedings of the Rapid Excavation and Tunneling Conference 2005, Hutton and Rogstad (eds.), p. 408–417, 2005.* Littleton: Society for Mining, Metallurgy, and Exploration, Inc.

Henn, R., Davenport, J., Tzobery, S., Bandimere, S., 2005, Additional Test Results for Comparison of Penetration of Grout Made with Various Ultrafine Cement Products, *Proceedings of the Rapid Excavation and Tunneling Conference 2005, Hutton and Rogstad (eds.), p. 1,039–1,050, 2005.* Littleton: Society for Mining, Metallurgy, and Exploration, Inc.

PRACTICAL CONSEQUENCES OF THE TIME-DEPENDENCY OF GROUND BEHAVIOR FOR TUNNELING

George Anagnostou

ETH Zurich

ABSTRACT

Time-dependent ground behavior may have important implications for the construction process or the life of a tunnel. The paper gives a concise overview of the influence of time on the deformations and the stability of underground openings, including topics such as face stability in soft ground or the time-development of pressure or deformations in squeezing or swelling rock. While keeping in mind practical questions of tunnel engineering, emphasis is placed on understanding the observed time-dependency, on the underlying mechanisms (such as porewater pressure dissipation or chemical processes) as well as on the respective inherent prediction-uncertainties and design-uncertainties.

INTRODUCTION

Ground often responds to tunneling operations with a considerable delay. The time-dependency of ground behavior can be traced back to three mechanisms: consolidation, creep and, in some rocks, chemical processes as well. It manifests itself in a variety of ways depending upon both the type of ground and the construction method, and may have important implications for the construction process or the life of a tunnel.

Starting with a description of the underlying mechanisms, a series of tunneling problems will be discussed in the following pages from the standpoint of time-dependency. The topics dealt with cover a wide range of tunneling conditions and practical questions, such as the stability of shallow soft ground tunnels, the importance of rapid excavation, the conceptual design of support and the risk of shield jamming in squeezing conditions, or the long-term behavior of swelling rock. The qualitative considerations will be illustrated by computational examples and substantiated by references to tunneling practice.

MECHANISMS

The time-dependency of low-permeability soft ground is due mainly to transient seepage flow processes that are triggered by the tunnel excavation and develop slowly over the course of time: The long-term deformations of the ground include, in general, changes to its pore volume and water content. The latter needs more or less time depending on the seepage flow velocity and thus on the permeability of the ground. In a low-permeability ground, the water content remains constant in the short term. Tunnel excavation generates excess pore pressures, however. As these are higher in the vicinity of the tunnel than further away, seepage flow starts to develop. So the excess pore pressures dissipate over the course of time, thereby altering the effective stresses and leading to additional time-dependent deformations ("consolidation"). The permeability of the ground is decisive for the rate of the pore pressure dissipation and thus for the time-development of the ground deformations or, if the latter are constrained by a lining, of the ground pressure. As in other geotechnical problems involving "unloading" of the ground (e.g., deep excavations), the short-term behavior (so-called "undrained conditions") is more favourable than the long-term one (so-called "drained conditions"). This is because excavation leads mostly to a volume increase of the ground. The short-term excess pressures are, consequently, negative (i.e., suctions) and increase the effective stresses temporarily. The resistance of the ground to shearing is, therefore, higher in the short term than in the long term.

Laboratory tests, field observations and theoretical considerations show that consolidation processes are also highly important for weak rocks prone to squeezing or swelling (Anagnostou 1993, Vogelhuber et al. 2004).

Another mechanism of time-dependency is associated with the rheological properties of the ground. So-called "creep" becomes evident if the ground is overstressed—particularly as the state of failure is approached. It is, therefore, of paramount importance for weak rock under high stress (squeezing conditions).

Many rocks are susceptible to physico-chemical actions. Rapid disintegration of serpentine rock, for example, has in some cases caused serious deformations shortly after excavation. Physico-chemical processes mostly evolve slowly, however, and affect tunnel construction rather rarely. Geochemical processes such as rock dissolution in karstic formations (leaching) or crystal-growth in anhydritic rocks can in some cases be accelerated by tunnel construction and have a big impact on the stability and serviceability of a tunnel during its life.

There are particularly large uncertainties concerning the long-term behavior of argillaceous rocks containing anhydrite. The strong swelling of these rocks leads to major heaves in the tunnel floor or, in the presence of an invert arch, to the development of considerable pressures, which may damage the lining or even heave the entire tunnel tube and the ground up to the surface. Swelling may even occur during the construction process, but the most serious problems are caused by swelling phenomena developing slowly over several decades, as they necessitate costly repair works and lengthy service interruptions. In spite of experience going back several decades with tunneling in anhydritic rocks, there are still large design uncertainties. An empirical design, based upon experience from projects carried out under similar conditions, is hardly possible at all as the intensity of the swelling often varies considerably over small distances within one and the same tunnel, even in the case of a macroscopicaly homogeneous rock mass with constant mineralogical composition. Additionally, the swelling rate usually fluctuates over the course of time, thereby making long-term extrapolations very uncertain. Field observations show that the rate of tunnel floor heave can be reduced by applying a counterpressure to the tunnel floor, but the relationship between long-term heave and support pressure, which is important for the conceptual design of tunnels in swelling rock (Kovári et al. 1988), is unknown. The extremely long duration of the swelling process, taking several years even under laboratory conditions at the scale of rock specimens, makes systematic field investigations and laboratory investigations very difficult. We are still far from understanding, let alone from quantifying, the mechanisms governing swelling, as a series of physico-chemical and seepage flow processes come into play (water adsorption by the clay minerals, chemical transport, dissolution of anhydrite and growth of gypsum crystals).

EFFECT OF THE ADVANCE RATE

In geological conditions with pronounced time-dependent ground behavior during construction (e.g., shallow tunnels through clay deposits or deep tunnels through weak rock), the advance rate greatly influences the development of ground pressure and deformation as the deformations due to creep or consolidation are superimposed in the region around the working face upon those occuring due to the three-dimensional redistribution of stress caused by the excavation.

Regarding consolidation processes, the higher the advance rate and the lower the permeability, the less will the pore pressures dissipate in the vicinity of the face and, consequently, the higher the will be shear resistance and the smaller the deformations. If, on the other hand, the permeability is high and the advance rate low, drained conditions, which are less favourable, will also prevail in the vicinity of the working face. Depending on the relation of advance rate to permeability, the ground response will be "undrained," "drained" or somewhere in-between. The ratio of advance rate to permeability governs the stability, the deformations and the ground pressure acting upon a lining or shield in the vicinity of the face.

Similar considerations apply for creep (Ghaboussi and Gioda 1977): the lower the advance rate, the bigger the deformations will be. In the borderline case of a very high advance rate, only small, elastic deformations develop in the vicinity of the face. Consequently, a high advance rate reduces deformations close to the face, whether the reason for the time-dependency is creep or consolidation.

STAND-UP TIME

The most frequently experienced time-effect in tunneling is related to the amount of time an opening can remain stable by itself, thereby providing a period for support installation. Among other criteria, such as the limitation of surface settlement in urban tunnels, the stand-up time is decisive for the type and location of support installation because the different support types need more or less time to be installed and to develop their bearing capacity (Figure 1). As support installation close to the working

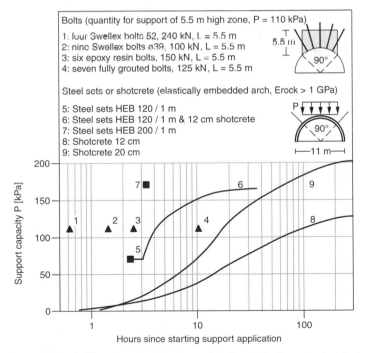

Figure 1. The capacity of different support types as a function of the time required for installation and strength development

face interferes with excavation, thereby reducing production rate, stand-up time may have a considerable impact on construction schedules, cost and equipment. This is particularly true for rock tunneling by open-type TBMs as low stand-up times may reduce utilization dramaticaly, depending on machine design and equipment (Beckmann & Krause 1982).

The stand-up time for rocks can vary from minutes to centuries depending on a number of factors such as joint conditions and the presence of water, the round length and the size of the opening (Schneider 2002). Possible manifestations of time-dependency in tunneling through hard rock, such as delayed detachment of blocks, are mainly due to clayey discontinuity infillings. The factors influencing stand-up time are understood in principle. Although stand-up time cannot be predicted by calculations or specific laboratory or field tests, the tunnel engineer's general professional experience and records from projects with similar conditions suffice in most cases for reliable predictions.

The time available for support installation in soft ground depends essentially on the clay content. Coarse-grained soils have zero stand-up time, thereby necessitating ground improvement by grouting or freezing or installation of support measures in advance such as forepoling. The behavior of silty ground is particularly tricky. When saturated, it resembles a slurry which flows immediately after being exposed by the excavation. When wet, it shows some cohesion due to capillary stress, but it dries with time to a cohesionless powder. In complex hydrogeological conditions, characterized by alternating soil layers of higher or lower permeability, all these forms may occur simultaneously at different locations of the exposed ground.

Related to the question of stand-up time is the question of whether the ground at the face can remain stable up to the next excavation round. The problem of face failure is important particularly for tunnels crossing soft ground and will be discussed in the next section. Face instabilities in hard rock tunneling are rather rare but may occur in the form of block detachments, depending on the joint conditions. Careful scaling of the face reduces the risks associated with unstable blocks that do not fail immediately with the blasting. Blocky rock having a low stand-up time can cause considerable problems particularly for mechanized tunneling (Kovári et al. 1993). Excessive deformations of the face, as observed sometimes in tunneling through squeezing rock (Lunardi 2000), may also lead to failure of the face.

SHALLOW SOFT-GROUND TUNNELS

On the Time-Dependency of Face Stability

The most serious risks in tunneling through soft ground are associated with a collapse of the tunnel face. In shallow tunnels the instability may propagate towards the surface, thereby creating a chimney and a crater on the ground surface. For the reasons mentioned above, the stability of the face depends greatly on time. A face that is stable during continuous excavation may collapse over the course of an excavation standstill after a period of time. Figure 2 shows, by means of a computational example for a shallow tunnel crossing silty clay, the safety factor of the face as a function of the applied support pressure. In the short term, the face is stable even without support. In the long term, a minimum support pressure of 110 kPa is needed for stability because the ground offers a lower resistance to shearing and it is additionally loaded by the seepage forces S generated by the flowing pore water. The stability-loss occurs more or less rapidly depending on the permeability of the ground. In a high-permeability ground, the unfavourable drained conditions would prevail right from the start (there is no distinct "short-term" behavior) and the face would fail immediately after excavation.

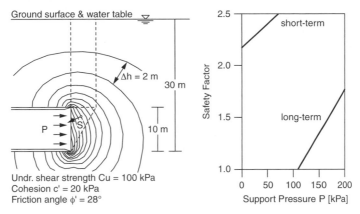

Figure 2.   The long-term hydraulic head around the tunnel and the safety factor of the tunnel face as a function of the support pressure

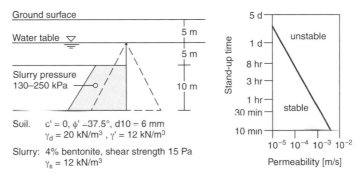

Figure 3.   The stand-up time of the tunnel face in slurry shield tunneling through extremely coarse-grained and poorly-graded ground as a function of permeability (after Anagnostou and Kovári 1997)

An Aspect of Slurry Shield Tunneling

Slurry shields provide support to the face by a pressurized bentonite suspension. The stabilizing effect of the suspension decreases with its infiltration depth into the ground. In order that the face remains stable, the infiltration depth should not exceed a critical value. Under normal operational conditions, a fine-grained and practically impervious layer (so-called "filtercake") forms on the tunnel face and inhibits infiltration. In extremely coarse-grained, poorly graded and highly permeable soils, however, face instabilities may occur (Babendererde 1991). Aggregates added to the slurry assist filtercake formation in coarse-grained soils but may present environmental or process engineering problems. Improved additives have been developed recently by Heinz (2006).

In the absence of a filtercake, slurry will displace the groundwater and, depending on its rheological properties, may reach critical depth after a period of time that is shorter or longer depending on the permeability of the ground. Consequently, for a given slurry composition, the permeability will govern the stand-up time of the face (Figure 3). The conditions during continuous excavation are more favourable since the removal of bentonite-saturated soil at the face partially compensates for the infiltration.

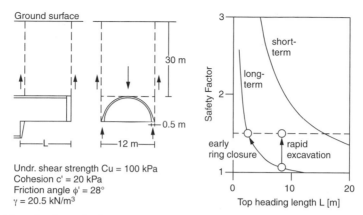

Figure 4.   The safety factor for the section of tunnel between the tunnel face and the bench excavation

The lower the permeability of the ground and the higher the advance rate, the smaller the infiltration depth will be. A rapid and continuous TBM operation is favourable also with respect to face stability (Anagnostou & Kovári 1997).

On the Timing of Ring Closure By an Invert Shell

Apart from face collapse or a failure of the unsupported crown, a series of other hazards need to be considered depending on the project conditions and on the excavation and support method. For example, when tunneling by the "top-heading and bench" method, a collapse up to the surface may also occur as a consequence of an insufficient bearing capacity of the ground beneath the footings of the temporary arch. Figure 4 shows the safety factor of the system as a function of the distance L between top heading and bench excavation. The smaller this length, the more pronounced the arching in the longitudinal direction and the higher the safety will be. Figure 4 illustrates a well-known fact from tunneling practice: in weak ground, rapid advance and closure of the ring near to the face improve stability conditions considerably.

According to the diagram, ring closure should be accomplished within a distance of one to several diameters from the face depending on how rapidly the ground responds to excavation. This difference is considerable from a constructional point of view. In complex soil deposits, the intensity of ground response and, according to Figure 4, the required timing of ring closure may vary within short distances. Full face excavation (in combination with systematic face stabilization) eliminates the respective construction schedule uncertainties. So, depending on the construction method, the ground characteristics may have a larger or a smaller impact on the tunneling works.

On the Time-Dependency of Surface Settlements

Field measurements show clearly that the settlements induced by tunneling through low-permeability clay deposits may increase for several months after excavation (O'Reilly et al. 1991). Groundwater recharge from the surface usually suffices in low-permeability ground for maintaining the watertable even if the tunnel lining is unsealed. Due to the drainage action of the tunnel, however, the pore pressures will decrease over the course of time. Since this happens within a zone which is considerably more extended than the influence-zone of the actual excavation, the settlement trough widens over the course of time. For collecting and preserving evidence in the case of urban tunneling through formations with pronounced time-dependent behavior, monitoring programmes should exceed the actual construction period and also embrace a wide zone alongside the alignment.

As previously mentioned, the higher the excavation rate, the lower the deformations of a low-permeability ground will be in the vicinity of the face. The importance of rapid excavation becomes evident when considering the fact that in shield tunneling the deformations of the ground ahead of the face and around the shield (gap closure) represent by far the main sources of volume loss and surface settlement in shield tunneling.

SQUEEZING ROCK

Squeezing, the phenomenon of large time-dependent deformations triggered by tunnel excavation, occurs mostly in weak rocks with a high deformability and a low strength, often in combination with a high overburden (Kovári 1998). It is also known that a high pore water pressure promotes the development of squeezing. Squeezing normally develops slowly, although cases have also been known where rapid deformations occur very close to the working face (see, e.g., Marty 1989, Grandori & Antonini 1994, Höfle 2001). As mentioned before, creep-processes and consolidation-processes are responsible for time-dependent increases in rock deformation or pressure, whereas it is hardly possible to distinguish these two mechanisms from another phenomenologicaly.

Here, the generation and subsequent dissipation of excess pore pressures will be considered as the dominant mechanism. In this case, the permeability of the ground governs the rate of pore pressure dissipation and, therefore, the intensity of squeezing. More precisely, where permeability is higher by a factor of ten, rock pressure and rock deformation will develop ten times faster. Permeability is a major source of prediction uncertainty: squeezing rocks, such as shales, mudstones or altered metamorphic rocks, exhibit very low permeabilities. The water inflows in tunnel portions with squeezing conditions are mostly negligible. The permeability coefficient k of a practically impermeable rock (i.e., in the range of $k < 10^{-9}$ m/sec) cannot be predicted reliably even in the specimen scale. Additional difficulties exist in heterogeneous rock masses with alternating layers of weak and hard rock, as the latter are often fractured and therefore increase the overall permeability. The practical consequences of the inherent prediction uncertainties associated with the permeability of the ground will be discussed in the next sections.

Conventional Tunneling

There are basically two design concepts for dealing with squeezing rock: the "resistance principle" and the "yielding principle" (Kovári 1998). In the former a practically rigid lining with a high load bearing capacity is adopted, which is dimensioned for the expected rock pressure. In the case of high rock pressures this solution is not feasible. By contrast, in the latter, rock pressure is reduced to a value that is manageable during construction by allowing a large convergence of the temporary lining. An adequate overprofile and suitable structural detailing of the lining permit the occurrence of rock deformations without damage to the lining, thereby maintaining the desired clearance from the minimum line of excavation.

The computational results of Figure 5 refer to two questions of practical interest: the time-development for rock pressure in the case of the "resistance principle" and time-development for rock deformation in the case of the "yielding principle." The curve P in Figure 5 on the left hand axis shows rock pressure developing within 2 weeks (i.e.,

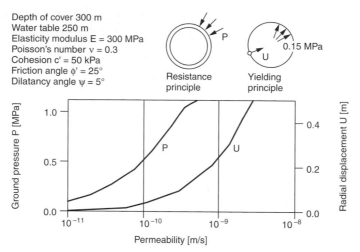

Figure 5. The squeezing pressure developing on a stiff lining within two weeks (curve P) and the respective deformation of a yielding support (curve U)

2–3 tunnel diameters behind the working face) as a function of permeability. For a permeability k in the range 10^{-11} to 10^{-9} m/s, the rock pressure amounts to 0.1–1.2 MPa. A light support suffices for a pressure of 0.1 MPa, while a heavy lining is required in order to sustain a load of 1.2 MPa—a huge difference from the constructional point of view. As a consequence of an unpredictable permeability variation, which can occur within a macroscopicaly homogeneous rock mass, the tunneling engineer would in one case encounter ground that is "competent," and in the other case ground that is "disturbed." The curve U in Figure 5 shows (on the right hand axis) the respective results for a flexible support. For the permeability range mentioned above, the radial displacement amounts to 0–0.25 m. The practical consequences of a poor estimate of convergence are modest compared to the consequences of a wrong estimate of the time development of the pressure acting upon a rigid support (damage to the lining and the need for re-profiling). So, depending on the construction method, the geological uncertainty may have a greater or lesser impact.

Mechanized Tunneling

Squeezing ground may slow down or even obstruct TBM advance. As tunnel alignments cannot always avoid difficult geological zones with sufficient reliability, planning a TBM drive in potentially squeezing ground involves the assessment of a series of hazards concerning the machine (blockage of the cutter head, jamming of the shield) or in the back-up area (inadmissible convergences of the bored profile, damage to the support). This analysis has to take into consideration the characteristics of the different TBM types (thrusting system, type of support, length of shield etc.), as TBM performance is the result of a complex interaction between machine, tunnel support and ground (Ramoni & Anagnostou 2006 & 2007).

When tunneling with a shielded TBM, some limited convergence can occur, due to the gap between the shield and the surrounding ground. If the convergences develop quickly, the ground closes the gap near to the tunnel face, thereby starting to develop pressure upon the shield. If the permeability is low and the advance rate high, the gap remains open for a longer period and, consequently, the pressure acting upon the

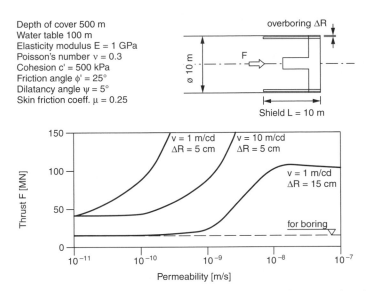

Figure 6.   The influence of ground permeability, of advance rate v and of amount of overboring ΔR on the thrust required in shield tunneling through squeezing ground

shield is lower or even zero. The thrust force needed to overcome shield friction during continuous excavation in given ground conditions depends on the length of the shield, on the amount of overboring, on the skin friction coefficient and on the advance rate. A reduction in the advance rate leads as a rule to an increase in the ground pressure, and thus to further deceleration or even standstill of the machine. The available thrust force depends on the installed thrust and on the capacity of the thrusting system. Figure 6 shows, by means of an example, the effect of permeability k on the thrust needed to overcome friction. The curves apply to advance rates v of 1 or 10 m/cd and an overboring ΔR of 5 or 15 cm. Accordingly, the thrust requirements can be reduced considerably by a major overboring—this comes, however, at the cost of possible steering difficulties and reduced production rates. Accordingly, if the permeability is higher than about 10^{-7}–10^{-8} m/s, the thrust requirements reach, in the present example, the limits of feasibility. In the case of a practically impermeable ground ($k < 10^{-10}$–10^{-11} m/s) the thrust needed to overcome friction is considerably lower even at moderate advance rates ($v = 1$–10 m/cd). In the permeability range of $k = 10^{-10}$–10^{-8} m/s, the results are highly sensitive to small variations in permeability, thus indicating a major source of prediction uncertainty as mentioned above.

The extrusion of the ground core ahead of the face represents another potential problem, as excessive deformations may lead to cave-in and blockage of the cutter head. Figure 7 illustrates the positive effect of a rapid excavation as well as the influence of rock permeability on the axial displacement of the face.

CLOSING REMARKS

Due to consolidation, creep or chemical processes, the behavior of certain soil or rock types is time-dependent. The time-dependency manifests itself in a variety of ways depending both upon the type of ground and the construction method, and may

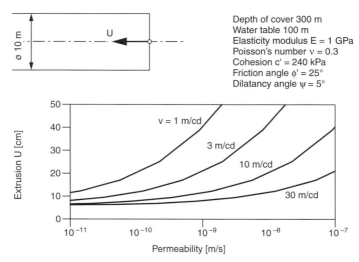

Figure 7. The influence of ground permeability and of advance rate *v* on the axial deformation of the tunnel face in squeezing ground

have important implications for the construction process or the life of a tunnel. Tunneling through anhydritic rocks is particularly demanding.

Theoretical considerations, underpinned by computational results, improve our understanding of the time-dependency of ground behavior observed in field, and explain the empirically established, positive effect of rapid excavation in the case of tunneling under adverse conditions, while also indicating the sources of inherent design uncertainties. Although the basic mechanisms are understood, reliable estimates of the time development of ground deformation or pressure are often impossible. Nevertheless, depending on the construction method, the impact of geological uncertainty on work schedules, construction time and cost can be lesser or greater for one and the same geotechnical situation. This underlines the importance of an adequate conceptual design and adequate contractual regulations.

REFERENCES

Anagnostou, G. 1993. A model for swelling rock in tunneling, *Rock Mechanics and Rock Engineering*, 26, 307–331.

Anagnostou, G., and K. Kovári 1997. Face stabilization in closed shield tunneling. *Rapid Excavation and Tunnel Construction*, Las Vegas, 549–558.

Babendererde, S. 1991. Kritische Betrachtungen zum Einsatz von Hydroschilden. In *Berichte des Int. Symp. Sicherheit und Risiken bei Untertagebauwerken* (eds R. Fechtig & K. Kovári), 47–51 (in German).

Beckmann, U., and Th. Krause 1982. Erfahrungen mit Tunnelbohrmaschinen in Stoerzonen—Einfluss auf Bohrgeschwindigkeit und Ausnutzungsgrad. In *ISRM Int. Symp. on Rock Mechanics related to Caverns and Pressure Shafts*, Aachen, 761–770.

Ghaboussi, J., and G. Gioda 1977. On the time-dependent effects in advancing tunnels. *Int. J. Num. & Anal. Meth. Geomech.,* Vol. 1, 249–269.

Grandori, R., and F. Antonini 1994. Double shield TBM excavation technique—Recent experiences and future development. *Felsbau* 12 (1994) Nr. 6: 490–494.

Heinz, A. 2006. Modifizierte Bentonitsuspensionen fuer geotechnische Bauverfahren in Boeden hoher Durchlaessigkeit. PhD Dissertation 16698, ETH Zurich (in German).

Höfle, H. 2001. Wasserstollen bei Bormio—Einsatzerfahrungen mit einer Teleskopschildmaschine. In 6. *Internationales Symposium Tunnelbau bauma* (München), 169–174 (in German).

Kovári, K., Amstad, Ch., and G. Anagnostou 1988: Design/Construction methods— Tunneling in swelling rocks. *Proc. of the 29th U.S. Symp. "Key Questions in Rock Mechanics"* (Eds. Cundall et al.), Minnesota, 17–32.

Kovári, K. 1998. Tunneling in squeezing ground, *Tunnel* 5/98.

Kovári, K., Fechtig, R. and Ch. Amstad 1993. Experience with large Diameter Tunnel Boring Machines in Switzerland. In *Burger (Ed.), Options for Tunneling, Dev. in Geot. Engng.*, 74, 485–496.

Lunardi, P. 2000. Design and construction of tunnels using the approach based on the analysis of controlled deformation in rocks and soils. *Tunnels and Tunneling Int. Special Suppl.*, pp. 3–30.

Marty, T. 1989. Tunnelboring in difficult rock conditions with a fullface machine as part of the Ilanz Power Station Project. *World Tunneling and Subsurface Excavation 2 Special edition*—April 1989: 14–20. London: The Mining Journal Ltd.

O'Reilly, M.P., Mair, R.J. and G.H. Alderman 1991. Long-term settlements over Tunnels. An eleven-year study at Grimsby. *Tunneling '91, Inst. of Mining and Metallurgy*, London, 55–64.

Ramoni, M. and G. Anagnostou 2006. On the feasibility of TBM drives in squeezing ground. *Tunneling and Underground Space Technology*, Volume 21, 3–4, 262. Proc. of the ITA-AITES 2006 World Tunnel Congress "Safety in the Underground Space." Ramoni, M. and G. Anagnostou 2007. The effect of advance rate on shicld loading in squeezing ground. *ITA-AITES World Tunnel Congress "Underground Space—the 4th Dimension of Metropolises,"* Prague (in Press).

Schneider, A. 2002. Sicherheit gegen Niederbruch im Untertagbau. Veroeff. Inst. f. Geotechnik, ETH Zurich, Vol. 219, Vdf Hochschulverlag AG, Zurich (in German).

Vogelhuber, M., Anagnostou, G. and K. Kovári 2004. Pore Water Pressure and Seepage Flow Effects in Squeezing Ground. *Proc. X MIR Conference "Caratterizzazione degli ammassi rocciosi nella progettazione geotecnica,"* Torino.

TUNNELING THROUGH FAULTS IN VOLCANIC ROCKS FOR THE CONSTRUCTION OF THE LA JOYA PROJECT (COSTA RICA)

R. Grandori

SELI SpA

L. Pellegrini

SELI SpA

ABSTRACT

The construction of the La Joya hydropower plant (Costa Rica) was awarded by the Spanish power company, Union Fenosa, to the Italian Joint Venture formed by the companies SELI SpA and Ghella SpA.

The project, located in a mountainous region of central Costa Rica, includes a 7.9-km-long head race tunnel having an excavation diameter of 6.18 m.

A double-shielded TBM was selected to perform the tunnel excavation, which involved mainly volcanic rocks in variable but generally very weak conditions.

The tunnel crossed long and continuous sections in completely weathered rock masses with no residual cohesion and with high water inflows.

The paper describes and analyzes:

- The special methods adopted to overcome these sections with the TBM including face stabilization by chemical grout and foams as well small by-pass tunnels in the crown combined with TBM advance/support

- The grouting design and the grouting method that was implemented to meet the waterproof design specifications and restore the original groundwater level in the mountain

- The additional geological investigations performed to improve the knowledge of the rock mass in front of the TBM and the outcomes of these investigations

- The performance of the TBM in different ground conditions and on average along the entire tunnel

Finally, the author's comment about the possibility of implementing and improving the technologies experimented with in this project in future similar applications.

INTRODUCTION

The Spanish Electric Company Union Fenosa (Client) has awarded the Contract for the construction of the La Joya hydropower plant to the Italian Joint Venture Ghella—SELI (Contractor)

The power plant is located on the Reventazon River in the Central Valley region of Costa Rica, and has a production capacity of 50 MW.

The construction contract included:

- One diversion structure,
- One water transfer tunnel,
- One surge reservoir,

- One pressure pipeline,
- One external power plant,

The 8-km-long water transfer tunnel has an excavation diameter of 6.18 m and is lined by 4–200 mm thick precast segments of the hexagonal honeycomb type.

The internal lining diameter is 5.6 m and the slope is 0.07%. The tunnel has been designed to operate under low pressure (max 1.4 bars) to transfer 52 m^3/sec of water from the outlet of the existing Chaci' power plant to the surge reservoir of the new La Joya power plant.

While Ghella Company took the responsibility of implementing the civil works, SELI Company supplied and operated the complete set of equipment for finishing the tunnel.

The equipment included a double-shield TBM, the TBM back-up, the rolling stock, and the other tunneling equipment.

GEOLOGY

The tunnel (see Figure 1) mainly crossed volcanic materials of the tertiary age with lavas, breccia and other volcanic deposits such as lahars.

The lavas showed a great variability in texture, crystallization and mechanical properties. The grade of fracturing was also very variable with the most fractured zones near the faults.

As a consequence of the rock mass properties, the permeability and behavior during excavation changed frequently along the tunnel drive, and long stretches of "bad ground" have been encountered.

The lahars were generally well-cemented and stable.

The tunnel cover varied from 50 m to 600 m and the water table level varied between 60 m above the tunnel crown up to a maximum of 120 m. Average water table level has been registered as approximately 80 m above the tunnel crown.

The actual geology encountered by the TBM during tunnel excavation was much different than the one foreseen by the tender documentation.

Table 1 shows the difference between encountered rock mass classes versus the predicted rock mass classes, according to Bieniawski classification.

The results show that ground conditions along the tunnel were much worse than predicted and very poor in general, with over 2,000 m in class V and 545 m in class Over V.

We classified as Over V a ground mass with very low cohesion, where parameters normally used to classify the rock could not even be detected and the stability time at the tunnel face was less than few minutes.

On the basis of the predicted geology, the Contractor chose the double-shield TBM Robbins 194-272 for excavation of the tunnel. This TBM was considered to be the proper one for the described situation. Despite expectations, the extremely adverse geology encountered as well as the high water inflows in specific locations required the Contractor to develop special procedures to overcome the poor ground sections and continue operating the double-shield TBM.

TBM PRODUCTION

The TBM specifications are shown in Table 2. The TBM performed well in classes III, IV, and V, achieving the advances shown in Table 3.

Figure 2 shows the TBM production time diagram.

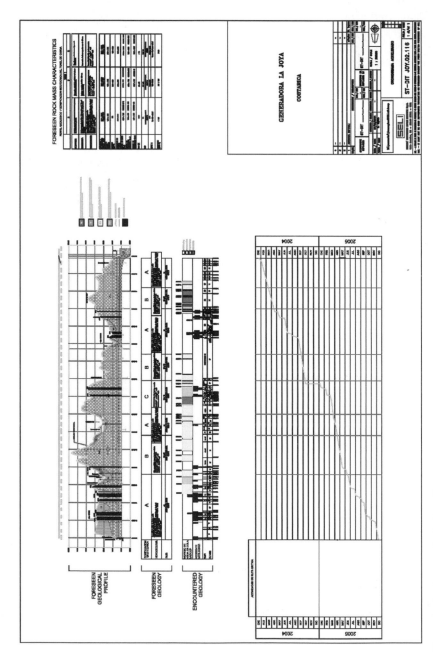

Figure 1. La Joya geological profile

Table 1. La Joya rock classes (Bieniawski classification)

RMR Rock Classes	Encountered Geology (meters)	Predicted Geology (meters)
II	0	950
III	1,058	2,450
IV	4,129	4,503
V	2,171	0
Over V	545	0
Total	7,903	7,903

Table 2. TBM specifications

Excavation diameter	6.18 m
Cutters	n.43×17" × 250 KN/cutter
Maximum operating cutterhead thrust	10,750 KN
Cutterhead drive	8×180 KW = 1440 kW
Cutterhead speed	0–7.2 RPM (variable frequency drive)
Max torque	3,440 KN (4,470 KN breakout)
Max main thrust	30.000 KN
Max aux. thrust	30.000 KN

Table 3. TBM production

Tunnel length in classes III–IV–V	7,593 m
Total working days	279 days
Daily average advance	27.2 m/day
Best daily advance	49.6 m
Best weekly advance	300 m
Best monthly production	892 m

Figures 3, 4 and 5 show TBM daily, weekly, and monthly advances along the tunnel, including the 545 m in Over V rock class sections.

From these graphs, it is seen that in rock class Over V, the TBM production drastically dropped from an average of 27.2 m/day to an average of 1.91 m/day. In these particular sections it was necessary to follow special excavation procedures that are described in detail in the following section.

By using these special excavation procedures, SELI was able to overcome all the fault zones, keeping the overall TBM advance rate above the average of 14 m/day. This was regarded as a very good achievement considering the actual geological conditions encountered during the excavation.

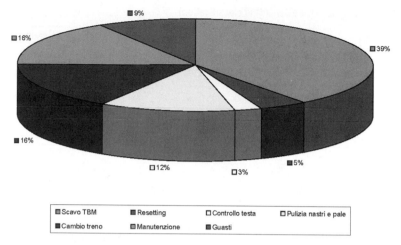

Figure 2. TBM production time diagram

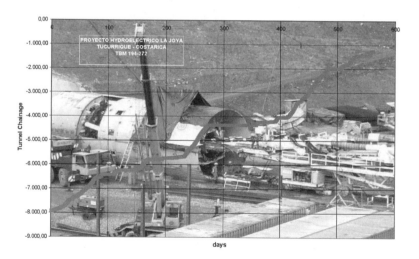

Figure 3. La Joya daily production diagram

TBM ADVANCE IN ROCK CLASS OVER V

The differences encountered between the predicted rock quality and the actual excavated rock forced the engineers of SELI to develop special procedures to minimize work stoppage due to the presence of an unexpected fault zone.

It is well known that being ready with all tools, materials and manpower is key to avoiding long and undesired stoppages.

For this reason, the TBM pilots have been instructed to react quickly in the presence of any sudden change of the most sensitive TBM parameters.

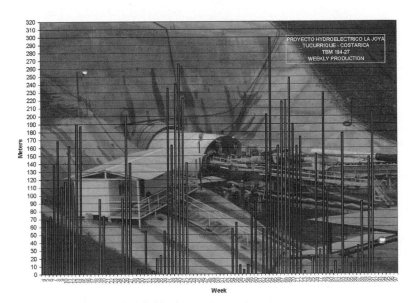

Figure 4. La Joya weekly advance report

Figure 5. La Joya monthly advance report

In the particular, the pilots should directly check the rock at the tunnel face when one of the following situations may arise:

- quick change of cutterhead torque, (i.e., increasing or dropping with respect to the average values),
- Increase of penetration rate at constant thrust force,
- Increase of resetting pressure,

Table 4. Rock mass data for class over V

Total extension of rock class over V	545 m
Number of interventions for ground stabilization	55
Total length of by-pass tunnels excavated by miners	490 m
Number of successful interventions of ground stabilization by the use of the only chemical injections	8
Number of successful interventions of round stabilization that required the excavation of by-pass tunnels associated to chemical grout injections	15
Quantity of PU resin used (chemical grout)	226,673 kg
Total number of TBM production days lost for chemical round treatment	14 days
Total by-pass excavation days	271 days
Total TBM production days lost for ground consolidation treatment	285 days
Average TBM production in fault zones	1.91 m/day

- Sudden change of size and aspect (i.e., colour, shape, water content) of mucked-out material,
- Increase of pressure in the hydraulic circuit of the extractor conveyor belt drive motor,
- Increase of excavated volume measured on muck cars,
- Sudden change of water quantity (increasing or decreasing).

Entering a fault zone with the TBM results in an over-excavation above and in front of the cutterhead. In such a situation it is extremely important to stop the TBM excavation at the proper time in order to perform stabilization work on the collapsed rock and to allow miners to work in a safer environment.

The stabilization work on the collapsed rock and the cavern generated in such a situation may be divided in two main categories:

1. Measures performed from inside the TBM, mainly consisting of ground treatment in front of the TBM with cement and chemical grouting.

2. Measures performed outside and around the TBM shields, mainly consisting of the manual excavation of a small by-pass tunnel above the shields in order to perform chemical grouting

Table 4 shows the main data for the 545 m of the Over V rock mass encountered.

Chemical grouting was performed using bi-component or water-reacting Poliuteran resin injected through ½" PVC pipes located in the body of the collapsed ground or in pre-drilled boreholes in the rock mass.

In the shorter fault zones it was generally necessary to treat the tunnel front with chemical grouting in order to improve the face stability, increase the self-support time at the face, and allow the TBM to restart the excavation on its own.

In the major fault zones, characterized by more than a few meters of extension, the most effective and convenient method, from a cost and time-lost point of view, was the excavation of the by-pass tunnel.

All the by-pass tunnels have been excavated at the top of the TBM shields and in association with the injection of chemical grout to enhance the stability of collapsing rock and improve safety for the workers.

Figures 6–11 illustrate the method adopted to create the by pass tunnels.

Figure 6.   Excavation of by-pass tunnel

Figure 7.   Forepoling

The by-pass tunnel is excavated at the top of the TBM shields. Access to the new working area is permitted through a window located in the TBM shield.

SELI's engineers have designed three special access windows strategically located in the top part of the TBM shield. Each window is a different distance from the cutterhead. Which of the windows has to be open in order to start excavating the new by-pass depends on safety and operational factors only.

All the windows allow easy access for miners and materials to the working area and can be connected directly to the extractor conveyor in order to facilitate muck out of the rock spoil.

The by-pass tunnel is excavated by miners using traditional pneumatic hand tools.

Stability is ensured by steel arch ribs, wire mesh and shotcrete.

Arch ribs are specially designed, trapezoidal section HEB steel profiles. All the arch ribs are closed at the foot and connected to the following one.

Figure 8. Inside by-pass tunnel

Figure 9. PU chemical grout

The by-pass tunnel is normally advanced up to the cutterhead to rescue the TBM and allow stabilization of the tunnel face and crown. The tunnel is eventually prolonged in front of the TBM until safe TBM start-up rock conditions are encountered.

If the prolongation of the by-pass tunnel in front of the TBM exceeds a certain length (i.e., 2–3 heights of the by-pass tunnel), then excavation of the by-pass tunnel is alternated with the TBM advance. With this solution, mucking out of spoils can be done through the cutterhead buckets, thereby improving the production cycle.

LINING

The tunnel lining, Figure 12, is made of n. 4–200 mm thick hexagonal precast concrete steel reinforced segments. Reinforcement consists of a steel mesh cage of approximately 60 kg of steel per m^3 of concrete.

Extrados diameter of the precast ring is 6 m, intrados diameter is 5.6 m, while the axial length is 1.6 m.

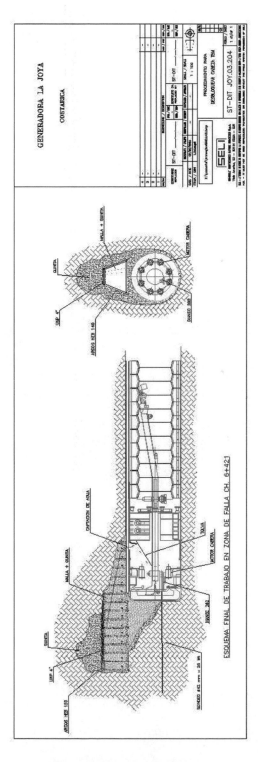

Figure 10. By-pass tunnel schematic

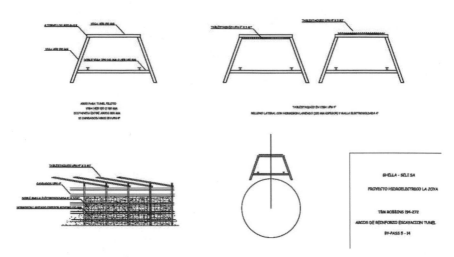

Figure 11.   Trapezoidal arch ribs

Figure 12.   Tunnel lining

The annular gap between the cut section and the lining extrados is filled by pea gravel at the moment of the extrusion of the precast ring from the TBM tail skin. All voids between pea gravel grains are grouted by cement injection just behind the backup tail.

The grout mix plant located at the tunnel portal has been designed to prepare a maximum of 15 m^3/hr of grout mixture, a/c = 0.75, and deliver a minimum of 4 m^3/hr at 8 km distance.

LINING WATERPROOFING

At the time of the tender, waterproofing of the tunnel lining was not required due to the fact that the tunnel draining capacity, according to the local water table, was considered negligible. Unfortunately, the actual behavior of the water table during tunnel excavation proved that the tunnel did have a strong impact on it. The drastic drop of the water table and the fact that several natural springs were dried out convinced the Client to ask the Contractor to modify the grouting procedures in order to waterproof the tunnel lining.

Lombardi Engineering (from Switzerland) was contracted to study the situation and to find, together with the Contractor, practical solutions to overcome the problem.

The new waterproofing procedure should combine with the actual hexagonal honeycomb lining available.

At the end of the investigation, a new maximum water inflow value and a new lining and grouting procedure were set:

- It was determined that 1 L/km/bar/sec should be the maximum acceptable water inflow in the tunnel. With this value in mind, it was believed that the water table would have quickly increased up to the original level. Considering the 80 m average water pressure and the 8-km-long tunnel, the formula resulted in a limit of 64 L/sec for the entire tunnel,

- Hydrophilic seals were installed in the segment joints,

- A special cement grouting procedure was developed, with 4 grout phases and grout pressures up to 5 bars measured at the packers,

- The first phase of the grout procedure, consisting of tunnel invert grouting, was always performed adding a special alkali-free accelerator to the grout mixture just at the packer position,

- In specific tunnel sections where the water inflows were still above the limit, a fifth grout phase was performed by drilling 4- to 2-m-deep holes in all the rings of a specific tunnel section and performing new cement grouting,

- Finally, in the few most critical sections, a sixth grout phase was performed, drilling 50 cm holes through the segments and grouting with water-active chemical resin.

At the end of the implementation of these special grout measures:

- The residual water inflows inside the tunnel were reduced to 36 L/sec, almost half of the admissible limit set,

- The ground water table in the mountain returned to its original level and all springs were active again with a few months.

Meanwhile, the 15 by-passes were filled with pea gravel and grouted with cement.

To enhance the quality of the ground in these specific sections, ground consolidation treatments have been performed. A path of 4 boreholes per ring, 2" diameter and 3 m deep were executed starting 5 rings before the beginning of the weak section and up to 5 rings beyond its end.

At the completion of all the works, a total of 16,000 ton of cement and 28,000 kg of special water-reacting resins were utilized.

GEOPHYSICAL INVESTIGATION

To predict with the most accurate precision the geological condition encountered by the TBM during the excavation of last part of the tunnel, campaigns of seismic and geo-electric investigations have been performed.

Overlapping the results of both investigations, and linking this information with the data collected during the excavation of the first part of the tunnel, the site geologist could develop a realistic model that helped engineers programming the execution of the works and the supply of consumable materials (Figure 13).

Both types of investigation were performed from the surface. Average coverage of the tunnel alignment in the area is approximately 160 m.

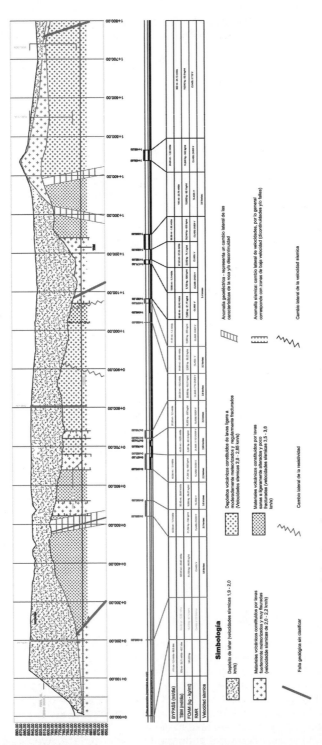

Figure 13.   Siesmic and geo-electric interpretation

In the seismic refraction investigation, geophones have been located with a path of 24 m from one to the next. The interpretation of the refraction data allowed the geologist to draw a new and well-detailed geological profile, locating the probable fault zones.

In the geo-electric campaign, the electrodes were located with a path of 40 m and the AB/2 distance has been set at 200 m. Interpretation of the ground resistivity recorded gave the geologist confirmation of the presence of fault zones, their extension, and other valuable information related to the alteration of the rocks and the presence of water.

CONCLUSION

The rock mass conditions encountered in the La Joya tunnel (Costa Rica—Central valley) were much worse than predicted and in general extremely bad.

The double-shield TBM selected on the basis of the geological data available at the beginning of the contract was able to advance with very good production in rock classes III, IV, and V. The good production achieved in these rock classes minimized disruption caused by encountering the poorer geology.

To pass through those specific sections where the rock mass has been classified as Over V (according to Bieniawski classification), special working procedures were developed to allow the TBM to by-pass the complicated zones and/or to reduce the environmental impact of the tunnel excavation.

Soil stabilization by the use of chemical foam injection and excavation of 15 by-pass tunnels, with the help of miners, helped the TBM to overcome various unexpected fault zones.

The special grouting measures developed by SELI engineers and performed on site were able to reduce water inflows inside the tunnel to acceptably low limits despite the high groundwater pressures and the high rock mass permeability.

The results of seismic and geo-electric campaigns executed in front of the TBM were well-correlated to the ground behavior and were a useful tool to anticipate critical geological conditions and predict TBM advance rates.

Despite the hydro-geological difficulties, the tunnel was completed in time to put the hydropower plant in operation on schedule.

TUNNELING THROUGH MOUNTAIN FAULTS

Don W. Deere

Deere & Ault Consultants, Inc.

ABSTRACT

Faults under high rock cover represent some of the most severe ground conditions a Tunneler can encounter. High ground water pressures and flows are usually associated with mountain faults. Ground behavior can range from blocky ground to flowing sands and squeezing clays. Sticking a TBM within a fault zone for many weeks, with the disastrous financial consequences, is still a common occurrence.

Interpreting the geology and detecting the location of a fault prior to penetrating it is a fundamental step to successful tunneling. Ground water control through pre-draining or pre-grouting is the second fundamental step.

This paper discusses techniques applied for detecting and crossing mountainous fault zones through the use of case histories across the globe.

INTRODUCTION

Traversing a mountain fault during tunneling often represents encountering treacherous conditions. These conditions include blocky, squeezing, and flowing ground often with flush water inflows of several hundreds of liters per second. The cost impacts of these significant geologic features on a project have seemed to increase over the last 20 years as more tunneling is done with TBM technology. It is not uncommon for projects to be delayed a year due to sticking the TBM in a fault zone with the resulting finger pointing between Owner and Contractor.

The American Geological Institute dictionary defines a fault as "A surface or zone of rock fracture along which there has been a displacement, from a few centimeters to a few kilometers in scale." For the Tunneler, the amount of relative displacement has little relevance; rather the important factor is the engineering behavior of the feature. For the purposes of this paper, faults also include foliation shears and shear zones.

This paper deals with mountain faults where conditions are at their worst due to the high cover over the tunnel. Mountain tunnels include projects that are typically deeper underground than the typical urban tunnel. A mountain tunnel can typically have rock cover of between 100 to 5,000 meters (328 to 16,400 feet) with ground water pressures of up to 75 Bars (1,088 psi). This is significant because these associated pressures are higher than what current TBM technology routinely deals with.

ENGINEERING GEOLOGY OF FAULTS

Faults are usually made up of a series of different zones that have resulted due to shearing displacement and water permeation. A simplified typical profile of a fault is shown on Figure 1.

In general the harder, stronger, and more massive the country rock mass, the more severe and complete the zone of shearing. Granitic rock masses (intrusive igneous rocks) display some of the most severe faults observed by the author. This observation was also made by Karl Terzaghi 60 years ago (Proctor and White, 1946) when he stated: "The intrusive rocks should be considered decidedly treacherous."

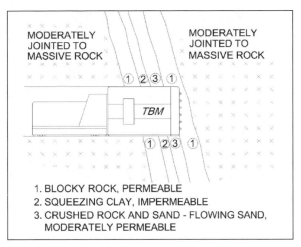

1. BLOCKY ROCK, PERMEABLE
2. SQUEEZING CLAY, IMPERMEABLE
3. CRUSHED ROCK AND SAND - FLOWING SAND,
   MODERATELY PERMEABLE

Figure 1.   Make-up of a typical fault zone

Blocky Rock

The boundary of a fault zone (either one side or both) is typically made up of a blocky rock zone. The blocky rock zone is fractured and usually permeable. Very high flush water flows can be associated with open jointing in the blocky rock zone. Often times joints bounding the blocks are slickensided resulting in low friction angles. The rock blocks can range from fresh to hydrothermally altered.

A very major problem with TBM tunneling in hard, very strong rock is boring through the blocky rock zone. Hard blocks loosen and dislodge from the face and can roll around in front of the cutter head causing severe vibration damage to the TBM head. Damages observed from blocky ground include broken cutters, as well as cracked and gouged machine heads, all due to point loading caused by the loose and dislodged blocks. This kind of behavior was observed in the Casecnan Hydroelectric Project in the Philippines, as well as the Manapouri Tailrace Tunnel No. 2 in New Zealand (Deere, et al., 2003). Both of these tunnel projects were excavated in granitic and metamorphic terrain with unconfined compressive rock strength approaching 200 MPa.

Squeezing Clay

It is common for the core of the fault to contain a zone, say 0.1 to 1 meter thick, of squeezing clay gouge. This zone is formed by tectonic crushing forces, as well as hydrothermal alteration. The clays are usually quite plastic and are typically bentonitic and chloritic clays. These clays tend to squeeze dramatically when excavated in a mountain tunnel under high cover because the unconfined compressive strength of the clay is much lower than the overburden (or other in-situ) stresses on the tunnel opening. These clays also can continue to squeeze due to swelling of the clay minerals. Swelling is usually a secondary phenomenon that occurs after the initial stress squeezing and the clay is disturbed and open to additional wetting fronts.

A photograph of squeezing clay encountered in a fault zone in the Peruvian Andes is shown on Figure 2. The clay that squeezed around the TBM dust shield appears on the lower left of the photograph. This squeeze occurred in a matter of seven or eight hours after excavation.

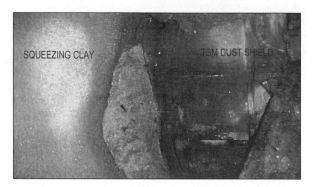

Figure 2.   Squeezing clay

Crushed Rock and Sand

Many major faults in hard rock contain a zone of rock that has been crushed by tectonic forces into sand and gravel size fragments. This material will turn into flowing sand conditions when subjected to a high hydraulic gradient. Sometimes the crushed sand has some cementation due to alteration, or it can be chemically intact as described by Terzaghi and then display zero standup time under high ground water heads.

Hydrogeology

The hydrogeologic setting typical of mountain faults is shown in cross-section on Figure 3. Faults and shear zones act as subsurface dams across them and as a drain in a parallel direction (Deere, D.U., 1973). The clay gouge zone described previously constitutes the natural low permeability barrier across the fault. The fractured blocky ground and crushed rock zones serve as drains parallel to the fault.

In unfaulted terrain with topographic relief, the ground water level typically mimics the surface topography, but in a subdued manner. This is shown on the left side of Figure 3 where as the ground cover rises, so do water levels but at a reduced comparative rate.

In faulted terrain, there can be abrupt changes in ground water levels (and pressures) on one side of the fault to the other because faults act as subsurface dams. Faults form boundaries of separate ground water "compartments" as illustrated on Figure 3. The treacherous conditions illustrated on Figure 3 are obvious. Where a fault is crossed in a mountain environment, changes of 500 meters in head across a 1 meter wide clay gouge zone presents extreme ground water gradients. These gradients can result in flowing ground and even catastrophic blowout due to these extreme seepage forces.

In addition to high pressures, high ground water flush flows are encountered once the clayey fault gouge is pierced due to the permeable nature of the associated blocky ground and crushed rock. Flush flows in fault zones can be as high as 200 liters per second (3,170 gpm). The flow rate often drops by 75 percent over a few days as limited storage in the ground water compartment is drained down.

The author has encountered high flush flows on projects crossing faults throughout the world. Photographs of a fault zone crossed in the Andes in Peru are shown on Figures 4 and 5. The fault zone collapsed on the TBM, primarily in blocky ground where the clay gouge zone was completely penetrated. Initial flush flows of 200 liters per second were recorded.

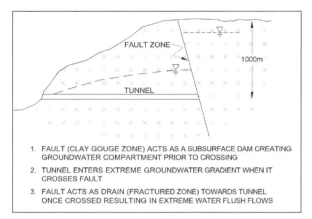

Figure 3. Hydrogeology of a fault zone

Figure 4. Collapse of blocky ground from fault zone

Figure 5. High water flush flow from fault zone; note water flowing over top of gripper pad

DETECTION OF FAULTS DURING PRE-CONSTRUCTION INVESTIGATION

A primary tool for detection of faults in geotechnical investigations is geologic mapping. The mapping includes both air photo interpretation and field mapping. Air photos are useful in detecting lineations that often times are faults. Core borings are the other primary tool for investigations. Most borings are drilled vertically, and although they provide good general information on the rock mass and water conditions, they provide very little data immediately on the tunnel alignment. Inclined core borings (45° to 90° from vertical) should always be used in conjunction with vertical borings in order to intercept steeply dipping rock structures.

Classical geologists are taught that faults appear as topographic lows reflected as stream valleys or saddles in a ridge. This is often the case, but the author has considerable experience where the observation does not hold true.

There are many case histories in tunneling where unknown and unexpected faults and shear zones with little topographic surface expression are encountered within the first kilometer (3,280 feet) of entry into a mountain front. These faults are usually never shown on published or project generated geologic maps, but are probably simply associated (a splay) with the more well known primary faults.

These unidentified faults entering a mountain usually have major impacts to the project because they are the first high pressure ground water compartment encountered within very poor quality rock. Problems occur because the tunneling contractor is usually not prepared for this unknown feature early on in the tunnel, and crews are still operating under a learning curve.

The most powerful tool for fault detection is horizontal core drilling. Horizontal borings should be drilled at or near the proposed tunnel crown and advanced from each portal as far as possible. A secondary benefit of a horizontal boring is it often provides for pre-drainage. Depending on rock conditions, today's technology allows for horizontal core boring lengths up to 650 meters (2,132 feet).

The base cost of horizontal core boring is typically double ($650 per meter or $200 per foot) than that of vertical core drilling (Ruen, 2007, personal communication). If tight survey control and steering are specified, the cost can be as high as four times that of vertical coring. However, the information provided is priceless as a continuous geomechanical log of the tunnel alignment can be prepared with a horizontal core.

The type of useful information given by a horizontal boring is illustrated on Figure 6. This graph was prepared for the Mariani Highway Tunnel investigation in Puerto Rico. The horizontal borings were geologically logged and a "Q" (Rock Mass Quality) was assigned. This graph provided an excellent tool for the contractor to plan his operation and there was good correlation between the actual support used and the Q value assigned from the horizontal core boring. A small fault (see right side of graph) that was not originally detected by field mapping or vertical borings was discovered and planned for based upon the findings of these horizontal borings.

DETECTION OF FAULTS DURING CONSTRUCTION

The most commonly used method for fault detection during tunnel excavation is the use of probe holes out in front of the face. Probes are usually drilled with a percussion drill rig and give an indication of soft rock versus hard rock and the presence of water. They are commonly drilled 25 to 50 meters (82 to 164 feet) out in front of the face.

Core drilling out in front of the face, because it is slower, more expensive and requires more room, is usually only utilized in select circumstances in civil tunneling to investigate an expected special feature or after problems such as a collapse have already occurred. Core drilling provides intact samples which gives much more reliable

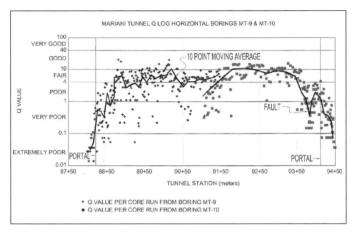

Figure 6.   Q log plot of horizontal borings

information than percussion probe drilling. In the underground mining industry, horizontal core drilling is a much more common practice since the drill rigs are on-site for mineral exploration.

Continuous pilot core drilling is currently being used in the high speed drill and blast excavation of twin access tunnels by PT Freeport Indonesia (Sinaga, 2006). These deep access tunnels into an ore body, with a total planned length of 18.6 kilometers, are using 650 meter (2,132 feet) long HQ core borings drilled from remucking alcoves along the tunnels. This innovative technique allows them to safely probe ahead of the face to warn them of bad ground and high pressure water compartments. They are able to achieve 30 meters per day of core drilling utilizing Long Year 75 hydraulic units (Barber, 2007, personal communication).

Seismic probing using the Amberg TSP (Tunnel Seismic Probe) is commonly used in some places such as Japan and Europe. The profile records reflection planes (joints, faults, contacts). It can typically provide information some 50 meters in front of the face. It relies on interpretive software for prediction. The author believes it currently works best when used in combination with drilled probe holes.

CONSTRUCTION METHODOLOGIES

The most common methods used for crossing faults include: (1) pre-drainage, (2) grouting, (3) pre-reinforcement, and (4) pilot drift hand mining. It is the author's experience that pre-drainage is a very cost effective fault handling methodology, yet this method appears to be underutilized in the business. If high water pressures and gradients can be reduced (Figure 3) then stand-up time can be increased. Efficiently removing flows from the working front also allows passage of the fault more quickly. Many contractors consider pre-drainage drilling as a time consuming nuisance, especially on TBM jobs. If planned for when ordering the TBM, drills can be installed on the machine that allow for drilling of drain holes while the TBM is cutting. Environmental constraints by regulatory agencies against temporary ground water level reductions sometimes prohibit the usage of pre-drainage. Fortunately, these constraints occur more commonly on urban tunneling under less cover than on the typical mountain tunnel.

Grouting can be quite expensive and time consuming and its effectiveness can be highly variable. Cement grouting of faults has had variable success, as it is difficult to permeate the crushed rock and sand zone or altered rock with even micro-fine cement. Grout takes are often excessive due to large grout travel and loss back to the tunnel. Chemical grouts, although costly are usually more successful in providing ground control because of better permeation and rapid set times. A case history of successfully tunneling through numerous faults in the Philippines using chemical grouting combined with pre-reinforcement resin dowels in the face was presented by Grandori (RETC 2001).

It is sometimes extremely difficult to even drill a drain hole, grout hole, or pre-reinforcement bolt hole through a fault as they collapse or squeeze shut. Self drilling rock bolt and drain hole continuous casing systems, such as provided by Alwag Techmo or Atlas Copco, should be specified and on hand for all mountain tunnel projects.

Even the biggest, and best planned and investigated projects can still have problems sticking TBM's in fault zones, and sometimes it is a simple matter of luck. The recent problems at the Gothard Deep Swiss Rail Tunnels were documented by the Discovery Channel. The Swiss specified drill and blast tunneling where the worst ground conditions were expected and TBM tunneling in better ground. The first TBM of the twin tunnels rapidly traversed an unknown fault while the sister TBM passing later was stuck for six months. The TBM was rescued utilizing a self-drilling pipe umbrella.

CONCLUSIONS

1. Fault zones continue to be one of the greatest challenges while driving a mountain tunnel due to squeezing, flowing, and blocky ground associated with high ground water pressures.

2. Detection of faults prior to penetration is the key to potentially safe passage.

3. Horizontal core borings are an underutilized powerful tool for fault detection. All mountain tunnels should utilize horizontal borings drilled at each portal as a standard investigative procedure. This procedure will detect the presence of splay faults commonly encountered when entering a mountain front. Pre-drainage of any splays encountered is a potential additional benefit of horizontal core borings.

REFERENCES

Barber, John, 2007, personal communication with PT Freeport, Indonesia.

Deere, D.U., 1973, "The Foliation Shear Zone, an Adverse Engineering Geologic Feature of Metamorphic Rocks," Journal of the Boston Society of Civil Engineers.

Deere, D.W., Keis, S., and Watts, C., 2004, "Manapouri Tailrace Tunnel No. 2 Construction—A Very Large TBM Tunnel in Very Strong Rock," North American Tunneling Conference, Atlanta, Georgia.

Grandori, R., 2001, "Manila Aqueduct (Philippines)—The Construction of the Umiray-Angat Tunnel Project—A Success of Organisation and Technology Against a Unique Combination of the Most Adverse Conditions Ever Encountered in Tunneling," RETC Proceedings.

Proctor and White, 1946, "Rock Tunneling with Steel Supports," published by Commercial Shearing, Inc.

Ruen, Arlan, 2007, personal communication with Ruen Drilling.

Sinaga, F., et al., 2006, "Geotechnical Concerns During the Development of the AB Tunnels in PT Freeport Indonesia," 4th Asian Rock Mechanics Symposium in Singapore, November.

5

Focus on Canada

Chair
M. Malott
McNally International, Inc.

DESIGN/BUILD AGREEMENT FOR THE NIAGARA TUNNEL PROJECT

John Tait

Hatch Mott MacDonald

Russel Delmar

Hatch Mott MacDonald

Harry Charalambu

Hatch Ltd.

Rick Everdell

Ontario Power Generation

ABSTRACT

Award of the Design/Build Agreement for construction of Ontario Power Generation's (OPG's) Niagara Tunnel Project was made after an intensive 8-month international procurement process. The 14.4-m diameter × 10.4-km long rock tunnel will divert 500 m^3/s of water from the Niagara River to the Sir Adam Beck hydroelectric generating complex. The procurement process and Design/Build Agreement included features to address the owner's requirements regarding risk allocation and cost and schedule certainty, while encouraging innovative ideas from the industry, both technically and commercially. The process included an international expression of interest and prequalification, followed by an invitation for design/build proposals to qualified proponents, and comprehensive proposal evaluations and negotiations with several respondent proponents. An honorarium was provided to the unsuccessful participants in the proposal stage. The resultant negotiated Design/Build Agreement includes Owner's Mandatory Requirements, negotiated contract terms and conditions, performance incentive bonuses and liquidated damages for schedule and water delivery performance, a multistaged Geotechnical Baseline Report, joint risk management, references to preexisting community impact agreements and measures to reduce impact on tourism. It also addresses regulatory approvals and permits, an owner controlled insurance policy that required reference to the International Tunneling Insurance Group (ITIG) Code of Practice for Risk Management as a condition of coverage, site specific safety, security and environmental plans, formal team building, a dispute review board, and escrow documentation.

INTRODUCTION

OPG is an Ontario-based electricity generating company whose principal business is the generation and sale of electricity in Ontario. OPG's focus is on efficient production while operating its generation assets in a safe, open and environmentally responsible manner. OPG's sole shareholder is the Government of Ontario.

Hydroelectric development at Niagara Falls started in the late 1800s to serve local users. Around 1900, six larger, commercial hydroelectric stations were constructed adjacent to the falls. These six Canadian and U.S. "Cascade" plants served more-distant customers through the first long-distance transmission lines and had combined capacity of about 500 MW. Operation of the last U.S. Cascade plant ceased in 2006. In the 1920s, the Queenston-Chippawa Power Development, later named Sir Adam Beck generating station No. 1 (SAB1), was the first Niagara River facility to utilize the hydraulic head available not only at the falls, but also through the rapids upstream and downstream over a (straight line) distance of about 8 km. For SAB1 approximately 600 m^3/s of Niagara River water was diverted through a 20-km open-cut canal from above the falls to a 400-MW powerhouse near Queenston. The 1950 Niagara Diversion Treaty established the future distribution of Niagara River flow between competing interests, confirming a priority for domestic and navigation purposes, stipulating a schedule of minimum flow over the falls for scenic purposes, and permitting remaining flow to be shared equally between Canada and the United States for power generation. The Sir Adam Beck generating station No. 2 (SAB2), including its diversion system comprising two 8-km diversion tunnels (600 m^3/s each) and a 3-km open-cut canal, and the companion Pump Generating Station (PGS), were constructed in the 1950s. In the early 1960s, coincident with closure of U.S. Cascade plants, the New York Power Authority constructed its Robert Moses generating station and companion Lewiston PGS to fully utilize the U.S. share of the Niagara River flow available for power generation.

OPG and its predecessor Ontario Hydro have been considering expansion of Ontario's Niagara River hydroelectric facilities over the past three decades. An environmental assessment (EA) for construction of two new diversion tunnels (2×500 m^3/s), an underground generating station (up to 3×300 MW) and transmission improvements in the Niagara peninsula was submitted in 1991. The EA received approval in 1998, including provisions for phased construction. Restructuring of Ontario's electricity market in 1998–2004 resulted in OPG's Niagara hydroelectric facilities becoming part of Ontario's Regulated Supply. This, combined with the growing need for renewable energy generation in Ontario, led to OPG's June 2004 decision to proceed with the project (Phase I under the EA approval).

OPG decided to utilize a design/build contracting approach to achieve owner and shareholder best value objectives including best practices in safety, environmental and quality management, appropriate risk allocation, cost certainty, schedule certainty, and early completion to replace energy production from the planned phase out of OPG's coal-fuelled generators. The target diversion capacity was established as a nominal 500 m^3/s. A 90-year design life was set as a minimum requirement. Project risks were to be owned by the party best positioned to manage those risks.

Table 1 summarizes the current and planned Niagara River water diversion volumes, generation capacities and average annual energy outputs. Since no additional generation equipment is being installed at the existing power plants, the estimated 14% increase in average annual energy output from the SAB complex will be the result of increased utilization of the existing hydroelectric generators due to the 27% increase in OPG's water diversion.

Figure 1 presents hydrographic data illustrating that Niagara River flow available to Canada for power generation exceeds OPG's current diversion capacity (one canal and two tunnels) about 65% of the time. With the addition of the new tunnel, the available flow will exceed OPG's diversion capacity only about 15% of the time.

Earlier OPG/Ontario Hydro engineering studies and commitments made through the EA approval process placed several constraints on the project configuration and implementation. Locations and conceptual designs for the intake and outlet structures were fixed based on physical constraints and hydraulic model studies. The tunnel horizontal alignment was required to substantially utilize the easement corridor established for the existing SAB2 tunnels. To minimize construction impacts on the City of

Table 1.   Water diversion volumes, installed generation capacity and yearly average outputs

Facility	In-Service	Diversion (m³/s)	Installation (MW)	Average Output (TWh/yr)
SAB1	1922	600	472	2.6
SAB2	1954	1,200	1,480	9.2
SAB PGS	1958	—	174	−0.1
Niagara Tunnel	2009	500	—	1.6
Total—Canada		2,300	2,126	13.4
Robert Moses	1964	3,000	2,400	14.2
Lewiston PGS	1964	—	300	−0.5
USA—Totals		3,000	2,700	13.7

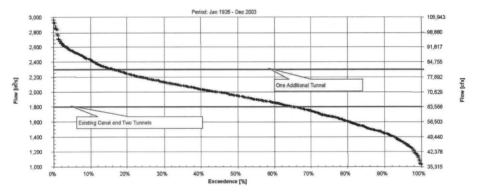

Figure 1.   Water available to Canada by treaty versus diversion capability

Niagara Falls and its tourism industry, there was a requirement that the tunnel be constructed by mechanical means from the outlet end, without any intermediate shafts,
and that construction traffic be limited to designated routes. The host rock, primarily
Queenston shale, had to be excavated and stored in a manner to facilitate its future
use by Ontario's clay brick industry. To minimize impacts on the Niagara River, work at
the intake area had to be accomplished within the flow, water level and ice management regime established by the 1950 Niagara Diversion Treaty and Directives of the
International Niagara Board of Control. To limit impacts on the ongoing hydroelectric
operations, the extent and duration of outages on the existing facilities required to
enable tunnel construction, needed to be minimized and well planned.

OPG's invitation for proposal outlined the key owner/shareholder objectives, constraints, mandatory requirements, functional requirements, key design parameters and
proposal evaluation criteria, while maintaining enough flexibility to capture contractor
experience and innovations.

Summary of Expectations and Constraints

From the preceding narrative, it can be seen that there were several overriding
expectations and constraints governing the procurement, design and construction of
the project. These can be summarized as follows:

Shareholder (The Government of Ontario) Expectations
- Quick delivery of additional renewable energy,
- Cost and schedule certainty.

Owner's (Ontario Power Generation) Expectations
- Best safety practices,
- Minimum environmental impact (consistent with the EA approval),
- Optimum utilization of available water (set by international treaty) in Niagara River,
- Nominal water delivery flow of 500 m^3/s,
- Cost and schedule certainty,
- Ninety-year design life,
- Risk allocation transfer to the most appropriate party,
- Minimum impact on current power production,
- Minimum impact on local communities,
- Minimum impact on local tourism.

Constraints
- Existing operating hydroelectric facilities,
- Existing (live) tunnels in the available corridor,
- Requirements of the Ontario Provincial Environmental Assessment Approval,
- Requirements of the 1950 Niagara Division Treaty and related agreements,
- Excavation method (TBM),
- No intermediate access shafts,
- Tunnel horizontal alignment
- Requirements governing disposal of excavated material.

Concept Scope of Work

The project comprises three major elements, namely, intake facilities, outlet facilities and the diversion tunnel.

Intake. The tunnel intake consists of an intake structure located beneath Bay 1 of the existing International Niagara Control Works (INCW) structure, situated about 1.5 km upstream from the Horseshoe Falls. This work will be carried out behind a cofferdam set in the river upstream from the Bay 1 gate.

Outlet. The tunnel outlet consists of an outlet structure and control gate discharging into a 390-m long canal connecting to the existing SAB diversion canal system near the PGS.

Diversion Tunnel. The 10.4-km diversion tunnel will be constructed as a two-pass tunneling system with boring taking place from the outlet canal and passing beneath the City of Niagara Falls to the intake excavation. Boring will be carried out by means of a 14.44-m diameter open gripper rock tunnel boring machine (TBM).

The final lined diameter of the tunnel will vary between 12.6 m and 12.7 m.

Figure 2 presents the elements and general layout of the work.

Figure 2.   Niagara Tunnel Project—concept elements and layout of the work

IMPLEMENTATION THROUGH DESIGN/BUILD AGREEMENT WITH CONTRACTOR

OPG selected a fixed price design/build format for the agreement between owner and contractor as the best means of achieving the following three objectives; (1) avoiding the risk of owner's design delays in what was to be an accelerated design and construction schedule with performance incentive bonuses and liquidated damages, (2) opening up the tunnel design and construction to the best ideas that the industry could offer constrained only by some specific and necessary Owner's Mandatory Requirements, and (3) delivering the maximum practicable water volume at the best price. These objectives would have been difficult to achieve had the owner's engineers fixed on a single preconceived design for a design/bid/build format. The above approach resulted eventually in receiving three different design concepts to choose from. OPG's proposal invitation did not specify any schedule requirements or a specific tunnel diameter other than to encourage a minimum flow delivery of 500 m^3/s. Proponents were required to offer their own targets for both. The invitation for proposal included information on how both flow and schedule would be evaluated by the owner to bring the proposals to a common basis. Having set those targets, they became contractual requirements in the final agreement against which bonus/liquidated damages calculations will be made. Also, the terms and conditions of the agreement were open to revision by the proponents and to subsequent negotiation between the parties prior to execution. In these ways, OPG was open to receive the ideas, methods and conditions that best suited all the parties to the task at hand and, in the process, should go a long way toward ensuring cost and schedule certainty.

OPG retained Hatch Mott MacDonald in association with Hatch Energy as its Owner's Representative to provide a concept design, contract procurement and contract administration services and full-time on-site monitoring and review of the contractor's work for compliance with the Design/Build Agreement.

Owner's Mandatory Requirements

The Owner's Mandatory Requirements (OMRs) are the minimum acceptable requirements for the design, construction, testing and commissioning for the performance of the work. These, in conjunction with the Summary of Work and Concept Drawings define the scope of the work to be performed under the Design/Build Agreement. Under the Agreement, the contractor is required to develop and detail all additional requirements and specifications needed to complete the work. Unlike other proposal assessment criteria described later, proposals were required to comply fully with the Owner's Mandatory Requirements and, as such, each proposal was required to pass a yes/no compliance gate before being considered for further evaluation. In addition to containing many of the expectations and constraints listed above the OMRs covered requirements with respect to environmental protection, safety by design and rules governing the disposal of excavated material.

Performance Incentive Bonus and Liquidated Damages on Schedule and Water Delivery Quantity

Having offered targets for schedule and guaranteed water flow delivery in their proposals, each proponent was evaluated on these offerings and was required to accept these targets in the Design/Build Agreement against performance incentive bonus payments for betterment and liquidated damages for failure to meet them These measures were designed to promote cost and schedule certainty.

The baseline schedule agreed between the owner and the contractor can be summarized as follows:

- Contract start—September 1, 2005,
- Commence tunneling—September 1, 2006,
- Substantial completion (water up) October 9, 2009,
- Final completion—December 8, 2009.

Multistage Geotechnical Baseline Report

Consistent with OPG's risk allocation philosophy, OPG assumed the risk associated with differing subsurface conditions to encourage the lowest fixed price by each proponent. Anticipated conditions were baselined by means of a multistage collaboratively developed Geotechnical Baseline Report (GBR). The first stage included a GBR in the invitation for proposals. This GBR, referred to as GBR A, baselined the conditions the owner expected to be encountered during construction of the tunnel and included a number of blank tables for completion by each proponent addressing their specific design and construction methodology elements and how these elements would address the identified subsurface risks. The second stage required each proponent to submit specific numbered responses to each of these GBR A tables in what was referred to as GBR B, included in their respective proposals. This process enabled OPG to (1) evaluate the comparative subsurface risk profile presented by each proposal, and (2) in the final stage, negotiate the final proposal specific GBR included in the Design/Build Agreement as the only basis for determining changes in or differing geotechnical subsurface conditions.

Risk Management

A third party facilitator was used to conduct expert workshops to analyze the risks inherent in the project. This initially produced a qualitative risk register that identified the risks that would need to be managed. Where appropriate some of these were then transferred to the contractor through the proposal invitation process and subsequently in the Design/Build Agreement. Later, many of these risks formed the basis of OPG's contribution to the joint risk register discussed later under the section on project insurance.

A second series of expert workshops were convened to populate a quantitative risk register from which the required value of project contingency was calculated.

Community Impact Agreement with Local Governments and Agencies

Prior to inviting proposals, OPG entered into a Community Impact Agreement with the local governments to address the potential effects of such a major project within their municipal boundaries. It was recognized early in the planning stages that the project might have significant impacts on the local urban and tourist areas. Agreement was reached to restrict truck movements to a few arterial roads away from the heart of the Niagara Falls tourist centre. A specified truck route, formally agreed with the City of Niagara Falls and the Regional Municipality of Niagara, was identified in the invitation for proposals as the only acceptable haulage route through the City. Proponents were required to incorporate these restrictions into their proposals and the identified truck route was subsequently carried forward into the Design/Build Agreement. OPG paid for the reconstruction and widening of the route accessing the outlet site entrance from the nearby major highway (Highway 405). Agreement was also reached with the Regional Municipality of Niagara to allow sewage from the project to be pumped into the local sewer system, eliminating the need for on-site holding or treatment facilities. Potable water is also provided from a City watermain.

These actions reflected OPG's desire to remain a good corporate citizen in a city and a region where it has carried out business for decades and helped to alleviate proponents' uncertainties over the costs of such services.

A Community Liaison Committee has been established comprising representatives from the three local host municipalities, OPG and the contractor. The purpose of the Committee is to keep the municipalities informed about the project and to provide an opportunity for the municipalities to identify any concerns they might have related to the project such that OPG can be proactive with any remedial action required. In addition, OPG has set up a 24-hour "1-800" complaints line to speedily deal with issues arising from project activities.

Reduced Impact on Tourism

During the proposal phase, proponents were reminded of the unique nature of the project. A major hydroelectric construction project was being contemplated in the middle of one of North America's busiest tourist attractions. Various constraints were introduced to cope with this situation, some of which are the same as those described in the preceding section to reduce the impact on the local host municipalities and the permanent residents. Other tourist-specific measures were also put in place. For example, the Niagara Parks Commission (NPC), an Ontario Crown Agency whose mandate is to preserve and enhance the natural beauty of the Niagara River corridor for the enjoyment of visitors, has ownership of the land adjacent to the INCW and the outlet. Agreement was reached to enable land to be leased from NPC for the intake construction yard. OPG has committed to minimize the effect of the project on NPC's tourist areas and activities. Access to the intake site has been routed away from the entrance to the NPC's People Mover parking facilities, a major tourist handling area adjacent to the tunnel intake. This also removes the potential for claims emanating from NPC interference with the execution of the work. Ongoing communications are maintained with NPC to ensure any issues can be addressed in a timely manner.

Through the Community Impact Agreement, OPG has also made a financial contribution to the City of Niagara Falls and the Town of Niagara-on-the-Lake to cover the cost of advertising and other communications about the project that will keep the population and tourism industry aware of what is going on and how it might affect them.

Provincial and Federal Approvals and Permits

OPG proactively obtained as many approvals as it could before award of the Design/Build Agreement in order to reduce uncertainty, ensure that the project would proceed in a timely fashion and keep the contract price as low as possible. OPG had applied and received approval for the project under the Environmental Assessment Act (Ontario) in 1998 and authorization under the Federal Fisheries Act in 1995. A number of conditions were imposed from both these approvals which required contractor input specific to its design and construction methods. Also, a number of other permits and approvals were required to be obtained by the contractor as a matter of course. The contractor's responsibilities for these conditions, permits and approvals were tabulated in the Design/Build Agreement and the contractor bears the risk of any delay in obtaining them.

Owner Controlled Insurance Policy

OPG, as part of the Design/Build Agreement, has provided Builders All-Risk insurance obtained through a consortium of major underwriters based in the United Kingdom. During the early discussions with the insurers, they insisted that, as a condition of providing coverage, the project must adhere to the provisions of the ITIG publication "A Code of Practice for Risk Management of Tunnel Works." OPG did not believe this code of practice was entirely applicable to the project due to the timing of the introduction of this code to the project and to North American-specific requirements. Following extensive discussions with the underwriters, a project-specific "Agreed Code of Practice for Niagara Tunnel Project" was developed and accepted. This project-specific code of practice, which is incorporated into the insurance policy, embraces all the key principles of the ITIG code and includes many of its principal requirements. An unusual requirement (at least for North America) of the ITIG code is that it compels the owner and contractor to closely and actively collaborate in the construction-phase risk management process. Our experience, to date, in this joint risk management process is positive.

Contractor's Site Specific Safety, Security and Environmental Plans

In order to be able to evaluate proponent's safety, security and environmental approach to the project, proponents were required to submit preliminary site specific (i.e., nongeneric) plans for safety, security and environmental protection with their proposals. In terms of the Design/Build Agreement, the contractor assumes the role of "Constructor" (as defined in the Occupational Health and Safety Act [Ontario]) for the project, except for a small portion of the project where the contractor was required to perform certain in-water elements of work, upstream from the INCW structure where the water-control gates must remain under the control of OPG at all times during execution of the work. For the limited time that in-water work was required in this area, OPG assumed the role of "Constructor" under a part-project designation approved by the Ontario Ministry of Labour. In terms of the Design/Build Agreement, the contractor was required to submit separate detailed site specific plans for safety, security and environmental protection for the project as well as the part-project elements of the work.

Formal Team Building Between Owner and Contractor

In order to optimize communications between the parties after award of the Design/Build Agreement and during the design and construction of the tunnel and other facilities, the Design/Build Agreement incorporates formal team building to cover the design/build contractor's key staff, its designers, subcontractors and other partners, as well as OPG senior management and staff, its owner's representative and other major participants. The initial session was formally facilitated at a third party location with subsequent interaction being regular but less formal. Additional formal sessions will be repeated at appropriate points as the project progresses through different phases of construction. The intent of this effort is to keep all communication channels open thereby reducing on-site disagreements and conflicts.

Dispute Resolution

The Design/Build Agreement requires the establishment of a Dispute Review Board (DRB). The DRB was set up soon after award with one member nominated by the contractor, one by the owner and the third (the chair) selected by the first two members. During the nomination process, both parties to the Design/Build Agreement had a limited right of refusal of the other's nominees. Since its inception, the DRB has met formally on site on a quarterly timescale and has been apprised of progress, changes to the contract and any issues arising. To date, no disputes have been referred by either party to the Design/Build Agreement.

In addition to having a DRB, a set of the contractor's back-up documentation for its proposal has been set aside in escrow. Access to these documents is only allowed by mutual consent of both parties to the Design/Build Agreement and only for purposes of dispute resolution.

THE PROCUREMENT PROCESS

Advertised International Invitation to Prequalify

In July 2004, an expression of interest (EOI) for the design and construction of the project was advertised in various key publications and on the internet. In response, 67 contractors and designers from around the world requested documents. Thirty of the respondents agreed to share their names with others. The EOI closed on September 9, 2004, with seven contractor/designer consortia having submitted responses. EOI evaluation was done by staff from OPG and the Owner's Representative in accordance with a predefined criteria and scoring methodology.

Prequalification of Applicants

The following EOI evaluation criteria were used to assess applicants for invitations to submit a proposal:

- Claims history,
- Environmental record,
- Financial capability,
- History with OPG,

- Performance on previous projects,
- Related construction experience on tunnel projects of similar complexity,
- Safety assessment,
- Tunnel construction experience in hard rock,
- Tunnel design experience.

Evaluation of a number of these criteria required making contact with references provided by the consortia with the objective of having a minimum of three references for each contractor and designer. Predetermined questions were formulated to maintain uniformity during contact with each of the provided references.

Following the prequalification process, four of the seven applicants were prequalified for invitation to submit a proposal.

Invitation to Prequalified Contractors to Submit Proposal

Invitations to submit proposals were issued to the four prequalified proponents at the start of a 5-month proposal period on December 22, 2004. Proposals from three proponents were received at the end of the period on May 13, 2005. Questions or queries from proponents during the proposal period were responded to in one of two ways—(1) any clarification regarding the invitation was provided in writing to all proponents, and (2) if the question or query necessitated a change to the Invitation, an amendment was issued to all proponents.

Evaluation of Proposals and Negotiation with Proponents

Evaluation of proposals was undertaken by a 30 person team comprising OPG management staff, Owner's Representative staff and external legal counsel, each of whom was required to sign a confidentiality agreement regarding proposal evaluation. The team was subdivided into commercial and technical evaluation teams. Details of the commercial proposal were kept secret from the technical team in order not to influence their technical evaluation. Code names were assigned to the proponents to preserve confidentiality and were used for all references made to proponents during the evaluation process.

The following predetermined criteria were established for proposal evaluation:

- Compliance with the Owner's Mandatory Requirements (Y/N gate),
- Proposed design and construction approach,
  - Cover of Summary of Work,
  - Draft drawings meet intent of concept drawings,
  - Preliminary designs meet requirements of invitation,
  - Proposed construction methodology and approach,
  - TBM equipment and operations,
- Response to Geotechnical Baseline Report,
- Price and other financial terms,
- Proposed project schedule,
- Tunnel flow guarantee quantity,
- Terms and conditions of the draft agreement,
- Risk management approach,
- Project team and key personnel,

- Preliminary project site specific safety, security and emergency response plan,
- Outline environmental compliance plan,
- Outline QA/QC program.

Relative scoring values for the criteria were not provided to the proponents.

All three proposals received were found to be compliant at the yes/no gate. As was expected by such an open proposal phase of the procurement, all three were significantly different from each other and all three had, to varying degrees, modified the terms and conditions of the draft agreement initially proposed by OPG. In order to produce an agreement acceptable within the expectations of the shareholder and the owner, that accommodated all of the constraints, and at the same time improve the design concepts to give a greater comfort level of achievement of the 90-yr design life mandated in the Owner's Mandatory Requirements, a hectic and detailed negotiation process began with all three proponents. During this process, the design details evolved and the terms and conditions of the agreement were modified to the satisfaction of both OPG and the proponents.

Recommendation to Owner to Enter into Design/Build Agreement (Contract)

Proposals were received on May 13, 2005. Negotiations commenced with all three proponents but were subsequently reduced to two and a preferred proponent was identified by the negotiating team in July 2005. The unsuccessful proponents were each paid an honorarium to offset the substantial costs of preparing their proposal and carrying out the necessary preliminary design work. Based on the recommendation from the negotiation team, project approval by the OPG Board and shareholder approval to finance the project were subsequently obtained and the Design/Build Agreement was signed by officers of OPG and Strabag AG August 18, 2005. Construction work commenced on site in September 2005.

CONCLUDING REMARKS

This paper serves to illustrate some of the key measures that were put in place by OPG and its Owner's Representative, Hatch Mott MacDonald in association with Hatch Energy, to address the unique aspects of procuring a construction contract for a large, technically challenging, water delivery tunnel in a major tourist area, spanning multiple municipal, regulatory agency and operational environments while satisfying shareholder and owner expectations. As of the date of writing this paper in January 2007, the project is tracking well, a key milestone was achieved exactly 12 months after contract award when the TBM commenced tunneling. Such a rapid start to tunneling is practically unheard of for such a large tunnel project.

ACKNOWLEDGMENT

The authors wish to thank Ontario Power Generation for its kind permission to publish this paper.

SAMPLE PROGRESS PHOTOGRAPHS

TBM nearing completion of assembly

TBM rear view before launch

Intake site showing marine construction

Intake cofferdam construction

Outlet site (RHS) and existing canals

ENVIRONMENTAL ASSESSMENT STUDY FOR EXTENSION OF SPADINA SUBWAY IN TORONTO

Verya Nasri

URS Canada, Inc.

Scott Thorburn

URS Canada, Inc.

ABSTRACT

The City of Toronto and the Toronto Transit Commission (TTC) have conducted an Individual Environmental Assessment (EA) for a 6.2 km, 4-station underground extension of the Spadina Subway from Downsview Station to Steeles Avenue. This Environmental Assessment Study was performed in three phases including routes and general station locations, alignments and station concepts and detailed assessment of the effects of the undertaking. URS Canada Inc. has led the consultant team for this Study. The anticipated total duration of the project is approximately 6 years and its estimated total cost is about $1.4 billion. This paper presents the planning process of the project, the environmental and engineering challenges, and the measures taken to address them.

INTRODUCTION

The City of Toronto and the Toronto Transit Commission (TTC) have conducted an Individual Environmental Assessment (EA) (in accordance with the requirements of the Ontario Environmental Assessment Act), for a 6.2 kilometre, 4-station underground extension of the Spadina Subway from Downsview Station to Steeles Avenue, with related commuter facilities (bus terminals, passenger pick-up and drop-off and commuter parking). This paper provides a summary of the EA Study, including background and context for the current Study, purpose and objectives of the EA Study and the Project, overview of the EA Study process, and key study findings and recommendations.

TTC and the former Metropolitan Toronto completed an Environmental Assessment Report for the Yonge-Spadina Loop Project. The EA established the need and justification for transportation improvements in northwest Toronto. The Study determined that extending and connecting the north ends of the Yonge and Spadina Subway Lines across Steeles Avenue was the preferred alternative. This recommendation was based on several alternatives including "doing nothing," roadway improvements, alternative technologies and modifications to the existing subway system. The Project became known as the Yonge-Spadina Loop.

In 1994, upon review of the TTC/Metro Toronto Environmental Assessment Report, the Minister of Environment and Energy authorized the extension of the Spadina Subway from Downsview Station to York University. At that time, it was anticipated that completion of the entire "Loop" from York University to Finch Station (via Steeles) would be many years into the future. Design and construction of the approved extension of the Spadina Subway did not proceed due to lack of funding.

Although the 1994 EA Approval remains in effect, its amending formula does not accommodate the consideration of alternative subway alignments and station locations. Therefore, a new EA has been undertaken to develop, review and analyze potential

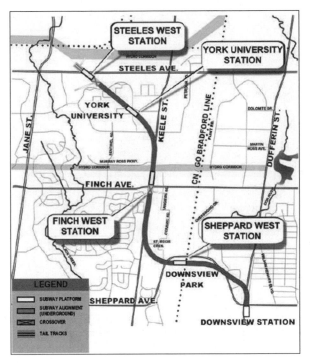

Figure 1.   The project

changes to the 1994 EA. The current EA Study builds on the Project need and justification established in the 1994 EA and the 2001 Rapid Transit Expansion Study. The new EA is being conducted to ensure the best alignment for the Spadina Subway Extension is chosen to capitalize on recent and future changes within the Study Area (Figure 1).

EA STUDY PURPOSE AND PROCESS

The purpose of the EA Study is to develop and analyze alternative alignments and station locations for the 1994 EA to extend the Spadina Subway from Downsview Station to York University and to study a further radial extension to a terminal station at Steeles Avenue. The EA Study has also determined environmental impacts and has developed mitigation measures to minimize any negative impacts.

Consistent with the 1994 EA, the objectives of the Spadina Subway Extension are to provide subway service to the Keele/Finch area, York University and a new inter-regional transit gateway and commuter parking facility at Steeles Avenue, provide improved connections between the TTC subway system and GO Transit, York Region Transit and other inter-regional transit services, support local population and employment growth up to 2031, in accordance with the land use and transportation policies of the new City of Toronto Official Plan and the York Region Official Plan, minimize any negative environmental impacts, and achieve reasonable capital and operating costs.

The Ontario Environmental Assessment Act requires the preparation and approval of a Terms of Reference as a prerequisite to the commencement of the Environmental Assessment Study. The Terms of Reference provides a plan for the preparation of the EA and a benchmark for the subsequent review and approval of the Study.

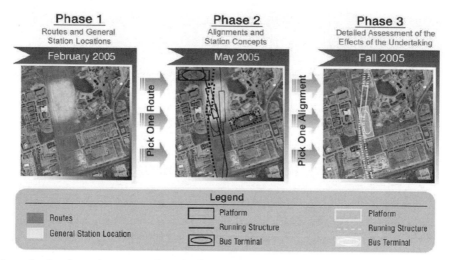

Figure 2. Spadina subway extension selection process

In 2004, TTC and the City prepared the Terms of Reference for the Spadina Subway Extension Environmental Assessment and retained URS Canada Inc. to conduct this EA, in accordance with the approved Terms of Reference.

The Canadian Environmental Assessment Act (CEAA) applies to federal authorities responsible for a decision or planned action, which enables a project to proceed in whole or in part. The requirements of the Act are "triggered" when a federal authority is the proponent of the project, provides funding or land to the project, or issues a permit, license or authorization (as prescribed in the Law List Regulations). Commencing with the Terms of Reference phase and continuing throughout the EA Study, the Study Team consulted with federal authorities (through the Canadian Environmental Assessment Agency) to identify any CEAA "triggers."

This EA Study was conducted in three phases, as follows:

Phase One: Routes and General Station Locations—This initial phase involved generating, evaluating and selecting alternative routes (i.e., corridors) and general station locations;

Phase Two: Alignments and Station Concepts—The second phase involved generating and evaluating more detailed alignments and station concepts within the technically-preferred route; and

Phase Three: Detailed Assessment of the Effects of the Undertaking—The last phase of the Study involved the detailed environmental assessment of the technically-preferred alignment and station concepts selected in Phase Two (Figure 2).

Public and stakeholder agency consultation is a key requirement of the Ontario Environmental Assessment Act. As such, a comprehensive plan was developed during the EA Terms of Reference and implemented during the EA Study. Accordingly, a dedicated telephone number (including TTY), fax, email address and web site were introduced during the Terms of Reference. Opportunities for public participation in decision-making at the three key milestones of the study included open houses (with commenting forms), interactive on-line commenting and facilitated workshops (both of which included detailed questionnaires).

A Technical Advisory Committee (TAC), consisting of senior staff of key stakeholder agencies, was established in early 2004 to provide ongoing advice and assistance to

TTC and the City of Toronto during the EA Study. TAC meetings were held on a monthly basis throughout the Terms of Reference and EA Study. In addition, contacts were established with a broad range of Federal, Provincial and Municipal agencies with a potential interest in the Project during the Terms of Reference stage. Stakeholder agency workshops, including participants from the TAC and a wide range of agencies, were held twice during the EA Study. In addition, ad-hoc meetings were arranged to discuss and resolve specific issues.

PHASE ONE: ROUTES AND GENERAL STATION LOCATIONS

Phase One of the Spadina Subway Extension selection process involved generation and evaluation of routes and general station locations followed by the selection of a preferred route and general station locations along that route. The preferred route and general station locations were then carried forward to Phase Two for further analysis. The Study Team defined routes as broad geographical corridors, within which a number of subway alignments may occur. General station locations were defined by the Study Team as broad geographical areas, within which, a number of station concepts may occur. The project objectives encompass transportation service, land use, socio-economic environment, natural environment and cost/revenue. Based on these project objectives, criteria were developed to guide the Study Team in the generation of routes and general station locations. The route generation criteria were released for review and comment during the first round of consultation.

Using the route generation criteria, the Study Team generated eight routes and general station locations. Seven key activities were conducted during the first phase of the EA Study, including a detailed inventory of existing and future conditions in the proposed Study Area, review and confirmation of the Study Area boundaries, review of alternatives to the Undertaking, development of alternative routes including general station locations, preparation of route evaluation criteria and indicators, public and stakeholder consultation, and evaluation and selection of the technically preferred route.

One of the main purpose of this EA is to identify existing conditions within the Study Area, including the existing transportation planning context, transportation systems, natural environment, socio-economic environment, existing and planned land use, and future transportation demands. This information is served as a baseline for the generation of alternatives, the prediction of condition changes/environmental effects and the identification of environmental protection measures.

The focal points of transit services within the Study Area are Downsview Station, the current terminus of the Spadina Subway, and York University, which is served by over 1,500 buses a day. Recent ridership trends show a significant increase in demand for bus service within southern York Region and to York University. In order to estimate future travel demand, the City of Toronto uses a transportation demand forecasting model, which converts predicted population and employment trends throughout the Greater Toronto Area into travel demand patterns for all modes including roads and transit.

Land use changes in the City of Toronto and York Region continue to result in greater demands on the existing transportation system. The Study Area is currently under the influence of a variety of planning policies, private sector redevelopment activities and major public sector investments. All of these planning documents, with their various levels of statutory approval and influence, support higher density redevelopment in anticipation of expanded rapid transit facilities. The Undertaking is needed to increase the overall person carrying capacity of the north-south transportation system that provides an essential interregional connection between the City of Toronto and York Region. The Undertaking will also provide a solution, which is consistent with provincial and municipal policies that promote more intensive land use patterns, densities and a mix of uses that encourage a balance of travel by all transportation modes.

During the first round of consultations, members of the public and stakeholder agencies were invited to review and comment on the inventory of existing and future conditions, review of alternatives to the Undertaking, proposed alternative routes and general station locations, and proposed route evaluation criteria and indicators. Based on these consultations, further details on local conditions were added to the Study Area inventory and several new or revised indicators were added to the route evaluation criteria.

Evaluation of Routes. A comprehensive list of evaluation criteria and indicators were developed by the Study Team in order to analyze and evaluate the alternative routes. These criteria were derived from the project objectives and included the following considerations: convenience for riders to walk to local stations, convenience for other modes of travel, conform with current approved planning documents, maximize redevelopment potential in support of the subway extension, maximize the potential to create a high quality urban/pedestrian environment, protect existing stable land uses, minimize the potential effects on important natural and cultural heritage features, minimize capital and operating costs of the subway extension, maximize revenue generated from the subway extension, maximize the subway extension in lands with no property costs to the Project, and quality of subway service.

The Study Team used two complementary evaluation methods to assist in the evaluation of alternatives: a qualitative method (Reasoned Argument Method) to evaluate the alternatives to select a preferred method of carrying out the Undertaking and a numeric method (Multi-Attribute Tradeoff System) as a test of traceability of decisions made.

Reasoned Argument Method (RAM) is the process of getting from one decision to the next through a series of logical steps. The method uses the measured amounts from the analysis step to rank alternatives from most preferred to least preferred. Indicators that generate no appreciable difference (i.e., are the exact same or are considered essentially equal) are identified as "not decision making criteria" and are eliminated from further analysis. Based on the preference by indicator, the alternatives are subsequently ranked at a criteria level, followed by the project objective level. This approach facilitates the creation of a list of advantages and disadvantages for each alternative. The relative significance of the effects is examined to provide a clear rationale for the selection of the preferred alternative.

The numeric method used was the Multi-Attribute Trade-off System (MATS). MATS-PC is a computer program designed by the U.S. Department of the Interior. It helps planners evaluate multi-attribute alternatives to reach a judgment of each alternative's relative worth or desirability. MATS-PC is an effective tool because it allows the application of sensitivity tests to determine how different weights and assumptions influence the results of the analysis. The output of the model is the composite of the analytical measurements and the weights assigned to each indicator. The preference is identified by the highest numerical output.

Following this process, Route 1 is found to be the most preferred in that it provides a cost-effective solution with the best connections between other modes, good service to each of the four identified catchment areas and strong support for future growth, while minimizing adverse environmental effects (Figure 3). Route 1 would include four following stations: Sheppard West Station, Finch West Station, York University Station, and Steeles West Station. During the second round of public and stakeholder consultation, the selection of Route 1 was supported by over 90% of respondents.

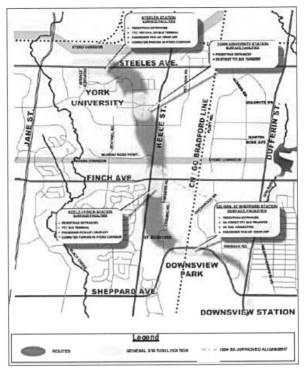

Figure 3.   Route 1

PHASE TWO: ALIGNMENTS AND STATION CONCEPTS

Phase Two of the Spadina Subway Extension selection process involved generation and evaluation of numerous alignments and station concepts within the Route 1 corridor and the selection of a preferred alignment and associated station concepts. The preferred alignment and station concepts were then carried forward to Phase Three.

Because all potential alignments would converge at the Keele/Finch intersection, Route 1 was divided into a southern section, from Downsview Station to Finch Avenue West and a northern section from Finch Avenue West to Steeles Avenue West. Alternative alignments were developed and evaluated for both sections. Any of the northern and southern alternatives could be combined to achieve the best overall Spadina Subway Extension alignment. Alternative alignments were developed using "alignment generation criteria," which were developed by the Study Team based on the overall objectives for the Spadina Subway Extension, subway operating requirements and TTC's engineering design criteria. Alignments were defined as 30-metre wide rights-of-way located within the technically-preferred Route 1. As shown in Figures 4a and 4b four alignments for the southern section and three alignments for the northern section of Route 1 were developed.

The alignment generation criteria were as follows: construct subway under road right-of-way to avoid disruption and minimize property acquisition, avoid petroleum storage facilities, avoid structures with deep foundations (buildings and existing bridges), construct below existing grade to minimize impacts to crossing roads and adjacent properties, avoid impacts to cultural and natural heritage features, avoid

Figure 4a.   Southern alignment alternatives   Figure 4b.   Northern alignment alternatives

stable residential areas, meet geometric design standards (absolute minimum radius = 300 m; desirable minimum radius = 700 m; minimum grade = 0.3%; maximum grade = 2.0%; minimum vertical curve K=60 m; all stations must be on at least 200 m of straight track), allow trains to operate at two-minute headways, provide cross-over and storage tracks for operational needs; and protect for further extension into York Region.

In support of the subway operations, "special track work areas" were identified based on the following subway operations requirements: provision for switching trains between northbound and southbound tracks at Finch West Station, to be used to provide reliable service or in emergency situations, which would require double cross-over tracks south of the station, turn back of trains at Steeles West Station terminus, requiring double cross-over tracks immediately south of the platform, tail-track structure north of Steeles West Station platform to allow full operating speed into Steeles West Station as well as to provide for temporary storage of trains, and the connecting track from Wilson Yard to Downsview Station, which has already been granted EA approved status (through the New Subway Storage and Maintenance Facility EA).

Evaluation and Selection of the Preferred Alignment. The Study Team developed a comprehensive list of indicators, which were refined based on public and stakeholder agency comments received during the second round of consultations. These indicators are based on the criteria listed below: potential for riders to walk to local stations, speed and comfort for subway passengers, convenience for transfers from buses (including Wheel-Trans and Mobility Plus) or rail to subway, convenience for access from other travel modes (taxi, bicycle, pedestrians (including ambulatory/nonambulatory disabled persons)), passenger pick up and drop off and commuter parking, flexibility for potential future subway extension into York Region, maximize redevelopment potential in support of the subway extension, maximize the potential to create a high quality urban/pedestrian environment, potential effects on natural heritage features, potential effects on geology, hydrology and hydrogeology, potential effects on socio-economic features, potential effects on pedestrian and traffic access/flow, potential effects on freight and rail passenger service and its signal systems at Sheppard West Station, potential effects on cultural heritage resources, potential effects on pipelines located in the Finch hydro corridor, minimize capital costs, minimize costs of property acquisition, and minimize net operating cost. Similar to the selection of the technically preferred route, the Study Team used both qualitative and numerical analysis methodologies to evaluate the alternative alignments and station concepts. Figure 1 presents the technically preferred alignment and station concept, which was endorsed by the public and the key stakeholder agencies during the third round of consultations.

PHASE THREE: DETAILED ASSESSMENT OF THE EFFECTS OF THE UNDERTAKING

The final phase of the study included public and stakeholder consultation, confirmation of the recommended subway alignment and station layouts, determination of environmental effects, development of mitigation measures for any negative effects, and commitments to future work to address any issues that are beyond the EA scope.

The environmental effects for the Spadina Subway Extension are classified as displacement of existing features by the subway, construction impacts, and operations and maintenance impacts. These impacts and proposed measures to mitigate any negative effects are summarized. During the EA process, partial takings have been identified for more than 30 properties. These include underground easements for the subway tunnels and partial surface takings for station entrances, commuter facilities (such as bus terminals and passenger pick-up and drop-off facilities) and ancillary facilities (such as Emergency Exit Buildings). TTC and the City of Toronto will conduct a Property Protection Study during the design of the Spadina Subway Extension, which will determine detailed property requirements, including temporary construction easements.

The City of Toronto (on behalf of TTC) would acquire properties and provide compensation through either a negotiated settlement or, in the event that expropriation is required, in accordance with the Ontario Expropriation Act. TTC and the City of Toronto are committed to the following process/principles for these impacted properties: early notification to property owners, ongoing meetings and discussions with property owners concerning property impacts to minimize property takings and identify mitigative measures, further investigations of alternative construction methods for the three-track structure, with the objective of minimizing direct impacts on these properties, while achieving reasonable Project costs, and uniform and equitable treatment, in accordance with the Ontario Expropriation Act.

For the TBM sections of the subway alignment, the construction impacts will be limited. Mitigation measures will be implemented to prevent or minimise or eliminate building settlement/structural stress due to tunneling. For stations, commuter facilities and sections of the alignment to be constructed by cut-and-cover method, mitigation measures will be implemented to prevent or minimise or eliminate the erosion and sedimentation due to excavation activities, effects of dewatering on aquifers and recharge/discharge areas, and building settlement/structural stress due to piling and dewatering.

Mitigation measures will also be implemented to minimize the disruption of existing vehicle circulation patterns due to road and lane closures and temporary traffic detours and diversions, impact on pedestrian circulation and safety due to road diversions and detours, noise and vibration generated by construction equipment, increase in air emissions including dust due to excavation activities and operation of construction equipment, impacts to local business operations (modified vehicle and pedestrian circulation patterns, reduced visibility of store fronts and signs, reduction in on-street parking, less convenient access to off-street parking, customer inconvenience due to temporary construction debris, noise and dust), and migration of mud and litter off site. Mitigation methods will include engineering studies and ongoing management and monitoring of construction activities.

Negative environmental impacts due to the operation of the Spadina Subway Extension and related commuter facilities would include potential traction power stray current impacts on buried utilities using metal piping, noise generated by subway, bus and private vehicle operations, vibration generated by subway and bus operations, potential electromagnetic interference created from the negative rail of the subway DC traction power system, and although the operation of the subway will result in overall reductions in air emissions, localized increases will occur due to increased bus traffic at Finch West and Steeles West Stations. Subway noise and vibration will be mitigated through the use of the double-tie track system, which has proven to be effective where implemented in the existing subway system. Stray current impacts will be mitigated through use of isolated and insulated power rails.

TTC will conduct further research and analysis for the construction of the Spadina Subway Extension including but not limited to the following activities: include noise, vibration and air quality monitoring and mitigation measures and construction site maintenance/upkeep requirements in construction contract documents, develop traffic, transit and pedestrian management strategies to be included in construction contract documents, prepare and implement tree and streetscape protection and restoration plans, undertake Designated Substances Surveys for any buildings or structures which require demolition and to reflect the findings in construction contract documents, develop procedures for disposal of excavated materials, including contaminated soils, in accordance with Ministry of the Environment requirements, prepare and implement a groundwater management strategy, prepare an erosion and sediment control plan, which complies with prevailing TRCA and City of Toronto water guidelines and require-ments, prepare an Environmental Management Plan including monitoring, triggers and contingencies in the event that further groundwater investigations indicate a potential adverse effect on the York University woodlots or other sensitive environmental fea-tures, arrange for a Stage 2 archaeological assessment to be conducted at areas where ground disturbance will occur during construction and which have archaeologi-cal potential, undertake buildings, structures and railway protection, monitoring and condition surveys, and undertake stray current protection and monitoring for pipelines and other utilities.

GEOLOGICAL CONTEXT AND CONSTRUCTION METHODOLOGY

The soil deposits in the Toronto region consist predominantly of glacial till, glaci-olacustrine, and glaciofluvial sand, silt and clay deposits. The soil deposits overlie the Georgian Bay Formation bedrock generally found about 50 to 75 m below the ground surface in the Study Area. The deposits encountered are grouped, in order from the ground surface down, as follows: Fill: typically ranging in thickness from about 1 to 5 m; Upper Till: with thickness from about 10 m to 25 m; Upper Sand/Silt Deposits: with thickness up to about 10 m; and, Upper Clay Deposits: typically ranging in thick-ness from about 5 m to more than 15 m. The lower permeability glacial till layers tend to impede groundwater flow whereas the interstadial deposits of silt and sand serve as local shallow aquifers.

Historically, three types of construction methods have been used for Toronto sub-way projects: TBM, cut-and-cover and open cut methods. In order to determine the environmental impacts arising during the construction of the Spadina Subway Exten-sion, preliminary investigations of the proposed construction methodology were con-ducted as part of the alignment development and evaluation process.

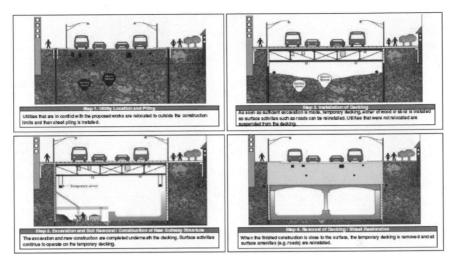

Figure 5.   Proposed top-down cut-and-cover construction method

Cut-and-cover method would be used for each of the four stations, the proposed cross-overs south of Finch West Station and Steeles West Station, and the tail tracks at the north-end of Steeles West Station (Figure 5). The remaining 4 kilometres of running structures would be constructed by TBM. Open cut method was also considered for the section of the alignment within the undeveloped Parc Downsview Park lands as an alternate construction method to TBM or cut-and-cover. This option was not carried forward due to potential long-term maintenance issues and concerns that an open cut configuration would be incompatible with short and long-term development plans.

Station concepts were developed based on travel demand forecasts, the proposed feeder bus network, future transit supportive development opportunities and TTC's engineering design criteria. These concepts included preliminary layouts of the subway platforms, station entrances, bus terminals, commuter parking, passenger pick-up and drop-off facilities and traction power substations (Figure 6).

Stations typically have a "centre platform" configuration in which passengers board and alight trains via a single platform between the two tracks. Centre platforms are preferred over side platforms because they allow greater utilization of vertical circulation, convenient cross-platform transfer where required and greater capacity to accommodate surges in traffic flow, especially during service interruptions. The concourse level is located directly above the platform and is connected to the platform through stairs, escalators and an elevator. The concourse permits transfers between TTC bus platforms at ground level and subway platforms.

Emergency Exit Buildings are used for the evacuation of subway train passengers in the event of an emergency and may be equipped with emergency backup power and ventilation equipment. In accordance with NFPA130, emergency egress from the tunnel shall be provided throughout the underground system so that the distance to an exit shall be not greater than 381 m. Therefore the maximum distance from emergency exit to emergency exit or emergency exit to station shall be 762 meters. Therefore, six Emergency Exit Building sites were identified for the recommended alignment.

To meet the traction power requirements for TTC's subway system, substations are typically 2.0 kilometres apart but cannot exceed 2.5 km in spacing. Since subway

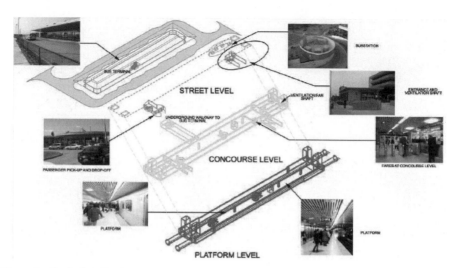

Figure 6. Typical station layout

stations require power for lights and equipment, TTC usually locates the electrical sub-stations near subway stations.

Ventilation shafts are incorporated into the subway station in order to balance air pressure within the tunnels and stations and to provide for emergency exhaust and fresh air supply in case of an underground fire. Ventilation fans can also be used to alleviate high summer temperatures in the underground stations.

Subway and Track Technology. TTC's subway cars have a length of approximately 22.8m and a width of 3.1m. The rated capacity for each subway car is 250 passengers. Trainsets of 6 cars result in a train length of approximately 137 m. Maximum operating speed is 80 km/hr. Trains are powered by electric motors, which utilize 600VDC current. Train operations, both locomotive control and opening/closing car doors, are manually controlled by on-board staff. Wayside signaling regulates the movement of trains along the line. Since this Undertaking is an extension of the existing 31 km Yonge-University-Spadina Subway, the current technology and operational requirements on the existing line will govern the operation of this Undertaking.

The track technology to be used is a combination of floating concrete slabs and double ties, which are designed to minimize the noise and vibration effects of subway operations to an acceptable level. The double tie trackbed system is designed to reduce vibration levels in the frequency range 30 Hz to 120 Hz by 14–16 dB in the box structure and by 12–15 dB in the tunnel structure. Sections of the Spadina Subway and the Sheppard Subway were built using such technology and have achieved desired results.

Project Cost and Schedule. The anticipated duration of a project of this scope would be approximately 6.5 years, including the following overlapping activities: design (2 to 3 years), construction (3 to 4 years), and testing and commissioning (1 year). The actual duration of the project would depend on project financing, property acquisition, construction methodology, contracting methods and procedures as well as other factors. The order-of-magnitude cost estimate for the 6.2 kilometre subway extension, including design and construction costs, property, yard improvements and vehicles, is $1.4 billion in 2005 dollars.

CONCLUSIONS

The presented EA Study developed and analyzed alternative alignments and station locations for the extension of the Spadina Subway in Toronto. It also determined environmental impacts and developed mitigation measures to minimize any negative effects. This paper provided a summary of the EA Study, including background and context for the current Study, purpose and objectives of the EA Study and the Project, overview of the EA Study process, and key study findings and recommendations.

REFERENCES

Toronto Transit Commission and the City of Toronto. 2004. Environmental Assessment Terms of Reference—Spadina Subway Extension Downsview Station to Steeles Avenue. Toronto, Ontario.

Toronto Transit Commission. 2001. Rapid Transit Expansion Study. Toronto, Ontario.

McCormick Rankin. 1993. Yonge-Spadina Subway Loop Environmental Assessment Report. Prepared for Toronto Transit Commission and The Municipality of Metropolitan Toronto. Toronto, Ontario.

NIAGARA TUNNEL PROJECT

Doug Harding

The Robbins Company

ABSTRACT

The Niagara Tunnel Project (NTP) is a 10.4 km (6.5 mi) long, 14.4 m (47.5 ft) bored tunnel that will run under the City of Niagara Falls from the Upper Niagara River to the Sir Adam Beck Power Station.

The completed project will enhance the capacity of the Sir Adam Beck Power Station by adding 500 m^3 (17,657 ft^3) of water per second through the tunnel.

The paper describes the project in which the world's largest hard rock TBM at 14.4 m (47.5 ft) in diameter is being used to excavate the tunnel. Features of the project include the use of a High Performance (HP) Main Beam TBM, state-of-the-art ground support and tunnel logistics systems. A continuous conveyor is used for muck transport from the TBM to the disposal area. The paper will also discuss the optimization of the TBM system as the tunnel advances through the varying geology.

In June 2004 the Ontario Government announced that Ontario Power Generation (OPG) had been given approval to proceed with the third tunnel under the city of Niagara Falls—the first two tunnels were constructed in the 1950s using drill and blast techniques.

The third tunnel, appropriately named The Niagara Tunnel Project (NTP), consists of a 10.4 km (6.5 mi) long TBM-bored, concrete-lined tunnel with a finished diameter of 12.5 m (41.1 ft).

In August 2005, OPG announced the selection of Strabag AG of Austria, one of the largest construction companies in Europe, as the contractor to build the NTP Project at a price of CAN $600 million.

In September 2005, Strabag A.G. selected The Robbins Company of Solon, Ohio to design, manufacture and deliver a new 14.4 m (47.5 ft) diameter High Performance state-of-the-art Main Beam Hard Rock TBM to be used in the excavation of the tunnel. In order to meet the aggressive construction schedule as requested by OPG/Strabag, the TBM needed to be supplied as "ready to bore" within a 12 month period after the contract was awarded to Robbins.

DESCRIPTION OF THE NIAGARA TUNNEL PROJECT (NTP)

The Niagara River is 59 km (35 mi) long and runs in a north-south direction from Lake Erie to Lake Ontario. It is an International Boundary between Canada and USA. The average flow of the river is estimated at 6,000 m^{3}per second (212,000 ft^3 per second).

Under the terms of a 1973 treaty all excess waters available for water diversion for power generation shall be divided equally between Canada and the United States. An exception allows Canada to divert an additional 142 m^3 (5,000 ft^3) of water per second from the Welland Canal or the Niagara River by a 1940 government agreement.

In order to best utilize the water available for diversion, the New York State Power Authority and Ontario Power Generation signed an agreement in 1965 to share generation capacity. Both power companies have rented available power generating capacity from each other to maximize utilization of their respective shares of water available for power production. This agreement allows each company to minimize the costs of equipment outages and to provide for the handling of ice or other power delivery problems. In

Figure 1. Tunnel route

simple terms, when Ontario Power Generation (OPG)–Sir Adam Beck Power Stations (Canada) has surplus water shares available for diversion and no extra generation capacity, they rent the generation capacity available at New York State Power Authority (NYSPA)–Robert Moses Power Station for power.

To meet increased demand, and to increase the power output of the Sir Adam Beck Project Station, a new tunnel to supply additional water capacity to the Sir Adam Beck power station and upgrading of the 16 existing generators will be done.

The new tunnel is known as the Niagara Tunnel Project, (NTP) and consists of boring a 14.4 m (47.5 ft) tunnel at a depth of 140 m (459 ft) below the city of Niagara Falls.

Currently 1,800 m^3 of water per second (63,566 ft^3 of water per second) are available to be diverted to the Sir Adam Beck Generating Stations for power production. The new Niagara Tunnel will allow an additional 500 m^3 (17,657 ft^3) of water per second when available to be diverted for power generation.

PROJECT ALIGNMENT

The new Niagara Tunnel will follow the same basic route as the existing two tunnels parallel to Stanley Avenue (see Figure 1).

The new tunnel starts on Ontario Power Generation property at Queenston with a −7.82% decline over a length of approximately 1,500 m (4,921 ft) reaching a depth of up to 140 m (459 ft) below the City of Niagara Falls. Here the tunnel proceeds with a relatively horizontal plane over a distance of approximately 7,400 m (24,278 ft). The alignment will follow a horizontal curve radius of over 1,000 m (3,281 ft) in length. The tunnel ends on the Niagara River at the International Water Control Dam located one mile upriver from the Horseshoe Falls with an incline gradient of +7.28% over the final 1,500 m (4,921 ft).

The inside diameter of the finished tunnel will be 12.5 m (41.1 ft) and will be lined with 50 cm (24 in) of un-reinforced concrete with double layer seal and pre-stressed injection concrete.

GEOLOGY

The geology is varied consisting of limestone, dolostone, sandstone, shale, and mudstone. The rock strength ranges from 15 to 180 MPa (2,100–26,000 psi), with

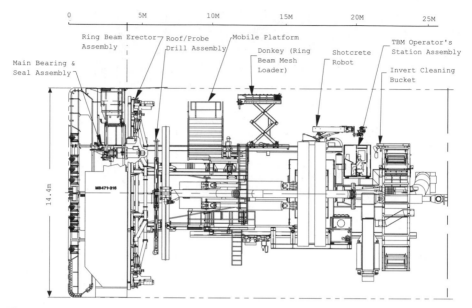

Figure 2.   TBM general assembly

most of the rock in the 100 to 180 MPa (15,000–26,000 psi) range. With the exception of sandstone, the geology is basically non-abrasive. Most of the muck (approximately 30%) removed from the tunnel will consist of Queenston Shale.

PROJECT APPROACH

For the construction of the NTP Project, Strabag A.G. purchased a new Robbins High Performance (HP) Main Beam TBM, and a new H.P. back-up system provided by Rowa Tunnel Logistics A.G. of Wangen, Switzerland.

The TBM Model 471-316, nicknamed "Big Becky," is the world's largest hard rock TBM ever manufactured. Design of the HP machine includes the use of 20" rear mounted cutters, high cutterhead power and state-of-the-art ground support equipment (see Figure 2).

Specifications for the TBM can be found in Table 1.

Cutterhead Design

The cutterhead design for the NTP project consists of a six piece bolted and doweled hard rock configuration, which includes 12 muck buckets with radial face and gage openings. Grill bars, abrasion resistant carbide buttons and abrasion resistant boltable bucket teeth are provided along the bucket openings. An abrasion resistant faceplate and gage plates along with periphery grill bars have been provided on the cutterhead structure. To avoid problems if sticky ground isencountered and to assist with the flow of the material and avoid plugging of the buckets, foam nozzles and rotary swivels have been provided. The finished weight of the cutterhead is in excess of 400 t (440 T) (see Figure 3).

Table 1. TBM specification sheet

TBM Type	Robbins HP 471-316
Year of Manufacture	2006
Machine Diameter (new cutters)	14.44 meters (47.2 Feet)

Cutters
Face
Center
   Number of Disc Cutters
   Nominal Recommended Individual Cutter Load
	
	Series 20 (508 mm)
	Series 17 (431.8mm)
	85
	35 t/cutter

Cutterhead
   Cutterhead Drive
   Cutterhead Power
   Expandable to 16 × 422 H.P.
   Cutterhead Speed
   Approximate Torque (low speed) 0–2.4 rpm
   Approximate Torque (high speed) 5.0 rpm
	
	Electric motors/safe sets, gear reducers
	6330 HP (15 × 422 HP)
	
	0–5.0 rpm
	18,800 kNm
	9,025 kNm

Thrust Cylinder Boring Stroke	(1729 mm) 68 inches

Hydraulic System
   System Operating Pressure at Maximum
   Recommended Cutterhead Thrust
   Maximum System Pressure
	300 HP (225 kW)
	275.7 bar (4000 psi)
	
	310 bar (4500 psi)

Electrical System
   Motor Circuit
   Lighting System/Control System
   Transformer Size
   Primary Voltage
	
	690 VAC 3-phase, 60 Hz
	120V/24 VDC
	4 × 1700 kVA + 1 × 1000 kVA
	13,800 V 60 Hz

Machine Conveyor Width	 54 inches (1370 mm)
TBM Weight (approx.)	1,100 metric tons, excluding drilling equipment

The cutterhead is equipped with 85 cutter discs, which include Robbins 20" wedge lock cutter assemblies with a nominal thrust capacity of 35 t (39 T)/cutter and an operating capacity of 50 t (55 T)/cutter. Overcut is provided by shimming of the outmost gage cutters should squeezing ground be encountered (see Figure 4).

Even though the Niagara geology is primarily soft rock, Strabag and Robbins agreed to provide the higher capacity cutters and 20" rings to reduce the need for cutter changes. In addition the 20" cutters and HP TBM configuration will allow the use of the TBM on future hard rock projects.

TBM ASSEMBLY

In order to comply with the aggressive construction program outlined by OPG/Strabag, the supplied TBM system had to be designed, manufactured, assembled and made "ready to bore" within 12 months after contract award.

The project team achieved this by the pre-assembly of the major critical components in a workshop and final assembly and commissioning of the complete machine at the project site. By doing this, in effect the "workshop" assembly was done at the

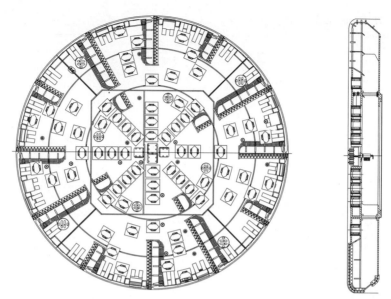

Figure 3.   Cutterhead configuration

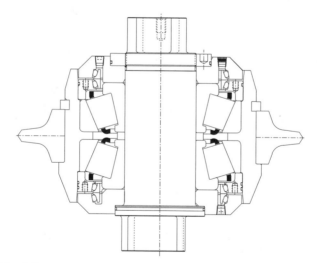

Figure 4.   Robbins 20" cutter assembly

jobsite using the operating personnel. Robbins supplied experienced supervision and specialty labor, while Strabag supplied the local labor.

This practice of jobsite assembly achieved a 12-month "ready to bore" schedule, which saved approximately 4–5 months on the TBM delivery schedule. In addition, there were project cost savings associated with labor and freight, as these operations only needed to be done once and not multiple times as with a workshop assembly (see Figure 5).

Figure 5.   TBM and backup assembly in launch pit

Table 2.   Rock support based on the type of geology encountered

Rock Type	Ground Support Required
Type 1	Steel wire mesh Shotcrete—50 mm thick
Type 2	Steel wire mesh Steel profile—UNP 100 Rock bolts—2.4 m long Shotcrete—70 mm thick
Typo 3	Steel wire mesh Steel profile—UNP 140 Rock bolts—3.6 m long Shotcrete—100 mm thick
Type 4	Steel wire mesh Steel profile—UNP 140 Rock bolts—3.6 m long Shotcrete—150 mm thick
Type 5	Steel wire mesh Steel beam—IPB 160 Rock bolts—6.0 m long Shotcrete—200 mm thick
Type 6	Steel wire mesh Steel beam—IPB 260 Shotcrete—100 & 200 mm thick

GROUND SUPPORT

The design concept of handling the ground support is to bring the primary support into the tunnel and handling of the support on top of the TBM. This allows the invert to be clear which allows free access of equipment for cleanup of the invert area. Contract requirements necessitated several different support systems based on the type of ground encountered. Rock support types ranged from Type 1 to Type 6 and are documented along with ground support methods in Table 2.

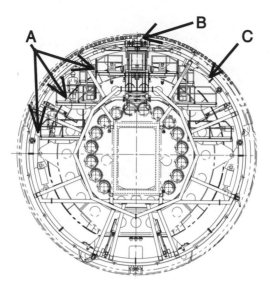

Figure 6.   Ring beam erector. Arrows B and C show the ring beam loader and ring beam erector assembly, respectively.

L-1 AREA ROCK SUPPORT

The L-1 area is located directly behind the TBM cutterhead support, which is approximately 4.1 m (13.5 ft) from the rock face. The installation equipment includes the following systems.

Ring Beam Erector

A rotary type ring beam erector with provisions to hydraulically lift the ring beam or channel section into place and hydraulically expand the steel sections against the bored rock is provided. The ring erector is located directly behind the TBM cutterhead support and allows placement of the ring beams under the protection of the roof shield fingers. The erector control functions are operated by a radio control system, which allows the operator the mobility to move along the top or bottom work areas as the ring is being erected (see Figure 6, arrow A points to the work platforms). Design and operation of the steel erector allows installation of the ring beams or channel sections during the mining stroke (see Figure 6).

Wire Mesh Erector/Material Handling Cart

A dual-function handling cart known as the "donkey" is located on the top section of the TBM main beam. The "donkey" transports the steel sections and wire mesh forward into the L-1 working area. Supply of the donkey includes a hydraulic lifting device to handle the wire mesh and steel sections to the crown where they can then be installed. Operation of the unit is by radio control and is independent of the boring stroke of the TBM.

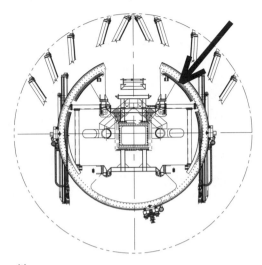

Figure 7. Roof drill positions

Rock Drills

Two Model 1532 hydraulic drills (provided by Atlas Copco of Stockholm, Sweden) were installed on 6 m (20 ft) long slides. The drills are installed on a rotary positioner, which allows independent operation of each drill. The positioner allows the various drill positions to be achieved in order to install the rock bolts per project requirements. Design of the system allows the bolts to be installed during the boring operation (see Figure 7).

Work Platforms

To assist the tunnel operating personnel in the installation of the rock support, there are various stationary and mobile work platforms located in the L-1 area. These platforms allow rock scaling, wire mesh and other ground support functions to be performed (see Figure 8).

Shotcrete Robot

Should shotcrete be needed in the L-1 area, a shotcrete robot has been installed and integrated into the work platforms. The robot has been supplied by Rowa (Wangen, Switzerland)/Meco (Sugarland, Texas, USA) and includes a boom to allow shotcrete coverage over a 180 degree section of the tunnel crown and at a rate of 20 m³ (700 ft³)/hour.

L-2 AREA GROUND SUPPORT

The L-2 ground support has been provided as part of the Rowa back-up supply. Included in the supply is the following.

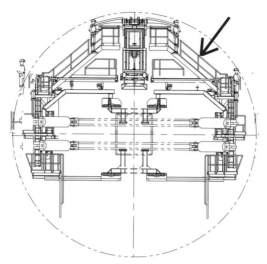

Figure 8.   Work platforms—arrow points to the work platforms, L-1

Rock Drills

To compliment the forward L-1 drills, two additional Atlas Copco Model 1532 hydraulic drills were installed on a rotary drill positioner to allow installation of 6 m (20 ft) long bolts.

Shotcrete Robots

Two remote controlled shotcrete robots were installed in the L-2 area. The units consist of a Meco Suprema unit, which allows 360-degree coverage. Each unit is independently controlled and has the ability to travel 8 m (26 ft) in the longitudal direction. The robots are charged by two Meco shotcrete pumps that deliver the shotcrete at a rate of 20 m (65 ft)/hour per pump.

Muck Haulage

Muck haulage is achieved by the use of a continuous conveyor system. As the backup is advanced, sections of conveyor are installed to allow continuous operation of the system. Muck is transported to the portal on the continuous conveyor where it is then discharged to an overland conveyor and to the storage area located adjacent to the jobsite.

PROJECT STATUS

As of drafting this paper, the TBM has advanced 250 m (820 ft) since start-up. Problems have been encountered due to higher than expected water inflows and handling of the water due to the 8.5% decline the TBM is excavating. The water removal system has been modified and the progress has increased to the expected advance rates.

As part of the boring process, Strabag A.G. will pour the final invert section on a bridge designed and supplied by BMTI, a sister company owned by Strabag A.G.

Once the TBM has advanced approximately 1.5 km (0.9 mi), the balance of the final concrete section will be installed on a secondary working bridge also supplied by BMTI. This system allows the final lining to be installed independent of the TBM boring operation.

CONCLUSION

The complete system used on the Niagara Tunnel Project has the capacity to achieve good advance rates and allow construction of the project per the required completion schedule required by OPG, the project owner.

The most important aspects to the project include the selection of a state-of-the-art TBM, ground support systems and backup system. The onsite assembly allowed for a quick start-up and cost savings. Strabag A.G.'s installation of the insitu lining concurrent with the TBM boring operations will allow the project to be completed per the aggressive construction schedule required by the owner. The assembly and technological innovations being used on the project have been successful to date because of proper up-front planning and partnering by the owner, contractor, and equipment supplier, as well as the dedicated personnel who operate and maintain the systems on a daily basis, and take great pride in the project.

ACKNOWLEGEMENTS

The author wants to thank Ernst Gschnitzer, Robert Goliasch, Christian Berger and Alex Herz of Strabag A.G. and Jost Wenk of Rowa A.G. for the information and assistance in providing data for this paper.

REFERENCES

Ontario Power Generation website: www.opg.com
Office of Public Affairs, Niagara Plant Group

PERMITTING NEW TUNNELS IN SOUTHERN ONTARIO, CANADA—PANACEA, PRAGMATISM, AND PROFESSIONALISM

Adrian Coombs

Regional Municipality of York

Paul Magowan

Marshall Macklin Monaghan Ltd.

Derek Zoldy

Earth Tech Canada, Inc.

ABSTRACT

Continued development and intensification are fuelling the urgent need for new trunk infrastructure within The Regional Municipality of York and throughout the Greater Toronto Area. Permitting of these projects is increasingly subject to public scrutiny to severely restrict dewatering through construction technologies and the requirement for a comprehensive and effective Environmental Management Plan.

Three current projects: one on the York Durham Sewage System and two in the City of Toronto

1. Conventional Design, Tender, Award: 4,150m (13,600 ft.) 2.7m (9 ft.) finished interior diameter soft ground tunnel for York Region's YDSS Interceptor Trunk Sewer

2. Negotiated Contract: 3,800m (12,500 ft.) 3.5m diameter (11.5 ft.) rock tunnels for Enwave's environmentally sustainable and innovative chilled water system in downtown Toronto

3. Conventional Design, Tender, Award: 2,200m (7,200 ft.) 3.3m diameter (10.8 ft.) rock tunnel for reinforcement of Hydro One's primary hydro feed in the downtown core of Toronto

These projects exemplify issues such as:

- EPB tunnel and sealed shaft technologies
- Substantial permitting requirements associated with dewatering
- Mitigation of rock creep
- Congested downtown rights-of-way
- Deep shafts and confined working compounds

Unique scheduling and cost benefits when Owner, Contractor and Consultant can work together cooperatively.

INTRODUCTION

Members of the general public, local residents and environmental interest groups apply pressure on the regulatory agencies that issue construction permits as well as on municipal and provincial politicians; to require additional studies, peer reviews, facilitated workshops and steering committees before municipal infrastructure can proceed to design and construction.[1] This allows cities and municipalities the opportunity to publicly demonstrate:

- How they propose to develop infrastructure that is located and fit for its intended purpose (appropriate technologies, appropriate pipe design criteria, consideration of topography, local versus municipal system),
- That the infrastructure is a timely response to approved, planned population and employment growth prediction,
- Stewardship of the environment,
- Recognition of existing legislation to protect valuable ground water resources, designated greenbelt areas and significant natural features,
- Provision of a forum for interested persons to learn the basics of responsible infrastructure management and advanced construction techniques.

All of these commitments come at a cost—to time, capital budgets and the construction industry. Delays in providing needed infrastructure can put existing sewers at risk of exceeding their capacity. In recent years the cost of York Region's capital delivery budget has escalated substantially, primarily to provide funding for a rigorous environmental and extensive monitoring program before, during and after construction. The heavy demand for construction technologies that respect and protect the environment puts a strain on the construction industry, materials and the work force, that is reflected in premium payments even when the construction is awarded to the lowest bid.

YORK DURHAM SEWAGE SYSTEM

Overview of the York Durham Sewage System (YDSS)

The YDSS is a wastewater collection system within the Great Lakes basin. It was constructed by the Province of Ontario in response to a 1965 decision that no additional sewage treatment plants could be built on the Humber, Don or Rouge Rivers. The concern at the time was that the assimilative capacity of receiving streams could be exceeded by continued local service. It fulfills some of the obligations imposed upon the Province of Ontario through the Canada-Ontario Agreement Respecting the Great Lakes Basin Ecosystem. What evolved from these initiatives was the most environmentally respectful system in the entire Great Lakes basin.[2] Because of the regulatory need to implement a sustainable YDSS, the initiation of the system was exempted from the full regulatory process of the day.

YDSS Interceptor Trunk Sewer

Currently, 94% of the Regional Municipality of York's existing population relies on sanitary services provided through the YDSS. The natural environmental, hydrogeological, physical and geomorphological settings as well as the current regulatory climate are collectively essential to the understanding of the project baseline and the potential impacts on the environment. For example:

Within the regulatory framework, the requirement for and siting of infrastructure is directed through the Provincial Places to Grow Planning Initiative; Greenbelt Plan for the Greater Golden Horseshoe; Oak Ridges Moraine Conservation Act; and the current York Region Master Plan update.

The project route in the Town of Richmond Hill extends along 19th Avenue from Yonge Street east to Leslie Street, and south on Leslie for a total of 4.4 km. The YDSS Interceptor Sewer is to be constructed in tunnel except for a 180 m section east of

Yonge Street by open cut. The tunnel section of the sewer will have a minimum internal diameter of 2.1 m with a minimum internal diameter of 1.65 m for the open cut. All construction has been accommodated under or along an existing Regional right-of-way.[2]

The paramount objective was through technology to protect and respect the Oak Ridges Aquifer Complex (ORAC) and the environmentally sensitive headwaters, watercourses and terrestrial features of the area surrounding the project alignment. For the purpose of the Environmental Approvals three general construction methods have been used: open cut where the water table lies below the trench invert; tunneling with a pressurized-face tunnel boring machine (Earth Pressure Balance (EPB) or Slurry Shield technology) for tunnel sections on 19th Avenue and Leslie Street; and sealed shaft techniques.[2]

This project must pass a rigorous review and evaluation from Conservation Authorities (CA) and the Ministry of the Environment (MOE). The construction technology and methods must also pass muster from other agencies when weighed against the importance of the terrestrial, ecological, aquatic and source water environments. Figure 1 displays the project location plan that passed the review of stakeholders and agencies.

TECHNOLOGY

The blueprint for construction technologies and methods has evolved by adopting the best of the old to advance often innovative construction techniques. The dual mandates for construction of the Interceptor by seeking alternatives to groundwater pumping and protecting the environment, is exemplified by the guidelines and constraints within the Minister of the Environment's letter of October 1, 2004. Abbreviated excerpts from four of the Minister's eleven requirements for the Interceptor include: an assessment of the impacts of the design and construction techniques on all local natural heritage features; monitoring and mitigation measures that address impact on long-term sustainability of groundwater and surface water resources in the watershed; cumulative impacts on the environment of dewatering activities on this and related sewer projects; independent peer review by appropriate experts of all technical studies.

Throughout the project was included a requirement to consult extensively with interested members of the public and stakeholders and to provide information and educational opportunities. Further, to document consideration of all concerns around the potential for damage to the function of the ORAC and dependent streams and wetlands in consequence of massive dewatering activities. The MOE adopted this pragmatic approach of specific guidelines and activities as the panacea to suit their permit approval process and to satisfy public and political concerns.

The geotechnical investigations (Geotechnical Investigation Data Report (GIDR) and the Geotechnical Baseline Report (GBR)) identified an upper till formation with a high water table and inclusions of saturated non-cohesive soils widely distributed within the till formation. Accordingly, the pressurized face tunnel method was adopted for tunneling while sealed shaft methodologies are mandated for all five shafts associated with the tunnel.[3]

The Minister's conditions of October 2004 for the YDSS Interceptor included conditions to be carried out prior to, and subsequent to, implementation of the sewer. These actions are particularly related to the assessment of any impacts on the environment resulting from construction dewatering activities. York Region takes the approach that, for this project, construction methods will be used that minimizes the use of dewatering.

The 19th Avenue and Leslie Street route and construction methodology was chosen to minimize overall construction impacts. Significant efforts have been made to

Figure 1. Project location plan

identify and mandate construction methods to limit the need to dewater; the number of dewatering locations; the volume of dewatering; and the duration of dewatering. Another feature of the project is the existence of a significant length of artesian conditions in conjunction with a limited thickness of confining till cap.[4]

For this project two Lovat Earth Pressure Balance (EPB) Tunnel Boring Machines (TBMs) (Model RME-129SEs) are being used to construct the 2.744 m internal diameter tunnels. The First EPB TBM will commence mining from Shaft 4 and mine eastwards for 3600 m. This TBM will pass through the ORAC and two intermediate sealed shafts during this drive. The second EPB TBM will mine from Shaft 4 for a single short drive of 533 m.

Although use of EPB and sealed shaft technologies comes at a known cost premium, the benefits far outweigh the risk of cost uncertainty and related impacts from dewatering to private wells and sensitive environmental features. These benefits include: minimized impact to the environment through the technologies of EPB—a one-pass system with the segmented rings (pipe) installed during excavation and sealed shaft construction methods to eliminate inflow of ground water; EPB technology has a growing presence in North America which is reflected in an increasing number of positive project outcomes; the MOE requires an application for a Permit To Take Water (PTTW) to address the worst case scenario of failure of proposed technologies, however, use of sealed shafts compliments EPB tunnel construction in water bearing soils

Figure 2.   Drilling at Shaft 2

to virtually eliminate the need for groundwater pumping, nevertheless a PTTW was granted by the MOE to York Region for such pumping, if required.[3]

The shafts installed in the aquifer are of sealed construction using slurry wall construction methods. With these two 25 to 28 m deep shafts, the inverts have been designed to resist uplift by the surrounding high water table. The contractor opted to use a thick concrete plug to resist the water pressure and rely on skin friction and mass, as opposed to a concrete slab keyed into the wall system. This method provides the safety bonus of not requiring divers to tie in the structural key. Two shafts outside the aquifer use gasketted liner plate and grouting around the shaft annulus to achieve the seal against ground water.[5]

The construction of the shafts by "sealed" construction methods has been mandated by this Contract eliminating the need to dewater.

Construction systems such as concrete diaphragm walls, interlocking piling systems and drilled secant pile shafts or ground treatment were all considered feasible. It was expected excavation of the lower portions of the shafts would be completed underwater to avoid the potential for heave or inflow at the base of the excavation. Since the shaft bottom slabs must be designed to resist hydrostatic pressures, these slabs will be installed with the shafts filled with water. These requirements were considered to be critical where the shafts penetrate the aquifer as is the case at Shaft 2 and Shaft 4 locations (Figures 2 and 3 depict construction at each shaft). Conceptual design of sealed shafts was based on the concrete diaphragm wall system.[3]

SHIFT IN RESPONSIBILITY

As indicated in Table 1, regulatory bodies now have to rely heavily on owners, their consultants and peers to identify, investigate and mitigate all construction risks and environmental impacts. To this end, agencies often dedicate in-house expertise to

Figure 3. Shaft 4

specific projects from the initiation of final design through final performance of the construction contract and often beyond. It's no longer the good old days; public and political pressures demand that governments change the way they do business and manage the environment:

> Decide that things are changing, and that "things ain't what they used to be and probably never was." —Will Rogers

Environmental Permitting

With the growing complexity, scale and scope of infrastructure projects, the onus of acquiring almost all permitting to address the health, safety and wellbeing of the public, workers and environment has shifted from the contractors to the owners and their consultants. The tunnel and shaft work within the ORAC must not detrimentally affect the groundwater or the water quality of area surface water. This means that applications are made by those most familiar with the need, the regulatory requirements and the resources to provide a thorough and meaningful appraisal of the project. It is now the owners' obligation to ensure contractors have the tools and understanding to comply with the new regulatory permitting requirements. Again, the provision of highly regulated and scrutinized projects comes at a premium cost both for the taxpayer and in extended time lines before commencement of permitted construction.[3]

One innovative initiative is the development and implementation of an Environment Management Plan Awareness and Compliance Training (EMPACT) for all project staff from the Region, the consultant and the contractor (including all site personnel from inspectors down to the machine operators). The Region's commitment is to ensure that the environment, public and worker health and safety are protected during the construction, operation and maintenance phases of the YDSS Interceptor project. The responsibility for environmental operating/work procedures and process is shared among all project members with a requirement for everyone to observe, identify and remediate any threat to the safety and wellbeing of the environment.

There is no single panacea or bandaid to enable the design and construction of new infrastructure. As an owner, York Region has developed an appropriately professional Regional specification for Environmental Management and Monitoring, based on the public interest in the Environment, availability of technology and the cooperative input of agencies.

Table 1. Shift in responsibility

Responsibility Issue	Owner		Contractor		Agency	
	Then	Now	Then	Now	Then	Now
Permit To Take Water (PTTW)	No direct responsibility	Applicant Responsible for regular reporting	Applicant Contractor retains consultant to perform field tests and report	Owner is Applicant Owner sets conditions in contract for access to Permit Owner retains consultant to perform field tests and report	Ministry of Environment (MOE) Issues PTTW to Contractor	Ministry of Environment (MOE) Issues PTTW to Owner
Permit for Discharge	No direct responsibility	Applicant Responsible for regular reporting	Applicant	Owner is Applicant and sets conditions for access to Owner Permit	Ministry of Natural Resources (MNR) issues Permit	Conservation Authority (TRCA) issues Permit with Department of Fisheries and Oceans (DFO) Letter of Advice
Environmental Monitoring of discharge	No direct responsibility	Applicant Responsible for regular reporting	Applicant, responsible for monitoring and periodic reports	Owner is Applicant Owner sets conditions in contract for access to Permit	DFO and MNR Lakes and Rivers Improvement Act	TRCA and DFO Lakes and Rivers Improvement Act
Environmental Management	No direct responsibility	Responsible for coordination and regular reporting	No direct responsibility	Responsibility extends down to individual workers, training of construction supervisors, project staff and resident inspectors	Reviewed and provided compliance reports and orders to Contractors	Current best practices at the time/primarily overseen by district offices of MOE, MNR, CAs and DFO

INTRODUCTION—ROCK TUNNELS, CITY OF TORONTO

Two further tunnel projects nearing completion in the downtown core of the City of Toronto have been successfully mined in shale. They include:

- Enwave Energy Corporation
  - Three separate, but linked, rock tunnels with three construction shafts
  - 3.5 m OD
  - 3.8 km total length
- Hydro One
  - One rock tunnel with two construction shafts
  - 3.3 m OD
  - 2.2 km total length

THE ENWAVE PROJECT

The Enwave project includes the construction of twin pipes, comprising a 1200mm diameter Cold Water Supply (CWS) and a 1200 mm diameter Cold Water Return (CWR), contained within a 3.5m OD rock tunnel. The project supplies environmentally sustainable chilled water to many of the high-rise office buildings, hospitals and condominiums in the downtown core of the City of Toronto.

The Hydro One project contains primary electrical power lines in a 3.3m OD rock tunnel. The tunnel connects two existing transformer stations and provides additional infrastructure strengthening to the primary hydro feed in the City's downtown core.

Who Is Enwave?

Enwave is one of the largest District Energy Providers in North America. They currently provide heating and cooling services to over 140 downtown office buildings.

What Is Deep Lake Cooling?

Deep Lake Cooling uses cold energy from Lake Ontario to cool buildings in downtown Toronto. Naturally cold water is drawn from Lake Ontario. The water is permanently at 4°C. Intake pipes are 83 m deep and the intake is 5 km off shore. The intake pipes were designed and built (by others) in conjunction with the City of Toronto. The City of Toronto uses the water, while Enwave uses the coldness.

Enwave utilizes heat exchangers to transfer the energy from the cold lake water to their closed chilled water supply loop. The energy transfer happens at the John Street Pumping Station.

Enwave has enough capacity to air condition ±100 office buildings (3.0 million square metres). Options to significantly increase their cooling capacity are currently being explored.

How Was Risk Managed for Enwave?

Uncertain Market. Initially there was an uncertain market for Enwave's ability to "sell" the concept of chilled water to Owners and Property Managers. The market was tested with a very limited first phase. When the market was proven to exist, the project was accelerated and expanded to meet the demand.

New "Utility." Part of the Owner's risk was that their chilled water system is considered a new utility in North America, with no track record to draw on for experience.

This risk was mitigated by the application of combined consultant and contractor knowledge. AWWA Potable Water Standards were applied and a Corrosion Specialist was engaged to monitor and address "stray" electrical currents from the nearby subway and streetcar systems.

Constructability issues were addressed by a thorough review of open cut vs. tunneling construction techniques. There was also an analysis of optimal tunnel shaft locations and an analysis of piping construction materials.

Cost and Scheduling Uncertainties. Cost and scheduling uncertainties were mitigated by teaming with a reputable contractor early on in the design process. A collaborative approach was used for:

- Scheduling
- Costing and
- Design

Rock Relaxation in Toronto Shale. Since the 1970s, the engineering and construction industries have become acutely aware of the rock relaxation that occurs in the Toronto shale after a tunnel is excavated. On occasion, in the past, the convergence has caused structural failure of the liner. However the risk was mitigated by monitoring the rock convergence and by specifying compressible foam lining where the schedule did not allow adequate time for the convergence to occur.

What Was the Consultant's role?

- Preliminary Engineering
- Design and Contract Administration for the extension of Enwave's Trunk Chilled Water System
- Extension of twin 1200 mm cement lined steel pipes in a 3.5 diameter tunnel under York, Wellington, Queen and Bay and
- Chilled Water Connections from trunklines to the highrise buildings

Unique Relationship Between Owner, Consultant and Contractor

The Original Tender only provided Outline Specifications and an Outline Project Description. "Indicative Bids" were solicited and C+M McNally were awarded the Contract.

Enwave, McNally & MMM continue to have weekly design/construction progress meetings. Construction contract extensions are now all negotiated. Currently there are three Master Construction Contracts with Change Orders issued when required.

THE HYDRO ONE PROJECT

The Hydro One tunnel is 2.2 km long and includes two construction shafts each 24 m and 29 m deep. The tunnel links two downtown substations. The tunnel route is under Front Street and under Union Station, the City's main subway and railway station. The tunnel also includes 2×90 degree tunnel bends, that were maintained within the public right of ways. This project was implemented with a conventional design, tender, construct process. DIBCO Underground is the contractor.

CONCLUSIONS—CITY OF TORONTO TUNNELS

Construction using tunneling technology in the downtown core of the City of Toronto, has greatly reduced the disruption to the pedestrians, traffic and business. After a hiatus of several years, the tunneling industry is once again alive and well in the Greater Toronto Area.

ACKNOWLEDGMENT

The authors would like to express their gratitude to Miss Anita Carley for her assistance in preparing this paper. Also, the authors would like to thank Mr. Steve Skelhorn of McNally Construction Inc. for his assistance and input into the preparation of this paper.

REFERENCES

[1] Vision 2026, York Region: Ontario's Rising Star, Towards a Sustainable Region, Fourth Annual Report on Indicators of Progress, Regional Municipality of York, Spring, 2006.

[2] YDSS Interceptor Sewer Study Report & Appendices, New Comparison of Alternative Route Alignments, 2005, Earth Tech in association with Alston Associates Inc., Beatty & Associates, Parish Geomorphic, and Michalski Nielsen Associates Limited.

[3] York Durham Sanitary System (YDSS) Interceptor Sewer on 19th Avenue & Leslie Street: Geotechnical Baseline Report, Richmond Hill, Ontario, 2006, Hatch Mott MacDonald in association with Earth Tech Canada Inc., Alston Associates Inc., W.B. Beatty and Associates Limited and Thurber Engineering Ltd.

[4] Chapman, L.J., Putman, D.F. "The Physiography of Southern Ontario" Third Edition, Ontario Geological Survey, Special Volume 2, Ministry of Natural Resources, Ontario, 1984.

[5] YDSS Interceptor Sewer Study, Environmental Management Plan, In Support of a Permit to Take Water Application and other Environmental Approvals for the 19th Avenue/Leslie Street Project, March 6, 2006, Earth Tech in association with Alston Associates Inc., Beatty & Associates, Parish Geomorphic, and Michalski Nielsen Associates Limited.

TUNNELING UNDER GLACIERS GALORE CREEK MINE ACCESS TUNNEL, BC, CANADA

Dean Brox
Hatch Mott MacDonald

Jason Morrison
NovaGold Resources, Inc.

Peter Procter
Hatch Mott MacDonald

ABSTRACT

The Galore Creek Mine project is one of the newest major mining projects under development in Canada. The project is located in a very remote area of northwestern British Columbia and is being developed by NovaGold Resources of Vancouver. The mine project comprises one of the largest and highest grade undeveloped porphyry-related gold-silver-copper deposits in North America. In order to access the mine site a new 120 km road is planned to be constructed including a 4.4 km mine access road tunnel. Glaciers with heavy snow pack cover approximately two-thirds of the tunnel alignment with a maximum cover of 1,200 m. The size of the access road tunnel is about 50 m^2 and has been designed as a large single lane road tunnel to accommodate the transport of large size mining equipment for mine start up and ongoing operations. The rock conditions along the tunnel alignment comprise very strong volcanic rocks. Key services for mining operations will be incorporated into the tunnel comprising diesel and power as well as the concentrate pipeline. Structural portal canopies have been incorporated into the tunnel design due to the risk of rockfalls and avalanches in the portal areas. The mine access road tunnel is a significant and critical-path component to the overall project to provide vehicular access to allow construction of the mine site. In order to meet the expected mine start up date the construction of the tunnel works are to commence prior to any other works at the mine site and will be constructed entirely by helicopter support. Geotechnical investigations commenced in the summer of 2006 with the removal of the overburden materials at the portals to assess bedrock conditions. Tunnel construction from both portals is expected to commence in the spring of 2007. The construction of the mine project is subject to and awaiting environmental approval in the spring of 2007. The mine access road tunnel will be the longest private road tunnel in North America. The key design and constructability issues for the mine access road tunnel project are presented and discussed.

INTRODUCTION

The Galore Creek Mine project is one of the newest major mining projects in Canada located in a remote area of northwestern British Columbia. The project is being developed by NovaGold Resources of Vancouver. The mine project comprises one of the largest and highest grade undeveloped porphyry-related gold-silver-copper deposits in North America. In order to access the mine site a new 120 km road will be required to be constructed including a 4 km road tunnel. The construction of the mine project is subject to and awaiting environmental approval in 2007. If constructed, the mine access road tunnel will be the longest private road tunnel in North America.

Detailed design for the mine access tunnel was completed in late 2006. The access road tunnel is about 50 m^2 and has been designed as a large single lane road tunnel to accommodate the transport of large size mining equipment for mine start up and ongoing operations. Key services for mining operations will be incorporated into the tunnel comprising diesel and power as well as the concentrate pipeline. The proposed tunnel will pass below two major glaciers with a maximum cover of over 1,200 m. A geotechnical investigation program commenced in the summer of 2006 with the excavation of overburden materials at the portals. Tunnel construction from both portals is expected to commence in the spring of 2007.

The tunnel was tendered in late 2006 and tender award is subject to environmental approval that is expected in March 2007. Notice to proceed is planned for March 2007 with early mobilization requiring the removal of significant snow in order to gain access to the portal areas to commence tunnel excavation. Excavation is planned to be undertaken from both portals in order to reduce overall project schedule which duration is estimated to be about 16 months.

The construction of mine access road tunnel at such a remote location introduces logistical challenges for the efficient mobilization as well as supply of labour, materials and equipment during construction required to meet a critical path construction schedule to allow for construction of the mine site.

PROJECT LOCATION, DEVELOPMENT AND LAYOUT

The project is located in a remote area of northwestern British Columbia, Canada within the historic Stikine Gold Belt. The mine property is approximately 75 kilometers northwest of Barrick Gold's Eskay Creek gold-silver mine and lies 70 kilometers west of Highway 37 and 150 kilometers northeast of the town of Stewart, British Columbia. The project area is characterized with steep sided valleys and glacier covered high mountain peaks up to 2,000 m in elevation with high snow pack. Figure 1 shows the typical glacial terrain of the project area.

The final feasibility study for the project was completed in October 2006 and estimated proven and probable reserves at 6.6 billion pounds of copper, 5.3 million ounces of gold and 92.6 million ounces of silver. Galore Creek also hosts estimated measured and indicated resources of 1.9 billion pounds of copper, 2.1 million ounces of gold and 24.5 million ounces of silver, with additional inferred resources of 2.4 billion pounds of copper, 2 million ounces of gold and 35.7 million ounces of silver. The ore deposit is planned to be mined by open pit methods over a period of about 20 years at a typical production rate of 65,000 tonnes per day. The estimated capital cost of the project is US$ 1.75 billion.

The key infrastructure of the mine project comprises a 120 km access road from near the Iskut River westwards via the More Creek Valley, a 4.5 km mine access road tunnel, an open pit, one of the world's highest tailings dam at 260m, and a remotely located filtration plant. Figure 2 shows the overall project location and layout of key infrastructure.

TUNNEL ALIGNMENT

The horizontal and vertical alignment for the mine access tunnel was defined by the selection of acceptable portal locations in terms of constructability. The tunnel portals were identified based on locations characterized with glacial overburden of shallow

Figure 1.   Glacial terrain of project area

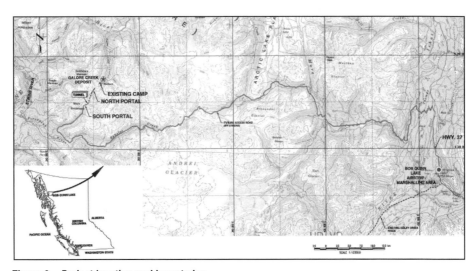

Figure 2.   Project location and layout plan

depth as indicated from seismic refraction surveys. The preferred location for the North Portal was at an elevation of about 960 m and out of the direct paths of potential rock-falls and avalanches from the overlying mountain slopes. The North Portal is located along the main north facing cirque slope surrounding the mine deposit. The portal slope extends from the base of the main Galore valley at an elevation of about 900 m up to a glacier covered mountain ridge at elevation of about 1,200 m. Figure 3 shows the general area of the North Portal.

The South Portal is located near the end of a major glacial valley upon the bench of a lateral glacial moraine at an elevation of about 1,080 m. Figure 4 shows the general area of the South Portal. The exact position of the South Portal was revised during the detailed design within Scott Simpson Creek valley to a preferred location of minimum overburden that would facilitate rapid preparation of the portal laydown area and

Figure 3.   North Portal area

Figure 4.   South Portal area

early excavation in bedrock for tunnel construction. Rockfall and avalanche hazards existed throughout the overall valley and no low risk locations were available to ideally site the South Portal.

The portal locations result in a horizontal alignment with a length of about 4,400 m and an overall downgrade northwards of 1% as shown in the plan and profile of Figure 5. The maximum elevation along the tunnel alignment is about 2,100 m located approximately at mid-way and provides a maximum cover of about 1,200 m. Approximately two-thirds of the tunnel alignment is covered by glaciers as can be seen in the orthophoto of the plan of Figure 5. A minimum cover of over 400 m is inferred below the glacier along the southern section of the tunnel alignment. A vertical alignment with a downgrade into the mine site from the South Portal was preferred for mine closure where groundwater inflows into the tunnel during operations would flow towards the main valley of the mine site and into the mine tailings area rather than require long term treatment of such flows into open stream courses.

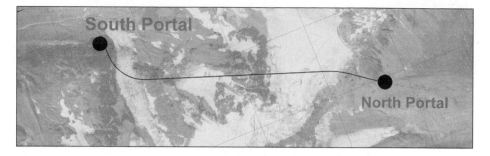

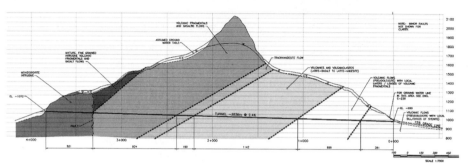

Figure 5. Tunnel alignment

MINE ACCESS ROAD TUNNEL CROSS SECTION

The cross-sectional size and shape was developed during the design based on providing internal vertical and horizontal clearances to allow for the passage of large size mining equipment for mine start up and subsequent operations. The dimensions of the largest envisaged mining equipment was 6 m wide by 4 m high. These dimensions in conjunction with allowances for tunnel support, excavation alignment control and rock deformation yielded tunnel excavation dimensions of 7.3 m wide by 7.0 m high. The tunnel cross section is shown in Figure 6.

A relatively flat arch shape was adopted for the tunnel roof in recognition of the hard rock environment and the expected ability to excavate and adequately stabilize the tunnel roof in a controlled manner without excessive overbreak. The floor of the tunnel incorporates a longitudinal side drain to convey expected groundwater inflows. A nominal road base has also been accommodated to provide a safe and low maintenance traffic surface that can be easily graded during mining operations.

GEOTECHNICAL CONDITIONS

Existing Information and Site Investigations

Numerous exploration boreholes have been completed within and in close proximity to the open pit deposit however there were no boreholes previously completed near the portals or along the proposed tunnel alignment. Regional geological mapping was undertaken as part of previous studies. In particular, detailed mapping has been completed in the project including the tunnel alignment by the B.C. Provincial GIS Geological Survey and digital data files of this mapping were provided. This information served

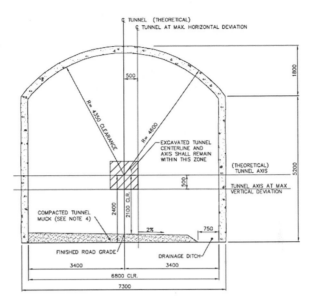

Figure 6. Mine access road tunnel cross section

to allow for the development of an original base geology map along the tunnel alignment for ground proofing. None of the regional faults that have been postulated in the
area of the open pit deposit are believed to extend near or across the tunnel alignment.

Detailed field mapping was completed along the tunnel alignment. Bedrock exposures were inspected and mapped for lithology and fracture information at the portal
areas as well as along the entire tunnel alignment. Limited exposures of representative
bedrock existed along the northern section of the tunnel alignment due to the presence
of glaciers and access was not possible along the central section due to the steepness
of the terrain. Following the field work a comparison of the field collected data was
made with the previously available base map to develop an updated geological plan
and profile of the tunnel alignment as shown as Figure 5.

Seismic refraction surveys have been completed at each of the portals to assess
the potential depth to and profile of the underlying bedrock. In view of the extensive
outcrops of representative geology along the surface of the tunnel alignment and the
absence of any foreseeable adverse geological formations along the tunnel alignment,
the required depth of drilling, and time/logistic challenges, investigation of the geotechnical conditions at tunnel elevation were not undertaken. Other geotechnical investigation work included petrographic analyses of selected rock samples and acid rock
drainage (ARD) and metal leaching (ML) testing.

Regional and Site Geology

The Galore Creek deposit area straddles the western boundary of the Intermont-
ane Belt with the Coast Belt, and is underlain by rocks of the Stikine Terrane (or
Stikinia, a composite allochthonous terrane composed of volcanic island arcs of mid-
Paleozoic to Middle Jurassic age, intruded along the western margin by Jurassic to
Paleocene plutons of the Coast Plutonic Complex).

The main rock types to be encountered along the tunnel alignment mainly comprise a repetitive package of moderately south-dipping volcanic/volcaniclastic bedrock

while syenite porphyry will be encountered near the North Portal and monzodiorite along the southern most section. The volcanic and volcaniclastic rocks along most of the tunnel alignment mainly appear to comprise (1) stratigraphically lowermost, plagioclase-bearing, dark green (chloritic) basalts with locally abundant bright yellow-green patches of epidote, (2) overlying pyroxene-phyric basalts (rarely observed in the tunnel area) and (3) an uppermost, thick, highly alkaline package composed of variably pseudoleucite-bearing flows and re-sedimented (volcaniclastic) rocks. However, along the proposed tunnel alignment, these units appear to be structurally repeated, presumably by folding and faulting, so that the basalts may occur at the highest elevations examined, and the volcaniclastic rocks at the lowest elevations examined as indicated in the geological profile of Figure 5.

The main rock types along the mine access road tunnel are fractured and can be characterized with several sets of fractures with different orientations. These sets of fractures will combine to form potentially unstable wedge blocks along the crown, haunches and sidewalls of the tunnel that will be required to be supported to prevent fall-out. The spacing of the rock fractures within the volcanics and granodiorites varies typically from 0.2 m to 0.5 m and from 0.1 m to 0.4 m respectively based on observations from surface outcrops. A total of five main fracture sets have been identified to be associated with the volcanics and four main fracture sets within the monzodiorites. Based on overall outcrop observations, the foliation/bedding (Fracture Set # 1) is the most prominent fracture set within the volcanics. A limited number of faults are inferred to be present along the tunnel alignment. A single major fault with a width of 15 m has been inferred to be present below the glacier along the south section of the alignment as shown in Figure 5.

Portal Conditions

Overburden materials are present at both portal based on observations and the results of the seismic refraction surveys. These materials at the South Portal comprise glacial tills and mixed talus to an estimated depth of about 12 meters. Figure 3 shows the overburden materials at the South Portal.

The North Portal is comprised of mixed talus material as shown in Figure 4. The existing rock slope immediately above the portal location is very steep and relatively good quality rock mass conditions. There are a number of potentially loose rock blocks along the upper portion and crest of the rock slope that will require local scaling and possible stabilization works to provide a safe working area below for the excavation of the rock face to start tunnel excavation.

Rock Mass Quality and Strength

Rock mass quality has been assessed using the Rock Mass Rating (RMR) rock mass classification system based on inspection of borehole core from the geotechnical drill holes completed along and near-by the tunnel alignment. The RMR rock mass classification system is based on five parameters describing rock conditions in terms of intact rock strength (UCS), rock quality designation (RQD), fracture spacing, fracture condition, and groundwater condition.

The rock mass quality of the main rock types along the tunnel alignment has been assessed based on outcrop mapping. Upon consideration of the rock mass quality data from the outcrops and recognition that additional minor faults and fracture zones can be expected, an overall summary of rock mass quality for the tunnel alignment is presented in Table 1. In general, most of the rock conditions expected to be encountered along the tunnel alignment can be described as "Fair" to "Very Good."

Table 1. Summary of rock mass quality

Rock Mass Quality	Rock Mass Rating Range	Estimated Percentage, %
Very Good	80–100	25
Good	60–80	30
Fair	40–60	25
Poor	20–40	15
Very Poor	0–20	5

The strength of the various rock types to be encountered along the tunnel alignment has been evaluated from a limited number of uniaxial compressive strength (UCS) tests completed on block samples collected at site. The laboratory test results are within the typical range for the volcanics and granodiorites. Based on a review of these testing results and field observations the rock strengths of the rock types are generally uniform and can be expected to vary only to a limited degree since there is no weakening effect due to significant alteration of the rock types along the tunnel alignment. It is particularly noted that the volcanic rocks can be generally characterized as extremely strong. The rock strength of the monzodiorite is estimated to be about 200 MPa and the rock strength of the volcanics about 260 MPa.

Groundwater Conditions and Predicted Inflows

The groundwater conditions and expected groundwater inflows along the tunnel alignment have been assessed based on consideration of the site topography, tunnel alignment geology and overall geological environment. The depth to the groundwater table is inferred to be relatively shallow along most of the tunnel alignment. Hydrostatic groundwater pressures can therefore be expected to vary accordingly with topography increasing to an inferred maximum of about 900 m along the central section. The maximum hydrostatic groundwater pressure below the glacier along the southern section may be as much as 275 m.

Groundwater inflows will emanate along open fractures and at major fault and/or fracture zones intersected along the tunnel alignment. Expected groundwater inflows along the tunnel alignment have been evaluated using an established empirical approach for both non-grouted and grouted fault zones conditions and assumed rock mass permeabilities typical for the rock types. Notwithstanding the assumed rock mass permeabilities for the groundwater inflow predictions, it is noted that higher rock mass permeabilities may exist within the rock conditions along the tunnel alignment due to the presence of open fractures at depth which is not uncommon based on past experience in similar geological environments. The predicted groundwater inflows along the entire tunnel alignment prior to breakthrough are 110 L/s without pre-excavation grouting and 55 L/s with pre-excavation grouting.

Rockfall and Avalanche Potential at Portals

The South Portal is located along the lateral moraine mid-way up along a major 300 m high slope. The existing slope comprises bedrock outcrops that have been subjected to alteration and weathering with the presence of some loose rock blocks. Rockfall and avalanche chutes are present along the slope which restrict rockfalls and snow to these locations however rockfalls and avalanches may manifest from higher areas and occur over wider areas around the portal. Rockfalls of varying size have occurred in the past along this slope. Rockfall protection measures comprising scaling and slope fencing will be prudent to install as part of the portal works to protect workers and equipment during portal and tunnel excavation. Given the high snow pack of the project area active full-time avalanche management is deemed necessary during portal and tunnel excavation to limit the extent of snow build-up for subsequent routine snow removal within a practical time frame.

The North Portal is also located near the toe of a major rock slope where potentially unstable rock blocks can be observed along the upper portions of the slope. Scaling and cleaning of the rock slope will be required to be completed as part of the portal works and rockfall protection measures comprising slope protection fencing will be prudent to install as part of the portal works to protect workers and equipment during tunnel excavation.

DESIGN AND CONSTRUCTABILITY ISSUES

General

During detailed design the following key issues were evaluated and addressed:

- Identification of preferred/acceptable portal locations
- Tunnel excavation and support
- Tunnel Operations Requirements—Refuge Bays and Ventilation
- Portal canopy requirements
- Advanced Construction of Portals

Preferred Portal Locations

The mine access tunnel provides the vital link connection from the Galore Creek valley at the mine site to the western end of the Scott Simpson valley. The western end of the Scott Simpson valley is characterized with steep sided natural rock slopes above which are covered with significant snow pack throughout much of the year as shown in Figure 6. Several narrow gullies are present along the eastern side of the valley and extend from the upper high slope areas down to the bottom of the valley. Rockfalls and avalanches are typically channelled along these gullies and presented challenges for the siting of the South Portal at the end of this valley. The final location for the South Portal was selected based on consideration of the pathways of rockfalls and avalanches, rock quality, and overall access road constructability restrictions. Similarly, the final location for the North Portal was selected based on consideration of the pathways of rockfalls and avalanches as well as the inferred shallowest depth of overburden materials.

Tunnel Excavation and Support

Given the relatively good quality, extremely strong rock conditions that are expected over most of the tunnel alignment, it is envisaged that the mine access tunnel will be excavated by full-face drill and blast methods. The tunnel cross section is approximately 50 m^2 and represents a moderately large cross section. However, this size of tunnel face area can easily be excavated with standard and modern tunneling equipment comprising a multi-boom jumbo drilling machine and muck haulage vehicles. The tunnel geometry may be modified from the current tunnel cross section subject to the availability and selection of the tunnel contractor's equipment and the preferred methods of tunnel construction.

Table 2. Tunnel support classes and percentage distribution

Tunnel Support Class	Description	Rock Mass Quality	Rock Mass Rating, RMR Range	Estimated Length, m	Estimated Percentage, %
1	Crown Pattern Bolting	Very Good	80–100	1,122	25
2	Crown Pattern Bolting	Good	60–80	1,346	30
3	Partial Pattern Bolting	Fair	40–60	1,133	25
4	Full Profile Pattern Bolting	Poor	25–0	673	15
5	Lattice Girders + shotcrete	Very Poor	0–25	224	5

Tunnel excavation is expected to proceed from both portals in order to minimize the duration of construction. Typical excavation advance rates of each heading can be expected to vary for this size of tunnel from about 1.0 m/day for the very poor ground conditions associated with major faults zones requiring the installation of high capacity support measures, up to 9 m/day in very good rock conditions with an overall tunnel excavation progress rate of about 6 m/day per heading.

Based on the expected rock conditions along the tunnel alignment and the assessment of tunnel stability, and in accordance with standard industry practice, an appropriate level of tunnel support will be required to be installed to maintain the long-term stability and safety of the tunnel given the operating requirements of the mine access tunnel and to minimize maintenance disruptions during operations.

With the likely fast-track nature of the overall project and critical schedule for the mine access tunnel, it is suggested that initial tunnel support systems comprising rock bolts, mesh, shotcrete, and lattice girders be installed as required concurrently with tunnel excavation and that final tunnel support comprising any additional rock bolts and any necessary shotcrete lining be installed after breakthrough and removal of the ventilation duct and all services. This tunnel excavation and support approach is considered to result in the shortest overall construction schedule without jeopardizing the integrity of the tunnel.

With the expected variable rock conditions to be encountered along the tunnel a series of five tunnel support classes will be required to be designed to cater for these conditions. Tunnel support classes are related to rock conditions as encountered defined in terms of rock mass quality using a recognized rock mass classification system. Rock mass quality data is collected as part of tunnel mapping during excavation. The required tunnel support for each excavation advance is evaluated based on consideration of the rock mass quality and geological mapping data as well as practical experience.

A typical series of tunnel support classes and their component descriptions for the proposed mine access tunnel is presented in Table 2. The estimated percentage for each tunnel support class is also presented in Table 2 and includes an allowance of Class 5 tunnel support for the intersection of the major fault below the glacier.

Tunnel Operations Requirements—Refuge Bays, Ventilation Jet Fans and Sealed Doors

The mine access road tunnel will be operated as a private use tunnel. The design incident scenario for the mine access road tunnel comprises a vehicle fire and the escape of maintenance workers dictates a fundamental requirement for emergency egress to fresh air. For a single tube tunnel this requirement is most commonly provided by self-contained refuge stations/safehouses located along the tunnel at appropriate intervals. The purpose of the refuge stations/safehouses is to provide a safe haven of breathable air and first aid supplies to any incident victims. The size of the refuge stations must be sufficient for the maximum number of expected people that could be impacted by a possible and representative incident in the tunnel.

While there is no recognized applicable regulation for the spacing of refuge stations for this type of mine access tunnel, it is deemed appropriate that the interval of the refuge stations can be greater than that required for normal underground mining operations where there are several workers located in close proximity (300 m), and greater than that required for a high traffic volume highway tunnel (250 m). For the selected representative safety design incident involving a major vehicle fire (20 MW), a medium duration travel time to the refuge station of no more than 8 minutes is considered to be appropriate based on a normal walking speed of 4 km per hour. In comparison, the maximum allowed travel time (NFPA 130) to a place of safety due to transit train fire in an underground transit station is 4 minutes. The Inspectorate of Mines of British Columbia has endorsed this overall approach.

Refuge bays/safehouses have been designed to be constructed at intervals of 500 m along the tunnel to provide emergency shelter in the event of a fire within the tunnel. The refuge bays will incorporate sealed fire proof doors, oxygen cylinders, carbon dioxide absorbers, water supply, blankets, first aid materials, toilet, a collapsible injury cot, and telephone communications in order to safely protect and to shelter 12 persons who may be traveling and/or completing maintenance work within the tunnel. A total of 8 refuge bays will be required. The refuge bays will be separated from the tunnel by an airlock entry system with double doors in each bulkhead. The toilet will be located in the airlock.

During normal operations the mine access road tunnel will be self-ventilating due to the difference in elevation of the portals. This natural ventilation may be affected by weather conditions at the portals. A mechanical ventilation system will be required to provide airflow when the natural ventilation is inadequate, or in the event of a fire.

Three banks of three jet fans each will be positioned within the tunnel along the northern (downhill) section and near the tunnel portals and activated in the event of unacceptable air quality as monitored by sensors and/or an incident/fire in order to regulate the air-flow and force/exhaust smoke out of the tunnel in the most appropriate direction depending on the location of the incident. In the event of a fire in the tunnel the ventilation system will be activated from the mine dispatch office following a telephone or radio call to mine dispatch from a person involved in the incident or following an alarm signal from an array of air quality sensors.

Additional fire, life and safety components that will be incorporated into the tunnel include radio communications, fire extinguishers, air quality sensors, telephones, reflective signage, and lighting. Finally, sealed mine-shop doors located inside each portal have been incorporated into the design of the tunnel that will be manually activated to allow passage through the tunnel. The doors will serve to prevent cold air entering the tunnel and causing freezing of groundwater inflows along the walls of the tunnel and will facilitate ventilation under emergency conditions.

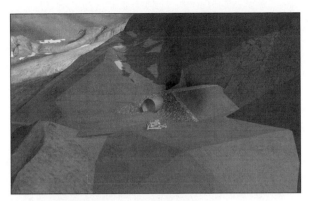

Figure 7. Portal canopy structure with backfill protection

Portal Canopy Requirements

Rockfall and avalanche hazards will exist during operations of the mine access tunnel. Portal canopy protection structures have been included at each portal to provide adequate protection of the tunnel against these natural hazards. The portal canopies will comprise pre-manufactured large diameter (9 m) metal culvert structures that are designed to accommodate variable impact loads from rockfalls and avalanches along the length of the canopies. The canopies have been designed to a length of 65 m that has been adopted based on consideration of rockfall and avalanches run-out analyses based on location-specific topographical and geotechnical parameters. Figure 7 shows a 3D image of an envisaged portal canopy including backfill placement around the structure.

Overburden Removal at Portals

In order to confirm the suitability of the selected portal locations, geotechnical investigation comprising the removal of overburden was undertaken during the summer of 2006 to expose bedrock. This work required the mobilization of large capacity earthworks equipment (D8 bulldozers and back-excavators) by a Chinook helicopter with a payload capacity of 10 tonnes. Geotechnical investigation at the North Portal required removal of talus slope debris down to bedrock. Geotechnical work that was undertaken at the North Portal laydown is shown in Figure 8. Excavation at the South Portal comprised establishment of temporary access up along the lateral moraine to the steep rock bluffs and removal of glacial overburden material to confirm the depth profile of the bedrock. This work will assist in the mobilization of large tunneling equipment by helicopter in the spring of 2007. Figure 9 shows the laydown area at the South Portal.

CONCLUSIONS

The Galore Creek mine access road tunnel comprises a very interesting tunneling project located in a remote area to be entirely constructed by helicopter support. The remote location introduces logistical challenges for the efficient mobilization as well as supply of labour, materials and equipment during construction required to meet a critical path construction schedule to allow for construction of the mine site. The mine access road tunnel will be the longest private road tunnel in North America.

Figure 8.   Overburden removal at the North Portal

Figure 9.   Overburden removal at the South Portal

ACKNOWLEDGMENTS

The authors gratefully acknowledge the permission of NovaGold Resources Inc. to publish this paper. Hatch Mott MacDonald of Vancouver completed the tunnel design. Dr. Craig Leitch provided specialist geological mapping services. Bruce Geotechnical Consultants (BGC) provided geotechnical input for portal rockfall assessment and mitigation measures. Snow Avalanche Services (SAS) provided specialist input for portal avalanche assessment and snow loading for the portal structural canopies.

6

Geotechnical and Ground Improvements

Chair
S. Boone
Golder Associates

BEACON HILL STATION DEWATERING WELLS AND JET GROUTING PROGRAM

Zephaniah Varley

Parsons Brinckerhoff

Richard Martin

Shannon & Wilson, Inc.

Red Robinson

Shannon & Wilson, Inc.

Paul Schmall

Moretrench American Corp.

Dominic Parmantier

Condon Johnson & Associates, Inc.

ABSTRACT

Sequential excavation of the Beacon Hill Station has been accomplished in highly complex and inter-layered sand, silt, and clay soils with multiple perched groundwater tables. Successful excavation has required the installation of a deep dewatering well system used to depressurize sand layers above and within the excavated tunnel faces. In addition, the presence of sand layers in the crown of the Platform Tunnels required jet grouting at depths of up to 49m (160-ft). This paper discusses the design intent, construction, and effectiveness of the dewatering system and the selective jet grouting of sand horizons and shaft break outs for the successful construction of the deepest SEM station excavation ever completed in the United States.

INTRODUCTION

The Beacon Hill Station and Tunnel project is a $280 million section of the Central Link Light Rail Line that extends for approximately 24 km (15 miles) between downtown Seattle and the SeaTac Airport. The project consists of an aerial station with 1.2 km (¾ of mile) aerial-guideway; twin 1.4 km (4,300-foot long), 6.4 m (21-ft) diameter running tunnels; 3 mined cross passages between the running tunnels; portal and tunnel approach structures with retained-fill and cut-and-cover structures; and a deep-mined underground binocular station with an entrance structure and surface plaza, station elevator access shaft and ventilation facility with a separate emergency egress and ventilation shaft at Beacon Hill.

STATION EXCAVATION DESIGN

The Beacon Hill Station will be the deepest station excavated in soft ground in the United States and will utilize the state-of-the-art Sequential Excavation Method (SEM; a.k.a. NATM in Europe) for construction of the station. The Station access shafts and head houses are supported with slurry walls, and the initial tunnel lining

346

for the underground Station excavation consists of lattice-girder reinforced shotcrete. See other chapters on the Beacon Hill Station for additional figures with tunnel labels and dimensions. Station excavation has progressed through a complex sequence of glacially overridden deposits consisting of very dense and hard clay, silt and sand, gravel and cobbles with multiple ground water levels in granular deposits, typically due to perched groundwater overlying clay and till units of cohesive soils. Deep jet grouting and a system of dewatering wells have been used to solidify and drain the larger granular deposits in advance of SEM mining. This paper describes the installation jet grout columns and deep dewatering wells, and describes the effectiveness of these two primary pretreatment tools with regard to SEM tunnel excavation to date.

SITE LAYOUT CONSTRAINTS

The Contract-defined, Noise Wall-enclosed Beacon Hill Station work site is a one-acre lot, and as is likely to happen on tight-schedule, tightly constrained work site, construction activities overlapped each other with regularity. Five months after Notice-to-Proceed there were exploratory drill rigs, dewatering well drill rigs, jet grout drill rigs and plant equipment, a Hydrofraise and a Clamshell slurry wall excavator, and slurry wall desanding plant and equipment all vying for space on site. The slurry wall excavators were alternating between deep shafts and shallower head house walls. Instrument drilling and utility work was also on-going both inside and outside the site, so initial jet grouting work began on the Northbound Platform Tunnel (NBPT) and proceeded into the adjacent El Centro parking lot in an area unto itself, while dewatering wells began along the Southbound Platform Tunnel (SBPT) and south of the Main Shaft. Original jetting limits also included work for the West Longitudinal Vent Adit (WLVA) west of the Main Shaft, however with slurry wall work concentrated at this same central shaft, WLVA jet grouting was deferred, and ultimately scaled back to just two rows of jet grout columns next to the Main Shaft for the WLVA break-out. Jet grout work in the intersection of 17th Av. and Lander St. was precluded from starting until late 2004, however with the discovery of sand in the platform tunnels jetting priorities shifted to the platforms to allow for TBM passage through the station. With all this on hand it was quite a task to orchestrate all the moving pieces and still get work underway. Figure 1 shows the congested site in plan view and Figure 2 shows an aerial view of the site looking southward during early jetting and dewatering well installation work.

PROBE AND INSTRUMENT DRILL HOLE PROGRAM

Due to real estate easement acquisition and design timing constraints, only a limited number of bore holes could be completed for the station footprint during the design and preconstruction phase prior to bidding. As a result of these timing constraints, and in particular due to the highly variable nature of the soils encountered in the pre-bid design phase (which included a test shaft on-site), an exploratory drilling program was added to the scope of the construction contract which encompassed both exclusive soil-sampling holes, and sampling of geotechnical instrument holes for ground condition monitoring around the BH Station. More than 40 instrument holes, 20 probe holes, and cuttings from 20 of the 39 dewatering wells were sampled and logged within the first 18 months of construction from the vertical street surface level alone. Horizontal in-tunnel cores and probe holes from SEM production mining have also been logged as SEM mining has progressed, however the vertical probe holes were the basis for the newly defined jet grout limits and the revised locations for additional dewatering and observation wells. A plan view of the instrument and bore holes is shown in Figure 3.

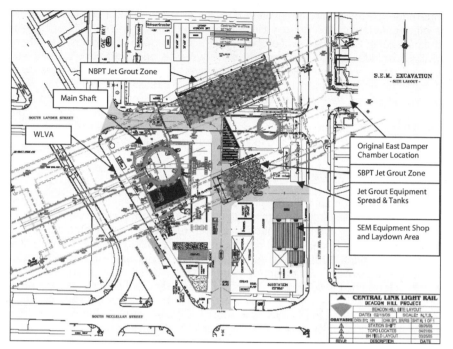

Figure 1.   BH site plan

Figure 2.   BH aerial photo

DEWATERING WELL INITIAL DESIGN

As part of the Geotechnical investigation for this contract, a test shaft was excavated at the location of the proposed access shaft to provide a detailed evaluation of ground conditions. The test shaft was excavated and lined with shotcrete and ring beams for excavation support, relying upon dewatering as an integral element of that scheme. Several non-cohesive granular soil zones were encountered which were not adequately dewatered and demonstrated significant instability in spite of well-points

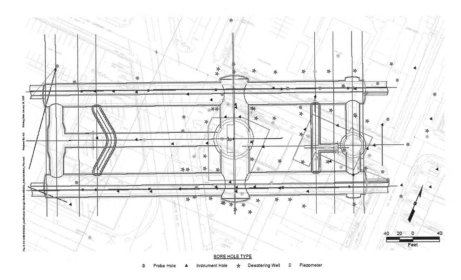

Figure 3. Instrumentation plan

and other drainage measures being implemented from within the shaft. The excavation of the test shaft by conventional methods was terminated at a depth of 30.5 m (100 ft) due to the unstable conditions, and further investigation was continued by large diameter drilling beneath. In response to the unstable ground conditions experienced in the shaft and the projected difficulties anticipated within the SEM mined tunnel, an initial array of 26 base and 12 optional vacuum assisted deep wells was laid out, heavily concentrating the dewatering effort immediately around the access shaft where the contract borings revealed non-cohesive granular soils, and the greatest potential for unstable conditions. The base 26 and additional 12 wells were designed with 9 m (30 ft) of well-screen between approximately 30–40 m (100–130 ft) depths, where the problematic soil material was anticipated.

In addition to the test shaft and initial exploratory boreholes, several aquifer pumping tests were performed in order to evaluate the hydro-geologic characteristics of this granular soil deposit and evaluate whether the material should be dewatered or modified. Well yields of approximately 75 lpm (20 gpm) were realized from the granular soils within the horizon of the northbound platform indicating permeable, non-cohesive ground which warranted jet grouting. The pump tests also indicated that there was communication between the multiple water bearing soil zones which reflected different groundwater table levels. This groundwater behavior was significant because the soils within the mined tunnel horizon were apparently recharged directly from the overlying soil zones.

DEWATERING WELL PRODUCTION WORK AND SYSTEM MODIFICATIONS

With the apparent source of recharge from overlying water bearing soil layers, there was a concern that even with the jet grouting, exterior water pressures of up to 1.7 bar (25 psi) could result in excessive external pressures on the SEM lining, as well as the potential for flowing sands through contacts between columns and through inadvertent windows in the jet grout. The originally contemplated array of 26 and 12 dewatering wells was expanded. In addition to the 18 of the first 26 wells that were already in place as part of the original scheme, an additional 21 wells (for a total of 39 on-site) were installed around the accessible areas of the whole mined station footprint to provide a more "global" or "areal" dewatering effort. These 21 new wells (some remaining original, some new/extra) were located based on localized, high pressure sand layers found in the exploratory holes, and were spaced to lower the water in the northbound platform sand to within 1.8 to 3 m (5–10 ft) of the bottom of the water bearing layer. The additional wells were constructed to depths of up to 58 m (190 ft) with screens as long as 30.5 m (100 ft) in order to tap and relieve any possible soil zones which might recharge the underlying water bearing soils in the tunnel horizon. A vacuum was applied to each well-head via dedicated piping from a surface mounted vacuum station. An additional 47 l/s (100 cfm) of vacuum capacity was provided for the expanded well system, which would have theoretically increased well discharge volumes quite substantially, however given the relatively low flow volumes for most of the wells installed to date has proved be an excellent backup device. The completed well locations and piezometers in Figure 3.

The initial 18 original "contract" wells were installed by mud rotary drilling techniques, using a polymer drilling fluid to stabilize the open 25 mm (10-in) drill-holes. The well installation technique was subsequently changed to a dual rotary, air flush technique to complete wells at greater depths as compared to the initial 18 wells, and to provide the Engineer with fail-proof quality assurance during well screen installation. With either drilling technique, well installation proved time consuming due to the very dense soil conditions. Each well required approximately four days for set-up, drilling, and well construction. Wells were toed a minimum of 1.8 m (5-ft) into competent clay and/or till layers, and cased with slotted 15 mm (6-in) Schedule 80 PVC well casing packed in filter sand, with bentonite chips or a neat cement grout plug at the bottom. Wells were furnished with 1.5 kW (2 hp) pumps capable of a maximum discharge of 115 lpm (30 gpm), with a 3 mm (1.25-in) vacuum assist line connected to one of two on-site vacuum manifolds.

Although the Station itself was to be constructed at considerable depth, a tremendous amount of underground activity had taken place within shallower depths. The system of deep wells had to mesh with the tiebacks installed for the shallower structures, the drilled barrel vault pipes over the crown of the larger SEM drifts, the jet grouting of the unstable soil zones, and of course the slurry walls and the footprint of the mined Station. Three wells were destroyed by tiebacks, four wells were damaged by the barrel vault holes, one well was grouted up, and seven additional wells were plugged by unknown means. Additionally, to protect the well system from the site activities, the well-heads had to be recessed in buried vault boxes and the well system discharge piping, vacuum piping, and electrical distribution was buried in trenches across the site. The entire well system discharge was directed to a storm drain connection on-site, however the plumbing also included piping through an automated pH adjustment (CO_2) treatment plant system for added environmental compliance assurance.

GROUNDWATER TABLE ELEVATION AND DRAWDOWN DATA

Upon startup of the dewatering system, an initial 11 dewatering wells were pumping at a combined rate of about 265 lpm (70 gpm), with most of the flow volume coming from 5 particular wells averaging 40 lpm (10 gpm) each. The remaining 6 wells produced generally less than 20 lpm (5 gpm) total. Subsequently, the remaining dewatering wells were installed and brought online, and the total flow rate gradually decreased over a 2-year period to about 40–80 lpm (10–20 gpm), as is the case at the time of writing.

Although individual well pump rates were generally low, the dewatering system has been extremely effective in lowering groundwater levels in the sand units near the

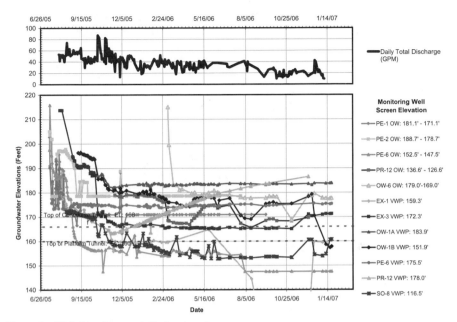

Figure 4.   Well drawdown and discharge volumes

crown of the tunnel. The dewatering system reduced groundwater levels to "steady-state" conditions over about a 4 to 8-week period. After about 3 months of dewatering system operation 8 of the 19 piezometers used to observe groundwater levels were 'dry'. Groundwater levels were lowered up to approximately 18 m (60 ft) prior to excavation of the station and tunneling. It is also noteworthy that when some of the higher producing dewatering wells were shut down, such as for maintenance, the nearby piezometers showed a fairly rapid rise in groundwater levels, some as much as about 1.8 to 3 m (5 to 10 ft) in a several-week period. Refer to Figure 4 for an illustration of well drawdown and flow volumes for the well system.

GROUNDWATER IN EXCAVATED SOILS

Sequential Excavation of the Station has encountered very little groundwater in the tunnels mined to date. The first sign of groundwater was found in the South Concourse Cross Adit (SCCA) eastern sidedrift top heading where the sidewall sprung a leak 4.6m (15 ft) into tunneling. This turned out to be connected to dewatering well SN-06 and was subsequently stemmed by capping the bottom 9m (30 ft) of well casing and moving the pump up over the crown elevation. Subsequent excavation in this same heading encountered first a trickle of water and then a small 50 mm (20 in.) diameter "dike" of flowing sand from above the crown that found its way through the protective barrel vault layer. The dike was plugged with wood-wool, welded-wire-fabric and shotcrete with the addition of grouted pipe spiles in the center top heading crown. Subsequent to that only minor pockets of water in the SBPT East, SBPT West Headwall, the East Transverse Vent Adit (ETVA) north of the Ancillary shaft, and the NBPT at the north and bottom edge of the jet grout zone have been encountered. Individual drain pipes and hoses have been plumbed into the shotcrete lining to drain these features, none of which have amounted to more than a few gallons per minute initially

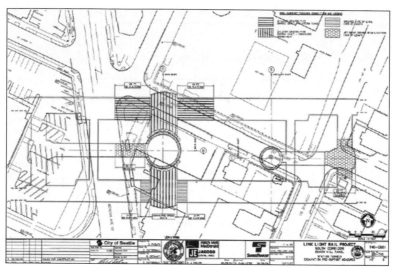

Figure 5.   Original jet grout zone

before decreasing to quarts and cups per minute. Waiting on standby, but yet to be used, vacuum assisted well points have been contemplated twice but deemed unnecessary after weekend gravity draining of various well points in the SCCA and the ETVA proved to be effective.

JET GROUT INITIAL DESIGN AND LAYOUT

Jet grouting was depicted on the Contract Drawings for the Main Shaft break-out for the CCAs, along the WLVA break-out extending westward from the Main Shaft, and for the East Damper Chamber adjacent to the Ancillary Shaft. Jet grouting was intended to "pre-stabilize potentially unstable water-bearing sand layers expected in the crown of…" these various tunnels [GBR 10.3.3] and to "allow excavation with minimal water ingress…" [Spec. 02343 Par. 1.01 B]. The jet grout zones originally specified in the bid documents for the WLVA and EDC consisted of stabilizing the entire section of the tunnel face with jet grouting. In addition to the predefined zones, additional areas of jet grout were to be identified based on the exploratory drilling program. The jet grout limits were described with a plan, profile and section view (grout pay envelope), however individual column spacing and target column diameter was left up to the Contractor to determine. The Contractor-designed layout was also to include angled holes with a potential deviation from vertical of up to 20 degrees for the WLVA below Beacon Ave. west of the Main Shaft. A plan view of the original contract jet grout zones is shown in Figure 5 and the revised final jet grout zones are shown in Figure 6.

JET GROUT TEST PROGRAM

Prior to beginning the production jet grouting initially planned for the WLVA and the Main Shaft Tunnel eyes, a test program was performed to verify the relationship between jetting energy and the resulting jet grout column diameter. On most projects with working depths less than 12–18 m (40-ft to 60-f), the pre-production test columns are typically jetted near the ground surface and exposed by excavation to allow visual

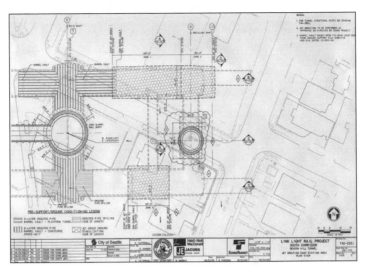

Figure 6. Revised jet grout plan

measurement and inspection of the columns to develop the jetting energy versus column diameter relationship.

Due to the jetting depths involved on this project, the use of shallow columns would have very limited applications to the production depths for two reasons. First, the near-surface soils in the available test area were very dense glacial tills with a substantial fines content that was not representative of the granular materials to be treated at depth. Second, the ground water pressure and the overburden stresses at a depth of 42.5–49 m (140-ft to 160-ft) differ so substantially from the near-surface conditions that an effective correlation can not be made between shallow test columns and the deeper production columns.

Normally, in cases where near surface ground conditions are not representative of the production work depths and where excavation and visual inspection of columns at the required depths is not practical, a pattern of columns would be installed at the working depth and cored at the intersections of the columns to verify closure of the columns. When coring is used to verify closure of a pattern of columns, there is less opportunity to minimize column overlap and refine the jetting energy. On this project, the Contractor decided to use a geo-physical testing method, "electronic cylinder testing," in conjunction with test column coring. The electronic cylinder test method attempts to determine the column diameter by measuring the difference in electric resistivity between a fresh jet grout column and native soils.

The resulting test program consisted of three shallow columns and four deep columns. The diameter of the shallow columns was measured using the electric cylinder followed by excavation and visual examination. The diameter of the deep columns was measured using the electronic cylinder method with limited coring to verify compressive strength. Due to the uniqueness of the electronic cylinder testing, the Contractor chose to perform a shallow test section to convince the project team of the usefulness of the geo-physical testing method by visually verifying the electric cylinder diameter projections on a shallow column. In the case of the shallow columns, visual observation was the primary method of measuring column diameter and the electric cylinder was the secondary means. Once the effectiveness of the testing method had been proven, it could then be used on deeper columns where the electric cylinder would be

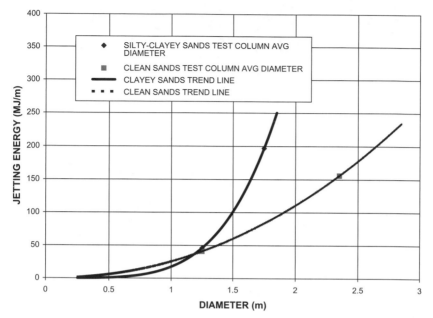

Figure 7. Jetting energy plot

the primary method of verifying the column diameter and coring would be the secondary means. The results of the shallow and the deep test program are presented below in Figure 7, with jetting energy per meter plotted against column diameter for the different soils encountered.

PRODUCTION WORK STARTING AT NBPT— PREDRILLED PATTERN DEVELOPMENT

After analyzing the results of the test columns to determine a reasonable range of expected jet grout column diameters for the target soils, a preliminary pattern of columns was laid out with a triangular spacing. At shallower depths, column layout pattern typically is based on expected diameter and the maximum drilling deviation measured or expected. The tolerance for potential gaps in the treatment is a function of whether the goal of the jet grouting is soil stabilization or groundwater control. Based on these three factors, a layout is chosen and the work typically proceeds without borehole surveys.

In the case of the depths involved at Beacon Hill Station, the potential drilling deviation associated with depths of 42.5–49 m (140-ft to 160-ft) would have resulted in an excessively tight column layout. Since the contract specifications required that every hole be surveyed, it would have been possible to base the layout on an expected maximum deviation and then perform some remedial work in the cases where the allowable drilling deviation was exceeded. However, it was determined that pre-drilling with hole surveying followed by adjustment of the column pattern and/or jetting parameters prior to performing the jetting would be a more efficient and controlled method of carrying out the jet grouting. In order to effectively utilize this method of pre-drilling, borehole surveying, and layout adjustment, a smaller working area within the larger column layout needed to be used. The actual working pattern within the global layout consisted

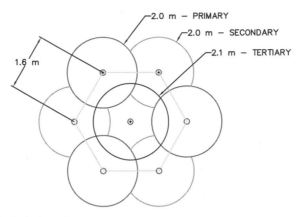

Figure 8.   Grout column layout

of a hexagonal pattern consisting of three primary columns and three secondary columns at the nodes of the hexagon and one tertiary column inside the six perimeter columns. Figure 8 shows a typical hexagonal pattern for the jet grout columns taken from within the global triangular-column spacing layout.

Downhole Surveys

The contract specifications required that all jet grout drill holes be surveyed. While there are numerous means of surveying open boreholes using magnetic based systems, there are essentially just two means of surveying a drill hole through ferrous metal casing or drill rods. One method consists of lowering a survey probe consisting of inclinometers and gyroscopes down a hole with a cable. The probe is then stopped at regular intervals to take measurements and the information is then downloaded and processed by a PC to generate a drill-hole profile consisting of discrete measurement points connected by straight lines. A second method consists of lowering a survey probe consisting of x–y inclinometers down a hole via a set of indexed rods. The probe is then stopped at regular intervals to take measurements and the information is downloaded and processed by a PC to generate a drill-hole profile. While both systems are essentially the same, the first utilizes a gyroscope to determine the orientation of the inclinometers at each measurement point while the second uses a rigid rod system to maintain the orientation of the inclinometer in the same direction during the survey. The Boretrak MKII rigid-rod system was used for drill-hole surveys at BH Station.

Pre-Drilling Methods

Jet grout hole pre-drilling was performed using a dual-rotary cased-hole drilling system. Although the use of mud-rotary drilling for the pre-drilling was considered, the use of the cased-hole drilling offered three advantages. First, the contractor was familiar with the drilling method from conventional micro-pile and tieback drilling such that the cost per foot for the cased-hole system was substantially lower than mud rotary. Second, the use of dual-rotary drill has the potential to produce consistently straighter holes than conventional mud-rotary methods. Third, the use of casing allowed the pre-drilling crew to immediately proceed to the next hole without having to wait for the drill-hole survey to be performed since surveying would be performed inside the casing and not through the drill rods. If mud-rotary without casing had been used, the drill crew would have had to wait for the survey to be performed through the center of the drill steel before tripping out the drill rods and moving to the next drill-hole.

Pre-Drilling Equipment. The pre-drilling was performed using a Klemm 806-4 drill rig with dual-rotary drilling motors and an air/water flush system for the drill cuttings. Pre-drill holes were cased with 15 mm (6-in) flush joint threaded casing in 2-m (6.5-ft) lengths, and the drill holes were advanced with tri-cone bits due to the presence of cobbles and boulders within the glacial till and outwash encountered in many of the pre-drill holes. The pre-drill holes were advanced to a depth of 2-m (6.5-ft) above the top of the jet grout zone in most cases. The holes were not advanced through the treatment zone to avoid potential communication of the high energy jets with the adjacent pre-drill holes. Once the pre-drill holes had been advanced to the depth of roughly 40m (130 ft), the holes were then surveyed and tremie-filled with a weak cement-bentontie grout to prevent the hole from collapsing prior to jetting and to avoid any potential for communication during jetting before removing the casing. Depending on site constraints, the pre-drilling crew was typically able to advance, survey, and tremie backfill approximately 3–4 holes per 10-hr shift.

Jet Grouting Methods

All jet grouting on the BH Station was performed using a double fluid jetting system. The double fluid system consists of a high-energy jet of fluid grout surrounded by an air shield to increase cutting efficiency. The grout mix consisted of water and cement with a specific gravity of 1.46 to 1.48. Jetting grout flows were on the order of 380 liters per minute (100 gpm) with line pressures of approximately 400 bar (5,800 psi) at the drill rig. Air pressure was varied with depth of treatment with a maximum pressure of 15 bar (200 psi) and a flow rate of 5200 liters per minute (185 cfm). The air pressure was set slightly above the fluid pressure of the column of jet grout spoils in the annulus of the drill hole above the treatment depth. By keeping the air pressure above the fluid pressure of the spoils at the bottom of the drill hole, the air shield is able to effectively increase jetting efficiency of the grout jet.

Jetting Parameters and Hole Deviations

Jet Grout Equipment. The jet grouting was performed using a Klemm 3012 which has a standard maximum single stroke drilling and jetting depth of 26 m (85 ft). The Klemm 3012 on this particular project was equipped with a three-rod magazine which increased the maximum drilling depth to 44 m (145 ft) although the rod magazine was replaced by manually adding rods with an auxiliary winch for safety reasons during the course of the project. Drill rods were 114-mm (4.5-in) in diameter due to the drill depths and the relatively high grout flow rate through the center annulus of the double system rods. The cement grout for the jetting was provided by a Recirculating Continuous Mixer (RCM), and the grout density was monitored with a Coriolis-type mass flowmeter. Once mixed, the grout was pumped at the high pressures required for jet grouting with a Soilmec ST-500J pump.

Jetting Parameters and Hole Deviations

As indicated previously, the procedure for performing the jet grouting consisted of pre-drilling and surveying the holes at the triangularly spaced locations and then constructing a plan view of the as-drilled hole locations at the top of the treatment depth. Once this plan view was constructed, the jet grout column layout and work sequence was then developed. In the cases where the drill-hole deviation precluded the use of the original design pattern, the column diameter and sequence was then adjusted to achieve the desired treatment based on the actual drill-hole pattern at the top of jetting column. When the drill-hole deviations were on the order of 0.5%, the only adjustment necessary was usually just a resizing of the target column diameter. However, larger drill deviations could result in altering the jetting sequence, jetting adjacent columns fresh-and-fresh, and possibly pre-drilling another hole. All abandoned holes were either jetted with minimal energy or redrilled and backfilled with cement grout to avoid leaving a potential preferential groundwater conduit into the future SEM tunnel horizon. The jet grout column diameters typically ranged from 1.8 m (6 ft) to a maximum of 2.6 m (9 ft), and it was often necessary to use fresh-and-fresh jetting sequences to create a continuous mass of soil-cement from the resulting drill hole geometry. The fresh-and-fresh method involved jetting two adjacent columns rather than following a primary/secondary/tertiary sequence with tertiary columns filling the remaining inter-section between already jetted (hardened) primary and secondary columns. In total 565 columns were pre-drilled with a redrill rate of less than 5% required due to vertical tolerances. As a result of the pre-drilling, surveying and re-drilling prior to jetting, the number of jet grout columns actually jetted was within 2% of the theoretical design quantity of 565 columns.

Production Quality Control

Wet Spoil Density and Compressive Strength Testing. During the initial test columns and the production jet grouting, the wet spoils coming from the annulus around the drill steel during the jetting were collected for testing. Immediately upon sampling, the density of the wet spoils were measured in a mud balance and com-pared to baseline density measurements as a measure of the efficiency of the jetting which is indirectly an indication of the column diameter and soil type assuming that the jetting energy is relatively constant. In the sands, more efficient jetting resulted in high spoil densities which is indicative of larger diameters while the same jetting energy in the silts and clays would produce lower spoil densities which was indicative of smaller diameters as expected.

The spoil samples were then cast into 75mm $\times$ 150mm (3×6-in) cylinders for cur-ing and compressive strength testing. Although the compressive strength measured by core samples was the specified acceptance criteria, the contractor took daily wet sam-ples to provide more timely feedback that the jet grouting was achieving the minimum specified strength of 3.5 MPa (400 psi). Due to the variability of the soils at the site and the tendency for the strongest core samples to survive the coring operation in tact with a minimum L/D of 2, it is difficult to make any specific observations on the relationship between core sample strength and wet sample cylinder strengths. However, in general the wet spoils sample strengths were typically in range of 5 to 7 MPa (600–800 psi), and the in-tact core samples ranged from 5 to 14 MPa (600–1,400 psi). This higher maximum strength from the in-situ samples is a function of the larger gravel aggregate which is less likely to be expelled with the spoils at the top of the drill hole, the fact that core samples were typically well older than 28-days prior to testing, and the effect of consolidation of the in-situ soil-cement due to the weight of the spoil column above the treatment zone during initial set and curing.

Data Acquisition. The Bi-Tronics data acquisition system on the Klemm 3012 drill rig provided a continuous readout of all drilling and jetting parameters in the cab of the drill rig. Both the drill operator and field engineering staff could monitor the grout den-sity, pressure, and flow; air pressure and flow; pull speed; rotation speed; and depth. All of these jetting parameters were also recorded and downloaded daily to create a graph-ical print out of each column jetted for review by the Contractor and submission to the Owner. The Bi-Tronics system was programmed to automatically control of the drill rod rotation speed and withdrawal speed during the jetting. Although the use of data acquisition for smaller projects is often cumbersome without replacing the need for a diligent operator, on a project of the size and nature of this the data acquisition records, and more importantly the daily review of those records was directly responsible for catching several field mistakes which had they gone unchecked could have had unfortunate consequences for the subsequent SEM tunnel operations in the station headings.

Quality Assurance

Core Drilling and Compressive Strength Testing. The contract required that a confirmation core hole and permeation test be performed for every 150 cubic meters (200 CY) of ground jet grouted. In order to prevent excessive core damage and increase recovery during coring, the in-situ soil-cement was typically allowed to cure and gain strength for 3 to 4 weeks following jetting before coring. The core locations were selected at both the intersection point and the mid-point of completed jet grout columns. Prior to mobilizing the core drill, cased holes were advanced by the pre-drilling operation and surveyed to insure that the core was taken from an appropriate location. Coring was performed using a triple tube PQ system which produces cores with a diameter of 85 mm (3.5-in). Of the thirty cores performed during the course of the project, the core recovery criterion was achieved in all but one of the cores in the targeted sands and gravels, however several of the cores contained sections of hard yet unjetted silts, tills and clay. Although it is difficult to truly verify the exact quality and strength of the in-situ material since the coring process tends to destroy and thereby remove from consideration weaker samples, the cores tested met the minimum strength criteria in all cases. Of the 360 holes over the NBPT only five required rejetting (two from the early learning curve and three as added insurance based on less-than-ideal core spoils recovery). None of the columns for the shaft breakouts or the SBPT required rejetting.

Falling Head Piezometer Test. The contract documents required that a falling head permeability test be performed in the treated jet grout zone to verify that the maximum permeability was less than 3×10^{-6} cm/sec. With the original jet grout work areas encompassing the entire height of the EDC, the method of assessing the results of the falling head test were more straight forward than in the station platforms where the jet grout treatment zone might only be 3 to 4 meters thick. In the end, the core holes themselves were typically used to run the permeability test. An inflatable packer was seated at the top of the borehole, and a falling head permeability test was run until a consistent flow rate was measured which indicated a steady seepage pattern. The results were then analyzed using a Hvorslev's equation for a well point filter at an impervious boundary. The equation that was used produced a range of permeability values depending on assumptions relative to horizontal and vertical permeability coefficients, and although it is difficult to justify or verify all the assumptions for this equation in this application, the calculated permeability ranges met the contract requirements and provided an indication of good quality (low permeability) jet grouted soil mass.

EXCAVATION OF TREATED SOILS

Sequential Excavation break out work for the WLVA from the Main Shaft began in late June 2005 exposing the finished jet grout product for the first time. Evidence of the jetted column was seen clearly two rows deep and with the exception of a very small unjetted pocket at the outskirt of the crown, the exposed ground stood up perfectly. In September 2005 CCA break out work from the Main Shaft began, and again a solid mass of jet grout was found behind the slurry wall across the entire opening. Platform tunnel jet grout was first encountered in June 2006 in the SBPT and again was found to be a competent mixture of jetted columns mixed with till and stiff clay that was too hard to penetrate by jetting. A Liebherr 932 excavator with a two pronged bucket attachment was used to excavate through a majority of the jet grout, however a milling road-header attachment was also used for about 20% of the SBPT by volume. Jet grout in the NBPT was encountered in November 2006 and is currently being mined (at the time of writing) with similar conditions to those found in the SBPT. In all the jet grouting program has provided a very competent soil stabilization mass, and has been mined with little incident and virtually no water influx to date.

CONCLUSION

The two methods of ground improvement, jet grouting and dewatering, combined with careful and systematic excavation and support methods in the Beacon Hill Station tunnels have proven to be very effective in limiting ground deformations and ground losses during station SEM excavation. Station SEM excavation has resulted in very small ground losses, of less than 0.1% of the excavation face volumes, corresponding to 0.8 to 1.6 inches of settlement, as measured in borehole extensometers with measurement rings located 1.8 to 3.6 m (5 to 15 ft) above tunnel crown. With the Station excavation over 80% complete, maximum surface settlements over the station are only 16.5 mm (0.65 in.). These surface settlements are a fraction of the over 100 mm (4 in) of surface settlement predicted using an assumed ground loss of 1%. Obviously the ground, both within and outside of the jet grouted and dewatered areas, has behaved better than anticipated, and resulted in much less ground loss than was predicted. However, much of the success in excavating the Station to date can be attributed to the proper and high quality field installation of these two ground improvement tools.

ACKNOWLEDGMENTS

The authors wish to thank Satoshi Akai, Steve Redmond & Rohit Shetty, Obayashi Co.

REFERENCES

Central Link Light Rail C710 Construction Contract Geotechnical Baseline Report (GBR), June 2004.
Central Link Light Rail C710 Construction Contract Specification Sections 02245—Dewatering Wells and 02343—Ground Treatment by Jet Grouting, June 2004.
Central Link Light Rail C710 Construction Contract Drawings Vol. 3, Book 1 of 6, June 2004.

GROUND FREEZING UNDER THE MOST ADVERSE CONDITIONS: MOVING GROUNDWATER

Paul C. Schmall

Moretrench American Corp.

Arthur B. Corwin

Moretrench American Corp.

Lucian P. Spiteri

Moretrench American Corp.

ABSTRACT

Moving groundwater is generally recognized as the most adversarial condition for ground freezing. Moving groundwater can occur due to natural or man-induced groundwater flow, but either way may result in freeze formation difficulties. History has shown that if the condition goes undetected, catastrophic failure of the frozen shaft or excavation can occur. The success of ground freezing under potentially flowing groundwater conditions depends on an evaluation of the likelihood for this condition; an accurate and early diagnosis, with a careful evaluation of freeze pipe temperatures and piezometer water levels; and timely reaction to mitigate the possible delays. Successful remedial measures have included the reduction of ground permeability by modifying the soil properties between the freeze pipes with various kinds of grouting, or increasing the freezing effect by adding freeze pipes and/or the use of liquid nitrogen as the freezing agent.

INTRODUCTION: THE NATURE OF THE PROBLEM

Excessive groundwater velocity can hinder the formation of a freeze. Movement of groundwater during freeze formation puts an extra heat load on the freeze pipes and the refrigeration plant, requiring more time to complete closure; that point in the formation of the freeze where complete interlocking of the individual frozen columns is achieved forming a continuous frozen wall. The extra heat load is a function of the groundwater velocity as well as the total amount of water passing through the "window," i.e., gap in the frozen wall. The freezing system, both the array of freeze pipes and the refrigeration plant, must have the capacity to remove the extra heat load introduced by the moving groundwater.

Where the velocities are excessive, ground water flowing past a single freeze pipe transfers the cooling effect down stream, which in plan view results in an egg-shaped formation of frozen soil around the pipe, growing more slowly on the upstream side. Even in a moving ground water environment, the egg-shaped cross sections can merge. However, the gaps which remain must accommodate a bigger share of the seepage flow; and, the fewer that remain; the more freezing effort will be required to achieve closure. Ultimately, perhaps only one "entrance" gap will remain on the upstream side of the flow, with a corresponding "exit" gap at the other side of the shaft and the freezing capacity applied between pipes may not be sufficient to close the gaps.

The critical groundwater velocity depends on several factors, namely the freeze pipe spacing, coolant temperature, the permeability of the soil, and the shape and size of the frozen structure. The threshold value for typical brine freezes, however, is generally recognized as, and has been frequently cited to be, 2 m (6 ft) per day. Velocities less than this value do not put an excessive heat load on a brine freezing system (Takashi, 1969; Andersland and Ladanyi, 1994).

A velocity of 1 m (3 ft) per day would be very large, and is rarely encountered under natural conditions. For example a clean medium to coarse sand with a hydraulic conductivity of 500×10^{-6} m/sec (1,000 gpd/ft^2) would require a hydraulic gradient of approximately 2.5% to generate natural groundwater velocities of 1 m (3 ft) per day. Such sands are not uncommon, but where they exist, the normal groundwater gradients are much flatter. Abnormally high, naturally occurring groundwater velocities are sometimes encountered due to external causes such as sudden rise or fall in the stage of an adjacent river.

PREVENTATIVE MEASURES DURING INSTALLATION

Quality Control

The first step towards problem prevention is proper quality control during the freeze installation. There are several essential specific measures that are imperative in order to achieve proper closure of a frozen earth cofferdam. These are as follows:

1. Confirm that the deviations of the drilled freeze pipes are within allowable limits. Excessive deviation between pipes may result in too large a gap between freeze pipes to achieve closure of the frozen wall in a reasonable period of time. The freeze pipes must be surveyed in order to confirm this.

2. A center relief well must be installed within the unfrozen core of the shaft to provide pressure relief as the freeze continues to grow inward and the encapsulated water expands with the phase change. This relief hole is also an absolutely essential instrument in that it will indicate when closure of the wall is achieved. If multiple aquifer zones exist, then multiple relief holes can be installed so that the response in each zone can be observed independently.

3. Piezometers must be installed outside of the frozen shaft to first measure groundwater gradients; secondly, confirm that the groundwater inside of the shaft has been isolated from the groundwater outside of the shaft.

Grouting Methods

Excessive groundwater velocities can be lowered to manageable values by reducing the hydraulic conductivity of the soil. Permeation grouting has been used effectively when the groundwater movement is occurring along preferential "paths of least resistance." Where warranted, permeation grouting of the soils can be performed concurrent with the installation of the freeze pipes or prior to activation of the freeze.

Boston Central Artery/Tunnel Contract 9A4

Groundwater movement was a major concern on Boston Central Artery/Tunnel Contract 9A4 which involved the jacking of three large tunnel boxes beneath the rail lines outside of Boston's South Station, and within a stone's throw of the Fort Point Channel. Early in the project, data logging instruments installed in several on-site observation wells indicated a groundwater fluctuation that mimicked the tide in the channel, suggesting that there was the potential for moving groundwater flow, particularly within the open rubble fills (Donohoe et al., 2001).

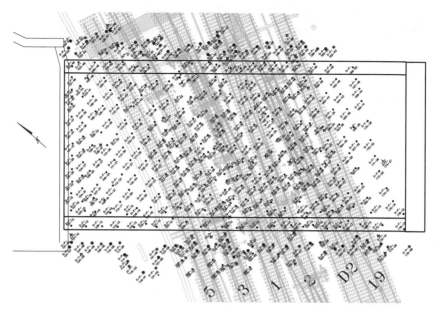

Figure 1.

It was imperative to ensure that the full area of each jack was hydraulically cut-off from the outside groundwater regime by continuity of the freeze along the outermost perimeter freeze pipes. Several measures were undertaken to mitigate the movement of groundwater so the freeze(s) would form on schedule, contiguously, and without gaps. The outside, or perimeter, freeze pipes were installed on tighter spacing so as to positively form an earlier perimeter closure to permit the formation of the internal mass freeze without groundwater movement. All of the perimeter freeze pipes were surveyed to confirm that the pipes were installed within tolerable deviations.

In addition, a grouting program was implemented during the installation of the perimeter pipes to detect and subsequently grout any voids encountered in the fill. The perimeter pipes were installed with a fluid-recirculating drilling technique which would provide an indication of ground hydraulic conductivity based on the fluid return (Donohoe et al., 2001). Where loss of recirculation fluid indicated voids, these holes were subsequently grouted. Additional pipes were installed where voids were encountered in the fill and where the pipe surveys showed unacceptable deviations with depth.

Figure 1 is a site plan of Box "D." The perimeter pipes are spaced closer than the interior pipes to ensure a perimeter closure with grouting of voidaceous ground and more concentrated perimeter freeze.

PREVENTATIVE MEASURES WITH FREEZE FORMATION— EARLY DIAGNOSIS AND TREATMENT

Where the geotechnical investigation suggests groundwater movement is likely, it is prudent to perform periodic temperature profiling of the freeze pipes. This is performed by measuring the static and stabilized brine temperature every couple of feet within each of the freeze pipes. Anomalous warm spots at the perimeter may be an indication of a window in the freeze wall. With typical freeze pipe spacing, temperature profiles can provide a very detailed overview of the perimeter ground temperatures and the progress of the freeze formation. External temperature monitor pipes, located outside of the freeze pipe circle, are also highly valuable. An external monitor that shows an anomalous cold spot will typically be an indication of an exit window. Likewise, an anomalous warm spot may indicate an entrance window.

In some cases the warm or cold spots are well defined and the location of the window can be precisely located. In most cases, however, the location of the window is not obvious. Where the window is well defined, the grouting or additional freezing effort can be applied with some degree of precision. Where the window is not well defined, the grouting or additional freezing effort must be applied to a broader area.

Where excessive groundwater velocities are encountered due to natural conditions, the soils are typically highly permeable gravels which are readily groutable with inexpensive and high viscosity clay-cement grout materials. Man-induced groundwater gradients are more difficult to address because the same excessive groundwater velocities (greater than 2 m/day) can be realized in lower permeability soils due to pumping from deep wells or other pumping devices. Less viscous, more penetrable and more expensive permeation grout materials must be used, typically injected using tube-a-manchette (TAM) pipes and double packers.

Kansas City Water Tunnel

The downshaft for the Kansas City Trans Missouri Tunnel provided an opportunity to monitor the closure of a freeze wall under anticipated, and somewhat controllable, man-made flowing groundwater conditions. A shaft was frozen near the Missouri River and 580 m (1900 ft) from a group of ground water extraction wells belonging to the Kansas City Water Department. A comprehensive study and hydrogeologic report was made available to contractors at the time of bid that predicted ground water flow averaging 0.9 m (3 ft) per day, with a maximum flow of 2 m (6 ft) per day.

Given the foreknowledge of moving groundwater, the freeze pipes were placed relatively close and additional refrigeration plants were provided to cool the brine several degrees colder than normal. Additionally, temperature monitoring devices were strategically placed to monitor the amount and direction of the anticipated eccentric growth of the freeze wall (Corwin et al., 1999).

About 14 days after ground freezing was initiated, anomalous temperature readings confirmed the movement of ground water towards the well field, which at the time was pumping 60 m^3/min (15,800 gpm). Within the upper 21 m (70 ft) of aquifer, temperatures, although unbalanced, were moving down at a satisfactory rate. In the lowest 9 m (30 ft), between 26 m and 35 m (85 ft and 115 ft) below ground surface, there was an unacceptably slow rate of progress. Closure of the wall was not occurring in the coarser sediments immediately above the rock from which the water supply wells were drawing.

On the 24th day of freezing, the water authority was able to tune back the pumping from the well group, reducing the flowrate to 17 m^3/min (4,600 gpm). A piezometer situated inside the freezewall had, up to that time, reflected general movements observed in the external piezometer, verifying that there was a connection via windows in the wall. A few hours after the reduction in ground water movement towards the well field, the water level in the center relief hole rose indicating closure of the freeze.

Shaft 29B, New York City Water Tunnel No. 3

The ability to control external man-induced groundwater gradients is not always possible, as was the case at Shaft 29B of New York City Water Tunnel No. 3. The tunnel alignment lays hundreds of feet below ground in the metamorphic, granitic rocks underlying the city. However, many of the vertical access shafts were sunk through

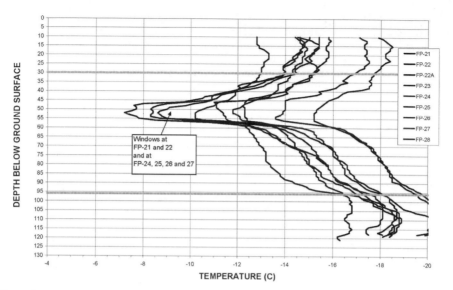

Figure 2.

glacial deposits and water-laden sands above bedrock. The shafts, in excess of 10.5 m (35 ft) excavated diameter, are sited in the dense urban environment which they serve.

Three of the five shafts recently frozen for the lower Manhattan leg of the tunnel showed evidence of groundwater movement. They were each grouted during the formation period to ensure closure within the pre-determined schedule. The most pronounced conditions were experienced at Shaft 29B. Groundwater was measured approximately 3 m (10 ft) below sea level due to a significant dewatering operation pumping several thousand gpm (L/min) of groundwater several blocks away. The most pronounced water bearing formation of the stratigraphy was a medium to coarse sand from water table to the top of rock at approximately 30 m (100 ft) depth. Although the ground was not nearly as permeable as the gravelly soils in which high natural groundwater gradients are typically observed, excessive velocities were realized due to the high-volume, man-made pumping operations.

Ground temperatures measured within the freeze pipes themselves indicated the presence of an upstream "entrance" and a downstream "exit" window due to the movement of groundwater through the shaft. Concurrently, piezometers throughout the site confirmed that the groundwater within the shaft was still in direct communication with the outside groundwater regime, and thus closure had not occurred (Schmall et al., 2006).

Temperature profiling data from a group of freeze pipes at shaft 29B is shown in Figure 2. The warm spot between 14 and 18 m (45 and 60 ft) depths reflects moving groundwater.

The ground freezing contractor installed a series of TAM pipes to permit discrete injections of grout at predetermined elevations. An ultrafine cement material was used to permeate the medium to coarse sands to successfully cut off the high velocity groundwater flow, and closure of the shaft was achieved. The success of freezing at this location, particularly in light of the moving groundwater situation, was due to an accurate diagnosis of moving groundwater and timely reaction to mitigate the possible delays (Schmall et al., 2006).

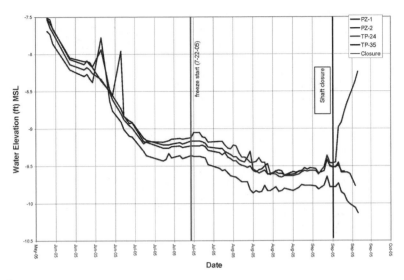

Figure 3.

Shaft 29B water levels measured in the center relief well and external piezometers are shown in Figure 3. Closure of the freeze is indicated with the pronounced rise in the center relief well level.

REACTIVE MEASURES—UNANTICIPATED CONDITIONS

Providence CSO Tunnel

In the opinion of the authors, the main access shaft of the Providence CSO project in Providence, Rhode Island has been the most challenging freeze closure to achieve to date. The shaft was 14 m (45 ft) in diameter, frozen into rock at an approximate depth of 52 m (170 ft). The stratigraphy consisted of a surficial fill, a 43 m (140 ft) thickness of alternating layers of glaciolactustine and highly permeable glaciofluvial deposits, over a thin mantle of glacial till and then rock. The groundwater table was shallow, approximately 4 m (12 ft) below the surface. The upper bound of the baseline soil permeabilities and the groundwater gradients as observed in a vast array of piezometers indicated the natural groundwater gradients were within acceptable limits.

The freeze was anticipated to close within six weeks of activation of the system. After almost eight weeks, although temperature monitors indicated cold ground temperatures and there were no other anomalies in the data, at 54 days, no indication of closure was indicated in the center relief hole. A complete round of temperature profiling of the freeze pipes was performed to evaluate the progression of the freeze and to determine why closure had not occurred. The profiling indicated temperature anomalies at shallow depth near the water table almost uniformly around the perimeter of the shaft. Closely spaced probes were subsequently advanced beneath the shaft collar to evaluate the advancement of frozen ground. The resistance encountered with the advancement of the probes indicated little frozen ground buildup at the water table in spite of ground temperatures between −12°C and −17°C (10.4°F to 1.4°F). Further investigation revealed a significant amount of free product from a 65 m³ (17,000 gal) gasoline spill which occured immediately at the location of the shaft rather than the

Figure 4.

dissolved phase of petroleum contamination anticipated. In response, a secondary series of shallow angled freeze pipes were installed, splitting the centers between the original freeze pipes. One week after the activation of the shallow secondary freeze pipes, probing confirmed closure of the freeze at the shallow depths. The diagnosis and full execution of the response required more than a month. The secondary ring of freeze pipes can be seen in Figure 4.

The temperature profiling performed on Day 54 also indicated slightly warmer spots on the frozen wall between depths of 27 and 38 m (90 and 125 ft) on the north side and depths between 46 and 49 m (150 and 160 ft) on the south side. These spots were not well defined or readily distinguishable. A grouting test program was commenced concurrent with the installation of the shallow secondary freeze pipes, however, low grout takes at high injection pressures suggested that high permeability ground and accompanying high groundwater velocities were not the cause of the problem. Temperature profiling on Day 83 indicated advancement and progressive growth of the freeze through some of the suspected warm spots.

No closure was observed, however, by Day 99 and a third round of temperature profiling of the freeze pipes was conducted. Based on the profiling data from Day 83, the windows, if they existed would be difficult to distinguish so the center relief hole was pumped concurrent with the profiling to "aggravate" the windows so they would be more distinguishable. This was considered an extreme measure because if a window existed, this pumping would cause the freezewall to erode in the vicinity of the window making it larger and more difficult to close; however, it was deemed necessary to understand the situation. The pumping from the center relief well revealed immediate communication with piezometers outside of the shaft in the glaciofluvial soils and rock, and a rapid water level rebound was observed in the relief well once pumping was ceased, both indications that the wall was not closed. Clusters of tube-a-manchette grout pies were installed at the north and south sides of the shaft where the windows were suspected and approximately 152 m³ (40,000 gallons) of bentonite cement grout was injected outside of the perimeter of the shaft. The grouting closed the windows after 18 weeks of freezing as observed by a sudden rapid rise in the water level in the center relief well.

Based on the typical limits of injectability of clay-cement grout into a granular material, the large takes of bentonite-cement grout realized within the upper and lower

Figure 5.

windows of the frozen wall reflected the presence of soils within these horizons of significantly greater permeability than indicated by the GBR. Ultimately, excavation of the shaft revealed significantly coarser and more permeable layers of clean gravel than baselined. Figure 5 shows a good example of open work gravel observed in the shaft. Additionally, several ungrouted boreholes were found within the shaft excavation, and observed flowing inside the excavation. The changed ground condition at the main access shaft was supported by the fact that two other frozen shafts, within 106 m (350 ft) of the main access shaft, were successfully closed in less than the calculated formation period utilizing the same design configuration, system operating parameters, and so forth, at fraction of the effort expended at the main access shaft.

The ultimate closure of the frozen wall would not have been possible without (1) the installation of the shallow secondary freeze pipes, and (2) grouting of the high permeability coarse glaciofluvial soils to reduce the groundwater velocities. The difficulties in closing the freeze in Providence were due to a combination of several unanticipated factors:

1. The presence of undissolved, free phase gasoline at the surface of the water table.

2. High permeability gravel layers and seams within the glaciofluvial deposits.

3. Flow of groundwater from the rock. Piezometric levels in the rock were higher than in the glaciofluvial materials. Several grout holes were improperly abandoned from an earlier grouting contract. During freezing, these grout holes acted as open conduits permitting free flow of water from the rock into the overburden. Water level reactions during and after freeze formation confirm that the center relief well acted as an interconnection between the source of water reaching it from the rock via the defective grout holes, and any gravel seams that it intercepted.

The windows in frozen walls are typically observed as a localized entrance window and a localized exit window. The Providence CSO main shaft was complicated by the concurrent action of no less than three contributing factors which required a sequential "process of elimination" in order to diagnose and rectify the problem(s). Grouting was ultimately the tool of choice, but could not be performed until the preceding shallow closure could be achieved and the groundwater flow through the shaft was distinguishable.

CONCLUSION

In theory and concept, the diagnosis of moving groundwater and subsequent closure difficulties is simple. In practice, a tremendous amount of experience and judgment is required. Expertise in groundwater movement and behavior as well as ground freezing is required to recognize anomalies in the data.

REFERENCES

Andersland, O.B and Ladanyi, B. (1994). *An Introduction to Frozen Ground Engineering*, Chapman & Hall.

Corwin, A.B., Maishman, D., Schmall, P.C. and Lacy, H.S. (1999). "Ground Freezing for the Construction of Deep Shafts." Proceedings of the Rapid Excavation and Tunneling Conference, Orlando, FL.

Donohoe, J.F., Corwin, A.B., Schmall, P.C. and Maishman, D. (2001). "Ground Freezing for Boston Central Artery Contract Section C09A4, Jacking of Tunnel Boxes." Proceedings of the Rapid Excavation Tunneling Conference, San Diego, CA.

Schmall, P.C., Mueller, D.K. and Wigg, K.E. (2006). "Ground Freezing in Adverse Geology and Difficult Ground." Proceedings of the North American Tunneling Conference (NATC), Chicago, IL.

Takashi, T. (1969). "Influence of seepage stream on the joining of frozen soil zone in artificial soil freezing." Highway Research Board Special Report, Transportation Research Board, Washington D.C.

INNOVATIVE RETAINING WALL WITH JET GROUT FOR NEW METRO TUNNEL UNDERNEATH AMSTERDAM CENTRAL STATION

B.J. Schat
ARCADIS

A.W.O. Bots
ARCADIS

O.S. Langhorst
Movares Nederland BV

ABSTRACT

Amsterdam City Council is building a new metro line, the North–South Line. The line passes under Amsterdam Central Station (a listed building) about 18m below street level. The retaining wall that has to be built underneath the existing building is an innovative construction consisting of two rows of steel piles interspersed with a solid mass of overlapping jet grout columns: a composite 'sandwich' wall. This paper gives an overview of the approach taken to designing (computational models) and building the retaining wall. Special attention is paid to the observational method used for the construction of the jet grout columns.

INTRODUCTION

In the centre of Amsterdam a lot of hard work is being done on the new North–South Metro Line, the metro connection to be constructed right under the centre of the city, including some historic parts. From the construction point of view, one of the complex parts of the underground metro is the intersection at the Central Station. Here, the tunnel has to be constructed underneath the existing station, which is a listed building. During construction the station will remain fully operational and the building may not be damaged. Figure 1 shows a bird's-eye view of the station and also indicates the future route of the North–South Line.

The station building stands on an artificial island that was built in the River IJ around 1875. It is a beautiful brickwork building, a site of national historic interest that was completed in 1889. The foundations underneath the building consist of about 8,700 wooden piles. In the middle of the station building is the central hall, directly under which the metro will be situated.

The layering of the ground underneath the station is indicated in a geotechnical longitudinal profile along the centre line of the future tunnel (see Figure 2). The ground level was raised from 5 m below NAP (the standard Amsterdam zero water level) to about 1 m above NAP with a soil fill of sand from the North Sea Canal with a low packing density. Below this, down to 15 m below NAP, there is a series of relatively weak layers, made up of River IJ clay and sandy clay layers—the former bottom of the River IJ. A thick layer of clay known as Eem clay is found from 28 m to 45 m below NAP. The Eem clay layer lies on a thick layer of glacial clay. The third sandy layer, with bearing capacity and high cone penetration resistance, is found at 56 m below NAP and lower.

Figure 1.   Station island and the North–South Line route

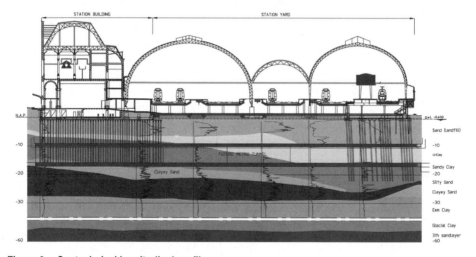

Figure 2.   Geotechnical longitudinal profile

DESIGN PRINCIPLE

The design of the new metro tunnel underneath the station is based on a combination of two well-known techniques used in tunnel construction: the excavation under
decking and the immersed-tunnel method. A channel, underneath the existing station,
is constructed using the first technique mentioned. The channel is connected to the
River IJ, which is at the back of the station. From the River IJ, a tunnel-element is then
floated in and immersed. A major advantage of this method is that there will be no
need for a dry trench to be created. As a result, the load on the trench walls will be limited to the load due to the ground pressure, instead of both ground and water pressure.

The special thing about the trench needed here is that the construction must be
placed starting from and underneath the existing station. While the trench is being made,
the load due to the original foundations of the station building has to be transferred to the
retaining walls too. For this purpose, a deck construction will be made between the existing building and the trench walls, which will transfer the loads from the station to the
retaining walls. The principle for the construction of the table is shown in Figure 3.

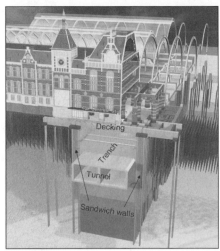

Figure 3. Visualization of the final situation

Trench Wall Underneath the Station Building

Several construction types were examined for the trench wall. In particular, the fact that the wall had to be constructed within the restricted space of the station hall and the presence of the wooden foundation piles meant that many conventional methods for the construction of such a wall could not be applied. The bending stiffness of the wall plays an important role too. The requirement that the station must not be damaged means that the bending stiffness of the wall must be as high as possible. This will limit settlement of the part of the station building behind the wall as a result of the walls deflection during excavation as much as possible.

Eventually, an innovative design turned out to be the solution: a wall made up of two rows of screwed steel piles with a body of jet grout between them. This is known as a sandwich construction, which is why this wall is referred to as a sandwich wall (SWW). The steel piles (round 457 with a wall thickness of 25 mm) are placed in a row 1 m apart, measured centre to centre; the rows are placed 2.5 m apart. This creates a wall that can take vertical as well as horizontal loads. The wall is constructed using techniques that have been applied before and which are eminently suitable for use in the immediate vicinity of existing foundations; the work has little effect on the building. These techniques are also excellently suited to being used in the restricted working space of the existing station.

Load Distribution in the Wall

The walls are intended to bear the load of the existing station vertically as well as to serve as a ground and water barrier for the trench to be dug between them.

The horizontal bearing capacity of the wall is derived from the interaction between the steel piles and the jet grout body. The steel piles on the inside of the wall act as the tension member; the grout acts as a pressure arch. To create this interaction, shear forces are transferred between the steel pilings and the grout by rings on the outside of the steel piles. The substantial stiffness provided by this construction is an important advantage: there is a lot of material in the outermost fabric of the wall and the substantial wall thickness will handle large internal lever forces.

THE GROUT IN THE SANDWICH WALL

The SWW is made up of steel piles and a grout body. The load-bearing and retaining functions of the entire wall depend particularly on the way the grout body is eventually created. Placing steel piles from an existing building close to existing foundations has been done many times before. This also applies to grout bodies underneath existing buildings. There is less experience with making a rather extensive and contiguous grout body between all the existing and newly placed elements. Numerous studies into the various alternatives for realising it were therefore performed beforehand.

The grout has to meet two principal requirements:

1. The grout must be able to transfer forces between the steel piles. To do this, at least half the circumference of the steel piles must be enveloped by the grout and the grout must be strong enough.

2. The grout must retain ground and resist water seepage; there may not be any grout-free conduits going through the wall.

Principle of Jet Grouting

To make a jet grout column, a drill string is first drilled into the ground as vertically as possible. Once the drill bit has reached the right depth, grout mortar is injected under high pressure from two horizontal nozzles, which are at the bottom side of the drill string. Then the drill string is simultaneously slowly rotated and pulled up while injecting the grout. This mixes the ground with grout in a circular shape, thus creating a cylindrical pillar of grout mixed with the soil previously present. This system is called the mono-jet system. To increase the diameter, it is also possible to add extra water or air as well as grout. This is called the bi-jet or tri-jet system.

With regard to making a single body from the various individual columns the grout column pattern, the combination of placements and diameters of the columns, strongly affects the final result. It is not merely that the columns have to be made between the earlier elements such as the wooden piles and the steel piles. There is also a considerable chance of columns not fitting together properly as a result of grout column deviations. The most important deviations that may occur are the following:

- A deviation in the placement of the column from the theoretical positioning
- Out-of-perpendicular positioning of the columns
- Deviations from the designed diameter
- Shadow effects, caused by the grouting tight up against an existing column or a wooden pile or steel pile
- Misdrilling: if drilling occurs in an already hardened column, no grout column will be made at all at this location

Figure 4 shows how the above-mentioned deviations may result in gaps in the entire wall. The occurrence of the above-mentioned deviations must therefore be avoided as much as possible. The following has been done to achieve this:

1. A field test was carried out at the site to gain experience with the local situation.

2. A column pattern has been developed that will make the greatest possible allowances for the occurrence of the various deviations.

3. Eventually the columns will be constructed according to the 'observational method.'

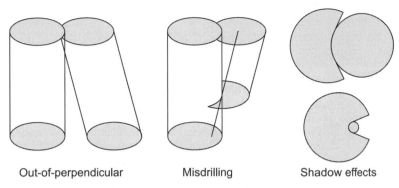

Out-of-perpendicular   Misdrilling   Shadow effects

Figure 4. Visualization of deviations that may cause gaps in the grout

JET GROUT FIELD TEST

Prior to the production of the jet grout columns within the SWW, test columns were made in the square in front of the station first. These test columns were constructed to gain experience with the local conditions, with the objective of finding the proper practical parameters for the required diameters. The best drilling method for drilling as vertically as possible was also examined. After construction, the test columns were excavated to a depth of about 7 m. This allowed the shape and diameter to be inspected visually. Several cores were also taken from the columns in order to determine their compression strength. A number of items relating to the design and execution process were adjusted after the results of the jet grout test.

To obtain the proper diameter and the necessary column strength, it is necessary to 'jet' the column in two stages. In the first phase, pre-cutting, the ground is cut up to obtain the right diameter. This stage means that a grout with a low density has to be used, since low density allows the energy to reach further, resulting in a larger diameter. At the second stage, post-cutting, the same section will be jetted a second time, using a higher density for the grout mixture. This will cause the cement content to increase, as a result of which the compression strength of the hardened column will increase.

Since the columns have to be made in two stages, it turned out to be necessary to divide the column into sections. Working on sections of about 6 m reduces the time needed between pre-cutting and post-cutting, avoiding having the grout stiffening too much after the first step, which would make the post-cutting stage useless.

The proper execution parameters for the diameters needed have been determined during the construction of the various test columns. The required diameter depends on the execution parameters used for jetting. Variations in pressure, nozzle diameter, rotation speed and withdrawal speed of the drilling head, the grout density and (where applicable) air pressure are possible. The execution parameters are determined as a function of the depth, ground condition and the desired diameter. This project works with the mono-jet system for the smaller diameters (800 to 1,200 mm) and with the bi-jet system, with air, for diameters 1,400 to 2,200 mm.

The large variation in the ground layers underneath the station makes it necessary to adjust the parameters of each column more than once during the jet grout process in order to obtain the same diameter along the height of a column. Each set of execution parameters can be converted into an energy level per metre of column inserted into the ground. To illustrate this, Figure 5 shows the various energy levels

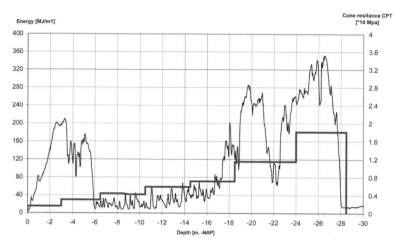

Figure 5.   Energy per metre of column, plotted against depth

used as a function of the depth for a column with a design diameter of 1 m. Some of the columns, including this one, needed as many as seven different sets of parameters. The same diagram shows the results of a CPT (cone penetration test).

JET GROUT PATTERN

The original design of the jet grout column pattern is given in Figure 6. Its basic principle is that the fit with the steel piles is created by the small columns of about 800 mm, called the edge columns, which are placed between the steel piles. The edge columns are made first, after which the space between the rows is filled with filler grout columns, always using a pair of two large columns. To make the columns match up properly with each other and with the steel piles, these columns were initially intended to be constructed 'wet in wet.' In the test this method was specifically examined by adding a dye to the second column with the objective of seeing (after excavating the columns) if this method could be used effectively. The time between constructing the first column and the second one turned out to be so large that the first column had already set too much, as a result of which the second column failed to reach the desired diameter. Based on these results, a new grout column pattern was looked for, which could be implemented without using wet-in-wet columns.

This new pattern is shown in Figure 7. It was decided to use various diameters for the filler columns, with the construction of the smaller primary filler columns first and the larger secondary filler columns afterwards. The space between the primary filler columns will remain free to such an extent that there will be sufficient space to do the drilling without touching the previously constructed columns while keeping the free space small enough that it can be filled with a single grout column. The location and tilt of the steel piles was known as this pattern was designed, since they had already been placed and measured with an inclinometer. This information was therefore included in the process of optimising the column pattern.

To determine if the pattern is sufficient, a Monte Carlo simulation model has been used, developed for the project by GeoDelft. This model, which models the various deviations stochastically, makes allowances for the occurrence of shadow effects and

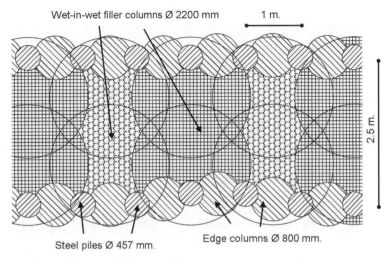

Figure 6.   Original design of the jet grout column pattern

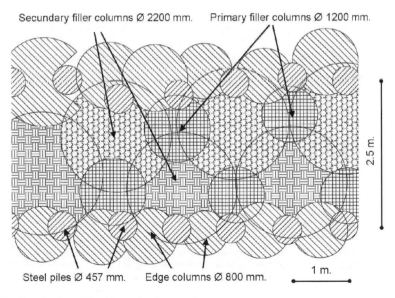

Figure 7.   New design of the jet grout column pattern

misdrilling. 10,000 simulations have been performed for each wall part and the chance of leakage through the grout body has been studied, as well as the extent of grout filling between the steel piles. Figure 8 gives an example of the geometry results of one of these simulations.

Appropriate adjustments of placement and diameter of the columns eventually determined the optimum design pattern.

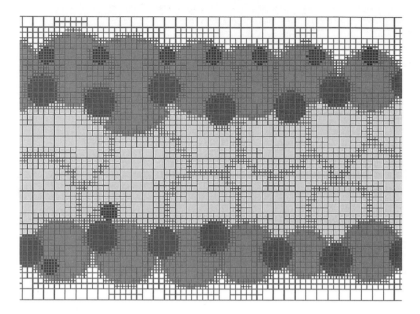

Figure 8.   Example of a Monte Carlo simulation

CONSTRUCTION ACCORDING TO THE OBSERVATIONAL METHOD

The construction of the jet grout columns in the SWW is a complex process that may generate a lot of deviations. Deviations of any individual column may render the entire wall system ineffective. That is the reason why the observational method principle will be followed during the construction. Taking measurements during the construction of the separate parts (jet grout columns) means deviations will be noticed as they occur. This will make it possible to prevent larger discrepancies by making adjustments during the process or, should this not be possible, by applying adjustments afterwards and adapting the design of the columns still to be constructed. Actively checking the quality of the parts (the jet grout columns) and adjusting them will in this way ultimately guarantee the quality of the entire product (the wall). Application of this method has resulted in an extensive measurement protocol being set up, to be used as the basis for adjustments to be made during execution. The quality actually achieved can be determined afterwards. As well as measurements of the jet grout process and its quality, several independent position measurement systems on the station building function as a check, by measuring the distortions in the building.

Organization and Planning

To make it possible to work according to the observational method, the executive organization has been adapted to allow the practical process to be steered suitably and effectively.

The construction of the jet grout columns in the SWW is distinguished by the fact that the principal, the contractor and the subcontractor involved are cooperating to optimise the quality of the SWW, because of the complex work and the complex method of execution chosen. The organization has been set up in such a way that decisions can be taken whenever needed to adjust the process and guarantee the quality of the individual columns and the entire wall. To this end, the executive organization has been split up into a steering group and a practical supervision team. The steering group controls the execution at the management level and also determines the parameters to be used for the columns to be constructed. The information needed for this includes what was learned from the analysis of information recorded about columns constructed earlier. During execution at the construction site, the practical supervision team is authorised to adjust the implementation process according to defined protocols. This may be necessary if deviations are measured or problems occur during the process. The principal, contractor and subcontractor are represented in both the steering group and the practical supervision team.

Both trench walls are made in two stages, since part of the station must remain available for travellers. Preparations for the southern parts of the SWW began in 2003. The steel piles were placed in 2004 and a start was made on the construction of the jet grout columns in the middle of 2005. The two southern wall sections were finished in the beginning of 2006. Preparatory work for the northern wall parts is currently being done.

Working Method

A plan of work has been drawn up for each column including all information relevant for its execution. The plan of work is like a scenario for the entire column, stating the parameters necessary for execution of each stage and including the measurements that need to be taken. The plan of work also contains the design information, a risk register and protocols in the event of deviations being observed.

Working according to the 'observational method' means that records and measurements from the construction of earlier elements are used to adjust the process for later elements. All measurement results are processed for each column and extensively analysed. The effect on the work to be done is determined for each column and measures can be taken if necessary. This principle will be used in two ways when making the SWW. On the one hand, based on diameter measurements and recordings of the earlier constructed columns, the actual jetting parameters (pressure, discharge, retraction speed, rotation speed) will be adjusted to match the designed diameter as closely as possible. On the other, the geometry of the grout columns (placement and design diameter) will be altered if an earlier column shows a discrepancy that may affect the surrounding columns still to be constructed.

Measurements and Recordings

A lot of measurements and recordings are performed. These measurements are mainly aimed at determining the inclination of the grout column and the diameter actually realised. An inclination measurement in the drill string is used to determine the tilt before starting the production of the column. The information about the inclination is worked out on site in a semi-3D drawing which indicates the direction of the column along its depth. Figure 9 gives an example of how this is done. If such a diagram indicates that the vertical deviation is too large, it may be decided to stop the column or to shorten it. It is also possible to adjust the execution parameters in such an event. Using the protocols provided beforehand, the sets of execution parameters can be adjusted to realise an altered diameter.

Diameter measurements are performed by a number of different methods. Hydrophone measurements were used to determine that the edge columns reached the steel piles; spin measurements (with an umbrella like device) were performed on a random

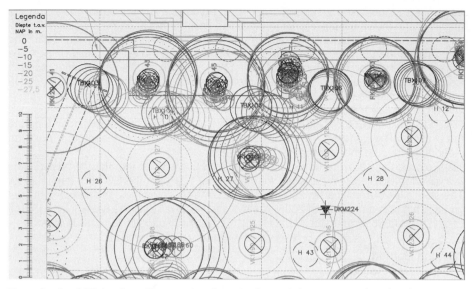

Figure 9. Semi-3D drawing with geometry of grout columns to be constructed or already constructed

basis. Return flow samples were continuously taken for all columns too. The density of the returning flow of material gives a direct indication of the diameter realised.

EVALUATION OF THE WALL

After the column is finished, all recordings and measurements during execution are collected and evaluated extensively. The 'as built' geometry of the column is plotted in a drawing, which indicates the position of the column every five metres (similar as Figure 9). For each column, all information is put together to determine if new adjustments will be needed for the columns yet to be created. If desired, the location of the columns yet to be made or their execution parameters can be adjusted.

When evaluating the entire wall, special attention is paid to the function of the wall as a system. The assessment focuses on the structural cohesion, ground density and water retaining function.

All 'as built' information for all columns constructed is put together to provide an overview of possible imperfections. Where serious imperfections are discovered, exploratory destructive drillings are performed. The resistance level observed when drilling helps judge whether there really is a gap. If a gap is actually drilled into, jetting may still be done using the drilling hole just created in order to fill the gap.

Cores will be taken from a random sample of the columns to check the strength of the grout. A core is taken along the entire length of the column. Various samples of the cores are sawn crosswise and their compression strength and tensile strength are tested. To determine whether the water-retaining function of the wall is sufficient, EFT (electro-flux tracking) is used to measure the damming function.

After all the columns were constructed, a Monte Carlo model (as also used for designing the grout column pattern) was used to calculate the 'as built' situation. This verification calculation provides insights into how the wall is filled with the grout. The

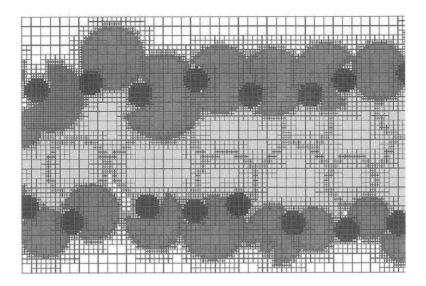

Figure 10. Example of a part of the 'as built' geometry of the grout columns

model determines the likelihood of a leak in the wall and shows the likelihood of grout filling in a cross-section. The probability of the presence of grout is plotted in colour shading in this cross-section. Figure 10 shows an example of the as built situation of one of the walls. Based on the output from this model, it has been determined that the likelihood of the wall being watertight is very high. The model also indicates that there is sufficient structural cohesion of the grout between the steel piles.

Based on the measurement results, the location and the angle out-of-perpendicular of the columns were entered deterministically into the model. Measurements of the diameter were also performed, but the diameter was not measured for every column. In addition, the diameter is not known at every depth even when actually measured. That is the reason why the diameter was entered in the model as the only stochastic variable. Input parameters for the diameter (distribution, average and standard deviation) were fitted to the available measurements.

CONCLUSIONS

According to all the measurements, the parts constructed to date are all watertight. The 'as built' information also backs up the assumption that the grout body is capable of transferring the forces between the steel piles.

Because of the phasing necessary to keep the station running, two wall sections still have to be constructed. Excavation cannot be started until these wall sections are finished. The final results of the grouting process will become visible during excavation.

BIBLIOGRAPHY

A.M.W. Duijvestijn, and B.J. Schat. 2001. *Een kanaal onder het Centraal* [In Dutch], Cement 2001 nr. 3, pp. 41–47. Stichting ENCI media: 's-Hertogenbosch.

J.M. van Esch, A.F. van Tol, H.R. Havinga, A.M.W. Duijvestijn, B.J. Schat, and J.C.W.M. de Wit. 2005. *Functional analysis of jet grout bodies based on Monte Carlo simulations*, 11th Int. Conf. on Computer Methods an Advances in GeoMechanics, Torino, Italy.

J.C.W.M. de Wit, P.J. Bogaards, O.S. Langhorst, R.D. Essler, J. Maertens, B.K.J. Obladen, C.F. Bosma, J.J. Sleuwagen, H. Dekker. 2006. *Uitvoering van de sandwichwand onder Amsterdam Centraal Station* [In Ducth], Geotechniek July 2006, pp. 54–61. Uitgeverij Educom BV: Rotterdam.

MONITORING USED AS AN ALARM SYSTEM IN TUNNELING

G. Borgonovo

Golder Associates

A. Contini

Golder Associates

L. Locatelli

Golder Associates

M. Perolo

Golder Associates

E. Ramelli

Golder Associates

ABSTRACT

This paper presents the ground settlement monitoring system used during the construction of two tunnels bored by two EPB TBMs, beneath the existing and active railway embankment in Bologna (Italy). The tunnels run close to large buildings and beneath bridges and other infrastructure.

Data are collected from remote total stations and from electrolevels installed directly on the railway tracks. The automated monitoring system acquires and analyses data then sounds an alarm in the event that there is a problem. Alarms are sent out in case of excessive tilt or settlement.

Data is provided through a dedicated web application for multiple-party access, to gather, manage and store monitoring data.

INTRODUCTION

In 2000, the joint venture of S.Ruffillo (Necso/Acciona Infraestructuras—Salini—Ghella) awarded the construction of one of the most critical sections of the new High Speed railway line between Naples and Milan. This section of the railway is to pass under the city of Bologna, starting from the S.Ruffillo quarter south of the city, to the new Central Rail Station of Bologna, located in the city's center (see Figure 1).

Golder Associates—Italy (Golder) was chosen to design and manage the monitoring system of lot n.5 which consists of:

1. Two Earth Pressure Balance (EPB) tunnels ("Pari" and "Dispari" tunnels), single track, 9.4 m diameter, 6 km length, 15 m interaxis; and

2. A NATM tunnel (115 m length) double track, 14 m wide, connecting the two tunnels to the Central Station.

The excavation was conducted mainly in heterogeneous alluvial and marine strata (see Figure 2) composed of sea clay and loose sandy deposits (Pliocene Clay and Yellow Pleistocene Sands) and of Savena river deposits, mainly gravel and sand strata with a high percentage of fines (lenses of clay and silt).

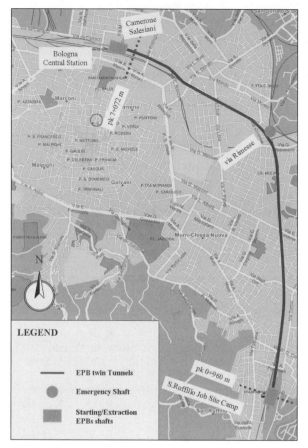

Figure 1.   Urban track of the tunnels

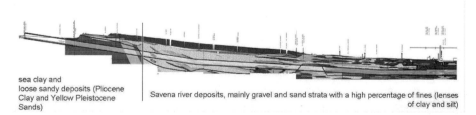

sea clay and
loose sandy deposits (Pliocene
Clay and Yellow Pleistocene   Savena river deposits, mainly gravel and sand strata with a high percentage of fines (lenses
Sands)   of clay and silt)

Figure 2.   Geological section along rail track

        The heterogeneity of the subsurface was a critical aspect of monitoring excavation
activities in regards to potentially rapid change in surface settlement and the TBMs'
behavior.

        The excavation of the two EPB tunnels started in July 2003. The original
monitoring plan was only designed to evaluate geotechnical sections for soil behavior
with the tunnel progress.

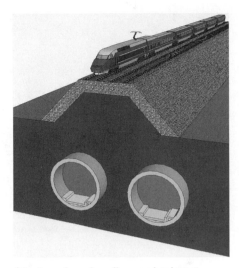

Figure 3. Cross section of the tunnels under railway embankment

Typical geotechnical sections were composed by in-place inclinometers, multi extensometers, and piezometers. All data readings were collected by dedicated data-loggers, interrogated with a server through a GSM system and were then available through a dedicated, protected web site named GIDIE (Golder Instrumentation Data Interpretation and Evaluation). The tunnel lining was also equipped with strain gauges to evaluate stress-strain behavior.

As the tunneling progressed under buildings, bridges and the existing Bologna-Firenze railway (the most important section or railway in Italy, connecting Rome with Milan, Figure 3) it became evident that the project needed a more complex monitoring system.

Golder proceeded with re-designing the original monitoring system in order to accomplish the main goals of monitoring both underground movement as well as the surface settlement. The final system was composed of the following:

- geotechnical monitoring;
- automatic topographic survey monitoring by means of total stations;
- TBM parameter monitoring; and
- monitoring of rail track geometry by means of electrolevels.

All the monitored data were collected through a dedicated Web server which allows users (Owner, Contractor, Costruction Manager, Designer) to access the monitoring data, query the database and display results by means of a dedicated WebGIS application.

On June 31st, 2006 the two EPB tunnels were completed (see Figure 4) and, at present, the NATM tunnel is in progress.

This paper focuses on the details of the monitoring system and its usage as an alarm system, therefore focusing on electrolevels and total stations.

Figure 4. EPB's breakthrough

Figure 5. Longitudinal electrolevel

INSTRUMENTATION

Electrolevels

As the EPB tunnels were driven under the active railway, there was a risk of causing a derailment on one of the busiest Italian railways. To reduce this risk, it was
decided to monitor the tracks inclination.

Railway displacements were monitored with a series of Electrolitic Tilt Sensors
(240). These sensors were installed onto the rail track in two different orientations:

- on the long axis of the rail track (Longitudinal Tilt, Figure 5); and
- on the long axis of the ties (Transversal Tilt, Figure 6).

The Longitudinal Tilt measured the variation of inclination of the rail track while
Transversal Tilt measured the twist of the rail track, which was defined as the difference of transversal inclination on a 3 meter section of rail. Tilt sensors were installed
and removed following the tunnels' progress underground.

Data from Electrolitic Tilt Sensors were acquired on a time interval of every 5 minutes and data were collected through short range modem connection, processed and
uploaded to the Web Server for real time data interpretation.

Figure 6. Transversal electrolevel

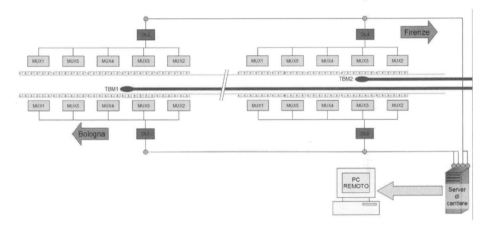

Figure 7. Modules' scheme

To follow the progress of the tunnels, tilt sensors were grouped in modules of 12 (6 longitudinal and 6 transversal), and were all connected to a dedicated multiplexer (see Figure 7). With a module it was possible to cover 54 m of rail track. It was necessary to use 20 modules to guarantee the necessary coverage (2 tracks with two EPB) with a maximum excavation speed of 30 m/day.

Total Stations

For the excavation of the two EPB tunnels, Golder designed a monitoring system for surface movements. The system was designed with total stations (TS—automatic theodolites) and topographical prisms installed on the railway embankment, on the buildings and existing structures (bridges and underpasses) that could potentially be affected.

Leica servo-controlled total stations, TPS System 1100—TCA 1101 Model with ATR 2 self-collimation devices were chosen for the monitoring.

The TSs were installed on dedicated tripods, adequately stiffened, powered by solar panels and connected to a GSM modem for data transfer (see Figure 8).

Figure 8. Typical TS installation

For every data cycle, at each monitoring point, the TS measured:

- slope distance;
- azimuth angle; and
- vertical angle.

After every cycle, the TS created an ASCII file with raw data for all the points' position.

The TSs were installed at the top of embankments so that they could "see" a large number of prisms. Because of this, the TSs were inside a potential subsidence basin and they had to be considered as mobile TSs. Therefore, before the calculated position of each monitoring point, it was necessary to identify any station movement.

Golder developed dedicated software, named GETS (Golder Elaboration of Total Stations), in order to process the measurement of topographical data. The software uses a statistical method for verifying the stability of the points measured by the TS. The most stable points are then used as reference points.

GETS chooses, among all the measured points, the ones showing the most homogeneous 3D displacement-vector (i.e., the 3D vector with the smaller variance). Such displacement vector is then applied to the TS position providing, for each cycle of reading by means of least square method, the actual displacement of the TS in a 3D space.

In spring of 2006, as the tunneling work proceeded, Golder was requested to design the monitoring and alarm system for two buildings during the completion of the NATM tunnel. It was easy to see that TS would be a good instrument to use for this task.

The Project

The NATM tunnel is being constructed under two buildings (see Figure 9) where the Italian railway service (FS) has their own offices. Golder is monitoring these building with 4 TSs and 79 prisms.

For each building, all prisms are monitored with a cycle of readings every hour and all the data are automatically downloaded and collected to a server that processes them. In addiction to the monitoring system designed for the EPB tunnels, Golder has

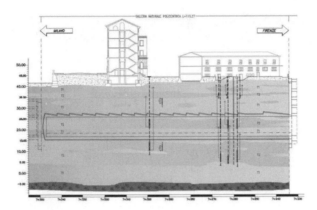

Figure 9.   NATM tunnel under buildings

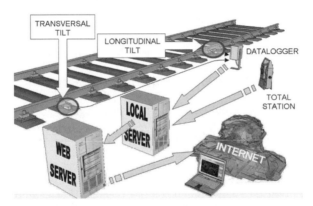

Figure 10.   Hardware configuration of monitoring system

developed an automatic alarm system that processes all data to evaluate two parameters (Beta and Inflection rate) to signal potential danger to the structures.

DATA MANAGEMENT

Data Loggers and Total Stations store the collected measurements in their internal memory.

A dedicated server can communicate with instruments via a LAN connection. If a LAN connection is not possible due to distance or access issues, a GSM is provided instead.

The server periodically downloads the new data from the instruments (see Figure 10). By the mean of a LAN it is possible to get data every few minutes. With a GSM connection, a lower frequency of downloading is allowed.

Software developed expressly for this system, checks that all the instrumentation is reachable and that every sensor is responding correctly. In the event that the server should encounter communication problems with an instrument or sensor, the software

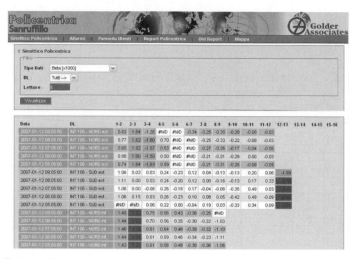

Figure 11. Data panel

will notify an analyst, by means of continuously placing phone calls to dedicated mobile phone until alarm is managed.

The software then calculates engineering data (mm, degrees, degrees/meter or else) from the electrical measurements it receives. When a sufficient amount of data is collected (i.e., no less than 5 data points) a statistical computations determines if the measurement is at a peak.

In the case of one peak measure, the event is logged but it doesn't raise any flag. In the event of many peak events on the same sensors a flag is recorded.

In the event that calculated measurements, such as track twist, gradient variation, beta or inflection ratio or other calculated parameters, are greater than prefixed thresholds, a new flag is arisen. Many thresholds are possible. For example: pre-attention (warning, something may go wrong), attention (some action must be taken), pre-alarm (warning to analyst), and alarm. Then the software analyzes the flags and it takes up the necessary actions to be submitted to the analyst. When the system creates a new alarm the server calls the analyst by mobile and repeated SMS are sent until alarm is taken into account by the analyst.

The analysts can log onto an Internet control centre (see Figures 11 and 12) from which they can examine all the raw data, calculated measurements, alarms and logs; in data table or graph form.

When a new alarm is created the analyst can check which measuring point created it, he can view the history of the measurements, and change the status of the alarm to "managed" (no new alarms will be created for the same threshold for that point, higher thresholds alarms are possible) or to "closed." If the analyst thinks that something significant has occurred, they would call the construction management to verify the data with manual instruments and take necessary safety measures (issue a warning to rail authority, etc.).

The control centre allows the user to schedule analysts' shifts. The server will call the analyst on duty, if the analyst on duty does not manage the alarm within a given amount of time, the server will then call a backup operator. In the event that this analyst doest not manage the alarm within a given amount of time, it will finally call all the operators in its list.

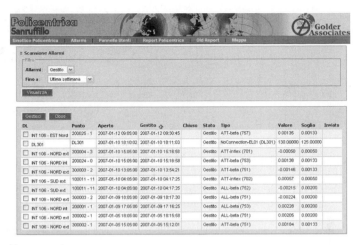

Figure 12.   Alarm panel

The server has a AMD Sempron™ Processor 3100+ with 1 GigaByte of RAM and two 160GB hard drives in raid1 configuration. It is equipped with Linux Operative System, Apache WebServer, MySQL Database server and all the software was developed using php and C++ languages.

ALARM SYSTEM

The alarm system is conceptually the same for both the rail tracks and for the buildings. It automatically retrieves data, processes the data by calculating the control parameters, and calls an analyst if an exceedance of a threshold limit occurs to decide if the exceedance warrents alarm procedures.

For rail tracks there are two parameters that can activate alarms: twist and gradient.

In building monitoring system both beta (angular rotation) and inflection ratio are compared with threshold limits and can create an alert.

Rail Tracks

The monitoring system allows a great number of readings and geometric parameters of the tracks that can be monitored continuously with measurements taken every 5 minutes.

In the event of a threshold limit exceedance (defined by railway regulations) the system activates the automatic alarm procedure and calls the analyst.

Geometric parameters analyzed by alarm system are:

- twist (on 9 m); and
- gradient.

Twist is calculated by dividing the difference between the transversal levels of two contiguous transversal electrolevels by their relative distance (measurement base of the twist) across two rails.

Transversal level is the measurement (in mm) of the difference in height between two contiguous rolling tables. It is the height of a right-angle triangle with 1,500 mm hypotenuse with a summit angle calculated as the angle between the rolling plane and an horizontal reference plane.

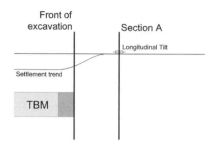

Figure 13. Longitudinal section undisturbed

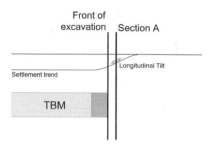

Figure 14. Longitudinal section next to excavation front

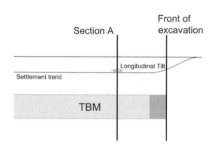

Figure 15. Longitudinal section, case "A"

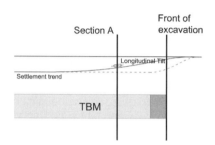

Figure 16. Longitudinal section, case "B"

Twist (expressed as ‰), is calculated on a 9 m base with the following formula:

$$\gamma_9 = \frac{(I_9 - I_0) \cdot b}{d}$$

where
I_0 = is the slope measured by the first transversal electrolevel (as mm/m)
I_9 = is the slope measured by the following transversal electrolevel (as mm/m)
b = is the rail interaxis (1,500 mm)
d = is the distance between two consecutive transversal electrolevels (9 m).

Therefore, by measuring transversal slope it is possible to calculate twist.

The Gradient is defined as the variation in time of the difference between two consecutive longitudinal slopes (i.e., the first derivative of the curve of the longitudinal slope measurement).

It was considered that these three parameters are the most important for the railway safety. Twist is considered the most dangerous geometric fault and can cause derailments. Longitudinal slope and gradient allow identification of rotation occurring on a surface passing through the tracks and the speed of the phenomenon.

Initially only a longitudinal section of the tunnel is considered. Figures 13–17 shows settlements trend versus advancement of the face of excavation. Before the arrival of the front, the longitudinal electrolevel, set in section A, is not affected by rotation (see Figure 13) As an excavation front advances, the electrolevel measures the inducted rotation (positive according to right hand law, Figure 14). When the front gets over section A there are three possibilities:

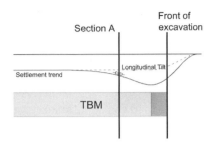

Figure 17.   Longitudinal section, case "C"

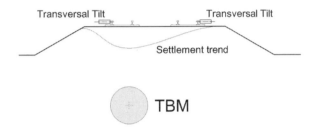

Figure 18.   Transversal section

1. if settlements measured ahead are similar to those in section A (Figure 15) electrolevel returns to the initial condition with null rotation
2. if settlements measured ahead are smaller than those in section A (Figure 16) electrolevel still measures a positive rotation
3. if settlements measured ahead are larger than those in section A (Figure 17) electrolevel measures a negative rotation.

When the excavation front passes through a transversal section, the electrolevel measures a rotation (in fact the excavation is external to rail tracks axis, see Figure 18). This rotation changes depending on drilling advance. Considering a portion of 9 m of tracks with two transversal (one at the beginning and one at the end) and one longitudinal electrolevels (at the centre of the portion, 4.5 m) it is possible to identify the rotation of the rolling plane given by the tracks. With a sequence of electrolevels it is possible to identify every movement of the tracks.

Beta and Inflection Ratio

Every hour, the topographic monitoring system verifies the attitude of the buildings related to all onsite operations (not just the tunnel boring itself but also the grouting process) connected to the construction of the tunnel. With a relevant number of prisms and the adopted configuration, onemeasure in an hour because the TS reading cycle takes 20–45 minutes depending on prism visibility.

In case of a threshold exceedance, the system activates the automatic alarm procedure and calls the analyst.

Topographic measuring targets are fixed on horizontal alignments along building facades. On GIDIE it is possible to see the movement (horizontal and vertical) of these targets in function of time.

Sagging measures allows the calculation of two parameters:

- Beta (angular rotation); and
- Inflection Ratio

Beta is calculated for every two contiguous targets and is given by dividing the difference of sagging between the two points by their relative distance.

Inflection ratio is expressed by the ratio between the distance of the rigid configuration of the building from its deformed configuration and the length of the building (or of its portion) affected by movements

Referring to Figure 19 it is possible to calculate the inflection ratio at every point (except the outermost points) with following method:

$$\text{Inflection ratio point B} = \max \left\{ \frac{\Delta_{B-AD}}{L_{AD}} ; \frac{\Delta_{B-AC}}{L_{AC}} \right\}$$

$$\text{Inflection ratio point C} = \max \left\{ \frac{\Delta_{C-AD}}{L_{AD}} ; \frac{\Delta_{C-BD}}{L_{BD}} \right\}$$

The software that elaborates all TSs readings provides all possible combinations of calculations along each wall of the building. For example having twelve targets means that 220 inflection ratios must be calculated. The alarm system uses these two parameters to determine possible damage to the structures.

ALARM PROCEDURES

Railway Tracks

In the case of a threshold exceedance, the alarm system automatically advises an analyst (or a group of specialists) that is constantly available. The analyst has different instruments to verify the cause of the alarm:

- time series of electrolevels data and parameters (see Figure 20);
- measurement of all the electrolevels in the alarm area;
- movement of the near topographic targets (data on GIDIE, Figure 21); and
- EPB parameters.

In 30 minutes, if the alarm is confirmed, the analyst has to inform the construction supervisor and the administrator of the railway. They will quickly meet to decide if it is necessary to stop the excavation and to further evaluate the risk to trains.

The redundancy of the system allows that the analyst can even evaluate if abort the alarm procedure in case other data confirm absence of risk and/or greater reliability.

The system controlling the railway tracks has been active for more than 6 months and in that time two alarms have been sent out. In both cases, the alarms were validated; however the presence of track ballast under the tracks caused the parameters to come back within threshold limits, in a few hours.

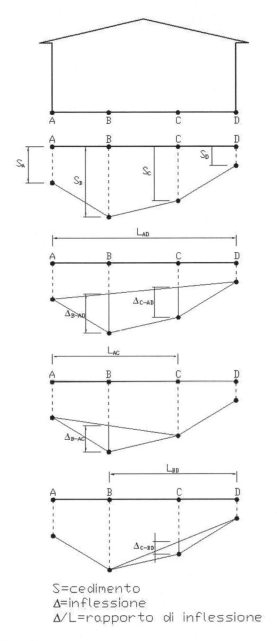

S=cedimento
Δ=inflessione
Δ/L=rapporto di inflessione

Figure 19.   Inflection ratio

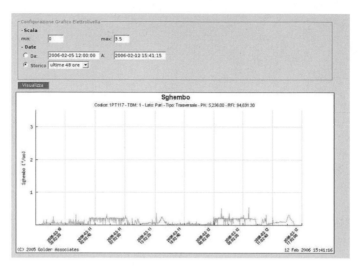

Figure 20. Twist versus time

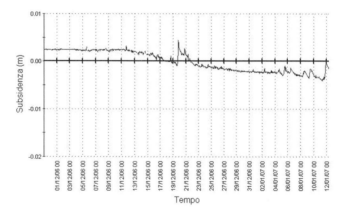

Figure 21. Graphic on GIDIE

Buildings

The TSs take nearly an hour to complete their reading cycle (measurement, data download, measurement calculation and publication on web), so the analyst has less data to evaluate alarms with. Because of this it is also very difficult to evaluate weather peaks are temporary or not.

For that reason it is important that analyst is kept informed of all the activities performed in the area and the behavior of every target. Inside the buildings, 68 structural electrolevels have been also installed allowing the analyst to have further data to evaluate alarms.

Since inflection ratio is a parameter whose theory is well known, but its evaluation is less intuitive than beta or differential settlement, more responsibility was charged to the analyst in determining the validity of alarms. In four months, the system automatically gave more than 100 alarms but the analyst only validated 11 of them (see, in

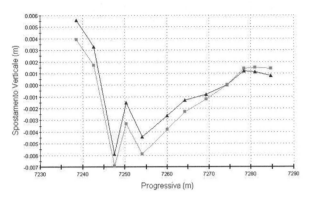

Figure 22. GIDIE, longitudinal section

Figure 22, a longitudinal section for an alarmed area) providing a valuable assistance to the site operation during grouting underneath the building.

After every alarm, the work supervisor was informed of the alarm and work was stopped three times to take precautions before proceeding.

CONCLUSIONS

Golder Associates designed, developed and deployed a monitoring system to prevent major infrastructure from damage from the construction of twin tunnels passing beneath the Bologna metropolitan area. The system allows for real time data interpretation of all data coming from total stations, geotechnical sensors, and electrolevels via the web, and provides an automated, real time alarm system.

The power of such a system allows the contractor as well as construction management to monitor, and therefore control, the effects that the excavation of the two tunnels has on the surrounding environment. The monitoring analyst becomes an important figure in the tunneling work, being responsible for validating and investigating alarms in real time and providing a safer and more efficient work procedure.

ACKNOWLEDGMENTS

The authors wish to thank Valentina Marchionni, Massimo Secondulfo and Fernando Minguez (S.Ruffillo s.c.a.r.l.) for their assistance in the setup of the whole system.

PRE-EXCAVATION GROUTING THROUGH WATER-BEARING ZONE UNDER HIGH PRESSURE UNDER EXTREME FLOW CONDITIONS

Robert Fu

East Railway Improvement Bureau, Republic of China

Adam Shang

Jines Construction Co., Ltd.

Alex Naudts

ECO Grouting Specialists Ltd.

ABSTRACT

The Yung-Chung-Tunnel in Taiwan, constructed for the Taiwan-East Railway Bureau, collapsed during construction when a water bearing, fractured marble zone was intersected. The inflow was approximately 4,000 liters/second. In spite of 30 relief holes the total inflow into the tunnel remained constant.

The pre-excavation grouting through the re-aligned tunnel had to be conducted under extreme flow-conditions. A hot bitumen grouting program was selected for technical and economical reasons out of 13 international proposals.

The technical challenges were:

- grouting under severe flow conditions
- high formation water pressures to be injected with hot bitumen at high temperatures and pressures

The hot bitumen grouting program resulted in a dramatic reduction in the hydraulic conductivity and stabilization of the formation and facilitated a problem-free excavation.

BACKGROUND

The New Yung Chung railway tunnel is located in North East Taiwan (Republic of China), along the line between Suau and Tonau, along the Northern Railway Route in South Ihran prefecture. The tunnel is approximately 4.5km long and is located directly to the west of the older Yung-Le tunnel.

The tunnel was excavated from both sides: from the north portal and from the south portal. The main tunnel is approximately 6.5m high and 5m wide. By-pass tunnels or "reconnaissance tunnels," approximately 4.5m high and 4m wide were excavated, parallel and in advance of the main tunnels, to explore the geology and to obtain information of the problems ahead of the main tunnel. The main tunnel was lined with in-situ cast reinforced concrete.

The tunnel was constructed using the drilling and blasting method. The construction of the tunnel was on schedule, until a huge water inflow occurred on 24 October 1998. The back of the tunnel collapsed and the inflow into the tunnel was in the order of 4m³/second. This event delayed the project by 3.5 years. By May 2001, more than 50 million m³ of water had rushed into the tunnel through the breach with no sign of slowing down or decreasing in pressure. The hydrostatic pressure in the drainage

Figure 1.   View of drainage tunnel with 36 pressure relief pipes

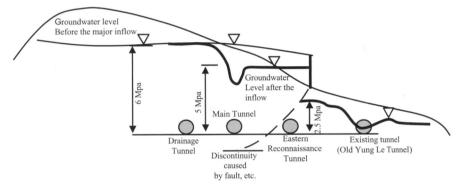

Figure 2.   Conceptual view of the water bearing mechanisim of the New Yung Chung Tunnel

holes, intersecting the flow path was exceeding 5 Mpa (750 psi). Rubble filled the tunnel over a length of 300m.

The water was running through a cavernous marble layer containing cavities and inter-crystalline porosity, extending to the west, from where the tunnel intersected this layer, to an area close to the peak of the mountain range. The average yearly precipitation in this mountain range is approximately 11,000 liters per square meter. During the subsequent investigation, it became clear that the marble layer acted as the "collector" for a vast area near the mountain top. The marble layer (locally referred to as Marble II) was sandwiched between shear zones of green shist with little strength. Ground squeeze and the presence of a vast, virtually unlimited reservoir of water under high pressure were the 2 main challenges to the tunneling operation.

An additional drainage tunnel (Figure 1) was excavated, from where 36 pressure relief holes, 20cm in diameter were drilled, intersecting the flow path well above the main tunnel. These drainage holes also produced approximately 3–4m^{3}of water/ second, but failed to significantly reduce hydrostatic pressure and flow through this water bearing formation (with reference to Figure 2). For this reason, this tunnel inflow was considered the "highest tunnel inflow in the world at the time."

The geology in and near the tunnel inflow area consists of amphibolite, green shist (including altered chert and "marble I"), marble ("marble II"), green shist and quartz mica shist (including altered chert). These layers dip upwards toward both the north and the south and their strike is from north-west to south-east. Marble II, the main conductor of the water, seemed to taper off near the point where it intersected with the

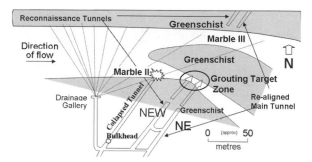

Figure 3.   Re-alignment of tunnels at the time of the grouting, early 2002

east planned tunnel alignment. Except for this point, this Marble II layer has good continuity, extending at least 1.2km to the west and 0.5km to the east.

The maximum thickness of the water bearing formation is 70m. Two major joints, which cross at right angles with each other are developed in the marble layers. One is running parallel with the shist and has been noticed in several outcrops, some near a cliff where it caused secondary breakage and at some spots forming porous coarse breccia. Along the already weakened line of joints and shistosity are ant-nest like, small leaching channels which are often filled with fine gouge or fine gravel. A lot of this infill material washed into the tunnel.

In order to complete the New Yung Chung tunnel, it was partly re-aligned and the north and the south faces were advanced to approximately 50m from the water bearing marble zone, which was approximately 20m thick, where the new tunnel alignment was going to intersect the water bearing zone. For the lay-out of the planned and existing tunnels, refer to Figure 3.

Since dewatering the formation was not working, the Taiwanese Railway Authority, assisted by an international panel of consultants, invited technical proposals from all over the globe, to construct a structural grout curtain, to be installed under flow conditions and under a hydrostatic pressure of 5 Mpa, that would facilitate safe tunneling through the water bearing area.

The technical proposal submitted by Jines Construction from Taipei, Taiwan, assisted by ECO Grouting Specialists Ltd. from Canada was selected from 12 other international proposals. The selection was based on technical merit, proven experience in this type of work and cost. The Jines group was also the only bidder who was offering a fixed budget price to perform this challenging pre-excavation grouting work and who guaranteed a successful outcome.

The remedial grouting job was awarded to Jines in November 2001 and construction of the grout curtain started in February 2002.

OVERVIEW OF THE PROPOSED GROUTING METHOD TO FACILITATE A SAFE TUNNELING OPERATION THROUGH THE WEAK WATER BEARING ZONE

A plan to facilitate safe excavation for both the reconnaissance and the main tunnels were tabled. It involved the installation of a series of heavy duty steel sleeve pipes, to be grouted with hot bitumen. The steel sleeve pipes were acting as grout delivery pipes and as a forespiling system.

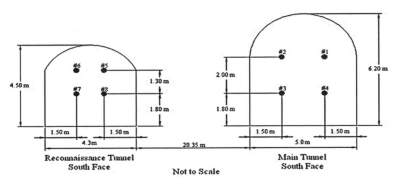

Figure 4.   Layout of primary grout holes for Phase I grouting

The plan proposed by the Jines team was serving two purposes:

- minimize the inflow of water into the new tunnel, in the water bearing zone when tunneling through it
- provide the necessary mechanical strength to the formation to minimize or prevent ground squeeze

It was envisioned that the grouting project would consist of 2, potentially 3 grouting phases. Contrary to what is commonly done in bitumen grouting projects, all grouting was to be performed with bitumen (no cement grouting). Phase I involved the following tasks and activities:

- Drill 8 cased holes (the primary grout holes) for the installation of the steel sleeve pipes: 4 per tunnel (refer to Figure 4)
- Perform the drilling and grouting activities from the south side, where a lot of space was available to accommodate this work
- Start the holes by installing heavy duty steel standpipes, grouted in place and tested to a pressure of 15 Mpa; attach a blow-out-preventor on each stand pipe, prior to drilling into the marble zone
- Drill a cased hole extending through the marble, through the blow-out preventor
- Install the in-hole grout delivery system, including the sleeve pipe in each grout hole, remove the casing and seal the annular space between the formation and the in-hole grout delivery system
- Design and install a hot bitumen grout plant in the tunnel, within 50m of the tunnel face, inclusive of a 40,000 liter heated and insulated holding tank for the hot bitumen. The grout plant had to be capable to inject 20m³ of hot bitumen per hour at pressures in the order of 7–12 Mpa. It involved 8 grout pumps acting in parallel, set up side by side (Figure 5).
- Construct the pipe lines to facilitate the pre-heating of the bitumen delivery system including the lines inside the tunnel and the in-hole grout delivery system
- Install thermocouples along the insulated grouting circuits
- Set up the bitumen delivery system:
  - from the plant, 300km away, to the job site, via mobile tanker trucks
  - The on-site bitumen receiving system near the portal of the tunnel including off loading facility, temperature re-boosting system and large (300m³) holding tanks (Figure 6)

Figure 5.   Grout plant in foreground, insulated bitumen holding tank in background

Figure 6.   Off loading facility and insulated bitumen holding tanks outside tunnel

> – Delivery of the hot bitumen to the holding tank, near the tunnel face at a rate of 20 m^3/hour—24 hours/day, via mobile tanker trucks

- Arrange for adequately qualified and experienced staff, consisting of local technicians and ex-pats to run this operation around the clock and inject in excess of 3 million liters of hot bitumen into the formation to create a grouted "donut" around the future tunnel trajectory
- Set up the grout assessment station in real time, based on the evolution of the apparent permeability of the formation (using grout as a test fluid)
- Provide the outline of the technical assessment for the first grouting phase, including accessible porosity of the formation, obtained grout spread and design for the second grouting phase as well as an assessment of all the components of the operation and make adjustments where necessary.

Phase II was to involve the following tasks and activities:

- Install 8 secondary grout holes, outfitted with the requisite in-hole grout delivery systems
- Perform the hot bitumen grouting
- Evaluate the results

Phase III (if necessary) was to drill and grout additional holes to ensure the future tunnel was completely protected by the grout curtain, having a thickness of at least one tunnel diameter.

HISTORY OF HOT BITUMEN GROUTING

By the end of the 19th century, grouting with hot bitumen for remedial repair work on dams and seepage control in rock tunnels was introduced. Bitumen was first used at European dams in Switzerland and France and later at dam sites in North America. There is documented evidence in the records of Puget Sound Power and Light that hot bitumen was used during the 1920s at Lower Baker Dam near Seattle, and on some dam grouting projects for the Tennessee Valley Authority (U.S.A).

The development of different types of environmentally friendly bitumen expanded its use in the grouting industry. Oxidized blown bitumen replaced the softer bitumen (typically used in road paving applications) and emulsions in grouting, and played an important role in marine, civil and mining applications, mainly for seepage control or prevention.

The use of hot bitumen was "rediscovered" during the early 1980s with the success of the Lower Baker Dam (USA) and the Stewartville Dam (Canada) grouting projects. Hot bitumen grouting made a remarkable comeback during the late 1990s. Projects in the Philippines, Brazil, Eastern Europe, New Brunswick (Canada), West Viginia (USA), and Wisconsin (USA) demonstrated that the application of bitumen technology is an efficient, economical and powerful tool to prevent or stop seepage and major leaks (Ref.2, Schonian, E. & A. Naudts, 2003).

The nature of most bitumen grouting projects typically involved emergency situations in which very serious water inflow problems needed to be solved. This has actually hampered the exposure of bitumen grouting in the mining and civil engineering world, since clients often do not wish that detailed information be disseminated on their misfortune or problem situation.

HOW HOT BITUMEN GROUTING WORKS

Dr. Erich Schonian, one of the first scientists to study the penetrability and behavior of hot bitumen, documented remarkable findings regarding bitumen penetration in cracks (Ref. 3, Schonian, E., 1999). As the bitumen comes in contact with water, the viscosity of the grout increases rapidly resulting in a lava-like flow. A hard insulating crust is readily formed at the interface between water and bitumen and shelters the low viscosity, hot bitumen behind it. The "crust" or "skin" is re-melted from within when hot bitumen continues to be injected.

When hot bitumen is injected into a medium with running water, it cools quickly at the interface with water. Steam is created at that point, decreasing the viscosity of the bitumen. The steam acts as an "air lift" drawing the bitumen into its pathway through small and large fissures or pore channels. The centre of the bitumen mass remains hot, and, as a result of the grouting pressure, continuously breaks through (re-melting) the skin formed at the interface of the bitumen and water. The faster the water flows, the faster the bitumen cools off. The skin prevents wash-out while the "sheltered" hot bitumen behind the skin behaves as a Newtonian fluid, penetrating in a similar fashion as solution grouts.

Because bitumen has good insulating characteristics, it can be injected for a very long time (days—even weeks) into the same grout hole without the risk of either premature blockage or wash-out. The width of the fissures accessible to hot bitumen depends on the duration of the grouting operation. The longer the grouting operation, the finer the apertures the bitumen will penetrate. Hot bitumen will penetrate fractures as small as 0.1mm as demonstrated during the Kraghammer Project in 1963.

When hot bitumen cools it is subject to significant thermal shrinkage. This phenomenon is partially overcome in smaller fractures if pressure continues to be applied and warmer bitumen pushes the cooling bitumen into the shrinkage gaps. Cement based suspension grout is often injected in conjunction with hot bitumen to compensate for the thermal shrinkage of the bitumen; to make the bitumen less susceptible to creep; and to increase the mechanical strength of the end product.

SPECIAL CONSIDERATIONS FOR USING HOT BITUMEN GROUTING

The equipment and set-up are generally more complex for bitumen grouting than for the applications involving regular cement based grouts or solution grouts. The operating temperature of the surface pipe system needs to be in the range of 180–225°C (356–437°F). Moreover, a supply of hot bitumen needs to be obtained and maintained at the requisite temperature. The bitumen should, ideally, be delivered to the site in heated and insulated bulk takers with the potential to boost or adjust the temperature on site in a custom build grout plant.

The piping system used during grouting to deliver the hot bitumen from the bitumen pumps to the sleeve pipe "stinger" located at the end of the bitumen grout hole, must either be pre-heated with hot oil, heat trace, or steam, potentially through a recirculation system. Additionally, the grout pipes must be insulated and equipped with temperature sensors and pressure gauges. The flow rate, total volume of grout injected, and grouting pressure must be monitored and recorded in real-time. This allows informed decisions to be made while the operation is in progress. The apparent Lugeon value is the permeability coefficient of the formation using grout as a test fluid (Ref. 5, Landry, E., Lees & Naudts, 2000). It is noteworthy that during the execution of a hot bitumen grouting program the apparent Lugeon value typically increases with time (contrary to cement grouting operations) due to the excellent penetration properties of the bitumen into the formation (which is warming up).

ENVIRONMENTAL ISSUES

The type for use in grouting is a "hard," oxidized, environmentally friendly type of bitumen with a high solidification point. Oxidized blown bitumen has a long history of successful use for lining (potable) water reservoirs in California (over 40 years) and in 1987 Washington and Oregon State wildlife authorities have used it for lining fish hatchery ponds.

Oxidized bitumen has proven to be in compliance with American Water Works Association (AWWA) standards for leachate resistance of materials for use in potable water applications. Indeed, it is now routinely used for water pipeline lining applications. It could be considered the most environmentally friendly grout presently available on the market.

THE ACTUAL GROUTING OPERATION AT
THE NEW YUNG CHUNG TUNNELS

The actual grouting work turned out to be a very challenging undertaking. Many setbacks had to be overcome. The first grouting phase took place in late February 2002 and the third grouting phase was completed by mid May 2002.

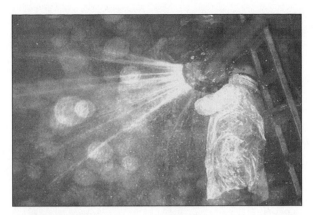

Figure 7. Drilling of cased holes for in-house grout delivery system

The drilling of the cased holes for the installation of the in-hole grout delivery system—including the sleeve pipe—was not an easy undertaking (Figure 7). In each hole, as soon as the marble zone was intersected, "all hell broke loose" and water inflows of more than a hundred liters/second under pressure of 5 Mpa gushed into the hole. Based on the outflow, the average permeability of the water baring marble formation was approximately 120–150 Lugeon (0.0013–0.0016 cm/s). All 8 primary holes were extended -albeit with great difficulty—to the far end of the marble zone. The in-hole grout delivery system consisted of 3 concentric heavy duty steel pipes (to facilitate preheating of the in-hole grout delivery system) ending into a sleeve pipe, located in the marble zone. Barrier bags were inflated with cement based grout around the in-hole grout delivery pipe and any residual seepage around these pipes was sealed using water reactive, hydrophobic polyurethane grout.

Each grouting operation started with the pre-heating of the insulated grout delivery circuits inside the tunnel. Hot oil was used for this purpose. After a few hours of pre-heating the grout lines and circuits, the in-hole grout delivery system—up to the sleeve pipe—was pre-heated followed by the actual bitumen grouting. Initially, a hybrid type of bitumen with a lower softening point was used in turn followed by the oxidized bitumen with a softening point of 95°C.

During the grouting of the first hole, it was quickly discovered that the 8 single action piston pumps operating at high pressures (10 Mpa) were not capable of operating more than 1 hour without the need to be repaired (Figure 8). The mechanics could not keep up with replacing the seals. It was also discovered that the temperature of the bitumen in the holding tanks outside the tunnel needed to be increased to 220 degrees Celsius to be able to conduct a permeation grouting program in this type of a formation. Eventually we ran out of spare parts for the grout pumps, before phase I could be completed. During phase I, only 6 of the 8 primary holes were grouted. The injected volumes in each of the holes fell short of the target volumes, necessary to create overlapping grout cylinders. The volumes injected varied between 17 and 68 cubic meters of bitumen per hole. All 6 holes came to refusal because of technical or logistical problems. The team needed to regroup at the end of phase I.

An additional 6 holes were drilled for the grouting of the second phase and the sleeve pipes installed. New, inexpensive pump seals were manufactured in Taiwan. A number of key-positions in the bitumen grouting operation were taken over by the local people. Phase II was very successful. The pump seals were standing up better under the high pump pressures (90–120 minutes in lieu of 60 minutes for the North America

Figure 8.   Single action piston pumps

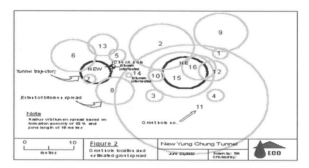

Figure 9.   Grout hole layout for the three grouting phases and theoretical grout spread

seals). The temperature of the bitumen was adequate (200–220°C). The mechanics managed to keep 5 out of 8 grout pumps operational at all times. Injection rates of 10–14 cubic meters per hour were maintained. The delivery of hot bitumen to the holding tank near the tunnel face was done safely. Seven of the 8 holes grouted during the second phase were brought to a natural refusal and each accepted between 80,000 liters and 420,000 liters of hot bitumen. Hole #11 did not refuse after 1,450,000 liters of hot bitumen had been injected during 4 consecutive days into this hole. The grouting mode was one of "permeation in conjunction with hydro-fracturing." At the end of the second phase, all the grouting information was assessed and 2 target areas were identified in which more grout needed to be placed (Figure 9). The average permeability of the formation (based on the water outflow) had been reduced to less than 5 Lugeon near the end of Phase II.

Based on the evolution of flow and pressure the apparent permeability of the formation (using bitumen as a test fluid) was calculated in real time and was graphically displayed on the computer-screen. It was important during this second phase of grouting to hydro-fracture the green-shist to minimize ground movement during subsequent excavation.

For phase III only two holes were drilled. The grouting was conducted flawlessly. Each hole required approximately 50 hours of continuous grouting. The bitumen

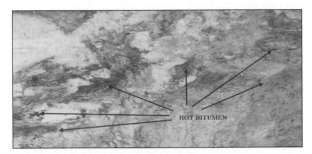

Figure 10.   View of tunnel face through Marble II formation

Figure 11.   Tunnel face entering the water-bearing grouted zone

placed during the second, but especially during the third phase filled up shrinkage gaps caused by the thermal shrinkage of the bitumen. During the 3 grouting phases, a total of 3,084,000 liters of hot bitumen were successfully injected. No bitumen washed out, in spite of the fact that water was still gushing into the abandoned and collapsed main tunnel, less than 100m away and in spite that the 36 drainage holes continued to produce 3–4m^3 of water per second.

At the conclusion of the third bitumen grouting phase, the residual permeability of the formation was less than 0.5 Lugeon. This value was obtained from drilling pilot holes prior to tunneling.

OUTCOME

A formidable and challenging grouting project made it possible to excavate—problem free—the reconnaissance tunnel and the main Yung Chung railway tunnel. Hot bitumen was observed to have filled even the smallest hairline fissures in the marble formation (Figures 10, 11 and 12). The residual inflow into the tunnel when crossing through the water bearing zone was in the order of a few liter per minute. The grout pipes acted very effectively as forespiling and prevented excessive squeezing of the green shist. Hydro-fracture planes filled with bitumen were observed criss-crossing the green shist. One of the most difficult bitumen grouting projects in history had been completed successfully. Hot bitumen grouting proved—once again—to be a formidable tool in the hands of experienced grouting practitioners.

Figure 12.   Sample of marble impregnated with hot bitumen

ACKNOWLEDGMENT

The authors would like to thank the staff of the East Railway Improvement Engineering Bureau, Republic of China, Jines Construction Co, Ltd., Taipei, Republic of China; and ECO Grouting Specialists Ltd. for their tremendous effort and dedication to bring one of the most challenging grouting projects ever undertaken, to a successful conclusion.

REFERENCES

Ref. 1. Van Asbeck, W.F. & E. Schonian, Bitumen in Wasserbau, Band 2 (Bitumen in Hydraulic Engineering, Vol 2). Huthig und Dreyer, Heidelberg, and Deutsche Shell AG, Hamburg 1968, pp. 323–326.

Ref. 2. Schonian, E. & A. Naudts, Hot Bitumen Grouting—Rediscovered, Bitumen, Heft 3 & Heft 4, Fall and Winter, 2003, pp. 118–123, 178–183.

Ref. 3. Schonian, E., The Shell Bitumen Hydraulic Engineering Handbook and CD (more text and Figures/Photos). Shell International Petroleum Company Ltd., London 1999

Ref. 4. Naudts, A & S. Hooey, Hot Bitumen Grouting: The antidote for catastrophic inflows, ASCE Conference, Grouting and Ground Treatment, Proceedings of the Third International Conference, New Orleans, 2003, Vol. 2, pp.1293–1304.

Ref. 5. Landry, E., D. Lees & A. Naudts, Dew Developments in Rock and Soil Grouting: Design and Evaluation, Geotechnical News, Richmond, British Columbia, September 2000 pp, 38–44.

PROGRAM-WIDE GEO-INSTRUMENTATION MONITORING FOR THE MTA EAST SIDE ACCESS PROJECT

Daniel A. Louis
URS Corp.

W. Allen Mawr
Geocomp Corp.

Jesus Schabib
Parsons Brinkerhoff

Dilip I. Patel
MTA Capital Construction

ABSTRACT

East Side Access (ESA) will extend Long Island Rail Road into midtown Manhattan via 11 km of tunnels, a new terminal station and associated rail support facilities. The hard-rock tunneling in Manhattan will traverse under Park Avenue and some of the most historically and commercially significant real estate in the world. The soft-ground tunneling in Queens will traverse beneath Amtrak's Sunnyside Yard and the Harold Interlocking, the busiest rail interlocking in North America. This paper details ESA's program-wide geo-instrumentation monitoring system for obtaining, processing and presenting all geotechnical monitoring data. Data streams will be real-time, such as from automated total stations, and manually read, such as from optical survey instruments and handheld dataloggers. All data will be input into a web-based Geo-Instrumentation Data Management System, which will be accessible to project stakeholders via ESA's intranet or via the Internet using remote access controls.

OVERVIEW—EAST SIDE ACCESS PROJECT

The Long Island Rail Road (LIRR) East Side Access Project, planned by the New York Metropolitan Transportation Authority (MTA), will connect the LIRR with Grand Central Terminal (GCT) in Manhattan. This will be the largest construction project undertaken by the MTA and will involve the multidisciplinary talent of planners, engineers, architects, contractors and operations specialists.

Improving commuter rail access between Long Island and the East Side of Midtown Manhattan has been under discussion for at least four decades. A new East Side terminal for LIRR was first considered as early as 1952. During the ensuing years, several studies examined and supported the concept of a new railroad terminal serving the East Midtown area. These studies resulted in the construction of the four-track, bi-level 63rd Street Tunnel and approaches between Manhattan and Queens. New York City Transit (NYCT) currently uses the upper level and the lower level was built for future LIRR use. In 1998, the MTA completed a Major Investment Study of the Long Island Transportation Corridor, resulting in the ESA Project. ESA will provide a higher level of service to those commuters who travel from Long Island into New York City, as well as relieving congestion at Penn Station and providing LIRR passengers direct access to East Midtown Manhattan.

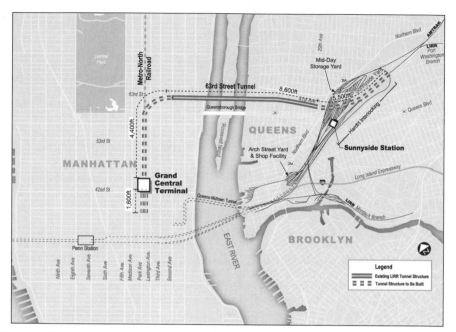

Figure 1. East side access plan

The ESA project alignment is depicted in Figure 1. The alignment begins with connections from the LIRR's Port Washington Branch and Main Line tracks within the Harold Interlocking in Queens. In addition to the main line track connections, lead tracks originate from the Sunnyside Yard and East Side Yard loop tracks.

From Harold Interlocking, the alignment proceeds through a set of four tunnels under Amtrak's Sunnyside Yard to a cut-and-cover tunnel section that begins at the edge of the East Side Yard. In the cut-and-cover section, the tunnels meet at a new Plaza Interlocking, pass under Northern Boulevard, and merge into two tunnels before entering the existing 63rd Street tunnel structure. The length of excavation from the Main Line and Port Washington tracks to the bellmouth is approximately 5,500 feet. From the bellmouth on the east end of the existing 63rd Street tunnel to its end at Second Avenue and 63rd Street in Manhattan, the distance is approximately 8,600 feet.

On the Manhattan side of the river, the alignment continues westerly from the existing 63rd Street Tunnel's terminus at Second Avenue and 63rd Street. These twin rock-mined tunnels pass under the NYCT's Lexington Avenue line and then turn south under Park Avenue beneath MNR's Park Avenue tunnels. The two tunnels proceed south under Park Avenue, passing under NYCT's 60th Street tunnel. At about 58th Street, each of the two approach tracks splits into a lower and upper track creating four tunnels stacked in a two-above-two configuration. These tunnels converge at two caverns, each of which accommodates an upper and lower diamond crossover. From that crossover cavern, each of four approach tracks splits again to service a total of eight tracks and four platforms under the lower level of GCT. The distance from the existing terminus of the 63rd Street tunnel to the new station beneath GCT is approximately 5,000 feet.

The new tracks and platforms will be located below GCT's lower level tracks in two twin-level rock caverns. Each platform will have stairs and escalators rising to four mezzanine-level cross-passageways above and perpendicular to the platforms. From

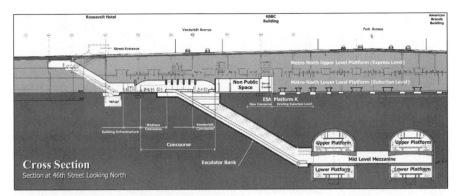

Figure 2.   Grand Central terminal cross section

these passageways, stairs and escalators will rise to a large concourse for the LIRR in the area currently occupied by MNR's Madison Yard. The concourse will include passenger amenities, such as ticketing and information booths, waiting room seating, retail shops, and required administrative and operational support spaces. A cross-section of the GCT caverns, passageways and escalator shafts is shown in Figure 2.

Daytime train storage will be provided for the LIRR trains serving GCT in the East Side Yard, which will be configured to accommodate 288 railcars, with the appropriate storage and running tracks. A vehicle maintenance facility, jointly funded by ESA and LIRR, has already been constructed at Arch Street Yard. To replace MNR's Madison Yard, a new facility has been constructed at Highbridge in the Bronx.

Current Status

The project has been under construction since 2001, with several smaller construction contracts being completed. These contracts have included the Highbridge Yard, Arch Street Yard, and other contracts aimed at readiness for the start of major construction. The large-scale ESA construction work began in 2006 with the award, mobilization and start-up of Contract CQ028—Queens Open-Cut Excavation and Contract CM009—Manhattan Tunnels Excavation. Together, these contracts comprise over $500M and will be in construction for several years.

In 2007, the project plans to award three additional major construction contracts. This new work is planned to include Contract CM019, which expands the current construction work in Manhattan; Contract CH053, which represents the first ESA construction contract in Harold Interlocking; and Contract CQ031—Queens Tunnel Excavation.

The Manhattan construction contracts involve a significant interface with Metro-North (MNR) railroad operations in GCT, some of which can be seen in Figure 2. The Queens construction contracts involve significant interface with both Long Island Rail Road (LIRR) and Amtrak. Figure 3 provides an overview of the major work in Queens, where many of the railroad and related interfaces can be seen.

These railroad interfaces introduce significant additional constraints to ESA construction, over and above those typically involved in tunneling and heavy construction in an urban environment. These added constraints have led the ESA team to implement a broad and rigorous program of geotechnical and structural instrumentation. This paper focuses on the rationale for selecting the geo-instrumentation and the monitoring systems to be employed.

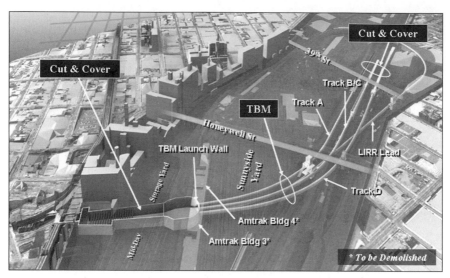

Figure 3.   ESA Queens and Harold interlocking construction

GEO-INSTRUMENTATION PLANNING

Geotechnical and structural instrumentation (i.e., geo-instrumentation) has become commonplace in large complex projects like East Side Access. This geo-instrumentation is a key part of the project's risk management approach. The principal reasons for employing geo-instrumentation on this project are as follows:

1. Provide an early warning of changing conditions or impending failure.
2. Reveal unknowns.
3. Reduce surprises.
4. Assess critical design assumptions and contractor's means and methods.
5. Alert of any potential damage to adjacent structures and facilities.
6. Monitor construction progress, operations and sequences.
7. Provide data to help corrective actions.
8. Measure, document and help to improve performance.
9. Inform stakeholders.
10. Satisfy regulators.
11. Reduce claims and litigation.
12. Advance state-of-knowledge.

A central theme in each of these reasons is the management of project risk. On East Side Access, this is a particularly important element of project management because of the complex interfaces with the key railroad stakeholders, namely LIRR, MNR, NYCT and Amtrak and the close proximity to the work of many major and historic structures.

The combination of a complex urban environment, numerous sensitive facilities above and adjacent to the work, diverse and often innovative methods of construction, and many involved stakeholders produce an environment where the potential consequences of adverse or unexpected performance are high. These consequences can involve money, time and political considerations. By using the geo-instrumentation to address the items listed above, the potential for adverse consequences can be effectively measured, understood and managed. Monitoring of geo-instrumentation is an essential part of ESA's risk management plans.

GOALS OF THE GEO-INSTRUMENTATION MONITORING PROGRAM

During preliminary engineering, the project began planning for a program-wide geo-instrumentation monitoring system for obtaining, processing and presenting all geotechnical monitoring data. As the design concept developed, the overriding performance-related concerns came from the potential for unexpected deformations caused by excavating the tunnels and station areas. The designer used their knowledge of the existing conditions and the engineering concepts to select and locate the appropriate type and number of instruments. The primary instrument systems planned for the project are as follows:

1. Optical Survey Points, including deep benchmarks, surface monitoring points, structure monitoring points and settlement/lateral displacement reference points.
2. Track Dynamic Profile Monitoring Points.
3. In-place Inclinometers to measure horizontal movements towards excavations, including those in soil/rock and placed within the slurry walls.
4. Horizontal In-place Inclinometers.
5. Multi-point extensometers to measure relative movements between points.
6. Tape Convergence Measurement Points within the tunnels.
7. Tilt-meters to measure rotation of building faces and foundations.
8. Liquid level systems for accurate settlement measurements along tracks.
9. Automated Total Stations with reflective prisms to measure Δx, Δy and Δz.
10. Crack Monitors to indicate crack growth in concrete and masonry, including grid crack gauges and vibrating wire crack gauges.
11. Vibrating Wire Strain Gauges and Loads Cells.
12. Seismographs to measure ground motions from construction activities.
13. Observation wells and piezometers to monitor alterations in groundwater levels.

As the plans developed for the type, location and number of required instruments, it became clear that the data collection and management activities had to be automated for the information to be of use in the rapidly-paced project. Some of the reasons for automating the monitoring processes are as follows:

1. Reduce time to obtain and evaluate data.
2. Provide automated warning of unacceptable performance in a short time.
3. Get data from remote, inaccessible, and dangerous locations.
4. Record more readings.
5. Manage data from thousands of points over duration of the project.
6. Take readings where rate of change is too fast for manual approaches.
7. Measure different parameters at the same time.
8. Improve reliability of data.

9. Provide immediate access to entire data base in electronic form to facilitate project wide evaluations.
10. Lower the total cost of monitoring.

Automated Geo-Instrumentation Monitoring

The key requirements for the automated monitoring system that developed over time on ESA are as follows:

1. Accept data from variety of instruments and systems installed by different contractors.
2. Manage data recorded as frequently at 1,000 times per second over a duration of up to 12 years.
3. Receive data in electronic formats and handwritten notes.
4. Convert instrument readings to meaningful engineering units.
5. Be reliably available 24 hours per day, 7 days a week for the duration of the project.
6. Combine data from TBM's with geotechnical data to aid in performance evaluations.
7. Isolate data by contract and also aggregate data across multiple contracts.
8. Provide messages by phone and email to a contact list within 5 minutes of an alarm condition.
9. Provide graphs and tables of the reduced data on the project intranet and the Internet in a format useful to engineers, contractors, the Owner and third parties.
10. Provide ability to select what information each party can access.
11. Provide backup and redundancy so no data are lost.

ESA Geo-Instrumentation Roles and Responsibilities

Because of the size and diversity of the project, the key responsibilities for the geo-instrumentation program underwent considerable study and discussion during the conceptual design and preliminary engineering phases. The result of this process is summarized below.

The project's geo-instrumentation specialist, Geocomp, is the core of the program and provides and maintains the data management system. The geo-instrumentation specialist is required to produce daily printed and electronic reports for the previous 24 hrs of readings, as well as weekly summary reports. This specialist is also responsible for notifications when data exceeds alert or threshold values; QA of instrumentation installation by Contractors and monitoring the Contractor's baseline activities. Despite the high level of automation of the data gathering, the geo-instrumentation specialist will maintain an on-site core team that will be available to respond to special situations and emergency requests.

The project's construction management team is responsible for managing the geo-instrumentation specialist, as well as the interface with outside stakeholders, including special requests for readings. The CM team will be the focal point for receiving and distributing periodic reports from the geo-instrumentation specialist, and for responding to issues related to the performance of the monitoring systems. The CM team will also review the need for action and coordinate responses to alert messages, as well as identify and manage necessary changes to construction operations.

The project's designer, a consortium of Parsons Brinkerhoff, STV and Parsons Transportation Group, has primary responsibility to specify (1) the types, locations and procedures for instruments; (2) the alert and threshold limiting response values; and (3) the frequency of monitoring. The designer also evaluates the need for changes when response values are exceeded and evaluates the need to modify monitoring frequency as appropriate.

On ESA, the construction contractors are responsible for the procurement, installation and baseline reading of specified instruments, as well as for maintenance, repair, replacement and re-baselining of damaged instrumentation. In response to monitoring data and performance evaluations, the contractor also implements necessary changes to construction operations to mitigate risks. The rigorous qualifications for the staff installing the geo-instrumentation typically dictate that most contractors will need to subcontract their instrumentation work to geo-instrumentation specialty firms.

During construction activities, the designer, CM team and contractor will review, analyze and evaluate measured performance. The Contractor remains responsible to use the instrumentation results for all issues related to safety of its construction activities.

THE GEO-INSTRUMENTATION MONITORING SYSTEM

The project will implement a full range of tools for automation of geo-instrumentation monitoring. These tools will include sensors, dataloggers, personal data recorders and hand-held readout devices. The project selected the geo-instrumentation specialist using a qualifications-based selection process. The selected firm, Geocomp Corporation, proposed to use their Web-based iSiteCentral system as the central component of the project's program-wide monitoring system. This system had been used on a number of projects with similar instrumentation requirements.

Figure 4 illustrates the layout of the iSiteCentral system as deployed for the ESA project. The system's primary infrastructure resides at Geocomp's headquarters in Boxborough, MA. Its central core is a Dell RAID-5 Server. An exact duplicate of this machine sits in another part of the building with an up-to-date copy of the entire system ready to kick in should the primary unit fail. In this manner, one can literally kill the primary server and the backup unit will automatically kick in without any interruption of service. A third identical unit resides in Geocomp's Atlanta office. It uses an entirely independent power system, Internet service provider and telecommunications system. This backup stands ready to kick in if service in Boxborough is lost.

These machines use the Microsoft 2003 Server operating system with MS SQL database. An outside group independently monitors these machines and notifies the Geocomp System Administrator immediately upon the failure of any of the three servers. Each server maintains a full copy of the database on three hard drives. The complete database is backed up weekly to tape, which is stored off-site in a rotating plan. The entire system sits inside a firewall to protect it from external assault.

Data comes into the central database via a communications control module (CCM). This module runs software processes to listen for an incoming call from a device on the project. An example would be a seismograph that has detected ground motions that exceed 0.5 in/sec. The instrument automatically makes a phone call to the CCM. The CCM figures out what type of device is calling and decodes the incoming data stream to get the information into a format that can be placed into the data base. Full data encryption and handshaking are used to prevent loss or scrambling of data. Any corrupt data packet is re-sent until the packet has good data. Should the communications link become lost during the transfer, the CCM re-syncs the link and completes the data transfer. The CCM can also call a device at the site and request a data download or a new reading.

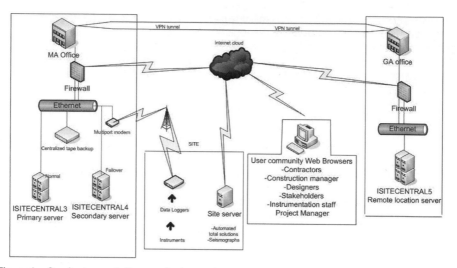

Figure 4. Geo-instrumentation monitoring system

The link between the CCM and devices at the site can be by Internet, cell modem, landline or radio. This capability facilitates data transfer from the many different data logging devices that the various contractors may employ. For example, the Contractor's specifications require that many of the instruments in GCT be connected to dataloggers that record their data onto computers which reside within the bowels of GCT. These computers have an Internet link. The CCM will constantly search these computers to pick up any new data to the central data base. Alternately a field technician may use a handheld device with a Web link to enter manually read data right at the instrument.

As data are loaded into the iSiteCentral database, each reading is checked using the pre-established Response Values for each instrument. If a Response Value is exceeded, the system goes to the alert messaging module (AMM) to issue alert messages. As soon as the data are loaded into the database they become available for viewing with a Web browser. Anyone given access enters a user name and password to see the opening screen for the system. Figure 5 illustrates this screen. Its contents depend on the access privileges the user is given. For example, a neighborhood group might see only the results of vibration monitoring in their community, while a system administrator would see many details for setting up and managing the system.

Alert messages are sent to previously established contact lists. The list is different for each alert level. The alert message includes the project name, the instrument ID that has alarmed, the current reading and the alert level that has been exceeded. Anyone on the call list can use a Web browser to acknowledge the alert message and prevent it from being resent multiple times.

This screen also shows a plan for an active part of the site with the locations of the instruments. Instruments which are currently installed are shown in color. The color indicates the Response Level at which the current instrument reading is at. Figure 5 shows eleven displacement monitoring points have been installed in the area covered by this plan and the current reading is in the Green Zone, i.e., below any Response Level. The symbol will change to yellow if the reading exceeds the Alert Level and red if the reading exceeds the Limit Value. The symbol turns blue if the sensor malfunctions and black if it totally dies. Placing the cursor over a symbol will open a box that shows the sensor ID and the most recent reading taken for the sensor. Clicking over

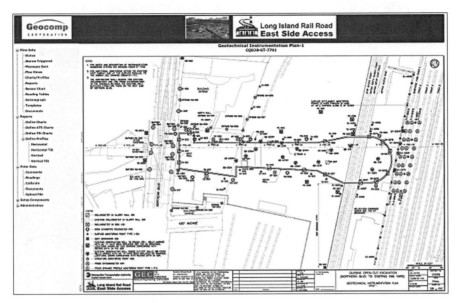

Figure 5. Contract-specific opening screen in iSiteCentral

the symbol will open a graph showing all data for the sensor. Figure 6 shows an example. Clicking over any data point will open the record for that data point and allow one to enter a comment about that specific data point. This facility provides a powerful way for anyone associated with the project to get a quick view of the status of the instrumentation; then drill all the way to the raw data with just a couple of clicks.

BENEFITS OF PERFORMANCE MONITORING

Experience on project's similar to ESA has shown that, in addition to reasons listed herein, geo-instrumentation performance monitoring can result in significant project cost savings by reducing damages and delays that would have been experienced had the monitoring not been in place or not been effectively used. On ESA, the use of the project's intranet will be beneficial to quickly disseminate information within the immediate project team. The Internet will be used to disseminate information to outside project stakeholders. As with geo-instrumentation monitoring systems on similar complex, urban projects, the use of these technologies will play a central role in managing, storing and making the monitoring data available on a current, controlled basis.

Looking forward, ESA will see a number of challenges to maximize the benefits of the geo-instrumentation monitoring systems. These challenges include prompt and accurate information for both internal and external stakeholders, quick response to safety concerns, and good coordination between all project participants. It will also be a major challenge to minimize false alarms from the warning system that result from failures in the measuring systems, physical damage to an instrument, and other unplanned occurrences.

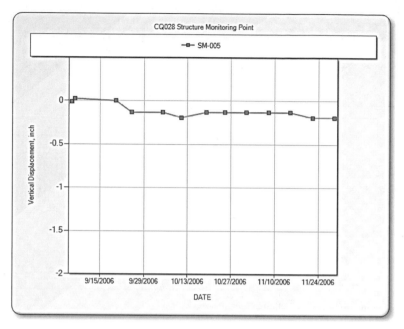

Figure 6.   Example data output screen

CONCLUSIONS

Performance monitoring with geotechnical and structural instrumentation is an essential component of the risk management program for the ESA Project. The ESA monitoring system is expected to provide warnings of unacceptable performance in time to reduce the consequences of that performance. Reducing probability and con-sequences of unexpected performance will help reduce risk. We anticipate the perfor-mance monitoring to return several times its cost by helping to manage risk.

The ESA geo-instrumentation performance monitoring program is automated as much as possible. This is being driven by lower technology costs and increased capa-bilities, the desire of owners to better control risks and more stakeholders who insist on being kept up to date with the impact of the work on their surroundings. Automated performance monitoring is expected to be an indicator of future performance and help reduce the probability of unacceptable performance. The ESA automated monitoring system will provide data several times a day at less cost than traditional manual moni-toring systems. Readings taken this frequently can indicate periodic changes caused by environmental conditions that otherwise appear as data scatter in manual data sets. By monitoring these periodic changes, we get confirmation that the automated system is functioning properly and we can remove the effects from the data set to get a true record of the facility performance. These benefits can greatly increase the reliability and believability of the data.

The Internet will play a central role in performance monitoring for ESA. Most of the project staff's use of the data management system is expected to be via Web browser. The geo-instrumentation specialist expects to perform all data collection, management and reporting with Internet-based tools. Selected data will be made available to third parties via Internet as a part of ESA's commitment to the neighbors and communities to minimize collateral damages from the work.

7

Innovations in Underground Construction

Chair
B. Fulcher
JF Shea Co., Inc.

ALTERNATIVE CONTRACTING METHODS

John Reilly

John Reilly Associates International

INTRODUCTION

Traditional methods of project procurement of have remained virtually unchanged in the U.S. infrastructure construction industry for more than 50 years. The traditional method, Design-Bid-Build (DBB), uses a two-step process that separates planning and design from construction. Design is generally done by the public agency and their consultants and the subsequent construction contract is procured from the private sector, almost always by a "low-bid" approach. This method has worked well for many projects, delivering quality work for the lowest bid price. This is particularly true for those projects where the work, and the public agency's management procedures, are well understood by the contracting community and there is relatively low uncertainty in the work or conditions, particularly geotechnical conditions for underground construction.

However, concerns exist with the DBB process. It is time consuming because all planning and design work must be completed prior to beginning any construction work and it is of concern for complex or large projects, plus those with significant risk, where the separation of design and construction, in conjunction with pressures resulting from the low-bid environment, often result in substantial cost and schedule increases. Frequently, these increases occur in a non-productive, inefficient, adversarial work environment leading to claims, disputes and costly litigation (Quick 2002).

Drivers for Change

Until the late 1980s, for the most part, methods of project procurement and administration in many countries were similar to those in the United States. Public transportation agencies retained tight control over the design and construction of projects. Prescriptive specifications and low-bid procurement methods were the public-sector tools of choice for procuring new works in both the United States and Europe. However, consideration of other methods was initiated due to problems such as:

- Cost and schedule overruns
- Reduced public funds
- Insufficient qualified staff
- Lack of innovation in addressing project needs
- Adversarial relationships
- Claims-oriented environments leading to disputes and litigation
- User frustration
- Political discontent

International Initiatives

In the late 1980s, agencies in other countries began to make significant changes to project procurement techniques. "Alternative" methodologies became primary contracting methodologies for many major projects in Europe, Asia and Australia with some significant changes for underground and tunneling projects (Martin & Leray 1996). In particular, European agencies appeared to be better at exploiting the efficiencies and resources that the private sector offers, through the use of innovative financing, alternative contracting techniques, design-build, concessions, performance contracting, and active asset management. These methods generally worked to create a team approach, establish an atmosphere of trust in the design and construction process, leading to better innovation and added value, and to allocate risk appropriately (Chambley et al. 2000). Many of these methods were reported to increase efficiency and add value through optimized work processes and increased innovation (FHWA 2002).

Initiatives and Approvals for Innovative Contracting in the U.S.—SEP-14 and SEP-15

The Transportation Research Board (TRB), in 1987, formed Task Force A2T51 to research Innovative Contracting Practices and to identify promising innovative contracting practices for further evaluation. In 1989 the Chairman requested that the Federal Highway Administration (FHWA) establish a means to evaluate some of the task force's more project-specific recommendations. Special Experimental Project 14 (SEP-14) was initiated on February 13, 1990, to allow to provide a means for evaluating some of the task force's more project-specific recommendations and allowed States to use alternative contracting techniques when using federal funds. In December 1991, TRB published the final recommendations of Task Force A2T51 in a benchmark document entitled Transportation Research Circular Number 386: Innovative Contracting Practices. While SEP 14 is still in use today to monitor innovative contracting methods, many innovative methods have become mainstream and do not require SEP-14 approval on Federal-aid projects. All initial innovative contracting practices allowed under SEP-14 (A+B bidding, design-build, best-value, etc.) have now been adopted as standard practices by FHWA, and are no longer considered to be experimental in nature.

In October 2004, FHWA initiated SEP-15, an experimental program to allow contracting agencies to explore alternative and innovative approaches to the overall project development process, focusing specifically on applications with public-private partnerships. SEP-15 is still quite new, but several projects have already been approved. It can form the basis of a new contractual approach such as GCCM or Alliancing, particularly if public-private partnerships are involved.

CONTRACTING METHODS—TRADITIONAL AND INNOVATIVE

The following contracting methods were considered in research for this paper. These are listed from established and well accepted—Methods 1 and 2—to more recent and innovative—Methods 3 and 4. With respect to innovation in contracting practices see Transportation Research Board 1991; Martin & Leray, 1996; Egan 1998; Smith 1997; FHWA 2002. The contracting methods being considered here are:

1. Design-Bid-Build (DBB)
2. Design-Build (DB)
3. General Contractor/Construction Manager (GCCM)
4. Alliancing/Relationship Contracting/Early Contractor Involvement (ECI)

TRADITIONAL PROCUREMENT METHODS

The following methods, DB and DBB, are well known to the underground construction industry—they are briefly treated here as a "baseline" from which new procurement methods can be evaluated.

DESIGN-BID-BUILD

The traditional method of public works procurement, in which the owner contracts for design by an independent architect or engineer (or in some cases accomplishes the work in-house) for a set of design drawings and specifications that are the basis of the contractor's bid. The architect or engineer (or the owner) certifies that the drawings and specifications meet all applicable requirements of building codes, engineering standards, and other design and safety standards. This bid package is then made available for all qualified bidders to analyze and provide a bid to the owner for construction of the work. Design of the work and construction of the work are accomplished by two independent, unrelated parties, contracted separately to the owner.

Design-bid-build (DBB) is not a rigid procurement method—it does allow for many variations that can enhance the owner's ability to obtain a quality product in a specified time and at a defined cost, for example A+B bidding and Incentive/Disincentive provisions (AASHTO, 2005). It does allow for the assignment of certain risk elements to the contractor, if appropriate. US Agencies have successfully used, and continue to use, DBB for the majority of projects.

DBB is the most basic U.S. contracting approach, best where there is clarity of the deliverable with a low probability of major risk or uncertainty which will cause significant changed conditions. Intended to create a clear and objective competitive environment, avoid problems of influence, collusion, corruption and/or bid-rigging. The intent is to provide taxpayers with the project "…at the lowest price that responsible, competitive bidders can offer."

Considerations

As noted in the introduction to this report, several concerns exist with DBB. It is time consuming because all planning and design work must be completed prior to beginning any construction work and it is of concern for complex or large projects plus those with significant risk—where the separation of design and construction, in conjunction with pressures resulting from the low-bid environment, often result in substantial cost and schedule increases. Frequently, these increases occur in a non-productive, inefficient, adversarial work environment leading to claims, disputes and costly litigation. Other concerns raised include:

- In order not to convey "unfair" advantage to one construction contractor, the design work is typically performed without contractor's input—therefore opportunity is lost for the construction contractor to shape and contribute to the design—with practical suggestions and methods that can add value.

- Design and construction must be performed sequentially, when savings of time might be achieved by concurrent organization of at least some early phases of design with phases of construction.

- The contractor's non-involvement in the design contributes to the baleful adversarial nature of the construction process. This often results in substantial cost and schedule increases during construction—changes that might otherwise be avoidable.

- Frequently, these increases occur in a non-productive, inefficient, adversarial work environment leading to claims, disputes and costly litigation

Additionally in DBB, as in DB, there is an inherent conflict embodied in the contractual provisions—as described in Quick's paper on Relationship Contracting (Quick 2002) and as noted following.

COMMENTS ON THE "LOW-BID" REQUIREMENT

In the conventional DBB model, the bidder who bids, say $17.432 million for the project, is awarded the contract over the second-low bidder who might bid $17.435 million. However, this model does not always result in the lowest cost for the public. The general assumption that the "low-bid" by the "lowest responsible bidder" results in best value and lowest cost is seriously open to question. First of all, the contractual environment is characterized by the ability of each party to treat the other party as an adversary—for their gain at the potential expense of the other—each party enters a contract at their own risk. To be the "low bidder," the contractor must do at least two things:

1. Determine his lowest cost to deliver the work specified at a minimum quality level.

2. Determine a strategy to bid that cost—or lower—to secure the work, with the expectation that any deficiencies in price can be made up in changes caused by new agency requirements, changed site or environmental conditions, defects in the design documents or other strategies that will accrue to his advantage.

This potential conflict in contract management and execution can be exacerbated by such problems as:

1. Interpretation of the contract—i.e., how are terms interpreted—do they mean that the owner (or the contractor) must bear a certain risk ? How is that known during the bidding phase? And therefore how can it be clearly priced?

2. How do contract terms deal with an event or consequence which has arisen from performance of the agency or contractor? Who is responsible for a breach of those terms? What did the parties really agree to—and how can this be priced or resolved in construction dispute?

3. Risk events which may occur and how they are treated under the terms of the contract—i.e., who has agreed to bear those significant (and usually unknown) risks? How can this be anticipated?

DESIGN-BUILD

Design-build (DB) procurement is actually one of the oldest methods of getting something built—in ancient times, the owner would hire a master builder to design and construct, without drawing a line between the two phases. Design-build is used extensively internationally and in the private sector in the United States. A large percentage of office buildings, factories, plants, systems and facilities in the U.S. and abroad, both private and publicly owned, are procured using design-build (USNCTT 1974; Design Build Institute of America 1995; Brierley & Hatem 2002; Molenaar 2003; Dornan et al. 2005; Warne & Schmitt 2005).

With a DB procurement, the contracting agency identifies the end result parameters and establishes the design criteria. The prospective bidders then develop design proposals that optimize their construction abilities. The submitted proposals may be rated by the contracting agency on factors such as design quality, timeliness, management capability and cost. These factors may be used to adjust the bids for the purpose of awarding the contract. The DB concept allows the contractor maximum flexibility for innovation in the selection of design, materials and construction methods.

Design-build has been utilized for a number of infrastructure projects across the U.S. and is frequently the base case for international project procurement. While not a "traditional" method in the U.S. yet, it has been shown to decrease time to completion, and projects done using design-build show less cost growth than those using DBB.

Results Reported

The California Design-Build Coalition, studied 21 projects across the country, ranging in size from $83 million to $1.3 billion (Warne, 2005). The purpose of this study was "to ascertain the performance characteristics of a population of design-build projects which would allow for an objective assessment of the effectiveness of this project delivery methodology." In this study:

- 76% of the projects were completed ahead of the schedule established by the owner.
- 100% of the projects were finished faster than if the design-bid-build were used.
- The cost growth of the projects in this study was less than 4% as opposed to an average of 5–10% characteristic (according to this study) of design-bid-build efforts.
- Many of the projects show a 0% cost growth rate largely stemming from the fact that many of these contracts are awarded on a lump sum basis requiring a contractor to deliver the project for a fixed amount of money.

Not all design-build projects are successful on all fronts, however. One difficulty is in estimating the final cost prior to inviting proposals. Normally, in DBB, cost estimates are done with design engineering 100% complete, just prior to soliciting for bid. In design-build that solicitation is done with design engineering between 10 and 30% complete, making the estimating task much more of a guessing game. However, the project cost is fixed through a lump-sum contract at 10–30% design and does not experience the cost growth that often occurs between 30% design and final bidding documents on a traditional DBB project. Another difficulty is that a design-builder will build what is required, but not necessarily what is desired. That is, in DBB, the public agency can, and does, modify design to include desirable changes right up to the time of bid, with the ability to incorporate input on aesthetics and other "should do" items. With DB, if the RFP does not specifically require it, it will probably not get done.

DISCUSSION—THE TWO MOST PROMISING INNOVATIVE CONTRACTING METHODS

The following describes the most recent and promising innovative contracting methods–Alliancing and GCCM—and discusses the benefits associated with each. The more usual contracting methods—DBB and DB—are not treated in detail since most readers are assumed to be familiar with them and many references are available (Brierley & Hatem 2002; Molenaar & Scott 2003; Warne & Schmitt 2005). Later in this paper there is a discussion on measures, currently under consideration (WSDOT 2007) that would incorporate the best attributes of Alliancing in current DBB and DB delivery approaches.

It should be noted that "best value" procurement can take many forms, and there is no one method that is inherently better than any other. Specifics depend upon project goals, laws, local preference, and the level of uncertainty or risk among other things. When using best value procurement, it is possible that the winning proposer might not be the lowest bidder, because of a better (for example more innovative, safer, or faster) method of completing the project. Also note that best value procurement can be used with several procurement methods such as DB or DBB.

GENERAL CONTRACTOR/CONSTRUCTION MANAGER

Also known as CM/GC and sometimes Construction Manager at Risk, with significant differences depending on the state and agency and what is permitted under legislation. This is a procurement method in which the design work is begun, either by the owner or a consultant, and a general contractor is engaged to work with the owner and designer to develop and deliver the project. This general contractor is selected based upon a combination of experience and price. This contractor may not actually construct elements of the project but is responsible for delivery of the project at a guaranteed maximum price (hence the term "at risk").

In general contractor/construction manager (GCCM), during the design phase, the contractor develops an independent cost estimate for construction. Working with the owner and designer, the contractor addresses areas of high cost and recommends changes. Once the design is complete, the contractor submits a Guaranteed Maximum Allowable Construction Cost (GMACC). The owner can accept this GMACC, and hire the contractor to perform the work, or the owner can reject the GMACC and find another contractor to do the work. Design risk is owned by the agency at design completion—that is, if there are major problems with the design causing construction claims, the owner is responsible—however, the contractor is responsible for completing the work at or under the GMACC. If a particular subcontract ends up costing more than estimated, the general contractor is liable.

Of specific interest to this conference, an initial application of GCCM for complex tunneling and underground construction has been the Portland, Oregon Portland Combined Sewer Overflow (CSO) program as described in the RETC 2005 proceedings (Gribbon et al. 2005) and in a memorandum (Reilly 2005). The Portland Tri-Met Light Rail extension to the Portland Airport used a similar approach.

An outline of the results of the Portland CSO GCCM project follows (the reader is encouraged to read the Gribbon reference)—in summary:

1. The basis of this contracting approach was the desire of the City of Portland to construct the project efficiently and effectively, avoiding the adversarial nature and the common disputes, claims and litigation of the conventional design-bid-build type contract. The City expected to pay a reasonable cost for the project and did not expect the contractor to fund or contribute monies to the project.

2. Experience with design-bid-build on previous contracts—even using recommended techniques such as partnering, geotechnical baseline reports, dispute review boards and escrow bid documents—had not been satisfactory.

3. There was a need to have the project participants, including contractor, focus on solving problems constructively—use that energy to add value to the project, not to waste it in adversarial contractual (claims) management, or posturing for litigation. How to capture that resource, that value? This leads to this type of contract.

4. The form of contract was open book cost reimbursement plus fixed fee to cover contractor overheads and profit. Fee was paid commensurate with percent complete of work.

5. Risk Mitigation for identified risks is included in the project. For risks that cannot be mitigated, a $17 million contingency was recommended for the West Side CSO Project and accepted by the City Council (the contract was $293 million). There were problems that may have required use of about ¾ of this amount, but the project was able to avoid use of the contingency through the collaborative process (a benefit of this contracting approach).

Benefits noted included:

1. Having the contractor on board during design—with input on how to construct, innovation, suggestions, value added and allow the contractor buy-in to the design and construction contract.

2. Changes handled quickly and in best interest of the job—no major paper work or delay for changes, no claims. Contract doesn't recognize differing site conditions (Type I) for prime contractor since direct costs were paid as reimbursable. Only Type 2 could increase fee if they impacted the critical path. (In general, there was no markup on extra work, unless it's Owner discretion and affects critical path).

3. Owner and contractor did not have to "take a position" or be adversarial, and therefore could resolve issues in the best interest of the work, which meant better objectivity and (team) alignment.

4. Owner is involved—on site, understands the work, is part of the process, can staff the work efficiently.

5. Bi-annual audits done, clearly—audit findings were very complementary to the work and process. Audit said contractor's cost control system was good.

6. Time to decision was quick

ALLIANCING OR RELATIONSHIP CONTRACTING

Alliancing (Henderson 1999; Reilly 2004; Ross 2003, 2005) requires that the owner, designers, contractors and suppliers work together as a single team, with contractually defined risk-reward (pain-gain) provisions. The owner(s) and the service providers (engineers, contractors) collectively assume responsibility for delivering the project against pre-agreed performance target outcomes.

After team selection, based on a determination of the best team to design and construct the project (but before a price is determined) the Alliance contractual agreement is executed. Then, owner, engineer and contractor develop a joint target cost estimate and schedule for the project, then proceed in to design and construct the project. An important part of Alliancing is that result of all outcomes—risks and benefits (pain and gain)—are shared based upon the agreement—there are no claims for delays or litigation within the alliance. All work is open-book and the team selects the best person for each a position, regardless of affiliation. Also called, or related to, Relationship Contracting (Quick 2002) and similar to Early Contractor Involvement in the U.K. (British Highways Agency 2004).

The characteristics of Alliance contracts include:

1. Parties are bound by one, relatively brief, agreement focused on the prime goals of the procurement

2. The owner, in some form, is contractually part of the alliance team

3. Obligations are stated collectively ("We shall…")

4. Costs are 100% reimbursed

5. Overhead and fee are negotiated and fixed (see below)

6. Share in profit or loss is related to meeting the defined performance goals (pain-gain structure)

7. Losses can be taken out of fee and overhead but not costs

8. Express commitment to resolve issues within the alliance without undue reliance on outside parties—no disputes or litigation within the alliance

9. 100 percent open-book accounting for all parties—requires trust, commitment and understanding

10. Decisions of the alliance board are binding and must be unanimous

11. Project is managed by an integrated management team, members are chosen on the basis of "best person for the responsibility" from any of the alliance partners.

12. Participants develop and commit to a set of "Alliance Principles," related to the key goals of the project

Volume of Work Under Alliancing in Australia

In Australia, from 1994–2005 Aus$4.7 billion of a total of Aus$37 billion in capital projects were underway or constructed using the Alliance process. This represents about 13% of the total and this percentage is growing.

Of relevance, in terms of adoption by public agencies for complex infrastructure and underground projects, is the recommendation of the Western Australia Commissioner of Main Roads advocating use of the Alliance process (Henneveld 2005) and the approval by the State of Victoria Treasury in their 2006, "Project Alliancing Practitioner's Guide" (Victoria Treasury 2006).

Benefits

Benefits were first reported from the UK offshore oil platform construction (see below). Subsequently, highway and infrastructure projects such as the Sydney Northside Tunnel Project (Henderson 1999); the UK Channel Tunnel Rail Link (Arnold & Myers 2000), Australian Naval ship procurement (Hill 2004) and the UK New Carrier Procurement have adopted the Alliancing method and many have reported successful results.

Significant benefits have been reported (Evans & Peck, 2003; Key 2004), for the Australian projects which include reduced cost, added value, improved schedule, higher levels of innovation, reduced or eliminated adversarial environment and the practical elimination of disputes, claims and litigation between owner, designer and contractor (Quick 2002).

Of note, the preface to the State of Victoria Treasury "Project Alliancing Practitioner's Guide" states:

> Project Alliancing is about providing better value for money and improved
> project outcomes through a more integrated approach between the
> public and private sectors in the delivery of infrastructure projects.
> Alliancing reflects a shift from more traditional procurement methods
> which focus on strict risk allocations, to a collaborative approach. This
> involves Government working with one or more service providers to align
> incentives and objectives, and manage project risks and issues.

Results Reported

The first alliance projects are reputedly the North Sea offshore oil and gas platforms, BP Hyde and BP Andrew Projects, constructed in the early 1990s. Results reported for these platforms were:

- £450m first estimate
- £370m sanction to proceed
- £290m Cost (22% savings)
- Completed 6 months ahead of schedule

In total, four U.K. offshore oil platforms aggregated savings of £550 million or 20% (NCE, 2000)

The Queensland Department of Main Roads, Port of Brisbane Motorway Alliance established a Target Cost Estimate (TCE) of Aus$112 million, representing a 30% probability cost outcome. The final result was Aus$8.3 million or 7.5% less than the TCE—even including an increase of scope of approximately Aus$7m, representing a total saving of Aus$15.3 million or 13.6% (Evans & Peck, 2003). The report notes additional value obtained through innovation, better design options and outcomes, better aesthetics, better "fit for purpose," better quality of workmanship and, a focus on innovation and improvement rather than pursuit of claims.

The U.K. Channel Tunnel Rail Link project reported results better than established targets for Safety—18% better; Process defects 45% better; Waste management 75% better; Complaints 63% better; Staff costs 26% better; and a 5 month reduction in schedule. (Halcrow, 2005)

Sydney Water reported, from a survey of 48 projects, that:

- 61% of Alliance projects exceeded expectations vs. 17% of non-Alliance projects,
- 72% of Alliance projects achieved lower cost than initial target/budget,
- 36% of Alliance projects were ahead of schedule compared to 10% for non-Alliance projects.
- The best Alliance was 35% ahead of schedule whilst the best non-Alliance was 10%.

Sydney Water reported an analysis for the Northside Tunnel Project (Evans & Peck, 2004) that compared the independent Target Cost Estimate (TCE) to the projected cost for Design & Construction using a Non-Alliance process, including the costs under that method for delays and scope changes actually experienced by the project. They concluded that the cost profile was, in summary:

- Budget Aus$451 million
- Final cost Aus$466 million (+3.3%)
- Design & Construct estimate Aus$483–Aus$507million (20% & 80% range numbers respectively)
- Design & Construct at completion Aus$567–Aus$573 million (20% & 80% range numbers respectively)

This is an actual cost of Aus$466 million for the Alliance and a probable cost (50% range) of Aus$571 million if using Design & Construct—a difference of Aus$105 million saved by Alliancing (+22%).

This analysis is retrospective and not absolute—however, even if the assumptions are overly optimistic for Alliancing and pessimistic for Design and Construct, the fundamental conclusion is that the Alliance process did deliver significant value in these circumstances.

Considerations Regarding Alliancing

Because the alliance agreement includes the owner, designer and contractor together concerns have been expressed about the lack of a contractual "arms-length" determination of cost and cost changes. Current legislation in the U.S. generally requires a fully-independent "low-bid" or "best value" determination—Alliancing therefore requires an independent validation of the contractual target cost, since Alliancing establishes the target cost only after selection of the contractor. The cost of setting up the alliance is also a factor—this is more efficient for larger projects and needs to be balanced against the potential benefits.

Additionally, resistance to consideration of Alliancing in the U.S. considering public law statutes and resistance from the construction community is a factor. Implementation of Alliancing in Australia was facilitated by a very corrosive disputes, claims and litigation environment in the mid-1990s—leading to a readiness for both owners and contractors to be willing to consider the Alliancing approach—which subsequently proved to be able to deliver increased value to both owners and contractors, as noted and referenced above (Reilly 2006).

Use of Alliancing in the U.S. and Canada is primarily by the oil and gas production industry. Specifically, in Alberta, adoption of Alliancing is primarily due to the increased value perceived by the owner, who is requiring this approach for the Calgary oil sands project (Reilly 2006).

INCORPORATING ALLIANCING AND GCCM METHODS IN DB OR DBB

In 2005 the Washington State Department of Transportation (WSDOT) initiated a study of Alliancing and GCCM (Reilly & Smith 2007) and their application to WSDOT projects, including the large, complex urban projects in the Seattle area. This possibility was further considered in 2006 and 2007 by an Expert Review Panel for the SR520 Lake Washington Floating Bridge Replacement Project which involves construction of large floating pontoons in a sensitive environmental framework (WSDOT 2007). The panel concluded that using Alliancing would be of value to WSDOT but that it might take time to secure legislative approval for the process. They concluded that it would be possible to implement most of the key features of Alliancing in a DBB or DB procurement under existing legislation.

Specifically they recommended that the contracting method should involve the owner, designer and contractor in meaningful roles, including from preliminary through final design, to ensure that the project could be built cost effectively and to achieve greater cost certainty. Additionally, the contracting method should effectively manage risk allocation and risk sharing, establish a means to resolve disputes effectively, accelerate schedule and produce best value for the taxpayer in terms of pricing, schedule and quality. All of these attributes or goals are at the core of the Alliance delivery process.

They concluded that DBB and DB methods are capable of greater flexibility than traditionally practiced. Especially in the context of a Public Private Partnership (FHWA 2004), the use of which is possible for the SR520 project, SEP-15 may allow use of the Alliancing and/or GCCM methods more directly.

SUMMARY AND CONCLUSIONS

1. Use of innovative contracting methods such as Alliancing and GCCM has the potential to delivery better value to owners for complex urban and underground projects in the U.S.

2. Such methods have been used and proven in international settings (Alliancing) and the U.S. (GCCM).

3. Studies documenting this added value are available

4. It may be possible to use the key attributes of Alliancing and/or GCCM in current DBB and DB contractual environments under existing legislative provisions and policy guidelines

5. Reference to the SEP-15 process may allow use of these processes for testing and development, particularly if Public Private Partnerships are involved.

REFERENCES

AASHTO, 2005, "Primer on Contracting for the Twenty-First Century." Report of the Contract Administration Task Force of the AASHTO Subcommittee on Construction, Fifth Edition Draft, American Association of State Highway Transportation Officials, Washington, D.C.

Arnold, J. & Myers, A. 2000, 'Mechanisierung der Neuen Osterreichen Tunnelbauweise beim North Downs Tunnel des Channel Tunnel Rail Link in England,' Osterreichischer Tunneltag 2000, Salzburg, 11 Oktober.

Brierley, G. & Hatem, D.J., 2002 "Design-Build Subsurface Projects" American Underground Construction Association/Zeni House Publications.

British Highways Agency, 2004, "Early Contractor Involvement: Contract Guidance Manual" ECI Model Contract—Issue 1 Revision 1, London, England.

Chambley, P.J., Bassford, C. & Tindall, S. 2000, 'The Project Team Approach to the Management of Risk for a Major Tunneling Contract at Kingston upon Hull,' American Underground Construction Association News, Vol 15, No. 3, Fall 2000 p 22.

Design Build Institute of America, 1995, Design-Build RFQ/RFP Guide for Public Sector Projects, 1995.

Dornan, D., Molenaar, K., Macek, N., and Shane, J. 2005. Study to Congress on the Effectiveness of Design-Build Project Delivery Relating to the Federal-Aid Highway Program—Final Report, Federal Highway Administration, Washington, DC, March 2005, 160 pp.

Egan, J., 1998, 'Report of the Construction Task Force to the Deputy Prime Minister, UK Rethinking Construction,' UK Department of the Environment, Transport and Regions, London, July.

Evans & Peck, 2003, "Alliance Learning Experience," Report for the Queensland Department of Main Roads, Port of Brisbane Motorway Alliance, September.

FHWA 2002, "Contract Administration, Technology and Practice in Europe" USDOT FHWA International Exchange Program, Report FHWA-PL-03-002, October

FHWA, 2004, "Report to Congress on Public-Private Partnerships." U.S. Department of Transportation, Federal Highway Administration, Washington, D.C.

Gribbon, P., Colzani, G. & McDonald, J. 2005 "Portland Oregon's alternative Contract Approach—A Work in Progress," RETC Proc. Ch. 2 pp 10–19.

Halcrow, 2005 "Channel Tunnel Rail Link—A Procurement Success," public release document.

Henderson, A., 1999 'Northside Storage Project,' Proc 10th Australian Tunneling Conference, Melbourne March, p 57. Alliancing.

Henneveld, M., 2005 "Getting Alliances up and Running," Strategic Decision-Making: Why Alliance Contracting Conference, Melbourne 24 August.

Hill, R. 2004 "Decisions relating to the future of the naval shipbuilding and repair (NSR)." ANZAC Alliance" Australian Defence Minister May 24.

Jones, D. 2000 "Alliancing and Other Forms of Co-operative Contracting with Government: A Local and International Viewpoint," paper presented at Alliancing Contracting in Construction Conference, 11–12 December, Sydney.

Key, J., 2004, "Do Alliance Projects offer Value for Money?" Study for Sydney Water, Currie & Brown, November.

Martin & Leray, 1996, "Negotiated compressed process for the world's largest privatized water supply system," Proc International Tunneling Assoc./North American Tunneling Conference, Washington D.C., April, Vol. II, pp. 667–670.

Molenaar, K. and Scott, S. 2003. "Examining the Performance of Design-Build in the Public Sector," Design-Build for the Public Sector—Chapter 3, Edited by Michael Loulakis, Aspen Business and Law, Aspen Publishers.

NCE 2000, New Civil Engineer Magazine, 3 February 2000, p 10.

NCHRP 10–49, 1999, "Improved Contracting Methods for Highway Construction Projects," final report, December.

Quick, 2002 "Introduction to Alliancing and Relationship Contracting," QLS/BAQ Symposium 2002—Session K, Construction Law—02 March 2002.

Reilly, J.J. 1999 "Successful procurement of large and complex infrastructure programs" Tunnels & Tunneling, North American Edition, June, p5

Reilly, J.J, 2004, "Alliancing for Underground Construction Projects," TBM Magazine, June.

Reilly, J.J., 2005, Memorandum on a meeting to review the implementation of Portland's innovative procurement method for the West (and East) side CSO projects, October 14.

Reilly, J.J., 2006, "Australia Alliancing Questions Summary," Memorandum prepared for the Washington State Department of Transportation, December.

Reilly, J.J. & Smith, R., 2007, Report "Alternative Contracting and Innovative Project Management" Washington State Department of Transportation, May.

Ross, J. 2003 'Introduction to Project Alliancing,' April 2003 Update, Alliance Contracting Conference, Sydney, Australia, 30 April.

Ross, J. 2005 "Project Alliancing Practice Guide," Project Control International Pty. Ltd.

Smith, R.J. 1997, 'Systematic evaluation, selection and implementation of better contracting practices,' Proc Tunnels for People, Ed. Golser, Hinkel & Schumbert, ISBN 90-5410-868-1, pp 757–762 Balkema Rotterdam.

Transportation Research Board, 1991 "Innovative Contracting Practices," Circular 386 National Research Council, Washington, D.C.

USNCTT 1974, 'Better Contracting for Underground Construction,' Standing committee No. 4 Contracting Practices, US National Committee on Tunneling Technology National Academy of Sciences, Washington D.C.

Victoria Treasury 2006 "Project Alliancing Practitioner's Guide."

Warne, T. and Schmitt, G. 2005. Design-Build Contracting for Highway Projects a Performance Assessment, Report prepared for California Design-Build Coalition, Tom Warne and Associates, LLC, S. Jordan, UT.

WSDOT 2007 "Special Projects Construction Site Report," Expert Review Panel, SR520 Project, January.

EVALUATING GROUND CONDITIONS AHEAD OF A TBM USING PROBE DRILL DATA

Steven K. Duke

GeoPentech

Jay Arabshahi

Metro Water District of Southern California

ABSTRACT

An economic, real-time probe drill monitoring system was developed for Metropolitan Water District's Arrowhead Tunnels project to quantitatively evaluate ground conditions ahead of a TBM. This system digitally records hydraulic feed, rotation, and percussion pressures and calculates drill advance rates during drilling. These data are computer synthesized to develop a digital representation of the ground ahead, and the resulting "picture" of the ground is distributed to the project team to assist with tunnel construction decisions (i.e., targeting grout holes, optimizing mine-cycle lengths, and avoiding becoming stuck). The analysis and interpretation procedure has now been calibrated and refined using over 5,000 m (3 miles) of geologic face mapping data.

INTRODUCTION

The geologic ground condition ahead of the advancing face is probably the most important factor to know during tunnel construction. Prior to construction, geologic features expected to be encountered during tunneling are typically estimated from widely spaced boreholes, surface exposures of geologic outcrops extrapolated to tunnel depth, or geophysical surveys. These initial exploration techniques are limited and often miss potentially hazardous ground conditions, especially in geologically complex areas. During construction, more reliable information relating to the ground ahead can be estimated from the drilling of probe holes ahead of the tunnel face. Observations of the drilling characteristics of the probe holes indicate changes in the strength of the ground, the presence of large discontinuities or cavities, and the presence of groundwater. In order to fully take advantage of the information provided during the drilling of the probe holes, an automated probe drill monitoring system was developed for Metropolitan Water District's (MWD) Arrowhead Tunnels project. This system continuously records probe drill pressure data during drilling and computer synthesizes the results to provide a look-ahead picture of the anticipated ground conditions.

The Arrowhead Tunnels are currently being excavated through highly variable ground conditions ranging from highly fractured to massive, weak to strong igneous and metamorphic rocks. Numerous faults and shear zones containing gouge and crushed rock are being encountered. High groundwater inflows with high hydrostatic pressures have also been encountered. For the Arrowhead Tunnels project, accurate and detailed evaluations of ground conditions ahead of the TBM are useful to identify hazardous ground, which could be problematic to excavate. The ground evaluations using the automated probe drill monitoring system assist the engineering and contractor teams with the daily planning of tunneling activities, including targeting grout holes, optimizing mine-cycle lengths, and minimizing the chances becoming stuck. This paper presents the probe drill methodology; setup of the probe drill instrumentation;

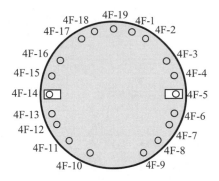

Figure 1. Hydraulic powered rotary percussion probe drill

Figure 2. Layout of probe hole ports through cutterhead

processing, analysis, and calibration of the probe drill data; and example probe data plots from the Arrowhead Tunnels project.

PROBE DRILL METHODOLOGY

For the Arrowhead Tunnels project, probes ahead of the tunnel excavation are typically advanced using hydraulic powered rotary percussion drills with 5-foot long drill rods and 2.25-inch or 3.75-inch diameter drill bits. The drilling involves rapid pounding of the drill bit as it is simultaneously rotated and forced into the ground with feed pressure maintained on the drill bit. As the rock is loosened by the action of the bit, the cuttings are expelled out the annulus surrounding the drill rod by the flow of water injected through the drill bit. When the entire drill rod has penetrated the formation, the head is disconnected and an additional drill rod is added. Two probes holes are typically drilled simultaneously by using two probe drills mounted on the left and right sides of the tunnel boring machine (TBM). A probe drill used for the Arrowhead Tunnels project is shown in Figure 1.

Probe drilling is predominantly performed through one of 19 ports located throughout the circumference of the TBM, as shown in Figure 2. From these locations, the holes are commonly drilled at 4-degree lookout angles through the cutterhead and into the excavation face. A minimum of two probes holes are typically drilled using a 2.25-inch bit to a distance between 24 m (80 ft) to 46 m (150 ft) ahead of the tunnel excavation and typically grouted. Depending on probe groundwater inflows and tunnel stability concerns, additional probes may be drilled and grouted. Sometimes, due to unstable ground conditions, probe drilling and grouting is performed in successive stages. In these instances, the initial stages of probe drilling are occasionally performed using the larger diameter, 3.75-inch bit.

During drilling, the probes are observed by tunnel inspectors. The inspectors note pertinent probing information, including (1) probe face station, (2) probe hole location, (3) probe bit size, (4) drill start and stop times to advance each 5-foot rod, (5) drill start and stop depths, (6) probe groundwater flow rate, and (7) other pertinent comments (i.e., hole caving, plugged or worn bit, or hanging steel). The inspectors note the drilling information on drill reports, and an example drill report is shown in Figure 3.

After all the probes holes have been drilled and grouted, excavation of the tunnel is performed to within typically 20 to 30-feet of the end of the drilled probes holes. This

EAST	WEST [x]		DRILL REPORT			IDR Page: 9 of 11	
Inspector: (Green)			Drillers Initials: RR		Ring No: 1705	Drill Page: of 1	
Shift: Day (Swing) Grave			Starter Bit:		Hole No: 4F11	VMT Station: 4638.15	
Day: M T W (Th) F Sat Sun			Production Bit: 2 1/4"		Category: S1	Face Station: 152+12 2043	

Steel #	Drill Time Start	Stop	Depth (feet) From	To	Pertinent Information — Note depth of formation changes, water loss or gain, cave, Etc.
1-2	2150	2158			Set up.
3	2159	2202	0	5	soft.
4	2203	2206	5	10	
5	2267.5	2210	10	15	
6	2211	2213.5	15	20	
7	2214.5	2217	20	25	
8	2218	2220.5	25	30	
9	2215	2223.5	30	35	
10	2225	2227	35	40	No Flow
11	2228	2230	40	45	No feed psi – No cave
12	2231.5	20345	45	50	
13	2235	38.5	50	55	
14	2239.5	2242	55	60	
15	2243	45	60	65	
16	2246	2249	65	70	
17	2250.5	53	70	75	
18	2254	56.5	75	80	

Figure 3.   Example inspector drill report

pattern of probe drilling and tunnel excavation creates a minimum overlap of 20 feet between successive setups of probe holes.

AUTOMATED PROBE DRILL MONITORING SYSTEM

Instrumentation Setup

The automated probe drill monitoring system for the Arrowhead Tunnels project was designed to be relatively inexpensive to setup and rugged enough to survive the harsh environment within the TBM. Instrumentation used by the probe drill monitoring system includes pressure gauges, a central datalogger collection and storage unit, and a modem connected to an outside phone line. It is noted that high-tech recording systems are commercially available for percussion drills; however, these systems have not yet been adapted for use in a tunnel. The manufacturer for the percussion drills on the Arrowhead Tunnels project indicated they might be able to adapt their existing drill recording system to work in the tunnel; however, the instrumentation costs would be expensive, on the order of ten times greater than the instrumentation costs for the current setup.

Pressure Gauges. Three pressure gauges are connected to each probe drill. These gauges are connected to the hydraulic lines that control the drill feed, rotation, and percussion pressures during drilling. To help protect the pressure gauges from the harsh TBM environment, the hydraulic lines used for the pressure gauge connections are plumbed to out-of-the-way areas within the TBM. Each pressure gauge is connected to a central datalogger unit with electrical cables for recording. While drilling, feed pressures typically range between about 1 kPa (25 psi) and 30 kPa (600 psi), rotation pressures typically range between about 55 kPa (1,200 psi) and 70 kPa (1,500 psi), and percussion pressures typically range between about 80 kPa

(1,700 psi) and 100 kPa (2,000 psi). When drilling is not being performed, these pressures stabilize to near zero.

Datalogger. The hydraulic feed, rotation, and percussion drill pressures for all the probe drills are digitally recorded and stored in a central datalogger unit located near the front of the TBM. The datalogger stores the drill pressure data, date, and time at 10 second intervals, whether or not drilling is being performed. Data are stored in a ring memory which overwrites itself after approximately five days. The datalogger has an LED readout screen mounted to the side of the datalogger box, which can display the datalogger time and most recent drill pressure measurements. The readout of the datalogger time is useful to help the inspectors synchronize their observed drill times to the pressure reading times recorded by the datalogger. The datalogger readout of the most recent drill pressure measurement is useful to help troubleshoot malfunctioning pressure gauges from within the tunnel.

Modem. Digital data from the datalogger system is downloaded via telephone modem out of the tunnel to an office location for processing. Probe drill data is setup to be automatically downloaded at 2 hour intervals; however, the most current probe drill data is always available to be downloaded at any time.

Data Processing and Analysis

Processing and analysis of the probe drill data is designed to be performed relatively quickly, usually within an hour or two after completion of drilling. Quick interpretation of the probe drill information is necessary so that it can be used in a timely manner to assist with tunnel construction decisions. The collection and analysis of the probe drill data is performed off the critical path of the construction activities, and therefore, does not create delays. Processing of the data is performed using various internally written and commercially available programs. Steps involved in processing the probe drill data are summarized below.

Drill Time to Depth Conversion. The drill feed, rotation, and percussion pressure data that are downloaded from the datalogger are recorded as a function of time at 10 second intervals. This drill pressure versus time data are converted to drill pressure versus depth using the drill time and depth information noted in the tunnel inspector's drill reports. In order to do this, the inspector noted drill start and stop times and depths for the drilling of each rod are input into an internally written program. The program matches the inspector noted times to times recorded by the datalogger and assigns the inspector noted depths to these times. Data recorded outside the drilling times noted by the inspector are removed, and data recorded during drilling are assigned interpolated depth values. The calculated drill depths are also converted to tunnel stationing taking into account the lookout angle of the drill holes.

Drill Rate Calculation. The drilling rate for each rod advance is also calculated. The drilling rate, in feet per hour, is calculated by dividing the length of the drill rod by the time duration required to drill the rod, as noted by the inspector.

Drill Data Assembly and Verification. The probe drill depth, tunnel station, date/time, pressures (feed, rotation, and percussion), and advance rate for each 10 second recording are assembled together with the face station, ring number, hole location, and drill bit size. An example clip of the processed data is shown in Table 1.

A plot of the drill feed, rotation, and percussion pressure and advance rate as a function of drill depth is produced as a visual check to evaluate input errors or pressure gauge malfunctions. An example of this plot is shown in Figure 4. Once the data are checked for errors, the processed data are combined into a single database for evaluation.

Table 1.   Example probe data set after assembly

Face Station	Ring No.	Hole Location	Bit Size (in.)	Drill Depth (ft)	Drill Station	Date/ Time	Feed (psi)	Rotation (psi)	Hammer (psi)	Advance Rate (ft/hr)
35927	1136	4F11	2.25	0	35911	1/19/05 4:51:03	185	1,313	1,760	18
35927	1136	4F11	2.25	0.05	35911.05	1/19/05 4:51:13	187	1,308	1,746	•18
35927	1136	4F11	2.25	0.1	35911.1	1/19/05 4:51:23	188	1,319	1,729	18
35927	1136	4F11	2.25	0.15	35911.15	1/19/05 4:51:33	186	1,295	1,730	18
35927	1136	4F11	2.25	0.21	35911.2	1/19/05 4:51:43	189	1,303	1,736	18
35927	1136	4F11	2.25	0.26	35911.25	1/19/05 4:51:53	188	1,315	1,739	18
35927	1136	4F11	2.25	0.31	35911.3	1/19/05 4:52:03	187	1,295	1,738	18
35927	1136	4F11	2.25	0.36	35911.36	1/19/05 4:52:13	188	1,298	1,733	18
35927	1136	4F11	2.25	0.41	35911.41	1/19/05 4:52:23	189	1,287	1,734	18
35927	1136	4F11	2.25	0.46	35911.46	1/19/05 4:52:33	178	1,244	1,752	18
35927	1136	4F11	2.25	0.51	35911.51	1/19/05 4:52:43	189	1,297	1,747	18
35927	1136	4F11	2.25	0.56	35911.56	1/19/05 4:52:53	189	1,295	1,756	18
35927	1136	4F11	2.25	0.62	35911.61	1/19/05 4:53:03	189	1,276	1,751	18

Probe FPI Calculation. The approximate ground strength is estimated using a combination of drill feed pressure and advance rate data. For the evaluation, the feed pressure is divided by the advance rate to calculate a field penetration index for the probes, or "probe FPI." In theory, stronger rock requires high feed pressures and low advance rates to drill, which creates a relatively high probe FPI; while, weaker rock requires lower feed pressures and higher advance rates to drill, which creates a relatively low probe FPI. Therefore, relatively high probe FPI is indicative of relatively stronger rock, and conversely, relatively low probe FPI is indicative of relatively weaker rock.

Rotation or Percussion Pressure Filtering. During drilling, the rotation and percussion pressures have been observed to be relatively constant, regardless of the rock strength; therefore, these parameters are not used in the calculation of the estimated ground strength. Although these data are not used to estimate ground strength, the rotation and percussion pressures are used to determine instances when the drill is not operating. As can be seen on Figure 4, processed drill rotation and percussion pressures go to zero in some instances, which indicate the drill was not operating. These instances usually occur near the beginning and end of a rod and are a result of synchronization differences between the inspector and datalogger times. Since the

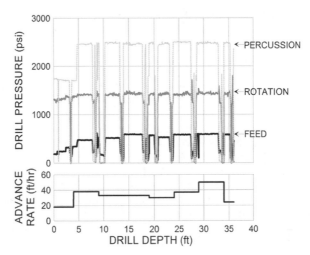

Figure 4.   Example drill pressure and advance rate data plot

drill is not operating during these times, these data are filtered out. Typically, the drill rotation pressure is used to filter the data; however in some instances, the percussion pressure data can be used as well.

Drill Bit Size Scaling. Occasionally different size drill bits are used to drill multiple staged probe holes. In these instances, the larger drill bit requires relatively more force to drill due to the larger surface area of the hole. Because of this, a constant scaling factor is applied to normalize the feed pressure data. For the Arrowhead Tunnels project, the holes were predominantly drilled using a 2.25-inch diameter bit; however, the initial sections of some holes were drilled using a larger 3.75-inch bit. A constant scaling factor was applied to the feed pressures used to drill the larger diameter holes to normalize these data to the 2.25-inch bit data. The scaling factor was calculated by comparing numerous adjacent 3.75-inch and 2.25-inch bit feed pressure data drilled in the same hole.

Caving Hole Adjustment. Occasionally, probe holes are hard to keep open during drilling due to caving conditions. Usually during these circumstances, the holes are drilled using relatively low feed pressures and extremely low advance rates due to the extra washing required to keep the hole open. Due to the extremely low advance rate under these conditions, the calculated probe FPI would incorrectly indicate very strong rock. Because of this, the advance rate is not considered when excessive hole washing is needed, instead only a normalized feed pressure is used.

Ground Evaluation Plot Creation and Interpretation. The resultant filtered and scaled probe FPI data are plotted spatially, as a function of tunnel station (x-axis) and hole location (y-axis), on an unrolled view of the tunnel and color contoured. The varying colors on the ground evaluation plot represent varying probe FPI amplitudes. For the Arrowhead Tunnels project, the probe FPI amplitudes have been correlated with geologic face mapping data, as discussed in the next section. The ground evaluation plot color-codes the various calibrated probe FPI value ranges that represent the anticipated rock strengths (i.e., weak, moderately strong, strong, and very strong). The ground evaluation plot is supplemented with other available data from the probe drill reports including groundwater observations and drill behavior observations to produce the final ground evaluation plot. The final ground evaluation plot is distributed to the project team via email to assist with tunnel construction related decisions. Although the probe analysis is semi-automated for quick turn around (usually 1 to 2 hours after completion of drilling), a human element is recommended for data interpretation, and it is not recommended to fully automate this system.

An example ground evaluation using the automated probe drill system is shown in Figure 5. Figure 5 shows a time-lapse of four plots that show the anticipated ground strengths as additional holes are progressively drilled through a zone containing weaker rock. The crown of the tunnel is located horizontally along the center of each plot and the invert is located horizontally along the top and bottom edges of each plot. The left springline of the tunnel is located on the upper half of the plot between crown and invert, and conversely the right springline is located on the lower half of the plot between crown and invert. For this paper, the estimated ground strength is represented by different shades of gray, where darker colors represent relatively weaker rock (lower probe FPI) and lighter colors represent relatively stronger rock (higher probe FPI). It is noted that the holes are drilled from left to right on the plots and the area immediately outside the hole locations are blanked to minimize excessive interpolation. Figure 5 also shows the locations of the drilled probe holes in tunnel cross-section view. As can be seen in Figure 5, the resolution of the anticipated ground condition plot improves as additional holes are drilled. The upper plot, which shows the initial two probe holes, indicates strong rock in the crown and a zone of weaker rock in the last half of the invert hole. The second plot shows the addition of two springline holes and suggests a weak zone on the left side and weak rock near the end on the right side probe. The next plot shows the addition of two more holes, six total, which further refines the location of the weak rock zones. The bottom plot shows the final interpretation of this area and is a compilation 11 probe holes. The anticipated zones with weaker rock are outlined on this plot.

Drilling Variables Not Always Considered. Many variables can affect the drilling behavior, and hence interpretations of the probe drill data. These variables may include operator drilling style (i.e., extra washing or long pauses while drilling rod), changes in water pressure washing through drill bit, drill bit condition (i.e., worn or plugged bit), or bent rods within drill holes requiring extra effort to drill. Although these variables are not always taken into consideration, they are annotated on the ground evaluation plots when noted on the inspector reports, and anomalous data are corrected or removed.

Probe Drill Calibration

The calculated probe FPI values have been correlated with the corresponding rock strength from geologic face maps logged by geologists on the Arrowhead Tunnels project. The purpose of this calibration was to estimate probe FPI ranges representative of different rock strength classifications (i.e., weak, moderately strong, strong, and very strong). This calibration is ongoing, and currently, probe FPI estimated strength ranges have been adjusted to approximate the rock strength logged in over 1,800 face maps. The resultant calibrated probe FPI strength ranges are color coded on the probe drill plots to visually show the predicted rock strengths.

A comparison of the average probe drill ground strength and the average as-built ground strength based on face mapping as a function of tunnel station is shown in Figure 6. This figure shows a section of the Arrowhead Tunnels that was excavated through ground having variable strength classifications, ranging from very weak to extremely strong. The normalized probe FPI scale corresponding to the face log strengths is shown on the right in this figure.

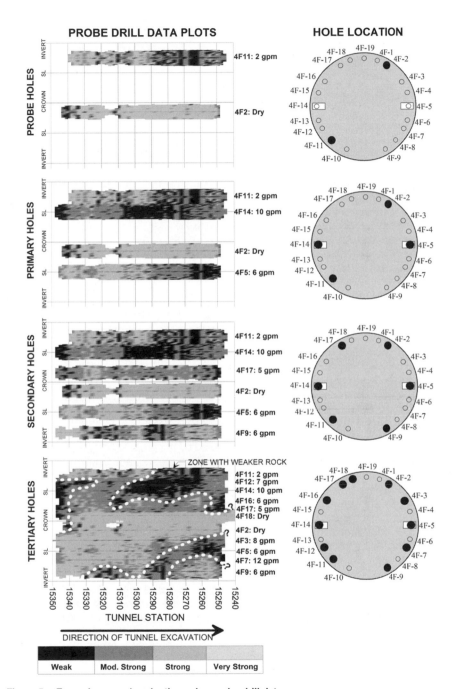

Figure 5.   Example ground evaluation using probe drill data

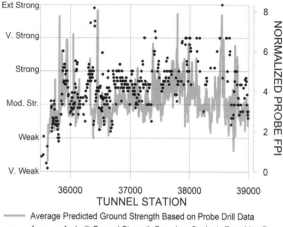

Figure 6. Comparison of probe drill strength and corresponding face map strength

AUTOMATED PROBE DRILL DATA EXAMPLES FROM THE ARROWHEAD TUNNELS PROJECT

To date, over 5,000 m (3 miles) of probe data for the Arrowhead Tunnels project have been collected and analyzed. The following are some examples of this data.

Examples Showing Weak Rock

Figure 7 shows an example of a relatively narrow weak rock feature that was first recognized by the automated probe drill data and subsequently mined. This figure shows the location of the weak feature as predicted by the probe drill data prior to excavation and the actual location of this feature based on subsequent tunnel geologic face mapping. The results of the geologic face mapping are shown on the histogram below the probe plot. The geologic face mapping histogram shows the percentage of rock mapped within the various strength categories, from weak to very strong, at various face stations. According to face map data, a relatively narrow zone of weak and sheared rock was mined through at the crown and right-side of the tunnel within the location predicted by the probe drill plots. The probe plot shows this zone to be approximately 5-feet thick at its narrowest, and this example illustrates the maximum resolution of the probe data, which is on the order of 5-feet.

Figure 8 shows another weak rock example that was initially predicted by the probe data and subsequently excavated. This example again shows the predicted weak zones on the probe plot and the subsequent actual conditions as documented by the geologic face mapping. As shown on the probe plot and on the geologic face map histogram, this section of the tunnel was excavated through two zones that contained predominantly weak rock. Beyond these two weak zones, the ground conditions improved to predominantly strong rock. The vertical dashed lines in Figure 8 delineate the start and end of the TBM excavation cycles. Figure 8 illustrates how the ground evaluation from the probe data can be used to optimize excavation cycle lengths to minimize the possibility of becoming stuck in weak rock zones. As shown in Figure 8, the initial excavation cycle (Cycle 1) was relatively short and completed to the beginning of the first weak zone. The next excavation cycle (Cycle 2) was relatively short

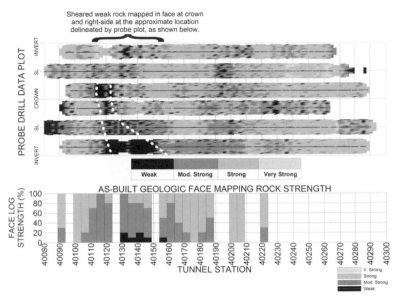

Figure 7.   Example probe drill plot showing narrow weak rock zone

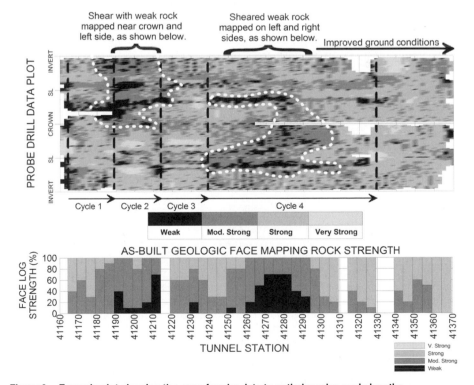

Figure 8.   Example plot showing the use of probe data to optimize mine cycle lengths

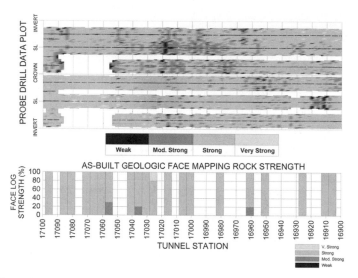

Figure 9. Example plot showing strong rock conditions

and completed through the first weak zone. Following this excavation cycle, Cycle 3 was also relatively short and was completed to the beginning of the second weak zone. The final excavation cycle (Cycle 4) was relatively long and was completed through the second relatively larger weak zone. Beyond the second weak zone, ground conditions improved to predominantly strong rock, and excavation cycle lengths were adjusted back to normal.

Example Showing Strong Rock

Figure 9 is an example probe ground evaluation plot showing favorable mining conditions. The probe plot in this example predicted predominantly strong rock and agrees with the subsequent face mapping data that is shown on the histogram at the bottom of the plot. Obviously, the ground evaluation information provided by the probe data in favorable ground conditions is not as important to assist with tunnel construction decision making, as during instances when the probe data suggests problematic ground conditions.

Example Showing Changing Ground Conditions

Probing cycles that contain multiple grout holes can create dynamic ground conditions ahead of the TBM. As holes are progressively grouted, ground strengths can progressively strengthen, and sometimes this ground strengthening is seen on the probe plots. Figure 10 shows two plots of probe drill data collected at the same face station. The upper plot shows the probes that were drilled during the initial stages of cement grouting. As can be seen on this plot, the probe data indicated a large zone of weak rock located predominantly on the right-side and extending to the left side. The lower plot shows only the holes that were drilled after the initial stages of cement grouting shown on the upper plot. As can be seen on the lower probe plot, it appears that the ground strength has slightly improved due to the efforts of the initial cement grouting, especially on the right side of the tunnel. Ground strengthening with time as holes are

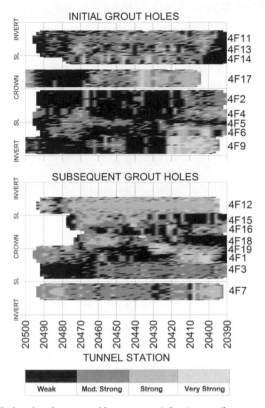

Figure 10.   Probe drill plot showing ground improvement due to grouting

progressively grouted is seen frequently on the Arrowhead Tunnels project, especially in areas of intense grouting.

SUMMARY

This automated probe drill monitoring system has proven to be a useful tool for evaluating ground conditions ahead of the TBM for the Arrowhead Tunnels project. This system was relatively inexpensive to setup and has now been calibrated with over 5,000 m (3 miles) of geologic face mapping data. The drill pressure data are analyzed off the critical path of the construction activities, usually within a couple of hours after the completion of probing. The resultant "picture" of the anticipated ground conditions is distributed to the project team to assist with tunnel construction decision making, especially in problematic ground. This automated probe drill monitoring system continues to be used on a daily basis for the Arrowhead Tunnels project.

EVALUATION OF GEOLOGICAL CONDITIONS AHEAD OF TBM USING SEISMIC REFLECTOR TRACING AND TBM DRIVING DATA

Takuji Yamamoto
Kajima Technical Research Institute

Suguru Shirasagi
Kajima Technical Research Institute

Koji Murakami
Kajima Technical Research Institute

Jozef Descour
C-Thru Ground, Inc.

ABSTRACT

For TBM tunneling no visual inspection of excavated ground is possible. This may hinder the high-speed excavation, particularly in complex ground conditions. The TBM Excavation Control System was developed for better prediction of geological conditions ahead of the tunnel face during excavation. The seismic reflector tracing produces volumetric images of anomalies in the ground in "near-real time." These images are verified using logging data from pilot boring and TBM driving data, to improve reliability and precision of ground mapping. Also good correlation was obtained between the rock strength index computed from TBM data, and one derived from volumetric seismic images.

INTRODUCTION

It is utmost important that the geological conditions along the route of excavated tunnel are known in advance both for the tunneling safety and for its construction economy. Typically seismic wave velocity and electrical specific resistance values are mapped from surface to generate a geological profile before the construction work commences. However, in a number of cases the predicted geological information does not sufficiently match the actual geological conditions, forcing design changes during tunnel excavation. This reveals shortcomings of the surface investigation methods (mainly Seismic Refraction and Resistivity Tomography surveys).

The core pilot drilling offers the most direct data ahead of the tunnel face. It is used mainly for parts of the ground identified as geologically defective by prior surface investigation. However, while the coarser geological conditions can be ascertained by drilling pilot boreholes, smaller structural features can be missed. Furthermore, the pilot drilling requires stopping the excavation. That is very costly as it disrupts tight construction schedules. Consequently, a strong demand exists for new survey techniques applicable from the tunnel itself that would be sufficiently reliable without disrupting the tunneling operations.

The new technique using reflected seismic waves is being pioneered in Japan since 1999. This technique is referred to as Three-dimensional Reflector Tracing or

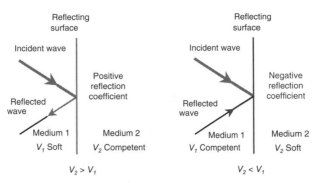

Figure 1.   Relation between rock quality and reflection coefficient

TRT. It offers a "near-real time, non-destructive method for 3D imaging of structural ground changes, while causing minimal interruption to the excavation process.

THEORY OF SEISMIC REFLECTOR TRACING

Structural anomalies in the ground are typically associated with a contrast in seismic impedance (the product of density and seismic velocity) in the rock mass. These contrasts occur at boundaries between geological layers or other structural discontinuities. The discontinuities of size comparable with wavelengths of incident seismic waves produce reflected waves (Figure 1) that can be detected by the array of sensors. The magnitude ratio between reflected and incident waves is called the reflection coefficient. The reflection coefficient for normal incidence of seismic energy onto a boundary between media 1 and 2 is defined by equation (1) (Watters 1978; Aki and Richards 1980):

$$w R = \frac{Q_2 - Q_1}{Q_2 + Q_1} \tag{1}$$

where:

R = reflection coefficient
$Q_1 = \rho_1 V_1$ = *seismic impedance in medium 1*
$Q_2 = \rho_2 V_2$ = *seismic impedance in medium 2*
ρ = rock mass density
V = seismic wave propagation velocity in the ground

Note that the seismic impedance for P-waves is proportional to the Young's modulus, and for S-waves it is proportional to the modulus of rigidity.

A transition from material with a lower seismic impedance to one with a higher value results in a positive reflection coefficient, and vice versa. Thus, features such as fractured zones within a more solid rock mass will give rise to reflections of reversed polarity.

The anomalies reflecting seismic waves are imaged by transposing time records for these waves back to the location of their origin, using the proper velocity model to convert time to distance, and then stacking reflected waveforms at that location (Ashida and Sassa 1993; Ashida 2001; Neil et al. 1999). Figure 2 shows the concept behind this operation. For each seismic source and receiver of known location the locus of all possible reflector positions for the same "equi-travel" time (Ashida and

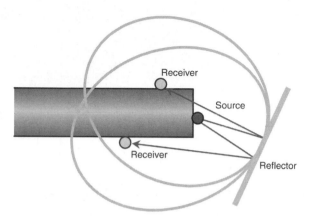

Figure 2. Principal concept for transposing reflected waves to the reflector

Sassa 1993) from source to reflector and back to receiver defines an ellipsoidal surface in three-dimensional space. For the proper velocity model the transposed reflected signals do match in phase, and add up forming a wave-like anomaly. That anomaly starts at the reflector surface, and often stretches behind that surface, away from the source-receiver pair. The polarity of the first peak of reflected waveform depends on which way the seismic impedance changes across the boundary of the reflector (1). The technique typically employs an average velocity measured for the direct waves traveling along the tunnel walls between sources and receivers.

A discrete image of reflective ground anomalies "ahead of the tunnel face" is calculated using specialized software, and for a grid of regularly spaced nodal points contained in a pre-defined rectangular survey volume.

MEASURING SETUP

The Three-dimensional Reflector Tracing (TRT) uses an array of 10 to 12 accelerometers, and a number of source points coupled in a predetermined pattern to the rock of the tunnel walls (Figure 3). Such an array of sources and receivers forms a 3D seismic transmitter and receiver "antennae" system around the tunnel walls. This system is capable of directional reception of reflected waves. However, its directional characteristics are the most uniform along and specifically ahead of the tunnel face.

Seismic waves are generated in the rock either by sledgehammer strikes, or by a magnetostrictive source driven by a swept frequency signal. Time records of seismic waves detected by accelerometers are collected as data files using a multi-channel, high sensitivity digital seismograph (Figure 4). Also, reasonably precise locations of source and receiver points are required for reliable imaging of the ground. Once collected, the data is processed to generate three-dimensional image of structural features in the surveyed ground volume. As the tunnel excavation advances, or when another zone of potentially unstable ground is approached, the array can be installed at the new location in the tunnel to collect another set of data, and to produce another image of the ground.

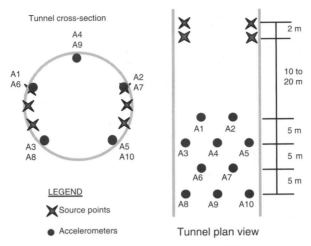

Figure 3.   Standard pattern of source and receiver locations for TRT seismic survey in TBM tunnels

Figure 4.   Seismic data acquisition system setup in a TBM tunnel

SEISMIC DATA PROCESSING AND ANALYSIS

The data processing starts from downloading the acquired data files (time records) along with coordinates of source and receiver locations into the data processing software.

Then a desired rectangular image block is defined to include zones of interest (Figure 5). It is followed by setting the number of nodal points in the block that determines the image resolution. The image block is typically plumb, and parallel to the tunnel route.

In the next step the frequency filters (Figure 6) are set to enhance the direct wave arrivals and to suppress possible noise so reliable direct wave arrival times can be picked (Figure 7). The difference between the arrival time, and the source trigger time defines the travel time for any source-receiver pair. Typically, with known locations of source and receiver points, an average velocity is assessed through linear regression for the time versus distance spread of picked travel times (Figure 8). This velocity is

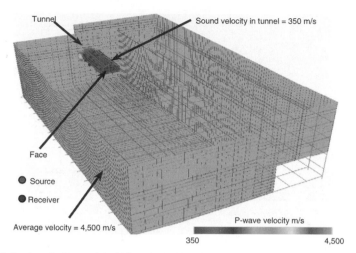

Figure 5.   Seismic velocity model within selected imaging block

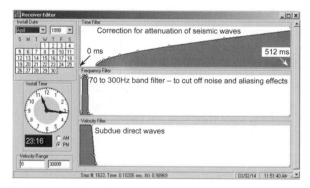

Figure 6.   Filters for data processing

used to build a velocity model within the selected image block (Figure 5). The model includes low velocity void (350 m/s for P-waves, and 0 m/s for S-waves) that represents excavated tunnel. The velocity model is a base for converting time to distance in order to generate a volumetric image of structural features that reflected seismic waves. The velocity model can be modified based on other geological or geophysical survey results thus improving accuracy for generated images.

Finally, three filters are set specifically for data processing (Figure 6). The time filter compensates for the amplitude decay of recorded signals due to attenuation (Neil et al. 1999; Descour et al. 2005). The frequency filter subdues the noise and limits the frequency range to avoid spatial aliasing related to nodal grid spacing. And the velocity filter suppresses amplitude of the direct waves within source-receiver array to enhance reflected waves. Figure 7 shows an example of a raw seismic record and the same record filtered for data processing.

The software calculates a reflectogram—a discrete image of reflectors within the image block following the procedure described in the Theory of Seismic Reflector

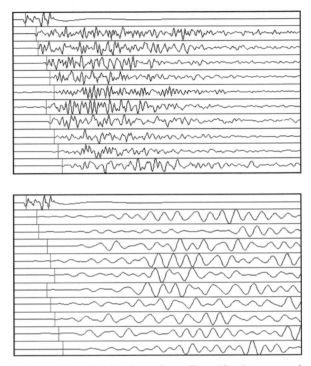

Figure 7.   Raw seismic waveforms (top) and waveforms filtered for data processing (bottom)

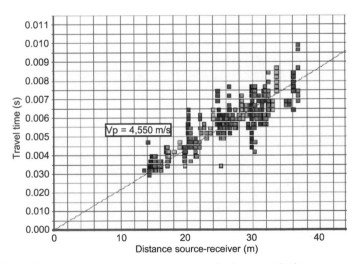

Figure 8.   Using linear regression to measure average seismic wave velocity

Tracing section. The data processing takes from a few minutes up to half-an-hour dependent on the total number of nodal points in the image block.

The data processing can be repeated after changing the velocity model and the filters until the reflectogram appears stable and balanced, and relatively free of noise. If any problem zones (faults, fractured rock) are anticipated based on other survey data, the image is used to verify the existence of these zones and their location.

The extent and hardness of reflecting boundaries is measured by normalized Reflection Magnitude (reflection number). Its range of −1 to +1 is proportional to the reflection coefficient as defined in section Theory of Seismic Reflector Tracing.

The reflected waves either retain their polarity for higher seismic impedance at the reflector boundary, or change their polarity if the impedance is lower (Figure 1), (1). A color code is used to distinguish between reflections from objects of higher impedance (orange to red—for harder rock), and objects of lower impedance (blue and purple—for softer, fractured, often water bearing rock).

To tie the reflectogram to the tunnel site, the topographic coordinates and the local tunnel chainage (STA) are added to the image background. The final analysis of structural features represented by their seismic reflections is done by inspecting two-dimensional sections through the image block, and analyzing the shape, extent, and alignment of contour images drawn for different values of reflection magnitude. To make this three-dimensional analysis easier, animation software allows rotating the reflectogram for better visual evaluation of features in the image.

EXCAVATION MONITORING AND CONTROL SYSTEM USING TBM DATA

The current TBM excavation monitoring and control system has two levels. In the first level the immediate control of the actual rock cutting operation relies on routine measurements during each excavation stroke. The second level incorporates the results of reconnaissance and pilot drill logging as forward geological information, the recent TBM data (such as thrust, torque, penetration rate, gripper pressure, weight of muck, rock strength and excavating energy) (Fukui and Okubo 1999), and the observation of tunnel walls. The acquired information is monitored at a site office as well as at the TBM drivers seat, and is used as a feedback for the next excavation stroke, and for selecting support patterns (Figure 9). The major components used for the TBM excavation monitoring and control system are briefly characterized below.

Drill Logging

The data from a hydraulic percussion drilling ahead of the tunnel face is entered into the equation for the Drill energy coefficient (2) (Aoki et al. 1990). This coefficient proportional to the hardness of the rock is used for evaluating ground conditions ahead of a TBM.

$$Ev = \frac{E \cdot N}{V \cdot A} \qquad (2)$$

where:

   Ev = drill energy coefficient
   E = percussion energy per one strike
   N = number of strikes per second
   V = drilling speed
   A = cross section of the drill hole.

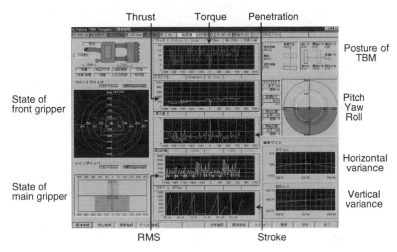

Figure 9.   Main TBM monitor

Ground Evaluation Using TBM Data

The TBM thrust, torque, penetration rate, and gripper pressure are automatically measured and used to calculate the rock strength index (3) (Fukui and Okubo 1999), and the excavating energy (4).

$$\sigma_{cF} = \frac{F}{C_\sigma \cdot Pe} \tag{3}$$

where:

σ_{cF} = rock mass strength
F = thrust
C_σ = constant
Pe = penetration rate

$$Ee = \frac{F}{A} + \frac{2\pi \cdot Nr \cdot Tr}{A \cdot V_t} \tag{4}$$

where

Ee = excavating energy
F = thrust
A = cross section of the tunnel
Nr = head revolutions per minute
Tr = torque
V_t = advancing speed

Belt Scale Readings

A scale is installed under the conveyor belt that carries muck out, and the weight of the muck and the speed of the belt are measured. The resulting muck flow readings (t/hour) are recorded continuously. If the readings show a sudden sharp increase, there is a possibility that the rock at and around the face is falling.

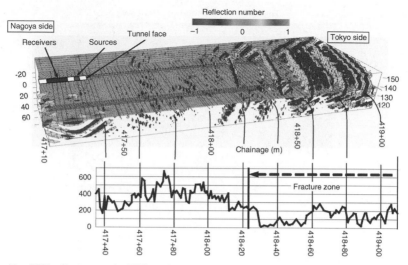

Figure 10.   TRT reflectogram for TBM cutting through predominantly mudstone layers in Site 1

CASE STUDY 1—SEISMIC REFLECTIONS FOR GROUND EVLUATION AHEAD OF TBM

Survey Purpose and Outline

A number of parameters used to monitor the performance of a TBM (driving data) correlate well with encountered geological conditions. Unfortunately, with exception of core drilling, this information is limited to the immediate proximity of the tunnel face.

The purpose of this study was to evaluate correlation between the driving data, and the ground characterization provided by TRT reflectograms in front of a TBM excavated tunnel for significantly different ground conditions. The results from two TBM driven Sites (1 and 2) are presented.

Survey in Site 1. The tunnel was excavated through mudstone-dominated layers. This ground had relatively low strength and deteriorated very easily when exposed. The results of TRT survey in this tunnel were compared with some of TBM driving data.

Survey in Site 2. The tunnel was excavated mainly through a competent and fresh andesite, separated by zones of tuff that were often fractured, with some cracks carrying water. In this site two overlapping TRT surveys were conducted and compared with the machine data.

Survey Results

Results in Site 1. The TRT reflectogram for Site 1 is shown in Figure 10. A relatively reflection-free zone in front of the tunnel is followed by an emerging series of reflectors starting around STA 418+30.

A cutter torque plot shown at the bottom was measured as the TBM was passing through. In this figure the second part of the torque plot also starting from STA 418+30 (fracture zone) shows significantly lower readings. Argillatious fractures were observed in this range which were caused mainly by the excavation. In overall, multiple reflectors shown in the reflectogram appear to correlate well with the range of fractured zone and with the lower torque values.

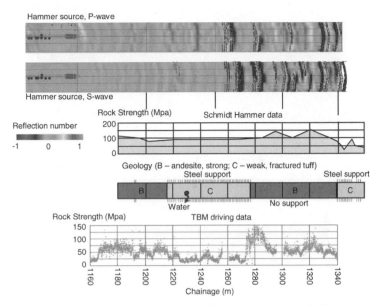

Figure 11.   TRT reflectogram for TBM excavating the first surveyed section in Site 2

Results in Site 2. Figure 11 shows a comparison between the reflectograms for the first survey (P- and S-wave for hammer source, and S-wave for swept frequency source), and a Schmidt Hammer and a TBM driving data plots for the same tunnel chainage.

The initial part of the P-wave reflectogram shows almost no reflections except a weak negative (blue) anomaly between STA 1220 and 1240. This anomaly appears to match the onset of a tuff zone and an inflow of water that occurred at STA 1230. A much stronger, generally positive (orange) anomaly shows first at STA 1260, and then at STA 1280. This last anomaly coincides with the end of the tuff zone.

The S-wave reflectogram shows a weak anomaly around STA 1340 that would match the start of a tuff zone. A weak negative anomaly at 1380 can be attributed to cracks carrying water at this chainage. This anomaly shows much stronger in a parallel reflectogram generated using a swept frequency source. Both anomalies are followed by smaller positive anomalies and a larger negative anomaly at approximately STA 1330 where the driving data shows a decline. The reflectograms for the second survey (for P- and S-waves) shown in Figure 12 are compared with a Schmidt Hammer and a driving data plots. A major negative anomaly around STA 1455 matches another tuff zone with fractures carrying water.

In general, both reflectograms appear to match major features in the ground as indicated by the Schmidt hammer and the driving data.

CASE STUDY 2—ROCK STRENGTH INDEX DERIVED FROM REFLECTION NUMBER

Velocity Assessment

According to experimental data, seismic wave velocity is proportional to the rock strength index. The TRT data from a TBM tunneling site was used to tie the two

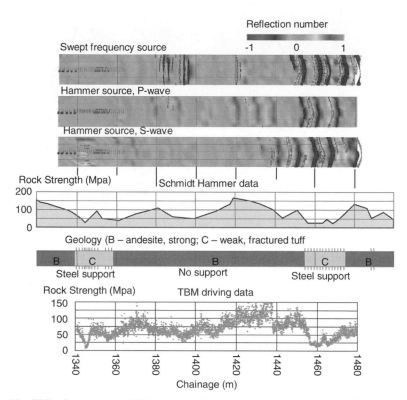

Figure 12.   TRT reflectogram for TBM excavating the second surveyed section in Site 2

parameters together and to investigate the level of correlation and possibility to use reflection numbers for the rock strength evaluation.

In numerical terms the presence of a reflective boundary is detectable as a change of values along nodal points in the image volume. In general, these values are proportional to the reflection coefficient and are referred to as reflection numbers γ. For the purpose of this analysis, and drawing from the formula for the reflection coefficient (1), the reflection number at the boundary between media k and k+1 was defined using a simplified expression (5) (Aoki et al. 2003):

$$\gamma_k = C \frac{V'_{k+1} - V'_k}{V'_{k+1} + V'_k} \qquad (5)$$

where:

V'_k and V'_{k+1} = approximated seismic wave velocities for media k and media $k+1$ respectively which are defined as $V' = C_oV$ where C_o assumed constant for this analysis; and,

C = functional related to the magnitude and polarity of incident waves and to the natural density change at the reflector boundary; it is also assumed constant and >0 for this analysis.

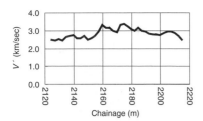

Figure 13. Assessment of seismic velocity along the tunnel alignment

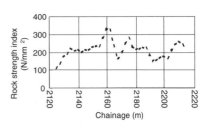

Figure 14. Rock strength derived from TBM driving data

The equation (5) led to an approximate velocity expression (6):

$$V'_{k+1} = \frac{1 + \dfrac{\gamma_k}{C}}{1 - \dfrac{\gamma_k}{C}} V'_k \tag{6}$$

Conversion of Velocity to Rock Strength Index

The reflection numbers obtained from TRT survey represent relative seismic impedance/velocity changes. The reflection numbers γ obtained from TRT survey at 2.5 m centers for consecutive sets of nodal points along investigated TBM tunnel alignment were converted to a matching set of V' values using equation (5). The C value was experimentally assessed, and the starting velocity V'_1 was set at 2.5 km/s as measured at the TRT survey site. The approximate velocity profile is shown in Figure 13. The rock strength index σ calculated from the TBM driving data along the same tunnel section is plotted in Figure 14.

The mean $m_{V'}$ and variance $s_{V'}$ for V' values in Figure 13, and the mean m_σ and variance s_σ for σ in Figure 14 can be expressed respectively as:

$$m_{V'} = \frac{\sum V'}{N}, \; s_V = \frac{\sum (V' - m_{V'})^2}{N} \tag{7}$$

and

$$m_\sigma = \frac{\sum \sigma}{N}, \; s_\sigma = \frac{\sum (\sigma' - m_\sigma)^2}{N} \tag{8}$$

where:
N = number of data points

Using these values, the rock strength was written as a function of V':

$$\sigma = m_\sigma + \sqrt{\frac{s_\sigma}{s_{V'}}} (V' - m_{V'}) \tag{9}$$

A comparison of both, the rock strength index for driving data, and the rock strength index derived from TRT reflection number is shown in Figure 15. The two sets of indexes match quite well. This match suggests that the rock strength can be predicted ahead of the tunnel face using TRT reflection numbers, particularly if the comparison between both indexes is updated and refined using incoming driving data as

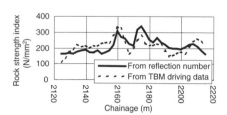

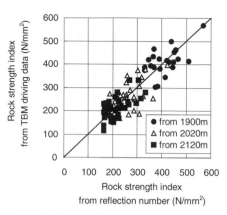

Figure 15. Comparison rock strength index from reflection number with one from TBM driving data

Figure 16. Correlation between rock strength index from TBM driving data and one from reflection number

the TBM advances through the TRT surveyed range. Furthermore, the rock strength around the tunnel can be estimated with better precision after the TBM excavation is completed over the range of TRT survey. The former gives useful information for the TBM driving, while the latter helps in the tunnel enlargement operation.

CORRELATION BETWEEN ROCK STRENGTH INDEX FROM TBM AND FROM REFLECTION NUMBERS

This investigation was conducted in a tunnel that was excavated using TBM as the pilot tunnel, followed by the tunnel enlargement operation. The diameter of the pilot tunnel was 16 m, and the total tunnel length was 4,450 m. As the tunneling progressed, three seismic reflective surveys, each covering approximately 100-m range ahead of the face, were conducted at three subsequent points: TD 1,900 m, TD 2,020 m, and TD 2,120 m. Using the approach discussed above, the results of TRT survey were converted to the rock strength index values and plotted against the strength index values from matching TBM data (Figure 16). The resulting correlation coefficient of 0.866 was quite high thus building confidence that the proposed method of using reflection numbers for evaluation of the rock strength ahead of TBM face is reliable and should be further explored and implemented.

CONCLUSIONS

This research has introduced to tunneling the Three-dimensional Reflector Tracing (TRT), a new ground imaging technique that is applicable from inside of any tunnel. This technique offers relatively inexpensive, non-intrusive, and near-real time means for seeing through the ground ahead and around an advancing tunnel in search for possible structural problems that can slow down or even severely disrupt the tunneling safety and operation.

This technique successfully integrates (1) the physical property of the ground change expressed by the reflection coefficient, and (2) the concept of equi-travel time surfaces that intersect in phase along the boundary of each sufficiently strong and large reflector. These surfaces can be calculated for each source-receiver pair of the proper size array coupled to the tunnel walls. It converts seismic waves reflected from structural boundaries in the ground into a reflectogram by transposing reflected waveforms to their origin and thus providing three-dimensional image of structural features in the ground.

In the authors' opinion this discussion of presented case studies for different geological and construction conditions leads to the following conclusions:

- In general, reflectograms generated by Three-dimensional Reflector Tracing (TRT) appear to correctly address major anomalies in the ground, particularly zones carrying water, as well as changes in the ground properties expressed by the value of seismic impedance;

- There is an apparent correlation between changes in the rock conditions that affect TBM performance, and at least the major structural features in the ground detected by TRT as seismic reflectors;

- Setting an optimum correction for wave attenuation appears still to be a problem as it is not only controlled by the contrast in seismic impedance, but also by the shape, size and orientation of wave reflecting anomalies with respect to the wavelength range of used seismic waves;

- Better feedback from surveyed sites could help improving efficiency and reliability of imaging the ground in front of advancing tunnel construction using the TRT technique;

- The rock strength index derived from digital ground images (reflection numbers) produced by Three-dimensional Reflector Tracing (TRT) appears to show good correlation with the rock strength index obtained from the TBM driving data acquired along advancing tunneling operation. This 3D rock strength index derived from TRT images can be continuously updated using incoming TBM driving data to improve volumetric prediction of the ground conditions in front of a TBM. It can also be refined after completion of the excavation over the range of prior TRT surveys to better characterize ground conditions around the pilot tunnel for safer and more effective tunnel enlargement operation.

REFERENCES

Aki, K., and P.G. Richards. 1980. Quantitative seismology. Theory and methods. Vol. 1. W.H. Freeman and Co., N. York.

Aoki, K., Inaba, T., Shiogama, Y., and Tezuka, Y. 1990. Evaluation of geological conditions by "Drilling Logging System" (in Japanese). *Proceedings of the 8th Japan Symposium on Rock Mechanics.* 67–72.

Aoki, K., Mito, Y., Yamamoto, T., and Shirasagi, S. 2003. Advanced support design system for TBM tunnels using the seismic reflective survey and TBM driving data. *Proceedings of the International Symposium on the Fusion Technology of Geosystem Engineering, Rock Engineering and Geophysical Exploration, Seoul, Korea, November 18–19.*

Ashida, Y., 2001. Seismic imaging ahead of a tunnel face with three-component geophones. *International Journal of Rock Mechanics & Mining Sciences* (38): 823–831.

Ashida, Y., and K. Sassa, 1993. Depth transform of seismic data by use of equi-travel time planes. *Exploration Geophysics* 1993 (23):341–6.

Descour, J., T. Yamamoto, and K. Murakami, 2005. Improving 3D imaging of unknown underground structures. In *Proceedings (CD-ROM) of FHWA Unknown Foundation Summit,* Lakewood, CO. November 15–16, 2005.

Fukui, K., and Okubo, S. 1999. Rock-properties estimation by TBM cutting force. *Proceedings of the International Congress on Rock Mechanics, France,* 12:1217–1220.

Neil, D.M., K.Y. Haramy, J. Descour, and D. Hanson, 1999. Imaging ground conditions ahead of the face. *World Tunneling,* 12(9):425–429.

Watters, K. 1978. *Reflection seismology. A tool for energy resource exploration.* John Wiley & Sons, New York.

GUIDANCE FOR PARTIAL FACE EXCAVATION MACHINES

Nod Clarke-Hackston
VMT GmbH

Jochen Belz
VMT GmbH

Allan Henneker
Surex Pty Ltd.

ABSTRACT

For the construction of tunnels and other underground structures, extraction of the exact amount of material is of paramount importance both economically and for engineering purposes. In the Sequential Support Method (NATM) immediate (sequential) and smooth support by means of shotcrete, steel arches, lattice girders and rockbolts, either singly or in combination are used; cutting of the precise (albeit sometimes of complex geometry) profile is an integral part of the method.

In order to save unnecessary excavation and provide better information to the machine operator, VMT GmbH has developed a system to support precise excavation of the tunnel profile when using roadheaders or other partial face cutting machines. This paper outlines the principles of this system with examples from Australia, Germany, Sweden and Spain and will cite examples of typical savings achieved.

HISTORY

Roadheaders were first developed for mechanical excavation of coal in the early 1950s. Today their application areas have expanded beyond coal mining as a result of continual performance increases brought about by new technological developments and design improvements. The major improvements achieved in the last 50 years consist of; steadily increasing machine weight, size, and cutterhead power, improved design of boom, muck pick-up and loading systems, more efficient cutterhead design, metallurgical developments in cutting bits, advances in hydraulic and electrical systems, and more widespread use of automation and remote control features. All these have led to drastic enhancements in machine cutting capabilities, system availability and service life. Roadheaders are the most widely used underground partial-face excavation machines for soft to medium strength rocks, particularly sedimentary rocks. They are used for both development and production in the soft rock mining industry (i.e., main haulage drifts, roadways, cross-cuts, etc.) particularly in coal, industrial minerals and evaporate rocks. In civil construction, they find extensive use for excavation of tunnels (railway, roadway, sewer, diversion tunnels, etc.) in soft ground conditions, as well as for enlargement and rehabilitation of various underground structures. It is chosen over other possible tunneling methods principally because of its application advantages which include free excavation area, considerable flexibility in regard to unforeseen changes in geological and hydrological conditions, flexibility in terms of cross-sectional changes, and rapid mobilization with readily available, unsophisticated and relatively inexpensive excavation equipment.

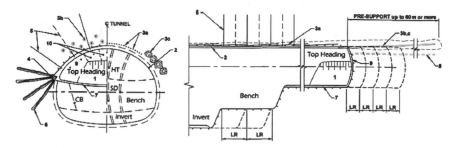

Figure 1.   Complex geometrical profiles

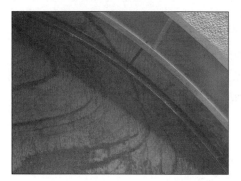

Figure 2.   Precise cutting for girder placement

Figure 3.   Initial shotcrete coating

GENERAL

All rock and soil has, to a greater or lesser degree, the inherent ability to support itself or stand freely for a certain amount of time. Strong rock can remain free standing for hundreds of years while the self-supporting ability of rocks and soils of weaker types is relatively short-lived and measured in minutes rather than hours, days or years.

For the construction of tunnels and other underground structures, extraction of the exact amount of material is of paramount importance both economically and for engineering purposes. In the Sequential Support Method (NATM) where immediate (sequential) and smooth support by means of shotcrete, steel arches, lattice girders and rockbolts, either singly or in combination are used; cutting of the precise (albeit sometimes of complex geometry) profile is an integral part of the method (Figure 1).

Extraction of too much material is costly and can affect the calculations for temporary support. Over extraction also results in the need for additional costly material to fill any void between the actually excavated ground and the designed permanent lining. Under excavation necessitates reworking of the area to enable preformed supports to be erected. The rate of excavation also has to be adjusted to the setting rate of the shotcrete (Figures 2 and 3).

When using a roadheader one of the main problems affecting the precise cutting of the face is visibility of the cutterhead. Frequently the amount of dust produced, or residual shotcrete in the air, is such that the operator is unable to see the position of the cutterhead (Figure 4). Marking of the face area to be excavated is time consuming for the survey crew and it takes place in an area where stability of the working area is at its potentially most vulnerable. Even if the machine is positioned correctly and has a

Figure 4. Poor atmospheric conditions

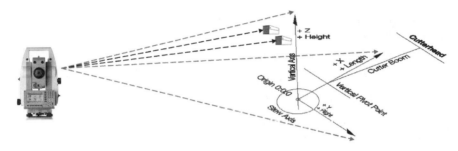

Figure 5. Determination of cutterhead position

limit on the movement of the cutterhead there is a tendency for the machine to move as a reaction to the cutting force applied by the machine.

In order to save this unnecessary excavation and provide better information to the machine operator, VMT GmbH has developed a system to support precise excavation of the tunnel profile when using roadheaders or other partial face cutting machines.

OPERATING PRINCIPLE

The operating principle is based on determination of the position and orientation of the body of the roadheader with respect to a Total Station mounted on the tunnel wall and sighting to two shuttered prisms located on the body of the machine. Rotation of the body of the machine is monitored by a dual axis inclinometer. Movement of the cutter arm and the position of the cutterhead itself is determined from sensors (Rotary and Linear encoders) mounted either in or on the arm. Values from these sensors are taken from the machines PLC or by means of direct output from the sensors (Figure 5).

There are two modes of operation: tracking mode allows fast and continuous distance measurements, lock mode allows the following of moving prisms. The continuous measurements are only interrupted by directional controls to the back target. The servo-motorized Total Station–Leica TCA 1203 is mounted on a bracket at the tunnel wall. Every 0.5 seconds the PLC sends data on the horizontal and vertical angles and the length of the cutter arm (kinematics of the road header). The position of the cutter head is calculated in the road header coordinate system and then transformed into the global coordinates of the tunnel project. Thus the cutter head position relative to the tunnel axis is known and may be continuously displayed graphically in the operator cabin, since the tunnel profile is maintained at right angles to the tunnel axis.

When the theodolite does a directional control check, the tracking mode is switched off and the standard measuring program is activated. This enables a higher accuracy of distance measurement can be obtained, but it is not possible to track the target. The average duration of the measurement cycle for the directional control is 25 seconds.

The data from the Total Station is transmitted to the Guidance PC on the road header via a pair of radio modems, one mounted next to the Total Station the other mounted on the body of the roadheader.

Sophisticated software enables the project surveyor to easily set the system parameters for the machine, enter the Designed Tunnel Alignment Data (DTA) into the program as well as each of the individual profiles to be excavated. A simple change in the profile icon enables the machine to rapidly change its excavation sequence in line with the project's requirements.

EQUIPMENT

The major components are mounted on the body of the machine itself. These include the Main Industrial PC, remote display and trackball control devise in the control cab, UPS, Central Box, 2× Shuttered Prisms, Dual Axis Inclinometer and Local Radio Modem. Additional components; the Leica TCA 1203 motorized Total Station, the Remote Radio Modem and the Backsight Target Prism are mounted on the tunnel wall. Power for the machine components is supplied from the machine via a UPS. The Total station and Remote Radio Modem are usually powered from the tunnel lighting circuit (Figure 6).

The Ruggedized Industrial PC, UPS, Central Box and Dual axis inclinometer are all securely mounted on the body of the machine in an area that gives the maximum protection to these components. The Shuttered prisms are typically mounted on the roof of the operators cab to give maximum viewability, however due consideration to the alignment of the tunnel should be taken into account when choosing these locations and to the sighting of positions for the Total Station. A continuous line of sight between the shuttered prisms and the total station is needed throughout the excavation cycle. The Machine body should be surveyed prior to installation of the shuttered prisms and inclinometer unit so that the main axis of rotation can be determined, the position of the prisms measured and the inclinometer be correctly zeroed.

The Shuttered prisms are standard survey prisms enclosed in a protective housing with a motorized protective shutter. The shutter, activated by the software of the SLS system, is only opened for the duration of the angle and distance measurement to the respective prism. This eliminates confusion of prisms or other reflective objects it also helps to keep the prism surfaces clean. The individual prisms can be connected by a bus system.

A fluid damped dual axis inclinometer for determination of inclination of the roadheader. The damping behavior of the inclinometer can be adapted to the vibration range of the mounting position, in addition the inclinometer is equipped with a temperature sensor for temperature compensation.

The main computer is a ruggedized industrial PC which is housed in a special Stainless Steel enclosure which is fixed on anti-vibration mounts between the housing

Figure 6.   System components

and machine body. The mounting location is chosen to give maximum protection for the unit. A UPS is supplied to maintain the appropriate power to the PC in the event of fluctuating power being available from the machine. The PC uses the latest Windows™ Operating system. A remote display for operator is installed in a prime viewing location in the operators cab. Easy access to functions is via track ball control. A remote USB connector is available for attaching a keyboard for initial data entry and for memory device attachment.

The Central Box interfaces between the various sensors of the system and converts these outputs for suitable entry into the Industrial PC. Control signals from the Industrial PC are also converted for transfer to the sensors such as the shuttered prisms.

Radio Modems are used to connect the components on the machine with those on the tunnel wall as the use of cables in such a consistently changing working environment would be problematic.

The Local Radio Modem is mounted on the machine and effects Wireless Data Communication between SLS-PC and Total Station whilst the Remote Radio Modem is mounted next to the Total Station with clear path for radio signals. Wireless reception is controlled by the SLS-software. The Remote Radio Modem receives the data from the Local Unit and transfers them to the Total Station. In case of power failure it changes automatically into battery mode and sends an alarm signal to the SLS-PC.

The principle measurement tool for the SLS-TM system is a Motorized Leica Total Station Model TCA 1203 with Automatic Target Recognition (ATR). Automatic fine pointing to the prisms speeds up measurements and improves productivity. A Standard, round retro-reflective prism is used as a rear backsight target. It is recognised and measured by the ATR target recognition unit of the Total Station and is used to check the bearing of the Total Station. A comparison between the Backsight Target and Total Stations' positions during each measurement cycle is used to give a warning in case of movement of either outside of any preset limits.

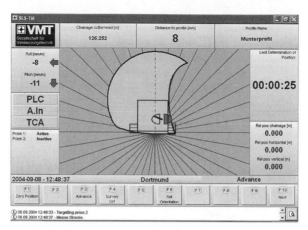

Figure 7.   Main operator display

SOFTWARE

The SLS-TM Roadheader Guidance System provides an automated means to minimize operating expenses, while maximizing the accuracy of cut and minimizing time. This is achieved through use of a highly unique, computer-driven machine operator's display screen (Figure 7).

The easy to use Windows™ software has been carefully configured to make data entry as simple as possible. Initial setup of the system is implemented through the System Editor data entry pages. Input of the Designed Tunnel Alignment (DTA) is achieved through the DTA Editor pages where the DTA, in point coordinate table format can be created from the principle geometric elements that make up the alignment in both horizontal and Vertical. From the Profile Editor each of the relevant profiles for the project can again be created from the geometric elements that make up the profile (Figure 8).

Where multiple profiles are to be excavated in rotating sequence they can all be stored in the main program. When the individual profile is to be excavated its profile is simply called up from the profile editor menu. Access to the various user levels is password protected. Several language options are available as well as Metric and Imperial unit options.

To aid the operator the background of the main graphical display is colour coded to indicate the time interval since the last valid survey reading. Although the measurement cycle can be set to an interval as short as every 20 seconds there may be times when a valid reading cannot be made, when this occurs a change in the background colouring is made to the operators display. The colours for the background are:

1. Within initial time interval   Grey

2. Time greater than first preset value   Yellow

3. Time greater than second preset level   Red

The typical time intervals being 3 minutes and 6 minutes.

A full dialogue of the survey positions is available from the survey position screens.

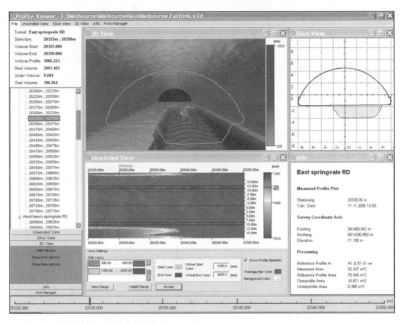

Figure 8.   Profile viewer—general display

DEVELOPMENTS FROM ORIGINAL SYSTEM

Since the basic system was created there has been a constant programme of developments, frequently inspired by the end users of the equipment.

All information that is monitored by the system is saved in the system database. This open platform enable customized reporting to be made and a comprehensive package of displays of the excavated sections of the tunnel including 3-D views and volumetric calculations have been created. This enables comprehensive reporting of the excavated profile to be created and presented to the owner in a clear concise manner. The full scope of the systems capabilities is still to be fully realised although its use in providing intermediate "as built" documentation is rapidly gaining popularity with the sites where the systems is in use.

EXTENSION INTO EXCAVATOR AND BOLTER GUIDANCE

Having seen the of the guidance control for the roadheader on the Eastlink project in Melbourne, Australia, the site personnel requested a system to be created for use on a Caterpillar excavator as the application was very similar to that of the roadheader. This was quickly and successfully implemented.

A further major concern on this project was for the safety of the survey personnel who had to enter the unsupported face area to mark out the positions for the drilling of the rockbolt support holes. Finding a way to accurately position the drill of the bolter without physically marking the surface for the various support classifications was achieved with and adaptation of the roadheader guidance system. Substantial costs savings were made due to the sequential nature of the use of the roadheader, for excavation and the Bolter for support which enabled the use of many of the main components of the system including the total station and wireless modems. To change

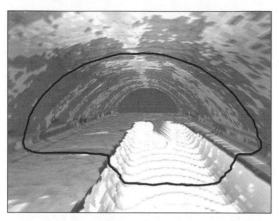

Figure 9.   Detailed view of section

Figure 10.   Guidance for bolter used in conjunction with Roadheader system

from one system to the other it is only necessary to switch off the system on one machine and initiating it on the next machine. The software automatically initiates a survey and calculates the machines position and profile section to be worked upon (Figures 9 and 10).

BENEFITS REPORTED BY END USERS

It has always been difficult to get precise figures from any project whilst production is still taking place, however the following favourable comments have been received from the site in Bilbao:

- "The operators like the system; they can cut a better profile faster than when it is done manually."
- "As it is no longer necessary to mark up the face by the surveyors there is a time saving of 1 hours per day when the machine is not in a nonproductive mode."

- "In one section where detailed records of were kept it was found that the more precise cutting achieved with the guidance system in use enabled a reduction from 600 m^3 to 423 m^3 in the amount of concrete in comparison to and identical section of tunnel excavated without the guidance system."

REFERENCES

Copur, H., Ozdemir, L. and Rostami, J., 1998, Roadheader Applications in Mining and Tunneling. Mining Engineering, Vol. 50, No. 3, March, pp. 38–42.

Chittenden, N., Müller H., 2004 New Developments in Automated Tunnel Surveying Systems. *Proceedings ITA World Tunneling Congress 2004*, Singapore.

Sauer, G., 1990 Light at the end of the tunnel. Construction Technology International in March 1990.

Smith, M., 2006 Face Drilling. Reference Editions Ltd. for Atlas Copco Rock Drills AB. 2006 Sweden.

White, A., 1993 The Installation of Automated Profile Guidance on a Dosco MkIIB Roadheader. *Paper Presented at Meeting of The Institution of Mining Engineers, May 13, 1993,* Nottingham.

INNOVATIONS IN UNDERGROUND COMMUNICATION INFRASTRUCTURES

Darrell Gillis

Consultant

INTRODUCTION

Recent accidents in the United States underground Coal Mines and the on-going carnage in Underground Chinese Coal Mines has shown the danger associated with the lack of communication infrastructures in an underground environment. The US Mine Safety and Health Administration (MSHA) have introduced legislation that states all Underground Coal Mines must have two-way voice and personnel tracking capabilities by 2009. If Coal Mining companies are being forced to deal with this problem Tunneling Companies will not be far behind.

This paper will examine current practices and best case technology to deal with these issues.

OLDER TECHNOLOGY

Fixed wire telephones using copper wire to hardened phones installed strategically throughout the underground workings. This only allows voice coverage to the points where phones are located and still requires the installation and ongoing expansion of the phone lines.

ULF (Ultra Low Frequency) technology transmits one way paging signals using a loop antenna on surface and frequently with a secondary loop being installed underground. The messaging system delivers a page message to workers that are in the coverage area. The main drawback of the system is that it is a one way signal with no way of confirming if the page was received.

Leaky Feeder is a 30-year-old technology that is almost standard in Hard Rock Mining and has been used in many Tunneling Projects. Leaky Feeder is an acceptable system for radio voice communications within 100 feet of line of sight of the co-axial cable. Leaky Feeder's drawback is the lack of speed (restricted to 9600 baud) for data. These systems have all the negative aspects that normal broadband coax networks have, plus they have the added drawback of being extremely susceptible to noise since they are designed to "pick up" any analog signal within their vicinity.

Today, Tunneling Contractors are looking at an infrastructure that will provide voice, data, video and tracking all on one encompassing backbone underground. Most projects currently use a combination of the above. Leaky Feeder or pager phones for voice, ULF for safety paging, copper or fiber optic cable for gas monitoring, seismic readings, CCTV (Video), ventilation control, TBM Data and PLC/SCADA controls.

SOLUTION

What is currently plugged into the back of your computer? It is not twisted pair or leaky co-axial cable. The solution is using the same technology that all businesses are using on surface for all Data and are beginning to use for Voice (VoIP). A solution for the all-encompassing backbone is Ethernet or a combination of Fiber Optic Cable and Ethernet.

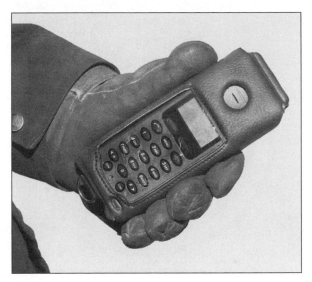

Figure 1. WiFi phone

The Ethernet System uses a backbone consisting of hybrid cable, switches and access points, the same system that all Companies to some extent use on surface to transfer data and control.

By installing a Fiber to Ethernet module the system becomes a wireless network with the addition of switches and access points. The switches also have numerous ports to extend in multiple directions or to hardwire existing equipment to the network such as hard-wired phones, PLC's, fan controls and gas monitors.

ETHERNET STANDARDS

The Ethernet Infrastructure is based on open, worldwide standards for networking. Backed by billions of dollars of conformance testing, along with a 15-year history of excellence the technology has proven itself to be superior in all industries. The System is based on the IEEE 802.3 Ethernet standard along with the IEEE 802.11b wireless standard which combines to offer total conformance and flexibility throughout the entire underground area.

The infrastructure carries all information (voice, video, data) in a digital format which makes it immune to noise, and carries this digital information in *Packets* which provides the system with error control, automatic retransmission of lost or degraded packets, destination addressing, and true multi-tasking which (as is the concept of any properly designed network) allows multiple devices to use the network infrastructure at the same time, as opposed to having to wait until the network is idle.

INFRASTRUCTURE

The IEEE 802.3 Ethernet standard specifies a Star Topology which ensures that every Client (personnel tag, wireless telephone, vehicle being tracked, ventilation fan, etc.) will have equal access to the network, and guaranteed bandwidth (Figure 1). In addition to those benefits this topology perfectly matches an underground drift topology.

Figure 2.   A hardened access point

Everywhere a tunnel splits off, the network can also be split off in several directions without affecting the rest of the Upstream network. This topology makes the it perfect for additions and changes as the tunnel expands since you can simply "Plug and Play" without reengineering the existing infrastructure.

Adding wireless capabilities onto the proven benefits of Ethernet creates unlimited possibilities for Resource and Operation Management. The backbone allows the project to extend all the benefits of Ethernet to your mobile Clients which might include: Personnel, machinery, sensors, etc. Since the Wireless and Ethernet standards are made to work together, you simply connect wireless Base Stations (or Access Points) to your wired network in those areas where you want wireless coverage. That's it! No special interface software or reengineering required.

VOICE

The VoIP phones allow clear digital voice communications over the access points (Figure 2). Minimum range line of sight from an access point is normally 1000 feet in underground conditions. They allow private phone conversations internally in the Tunnel as well as phone conversations outside the Tunnel using your existing Surface Phone Connection. Features include voice mail, all page, as well as built-in tracking capabilities. Each time a worker passes an access point the unique address assigned to the phone is registered with the access point and transmitted to surface.

FREQUENCIES

Most large Tunnel projects take place in an Urban Environment. This becomes a large problem when Tunneling Companies seek to apply to the FCC for licensed frequencies. Radio systems such as Leaky Feeder require licensed frequencies in the 100 or 500 MHz Range for both Surface and Underground areas. These frequencies are becoming very difficult to find and license. Using Voice over Internet Protocol (VoIP) does not require FCC licensing.

PROCESS CONTROL AND FEEDBACK

Vendors have designed industrial quality control units which can interface to everything from ventilation fans to pumps, doors, and other machinery. These control units come with either Analog or Digital (or a combination of both) Inputs and Outputs to turn units ON or OFF as well as provide instant feedback. They can also respond to alarms and take appropriate action to contain emergencies as opposed to being "reactive only." Each control unit is modular, meaning that it starts with the basic microprocessor and Ethernet interface,then the rest is built to the specifications depending on what you are controlling. They use very low power and are properly isolated from your high-powered equipment by using a "logical voltage" to trigger events. Much like a Relay, these units can turn ON a 440 Volt 50 Amp pump by simply toggling a 24VDC logical voltage. Anything that can be manually operated can be activated with a Process Control unit.

VIDEO

The benefits of Ethernet technology really come through when you need multiple control stations monitoring the same section of the Tunnel. Using an Ethernet camera allows many users to view the same live video on their computers simultaneously—even from totally different geographic locations. Also, the cameras themselves have become commodity items with most home computers using a low end Ethernet Camera.

COMMUNICATION WITH THE TBM

One of a Tunnel Contractors single biggest asset is the Tunnel Bore Machine working underground. Ethernet can "enable" your machine to talk to your control room by installing a wireless device onto the machine. Then the TBM can upload all of the status and diagnostic information to the Surface. This not only eliminates the "paper trail" at the end of each shift, but it gives you up-to-date information so that you can make decisions about your machines before they break down.

RESOURCE TRACKING

There is an obvious advantage to knowing where your people, machinery, and other resources are. Every wireless device that talks across the Ethernet network can be located instantly and, if needed, can be continuously tracked as it moves through the Tunnel. Imagine how much quicker, and more accurate it would be if you had an up-to-the-minute database keeping track of all that information for you. Or if you have a "Man Down" and he can't alert you of his location you can find him in a fraction of a second with Tracking Software (Figure 3).

Knowing where your vehicles are will also improve production, and ultimately save money. When vehicles are idle (when they are scheduled to be working) the system will tell you and you can find out why. This allows you to redeploy other machines to keep your production going, and get the maintenance scheduled all in the same minute.

SENSORS

Literally anything can be made to communicate across an Ethernet network— even your Guidance and Ground Control sensing equipment. There is a multitude of Vendors manufacturing sensors for underground applications, and almost all of them

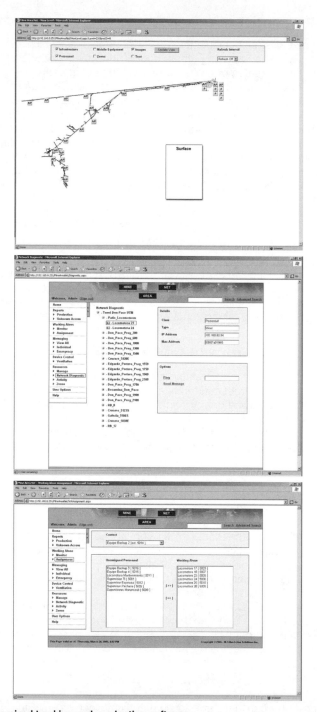

Figure 3.   Customized tracking and production software

record their information in some kind of "Datalogger." Instead of deploying a person underground to periodically take these readings you can connect the Dataloggers to the Ethernet Network and obtain the readings delivered right to your computer in a fraction of a second.

EMERGENCY

To provide instant voice and data communications at the face it is possible to use battery powered portable access points to reach the face without the permanent infrastructure having being installed. The access points also have battery back up in the case of power outage or emergency situations.

CONCLUSION

It is time that Tunneling Contractors look above and see the off the shelf technologies that every other industry is using for voice and data. Tunneling equipment has made great strides in the last 20 years but the infrastructures to support, track and maintain this equipment has not. It is time for reliable, clear communications for the workers underground. There is a clear economic reason for a phone being installed at every desk. Does it not make sense to have the same underground? In case of emergency does the Contractor not want to talk to the workers, and if not able to communicate (if the infrastructure is lost in a fire or an explosion) know their last tagged location in the Tunnel?

REFERENCES

Ben Bjorkman, *State of the Art Underground Communications*
C. Mallett, G. Einickle, P. Glynn, CSIRO Exploration & Mining, *Mine Communications & Information for Real Time Risk Analysis*

LOS ANGELES METRO GOLD LINE EASTSIDE EXTENSION— TUNNEL CONSTRUCTION CASE HISTORY

Brett Robinson

Traylor Frontier Kemper JV

Christophe Bragard

Traylor Frontier Kemper JV

ABSTRACT

This paper reflects the challenges encountered during the construction of the light rail tunnels under Boyle Heights, in Los Angeles, California.

Equipped with two state-of-the-art EPB TBMs, the construction team undertook the mining of twin bored tunnels in the Los Angeles alluvium. The gassy underground conditions dictated many aspects of the design of the tunnels, as well as the tools and means used for their construction. This paper will detail the challenges encountered during construction, as well as the means, methods and logistics used on the project. Discussion includes: EPB TBM machines with some very unique/specific features; complex assembly sequence due to limited shaft access; final precast concrete liner with double gaskets; NATM cross passage excavation in soft ground and below the water table; compensation grouting; mucking shaft arrangement.

INTRODUCTION

The Metro Gold Line Eastside Extension Project (MGLEE) is a six-mile, dual track light rail system, with eight new stations and one station modification. The system originates at Union Station in downtown Los Angeles. East of the Los Angeles River, near 1st and Utah Streets, the alignment transitions to tunnel for approximately 1.7 miles, and then continues beneath 1st Street to underground stations at 1st Street and Boyle Avenue and 1st Street and Soto Street. The alignment returns to the surface near the intersection of 1st and Lorena Streets (Figure 1).

The MGLEE project was awarded to Eastside LRT Constructors (a Joint Venture comprised of Washington Group, Obayashi Corp. & Shimmick Construction Co.) in July 2004, and Traylor—Frontier Kemper Joint Venture (TFK) were awarded the subcontract to construct the tunnels and cross passages.

GEOLOGY AND ALIGNMENT

Geology

The geology for most part was a mixed percentage of sandy clays and clayey sands. At no point did we have pure sand or pure clay. This combination of ground types, was actually quite good. The clayey ground had enough sand in it to prevent the muck from getting to sticky, and we rarely plugged the cutterhead or screws. The sandy ground had enough clay content to give it some body, and didn't allow the sandy ground to pack, or lose all its water content. After traveling through 200ft (61m) of screw conveyors, the muck was very homogeneous and quite "fluffy." We rarely reached the earth pressure levels we had anticipated at bid time.

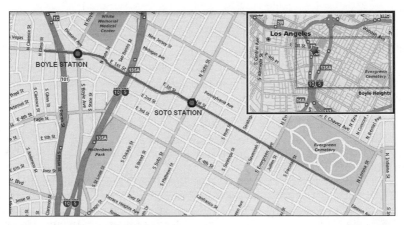

Figure 1.   Location of the tunnel project

The alignment traveled thru some contaminated zones, where muck was segregated and taken to a specific Dump Site for processing. The first zone we encountered happened to be in an are where we needed to change cutters. We tried on three occasions to enter the cutterhead. The first two times, we had to abandon the attempt due to high levels of benzene. Other than this instance, the contaminated ground never played any significant role. As you will read later, we had prepared ourselves well for this.

Alignment

The two tunnels were bored approximately 40ft (12m) apart (center to center). The tunnels were mined in a west to east direction. The south lying tunnel was for the East Bound (EB) train, and the north lying tunnel was for the West Bound (WB) train. For simplicity, the tunnels were refered to as the EB and WB tunnels. To confuse the matter even more, both tunnels were mined from the Boyle Station (an underground station constructed by ELRTC) in an easterly direction to Soto Station (the second underground station constructed by ELRTC), and then further east to the East Portal, where the subway trains will eventually enter to the surface. The first reach of tunnel (from Boyle Station to Soto Station) was approximately 2,850ft (869m), and the second reach (from Boyle Station to East Portal) was approximately 4,125ft (1,257m).

The total alignment contained five major horizontal curves, the tightest being 800ft (244m) radius. The vertical alignment contained five vertical curves. The maximum grade traveling in the direction of TBM mining (East) was 5% at the East Portal, and the maximum grade traveling away from the TBM mining was 2%, which was close to the Boyle Station.

In some places, the lowest cover to the street or in some cases, buildings above the tunnels was 35ft (11m).

For more details about the geology, refer to "Planning and construction of the MGLEE Tunnel" (Elioff et al.).

PRE-GROUTING AND COMPENSATION GROUTING

Chemical Grouting

Various zones along the alignment were specified to be chemically grouted:

- the cross passage zones between the tunnels
- the break-in and break-out zones for the TBMs at the stations
- some settlement sensitive zones: a freeway bridge and a water main.

Cross passages grouting will be discussed in more detail further in this paper.

The freeway and water main grouting was only an extra measure to mitigate for settlement risk.

The break-ins and break-outs proved to be efficiently stabilized and worked very well together with the caution and methodology adopted by the TBM's engineering practice.

Chemical grout was performed by Hayward Baker using conventional methods. A triangular pattern of holes was drilled 5 feet, 7 feet or 8 feet apart, depending on the application and the ground. Then using different port sleeves, grout was injected into the soil in a specific sequence, first using odd ports then even ones and making sure to "push" the water away from the grout instead of trapping it in the heart of the grout zone.

The grout was injected until one of the following conditions was met:

- Design volumes were achieved (these depend on the porosity of the ground)
- The ground did not "take" anymore grout (<0.5 gal/min under a pressure of 2psi/ft of overburden)
- Surface or utility heave was encountered (max ½"). This condition occured only once at a surface street during injection in one of the break-out zones.

Compensation Grouting Set-Up

In order to allow for quick reaction in case of buildings settlements resulting from the TBMs operation, MTA had required the installation of a network of compensation grouting pipes under the buildings that could potentially be affected by the path of the TBMs. Instead of a very complex series of pipes that would be complicated to manage and would tremendously impact traffic and businesses, Hayward Baker simplified the layout from 210 short range drill holes to a series of 33 horizontal drill holes up to 428 lft in length.

Such long horizontal drilling had, according to the authors, never been performed in the Western United States. Such achievement was even more challenging since the holes did not surface on both sides.

The 6" holes were drilled using a Ditch Witch horizontal drilling system with bentonite. On the surface an HDD tracking unit monitored and steered the bore hole. Cement/bentonite slurry was injected in the hole as the drill was pulled back. Then a 2" schedule 80 PVC or steel pipe (both were used) was pushed into the hole. Because the holes were blind drilled, there was no possibility of pulling the pipe from the other side. The pipes were provided with sleeve ports every 2 to 2.5 feet. Using the primary ports, the ground was preconditioned (also called soilfrac): grout was injected in the soil until either theoretical volume was attained or movement of the buildings was detected. The secondary ports remained free in order to limit and counteract any potential settlements that might occur during the TBMs' operations.

Survey of Buildings

For monitoring of the buildings, Hayward Baker also took the challenge and set up a real time automatic survey system. Small reflectors were installed on the roof corners of the buildings. A total station instrument was programmed to measure those reflectors constantly. The information was then communicated to the network and analyzed with only a couple of minutes delay by their engineers.

In order to understand the movements properly, a pre-study was conducted and corrections were made in order to take into account temperature variations that affected both the position of the total station and of the reflectors. Some smaller factors such as weather changes were not filtered but could be understood by the field engineers. A grouting crew was always available in case compensation injections were needed.

Settlement tendencies were noticed but none presented values above the minimum action level of 0.2 inches. The values were so small in fact, that only by means of a graphical representation of the position on the reflectors over a longer period of time (a few days) could some slight movement actually be seen.

INITIAL PLANNING AND INTERFACE ISSUES

As a subcontractor only in charge of tunnel construction, we (TFK) were required to interface with ELRTC on general construction activities on the complete project. The major clashes were in the stations, where ELRTC wanted to get as much work as possible completed before we commenced TBM installation, but also wanted to be sure not to hold us up.

In the initial scheme of work, at both Boyle and Soto Stations, ELRTC would construct the station ground support and TBM thrust reaction foundation, we would come along and mine the tunnel, and then afterwards, ELRTC would commence final concrete work and fit out of the station. ELRTC were very keen to install some of the final concrete before we started mining, so that they could gain some time on their schedule. The process was an iterative one, with give and take on both sides.

Boyle Station

Installing the final invert ahead of our operations, provided a couple of advantages to ELRTC in addition to the gain in schedule. Installing the final invert before we started, saved ELRTC from having to install a 2ft (0.6m) thick reinforced sub-invert slab throughout the 400ft (122m) long, 50ft (15m) wide excavation, the purpose of which was to support the toe of the shoring. Also, having a final invert eliminated ELRTC's need for a special purpose thrust reaction foundation for the TBM. For TFK, the final invert allowed the TBM Trailing Gear to be close to the correct elevation for mining, without the need for elevated track. If the original sub-invert were only provided, the TBM Trailing Gear for each TBM would have had to be elevated almost 56in (1.4m). The down-side to TFK with the final invert plan, was that the TBM Trailing Gear needed to start-up in a "snakey" fashion, twisting in a 700ft (213m) radius S-curve to avoid the rebar and concrete of the final station walls.

Due to some minor details that concerned both parties, it was decided to install the final invert 12" (0.3m) low, which coincided exactly with track level in the tunnel. The final 12" of concrete, with drains and pedestals for track, would be installed later.

In order for ELRTC to gain as much as possible on schedule, an additional access shaft was installed at the East "hammerhead" of the station (the enlarged area where the TBM shields were assembled and launched). Initially, ELRTC was going to construct a small shaft and join it to the main excavation, but this did not allow TFK sufficient space to install the TBM shields, which was a necessity if the final concrete were

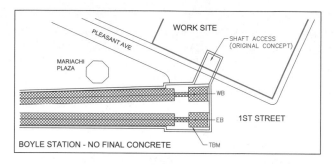

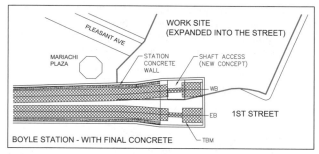

Figure 2.   Original plan versus the new adopted start-up plan for the TBMs in the Boyle Station

poured, as the final concreted station would not have provided enough room to travel the shield pieces through the station. TFK came up with a solution to expand the work site into the street so that street decking could be removed providing an opening large enough for TBM assembly. This solution required cooperation from LACMTA, as well as the community and other City agencies. The team was able to make this happen, and a new shaft was created (Figure 2).

Installing the final invert in time before the TBM arrived appeared to be achievable. The concrete walls and first level ceiling were a little in doubt. Either way, TFK had devised a means to start-up the TBMs if all the concrete were installed, just in case ELRTC was able to achieve getting that far. In the end though, ELRTC was only able to pour the invert (Figure 3).

Soto Station

The need for final concrete at Soto Station, before we reached there with the TBMs, was actually the first issue we discussed with ELRTC, and solutions developed at Soto Station were adopted at Boyle Station.

We developed a plan similar to the plan at Boyle Station, which allowed ELRTC to pour the final concrete invert, 12" (0.3m) low for the entire station, except the West hammerhead sections where the TBMs entered the station, and at the East end where TBMs started mining again.

Final concrete walls could be poured, but not the ceiling, as headroom was required for the TBMs to travel through the station.

The advantages for both parties were the same as they were at Boyle Station. For TFK, the final invert meant there did not need to be any elevated track section through the station once the TBM had traveled thru, but this came with some payoff. Like with Boyle Station, the TBM Trailing Gear would need to start in a tight S-curve, when mining

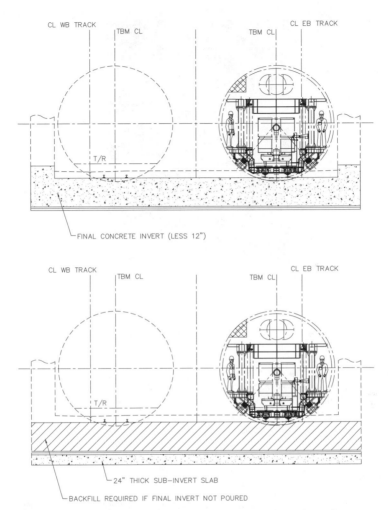

Figure 3.   Difference of plan versus the new adopted invert layout in Boyle station

commenced. But the most significant obstacle was that the TBM shields, once they broke thru into the station, had to be lifted 30in (0.76m) before they could be jacked to the far side of the Station for startup, and then, once on the other side, the TBM shields had to be lowered 30in (0.76m). This was referred to the "Soto Hop" (Figure 4).

Like Boyle Station, ELRTC was only able to install the invert.

Work Site Layout

The deal to allow ELRTC to proceed with pouring the final invert, came with costs as well as benefits to TFK. TFK agreed to delay their schedule 19 calendar days to give ELRTC the necessary time to complete concrete work, and TFK agreed to absorb the overhead costs for this delay. In turn, ELRTC allowed TFK to operate their tunnel activities out of a second work site which included the new created shaft.

The main work shaft, Bodie Shaft, (named after an adjacent street) was used to install the TBM Trailing Gear and then afterwards, for the removal of muck during mining.

Figure 4. Photograph showing the jacking of the 500-ton WB TBM up 30 inches (0.76m)

The new created shaft, Bailey Shaft (also named after an adjacent street) was used to install the TBM shield components and then afterwards, for tunnel liner segment and utility handling. The grout batchplant was assembled on the new work site to divert all grout material trucks away from the mucking activities.

The realization of the benefits of splitting the tunnel mucking from the rest of the materials handling activities is better realized in the next section. One major benefit was that it eliminated many safety hazards. The other benefit was that it allowed the mucking operation to have it's own dedicated work area. Trucks, for mucking, potentially needed every 3 minutes, now did not have to negotiate around other trucks arriving with deliveries.

MUCKING SYSTEM

The tunnel is basically a factory. There are many driving factors in the tunneling process that dictate the progress of tunnel mining. In order to optimize the tunnel mining rate, the individual processes must be analyzed separately to ensure they will meet the mining requirements.

What sets the limits for tunnel mining rate? One could say the TBM speed, but then on the other hand, if you can't get rid of the muck produced by the TBM, then this will essentially dictate the mining rate.

All of the relevant factors must be analyzed in order to fully plan and prepare the job.

But simply, the TBM mines, it generates muck that first needs to get out of the tunnel, and then the spoil needs to be removed from the site.

On most jobs you hope you can buy a TBM that goes as fast as possible, and then everything else is sized to work at least as fast. Getting muck removed from the tunnel and then from the job however, turned out to be the driving force that would dictate the tunnel progress on this job. The most critical process on the job was the off haul by trucks.

To meet our schedule, we were required to attain an average mining rate of 50ft per day per tunnel. This is the average rate, so some days we might have 0ft mined

tunnel diameter	mine rate	volume of muck		weight of muck	truck loads	trucks/hr required	truck loading time required	truck loading time required
		(1 tunnel)	(2 tunnels)	(150pcf)	(20ton)	(in 12 hrs)	(1 loader)	(2 loaders)
ft	ft/day	bank cft	bank cft	tons	trucks	trucks/hr	minutes	minutes
21.5	100	36,305	72,610	5,446	272	23	2.6	5.3

Figure 5.   Summary of major criteria for muck removal

and some days we might have 100ft mined. We needed to be sure we could cope with the peak (100ft per day/tunnel), to meet our average rate (Figure 5).

Some other factors that affected muck removal were the Dump Sites themselves, and also contaminated muck. Some Dump Sites were only open at certain times, and some Dump Sites were selective about what muck they would accept. Muck from an EPB TBM can often be sloppy. Also, the tunnel alignment passed through some contaminated ground. This contaminated muck needed to be segregated from the other tunnel's muck. A two bay, large Muck Bin was planned for site, which would provide the means to load two trucks at a time, but would also allow muck segregation.

The next critical process was the muck haul out of the tunnel. It is simple enough to figure out cycle times for trains to ensure the train is able to travel out, drop off full cars, pick up new empty cars, and get back to the TBM by the time a ring build is complete, so that the TBM is able to produce the peak 100ft per day. The complication comes when you add another tunnel to the picture, and must funnel all of the train traffic and unloading of trains into one place.

The process of removing muck from the tunnel consists of two basic processes:

1. Muck car dumping—The full muck cars are cycled around and emptied, usually by a crane lifting them and dumping the contents in a temporary muck storage bin on the surface.

2. Train travel—The train travels from TBM to shaft (or portal) where full boxes are exchanged for empty boxes, and then the train travels back to the TBM.

Each process had to meet the 100ft/day/tunnel requirement.

For the dumping process, there were potentially 7 cars for every 5ft of tunnel mined that required cycling, lifting, and dumping. For two tunnels, that meant there were 280 muck boxes that needed to be cycled per day. For two 8hr shifts, that equates to dumping 18 muck boxes per hour (one every 3.3 minutes). To empty the muck boxes in this efficient a manner, three things had to work well:

1. the exchange of full cars with empty cars had to be quick

2. the indexing of full cars to the mucking crane had to be quick

3. the connection of the crane to the muck box, lifting, dumping, and returning the empty muck box to the bottom had to be quick

Plans were made to install an elaborate, but efficient track system in the shaft to enable quick switching of trains from track to track. The track system centralized the actuation of the points of 11 switches, so that this could be done quickly and safely (Figure 6). This took care of No. 1.

An automated car indexing system designed and built by Construction Tunneling Services (CTS) would take full muck cars, one at a time to the crane loading area, and then return the empty muck car (Figure 7). This took care of No. 2.

A self engaging and disengaging bridle system, the "Fikse Grab," which has been used previously on Traylor tunnel jobs was utilized. This system is quick, reliable and

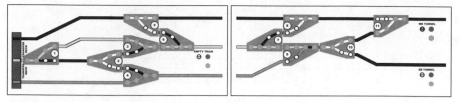

Figure 6. Control console for the main switch yard

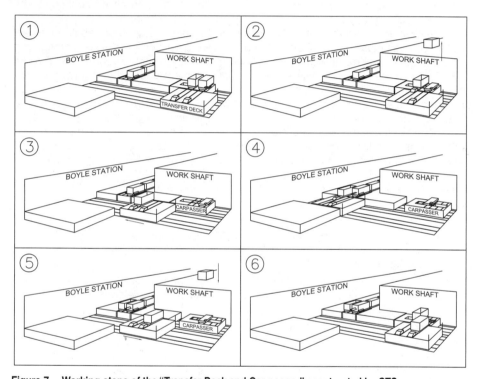

Figure 7. Working steps of the "Transfer Deck and Car-passer" constructed by CTS

safe. A new Liebherr HS 895 HD Crane was used as the lift crane. These both took care of No. 3.

Train travel was just as important. Grades in the tunnel created two problems:

- A 40ton locomotive may not have the weight to pull the full train up the tunnel, and wheels may slip.

- Locomotives for the tunnel were required to meet special requirements of MSHA (Mine Safety & Health Administration), and the only engines approved for use on new locomotives were low on horse power. Train speed therefore becomes a factor.

To overcome these problems, plans were made to ensure the rolling resistance on the muck cars was minimized, by installing all new bearings on wheels, and locomotives were configured in such a way that they could be joined together in tandem.

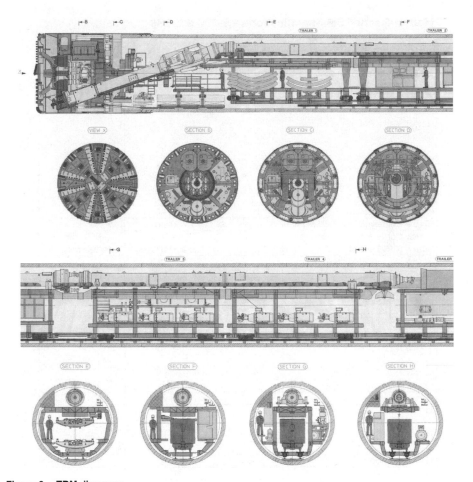

Figure 8. TBM diagrams

TUNNEL BORING MACHINES

Traylor's previous tunnel project in Los Angeles, the North East Interceptor Sewer (NEIS) was nearing completion when Tunnel Boring Machines (TBMs) for MGLEE's were selected. After a detailed review of TBM proposals, Herrenknecht USA received the order for two Earth Pressure Balance (EPB) TBMs (Figures 8 and 9).

Experiences from NEIS influenced the design greatly, and consequently, ventilation, TBM Muck Handling and the Backfill Grout System were all taken to a new level.

Ventilation

After experiencing numerous tunnel evacuations on NEIS due to gas, a major goal for MGLEE was to aim for "no tunnel evacuations." A tunnel evacuation due to gas is a dangerous circumstance, and also a costly one, as a day or so of production is usually lost to the job, waiting on OSHA investigations and corrective actions.

For MGLEE, being a "gassy" tunnel like NEIS, ventilation needed to be seriously analyzed. In most cases, a TBM is designed to fit operational systems and one of the

Herrenknecht TBM Specifications - S-297 & S-298

Description	Imperial Units	Metric Units
TBM Cutterhead Diameter	257.6 in	6544 mm
	21.5 ft	
Cutter Drive Type	hydraulic	
Cutterhead Torque	3,208,000 lb.ft	4377 kNm
Cutterhead RPM range	0 to 3 rpm	
Cutter Type	double disc monoblock	
Cutter Size	17 in	432 mm
No. Cutters	18 no.	
No. Openings	8 large	
	8 small	
No. Injection Ports	6 no.	
No. Wear Detectors	3 no.	
Screw Conveyor Qty	3 no.	
Screw Conveyor Diameter	35.4 in	900 mm
Screw No.1 Length	46.6 ft	14200 mm
Screw No.1 Inclination Angle	18 deg	
Screw No.1 Drive Type	external - hydraulic	
Screw No.1 Drive Torque	150,000 lb.ft	200 kNm
Screw No.1 Speed	0-8 rpm	
Screw No.2 Length	74.5 ft	22700 mm
Screw No.2 Drive Type	external - hydraulic	
Screw No.2 Drive Torque	150,000 lb.ft	200 kNm
Screw No.2 Speed	0-8 rpm	
Screw No.3 Length	81.0 ft	24700 mm
Screw No.3 Drive Type	rear - hydraulic	
Screw No.3 Drive Torque	150,000 lb.ft	200 kNm
Screw No.3 Speed	0-8 rpm	
TBM Shield Diameter	256.5 in	6514 mm
	21.4 ft	
TBM Shield Length	30.3 ft	9230 mm
TBM Weight	500 ton	454 tonne
TBM Total Length	394 ft	120 m
Thrust Jack Qty	16 pairs	
Thrust per Jack Shoe (Jack Pair)	294 ton	267 tonne
Total Thrust	4714 ton	4277 tonne
Thrust Pressure Groups	6 groups	
Thrust Jack Stroke	86.6 mm	2200 mm
Stroke Measuring Device Type	temposonic	
Stroke Measuring Device Qty	4 no.	
Tunnel Liner Width	60 in	1524 mm
Articulation Type	active	
No. Articulation Jacks	12 no.	
Articulation Pressure Groups	3 no.	
Stroke Measuring Device Type	temposonic	
Stroke Measuring Device Qty	4 no.	
Erector Type	ring	
Segment Pickup Mechanism	vacuum	

Figure 9.   TBM specifications

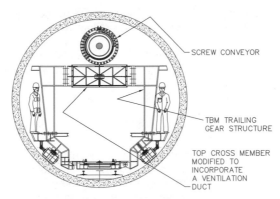

Figure 10.   Auxiliary ventilation system on the TBM trailing gear

last considerations is the installation of the auxiliary duct. The contract required the tunnel ventilation capacity to be 100,000cfm (2,830m^3/min). 72in (1.8m) rigid steel duct was used in the tunnel, and a "bomb-bay" (or "clam shell") was designed to load 30ft (9.1m) duct lengths at the end of the TBM Trailing Gear. The large duct, although an expensive investment up front, required only one 200HP vent fan to provide the required flow for the complete tunnel, so power savings were significant. The other advantage of the large duct was that it provided the means to increase ventilation flow if required.

The TBM Trailing Gear was configured to accommodate 100,000cfm up to the point where muck was discharged (approximately half way), and then 50,000cfm (1416m^3/min) for the zone from the muck discharge to the TBM shield. This was not so difficult to accommodate for the rear half of the TBM Trailing Gear, but up front, the Trailing Gear was more congested with a screw conveyor, hydraulic systems, and the like.

Herrenknecht designed a special cross section for the forward section of the TBM Trailing Gear decks which turned the top cross member of the decks into a duct (Figure 10).

TBM Mucking System

As mentioned, EPB machines were selected. One of the biggest headaches when mining an EPB tunnel is handling muck, or managing the transfer of muck from the screw conveyor to the belt conveyor. In San Diego, on the South Bay Ocean Outfall (SBOO), we used a continuous screw all the way to the discharge. SBOO had its own challenges due to extremely high pressures and difficult ground, but the tunnel alignment was straight. On NEIS, we felt a bit better about the ground conditions, and initially planned to do the same as SBOO and use a long screw, but voted against it due to the curves in the tunnel. On NEIS we had a second screw, but still there was a transfer point, and in hindsight, we felt pehaps that the long screw may have been the better option after all.

So, along comes MGLEE, with curves, gas, and low ground cover. A job that certainly would not tolerate any ground support failures in light of previous Metro tunnel projects.

We used the experience gathered from the Mitsubishi TBM on SBOO, along with the experience we gathered from Lovat TBMs on local ground conditions on NEIS, and worked with Herrenknecht to develop a long screw conveyor concept. The size of screw, the operational speeds and power of the screw, as well as the drive concept, were developed jointly.

A system of three interconnected, individual screws were chosen. On SBOO we had often jammed the rear long screw, so we felt it may be wise to split this into two. The outlet of the first screw fed directly into and second screw, and the outlet of the second fed directly into the third screw. This was something Herrenknecht had not done before.

All screws were the shafted type. On SBOO we had a ribbon and a then a shaft which created numerous headaches when we reached boulders, as the ribbon could digest anything, but the shaft was more selective (Jatczak et al.)

With this system, muck could be contained the whole way until it was discharged into the muck box, 200ft (61m) from the mined face. This reduced muck spillage as there was no transfer to a belt conveyor, plus, there was no belt conveyor. The safe part about the system was that gas, if trapped within the mined muck, was contained until it reached the discharge point, and at this discharge point, we had 100,000cfm of ventilation.

The tunnel was never evacuated due to gas for the entire time we were mining with the TBMs.

Backfill Grout System

The Los Angeles Metro Transit Authority (LAMTA) had specified we use "Tail Shield Grouting." The specification was written in a manner that was conducive to "mortar-type" grouts. It required us to grout using 6 ports thru the Tail Shield.

We had fantastic success with an "accelerated-type" grout on NEIS injected through the tunnel liner, but this system, unlike the "mortar-type" required a special back-flushing system if the grout were injected through the Tail Shield. The "accelerated-type" grout technology we had adopted, was based on the backfill grout technology used on numerous projects in Japan. In Japan, grout is typically injected only in the upper part of the shield, typically at the 2 o'clock and 10 o'clock location.

Some important positive factors about the "accelerated-type" grout:

- It requires much less energy to force into the void, which puts far less strain on the segments. The "accelerated-type" grout is much more liquid and can travel and penetrate the void space more effectively, and with much less pressure.

- Since the pumping pressure is less, and also since the pumps used are smooth flow, the control of pumping pressure is much more effective compared to mortar grouts, which require piston type pumps that operate at high pressure.

- The grout, since its sets quickly, is also able to support the tunnel liner almost immediately. With some mortar grouts, the set time can sometimes be so slow that the tunnel lining can later move or distort.

One negative of this system is that it requires some rather sophisticated controls, and sophisticated systems to keep it working and cleaned. However, most of the difficult learning with this grout system occurred on NEIS.

Negotiations were made with the LAMTA to veer away from what was written in the contract, upon the understanding that the alternative system (2 injection ports only) had some advantages, but it had to be proven effective. Grout volumes and pressures would be recorded in "real time," and could be viewed throughout the job by operators, inspectors, and management. Check holes could also be drilled in the segments at agreed locations to physically inspect whether the void was full.

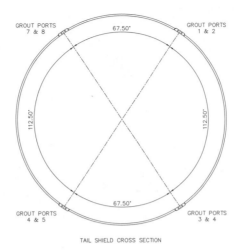

TAIL SHIELD CROSS SECTION

Figure 11.   Grout ports location

As mentioned, the "accelerator-type" grout, when injected through the Tail Shield, requires a back-flushing system. The accelerated grout, if not cleaned out of the injection ports, will setup, and the port will block. Our accelerated grout sets within 10 to 15 seconds, so it is important to have a quick and reliable system to clean out the injection pipes.

The back-flushing concept we incorporated was copied from what is typically done in Japan on TBMs that require Tail Shield Grouting. In Japan, the TBMs Tail Shield are built with two bulges on the top (at 2 o'clock and 10 o'clock) to house the back-flushing systems for each port. The shield has special static ripper teeth placed up near the cutterhead which plow a clear path for the bulges. We took this system that is normally housed within the bulges, re-engineered it, and directed Herrenknecht to install it within the Tail Shield skin—and have no bulges (or nearly none). The Tail Shield skin was 1" (25mm) thicker locally where the back-flushing was located. The thicker skin locally was not greater than the over cut of the cutters, so no special ripper teeth were required.

The Tail Shield was fitted with a pair of grout ports in four locations (Figure 11). The pair of ports provided for an operational port, and a backup port if the operational one were blocked. The ports in the lower part of the Tail Shield were not intended to be used, but were incorporated to satisfy the requirement of the LACMTA specification.

The grout consists of two parts; a stabilized cementitious part A, and an accelerator part B. These two parts are pumped in a controlled manner to a mixing nozzle which is located just before the grout ports housed in the Tail Shield.

The grout port actually consists of three ports that all join at the exit point. In the center port, there is a piston/poppet that is extended and retracted. On one side is the grout port, where accelerated grout is pumped. On the other side, is a high pressure water line that supplies the back-flushing water. When the poppet is in the "retracted" position, the grout port is open to the grout void, grouting occurs, and the high pressure water port is blocked. When grouting is complete, the poppet extends, blocks the port to the grout void, and at the same time connects up the high pressure water port with the grout port, and the grout port is back-flushed with high pressure water. The sequencing of all this is controlled automatically by a PLC. As well as this, the pumping

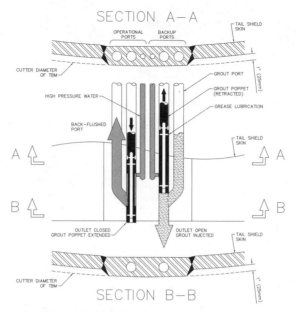

Figure 12.   Configuration of the grout "back-flushing" system

is controlled by a PLC which maintains the highest flow rate of grout possible within the required pressure limits (Figure 12).

SCHEDULE AND PRODUCTION

Schedule

The contract for MGLEE was awarded on June 1, 2004. TFK was immediately awarded the subcontract for the tunnel portion of the project, and planning began (Figure 13).

Additional Delays

After agreeing early on in the planning stages to delay TBM delivery 19 calendar days, TBM delivery was delayed a further 25 calendar days by ELRTC, to provide them enough time to complete concrete work in the Boyle Station. It was taking longer than planned. In addition to this, shoring loads were higher than had been anticipated, and additional struts were required. TFK incurred an additional delay of 52 calendar days, but the TBM had already left Herrenknecht AG in Germany, and there was no stopping it. The total delay to TFK was 96 calendar days, 19 of which TFK had agreed to absorb the overhead costs.

TBM Assembly and TBM Start-Up

On September 21, 2005, both TBMs arrived on the same ship at the Port of Los Angeles and shipment to the work sites commenced. Due to the delay, a third site was used to take care of the overflow.

Metro Gold Line Eastside Extension

TIMELINE OF TUNNEL ACTIVITIES

Activity Description	Start Date	Finish Date	2004	2005	2006	2007
Notice to Proceed	06/01/04					
TBM Assembly & Delivery to Site	07/21/04	09/21/05				
Site Establishment	03/01/05	09/06/05				
Delay - by ELRTC	09/06/05	11/21/05				
Assemble and Test TBM's	11/21/05	04/27/06				
Mine EB Tunnel	02/23/06	11/14/06				
Mine WB Tunnel	04/27/06	12/09/06				
Remove TBMs at East Portal	11/14/06	01/26/07				
Construct Cross Passages and Sumps	10/01/06	06/07/07				
Final Concrete	01/28/07	06/21/07				
Final Completion		08/24/07				

Figure 13.   Basic summary schedule of key activities

TBM Trailing Gear decks were pre-assembled on the surface and lowered into the main work shaft (Bodie Shaft) by a Liebherr 895 Mucking Crane. There were 8 decks per TBM and the heaviest deck was approximately 150,000lb (68 tonne). In the bottom of the Bodie shaft, the Transfer Deck and Carpasser (later to be used for mucking) were joined together to make a 40ft long traveling car. Each TBM Trailing Gear deck was lowered on to the traveling car and then moved approximately 70ft (21m) to align with tracks located in the station. The decks where pushed off by a locomotive.

We were given access to the Bodie Shaft a little earlier than the Bailey Shaft, so TBM Trailing Gear was almost completely installed in the station, for both TBMs when we finally received access to the Bailey Shaft on November 22, 2005. Bailey Shaft consisted of a 20ft × 24ft (6.1m × 7.3m) access hole in the street decking. The station excavation ran east-west, with the EB TBM being on the south side and WB TBM being on the north side. The access hole (Bailey Shaft) was on the north side, so the EB TBM shield had to be built first, slid to the south and then joined to the TBM Trailing Gear waiting behind it in the station.

The TBM shield weighed approximately 500ton (454tonne) once completely assembled. Cradles mounted on Hilman rollers facilitated in the mating of each shield component in an east-west direction, and then the rollers were rotated, to allow the assembled TBM to move south. The Thrust Frame was also lowered and attached to a cradle, and this helped support the No.1 Screw as the EB TBM traveled south. For the WB TBM, the process was almost the same, although no traveling north or south was required once the complete TBM was assembled.

Each TBM shield took 6 working days to complete, working two 10 hour shifts. The heaviest shield section was the Forward Shield which weighed approximately 293,000lb (133tonne).

Production

The EB TBM started mining on Feb 23, 2006, approximately 16 calendar days late. The WB TBM started mining on April 27, 2006, approximately 52 calendar days late (Figure 14).

Following are some milestone dates for each TBM at Soto Station and the East Portal.

- On June 10, 2006, we stopped mining for a week to install the switch yard in Boyle Station.
- EB TBM "Break Thru" at Soto Station on July 21, 2006
  WB TBM "Break Thru" at the Soto Station on August 24, 2006.

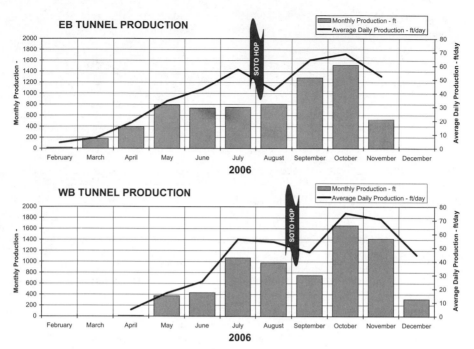

Figure 14.	Monthly summary of TBM production

- EB TBM "Start-Up" at Soto Station on August 7, 2006. "Soto Hop" took 5 work days, and maintenance took 5 work days.
  WB TBM Start-Up" at Soto Station on September 8, 2006. "Soto Hop" took 5 work days, and maintenance took 5 work days.
- EB TBM "Break Thru" at East Portal on November 14, 2006. 11 days late.
  WB TBM "Break Thru" at East Portal on December 9, 2006. 3 days late.
- Best day (2 × 10 hour shifts) on the EB tunnel was 90.5ft (27.6m).
  Best day (2 × 10 hour shifts) on the WB tunnel was 95.2ft (29.0m).
- Best week on the EB tunnel was 376.3ft (114.7m)
  Best week on the WB tunnel was 425.4ft (129.7m)

Mining on the EB TBM gained 5 days on schedule, and WB TBM gained 49 days on schedule.

Two 8 hour shifts were used on both TBMs up until July 10, 2006. From that date forward, two 10 hour shifts were used on both TBMs.

Cutters and Cutter Changes

Cutters used were manufactured by Herrenknecht USA (HK-USA) and Construction Tunneling Service (CTS). Cutters were monoblock, 17" (432mm) double disc with tungsten carbide button inserts on the rings as well as hardfacing.

The HK-USA cutters were dressed on the EB TBM and the CTS cutters were dressed on the WB TBM. The CTS cutters seemed to perform marginally better than the HK-USA cutters, but it is very difficult to analyze this, since the WB TBM traveled faster. Both cutter designs were far superior to what we had used in the past. In fact, in the second reach of tunnel from Soto Station to East Portal, the cutters lasted the entire 4125ft (1257m) length of the tunnel for both TBMs.

On only one occasion did we need to enter the cutterhead to replace cutters. Early on, approximately 300ft into mining, the EB TBM hit some heavily cemented Alluvium material and TBM penetration was extremely slow. The ground consisted of boulders cemented in a very hard sandy/clayey matrix. Approximately 500ft through this zone, we entered the cutterhead and changed out cutters. The lower part of the face was extremely hard and the top of the face was extremely soft. The hard section in the bottom of the face needed to be spaded with a jack hammer, as it resembled a weak concrete. The top you could easily pick away with your fingers.

Although the cutters were not completely worn, they had worn remarkably for such a short distance, so we changed out everything (including bucket lips and bucket teeth) to new. The extent of the zone, or for that matter, the existence of the zone, seemed difficult to pin point when looking at the boring logs and even when a visit to the lab to view the bore samples did not help us understand the situation. Also, the tunnel was decending deeper and water seepage was already evident when we entered the cutterhead. We had no choice but to make the cutterhead as new as it could be, as the tough ground may have been with us for a while, and we certainly didn't want to have to use our compressed air facilities. As it turned out, we had only 200ft of the hard ground left. We persisted through this zone for approximately 700ft in all, which caused us to fall further behind on schedule.

The WB TBM, however, managed to get through this zone a little quicker, which is one reason why it gained so much time on the EB TBM.

SITE NETWORKING OF INFORMATION

Typical now on most tunnel projects that Traylor manages is an elaborate network of communications between surface and tunnel equipment, with links to remote offices and Engineers homes. All is done with fiber optic, IP addresses, Internet, VPNs and the like.

On the Los Angeles Metro tunnel project, the project has a network that links all operator interfaces, office computers, printers and PLC's (Programmable Logic Controller's) together. The network starts at each tunnel, links to surface equipment and then connects to office networks on site, to the Main Office (approximately 3 miles away), to multiple MTA offices (up to 2 miles away) to Engineers laptops at home or abroad (in some cases 100+ miles away), and also to firms overseas.

On each TBM, there are 4 operator interfaces (industrial PC's with touch screens) and 2 PLC's. On the surface and in the shaft there is the Grout Batchplant, which has 1 PLC and two operator interfaces. There is a TBM Cutterhead Foam Batchplant, which has 1 PLC and 2 operator interfaces, there is the automated Mucking Transfer Deck which has a PLC and operator interface, and there is a Water Cooling Tower that includes a PLC.

The site office has 2 servers and 4 desktops, for data monitoring and recording. There are 4 remote desktops in MTA offices used for data monitoring and recording. The Traylor Joint Venture Main Office has 2 servers and 2 desktops dedicated for data monitoring and recording.

The network is relatively simple. On site, the tunnels are connected to the surface via a fiber optic cable. From the site, VPN's (Virtual Private Networks) link the site to the remote locations using the internet. This technology is not new. The most expensive cost is the fiber optic cable. File transfers from office computers on site to the TBM computers are extremely fast. This has saved an enormous amount of time when new programs are added, or reloading of software is required. All can be done through the network, and no floppy drives or CD players are required down in the tunnel.

This elaborate system may appear at first glance to be overkill, but when one looks at the ease with which the job can be monitored and managed from more than one location, it provides a means for many people to know what is happening, and provides a means for many people to gain access to areas to troubleshoot, program, or teach operators how things work remotely.

For example, having all these computers linked via Ethernet, allows an Engineer at home, late at night, to log in, dock into a computer that an operator is using, and help that operator with a problem. Or, if there is a programming glitch, the Engineer can log in and re-program from home. The alternative is that the job stops, waits for the Engineer to drive to the site, travel to the TBM and then solve the problem. When TBMs start up, there are always many problems to solve. With our system, the problem can usually be solved in minutes.

A telephone line is carried over the fiber optic cable to allow the operator to dial out, or for someone to dial in.

CROSS PASSAGES

Ground Conditions

At bid time, TFK had envisioned to excavating the cross-passages using classic excavation methods. The ground would be chemically grouted (per contract) and the excavation would be preformed with standard means (ribs and lagging system) (Figures 15 and 16).

A closer look at the soil borings and some additional soils investigation showed that the ground was not ideal for this excavation method. The geology consists of an accumulation of non-homogeneous sediment layers that vary from medium dense clays to lose silts and sands. The layer thicknesses vary from 1 ft to 20 ft, making field analysis quite difficult. A lot of the soils showed to be non groutable, yet non clayey, meaning that TFK could not rely 100% on the chemical grout or cohesion of the ground for stability of the soils. Most cross passages are located under the water table, and up to 1.5 bar of hydrostatic pressure is expected in the excavation. For environmental reasons no major dewatering system is allowed on this portion of the work.

Both TFK and MTA wanted to avoid high risk excavation for various reasons with safety of the public and of the workers being the major one. TFK decided to consult Austrian based Beton Und Monierbau (BEMO), a decision which MTA later supported. NATM excavation is BEMO's field of expertise. BEMO's experience on similar cross passages projects in soft ground using NATM methods had proven to be safe and efficient.

Grouting Zones

TFK also expressed concern about the size of the chemically grouted zones around the cross passages and the possibility of uplift and bulking. TFK and MTA decided jointly to re-design each grouting zone individually. Extra borings were performed and a couple of pressure meter test were undertaken in similar chemical grout plugs to determine the E-modulus of the grouted soil (the break-ins and break-out for the TBMs at the stations had already been chemically grouted). This E-modulus would be key in determining the time response of the grouted ground and therefore the shotcrete strength development needed to stop those movements. Together with Hayward Baker input and analysis, BEMO and TFK analyzed each cross passage separately as

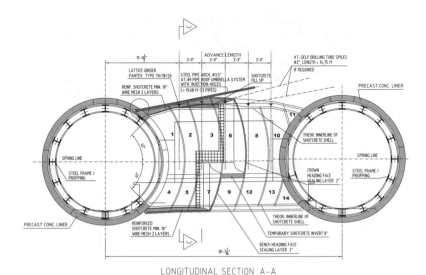

LONGITUDINAL SECTION A-A

Figure 15. Cross passage sequential excavation method

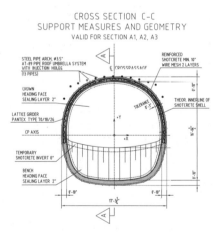

CROSS SECTION C-C
SUPPORT MEASURES AND GEOMETRY
VALID FOR SECTION A1, A2, A3

Figure 16. Cross section of cross passage excavation

best they could, determining which layers would be groutable and how far from the excavation profile the ground should be grouted.

It appeared that many layers would be hardly groutable, some being clayey but others being silty and non-coherent which would lead to challenges in ensuring ground stability.

Chemical grout was performed by Hayward Baker with conventional methods.

NATM Design

BEMO developed an NATM design to best fit the 6 cross passages along the alignment. Two different egg-shaped cross sections (due to different cross passages sizes) were developed.

To determine the support required, the ground loads were simulated in a 3D model at different time steps of the process taking into account a layered simulated ground structure. Heavy "squirrel" cages (each 70,000 lbs) were installed in the tunnel to transfer the loads of the opening to adjacent segments. The excavation support in the cross passages consisted of 10 to 12 inches of reinforced shotcrete applied in a specific sequence (Fiber shotcrete was not used due to the heavy loads expected).

Aside from design calculated support, extra measures were taken to ensure a safe excavation such as a system of pipe umbrella, lattice girders and optional spiles and face bolts.

The work proceeded in a specific sequence:

- Probe holes drilled on both sides of the excavation to ensure ground quality and to potentially drain the ground around the excavation.
- Installation of a pipe umbrella consisting of 11 heavy steel pipes, 20 ft long and 3.5" in diameter.
- Installation of squirrel cages in both tunnels
- Removal of the segments with concrete chain saws and jack hammer
- Sequential excavation in 3'-9" increments, top-heading, bottom-bench until reaching the other tunnel.
- Removal of the segments in the other tunnel

Dry shotcrete was chosen for various reasons, a few being that shotcrete would always be available if needed, dry shotcrete has a flashing property that can help stabilize wet ground faster, logistics and costs (only about 800 CY of shotcrete total). Shotcrete is directly applied to the excavated ground, including a sealing layer on the excavation face and a temporary invert during the top excavation.

Equipment

Once again, gassy conditions dictated the selection of the equipment for this task.

TFK possessed several hydraulic drills and proceeded to do some testing for the probe holes. The standard Symmetrix-T drill bits with an hydraulic rock drill were proven to be most efficient, although TFK designed an alternative system for more clayey zones if needed.

Hydraulic chain saws for concrete segment removal, although sometimes costly proved to be most flexible and easiest to use.

A special electric excavator (the Butor) was purchased from Europe and modified for use in Class1-Div2 environment. The excavator is small in size but has adequate power. Also, the boom is provided with a knuckle that allows fine detailing of the excavation. A standard Cat mini excavator was also used, which proved to be more flexible and more "operator friendly."

The shotcrete installation consisted of REED shotcrete pumps with air motors that were fed with bulkbags of shotcrete through a hopper. The material was then projected through a two inch hose with a hydronozzle assembly system and a straight fiberglass nozzle tip. The material was pre-dampened at the first water ring, flowed through 12 ft of hose then gets mixed with an accelerator enriched water, right at the nozzle. The challenge was to obtain the required 4,200 psi at 3 days while using enough accelerator to allow for very early strength too. TFK undertook a lot of testing to fine tune their installation. All aspects needed to be looked at to ensure best results: nozzleman training, water rings types, shotcrete pump maintenance, water adjustment, accelerator adjustment, shooting technique, hose lengths, air pressures, cleaning of equipment. BEMO's personnel proved helpful in this process.

Figure 17.   Cross passage 2 excavation

Challenges Encountered

Cross passage 1 was expected to be an easy task. The first surprise came fast when TFK encountered the presence of "methane" gas when drilling the probe holes. This needed special attention and once again strong ventilation proved to be the most effective solution. Local presence of gas was detected during the excavation but the ventilation diluted the gasses so fast that they were never reached any serious concentration.

The ground quality also proved to be surprising. The bottom half of the excavation was in hard grouted sand below the water table and required jack hammering for digging. The top half of the excavation was in loose silty sand, moist but not saturated. After a few excavation increments, the crown profile was not protected by the pipe umbrella anymore and the silty sand started to ravel. A quickly applied flash coat of shotcrete and a lot of spiles took care of the problem.

Cross passage 2 was peculiar for it was not chemically grouted. Hayward Baker had classified the ground as non-groutable, therefore trying to grout it would have been a waste of time and money. The ground was expected to be mostly cohesive, although some layers were of uncertain quality. The probe holes proved the crown to be clayey and stiff while the bench showed some permeability. Extra drain holes were drilled to relieve the excavated faces from any water pressure but not vacuum pumping was performed to, as much as possible avoid disturbing the ground. Water flows remained minor. The excavation went smooth and no major challenge was encountered (Figure 17).

The next four cross passages are still to be excavated at the time of writing this paper. Some probe holes already proved that it will not be easy. Unexpectedly large water flows were detected in some of the probe holes as well as presence of hydrogen sulfide and methanol, which may bring other challenges in the excavation method.

FINAL CONCRETE

Final concrete work in the tunnel consists of pouring an invert slab and a walkway (Figure 18). Concrete was pumped using a Putzmeister Pump model BSA 14000 from the surface through 5" (127mm) slick line over 1,000ft (305m). The concrete pump was typically located on the work sites available, but where distances were excessive, a 15" (381mm) diameter hole drilled from the street to each Cross Passage Sump, were utilized.

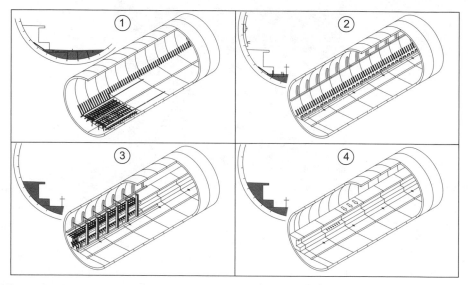

Figure 18.   Typical arrangement of the invert and walkway installation

CONCLUSION

The previous tunnel project in Los Angeles, NEIS as well as the new Metro tunnel job have been great successes for Traylor. These jobs have helped cement Traylor Bros, Inc as a preferred contractor, having a great reputation in the City of Los Angeles for future work. The job has performed with an unprecedented safety record, and has developed new project personnel to high levels of expertise and competence.

REFERENCES

Choueiry, Elioff, Richards and Robinson, "Planning and Construction of the Metro Gold Line Eastside Extension Tunnels, Los Angeles, CA," Proceedings, RETC 2007.
Jatczak and Robinson, "South Bay Ocean Outfall Case history," Proceedings, RETC 1999.

8

Mega Projects

Chair
B. Hansmire
Parsons Brinckerhoff

DESIGNING THE ALASKAN WAY TUNNEL TO WITHSTAND A 2,500-YEAR EARTHQUAKE

Celal Kirandag
Parsons Brinckerhoff

James R. Struthers
Washington State Dept. of Transportation

Gordon T. Clark
Parsons Brinckerhoff

INTRODUCTION

Project Background

The Washington State Department of Transportation (WSDOT) hired the Seattle office of Parsons Brinckerhoff to lead a team of consultants to produce an Environmental Impact Study and perform conceptual and preliminary engineering. The purpose of the Alaskan Way Viaduct and Seawall Replacement project is to replace the existing earthquake damaged and aging viaduct and sea wall, adjacent the downtown waterfront of Elliott Bay, with structures that have improved earthquake resistance and that maintain or improve mobility and accessibility for people and goods along the existing corridor. With an estimated replacement cost over $4 Billion and construction duration exceeding 8 years, it is one of the largest transportation design projects on the West Coast of the United States today. Seattle, Washington is a city of over 2 million people in the Northwest corner of the United States. Ever since the Klondike Gold Rush in 1897, Seattle has been considered a departure point to Alaska. Because of this, the street along the waterfront was named the Alaskan Way in 1930.

Alaskan Way Viaduct and Seawall

The Alaskan Way Viaduct (AWV) carries State Route 99 along the shoreline of Elliott Bay through downtown Seattle and is a vital part of Seattle's highway system. The viaduct carries approximately 110,000 vehicles per day or 25% of all north-south traffic through Seattle (Figure 1). It is 3.4 km (2.1 mile) long with 3 lanes in each direction and was constructed of cast-in-place reinforced concrete between 1952 and 1956.

The wear and tear of daily traffic, salty marine air, and several earthquakes have taken their toll on the facility. Studies in the mid-1990s provided early evidence that the viaduct was nearing the end of its useful life. The viaduct's increasing age and vulnerability was apparent by crumbling and cracking concrete, exposed rebar, weakening column connections and deteriorating railings.

In early 2001, a team of structural design and seismic experts began work to determine whether it was feasible and cost-effective to strengthen the viaduct by retrofitting it. In the midst of this investigation, a 6.8 magnitude earthquake shook the Puget Sound region (the Nisqually Earthquake, located 56 km (35 miles) from Seattle and deep below the surface) caused moderate damage to the reinforced concrete structure. WSDOT closed the viaduct for inspection and repairs intermittently for several days over a period of several months.

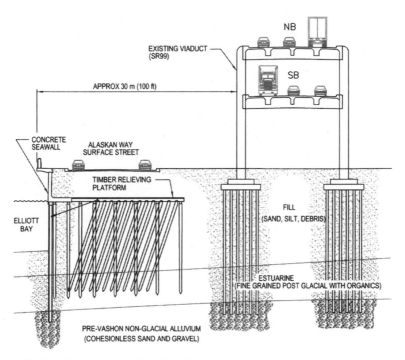

Figure 1. Section of existing seawall and viaduct

Seattle Seawall

The Seawall was built in phases between 1906 and 1934 with numerous repairs and replacement of sections since then. The seawall configuration varies along the length depending on when it was built and how deep the water is at the foot of the wall. In the south, there are several sections of massive concrete gravity wall interspersed with pile supported sidewalk. In the middle and northern portions the seawall structure employs a relieving platform to stabilize the ground supporting the surface road during the daily tidal fluctuation of 2.5 to 4m (8 to 12 ft). See Figure 2 for an illustration of the gravity type seawall. The seawall extends along Seattle's waterfront and supports the soil and many subsurface utilities as well as the foundation of the Alaskan Way Viaduct. See Figure 1 for an illustration of the relieving platform.

Shortly after the Nisqually Earthquake, a 30 m long by 3 m wide (100 ft by 10 ft) section of the Alaskan Way surface street settled, raising concern about the structural integrity of the Alaskan Way seawall. If the seawall were to fail, sections of the viaduct, the Alaskan Way surface street and adjacent structures and utilities could collapse or become unsafe.

Cut and Cover Tunnel

A 1.6 km (1 mile) cut-and-cover tunnel is the most favored of the alternatives under consideration by City officials because it would eliminate the visual dam and noise of the existing viaduct between downtown Seattle and the historic wharfs and tourist attractions along Elliot Bay. At this time, a "preferred alternative" has not yet been selected. The design of a typical cross section of the replacement tunnel along

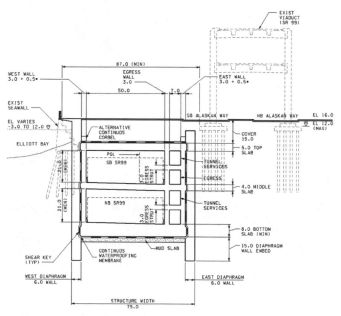

Figure 2.   Cross section of stacked waterfront tunnel

the waterfront includes double stacked box sections with transitions to side-by-side grade roadways and aerial structures at either end. The tunnel boxes are flanked by large secant pile walls that extend below the bottom slab into competent soils. Inside these walls is an inner tunnel box that houses the traffic lanes, system corridors, and egress space. Each roadway in tunnel is designed for three lanes of traffic and is approximately 50 feet wide and 21 feet high. The western edge of the tunnel would serve as a replacement for the existing sea wall. See Figures 2 and 3.

DESIGN CONSIDERATIONS

Seismic Design Criteria

The seismic design of the project is based on two levels of design earthquakes with explicit design checks to make sure that "life safety" performance objectives and damage states are met as set per MCEER/ATC-49 Guidelines. The two levels of design earthquakes are:

- **Expected Earthquake (EE).** Lower level design event that has ground motions corresponding to 50% of probability of exceedance in 75 years (i.e., return period of 108 years)

- **Rare Earthquake (RE).** Upper level design event that has ground motions corresponding to 3% of probability of exceedance in 75 years (i.e., return period of 2,475 years)

In the concept of selected "life safety" performance objective, one may expect "significant disruption" following an RE event but "immediate" full access to normal traffic should be available following an inspection after an EE event. Although "significant" inelastic response is allowed in RE event, "minimal" or no damage should occur in EE event.

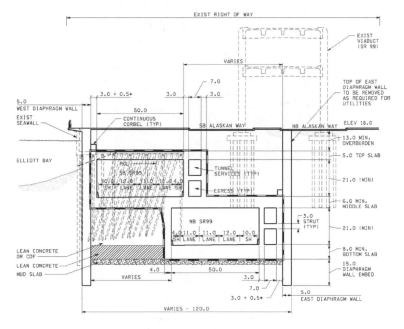

SECTION NEAR STATION 230+00 - UNIVERSITY ST

Figure 3. Cross section of transition waterfront tunnel

Geology

Seattle is located in the central portion of the Puget Sound lowlands bordered by the Cascade Range to the east and the Olympic Mountains to the west. The regions geology and geography are dominated by scour and deposition from Pleistocene glaciations. Pleistocene continental ice moved in a generally southward direction from British Columbia into the Puget Sound Basin. A series of at least five major advances produced a complex sequence of lacustrine deposits, advance outwash, glaciomarine drift, till, and recessional outwash. The most recent ice retreat left behind a series of north-south trending ridges separated by glacially scoured troughs. The troughs are now filled with marine waters (Puget Sound) or fresh-water lakes (Lake Washington and Lake Union). Erosion of the ridges has filled the troughs with silt, and organic material. Leveling of Denny Hill between 1906 and 1930 placed significant amount of spoil in and along Elliott Bay adjacent downtown Seattle. Below these unstable materials, geologic exploration indicates sand, silt and clay at depths of 30 m (100 ft.) below the surface. The cut-and-cover tunnels will be bedded in this glacially consolidated material.

Holocene deposits range from approximately 9 to 18 m (30 to 60 feet) and consist of fill and Estuarine deposits. The fill ranges in thickness from 5 to 13 m (15 to 40 feet) and consists of sand, silt, and clay with some cobbles and boulders mixed with debris from the Great Seattle Fire of 1889. These materials are generally considered to be susceptible to liquefaction during seismic events. The Holocene soils are underlain by non-glacial material generally consisting of sand and gravel that extend to the maximum depth explored of 80 m (250 feet).

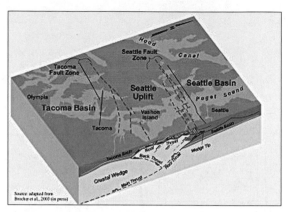

Figure 4.   Seattle fault

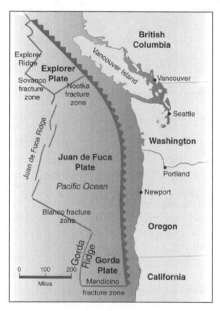

Figure 5.   Cascadia subduction zone

Seismic Setting

The project lies within the moderately seismically active Puget Sound Lowland. In this region, seismicity is characterized by common low to moderate-sized earthquakes with occasional strong earthquakes documented both historically and in the geologic record. Seismic sources typically considered in design within this region include the Seattle Fault (Figure 4) and the Cascadia Subduction Zone (Figure 5).

As described earlier, structural design criteria for the project included consideration of both 108 and 2,475-year ground motion levels. Design ground motions for both hazard levels were developed using probabilistic seismic hazard analysis (PSHA) performed by Shannon & Wilson, Inc. (2004). Design of the tunnel for the approximate 2,500-year ground motion indicated that the primary contributor to seismic hazard for this level of risk is the Seattle Fault Zone which lies in the immediate vicinity of the project area. Although the Cascadia Subduction Zone is capable of generating strong earthquakes (in excess of moment magnitude 8.5), the large site to source distance results in a relatively low contribution to the 2,475-year return period ground motion at the site.

Geotechnical Investigation and Analysis

Due to large size of the project and unfavorable geotechnical conditions within the project corridor, extensive geotechnical investigations were conducted during preliminary project design. To date, over 150 geotechnical borings and monitoring wells have been installed at locations along the length of the project. The primary purpose of these borings was to collect information on the disposition and strength of subsurface soils to be used in geotechnical design of project structures, the borings facilitated geotechnical testing that included pumping and slug tests, tidal monitoring, collection of samples for standard geotechnical and cyclic shear testing, and down-hole shear-wave velocity testing at 8 locations.

Data gathered during the geotechnical investigation was used as the basis for generating site specific ground response parameters for both the EE and RE. The development of these ground motions included consideration of soil liquefaction, basin response, non-linear soil response, and fault directivity effects. Figure 5 shows the recommended response spectra developed for the project. These spectra formed the basis for the development of time histories used in subsequent geotechnical and structural analysis of the tunnel.

Development of design earth pressures for use in structural design was performed using a FLAC finite difference model. In this modeling, time-history analysis was used to model soil-structure interaction and develop a seismic earth pressure increment for use in modeling the structural elements of the stacked tunnel. Soil-structure modeling considered the effects of pore-pressure generation, soil liquefaction, and permanent displacement of soil on the seaward side of the tunnel structure.

Liquefaction

Geotechnical studies conducted for the project indicate that the viaduct alignment is underlain by loose fill soils with a median Standard Penetration Test (SPT) blow count of about 10. Analyses indicate that these soils have a high susceptibility to liquefaction under moderate to strong earthquake ground motion. Geotechnical studies conclude that widespread liquefaction could occur along the viaduct for peak ground acceleration of 0.16g. The findings also indicated that widespread liquefaction would likely result in failure of the Alaskan Way seawall, and lateral spreading of the adjacent soil. Current seismic design criteria call for no collapse in the event of peak ground accelerations of 0.76g. Project design efforts have focused on mitigating the impacts of liquefaction and lateral spreading through design of the proposed tunnel so that it structurally serves as a replacement for the existing seawall system, alleviating the need for extensive outboard ground improvement.

STRUCTURAL REQUIREMENTS

Displacement

The control of wall displacements both during construction and in final configuration is a very important design requirement. Ground stability and control of settlement of adjacent structures during excavation and erection of the tunnel is very much a function of wall stiffness. Excessive inward wall displacements can cause serious settlement issues for existing nearby structures. Close proximity of viaduct footings and nearby buildings necessitated a very rigid excavation support wall system to minimize these effects. In case of RE seismic actions, the integration of these rigid excavation support walls with the tunnel box itself will help increase lateral stiffness of the overall structural system and help reduce racking deflections. For this reason, shear keys interacting in between tunnel walls and diaphragm walls are introduced.

Water Tightness

The top of roadway of the southbound tunnel is over 15m (50 feet) below both mean sea level the landside groundwater table. This required the development of a robust waterproofing system and a structural limit on crack width during severe loading such as the RE seismic event. Several waterproofing systems are currently under consideration and even at this early stage are being accounted for in both the space allocation and construction budget. It is currently thought to fully "tank" the tunnels on the interior of the secant pile walls. The secant pile walls will resist earth pressures and the inner box will resist the water pressure.

Ductility

Based on the performance objective of the seismic code for the RE event, permanent offsets are allowed due to inelastic behavior of concrete but no collapse should occur. Therefore, the concrete members comprising lateral stiffness of the complete structure must be designed to be ductile enough to allow this deformation without having excessive lateral displacements of the tunnel walls. In order to account for ductile behavior of the concrete members in the structural analysis, reduction factors were applied to moments of inertia to account for flexural cracking of members in the RE event. A balancing of criteria for ductility to reduce the magnitude of structural response forces and minimizing of crack width to mitigate leakage is being done. The calculation of crack widths and overall stiffness of the tunnel is an important factor in design and configuration of reinforcement as well as member sizes.

Dewatering

The close proximity of Elliott Bay and high groundwater table pose some of the greatest risks that will be encountered during construction. Mitigation of this risk during the design phase required looking at various types of construction approaches including massive jet grouting, a thick base slab with tie-downs, and even trying to construct the tunnel in the wet. It was ultimately decided that the best approach was to utilize dewatering wells to lower the water table and depressurize deeper permeable soils in between the outer secant pile walls. There was considerable concern regarding the dewatering of the fill material and the potential for negative effects on adjacent buildings. It is thought that the use of injector wells will minimize settlements of existing buildings due to draw-down of groundwater table. During excavation in-between the

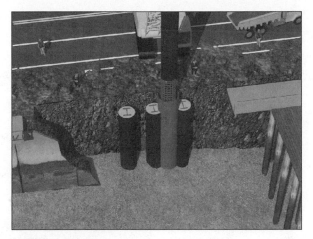

Figure 6. Secant pile wall construction

secant pile walls, constant monitoring of both groundwater level and ground movement will be performed.

Temporary Support and Staging

Temporary bracings and tiebacks will likely be used to support excavation support walls during a cut-and-cover tunnel construction. A sequential excavation procedure was applied such that locations of cross lot bracings and tiebacks would be compatible with the final tunnel geometry. Unsymmetrical lateral loads will result from soil on one side of the tunnel and Elliott Bay on the other. These large unbalanced earth pressures and hydrostatic pressure were taken into account in the analysis of these sections. This unbalanced loading posed the greatest impact on the transition sections (Figure 3). The displacement patterns of each wall were unique for each stage of construction. A series of sensitivity analyses were carried out to determine optimum locations of bracings and tiebacks considering the forces in them and the moments that developed in the diaphragm walls.

Secant Piles

Two secant pile diaphragm walls (Figure 6) will be required to construct the cut-and-cover tunnel to replace the Alaskan Way Viaduct. These walls while supporting open excavation in construction of tunnel will provide additional lateral resistance to the final completed tunnel structure during large seismic loads. The secant pile wall on the bay side of the tunnel will replace the existing seawall and together with a waterproof membrane will create an impermeable barrier between the excavation and Elliott Bay. This barrier will also reduce the amount of debris and contaminates that could potentially find their way into Elliott Bay. The secant pile wall on the landward side of the tunnel will provide both an effective groundwater cut-off wall as well as bracing the ground during excavation. Ultimately the secant pile walls on each side of the tunnel will be tied together to become part of the structural system to resist seismic loading. Currently, the secant pile walls have been designed with a 1.5 m (5 feet) diameter and overlap each other by 0.3 m (1 foot) on each side. Every other secant pile will have a rolled wide flange section to increase the flexural capacity needed to resist seismic actions. Intermediate piles will have a rebar cage for reinforcement.

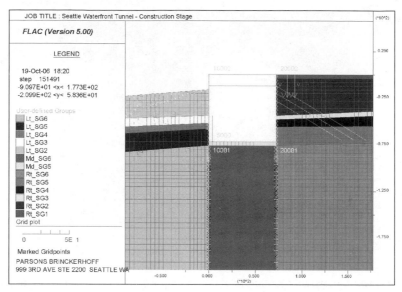

Figure 7. 2D FLAC model of tunnel excavation

Jet Grouting

For the most part, the base slab of the stacked waterfront tunnel will rest on competent glacial soil with the exception of a transition zone on one end which will not extend deep enough to reach that layer. In this region, jet grouting is proposed to create vertically supporting contact zone between the bottom slab and glacial soils and lateral support for diaphragm walls.

Structural Analysis

A series of structural analyses utilizing finite element method and finite difference method have been performed to understand the response of structure to static and dynamic forces during construction and during its useful design life. For each different construction activity, a non linear dynamic soil-structure interaction (DSSI) analysis was carried out with FLAC software to estimate the maximum forces that would develop in the bracing system and tiebacks (Figure 7). An analysis was also performed on critical stages to prevent catastrophic failure that might happen during construction according to AASHTO LRFD standard. Special meshing subroutines were developed to allow the 3D FLAC models to account for the complex glacial geology (Figure 8).

A separate DSSI analysis was also performed using FLAC on the completed stacked tunnel configuration to determine the maximum earth pressures that could occur during EE and RE seismic events. These DSSI analyses also considered possible liquefaction effects. Using these apparent earth pressure diagrams and vertical acceleration effects, the finished structure has been analyzed and designed in accordance with the specified load combinations given in AASHTO-LRFD.

Preliminary Results

Current design analyses showed that 1.5 m (5 ft) diameter secant piles performed well for both sequential construction staging and for final configuration loadings including

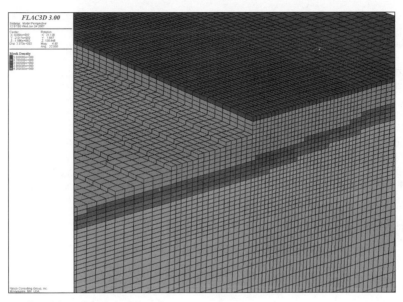

Figure 8.   3D FLAC model of glacial geology

EE and RE seismic actions. Although the use of shear keying action between inner box walls and diaphragm walls is not yet validated because of the difficulties involved in construction of it, there are other options being investigated. The diaphragm wall displacements occurring during RE seismic actions are in the order of 10 to 12 cm (4 to 5 inches). So far these numbers appear to be acceptable. As the design progresses, a more detailed non-linear structural analysis will be done to evaluate crack size opening. This analysis will take into account the effect of axial load in determining water tightness of the system.

REFERENCES

Gleaton, S.J., Clark, G.T., 2003, Tunnel Alternatives for the Replacement of the Alaskan Way Viaduct: Proceedings of Rapid Excavation and Tunneling Conference, SME, p 1301–1311.

Applied Technology Council (ATC), 2003, Multidisciplinary Center for Earthquake Engineering Research Design of Highway Bridges, (MCEER), 49.

Blakely, R.J., Wells, R.E., Weaver, C.S., and Johnson, S.Y., 2002, Location, structure, and seismicity of the Seattle fault, Washington: Evidence from aeromagnetic anomalies, geologic mapping, and seismic-reflection data: Geological Society of America Bulletin, v. 114, p. 169–177.

Gower, H.D., Yount, J.D., and Crosson, R.S., 1985, Seismotectonic map of the Puget Sound region, Washington: U.S. Geological Survey Miscellaneous Investigations Series Map I-1613, scale 1:250,000.

Johnson, S.Y., Potter, C.J., Armentrout, J.M., and others, 1996, The Southern Whidbey Island Fault, an active structure in the Puget Lowland, Washington: Geological Society of America Bulletin, v. 108, p. 334–354.

Johnson, S.Y., Dadisman, S.V., Childs, J.R., and Stanley, W.D., 1999, Active tectonics of the Seattle Fault and Central Puget Sound, Washington—implications for earthquake hazards: Geological Society of America Bulletin, v. 111, No. 7, July, p. 1042–1053.

Johnson, S.Y., and others, 2003, Evidence for one or two late Holocene earthquakes on the Utsalady Point fault, Northern Puget Lowland, Washington: Abstracts with Program, Annual National Meeting, Geological Society of America.

Shannon & Wilson, Inc., 2004, Seismic Ground motion study report, SR-99, Alaskan Way Viaduct Project, Seattle, Washington: Report prepared by Shannon & Wilson, Inc. Seattle, Wash., 21-1-09490, for WSDOT, Federal Highway Administration and City of Seattle, October, 2 Volumes.

Stanley, D., Villasenor, A., and Benz, H., 1999, Subduction zone and crustal dynamics of Western Washington: a tectonic model for earthquake hazards evaluation: U.S. Geological Survey Open-File Report 99-311 on-line edition, 66 p.

KÁRAHNJÚKAR HYDROELECTRIC PROJECT, ICELAND
EXTREME UNDERGROUND CONSTRUCTION

Johann Kröyer

Landsvirkjun

Bernhard Leist

Lahmeyer International

Holger Evers

Mott MacDonald

William D. Leech

Underground Plus Construction Services, Inc.

ABSTRACT

Landsvirkjun—the national electric power company of Iceland—is building the Kárahnjúkar Hydroelectric Project in eastern Iceland. Two glacial river systems will be dammed and channeled through a series of tunnels into a 690 MW underground powerhouse. The Project has a total of 73 km of tunnels to be excavated of which most have been completed. From three access adits, three large diameter (7.2 m to 7.6 m), main-beam, gripper-type, Tunnel Boring Machines (TBMs) are being used to complete 50-km of the waterway tunnel. Using the TBMs was a first in Iceland to excavate the complex Icelandic basalt formations, which consist of inclined layers of hard basaltic flows with weaker scoriaceous and sedimentary interbeds. Faults, fissures and dykes that frequently cross the lava flows posed both geologic and hydrogeologic difficulties to the TBM tunneling. This paper describes the extreme challenges that the men and equipment had to endure to complete the tunneling, including the harsh winter weather on the surface and the cold, abundant ground water inflows within the tunnels.

INTRODUCTION

By their very nature, hydroelectric power projects are frequently located in remote and mountainous regions, often with inhospitable climates. Where topographic and/or hydraulic conditions dictate, they can also feature extensive tunnel drive lengths to maximize the output of energy. Bored tunneling options have the potential to reduce the overall construction duration for the execution of long tunnel drives.

The extensive tunneling layout for the Kárahnjúkar Hydroelectric Project required a range of geological and hydrogeological challenges to be overcome to bring the project to a successful completion. The inhospitable climate during the long winter months posed additional challenges for both men and equipment working on the Project. An overview of some of these is given in this paper.

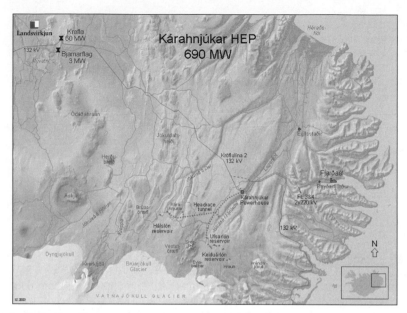

Figure 1.   Project location

PROJECT DESCRIPTION

Scope of the Project

The Kárahnjúkar Hydroelectric Project in Eastern Iceland will harness the might of two glacial rivers north of Vatnajökull, Europe's largest glacier. Six 115 MW Francis turbines, operated under a gross head of 600 m and with a total discharge capacity of 144 m³/s, will generate 4,600 GWh of energy per year to a new aluminum smelter in Reyafjördur, a town in Iceland's East Fjords.

The main storage reservoir for the seasonal flow (the Hálslón reservoir) will hold 2,100 million m³ and is located on the western river system confined by three dams along its northern perimeter. The main dam (the Kárahnjúkar Dam) is a 200 m high concrete-faced rock fill dam (CFRD), currently the highest of its type in the world, incorporating 9 million m³ of rock fill. The dam is constructed across a 50 to 90 m deep canyon with vertical cliff walls. Two earth-fill saddle dams with a central core of moraine (of 68 m and 25 m height respectively) are situated on either side. The second stage of the project will utilize the flow of the eastern river system from the Ufsalón reservoir, and will include three more earth fill dams, see Figure 1.

The underground works of the Project comprise a total of 73 km of tunneling, several shafts as well as two large underground caverns to house the power turbines and transformers. The Headrace Tunnel (HRT) linking the Hálslón reservoir to the power house includes some 50 km of excavation, mainly by open, main-beam, gripper-type, TBMs of diameters 7.2 and 7.6 m, all from the manufacturer, Robbins.

An overriding challenge to the successful outcome of the Kárahnjúkar Hydroelectric Project has been to coordinate the tunneling works with the other key project components—e.g., the dams, pressure shafts, powerhouse, transmission lines, and the smelter—and to manage their interfaces. The construction of the Headrace Tunnel by

Figure 2. Project data

three TBMs through the heterogeneous geology to a tight schedule has been central from the start to the project's success.

The geographic remoteness as well as the topographic and extreme climatic conditions, two degrees below the Arctic Circle, makes the biggest construction scheme in Iceland's history even more demanding.

The principal project data is summarized in Figure 2.

Construction Program and Project Members

Construction started in spring 2003 and the Headrace Tunnel is due to be ready for water transfer in the second quarter of 2007. Power generation will commence during mid 2007, although work on the second stage extension of the project for dams on the Jökulsá in Fljótsdal River and diversion to the main headrace tunnel will continue into 2008. The main contracts for the Kárahnjúkar dam (contract KAR-11) and the Headrace Tunnels (contract KAR-14, including a 10 km TBM drive for the Jökulsá diversion) were awarded to the Italian contractor, Impregilo. The powerhouse has been constructed by a joint venture led by the German contractor, Hochtief. Icelandic contractors have played a significant role in the overall project—Ístak is a partner in the joint venture with Hochtief—Sudurverk is constructing the saddle dams for Hálslón reservoir—Arnarfell is excavating the drill and blast sections of the Jökulsá diversion tunnel.

The designers for the underground works is a group of international and Icelandic consultants, with Electrowatt (now Poyry Energy Ltd.) of Switzerland responsible for the tunneling works, and the construction supervisors on site for the tunneling is another joint venture of seven companies led by the British firm Mott MacDonald.

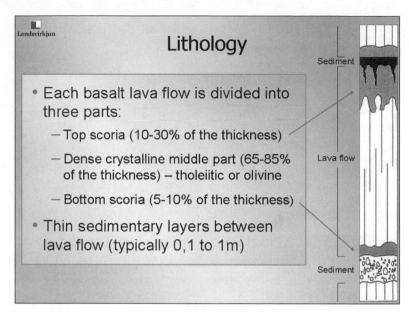

Figure 3.   Lithology

Geological Conditions

The Headrace Tunnel, located between 100 and 200 m below the ground surface, passes through sequences of shallow-dipping basalt flows. The bedrock was formed during the last 0.1 to 10 million years and consists of a lava pile almost 3 km thick. Scoria are normally found at the top and bottom of each basaltic flow, forming the contact zone to layers of glacial sedimentary rock which developed in between volcanic eruptions. Figure 3 shows a typical sequence.

The lithology generally dips at 3 to 5° towards the west. Two relatively dense subvertical fault, fissure and dyke systems are apparent striking in a northerly direction. The water table is generally close to the surface and above the Headrace Tunnel. Precipitation in the Icelandic Highlands is generally low, but the melting of the glacier causes a variation in the natural ground water level.

The Scoria is characterized by its highly chaotic nature. Similar to the sedimentary rocks, its layers are variable in thickness along their extent. The strength varies over a wide range, but is significantly lower than the basaltic rock. To the west of the project area formations such as pillow lava, pillow breccias, tuff breccias and tuffs are found and these formations are referred to as Móberg, a generic Icelandic term for lava flows and pyroclastic rocks that were formed under water or glaciers.

Table 1 contains a summary of geological parameters for the majority of rocks encountered.[1]

Pre-Investigations

Engineering geological investigations were carried out in the Project area in various phases between 1970 and late summer 2001. Geological maps covering the whole Project area are available. Aerial photos in the scale of 1 to 36,000 and 1 to 40,000 were used to trace tectonic features. Cored drill holes were available along the headrace tunnel alignment at a general spacing of 1.5 to 3 km. Longitudinal sections

Table 1. Summary of geological parameters

| Rock Class | Bore Class | Rock Type | UCS (MPa) | | E-Mod. (GPa) | Joint Spacing (m) |
			Normal	Maximum		
Sediments	H1	Tillite	40–100	150	8–12	0.5–1
		Conglomerate	20–80	110	2–14	0.5–1
		Sandstone/Siltstone	20–80	110	2–14	1–10
Scoria/Móberg	H2	Scoria	30–80	150	8–12	—
		Móberg	50–200	250	8–12	0.05–0.3
Basalt	H3	Basalt	100–250	350	15–30	0.3–3
		Andesite	150–300	400	15–20	0.1–5

were prepared using the geological maps, detailed mappings of profiles across the lava pile, and the results of drill holes performed along the route. These sections indicated available information on the geometry, geology, rock parameters, ground water conditions and potential hazard assessments along the tunnel alignments. However, factors such as extrapolations from the exploratory holes, variations in the general dip of the lava suite and numerous sub-vertical faults, fissures and dykes limit somewhat the accuracy of the longitudinal sections. Variations from given ranges could therefore be expected.

The tectonic lineaments apparent from the aerial photos were described as normal, sub-vertical features with an observed down throw or displacement of between 1 and 40 m and an observed spacing in the range of 200 to 250 m. The faults and fissures, particularly those caused by higher displacement, are generally of a width of between 0.5 and 2.0 m. They are characterized by sheared, crushed rock with a grain size distribution ranging from blocks down to clay. Both well-cemented crushed material as well as weathered and loose material was observed in the faults and fissures during tunnel excavation.

TBM EXCAVATION

Hazards, Machine Selection, Layout

The combination of highly variable rock conditions (see Table 1), with a shallow-dipping lithology, was known to lead to 'mixed-face conditions' over potentially long stretches during excavation.

The joints, fissures and faults provide a major pathway for ground water seepage, along with the contact zones of the geological strata, and give rise to the risk of potentially high and sudden water inflow during excavation. The water table close to surface level could result in hydraulic pressures of potentially 20 bar at tunnel horizon with infinite recharges from the various streams and lakes in the Project area. The seasonal melting of the glacier was expected to result in a time-dependent variation of the natural water table and potentially variable inflows into the tunnel over time. In general, only very limited experience from other projects regarding the ground behavior during tunnel excavation could be relied upon.

After careful evaluation of the rock conditions and the associated risks, the suitability of mechanical excavation by open, gripper-type, TBMs was considered, and finally chosen for the excavation for approximately 90% of the Headrace Tunnel. Not

Table 2.   TBM specifications

Manufacturer/model	Robbins, model 236	Number of cutters	46/48 (back-loading)
Type	Open gripper TBMs	Cutter size	19" (483 mm)
Excavtion diameter	7.2/7.6 m	Maximum cutter load	311.4 kN
Maximum thrust	14,432 kN	Cap. machine c/v	1,000 t/hr
Inst. power	10 × 300 kW		
Cutterhead speed	0–8.3 rpm 3,490 kNm (@ 8.3 rpm)		
Torque	6,275 kNm (@ 4.6 rpm)		

only would the higher advance rates (estimated to be four times the rate of conventional drill and blast methods) benefit the overall construction schedule, but also the smooth excavation profile from the TBM minimize hydraulic losses during operation and reduce requirements in terms for temporary and permanent rock support.

The machines were designed and specified for the anticipated ground conditions to deal effectively with both the anticipated hard rock and mixed-face conditions, in which rocks of strength ratios as high as 1:15 would be encountered. The TBMs are sized at 7.2 in diameter upstream and 7.6 m downstream of the Jökulsá bifurcation, and equipped with large 19 inch back-loading cutters (total number of 46 and 48 respectively) to ensure maximum penetration. The specifications of the TBMs are summarized in Table 2.

The contract start-up activities for the portal areas and access adits were excavation by conventional drill and blast methods to allow mobilization and on-site assembly of three TBMs. The location of the access adits was chosen to divide the Headrace Tunnel into approximately equal lengths, to limit the actual drive lengths of the adits themselves, and situated in local depressions to allow the surface works to remain unaffected by winter weather conditions as far as possible, see Figure 4.

Launching

TBM 3, with a diameter of 7.2 m, was launched in April 24, 2004 from Adit 3, thus becoming the first "tunnel boring machine" to break into Icelandic rock. TBM 2, with the same diameter, started shortly afterwards from Adit 2 on July 26, 2004; TBM 1, with a diameter of 7.6 m, finally commenced excavation from Adit 1 on September 21, 2004.

All TBMs were driving upstream at gradients of generally less than 1% to allow gravity drainage of the water from the excavation face towards the portals. Mucking of excavated material was achieved by a fixed conveyor with the TBM (1,000 t/hr capacity) feeding into an extendable, continuous tunnel conveyor system to the portal (600 t/hr capacity). From the portal it was envisaged that surface conveyors transport the muck directly to the designated spoil areas, but after increasing downtimes during the first winter it was permanently reverted to conventional mucking by trucks from the portal.

Excavation Classification

For tendering and payment purposes, a combination of two factors is used for the excavation classification.[2] The geology along TBM driven tunnels is allocated into 'Bore Class' groups, to define encountered rock materials in terms of their boreability and their associated risk for advance such as cutter wear and cutterhead clogging. Three homogeneous bore classes H1, H2 and H3 were defined (see Table 1). For mixed-face conditions, reduced thrust and/or revolutions were expected to limit the wear on the machine cutters, which would potentially result in reduced penetration.

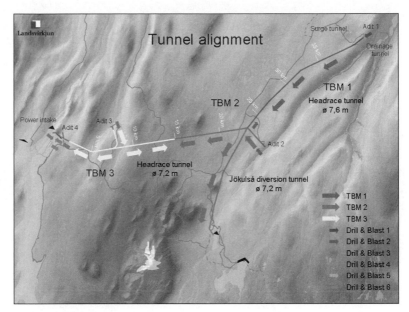

Figure 4. Tunnel alignment

Therefore two mixed-face bore classes M1 and M2 were defined alongside the homogeneous bore classes.

The second factor, covering hindrance of the TBM advance resulting from the installation of support in the machine zone, is reflected by 'Excavation Classes.' Classes I to III are defined either by the number of rock bolts or dowels installed per tunnel meter and/or the surface area of applied shotcrete. Class IV includes heavy steel-arch support with or without spiling, steel lagging or backfilling. Negligible or no hindrance to the TBM excavation was foreseen in Classes I and II, whereas Classes III and IV would cause considerable or major hindrance to the TBM advance. Figure 5 shows the typical support amount/extent for each excavation class. The variations of bore and excavation classes results in a matrix of 20 possible combinations.

It is noted that, in line with European industry practice, the hindrance caused by handling groundwater is dealt with separate from the bore and excavation classes.

Rock Support

The rock reinforcement mainly consists of steel-fiber reinforced shotcrete (fiber dosage 40 kg/m^3) of up to 250 mm thickness and mortar-embedded rockbolts (25 mm, generally 3 to 4 m length). Where required steel ribs are used as heavy and immediate support. The rock support serves both for temporary support for safe working and as permanent support ensuring long-term stability and to provide protection during operation against deterioration and erosion of weaker sedimentary layers over time when subjected to constant flow velocities of about 2–3 m/s. Other, less frequently used support types include wire mesh, steel lagging, straps, half-ribs as well as support ahead of the face by spiling and self-drilling anchors. The rock support was instructed based on a project-specific risk matrix.

The lower sections of shotcrete, below the spring-line, and the required lengths of invert concrete have generally not been installed from the TBM as originally planned,

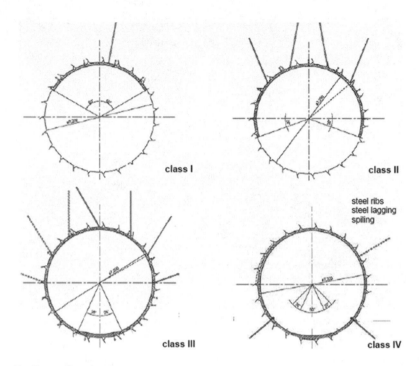

class I

class II

class III

class IV

steel ribs
steel lagging
spiling

Figure 5. Excavation classes

but moved to the post-excavation finishing works to be performed along with post-grouting and any rock support which has not been installed during excavation.

Performance

The actual distribution of bore classes encountered has been approximately 22/16/26% for bore classes H1to H3 and 13/23% for mixed-face bore classes M1 and M2, comparing well to the estimated distribution of 23/10/28% and 18/21% respectively.

The mixed-face conditions did occur as anticipated, but did not require overly cautious operation of the TBMs and the cutter wear was not significantly affected by the mixed-face conditions.

Approximately 75% of the encountered length of bored Headrace Tunnel has been designated as Excavation Class I (67% is without any support and 7% with limited support, see Figure 5). For the remainder, about 24% of the tunnel was split almost equally between Excavation Classes II and III, and about 2% was in Class IV. This compares reasonably well to the predicted distribution of 67/21/11/1% for Class I to IV.

To date an average of about 0.3 m^3 of shotcrete has been applied per linear bored tunnel meter. Some 3.8 km of invert concrete, equivalent to approximately 10% of the total TBM drive length, remains to be placed. A total of 550 steel ribs have been installed, about 50% in concentrated stretches, e.g., around Chainage 36+400 where weak sedimentary layers, prone to slabbing, occurred in the tunnel crown over a length of approximately 300 m, and locally from Chainages 13+000 to 14+350 where closely fractured, brittle andesite required early support. Another area of heavy steel rib support on the TBM2 drive is discussed further.

In general, the performance of the TBMs has been very satisfactory considering the highly variable geology along the Headrace Tunnel alignment. Both extremely strong basaltic conditions as well as moderately strong sedimentary layers have been excavated without any major problems to the machines or effects to their operation. The net penetration was generally between 2.5 to 3.5 m/hr. The predicted penetration rates have been largely in good agreement with predictions.

The cutter downtime, taking 3.6% of production time, has been lower than predicted, and the machine availability varied between 92 and 95% for the three TBMs. Generally, the largest down-time increases incurred were due to the conveyor system and to lengthy durations needed for rock support (see also Figure 6).

The best daily, weekly, monthly records achieved were 92.1 m (TBM2, June 09, 2006), 326 m (TBM3, week 24 of 2004) and 1.192 m (TBM3, August 2004).

SELECTED CHALLENGES DURING TBM EXCAVATION

Heavy Water Inflow at TBM3 Upstream

Following initial and continuously good performances of 150 m/week in the upstream direction, TBM3 stopped for the first time on January 15, 2005 to carry out pre-grouting to reduce water inflow. Over the following six months (or 1,500 m), the more permeable Móberg geology being intersected required extensive pre-grouting.

A total of 22 grouting arrays were drilled and water-cement grout (generally 1:1 to 1:2) injected at water pressures of up to 15bar and at water temperatures between 2–4°C. A total of 876 tons of cement or 486 m^3 of grout were injected; the last array alone took 157 tons of cement (or 86 m^3 of grout), before the decision was made on July 7, 2005 to stop the TBM drive short of its foreseen length, disassemble the TBM, and turn it around to excavate in the downstream direction from the adit junction. The remaining 1,119 m of the foreseen TBM upstream drive was then performed by conventional drill and blast excavation.

The progress of the TBM during the grouting works had dropped from an average 25.5 m/day before any pre-grouting was carried out to an average 10.4 m/day until stoppage. The utilization decreased by 40% and the downtime for probing increased more than threefold.

The water flow for the entire drive increased during the month of January 2005 from roughly 250 L/s to 500 L/s. The pre-grouting was only partially successful in reducing the water influx, and the water flow continued to rise reaching about 850 L/s when the TBM was stopped.

As an interesting note, the Móberg formations did not require significantly more rock support, the allocated time for rock support fell by 18% (however the pre-grouting allowed at times for concurrent activity of rock support), thus the pre-grouting was predominantly performed to keep the water inflow at manageable levels.

Major Fault Zone at TBM2

At the end of April 2005, TBM2 was driving upstream from Adit 2 and encountered a set of three major fault zones over a 50 m tunnel length resulting in stoppages and very little progress for six months between May 2005 and October 2005 (see Figure 7). Up to this time the TBM had performed well and had excavated 5,000 m in eight

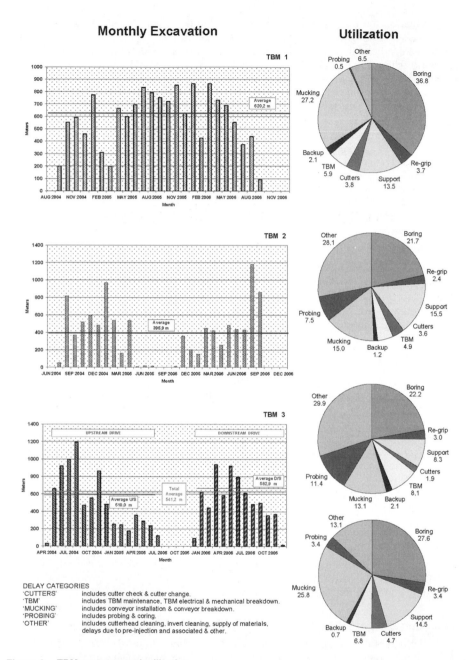

Figure 6.   TBM progress and utilization

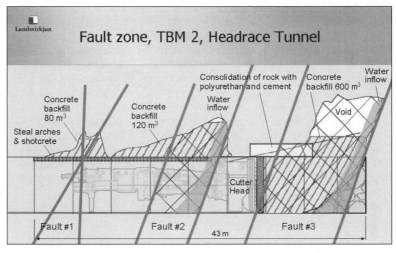

Figure 7. Fault zone TBM2

months without major problems and with production figures of more than 200 m/week in generally good rock conditions. By the end of April 2005, at Chainage 19+640, the conditions started to deteriorate rapidly and Excavation Class IV support with steel arches began to be installed. The ground was particularly blocky with clay infilling in joints and needed immediate shotcrete application behind the cutter head. The excavation was advanced with steel arch spacing reduced to 0.4 m, steel lagging behind the ribs in the crown and side walls and steel spiling ahead of the face. In the first week of May 2005 the first void became visible behind the cutterhead, and ground loss at the crown continued and for which the spiling measures were no longer adequate. Rapid reacting polyurethane foam was then promptly used for stabilizing the area. At Chainage 19+620, when a sudden 150 L/s inflow of groundwater exacerbated the situation by causing more ground loss, the cutterhead rotation was temporarily lost.

After removal of the previously installed steel arches, the TBM was reversed 10 m back. The front of the cutterhead was protected with timber planks and polythene sheeting and a concrete plug installed to stabilize the fault zone.

By the third week of May 2005, 620 m³ of concrete had been used to fill about 80% of the void ahead. However, wash out of the concrete by the ingress of water was very severe and, despite not having filled the critical roof section, the concreting operation was suspended. Drainage relief holes were drilled and polyurethane foam was used to seal off the water influx and to enable the resumption of concrete backfilling and further concrete backfill/grouting operations were attempted from the TBM in order to consolidate an area above the cutter head. In the first week of June 2005, the TBM started re-excavation through the concrete plug reaching Chainage 19+614 by the middle of June 2005.

Additional core drilling indicated a further and wider fault zone lying ahead between Chainage 19+610 and 19+600. Initially, drilling for pre-grouting and consolidation ahead of the face was attempted using self-drilling, injectable anchors, but the attempts were limited by the poor rock conditions and material wash out. Finally, the TBM was cautiously advanced with steel arches at close spacing, steel lagging support, and pre-grouting at an average of 1 m/day.

On July 21, 2005, at Chainage 19+597, when the cutterhead had encountered partially competent rock at the face, the tunnel conveyor broke down and repair works needed almost 2 days. The continued collapse of ground during this stoppage blocked the cutterhead and again rotation was lost. Attempts to restore the rotation in-situ failed. Consolidation grouting operations, reversing the TBM with removal of short sections of steel arches, and attempts to restore the cutterhead rotation continued throughout July and August 2005. Spiling and grouting through 90 mm diameter, 12 m long, steel pipes (System Bodex) were made. The TBM was eventually reversed to Chainage 19+612, the cutterhead re-dressed and rotation restored. The TBM was then moved about 3 m forward under umbrellas of self drilling bolts and pre-grouting and a second concrete bulkhead was constructed to allow backfilling of the void. In early October 2005, the void was backfilled to 3 m above the crown level mainly using lightweight concrete with foaming agent. Subsequently re-boring through the concrete plug commenced and the TBM advanced to the originally achieved face at Chainage 19+597. Normal TBM production could be resumed by the end of October 2005.

MITIGATION MEASURES

Investigations During Construction. Concurrent to the works to overcome the challenges on the TBMs, further geological investigations were carried out to assess the likelihood and frequency of further major features to be encountered.

Three tests with a tunnel seismic prediction (TSP) system were performed underground in various TBM tunnels, but the results were found to be too inaccurate to be useful and the exercise was not continued.

In addition, geophysical exploration from surface level by means of refraction seismic and ground magnetic surveys were undertaken along the tunnel alignment covering 1,000 m ahead of the actual face and 500 m of already excavated tunnel for comparison and correlation purposes. The rock cover in this area ranges between 180 and 200 m.

The results of the refraction seismic survey showed considerable variations in the seismic velocity but the interpretation of fault zones was found to be very difficult and inaccurate and the refraction seismic survey has not been continued.

The magnetic data was sampled on lines at 25 m intervals up to 200 m left and right of the tunnel route. Magnetic anomalies can arise from several factors, that is mainly from dykes and normal or reverse faults. Due to the absence of dikes in the area investigated, the local magnetic anomalies could be related to fault zones and the correlation to already encountered features was found to be quite accurate. The field work was completed in five days and interpretation reports were available after four weeks with preliminary results submitted for information earlier. The magnetic survey was extended a further 1.600 m ahead. It proved to be very useful to predict potential problem zones and supported the accuracy of time schedules and forecasts for planning purposes.

Additional Adit. When TBM3 was reversed following excavation in the upstream direction, the remainder of the drive (1,119 m) towards the Power Intake was to be excavated by conventional drill and blast means. In order to allow an unhindered TBM excavation in the downstream direction, as well as to relief the overall construction schedule, a new 400 m long adit (Adit 4) to the Headrace Tunnel was introduced, which was originally not foreseen (see also Figure 4). From Adit 4, the Headrace Tunnel was advanced in both directions. This additional access has also become of significant benefit with respect to accelerating the finishing works in this section of the Headrace Tunnel. In addition, Adit 4 allowed for further contingencies for pre- and post-grouting to be undertaken where required.

Vertical Pump/Concrete Shafts. To manage the high water flows in the HRT3 drive, and to facilitate the work progress, two 80 m vertical boeholes were drilled down from surface level about mid-way between Adit 3 and Adit 4. Two heavy duty pumps (type KSB) of 400 L/s capacity were installed diverting most of the water to surface level. This pumping facility will also be of help for the future post-grouting and finishing works activities, and will finally be used as air vent shafts.

Another 170 m deep, $\varnothing$ 600 mm borehole has been constructed to aid concrete supply to the lining works at the Jökulsá Valve Chamber. It was initially foreseen that the concrete would be supplied via Adit 2, but the delays to the dismantling of TBM2 as a result of the encountered fault zone, as well as the associated delays to the finishing works for HRT1 and HRT2 (all be supplied through Adit 2) favored a separate supply line to reduce the logistical problems and potential impact to the constrained program.

CONCLUSION

On the Kárahnjúkar Hydroelectric Project, TBMs have been used for the first time in the complex Icelandic conditions consisting both of extremely hard rock as well as extensive sections of mixed-face conditions.. Despite substantial geological and hydro-geological challenges, including faults, open-joints, fissures and dykes as well as high, localized groundwater inflows, the Project has demonstrated both the suitability of TBMs tunneling methods and the TBMs potential for high rates of advance. The Project also justified the selection of TBMs for the harsh Icelandic environment where freezing temperatures are encountered for most of the year. However, the Project has shown the associated logistics to be pivotal to continuously achieve the benefits of lower costs and schedule durations. This is especially crucial for the systems and equipment for shotcrete application, rock bolt installation, pre-grouting, and mucking or spoil removal that need to be carefully selected and specified to suit the underground excavation methods.

The data and statistics contained in this paper have been compiled from site records.

REFERENCES

[1] *Kárahnjúkar Hydroelectric Project, Contract Documents KAR-14, Appendix A to Part II, General Information* Geological Report; January 2003.
[2] *Kárahnjúkar Hydroelectric Project, Contract Documents KAR-14, Volume 2, Part IV Technical Specifications.*

THE KENSICO-CITY TUNNEL WILL ADD RELIABILITY TO NEW YORK CITY'S WATER SUPPLY SYSTEM

Kevin Clarke

New York City Department of Environmental Protection

Eric Cole

Earth Tech

Heather M. Ivory

URS Corp.

Verya Nasri

URS Corp.

BACKGROUND

Development of the New York City Water Supply System

The New York City water supply system serves a population of almost nine million, including residents in neighboring counties. Peak day demands on the system are currently running at about 70 m³/sec (about 1,600 mgd), which is significantly lower than in the 1990s. This trend of reducing demands over the past 25 years, which has been experienced in other major urban centers in the US, is mainly the result of conservation measures and increased cost consciousness on the part of water users. In the long term, demands are expected to once again follow an increasing trend with population growth, so that by the end of the planning horizon of these studies, the year 2045, the peak day demand is expected to increase to about 85 m³/sec (2,000 mgd). The system is served largely by surface water sources, regulated by a system of 19 reservoirs and three controlled lakes, which are connected to the City by a series of aqueducts, the longest of which is about 87 km (54 miles). The total storage in the system amounts to about 560 billion gallons, providing a system firm yield of about 1300 mgd. The Delaware system provides about 50% of the City's water, the Catskill about 40%, with the remaining 10% provided by the Croton. The system as a whole, which is shown on Figure 1, has been built up over the past 150 years, and is truly a marvel of engineering.

The Catskill and Delaware aqueducts both discharge into Kensico Reservoir, which is a daily balancing reservoir located in Westchester County about 24 km (15 miles) north of the New York City line. The two aqueducts extend south for a further 14 miles to combine again at Hillview Reservoir in the Bronx, which provides hourly balancing storage for New York City system. The Croton aqueduct runs nearby as shown on Figure 2, but does not connect into the two reservoirs. From Hillview Reservoir the water is supplied to the city via three main distribution tunnels, known as City Tunnels 1, 2, and 3. City Tunnel No.1 was developed in the 1920s as part of the Catskill system. City Tunnel No. 2 followed in the 1960s as part of the Delaware system. City Tunnel No. 3 (CT3) is partly under construction and partly in operation, and is being implemented to permit the other distribution tunnels to be taken out of operation for the first time for inspection and maintenance. Presently CT3 is supplied with water from Hillview Reservoir via a large underground valve chamber, the Van

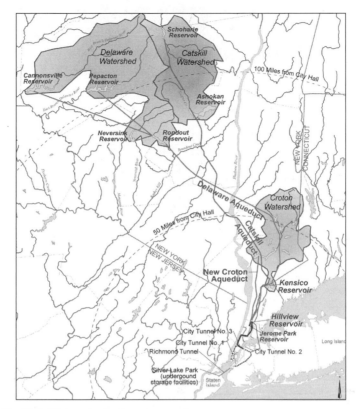

Figure 1. New York City's watersheds and reservoirs

Cortlandt Valve Chamber (VCVC), which was completed in the 1980s as part of the development of CT3.

It will be appreciated that the conveyance corridor between Kensico and Hillview reservoirs, which carries about 90% of the water supplies to New York City, is a critical reach of the conveyance system. Both the Catskill and Delaware aqueducts have been in operation for many decades with no opportunity for inspection and repair. It is for this reason that NYCDEP has embarked upon this planning study into a third water tunnel downstream of Kensico Reservoir, referred to as the Kensico-City Tunnel or simply the K-CT. The K-CT will connect Kensico Reservoir with the VCVC, and will be about 24 km (15 miles) long.

Filtration Avoidance

One of the unique features for such a large water supply system is that 90% of the supplies, namely the Catskill and Delaware (Cat/Del) water, are unfiltered. The supplies from the Delaware and Catskill catchments are of high quality and NYCDEP has gone to great lengths to protect these headwaters, thus avoiding the need to add costly filtration facilities. The United States Environmental Protection Agency (USEPA) has the responsibility of enforcing national drinking water standards, and monitors unfiltered supplies, such as that for New York City. Over the years USEPA has issued "Filtration Avoidance Determinations" (FADs) for the City's water supplies, and one of

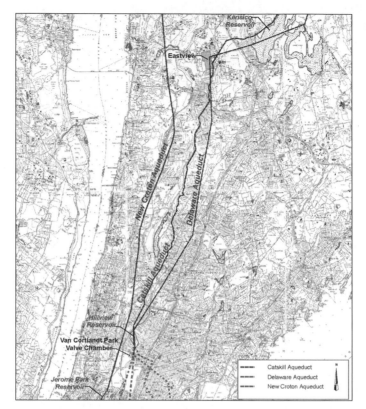

Figure 2. Existing conveyance systems in project area

the conditions of the latest FAD was that NYCDEP must implement a UV facility for the Cat/Del water. UV disinfection alters the DNA of microorganisms such that they are no longer able to replicate and cause illness.

The UV facility for the Cat/Del water will be located on a city-owned tract of land referred to as the Eastview Site, which is located adjacent to the aqueduct alignments, as shown on Figure 2. Construction of this UV facility, which with a treatment capacity of just over 85 m^3/sec (2,000 mgd) will be by far the largest such facility in the US, has already commenced and it is estimated that it will be completed in about 2010, well in advance of the K-CT. The UV facility at Eastview thus forms part of the existing system as far as planning of the K-CT is concerned. It is the policy of NYCDEP to avoid filtering the Cat/Del water, nevertheless should filtration be necessary sometime in the future, the filtration plant would also be located at the Eastview Site.

The Conceptual Plan for the K-CT

Aqueduct and tunnel planning and construction for the New York City system has evolved over the past 100 years. The first major aqueduct, the Croton aqueduct, was of cut-and-cover masonry construction. This was followed by the Catskill Aqueduct, which is an open-surface flow concrete aqueduct constructed largely by cut-and-cover techniques. In contrast the Delaware Aqueduct was constructed as a deep rock tunnel using drill and blast excavation methods. Placing the tunnel at depth reduces

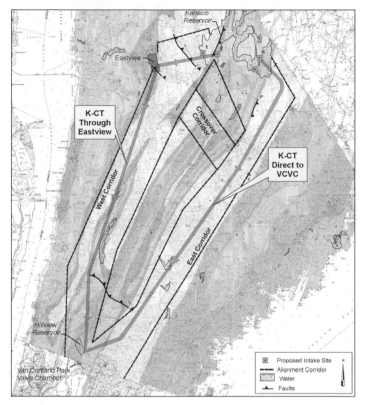

Figure 3. Project area geology

the environmental impact and takes advantage of better rock conditions. All three of the City Distribution tunnels were constructed as deep rock tunnels, the first two by drill and blast methods, and the third and most recent, CT3, using TBMs. The K-CT will also be a deep rock tunnel, which for most of its length will be excavated by TBMs.

The K-CT was originally conceived to be part of the CT3 development, and in the early years of planning it was considered that the tunnel should pass directly from Kensico Reservoir to the VCVC, bypassing Hillview Reservoir. The route selected was along a broad band of Manhattan Schist located to the East of the project area, as shown on Figure 3, which is a summary geological map of the project area. The introduction of the Eastview UV facility into the system, has prompted consideration of an alternative alignment that would pass through the Eastview site, as shown on Figure 3, traversing much more complex geological conditions. For the more easterly alignment, it would be necessary to construct an additional UV facility for the K-CT, located at the intake site, which is the only place where the City already owns significant areas of land. The question thus becomes "can the probable lower cost of tunneling along the Eastern alignment offset the cost of the additional UV facility?"

KENSICO—CITY TUNNEL AND SYSTEM RELIABILITY

Critical infrastructure features for large urban centers in the US are a cause for concern in these security-conscious times, and New York City water supply system is no exception to this. With virtually all the City's supplies conveyed by three main aqueducts, the loss of any of these aqueducts would clearly have a profound impact. This is particularly true in the case of the Delaware Aqueduct which carries about half the peak daily demand of the system. This concern for the potential vulnerability of the system, together with the significant age of the existing aqueducts has made the implementation of the K-CT a high priority for NYCDEP.

The K-CT Will Improve the Reliability of the Main Conveyance System

Adding additional conveyance capacity downstream of Kensico Reservoir will enable the existing aqueducts to be taken out of service in turn, and with appropriate maintenance brought back to high levels of reliability. This will require a defined program of planned outages of each component of the system, namely the Croton, Catskill and Delaware aqueducts, which could extend over a 20-year period in all. As a minimum, the capacity of the K-CT should be sufficient to allow the largest existing component of the conveyance system (in this case the Delaware Aqueduct) to be taken out of service for inspection and repair. However, during any planned outage of a specific component, there is always the possibility of other components suffering unplanned outages, and the capacity of the K-CT would need to be larger if such unplanned outages were also to be covered. Planning studies into the optimum sizing of the K-CT thus involve weighing up the additional costs in increasing the capacity of the K-CT against the benefits of increased levels of system reliability.

A reliability model was developed for the conveyance system downstream of Kensico Reservoir, which enables overall system reliability in meeting a specified demand to be computed taking into account all combinations of planned and unplanned outages of the system components. The model required assessing levels of reliability of the existing aqueducts, a process that involved close consultation with the operations staff of DEP. The model will be applied to each year of projected demands following the year of implementation of the K-CT, taking into account a defined schedule of planned outages for each component of the system. The model also computes the probable deficiency in meeting target demands, for various K-CT capacities, and by placing an economic value on un-served water demands, the optimum capacity of the K-CT can be assessed. One of the basic inputs to the reliability model is the cost of the K-CT as a function of discharge capacity and one of the early steps in project screening was to develop such cost functions. Typical outputs from the reliability modeling for a particular planning scenario are shown on Figure 4. As indicated a minimum present value scheme can be identified in this way.

Planning the K-CT for Extreme Events

In addition to outages of the existing aqueduct as discussed above, planning of the K-CT also recognized that other extreme events are possible that could have a serious impact on water supplies. One such extreme event would be the complete failure of Hillview Reservoir, which would make the Delaware and Catskill aqueducts inoperable downstream of the Eastview site. Connecting the K-CT into the VCVC and bypassing Hillview Reservoir would go a long way in maintaining supplies to New York City, even in this extreme event. This would mean that the K-CT would supply CT3 only, with the other distribution tunnels out of service. Computer studies are being undertaken to determine the extent to which the CT3, along with the Croton aqueduct could meet the demands on the City's distribution system. Another extreme event is serious damage to the facilities at the Eastview site, again with the possibility of interrupting flows in the Delaware and Catskill aqueducts. For this reason several of the options for the K-CT include facilities that will enable the Eastview site to be bypassed.

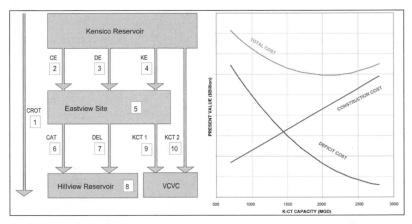

Figure 4.   Reliability model

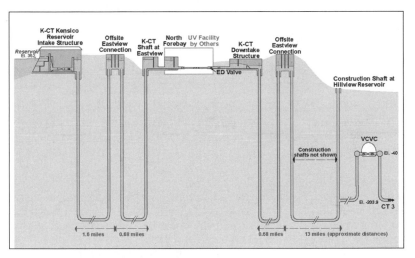

Figure 5.   K-CT hydraulic profile through Eastview site

INTEGRATING THE K-CT INTO THE SYSTEM

Connecting Structures and Shafts

For several reasons, mainly geological, the K-CT will be a deep tunnel, varying in depth from about 400 to 700 feet below ground level. In addition to construction shafts, several vertical water shafts will be required to connect the K-CT into the various surface and subsurface components of the conveyance system. This is best illustrated by Figure 5, which is a schematic profile for the K-CT options that pass through the Eastview site. Starting at Kensico Reservoir, a new intake structure will be required with a connecting shaft down to the tunnel. In the vicinity of Eastview the vertical connecting shafts increase in complexity. The option illustrated on Figure 5, includes an off-site bypass arrangement included for security purposes, which would allow the K-CT to continue supplying New York City with water, even if the whole of the Eastview facilities

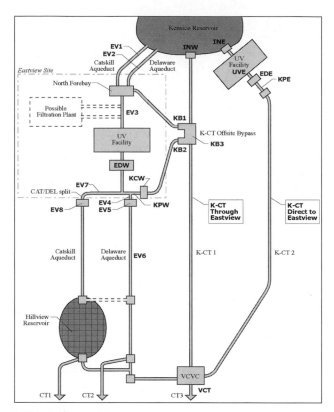

Figure 6.   The K-CT hydraulic model

were out of operation. For this option, a further 6 vertical water connection shafts would be required. Some of the seven connecting shafts in the Kensico/Eastview reach would also be used as construction shafts, and would be constructed to a larger diameter for that purpose. In the reach between Eastview and the termination point of the K-CT at the VCVC, two or three construction shafts would be needed, depending on how the project is broken down into construction contracts. Most likely these construction shafts would be subsequently converted to distribution shafts to supply water to Westchester County. Finally at the VCVC itself, a vertical water connecting shaft would be required to connect the K-CT with the stub tunnel that has already been constructed at the VCVC to receive the K-CT.

Hydraulic Modeling

Once in place the K-CT will operate in conjunction with the existing Delaware and Catskill aqueducts, forming a fairly complex hydraulic system, which includes the constantly varying Hillview Reservoir and the complicated VCVC structure with its 17 lateral flow conduits. This hydraulic system, which is shown schematically on Figure 6, was modeled using H2ONET software, so that the impact of varying the size of the K-CT tunnels could be analyzed. Typical results from the H2O modeling are shown on Figure 7, which gives the required diameters of the K-CT for various discharge capacities and operating scenarios. One of the interesting additional outputs of the modeling

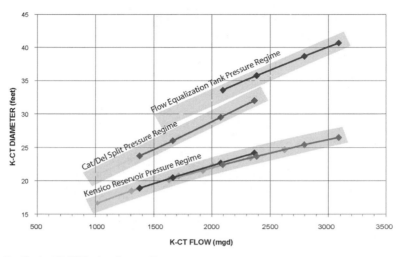

Figure 7. Typical K-CT hydraulic results

is the storage requirements at Hillview Reservoir for various target demands, and the extent to which one compartment only of Hillview Reservoir would be sufficient to meet the hourly variation in the demand on the system. NYCDEP is currently considering various strategies for covering the 90-acre Hillview Reservoir, and these results could influence the staging of this high cost cover. Another interesting outcome of the hydraulic modeling is the potential for operating the K-CT as a separate hydraulic system from the Delaware and Catskill aqueducts and supplying water at higher pressure to the New York City distribution system.

Several sites for the intake structure on Kensico Reservoir are under consideration as shown on Figure 8. One of the factors that could influence the choice of intake site was the impact of intake location on the quality of the water withdrawn from the reservoir. This is potentially a very important factor for the unfiltered New York City water supplies. To investigate this, two and three-dimensional water quality models of Kensico Reservoir were developed. The grid for the three-dimensional model is shown on Figure 9. These models have been successful in quantifying the impact of adverse water quality events in the incoming flows to Kensico Reservoir, on the quality of the water that can be expected at the various intake sites, and have influenced the conceptual design of the intakes themselves. It has been decided that the intakes should have the capability to draw water from the reservoir at several levels.

MINIMIZING THE COST OF THE K-CT

The K-CT, which by itself under certain operating conditions, will be required to supply water to the 9 million plus customers of the system, will be a very large capital investment, on the order of 4 to 5 billion dollars. NYCDEP wishes to minimize the cost of the K-CT, without compromising on system security and reliability, and without compromising water quality. The biggest single factor affecting project cost is the cost of the tunneling, which in turn will be influenced most significantly by the geological conditions and the characteristics of the various types of rock along the prospective tunnel alignment.

The project area from the Kensico Reservoir and Eastview sites to the north, down to the Hillview/VCVC location to the south, is about 15 miles in length and about

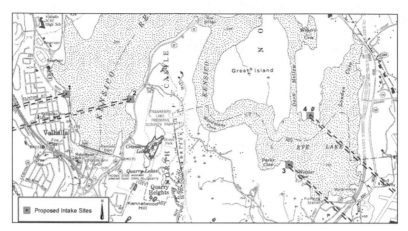

Figure 8.   Layout of intake sites

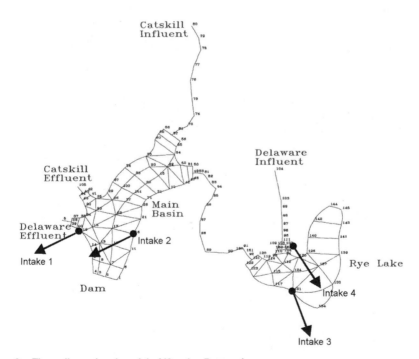

Figure 9.   Three-dimensional model of Kensico Reservoir

5 miles wide. There is a multiplicity of possible tunnel alignments in this area, each traversing different geological formations. This is illustrated on Figure 4, which is the geological map of the project area. The project area is made up of metamorphic rock types consisting of bands of schists and gneiss with intermittent bands of marble, all of which are orientated in a general NNE-SSW direction, which essentially will be the overall orientation of the K-CT.

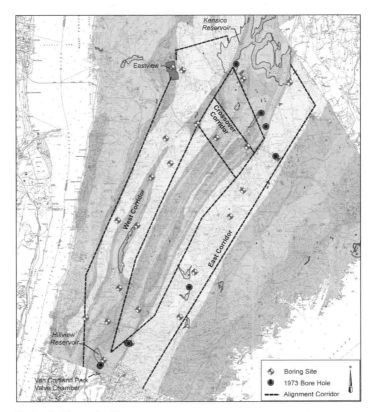

Figure 10. Geological investigations

Site Investigation Strategy

At this stage in the planning process it would not be cost effective or appropriate to undertake a site drilling program tied to a particular tunnel alignment. The strategy that has been developed is to undertake the site investigations in two stages. The Stage 1 investigations are designed to support review of alternatives, leading to a decision on which alignment to select for the K-CT. Stage 2 investigations will provide the basis for the design of the selected project and provide information for bidders. Stage 1, which has now been completed, consists of geological reconnaissance, review of all available previous subsurface data in the project area, particularly the experience derived from the excavation of the Delaware tunnel which was constructed in the 1960s, and finally by a broad brush site drilling program to investigate the typical characteristics of the various rock formations at depth. The program included 21 borings with extensive down-hole and laboratory tests. Also included was a re-logging and testing of cores from 7 exploratory holes that were drilled in the 1970s as part of an earlier planning study for the K-CT. The location of all these holes is shown on Figure 10. The tests on the cores covered the usual range of tests, but in addition included a range of tests performed by SINTEF in Norway to determine the TBM characteristics of the rock types encountered.

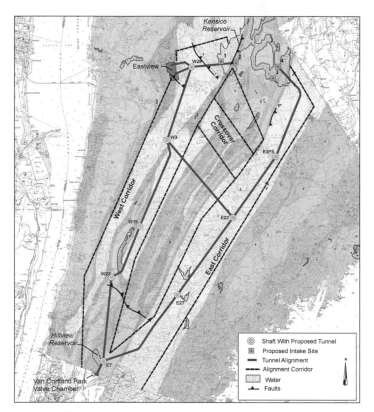

Figure 11.   Typical K-CT alignments

Tunnel Alignment and Profiles

A typical family of tunnel alignments that have been examined is shown on Figure 11. It will be noted that one of the alignments passes through the Eastview site on its way to the VCVC, and another passes directly from Kensico Reservoir to the VCVC. Also shown on Figure 11 is the alignment of the Delaware Aqueduct, which has had a large impact on the selection of tunnel alignments and profiles for the K-CT options passing through Eastview. The Delaware Aqueduct was located in a fairly wide area of schist for the upper half of its alignment, and then follows a moderately broad band of gneiss down to Hillview Reservoir. The invert of the aqueduct is at about El –200, which places it at a depth of between 300 to 600 feet below ground level. The Delaware Aqueduct is a vital component of the water supply system for New York City, and it was determined early in the planning studies that a conservative spacing (in the order of 150 to 200 feet) must be maintained between it and the K-CT.

The rock tests have indicated that TBM penetration rates can be expected to be relatively low in the hard schist and gneiss formations, and relatively high in the marbles. The latter however are more fractured and weathered even at depth, particularly at some of the boundaries with the schist formations. It is estimated that extensive support measures will be required in the marbles for the large diameter tunnels under consideration, and therefore overall TBM advance rates can be expected to be low in these formations, despite their favorable penetration rates. In contrast, past TBM experience in the schist and gneiss rocks in the New York City area has shown that reasonably high

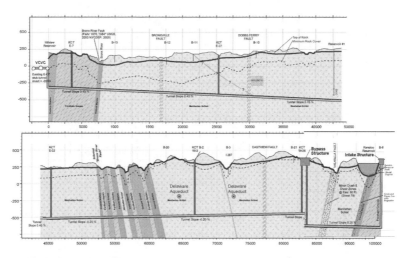

Figure 12.   Typical tunnel profile

consistent advance rates can be achieved despite the hardness of the rock. In the construction of City Tunnel No. 3 in the Manhattan Schist, average advance rates over a five mile TBM drive of almost 100 feet per day were achieved. While this is by no means record breaking, it is respectable. From the geological viewpoint, the strategy in selecting tunnel alignment options therefore has been to follow the wider bands of the schist and gneiss formations, rather than the relatively narrow bands of marble.

Identification of acceptable construction shaft sites has also been a significant factor in tunnel alignment selection and in the breakdown of the project into construction contracts. Westchester County is one of the most valuable areas of real estate on the East Coast. The area is fully developed, and contains many areas of high value housing, numerous select golf courses, and highly appreciated park land. Locating TBM mucking shafts that would be in continuous operation for several years, in this area is particularly challenging. The strategy adopted has been to locate mucking shafts where possible on DEP owned land (at Kensico and Hillview reservoirs and at the Eastview site), and to minimize the number of construction shafts that would require purchase of private land. An extensive screening process was undertaken of potential shaft site areas, and over 70 possible sites with a minimum area of 2 acres were identified. These were further screened until about 12 promising shaft sites in each of the tunnel corridor were identified for the alignment study. From these, shaft sites have been selected that would limit the TBM drives to about 5 miles or less.

The tunnel profiles were established to ensure minimum rock cover in relation to the tunnel internal water pressure, to take maximum advantage of higher rock quality at depth, to maintain an acceptable spacing from the existing Delaware Aqueduct, and to provide reasonable drainage gradients. The plan and profile of one of the typical tunnel options under consideration is shown on Figure 12.

Development of Comparative Tunnel Costs

The development of tunnel costs as a function of tunnel diameter, for the various alignments is probably the single most important aspects of this planning study. The crucial aspect of this is to reflect as realistically as possible the impact of the geological conditions along the various alignment options. A two-pronged approach has been

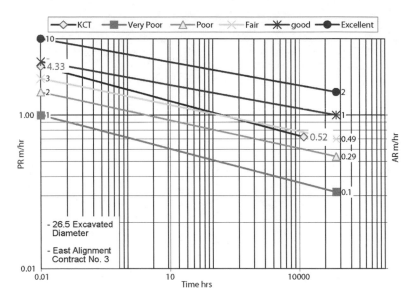

Figure 13.   Typical TBM penetration and advance rates

taken, involving (1) an in-depth review of TBM tunnel operations costs in similar geo-logical conditions in the New York City area, and the US in general, and (2) the devel-opment of resource based cost estimates of the type used by contractors for bidding purposes. The latter approach formed the base case, with the data from previous TBM tunnels in the New York City area used as a reality check.

Resource-based estimates required assessment to be made of construction equipment costs (most notable the TBM), material costs (mainly for temporary support measures, and for the final in-situ concrete lining), and the cost of the work crews for TBM operations, disposal of excavated materials, placement of temporary support, and placement of the permanent concrete lining. Models of this type were developed for a range of tunnel diameters. The critical factor affecting tunnel costs is the assump-tions made on advance rates in the various rock types. The advance rates used were developed using the Q-method of rock mass classification, and Q_{TBM} for estimation of advance and penetration rates, making use of the known geological conditions (partic-ularly with respect to rock weathering and fracturing), experience from other tunnels (back analysis into Q_{TBM}), and the results of the rock core testing. The Q-System was developed primarily from analyses of tunneling experience `in metamorphic rocks, which are the dominant rock type in the K-CT project area. Estimated daily progress rates varied from as high as 150 feet per day down to 10 feet per day, depending on the geological conditions in the various rock types along the tunnel alignments. Tensile and compressive rock strength characteristics are used in Q_{TBM}, which can account for the significant impact of gneissic layering on TBM advance rates that have been expe-rienced in past TBM tunneling in the New York City area. The relative levels of quartz and garnet concentrations are also considered with regard to penetration and advance rates. Typical advance rates and penetration rates assessed for one of the K-CT con-tracts are shown on Figure 13.

The cost estimates that have been developed for each of the TBM contracts, include the construction shafts required for that contract. This is a work in progress, but average tunnel costs for the various contracts are falling in the range of $12,000 to $25,000 per foot of tunnel, excluding contingencies and cost escalation.

THE NEXT STEP

At the time of writing of this paper, finalization of the comparative costs for the various components making up the complete K-CT development options were well advanced. The next step in the screening process will be to apply the reliability model in conjunction with the cost functions for the K-CT to determine the optimum size of the tunnel. Thereafter economic comparisons will be made of the tunnel alignment options leading to the selection of a tunnel corridor and preferred alignment. Once the tunnel corridor has been selected, the planning process will continue with public outreach programs, preparations for land acquisition, and the planning for the EIS scoping meeting. In parallel, the K-CT in the selected corridor will be optimized with respect to final alignment, shaft site locations, intake site choice and options for improving the security and flexibility of the selected scheme.

TUNNELING FOR THE EASTLINK PROJECT, MELBOURNE, AUSTRALIA

Harry R. Asche
Connell Wagner Pty Ltd.

Edmund G. Taylor
Thiess John Holland JV

P. Scott Smith
Connell Wagner Pty Ltd.

Peter Campi
Connell Wagner Pty Ltd.

Ian P. Callow
Australasian Groundwater and Environmental Consultants Pty Ltd.

ABSTRACT

The US$1,850 million EastLink project is a 39km tollway and includes twin 3-lane, 1.5km long, undrained driven tunnels. The hard rock tunnels are being built to avoid passing the motorway through the Mullum Mullum Valley and so to help preserve the unique flora and fauna in the area.

This paper describes several of the key issues in the tunnel design and construction, including:

- the integration of the temporary support design with the Contractor's tunnel excavation and support procedures, including sharing of responsibility between the Designer and Contractor for support type selection and the production of a project specific rock mass classification system to enable more objective determination of ground conditions.

- detailed hydrogeological assessments during the design and construction phases to allow confirmation of design loads for the tunnel lining design and the various suite of waterproofing tools in order to meet the project inflow requirements.

- development of a pre-cast invert solution to the main tunnel secondary lining to enable faster lining construction and to minimize risks to the tunnel membrane associated with fixing reinforcement and construction traffic.

THE EASTLINK PROJECT

Project Overview

The US$1,850 million EastLink project is a 39km tollway being constructed in Melbourne, Victoria, Australia. The tollway links the Eastern Freeway in the north-east of Melbourne to the Frankston Freeway in the south-east. Along the alignment the tollway includes tunnels, 88 separate bridges, 17 major interchanges and a dedicated Shared

534

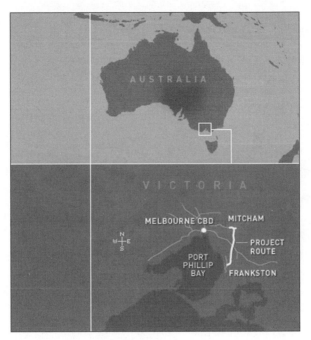

Figure 1.   EastLink Project in Melbourne, Australia

Use Path for pedestrians and cyclists. Figure 1 shows the overall EastLink project in relation to the city of Melbourne.

The State Government of Victoria, through the agency of the Southern and Eastern Integrated Transport Authority (SEITA, www.seita.com.au), has set up the project as a Build, Own, Operate and Transfer (BOOT) scheme. The proponent, ConnectEast (www.connecteast.com.au) is arranging financing, design and construction of the facility, maintenance and tolling.

ConnectEast have let a Design and Construct Contract (known in the US as a Design-Build Contract) to the Thiess John Holland Joint Venture (TJH, www.thiessjohnholland.com.au) as the Principal Contractor for the Project. CW-DC Pty Ltd, a wholly owned subsidiary of Connell Wagner, has been engaged by TJH to provide the detailed design documentation for the civil works of the northern 6km of the Project, including the twin tunnels.

The Driven Tunnels

The project includes twin 3-lane, 1.5km long, undrained driven tunnels built to avoid passing the motorway through the Mullum Mullum Creek Valley and so to help preserve the unique flora and fauna in the area. Figure 2 shows the tunnels in plan.

The 15.8m wide by 11.9m high driven tunnels are excavated through ground conditions with varying strengths from 3.5MPa to 200MPa, through the cross bedded bands of siltstone and sandstone. The concrete secondary lining for the tunnels has been designed to carry the full hydrostatic pressure and the long term ground loads. The temporary support for the tunnel is provided by a suite of five support types of increasing stiffness of rockbolts, shotcrete through to steel sets.

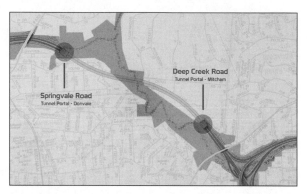

Figure 2. EastLink Tunnels plan

The majority of the main heading drive for the tunnels has been excavated using roadheaders with drill and blast techniques used through the areas of harder rock. Excavation of the bench and invert sections used roadheaders, rock hammers and drill and blast techniques.

The tunnel includes several key fire life and safety aspects, including:

- a 1,200mm walkway at roadway level located in front of the road side barriers;
- egress cross passages at minimum 120m centres;
- a deluge system throughout the tunnel; and
- monofilament polypropylene fibres within the concrete lining.

EXCAVATION AND TEMPORARY SUPPORT PROCEDURES

Legal Framework

There have been two major incidents within Australian hard rock tunneling in recent years. The first resulted in a tragic loss of life due to a rockfall, the second a major collapse to the surface which severely damaged an apartment block. These incidents have highlighted the complexities of Design and Construct Contracts with respect to safety. At the same time, Victoria's Occupational Health and Safety legislation has been recently strengthened (as has similar legislation in the other Australian states,) adding new responsibilities for designers and contractors.

In all tunnel contracts there is a tension between the designer and the contractor with respect to safety and support. The designer must attempt to foresee the ground conditions and to provide sufficient support elements for the proper support of both pervasive and individual ground conditions. The contractor must implement these support systems but wishes to do so on his terms, to maximize the benefits of his own equipment and knowledge.

In a traditional contract (where the designer is employed by the Principal), the designer must provide direction in situations:

- where the contractor is unsure of how to proceed safely, or
- where the designer observes that the ground response is not as expected,

Under these circumstances, the responsibility for safety and support is clearly with the designer to provide the correct direction, and with the contractor to follow that direction.

In a Design and Construct Contract, the designer is employed by the contractor. The designer cannot contractually direct, he may only advise. On the contractor's side, to the same extent that the contractor follows advice from the designer, the responsibility remains placed upon the designer. However, these lines of responsibility can quickly become obscure.

In the EastLink project, the contractor (TJH) and the designer (CW-DC) have carefully set out to make the responsibility for safety and support clear between the parties. This has been achieved firstly in the specifications and work method statements, and then in procedures. In addition to clarifying responsibility, the procedures have been designed to improve communication between all people in the tunnel with benefits for safety and progress.

The procedures included the following:

- daily support reviews,
- daily tunnel inspection checklists,
- ground awareness inductions and daily tunnel face photos, and
- weekly monitoring reviews.

Daily Support Reviews

Every day, or more often if required, a meeting between the designer and the contractor was held to review the support at every face and to adjust the support being installed where necessary. The fundamental principle agreed by the parties was that an increase in support was encouraged as a unilateral decision by the face crew, but that a reduction in support required agreement from all parties at the daily support meeting.

The process was embodied in a form which acted as both the agenda and record of the review meetings. At the same time, it was important that the form was easy to use and provided information clearly to all using the system, especially the construction crews at the face.

The content of the form was refined through a series of reviews between CW-DC and TJH to ensure that the form was accepted and owned by all parties. It was recognised that the form was a "look ahead" rather than a "QA" document recording the installed support. However, the review process included a prompt to consider the installed support in the light of the ground conditions exposed in the recent advances, as these might provide additional information on features not exposed earlier. The form was designed to:

- be limited to a single page;
- record the review items listed in the project specification, reflecting the design requirements;
- record the agreed support type as one of a standard set, with provision for noting additional requirements if needed;
- comprise checklists that allowed for easy recording of the reviewed items by circling responses;
- retain flexibility to allow its use for support determination in heading, cross passage or bench; and
- record the reviewers and their responsibilities in the process.

Daily Tunnel Inspection Checklists

To assist in the effective implementation of the support review process and to avoid complacency during site inspections, CW-DC developed two daily checklists, the Tunnel Site Visit Checklist and the Daily Tunnel Support Review form. These checklists were a designer initiative; they formed the key designer inputs into the daily support review process, including:

- a review of the geological model;
- a review of the design documentation, showing support restrictions, monitoring requirements and information on anticipated conditions;
- a review of the lead heading support and conditions when considering the trailing heading;
- a visit to the tunnel for observation of the ground conditions and support behavior; ands
- a review of the monitoring data.

The checklist included a record of the pre-visit risk assessment. This comprised a review of the tunneling records or contact with senior staff of the Contractor. It alerted CW-DC staff entering the tunnel to any potentially hazardous conditions or new activities in the tunnel. It also served as a briefing on the current activities and any incidents or exposed conditions since the previous visit.

The body of the form provided prompts for observing the rock conditions and support in the tunnel during the visit. These observations were recorded in a manner that matched the inputs to the daily support review. An area for sketches allowed for recording the layout of features in the face and the backs, a clearer record than a description in words.

Ground Awareness Induction and Daily Tunnel Face Photos

The conditions at the EastLink tunnel were different to other tunnels in Australia, and particularly different from conditions in Sydney where many of the tunnelers had recently worked. TJH requested CW-DC to provide training which dealt with the geological issues specific to these tunnels. In addition, CW-DC and TJH geologists provided daily reminders of the issues for the benefit of the workforce:

- a ground awareness induction was set up to educate the tunnel workforce. The induction session involved a video followed by a PowerPoint presentation. These sessions were aimed at inexperienced underground workers or experienced workers who had not worked in this type of ground before. Updates of these sessions were undertaken throughout the year.
- on a daily basis, annotated digital photographs were presented to the tunnelers at their toolbox meetings. The photos provided geotechnical information in areas of potential concern and an overall general description of the tunnel faces. On occasions, photographs were combined with some basic analytical information regarding potentially poor ground or likely rock wedges. Refer to Figure 3 for an example.

Weekly Monitoring Reviews

Each week, all the monitoring results were presented and discussed by the parties. The monitoring included:

- extensometer readings,
- settlement arrays,

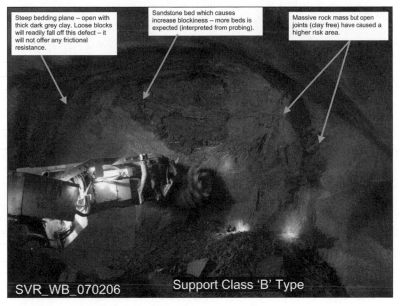

Figure 3.   Example of annotated face photographs

- convergence arrays,
- groundwater monitoring bores,
- flow measurements in Mullum Mullum Creek.

This meeting was usually attended by the external review parties representing the interests of ConnectEast and SEITA. The monitoring results confirmed the visual inspections made of the tunnels by the parties and confirmed that the construction was meeting the design expectations.

The process described above has been a success, and both tunnel headings broke through in late 2006.

HYDROGEOLOGY

The section of the project which is now constructed in tunnel was the subject of an Environmental Impact Study (EIS) in 1978, and approval was granted to build a free-way on the surface. However, in the years after the EIS was approved, public interest in maintaining the native bushland along Mullum Mullum Creek led to the construction of the road tunnels. The tunnels are therefore intimately linked to Mullum Mullum Creek. A key issue in the design and construction of the tunnels has been the understanding and modeling of the hydrogeology, as well as the design and implementation of the waterproofing and the secondary lining.

The tunnel alignment chosen involves a downgrade to reach acceptable cover at the creek crossing, followed by an upgrade to return to the surface. At either side of the creek crossing the tunnel passes beneath hillsides.

At concept design stage, some issues relating to hydrogeology were not well known. Issues which have emerged and which are now better understood include:

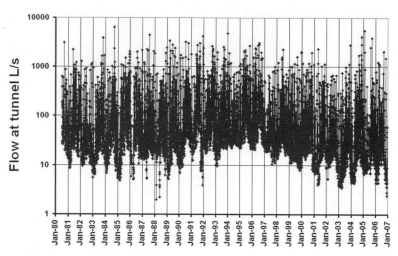

Figure 4.   Mullum Mullum Creek flows

- the hydrology and hydrogeology of Mullum Mullum Creek, and how this relates to the environment,
- the groundwater chemistry,
- the hydrogeological modeling of a tanked tunnel, and
- waterproofing and structural design of the secondary lining.

Hydrology and Hydrogeology of Mullum Mullum Creek

Mullum Mullum Creek is a tributary of the Yarra River. Set in the slightly more hilly terrain to the east of Melbourne city, the creek catchment experiences an average rainfall of about 860mm (34 inches) per annum. The rainfall, like that in Australia generally, is not constant, but varies significantly each year. In addition, the temperature and the consequent evaporation characteristic varies significantly seasonally, as well as yearly.

A gauging station measuring the daily flow in Mullum Mullum Creek is located downstream of the tunnel crossing and has operated since 1980. The measurements (adjusted to account for the upstream location of the tunnel) are shown in Figure 4, with a log scale for flows to denote the occurrence of low flows.

During December to March (summer) the rainfall is more frequently a heavy storm downpour (with higher than usual runoff) and there is an increase in evapo-transpiration due to higher temperatures. As a consequence the low flows reduce. The records show significant variation over two and a half decades. The years from 1993 to 1999 were exceptionally wet with flow never falling below 10 L/s, whilst the years from 2001 to present have been exceptionally dry, with flows falling below 5 L/s over summer in most years.

From a hydrogeological perspective, the issue is that water that may flow into the tunnel is water derived directly from the creek, or is water which may have entered the creek but which has been intercepted by the tunnel.

Mullum Mullum Creek runs through a landscape composed of Silurian rocks of the Anderson Creek Formation. This formation of interbedded mudstones and sandstones is gently folded and sheared in places. Some dykes and sills of basaltic material have been injected into the formation, usually at pre-existing weak zones. The rock is

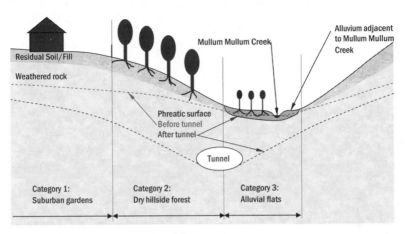

Figure 5. Hydrogeological conceptual model

weathered towards the surface. At the surface, the material may have weathered to a residual soil, with some fill or colluvium also present. In the valley of Mullum Mullum Creek, there is a small pocket of alluvium up to 4m deep, several tens of meters wide. In the slopes on either side of the creek, there is a native forest and higher up the slopes, houses sit in lots of relatively large size with gardens around each house.

Figure 5 shows a conceptual model of the hydrogeology before and during tunneling construction. Three ecosystems are identified:

1. Suburban gardens which sit high on the hillside. The plants are fed by rain and by irrigation systems. The roots of the plants sit in the unsaturated zone, above the water table. During tunnel construction it is expected that the water table will be pulled down to the tunnel level, but this is not expected to affect these gardens.

2. Hillside forest. The plants are native trees which sit above the water table and derive water from the unsaturated zone. Again it is not expected that these trees will be affected during construction. However, this ecosystem is monitored by the contractor (and to date no impact due to tunneling has been observed).

3. Alluvium at Mullum Mullum Creek. The plants are native plants and shrubs which sit in the water table. During construction a double phreatic surface has developed with the alluvium recharged from the creek water. These plants are being monitored (and currently showing no impact due to tunneling).

Groundwater Chemistry

The groundwater chemistry was an important driver for the project—it was a key factor in the decision to construct a fully tanked tunnel. CW-DC engaged an expert in the field of groundwater attack on concrete structures, Professor Brian Cherry of Monash University. Professor Cherry's work is discussed in summary here. Two main points are important for tunnels in groundwater.

Firstly, in order to understand the groundwater chemistry, it is very important to sample properly. Unless special care is taken, the samples are an uncontrolled mixture of groundwater from all levels of the borehole, as well as drill and flush water in the borehole. With such a sample, useful chemical analysis is impossible. Cherry (1993) describes in detail the necessary procedures for taking a useful sample.

Secondly, amongst many components of the groundwater, the effect of dissolved CO_2 can be critical for a tunnel in contact with the groundwater. Two situations can be distinguished.

1. Under certain circumstances, the CO_2 can attack the concrete itself:

$$CaCO_3 + H_2CO_3 \rightarrow Ca(HCO_3)_2$$

    In addition under the same circumstances, similar reactions will occur which can be deleterious to reinforcement.

2. In the reverse situation, the CO_2 will combine with Calcium or Magnesium salts to form an insoluble precipitate:

$$Ca(HCO_3)_2 + Ca(OH)_2 \rightarrow 2\ CaCO_3$$

$$Mg(HCO_3)_2 + Ca(OH)_2 \rightarrow CaCO_3 + MgCO_3$$

    The propensity for these reactions to occur are increased with a reduction of CO_2 partial pressure as the groundwater is released from pressure, and also with an increase in pH due to contact with non-hydrated cement, such as in no-fines concrete.

Distinguishing between the potential to attack the concrete or to clog drainage systems is made using the Langelier Saturation Index (LSI) which is based on the pH of the groundwater compared to the pH of saturated Calcium ions.

On the EastLink project, the potential to clog was found to exist in most of the boreholes. A practical demonstration of the effect was made by passing groundwater sampled from one of the boreholes through a small sample of no-fines concrete. The flow completely blocked up within a few days by the deposition of insoluble precipitate of Calcium and Magnesium Carbonate. In practical terms this meant that a drained tunnel with a flat invert would require significant maintenance to avoid clogging of the drainage systems and that the invert would be potentially subject to uplift forces.

Hydrogeological Modeling

The modeling methodology used was typical for tunnel projects:

- choose the model boundaries and input the topography (refer Figures 6 and 7)—note that the model covers an area of 5.6 km^2 approximately;
- extrapolate the known head data to provide initial conditions for the model;
- adopt a (continuous and uniform) recharge rate;
- calibrate the model for hydraulic conductivity at steady-state against the known groundwater heads (refer Figure 8);
- model the tunnel to produce predictions of impact on groundwater.

During the design of the secondary lining, the modeling assumptions were subjected to some scrutiny, given that the output of the modeling included predictions of hydrostatic head on the structure. It is noted that the adopted unfactored design head onto the tunnel was below the ground surface.

Particular issues which were identified included:

- Choosing the appropriate initial conditions for the model. In the EastLink project most of the head data was collected during dry years. Therefore a correction was made for the initial conditions based on a 2D transient model using all rainfall data from 1930 onwards.

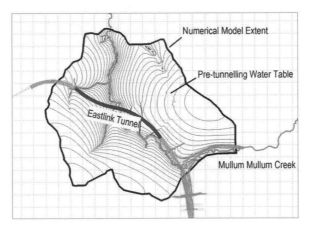

Figure 6. Hydrogeological model—plan

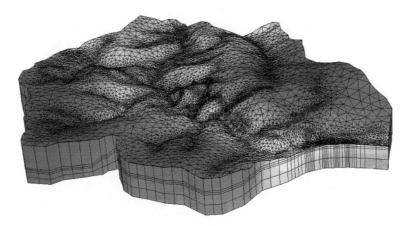

Figure 7. Hydrogeological model—oblique

- Choosing the appropriate recharge rate. This rate is a simplification of a very complex process, therefore three recharge values were chosen (dry, wet, and after discussion, exceptionally wet). The values adopted are given in Table 1.

- Choosing the appropriate storativity value for the model. This parameter does not affect the initial steady-state calibration, but is a key parameter for assessing transient effects and had to be calibrated following excavation of the tunnel and measurement of groundwater drawdown compared with time and assessment of the measured inflows into the tunnel. The value finally adopted was 0.025% which is a number which is relatively low but compatible with the fracture spacing and packer test results. A typical transient model result is shown in Figure 9, compared with measured values.

- Modeling longitudinal flow along the tunnel. This issue turned out to have the greatest impact on the final groundwater table. Initial modeling adopted the simplistic view that the completed secondary lining could be represented as an

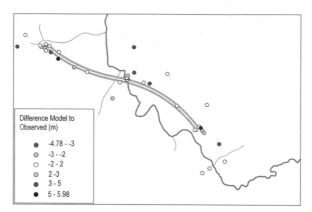

Figure 8. Model calibration at monitoring bores

Table 1. Recharge rates for EastLink modeling

Scenario	Recharge as % Annual Rainfall (860 mm/a)	Recharge to Catchment (5.6 km²)
High Recharge	5.7	8.7 L/s
Low Recharge	2.3	3.5 L/s
Very High Recharge	5.7% of 1,300 mm/a	13.1 L/s

impermeable block of rock. However, the structural analyses of the secondary lining showed that invert deflections create a significant gap, along which water will travel. This effect was analysed in the model by the incorporation of a 2D planar element which was positioned at the tunnel invert. To mitigate the longitudinal flow and also allow improved recovery of the groundwater table, external compartments were positioned at three discrete locations along the alignment. At the locations of these external compartments, the planar element was interrupted. Figure 10 shows the expected steady state groundwater table along the tunnel. Note how the longitudinal flow affects the groundwater table.

WATERPROOFING APPROACH

The project requirements for waterproofing were very specific and required an undrained tunnel fully tanked using a membrane system. There was also an upper limit of 1 L/s placed on groundwater inflow into the total tunnel system.

It is within this context that the waterproofing details of the tunnel are based.

Construction Risk Assessment

As a part of the constructability input into the design, a detailed construction risk assessment was undertaken by the contractor to highlight the potential areas of risk in achieving these stringent project requirements. The resulting design was then developed with the following philosophy.

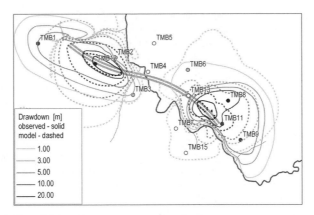

Figure 9.   Transient model drawdown versus measured values

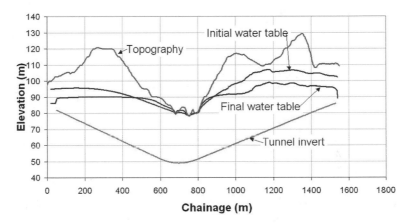

Figure 10.   Predicted groundwater head along tunnel alignment

With the design specifying a continuous, fully circumferential and impermeable membrane as the primary waterproofing medium, there was a strong focus on ensuring membrane integrity at all times.

Surface preparation of the shotcrete placed as a part of the primary support was the initial focus. With the tunnel action under full hydrostatic head, there are large forces exerted from the lining to the rock in the haunch and crown areas, consequently sandwiching the membrane. Therefore the shotcrete finish and profile as well as the geotextile protection layer were given a high priority.

Tunnel membrane installation and welding techniques are well established so the design and construction focus was on quality control, the use of experienced labor and the use of project specific installation gantries.

Following membrane installation, the design and construction methods were developed to ensure minimal opportunity for damage to the membrane. The tunnel structure required reinforcement to the invert and lower haunches. To minimize potential damage to the membrane from construction traffic in the invert and from the placement of reinforcement to the invert and haunches, two key initiatives were developed.

Precast tunnel segments were designed for placement in the invert. This greatly reduced the need to traffic over the membrane as well as eliminating the placement of reinforcement in this zone. The segments were trafficable once they were fully grouted in place. (Refer to the next section for more details).

With the invert segments in place, haunch reinforcement could be positioned from the starter bars extending from the segments. A specially designed and manufactured reinforcing mesh was developed to meet the structural requirements. The mesh was rolled to suit the tunnel curvature which, in conjunction with the mesh concept, simplified the installation and so minimized the potential damage to the membrane.

The final aspect to the construction risk assessment was recognition of the importance of the hydrogeological regime. Observations of groundwater inflow into the tunnels during the excavation phase combined with an understanding of the impact on groundwater levels enabled sections of the tunnel to be defined as "wet." Such wet areas were then considered to be higher risk zones for potential tunnel leakage and so a suite of contingency (or supplementary) waterproofing details were prepared for use as determined by the contractor in these areas.

Groundwater Recovery and Operational Impacts

Following completion of the tunnel lining, the groundwater will recover from the drawdown state that existed during construction. The tunnel structure becomes fully loaded hydrostatically and as a result will deflect. It is anticipated that the invert may deflect up to 15mm in the long term.

To limit the longitudinal flow, external compartments or flow barriers were designed to seal this flow gap around the tunnel. The compartments comprised rear guard water barriers cast into the rock with a membrane flap welded between them and the tunnel membrane. The flap is designed to accommodate movements from the tunnel lining deflecting under load and from long term creep. To ensure the water path is not short circuited through the rock, pressure testing was carried out in the vicinity of these external barriers and where the ground was found not to be tight, grouting was carried out.

Despite design and construction best efforts, there is the potential for groundwater to penetrate the membrane and leak into the tunnel. To both isolate potential leaks and to protect the tunnel lining, various additional measures are incorporated into the design. Internal compartments are created at the membrane-to-concrete interface by use of fully circumferential and longitudinal rear guard waterstops welded to the tunnel membrane. The joints between precast invert segments and between segments and the cast in-situ arch are sealed by use of hydrophilic waterstops and low-shrink grout. The internal compartmentalization therefore allows any operational rectification of leaks to be targeted to a specific segment of the tunnel.

Waterproofing Details

A "tool box" of waterproofing design details were developed that comprised the "standard waterproofing solution" to meet the project specifications and operational requirements as well as "supplementary waterproofing solutions" for use by the contractor as dictated by the construction risk assessment (as discussed above).

The standard solution comprised:

- geotextile fleece, 500g per m^2 to the full tunnel perimeter;
- 2mm thick PE membrane to the tunnel arch;
- 3mm thick PE membrane to the invert;

- continuous longitudinal rearguard waterstop above the invert/arch construction joint;
- radial rearguard waterstops at 40m maximum centres;
- hydrophilic waterstops to joints between invert segments and between invert segments and the arch; and
- a minimum number of external compartments that enable reasonable recovery of the groundwater regime.

The supplementary details, provided for use by the contractor to supplement the standard details, as a line of defence to minimize risk in areas of potential high inflow (or "wet" zones) are:

- additional external compartments to isolate "wet" from "dry sections of the tunnel;
- reduced spacing of internal compartment rearguard waterstops;
- inclusion of re-groutable tubes at compartment waterstops; and
- the provision of a double membrane sandwich with groutable systems to the invert.

Construction Phase Assessments

The hydrogeological assessment (as discussed in the earlier section on Hydrogeology) was based upon a model calibrated against observed geology, groundwater movements and inflow into the tunnel during the excavation phase. On-going groundwater monitoring enabled the best possible understanding of the groundwater regime and this knowledge was used to both determine the best possible location for external compartments and also clearly define the "wet" areas of the tunnel.

The process to establish the minimum number and location of external compartments was to run the hydrogeological model under transient conditions and compare the projected groundwater recovery to the pre-existing conditions. The location of the compartments was fine tuned by a detailed review of the mapped geological conditions in the tunnel, focusing on areas of "tight" rock (minimal jointing and absence of potential interconnection features).

During and following excavation, the tunnel was mapped and monitored for groundwater inflow. These observations enabled the tunnel to be divided into "wet" and "dry" areas. At the time of writing, this information has not yet been translated into what additional or supplementary waterproofing measures the contractor may employ.

THE PRECAST INVERT

The Drivers—Program, Leak Prevention

The EastLink tunnel is being delivered by a BOOT scheme in which the private sector provides financing to build the project in return for toll revenue. This delivery method emphasises speed of construction; the financial outcome is better if the toll revenue arrives earlier.

During early studies, the time taken to construct the invert arch was identified as a critical aspect for the tunnel program. The decision was therefore taken to use a precast invert section in the secondary lining. This decision has brought with it one major benefit, as well as some significant challenges.

Initial thinking involved a precast invert with gasketed seals, similar to a bolted segmental lining. However, it proved impossible to design a connection between the segmental invert and the waterproof membrane in the crown. The decision was then

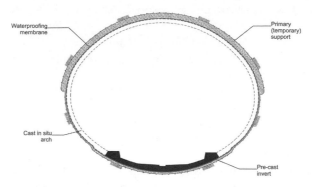

Figure 11.   Precast invert and secondary lining

made to proceed with a full circle of membrane and to place the precast invert inside the membrane. Figure 11 shows the precast invert.

Schockemöhle and Heimbecher (1999) review the success of membrane water-proofing tunnels with invert arches. The German experience is that most membrane damage occurs when invert reinforcing steel comes accidentally in contact with the membrane; the risk being greatest at the bottom of the tunnel. This damage is usually extensive and that without considerable repairs, the waterproofing will not work. (The second most common cause of membrane damage is due to voids at the crown of the tunnel.) The major benefit of using a precast invert segment is that the segment protects the area of greatest risk.

The challenges which are being overcome in the EastLink tunnel include:

- placing the precast units onto membrane in such a way that the membrane is not damaged;
- managing the joint between the precast units and the cast-in-situ lining; and
- providing internal and external compartmentalization of the membrane with a precast unit.

Design Issues

The principal load on the lining is the hydrostatic load caused by the recovery of the groundwater following completion of the waterproofed secondary lining. Examination of Figure 10 shows that the hydrogeological model, even when calibrated to a pre-construction steady state, can include uncertainty. Consideration of other effects such as the variation between wet and dry years has highlighted the probabilistic nature of the groundwater loading. These effects have been taken account of in choosing an appropriate load factor in the structural analyses.

Other loads which have been taken account of in the design include:

- the transient loading as the water table rises;
- the effect of a possible leak around the perimeter to create asymmetric loading;
- the effect of construction loading including trucks, concrete formwork, fill placing;
- the invert fill and roadway pavement;
- the gap between the secondary lining and the rock, filled with the compressible geotextile fleece and membrane;

Figure 12.   Blinding beneath precast invert

- the effect of different concrete moduli depending on the load and time; and
- for the precast invert—casting and handling loads.

Following assessment of each of the design load cases and several iterations regarding the shape of the tunnel invert, the final design solution adopted included:

- reinforced 50MPa concrete invert segment complete with concrete corbel as a base for the arch formwork and starter bars into the arch pour;
- lightly reinforced 40MPa concrete lower haunch areas with the reinforcement extending from the precast invert segments to approximately roadway level; and
- unreinforced 40MPa concrete tunnel arch.

A particular issue for the reinforcement design was the difference in shrinkage between the precast invert modules and the cast-in-situ lining. This differential shrinkage occurs between each arch pour and was controlled with additional longitudinal reinforcement near the invert segment to arch construction joint.

Construction Details

The main issue for the construction is the placement of the precast invert such that the membrane is not damaged whilst maintaining the tight placement tolerances set by the designer. Two issues must be considered, damage to the membrane when placing the unit, and avoiding air voids in the grout under the units so that the membrane has no place to expand and tear due to the decreased section thickness.

The process adopted starts with the production of concrete invert blinding to within 10mm of the theoretical position. This is achieved by using curved steel screed formers, placed by survey, in a hit and miss sequence (Figure 12).

The precast invert (Figure 13) is then placed on groutable pads on either side of the units. These pads are made from polythene and are a flat rectangular profile normally used in strip drains, with a water path in all directions. This detail allows the free flow of grout to allow the complete filling of the underside of the profile.

A low shrink grout is then pumped into the void beneath the segments through pre-formed grouting holes. Due to the curved profile of the segments and the longitudinal grade of the tunnel the grouting process can only be undertaken up to four segments at a time if grouted in a single stage (Figure 14).

Figure 13. Precast invert units   Figure 14. Precast invert units located in tunnel

Due to the complex nature of this construction, detailed full scale trials were undertaken in a purpose made structure outside the tunnel portals. The structure built to replicate the tunnel invert profile and grade and allowed repetitive trials for the placement and grouting of the precast segments and enabled the optimization of the contractor's work methods and a training ground for the tunnel workers. Only on successful completion of each of the aspects of the placement and grouting process did production within the tunnel commence.

Joints between the precast units have a shear key profile to lock the units together. The bottom of the shear key joint included a seat for locating a triangular section of hydrophilic material. The rest of the joint is grouted up with low shrink grout.

REFERENCES

Cherry, B.W., 1993. "Cathodic Protection of underground reinforced concrete structures" B.W. Cherry, in "Cathodic Protection Theory and Practice" edited by V. Ashworth and C. Googan, Ellis Horwood, pp 326–350.

Schockemöhle, B. and Heimbecher, F., 1999. "Stand der Erfahrungen mit druckwasserhaltenden Tunnelabdichtungen in Deutschland," Bauingenieur, Vol. 74, No 2, February 1999, pp 67–72.

WATER TUNNEL IN LIAONING PROVINCE, NORTHEAST CHINA

I. Michael Gowring
Consultant

Zhong Gang
Department of Water Resources of Liaoning Province

Chen Yong Zhang
Design Research Institute of Water Resources and Hydropower

Li Jiuping
Liaoning Runzhong Water Supply Co., Ltd.

ABSTRACT

The Liaoning Dahuofang Reservoir Water Transfer Project is an important development for the Department of Water Resources of Liaoning Province in northeastern China. The 85.3 km water conduit will transport water from the Hun River downstream of the Huanren Reservoir to the outlet on the Suzi River leading to the Dahuofang Reservoir. The additional water is required to sustain the growing cities of Fushun, Shenyang, Liaoyang, Anshan, Panjin and Yingkou. The project is divided into seven contracts. Three TBM contracts comprise almost 75% of the work and the others are drill and shoot. The focus of this paper is about the creditable TBM performances achieved and some of the problems encountered by Chinese contractors; notably two having had no previous TBM experience.

REGIONAL DEVELOPMENT

Liaoning Dahuofang Reservoir Water Transfer Project is a gravity diversion scheme to transfer water from the Hunjiang catchment area in Huanren County, Liaoning Province through a tunnel to the Suzi River in Xinbin County. The additional water then flows in the Suzi River to the Dahuofang Reservoir from where it is regulated to provide water to six cities in the central area of Liaoning Province. The reservoir is in the heart of the grain producing area of Liaoning Province in addition to being close to a heavy industrial region which produces iron and steel, coal, petroleum derivatives, chemicals, electric power, machinery and building materials. The region has a developing economy and a dense population, but it lacks water resources. A 544m^3 per capita of water resource availability and a utilization ratio of 81.9% makes the area one of the most severely lacking in water resources nationwide. In addition, the water quality of Hunhe, Taizi and Daliaohe rivers is seriously polluted, the ecological environment is deteriorating and other problems such as land subsidence are occurring as the result of lowering of the water table due to widespread pumping of underground water. This project came to fruition to strengthen the local economy by providing additional water for industry, agriculture and urban development. The adopted solution was to transfer water from the mountainous area in the upper reaches of the Hunjiang River, which is sparsely populated and enjoys a per capita water resource availability of 4187m^3 and a utilization ratio of only 9.39%. Figure 1 shows the location of the project.

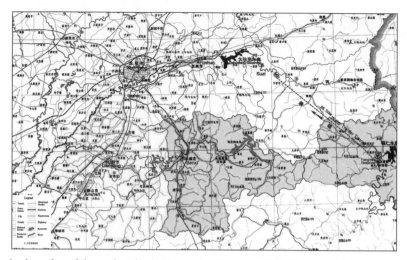

Figure 1.   Location of the project. North Korean border, upper right; Shenyang, upper left.

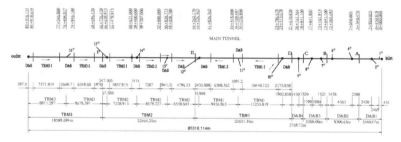

Figure 2.   Schematic layout of Dahuofang water diversion project

PROJECT OVERVIEW

The project comprises an intake structure, the main diversion tunnel, an outfall and a dissipating structure. The total tunnel length is 85.31km, with an excavated diameter of 8m. The designed open channel flow within the tunnel is 70m³/s, with allowance for a peak of 77m³/s. The gradient is 1:2380. The tunnel construction was divided into seven contracts, three utilizing TBM excavation and the remaining four being drill + shoot methodology. In addition to the inlet and outlet portals, there are 14 intermediate adits used for construction access. A schematic arrangement of the tunnel and adits is shown in Figure 2. A summary of the contract limits and the contractors is shown in Table 1.

The access adits are of varying lengths and gradients. In the TBM sections, the adits are used to move the operation forward as the excavation progresses, to reduce the distance for muck removal and the travel time for people and materials to and from the face. The steeper adits are serviced by a rail haulage way with motive power provided by a surface winch. Table 2 summarizes the adit dimensions. The adits with larger cross sections have been designed to accommodate the installation and removal of the TBMs.

Table 1. Summary of contract limits and contractors

Contract	Contract Limits (stations) Start	Finish	Total Length	% of Total	Contractor
D&B1	0 + 000.0	5 + 340.5	5,340.5	6.3	China Railway 13th Bureau Co. Ltd
D&B2	5 + 340.5	13 + 641.1	8,300.6	9.7	China Railway 13th Bureau Co. Ltd
D&B3	13 + 641.1	18 + 841.1	5,200.0	6.1	Sino Hydro Engineering Bureau 6th
D&B4	18 + 841.1	21 + 610.8	2,769.7	3.2	Liaoning Water Conservancy Bureau (LWCB)
TBM1	21 + 610.8	44 + 262.2	22,651.4	26.6	Beijing Vibroflotation Engineering Co. (BVEC)
TBM2	44 + 262.2	66 + 726.4	22,464.2	26.3	China Railway Tunnel Group (CRTG)
TBM3	66 + 726.4	85 + 316.1	18,589.7	21.8	Liaoning Water Conservancy Bureau (LWCB)
Total (m)			85,316.1		

Table 2. Adit dimensions

Adit No.	Length (m)	Decline %	Dimensions (m)
1	277.9	4.5	5 × 5
2	796.2	14.0	5 × 5
4	843.0	46.6	5 × 5
6	377.1	24.9	5 × 5
7	519.1	20.8	5 × 5
8	925.6	12.0	5 × 5
9	1,108.7	8.0	5 × 5
10	1,573.7	10.0	6.6 × 6
11	1,004.7	21.5	5 × 5
12	2,606.7	12.0	6.6 × 6
13	1,371.3	16.9	5 × 5
14	1,283.2	11.6	5 × 5
15	1,776.2	9.0	6.6 × 6
16	642.4	10.7	6 × 6.6

The agency responsible for the project is the Department of Water Resources of Liaoning Province. The construction management is undertaken by the Liaoning Runzhong Water Supply Company Limited. The Water Supply Company is an "arm's length" subsidiary of the Department of Water Resources. The Department, a Government Agency, is prevented by law from managing construction work. The design is performed by the Design and Research Institute of Water Resources and Hydropower Group of the Water Conservancy Ministry of China for Liaoning Province. The Department of Water Resources is based in Shenyang. The main construction office for the Water Supply Company is in Huanren which is close to the intake, with a subsidiary office at Yongling towards the outlet.

This is an ambitious project, even by world standards. TBMs are not new in China, but the related management skills and construction workforce proficiencies are still developing. Construction started in September 2002 and project completion was originally scheduled for December 2008. Overall progress has been less than scheduled

Table 3. Rock classifications in tunnel

Rock Class	Rock Characteristics	Rock Mass Index T	Stress Ratio S	Type of Support
II	Basically stable. Rock is integrally stable, without plastic deformation. Blocks may fall locally	85≥T>65	>4	Random rock anchors or thin layer of shotcrete. For large span, pattern rock anchors with mesh reinforced shotcrete
IIIa	Rock is integrally stable, generally without plastic deformation. Local stability is poor with blocks falling and walls caving	65≥T>55	>2	Pattern rock anchors. Mesh reinforced shotcrete, minimal thickness
IIIb	Poor local stability with the likelihood of plastic deformation. Rock falls may occur without support although softer rock without discontinuities may have a limited stand up time	55≥T>45	>2	Pattern rock anchors. Mesh reinforced shotcrete, increased thickness. For 20–25m spans, place concrete lining
IV	Unstable. Very short stand up time. Large scale deformations and failures will occur without support	45≥T>25	>2	Initial support: Pattern rock anchors and mesh reinforced shotcrete. Permanent support: Concrete lining
V	Extremely unstable. Rock will not stand unsupported and deformations will occur	T≤25		

and although mitigating measures have been adopted, the currently projected completion date is April 2009. The final cost of the project is expected to be approximately RMB 5 billion ($654 million).

A follow on project (Stage 2) is in the planning stage. This work will consist of a network of pipelines and distribution structures to convey water from the Dahuofang Reservoir to the aforementioned cities and surrounding districts. There will be a total of 260km of spirally prestressed concrete pipes ranging in diameter from 3.2m to 1.4m.

DESIGN OVERVIEW

The tunnel route is located in an area of low mountains, passing under about 50 peaks, the maximum height being 885m, although most are within the 500–650m range. The depth of cover varies between 52m and 600m; 8% is less than 100m; 55% is between 100m and 300m; and the remaining 37% is in excess of 300m. The tunnel route passes through Archean, Proterozoic and Mesozoic strata, the rock being mostly medium hard to slightly weathered. A total of 29 faults have been identified along the tunnel alignment, of which 5 are considered to have an influence on the design.

The classification of rock is based on the method used in the *Code for Geological Survey in Water Resources and Hydropower Project Ref: GB50287-99*. The Rock Mass Index "T" is derived from (1) the uniaxial saturated compressive strength, (2) the seismic velocity coefficient, (3) the condition and spacing of the joints, (4) the water inflow and (5) the strike of the main joint. Table 3 identifies the rock classifications that apply to the tunnel.

Table 4. Distribution of rock types and classifications

Station	Length of tunnel (m)	Length of various classes of rock sections (m)					Stratum Code	Rock Description
		II	IIIa	IIIb	IV	V		
0+000–1+060	1,060		520	250	260	30	K_1l	Volcanic breccia, tuff
1+060–2+050	990			830	140	20	Mr_2^1 / Pt_1gx	Compound granite, marble
2+050–12+420	10,370	5,270	2,930	1,770	360	40	Qny	Quartz sandstone
12+420–20+320	7,900	1,210	700	2,100	3,490	400	Pt_1d/ Pt_1gx	Marble, granulite
20+320–30+045	9,725	3,345	1,090	4,880	380	30	$\xi\pi_5^{3(1)}$	Syenite
30+045–32+855	2,810		2,800			10	Mr_1	Compound granite
32+855–42+560	9,705	6,020	2,145	1,320	200	20	M_1	Migmatite
42+560–49+040	6,480		3,565	2,715	185	15	K_1xl/ J_3x	Brecciated lava, andesite, tuff
49+040–66+760	17,720	8,700	3,260	4,330	1,320	110	M_1	Migmatite
66+760–69+330	2,570			2,040	520	10	K_1xl	Tuff
69+330–75+450	6,120	1,800	990	2,970	340	20	M_1	Migmatite
75+450–85+315	9,866		1,950	6,653	1,243	20	K_1xl	Brecciated lava, andesite, tuff

The distribution of rock types and classifications along the tunnel alignment is summarized in Table 4.

Figure 3 illustrates the initial and final ground support specified for the various rock classifications. It may be noted that the minimum support required is random rockbolts and 80mm shotcrete lining. Also, as a minimum, a concrete invert is maintained throughout the entire length of the tunnel. The requirement for a full circle shotcrete lining in the most competent rock is a departure from the practice in other countries, at least in North America. Furthermore, the shotcrete is placed immediately behind the TBMs during the course of excavation. In North America there are many miles of water tunnels under pressure, and more specifically open flow, in unlined competent rock that have provided years of satisfactory service. A second observation is that all the rockbolts are identical—22mm dia. × 2m long, for all ground classifications. They are more accurately described as untensioned rock dowels. The rock dowels are installed with cement grout capsules which require one day to set up and acquire a working strength. The outer ends are pre-bent to a "J" configuration. There is no anchor plate or nut. A final point of note is the size of the I-16 steel ring beams. This is a slender section of $160 \times 88 \times 20.5$kg/m with a section modulus of 141cm^3. In North America, an appropriate steel section for this diameter of tunnel would be a W6 × 25 wide flange ($150 \times 150 \times 37.2$kg/m) which has a modulus of 274cm^3.

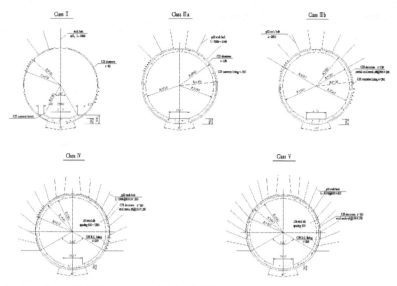

Figure 3.   Structure of initial support and lining in TBM

TBM PROCUREMENT AND SPECIFICATIONS

Three TBMs were furnished for three separate contracts on this project. Two machines were designed and furnished by The Robbins Company of Solon, Ohio, USA and the third was supplied by The Wirth Company of Erkelenz in Germany. The Robbins machines and back up were manufactured in Dalian, Liaoning Province. The Wirth machine was remanufactured in Erkelenz. The back up was designed by Rowa of Wengen in Germany and fabricated and assembled in China with components supplied from Europe. This machine was originally built in 1992 for the 12 km T4 section of the Vereina tunnel in Switzerland.

The Robbins machines are identical. They were manufactured by Dalian Huarui Heavy Industry International Co. Ltd., (DHI). DHI did not have previous experience with TBMs, but they were a logical choice as a manufacturer of heavy equipment such as steel mill machinery, coal mining equipment, and heavy bridge and gantry cranes. Other factors included the encouragement by the Water District to use a Liaoning based facility for manufacture and the availability of good transportation to the project site. Robbins provided significant supervision at DHI. The senior shop manager and two shop technicians from Robbins USA were seconded to Dalian for the entire duration of manufacture. Their function was to provide quality assurance and to monitor the schedule. Four engineers and eight mechanics from Robbins China worked at DHI for the entire duration. During final assembly and commissioning, a hydraulics technician, an electrician and two field service supervisors provided expertise. In addition, Robbins USA engineers and the project manager provided support as required.

A comparison of the basic TBM specifications is summarized in Table 5.

Table 5. TBM specifications

Description	Unit	Robbins	Wirth
Machine Diameter (nominal)	m	8	8
Number of Cutters	no.	51	59
Diameter of Cutters	in.	19	17
Maximum individual cutter load	kN	311	240
Cutterhead Operating Thrust	kN	17,105	16,500
Maximum Operating Thrust	kN	22,934	21,000
Maximum Hydraulic Pressure	bar	345	325
Number of Drive Motors	no.	10	12
Power/Drive Motor	kW	300	280
Total Cutterhead Power	kW	3,000	3,360
Cutterhead Speed (constant torque)	RPM	0–4.63	0–7.4
Cutterhead Speed (constant power)	RPM	4.63–6.93	
Number of Thrust Cylinders	no.	4	8
Stroke	m	1.83	1.8
Primary Voltage	V	10,000	10,000
Secondary Voltage	V	690/380	690/380
Conveyor Capacity	m^3/hr	1,110	780

TBM CONTRACTS

Summary of the Work

The schematic layout in Figure 2 provides an indication of the limits of the three TBM contracts. Each TBM contractor is responsible for a defined length of the main tunnel, which in addition to the TBM excavation, includes construction of the access adits, underground work areas and surface maintenance facilities including shops, warehouse, office and camp. Each contract spans at least 2 adits and in the case of TBM2, the contract extends to 3 adits. The adit spacing is typically 10km or less. This distance is considered the optimal maximum for an efficient tunnel operation. As each TBM holes through to the next adit, the entire operation is jumped forward, leaving the completed reach available to complete the permanent lining. A noteworthy feature common to all the TBM contracts is the muck removal by continuous conveyor. The reaches as shown on the schematic have not necessarily been excavated by TBM. In fact for reasons discussed later, excavation of a significant length of tunnel has been substituted by drill + shoot methods. Another noteworthy feature is the sophistication of the chambers at the foot of Adits 10 and 15 that are equipped with overhead bridge cranes for underground assembly of the TBMs. These cranes continue to be used for materials handling during the tunnel operations. See Figure 4.

TBM1 Contract encompasses 22,651m of tunnel. BVEC is the contractor. The contract calls for two reaches to be excavated by TBM—11,256m and 9,456m. The work started at Adit 10 with the mining progressing downhill. Adit 11 is the intermediate adit and the machine is due to be removed through Adit 12.

TBM2 Contract encompasses 22,464m of tunnel. There are three reaches— 5,982m, 8,579m and 6,559m. Contractor CRTG started at Adit 15, mining uphill. The

Figure 4. Adit 10 TBM assembly and materials handling chamber

intermediate adits are Adit 14 and Adit 13 with a planned TBM recovery at Adit 12 at the end of Reach 3.

TBM3 Contract starts at the outlet portal, without the need for an access adit. LWCB is the contractor. The contract calls for two reaches of 8,911m and 9,678m for a total of 18,589m. The intermediate adit is Adit 16 and the mining continues uphill towards Adit 15 where the machine will be recovered.

Equipment

This section discusses the differences in approach relating to the equipment used by the TBM contractors.

TBM1 Contractor BVEC uses a Robbins TBM with Robbins backup. The tunnel and adit conveyors are Robbins supplied. The tunnel conveyor has an advancing tail-piece mounted on the backup with the storage magazine located at the foot of the adit. The Chinese made conveyor belt is steel cored. The uphill (loaded) conveyor requires a 600kW (2 × 300kW) main drive at the head drum and 3 intermediate 600kW booster drives in the tunnel. The conveyor capacity is 800t/hr, equivalent to 5.85m of TBM advance. All the contractors have become proficient at splicing the belt using their own forces. A typical belt splicing time including curing is about 15 hours. The adit conveyor is 1,629m long with one 600kW drive unit. The adit gradient is 10%. Outside the adit there is a muck discharge station that loads directly into trucks, without any surge pile except for an emergency. For most of the time, the trucks are able to keep up with the TBM progress. Relatively small battery electric locomotives (8–10t) are used to service the tunnel. The battery charging station is located underground in the back drive at Adit 10. Shotcrete is batched underground and transported to the backup in open top cars with spiral agitator paddles. The shotcrete pump and robot system on the backup is of Meyco supply. The Robbins backup straddles the mainline track and runs on jump rails. The shotcrete cars have to be hoisted from the mainline for discharge into the shotcrete pump on one side of the backup. BVEC has a cutter maintenance contract with Robbins. This work is done at a shop off site at Muqi which also services TBM3.

TBM2 Contractor CRTG utilizes a Wirth TBM with a Rowa backup. The tunnel and adit conveyors are supplied by REI of France. The conveyor capacity is 750t/hr with 900mm belt width. The tunnel conveyor has a single 800kW drive (4 × 200kW). The

Figure 5.   TBM3 at downstream portal

storage magazine has 600m capacity. The muck discharge station at Adit 15 is similar in concept to TBM1. The tunnel locomotives are Schöma diesels, approximately 18t. Shotcrete is batched at a surface plant into frame mounted diesel powered Moran-type drum carriers which are trucked down the adit and transferred to rail cars. The Rowa backup has a raised deck onto which the shotcrete cars can be driven for direct discharge into a Sika pump and robot system. CRTG has a cutter repair shop on site and uses its own workforce to maintain cutters.

TBM3 Contractor LWCB has a Robbins TBM and backup that is identical to TBM1. There is no adit at the outlet portal. Robbins supplied the 693m long transfer conveyor linked to the discharge conveyor supplied by the contractor that runs along the embankment of the Suzi River. The conveyor discharges to a stockpile. The tunnel conveyor has a single 900kW (3 × 300kW) head drive and a 600m storage magazine. The service locomotives are battery electric and although small, have proven to be adequate for the task over the full distance of Reach 1. Battery charging is done within a building at the portal site. Shotcrete is made at a plant at the portal site and transported to the back up in open top rail cars of the same design as used in TBM1. LWCB has a cutter maintenance contract with Robbins. Figure 5 shows TBM3 assembled at the portal.

Commentary

It should be noted that two of the contractors, BVEC and LWCB, had little or no previous experience with TBM operation, although CTRG has worked on a number of TBM contracts before. Certain key people with TBM experience were hired by BVEC and LWCB for this project. The principal author of this paper was initially engaged to offer construction advice about the tunnel operations. On subsequent visits the improvements to the work practice became increasingly evident. In some instances, personnel changes were made to strengthen the expertise within management. Contractors in China are experienced and proficient in conventional drill + shoot tunneling and produce very high quality work. The transition to TBM tunneling is an evolving procedure, as in any country. To the outside observer, however certain work practices seem arduous, although in China they are commonplace. For example, there is nothing unusual about a 24-hour 7-day work schedule in tunneling. What is unusual is that the workforce, living in camp, has no regular schedule for time off. Employees who need time to travel and visit their family will make individual arrangements with their foreman and their absence is covered by other people on the crew. If there is any downtime on the machine, scheduled or unscheduled, the employees who are not required for the repair work will take time off until production resumes. Typically there are two production shifts and an overlapping maintenance shift. Another practice which seems arduous to the outside observer is that in order to provide more rest time for the employees there are three crews to cover the two production shifts. The crews rotate so that after coming off shift they are not due back to work until two shifts later, resulting in a constant dayshift/nightshift regime.

One of the problems that hampered early production was crews' lack of understanding about the need for containment of bad ground, particularly in the case of TBM3. Main beam machines do not perform well in broken ground and special precautions need to be taken. The philosophy early on was to allow the ground to collapse and take remedial action later, causing endless problems. The importance of standing the ring steel square to the tunnel axis, expanding the rings, providing longitudinal restraint and blocking the rock was emphasized. In fairness to the tunnel crews, certain factors contributing to these problems should be noted. The slender ring beams with narrow flanges make ground support very difficult. The J-type rock dowels without nuts or anchor plates and a cement grout that takes 24 hours to gain strength could not be used to provide immediate anchor for steel strapping or steel wire mesh. The ring beam erectors needed modification and access platforms had to be added to the machines. A mishap occurred on TBM3 when the sidewall broke away allowing the grippers to over-extend causing the bolts to shear at the spherical connection to the gripper shoe. This type of failure is not uncommon on main beam machines, and serves to illustrate importance of understanding the limitations TBMs and anticipating problems which only develops with experience. As time progressed, the crews gained experience and confidence in overcoming day-to-day problems.

Progress

Considering the remoteness of the region, inexperienced work crews and that two of the three contractors had virtually no previous TBM experience, the project schedule is ambitious. The projected rate of progress for TBM1 was 697m per month. Assuming an effective working year of 350 days, this represents an average of 23.9m per day. There are a number of factors contributing to the deteriorating schedule. The Robbins machines were late in being delivered. There were reasons for this which included quality control problems at DHI and some contractual issues between the companies that had to be resolved during the course of manufacture. It is perhaps worth mentioning that Liaoning Province now has a facility that is trained for the production of future TBMs, of which there will be many in China. The Wirth machine, although delivered on schedule, has not been trouble free. Early on in the first reach it was discovered that there were main bearing problems. As a result, the rate progress was limited by reduced thrust, until the machine reached Adit 14 where the bearing was changed. The down time for refurbishment at Adit 14 was more than 3 months, considerably longer than scheduled. Adding to these delays, the contractors have been unable to achieve the projected rates of progress. Figure 6 illustrates the monthly progress records of all the TBMs. There have been some impressive performances. Both TBM1 and TBM3 have progress records in excess of 1km in one month. The best month's performance for TBM2 has been 732m. The total excavation distances, measured at the end of December 2006, were TBM1—10,438m; TBM2—10,133; TBM3—7,624m. It

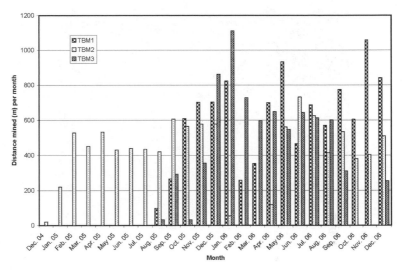

Figure 6. Comparison of TBM progress

is worth noting that in November 2006, when TBM1 achieved 1,058m, the daily utilization was in the order of 16 hours.

The measures adopted to recover the project schedule have been to substitute sections of tunnel originally planned to be excavated with TBMs for drill + shoot methods. This procedure although slower, allows concurrent working to make up time. For example, when Adit 16 was developed to receive TBM3, some 2,600m of main tunnel had been excavated before the TBM holed through. Additionally, the entire 6.5km long Reach 3 for TBM2 will now be excavated by drill + shoot methods. There may be an added benefit to this approach because Reach 3 has some of the worst ground in the TBM2 Contract. The TBM contractors are responsible for 74% of the tunnel length. The projected amount to be excavated by TBM was 72%. With the remedial recovery measures in place, this amount will be reduced to less than 60%.

CONVENTIONAL TUNNELING

As originally designed, excavation by drill + shoot methods was confined to the upstream section (approximately 25%) of the tunnel. The reason for this decision was because the more difficult geological conditions in this area are less suitable for TBM excavation. For reasons already discussed, the total length has increased for this type of construction. Local contractors are proficient at this work using simple equipment. The quality of the product, particularly the concrete lining is excellent. This section of the work is divided into four contracts and is accessed from seven adits. Some of the work is challenging, having encountered significant water inflows that required grouting.

The face is drilled from a gantry platform with decks using jack legs; see Figure 7. In Class II ground the round length is about 3.5m. In Class IV and V ground the round length is reduced to 1.5m. The muck is loaded with side dump loaders into trucks where there is rubber tired access. The steeper adits have a winch haulage system and rail equipment. At Adit 6 there is a storage bin for the muck at the bottom of the decline. For headings with truck access the cycle time is approximately 18 hours and about 25 hours where there is winch access.

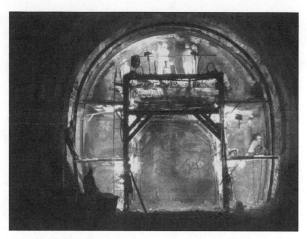

Figure 7.   Drill + shoot platform jumbo

Concrete is typically placed in forms 12m long. The lining contains bar reinforcement and the construction joints have water stop. Concrete is pumped through the form. The finished walls are typically very high quality and the need for concrete repairs or patching is rare.

CONCLUSION

This is an important development for northeast China. The contractors and management have stepped forward from an established conventional tunnel industry to newer methods, using a mainly local workforce and adjusting to the challenges without the need for foreign joint venture partners.

ACKNOWLEDGMENTS

The authors would like to thank Biyue Li, Steve Smading, Dave Fisher and Angelo Cinarelli of the Robbins Company, Heinz Küppers, Detlef Jordan and Ron Kurta of the Wirth Company and their associates for providing data and photographs for use in this paper and for the subsequent presentation at RETC.

The principal author would also like to thank Zhu Yun Chan, interpreter and civil engineer, for his invaluable assistance during several visits to the project and for translating some of the data used in this paper.

9

New and Future Projects I

Chair
M. McRae
Jacobs Associates

THE DULLES CORRIDOR METRORAIL PROJECT— EXTENSION TO DULLES INTERNATIONAL AIRPORT AND ITS TUNNELING ASPECTS

J. Rudolf

Bechtel Infrastructure Corp.

V. Gall

Gall Zeidler Consultants, LLC

ABSTRACT

The Virginia Department of Rail and Public Transportation and the Metropolitan Washington Airports Authority are undertaking the extension of Washington Metropolitan Area Transit Authority's Metrorail service to Washington Dulles International Airport and beyond to Route 772 in Loudoun County, Virginia. The roughly 37 kilometers long, double track alignment involves two 700 meters long single track, soft ground NATM tunnels at Tysons Corner, two 3.3 kilometers long single track rock TBM tunnels at Dulles Airport and one 25-meter deep station at the airport to be constructed by NATM in sedimentary rock. The design-build project is being implemented in a Public-Private-Partnership. A joint venture of Bechtel and Washington Group International has concluded the preliminary engineering and construction is scheduled to start in late 2007.

INTRODUCTION

The Dulles Corridor Metrorail Project will extend Washington Metropolitan Area Transit Authority's (WMATA) rail services from the Metrorail Orange Line in Fairfax County, Virginia to Route 772 near Ashburn in eastern Loudoun County, Virginia. This corridor encompasses several activity centers including Tysons Corner, Reston, Herndon, and International Airport Dulles (IAD) as well as emerging activity centers in eastern Loudoun County. The proposed project alignment within the Dulles Corridor is displayed in Figure 1.

Rapid Transit for the Dulles Corridor was initially explored in the 1950s as part of the planning process for Dulles Airport. At that time it was decided to reserve the median of the Dulles Airport Access Highway for future transit access to the airport. Preservation of this median allows the alignment to be at grade for most of its length within the corridor. Since the initial planning, the need for transit in the Dulles Corridor had been studied and although rail transit in the corridor was not part of WMATA's originally adopted rapid transit system, rapid transit service for the corridor remained a local and regional goal (Schrag, 2006).

The strong growth of the activity centers within the corridor in particular during the1990s and 2000s that continues today has led to momentum for Metrorail in the Dulles Corridor. Current and projected, regional growth data exemplify the need for rapid transit and its timely implementation (Dulles Transit Partners, 2006):

- Tysons Corner is the largest employment center in Virginia with 115,000 jobs and close to 4 million square meters of commercial space.
- Reston/Herndon is home of 70,000 jobs and 2.7 million square meters of commercial space.

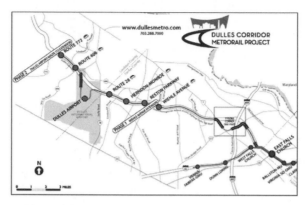

Figure 1. Dulles corridor Metrorail project

- In Fairfax County employment is expected to increase 63 percent in the next 20 years.
- Loudoun County grew by 49 percent in the last 5 years and is currently the fastest growing county in the nation.
- In the last nine years traffic on the Toll Road in Loudoun County has increased from 50,000 to 90,000 cars per day.
- Dulles International Airport employs more than 19,000 people and serves 27 million passengers per year and presently is being expanded and modernized. Modernization includes a new underground automated people mover system with multiple stations at main and mid terminals.

Regional growth and progress result however in urban and social challenges:

- The Washington, DC region has the 3rd worst congestion in the US.
- The annual delay amounts to 69 hours per traveler resulting in a "congestion cost" of US$2.5 billion per year.
- 5 of 8 main roads in the corridor will be gridlocked by 2010.

The implementation of the project began with Preliminary Engineering in 2004 under a public private partnership agreement between Virginia Department of Rail and Transportation (DRPT) and the joint venture of Bechtel and Washington Group International referred to as Dulles Transit Partners (DTP). Other funding partners in financing the project and approving the preliminary engineering effort are the Federal Transit Administration (FTA), the Metropolitan Washington Airports Authority (MWAA), County of Fairfax, Loudoun County, the towns of Reston and Herndon and WMATA as the technical reviewer who will operate the system. At the end of 2006, ownership of the project was essentially transferred from DRPT to MWAA.

The Dulles Airport extension, to be known as the Silverline once completed, will significantly increase the length of the existing Metrorail system. The original system as conceptualized in the 1960s included 166 kilometers ("103-mile system") and was designed and built between 1969 and 2001. Additions including the Largo Line were accomplished between 2001 and 2004 extending the total system length to about 171 kilometers. The planned extension to Dulles Airport and into Loudoun County when fully completed will constitute an addition of some 23% in length.

Table 1.   Current Metrorail system

	Double Track Length (km)	Number of Stations
System wide		
Subway Tunnels Includes cut-and-cover construction	80.55	47
Surface	70.41	32
Aerial	14.84	7
Metro System (Total in 2001) Without Largo segment	165.79	84
By Jurisdiction		
District of Columbia	61.64	40
Maryland	61.55	24
Virginia	47.43	20
TOTAL Metro System With Largo segment added in 2004	170.62	86

Figure 2.   Metrorail system map

WMATA METRORAIL SYSTEM AND TUNNELING EXPERIENCE

WMATA's existing Metrorail system is displayed in Figure 2. A summary of the existing WMATA Metrorail system components is provided in Table 1 followed by a summary of WMATA's tunneling experience of the three decades between the early 1970s through the beginning of 2000. This experience summary is based on the main author's involvement in the construction of the Washington Metrorail System for 30 years where in particular he served as the WMATA Chief Civil/Structural Engineer between 1985 and 2003.

Tunneling Experience

WMATA's more than 80 Kilometers of subway construction provides many examples of tunneling methods and types of tunnel construction and displays a continuous development of tunnel design and construction methodology spanning some 30 years.

In the 1970s WMATA had employed tunneling methods nowadays considered an "old-standard." Soft ground methods involved mandatory dewatering for tunneling with open face digger shields, breasting and temporary support by steel ribs and lagging. These soft ground tunnels were designed for loading conditions assuming a load equivalent to full overburden. Consequently, the final tunnel lining was a rigid, heavily reinforced cast-in-place concrete structure with PVC waterstops in contraction joints as the only means of positive waterproofing. Such construction was used on the Inner City A-Redline, D-Orangeline, and Outer G-Blueline. During that time there are examples of utilizing cast iron bolted segmental linings with lead waterproofed joints between the liner segments. Cast iron linings were used for the Potomac River Tunnel on the C-Orangeline and the Waterfront Tunnel on the F-Greenline. Immersed ("sunken") tube construction was used across the Washington Channel leading toward the bridge across the Potomac River (L-Yellowline to Virginia).

For tunneling in rock, drill-and-blast methods were used for excavation with steel ribs and cribbing as temporary support followed by cast-in-place reinforced concrete for final tunnel support. During this period WMATA already used a modern, gripper-type rock TBM when good bedrock conditions were present, with cast-in-place reinforced concrete lining as final tunnel support. An example of such TBM tunneling is one section on the A-Redline. For the construction of a large, approximately 20-meter wide and 16-meter high mined station vault in rock, drill-and-blast methods were used for excavation. First, a pilot tunnel was employed in the crown followed by mining of a multiple drift cavern excavation. Support was by heavy rock bolting and massive steel ribs embedded in shotcrete for both temporary and permanent support. The final structure was established as an independent architectural segmental, pre-cast concrete structure erected within the mined vault. For the design of the permanent support in rock some arching effect was considered. Tunnel construction on the A-Redline under Connecticut and Wisconsin Avenues features examples of such rock tunneling to construct station vaults.

In the 1980s soft ground tunneling was accomplished using sophisticated Earth Pressure Balance Machines (EPBM) with a single pass segmental, pre-cast concrete lining with gaskets, both fabricated with tight tolerances. The tunneling was performed under the Anacostia River in adverse ground conditions with about a 3-bar hydrostatic pressure. A very successfully waterproofed tunnel was achieved largely as a result of well-designed and tight tolerances that were required for segment construction and gasket fabrication. A finger type shape, dense (closed cell), neoprene gasket was developed and tested during construction. This EPBM tunneling was used also on two different sections under M Street, namely Sections F3a and F3c on the Greenline. Even in most difficult conditions as under the Anacostia River tunneling was successfully completed with no water leaks through the joints, which are still dry after over 20 years in service. Dry tunnel conditions depended on a proper design and tight tolerances of the pre-cast concrete segmental lining and the gaskets, including appropriate testing that was explicitly specified. Successful installation of bolted segments with a hard gasket depends on having appropriate ring erection equipment and on contact grouting within the time specified to avoid squatting and misalignment (see Considerations for One-Pass Tunnel Lining Design Under High Hydrostatic Pressure). The contractor experienced difficulty in closing the rings due to lack of an erector ring offered by the machine manufacturer, however the final result thanks to every ones effort was a successfully constructed and watertight tunnel.

On other Metro sections open face TBM tunneling was utilized. On one section with a low hydrostatic pressure compressed air was employed to control ground water inflow. On another section with an open face TBM systematic dewatering was performed. Both open face TBM drives utilized a one-pass segmental, gasketed, pre-cast concrete lining which also was successfully installed and remained fairly dry after construction. After extensive material testing soft gaskets were used for these applications.

Also in the 1980s WMATA allowed new, at that time progressive, tunneling and waterproofing approaches. Consequently, in 1984 WMATA accepted the use of NATM rock tunneling proposed by the contractor in a Value Engineering Change Proposal (VECP) framework. This was the first application of a dual lining NATM system with PVC waterproofing in the US. It was utilized for running tunnels and station construction on the B-Redline to Wheaton, MD. The design considered arching of the surrounding ground and interaction between ground and the initial lining. Un-reinforced, cast-in-place concrete lining was used for final support. The design was conducted utilizing the German DIN Code, as the ACI Code had no provision in this application for plain (un-reinforced) concrete. Tunnel and station waterproofing was by an "umbrella type" PVC membrane with fully immersed sidewall drains located in the invert and on both sides of the tunnel arch. This system resulted in completely dry tunnels and station in contrast to the A-Redline rock tunnels experiencing persistent leaks. At the end of the 1980s and at the beginning of the 1990s NATM tunneling was used again, but this time in soft ground conditions for running tunnels and complicated split station vault construction at Fort Totten on the Greenline. The Fort Totten station of Section E5a was built using five different drifts. The first center drift was excavated for the installation of a column line located in the middle of the future station platform. A heavy pre-support by 25-meter long secant jet grouted piles installed from the portal wall to form a crown arch was specified. This roof pre-support was to be followed with overlapping forepoling sheets. In lieu of the jet grouted arch the contractor provided a heavy temporary portal wall and used soil stabilization by micro-fine cement grouting in combination with overlapping sheeting driven above lattice girders using a hydraulic ram. Both, the station and adjacent NATM running tunnels were fully encased by a PVC membrane for waterproofing.

Soft ground NATM tunneling was used again in the mid-1990s to create the connection to the Fort Totten Station. This involved tunneling under the historic Rock Creek Cemetery by employing dewatering from inside the tunnel using vacuum lances, since access to the cemetery was excluded for construction purposes. A grouted arch as a crown pre-support was used to control surface settlements. The grouted pipes were installed by "directional drilling" methods for approximately 250 meters under the Rock Creek Cemetery from a shaft at New Hampshire Avenue. The tunneling operation and the results were very successful, limiting surface settlements to below 1.5 cm. This section E4b was part of the Mid-City E-Greenline.

Also in the 1990s WMATA adopted a cost effective "Two-Pass" lining system for the circular soft ground tunnels excavated by the open face digger shield method introduced by contractors through a VECP on the Outer E-Greenline tunnels (Sections E6e and E8a) which were originally designed and specified for TBM tunneling with a single pass lining and for NATM mined tunneling, respectively.

The two-pass lining system consist of 1.2 meter wide and typically 23 cm thick initial pre-cast reinforced concrete lining segments that form a ring installed within the shield tail which is then shoved out of the tail. The ring is then expanded against the ground using 100-ton capacity jacks at 10 and 2 o'clock locations. After expansion is achieved, steel struts ("Dutchman") are inserted and grouted in place to form the structural initial lining ring. Once the tunnel opening was supported by the initial lining and monitoring indicated acceptable ring stability, PVC waterproofing membrane installation followed fully encasing the tunnel. Subsequently, a plain, cast-in-place final lining is installed, typically 42 cm thick.

Apart from the need for dewatering, the digger shield method also required the use of ground modification techniques such as chemical grouting applied systematically from the surface prior to tunneling. Examples are the 14th Street tunnels and the under/over tunnels at Park Road, both part of the Mid-City E-Greenline in Washington, DC. Construction of these tunnels started in 1994. The roughly 35 meter deep under/over tunnels were first partially dewatered following the owner's designed drawdown system using deep wells. This system was only partially effective above the tunnel invert and was followed by an extensive chemical grouting program using sodium silicate to stabilize mainly sandy ground that existed across the tunnel profile and above the tunnels. Following this grouting program, although costly, tunneling was accomplished very successfully in this urban setting.

The design of the two-pass lining system utilizing an expanded pre-cast initial liner in soft ground at that time generally assumed the initial liner to be sacrificial or "throwaway" and temporary in nature. However, WMATA changed the design philosophy and accounted for the structural capacity of the initial lining in the design of the final lining support system. The premise for this assumption and adaptation of design philosophy was that the initial support created a solid, closed concrete ring. This consequently excluded the use of wooden wedges between segments. Further, the pre-cast lining was required to be fully stabilized before the final concrete lining was cast in place. The final liner was designed taking the combined support of both liners into consideration. Using soil-structure interaction and assuming flexibility of the initial lining the liners were designed for "Short Term Loading" and all WMATA loading combinations including full hydrostatic pressure acting on the final lining for "Long Term Loading." Using these assumptions the initial pre-cast and the final cast-in-place linings share the long-term loading combination. This allowed the use of an un-reinforced, cast-in-place final concrete lining. For the initial liner segments installed as expanded rings, success depended upon dewatering, chemical pre-grouting, and immediate expansion by jacking of the segments against the ground.

Depending on the nature of the soils, the ground water level and the difficulty in dewatering, such as from aquifers of artesian nature, it was necessary to use EPBM technology again. In such instance the initial liner was of a non-expansion type, consisting of lightly bolted segments similar to those in single-pass installations but with temporary, soft gaskets designed for partially dewatered conditions. Upon initial lining installation, a PVC waterproofing system was installed followed by an un-reinforced cast-in-place concrete lining. This method was referred to as "Modified Two-Pass" system. Such systems were used on the Outer F-Greenline, Sections F6a and F6c at Suitland Parkway Line to Branch Avenue. Here, the two-pass lining system was used for the first time with the EPBM tunneling method on the WMATA system. In this application the usual rings of four (4) reinforced concrete segments with a key segment are only lightly bolted in the longitudinal joints. Sponge type gaskets in joints and the initial liner are designed for temporary hydrostatic pressure as the final waterproofing is achieved by the PVC membrane installed around the entire lining circumference. This system is obviously more costly, but was necessary to overcome the most adverse ground and water conditions where full dewatering was not allowed due to environmental concerns. For the initial lining installation success depended upon water control, proper erection systems, and accomplishing contact grouting immediately behind a sealed tail of the TBM shield (Rudolf, 1997).

Considerations for One-Pass Tunnel Lining Design Under High Hydrostatic Pressure

From the lessons learned on the WMATA system in the past three decades a number of important considerations for the design of one-pass tunnel linings for soft ground under high hydrostatic pressure can be derived. These are equally applicable to design-bid-build and design-build type project delivery methods and applicable regardless whether the specifications are of a method or performance type and summarized below.

Segment Considerations. Apart from proper selection of segment geometry with regard to the shape (rectangular or trapezoidal), the number of segments in a ring with key segments at crown level, tapering and thickness, adequate strength for ground and construction loadings leading to segment thickness and being able to accommodate gasket pockets, grooves, bolts and packing materials it is important to:

- Specify tight tolerances for segments and very tight tolerances for gasket pockets. The British Tunneling Society's recommendations for tolerances for the fabrications of special segments are suggested (British Tunneling Society, 2000).

- For pre-cast concrete segments specify high strength and high performance concrete, typically 42 MPa (6,000 psi) to 50 MPa (7,000 psi) concrete reinforced to withstand temporary and long-term loadings and appropriate loading combinations depending on type of soil and overburden as well as construction loading from longitudinal and circumferential forces including handling and bolting.

- To assure a high quality product and the tunnel's longevity all aspects of the fabrication and installation must be rigorously controlled prior to and during construction.

Gaskets and Packing Considerations

- Select a high quality gasket material with high permanent resilience (stress relaxation) and specify a thorough material testing. Note that the "finger shaped," closed cell-hard 1⅞" wide neoprene gasket and 30-mil neoprene packing on each segment are still performing well in the Anacostia River Tunnel, although the industry nowadays prefers EPDM materials.

- Specify minimum gasket width considering possible segment offset due to installation. A width of 45 millimeters (1¾") is preferred for a 13 millimeter (0.5") offset. To prevent overfill specify that for any condition the total cross sectional area of a "hard" gasket shall not exceed 95% of the total area of the pocket between segments, calculated when the faces of the segments (including the neoprene packers) are in a full contact.

- Specify a Minimum Working Pressure (hydrostatic pressure × safety factor) and a Maximum Pressure to fully compress the gasket in the confined pocket. For the Anacostia River Tunnel the working pressure was 1.4 MPa (200 psi) and the maximum pressure was 2.8 MPa (400 psi).

Liner and Gasket System Testing

- Specify Stability Testing for water tightness to withstand the minimum working pressure without leaks.

- Specify Load Deflection Testing to measure force closure required to fully compress the gasket confined in the pocket.

Note that prior to the design of the first single-pass WMATA tunnel with reinforced concrete pre-cast segmental tunnel linings an extensive testing of segments for strength in conjunction with the gaskets was performed by Prof. S.L. Paul for WMATA's General Engineering Consultant, DeLeuw, Cather and Company (Paul, 1978 and 1984).

It is recommended that a few gaskets be pre-fabricated to the detailed dimensions with minimum and maximum tolerances. Both types should be tested in the specified pockets (between steel plates separated by packing). This should be done in a laboratory environment similar to the Stress Relaxation Test. These requirements should be specified to verify minimum and maximum compression pressures.

These tests carried out in a laboratory environment should determine the maximum and minimum pressures obtained for the extreme values of specified tolerances so they are not exceeded in either direction. The larger gasket, i.e., maximum tolerance, should be placed in the smallest pocket with the minimum tolerance and load tested by a Load Deflection Test to verify the maximum pressure at closure. The gasket with the minimum tolerance shall be placed in the largest pocket with the maximum tolerance and tested for leakage by a Leak Test performed by squeezing to closure at the minimum pressure specified. This test shall include the compression packing material. This test should further consider an offset of 10 millimeters (0.4") to 13 millimeters (0.5") in horizontal direction.

Erection of Ring and Segment Bolting Considerations

- The lining erector shall be composed of a full erector ring and erector arm capable of squeezing the gaskets with a packer in the gap between segment faces without fully relying on bolts. The segments and erection ring and erection arm must be compatible with the TBM and the liner system to ensure safe and efficient segment installation and ring closing.

Contact Grouting Considerations

- Require immediate grouting behind the wire brush seal in the shield tail.
- Require grout sufficiently stiff to provide immediate passive reaction to limit liner ring squatting.
- Require full grouting of each ring after ejection from the shield tail before the next excavation cycle begins.
- Specify the allowed over excavation and maximum annular space between the outside surface of the segments and the excavated ground surface.

Other Considerations

- Require the contractor to lay out corrective methods due to misalignment which would involve bolting to the adjacent ring.
- Specify the allowed over excavation and maximum annular space between the outside surface of the segments and the excavated ground surface.
- Require a jacking ring or shoes with pads that will equally distribute the jacking force to the liner.
- Specify the maximum jacking force that can be applied to the liner without damaging it.
- Specify requirements for installation and instrumentation and monitoring.

These considerations and suggested requirements are to ensure that the contractor can achieve the specified structural performance of the tunnel over its design life and beyond and in particular its water tightness without costly changes during construction and/or costly repairs and post construction grouting to restore the tunnel water tightness. These also facilitate use of adequate construction techniques to prevent excessive ground loosening, development of voids, inadequate backfill grouting, and excessive liner distortion. Construction methods must facilitate a quick development of a passive reaction with the ground to limit displacement as otherwise the liner can distort beyond the design limit (Kaneshiro & Navin, 1996).

DULLES CORRIDOR METRORAIL PROJECT

The roughly 37 kilometers long guideway alignment will be constructed in two phases. The Phase I segment is 19 kilometers long and involves five stations (two at grade and three elevated) and is scheduled to be operational by 2012. Phase II will extend rail to Dulles International Airport and beyond to a terminus station in Ashburn, Virginia. This alignment is mainly located at grade and on aerial structures within the median of the Dulles Access Road and the Greenway, a six lane highway. The airport area alignment segment and the metro station in front of the Airport Terminal will run deep underground in fairly competent rock conditions and will be constructed using TBM tunneling and NATM station mining. This second Phase is scheduled for completion in 2015. This description concentrates on the tunneling aspects of the project at Tysons Corner (Phase I) and at Dulles Airport (Phase II). The preliminary engineering of Phase I essentially followed the general plans of the Locally Preferred Alternative (LPA) selected by WMATA and approved by other Agencies out of many alternate alignments studied including a long tunnel at Tysons Corner with underground stations. The LPA as portrayed in the approved Final Environmental Impact Statement (FEIS) is designed mainly as an aerial guideway with a short tunnel through Tysons Corner.

Late in the preliminary engineering of Phase I WMATA, in conjunction with a Spanish contractor and an Austrian design group strongly supported by a local developer, proposed an all-underground option for the roughly 6.0 kilometers long segment at Tysons Corner. The envisioned tunnel would have been a large bore, 12 meter diameter or more TBM driven tunnel to accommodate two over/under tracks and stacked station platforms. It was based on a deep tunneling experience gained at the Barcelona Light Rail system recently constructed (Della Valle, 2002 and 2005). Despite support of an underground option by many parties involved, its realization was found to cost from US$250 to over $800 million more, based on various estimates, than the mostly elevated and partially at-grade alignment including the 700 meter long twin single track NATM soft ground tunnels. In reality, the large bore is four times larger in volume than one single track tunnel and two times larger than two single track Metro tunnels. There would be even a higher factor than two when comparing the concrete volume installed in the large bore vs. two single track tunnels. The large bore presents more risk than the excavation of two significantly smaller single bores, particularly when driven through mixed ground conditions with shallow soft ground cover. At several locations the proposed alignment indicated less than ½ tunnel diameter of mainly weak soil or fill cover. With the large bore extensive and deep excavations still would be needed for station entrances, ventilation and emergency access/egress. The large tunnel bore alone would have required handling of approximately 2.2 million cubic yards of excavated material. These facts indicate the trend towards much higher cost of the tunnel, which would be difficult to compare with an aerial and at-grade alignment.

Furthermore the large diameter tunnel option proposed throughout the entire Tysons Corner segment would have significantly deviated from the NEPA selected and approved alignment as portrayed in the FEIS and the preliminary engineering documents. This new tunnel concept would have therefore involved another environmental approval process, and additional geotechnical studies to be followed by a new preliminary engineering. This in turn would have resulted in a project delay of some 2.5 to 3 years. The additional projected cost for the tunnel alternative would have practically led to the loss of funding by the Federal Transit Administration (FTA) and substantially delayed the project. These factors and the fact that traffic congestion relief would have been postponed by another up to three years made the decision to move forward on the all-tunnel scheme very problematic. Supported by federal officials and local congressmen Virginia's Governor Timothy M. Kaine reaffirmed the Commonwealth's selection of the aerial alignments through Tysons (MacGillis, 2006), and DTP resumed design work on the original Phase 1 project alignment.

Soft Ground NATM Tunneling for Phase I

The mined tunnel segment includes twin single track NATM tunnels at a length of 700 meters each and an emergency cross-passage. Short cut-and-cover sections will be utilized at the portals. These tunnels will be constructed in soft ground and will be located adjacent to existing structures and utilities that are sensitive to ground movements.

The soils encountered along the tunnel alignment include mainly residual soils and soil like, completely decomposed rock. The residual soils are the result of in-place weathering of the underlying bedrock and are typically fine sandy silts and clays, and silty fine sands. According to project classification the residual soils are identified as Stratum S which can be divided into two substrata based on the consistency and the degree of weathering. The upper substratum, S1, typically exhibits lower N-values (averaging 16 bpf or less) and has higher fines content. Typical USCS classifications are ML, CL, and/or SM. Within the tunnel alignment, the thickness of substratum S1 varies considerably, from 0–2 feet to almost 30 feet. The lower substratum, S2, is similar to S1, but typically exhibits higher N-values (averaging 16 bph or greater) and is made up of more granular particles. Its thickness within the tunnel alignment ranges from 4 feet to 60 feet. Substrata S1 and S2 will be the predominant soil types encountered during tunnel construction with tunneling within the S1 stratum mainly near the portals and stratum S2 where the tunnel is located deeper in the mid portion of the alignment. Only to a limited extent where the tunnel is deepest will tunneling encounter decomposed rock referred to as D1 in bench and invert. The decomposed rock is a soil like material but has higher blow counts with N-values between 60 bpf and 100 bpf. Ground water at portal locations is generally at invert elevation, in mid-point of the tunnel alignment it rises up to the tunnel spring line.

Prominent building and infrastructure elements located in the tunnel's vicinity include an underground parking garage at a distance of some 8 meters from the outbound tunnel wall and bridge piers of the Route 123/Route 7 overpass, at a clear distance of approximately 15 meters from the inbound tunnel, as well as International Drive, a six-lane divided highway located about 4.5 meters above the future tunnel crowns. Deepest overburden cover exists at about mid-point of the alignment with nearly 12 meters. At the west portal and in the center of Route 7 the overburden cover is just 4 meters. A section indicating geology, arrangement of tunnels near the parking garage and local roadway is shown in Figure 3.

Because of the shallow depth, the prevailing soft ground conditions, the relatively short tunnel length, and the need to control settlements the NATM has been chosen as the preferred tunneling method. To enhance stand-up time of the soils and minimize settlements a single row of a grouted pipe arch umbrella will be utilized for the entire length of the tunnels. This will be sufficient for pre-support where the overburden is greater and surface structures are less sensitive. An additional row of pipe arch umbrellas, using closely spaced approximately 150 mm diameter sleeved steel pipes (tube-a-manchette) will be used on the first 100 meter length at both portals where tunneling is shallow with less overburden. The pipes will be installed at 30-cm centers around the tunnel crown. Figure 4 displays the double row pipe arch umbrella above a typical single track NATM tunnel with shotcrete initial lining, closed PVC membrane waterproofing system and a cast-in-place concrete final lining.

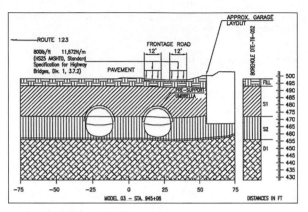

Figure 3.   Arrangement of soft ground NATM tunnels

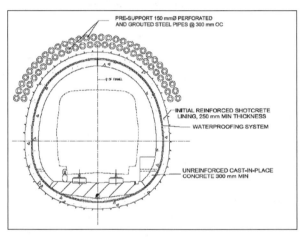

Figure 4.   NATM tunnel with pipe arch pre-support

Rock TBM and NATM Tunneling for Phase II

The underground segment of Phase II lies within Dulles International Airport property with the metro station referred to as Dulles Airport Station just north and in front of the main terminal. The main terminal has considerable traffic and existing infrastructure with much of the project area having a high concentration of existing utilities. The underground structures include twin single-track TBM tunnels, emergency cross passages, shafts and two mined caverns for the Dulles Airport underground station. These underground openings will be located below existing structures and utilities that are sensitive to ground movements. The host geologic formation for tunneling will be generally competent bedrock whereas the over burden includes fill, residual soils, and decomposed rock.

The principal bedrock unit at the project site is the Balls Bluff Formation, which generally consists of interbedded mudstone and siltstone with lesser amounts of claystone and sandstone. These lithologies are described as micaceous or calcareous, with varying degrees of weathering and alteration. Where present, the bedding of this

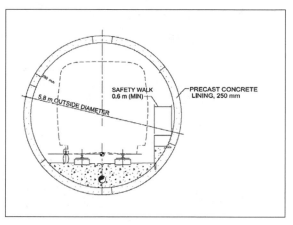

Figure 5. Typical TBM tunnel section

formation is generally well developed, ranging from laminar (beds less than 15 mm thick) to medium bedded (beds from 20 cm to 60 cm thick).

The bedrock is occasionally to moderately jointed and the prevailing bedding planes dip at an angle of about 15° to 30° to predominantly the west. Occasional zones of highly fractured rock intercept the rock mass. While the siltstone bedrock represents a favorable tunneling medium for both TBM and road header excavation ground control and support measures have to account for the jointing and bedding planes that, if left unsupported, may develop blocks and wedges with the tendency to fall-out or slide into the excavation.

The TBM tunnels have an approximately 6 meter outside diameter and are about 3.3 kilometers long each. The tunnels will be constructed by either a shielded rock TBM using a single pass, pre-cast concrete, gasketed lining or a rock gripper type TBM with an initial rock support followed by installation of a PVC membrane waterproofing and a final cast-in-place concrete lining. Figure 5 displays a typical, single pass lining cross section for the TBM tunneling.

The mined portions of Dulles Airport Station will be constructed using NATM techniques with excavation to be carried out by road headers. Initial support will consist of rock reinforcement and shotcrete lining. All mined station and associated structures will be waterproofed using an open, "umbrella type" waterproofing system with sidewall drain pipes. The station platform is about 25 meters below the ground surface. To allow for a twin station tunnel configuration, where there are two parallel station vaults, the centerline track-to-track distance is 28 meters. Both station platform tunnels are 183 meters (600 feet) long and unobstructed by vertical circulation. The station platforms are connected with cross-passages between the station tunnels. Access to the platforms is provided by a central access structure located between the two station vaults. Figure 6 displays a typical station tunnel configuration at the central cross passage with 5.2 meters wide platforms.

All station construction will be mined except for the mezzanine and ancillary rooms, which will be constructed using cut-and-cover techniques. Mined station construction has been selected to minimize disruption to airport activities. Surface disruptions will therefore generally be limited to Mezzanine and ancillary room construction using shallow (±8 meters) cut-and-cover excavation while maintaining airport pedestrian circulation above, except for the time period when the mezzanine box will be connected to an existing pedestrian tunnel "Node" that will provide Metrorail

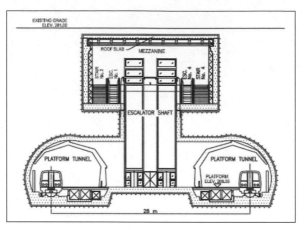

Figure 6.   Station typical structural cross section

Figure 7.   Dulles Airport Metrorail station at Dulles International Airport

Station access. Figure 7 displays a composite section of the main terminal, walkways and Metrorail underground station.

An architectural rendering for the station tunnel configuration is shown in Figure 8. Figure 9 displays the underground alignment at Dulles Airport.

IMPLEMENTATION

Public Private Partnership (PPP)

The project is being implemented in a Public-Private-Partnership under the Public Private Transportation Act (PPTA) an innovative project delivery framework as established by the Virginia Department of Transportation (VDOT) in 1995. Its implementation is in accordance with the guidelines as amended by the General Assembly in 2005 (The Commonwealth of Virginia, 2005). The essential goals of the PPTA are to encourage investment in the Commonwealth by creating a more stable investment climate and increasing transparency in a competitive environment and public involvement in the procurement process. According to the guidelines the private entity charged with project implementation is required to provide certain commitments or guarantees and enters into a negotiated risk sharing. Development of the Dulles Corridor Rapid Transit Project is an example of a PPP, where a private consortium facilitates public financing for the project and provides its full development in exchange for a negotiated Design-Build contract of the facilities. Per the terms and conditions of the comprehensive agreement, a firm fixed price (FFP) for construction is submitted to the client. This FFP is a detailed (bottom-up contractor's estimate) Design-Build proposal, which is then

Figure 8.   Station tunnel rendering (by diDomenico+Partners, architectural design consultant)

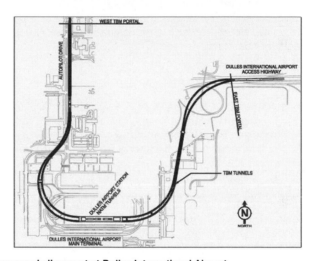

Figure 9.   Underground alignment at Dulles International Airport

negotiated on an open book basis before Final Design and Construction starts. (Martinez, 2006).

Design and Construction

The project is being realized under a design-build contract. The proposed design-builder, Dulles Transit Partners is required to initially develop preliminary engineering for the rail project. The cost for the preliminary engineering is shared between the design-builder and the project partners, DRPT, FTA, MWAA and the counties of Fairfax and Loudoun. The preliminary engineering then forms the basis to develop a fixed firm price by the design-builder. To maintain previously established budget limits this results in design challenges and the need to optimize design and construction methods to build to budget. Consequently, many design iterations are required during preliminary engineering. The design and construction team constantly weighs the benefits of underground space to keep everyday routines undisrupted versus its increased cost when compared to at grade and above ground construction.

Value Planning (VP) and Value Engineering (VE) exercises are a central activity of the design development in pursuit of the most economical approach with least impact on the surroundings. In Phase I these exercises led to a series of transformations of the underground segment at Tysons Corner. This alignment was initially envisioned as a deep, 1.6-kilometer long twin TBM tunnel scheme in mixed ground conditions with high hydrostatic head and a roughly 24 meter deep underground station constructed by cut-and-cover methods within the Route 7 road lanes, a busy traffic artery. As a result of Fairfax County requirements the alignment was moved to the median of Route 7. During the cost reduction process that was mandated by the client a rigorous analysis of construction cost on alternate alignments was performed. This analysis favored the implementation of the short NATM tunnels with a quasi at-grade station within the median of Route 7 at a cost saving of roughly US$200 million. In Phase II the VP exercises led to selection of a deep TBM tunneling and NATM station construction in rock instead of a cut-and-cover excavation for station and running tunnel construction originally depicted in the FEIS. Since the rock formation at the Airport is close to the surface this selection resulted in considerable cost and schedule savings. This construction will also considerably reduce impacts on the Airport operation. VE exercises, which are to follow, will search for further cost reductions; if successful these will become a new basis for construction.

REFERENCES

Capital Transit Consultants, (2002). Engineering Design Report, Dulles Corridor Rapid Transit Project, Virginia Department of Rail and Public Transportation and Washington Metropolitan Area Transit Authority.

Della Valle, N., (2002). Barcelona's New Backbone Runs Deep, Tunnels and Tunneling International. March 2002.

Della Valle, N., (2005). The Barcelona TBMs' Learning Curves, Tunnels and Tunneling International. February 2005.

Dulles Transit Partners, (2006). Project Update Presentation, March 2006.

Kaneshiro, J.Y. & Navin, S.J., (1996). Unique Precast Concrete Segmented Liner for South Bay Ocean Outfall Project, 1996 North American Tunneling (NAT) Proceedings, Washington, DC.

MacGillis, A., (2006). No Tunnel for Tysons, The Washington Post, September 7, 2006.

Martinez, J., (2006). Innovative Contracting—Dulles Metro Public Private Partnership, Presented to ASCE Geo-Institute Workshop.

Paul, S.L., (1978). Load Testing of Precast Concrete Tunnel Liner and Stability Tests, Technical Report Prepared for DeLeuw Cather & Company.

Paul, S.L., (1984). Hydraulic Pressure and Stiffness Tests of Gasket for Precast Concrete Tunnel Liner Segments, Technical Report Prepared for DeLeuw Cather & Company.

Rudolf, J., (1997). Waterproofing and Its Effects on the Design and Construction of WMATA Subway Tunnels. Proceedings, 1997 APTA Rapid Transit Conference. Washington, DC.

Schrag, Z.M., (2006). The Great Society Subway: A History of the Washington Metro, The Johns Hopkins University Press, Baltimore, MD.

The British Tunneling Society and the Institution of Civil Engineers, (2000). Specifications for Tunneling, Thomas Telford Publishing, Thomas Telford Ltd., 1 Heron Quay, London E14 4SD.

The Commonwealth of Virginia, (2005). The Public-Private Transportation Act of 1995, (as Amended, Implementation Guidelines, Revised October 31, 2005).

DURBAN HARBOUR TUNNEL—FIRST USE OF A SLURRY TUNNEL BORING MACHINE IN SOUTH AFRICA

J. Andrew Hindmarch
Mott MacDonald

A.L. Griffiths
Goba (Pty) Ltd.

Andrew K. Officer
Goba (Pty) Ltd.

Gert Wittneben
Hochtief-Concor JV

ABSTRACT

Durban Harbour in South Africa is one of the continent's busiest ports however the harbor entrance is currently too small for the new generation of container ships. Therefore it is to be deepened and widened by the National Ports Authority in the near future. An existing immersed tube tunnel carrying major sewer, water and other utilities under the harbor first needs to be removed and replaced by a deeper 4.4 m ID services tunnel. Construction of the deeper tunnel saw the first use of a slurry machine in South Africa and was required to negotiate grades up to 20%.

PROJECT DESCRIPTION

The city of Durban lies on the eastern coast of South Africa in the KwaZulu-Natal province on the Indian Ocean, and has a population of about 3.3 million. The city is a thriving business and tourist destination. The harbor is in the natural location of Durban Bay with the area of highland called 'The Bluff' on the south side and the city business district to the north (Figure 1). It is the busiest port in Southern Africa, operating 24 hours a day, 365 days a year. The harbor entrance has a navigable channel of 120 m and can accommodate vessels with a maximum draught of about 12 m.

The current harbor entrance configuration was established in the late 1800s and for the larger ships is quite difficult to navigate during heavy weather with a risk that a ship could breach and thereby block the entrance. The harbor is also not able to accommodate the new generation larger ships. Consequently, the National Ports Authority intends to widen the harbor entrance by relocating the breakwater (North Pier) and deepen the channels within the port to 18 m at the ocean side and from about 13 m to 16 m at the port side (Figure 2).

At present a number of water, sewer, power and communication utility services are carried across the harbor entrance in an existing 350 m long, 3.6 m diameter immersed tube tunnel. The major services are twin 1,000 mm diameter wastewater pipelines and a 450 mm diameter potable water pipeline. This tunnel was built in the 1950s and is located at a shallow depth beneath the harbor. Before the widening and deepening of the harbor entrance channel can take place it is necessary to divert

Figure 1. The city of Durban and harbor

Figure 2. Durban Harbour showing proposed widening of the entrance

these existing services into a deeper bored tunnel, which requires the first use of a full face slurry tunnel boring machine (TBM) in South Africa.

Once the services have been relocated to the new tunnel, the services in the existing immersed tube tunnel will be stripped out and then the tunnel itself will be removed.

Though initially it was planned to construct a second tunnel using the same TBM and linings to facilitate the installation of a pedestrian crossing or light rail system across the new harbor entrance, it was decided that this tunnel would not be built at this stage.

The lead consultant for the project is Goba (Pty) Ltd of South Africa with the work for the replacement services tunnel commissioned by eThekwini Municipality Water and Sanitation Department. Mott MacDonald in the United Kingdom provided specialist advice on the tunneling for the project.

After a five month tender period the tunnel contract was awarded to Durban Harbour Tunnel Contractors (a Joint Venture between Hochtief Construction AG of Germany and the South African construction company Concor).

Though the project originally called for a 4.5 m id tunnel, it was recognized that to meet the tight construction schedule a secondhand TBM would be required. Thus to maximize the availability of existing plant the Tender document permitted contractors to offer variants wherein the diameter could be between 4.4 m and 5.0 m ID.

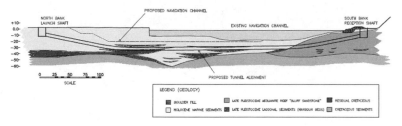

Figure 3. Tunnel long section and geology

ALIGNMENT

The proposed profile of the new harbor entrance largely governs the vertical alignment of the tunnel. However, commercial and residential development constraints on the north side and topographic constraints (the Bluff) meant that deep shafts would be needed for TBM launch and reception. To minimize the depth of the shafts, the position of the tunnel under the shipping channel was as shallow as considered prudent. Studies were carried out to determine the maximum grade that could practically be negotiated by a TBM (and any associated increased tunneling cost), which could then be compared with the cost of tunneling at lesser gradients but with increased shaft depths and associated work necessary to bring the services to surface. These studies included research on precedent experience on other projects worldwide and discussions with TBM manufacturers.

This led to the adoption of the unusually steep gradients of 20% on the incline and decline of the tunnel for about 100 m at each side with a 300 m section at 0.5% grade across the main part of the harbor entrance. The vertical curves between the inclined and subhorizontal sections are on a 300 m radius curve.

There was little option to achieve a shallower gradient on the south side of the harbor, other than deeper shafts, as there was very little flat ground before the terrain rises steeply up 'The Bluff.' Additionally the position of the TBM reception shaft on the south side of the harbor required sufficient space for a working construction site. Though there was scope for reducing the gradient of the tunnel at the north side of the harbor entrance by increasing the length of the tunnel, this would have disrupted development at this side along Point Road. A shallower gradient was possible on the south side of the harbor but this would have significantly increased the shaft depth required.

Figure 3 shows a section through the harbor channel showing the existing and proposed entrance channels. The cover for the replacement bored tunnel below the existing seabed was chosen as a minimum of 9m (two tunnel diameters). After the dredging of the deepened harbor channel (after tunnel construction) the minimum cover chosen was 5 m, which occurs at a pinchpoint on the north side of the harbor between the new harbor channel and 20% gradient of the new tunnel. These depths also allow for any over dredging when the harbor widening works take place. Though the removal of the extra cover over the new tunnel would increase the risk of flotation, there was still a sufficient resistance to flotation with the reduced cover.

For the horizontal alignment of the tunnel, the diversion of the existing services across the harbor was from a location clear of the proposed widened harbor at Point Road. Hence, the alignment of the new tunnel is to the west of the existing immersed tube tunnel to pick up the wastewater pipelines where they currently run to the existing tunnel on the north side of the harbor (Figure 4). On the south side, the pipelines will exit out of the tunnel to tie into the line of the existing pipelines running round the 'Bluff.'

Figure 4. Connection of existing services through new tunnel

The launching of the TBM was from within a slurry walled shaft on the north side of the harbor and reception on the south side was also into a slurry walled shaft located near an old whaling station slipway.

GEOLOGY AND SETTING

Eleven boreholes were drilled as part of the geological and geotechnical investigations, two of which were within the existing shipping channel. Figure 3 shows the idealized geology of the harbor which is largely Holocene and Pleistocene marine and lagoonal sediments. This is mostly classified as dense fine cohesionless sand with a clay/silt content in the order of 10%. However, lagoonal sediments comprising mostly silt and clay were identified which could occupy up to 50% of the tunnel face. Approximately 420 m out of the 515 m total tunnel drive would be through these sediments. The strata changes to aeolianite reef sandstone on the south side of the harbor forming the harder geology of the 'Bluff' and the tunnel drive would pass through this geology for the remainder of the drive. The sandstone on the south side was uniformly grained but heavily laminated with a strength varying between 0.5 and 50 Mpa.

A full head of water pressure of around 3.5 bar was also expected.

DESIGN

A feasibility study for the project was undertaken in 2003/2004 and several options for carrying the services across the harbor entrance were evaluated. These included shallow and deep bored tunnels, immersed tube tunnels, microtunneling and directional drilling. A shallow bored tunnel was selected as the preferred option for the new crossing on a balance of utility and cost.

Detailed design of the permanent works was carried out by Goba (Pty) Ltd of South Africa, in association with Mott MacDonald in London (design of the tunnel lining and advice on tunneling methods and specifications), Wilson and Pass Inc. in Durban (initial design of the shafts and review of the Contractor's shaft designs during implementation), and Drennan Maud and Partners in Durban (geotechnical investigation along the tunnel alignment).

The TBM launch and reception shafts would not form part of the permanent works and so responsibility for the detailed design of the shafts during implementation was that of the Contractor. This allowed the Contractor to choose a configuration to suit his methods of construction and anticipated loading by the launch and reception of the tunnel boring machine. Jones and Wagener of Johannesburg, South Africa carried out this detailed design on behalf of the Contractor.

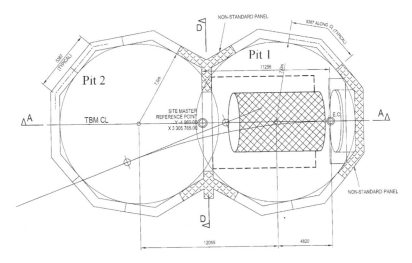

Figure 5. North shaft plan

The tunnel lining was designed for its ultimate service loading condition by Mott MacDonald. However, the Contractor was responsible for ensuring that the lining design was sufficient to withstand the loads that would be exerted during manufacturing, transport, installation and advance of the TBM. This required considerable liaison between Mott MacDonald in London and Hochtief in Germany.

Tunnel Lining Design and Manufacture

Though a 4.5 m id tunnel was initially specified, it was known early on in the project that a suitable Herrenknecht mix-shield slurry TBM with a 4.4 m ID lining had been used on the recently completed Kai Tak Transfer Scheme for the West Kowloon Drainage Improvement in Hong Kong. The 4.5 m id lining design was therefore based on this 4.4 m ID lining, thus allowing for minimal redesign if the Kai Tak lining molds were obtained. Fortunately, all four tenderers for the Durban Harbour Tunnel included the Kai Tak TBM in their bids and as it turned out, both the TBM and lining molds from the Kai Tak project were procured for the Durban Harbour Tunnel.

The 4.4 m id lining was a universal tapered ring, nominally 1,200 mm long. There were three ordinary plates, two top plates and a key which could be placed in any one of 16 positions at 22.5° round the tunnel circumference to match the TBM ram spacing and with a taper of 28 mm for a minimum 250 m radius curve.

Shaft Design

The North Shaft was designed as a twin cell structure, with a central diaphragm wall to transfer loads exerted by the surrounding ground and external water table. This wall had an opening allowing installation of the TBM shield and first gantry, to enable launch of the TBM. Pit One was designed to accommodate the TBM support works during tunneling, while Pit Two allowed the Contractor to commence early with construction of the cut and cover civil works, once the TBM and its backup had moved into the tunnel (see Figure 5).

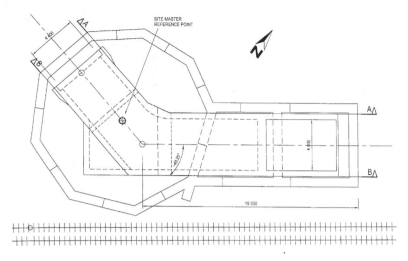

Figure 6.   South shaft plan

The jet grouted base plug initially offered by the contractor at the time of tender was changed back to a 3.5 m thick unreinforced mass concrete plug.

The South Shaft required a different configuration, incorporating a single cell reception shaft and rectangular inclined shafts to accommodate the cut and cover civil work constructed at an angle to avoid the railway tracks running immediately adjacent to the site as well as the steep rise up the Bluff (see Figure 6).

The launch and reception eyes in the North and South shafts were constructed using glass fiber reinforcing bars, to enable the TBM to pass through the eye walls without damage to the cutterhead (see Figure 7).

SHAFT CONSTRUCTION

The shafts were constructed using under-slurry diaphragm wall techniques, using the following procedure:

- Guide walls were constructed at ground level, on the required configuration of the shaft, to ensure the correct alignment of the excavation equipment.
- Alternate primary panels were excavated using a combination of Continuous Flight Auger (CFA) and mechanical grab. Panels were overexcavated by about 0.5 m to ensure fit of the reinforcing cage. All excavations were carried out under bentonite slurry to maintain stability of the sides of the slot beneath the water table.
- Steel stop-ends were installed at each end of the primary panel.
- Reinforcing cages were then lowered into the slot, incorporating 75 mm roller blocks to ensure the correct cover and to allow the cage to move freely down the slot (see Figure 8).
- Concrete was placed from the bottom of the excavation using a tremmie pipe, with an insertion depth of about 500 mm being carefully controlled through constant sounding of the concrete surface.

Figure 7. Soft eye formed using glass fiber bars

Figure 8. Shaft diaphragm wall reinforcing panel being lifted into position

- Secondary panels were then excavated and cast as infill sections following a similar procedure.

On completion of the diaphragm walls, the shafts were then excavated using a combination of:

- Dry excavation to a point where either the groundwater inflow exceeded pump capacity or the upward pressure of the groundwater resulted in instability of the base of the shaft, whichever occurred sooner
- Underwater dredge pump excavation in softer material
- Underwater mechanical grab excavation
- Suction airlift operated by divers for the final trimming of the base of the excavation prior to casting the base plug

The base plugs were then cast underwater using a high slump pump mix, incorporating an anti-waterlogging admixture to prevent washout and dispersion of the fresh concrete underwater. The casting operation was carried out from a single point using truck mounted concrete pumps, and divers to ensure embedment of the pump line by about 300 mm at all times. Dips taken during the casting process showed good flow of the concrete horizontally, and cores taken after dewatering showed little or no segregation.

After sufficient strength gain on the plugs, the shafts were dewatered. A problem was encountered on the North Shaft shortly after dewatering, with a leak coming from under the base plug of Pit Two through a joint on the intermediate cross panel. The leak caused sand washout and a sinkhole to form outside the shaft adjacent to the tower crane. The shaft was immediately flooded for stability. Extensive grouting underwater from both Pits One and Two was carried out using a combination of cementitious and polyurethane grout before the shaft was again dewatered.

Following completion of the North Shaft and construction of the TBM support works, the TBM was established at the base of the shaft (see Figure 9).

TUNNEL CONSTRUCTION

The limited space available for the construction site on the south side of the harbor meant that the TBM was launched from the north side with its easier access and sufficient land for the TBM ancillary equipment and segment casting and storage yard.

Figure 9.   TBM established in the north shaft prior to launch

Tunnel Boring Machine

It was a requirement of the project that only a slurry TBM was considered for the tunnel excavation. This was due to the ground conditions expected along the tunnel drive and the high water pressure of up to 3.5 bar. Additionally the removal of excavated material on the 20% slopes by either muck wagon or conveyor will have been difficult with slippage of excavated ground likely using a conveyor method of muck removal. Rail methods of muck removal will have required winches to haul trains up the steep slopes.

The project used the 5.17 m diameter Herrenknecht mix-shield TBM previously used on the Kai Tak Transfer Scheme in Hong Kong. This was the fourth project the TBM had constructed. The TBM was renamed 'Nomfundo' after the Durban Mayor's wife.

The cutterhead consisted of 50 scrapers for the sand and clays and 24 disc cutters capable of excavating through the sandstone on the south side of the harbor (see Figure 10).

The TBM operated as a mix-shield slurry TBM with the excavated material being mixed with a bentonite slurry pumped through a pipeline to the surface where it was separated using a screening and sand/mud separation centrifuge plant.

Due to the steep 20% decline and inclines in the tunnel, a trackless system of transport was developed for the operation of the TBM that was specially designed to carry the tunnel segments and grout mix and fit under the backup sledges. This Multi-Service Vehicle was specially designed and fabricated by Techni Métal Systèmes (TMS) in France for the project, and tested using simulated conditions for full load operations on a 20% slope (see Figure 11).

Performance of TBM and Challenges

Limited Advance Thrust in Soft Material on the Down Drive. In the initial stages of the down-grade drive, a combination of the soft material being excavated and the gradient of the drive necessitated low thrust pressures to prevent the TBM "running away." Dipping or diving of the cutterhead was also experienced. The consequence was some difficulty in steering the TBM since it was hard to develop the differential thrust needed to keep the front of the shield up. Also the lower thrust on the

Figure 10. TBM cutterhead

Figure 11. Multi-service vehicle

installed segments gave some problems with ensuring good seating of the rings during the advance.

Thrust Ram Alignment in the Vertical Curve. A steep down grade inevitably results in a vertical curve to bring the tunnel back to subhorizontal. One problem encountered here was the angle of the line of thrust from the TBM rams on the leading edge of the built ring. Some damage was picked up on the internal and external peripheries of the leading edge, thought to be due to eccentric loading of the thrust rams with the changed angle as the TBM advanced through the curve. This was solved through the repositioning of the thrust rams and shoes at the midpoint of the drive. The sideways movement of the ram was quite noticeable as soon as the thrust load was released.

Advance Rate. An average advance rate of about 12 m per day (or ten rings per day) was originally planned for the tunneling operation, working on a 6 day week, 24 hours per day.

However, problems such as the leak in the North Shaft resulted in a delayed start to tunneling. This led to the decision by the Contractor to change the tunneling operation to a 7 day week in order to recover. This was achieved using three teams working two 12 hour shifts, on a 10/5 day rotational basis.

With a tunnel of only 515 m length, the challenge would always be to get past the learning curve as quickly as possible. In the case of the Durban Harbour Tunnel, this learning curve took place on a 20% down-grade in soft ground conditions—a tall order. The initial rate of advance was slower than originally planned, which persisted through most of the down-grade drive. The rate of advance started to recover when the TBM moved into the subhorizontal drive, but was then affected by encounters with significant quantities of clay, in two main areas, largely as a result of interventions required to clear blockages and problems caused at the separation plant. However, once the clay areas had been negotiated, the rate of advance through the subhorizontal section and on the 20% up-grade drives were essentially as originally planned, albeit on an extended shift basis (see Figure 12).

Analysis of the total downtime experienced on the TBM drive is summarized in Table 1. A significant amount of time was lost with problems experienced with TBM faults and with the annulus grout system, the latter being mainly a large number of blockages in the grout lines. Segment damage, repair and removal was a significant cause of downtime in the early stages, but this reduced as the teams became familiar with the plant and conditions under which they were operating.

Water inflows occurred on a few occasions, as pressure was exerted on the lining when the ring moved past the rear of the shield, causing the key segment to "squeeze"

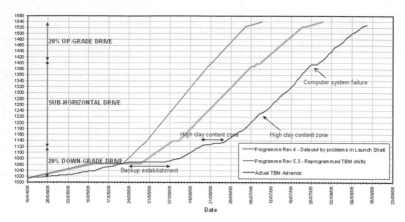

Figure 12. Time-chainage plot

out of position, thus compromising the gasket seal. This required additional restraint of the key block to prevent this movement.

Clay Conditions on the Tunnel Alignment. The geotechnical report based on the investigation carried out along the tunnel alignment indicated the presence of clay lenses intersecting the face of the tunnel over a length of approximately 150 m. While not extensive enough to warrant concern with regard to the selection of the TBM, the clay did cause a number of difficulties during the excavation process.

The excavated material, in this zone, being a very stiff to hard clay, showed a tendency to clog the access chamber behind the cutter head. This blocked the slurry level sensors, and also interfered with the slurry/air interface in this chamber, both of which are crucial to the control of slurry pressure during the drive. This necessitated a number of interventions to clear clay away from the sensors and to unblock flushing lines.

The amount of clay encountered also posed a problem for the separation plant. This was exacerbated by difficulties in determining which combination of flocculent and coagulant would be most suitable for the combination of bentonite and natural clay being encountered.

Annulus Grout. On the 20% decline, control of the annulus grout system proved to be absolutely critical. The TBM, being an older machine, was not fitted with a skirt at the rear of the shield to prevent backflow of annulus grout along the shield. On the down-grade drive, the grout showed an increased tendency to flow forward along the shield.

Problems were experienced with very high thrust pressures being required on the middle part of the down-grade drive, which resulted in a loss of directional control with the inability to develop a differential thrust. This was thought to be due to the presence of annulus grout which had accumulated and hardened on the outside of the shield. Various attempts were made to dislodge this grout, including the fitting of form vibrators to the inside of the shield, which generally proved unsuccessful.

It was also thought that the accumulation of annulus grout on the shield may have blocked the articulation gap, again hampering the ability to articulate the shield and steer effectively. This was partly alleviated by continuous maneuvering of the shield to remove grout from the articulation joint.

Eventually, an intervention was carried out and the gauge cut was increased by 20 mm to reduce the friction on the shield. This allowed the advance to proceed using normal thrust pressures and with improved directional control.

Table 1. TBM performance analysis

Downtime Description	Total Downtime (hours)
TBM Faults	253.75
Ring Erector	41
Segment Feeder	4.25
Slurry Pump	33.25
Slurry Pipeline	63
Waste Water Pump	25.25
Grouting System	246.75
Water Supply	1
HP Air Supply	0
HV Electrics	10.5
LV Electrics	12
Ventilation	1.5
Multi-Service Vehicles	6.25
Survey	4.5
Guidance System	0
Segment Repair/Removal	121
Surface Operations	0
Accidents	0
Extend Pipelines	13.5
Extend Cables	2.5
Control Cabin	0
Water Inflow	78.75
Planned maintenance	269.25
Separation Plant	41.75
Miscellaneous	140.75
Total Downtime Hours	**1,370.50**
Total Working Hours	**2,519.50**
Total Production Time	**1,149.00**
TBM Availability	**46%**

The main effect of this problem with directional control was a deviation from the designed alignment of some 450 mm horizontally and vertically. While this was out of specification, the decision was made not to force the TBM immediately back onto the designed alignment, as this could have resulted in damage to the segmental lining with high thrust pressures and possible contact of the extrados of the lining with the trailing edge of the shield. Instead, a gradual approach back to the design alignment was adopted, and a correction was made to the VMT guidance system. The principal concern at this stage was with the vertical curve at the bottom of the down-grade, required to bring the tunnel onto the subhorizontal section for the crossing of the harbor channel. However, the TBM was brought onto the correct alignment at the commencement of this vertical curve.

Circumferential Joint Dowels. The circumferential and radial joints of the precast concrete segmental lining were fitted with dowels and bolts respectively. The circumferential joint dowels were installed using pneumatic spanners immediately prior to erection of each segment, as part of the ring build process.

Problems were initially noticed during the installation of the first rings in the tunnel, with severe spalling around the positions of the dowel sockets. On close inspection of the inserts, once the dowels had been fitted immediately prior to segment erection, hairline cracks were noted in the segment trailing edge face, running through the socket and extending to the inner face of the segment.

On investigation, it was found that the dowels had a nominal diameter greater than that of the inserts. This resulted in compression of the plastic insert on insertion of the dowel, which caused stressing of the concrete around the socket, effectively setting up lines of pretensioning through the segment concrete. During installation of the segments, the slightest misalignment of the segment caused additional stresses on the dowels, which resulted in failure of the concrete around the dowel.

Consideration was given to reaming the inserts to allow the dowels to be installed with less force and to reduce any pretensioning of the concrete. However, the fit of the dowel in the socket is critical to the resistance by the segments to the thrust exerted by the sealing gasket, and to ensuring that the segments remain locked in during installation. Site trials were conducted to test the effect of reaming the sockets on the pullout force of the installed dowel. It was determined that limited reaming would ensure adequate lock-in force and this solution was implemented successfully on site.

Interventions. A total of 12 No interventions were carried out during the tunnel drive. At the start and end of the drive, interventions were carried out to check the condition of the cutterhead after and prior to driving through the shaft walls at the launch and reception pits respectively.

On the down-grade drive, the problems with the accumulation of annulus grout on the tailskin necessitated interventions to investigate possible causes and ultimately to increase the gauge cut.

The remaining interventions were necessitated by the build up of clay around the slurry level sensor and on the slurry/air interface, which disrupted the control of the slurry level and pressure. No interventions were required during excavation of the subhorizontal section or during the upward inclined drive in the sandstone, apart from that required to check the cutterhead condition prior to cutting into the reception shaft.

All interventions were supervised by qualified divers in control of the airlock operation and included in the entry team to work under compressed air conditions.

Slurry Treatment Plant

The slurry treatment plant was provided by Piggot Shaft Drilling (PSD) from the UK. The principal components of the package separation plant consisted of:

- One SD600DP declined deck primary shaker (removal of oversized solids and clayballs)
- Two SM450PSDP desander/desilting units (separation of coarse particles including sands)
- Two S3 centrifuges (fine particle separation)

The main plant was capable of operating at a capacity of approximately 600 m^3/h, with the twin S3 centrifuges able to operate at around 15 m^3/h each.

The process also incorporated a combination of flocculants and coagulants to enhance the separation of fine particles. Some problems were experienced with the

Figure 13.   View along tunnel alignment
showing segment production area

Figure 14.   Typical segment in finishing yard

reaction of the flocculant with the natural clays and various chemicals were tested
before achieving the most favorable combination.

The clay encountered on the tunnel alignment caused some problems with the
separation plant, with the capacity of the hydrocyclones on the desanding units being
exceeded, resulting in overtopping. The amount of clay being dispersed into the slurry
also resulted in high mud weights, which required extensive cleaning through the
separation process.

TUNNEL LINING PRODUCTION

The manufacture of the lining segments took place on site close to the North Shaft
(see Figure 13). With the limited production requirement for the short length of tunnel,
and the generally favorable weather in Durban, the use of an indoor production facility
was not felt to be warranted. However, portable screens were provided for the molds to
give protection against sun, wind and rain. The molds were retrofitted with form vibrators to assist with compaction of the concrete.

Prior to the commencement of production of segments, a check was carried out
on the molds using a three-dimensional photogrammetric digital survey. From this survey, digital models of each mold were produced to 0.1 mm accuracy to confirm compliance with the required accuracy specified.

Extensive trials were carried out on site to determine the optimum consistency/
slump for the specified 50 MPa mix, and to obtain the best sequence of casting,
opening the molds and floating. Some problems were experienced with cracking of the
extrados surface, thought to be due to the additional working of the concrete to achieve
a steel float finish. The decision was taken to allow a wood float finish to the extrados,
which reduced the incidents of cracking. A possible effect of this change was
increased wear on the wire tailskin seals, but the Contractor felt that this would not
pose a problem over the short length of tunnel.

Steel reinforcing cages amounting to 465 kg per ring were preassembled in the
segment manufacture area. Cover to the steel was 50 mm.

Segments were demolded at approximately 18 hours using a vacuum lifting
device, and removed to a finishing yard where they were stored on specially fabricated
steel frames (see Figure 14). At this stage minor repairs and crack sealing were
carried out, and the EPDM Phoenix gasket was fitted. The segments were then moved
to a storage area, and stacked in complete rings. Each segment was also given a

unique code, identifying the date of casting, the segment type and the ring number. This code was logged through the life of the segment, including all pre and post installation repairs.

During the tunneling operation, a day-yard was established, where complete approved rings were stacked ready for installation on the multi-service vehicle for transport to the TBM. At this stage the secondary hydrophilic gasket was fitted to the segments, and all dowel inserts were drilled out as required following the site trials previously described.

REMOVAL OF EXISTING IMMERSED TUBE TUNNEL

The existing/immersed tube tunnel is approximately 310 m long, and comprises a 3.6 m ID reinforced concrete pipe with a 610 mm wall thickness. Removal has yet to be carried out, but the following possible methods of removal have been identified:

1. A reversal of the original immersed tube installation method, involving installation of internal watertight bulkheads, exposure and cutting at original joint positions, adding buoyancy via airbags, and lifting and towing to the approved disposal area out at sea.

2. Flooding and cutting the tunnel into lengths that can be lifted and handled by the equipment available in Durban, followed by marine transportation to the approved disposal area.

3. Breaking up the tunnel insitu, and removal by grab dredger or similar to barges for marine disposal.

At present it is considered that the second option is the more practical as follows:

- Sand that has accumulated either side of the immersed tube tunnel pipe is removed by dredger to the level of the new channel. The tunnel was originally positioned on concrete pedestals spaced approximately 66 m apart prior to backfilling. A suction pipe will be used to remove the sand bedding under the tunnel such that it is left, as far as possible, standing only on the cradles.

- Buoyancy bags each capable of lifting 5 tons are available in Durban. A 20 m length of tunnel, which has a submerged weight of about 40 tons, would thus require eight or so bags for flotation without the need to position bulkheads for additional internal buoyancy.

- Each section of tunnel between pedestals will be cut into 20 m sections leaving a short section remaining on the pedestal. A diamond wire cut or a disc saw could be used and both are expected to cut through the reinforcing steel without problem. Wire saw cutting would be from the outside whilst disc saw cutting would be from the inside out.

- Holes will be punched or drilled through the tunnel crown, and chains would then be attached to a strong-back inside the tunnel and to the buoyancy bags above the section. The bags would then be inflated one by one until the section lifts clear of the channel bed.

- The tunnel sections will then be towed out to sea for disposal. Alternatively, they can be beached in the harbor and broken up into smaller sections which could be dumped in landfill or used for new sea wall protection.

- The short section remaining on the concrete pedestals and the pedestals themselves can be drilled through and lifted with bags in a similar fashion or they can be cut into smaller pieces with a top down wire cut if required.

Notwithstanding these factors, selection of the preferred method of removal will be left to the successful tenderer on the Durban Harbor Entrance Widening and Deepening contract. Whichever method is selected, a critical aspect will be to ensure that the operation does not result in any disruption to the navigable depth and width of the demarcated navigation channel.

SUMMARY AND CONCLUSIONS

The Durban Harbour Tunnel Project was the first use of a slurry TBM in South Africa and demonstrated how the difficulties of tunneling on a steep gradient of 20% were able to be overcome. The successful completion of the tunnel means that the proposed widening of the harbor entrance can now proceed as planned.

The paper discusses tunnel design and construction issues with regards to harbor crossings and tunneling with closed face tunneling machines on steep gradients. It is rare for tunnels with such steep gradients to be constructed by TBMs with segmental lining and therefore will be of interest to tunnel designers, owners and precast segment manufacturers. With the arrival of larger container ships, the issue of harbor widening to accommodate these ships and diversion of services will be of relevance internationally. It may also become increasingly necessary to negotiate steep gradients with closed face tunneling machines as underground space in urban environments becomes more constrained.

ACKNOWLEDGMENTS

The authors would like to thank Ethekwini Municipality Water and Sanitation Department for their permission to publish this paper. Thanks also to Yogini Vimalanathan of Mott MacDonald for her helpful review comments.

JOINT WATER POLLUTION CONTROL PLANT TUNNEL AND OCEAN OUTFALL PROJECT

Steve Dubnewych

Jacobs Associates

Jon Kaneshiro

Parsons Corp.

Calvin Jin

Sanitation Districts of Los Angeles County

ABSTRACT

The Joint Water Pollution Control Plant (JWPCP), operated by the Sanitation Districts of Los Angeles County (the Districts), treats wastewater generated by over 3 million people and processes wastewater solids generated by over 5 million people. The JWPCP Tunnel and Ocean Outfall Project, if constructed, would be one of the major marine outfall projects in the world. The new tunnel and ocean outfall system would provide relief to the existing outfall and allow inspection, maintenance, and repair of the existing tunnel and outfall system, portions of which were constructed as early as the 1930s.

Geologic conditions are challenging along the alignments within the study area being considered. Conditions include saturated alluvial soils, weak sedimentary rock, mixed face conditions, and squeezing ground conditions. Other significant challenges include the potentials for encountering up to 11 bars of water pressure, crossing seismically active faults, and encountering gassy and contaminated ground conditions. This paper discusses the details of the proposed project and the associated challenges.

INTRODUCTION

The Districts are 24 independent special districts that serve approximately 5.3 million residents in Los Angeles County. Seventeen of the districts that provide sewerage to metropolitan Los Angeles are signatory to a Joint Outfall Agreement that provides for a regional, interconnected system of facilities known as the Joint Outfall System (JOS). The JOS serves an area that encompasses 73 cities, unincorporated territory, and parts of the City of Los Angeles. Figure 1 shows the area served by the JOS. The JOS provides sewage treatment, reuse, and disposal for households, businesses, and industrial customers. It includes seven treatment plants, the largest of which is the JWPCP, located in the City of Carson. Currently, secondary effluent from the JWPCP is conveyed through two parallel tunnels of 2.4 and 3.7 m (8 and 12 ft) in diameter. The tunnels interconnect at a manifold structure at Royal Palms State Beach on the Palos Verdes Peninsula, from which two operational seafloor outfalls extend offshore.

During a storm in January 1995, which was particularly severe in the southern portion of Los Angeles County, discharge through the tunnel and outfall system reached the maximum hydraulic capacity. The new tunnel and ocean outfall system, if constructed, would provide additional capacity and long-term redundancy, and would allow inspection, maintenance, and repair of the existing tunnel and outfall system. The

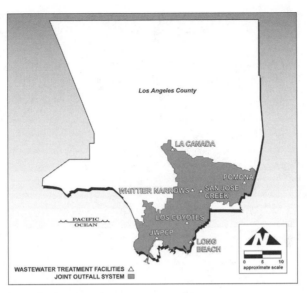

Figure 1. District's service area

peak flow for the new JWPCP Tunnel and Ocean Outfall will be determined as a part of the JOS Master Facilities Plan and may be up to 5.68 million m³/day (1,500 million gallons per day). The new tunnel and ocean outfall would begin at the JWPCP shaft site. From the JWPCP shaft site, an onshore tunnel will be constructed in a southward direction to either an onshore shaft or if an onshore shaft site cannot be established the tunnel will continue offshore. The offshore alignment will begin with a tunnel which will either continue offshore to a terminal diffuser riser array, or to an offshore riser connected to sea-floor pipelines and conventional pipeline type diffusers.

In June of 2006, feasibility studies and preliminary engineering for the JWPCP Tunnel and Ocean Outfall Project was awarded to the Parsons Corporation in association with Jacobs Associates, who are leading the underground structures design efforts.

EXISTING COMPONENTS

The JWPCP provides treatment for approximately 70% of the wastewater produced in the JOS, and treatment and processing for all of the solids produced. Secondary effluent is conveyed by parallel tunnels of 2.4 and 3.7 m (8 and 12 ft) in diameter (see Figure 2). The tunnels extend from the JWPCP to the manifold structure located at the Royal Palms State Beach. The effluent is discharged to the Pacific Ocean through ocean outfalls of 2.29 m (90 in.) and 3.05 m (120 in.) in diameter that extend 2.4 km (1.5 mi) off the Palos Verdes Peninsula to a depth of approximately 60 m (200 ft). There are also two standby outfalls that are 1.5 m and 1.8 m (60 in. and 72 in.) in diameter that are used only during peak storm flows. The maximum hydraulic capacity of the treatment plant and tunnel and outfall system is 2.5 million m³/day (670 mgd) with pumping. The average daily flow is approximately 1.25 million m³/day (330 mgd) with a dry weather peak of 1.63 million m³/day (430 mgd). The 2.44-m (8-ft) tunnel has been in operation since 1937. The 3.66-m (12-ft) tunnel has been in operation since 1958, respectively. Both tunnels are required to be in service at all times and have not been inspected since 1958.

Figure 2. Study area

PROPOSED STUDY AREA

All proposed alignments for the JWPCP Tunnel and Ocean Outfall start at the JWPCP and extend south to a point on the continental shelf. The onshore study area (see Figure 2) includes multiple alignments from the JWPCP to the shoreline—from White Point on the Palos Verdes Peninsula to the far eastern boundary of the Port of Los Angeles. The offshore study area extends from the shoreline, as defined above, southwards to the San Pedro shelf.

SITE CONDITIONS

Geologic Setting

The proposed study area lies along the southwest boundary of the Los Angeles Basin and straddles the paleo-Los Angeles River and two prominent geomorphic features, the Palos Verdes Hills and the Newport-Inglewood Uplift The dominant structural features are the Palos Verdes fault, which forms the northwest-trending Gaffey syncline and anticline. The other pervasive feature of the area is that it is underlain by the Wilmington Oil Field, which consists of a well-developed, northwest-trending anticline structure of oil-bearing rock. The stratigraphy of the Los Angles Basin and the Palos Verdes Hills is summarized as follows.

- Quaternary age: Surficial deposits include recent sediments consisting of fills, alluvium, sand dunes, and terrace deposits. These are underlain by Pleistocene sediments including the Lakewood Formation and the San Pedro Formation, which consists of Palos Verdes sand, San Pedro sand, Timms Point silt, and

Lomita marl. These formations are primarily unconsolidated sediments to very weak rock.

- Tertiary age: Pliocene sediments include the Pico Formation and the oil-bearing Repetto Formation, which are underlain by the Miocene-age Monterey Formation. The Monterey Formation consists of the Malaga Mudstone, Valmonte Diatomite, and Monterey Shale. These formations are primarily weak rock to moderately strong rock.

- Jurassic age: Basement rock is the Catalina Schist. This is a metamorphic hard rock, which varies from moderately strong to very strong.

Seismic Setting

The seismicity of the region is dominated by the intersection of the northwest-trending San Andreas (right-lateral, strike-slip) fault system and the east-west trending Transverse Ranges (vertical, reverse-slip or left-lateral strike-slip) fault system. Both systems are responding to tectonic strain related to the relative motions of the Pacific and North American tectonic plates. The effects of these movements include mountain building, basin development, widespread regional uplift, and earthquakes.

Forrest et al. (1997) note that the recurrence interval for a magnitude M 7+ event is between 400 and 1,000 or more years, based on empirical relations of length and slip rate. M6.5 earthquakes are expected every few hundred years, while M7 earthquakes every thousand years. Forrest et al. also note that small earthquakes in the proposed study area do not appear to be as frequent as on other faults within the Los Angeles Basin. At least two M2+ earthquakes have occurred every year since 1980. In 1988, there were over 25 M2+ events, but activity has decreased since then. There have been fewer than 15 M3–4 events since 1944. Most of the activity is about 8 km (5 mi) deep, with a second horizon of activity about 3.2 km (2 mi) deep. The maximum depth of activity is about 14.5 km (9 mi).

Potential Fault Crossings. A significant feature of the regional tectonic effects is the Palos Verdes fault, which is about 100 km (62 mi) long and extends northwestward into Santa Monica Bay and southeastward off the shore of the Port of Los Angeles harbor. The fault may feed into the Coronado Bank fault (San Diego) and extends as far south as Ensenada, Mexico. The Palos Verdes fault originated as a subduction zone. Then, the tectonic regime changed from a convergent plate margin to a transform or strike-slip margin. It became a right-lateral strike-slip fault somewhere between 3 and 16 million years ago (mya) (in the middle Miocene to early Pliocene age). From 1.5 to 3 mya (the middle to late Pliocene age) the southwest side of the fault was uplifted by reverse motion along it, forming the Palos Verdes Hills. South of San Pedro, strike-slip motion remained dominant during this time.

For its 2020 Expansion between Terminal Island and beyond the middle breakwater, the Port of Los Angeles conducted eExtensive geotechnical and geophysical investigations (Fugro-McClelland, Inc., 1992). These showed that the Palos Verdes fault is a series of several segments with two known prominent left steps of 200 to 250 m (660 to 820 ft) located north of the breakwater, and a second step of 100 to 700 m (325 to 2,300 ft) located south of Terminal Island. The active faulting was shown to be concentrated in a zone 100 to 300 m (325 to 1,000 ft) wide. The fault dips slightly to the east at the surface. The southwestern bock is uplifted consistent with the Palos Verdes peninsula. During the 2020 geotechnical investigations, two ancestral channels of the Los Angeles River were discovered and were offset by the Palos Verdes fault. From these investigations, the average strike slip-rate was determined to be 3 mm/year (0.1 in./year). The horizontal to vertical displacement ratio was determined to be 8:1.

There are several other significant faults and splays that would be of concern in the proposed alignment corridors; these include the Cabrillo fault and possibly the THUMS-Huntington Beach blind thrust fault.

Groundwater Conditions

On shore, toward the JWPCP in the City of Carson, groundwater is about 16 to 20 m (54 to 63 ft) below the ground surface, which is about 7.5 to 9.5 m (–25 to –31 ft) below mean sea level. Perched water may exist in these Quaternary deposits. There are four regionally extensive hydrogeologic units within the Quaternary deposits. These may cross alternative tunnel corridors. From youngest to oldest they are the Gaspur aquifer, Gage aquifer, Lynwood aquifer, and Silverado aquifer. As the corridor moves to the edge of the harbor, the groundwater level is likely to be just 3 to 6 m (10 to 20 ft) below the ground surface or at mean sea level.

Off shore, alignment corridors which are expected to be in flat lying sedimentary deposits, groundwater and hydraulic pressure will be reflective of sea level. Alignment corridors, which are in competent rock, the groundwater inflow would be limited to joints, fractures, or interbedded laminations and would not likely come through the rock matrix. More copious inflows are to be expected with the very block and seamy rock and the fractures associated with major shear zones and/or faults.

Oil and Gas

The study area lies along the southwest boundary of the Los Angeles Basin. This area is underlain by the Wilmington Oil Field, which consists of a well-developed north-west-trending anticlinal structure of oil-bearing Miocene and Pliocene shale and sandstone overlying crystalline bedrock (see Figure 3). In the 1930s, the Wilmington Oil Field produced more than any other field in the Los Angeles Basin. The basin had experienced general subsidence of more than 5.5 m (18 ft) due to oil withdrawal. Consequently, there is strong potential that gaseous tunneling conditions will occur in some alignments.

PROJECT COMPONENTS

The project would convey flows from the JWPCP to a point offshore. The point would be sufficiently far off shore to achieve the desired initial dilution and dispersion characteristics. The terminus of the outfall has not yet been identified. The offshore tunnel would connect to a riser, which would lead to either: (a) pipelines and ocean floor diffusers, or (b) riser/diffusers. A brief description of each major component follows.

Shafts

The project would include a minimum of one shaft to provide access to the tunnels during construction. After construction, the shaft(s) would be completed as permanent structures to allow access to the tunnels and/or house permanent tunnel maintenance and operation facilities. Based on the current understanding of the geotechnical and groundwater conditions, the groundwater control requirements, the anticipated shaft depths, and the construction risk factors, three excavation support systems are considered appropriate for this project: diaphragm (slurry) walls, secant pile walls, and ground freezing. All three methods are essentially watertight systems that effectively control groundwater without the need for dewatering.

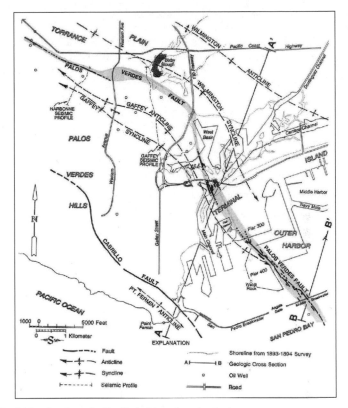

Figure 3. Geologic structures in the Port of Los Angeles harbor area

Other excavation support systems, such as deep soil mixing, interlocking steel sheet piles, or sequential excavation with shotcrete and rock reinforcement (in sedimentary rock only), may be considered appropriate, depending on the depth of the tunnels and the findings of geotechnical investigations conducted at each shaft location.

The internal dimensions of each shaft will be based on an evaluation of construction requirements and the area needed for permanent facilities. Construction requirements will vary based on the shafts use during construction. For example, main construction shafts may range from 15.2 to 18.2 m (50 to 60 ft) in diameter, while receiving shafts may range from 7.6 to 10.7 m (25 to 35 ft) in diameter. If a mining shaft is used to support two tunnel drives it will require an even larger footprint than noted above. Shafts may approach a depth of 200 ft or more in certain alignment corridors. The final depth will be determined based on geologic conditions, engineering analyses, and the need to avoid conflicts with existing underground utilities and surface structures.

Tunnels

The tunnels would be comprised of an onshore tunnel and an offshore tunnel. The onshore tunnel would have a length ranging from approximately 10 to 13 km (6 to 8 mi), depending on the final alignment alternative chosen. The offshore tunnel would have a maximum length of approximately 18 km (11 mi). If a tunnel/seafloor pipeline combination was found to be more advantageous, these lengths would be reduced.

The excavated tunnel diameter would be based on the peak flow, but could be up to 6.7 m (22 ft).

Ground, groundwater, and gaseous tunneling conditions dictate: a precision, pre-cast-concrete, gasketed, segmented tunnel liner; a controlled face; and soft ground TBM methods with rock excavation capabilities. A pressure-face TBM would be of par-amount importance to this project due to high groundwater pressures and the varying permeability and strength of the ground units along the alignment corridors. The two methods that will be studied in detail are Slurry TBM and Earth Pressure Balance TBM (EPB TBM). The choice of a Slurry TBM or EPB TBM, or even a hybrid EPB TBM/Slurry TBM will be based on the following factors:

- Soil/rock type
- Ground permeability
- Ability to withstand maximum groundwater pressure up to 11 bars
- Provisions to allow hyperbaric interventions
- Ability to install segmental lining to the required tolerances
- Ability to condition the spoils effectively for transport and disposal
- Ability to operate in gassy ground conditions

The design will ensure that adequate primary components and backup systems are incorporated into the TBM design so that major problems (e.g., delays and cost overruns) are avoided.

Riser

The tunnel is connected to the ocean floor via a vertical connection known as a riser (see Figure 4). Depending on the type of diffusers, the geologic conditions, and ship traffic considerations, the connection can consist of a single riser or multiple smaller risers. A few tunnel outfalls, such as the Boston Harbor outfall, have a diffuser section consisting of a number of vertical risers each connecting the tunnel to a flow distributing head. The head is a rosette that jets out the outflow in a circular pattern through several nozzles. Most other tunnel outfalls discharge the effluent through a tra-ditional diffuser pipe built directly on the ocean floor or protected in an excavated trench ballasted with riprap rock armor. The San Diego outfalls, Singapore Outfall, and San Francisco Southwest Outfall are examples of this type of system.

Construction of a large riser in the ocean is a delicate and risky operation. Gener-ally, the construction of the riser proceeds independently of and prior to tunnel con-struction. When both the riser and tunnel are completed, the connection between the two is made from the tunnel. This is the riskiest part of the operation, and its success is determined in large part by the accuracy of the planning during the design. The design has to address all that can go wrong with the different methods as applied to the spe-cific site geology, and a procedure that has built-in safety redundancy must be selected.

Diffusers

There are two types of diffusers possible for this project:

- A system consisting of several small risers connected to the tunnel at the bot-tom and terminating with a flow distribution cap (e.g., Boston Harbor outfall)
- A conventional diffuser pipe built on the ocean bottom (e.g., South Bay Ocean Outfall).

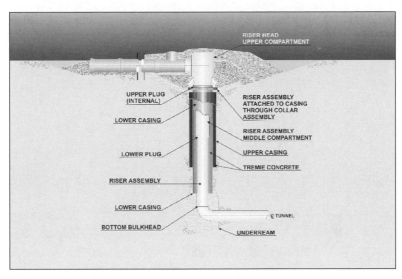

Figure 4. Riser shaft assembly

Conventional diffusers are considered more economical if geologic and ship traffic conditions are favorable. Also, they have better dilution performance characteristics and can be repaired more easily than non-traditional diffusers. However, conventional diffusers still need to be protected from anchors and cables (e.g., by burial and riprap armor protection and by protective structures surrounding the discharge heads). If a conventional diffuser was used, it might consist of a single pipe in line continuing along the tunnel alignment or in a different direction to optimize site conditions and dilution characteristics of the effluent. The diffuser would start either at the top of the tunnel riser or from the seafloor pipeline. If the JWPCP outfall system were designed for the larger spectrum of design flows, sea floor pipelines and conventional diffusers would need to be of very large diameters to accommodate the larger-than-average flows. The diameters that would be required are not standard sizes in the pipe and marine industry. As an alternative to a single large-diameter diffuser, two smaller diffuser sections could be connected to the riser head or a seafloor pipeline, each discharging half of the outfall flow. The total length of the diffuser would be the same in both cases, but there would be some savings in terms of the constructability, because smaller pipe sizes would be used. In addition, operation and maintenance flexibility would be improved with two diffusers, as one section could be closed for maintenance while the second remained in operation.

DESIGN CONSTRUCTION CHALLENGES

The entire study area crosses a complex geologic zone of structural folding and faulting. The alignments will be evaluated based on: the potential for expansive clays, squeezing ground, crossing a potentially active fault, hard rock, hydrocarbons, and high external water pressures.

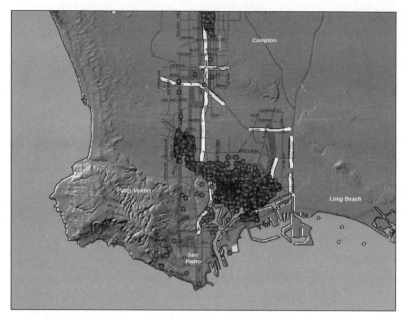

Figure 5. Abandoned oil wells

Methane and Hydrogen Sulfide Gas

The study area traverses a region known for hydrocarbon deposits. The final alignment may pass through ground that is potentially gassy, gassy, and possibly extra-hazardous. The gassy designation is applied to tunnels with flammable gas accumulations greater than 5% Lower Explosive Limit (LEL). The extra hazardous designation is applied to tunnels with 20% LEL.

In addition to methane gas, hydrogen sulfide gas may also be encountered. Hydrogen sulfide has an odor similar to rotten eggs, and the threshold of smell is less than one part per million.

In these conditions, a Slurry TBM will be advantageous since it operates in a "closed circuit," minimizing workers' exposure to gas underground. An EPB TBM would discharge the muck into muck cars underground and transport it by rail to the surface. During transport, gases could be released and could be dangerous. A number of provisions would need to be in place to mitigate dangers, such as increased ventilation and restrictions on electrical equipment. An EPB TBM could be outfitted with a "closed circuit" muck removal system similar to a slurry system to prevent discharge of gas into the tunnel environment.

In addition to methane and hydrogen sulfide gases, records from the State of California's Department of Conservation Division of Oil, Gas, and Geothermal Resources indicate that numerous active and abandoned oil wells are located along all of the alignment corridors (see Figure 5). These wells were drilled for exploration and production purposes in the Wilmington Oil Field. The active production zone is thousands of feet below the ground surface. However, the well casings would be obstacles for TBMs.

Table 1.   Recommended pressure working range for hyperbaric interventions on TBM

Environment	Recommended Working Pressure Range (bars)
Compressed Air	0 to 3.6
Mixed Gas	3 to 8
Saturation	4.5 to >45

TBM Interventions

Groundwater pressure of up to 11 bars could be encountered, depending on the chosen alignment. To access the cutterhead for maintenance, compressed air pressure could be required—if good ground conditions could not be found—to maintain stability and prevent groundwater inflow at the tunnel excavation face. Work performed in a compressed air environment above 3 bars requires the use of mixed gases and/or saturation diving techniques. Typical working ranges as noted by Holzhauser, et al. (2006) for use of compressed air or mixed gases and saturation diving are summarized in Table 1.

It should be noted that Cal/OSHA Tunneling Safety Orders do not cover compressed air work above 3.5 bars. This is to discourage the use of compressed air tunneling and to control the safety of operations by reviewing each job under a variance request. Every attempt would be made to identify favorable geology for cutterhead maintenance so that work could be performed in free air. Alternatively, ground freezing or pre-excavation grouting ahead of the TBM could be used to create a safe haven for tool change interventions.

Mixed Face Conditions

Geotechnical conditions within the study area are very complex and vary significantly. Ground conditions include saturated Quaternary deposits of the Los Angels Basin, folded and faulted Miocene-aged sedimentary rocks, localized intrusions of hard rock lenses, and high points of very hard Catalina Schist.

Due to the potentially long tunnel lengths and the variability in geotechnical conditions within the study area, it is unlikely that mixed face conditions could be avoided. A particularly undesirable condition is a mixed face condition where the face is in both rock and soft ground or several materials of widely-differing density and hardness. When these conditions persist, problems may develop. Problems may include difficulty steering the TBM and maintaining constant face pressure, higher cutter wear, and reduced progress rates. As the feasibility studies are completed, tunnel alignments will be optimized as much as practical to be favorable to TBM excavation methods.

Squeezing Ground Conditions

There is a potential for squeezing ground conditions, particularly in the Palos Verdes fault zone. This phenomenon was confirmed during the construction of the original tunnel, where squeezing ground conditions were encountered in four separate zones along the tunnel (Schultz 1937). This ground condition is problematic not only for tunnel excavation, but for lining design.

Squeezing ground usually occurs when two ground conditions exist: high in situ stresses and clay or weak rock. Squeezing is a serious problem when the magnitude is such that friction on the sides of a TBM is high enough to resist the forward movement of the TBM. Special measures would be necessary to minimize the effects of potential squeezing, such as: increasing the overcut, lubricating the TBM shield skin, reducing the TBM shield length, limiting TBM stops at critical stations, and monitoring tunnel deformation and earth pressure.

CONCLUSIONS

The JWPCP Tunnel and Ocean Outfall Project would involve many engineering and construction challenges. As discussed above, the work would encounter diverse and variable ground conditions, such as saturated alluvial soils, weak sedimentary rock, mixed face conditions, and squeezing ground conditions. High external water pressures, gassy ground conditions, and fault crossing(s) are also anticipated. As the feasibility study and design progresses, some of the design and construction concepts described herein will be modified. Planning, environmental documentation, geotechnical investigation and project design would last approximately six years. The feasibility study is to be completed in September 2008, with preliminary design anticipated to be complete in September 2009 and final design in 2012.

ACKNOWLEDGMENTS

Thanks are due to the design team: G. McBain, L. Meiorin, C. White, M. McKenna, S. Klein, and B. Schell, and to the Districts' personnel leads: M. Selna, T. Sung, A. Howard, and D. Haug.

REFERENCES

City of Los Angeles, Bureau of Engineering, Department of Public Works. 2002, *Oil Wells*, Prepared by GIS Mapping.

Earth Mechanics, Inc., 2006, Port-wide ground motion and Palos Verdes fault study, Port of Los Angeles, California: Consultant's Report to the Port of Los Angeles, San Pedro, CA., Final Report, EMI Project No. 02-131-11, Dated December 22, 2006.

Forrest, M., T. Rockwell, T. Henyey, and M. Behthien, 1997, The Palos Verdes Fault Guide, Shattered Crust Series #2, Southern California Earthquake Engineering Center, 60p.

Fugro-McClelland, Inc. 1992, Final Soils Report, 2020 Plan Geotechnical Investigation, Port of Los Angeles (Volume 1), Unpublished report prepared for the Los Angeles Harbor Department, FMWI Project No. 0901-2027, December 31.

Holzhauser, J., Hunt, S.W., Mayer, C. 2006, Global Experience with Soft Ground and Weak Rock Tunneling Under Very High Groundwater Heads, North American Tunneling, Chicago, IL.

Schultz, J.R., 1937, Geology of the Whites Point Outfall Sewer Tunnel, Thesis for PhD Degree, Institute of Technology, Pasadena, CA.

NEW YORK CITY'S HARBOR SIPHONS WATER TUNNEL

Michael S. Schultz

CDM Inc.

Colin Lawrence

Hatch Mott MacDonald

ABSTRACT

Due to the planned deepening of the NY Harbor, NYCEDC/NYCDEP is planning to replace two existing water lines between Brooklyn and Staten Island with one deeper water pipe. The crossing, just north of the Verrazanno Bridge is 1.75 miles long. The new pipeline will be tunneled using gasketed concreted segmented liners. A pressurized face machine will be used through the soft ground. This will be the first sub aqueous application using this technology in New York City. This paper discusses the planning, design and risk issues associated with the crossing and the challenging soil conditions that resulted in the decision to use a pressurized face machine.

INTRODUCTION AND BACKGROUND

The Port of New York and New Jersey is one of the most heavily used transportation arteries in the world, handling nearly 40 percent of the North Atlantic shipping trade and directly providing more than 229,000 jobs to the local economy. In 2004, $100 billion worth of consumer goods ranging from cocoa and orange juice, to automobiles and machinery moved through the Port. In order to accommodate future cargo volumes in the Port that are expected to double over the next decade and possibly quadruple in 40 years, deeper shipping channels are needed to provide access for a new generation of cargo mega-ships with drafts exceeding 45 feet. Current channels within the Harbor range in depths of up 45 feet, thus preventing carriers from suing these ships, or drastically reducing their cargo capacity in order to operate safely within the Harbor.

The Port Authority of New York and New Jersey (PANYNJ) in cooperation with the United States Army Corps of Engineers (USACE) is undertaking a comprehensive Harbor deepening program. There are 2 existing siphons, 36-inch and 42-inch diameter water transmission pipelines, that serve as a backup source for potable water supply for Staten Island. The existing siphons originate in the Fort Hamilton area of Brooklyn, and terminate in the Stapleton and Tompkinsville areas of Staten Island. The existing siphons require replacement to accommodate proposed dredging, to 50 feet below mean low water, of the Anchorage Channel, a primary navigation channel leading to the Port of New York and New Jersey (see Figure 1).

The CDM/HMM Joint Venture (JV) has been engaged by the New York City of Economic Development Corporation (NYCEDC) to provide engineering services in connection with the replacement of two existing siphons, which cross the New York Harbor between Brooklyn and Staten Island. The NYCEDC is managing the project on behalf of the New York City Department of Environmental Protection (NYCDEP), owner of the siphons.

The NYCDEP identified a preliminary horizontal alignment for the replacement of the proposed siphon(s). On the Brooklyn side, the siphon would begin in Shore Road Park, which is located between Shore Road and the Belt Parkway, near Fort Hamilton

605

Figure 1.   General location plan

High School. This park is owned and maintained by the New York City Department of Parks and Recreation. On the Staten Island side, the siphon would terminate on a vacant site located at the intersection of Murray Hulbert Avenue and Front Street. This property is currently owned by the City of New York.

Preliminary studies conducted by the JV included an assessment of alternative methods of constructing the siphon(s) across the Harbor as well as an evaluation of constructing a single larger pipe versus replacing the two single pipes.

The alternatives considered can be broadly classified into "dredged trench" and "tunnel" methods. A "dredged trench" alternative would generally involve dredging of a trench across the bottom of the harbor and laying a pipeline, underwater, in the trench. A "tunnel" method would involve construction of a lined tunnel utilizing a tunnel boring machine (TBM) and installation of a pipeline inside the tunnel. Other alternatives, including micro-tunneling and horizontal directional drilling, were considered but ruled out as technically infeasible. The initial assessment also considered alternative pipeline materials and sizing of the pipeline.

GEOLOGIC CONDITIONS

Regional Geology

The project site lies within the Harbor River Basin on the border between Staten Island and Brooklyn, New York within the Coastal Plain Physiographic Province. The region is generally characterized by thick glacier derived sediments overlying sedimentary and metamorphic rock formed during the Cretaceous period.

The following geological deposits are in the vicinity of the project area (D.W. Fisher et al., 1970; and C.A. Baskerville, 1994):

- **Recent Deposits (al).** Variably graded fine sand, silt to clay and organic sediments deposited through the Holocene Epoch.

- **Raritan Formation (Kr).** Raritan overlies Newark Basin and older rocks, representing the beginning of a series of major transgressions and regressions of the seas during Cretaceous time. The Raritan consists of clay, sand, lignite, and gravels representing progradational alluvial plain, coastal and near shore marine environments.

- **Hartland Formation (OCh).** Fine- to course-grained, gray-to-tan-weathering, quartzofeldspathic, muscovite-biotite-garnet schist...of Middle Ordovician to Lower Cambrian age.
- **Manhattan Schist (Cm).** Gray, medium- to coarse-grained, layered sillimanite-muscovite-biotite-kyanite schist and gneiss...of Lower Cambrian age.

Local Geomorphology

The terminal moraine or the leading edge of the most recent (Wisconsin-age) glaciations terminates approximately at the project site (D.H. Cadwell, 1989). The surficial soil on both banks of the Hudson River in the vicinity of the project is mapped as till moraine; however, on the Brooklyn side, the alignment lies near the confluence of the till moraine (locally known as Harbor Hill) and an outwash plain. The moraine material is described as variably sorted, relatively permeable and a near glacier deposit that may contain ablation till (usually 10 to 30 meters thick). The glacial outwash is defined as a course to fine gravel with sand deposited in a pro-glacial fluvial environment. The material is well rounded and stratified with generally finer textures away from the glacier parameter. Both the outlined surficial deposits were formed during the Pleistocene.

The lower New York Harbor area has become a tidal estuary over the past 20,000 years due to the Holocene sea level regression. The surficial soils of the inner harbor channel in the project area were deposited during this epoch and generally consist of black organic silt, fine grained silty sand, silt and clay. The dense soils at depth were likely deposited prior to the Wisconsin glaciations, during the Pleistocene.

GEOTECHNICAL CONDITIONS

Phase I Program

The JV conducted subsurface investigations both on land and over the water to collect the information necessary for environmental testing, for design and to prepare the geotechnical reports, including the Geotechnical Baseline Report (GBR), and to provide information for contractors during bidding. The program was conducted in phases, with Phase I (marine only) occurring during the initial assessment as described above. Phase I marine boring program included 7 soil borings spaced along the approximately 1³/₄ miles of the alignment. These borings were performed for both environmental testing and for the preliminary assessment and design. The environmental testing was critical to assess the feasibility and costs for ocean disposal of dredged material for the dredged crossing alternative.

Phase II Program

The Phase II subsurface investigation program included both marine and land borings, as well as geophysical survey, pressure meter and field vane shear testing, and a geotechnical laboratory testing program. The program was not yet complete at the time this paper was prepared for publication. The program elements are summarized in Table 1 and Figures 2 and 3 show pictures taken during the marine investigation.

SUBSURFACE STRATIGRAPHY

The available subsurface data gathered during the multiphase geotechnical investigation program indicate complex and variable subsurface conditions along the proposed tunnel alignment. The subsurface soil stratigraphy along the proposed siphon

Table 1. Subsurface investigation summary for Phase II

Marine Program
Phase I Borings—7
Phase II Borings—27
Pressure Meter Test Locations—6
Field Shear Vane Test Locations—4
Geophysical Reflection Survey
Staten Island Program
Phase II Borings—29
Pressure Meter Test Locations—2
Piezometer/Monitoring Wells Installed—14 (7 at Shaft)
Brooklyn Program
Phase II Borings—12
Pressure Meter Test Location—1
Piezometers/Monitoring Wells Installed—8 (5 at Shaft)

Figure 2. Positioning of spud barge during marine drilling

alignment can be categorized in six primary stratagraphic units: (1) Fill, (2) Dense Granular Soil, (3) Recent Harbor Sediment, (4) Marine Clay/Silt, (5) Marine Sand/Silt and (6) Decomposed to Weathered Rock.

- **Fill.** The granular fill typically consists of miscellaneous sand and gravel with variable amounts of silt and clay and debris. The SPT N-values vary significantly in the fill material, ranging from weight of rod (WOR) to over 50 blows per foot (bpf). The fill is typically encountered in the upper 10 to 25 feet at the land-side portions of the alignment on Staten Island and Brooklyn.

- **Granular (Glacial) Soil.** The granular (glacial) soil typically consists of dense to very dense fine to coarse sand with variable amounts of silt and gravel. The granular soil typically has SPT N-values greater than 30 (bpf), though lower N-values are encountered near the surface of the unit. The dense granular soil

Figure 3.   Marine boring operation with Verrazano-Narrow Bridge in background

is likely glacial in origin (glacial till and outwash deposits) and is encountered beneath the granular fill at the landside portions of the alignment on Staten Island and Brooklyn. This deposit also extends beneath the Harbor and underlies the more recently deposited marine sediments in the Harbor.

- **Recent Harbor Sediment.** The recent harbor sediment typically consists of very soft to soft organic silt and clay. The undrained shear strength of the deposit typically ranges from 100 to 250 pounds per square foot (psf). The recent harbor sediments are typically encountered in the upper 10 to 30 feet, beneath the harbor bottom on the Staten Island side of the alignment. The thickness of this unit thins to the east and is non existent approximately half way across the harbor. In addition, a relatively thin surficial layer of the recent harbor sediment is present at the eastern end of the harbor crossing near the Brooklyn seawall. The harbor sediment was also encountered as a thin, locally present, organic silt/clay layer where the existing fill was placed directly over the recent harbor sediment at some locations on Staten Island.

- **Marine Sediment.** The marine sediment encountered beneath the majority of the harbor crossing primarily consists of two sub-units (1) Silt and Clay and (2) Sand and Silt.

  - **Marine Silt and Clay.** The marine silt and clay typically consists of soft to medium stiff slightly organic, plastic silty clay. The undrained shear strength of the unit typically ranges from about 350 psf at the surface of the unit to 900 psf near the bottom of the unit. The marine silt and clay was encountered beneath the recent harbor sediment and overlying the granular soil and decomposed bedrock on the Staten Island side of the alignment. The thickness of the unit ranges from about 20 feet at the Staten Island side of the alignment to as much as 70 feet in the western third of the alignment. The thickness of the unit gradually tapers to the east where it pinches out approximately two thirds of the way across the alignment.

  - **Marine Sand and Silt.** The marine sand and silt typically consists of very loose to loose silty fine sand and fine sandy silt. The unit typically has little to no plasticity with SPT N-values typically being WOR. Undrained shear strengths in the primarily silty material range from about 700 to 1,100 psf.

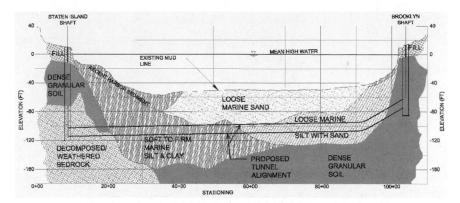

Figure 4. Generalized subsurface conditions along tunnel alignment and shafts

The unit is encountered approximately one third of the way across the alignment from Staten Island and comprises the upper 45 to 70 feet of the sediment from that point to the east where it overlies the dense granular soil on Brooklyn side.

- **Decomposed/Weathered Bedrock.** The decomposed/weathered bedrock typically consists of decomposed to highly weathered, extremely to moderately fractured, coarse-grained mica Schist. The RQD values typically ranged from 0 to 50 percent with unconfined compression strengths ranging up to 20,000 psi. The decomposed/weathered bedrock is encountered beneath the dense granular soil on Staten Island, with the rock surface dipping to the east.

The relationship and existence of these six stratagraphic units varies significantly along the length of the alignment as shown in Figure 4.

As shown in the Figure 4, the currently proposed siphon tunnel alignment, starting from Staten Island, passes through about 700 feet of decomposed mica schist, then transitions into an approximately 1,000 foot long mixed face condition of decomposed rock and granular soil before transitioning into the marine silt and clay for 3,000 feet. Approximately half-way across the alignment the tunnel transitions into the marine sand and silt unit then enters the dense granular soil around the Brooklyn seawall.

DESIGN

Preliminary Study

As discussed above, a preliminary study was undertaken to select the method for crossing the harbor. Several construction methods were considered, including micro-tunneling, horizontal directional drilling (HDD), a dredged pipeline, and conventional tunneling using a tunnel boring machine (TBM). After an initial screening, micro-tunneling and HDD were determined not to be feasible options given the length of the crossing, installation and operational requirements for the transmission main, and construction issues.

A comprehensive evaluation of the dredged crossing versus the conventional TBM tunnel crossing was performed. The study compared both methods against several criteria including: schedule; constructability; construction risk; environmental impacts; cost; compatibility with possible future projects and navigational impacts. Ultimately the TBM tunnel crossing was selected as it could most likely be completed sooner, with substantially less risk of delay, compared to a dredged trench crossing. A tunneled installation will also have significantly less impact to the marine environment and to harbor navigation than a dredged trench installation.

Horizontal Alignment

The horizontal alignment required connection of a water transmission pipeline in a tunnel that connected to the existing landside water distribution networks of both Staten Island and Brooklyn. The shortest route between the shaft in a vacant area at Murray Hulbert Avenue in Staten Island and Shore Road Park in Brooklyn is a straight alignment; this was the initial assumption for the conceptual design. From a starting point in Staten Island, the horizontal alignment extends under Murray Hulbert Avenue and the adjacent western New York Harbor Bulkhead. The majority of the alignment crosses underneath the New York Harbor, more specifically the Stapleton Anchorage, Anchorage Channel and Bay Ridge shipping channels. Towards Brooklyn the alignment crosses the eastern NY Harbor Bulkhead, extending under the Belt Parkway and terminates at the shaft in Shore Road Park.

During the development of the design, available construction records for old piers (circa pre-1940) that use to exist on the Staten Island shoreline, in the vicinity of the proposed alignment were reviewed. Piers 8 & 9 were removed in 1983 as part of the New York Harbor Collection and Removal of Drift Project. Although a large number of piles were removed as part of this earlier project, where difficulty was encountered during extraction the piles were cut off at the channel bed.

The estimated pile toe elevations were reviewed based on historical information and archive design drawings of the piers. Based upon the design drawings, the pile toe elevations vary from −46.33 ft to −91.33 ft (NGVD29). Unfortunately, insufficient records exist from as-built information to accurately determine which of the piles remain intact below the channel bed.

Marine borings taken during the site investigation for this project identified a pronounced increase in depth of low strength, weight-of-rod soil conditions toward the eastern end of the abandoned piers. This information increased the risk that the as-built piles for the abandoned piers were driven through the weight-of-rod soil to a deeper toe elevation than that indicated on the original pier design drawings.

The combination of abandoned piles being cut off at the channel bed and the likelihood of deeper piles being driven to refusal through weight-of-rod soil conditions toward the eastern end of the piers raised significant concern that the existing piles could conceivably extend down to the conceptual tunnel alignment. The possibility of encountering timber piles in this reach during tunnel excavation was identified as being a high risk for the project with regard to tunnel delay and costs. As a risk mitigation measure, the horizontal alignment was modified to divert the tunnel around the area of concern thereby reducing the risk during construction. A horizontal curve of 5,000 ft radius was applied to the horizontal alignment to skirt outside of this identified zone, returning to a straight alignment at a tangent point to allow a direct route to the Brooklyn shaft.

Vertical Alignment

The main controls for establishing the vertical alignment included:

- Minimizing the depths of the shafts at Staten Island and Brooklyn.
- Avoiding existing structures such as the harbor bulkheads and existing piles. As discussed above, although the encroachment of existing piles into the tunnel

horizon impacts the vertical alignment, by adjusting the horizontal alignment the risk was significantly reduced.

- Setting the tunnel alignment below the established future harbor dredging limits. In order to address the fundamental need for the project, the future dredged depth of the New York Harbor to allow access for post-Panamax vessels results in an assumed dredging limit at elevation −65.9 ft (NGVD29). In addition, as the project crosses a main marine anchorage area, a disturbance zone of 10 ft (3m) for potential disruption due to ship anchor penetration has been allowed for.

- Providing sufficient cover to allow buoyancy to be balanced both during and after construction. As tunneling will take place prior to the harbor dredging above, two scenarios were checked for buoyancy forces. The first scenario is during tunnel construction with existing overburden to the channel above remaining intact, the tunnel is assumed to have a segmental lining without any other contributing ballast against buoyancy; The second scenario is post tunnel construction with the dredged overburden removed, the tunnel is assumed to have a segmental lining and also have an internal steel water transmission pipeline grouted in place to act against buoyancy.

- Optimizing the alignment at the most suitable soil conditions anticipated at the tunnel horizon with respect to soil characteristics and pressures.

- Accommodating the client's operational and maintenance requirements.

Tunnel Diameter

The TBM excavated diameter was determined from a combination of the internal pipe diameter, sufficient space for placement and jointing of the internal water transmission pipeline, the required thickness of the pre-cast concrete tunnel lining and the outer excavated annulus.

Although a 10-foot diameter TBM would theoretically physically accommodate the 6-foot diameter steel transmission pipe it was considered prudent to assume a 12-foot (3.66m) excavated diameter to provide sufficient internal space given the length of the tunnel at 9,460 feet, and for tunnel logistics, TBM operations and productivity. A cross-section showing the main dimensions and tunnel configuration is shown in Figure 5.

Transmission Pipeline Within the Tunnel

The new siphon will serve as a backup water supply to the existing 10 ft (3 m) Richmond Tunnel whenever the tunnel is taken out of service. As required by NYCDEP, the new siphon shall be designed for a maximum flow rate of 570 million liters per day (150 MGD). Although a number of different pipe materials were considered, only three different pipe materials, Fiberglass Reinforced Pipe (FRP), Prestressed Concrete Cylinder Pipe (PCCP) and welded steel pipe, were available at 72-inch (1.84m) diameter and considered suitable for the project.

After operational, maintenance, construction logistics, and overall life-cycle performance of the pipeline, welded steel pipe was selected as the most appropriate for installation within the tunnel.

Tunnel Boring Machine (TBM)

The Harbor Siphon project is in a prominent location at the marine gateway to New York Harbor. As the location is one of the busiest reaches of waterway within the region, any tunneling method should serve to minimize the potential for disruption to marine traffic. Ground conditions identified by the marine borings performed to date

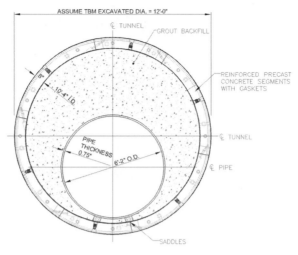

Figure 5.   Cross-section of tunnel

suggest a significant variation in both clay and sand in association with full hydrostatic conditions along the alignment. The ground conditions identified so far in conjunction with the limited cover to the channel and the sub-aqueous setting of this tunnel will require a tunnel machine with the ability to provide full closed face support on demand.

Given the nature of the ground conditions, a pressurized face tunnel boring machine has been identified for tunnel excavation. As the geotechnical site characterization program is not yet complete, the type of pressurized machine has not been specified, although at this stage the selection of the pressurized face machine is likely to be left for the contractor to determine.

Pre-cast Segmental Lining

There were a number of controls identified during the development of the design that led to the adoption of a pre-cast concrete lining. In determining that a pressurized face TBM will be required due to the ground conditions expected in turn requires a lining that can be immediately erected and provide both ground support and a seal against water infiltration. In terms of ground support, a one-pass lining provides desirable schedule benefits on a tunnel drive of this length.

The corrosion protection provided by the segmental lining to the internal steel transmission pipeline was also desirable due to the nature of the ground conditions. As corrosion of the pipeline was an identified risk, the use of segmental lining proved to be a risk mitigation measure to improve the long term durability of the pipeline.

The design of the segmental lining for the Harbor Siphon project was controlled by the required tunnel geometry, the expected ground pressures, the anticipated TBM advance forces, the water inflow criteria, as well as handling and installation issues.

As shown in Figure 6, a typical project ring consists of 6 segments, where the average segment length is 56.5 inches (1,435 mm) and the internal diameter is 124 inches (3,150 mm). Due to the nature of the ground conditions and the requirement to follow a curved alignment, a universal pre-cast ring has been designed. Both conventional steel reinforcement and fiber reinforcement were considered during development of the preliminary design, however, due to the anticipated low horizontal

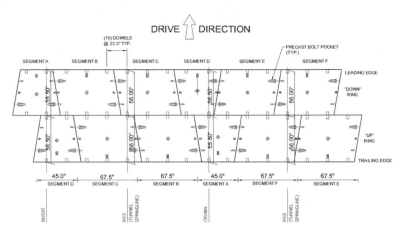

Figure 6.   Six segmental tunnel lining layout

in-situ stresses and resulting bending moments in the lining, it was determined that conventional steel reinforcement should be adopted in the design.

Launching and Reception Shafts Design

Given the proximity of the connection points to the existing water distribution network and the Richmond Chlorination Station, a vacant plot of land on Staten Island provides a suitable location for the launch shaft for the tunnel crossing. At the Brooklyn side, available shaft sites are at a premium in this heavily developed part of Brooklyn. A portion of an existing park was identified for the reception shaft, the park being bounded by the Belt Parkway to the west and Shore Road to the east, which serves an adjacent school and a residential neighborhood of Brooklyn known as Bay Ridge.

The minimum effective internal diameter for the Staten Island TBM launch shaft is 28 feet (8.5m) and for the Brooklyn TBM reception shaft is 24 feet (7.3m).

Although a number of methods were considered for shaft construction, after consideration of the ground conditions at both sites; the need to control groundwater and prevent surface settlement; the possibility of abandoned obstructions within the shaft excavation; the need for maintenance access during operations; mitigating noise and vibration during construction ; and the proximity to surface structures such as the Belt Parkway, residential housing and the Fort Hamilton School in Brooklyn, two methods of shaft excavation were identified. The two shaft methods chosen for design were permanent slurry walls and ground freezing.

RISK MANAGEMENT

The design team has used a dedicated risk management approach to identifying and analyzing the project risks during the design phase. A series of qualitative risk workshops have been performed with all the main stakeholders involved with the design of the project. As a result of the workshops, some 40 risks were identified and mitigation measures were also developed for implementation in the design process. The results of the workshops were captured in a risk register. As part of the design team's quality driven approach the risks are routinely revisited on a routine basis and updated, assessed and mitigated as required. This has enabled the design to be fine tuned in a proactive efficient way and allowed all identified risks and mitigation measures to be addressed on a regular basis with the owner at monthly progress meetings.

As is commonly the case with underground projects, tunneling is potentially a very risky business. The risk profile developed at the beginning of the design for the Harbor Siphon Project was determined to be high. Through the course of the design the profile has been trending towards the medium to low risk profile. A few examples of the typical risks identified include: threats to the schedule completion date; construction cost over-runs; potential causes of delay to TBM operations during construction; and delays in obtaining certain permits.

CONCLUSION

Historically, the soils underlying the Hudson River have proven a difficult barrier to negotiate for past tunnel projects such as the Lincoln Tunnel, the Holland Tunnel and the PATH Tunnels. Tunneling technology back then used compressed air, hand mining and cast iron segmental lining successfully performed but achieved with a cost to human life. In modern times the use of pressurized face technology to provide more efficient, cost effective and safer means of tunnel excavation provide an opportunity to construct a new crossing. The New York Harbor project represents one of the greatest tunneling challenges performed to date in the New York area. It will be the first use of pressurized face conventional tunneling technology ever to be undertaken in New York City and will provide significant experience and insight that may benefit several other projects currently planning to use this technology within the city.

Most importantly, the Harbor Siphon project will enable dredging of the New York Harbor which in turn will enable some of the world's largest vessels to access New York and New Jersey. This forward planning will ensure that the region's economic vitality continues.

ACKNOWLEDGMENTS

The authors wish to express their gratitude to the following individuals for their assistance and support: Jawad Assaf (NYCEDC); Nazir Mir (NYCEDC); Magdi Frag (NYCDEP); Jim Garin (NYCDEP); Odd Larsen (NYCDEP); Nick Barbaro (NYCDEP); Tom Costanzo (PANYNJ); Jim Cleary (PANYNJ); Paul Paparella (HMM); David Watson (HMM); Shimson Tzobery (HMM); Roger Howard (CDM); Mahammad Jafari (CDM) and David Tanzi (CDM).

REFERENCES

Baskerville, C.A., 1994. U.S. Geological Survey, "Bedrock and Engineering Geology Maps of New York and Parts of Kings and Queens Counties, New York, and Parts of Bergen and Hudson Counties, New Jersey," Miscellaneous Investigation Series Map I-2306 (Sheet 1 of 2).

Cadwell, D.H., 1989. University of the State of New York, "Surficial Geologic Map of New York: Lower Hudson Street," N.Y.S. Geological Survey.

Fisher, D.W., et al. 1970. University of the State of New York, "Geologic Map of New York: Lower Hudson Street," New York State Museum and Science Service Map and Chart Series No. 15, N.Y.A. Geological Survey.

PRELIMINARY DESIGN OF THE CALDECOTT FOURTH BORE

Bhaskar B. Thapa

Jacobs Associates

Michael T. McRae

Jacobs Associates

Johannes V. Greunen

Jacobs Associates

ABSTRACT

This paper describes the preliminary design of the fourth bore of the Caldecott Tunnels along State Route 24 in Oakland, California. The paper focus is on two major aspects of the project: initial support system design and seismic design. The initial support system was designed based on the New Austrian Tunneling Method (NATM). NATM provides the required flexibility to accommodate the variable ground conditions in the weak, folded, and faulted rock along the alignment. The seismic design provides a final lining system that meets serviceability requirements for a 1,500 year seismic event with an estimated peak ground acceleration of 1.2g.

INTRODUCTION

Project Background

The existing Caldecott Tunnels consist of three bores along State Route 24 (SR 24) through the Berkeley Hills in Oakland, California. The California Department of Transportation (Caltrans) and the Contra Costa Transportation Authority (CCTA) propose to address congestion on SR 24 near the existing Caldecott Tunnels by constructing a fourth bore that will provide two additional lanes. The length of the proposed fourth bore is 1,036 meters (3,399 feet). The project will include short sections of cut-and cover tunnel at each portal; seven cross-passageway tunnels between the fourth bore and the existing third bore; and a new Operations and Control Building.

The fourth bore includes two 3.6-meter (12-foot) traffic lanes and two shoulder areas that are 3 meters and 0.6 meter (10 feet and 2 feet) wide. The horseshoe-shaped mined tunnel is 15 meters (50 feet) wide and 9.7 meters (32 feet) high. The tunnel includes a jet fan ventilation system, a wet standpipe fire protection system, and various operation and control systems including CCTV monitoring, heat and pollutant sensors, and traffic monitoring systems.

GEOLOGY

Major Geologic Formations and Structure

The geology of the alignment is characterized by northwest-striking, steeply-dipping, and locally overturned marine and non-marine sedimentary rocks of the Middle to Late Miocene age. The western end of the alignment traverses marine shale and sandstone of the Sobrante Formation. The Sobrante Formation includes the First

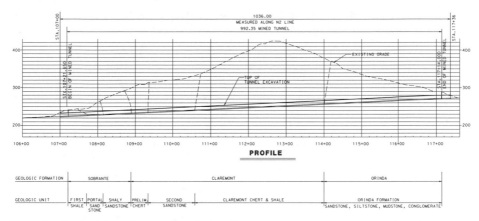

Figure 1.   Geologic formations and geologic units

Shale, Portal Sandstone, and Shaly Sandstone geologic units as identified by Page (1950). The middle section of the alignment traverses chert, shale, and sandstone of the Claremont Formation. The Claremont Formation includes the Preliminary Chert, Second Sandstone, and Claremont Chert and Shale geologic units (Page, 1950). The eastern end of the alignment traverses non-marine claystone, siltstone, sandstone, and conglomerate of the Orinda Formation. Major formations and geologic units within these formations are shown Figure 1.

The regional geological structure of the project area has been characterized as part of the western, locally overturned limb of a broad northwest-trending syncline, the axis of which lies east of the project area. The fourth bore alignment will encounter four major inactive faults, which occur at the contacts between geologic units. These faults strike northwesterly and perpendicular to the tunnel alignment. In addition to the major faults, many other weakness zones will be encountered away from the major faults, such as smaller-scale faults, shears, and crushed zones.

West of the fault contact between the Preliminary Chert and Shale and the Second Sandstone, the bedding encountered in the fourth bore generally dips predominantly toward the northeast. East of this fault contact, the bedding dips to the southwest. Several joint sets occur within each geologic unit, and random joints occur in almost any orientation in all geologic units. Intrusive sandstone dikes and hydrothermally altered diabase dikes occur most frequently in the Claremont Chert and Shale, but may also be encountered less frequently in other geologic units.

The structure of the rock mass units along reaches of the alignment varies from being blocky in the best ground, down to a disintegrated or crushed condition in the poorest quality rock. Average values of unconfined compressive strengths varies from 9.6 MPa (1,400 psi) to 48.5 MPa (7,000 psi) in the various geologic units along the alignment Rock Mass Ratings (Bieniawski 1989) vary between 20 and 65 along the alignment.

Seismicity

The San Francisco Bay Region is considered one of the more seismically active regions of the world, based on its record of historical earthquakes and its position astride the tectonic boundary between the North American and Pacific plates. During the past 160 years, faults within this plate boundary zone have produced numerous small-magnitude (M<6), and more than a dozen moderate- to large-magnitude (M>6), earthquakes affecting the region. Major faults that comprise the 80-kilometer-wide plate boundary within the San Francisco Bay Region include the San Gregorio, San Andreas, Hayward, and Calaveras faults.

The active Hayward fault, located 1.4 kilometers (0.9 miles) west of the Caldecott Tunnel, is the closest regional fault to the project site. The southern segment of the Hayward fault produced the 1868 Haywards earthquake of estimated magnitude 6.8 that was accompanied by 30 to 35 kilometers (19 to 22 miles) of surface faulting. The most recent large earthquake on the northern segment of the Hayward fault is estimated to have occurred between 1640 and 1776. Presently, both segments of the Hayward fault exhibit tectonic fault creep at rates of up to 1 centimeter/year (0.4 inches/year).

GROUND CLASSIFICATION

The design and construction of the fourth bore is based on the sequential excavation method (SEM), also called the New Austrian Tunneling Method (NATM). The ground classification process was twofold: identification and characterization of Rock Mass Types (RMT) along the alignment having similar mechanical characteristics, and identification of ground classes based on similarity of anticipated ground behaviors of each RMT in response to excavation.

The identification of RMTs was based on the distribution of geological characteristics and relevant geotechnical parameters. The alignment was divided into RMTs based primarily on lithology, fracture density, discontinuity properties and unconfined compressive strength (UCS). Mechanical properties were determined for each of the RMTs along the alignment and ground behaviors were evaluated considering the identified boundary conditions.

The RMTs were then grouped into four ground classes based on the similarity of anticipated ground behaviors in response to excavation. An appropriate support category was then developed for each ground class. For example, Ground Class 1 comprises all RMTs along the alignment that require Support Category I. Ground Class 2 correlates to Support Category II, and so on. Individually, the support categories address sets of similar ground behaviors, and as a whole they address all anticipated ground behaviors along the alignment. The ground classes were the basis of design for the initial support categories.

Ground Classes

The actual ground classes along the alignment will be determined during construction based on probe drilling ahead of the lead drift, geologic mapping of the tunnel, and tunnel monitoring. The ground classes encompass a broad range of rock properties as shown in Table 1.

EXCAVATION AND INITIAL SUPPORT

NATM excavation sequences and support designs were developed for the four support categories that correspond to the four ground classes described above. This section describes:

- The excavation sequence
- The main analyses performed to determine support element requirements within each support category

Table 1. Ground classes

Ground Class	Rock Mass Description
1	Blocky rock masses with poor to good discontinuity conditions and weak to medium strong intact unconfined compressive strength
2	Very blocky to blocky/disturbed/seamy rock masses with poor to good discontinuity surface conditions and weak to medium strong intact rock unconfined compressive strength
3	Blocky/disturbed/seamy rock masses with poor discontinuity surfaces, or disintegrated rock with poor discontinuity conditions
4	Disintegrated First Shale rock mass at the west portal with poor to fair discontinuity conditions and weak to very weak intact rock strength.

- The major support elements and support selection considerations used to develop the support categories
- Construction monitoring to be used during application of the designed support categories

Excavation Sequence

The overall excavation and support sequence consists of a top heading and bench. The top heading excavation will be accomplished using a single drift with a sloping core for face support. The bench excavation will be done in one or two stages depending on the support category and the lag maintained between the top heading and bench. A minimum lag between the top heading and bench is required to ensure equilibrium of the top heading under biaxial loading before additional loading is introduced as a result of the excavation of bench drifts. Drift advance length is primarily controlled by anticipated ground stand-up time and the size of the drift.

Analysis for Determination of Support Requirements

Convergence-confinement analyses were performed using Fast Lagrangian Analysis of Continua (FLAC 5.0, Itasca, 2005) to determine the required thickness of shotcrete lining and the length of rock dowels in the four support categories. The FLAC analyses simulated the excavation sequence and installation of perimeter rock dowels and several lifts of shotcrete. In addition, the models simulated the strength/stiffness gain of the shotcrete with time. The models were used to estimate the moments and thrusts that develop in the shotcrete lining and these results were plotted on thrust-moment interaction diagrams to verify that the loads are less than the capacity of the lining (see Figure 2).

The FLAC models incorporated an elastic-plastic material model to simulate the inelastic behavior of FRS. Each beam element was assigned a tensile and compressive strength, which, along with the section geometry, defines an interaction diagram. At each time step, the axial forces and moments are computed for each beam element and these forces and moments are compared to the capacity envelope. If the axial forces and moments fall outside the interaction diagram, the axial forces and moments are adjusted to return the values to the capacity envelope. This approach effectively limits the tensile stress that will develop in the shotcrete lining and permits plastic rotations and deformations to develop if the tensile capacity is reached. To assure that the

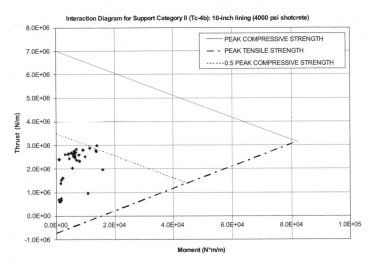

Figure 2.   Moment —thrust diagram

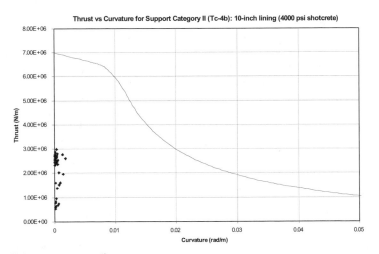

Figure 3.   Thrust—curvature diagram

section remains structurally viable, rotations that develop in the lining are plotted on thrust-curvature diagrams to verify that they are within allowable limits (see Figure 3).

The elastic modulus of shotcrete used in the FLAC analyses varied from 5,000 to 15,000 MPa (725 to 2,175 ksi) to account for creep in early age shotcrete (John and Mattle, 2003). The lining is initially assigned an elastic modulus of 5,000 MPa (725 ksi) immediately after installation. The elastic modulus is gradually increased to 15,000 MPa (2,175 ksi) for hardened shotcrete, as the excavation progresses, to simulate the stiffness increase of the shotcrete as it increases with time.

FLAC3D models of the full NATM excavation and support operation (see Figure 4) in each support category were used to estimate the amount of relaxation that occurs in the ground ahead of drift headings, evaluate face stability, estimate the required bench lags,

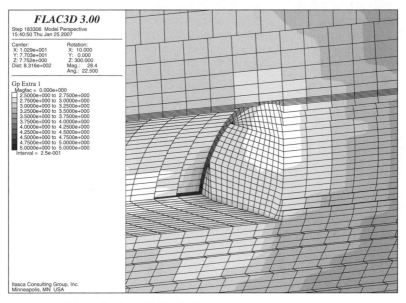

Figure 4.   Face stability evaluation with FLAC3D

and evaluate spile performance. The methodology used to evaluate relaxation ahead of the face is shown on Figure 5. Results from the FLAC3D analyses were also used to cross-check FLAC2D results. Finally, keyblock analyses were performed to determine support requirements to prevent block failure at the tunnel face and along the perimeter.

Initial Support Design

Support elements planned include plain and fiber-reinforced shotcrete (FRS), lattice girders, fast-setting cement-grouted rock dowels, fiberglass rock dowels, self-drilling and grouted spiles, injection spiles, and self-drilling grouted pipe spiles. The FRS will have a strength of 27 MPa (4,000 psi) in all support categories. Self-drilling spiles are used because drill holes are expected to be unstable where spiles are required.

Rock reinforcement, consisting of cement-grouted dowels, will be used in all support categories. In some of the more adverse rock, self-drilling cement-grouted dowels will be required. The top heading tunnel perimeter support will be installed within three rounds of the working face.

Spiles will be installed as part of Support Categories 1 through 3 so as to minimize overbreak and maintain the excavation profile line. In Support Category 4 that includes a pipe canopy, the excavation profile is flared along the inclination of the pipe canopy, to provide adequate clearance for the drill boom used to install the pipes.

Drainage holes and probe holes will be drilled ahead of the top heading to control the impact of water inflows on the stability of the ground around the tunnel excavation, and to identify adverse ground conditions ahead of the face.

Shotcrete thicknesses for the four support categories range from 203 millimeter (8 inch) to 304 millimeter (12 inch). Rock reinforcement will consist of 4-meter-long (13 foot) and 6-meter-long (20 foot) rock dowels. A sloping core is used for face support in all support categories. Systematic pre-support and lattice girders will be required in Support Categories 2–4. Maximum advance lengths for the four support categories range from 1 meter (3 feet) to 1.8 meter (6 foot).

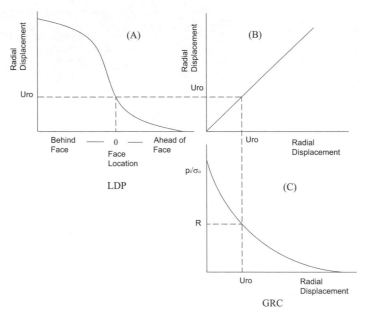

Figure 5. Schematic of estimation of ground relaxation factor using FLAC3D results

Monitoring

An initial support and ground response monitoring program will be implemented during construction to verify that the performance of the initial support systems is within the anticipated range. The monitoring data will also be used as supplemental information to facilitate the selection of appropriate support categories during construction, and to help determine where additional support measures are needed. The monitoring instruments will measure displacements and loading of the shotcrete lining at points around the perimeter of the tunnel, and will monitor ground movements within the rock mass in the rock pillar between the third and fourth bores near the portals as well as within the slopes adjacent to the portals.

The tunnel monitoring program will use monitoring bolts and pressure cells to measure deflections and stresses in the shotcrete lining. Sets of deflection points and load cells will be installed together at defined points around the circumference of the shotcrete lining. Five monitoring types, defining the instrument locations along the tunnel lining in a cross-sectional view, will be used for the range of support categories along the main tunnel alignment and in the cross-passageways.

FINAL LINING AND SEISMIC DESIGN

Final Lining System

The Caldecott fourth bore uses a double lining system consisting of an initial support system (discussed above) and a cast-in-place reinforced concrete final lining (Figure 6). A waterproofing membrane with a geotextile backing layer for drainage will be installed between the initial support and the final lining. The initial support system is designed to carry the ground loads that develop during construction while the cast-in-place reinforced concrete final lining is designed to carry long-term ground loads and any additional loads

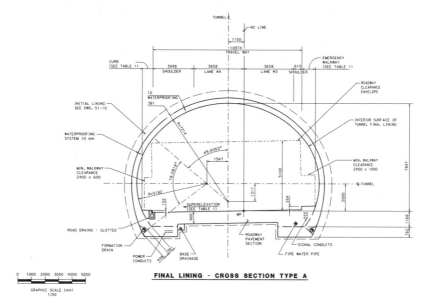

Figure 6. Final lining

due to interior finishes or equipment anchored to it. The final lining will also accommodate seismic deformations and provide a durable and sound tunnel lining.

Loads and Load Combinations

The final lining will support its own dead load, ground loads from load sharing with the initial support system, rock wedge loads supported by the initial supports during construction, and seismic deformations. Two critical load conditions have been identified. The first load combination could occur during the first few years after completion of construction before any ground loads are imposed on the final lining. This load combination consists of the final lining dead load and seismic deformations. The second critical load combination combines maximum ground loads with seismic deformations.

Ground Loads from Load Sharing

The initial and final linings will function as a combined support system in the long term. Over time, after the completion of construction, a portion of the ground load carried by the initial support system will be transferred to the final lining due to deterioration of the initial support system rock dowels and shotcrete. Analyses were performed to assess the effect of the degradation of the initial support and to determine the part of the ground load that will be transferred to the final lining. The analyses assumed the dowels deteriorate completely in the long-term and that the modulus and strength of the shotcrete degrade to approximately 60% of the original design values. The initial shotcrete lining is also assumed to have no flexural capacity in the long-term due to possible deterioration of any reinforcing embedded therein. The results indicate that the final lining will attract a maximum of approximately 50% of the ground load supported by the initial lining. The final lining was conservatively designed to support ⅔ of the ground load supported by the initial lining.

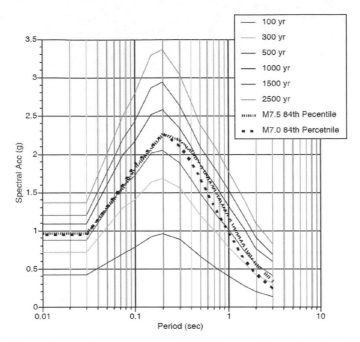

Figure 7. Uniform risk equal hazard spectra and MCE 84th percentile spectra at the Caldecott fourth bore

Seismic Demand

General Performance Requirements. In accordance with general Caltrans practice for "important" facilities on lifeline routes such as SR 24, the seismic design for the tunnel is based on the Safety Evaluation Earthquake (SEE) and a lower-level Functional Evaluation Earthquake (FEE). The project uses a 1,500-year return period for the SEE event and a 300-year return period for the FEE event.

The performance requirements for the SEE are that the fourth bore will be open to emergency vehicle traffic within 72 hours following an SEE. Performance requirements for the FEE are that the fourth bore remains fully operational and experiences minimal, if any, damage.

Seismic Hazard Analysis. Deterministic seismic hazard analysis (DSHA) and a probabilistic seismic hazard analysis (PSHA) were used to characterize the seismic hazard at the project site (EMI, 2005). While numerous faults have been identified in the Bay Area, the Hayward fault was found to be the controlling fault because of its close proximity to the Caldecott Tunnels. Figure 7 shows uniform risk equal hazard spectra developed from the results of the PSHA and DSHA analyses.

Ground Motion Characterization and Wave Scattering Analysis. Based on site specific rock acceleration spectra, three sets of time histories were developed for each of the ground motion events (i.e., SEE and FEE). Wave scattering analyses using these time histories were performed to evaluate the effects of seismic wave propagation and to estimate the ground distortion around the tunnel lining for each time step of the input motion. These displacement time histories were used as input for the pseudo-static time history analysis described as follows.

Tunnel Final Lining Seismic Analysis. Ground shaking and the associated ground deformations are the primary seismic design issue for the fourth bore. Three types of lining deformation were evaluated due to ground strains caused by wave propagation: longitudinal axial compression and tension, longitudinal bending, and ovaling or racking of the cross section. Two types of analyses were used to assess the behavior of the fourth bore lining due to longitudinal and racking seismic deformations. The first method uses closed-form solutions (Hashash et al., 2001 and Penzien, 1998 & 2000). The second is a state-of-the-art numerical method that uses beam-spring and beam-continuum models to perform pseudo-static time history analyses using the results of the scattering analyses described above. Two types of numerical models were used to calculate lining strains, stresses, and forces: 2-D SAP2000 (CSI, 2005) beam-spring models with nonlinear support springs (gap elements) to model ground behavior; and 2-D beam-continuum models using both FLAC (ITASCA, 2005) and ADINA (ADINA R&D Inc.) with elastic continuum elements to model ground behavior. Both methods were used to calculate strains, stresses, and forces in the fourth bore lining and cut-and-cover structures, and to ensure that the results were within acceptable stress and ductility limits.

Final Lining Design

Critical cross-sections in each support category were evaluated to determine the ability of the final lining to support the load combinations referenced above. Results of the analyses indicate that a 381 millimeter (15-inch) final lining with 35 MPa (5000 psi) concrete can support the ground loads and accommodate the seismic deformations. This final lining thickness was selected for constructability and is controlled by the thrust resulting from the ground loads in the high cover section of the alignment. Two layers of reinforcing will be used for the final lining to meet Caltrans criteria. The seismic demands, although very high, do not control the thickness of the final lining.

CONCLUSIONS

Preliminary design of the 15-meter-wide two-lane Caldecott fourth bore included identification of four ground classes that are expected along the alignment. Four corresponding excavation and initial support categories have been developed for NATM construction of the mined tunnel. Support elements include plain and fiber-reinforced shotcrete, lattice girders, fast-setting cement-grouted rock dowels, fiberglass rock dowels, self-drilling and grouted spiles, injection spiles, and self-drilling grouted pipe spiles. Shotcrete lining thickness in the four support categories ranges from 200 millimeters (8 inches) to 300 millimeters (12 inches).

The final lining will support dead load, ground loads, rock wedge loads, and seismic deformations. The design analysis shows that a 381 millimeter (15 inch) final lining with 35 MPa (5,000 psi) concrete can support the ground loads and accommodate the seismic deformations. Seismic demands do not control the thickness of the final lining, despite the close proximity of the project to a major active fault and seismic design criteria corresponding to an earthquake with a 1,500-year return period and a peak ground acceleration of 1.2g.

ACKOWLEDGMENTS

The authors would like to acknowledge Geomatrix Consultants for their work on the site geology, ILF Consultants for independent reviews of the initial support designs,

Earth Mechanics Inc. for their work on seismic hazard and wave scattering analysis and SC Solutions for their work on seismic demand analysis.

The contents of this paper were reviewed by the State of California, Business, Transportation and Housing Agency, Department of Transportation and the Contra Costa Transportation Authority. The contents of the paper reflect the views of the authors who are responsible for the facts and accuracy of the data presented herein. The contents do not necessarily reflect the official views or policies of the State of California or the Contra Costa Transportation Authority. This paper does not constitute a standard, specification or regulation.

REFERENCES

Page, B.M. 1950. Geology of the Broadway Tunnel, Berkeley Hills, California, Economic Geology, Vo. 45, No. 2.

John, M. and Mattle, B. 2003. "Shotcrete Lining Design: Factors of Influence." RETC Proceedings.

Itasca Consulting Group Inc., 2005. Fast Lagrangian Analysis of Continua (FLAC) Version 5.0, Minneapolis.

Hashash, Y.M.A., et al., 2001. Seismic Design and Analysis of Underground Structures, Tunneling and Underground Space Technology 16, pp. 247–293, Elsevier

Penzien, J., Wu, C., 1998. Stresses in Linings of Bored Tunnels, Int. J. Earthquake Eng. Struct. Dyn., 27, pp. 283–300.

Penzien, J., 2000. Seismically Induced Racking of Tunnel Linings, Int. J. Earthquake Eng. Struct. Dyn 29, pp. 683–69.

SAP 2000, Computer and Structures Inc., Berkeley, California, 2005.

ADINA, ADINA R&D, Watertown, Massachusetts.

Earth Mechanics Institute, 2005. Seismic Hazard Report and SEE Ground Motions, Caldecott Improvement Project.

Bieniawski, Z.T. 1989. Engineering Rock Mass Classifications, Wiley, New York.

UPPER ROUGE TUNNEL CSO CONTROL

Wern-ping Chen

Jacobs Engineering

Mirza M. Rabbaig

Detroit Water and Sewerage Department

Robert Barbour

Jacobs Engineering

Jean Habimana

Jacobs Engineering

Jinyuan Liu

Jacobs Engineering

ABSTRACT

To mitigate and control the combined sewer overflows from 17 outfalls in the City of Detroit, three outfalls in Dearborn Heights and eight outfalls in Redford Township, the Detroit Water and Sewerage Department has completed the preliminary design of a storage tunnel in 2005. In 2006, a final design has revised the alignments and the inside diameter (ID) of the tunnel. The revised tunnel system includes an 11.28-km (7-miles) 9.14-m (30-ft) ID tunnel, a 24.4-m (80-ft) ID pump station, two work shafts about 15.2-m (50-ft) ID each, and 14 drop shafts with various IDs. This paper discusses the latest development of this tunnel system.

INTRODUCTION

Combined sanitary and storm sewers, a small portion only, serve the majority of the Rouge Valley watershed, including Detroit and most of the western suburbs. The Northwest Interceptor (NWI), which extends between Eight Mile Road and Ford Road and roughly parallel to the Rouge River, intercepts sanitary sewage and some wet weather flows from trunk sewers in the tributary areas for transport to WWTP for treatment. During extreme storm event, when the capacity of the NWI is exceeded, the excess combined sewer and overflow (CSO) overflows into the Rouge River through 17 outfalls to local rivers. The purpose of the Upper Rouge Tunnel system (URT) is to capture and significantly reduce the uncontrolled CSO from these 17 outfalls, in the City of Detroit, three outfalls in Dearborn Heights, and eight outfalls in Redford Township. When implemented, the URT storage capacity would result in less than one overflow per year, on average, into the Rouge River. It will capture and store the combined sewage, and later, within 48 hours, discharge the combined sewage back to the interceptor when capacity is available and then to the Detroit Wastewater Treatment Plant for treatment and release. The owner of the URT system is the Detroit Water and Sewerage Department (DWSD).

The preliminary design of the URT was completed in November 2005. The notice to proceed (NTP) of the final design was issued in mid February 2006. This paper updates the URT design developments since the completion of its preliminary design.

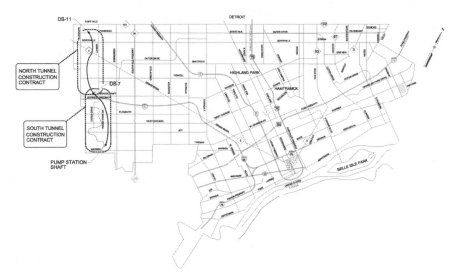

Figure 1. Final URT site location and plan

Preliminary Design

The Upper Rouge CSO Tunnel Basis of Design Report (BOD), completed in April 2005, summarizes the major components at the end of the preliminary design of the URT system, which include:

- Diversion chambers to intercept overflows and direct them toward the tunnel
- 14 drop shafts that convey the intercepted flow from the surface down to the tunnel level
- An approximately 11.27-km-long (7-mile) 201-MG capacity tunnel that generally follows the Rouge River's course from Warren Avenue north to Pembroke Avenue to store the captured flows
- A 100-MGD dewatering pumping station located just northeast of the intersection of Warren Avenue and W. Outer Drive that is capable of emptying the URT in 48 hours
- A 975-m-long (3,200-ft), 1.52-m (5-ft) diameter inverted siphon that carries flow by gravity from the URT pump station back to the DWSD sewer system at the conclusion of the wet weather event

The proposed URT alignment is mostly in the property of the City of Detroit. The project site is about 16-km (10-mile) northwest to the City of Detroit. Figure 1 shows the final design URT site location and plan, which follows almost exactly the footprint of the preliminary design alignment, except modifications in associated with the Drop Shaft 7 site relocation as discussed later in this paper.

Two tunnel vertical alignments were evaluated in the preliminary design phase. The first alternative is a split tunnel alignment concept, where the southern part of the system, South Tunnel, has a tunnel diameter of 10.5 m (34.5 ft) and the northern part of the system, North Tunnel, has a tunnel diameter of 6.09 m (20 ft). In this scheme, a transfer shaft is located at the interface between the South and North Tunnels. The second alternative is a single tunnel bore alignment concept, with a uniform finished

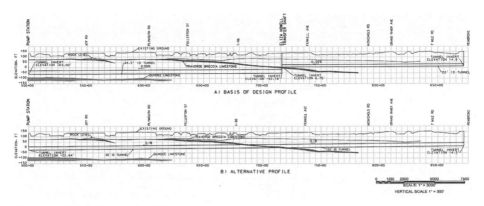

Figure 2.   URT BOD tunnel profiles

Table 1.   Key features of selected BOD alignment

Feature	
Total length	11,320 m (37,140 ft)
North tunnel length	4,724 m (15,500 ft)
North tunnel slope, %	0.05%
North tunnel finished diameter (ID)	6.09 m (20 ft)
South tunnel length	6,596 m (21,640 ft)
South tunnel slope, %	0.05%
South tunnel finished diameter (ID)	10.5 m (34.5 ft)
Number of drop shafts	14

diameter of 9.14 m (30 ft). Figure 2 shows the preliminary design tunnel alignment alternatives.

The split tunnel concept was selected as the recommend tunnel vertical alignment at the end of the preliminary design phase. Table 1 summarizes the key URT features of the selected alignment.

The primary reason for selecting the split tunnel concept was to avoid tunneling through Traverse Breccia Formation, which is considered heterogeneous with potential of encountering high groundwater infiltration during tunneling.

Final Design

DWSD selected JE Associates, Inc. (Jacobs Engineering, Detroit) and a team of support firms to provide final design and construction management for the URT system under DWSD Contract Number CS-1431. Table 2 presents the URT final design team and the responsibility of each firm.

The NTP for the final design was issued to Jacobs on February 16, 2006. The initial task of the scope of work was to review the BOD, prepared under DWSD Contract Number CS-1402, and to issue a Supplemental Basis of Design Report (SBOD) to summarize recommended basis for final design. Jacobs issued a SBOD on September 2006. Most of the design details provided within the BOD have been accepted and remain unchanged. The major recommendation of the SBOD was the selection of a

Table 2. URT final design team

Firm	Responsibility
Jacobs Engineering	Prime designer; also responsible for the South Tunnel and North Tunnel designs
Arcadis	Tunnel Design Support, Pump Station Design, Value Engineering, Environmental Investigations, Community Outreach
Berg Muirhead & Associates	Community Outreach
Hinshon Environmental Consulting	Permits and Regulatory Requirements
Multi Tech Resources, Inc	Electrical Design
RI Geotechnical, Inc	Geotechnical Review
Sigma Associates	Architectural. HVAC, I&C, Near Surface Facilities Design
Somat Engineering, Inc	Geotechnical Investigations
Stout Risius Ross	Appraisals
Tucker, Young, Jackson, Tull (TYJT)	Near Surface Facilities Design, Procurement Support
Wade Trim	Near Surface Facilities Design, Easement and Land Acquisition, Hydraulics
Williams Acosta, PLLC	Legal Support

single bore tunnel alignment with an ID of 9.14-m (30-ft) and the relocation of the site of the transfer shaft, Dropt Shaft 7 (DS-7 Shaft), as explained later in the paper. The total length of the proposed main tunnel, in final design phase, is about 11,278 m (37,000 ft).

Details of the URT system updates, construction contract packaging, design and construction considerations, and design and construction schedules are addressed, in this paper, after the Geology and Subsurface Exploration Program section.

GEOLOGY AND SUBSURFACE EXPLORATION PROGRAM

Geological Setting

The project site is located on the southeasterly edge of the Michigan Basin, a structural basin nearly circular in shape, in which layers of sedimentary rock slope or dip inward from the rims toward the central area of the Southern Peninsula. The basin itself is gently inclined and concentric where a rock formation outcropping along the margin of the state becomes more deeply buried by progressively younger formations toward the center of the basin.

Geology in the vicinity of the project consists of unconsolidated soil deposits approximately 9 to 30-meter (29.5 to 98-feet) thick, comprising fill, alluvial, and undifferentiated glacial deposits. The underlying bedrock consists of sedimentary rocks, comprising shale, limestone, and dolomite. In the northern reach of the proposed tunnel alignment, the soil is directly underlain by the Antrim Shale formation. This rock comprises a relatively hard, brittle, dark brown to black bituminous shale that has a history of producing methane gas. Further south along the alignment, the formation directly under the glacial drift makes a transition to the Traverse Group. This formation is informally divided into five sub-units: Traverse Dolomite, Traverse Breccia, Traverse Limestone, Traverse Limy Shale, and Traverse Basal Limestone. Like Antrim Shale, the

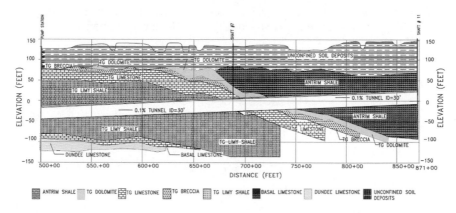

Figure 3.   Geological profile along the proposed tunnel alignment

Traverse Group has a potential to produce methane gas and can also produce minor quantities of hydrogen sulfide gas. Isolated occurrences of crude petroleum and groundwater under an artesian pressure head have been observed in the Traverse Group rocks. Dundee Limestone is observed below the Limy Shale in the southern part of the proposed tunnel alignment; however, it is well below the selected vertical tunnel alignment. The Dundee Limestone, which historically is an oil and gas producer and is occasionally highly fractured upper reaches, often contains groundwater under artesian pressure. Groundwater from the Dundee Limestone can contain dissolved sulfide and methane that are released as hydrogen sulfide and methane gas upon exposure to the atmosphere; therefore, it is purposely to select the vertical tunnel alignment to avoid the encountering of Dundee Limestone Formation. Figure 3 shows the geological profile along the proposed tunnel alignment.

Subsurface Exploration Program

Fifty-six (56) borings, varying in depth from 14 to 77-meter (46 to 253-feet), were performed from 1997 to 1998 during a feasibility study conducted under DWSD Contract No. CS-1283 by NTH Consultants, Ltd. Additional 55 borings, varying in depth from 9 to 78-meter (30 to 256-feet), were performed from 2004 to 2005 during the preliminary design, under DWSD Contract No. CS-1402, by Parsons Brinckerhoff Michigan, Inc.

During final design, an additional exploration program was conducted by Jacobs Engineering with Somat Engineering, under DWSD Contract No. CS-1431. This program includes four phases:

- Phase 1 includes ten (10) inclined borings, drilled along the tunnel alignment and through Traverse Formation rock to examine the possibility of encountering vertical or near vertical discontinuities along the revised tunnel alignment. These discontinuities, if exist, may provide paths for groundwater inflow during the tunnel construction. These borings are inclined to 25 degrees from vertical. Packer tests and acoustic teleview were performed to estimate hydraulic conductivity and beddings/fractures and their orientations within the rock mass.
- Phase 2 includes 85 shallow borings to support the design of the near surface structures associated with the pumping station and the 17 outfall structures.

- Phase 3 includes two (2) horizontal borings with a total length of 1,910 meters (4,000 ft) at the tunnel level. The locations of the horizontal borings were selected such that they will go through most of the anticipated geologic formations to be encountered during tunneling, expect Antrim Shale, which consists of more uniform and less controversial geological properties. Acoustic teleview, packer tests, and hydraulic fracturing test are performed.

- Phase 4 includes ten (10) deep boreholes, up to 76 m (250 ft) and mainly drilled along adits and in major shaft sites. Packer tests were performed full length in rock for these borings.

URT SYSTEM UPDATES

This section updates the latest design developments for components of the URT system as well as recommendations made in the SBOD, which forms the basis for final design.

Tunnel Diameter

Based on geological and hydrogeological parameters from previous studies (including Geotechnical Data Reports from CS-1283 and CS-1402 contracts, Geotechnical Interpretative Report from CS-1402 contract, and the subsurface exploration program for CS-1431 contract) along the proposed URT alignment and the latest subsurface exploration programs, it is considered that a single bore tunnel alignment is feasible. A single bore tunnel, in lieu of the split tunnel system, will also ease the tunnel system operations, as it eliminates the requirement of the transfer shaft and raises the pump station invert by approximately 6 meters (20 feet). The required ID of a single bore tunnel to provide the same storage capacity as that concluded in the BOD is approximately 9.14-m (30-ft), which becomes the basis for final design.

Tunnel Alignments

Horizontal Alignment. Because of a potential industrial building development in the Eliza Howell Park, the DS-7 Shaft site, as selected in the BOD phase, is not available for the URT project. During the SBOD phase, two potential sites within the park were identified for consideration. The first site is located at the northern end of the Park near Murphy Middle School (23901 Fenkell Street, Detroit). The second site is located at the southern end of the park near Interstate highway I-96. The selection of the appropriate site and corresponding adjustment of the tunnel horizontal alignment were based on the following considerations:

- The length of tunneling through the Traverse Breccia formation to be as short as possible

- The approximate starter tunnel length of 91.4-m (300-ft) for a tunnel boring machine (TBM)

- A tangent line, as long as feasible, at the beginning of the tunnel alignment to facilitate the tunnel excavation logistic

Considering the geological condition, environmental and local community impacts during construction, the construction schedule, and packaging impacts, the second site, at the southern end of the park, is preferred for DS-7 Shaft site, as shown on Figure 4. The tunnel horizontal alignment starting from the revised DS-7 Shaft location going north is a tangent line about 412-m (1,350-ft) long, followed by an approximate 914-m (3,000-ft) radius curve.

Figure 4. Revised tunnel alignment at DS-7 site

Vertical Alignment. The vertical alignment of the tunnel for final design is primarily determined by:

- Rock cover over the excavated tunnel crown—approximately one excavated tunnel diameter

- Tunnel slope—adequate to create scouring velocity to reduce grit accumulation during system operation

- The life cycle cost of the operation of the tunnel pump station—which prefers a shallower pump station to be cost effective for construction and operation

The shallowest rock cover over the URT tunnel is in the northern part of the project, within a reach approximately from STA 798+00 to the end of the tunnel at STA 870+20. The deepest top of competent rock elevation is identified approximately at elevation 16-m (52.6-ft). However, the majority of the top of competent rock elevations within this reach are observed at elevations at or above 17.4-m (57.0-ft). It is therefore selected the excavated tunnel crown elevation to be 7.4-m (24.2-ft) at the northern terminal end of the URT tunnel, to maintain approximately a rock cover of about one excavated tunnel diameter for most portion of the tunnel. One exception occurs nears STA 850+00, where the shallowest rock cover, approximately 8.5-m (28 ft), is encountered. In general, the depth of the tunnel is about 46-m (150-ft) below grade.

The BOD recommends a vertical slope of 0.05% for both the South and North Tunnels. Based on a hydraulic analysis of tunnel slope conducted by Arcadis on 2006, considering the ease of grit removal and maintenance of the system, a slope of 0.1% is considered adequate.

Based on the recommended 0.1% tunnel slope, the tunnel invert elevation at the proposed Pump Station is approximately 6-m (20-ft) higher than that proposed in the BOD. This scheme results in allow fewer pumps while still maintains the capacity to dewater the tunnel within 48 hours. Raising the pump station invert elevation is also beneficial to the operation and maintenance of the pump station within its design life.

CONSTRUCTION CONTRACT PACKAGING

Construction packages include two (2) tunnel contracts, one (1) Pump Station finish contract, four (4) near surface facility contracts, and one (1) near surface pipeline contract. The two tunnel contracts, South Tunnel and North Tunnel of about 5.6-km (3.5-mile) each, are separated at the DS-7 Shaft, which is near the mid point of the URT tunnel system. The South Tunnel contract includes the excavation of the Pump

Station Shaft, the excavation and final lining of the South Tunnel, and excavations and final lining of connecting adits and deaeration chambers south of DS-7 Shaft. The North Tunnel contract includes the excavation and final lining of DS-7 and DS-11 Shafts and all connecting adits and deaeration chambers north of DS-7 Shaft. Each near surface contract includes the excavation of drop shaft to the top of deaeration chamber for that specific site.

Each of the two tunnel packages has each own distinct geological features. The South Tunnel will fully encounter in Limy Shale. The North Tunnel will encounter about 1,220-m (4,000-ft) of Traverse Formation at the beginning of the tunneling and will then fully into Antrim Shale for its northern reach.

TUNNEL DESIGN AND CONSTRUCTION CONSIDERATIONS

South Tunnel

Key considerations for the South Tunnel are associated with tunneling through the Traverse Limy Shale, namely, its slaking propensity for degrading into generally non-durable residual material upon exposure to air and its fissile nature with clay filling bedding joints that disk when extracted from the ground and dried. Other issues for tunneling in Limy Shale are the bearing capacity of the Limy Shale for TBM grippers if an open face TBM is used and the feasibility of rock bolts as temporary support in Limy Shale with regard to its fissile nature. FLAC Ubiquitous model has been used to simulate the lamination behavior of the Limy Shale. It is found that the behavior of the Limy Shale after excavation, if not sealed, would potentially have a sudden progressive failure mode through overstressed laminated layers. The overstressed, plastic/yielding, zone would then propagate upward creating a cone shape failure zone above the crown. If sealed appropriately, relative to the time of opening, the Limy Shale does not require significant initial support. The best countermeasure for the construction in Limy shale is the immediate sealing, by shotcrete or other sealing material, of its exposed surface; however, this approach may be difficult when employing TBM excavation. The rebound of the shotcrete or sealing material immediately behind the cutterhead will slow down the TBM utilization and is not recommended by most TBM manufactures. The current design philosophy has adopted two tunnel lining alternatives. The first alternative includes precast concrete expanded segment as initial tunnel support and cast-in-place concrete as tunnel final lining. The second alternative includes one-pass segmental concrete lining to serve as both initial tunnel support and final lining. Both alternatives will be designed and presented in contract document for bids.

North Tunnel

As mentioned earlier, the North Tunnel will be driven through Traverse Limestone, Traverse Breccia, and Traverse Dolomite for about 1,220-m (4,000-ft) and then in Antrim Shale for the reminder 5,500-m (18,000 ft). For this tunnel reach, key construction considerations are associated with groundwater inflows and hazardous gas, methane and hydrogen sulfide at the interface between different geological formations. Vertical to sub-vertical joints have been reported within the Michigan Basin and may bring significant groundwater inflows during construction, though they has not been found from subsurface exploration programs conducted for this project, at the time when this paper is drafted. The current design philosophy for tunnel support and final lining employs two-pass lining system, which includes immediate support type rock dowel (such as swellex), steel strap, and steel rib and lagging. It is anticipated that, in

Antrim Shale, shotcrete sealing to seal off the exposed surface may be required to avoid the potential of rock mass deterioration when it is exposed. This thin layer of shotcrete could be applied behind trailing gear to avoid its interference with TBM excavation at the tunnel face.

Construction

Rock tunnel and shaft construction in southeast Michigan has a history of problems associated with encountering explosive gas (methane), hazardous gas (hydrogen sulfide) and high groundwater inflows. It is anticipated that the TBM and its supporting systems will be required to conform to the OSHA 'Potential Gassy' tunnel classification requirements. Safety requirements in constructing the URT project will follow MIOSHA regulations.

SCHEDULE

The design of the two tunnel packages is expected to be completed at the end of June 2007 and all other near surface and pipeline design packages to be completed around August 2007. Tunnel construction is expected to be commenced in early 2008 and the overall program is expected, at the time when this paper is drafted, to be completed in 2014. The logistic of the near surface packages will facilitate the construction of the two main tunnel contracts such that the construction of the drop shaft for each site will be completed prior to the construction of its associated deaeration chamber.

REFERENCE

Parsons Brinckerhoff Michigan, Inc., 2005. Upper Rouge CSO Tunnel Basis of Design Report. Prepared for the City of Detroit, Water and Sewerage Department, April 2005.

Jacobs Engineering, 2006. Upper Rouge Tunnel CSO Control Project Supplemental Basis of Design Report. Prepared for the City of Detroit, Water and Sewerage Department, September 2006.

Arcadis, 2006. Upper Rouge Tunnel CSO Control Project Hydraulic analysis for Tunnel slope. Prepared for the City of Detroit, Water and Sewerage Department, May 2006.

10

New and Future Projects II

Chair
S. Hunt
CH2M HILL

THE DC WASA ANACOSTIA RIVER CSO CONTROL TUNNEL PROJECT

Larry Williamson
CDM/HMM JV

Leonard R. Benson
District of Columbia Water and Sewer Authority

David C. Girard
CDM/HMM JV

Ronald E. Bizzarri
District of Columbia Water and Sewer Authority

ABSTRACT

On the east side of Washington DC, the DC Water & Sewer Authority is developing a Facility Plan for the Anacostia River CSO Control Projects, including 8 to 12 miles of CSO storage/conveyance and flood relief tunnels and associated facilities. The project includes major technical and non-technical challenges, including: completion under a Federal Consent Decree schedule; large diameter (estimated 25 to 35 ft) soft ground tunnels and shafts up to 200 ft depth; operations without use of active flow control systems; and close coordination with several major public and private construction and development projects that are being advanced in the project area.

INTRODUCTION AND BACKGROUND

Washington DC is undergoing a significant increase in property and infrastructure development, particularly on the east side of our nation's capital city along the Anacostia River. These development projects include several major private and public property developments, several major roadway and bridge improvement and replacement projects, and an approximate $2 billion program for control of combined sewer overflows (CSO's). This paper will outline the scope and planning activities for a portion of the CSO control program, specifically the Anacostia River Projects CSO tunnel facilities. This project is one of the initial components, and the most sizeable component, of the overall Long Term CSO Control Plan (LTCP) being implemented by the District of Columbia Water and Sewer Authority (WASA). The LTCP is being performed under a Federal Consent Decree agreement that requires all facilities to be completed by March 23, 2025 under a 20-year program. The Consent Decree also specifies several interim milestones for completion or start of various activities, including two significant milestone dates for the Anacostia River Projects—placement of the Anacostia River Tunnel (ART) into operation by March 23, 2018 and the placement of the Northeast Boundary Tunnel (NEBT) into operation by the March 23, 2025.

The project will include several miles of deep tunnels for storage and conveyance of combined sewer flows plus associated flow diversion structures, consolidation piping and/or tunnels, and shafts for construction, inflows, overflows, maintenance and air

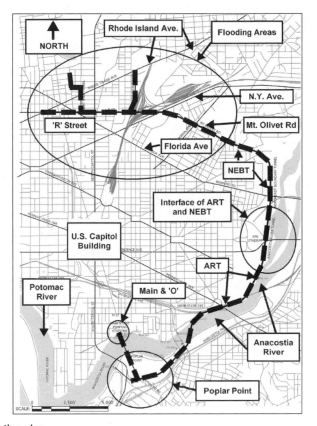

Figure 1. Location plan

release. This paper is being written at the conclusion of Phase I of the facility plan program and the initiation of Phase II, which will encompass full development of the facility plan; performance of an initial subsurface investigation program; completion of design to an advanced conceptual level; establishment of design, construction and operational standards and criteria for the project; and identification of a project implementation schedule and contract packaging for final design and construction.

The facility plan is being developed by CDM/HMM, a Joint Venture (CDM/HMM), which is comprised of partner firms Camp Dresser & McKee and Hatch Mott MacDonald, as consultant to WASA. During Phase I, accomplishments included: mobilization of the project team; gathering of existing data for the project; identification of alternative tunnel alignment corridors; performing two initial borings to depths of 250 feet; planning and procurement of a subcontractor for Phase II subsurface investigations; initiating coordination activities with other projects and agencies that will be performing development and/or construction within the project area; performing initial hydraulic operational model analyses; and preparing several technical reports to outline the issues and objectives to be addressed during Phase II services.

As previously noted, the project includes several miles of tunnels and associated facilities. Major components include:

- Approximately 8.7 miles of predominantly soft ground tunnel with bored diameters ranging up to 28 feet to provide a total storage volume of 126 million gallons (see Figure 1),
- A new pumping station for dewatering the tunnel and conveying of stored volumes to the flow treatment facilities,
- Drop shafts (approximately 14 are anticipated, at present), diversion structures, consolidation and conveyance conduits (pipelines/smaller diameter tunnels) to intercept and convey inflows to the storage/conveyance tunnels, as well as associated air release and maintenance access shafts,
- Overflow structures, including shaft(s), at the downstream and possibly mid-tunnel locations to facilitate operations of a system without an active flow control system, and
- Construction shafts as required to facilitate launching and extractions of the tunnel boring machines; and perform tunnel construction activities.

EVALUATIONS AND SELECTION OF TUNNEL CORRIDORS

The tunnels will be predominantly excavated and constructed in soft ground conditions consisting of water-bearing Cretaceous-age clays, silts and sands. They will be lined with precast concrete segmental liners coincident with tunnel excavation. There is potential for some rock or mixed-face tunneling near the northern terminus of the tunnel system; however, studies and investigations are underway in attempt to raise the tunnel vertical alignment so as to complete all tunneling within soft ground strata, if feasible. These studies include analyses of the hydraulic operational requirements to achieve sufficient collection of stormwater flows to mitigate flooding problems in these areas and divert excessive flows from the existing sewer system during rain events; as well as subsurface investigation programs. To-date, no borings have been performed in the potential rock or mixed-face areas; however, preliminary geophysical surveys have provided some limited data on the potential soil/rock interface in these areas.

Several segments of tunneling are anticipated. The initial segment, called the Anacostia River Tunnel (ART), will start in the vicinity of Poplar Point which is located at the South Capitol Street Bridge, also known as the Frederick Douglas Bridge, on the Southeast bank of the Anacostia River. This site is also the proposed location of the tunnel system pumping station for dewatering the tunnel, the downstream tunnel overflow structure, and the Fort Stanton Interceptor and drop shaft, unless individual drop shafts along the tunnel can provide the same operational hydraulic performance. The ART is then planned to progress northward to a terminus presently anticipated to be just northeast of the existing RFK Stadium. The ART would first cross under an existing underground WMATA transit line proceeding northeast under the Anacostia River amid a myriad of piers, piles and foundations associated with past, present and future highway and railroad structures and riverbank retaining structures. This segment of tunnel will require drop shaft(s) along the east side of the river to pick up CSOs in the Fort Stanton area and drop shafts along the more highly developed west side of the river where most of the CSO outfalls are located. Consolidation conduits by pipeline and/or tunnel will also be required, specifically along the west side of the river where three CSO outfalls, presently discharging at marine sites, are required to be eliminated and their respective flows diverted elsewhere.

Also to be constructed from Poplar Point is a branch tunnel connecting to the existing WASA Main and "O" Street Pumping Stations, located almost directly across the river from Poplar Point. The Main and "O" Street site is an extremely congested site with a large CSO outfall. It is located near the terminus of the old Washington Canal and is a convergence location for several major sanitary and combined sewers from which flows are collected and conveyed in siphon lines across the river to Poplar Point. The extent of the converging wastewater infrastructure and confined nature of the site will present several challenges to design and construction of flow diversion and shaft structures. To further complicate design and construction of new facilities at this location, the site is bounded by the Anacostia River to the south, the new Washington Nationals baseball stadium and associated property development to the west, and the new Southeast Federal Center to the north and east.

The second major tunnel to be constructed, which is termed the Northeast Boundary Tunnel (NEBT), will originate at the terminus of the ART and then progress in a northerly direction to Mount Olivet Road and then under New York Avenue and the adjacent railroad yards, then westerly along "R" Street, NE and NW, terminating on the western side of Florida Avenue. This segment will also include connecting conduits to identified surface flooding zones in the northern reaches of the project area. The primary functions of the NEBT are to provide flood relief in several identified areas and to intercept and convey stormwater flows before they become CSOs. Due to on-going and future Phase II studies, the alignments and configurations of the tunnels and connecting pipelines west of New York Avenue are the most likely components to incur changes in alignment, sizing and construction methodology. These Phase II studies include the subsurface investigation and field surveys that will provide new information on subsurface and site conditions as well as additional hydraulic modeling studies on the sewer collection system operational performance.

To date, approximately 12 alternative horizontal alignment corridors have been studied for the ART and approximately 6 alternative corridors have been studied for the NEBT. Alternatives were short-listed based on several considerations, including: ability to satisfy operational hydraulic requirements, maintain the tunnel corridors within public lands, ability to avoid subsurface obstructions, and ability to mitigate and manage risks associated with underground construction. Cost and schedule for the various alignments were also considered; however, cost and schedule did not appear to provide any significant degree of differentiation between alternatives at the time of assessment due to the accuracy of the estimates we were able to develop based on the level of engineering and data analysis that had been performed under Phase I of our facility plan effort. Updated cost estimating and establishment of the program schedule will be accomplished under Phase II of the facility plan effort.

Tunnel sizing will be determined based on the minimum 126 million gallon storage requirement of the Consent Decree, additional hydraulic modeling analyses, and constructability considerations. At present, the main tunnels are anticipated to be, on average, approximately 26 feet inside diameter with branch tunnels on the order of 15 feet I.D. Phase I studies considered larger diameter tunnels up to 40 feet or greater I.D. for the ART as potential measures to reduce construction cost and mitigate hydraulic surges during tunnel filling.

Depth of tunnel is governed by several factors. Anticipated depths of the LTCP schematic alignment and other early tunnel alignment corridors were on the order of 150 to 180 feet below ground surface. These alignments were largely governed by at least one under-crossing, and potentially multiple under-crossings, of WMATA rapid transit lines. The schematic alignment corridors also passed under several major buildings and utilities while transitioning between the tangent sections under major streets and avenues. The LTCP schematic alignments also had the NEBT initially located beneath a major sewer line, the Northeast Boundary Trunk Sewer (NEBTS) under Florida Avenue so that flows from the existing trunk sewer could be off-loaded into the new tunnel system periodically along it's length. For the NEBT, subsequent hydraulic modeling analyses indicated that sewer system flows could be intercepted before they reached the NEBTS by moving the NEBT to coincide approximately with Mount Olivet Road and 'R' Street NE and NW. This revised alignment provided additional benefit by eliminating or minimizing side tunnel connections to several of the identified flooding areas. Accordingly, the same hydraulic performance and improvements to the existing sewer system infrastructure could be achieved by the 'R' Street alignment while also eliminating a major risk of tunneling under a major utility for a great length.

Alignment studies along the ART also resulted in some opportunities for risk reduction and decrease in tunnel depth. The project team investigated potential corridors that could more closely follow park land along the east side of the river before traversing to the west side for pick-up of the major CSO flows. It was found that the tunnel depth could be significantly decreased because the under-crossing of the underground WMATA rapid transit line in the vicinity of Poplar Point could be accomplished at a higher elevation as the rapid transit line approached the Anacostia transit station southeast of the Anacostia River. Consideration of buoyancy and depth of the river channel then became a critical factor for the vertical alignment of the tunnel. Further investigations of historical records from several sources also resulted in a better understanding of the influence on the vertical alignment of past, present and future foundations and construction excavation support systems used for the transit lines, bridges and shoreline retaining structures along the proposed tunnel corridors. Horizontal and vertical alignment studies have not been concluded as of this writing; however, it does appear that an approximate decrease in depth as much as 80 feet may be achievable as compared to the earlier schematic layouts of the proposed tunnels. This results in a moderate savings in future operating costs due to a higher wet well elevation for the pumping station; however, more significant than long-term cost savings, this decrease in depth may provide a betterment for tunnel construction because tunneling would occur under lower groundwater pressures. It should also be again noted that the horizontal and vertical alignment studies to-date have been performed without benefit of new borings along the selected alignment corridors. Accordingly, additional adjustment may be made upon obtaining additional data on subsurface conditions in an effort to locate the tunnel in the most favorable soils for constructability and risk management considerations.

SUBSURFACE INVESTIGATION PROGRAM

The project area is located to a great extent along the lowlands of Anacostia River for the ART and the initial (southern) reaches of the NEBT and in more upland areas of the Northeast and Northwest quadrants of the District of Columbia for the upstream (northern) reaches of the NEBT. Prior to performing any subsurface investigations along the alignment corridors, existing geotechnical information for the project area was gathered from various sources, including data from public and private developments and construction projects and the tunnel projects that were completed for construction of the WMATA rapid transit system. Unfortunately, the vast majority of the existing borings were not completed to the depths anticipated for the CSO tunnel, and therefore, a comprehensive investigation program was planned for the facility planning phase of the project.

During Phase I of the facility planning effort, two (2) borings were performed in the vicinity of WASA's "O" Street pumping station. These were demonstration borings that were completed early in the program to evaluate the ability to perform soil borings and obtain soil samples at the anticipated depths of the CSO tunnel. Borings were advanced to depths of 250 feet using both conventional and sonic drilling equipment. Both methods performed adequately, and the lessons learned from the two demonstration borings were carried forward into the planning for a more extensive boring program during Phase II of the facility planning effort.

The Phase II effort will consist of thirty (30) borings that have been located at the presently anticipated shaft sites and along the proposed alternative tunnel corridors that remain under consideration. Twenty (20) borings will be performed using conventional drilling equipment and ten (10) will be accomplished using sonic drilling equipment. Two conventional borings are planned to be performed within the river. The conventional soil borings will obtain geotechnical samples at frequent intervals within the prescribed zone of the soil column that is most important to tunnel design. The sonic borings will obtain a continuous soil sample column for the full depth of the boring. In addition, associated laboratory and in-situ testing will be performed.

The boring program as presently planned consists of a reduced number of borings from that initially contemplated during the Phase I facility planning efforts. WASA has requested this reduction pending resolution of an outstanding legal challenge relating to allowable discharges to the receiving waters from remaining CSOs approved in the LTCP.

Other activities included in the facility planning effort include geophysical surveys, as previously noted, in the areas of potential mixed-face tunneling, as well as groundwater monitoring. Seismic reflection and refraction surveys have been performed in the vicinity of the potential tunnel alignments from approximately the intersection of Mount Olivet Road and West Virginia Avenue across New York Avenue and west along 'R' Street, NE to approximately Florida Avenue. While the results of these surveys are preliminary at this point in time and have not been confirmed by borings, they appear to reveal an undulating bedrock profile that generally mirrors the surface topography.

As stated, groundwater monitoring will also be performed via instrumentation maintained within a number of the borings subsequent to the drilling and sampling activities. A Geotechnical Data Report will summarize the findings of the subsurface investigation program for use in Final Design and Construction stages of the project.

TUNNEL DESIGN AND CONSTRUCTION CONSIDERATIONS

Alignment corridor studies to-date have been developed based upon presently known data and operational requirements as well as constructability and risk management considerations. The tunnels will be constructed by tunnel boring machine (TBM); however, the selection of tunneling equipment can not be better defined until the Phase II subsurface investigation program has been substantially completed. Pressurized Face Tunnel Boring Machines are anticipated to be utilized, including potentially Earth Pressure Balance Machines or Slurry Face Pressure Tunneling Machines.

As geotechnical data and associated lab and in-situ testing data are obtained and analyzed by completing the Phase II borings, the type of tunneling equipment appropriate to the anticipated subsurface conditions can be better defined. The project intent is to recommend a machine type and an associated vertical alignment that can best apply a consistent application of the same equipment thru the most favorable soil conditions as may be assessed from the boring data, with consideration of risk assessments and construction contract staging and packaging. The opportunity to use the same tunneling equipment on multiple tunnel drives is a factor under consideration.

Tunnel design will also consider minimum turning radius appropriate for the respective machine size and lining design being contemplated as well as other construction driven considerations such as: available construction staging areas; mining shaft area layout, access routes for material handling, protection of structures, avoidance of subsurface obstructions; need for ground improvement; and potential need for interventions and/or recovery shafts in event of machine failure. As previously noted, the proposed alignments also seek to maximize use of public ways and property and avoid acquisition of private property or easements as much as practicably feasible.

Available staging area is also of consideration. The alignment corridors are coincident with, or in close proximity to, several major development and infrastructure projects. Many of these projects will be under construction during the same time frames as construction of the Anacostia River Projects. Accordingly, close coordination with other public agencies and private developers regarding construction staging and mitigation of adverse construction impacts on the communities is necessary. WASA and the facility planning team have established working relationships with many of these parties, which will continue thru the remainder of the project.

Coordination requirements and allocation of staging areas along the river are most critical to the project. The most congested site will be the Poplar Point site where several major projects are being planned simultaneously. Besides the WASA CSO facilities, these projects include the replacement of the South Capitol Street Bridge (also referred to as the Frederick Douglas Bridge) and the adjoining Route 295 interchange by the District Department of Transportation; and the development of the Poplar Point site for recreational, memorial, residential and commercial retail use by the Anacostia Waterfront Corporation (AWC). AWC is an agency created by the D.C. Government to sponsor development of properties along the river for public and private use. They are presently developing plans for the site for public use and private development. These plans are intended to provide beneficial use, contribute to improvements to the neighboring communities and enable better access and use of the Anacostia River. The Poplar Point property and other parcels along the waterfront are the subject of a land swap between the District of Columbia (DC) and the federal government.

Much of the property along the river is under the jurisdiction of the National Park Service (NPS), which issues permits for construction activities, including borings, within their lands. Permits for construction are also anticipated from the Army Corps of Engineers, various DC agencies, WMATA, Railroad entities, and others. The requirements for these permits, as they are acquired, will be incorporated into the Project design standards and requirements and the respective construction contract documents.

As the southern terminus of the ART, the Poplar Point site is a critical area for the Anacostia River CSO facilities. All mining and tunnel construction activities for the ART and the Main and 'O' Branch Tunnel will be performed at this site, as well as the construction of the Poplar Point Pumping Station, the downstream overflow structure, and the Fort Stanton Interceptor facilities, which may include near surface pipelines or tunnel, diversion structures and drop shaft(s). In discussions with interfacing agencies, a staging area requirement of approximately five acres has been quoted as the minimum requirement for the CSO facilities construction. This will be reviewed and considered as all projects are advanced to more definitive degrees. Selection of type of tunneling equipment will also be a factor in defining minimum staging area requirements; for example, requirements for a slurry plant if a Slurry Face Pressure Tunneling Machine is recommended.

DESIGN AND CONSTRUCTION OF SHAFTS, SHAFT-TUNNEL CONNECTIONS AND DIVERSION STRUCTURES; AND ASSOCIATED RELATIONSHIPS WITH TUNNEL SYSTEM HYDRAULICS

The design of drop shafts, their connection to the tunnels and associated hydraulic structures for diversion of flow and for deaeration are of significant importance in configuration of the facilities. Many previous CSO tunnel projects have been constructed as deep rock tunnels where deaeration may be achieved through enlarged chambers at the entrance point for flows into the tunnel, as prescribed by the conventional Jain & Kennedy models. Whereas the Anacostia River Projects CSO tunnels and shafts will be constructed in soft ground, the design and construction considerations for incorporation of a deaeration chamber between the drop shaft and the tunnel, and for shaft/tunnel connection become more difficult and complex. This has an impact on determination of horizontal alignment of the tunnel as well due to the offset of the shaft from the line of the tunnel, further complicating the designs of shafts and tunnel alignments while attempting to maintain project facilities within public ways.

Constructability of the deaeration component may be addressed by either development of a configuration where deaeration may be achieved within the shaft structure or by application of the conventional Jain & Kennedy models and constructing the deaeration chamber under conditions where the ground has been improved by grouting or by ground freezing. The later option can not be fully evaluated until Phase II geotechnical studies have been performed, and the former option requires that additional hydraulic computational and physical model analyses be completed to assess the operational performance of such a configuration. With regard to the shafts themselves, use of slurry walls and/or ground freezing excavation support systems will be evaluated as the conceptual design is advanced and subsurface and site conditions at the respective shaft locations are better known.

As of the writing of this paper, hydraulic studies have been limited to those performed relative to the existing and future sewer infrastructure system to define the peak inflow parameters for the CSO tunnel system, as well as basic modeling of the tunnel system hydraulic performance utilizing the computed peak inflow rates. Surge analyses have not been performed to-date. More extensive hydraulic modeling of flows and surge analyses will be performed during Phase II of the facility planning effort, including consideration of a physical model for better assessment of hydraulics within the drop shaft structure and its interface with the tunnel. Determination of operational design parameters for energy dissipation, mitigation of surge impacts, deaeration and release of air during filling of the tunnel system will be primary objectives of the Phase II hydraulic modeling studies.

With regard to diversion structures and conveyance of flows from the existing collection systems and the drop shafts, a tangential approach per the Jain & Kennedy model is anticipated for most if not all drop structures. This configuration will be evaluated further for each respective drop shaft location subsequent to additional hydraulic analyses, advancement of tunnel design and acquisition of site characterization information. Another unique feature of the Anacostia River Projects CSO control system will be the application of passive flow control systems for diversion of flows into the storage tunnels, overflows, and control of flows within the system. Flow metering and event notification systems will be utilized; however WASA's objective is to not use active controls of any kind, such as gates, to control flows while the system is filling during a storm event. Isolation gates of some form, however, will be incorporated to enable isolation of flows for safe inspection and maintenance of the tunnels, shafts and associated facilities.

Alternative types of drop structures, including vortex, helicoidal, plunge and baffle drops were assessed during the Phase I evaluations with regard to relative hydraulic performance, flow capacity, cost, durability and maintenance, and constructability. At present, a vortex type drop shaft is envisioned, subject to additional hydraulic studies during Phase II of the facility planning effort.

Table 1. Consent decree milestones

Project	Start Design	Start Construction	Place in Operation
Facility Plan	September 23, 2005	Not Applicable	September 23, 2008*
Anacostia River Tunnel	March 23, 2009	March 23, 2012	March 23, 2018
ART Outfalls Consolidation	March 23, 2013	March 23, 2016	March 23, 2018
NE Boundary Tunnel	March 23, 2015	March 23, 2018	March 23, 2025
NE Boundary Side Tunnels	March 23, 2019	March 23, 2022	March 23, 2025
Separate CSO 006	March 23, 2006	March 23, 2008	March 23, 2010
Fort Stanton Interceptor	March 23, 2013	March 23, 2016	March 23, 2018

* Requires WASA to submit a report and detailed implementation schedule to the United States Environmental Protection Agency (EPA).

FUTURE PROJECT ACTIVITIES AND PLANNING

As previously stated, the Anacostia River Projects CSO Control Program is being performed in accordance with a federal Consent Decree agreement between the DC Water and Sewer Authority, the District of Columbia, and the United States via the U.S. Environmental Protection Agency and other plaintiffs. The Facility Plan now under development will more fully define the project in the upcoming months through more extensive site characterization, geotechnical investigations, engineering analyses and coordination with others. Several key Consent Decree schedule milestones are outlined in Table 1 for reference. WASA and the Facility Plan team look forward to a successful completion of this very important and environmentally beneficial project.

THE LAKE MEAD INTAKE 3 WATER TUNNEL AND PUMPING STATION, LAS VEGAS, NEVADA, USA

Michael Feroz

Parsons Water and Infrastructure

Marcus Jensen

Southern Nevada Water Authority

James E. Lindell

MWH/CH2M HILL JV

ABSTRACT

The Southern Nevada Water Authority (SNWA) operates two water-intakes in Lake Mead, located 20 miles east of Las Vegas, Nevada. Severe drought has caused decline in lake level by 85 feet since 2000 and future declines are expected. SNWA will construct a third deep-water intake to ensure existing water system capacity. This paper describes expected construction challenges for the new project that consists of: an intake structure and tunnel beneath the lake, pumping station, pipeline to connect existing treatment facility, and interconnecting underground adits. The project will be completed in the year 2012 for total estimated cost of approximately $600 million.

INTRODUCTION

The Southern Nevada Water Authority (SNWA) presently operates two water intakes and pumping stations at Saddle Island on the west shore of Lake Mead, approximately five miles northwest of Hoover Dam and approximately 20 miles east of Las Vegas in the Lake Mead National Recreation Area (Figure 1).

Severe drought has caused declining water levels in Lake Mead during recent years. The Lake elevation was approximately 1,127 feet (above mean sea level [amsl]) in August 2006 and the lake is expected to decline even more over the next several years. In order to protect the existing water system capacity against the potential inoperability of Intake Pumping Station No. 1, should water levels in the lake fall below 1050 amsl, SNWA is planning the construction of Lake Mead Intake No. 3 Project. This Project will consist of a deep-water intake, a pumping station and a large-diameter (~20 feet) water supply tunnel between the intake and pumping station, and will serve the existing Alfred Merritt Smith Water Treatment Facility (AMSWTF). The tunnel and intake will be constructed beneath Saddle Island and Lake Mead, terminating approximately 18,000 feet northeast of Saddle Island (Figure 2).

The intake will be north of the Las Vegas Wash, in the Boulder Basin of Lake Mead, and will be designed for a capacity of 1,200 million gallons per day (mgd) at an intake elevation of 860 feet amsl. The intake will be a lake tap structure with a riser structure above the lake bottom and a shaft below the lakebed that connects the riser to the intake tunnel. The riser structure will provide for a future extension to deeper waters northeast of the planned location.

Pumping station alternatives are being considered on the northern half of Saddle Island. Existing Intake Pumping Station No. 1 (IPS-1) and Intake Pumping Station No. 2 (IPS-2) are on the southern half of Saddle Island. The proposed IPS-3 will be

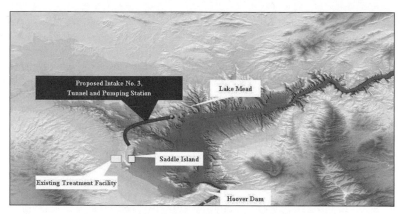

Figure 1. Proposed project site in Lake Mead relative to Hoover Dam

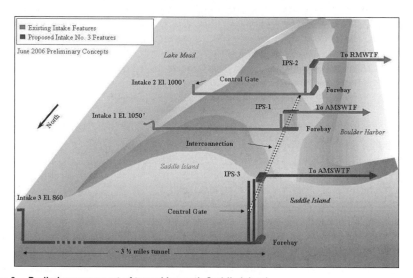

Figure 2. Preliminary concept of tunnel beneath Saddle Island

designed for a pumping capacity of approximately 600 mgd. Water will be pumped from IPS-3 to AMSWTF via a cut-and-cover pipeline across Saddle Cove. A connecting tunnel will also be provided between IPS-3 and IPS-2 to allow IPS-2 to draw water from Intake No. 3 in the event that Lake Mead falls below elevation 1000 feet (amsl). Provisions for a future 600-mgd pumping station on northern Saddle Island will be included in the design of IPS-3.

   Pumping station alternatives are being considered on the northern half of Saddle Island. Existing Intake Pumping Station No. 1 (IPS-1) and Intake Pumping Station No. 2 (IPS-2) are on the southern half of Saddle Island. The proposed IPS-3 will be designed for a pumping capacity of approximately 600 mgd. Water will be pumped from IPS-3 to AMSWTF via a cut-and-cover pipeline across Saddle Cove. A connecting tunnel will also be provided between IPS-3 and IPS-2 to allow IPS-2 to draw water from Intake No. 3 in the event that Lake Mead falls below elevation 1000 feet (amsl).

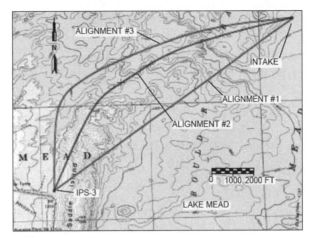

Figure 3. Potential tunnel alignments

Provisions for a future 600-mgd pumping station on northern Saddle Island will be included in the design of IPS-3.

This paper describes the preliminary investigations performed for the design and construction of the Intake tunnel. Basis for selection of the preferred tunnel corridor and a summary of performance requirements for the tunnel boring machine are discussed. The Intake Tunnel and the Intake Structure will be built using the Design-Build contracting method. The Pump Station and appurtenant structures will be built using the Design-Bid-Build contracting method.

INTAKE TUNNEL DESCRIPTION

The intake tunnel will connect the pumping station and the new intake riser. Elements that determine the location and alignment of the intake tunnel are:

- Location of the intake riser
- Location of the proposed pumping station (IPS-3)
- Hydraulic requirements
- Topography
- Ground conditions
- Tunneling technology

The tunnel will be constructed from a shaft near IPS-3 with an uphill gradient to the connection with the intake riser. The topography between the intake riser and pumping station is highly variable, as confirmed by pre-lake impoundment photographs and the results of the geophysical, bathymetric, and geotechnical field investigations. The most significant topographical feature between the intake riser and pumping station is the submerged continuation of the Las Vegas Wash. The intake tunnel needs to cross the submerged wash with adequate ground cover, creating a vertical constraint for the intake tunnel alignment. Three horizontal tunnel alignments were considered, as shown in Figure 3.

The geotechnical investigations and laboratory testing necessary to refine route alignment are ongoing; however, the intent is to select an alignment that has the least risk associated with tunnel construction. The potential tunnel alignments and tunneling issues are described later in this paper.

Alignment No. 1, the East Tunnel Corridor, is the shortest distance between the intake and the pumping station sites. The East Tunnel Corridor includes a tunnel section under the center and east side of Saddle Island, a long reach under generally deep water northeast of Saddle Island, and a crossing of the valley of the Las Vegas Wash near its original confluence with the Colorado River and under the ridge between the wash and the alluvial fan.

Alignment No. 2, the Center Tunnel Corridor, has a relatively long reach under Saddle Island and a long reach under the now submerged ridge north-northeast of Saddle Island. The corridor crosses the Las Vegas Wash at the downstream end of the canyon. The corridor then converges with the East Tunnel Corridor under the ridge to the alluvial fan.

Alignment No. 3, the West Tunnel Corridor, involves a short west-side departure from Saddle Island and a generally large radius curve west of the submerged ridge. It crosses the Las Vegas Wash in the upstream portion of the canyon. It converges with both the East and Center Tunnel Corridors under the ridge to the alluvial fan.

This Project will be the deepest sub-aqueous tunnel constructed with a pressurized-face TBM in the world to date. The current lake level is 1,128 feet amsl; however, the lake has the potential to rise to 1,229 feet. With a tunnel elevation of approximately 650 feet amsl, this Project will have a current hydrostatic pressure of approximately 14.3 Bar, but with a potential pressure of approximately 17.4 Bar. Tunneling in pressures of this magnitude in closed (pressurized-face) mode presents significant risks. Mitigation of these risks involves developing alignments and profiles that minimize this pressure, and tunneling in geologic units that minimize the amount of pressure necessary for stability and water control. These issues are discussed later in this paper. Based on current evaluations, the West Tunnel Corridor appears to offer the best potential to minimize the potential number of cutterhead interventions in hyperbaric conditions. The West Tunnel Corridor also appears to offer the least adverse geological conditions and shortest construction period. Consequently, the West Tunnel Corridor should offer the lowest relative risk and lowest construction cost. This should not be interpreted that there will be no risk or challenges during construction.

PRELIMINARY ASSESSMENT OF SITE CONDITIONS

The geotechnical investigation of intake tunnel alignments to date has included a desktop study of literature, field geologic mapping, bathymetric surveys, reflection seismic testing, and borings. The literature provides geologic information about origin and structure of the metamorphic bedrock of Saddle Island and areas northeast of Saddle Island that are now under Lake Mead. Field geologic mapping provided information about the three major geologic elements of Saddle Island, namely the lower plate, the upper plate, and the intervening detachment fault. The reflection seismic test program involved seismic lines on Saddle Island and seismic lines in Lake Mead within the corridor defined by the three tunnel alignments. The seismic results show two and three geologic units of different properties to a reported depth of 600 feet.

The geotechnical program to date has involved 16 fully cored land borings, between 200 and 700 feet deep and 33 borings in Lake Mead with depths below mudline ranging from 80 to 512 feet. Most of these borings were logged by temperature, natural gamma, optical, and acoustic devices and were water pressure tested. Core samples have been subjected to point load, index testing and laboratory testing to determine physical and mechanical properties of the materials.

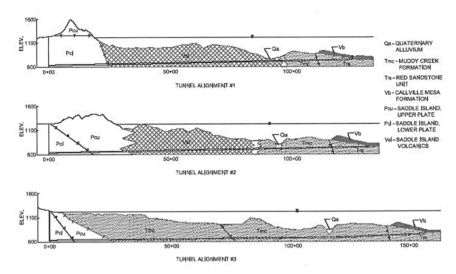

Figure 4. Geological profiles along the three tunnel alignments

ANTICIPATED GROUND CONDITIONS

The main geologic formations encountered along the tunnel alignments are shown in Figure 4 and summarized in Table 1.

A preliminary assessment of the geology and expected ground behavior in the East, Center, and West Tunnel Corridors is shown. The preliminary opinion of lateral extent and expected behavior of the lower and upper plate metamorphics, weak and strong sedimentary rocks, and volcanic rocks along the tunnel corridors is based on limited subsurface information. These preliminary assessments provide a framework on which to evaluate the relative ground conditions along the three tunnel corridors. The main strata are terrestrial sedimentary rocks and volcanic rocks, which are likely to vary both laterally and vertically. Both units lack the "key beds" that assist in the stratigraphic definition of marine rock units. The site does not provide exposures of these units to allow correlations between the borings.

TUNNELING ISSUES

Issues Identification

Critical issues for the intake tunnel include design and operational criteria, ground conditions, constructability, and contractor means and methods. Methods of construction are typically determined by contractors for tunnel projects, with significant corresponding influence on cost, schedule, risk, and related construction impacts.

Tunnel Configuration

A preliminary assessment of the hydraulic performance within the intake tunnel has been performed. To minimize abrasion, head losses, and surge capacity, maintaining a velocity below 6 feet per second (fps) at 1,200-mgd flow was considered preferable in a concrete-lined tunnel. A review of the construction cost and present worth operational costs indicates that at 600 mgd, an 18-foot-inside-diameter (I.D.) tunnel

Table 1. Evaluation of tunnel corridors (anticipated conditions)

Geologic Unit	Lower Plate Metamorphic	Upper Plate Metamorphic	Weak Sedimentary	Strong Sedimentary	Volcanic
Typical Rock Condition	Strong, hard metamorphic rock with widely spaced very hard igneous intrusions	Strong to weak metamorphic host rock with closely spaced very hard igneous intrusions	Massive, dense, generally low-strength mudstone, siltstone, sandstone, conglomerate/breccia	Massive, dense, medium-strength sandstone, conglomerate/breccia	Very weak andesite, diorite, and basalt mainly hydrothermally altered
Typical Groundwater Condition	Dry except for high-pressure inflows at major discontinuities	Mostly wet because of high-pressure inflows at major discontinuities	Mainly dry, wet where the unit is porous	Mainly dry with wet zones where joints connect to the lake	Mainly dry except in non-altered zones where the rock is jointed
Abrasiveness	High	Very high	Low	Medium	Medium
Face Stability	Excellent; Face condition—firm	Good to fair where rock is weak and jointed; Face conditions—firm and raveling	Generally good, fair where ground is weak, poor where ground is wet;Face conditions—firm, raveling and flowing	Excellent; Face condition—firm	Fair to poor depending on degree of alternation and water; Face conditions—raveling and flowing
Risk of Hydrostatic Pressure over 10 Bar	High risk near and under Lake Mead	High risk near and under Lake Mead	Low	Low	Medium
Water Pressure Tests, cm/sec, (high/low)	>3.E-04/2.E-06	1.E-04/3.E-06	4.E-05/7.E-07	9.E-05/7.E-07	2/E-03/2.E-06
Point Load Test, psi (high/low)	45,000/<700	49,500/700	2,400/0	13,500/0	37,800/<700
East Tunnel Corridor—Alignment No. 1					
Approx. total length (ft)	3,000		3,000		7,000
Center Tunnel Corridor—Alignment No. 2					
Approx. total length (ft)	2,600	2,600	3,100	1,500	3,800
West Tunnel Corridor—Alignment No. 3					
Approx. total length (ft)	1,000		9,500	5,500	

has a cost benefit of approximately 1 percent and at 1,200 mgd, a 21-foot-I.D. tunnel has a cost benefit of 1 percent. Therefore, for a single intake tunnel, an I.D. of approximately 20 feet is the preferred size with a resultant head loss of approximately 5.1 feet.

Tunneling Methods

Excavation methods for tunnels in rock include:

- Drill and blast
- Mechanical excavations (e.g., roadheader, hydraulic breaker)—staged excavation (e.g., bench and heading)
- TBM—full-face excavation

For shorter drives, in appropriate ground conditions, drill and blast or mechanical excavation is more effective as the equipment can be mobilized much more quickly. Drill-and-blast methods were used to construct the Lake Mead Intake No. 2 Project; however, significant grouting for control of water was required. Drill-and-blast methods are appropriate for excavating the construction shaft, TBM launch chamber, and tail tunnel required for the Intake No. 3 Project. The time required for constructing a tunnel by drill and blast or mechanical excavation can be reduced only if the work is executed concurrently from multiple working faces, which, in a linear tunnel, requires multiple points of access into the tunnel. None of the intake tunnel alignments offers cost-effective intermediate access points. Drill and blast or mechanical excavation for the intake tunnel requires significant grouting of poor ground and faulted zones to prevent high-pressure water ingress flooding the tunnel.

With a shielded TBM, a bolted and gasketed precast concrete lining may be installed within the tail of the shield and thereby cut off groundwater (and provide the lining) as the tunnel is advanced. TBMs also allow more groundwater control alternatives at the heading including pressurized-face tunneling, advance probing and grouting, and open-mode tunneling with controlled seepage. These water control features make shielded tunneling with a one-pass bolted and gasketed precast concrete segmental lining a better choice than the other methods.

For long tunnels, TBMs are generally considered more cost-effective than drill and blast or mechanical excavation because of their higher advance rates. Long-term average TBM advance rates are expected to be more than double that of non-TBM tunneling methods and provide a significant schedule and cost advantage. A TBM typically takes approximately 18 months to manufacture and mobilize, but the higher advance rate results in a shorter overall construction schedule and cost. A TBM is considered the only feasible method to construct the intake tunnel within the requirements of the project. The tunnel length with a single access shaft on Saddle Island supports the use of a TBM for construction of the intake tunnel.

TBM CONSIDERATIONS

TBM Operating Modes and Configurations

TBM performance is directly related to the selection of equipment to match ground and groundwater conditions within which the TBM is to operate. Since modern TBMs were introduced in the 1950s, continuous developments in design and experience have improved TBM performance. TBMs are manufactured in a variety of configurations to operate in a wide range of ground and groundwater conditions. No single TBM configuration has yet been developed that will perform at peak efficiency in all potential ground conditions. Therefore, whichever TBM configuration is selected, progress rates will depend on the ground and groundwater conditions encountered.

Open Mode

In open mode, the ground at the tunnel face (soil or rock) has no pressurized-face support. This mode requires stable ground and manageable groundwater seepage rates. Unstable ground is improved with permeation grouting (or dewatering or ground freezing) in advance of mining to improve stability and reduce groundwater inflows to manageable or permitted levels. Ground improvement measures ahead of the TBM are used to advance an open-face TBM through intervals of unstable ground, but reduce the rate of progress significantly. Spoil removal within the TBM is by belt conveyor or screw conveyor.

An open-face TBM provides limited or no support to the ground at the face during excavation. Open-face TBMs have limited ability to safely advance at maximum speed through unstable ground conditions or sustained high water inflows. Open-face TBMs can be configured with a single or double shield for added protection and improved production rates in poor-quality ground, but they are generally employed to excavate tunnels in competent and relatively impermeable ground.

Closed Mode

Closed mode, or pressurized-face mode, is defined as direct full contact between the ground at the tunnel face and the excavated material (spoil), spoil paste, or slurry in front of the TBM cutterhead. This provides positive support to the tunnel face and balances the hydrostatic groundwater and earth pressure. Spoil is extracted by a screw conveyor or slurry lines and pumps.

The two most common closed-mode TBMs are the EPBM and the STM. Less commonly, these two types can be combined in one TBM. Additionally, operating in partially closed/partially open mode with a closed-mode TBM is possible by excavating with a partially full pressure chamber or at pressures less than external hydrostatic pressure.

TBMs have been developed with the capability to operate in both open and closed modes. Dual-mode (also referred to as convertible) TBMs are appropriate for tunnels in predominately good ground that can be excavated using open-mode operation with relatively short reaches of ground that require closed-mode operation. In some cases, dual-mode TBMs can advance only in open mode, and closed mode is used to stabilize the face in poor ground while the TBM is stopped for ground improvement activities. The appropriate mode of operation is selected for the soil or rock conditions. This allows production rates to be optimized for a range of ground/groundwater conditions. Dual-mode TBMs include Open Mode/EPBM, Open Mode/STM, and EPBM/STM. Typical advance rates in the appropriate ground conditions will be higher in open mode than in closed mode. Spoil transport from the heading may change from belt conveyor or screw conveyor in open mode to pressurized screw conveyor or slurry lines in closed mode. The Hallandsas Tunnel in Sweden is being constructed using an Open Mode/STM designed to operate in a maximum pressure of 13.5 Bar.

Based on a preliminary assessment of EPBM, STM, and open-mode technologies and precedent practice, a STM appears to be more readily adapted to tunneling in variable rock conditions of the site at higher pressures. A dual-mode TBM allows improved rate of production in favorable ground conditions.

TBM Health, Safety, and Environment

A critical issue for tunnel construction is air quality in underground spaces. Underground ventilation systems that consist of portal fans, ducts, and booster fans in the tunnel provide sufficient air flow to dilute and remove carbon monoxide and other toxic gases and maintain an appropriate level of oxygen at all times. Ventilation systems are designed in consideration of the maximum number of workers and diesel-powered equipment expected in the tunnel. Air quality can also be adversely affected by high concentrations of naturally occurring toxic gases in the ground and groundwater. The two most likely toxic gases to be found underground are methane (CH_4), which is lighter than air, and hydrogen sulfide (H_2S), which is heavier than air.

Tunnel air quality and gas concentrations will be continuously monitored throughout the tunnel and on the TBM during tunneling. TBM-mounted gas monitoring devices will give early warning of potential hazards and may be connected to safety interlocks for automatically shutting down the TBM. Supplemental ventilation can be used to dilute potentially hazardous concentrations of gases. Gas infiltration to the tunnel should be minor and readily handled by the ventilation system; however, higher concentrations of H_2S in the Upper Diamond Fork tunnel in Utah caused the contractor to abandon the unshielded rock TBM and seal and abandon the upper reach of the tunnel.

A conservative approach requires the contractor to provide a TBM and equipment rated for a "gassy" environment (Class I, Division II—explosion proof) to prevent delays and costs in the event of a later re-classification.

Cutterhead Design

The maintenance of face excavation tools is essential for optimizing progress and protecting the TBM cutterhead. Estimates of tool wear must be used with caution as the highly variable ground conditions make it difficult to reliably estimate tool changes for the total tunnel length.

The number and means of interventions to replace and maintain cutting tools will directly affect the risk profile of the Project, including bid and completed construction costs. During construction, instrumentation should be used to remotely monitor tool/face wear, and progress rates should be monitored to establish the appropriate time to change tools. A mechanical method to support the ground at openings in the cutter face during interventions is preferred.

The cutterhead should be plated with abrasion-resistant material and have discrete openings rather than an open spoke design. The bucket openings should be adjustable by flood doors or face plates without intervention to allow partial or full closure to support the face and restrain ground movements during manned intervention in unstable ground conditions. Good abrasion resistance is required on all faces in contact with excavated material.

The largest potential cost of increased tool wear and replacement is associated with manned interventions for cutter changes under pressurized-face conditions.

TBM manufacturers have produced TBMs with cutterheads that can facilitate tool changes at high water pressure by the use of physical isolation systems instead of compressed air. These systems could reduce cost, schedule, and health risk to maintenance personnel. The ability to change cutter tools in poor ground with high water flow and pressure without hyperbaric manned entries should be investigated further.

EXCAVATION FEASIBILITY ASSESSMENT

Open Mode

Open-mode tunneling with pre-excavation grouting has been used with limited success on projects world wide. Pre-excavation grouting is useful for advancing a productive open-mode TBM through short reaches of difficult ground.

Results of the preliminary geotechnical site investigation indicate that most of the Muddy Creek Formation and the Tertiary volcanics have few to no fissures and have little intergranular porosity. Open-mode TBM operation may be practical if these materials are naturally tight and stable. If a stable face exists, but groundwater inflow is a concern, formation drainage/de-pressurization, pumping, and discharge are often less expensive than pre-excavation grouting (advance probing and grouting) and its feasibility should be assessed.

Closed Mode

A significant number of tunnels have been successfully excavated in closed mode at sustained pressures up to 4 Bar with brief periods at pressures up to 7 Bar. Closed-mode tunneling at pressures above 11 Bar has not yet been attempted.

INTERVENTIONS FOR CLOSED-MODE TBMS

A major consideration in a TBM operation is the repair and maintenance of the cutterhead. Accessing the TBM pressure chamber becomes increasingly more difficult when ground conditions deteriorate and external pressures increase. Some machines are designed to convert from open mode to pressurized closed mode and resist high pressures while stopped for pre-excavation grouting, but they cannot tunnel in closed mode.

New technologies permit cutter replacement without exposure to a pressurized environment, but these are typically only available for larger TBMs. These technologies need further investigation.

Interventions for repair and maintenance may be planned or unplanned. Planned interventions will take place at predetermined times or locations. Unplanned interventions will take place when favorable ground conditions are encountered or if reduced progress is due to mechanical deterioration (or breakdown) of cutters or the cutterhead.

When an intervention is required while tunneling in closed mode, the intervention will be conducted under one of the following excavation chamber conditions:

- Free air with cutterhead openings open (requires stable exposed rock, tunneling to pre-existing stable zones, or creating stable zones by ground improvement to reduce or eliminate the need for compressed air)

- Free air with cutterhead openings closed by face plates or flood doors (requires generally stable rock, which provides ground support but does not prevent groundwater inflow or piping of fines)

- Compressed air up to about 4 Bar partially balancing exterior hydrostatic pressure (requires competent ground with risk of only small to moderate inflows and instability under free air)

- Full compressed air balancing exterior hydrostatic pressure with workers using mixed-gas compression and decompression from free air (allows only short periods of work under pressure) or using mixed-gas compression and partial decompression from a pressurized shuttle or habitat called saturation intervention (allows longer periods of work under pressure)

GROUND IMPROVEMENT

The primary objectives of ground improvement are to stabilize the ground and limit groundwater inflow. Ground improvements can be performed routinely as part of normal tunneling activities or can be used to establish safe zones for interventions to perform routine maintenance or unplanned repairs.

PRE-EXCAVATION GROUTING

Grouting Systems for Ground Improvement

Grouting may be used to improve the condition of the ground (permeability, stability, strength) to permit progress and planned interventions and facilitate unplanned interventions. For the Intake No. 3 Project, ground improvement grouting may be completed from Saddle Island, the lake surface, the TBM launch shaft, or the TBM.

A grouting specialist examined core samples available in April 2006 from the metamorphic rocks of Saddle Island and the water borings in the weak and strong sedimentary rocks, the altered volcanics, and the basalt. Based on a single examination of these cores, he concluded that the metamorphic rocks, the jointed strong sedimentary rock, and the basalt can be grouted; that grouting can significantly control water movement through these rocks; and that grouting can improve the stability of these rocks. He also concluded that grout will not likely penetrate the fine-grained weak sedimentary rock or altered volcanic rocks and improve these materials for water control or stability.

For planned interventions, pre-excavation grouting is significantly more effective than post-excavation grouting; therefore, except in an emergency or breakdown, the grouting should be done before the arrival of the TBM. Control of remaining inflow is critical in weak or erodible ground because at 13 to17 Bar pressure, the flow will rapidly break down soft and poorly cemented ground.

Grouting from the Lake Surface

Grouting from the surface of the lake could allow the pre-excavation grouting to be completed independent of the tunneling operation and, therefore, in advance of the TBM. However, even with accurate drilling, the potential ineffectiveness of the grouting will be uncertain. The cost of pre-excavation grouting from the lake surface will be significant.

Grouting from the TBM

Grouting from the TBM is typically limited to face grouting through openings in the cutterhead and perimeter grouting through the TBM shield. However, restricted access limits the number of locations that a grout hole can be positioned. Successful grouting from the TBM requires that the machine be designed to facilitate drill positioning and connection to grouting equipment. Grouting through the tunnel face is time-consuming and will significantly increase the duration of tunnel construction.

Pre-excavation grouting is likely to be effective in the basalt at the riser shaft site, in the metamorphic rock east and north of Saddle Island, and in the strong sedimentary rock. Pre-excavation grouting is unlikely to be effective in weak sedimentary rock and volcanic rock. The tunnel alignment should reduce reliance on pre-excavation grouting. The risk and consequences of grout loss into the lake should be evaluated.

DRAINAGE AND PUMPING VERSUS PRE-EXCAVATION GROUTING

There may be advantages to using drainage/dewatering to divert water from the cutterhead or reduce the hydrostatic pressure for manned interventions. Drainage to improve tunneling progress is only applicable for open-mode operation. TBM production during open-mode operation will be adversely affected by inadequate or incomplete drainage. The cost of drainage and pumping to improve tunneling conditions can be significant. Provisions for drainage and pumping groundwater during open-mode operation of the TBM should be considered in TBM design.

Ground Freezing

Artificial ground freezing can be used to temporarily create a strong and impermeable block of ground ahead or around the TBM in otherwise saturated and unstable ground. Ground freezing can be used to permit safe access to the outside face of the cutterhead. Ground is frozen by a chilled brine circulating through a pipe system drilled ahead of and around the TBM cutterhead. Installation of the pipe system and development of the freeze takes considerable time. In areas of high water flow velocity or high ground temperatures, achieving a freeze can be difficult or impossible. It is also the most expensive and time-consuming method of ground treatment.

INITIAL GROUND SUPPORT AND LINING

Conditions anticipated for the Intake No. 3 tunnel include intervals of unstable ground and high water pressures and flows. Tunneling through these unstable conditions requires installation of an initial ground support system that can isolate high-pressure groundwater during construction and a future tunnel dewatering-inspection-repair event.

Traditional rock tunnel initial support systems of either rock bolts and shotcrete or structural steel rings and lagging will not be effective at controlling groundwater inflows and providing tunnel opening stability for the anticipated conditions. The only proven initial support system suitable for the intake tunnel is a pre-cast segmental concrete lining system installed immediately behind the TBM as the tunnel is constructed. Two pre-cast systems were considered:

- A two-pass system consisting of temporary pre-cast expanded segmental rings as initial support, followed by a permanent cast-in-place concrete lining
- A one-pass bolted and gasketed pre-cast segmental ring system

Two-pass lining systems can provide added structural capacity and improved hydraulic characteristics. However, a two-pass lining requires as much as 12 months to place the cast-in-place concrete inner liner.

Pre-cast bolted and gasketed segmental liners have adequate structural capacity and hydraulic characteristics for the intake tunnel. This system has been used for many projects with similar difficult ground and high head conditions. Based on an assessment of cost, schedule, durability, water-tightness, and safety, the one-pass bolted and gasketed pre-cast segmental lining system is recommended for the intake tunnel.

SPOIL DISPOSAL

Excavated spoil from the tunnel driven in closed mode will include materials other than natural rock. Dumping the spoil in an area that may become saturated or submerged by Lake Mead will present difficulties in containing or treating release of spoil-conditioning agents. Several conditioning agents are reported to be biodegradable and non-toxic; however, the specific chemical characteristics of potential conditioning agents are not known. It is not possible to determine with confidence that the conditioning agents will have negligible effect on Lake Mead or the shoreline. Therefore, material that contains spoil-conditioning agents should not be disposed of where it may have potential to contaminate the Lake Mead National Recreation Area.

ALIGNMENT EVALUATION

Development of Preferred Tunneling Conditions (Table 2)

The basic tunnel alignment for the Project is controlled by three key locations.

- TBM construction access shaft
- Intake riser
- Crossing at the Las Vegas Wash

The construction and intake riser locations are determined to a greater extent by project requirements and geology than by tunneling requirements and do not constrain the tunnel alignment between them. To develop a preferred alignment between these three points, a number of factors need to be considered to minimize risk. It is necessary to consider the following:

- Horizontal alignment and length of tunnel
- Vertical alignment required to tunnel under the Las Vegas Wash
- TBM design and operation considerations
- Health, Safety, and Environmental (HSE) concerns associated with TBM operations
- Cost and Schedule

TUNNEL HORIZONTAL ALIGNMENT

The geotechnical site investigation has developed preliminary information on the three alignment alternatives shown in Figure 3. Horizontal tunnel alignments can be compared using the preliminary assessment of ground conditions.

During detailed design, the tunnel alignment will be refined to:

- Maximize the lengths of straight tangents
- Minimize the number of curves
- Use a minimum curve radius of 1,000 feet
- Avoid the detachment fault or cross as perpendicular to minimize the length of tunnel in the fault

The Las Vegas Wash crossing should be sited in the cemented conglomerate.

TUNNEL VERTICAL ALIGNMENT

The depth of the intake tunnel is a primary factor in assessing constructability and operational performance. Preliminary alignment concepts included both deep and shallow profiles. Evaluating these options requires an assessment of cost, schedule, impacts, and risk. The risk of encountering poor-quality ground conditions is a critical differentiator for the alternatives and requires careful consideration during detailed design based on the best available geotechnical information.

Table 2.   Preliminary assessment of preferred tunneling conditions

Design Considerations	HSE Impacts	Cost Impacts	Schedule Impacts
Avoid high permeability rock	Reduced potential for compressed-air interventions	Lower pumping cost during construction. Cost of face interventions decreases.	Less potential for slow production
Avoid rock with potential hydraulic connectivity to lake	Reduced pressure if hyperbaric intervention is required	Cost of face interventions decreases	Less potential for slow production
Avoid rock that may be unstable	Reduced potential for compressed-air interventions. Reduced potential for collapse during open-mode.	Open-face working decreases cost. Cost of interventions decreases. Reduced cost of connection to intake riser.	Open-face working is faster. Interventions may be faster. Less potential for slow production.
Minimize depth	Reduced pressure if hyperbaric intervention is required. Increased separation from potential underlying igneous intrusions.	Lower shaft construction cost	Less potential for slow production. Shaft construction periods reduced. Shorter decompression period.
Minimize length	Shorter exposure time to risks	Cost is generally related to length in comparable ground conditions	Schedule is generally related to length in comparable ground conditions
Minimize number of curves	N/A	Increased production and fewer special segments reduce cost	Potential faster production
Use a minimum curve radius of 1,000 feet	N/A	Increased production and fewer special segments reduce cost	Potential faster production

A preliminary evaluation of the geotechnical investigation indicates that ground conditions will not improve with increasing tunnel depth. In many geologic settings, a greater depth of cover improves the ground conditions because of a reduction in the degree of weathering. At this site, weathering is not the key to rock quality except in the first few hundred feet of depth on Saddle Island. The quality of material at depth along the alternative tunnel corridors is mainly controlled by the degree of alteration in the case of the volcanics, and the degree of cementation in the case of the weak sedimentary rocks. An increase in the tunnel depth of tens to several hundred feet is unlikely to site the tunnel in a different geologic condition. A vertical profile that provides a minimum safe separation of the tunnel from the submerged valley of the Las Vegas Wash is preferred. The alignment should be as high as reasonably practicable to minimize the hydrostatic pressure at tunnel level and reduce the cost and schedule of shaft and intake riser construction.

PRELIMINARY SUMMARY ASSESSMENTS

Drill and blast offers the greatest flexibility for the construction of short tunnels in the Pre-Cambrian lower plate rock that are anticipated at the base of the construction shaft. Based on the alignment and anticipated geologic conditions, a transition from drill and blast to mechanized excavation will be adopted for the balance of the intake tunnel. The most logical scheme would be to use a single TBM launched from a pre-excavated tunnel in good quality rock under Saddle Island.

If a significant amount of Tertiary age deposits as previously described are encountered, the strength of these materials and their permeability will determine whether the TBM is operated in a Closed or Open Mode. If continued geotechnical investigations prove out the presence of a significant amount of Muddy Creek Formation is to be encountered, conditions will favor a dual mode TBM capable of converting from Open to Closed (Slurry) mode. The slurry tunneling approach is suitable for all of the anticipated ground types but a dual mode convertible TBM is better suited for achieving optimum advance rates in Open Mode where Closed face tunneling is not required. Pre-excavation probing would be required while in Open mode and while investigating safe intervention zones during Closed Mode tunneling. The TBM will be abandoned underground and limited equipment will be salvaged. Detailed specific recommendations will be developed during future project development as more geotechnical investigation is completed.

Preliminary assessment of TBM requirements include:

- The dual mode TBM should be equipped with a crusher to handle large clast sizes (i.e., cobbles and boulders)
- Plan to pre-qualify for dual mode or Slurry TBM experience and encourage North American tunneling contractors to team with other (overseas) tunneling contractors to obtain necessary slurry TBM experience
- The alignment and profile should be optimized to minimize the anticipated time spent in closed mode to improve schedule and reduce costs.
- Undetected faults and shears between borings will always remain as a potential risk and will have to be addressed with contingency planning.
- Initiate discussions with safety authorities on the use of hyperbaric interventions
- Contact three or more TBM manufacturers regarding their ability to design and manufacture closed face TBMs suitable for up to 17 Bar hydrostatic pressure

The West Tunnel Corridor appears to offer the best potential to minimize the potential number of cutterhead interventions in hyperbaric conditions. The West Tunnel Corridor also appears to offer the least adverse geological conditions and shortest construction period. Consequently the West Tunnel Corridor should offer the lowest relative risk and lowest construction cost. This should not be interpreted that there will be no risk or challenges during construction.

Once the alignment endpoints are established based on the results of ongoing geotechnical site investigations, consideration of construction methods, limitations and capabilities in relation to anticipated ground conditions should be primary criteria for selection and refinement of the preferred alignment and profile. Due to the depth of the tunnel and expected ground conditions, hyperbaric manned interventions may be required. These interventions must be feasible on the adopted alignment and profile from both regulatory and technical viewpoints. Impacts to cost, schedule, the environment and lake recreation should be considered during detailed tunnel alignment development.

ACKNOWLEDGMENTS

The authors are thankful to the Southern Nevada Water Authority, their preliminary design engineering team (MW/HILL Joint Venture), and Parsons Water Infrastructure team's Program Managers for their continued support during the Project. A special thanks to Chris Ottsen and David Whittaker for their technical input to the geotechnical and tunneling evaluation programs for the Project.

REFERENCES

Holzhauser, J., Hunt, S.W., & Mayer, C., 2006, Global Experience with Soft Ground and Weak Rock Tunneling under Very High Groundwater Heads, North American Tunneling Conference, Chicago, Ill.

Herrenknecht, M., 2005, Speech at 2005 Rapid Excavation and Tunneling Conference, Seattle, WA.

Herrenknecht, 2006, personal communication.

Le Pechon J.C., Sterk W, Van Rees Velinga T.P Saturation Diving for Tunneling Operations, AITES-ITA, 2001.

MOVING FORWARD ON NEW YORK'S SECOND AVENUE SUBWAY

Anil Parikh

MTA Capital Construction

Geoffrey Fosbrook

DMJM+Harris•Arup JV

Donald Phillips

DMJM+Harris•Arup JV

David Caiden

DMJM+Harris•Arup JV

Jon Hurt

DMJM+Harris•Arup JV

Jaidev Sankar

DMJM+Harris•Arup JV

ABSTRACT

This paper will cover the progress being made on Phase 1 of New York's Second Avenue Subway line. It will include a description of construction progress on the first contract to be let, which consists of a 248 m (815 ft) long launch box, 3.9 km (2.4 miles) of TBM tunnels and a tunnel sump; the packaging strategy for the remaining work; and the progress through Final Design of the remaining packages, including the two large span rock cavern stations.

INTRODUCTION

Plans for the Second Avenue Subway have been around since the 1920s, with some construction actually starting in the 1970s. Since 2002, MTA New York City Transit (NYCT) has been working with the DMJM+Harris•Arup Joint Venture (DHA) to progress the design of the Second Avenue Subway. In 2007, the first contract within Phase 1 of the Second Avenue Subway is anticipated to be awarded, with bids having been opened just as this paper was submitted.

The history of Second Avenue Subway and descriptions of the entire 13.6 km (8.5 mile) long route from Hanover Square in Lower Manhattan to 125th Street are well covered in other papers (Caiden et al., 2006; Parikh et al., 2005). This paper concentrates on Phase 1 of the Second Avenue Subway, which will connect into the existing New York City Transit (NYCT) subway system at 63rd Street and run north under Second Avenue. Three new stations will be added at 72nd, 86th and 96th Streets. The layout of Phase 1 is shown in Figure 1. There will be a total of four phases to construct the entire Second Avenue Subway. The project is being funded by a combination of State and Federal funding.

663

Figure 1. Second Avenue Subway Phase 1

By extending the 'Q' train via the existing station and stub tunnel at 63rd Street, Phase 1 of the Second Avenue Subway provides early revenue service, with ridership expected to be over 200,000 weekday riders when operational.

The paper describes each of the major civil works contract packages currently envisaged for Phase 1, describing the work included in each contract, the current status, and any particular challenges associated with the work. In addition to the civil works packages, there will be contracts not described in this paper to install the track, signals and communications systems.

TBM CONTRACT

Description

This Contract (MTA Contract No. C26002) primarily consists of the construction of a TBM Launch box within the limits of 2nd Avenue between 92nd to 95th Streets in the Upper East Side of Manhattan, two circular shafts at 69th and 72nd Streets, and the construction of two TBM bored tunnels extending from 92nd Street to 63rd Street.

TBM Launch Box. Due to the congested nature of the Upper East Side and the desire not to acquire adjacent properties, it was difficult choosing a location for the TBM launch and support operations. After considering various options, it was decided to launch the TBM from a location that will eventually form the southern portion of 96th Street Station extending from approximately mid-block between 91st and 92nd Streets to 95th Street situated within the right-of-way of 2nd Avenue, as shown in Figure 2. This location has the best geological conditions for launching the rock TBM within the future location of 96th Street Station. The approximate length of the launch box is 248 m (815 ft), its width varies between 20 m (65 ft) and 24 m (80 ft) and it is approximately 18 m (60 ft) deep. The geology of the launch box varies between soil, rock/soil interface and rock. The top of sound rock is higher in the southern end of the launch box and drops off towards 93rd Street.

In the northern portion of the launch box where the walls are founded on soil, it was decided to use slurry walls, which will provide the desired cut-off of groundwater flow through the walls. The slurry walls will also form the permanent station walls and are reinforced with rebar cages. Couplers will be used to connect to the future station invert, mezzanine and roof slabs.

At the southern end of the launch box, part of the excavation will be through rock, and the Temporary Support of Excavation (S.O.E.) walls in this section are the design responsibility of the Contractor. The design must comply with noise, ground movement and vibration criteria given in the Contract documents to limit disturbance and damage to adjacent structures and utilities. The suggested design included in the Contract documents shows a secant pile wall system that is toed into rock. This system was chosen

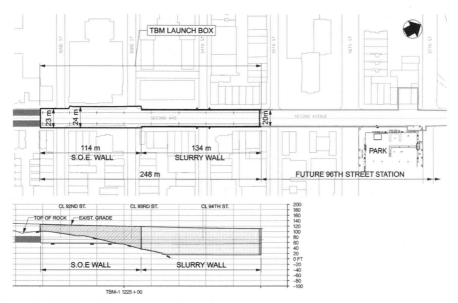

Figure 2.   TBM launch box plan

because it minimizes ground movements and is able to minimize groundwater infiltration at the toe. This type of wall also offers the flexibility to allow installation around existing utilities. The permanent station walls in this section will be reinforced cast-in-place concrete and will be installed as part of the 96th Street Station contract.

The TBM Launch box is decked over to allow work within the launch box as traffic flows above. The access through the deck to the launch box is achieved through openings in the deck located on the east side of 2nd Avenue allowing four lanes of southbound traffic on the west side. The design of the deck is the Contractor's responsibility.

A park between 96th and 97th Streets on the east side of 2nd Avenue has been acquired by the MTA for use during the construction of both the TBM Contract and the 96th Street Station Contract, which will be made available to the Contractors. In the TBM Contract, the Contractor will construct the TBM power substation within the limits of the park.

Circular Shafts. Two circular shafts, located at 69th and 72nd Streets, are included in this contract to provide construction access to the future 72nd Street Station site. The early construction of these shafts allows a shorter overall schedule for the construction of Phase 1. Both shafts are nine meter (30 ft) internal diameter and are concrete lined. The support of excavation walls are designed by the Contractor, with secant pile walls being shown in the Contract documents as part of a suggested design. The top of sound rock at both shafts is approximately three to four and one-half meters (10 to 15 ft) below existing grade.

The invert level of the two shafts is approximately the elevation of the top bench of the future 72nd Street Station cavern which will be mined by drill-and-blast. The depth of the shaft at 69th Street is 15.5 m (51 ft) and the depth at 72nd Street is 12 m (40 ft).

Utilities. Extensive utility relocations are required for the construction of the TBM Launch Box and the two access shafts at 69th and 72nd Streets. On-going coordination with the utility companies, including Con Ed, Verizon/ECS, and DEP, has resulted in proposed relocations to facilitate the work and is shown in the Contract Documents. The staging shown for the utility relocations attempts to relocate utilities in one move.

Therefore, many utilities were relocated to the space under the sidewalk between the support of excavation wall and the existing buildings. The remaining utilities are supported from the deck in the locations that they will eventually be restored to once the 96th Street Station construction is complete and the street restored.

TBM Tunnels and Sump. Two TBM bored tunnels, which will become the running tunnels for the new subway line, will run from 92nd to 63rd Street on the west side of 2nd Avenue and from 92nd to 73rd Street on the east side of 2nd Avenue. The total length of the TBM drives is approximately 3,900 m (12,800 ft).

The geologic setting of the tunnel drives consists mainly of the Precambrian Manhattan Schist formation consisting of strong and abrasive crystalline quartz, mica and feldspar with abundant garnet, amphibole rich zones and pervasive alteration. The typical strength of the rock is between 34.5 and 83 MPa (5,000 and 12,000 psi). In addition, the geotechnical investigation has also identified major faults near 64th, 68th and 90th to 92nd Streets.

The first TBM drive starts from the southern end of the TBM launch box under the west side of 2nd Avenue. The drive continues down 2nd Avenue to 69th Street where the TBM shifts towards the middle of 2nd Avenue, then continues south to 65th Street. At 65th Street, the TBM turns west following a 188 m (616 ft) radius curve until it holes through at the existing rock chamber at 63rd Street Station. The TBM is then disassembled and removed back through the TBM Launch Box. The second TBM drive starts at the southern end of the TBM launch box and proceeds south under the east side of 2nd Avenue to 73rd Street, where the drive ends and the TBM is disassembled and removed through the TBM launch box. This drive is essentially straight.

The finished diameter of the TBM bored tunnels is 6 m (19 ft-9 in) and was determined by car clearances, tunnel utilities and signals, and fire/life safety requirements. The contract was initially advertised with a precast concrete (62 MPa/9,000 psi) segmental lining 250 mm (10 in) thick. The excavated diameter was set to 6.75m (22 ft-1in). Subsequently the option of a cast-in-place concrete (34.5 MPa/5,000 psi) option was included based upon requests by the Contracting community.

As the TBMs proceed under 2nd Avenue, they will pass through areas where future caverns will be built in subsequent contracts. In these sections of the TBM tunnels the final concrete lining will not be installed. Instead, the Contractor will install initial support only with support types and assumed locations given in the Contract documents (although actual field conditions may dictate variations in the locations). The tunnels left with only initial support comprise approximately 35% of the total length of the TBM tunnels.

To comply with the requirements of the EIS, trucking from the TBM Launch Box is prohibited at night. However, in order to meet the schedule, the TBM operation must go on 24 hours a day. This will require the storage of night shift muck underneath the deck in the TBM launch box to be trucked during the day shifts.

In addition to the TBM tunneling, a sump will be mined under 78th Street and will provide temporary drainage during the interim construction phase and final drainage when Phase 1 is operational. Discharge will be via a connection to the existing sewer system. This eight meter (25 ft) wide, nine meter (28 ft) high and 18 m (60 ft) long sump will be mined between the TBM tunnels by drill-and-blast with initial support of shotcrete and bolts. The final lining of this sump is a reinforced cast-in-place concrete liner with membrane waterproofing.

Contract/Schedule. The duration of the TBM Contract is 40 months with Milestone 1 at 33 months from Contract Award. Milestone 1 includes turning over the first TBM tunnel south of 73rd Street, the shafts at 69th and 72nd Streets and the restoration of traffic on the deck between 93rd and 95th Streets. This will allow the award of the 96th Street Station and 72nd Street Station Contracts.

The contract has been let as a low bid lump sum contract with unit pricing for fissure grouting events ahead of the TBM and includes a Geotechnical Baseline Report to aid in the event of claims resulting from Differing Site Conditions. Because this Contract is on the critical path of the Phase 1 schedule, an early completion incentive is included for Milestone 1.

Status

In October 2006, MTA Capital construction advertised the construction contract for the TBM Contract of the Second Avenue Subway. At the opening of the bids on January 18, 2007, the apparent low bidder was S-3 Tunnel Constructors (Skanska/Schiavone/Shea).

Challenges

Second Avenue is a major southbound route consisting of 6 lanes of traffic and twenty foot sidewalks and is lined with buildings of varying type, use and size. Access must be provided to these buildings at all times during construction for businesses, residents and emergency personnel. While the work for Phase 1 progresses, four lanes of traffic and a minimum of 1.5–2 m (5–7 ft) sidewalks must be maintained. In addition, a path must be clear at mid-block within construction work zones to provide access by emergency personnel to buildings.

The space underneath Second Avenue contains a myriad of utilities. The relocation of these subsurface pipes and conduits is necessary to construct the TBM Launch Box. The staging of these relocations to accommodate the construction of the support of excavation walls will require up front planning and coordination with the utility companies.

In addition to the challenge utilities pose, many buildings along 2nd Avenue have basements used for storing retail supplies, restaurant supplies and storage for residents. Some of these basement areas, however, extend beneath the sidewalks along 2nd Avenue. These existing encroachments have been condemned to allow work on the Second Avenue Subway to begin because the space underneath the sidewalk will be used to relocate utilities. These vaults are supposed to be filled in by the building owners prior to construction. However, if the vaults are still in place after construction starts, the Contractor will have to remove them. These vaults are of particular concern when constructing the TBM Launch Box for the TBM Contract and also the 96th Street Station.

Because of the urban environment of the Phase 1 area, limitations are set forth in the Contract Documents for work hours, noise and vibration during construction to limit damage to adjacent buildings and to minimize the impact on the area's residents.

96TH STREET CUT-AND-COVER STATION

Description

The construction contract for 96th Street Station is planned to include the installation of slurry walls and excavation of the remainder of the cut-and-cover station box a further length of 290 m (954 ft) from the northern end of the TBM launch box to the existing section of tunnel at 99th Street. In addition, this Contract includes the construction of the permanent station structure, ancillary buildings and entrances and the fit-out of the station. Refurbishment and modifications to the existing tunnel structure from 99th to 105th Streets is included to allow for their future use for train storage. Also part of this Contract is placement of the concrete invert and duct benches of the TBM

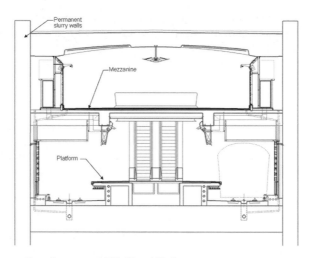

Figure 3.   Cross-section of proposed 96th Street Station

tunnels between 96th Street Station and 86th Street Station, and the mechanical, electrical and plumbing fit-out of these tunnels.

A cross-section of the proposed station is shown in Figure 3. The structure will consist of a reinforced concrete cut-and-cover box, formed with permanent slurry walls. The low rock elevation previously described at the northern end of the launch box structure remains over the entire length of the rest of the station structure, with all excavation taking place in soft ground consisting of fill, over a three meter (10 ft) thick deposit of organic materials, over sand or silty sand, over varved silt and clay. Entrances and ancillary buildings will require underpinning and other structural modifications to existing buildings.

The existing tunnel structure shown in Figure 4 was built in the 1970s as part of an earlier attempt to build the Second Avenue Subway. It runs between 99th and 105th Streets, and consists of a cut-and-cover box formed in the traditional New York subway manner with steel bents at one and a half meter (five foot) centers encased in concrete. The tunnel is in good structural condition and work will be limited to construction of duct benches and installation of a central blockwork wall to separate the ventilation of the tunnels.

Status

The 96th Street Station contract package is the most advanced of the station packages in the Final Design process. It is intended to be awarded while the TBM contract is ongoing, once the construction of the section of the launch box between 93rd and 95th Streets is completed. With this section of the launch box decked over, this will allow a separation of two blocks between the two contractors, which is sufficient to allow the 96th Street Station contractor full flexibility in traffic management.

Challenges

96th Street Station faces the same challenges as the TBM contract in dealing with the dense urban environment. Management and protection of traffic, business and residential access and utility diversion and relocation will be a major focus during the initial phases of slurry wall construction and excavation. In addition, working under and

Figure 4.   Existing tunnel structure between 99th and 105th Streets

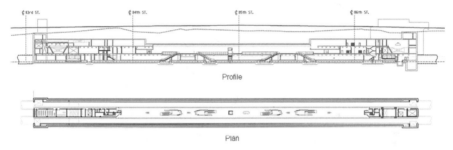

Profile

Plan

Figure 5.   Plan and long section of proposed 86th Street Station

around existing buildings to form the entrances and ancillary buildings will be complex. Minimizing the impact to the residential and retail areas surrounding the contract will require a significant management effort to minimize noise, dust, obstruction and other inconveniences.

86TH STREET AND 72ND STREET MINED STATIONS

Description

Mined stations were selected at 86th and 72nd Streets to take advantage of the high elevation of the top of rock, and minimize the surface disruption to utilities, traffic and the surrounding neighborhood.

86th Street Station is a two track station with a 22 m (71 ft) wide cavern. The cavern height varies between 15.5 m (51 ft) in the platform areas and 19 m (62 ft) in the ancillary areas at the ends of the station. At each end of the station there is a short length of cut-and-cover box. There are a number of penetrations into the cavern for entrances and connections to ancillary areas. The entrances emerge at street level in lower levels of existing buildings. A plan and long section of the station is shown in Figure 5.

72nd Street Station is a three track station. While three tracks are not required in Phase 1, in later phases trains leaving 72nd Street will either head south down Second Avenue or west to 63rd Street, and the three platforms will provide maximum operational

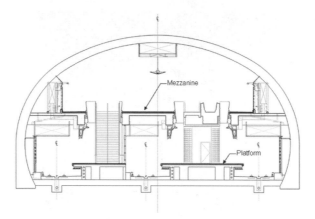

Figure 6.   Cross-section of proposed 72nd Street Station

flexibility. 72nd Street Station cavern is 29 m (95 ft) wide and 15 m (49 ft) high. There are no cut-and-cover sections in this station, with all entrances and ancillary connections contained in mined tunnels. A cross-section of the station is shown in Figure 6.

The general subsurface profile for the 72nd St. and 86th St. station caverns consists of miscellaneous granular fill over decomposed rock. Top of sound rock is typically less than 10 m (33 ft) below ground surface and consists of quartz mica garnet schist intercalated with thin zones of intrusive pegmatite or non-foliated granofels.

Status

Both stations are in the final design process. It is intended to award the 72nd Street Station Contract once the first TBM drive through the station is completed (as described above, the second TBM drive stops to the north of the station). The Broadway Connection, described in the following section, will also be part of this contract. The 86th Street Contract will be awarded such that blasting work for rock excavation will not start until the second TBM drive and work on the tunnel sump in the TBM contract are completed. This is to avoid blasting work being carried out above the TBM tunnels while they are still being used for other construction.

Challenges

In addition to sharing many of the challenges already described for the previous contracts, the mined stations pose technical challenges. The requirements to maintain a shallow alignment to maximize the attractiveness to passengers, provide a station platform with a good level of service and to minimize property acquisition for ancillary space have resulted in wide caverns with shallow rock cover. Providing an efficient and safe design for the construction of the caverns is a challenge that is described in detail in other papers (Desai et al. 2007, Desai et al. 2005). During final design, Itasca Consulting Group has been subcontracted to provide three dimensional discrete element modeling using 3DEC in addition to the two dimensional UDEC modeling being performed by the DHA design team and the empirical systems that the cavern design is based upon.

Access for mining will also be relatively limited. Typically only two openings, approximately nine by nine meters (30 by 30 ft) in size, will be available for each station cavern

during much of the mining process, at least until the excavation of the entrances has proceeded far enough to provide additional access.

BROADWAY CONNECTION

Description

The Broadway Connection, also known as the G3/G4 tunnels after track designa-
tions, is a two-track spur off the main Second Avenue alignment that connects into the
existing station at 63rd Street. At the point of departure from the north-south Second
Avenue alignment the two tracks run side by side, but as they curve under 63rd Street
one track drops below the other, to tie in with the existing station at 63rd Street, which
has stacked platforms and is oriented east-west.

As part of the TBM Contract, the first TBM drive will include the Broadway Connection and run along the alignment of the upper tunnel. The remaining excavation is
anticipated to be carried out by drill-and-blast and will be included in the 72nd Street
Station contract.

Status

The design of the G3/G4 tunnels is being progressed in parallel with the final
design of the 72nd Street Station.

Challenges

The track alignment in this area is very tight and is constrained by the existing
63rd Street Station, other existing NYCT tunnels, and provision for the future phases of
Second Avenue extending to the south. The configuration of the tunnels has been
rationalized as much as possible to maintain consistent internal tunnel cross-sections.
However, some excavation sequences will be complex with rock pillars as narrow as
1.2m (4 ft) needing to be maintained.

63RD STREET STATION REFURBISHMENT

Description

The new Second Avenue line will connect into the existing NYCT system at Lexington Ave/63rd Street Station. The existing station was opened in 1989 and consists
of two levels with two tracks on each level. Currently only one track is used on each
level, serving the 6th Avenue 'F' line that runs east-west to Queens via Roosevelt
Island. The other tracks, currently inaccessible to the traveling public, are used as storage tracks at the northern end of the Broadway 'Q' line. Figure 7 shows a cross-section
of the existing station.

The proposed work at 63rd Street Station involves opening up the sections of the
station that are currently closed to provide access to the tracks that will connect to the
Second Avenue line via the Broadway Connection. Additional ventilation, other systems and new entrances will be added.

Status

The final design of 63rd Street Station is ongoing, although construction is not currently planned to start until later in the Phase 1 construction period.

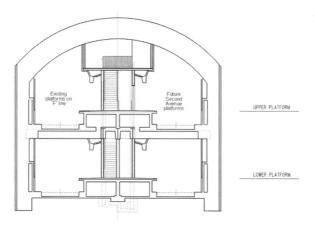

Figure 7.   Cross-section of existing Lexington Avenue/63rd Street Station

Challenges

The 63rd Street Station shares some of the difficulties with property impacts and lack of available working space at street level that the other contracts have. It has the advantage of not being on the critical path of Phase 1. It is, however, an integral part of a structure containing an operating subway system, and the need to maintain a safe environment for the trains and traveling public will add constraints to the construction process.

CONCLUSIONS

Construction of the Second Avenue Subway is due to commence in 2007, over 80 years after the line was first planned. There are significant construction difficulties, many resulting from the mass of subsurface utilities, the need to maintain traffic flows and the densely populated neighborhoods. Construction of a large cut-and-cover box and two major caverns will be significant challenges that will need to be overcome in the next few years.

Views and/or opinions expressed in this paper do not necessarily reflect the views and opinions of the MTA and/or its Agencies.

REFERENCES

Caiden, D., Sankar, J., Parikh, A. and Redmond, R. 2006. Rock Tunnels for the Second Avenue Subway. North American Tunneling Conference (NAT), Chicago, IL.

Desai, D., Lagger, H., Rossler, K. and Stone C. 2007, New York Subway Stations and Crossover Caverns—Update on Initial Support Design. RETC, Toronto.

Desai, D., Naik, M., Rossler K. and Stone, C. 2005. New York Subway Caverns and Crossovers—A tale of trials and tribulations. RETC, Seattle, WA.

Parikh, A., Fosbrook, G. and Phillips, D. 2005. Second Avenue Subway—Tunneling Beneath Manhattan. RETC, Seattle, WA.

PLANNING AND DESIGN OF THE
A3 HINDHEAD TUNNEL, SURREY, UNITED KINGDOM

Tom Ireland

Mott MacDonald

Tony Rock

Mott MacDonald

ABSTRACT

The proposed 1.83km long twin bore A3 Hindhead road tunnel planned for construction in 2008 is being delivered under a Highways Agency Early Contractor Involvement (ECI) contract.

The proposed construction methodology is progressive mechanical excavation of a horseshoe shaped tunnel with shotcrete installed close behind the face using robotic spraying equipment. The tunneling medium is a 'soft rock' sandstone for which there was little previous tunneling experience, requiring extensive geotechnical investigations and an innovative approach to lining design. A 240m long portal section though sandy material utilising a steel pipe umbrella is also required.

This paper describes the development of the tunnel planning and design including the most recent developments in UK road tunnels, and the innovation included in the 'permanent' primary lining design.

INTRODUCTION

The A3 Hindhead project located in Surrey, UK is a 6.7km dual carriageway trunk road that includes a 1.83km tunnel being delivered under a Highways Agency Early Contractor Involvement (ECI) contract. This paper describes the development of the tunnel design through the planning and design phases including vertical and horizontal alignment, tunnel cross section, ground support measures, and the design of the tunnel structure to survive fire loading.

The tunnel is to be constructed using the Sequential Excavation Method (SEM) and the various support types and additional support measures are described and the design approach and methodology including details of the numerical methods of analysis are discussed.

PROJECT DESCRIPTION

The A3 Hindhead project is one of the schemes in the UK Government's Targeted Programme of Trunk Road Improvements. The project will complete the dual carriageway link between London and Portsmouth and remove a major source of congestion, particularly around the A3/A287 traffic signal controlled crossroads. Refer to Figure 1 for location details.

The project will deliver quicker, more reliable journeys on a safer road, and remove much of the present peak time "rat-running" traffic from unsuitable country roads around Hindhead. The centre of Hindhead will be freed from the daily gridlock that blights the area, with the result that the project will bring benefits to road users, local residents, and the highly prized local environment.

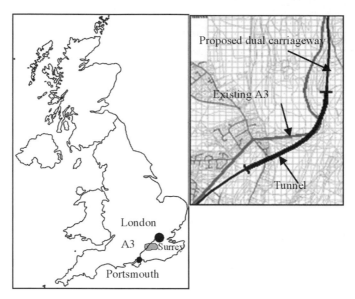

Figure 1. Project location

Planning Phase

The A3 Hindhead project has been planned since 1983 when it was included in the Government's Trunk Road program. The challenge for this project was balancing environmental impact with user economic benefits. Striking the right balance between these competing issues has taken over 20 years to resolve. Almost all the scheme lies within an 'Area of Outstanding Natural Beauty' (ANOB) with Site of Special Scientific Interest (SSSI) that are part of the Wealden Heaths SPA—an EU Designation that prohibits development except for "no choice" and national economic or safety considerations. The route passes through National Trust owned land which is classified as "in-alienable," that is, if National Trust so chooses, it can only be compulsorily purchased with the approval of the UK Parliament.

In the 19 years since the project was first mooted, and the commencement of the current phase of project development in October 2002 there has been a sea change in planning and approval thinking whereby economic benefits originally had primary importance and now environment damage mitigation is uppermost. This change in planning approach with time is shown on Figure 2. The key to a successful planning process in this case was to provide a tunnel under the environmentally sensitive areas, and to close the old A3 thereby reuniting sections of the SSSI/SPA. Other environmental benefits included the removal of intrusion and severance within the AONB, and the removal of air, noise and light pollution.

A significant change in approach resulted from the adoption by the Highways Agency of a partnership approach for scheme development. This was manifested on the A3 Hindhead project by the establishment of a Project Advisory Group (PAG) incorporating all the key stakeholders including the protectors of the environment such as National Trust, English Nature, Countryside Agency and English Heritage. The objective of the PAG was "To assist in developing the tunnel scheme to minimize its impact on the built and highly prized natural environment, and one that is broadly acceptable

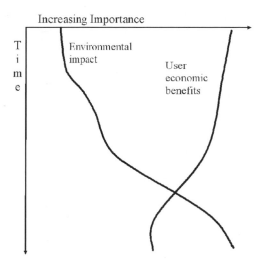

Figure 2.   Change in approach to development of A3 Hindhead Scheme

to the local community, while ensuring that all impacts have been addressed." Table 1 outlines the project development timetable since inception in 1983.

Early Contractor Involvement (ECI) Contract

The Highways Agency introduced the principle of "Early Contractor Involvement" in 2001. This new form of procurement is concerned with bringing suppliers and designers together much earlier in scheme conception than previously occurred, allowing them to work together more closely. This allows more scope for innovation, improved risk management, better forward planning of resource requirements and minimization of long term environmental impacts, improved consideration of buildability and health and safety, shorter construction periods and reduced environmental impacts during construction. Overall, the early creation of delivery teams clearly offers the opportunity for better value and improved performance.

GEOLOGY

Overview

The geology of the Hindhead area comprises a sequence of fine grained sedimentary deposits laid down during the Lower Cretaceous period in near shore transgressive marine conditions on the margins of the subsiding Weald Basin. The tunnel is within the Hythe Beds—a 90m thick sequence within the Lower Greensand Series formation.

The Hythe beds are variably sorted, highly glauconitic, variably bioturbated and cross-bedded sands and sandstones.

The Hythe bed unit is divided into six litho-stratigraphic subdivisions, through four of which the tunnel passes.

Table 1.   Project development timeline

Date	Activity
1983	Enters Trunk Route Program
1987	Single Route Consultation
1988	Red Route confirmed as preferred
1992	Second public consultation
1993	Modified yellow route with bored tunnel announced as preferred route
1995	Work suspended
1998	Roads review—Road Based Study into tolling announced
1999–2000	Tolling Study
2001	Enters Targeted Programme of Improvements (TPI)
10/2002	Early Contractor Involvement (ECI) contract awarded
9/2004–2/2005	Public Inquiry
8/2005	Inspectors Report received
9/2006	Secretary of State decision
1/2007	Commence construction
7/2011	Open Tunnel
3/2012	Complete Scheme

Tunneling Conditions

The tunnel at the southern end passes through units Upper Hythe A and B which are similar units with an increasing number of sandstone bands with depth, described as "medium dense thinly bedded and thinly laminated, clean to silty and clayey fine and medium SAND with subordinate weak to strong sandstone, cherty sandstone and chert."

The majority of the tunnel passes through the more competent Upper Hythe C and D, and Lower Hythe A units, described as "Weak, locally very weak to moderately strong, slightly clayey fine to medium SANDSTONE with occasional thin beds of clayey/silty fine sand."

The remaining unit is Lower Hythe B which has been avoided by the tunnel as clays and sand become dominant in the lower half of the unit.

The sandstone within Upper Hythe C/D and Lower Hythe A has typical UCS values of between 2 and 5 MPa and is heavily fractured with six joint sets including the sub-horizontal bedding with mean fracture centers varying between 190 and 815mm.

The tunnel also intersects a small number of discrete thin bentonitic Fuller's Earth Beds, formed from air- and water-bourne volcanic ash, including crystal and lithic tuffs.

The tunnel is above the historically observed water table, with the maximum predicted water table exceeding the invert level in only one location by a depth of less than 1m. Refer to Figure 3 for geological longsection.

Ground Behavior Model

A challenge for this project was to define a ground behavior model in an unusual material that had not been tunneled previously. The difficulties in interpretation stem from the weak to very weak nature of the sandstone material in combination with the content of up to 20% interbedded soil layers. A recurring challenge in tunneling is to determine the rock mass strength and stiffness, with empirical methods such as GSI,

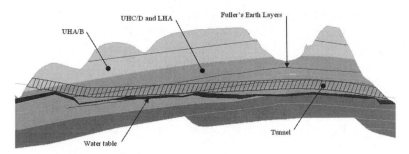

Figure 3.   Geological long section and vertical alignment (1H:4V)

RMR or the Q-method (Bieniawski, 1984) often used. The strength and stiffness relationships that are the cornerstone of these methods are generally determined from
data from significantly stronger rocks than the 2–5MPa sandstone and do not take
account of the influence from the soil layers leading to a significant overestimate of
stiffness.

An extensive geotechnical investigation was undertaken with sonic testing, pres-
suremeters and triaxial testing all used to determine the Elastic Modulus of the rock
mass. Pressuremeter testing was found to be the most reliable with the sonic testing
over-estimating the stiffness, and the triaxial testing surprisingly under-estimating the
stiffness in a number of cases. This was thought due to the difficultly in finding a
300mm long specimen for testing in a material with 6 joints sets and an average bedding spacing of 190mm. Fortunately some good quality rock joint shear box testing
was undertaken that provided a lower bound for strength and stiffness interpretations.

The interpreted model used in design was a small strain stiffness model ($E_{(\varepsilon a)}$)
that varied the stiffness of the rock mass with strain, and a Mohr-Coulomb strain softening model used for strength. Due to the large variation in cover from 16m to 58m the
rock mass stiffness was also related to depth (α) in order to take advantage of the positive effect of larger insitu stresses at depth.

ALIGNMENT

The horizontal alignment for the tunnel was determined based on road design
considerations and environmental constraints resulting in a reverse curve through the
tunnel with a minimum radius of 1,050m.

The vertical alignment was determined based on geological constraints with the
desire to minimize the length of tunnel through the sand at the southern end, to keep
the tunnel above the water table and also to maximize the vertical clearance to the
Lower Hythe B material which has insufficient strength to carry horizontal stresses
around the tunnel opening. The tunnel passes beneath the Devil's Punch Bowl which
is a re-entrant, primarily spring-sapped valley system with erosion feeding backwards
from the Hythe Bed/Atherfield Clay interface at the valley base. The crossing of the
punchbowl provides a cover constraint to the tunnel, and the cover changes rapidly
from around the minimum cover of 16m to the maximum cover of 58m within a horizontal distance of 130m.

The decision to follow the optimal tunnel material and to avoid the softer Lower
Hythe B material results unfortunately in a low point within the tunnel.

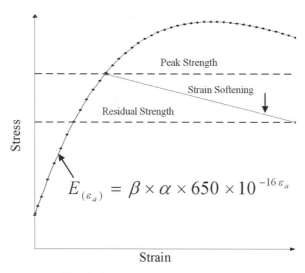

Figure 4. Ground model used for design

DESIGN

Cross Section

The Hindhead tunnel layout comprises twin 2-Lane bores with cross passages at 100m nominal centers. Refer to the typical cross section in Figure 5. Each bore has two 3.65m lanes, with full batter curbs and 1.2m wide verges on each side of the tunnel. The verge width is sufficient to allow for sight-lines due to the horizontal curvature of the tunnel, to accommodate electrical services and also to provide wheelchair access to the cross passages and emergency points at 100m nominal centers along the tunnel.

The vertical traffic gauge provided is 5.03m with an additional clearance of 250mm to the Equipment Gauge to allow for flapping tarpaulins and other transitory gauge infringements.

A continuous drainage system is utilized, located beneath the curb and verge alongside the cable duct bank. Other services such as the fire main, high voltage cables and pump mains are buried beneath the carriageway, with jets fans, lighting and communication cables contained within the crown.

These requirements result in a horseshoe shaped tunnel structure with an internal diameter of 10.6m and an excavated diameter of 11.6m.

Fire and Life Safety Provisions

The tunnel has cross passages at 100m nominal centers to allow an emergency escape to the non-incident bore. Cross passages include fire hydrants, dry pipe connections, fire extinguishers and emergency telephones. Emergency Points (EP's) are also provided at 100m nominal centers located at the mid-point between cross passages. Each EP has an emergency telephone and fire extinguisher.

A longitudinal ventilation system comprising 20 jet fans per bore is provided for smoke control. Consideration was given during the preliminary design phase to the inclusion of a Fire Suppression system, however it could not be justified on cost benefit

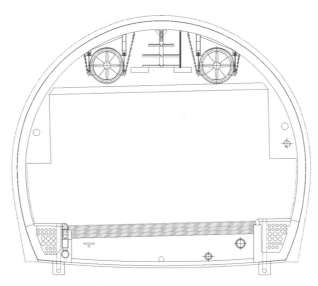

Figure 5. Typical tunnel cross section

grounds. The decision was taken to make space provision for the future installation of a fire suppression system, should further evidence of the benefits become available or standards and or technology change.

Design for Maintenance

Due to the comparative life cycle cost of managing tunnel assets and increasingly stringent health and safety regulations, the design team focused on minimization of whole life costs and the development of corresponding design details. Design for maintenance included:

- Minimization of unplanned maintenance interventions
- Consideration of the safety of maintenance
    - deletion of equipment where possible
    - selecting materials with the longest design life
    - Provision of safe maintenance access
    - Provision of replacement options for all infrastructure and equipment
- Maintenance risk assessments

Specific design initiatives incorporated in the works included:

- Provision of spare HV conduit and blind pits to allow replacement of life-expire HV cables in routine closures
- Modular hydrant connections allowing replacement in routine closures without disturbance of cable ducts and other services
- Relocation of in-tunnel sump to outside the tunnel by using a directional drilled gravity drain allowing maintenance access to sump without a tunnel closure

TUNNEL EXCAVATION AND SUPPORT

The presence of the sand layers, in one location up to 2m thick, led to the selection of the Sequential Excavation Method (SEM) whereby shotcrete is sprayed at the face following each excavation advance also known as Sprayed Concrete Lining (SCL) in the UK. Standard hard rock tunnel support techniques such as pattern bolting were not considered suitable due to the sand layers and the very low bond stress negatively impacting the effectiveness of rock reinforcement.

Construction Sequence and Support Types

Four basic support types have been designed for the standard tunnel cross sections with minor variations required at cross passage junctions and Emergency Point niches. There is one main support type for the sandstone section, with three support types covering the section through sand and the transition from sand to sandstone.

Excavation and support types are specified based on tunnel chainage and have been designed to cover all expected ground conditions. It is not proposed that support types be selected based on geological inspection. The Hythe beds have 6 joint sets and an average joint spacing of less than 200mm. The UCS values are typically 2–5MPA. This material is expected to act as a continuum, and given the heavily fractured nature of the material, and presence of sand layers, meaningful variations in rock quality are expected to be difficult to detect. A suite of 'additional measures' discussed below have been designed to account for any local stability issues. Geological inspection and mapping of the open excavation, and monitoring results will be used to determine the advance length that may vary between 1m and 2m, with 1m advances specified at critical locations such as beneath surface structures and roads.

Support Type 1

At the northern end of the tunnel (Chainage 3120 to 4650), excavation is in rock (UHC/D, LHA) and Support Type 1 is specified throughout. The tunnel is generally excavated with a full face heading followed at a distance by the bench excavation. Due to the generally stable nature of the ground and tunnel location above the water table, a closed invert is not required and the horse shoe shaped primary lining is supported on elephants feet. Refer to Figure 6 for details of the primary lining in the sandstone section.

Additional Support Measures

In addition to the sequences and support requirements for each of the support types, contingency 'additional support measures' have been specified. The requirement for the contingency measures will be triggered by geological inspection and mapping, and monitoring results, and will include:

- **Spiling.** Self drilling GRP tubular spiles will be installed when ground conditions result in excessive overbreak, or instability in the crown. Spiling is detailed as mandatory for approximately 30% of the Support Type 1 excavation in areas of known potential crown overbreak such as where the tunnel crosses Fullers Earth bands, and where a 2m thick layer of sand and shattered rock intersects the crown.

- **Additional Face Support Measures.** Geological inspections will inform the selection of additional face support measures such as sealing layers, face support wedge or face dowels.

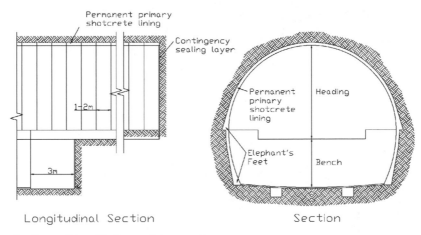

Figure 6. Support Type 1 in the sandstone section

- **Probe Drilling.** Probe holes drilled ahead of the face have been specified for the entire length of Support Type 1 to relieve any potential hydrostatic pressure from perch water ahead of the face. If water is detected in the probe holes then additional holes will be drilled to drain any water.

- **Grout Stabilization.** Microfine cement or chemical grouts will be used stabilise running sand bands, or other local areas of instability caused by sandy layers. It is proposed to conduct site trials prior to construction to determine the optimum method and materials for stabilization of sandy materials.

- **Invert Strut or Rib Bolts.** Convergence monitoring will be undertaken during construction with and a three stage trigger limit system implemented. Unplanned convergence resulting from worse than expected ground conditions will be addressed by installation of an invert strut at bench level, and in the unlikely scenario that convergence of the heading occurs then GRP self drilling rib bolts shall be installed.

Support Types 2–4

At the South end of the tunnel (Chainage 2880 to 3120 (m001)), excavation is in sand (UHA/B) and support types 2, 3 and 4 are specified. The excavation will be carried out on dayshift only due to constraints on working hours and is made stable with the use of a steel pipe umbrella and face dowels. The pipe canopy comprises 12m long 114mm diameter tubes at 400mm centers with an overlap of 4–5m. The advance length is a maximum of 1m for these support types.

Support Type 2 has sandy material (UHA/B) in the heading only, with the heading elephants feet supported on the sandstone material (UHC/D). This means self-drilling GRP face dowels are required in the heading only, and the heading can advance ahead of the bench. The face dowels are 12m long with a 4m overlap and are installed with the same drill jumbo used to install the pipe canopy.

Support Type 3 has a full face of sandy material with the elephants feet of the bench supported on the sandstone material. As the heading elephants feet are not supported on sound material, the heading must be advanced with the bench, with a 2m separation provided to maintain face stability. Face dowels are required for both the heading and the bench (see Figure 7).

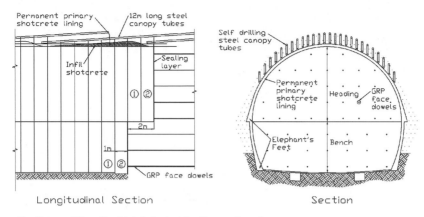

Longitudinal Section   Section

Figure 7.   Support Type 3, which is typical for the sand section

Support Type 4 has a full face of sandy material that extends below the tunnel, and therefore a closed invert is required. The heading must be advanced with the bench, with the invert closed a maximum of 6m behind the face.

Design Approach and Methodology

The excavation sequences outlined above are designed to control strains in the ground so that as much as possible of the ground load bearing capacity is used and the strains are maintained at levels that minimize yielding.

The main design approach was to utilize numerical methods of analysis (FLAC Version 5.0) to model in 2-D the different excavation stages and predict the performance of the linings in terms of the stresses and corresponding deformations. The model included the non-linear small strain stiffness material model and Mohr-Coulomb failure criterion with strain softening to residual strength parameters.

A modelling simplification inherent in 2-D modelling is the relaxation of the ground ahead of a excavation face, a largely 3-D effect and the time dependent development of lining strength and stiffness. In order to determine the appropriate value for these ground conditions we have utilised a 3-D numerical model (FLAC Version 3.0) which can accurately model the ground relaxation based on the proposed construction sequence and also the time dependent strength and stiffness development of the shotcrete.

For the 2-D model we used the 28 day stiffness value in accordance with Eurocode 2 divided by creep and shrinkage factor of 2. This factor reduces the stiffness of the spayed concrete lining to take account of the creep and relaxation that occurs when loading during early age (John et al. 2003). The 2-D model was then calibrated against the 3-D modelling to determine an equivalent relaxation of 60%.

In addition to providing calibration of the relaxation figure, the 3-D FLAC modelling was also used to model stability of the face, early age capacity of sprayed concrete lining near the face, and the effect of variations in advance length.

The results of the analysis were checked in accordance with Eurocode 2 to ensure that the design complied with the Ultimate Limit State.

A principal innovation with the support measures is the design of primary lining as permanent. This is possible due to a number of advances in tunneling technology in recent years. Firstly, non-alkaline accelerators are now available with no loss in shotcrete strength with time. A recent innovation is the use of 3-D scanning survey equipment that provides excellent shape control for both excavation and spraying, and allows shotcrete lined tunnels to be constructed without lattice girders. This technique has been recently used successfully for the Heathrow T5 project (Williams et al. 2004). Historically the inclusion of lattice girders meant the primary lining had to be considered temporary due to the corrosion potential of the steel lattice girder within the primary lining. Spiling is envisioned in several locations due to adverse soil layers. This will be carried out with self-drilling Glass Reinforced Plastic (GRP) dowels, again with no adverse durability issues. The sprayed concrete will be reinforced with steel fibers as is required for safe installation, however the design does not rely on the flexural capacity of the steel fibers, and the lining is designed as plain concrete. This is possible due to the curved shape of the section with all moments resisted by axial forces within the lining.

SECONDARY LINING DESIGN

A secondary lining is provided to support the proposed sheet waterproof membrane, and also to provide fire resistance to the tunnel. The secondary lining is constructed from plain concrete, with all tensile loads in the lining resisted by the tensile capacity of the concrete. The main issue is to minimize the heat of hydration and shrinkage, and this is achieved with a 35% Pulverized Fly Ash (PFA) cement replacement mix design with low shrinkage. Other important design issues include having a maximum fresh concrete temperature of 20°C, pour lengths of 12m and controlled curing conditions.

Fire resistance is achieved by adding 1–2kg/m^3 of polypropylene fibers to the concrete mix in order prevent explosive spalling. The precise dosage of fibers will be confirmed by fire testing.

CONCLUSIONS

The development of the planning and design of the A3 Hindhead project has been outlined in this paper. The main conclusions are:

- Projects through environmentally sensitive areas need to provide special mitigation measures in order to achieve planning consent.
- Tunnel details have been developed to avoid non-routine maintenance closures through the life of the tunnel.
- SEM design has been undertaken utilizing a combination of 2-D and 3-D numerical methods of analysis which included a non linear small strain stiffness ground model.
- The design of the tunnel support includes a permanent primary lining possible due to:
  - Availability of non-alkaline accelerators
  - Use of 3-D laser scanning survey technology for control of robotic spraying equipment to achieve tight shape tolerances avoiding the need to use lattice girders
  - Use of self-drilling GRP spiles in areas of poor ground.

ACKNOWLEDGMENTS

The authors wish to thank Paul Arnold of Highways Agency for his constructive comments, and Paul Hoyland and Roger Bridge of Balfour Beatty who have provided valuable input in developing the bored tunnel design.

REFERENCES

Bieniawski, Z.T. 1984. Rock Mechanics Design in Mining and Tunneling. Rotterdam: A.A. Balkema.

British Tunneling Society. 2006. Occupational Exposure to Nitrogen Monoxide in a Tunnel Environment. London. http://www.britishtunneling.org

John, M., Mattle, B. 2003 Shotcrete Lining Design: Factors of Influence. Rapid Excavation and Tunneling Conference, pp. 726–734

Williams, I., Neumann, C., Jäger, J. & Falkner, L. 2004. Innovativer Spritzbeton-Tunelbau für den neuen Flughafenterminal T5 in London. In Proc. Österreichisher Tunneltag 2004, Austrian Committee of the ITA, Salzburg, pp. 41–62. Salzburg: Die SIGN Factory.

PLANNING AND DESIGN OF THE BAY TUNNEL

R. John Caulfield

Jacobs Associates

Victor S. Romero

Jacobs Associates

Johanna I. Wong

San Francisco Public Utilities Commission

ABSTRACT

The existing Hetch Hetchy water system, owned and operated by the San Francisco Public Utilities Commission (SFPUC), serves 2.4 million people within the San Francisco Bay Area. The system has existing 1920s-era pipelines crossing under the Bay that are vulnerable to seismic damage from the nearby San Andreas and Hayward fault zones.

The SFPUC plans to replace the pipelines with the first TBM-driven tunnel underneath the San Francisco Bay. The tunnel will be constructed using an Earth Pressure Balance machine. The new tunnel will pass under environmentally sensitive habitats and will be approximately eight kilometers (five miles) long with no intermediate shafts. The project challenges will include tunneling through a prominent buried bedrock ridge that transects the alignment and through soft soils under the Bay. Seismic performance criteria include maintaining service flows after a large earthquake on one of the active faults in close proximity to the project site.

BACKGROUND

The San Francisco Public Utilities Commission (SFPUC) manages a large water supply system of reservoirs, tunnels, pipelines, and treatment systems that stretches about 320 km (200 mi.) from the Hetch Hetchy Reservoir in Yosemite National Park to the City of San Francisco. The water comes from a protected watershed in the Sierra Nevada Mountains, and is some of the purest in the United States. The system is also economical, as it is almost entirely gravity-fed.

The SFPUC is the third-largest municipal utility in the State of California and serves 2.4 million residential, commercial, and industrial customers in the greater San Francisco Bay Area. Approximately one-third of the SFPUC's delivered water goes to retail customers in the City of San Francisco, while wholesale deliveries to numerous other suburban agencies in Alameda, Santa Clara, and San Mateo counties comprise the other two-thirds of deliveries. In November 2002, the SFPUC launched a $4.3 billion Water System Improvement Program (WSIP) to repair, replace, and seismically upgrade the system's aging pipelines, tunnels, reservoirs, pump stations, storage tanks, and dams. The WSIP is funded by a bond measure approved by San Francisco voters. The program will achieve key service and operational goals related to seismic recovery, water quality, drought reliability, and sustainability. More than 75 projects in San Francisco and the surrounding region will be completed by the end of 2015.

The SFPUC found two major pipeline arteries in the system to be particularly vulnerable in a seismic event. The replacement or upgrade of these pipelines is a key

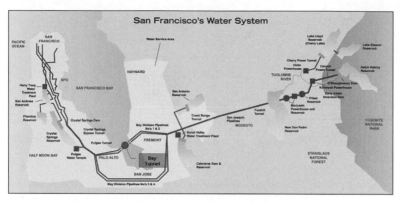

Figure 1. Schematic diagram of SFPUC's water distribution system

Figure 2. Bay Division Pipelines 1 and 2 at shoreline in East Palo Alto

element of the WSIP. The two important pipelines are Bay Division Pipelines 1 and 2. These aging pipelines travel from the City of Newark, under the southern portion of the San Francisco Bay, up into the Western Bay mudflats, and then across a pile-supported trestle into the City of East Palo Alto. (see Figure 2)

The SFPUC commissioned initial engineering studies to evaluate various replacement and upgrade options. Options included upgrading the existing pipelines, replacing them with cut and cover and submarine pipelines, and building a new tunnel. The studies showed that the preferred solution was to construct a new tunnel under the Bay.

The tunnel will be constructed while the existing pipelines remain in service. The existing pipelines, and a new pipeline to be constructed, will be tied in to the tunnel at either end during short outage windows. The shorter windows will limit system disruption and avoid customer supply issues.

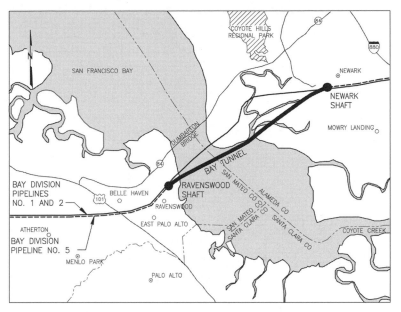

Figure 3. Proposed Bay Tunnel horizontal alignment

FACILITY LAYOUT

The Bay Tunnel will be located between two shafts. The presence of environmentally-sensitive habitats on the Bay margins resulted in the need for an 8 km (5 mi)-long tunnel with only launching and receiving shafts and no intermediate construction shafts. These two shafts will be located on properties owned by SFPUC in the City of Newark (Newark shaft) and the City of East Palo Alto (Ravenswood shaft), as shown on Figure 3. The shafts are expected to be constructed using diaphragm slurry wall construction methods with tremied structural concrete slabs tied into the shaft bottoms. The TBM launching and construction shaft located at the Ravenswood site will be approximately 17.7m (58 ft) in diameter and 33.5m (110 ft) deep. The receiving shaft located at the Newark site will be approximately 8.5m (28 ft) in diameter and 22.5m (74 ft) deep.

The preferred horizontal alignment was chosen to meet identified design and constructability criteria. Criteria included providing straight departures from the shaft sites and maintaining minimum 305m (1,000-ft) radius curves. Pile supported structures planned along the shoreline of the adjacent rail corridor made it desirable to cross this corridor onshore and at some distance from the shoreline on the Ravenswood side of the Bay. In the Newark area, a horizontal curve was placed in the alignment to avoid schedule delays that might be caused by passing under a parcel owned by a private party that was unwilling to grant an underground easement. The total length of the preferred alignment is 7,994m (26,226 ft).

The preferred vertical alignment was based upon a number of different design and constructability criteria, including maintaining minimum clearances under the bottom of the San Francisco Bay and keeping the tunnel as shallow as possible to reduce construction costs. It was also necessary to keep the tunnel alignment under the Young Bay Mud deposits to avoid negative impacts to seismic performance.

Table 1. Description of anticipated geologic units

Formation	Description
Young Bay Mud	Soft to medium silty clay
San Antonio Formation (also described as the Merritt-Posey-San Antonio Formation)	Interlayered medium stiff to hard silt and clay, with loose to dense sand lenses
Old Bay Clay	Stiff to hard silty clay with dense sand lenses
Franciscan Complex	Sedimentary and low-grade metamorphic rock, highly weathered

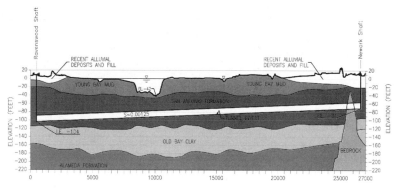

Figure 4.   Geologic profile along the proposed Bay Tunnel alignment

GEOLOGIC CONDITIONS

Site Geology

The initial phase of the field geotechnical investigation was completed in 2006. The field investigations included a series of marine-based borings across the San Francisco Bay, and land-based borings and cone penetrometers on the Newark and Ravenswood site locations. In addition to the conventional exploration, an extensive marine- and land-based geophysics program was performed along the majority of the proposed alignment corridor to help the design team make interpretations between boreholes. Suspension logging was also performed in selected boreholes at the Ravenswood and Newark sites. As a result of the field investigation program, four primary geologic units were identified as underlying the proposed tunnel alignment corridor. These geologic units are summarized in Table 1.

After the initial geotechnical investigation and associated laboratory testing were complete, a long section plot was generated showing the stratigraphy of the proposed alignment. The anticipated geology based on the results of this stratigraphy plot is summarized on Figure 4.

Groundwater

The geotechnical land-based investigation included the installation of multilevel, vibrating-wire piezometers to monitor groundwater pressures near the two shaft sites. Piezometer readings indicate that the groundwater pressures are generally consistent at 1 to 3.4m (3 to 11 ft) below the ground surface and are influenced by tidal variations.

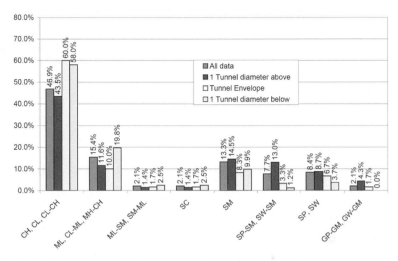

Figure 5. Distribution of USCS soil types along the proposed tunnel alignment

Readings between piezometer levels indicate that there is some hydraulic conductivity between the individual geologic units.

Anticipated Tunneling Conditions

Geotechnical evaluations resulted in the tunnel being situated in the San Antonio Formation to optimize tunneling conditions, depth, and seismic performance. The tunnel section was superimposed on the geologic profile such that it was possible to locate localized intersections of the stratigraphy within the tunnel envelope. Using the tunnel envelope stratigraphy, including one tunnel diameter above and below, it was possible to develop a preliminary assessment of the distribution of soil types anticipated to be encountered along the alignment. This allowed the design team to make a recommendation about what tunneling methods would be most appropriate.

Laboratory testing, including sieve analyses and hydrometer testing, was done on selected soil samples to determine the gradation curves for the various soil types sampled in the borings. This data, along with Atterberg Limits testing and visual/manual soil classification were used to classify soil with the Unified Soil Classification System (USCS).

The San Antonio Formation consists of interbedded clays, silts and sands. The entire tunnel alignment is under the water table, subject to approximately 3.2 bars (48 psi) of hydrostatic pressure. To select and specify a TBM, it was important to quantify the amount of soil with a high fines content (clays and silts) and the amount of soil comprised of coarse materials (sands and gravels). It was also important to characterize the stratigraphic distribution of the soil types, particularly for poorly-graded sands and gravels, which can create problems for some TBMs if they are encountered in excessively thick or laterally continuous layers rather than isolated lenses.

Some basic statistical analysis was performed to synthesize the data and draw some preliminary conclusions regarding soil type distributions. Figure 5 summarizes the results of the statistical analysis. This table includes data from one tunnel diameter above and one diameter below the proposed tunnel alignment in order to provide a

Table 2.   Review of squeezing conditions in clay and silt deposits along the proposed tunnel alignment

Degree of Squeezing	Probability of Occurrence Based on Number of Test Samples
Minor	29%
Moderate	52%
Severe	13%
Extreme	6%

more complete picture of the kinds of soils expected and the variability which may exist within the tunnel horizon.

Within and near the tunnel envelope, the majority of materials are low to high plasticity clays. Smaller amounts of silty and sandy materials are present. The percentage of fine grained material (relative to coarse grained material) tends to increase with tunnel depth. However, within the tunnel envelope a significant amount of coarser sands are also present. Stratigraphic thicknesses were also analyzed and it was concluded that when the excavation does encounter the coarser grained materials, much of the time it will encounter a full-face of coarse materials, as opposed to a mixed-face of coarse and fine materials.

The stability behavior of clay and silt deposits was reviewed by computing the overload factor (also known as the stability factor) along the proposed alignment depth. The overload factor (OF) was calculated using the following equation:

$$OF = \frac{P_z - P_a}{S_u}$$

Where

P_z = the total vertical pressure at tunnel depth

P_a = assumed to be atmospheric pressure (for an unsupported face)

S_u = the undrained strength from unconsolidated-undrained triaxial tests

A breakdown of the degree of squeezing behavior within the clay and silt deposits is summarized in Table 2.

It should be noted that the above analysis assumed an unsupported face, which will not be the case with the Bay Tunnel, since pressure-face methods will be specified. However, this analysis illustrates why pressure-face methods are required for most of the excavation. Because approximately 70% of the tunnel could experience moderate to extreme squeezing conditions, it was recommended that control and regulation of face pressure be maintained through pressurized mode excavation. The majority of extreme squeezing and severe squeezing behavior occurs at shallower depths and the degree of squeezing tends to decrease with depth. This is because the shear strength of the material increases significantly with depth, and governs the degree of squeezing.

The pressurized-face TBM handles squeezing ground conditions better than other methods since face pressures are balanced, preventing the soil at the face from squeezing. However, the main area of concern in the Bay Tunnel is along the body of the TBM, where high friction forces may occur due to the relaxation of the material into the steering gap created by the cutterhead overcut. Solutions for these conditions include using shield lubrication with bentonite, limiting work stops in high risk areas, making provisions for sufficient overcut on the cutterhead and providing additional thrust capacity.

In addition to the potential squeezing behavior of the cohesive fine grained materials, the silts and sands present along the alignment could present running or flowing ground conditions if the face is not pressurized during excavation.

In addition to the San Antonio Formation, the tunnel excavation is expected to encounter a short section of highly weathered Franciscan Formation bedrock. The bedrock is expected to consist of highly weathered sandstone, shale, serpentinite, and other rock types typically associated with the Franciscan Formation.

SEISMIC CHARACTERIZATION

Characterizing the potential seismic impacts on the facility and developing seismic design parameters were critical activities in the design process. The SFPUC expects the water system to be fully operational following a major seismic event.

The proposed Bay Tunnel is located in a seismically active area—central coastal California. Although the proposed alignment does not cross any identified active faults, it is situated between the San Andreas and Hayward faults, which are capable of generating large (M [moment magnitude] ≥7) earthquakes. There are also numerous smaller active faults within 50 km (30 mi) of the proposed alignment. A total of 15 earthquakes of M ≥ 6.0 have occurred in the San Francisco Bay region between 1850 and the present. These include the 1906 M7.9 San Francisco earthquake and the 1989 M6.9 Loma Prieta earthquake. The Bay Tunnel project is expected to experience significant ground shaking during its design life of 100 years.

As required by the SFPUC's general seismic design requirements, the Bay Tunnel will be designed for ground motions that will have a 5% probability of exceedance in 50 years (975-year approximate return period).

As part of preliminary design studies, a probabilistic seismic hazard analysis (PSHA) for ground shaking along the alignment was performed. The purpose of this evaluation was to estimate the levels of ground motions at a specified exceedance probability. Deterministic scenario ground motions were also calculated and compared to the probabilistic ground motions. An evaluation of liquefaction was also performed.

ENVIRONMENTAL CONSTRAINTS

The San Francisco Bay margins consist of environmentally sensitive mudflats, salt marshland, and vernal pool habitat. The Don Edwards Wildlife Refuge, the first urban National Wildlife Refuge established in the United States, is located along the eastern portion of the project site. Many endangered species, like the California clapper rail and salt marsh harvest mouse, live in the wildlife refuge. One of the key motivations for the proposed tunnel system was the benefit of putting the system underground so that environmentally-sensitive areas would not be disrupted. In fact, the geologic investigations for the project were very difficult to permit and to obtain access for field work. Only specific windows of opportunity were available for the investigations and many constraints were placed upon the work operations.

Another environmental issue at the project site is hazardous waste contamination. Within the area surrounding the Newark receiving shaft, elevated levels of chlorinated volatile organic compounds in the groundwater have been identified. The groundwater in this area will require pretreatment and special handling prior to discharge. The soil will require special handling and disposal.

Elevated levels of chrysotile asbestos were also found in some of the samples from the Franciscan Complex bedrock materials that were encountered. Portions of the materials excavated from this area will require disposal as hazardous waste.

SHAFT AND TUNNEL CONSTRUCTION METHODOLOGY

The Bay Tunnel will be constructed as a two-pass system. The first pass will involve a TBM-excavated tunnel, with initial ground support consisting of a bolted and gasketed, precast concrete segmental lining erected immediately behind the TBM. The final lining will consist of either a welded steel pipe or reinforced concrete cylinder pipe, either 2.74 or 3.05m (108 or 120 in) in finished diameter. Once the final lining is installed inside the tunnel, the annular space between the outside of the pipe and the initial support will be backfilled with cellular concrete.

The shafts on this project are required to be watertight to reduce the risk, particularly at the Newark site, of disturbing contaminated groundwater plumes. Excavation support systems that were identified as potential alternatives included secant piles, slurry walls, and ground freezing. Each alternative was evaluated for construction cost and schedule, constructability of the wall to the required depth, water-tightness, and continuity of the wall. The diaphragm slurry wall option is best suited for construction of the Ravenswood and Newark Shafts. Diaphragm slurry walls provide maximum safety against leakage through the wall and can be constructed without any "windows." The stiffness of the wall also helps to reduce adjacent ground settlements resulting from wall deflection. Once tunnel excavation and final lining installation are complete, a steel riser pipe will be installed in the shafts and the annular space backfilled with a combination of concrete and controlled low strength material.

Due to the large percentage of silts and clays and the cohesive nature of the soils, an EPB TBM is the most appropriate mechanical excavation method for the tunnel. As can be noted in Table 2, the majority of materials encountered within the borings at tunnel level are fine grained and cohesive material. The remainder will represent either a full-face or mixed face of sandy soil. In conditions where sand and clay are present in the face, the cutterhead will tend to mix the soils and prevent sand under groundwater pressure from becoming uncontrollable. The coarser grained sandy materials are expected to require ground conditioners in some combination of foam, polymers and bentonite. In most cases it is expected that ground conditioning injected from the TBM cutterhead will cope with most, if not all, of the materials expected. The highly weathered bedrock materials are also expected along a short section of the alignment. The TBM cutterhead will be outfitted with appropriate excavation tools to handle these materials when they are encountered.

PROJECT SCHEDULE

The final design for the Bay Tunnel is currently underway and is expected to be complete in November 2007. The project environmental documentation is being performed in parallel with the design and is expected to be completed in 2008. The project is expected to go to construction in July 2009, with an anticipated completion date of August 2013.

SUMMARY AND CONCLUSIONS

The proposed 8 kilometer (5 mile) long Bay Tunnel is a critical lifeline water supply facility for the City of San Francisco and the San Francisco peninsula communities. It will be the first tunnel excavated by a TBM under the San Francisco Bay. The proposed tunnel is located in an area of high seismicity and will need to remain serviceable after a large earthquake on nearby active faults. It is expected to encounter soft sediments, primarily sandy and silty clays, that will be well suited to the Earth Pressure Balance excavation methodology.

ACKNOWLEDGMENTS

The authors wish to acknowledge the support provided by the San Francisco Public Utilities Commission and Jacobs Associates for the preparation of this publication.

REFERENCES

Jacobs Associates and Fugro West, December, 2006. Bay Division Pipeline Reliability Upgrade (CUW36801), Bay Tunnel Project, Draft Geotechnical Interpretive Technical Memorandum.

Jacobs Associates and URS Corporation, November, 2006. Bay Division Pipeline Reliability Upgrade (CUW36801), Bay Tunnel Project, Draft Geotechnical Data Report.

San Francisco Public Utilities Commission. May, 2006. Bay Division Pipeline Reliability Upgrade, Phase 3, Conceptual Engineering Report. Prepared by Engineering Management Bureau and Project Management Bureau and Parsons, Project CUW36801, Control No. 201441.

ROCK ENGINEERING ASPECTS OF DESIGNING LARGE-SPAN CAVERNS AT DEPTH IN THE DEEP UNDERGROUND SCIENCE AND ENGINEERING LABORATORY

C. Laughton

Fermi Research Alliance

ABSTRACT

Construction of a Deep Underground Science and Engineering Laboratory (DUSEL) is proposed under the auspices of the US National Science Foundation (NSF). DUSEL will provide scientists and engineers with a research facility capable of supporting a broad spectrum of fundamental and applied research deep in the earth's crust. Research partners are physicists, biologists and geoscientists. DUSEL research calls for the construction of a network of shafts, tunnels and caverns that will provide laboratory facilities at depths of up to approximately 2 kilometers. The Project is still in the planning stages but, the initial experimental program is taking shape and conceptual designs have already been developed at two hard rock mine sites.

This paper discusses the overall program and key engineering tasks being developed to support the construction of deep, large-span caverns. These caverns will house a new generation of physics experiments. As currently scoped, the experiments call for the excavation of permanent caverns with spans of up to 60 meters, sited at depths in excess of a kilometer.

INTRODUCTION

The DUSEL is a multidisciplinary endeavor that will provide a diverse program of scientific and engineering research in a deep-earth setting. A core partnership of physicists, biologists, geoscientists and engineers is collaborating on designs that can meet a broad spectrum of research objectives within the borders of a shared-site, for an affordable price and acceptable level of risk.

The DUSEL partners are all seeking long-term occupancy in dedicated, deep laboratories in order to conduct pioneering research in their respective fields. Physicists require substantial overburden thicknesses in order to shield their sensitive particle detectors from the deleterious impacts of cosmic rays. Biologists need access to depth in order to access and study the unique, long-sequestered life forms that survive within the extreme conditions of stress and temperature prevailing deep underground. Geoscientists have an imperative for deep earth access in order to better understand the complex and dynamic processes that interact to control stress distribution, ground movement, fluid flow and mineral deposition within a large-scale geologic context.

Engineers have their own distinct reasons for conducting research underground. Unlike other engineering disciplines, where the construction materials are manufactured or selected to specification, the miner must deal with what he or she finds at the heading. Often times the materials encountered and their behaviors upon excavation are far from those anticipated by the designer. Poor project performance or outright failures, associated with collapse, water inrush, draw-down and contamination and/or conditions or behaviors resulting in a loss in productivity continue to plague the industry. In general, the underground option continues to carry a higher price tag and an elevated level of

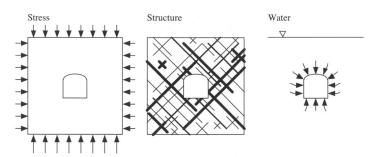

Figure 1. Common factors determining rock mass behavior in hard, blocky rock masses

contingency compared to surface-based alternates. Despite these cost handicaps, the subsurface option is increasingly selected. Societal pressures to preserve surface space and limit environmental impacts, within both urban and rural settings, are increasingly driving sponsors to consider the adoption of a subsurface solution. As more facilities are sited underground, the demand for research to develop safer, more economic and reliable underground design and construction practices, is becoming ever more acute.

THE CANDIDATE DUSEL SITES

In 2005, NSF short-listed two sites for conceptual study; they are the Henderson Mine, Empire, Colorado and the Homestake Mine, Lead, South Dakota. NSF provided funding to the two site teams to develop site-specific 100-page Conceptual Design Reports (CDR). These CDRs were delivered to the NSF in June of 2006. More detailed CDR's (250-page limit) from the Henderson and Homestake teams and other interested in proposing were provided to NSF in early 2007. These proposals will be peered-reviewed with a siting decision targeted for later in the year. A construction start with NSF funding is currently projected for the end of the decade.

The Henderson DUSEL site would be located adjacent to an operating molybdenum mine in the front range of the Rocky Mountains some 80 km West of Denver in Colorado. This site would be accessed and serviced through a combination of existing and purpose-mined shaft and decline tunnel excavations. The Henderson DUSEL would be sited in granite. The Homestake DUSEL site would be located within the boundaries of the closed Homestake gold mine, located on the eastern flank of the Black Hills, some 60 kilometers to the northwest of Rapid City, South Dakota. This site would be largely accessed and serviced through existing excavations. The Homestake DUSEL would be sited in meta-sedimentary and meta-volcanic rock units.

To date only a limited amount of investigative data has been assembled to support rock mass characterization at the two sites. However, based on the general descriptions of the sites presented by Kutcha & Golden (2005) and Roggenthen (2005), it appears that the rock masses at both sites can be broadly categorized as "hard and blocky."

In hard-blocky rock masses such as those present at the Henderson and Homestake sites, stress, structure and water are all potentially attributes that could influence or control excavation stability at depth below the water table, as shown schematically in Figure 1. These same three attributes are also likely to be prime research targets for the geo-scientific and rock mechanics communities.

With particular respect to rock engineering research, local site conditions should be conducive to the study of a number of important behaviors, related to topics such as "rock under stress" and "fracture-based aquifer flow." In situ stress levels at both proposed DUSEL sites should allow for the observation of fractured rock mass behavior over a wide range of stress conditions. The deeper locations should readily permit the long-term study of isolated rock mass pillars or blocks under conditions of yield. Although significant local transient inflows may be anticipated when undertaking new excavations below the water table, steady state water inflow rates are likely to be moderate at depth. Water inflows at both mine sites are moderate (2,000–4,000 L/min). However, it is anticipated that water quantities will be adequate to support key tests and experiments at both sites. DUSEL is likely to offer opportunities to perform experiments in aquifer engineering, under controlled conditions, and over a range of in situ scales.

The shared needs of the underground engineers and geo-science researchers to characterize the key factors of stress, structure and water will strongly overlap and should offer outstanding opportunities for cross-cutting advances in investigative and characterization technologies. In addition, opportunities to improve the engineer's fundamental understanding of the mechanisms that control the ground and hydrogeologic response to excavation are likely through the performance of "coupled-process" experiments that combine observations of fundamental geo-scientific, rock mechanics and engineering parameters.

PERMANENT LABORATORY FACILITIES

Although the scope of the individual facilities is still to be defined a few potential cavern sections that may be excavated as part of an initial scope-of-work are shown for discussion purposes in Figure 2. Each cross-section in the figure represents a permanent underground laboratory housing that would be constructed on an intermediate or deep level (~1 to 2 km deep) at the DUSEL site. Under this scenario, chambers and caverns, ranging in span from roughly 5 to 20 m would be constructed. Some caverns will accommodate the headroom required for overhead cranes and, in one case, a large drill-rig. The drill-rig would be used to undertake deep coring and retrieve biologically pristine samples from zones, under conditions of high stress and temperature, at depths of 3 km and beyond.

Some smaller experimental rooms (span 3~5m) may also be built but are not shown in the figure for reasons of clarity. These may be built adjacent to or within water bearing or fault zones which might be identified as "targets of opportunity" by the geoscience researchers during the investigation or excavation phases.

The final siting and dimensioning of the laboratory excavations and the alignment of connecting tunnels and shafts will be performed to satisfy both site-wide (e.g., environment, safety, health) and experiment-specific (e.g., dryness, cleanliness, space, electrical, mechanical) requirements. The alignment and siting of many underground structures is rigidly constrained in two or three dimensions. These constraints are in-place to connect specific locations, stay within existing rights of way or owned-land footprints and avoid existing natural or man-made features. DUSEL planners will have a good deal more layout flexibility. This flexibility should allow designers to optimize stability, relative to key factors such as stress and structure by adjustment of opening parameters such as location, alignment, orientation, spacing and sectional shapes.

Firstly, the rock mass volumes contained within the proposed sites' boundaries are large and there will be significant, early opportunities, through site investigation, to identify, delineate and avoid or, at least, minimize the negative impacts of mining and supporting excavations within or adjacent to poorer quality, (shear, faults, fractured zones, across contacts, within altered contact margins, dykes and other intruded structures) or more highly-stressed volumes prone to squeeze or burst behavior.

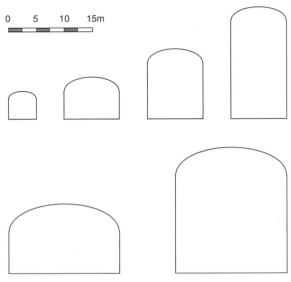

Figure 2.   A generic suite of permanent laboratory cavern cross-sections for the initial phase of DUSEL

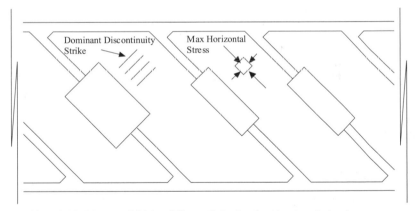

Figure 3.   Hypothetical layout—initial stability optimization for stress and structure

Secondly, with increased investigation, there will be opportunities to identify and reserve the "best rock mass volumes" for the construction of the more challenging excavations, such as the large-span caverns discussed later in this paper.

Thirdly, once potential sites are identified and a modicum of site-specific investigation has been performed to validate key stress and geo-structural parameters, there may be further opportunities to improve the self-supporting ability of the excavation through further refinement of the individual excavation designs.

The suites of permanent laboratory excavations constructed for each phase of DUSEL are likely to be clustered to facilitate the sharing of infrastructure. A sample set of clustered excavations, or campus, is shown in Figure 3. An en-echelon layout is preferred to facilitate delivery of larger experiment support elements, thus reducing the need for underground assembly. The layout also provides for an initial optimization of

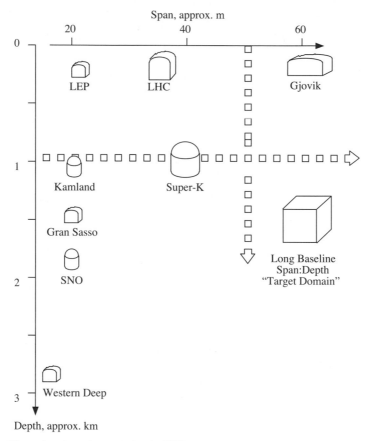

Figure 4.   Dimensional requirements for the UNO cavern

the excavations relative to the orientation of the maximum horizontal stresses (trending NW-SE, with respect to page North) and dominant planes of weakness (striking NE-SW). This optimization necessarily references regional rather than site-specific data sets and further investigative work will be required to evaluate conditions at the selected sites before layouts can be finalized.

At this conceptual stage, a twenty meter span is identified as an upper limit for the span of laboratory caverns. This span was considered to be consistent with cost-effective hard rock practices. However, later construction phases could include the construction of very large span cavern such as that proposed by the UNO collaboration (Jung, 1999) to house a major new physics experiment.

UNO is one of several physics proposals on the table that would house a Large-Deep cavern for a physics detector. Although this proposal requires a cavern that is somewhat larger than any other cavern mined for physics purposes it is by no means the first rock cavern built for physics. Over the last 30 years, the physics community has constructed a number of such facilities worldwide. Figure 4 plots-out in span:depth space some existing purpose-built caverns at the following underground complexes including:

- Large Electron Positron, LEP (Laughton, 1990),
- Large Hadron Collider, LHC (Wallis, 2001),

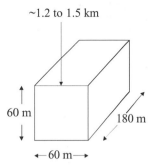

Figure 5.   Large physics caverns span versus depth

- Gran Sasso (Ianni, 2005),
- Super-Kamiokande (Yamatomi et al., 1995),
- Sudbury Neutrino Observatory, SNO (Zsaki & Curran, 2002).

A number of caverns have been constructed at the European Particle Physics Laboratory (CERN). This site straddles the Franco-Swiss border on the outskirts of Geneva, Switzerland. Underground structures at CERN include four 21-m span caverns constructed for the Large Electron Positron (LEP) particle accelerator and two 35-m span caverns constructed for the Large Hadron Collider (LHC) accelerator.

The largest and deepest domed physics caverns plotted in the figure were mined at the Kamioka and Creighton mines in Japan and Canada respectively. Here, 40 and 20 m span caverns were excavated in hard blocky rocks at depths of approximately one and two kilometers respectively. An underground laboratory has also been built at Gran Sasso under the Italian Appenine mountains adjacent to a road tunnel. Here, three 20 meter span caverns were constructed in a dolomitic rock mass.

To place these physics caverns in a broader industry perspective, "reference caverns" (black squares) have been added on to the plot to indicate an approximate envelope for the state-of-the-art in permanent cavern construction. The largest span excavation in Figure 4 is the 60m-span Gjovik Hall, Norway. This cavern was mined at relatively shallow depth (25–50m), in hard metamorphic rock, under favorable stress-field conditions (Meland & Broch, 1994). The deeper excavation plotted has a span of roughly 17m. It houses a hoist system at a depth approaching three kilometres in the Western Deep mine of South Africa (Hoek & Brown, 1980).

As currently specified, the construction of a cavern such as UNO could be considered at or slightly beyond the state-of-the-art. However, similar spans have been achieved at depth in the mining and oil storage industries and larger rock spans can be found in natural cave systems. Given the existence of such rock spans, some of which have been freestanding for significant periods of time, there appears to be no intrinsic reason why a 60 m-span permanent cavern cannot be mined and supported safely at depth, at either of the DUSEL sites (Figure 5). However, in the case of DUSEL, strict stability requirements for multi-year occupancy must be reliably met. The design and construction of such a cavern will be a major undertaking.

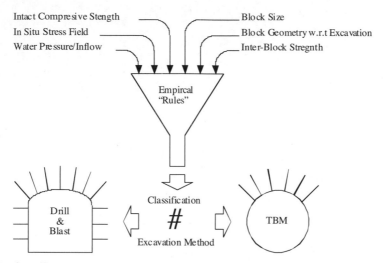

Figure 6. Classification process—from rock mass factors directly to rock support

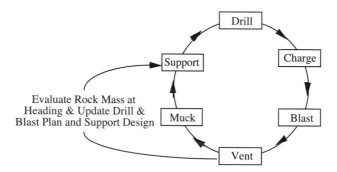

Figure 7. Design-as-you-go—facilitated by real-time feedback on rock mass behaviors

CHALLENGES AND OPPORTUNITIES FOR LARGE CAVERN ENGINEERING

Although there may be no intrinsic reason why large-span caverns cannot be constructed at depth at a hard rock DUSEL, the combination of span and depth are without precedent and particular attention will need to be paid to the evaluation of site conditions in order to predict the potential for adverse ground behaviors. In particular, the key parameters of in situ stress and structure will need to be accurately characterized and realistically modeled in order to develop a reliable prediction of the ground response to the engineering processes (rock treatment, excavation and support) (Figure 6). It will be critical that all instabilities driven by gravity and/or in situ stress over the short and long-term are properly mitigated during the construction phase of the project. Once excavation is complete, rehabilitation of subsequent instabilities occurring at height on the sidewalls or crown, up to 60 m above the invert, will be difficult to physically accommodate. Instability occurring after installation of the detector has begun would be even more problematic.

Given the important repercussions that any post-excavation instability could have on the experimental program an in-depth study that includes an analysis of all potential instability mechanism is merited. At depth underground, where stress or gravity can act singly or in combination to drive failure of intact or fracture rock, the use of classification systems and/or continuum models as basis for design may prove inadequate. Stress-driven failures may be under-represented within the case history data bases that under-pin common classifications systems, as indicated by the absence or limited weighting of stress-related parameters within the systems themselves. Similarly, continuum model, by definition, cannot explicitly account for the potential of block or wedge fall-out around the opening. This is not to say that classifications and continuum modeling should not be exercised, indeed their application may have particular merit during the preliminary stages of a project, where site-specific data is minimal and there is a need to perform initial assessments of potential sites. However, for such a large structure in hard, blocky rock where both scale dependent (stress) and scale independent (discontinuum) parameters could shave a significant impact opening stability, the final design will need to be based on models that explicitly take account of stress and structure.

Of course there are limits to the level of accuracy with which the rock mass can be characterized and modeled and its response to excavation and support predicted. Given these limits, the size of the excavation and the consequences of instability, there will be a need to verify the modeling accuracy by the use of instrumentation and moni-toring of the rock mass throughout both the construction and operational periods. Design-predicted values and geological and engineering models could all be cali-brated, or "ground-truthed," against excavation realities, with differences studied and model improvements made in real-time, as shown in Figure 7. Such round-to-round flexibility in support design would allow the engineer to respond immediately to the observed rock mass conditions and monitored behaviors at the heading.

CAVERN ENGINEERING RESEARCH

There has been a significant trend towards the adoption of the underground solu-tion. However, technological progress to match this trend has been relatively slow. Underground, at a production heading, engineers are rightly reluctant to try out unproven technology where risks of critical path delay/cost-overrun, to the contracted work or over-all project, commonly outweigh any time/cost benefit of a successful implementation. DUSEL can provide test-bed facilities where new methods and means can be proved-out away from the production heading in a dedicated tunnel or chamber. Having access to DUSEL, would allow researchers to "try things out" underground in a "forgiving envi-ronment" where the financial consequences of a long learning curve or even complete failure are minimal. Activities that could benefit from DUSEL access include equipment trials (alternate excavation and support systems etc.), material testing (super-skins or Thin Spray-On Liners, TSL, petro-membranes etc.), safety demonstrations (fire handling etc.) and personnel training (education, safety, rescue, equipment, etc.). Proof of princi-ple needs to have been demonstrated before a research product can seriously be con-sidered for inclusion within the scope of a construction contract.

As a part of this discussion on rock engineering research it should be noted that some underground sites have already been established that fulfill the role outlined above for individual universities, private companies or are used by national organiza-tions for highly function-specific uses. However, many of these sites are costly, single-purpose laboratories focusing on topics of national interest such as defense or nuclear waste storage, or the particular interests of a university department or private firm. These laboratories, generally cannot match the depth, space, infrastructure and tech-nical support or the economy of scale that a fully-developed DUSEL will be able to pro-vide. When opened for business, DUSEL is likely to attract a wide range of research projects from industry and academic consortia.

CONCLUSIONS

Creating a deep facility where space, infrastructure and operating costs can be shared amongst multiple users makes good financial sense. The science and engineering arguments for DUSEL are compelling and the inter-disciplinary nature of DUSEL is proving valuable with each partner able to draw upon the strengths and experiences of the other. Program development and conceptual design work for Henderson and Homestake are in a preliminary stage, but it is already clear that either of the two candidate sites would provide the research community with a first rate site for undertaking a broad spectrum of research goals.

From an engineering perspective, despite the growing demand for subsurface space, major engineering obstacles to increased underground use remain to be overcome. Unlike engineered materials such as concrete and steel, properties of rock vary markedly in space and time and cannot be accurately predicted. Resultant uncertainties in the underground engineering process can lead to major increases in cost during construction and make the underground option a risky proposition. Research is needed to develop new technologies that can accurately predict rock conditions and take full advantage of the underground dimension.

Research performed at DUSEL relative to large cavern construction can support improvements in the characterization and engineering process in hard blocky rock masses. Indeed, DUSEL represents a unique opportunity to conduct fundamental and applied research that can directly address industry needs in design and construction. If properly managed, DUSEL could be a real catalyst, driving major advances in underground engineering. Research deliverables can contribute strongly to a more fundamental understanding of the rock mass as a construction material and can provide a fast-track proving ground for underground technology.

REFERENCES

Hoek, E. Brown, E. (1980). Underground Excavations in Rock. Institution of Mining and Metallurgy.

Ianni, A. (2005). Status and Future Prospects of Gran Sasso. Next Generation of Nucleon Decay and Neutrino Detectors. Aussois, France. 2005.

Jung C K. (1999). Feasibility of a Next Generation Underground Water Cherenkov Detector: UNO. Proceedings of the Next Generation Nucleon Decay and Neutrino Detector Workshop. Stony Brook, New York USA.

Kuchta, M., Golden, R. (2004) Summary of Candidate Site Geology etc.—Henderson. Blacksburg.

Laughton, C. (1990). Support of the L3 Experimental Hall. International Symposium on Unique Underground Structures, Denver, Colorado.

Meland, O, Broch, E. (1994). "Underground openings for public use—Some Results from the Research Program for the Gjovik Olympic Mountain Hall." In Abdel Salam (ed). ITA Conference Tunneling and Ground Conditions. 49–57.

Roggenthen, W. 2004 Summary of Candidate Site Geology etc.—Homestake. Blacksburg.

Wallis, S. (2001). "Great Excavations." Civil Engineering Magazine. August. 34–41.

Yamatomi, J et al. 1995. Waste-less Mining—the Super-KAMIOKANDE and Subsurface Space utilization at Kamioka Mine, Japan. Proceedings of the International Society of Rock Mechanics Congress. Tokyo. 1,649–1,656.

Zsaki, M. & Curran, J.H. 2002. Parallel Computation of field Quantities in an Underground Excavation Analysis Code. In Hammah et al. NARMS-TAC. Toronto. 671–677.

11

Pressurized Face Tunneling I

Chair
M. Roach
Traylor Bros., Inc.

THE BIG WALNUT—A PLANNED APPROACH

Steve Skelhorn

McNally/Kiewit JV

Tom Szaraz

McNally/Kiewit JV

ABSTRACT

Part of a major sewer expansion for the City of Columbus, Ohio. The Big Walnut Outfall Augmentation Sewer (BWOAS) consists of 4,000m (13,200 ft) of 3.66m (12 ft) diameter tunnel passing mainly through glacial deposits with a water head of up to 2 bar. The project involves many new innovations, including:

- Slurry wall work shaft with excavation below the water table;
- Hybrid universal rings combining traditional steel reinforcement with fibers;
- New, full EPB Lovat TBM;
- Two component grout system pumped to the TBM from the surface; and,
- Linabond corrosion protection liner.

The paper looks at the planning approach adopted to meet the various challenges presented, combined with an overview of progress to date.

INTRODUCTION

Located on the south side of Columbus Ohio, BWOAS is the second stage of the Clean Rivers project for the city of Columbus. Designed by a team lead by URS Corporation with Lachel Felice & Associates and H.R Gray as sub consultants. The tunnel was specified as a closed face pressurized TBM drive lined with bolted and gasketted pre-cast cast concrete segments (see Figure 1).

The alignment parallels Alum Creek Drive, the centerline of the tunnel being within 16m of the west edge of pavement. The tunnel crown varies from approximately 14m to 20m below grade (see Figure 2).

The contract documents specified EPB tunneling and a number of measures to control the work, including: restriction of de-watering; sealed slurry walls or auger casings shafts; a TBM incorporating the latest enhancements to control and monitor settlement; and an extensive array of monitoring points.

This paper focuses on the TBM tunnel, examines the decisions made at the planning stage to improve constructability, and concludes with an outline of how the TBM performed.

GEOLOGY

A Geotechnical Baseline Report was issued with the contract documents. Generally, the expected stratum consists of glacial on post glacial deposits with up to 2-bar water head. Cobbles and boulders are expected along the entire route. The first 40% of the tunnel is located within a competent till layer, after this water charged sands and gravels are expected.

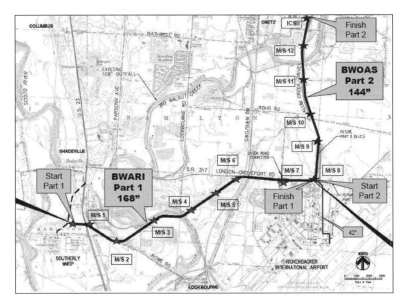

Figure 1.   General layout

Figure 2.   Alum Creek

TBM

The TBM selected was a Lovat full face EPB machine. The 4.26m (168") diameter TBM is the latest generation of Lovat machines and incorporates many of the latest innovations for tunneling, most of which were required to meet the specifications.

Figure 3.   Lovat RME-169 TBM

Cutter Head

The machine is equipped with a soft ground cutting head incorporating ripper teeth, scraper teeth and disc cutters. Wear protection is provided with Trimay plating over the entire face. Drive is by a VFD consisting of four 200kW (265Hp) electric motors. A small diameter bearing was initially proposed to allow the screw conveyor to be mounted at the bottom of the head; however, concerns were raised regarding the potential to jam on boulders. Consequently, a larger diameter bearing was selected with the screw mounted approximately one quarter of the way up the face (see Figure 3).

Segment Erector

Within a 3.66m tunnel, space is limited especially around the ring build area. The preference during design was for a ring erector mounted around the screw conveyor. Recent developments have provided a ring type erector mounted to the bulkhead at the rear of the stationary shell. This style of erector, utilized previously on a Lovat for the NEIS project in California, provided greatly improved clearance within the ring build area and provided a larger window for the guidance laser.

Screw Conveyor

To meet the specified positive displacement screw discharge system, a twin screw was adopted. The rationale being that the two screws can be rotated at different speeds, controlling face pressure with the forward screw and belt discharge with the rear. The double screw provides a secondary advantage in that the screw discharges to a horizontal section of belt, mitigating the potential for spillage.

Grout System

A two component grout system was selected for this project. Retarded grout is pumped to a holding tank on the TBM allowing an unlimited supply of grout not dependent upon the train. Additionally, the omission of a grout transfer car allows a shorter train and eliminates the need to split the train under the TBM gantry during the mining phase.

The grout was developed by Master Builders and consists of a cement/fly ash blend with a chemical retarder and a viscosity modifier added at the batching plant. The retarder volume can be modified to adjust the initial set from a few hours to several days. Bentonite was subsequently added to the mix to mitigate the drop-out observed in the pump lines. At the point of injection an accelerator (sodium silicate) is injected into the grout mix.

Grout is batched in 0.3 m^3 and transferred to a 1.5 m^3 cubic yard holding tank for transfer to the tunnel. A three stage Moyno pump transfers the grout through a 50mm (2") line to the TBM.

At the TBM, holding tanks are provided for the grout and accelerator. The theoretical grout volume is 1.5m^3 per ring; a 3m^3 tank was supplied with the TBM. Two Moyno pumps are used to inject the grout, coupled to two peristaltic pumps to dose the accelerator. The injection system is controlled by a PLC integrated with the TBM.

Ground Conditioning

A ground conditioning system is included on the TBM, and consists of four independent pumps and five injection ports ahead of the cutter head. The system allows foams, polymers or blends to be readily injected through any of the ports. For the first section of tunnel, hard till was encountered, requiring foam rates of up to 80% (foam injection ratio). After this the foam rate is expected to be much less, but will need the addition of polymers to control the wet sands and gravels.

Guidance System

A tacs guidance system was purchased with the TBM. The unit incorporates a Leica 1103 motorized theodolite, and the latest software allowing TBM progress to be plotted in real time on an imported AutoCAD drawing of the project plan, profile or geotechnical profile.

The unit was required to meet the class one div 2 electrical specifications, tacs supplied both the tacs control computer and the Leica enclosure in explosion proof boxes (see Figure 4).

The tunnel alignment consists of 15 250m radius curves. This presents potential problems with the line of sight for the laser and requires the laser to be moved forward every 9 rings during transition of the curves.

TUNNEL RINGS

The contract specified six-piece bolted segmental rings consisting of 3 parallel plates, one key and two top plates. During bid stage it became clear that a universal style ring would be preferable because of the requirement to line the tunnel on completion. Fewer bolt pockets were preferred to minimize irregularities which must be filled prior to affixing the liner. To minimize the bolt pockets, it was elected to use dowels rather than bolts for the circumferential joints. A universal ring, by virtue of each plate being a trapezoid, allows for much easier building with a dowel.

Figure 4.   Tacs guidance system

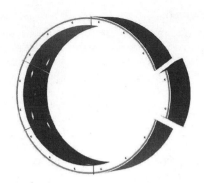

Figure 5.   Tunnel ring

A six piece universal ring was selected consisting of four 67 ½ degree rhomboid and two 45 degree trapezoid segments. The design incorporates a 38mm taper to the ring and allows the ring to be rotated through 360 degrees about the tunnel axis. A ring width of 1.5m was selected. Although the width creates limited clearances for transport under the TBM gantry, it reduces the total number of rings, further minimizing the amount of patching required.

The universal design allows every ring to be delivered to the TBM in the same order. Ring building always commences with the same 45 degree trapezoid counter—key plate. On occasions this results in the first plate being built above spring line and the final key plate inserted below, however the versatility of the erector allows this system to work without any problems (see Figure 5).

Ring design was carried out by Hatch Mott Macdonald (Toronto, Canada) with assistance from Chris Smith of CRS Consultants (UK). During the initial stages of design, the option of using steel fibers was explored. Based on success in European tunnels, the use of fibers simplifies the segment casting process and also provides a tougher ring that is less susceptible to superficial damage during handling. Unfortunately, the 1.5m width of the ring, combined with the geometry of the TBM and the taper angle of adjacent plates eliminated the possibility of using full fiber rings due to torsion stresses. However, Hatch were able to design a hybrid ring consisting of a reduced re-bar cage and fiber reinforcement. This provided the advantage of reduced handling in the segment casting plant, combined with the increased durability form the fibers (see Figure 6).

CBE custom built sufficient forms to allow NASCO to cast 16 full rings each day, working on a two shift basis. Steam curing allowed for rapid turn round on the forms. Using a single type universal ring allowed the segments to be stripped from the forms and stacked in half rings. These stacks remained intact until delivery to the segment unloading station on the TBM, minimizing the potential for damage during handling (see Figure 7).

SHAFT STRUCTURES

To safeguard the ground around each of the shafts, a jet-grout block was specified. This consists of a block of grout around the two tunnel eyes at shaft 8 and a block encasing the bases of the intermediate shafts.

Figure 6. Test ring erected

Figure 7. Nasco casting yard

Figure 8. Slurry wall shaft

Slurry Wall Shaft

Shaft 8 is a 12m diameter, 25m deep shaft with a 4.5m thick base plug. The slurry wall design incorporated soft eyes (fiberglass re-bars) at the location of the tunnel and included box outs to provide keyways for the base slab.

On completion of the slurry wall, excavation was carried out underwater by clam bucket in a flooded shaft. For the box out connections, it was necessary to employ divers to access the base of the shaft (see Figure 8).

On completion, the shaft base was tremmied in place before pumping out the shaft.

Auger Shafts

These involved 3m diameter future access shafts augured within a steel casing. Shaft depths were generally around 20m (60 feet). All shafts were augered to tunnel axis before backfilling with uncompressible fill to just above the tunnel. The fill was required to provide a seal for grouting the casing. On completion the fill was removed ready for the TBM entry.

Figure 9. TBM launch

LAUNCH PROCEDURES

Due to the offset of the tunnel in the shaft, the effective launch window was only 10m. While the TBM could be installed in this space, there was insufficient room to enable the main gantry components to be connected.

The use of an umbilical system to connect the TBM was discounted due to the cost. The TBM was installed in sections and launched with a short screw extending the trailing gear and conveyors as the tunnel advanced (see Figure 9).

With over 70 electrical connections between the TBM it was necessary to suspend the tail can from a "shelf" within the shaft. This allowed the cables to be passed through the can negating the need to reconnect them.

DATA ACQUISITION

As with most recent tunneling contracts, data acquisition was a major component of the owner's pre-requisites. The TBM is equipped with a PCL controller which reports to a data logger continuously. This includes most of the TBM operational data as well as inputs from the guidance system.

The data is transmitted by modem to surface computers and is downloaded to a VPN. This enables real time data viewing over the internet. The information is shared with the owner and the TBM manufacturer who can greatly assist in troubleshooting.

PROGRESS TO DATE

The BWOAS contract was awarded to the McNally/ Kiewit JV in October 2004. The machine was launched in August of 2005. The initial breakout through the jet grout zone allowed a non-EPB launch; however, the reduced clearance within the shaft proved difficult. In September the first rings were built progress was very slow as it was necessary to stop every 6 m to install the next gantry section. It was not until January 2006 that the entire TBM was fully buried and the trailing gear hooked up.

The tunnel was stopped in October 2006 as per the contract to allow the contractor from Stage 1 one to remove their TBM from the drive shaft. Mining resumed on December 4, 2006; at the time of writing (Jan 2007), the tunnel is 42% completed and 1097 rings have been constructed.

There have been some teething problems with the initial launch, some more problematic than others; however, most of these were ironed out as the tunnel advanced. In

Figure 10.   TBM within shaft

terms of the major components, most of the systems are now performing well. The double screw enhances the operator's control of EPB. The foam systems versatility allows the operator to control the EPB and the quality of the muck. The guidance system has created some issues, this involves the reduced laser window requiring the laser to be moved very close to the rear of the TBM. Additionally some movement of the rings, although minor, was affecting the operation of the system.

The ring type erector, although not favored by the operators at the onset is performing well. The operator's line of sight is limited by the erector shoe; however the increased clearance through the can greatly enhances space in the build area.

The VFD drives took a while to setup and balance; however these are now performing well.

Average mine times are now between 8 and 15 minutes with ring build times around 20 minutes (see Figure 10).

CLOSING REMARKS

This project incorporates the state of the art systems for TBM tunneling, VFD drives, fiber reinforced concrete segments, enhanced data recovery and a fully computerized TBM. Unfortunately, as with all new innovations, there are new learning curves.

To date, the crew has mastered all the systems and tunneling is progressing as expected.

CONSTRUCTION OF THE BIG WALNUT AUGMENTATION/RICKENBACKER INTERCEPTOR TUNNEL (BWARI, PART 1) —COLUMBUS, OHIO

Jeremy P. Theys

Traylor Bros., Inc.

Mina M. Shinouda

Traylor Bros., Inc.

Gary W. Gilbert

City of Columbus, Department of Public Utilities

Glen D. Frank

H.R. Gray

PURPOSE OF THE PROJECT

The Big Walnut Augmentation Rickenbacker Interceptor project began in 1995 with the preparation of the Scope of Services for the design of the Rickenbacker Interceptor, with service to and around the Rickenbacker Air Force base. The air base was soon to be placed under civilian control through a Port Authority, with an expected influx of businesses and services, requiring increased sanitary service capability. Due to political differences regarding annexation and other issues, the project was placed on hold. In mid 1996 the Scope of Services was rewritten with more emphasis on a tunneling option, but still only addressing the air base and surrounding tributary area sanitary service requirements.

The consultant selection process began in late 1996 with the Request for Statements of Qualifications issued in December. Three firms were selected to furnish proposals based on the revised Scope of Services that now included requirements for hydraulic modeling of the service areas and the Big Walnut Outfall (BWO) and an expanded soil investigation requirement. All three firms were requested to make presentations to the selection committee, with the committee selecting URS Greiner for further consideration and negotiation. A contract was entered into in August of 1997 with URS Greiner. The major sub-consultants were Lachel and Associates, Inc., H.R. Gray and Associates and Solar Testing Laboratories, Inc.

The project began with geological investigations, preliminary soil borings, well investigations, tributary area evaluations, hydraulic modeling of the BWO system and research into other projects planned for the Rickenbacker airport and surrounding area.

The hydraulic modeling of the BWO and the proposed Rickenbacker Interceptor confirmed that the BWO suffered wet weather surcharging and would require augmentation to properly convey the future flows to the Southerly Wastewater Treatment Plant (SWWTP). The project scope expanded to include evaluation of both the Rickenbacker Interceptor and the Big Walnut Augmentation projects. Both projects previously identified in the Capital Improvements Program. The work on the expanded scope for the Big Walnut Augmentation and the Rickenbacker Interceptor began with the evaluation of a parallel 2.75 meter (108") diameter augmentation sewer from SWWTP to S.R. 317 and Parsons Avenue, then continuing east along S.R. 317 as the Rickenbacker Interceptor. The parallel 2.75 meter (108") sewer had been planned for in the design of the BWO and a 30.5 meter (100') easement obtained for its construction and the installation of the second pipe.

Well pumping tests were performed near S.R. 317 and Parsons Avenue and the results entered into modeling performed by the U.S. Geological Services and URS Greiner's Woodward Clyde. The model indicated significant effect on the City's Division of Water well field to the north of S.R. 317.

During these evaluations, the Ohio EPA began implementing their anti-degradation rules that made permits for stream crossings much more difficult to obtain. The BWO had several stream crossings of the Big Walnut Creek, making a parallel augmentation sewer more difficult to permit.

This combination of design constraints moved the Division of Sewerage and Drainage (DOSD) to evaluate a southerly route for the project. The southerly route (present location) was evaluated with the determination that the sewer would need to be deeper to provide adequate cover for installation by tunneling methods. The route and size would provide a larger tributary area, serve as a relief (augmentation) sewer for the BWO, provide wet weather storage until build out to minimize bypassing at the SWWTP and eliminate the need for two pumping stations.

The geological investigations of the southerly route, well data, soil borings and other design considerations produced two major concerns for the design and construction; (1) vast quantities of water to handle if dewatering and (2) the presence of boulders in the soil matrices. Public meetings were held with the residents of the area and their concerns centered on the project's effects on their wells, as previous construction work at SWWTP had affected wells in the area due to the required de-watering.

The design team addressed the major concerns, in part, by requiring either an Earth Pressure Balance Tunnel Boring Machine (EPBM) or a Slurry Boring Machine to preclude the need for dewatering along the selected route. The project was advertised on May 17, 2003 and bids received on July 16, 2003. Jay Dee/Michels/Traylor Joint Venture was the low bidder. The Notice to Proceed with the work was issued on December 3, 2003.

THE CONSTRUCTION MANAGEMENT TEAM (CMT)

The BWARI Construction Management Team (CMT) serves as the representative for the City of Columbus during the construction of the BWARI tunnel and associated facilities. The CMT is a direct descendent of the consulting team responsible for the design of the project and all of the entities that were primarily responsible for the design are involved in the day-to-day administration of the construction contract.

The design team was led by the Columbus, Ohio office of URS Corporation, who was supported primarily by two subconsultants: H.R. Gray, based in Columbus, Ohio and Lachel Felice and Associates, Inc., based in Golden, Colorado. URS led the design effort, and performed the alignment planning and easements, hydraulic design and preparation of drawings and specifications not associated with the tunnel or shafts, Lachel Felice and Associates supervised the geotechnical investigations, prepared the Geotechnical Data and Baseline Reports and designed and prepared contract drawings and specifications for the tunnel and shafts. H.R. Gray prepared the administrative portions of the contract, the engineer's estimate, and performed constructability review.

The CMT is made up primarily of the same group of consultants with H.R. Gray leading the overall administration of the contract and management of the construction project, and URS and Lachel Felice and Associates jointly serving as the resident engineer on the project. H.R. Gray processes and manages all of the correspondence between the contractor and the owner's representatives, including; submittals from the contractor and the subsequent review comments from the engineer, requests for information from the contractor, progress payment applications from the contractor, etc. In addition H.R. Gray prepares and administers all of the meetings required between the contractor and representatives of the owner, including; coordination meetings, monthly progress meetings, quarterly dispute review board meetings and hearings, as well as any special meetings that might be required as issues come up during construction. URS and Lachel Felice and Associates provide field representatives, reviews submittals, and provide recommendations to the CMT on construction issues to ensure that the design philosophy is maintained throughout the construction of the BWARI facilities.

PROJECT PHYSICAL AND GEOLOGICAL SETTING

The geologic setting for the BWARI project is the Till Plains Physiographic Province (ODNR), which is characterized by subdued topography and low relief. The bedrock underlying the area is a sedimentary sequence of limestone and shale of Devonian age. The unconsolidated materials above the bedrock, through which the project will be constructed, are glacial and post-glacial soils, consisting of both glacial outwash and glacial till soils.

The glacial deposits covering this area were deposited by the continental ice sheets that repeatedly covered the region over the last 2 million years and by water flow resulting from melting of these glaciers. Although several glacial epochs occurred during this time period, the most recent, termed Wisconsinan, is the most significant in regards to the deposition of materials expected to be encountered by project construction activities. As it is currently understood, the Wisconsinan glaciations included at least two and potentially three ice advances and retreats in the project area. The oscillation in the deposition environment has resulted in complex subsurface conditions, including:

- Groundwater levels of 30 to 50 feet above the alignment
- Relative clean sands and gravels that would tend to flow in the presence of groundwater
- Numerous cobbles and boulders throughout the alignment
- Extensive layering of soils of greatly differing grain size and permeability

An extensive geotechnical investigation program was performed to delineate the geologic conditions and included soil borings an average of 500 feet apart along the alignment. Innovative methods were required to evaluate for the presence of boulders, which are difficult to evaluate with typical geotechnical drilling methods. These methods included large diameter (36-inch) auger borings, rotosonic borings, counting boulders during several days' production at a sand and gravel mining operation near the tunnel route, and a survey of the boulders present on the gravel pit location. The scope and results of this investigation are described in a paper previously presented at the 2001 RETC (Frank & Chapman 2001).

Risks associated with these geologic conditions were allocated in accordance with current engineering and contractual practice for tunnel projects. This included clearly describing the expected conditions in a Geotechnical Baseline Report (GBR), including definitive estimates of the number of boulders and cobbles expected. A Geotechnical Data Report (GDR) was also included that contained the field testing results.

Boulder and Cobble Study

The overall goal of the boulder and cobble study was to produce a baseline as described in Table 1.

Table 1.   Boulder and cobble study results

			Boulders						
	Cobbles	Small Boulders	Obstructions						
Cobble or boulder size in feet	0.5–1	1–1.5	1.5–2	2–3	3–4	4–5	5–6	>6
Number expected in shafts and open cut (10,000 cyds)	14,250	616	38	10	1	1	0	0
Number expected in tunnel (10,000 ft of tunnel)	113,000	3,146	415	126	16	4	1	1
Number expected in 500 lineal feet of microtunnel	460	21	5					
Number expected in 200 lineal feet of bore and jack	184	8	2					
Maximum for shafts and open cut*	285	18	2					
Maximum for tunnel (per 50 lineal feet of tunnel)†	11,300	300	50					
Total number of obstructions expected in 10,000 lineal feet of tunnel = 563 Average number of lineal feet of tunnel between obstructions = 18								

* Occurrences are for volumes of 100 square feet in horizontal area and 5 feet in vertical thickness. (Areas of highest concentration are expected to encompass 10,000 square feet.)
† No more than 5 obstructions will be encountered simultaneously.

This baseline table gives a very precise and statistically defendable estimate of the abundance and size distribution of the "very large fraction," those particle sizes likely to pose significant challenges to a pressure balance tunneling machine. The baseline was developed utilizing a mathematical model approach outlined in a paper previously presented at the 2005 RETC (Frank & Chapman 2005). The basic tenet of this boulder base lining method is that the abundance and size distribution of the large cobbles in a geologic formation (which can be sampled by large diameter borings) can be extrapolated out to give an estimate of the abundance and size distribution of boulders in the same geologic formation.

The accuracy of the boulder baseline and the accuracy of the underlying estimating method are hard to determine. From the standpoint of the owner, the boulder baseline set in the contract documents should be considered a success, as the tunneling was completed on schedule and below the budgeted cost, however there was no way to count the boulders that were encountered by the earth pressure balance tunneling machine. Boulders were counted on one section of hand mining that was performed approximately 10 meters (30 feet) above the main tunnel. The results indicated that the baseline estimate was approximately 300 percent (expected 627, encountered 234) higher than the count for large cobbles, 600 percent (expected 36, encountered 6) higher than the count for small boulders, and 20 percent (expected 5, encountered 6) lower than the count for obstructions (clasts that would not fit through an 18-inch grid). One could interpret this as a safety factor of 3 on large cobbles, 6 on small boulders, and 0.8 on larger boulders. This data, taken as a whole, appears gives a fairly accurate estimation of the actual conditions encountered.

Geologic Conditions Encountered

The reality of tunneling with an earth pressure balance tunneling boring machine dictates that the ground is never observed in situ, and boulders in particular are significantly altered during the excavation process. Therefore the accuracy of the anticipated subsurface conditions is very difficult to ascertain, and the ground description provided in the GBR can not be proved to be accurate. In general one must assume that if the tunneling method that was properly designed and implemented based on the ground description works (installs the tunnel at a cost that both parties can live with), then the ground description must have been accurate. Implicit therefore, is that if a tunneling method that was properly designed and implemented based on the ground description does not work, then the ground description must have been inaccurate.

Based on the criteria above, the site characterization for the BWARI tunnel project was accurate, as the tunneling method employed was successful. The data gathered while not completely supporting the site characterization given in the GBR, does not necessarily refute the accuracy of the GBR either. Overall the muck samples taken every 100 feet of tunnel advance show approximately half as much clean gravel and three times as much dirty sand and gravel as was expected. This is not surprising since the average stratigraphic layer thickness in the project area is about 1 meter, and the muck samples would be a combination of all of the geologic units at the face.

BWARI PROJECT CONSTRUCTION

Project Construction Team

The BWARI Project was constructed by a Joint Venture between Jay Dee Contractors (Livonia, MI), Michels Pipeline (Brownsville, WI), and Traylor Bros., Inc. (Evansville, IN) (referred to as JDMT). The management and construction of the project was performed by Jay Dee Contractors, Inc., with engineering assistance from Traylor Bros., Inc.

Tunnel Alignment

The total length of the BWARI tunnel was 6.3 kilometers (20,770 feet). The tunnel alignment runs primarily through the rural areas on the south side of Columbus near the town of Lockbourne, OH. The alignment runs through farmland but also crosses under a cemetery, an active gravel quarry, a river, a 4-lane highway, and 3 road crossings. There are no significant buildings or structures along the tunnel alignment. The initial alignment from URS was comprised of curves on a 244 meter (800 feet) radius. The easement for the project was 18.2 meter (60 feet) wide, with the tunnel centered on the easement. JDMT studied the alignment and proposed a redesigned alignment that utilizes relaxed curves on a 366 meter (1,200 feet) radius. This revised alignment fit entirely inside the existing easement and provided many benefits during construction. This alignment permitted the use of less taper on the precast concrete segments and allowed for an easier steering design for the tunnel boring machine components. The relaxed radius of the curves also shortened the overall tunnel alignment by approximately four meters. The proposed change was acceptable to the City and JDMT went forth in the design of the Tunnel Boring Machine (TBM) and the precast concrete segments using this revised alignment. The tunnel alignment is shown in Figure 1.

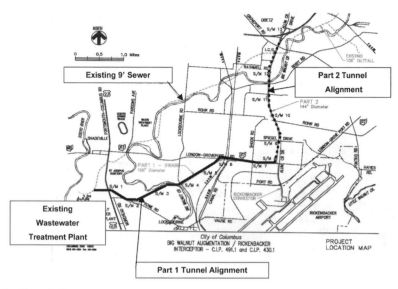

Figure 1.   Tunnel alignment

Main Shaft Design

The main shaft site is located in a farm field adjacent to the Southerly Wastewater Treatment Plant. The SWWTP and adjacent U.S. 23 Highway were constructed at the 100-year flood elevation of the Scioto River (694 feet). The Scioto River runs past the jobsite to the West by about 400 meters (0.25 miles). Due to the typically wet spring weather in the Columbus area, the river occasionally overflows its banks. Within the first twenty days on site, the river flooded, resulting in a foot of water covering the area. This led to delays in the initial surface site development. Once this flooding potential was realized, the main shaft was redesigned. The initial shaft support design was comprised of ribs and lagging for primary support of a circular opening. The redesigned shaft was a rectangular sheet piling shaft using internal steel walers for support. This revised construction shaft allowed for flood protection of the TBM and the completed work throughout the project. The geometry and size of the new, larger shaft allowed for the potential of building a larger final access shaft for equipment access to the tunnel to perform sewer maintenance. The construction shaft that was used is 13.7 m × 7.3 m (45 feet long × 24 feet wide). The inside clearance is 12.5 m × 9 m (41 feet × 29.5 feet). This clearance was designed to allow the tunnel boring machine components to be lowered through during the launching process.

Cast in Place Box Tunnel Construction

The project included construction of a 68.5 meter (225 feet) long cast in place concrete box tunnel with cross-sectional dimensions of 4.26 m × 4.41 m (14 feet × 14.5 feet). We recognized the value of having this tunnel section constructed early so it could be utilized as a tail tunnel for launching of the tunnel boring machine. The box tunnel was constructed in an open cut excavation. A fairly large dewatering system was installed and was capable of pumping approximately 10,000 gallons per minute and was run for four to five months before the excavation could proceed to full depth. The excavation and construction of this box tunnel was subcontracted to the adjacent contractor of the S-65 Pump Station Project, C.J. Mahan Construction, Inc. Once the box tunnel was completed, the main shaft could be constructed. The main shaft had a 1.52 meter (5 feet) thick concrete mat poured for the base; the steel sheet piling was installed on top of the base. Once the steel sheet piling was erected and the steel walers were welded and bolted in place, the lower portion of the shaft was backfilled with 1.0 MPa (150 psi) compressive strength flowable fill. Compacted excavated material was used for the remainder of the backfilling process.

Jet Grouting

A jet grout block was required to be installed at each access and drop manhole location to reduce permeability and increase the compressive strength of the soils. The original design of the access and drop shafts for the project consisted of a 1.82 meter (6 feet) internal diameter manhole at all locations and a 0.91 meter (3 feet) diameter drop pipe at manholes #4, 5, 6, and 7. The layout of the manholes detailed the drop shaft on the upstream side of the access shaft, with each shaft encapsulated in a separate jet grouted block. The grout block was to be installed at a specific depth to encapsulate the tunnel and the bottom of the shaft at their intersection point with the intention of providing a dry environment during the performance of the shaft/tunnel connection work.

Redesign of the Jet Grout Block. We analyzed this design and determined that there would be a few advantages to changing the orientation of the manholes for both the construction activities as well as for the life of the system. We concluded that if the location of the drop manhole was to be downstream of the access manhole the inspection of the live sewer would be easier. An inspection of the incoming flow from the drop sewer could be viewed from the bottom of the adjacent upstream access manhole without the inspectors being in the path of the incoming flow. In live conditions the drop shafts may provide large amounts of flow which may travel up the tunnel walls for a short section of the tunnel upon entry into the main tunnel due to the velocity. Another advantage of changing the orientation of the manholes was that both shafts could be located entirely in one single jet grouted block. This jet grouted block was designed to be large enough to provide a stabilized zone of ground, in which the TBM could tunnel into and seal off any ground water that may be present outside the shields (See Figure 2). This would allow free air cutterhead maintenance from the access shaft locations. This design was critical to the success we experienced in the cutterhead interventions throughout the project.

The details of the modified jet grout block are shown on Figure 3. The jet grout block was composed of an array of jet grout columns formed in rows adjacent to each other and spaced at 1.06 meters (3.5 feet) center to center. The rows of columns were staggered to achieve the best interlocking interface. The jet grout columns were terminated at the circumferences of the access and drop shaft. No columns were installed inside the perimeter of the shafts to eliminate any unnecessary strengthening of soil that would be drilled at a later stage, and also to enable an easier drilling process of the shafts. The columns installed around the perimeter of the shafts were drilled much taller than the typical block columns in an effort to further reduce the permeability of the soil at the location of the shaft to tunnel connection, and to create a strengthened soil curtain around the shaft to prevent soil collapsing during shaft drilling. These perimeter jet grout columns were extended below and above the jet grout block. The lengths of these columns differed at each manhole location depending on the ground conditions.

Installation. Due to the presence of cobles and boulders in the ground, it was decided to pre-drill pilot holes at the location of the jet grout columns. This process enabled the location of any obstructions that may damage jet grouting equipment or

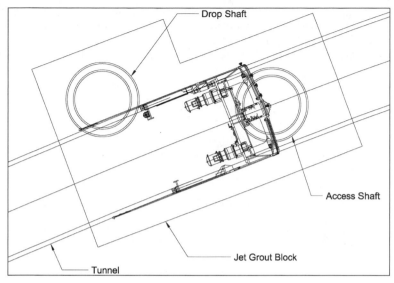

Figure 2.   TBM in jet grouted block

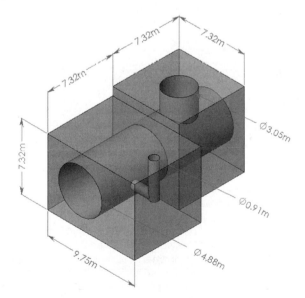

Figure 3.   Modified jet grout block

create "shadows" of un-grouted soil on the back side of the obstructions. If obstructions were encountered, the spacing of the columns could be adjusted in order to provide a thoroughly grouted area. The pilot holes were drilled using a Down-The-Hole hammer (DTH). The jet grouting rods and monitor were then inserted into the pilot hole and advanced to design depth using a rotary method with external water flush. Upon reaching the design depth, cement grout was injected, at the set injection parameters, from bottom to top of the treatment zone. The rig was then moved to the next adjacent hole and the process was repeated. The injection parameters were determined from a test program that was performed at the jobsite location where the jet grouting was to be executed. The injection pressure was 500 bars, air pressure was maintained at 8–10 bar pressure, and the injection rod rotation was set at 10 rpm. The withdrawal rate was 150 mm (6 inches) every 30 seconds.

For the most part, the jet grout blocks performed a decent job of reducing the permeability of the soil at the shaft locations. There were two locations in which we found a large amount of groundwater when performing cutterhead maintenance. In these locations, the jet grouting did not perform to the level that was expected. Additional grouting work was required to reduce the permeability to a low enough level to make the tunnel/shaft connections.

DRILLED SHAFTS

The project included installation of six access manholes and four drop manholes, not including the entrance and receiving shafts at both ends of the tunnel. Manhole #2 is a Y-shaped structure with a stub-out for a future connection and it is still under construction at the time of writing this paper. The shafts for the manholes were excavated using the cased augur shaft method. Corrugated metal pipe (CMP) was used as the casing for the shafts. All the shafts, with the exception of access shaft #3, were drilled in two stages. Access shaft #3 was drilled in one stage. The first (top) stage was drilled dry, and the second (bottom) stage was drilled in the wet. Drilling was performed in two stages due to the limit on the length of the shaft casings that could be shipped using standard means, which is 12.2 meters (40 feet). All the shafts with the exception of drop shaft #6 had upper and lower CMP casings of 3.66 meter (12 feet) diameter and 3.05 meters (10 feet) diameter, respectively. While the lower part of drop shaft #6 was similar to the rest of the shaft, it was cased using a rib and lagging method with a diameter of 4.88 meters (16 feet) at the top section. This was necessary as it was to be used as a mining shaft for the shallow sewer connected to it. Details about the shallow sewer are presented at a later portion of this paper. The typical cross-section of the shafts is shown in Figure 4.

Construction

The upper portion of the shaft was drilled to a diameter of 4.06 meters (5 feet 4 inches), leaving a gap of approximately 200 mm (8 inches) around the CMP casing, which was filled later with grout. After reaching the desired depth, the CMP casing was installed and grouted in place. As per the contract specifications, dewatering was not permitted in most of the sites. Thus, dry drilling was limited to few feet below the existing ground water table which in turn dictated the length of the upper casing. The top casing was also extended above ground surface to a level above the 100-year flood elevation to secure the shaft from potential flooding. This also acted as a protective ledge around the shaft for safety during construction.

After installing the upper casing, the shaft was then drilled to the design depth at a diameter of 3.35 meter (11 feet). A gap was left of approximately 152 mm (6 inches) for grout. The 3.05 meter (10 feet) diameter can was then installed and grouted in place, with the bottom about 457 mm (18 inches) above the crown of the tunnel. An overlap of 610 mm (2 feet) between the upper and lower casing was maintained in all the shaft locations. All the shaft excavations were completed prior to tunnel boring machine reaching each specific location. This enabled the access shafts to be used for cutterhead access during maintenance periods.

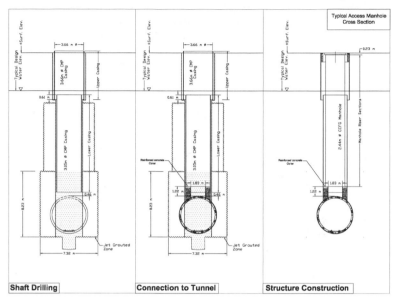

| Shaft Drilling | Connection to Tunnel | Structure Construction |

Figure 4. Typical cross-section of drilled shafts

Access Manhole Connections to the Tunnel

As mentioned before, the access shafts were used during maintenance periods on the cutterhead throughout the tunneling operation. After the tunnel boring machine was advanced a distance in which the trailing gear was clear of the shaft location, the connection to the tunnel was executed to facilitate usage of the shaft for tunnel ventilation. In order to make the cut in the tunnel liner, it was mandatory to develop a permanent structure element that would bear the tunnel liner loading after the opening is made. A cast in place reinforced concrete collar was proposed to be poured at the shaft/tunnel connection before cutting the liner (see Figure 4).

The contract specifications required the inside diameter of the access manholes to be at least 1.83 meters (6 feet). The inside diameter of the collar was held at the same diameter. The outside diameter of the collar was limited to a maximum of the 3.05 meter (10 feet) diameter of the CMP shaft casing. The height of the collar was 1.22 meters (4 feet), above the crown of the tunnel. To install the collar, the top of the tunnel was uncovered and cleaned from inside the shaft. The tunnel liner surface was roughened and the collar reinforcement was installed in place. The collar was anchored to the liner with steel dowels that were installed around the perimeter of the collar. The collar was then cast in place and given adequate cure time before making the connection to the tunnel liner. The cut was made using hydro-blasting methods. As part of the concrete collar design, a steel sleeve will be installed inside the collar with a length covering the full height of the collar plus the thickness of the liner. This sleeve will be set in place and field cut to match the inside diameter of the tunnel. The sleeve will be epoxy grouted to the concrete behind it.

Final Manhole Risers

Fiberglass pipe manufactured by HOBAS Pipe Inc. was chosen to be used as the manhole riser sections. The Hobas pipe sections had a nominal diameter of

2.44 meters (8 feet), and were custom made in lengths totaling the specific height of each manhole. The pipe sections were connected with fiberglass FWC couplings. Each manhole was equipped with a retractable ladder that will be used for tunnel inspection during the service life of the tunnel. The guides and hangers of the ladder were installed in the Hobas pipe before it was set in the shaft. The Hobas pipe sections were stacked in the shaft on top of the concrete collar and the annulus gap was grouted. Final fiberglass landings are installed prior to turning the shafts over for final service.

Drop Manhole Design

The drop manholes are still under construction at the time of writing this paper and the design includes four components. The four components are the connection to the tunnel, the drop pipe, the transfer of the incoming flow to the drop manhole, and the top manhole riser section. The connection to the tunnel will be made at the spring line of the tunnel. A reinforced concrete collar, similar to that used in the access manholes, will be utilized to bear the tunnel liner loading. This concrete collar will be anchored to the precast concrete segment tunnel liner and will be cast in place prior to cutting the tunnel liner. This allows immediate transfer of the tunnel liner loading to the reinforced concrete collar. This avoids any temporary instability that may occur when the tunnel liner is cut and removed. The tunnel liner cut will be performed from inside the tunnel.

The drop pipe is the portion from the tunnel spring line elevation to the incoming inlet pipe. A 914 mm (36 inch) Hobas pipe will be used as the drop pipe for the manholes. At the bottom, a 914 mm (36 inches) Hobas pipe will be utilized to connect the drop pipe to the tunnel. The connecting pipe will be installed inside the concrete collar and attached with a 90 degree Hobas pipe elbow to the drop pipe. This connecting pipe will be at a 90 degree angle to the vertical and oriented at 75 degrees to the alignment of the tunnel in the horizontal plane directed downstream. The bottom of the drop pipe and the connecting pipe will be encased in concrete, and the remaining length of the drop pipe will be encased in grout. Typical cross section and plan view of the drop manholes is shown in Figure 5.

A tangential inlet will be utilized to transfer the incoming flow from the inlet pipe to the drop pipe. This tangential inlet will be a cast in place concrete open channel having a tapered rectangular cross section. This open channel will have a sloping base and will have a width equal to the diameter of the inlet pipe at the inlet pipe end, and a width of 305 mm (12 inches) at the drop pipe end. Figure 5 shows the details of the tangential inlet. This open channel will be tangent to the drop pipe, consequently allowing the incoming flow to have a helical action going down the drop pipe to reduce velocity.

The top manhole riser section is the portion of the manhole from the ground elevation to the inlet pipe. This portion is composed of 2.44 meters (8 feet) diameter Hobas pipe sections similar to those used in the access manholes with the exception of a "doghouse" type cut at the bottom. This "doghouse" cut is made to accommodate the incoming flow pipe that will be in place during installation of the manhole riser sections.

TUNNEL PRECAST CONCRETE SEGMENTAL LINER

Contract Guidelines

The tunnel was lined with precast concrete segments having a final inside diameter of 4.26 meters (14.0 feet). These precast concrete bolted, gasketed segments acted as the final liner rings in the tunnel. The original design was to utilize tapered 1.21 meter (4.0 feet) long rings; each ring comprised of six segments each having at least one grout/lifting hole per segment. The tapering in the ring was to provide

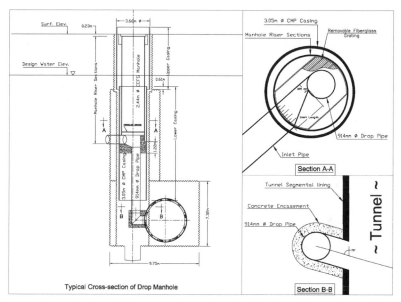

Figure 5. Typical drop manhole layout

changes in tunnel alignment and elevation. The contract called for 28 day compressive strength of 41 MPa (6,000 psi) and stripping strength of 17 MPa (2,500 psi). The curing of the segments was required to be done on two stages, primary and secondary curing. The segments were primary steam cured while still in the molds until the minimum stripping strength was attained. The segments were then moist cured for a minimum of five days.

Change in Ring Length

There were many reasons to change the geometry of the rings to utilize a universal type ring having a length of 1.52 meter (5.0 feet) in lieu of the 4.26 meter (4.0 feet) length design provided in the contract. Using a 1.52 meter (5.0 feet) ring resulted in less total rings, and allowed for roughly twenty percent fewer joints in the tunnel liner that could lead to possible leak locations. Furthermore, the smaller total number of segments to be erected resulted in less labor costs due to twenty percent less rings that had to be erected. Using a universal type ring also facilitated employing one type ring that can be installed in multiple arrangements to allow the liner to be steered in any direction forming the curves in the tunnel alignment.

Design of the Universal Ring

A joint venture between CSI and Hanson pipeline (CSI Hanson, J.V.) was formed to manufacture the segments, which in turn employed Hatch Mott MacDonald (HMM) to perform the design of the tunnel lining. The lining ring chosen was comprised of tapered segments which included four 67.5 degree segments and two 45 degree segments. The nominal length of each ring was 1.52 meters (60 inches), having a taper of ±19 millimeters (0.75 inches). Thus the ring was 1.54 meters (60.75 inches) long at the widest spot and 1.50 meters (59.25 inches) long at the thinnest location. The lining thickness was 229 mm (9 inches). Figure 6 shows a cross-section of the universal tunnel ring.

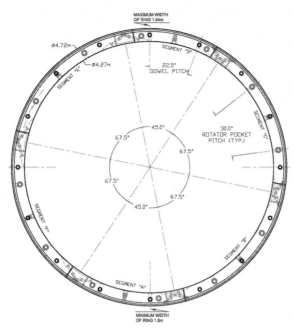

Figure 6.   Universal ring cross-section

Each segment was provided with an EPDM gasket to resist water ingress. The individual segments were connected by self-locking dowels along the circumferential joints and by steel bolts with plastic inserts along the radial joints. The main purpose of these connectors was to ensure that the gaskets remained compressed throughout all the phases of the ring erection and the erection of the subsequent rings. The dowels also minimized relative segment movements due to unbalanced loads during the installation and grouting processes. Finally, a non-return valve was threaded into the grout ports to prevent grout inflow during tunnel backfill grouting.

Special Segments for the Settlement Sensitive Zones

Although the tunnel runs through farmland with no significant buildings or structures nearby, there were three settlement sensitive zones along the alignment. A four lane highway (U.S. 23), a dual railroad line crossing, and a 2-lane highway (S.R. 317) were critical surface areas. Special segments with two extra grout holes per segment (except for the key and counter-key pieces) were used at these locations. The additional grout holes were used to optimize the grouting pattern for settlement control. Segments with extra grout holes were also used at the access and drop manhole areas to enhance the grouting capabilities at these locations.

Big Walnut Creek Crossing

The tunnel alignment crosses under the Big Walnut Creek, at which location the crown of the tunnel is roughly 2.1 meters (7 feet) below the creek bottom. Due to the shallow ground cover at this area, possible water pressure surcharge of the tunnel may occur which mandated reanalyzing the segmental lining at this location to handle the additional internal forces. It was found that the segments at this location had to be

Figure 7.   CSI/Hanson segment casting facility

equipped with additional embedded galvanized plates under the bolt head. Moreover, steel bolt insert threads had to be used instead of the plastic and higher strength bolts were also utilized. A protective concrete slab was cast in place at the bed of the creek over the tunnel envelope. The slab was necessary to prevent buoyant uplift of the tunnel and also provide long-term protection for the riverbed and the tunnel from scouring the river channel. The concrete slab was installed prior to excavation of the tunnel.

Precast Concrete Segment Casting Facility

The concrete segments were manufactured at the Hanson plant located at a close proximity of the job site, utilizing the "carousel system." The carousel production process used 48 segment moulds, eight sets of each six separate pieces of the ring. The steel moulds were oriented with the intrados of the segments facing down. The moulds moved along the production line from the first station where the segment was demoulded to the last station where the extrados surface of segment is trowel finished. The moulds were then transferred alternating to two primary cure lines. The moulds indexed in the opposite direction through the temperature zone steam kiln to emerge at the opposite end five hours later, where they return back to the production line to begin the casting process again. The kiln had a total capacity of 48 moulds on two tracks. The casting facility is shown in Figure 7.

At the start of the production line, the segments were removed from the moulds using an overhead crane equipped with a rotator beam. The segment was then rotated 180 degrees (intrados facing up), and moved to the segment gasket installation and finishing area, where it was stacked on dunnage as full rings in sequential order. As each segment was stacked, adhesive was applied to the gasket grove and the appropriate gasket was fitted in place. Once the ring stacks were complete, they were transported to the secondary cure area for a period of five days before being transferred to the stock yard where they remained until delivered to the job site. The production line was fully automated, after the first stage where the segment was removed from the mould, the mould automatically moved down the production line track to the next station to be cleaned and oiled. The mould was then bolted together, mould embedment parts were fitted, and the reinforcement cage was set in the mould. At the next station, the mould was gauged and checked for embedment content to ensure it passed quality control

approval before mould covers were bolted on, and the mould was advanced to the concreting station. At the concrete filling station, the operator hydraulically clamped the
mould to the vibration table and started filling the mould with concrete. As the concrete
was fed into the mould, the vibration table was activated and filling continued until the
concrete was properly consolidated. The mould was then unclamped and lowered on the
production track to proceed forward to the next station. After the concrete had stiffened
sufficiently, the mould covers were unbolted and raised to allow a rough trowel of the
extrados face of the segment. The moulds continued along the production line until the
concrete received the final trowel finish. Moulds were then advanced to be placed automatically on alternate kiln lines were they stayed until the concrete reached its stripping
maturity approximately five hours after the production cycle had begun. The reinforcement cages comprised of steel ladders, curved bars, and fabric mats which were all
assembled together by tack welding at the CSI/Hanson segment plant. The cage assembly stations were located next to the production line and were connected by an overhead
crane delivery system.

TUNNEL BORING MACHINE

The tunnel boring machine selected for the project was an Earth Pressure Balance TBM built new for the project from LOVAT, Inc and is a Model RME192SE
(Figure 8). The TBM was named "Mary Margaret," named after the Project Manager's
mother. The contract specifications mandated using either a new Earth Pressure Balance or new Slurry Tunnel Boring Machine. Throughout the bidding process both types
of machines were considered, but the advantages of using an Earth Pressure Balance
TBM to excavate the cobble and boulder laden ground were the deciding factor. There
was estimated to be over 226,000 cobbles 305 mm (1 foot) in diameter and smaller
during the drive and the EPB Tunnel Boring Machine could easily digest those rocks
whereas a Slurry TBM would have to grind up that entire quantity. Furthermore,
according to the GBR, we were to expect roughly 7,000 boulders over 305 mm (1 foot)
in diameter along the tunnel alignment. This amounts to one boulder expected every 3
feet along the tunnel drive. The primary emphasis in the design of the TBM was to
excavate the cobbles and boulders in an efficient manner. A more detailed paper on
the TBM design and capabilities, with the focus strictly on excavating the cobbles and
boulders, is printed in this RETC volume (DiPonio, Chapman, and Bournes, 2007).

TBM General Arrangement

The BWARI tunnel had an excavated diameter of 4.89 meters (16.04 feet). The
overall length of the tunneling machine was 97.8 meters (320 feet). The cutterhead
and shield sections made up the first 9.6 meters (31.5 feet), while the muck discharge
areas and trailing gear comprised the rest. The general arrangement drawing of the
tunnel boring machine is shown in Figure 9.

Many design characteristics of the tunnel boring machine were chosen based on
the results of the Northeast Interceptor Sewer Project (NEIS) and North Outfall
Sewer—East Central Interceptor Sewer (NOS-ECIS) project, both of which were completed in Los Angeles, CA using Lovat tunnel boring machines during the design
phases of the machine for the BWARI project. Due to problems experienced on the
NEIS project with unbalanced trailing gear riding up the walls of the tunnel as the
machine advanced, we chose a trailing gear design that was rail mounted. This decision
complicated matters somewhat due to the fact that we needed a ramp at the rear of the
TBM as well as a dedicated rail installation area in the gantry sections. Due to the size
of the tunnel, the increased size available over the smaller tunnels allowed the use of

Figure 8.   TBM during final testing

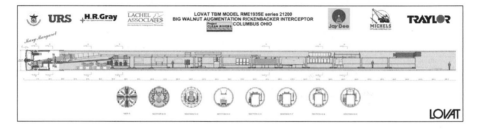

Figure 9.   TBM general arrangement

this type of trailing gear. The use of a solid deck through the trailing gantries provided a clean, easy accessible work area as well as increased visibility and an easy base for cleanup of spilled muck from the conveyor belt. The solid deck also provided a more stable structure over the open floor option. A similar backfill grouting system was also chosen as was used on the NEIS project based upon the success of those tunnels.

The TBM was designed to mine under earth pressures ranging up to 3.0 bars, which exceeded the maximum expected earth pressures expected along the tunnel alignment which were up to 2.0 bars. Average earth pressure throughout the project was roughly 1.4 bars.

TBM Cutterhead

The mixed face cutterhead of the TBM has a diameter of 4.89 meters (192.5 inches) and was equipped with 32 ripper teeth and six 30.4 cm (12 inch) diameter disc cutters as the initial startup configuration. Thirty one of the locations utilize adaptor boxes and can be replaced to use disc cutters if needed. Five of the ripper locations around the perimeter of the cutterhead were not interchangeable with disc cutters. The outer most perimeter ripper tooth was plumbed to a hydraulic system and used as a wear indicator. The cutterhead is a single piece design with four main spokes and four blinds. The eight mucking locations can all be closed off with flood doors to provide face isolation and ground control. There were three sets of grizzly bars in which the outer two were used during tunneling. The inside smaller set was left off in order to provide a free path for the muck to enter the plenum without interference. Replaceable scraper teeth, 88 in total, lined the spokes on both sides. A replaceable center nose cone with carbide teeth protected the center portion of the cutterhead from wear. The cutterhead also featured five soil conditioning ports evenly spaced across the TBM face for injection of foam. The entire cutterhead face and rim was plated with 19 mm (0.75 inch) thick abrasion resistant chromium carbide plating, to prevent accelerated wear of the structural steel. Additional cross-hatched hard facing was installed on the perimeter of the cutterhead which experiences the most wear. Two probe drill ports were also available if any probe drilling was deemed necessary during the tunnel drive. The probe drill capabilities were never used during the project.

Main Drive System

Due on recent successes of the Lovat EPB tunnel boring machines in Los Angeles, it was recommended that the BWARI TBM was powered by electric drive motors controlled by Variable Frequency Drives (VFDs). These electric motors were chosen based on the need to have as much horsepower and torque available in order to break any boulders encountered. The inefficiencies of hydraulic motors and the heat generation that results from those inefficiencies were not desirable. The VFD drives chosen, as well as most of the electrical gear on the TBM was manufactured by Cutler-Hammer. The use of Cutler-Hammer electrical products was chosen as they are based in Ohio and could provide immediate support in case of problems.

The eight electric main drive motors, each 113 kW (150 Hp) provided a combined 894 kW (1,200 horsepower) to the cutterhead and were water cooled. The maximum torque was 561 metric tons (4.06 × 10E6 ft. lbs) at 1.56 rpm. The peak starting torque was 701 metric tons (5.06 × 10E6 ft. lbs). The electric main drive motors proved to be very worth as they had no difficulty excavating the cobbles or breaking the boulders during the tunnel drive. Tuning of the variable frequency drives took a fairly long time at startup. Electrical assistance from Lovat technicians was critical to minimizing the down time.

Main Bearing Design

The bearing chosen for the TBM was a triple roller design. By contract specifications it had a 10,000 hour design life and was also replaceable from within the tunnel. The bearing seal system was manufactured by Lovat and Merkel, and was a positively pressurized automatic PLC controlled system that used the EPB sensors for pressure inputs. A very large bearing was chosen in order to provide high torque and advance capabilities to the tunnel boring machine in order to dispose of the cobbles and boulders.

Screw Conveyor Design

Due to the large amount of cobbles and boulders present in the glacially deposited soil, a very large screw conveyor was necessary. A 0.91 meter (36 inch) diameter screw conveyor was chosen due to the fact that a 292 mm (11.5 inch) rock can pass through the screw conveyor without blocking. The total length of the screw is 12.2 meters (40 feet), and uses an externally mounted hydraulic drive motor for power. The speed capability was 0–18 rpm and was variable and reversible. The rough capacity of the screw conveyor is 300 cubic meters per hour at maximum revolutions per minute. The shaft of the screw was retractable so it could be completely withdrawn from the plenum if needed. Two caliper doors were mounted inside the plenum and could completely seal the screw conveyor opening if ground conditions mandated. There was a discharge chute and sliding guillotine door at the rear of the conveyor housing to control the discharge of the sloppy muck. In anticipation of large quantities of cobbles, the front portion of the screw was replaceable and was also plated with abrasion resistant chromium carbide as well as lining the front of the auger housing. Two earth pressure sensors were located inside the screw housing to monitor conditioning of the soil. Two inspection hatches were also installed in critical locations. A third inspection hatch was installed during the tunnel drive at the screw box joint between two sections of the screw. A rock lodged in this location during the early stages of the mining process.

The location of the screw conveyor front is partially up the face of the plenum chamber bulkhead. This location was chosen because the TBM needed a very large bearing in order to provide the thrust capacity to break large obstructions that may be encountered. The location of the screw tip off the bottom also provides protection in case cobbles and boulders concentrate at the bottom of the plenum between mining pushes. This geometry also provides a flatter angle to the screw conveyor, which allowed a flatter slope for the secondary belt conveyor. This flat slope aids in the disposal of wet material.

Soil Conditioning System

The cutterhead included five soil conditioning ports in which to inject soil conditioning materials through. Four of the ports were plumbed to the foam injection system, while the perimeter port was plumbed to a bentonite slurry injection system. Each foam port had its own dedicated pumping system. This system allowed the use of different conditioners through each port as well as different flow rates. This capability was very important due to the changing conditions experienced along the tunnel drive. The Geotechnical Baseline Report (GBR) for the project provided very detailed geologic conditions throughout the alignment. Using the results of the preconstruction borings that were taken, we developed a chart that lists the anticipated Foam Injection Ratios (FIR) necessary at each location along the alignment. To calculate the values, we followed the methodology developed by the Rheological Foam Shield Tunnel Association of Japan. This methodology was also used and further developed by Traylor Bros., Inc during the construction of the South Bay Ocean Outfall in San Diego, CA as described in a 1999 RETC paper (Williamson, Traylor, Higuchi, 1999). The chart in Figure 10 shows the calculated foam injection ratio for each portion of the alignment. A commonly accepted upper limit for the potential use of foams as the sole method of soil conditioning is in the fifty to sixty percent range.

The alignment of the BWARI tunnel includes zones of soil comprised solely of clean gravels and cobbles without the presence of any fine material. It was immediately recognized that the highly permeable zones of glacially deposited gravels would take very large quantities of soil conditioners in order to excavate them in a controlled manner. The capability of injecting bentonite slurry as well as polymers was critical to the mining success.

Roughly fifteen percent of the tunnel was comprised of hard pan which required foam or water in order to excavate. We found that the injection of large quantities of water was the most successful mining method, while a very small amount of foam was added for lubrication. Roughly sixty percent of the tunnel alignment zones consisted primarily of sandy material which would flow relatively easy but required foam in order to provide lubrication and to provide the proper fluidity to the muck in order to balance the earth pressure. The remaining twenty five percent of the tunnel alignment was made up of clean gravels or cobbles that required large amounts of foam as well as

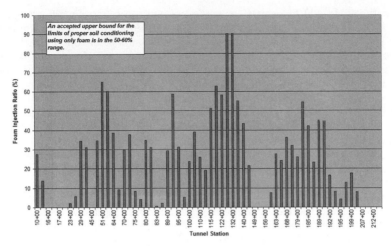

An accepted upper bound for the limits of proper soil conditioning using only foam is in the 50-60% range.

Figure 10.   Calculated foam injection ratio along the tunnel alignment

additional polymers or bentonite. These highly permeable zones were very difficult to advance through and maintain consistent earth pressure. We found the injection of large amounts of foam and roughly three percent (by volume) of polymer (Soilax P) would provide a mass that had the proper consistency to excavate. Early on in the drive we tried injecting bentonite slurry mixed with polymer and experienced poor results. The most successful conditioning method was found to be injecting the polymer through a bulkhead port into the plenum adjacent to the screw conveyor pickup location. The injection was handled by a pneumatic pump close to the bulkhead wall that was fed by a small bucket of polymer.

Generally in the silty sand zones a minimum of twenty percent foam injection ratio was maintained. This foam ratio was found to provide a good lubrication to the soil and aid in flushing the material towards the perimeter of the TBM where the scrapers would bring the material into the plenum. The foam provided critical lubrication to prevent wear to the steel components of the cutterhead and screw.

The bentonite injection system was used throughout the initial twenty percent of the tunnel alignment. The bentonite slurry was pumped using a large Moyno pump to distribute the slurry around the perimeter of the TBM through the outer most soil conditioning port on the cutterhead. The purpose was to provide a material around the TBM as it advanced that would cut down on the shield friction generated during the mining advance of the TBM. We visualized that the bentonite slurry would also provide some waterproofing capabilities to cut off ground water between the outside of the shields and the in-situ soils. Partway through the drive we found that the swivel had been damaged and up to 213 meters (700 feet) of the tunnel was mined without using the bentonite slurry. No noticeable difference was found in the mining rates or pressures, so the system was not re-enabled for the remainder of the tunnel drive.

Backfill Grouting

As the TBM advances, the annular gap between the extrados of the segments and the soil had to be filled with backfill grout at a rate necessary to completely fill the void as the tunnel advanced. This mandated having an adequate, and possibly variable supply of backfill grout at the heading at all times. A two-part grout mix was chosen and was stored in a large quantity grout reservoir mounted on the TBM trailing gear. Due to the long supply lines, unpredictable nature of tunneling, long maintenance periods, and weekends/holidays, the grout needed to remain liquid in the holding tank and delivery lines for a long period of time before being injected into the ground. It was predicted that a gel time of ten days would be sufficient. The first component, Part A (grout) was comprised of cement, fly ash, bentonite, water, and stabilizer. The second accelerator component, Part B, was sodium silicate.

The two-component backfill grout was pumped from the surface into holding tanks on the TBM trailing gear. The grout holding tank had a 6,000 liter capacity and included three agitation mixers. The accelerator tank had a 1,200 liter capacity. All liquid tanks on the trialing gear had capacity for more than two complete mining advances in case of problems with the delivery systems from the surface. The grouting process was not tied to the muck removal train so there was time savings realized during each mining cycle. In the highly permeable tunnel zones, it was also envisioned that these areas may require more grout to be pumped if it traveled away from the tunnel annulus into the adjacent soils. The extra grout quantities on the trailing gear provided backup in case this situation ever materialized. With a standard grouting setup using grout cars for supply, the volume available is always limited.

The grout pumps used for backfill grouting were peristaltic pumps. The pumps were highly reliable and the only maintenance needed during the complete mining process was a single change of the internal hoses. The pumps were VFD controlled via the PLC systems on the Tunnel Boring Machine trailing gear. A remote Allen Bradley panelview was located on the first trailing gear gantry so the system could be monitored by the operator which had a clear view of the grouting packers. Dual pumping systems were installed on the machine in order to provide redundancy in case of a pump failure. Each system was sized in order to provide complete filling of the annular void between the extrados of the concrete segments and the ground as the TBM advanced. The grout was injected through 2 inch diameter grout hoses and was pumped through custom grout packers threaded into the grout port located in the center of each concrete segment. The grout accelerator was pumped through one inch diameter hoses and mixed with the grout in a static mixer, part of the custom grout packer. Flowmeters and pressure sensors were installed on each line to provide monitoring and logging capabilities to the system. For each mining advance, 1,892 liters (500 gallons) of grout was injected behind the ring for the 1.52 meter (5 feet) advance. This volume used was the theoretical volume between the extrados of the concrete segments and the excavated ground, plus an additional ten percent. The grouting system had a preset maximum pumping pressure cut off level of 413 kPa (60 psi) to prevent point pressure loading of the segments to resist ring deformation. Under standard conditions, the two grouting systems would pump the backfill grout in the closest grout hole to springline on each side of the tunnel. The ring grouted through was the closest ring to the tail seal brushes, which provided the opportunity to fully fill the annulus as the machine advanced. Once mixed, the grout has a gel time of 40–45 seconds, but remained movable by pressure for an additional twelve to eighteen minutes. This allowed enough time for the grout to fill the entire void behind the liner. Final set time was in the range of 40–45 minutes. The accelerator was injected at roughly a ten percent ratio to the grout, but was adjusted if the grout was seen flowing into the tail shield. The compressive strength of the two component grout averaged 190 psi, well above the contract specified 28-day strength of 1.0 MPa (150 psi). Throughout the mining, ring diameters were surveyed and recorded. The grout performed very well and provided the support to keep the rings within 10 millimeters of perfectly round on average.

Compressed Air Capabilities

The contract specifications mandated the capability of using a compressed air system in order to perform cutterhead maintenance interventions without surface settlement or depressurization of the ground water table that would impact local residential water wells.

Early on in the project, we proposed the use of a compressed air method that uses the complete tunnel a compressed air chamber, instead of only using the TBM plenum as a compressed air chamber. The system would utilize a large compressed air facility on the surface to supply large volumes of air to a compressed air bulkhead that would be installed inside the tunnel. The plant has capabilities to bring the complete tunnel up to one bar pressure within a 24 hour time span.

There are many reasons why this compressed air method was chosen for this project. Due to the desire to have a large screw conveyor on the TBM, the forward shell area is very cramped. A standard compressed air lock as supplied by Lovat, Inc. would take up a lot of valuable space and would limit the size of the essential components of the TBM, namely the screw conveyor location and size, as well as the number and dimensions of the main drive motors. Another factor that was considered in the analysis is the very permeable soils that the tunnel alignment passes through, along with the fairly low amount of ground cover above the tunnel. The permeable sands and till along with roughly eight to ten meters of ground cover above the tunnel for the first half of the alignment, there is potential for massive amounts of compressed air loss through the ground. When using a standard compressed air system (man lock and high pressure plenum), the volume of compressed air is very small. This low volume of air does not allow for large quantities of air to escape through the ground, and maintain a consistent pressure in the work area. The only way to provide a consistent level of air pressure is to provide a massive air lock in which the losses through the ground account for a very small percentage of the overall volume. Using a large lock, maintaining a consistent air pressure is more easily achieved and the workers can perform the cutter changes in a safe environment.

Tunnel Guidance System

The guidance system used on the project was a ZED 261 tunnel guidance system, manufactured by ZED Tunnel Guidance Ltd. The ZED 261 tunnel guidance system was used to accurately monitor the position of the tunnel boring machine during excavation. The system provided the operator with details of the TBM deviations from its planned alignment so that corrections could be made quickly. Communicating with the tunnel boring machine PLCs at all times, real-time alignment and steering data is continuously provided to the operator. Per the project specifications, this data was also logged throughout the mining process. The guidance system proved highly accurate, due in large part to a very knowledgeable surveying team. The tunnel remained within 305 mm (1.0 feet) of planned tunnel centerline and 102 mm (4 inches) from planned grade throughout the drive.

Ring Build System

To avoid damaging segments or becoming "iron bound," careful consideration to ring positioning has become a critical part of the process. Previously this responsibility fell upon the shift engineer. It was performed at the completion of each advance and was typically accurate for only the next 2–3 rings. With today's electronic advancements in TBMs and the desire to increase production, a computerized ring building program was desired to perform these calculations and predictions with consistency and speed. The system chosen for this project was the SLS-T system, built by VMT GmbH. The system was fully compatible with the ZED guidance system and the Lovat TBM systems. The ring sequencing software is capable of determining ring type and general orientation numerous rings in advance. An orientation recalculation was performed just prior to building the ring, at the end of each tunneling advance. The VMT and ZED systems share a common data table, ensuring seamless integration of the two systems. When the TBM deviated from the tunnel alignment, the VMT ring build module calculated the "correction curve" necessary to bring the TBM back to planned line and grade. This calculation was based on the maneuverability capabilities of the TBM, and the ring geometry. All correction curves were calculated to bring the tunnel tangentially back to the tunnel axis.

Excavated Muck Handling System

The discharge of the excavated material was performed via the screw conveyor and a belt conveyor. The 12.2 meter (40 feet) long screw conveyor was the primary device to discharge the material from the plenum. The screw conveyor discharged the muck onto a 0.91 meter (36 inch) wide conveyor belt that was approximately 45 meters long. The conveyor belt, using drive units on each end, had the capabilities of moving the sloppy muck at high speeds (up to 100 meters per minute). This was very important when excavating the wet muck as the initial section of the conveyor belt was inclined. Without the high speeds, the wet muck would have slipped down the incline and piled up in the segment unloading area. At the rear discharge, there was a polyurethane belt scraper which helped keep the belt clean. As the TBM advanced, the muck would discharge into 10.7 cubic meter (14 cubic yard) capacity muck boxes, of which five were included on each muck train. The muck boxes were separated into three sections by the use of internal baffles. The baffles prevented the spillage of muck when the train stopped.

Mining Train

For the construction of the tunnel there was a single 36 inch gauge rail installed. A moveable California switch was installed at the halfway point for passing of trains. A second California switch was purchased but was not installed as it was not needed. A stationary switch and dual rail lines were installed inside the 68.5 meter (225 feet) long square cast in place box tunnel that was situated to the rear of the main mucking shaft. Sixty-six lb rail was used throughout the entire project. The rail was mounted to custom steel ties that were bolted directly to the precast concrete segments. This type of solid rail mount was chosen in order to provide a safe rail system to operate the trains at high speeds. The typical muck train consisted of a single fifteen ton locomotive, five 10.7 cubic meter (14 cubic yard) capacity muck boxes, a flat car, and two segment cars. Five muck boxes were not needed, but provided extra capacity in order to not overfill the boxes which would result in spillage of the sloppy muck.

SURFACE FACILITIES

Main Site Development

The main site design was changed from what was originally planned due to the flooding risk. We recognized the importance that all critical equipment and facilities needed to be located above the flooding elevation for the duration of the project. The essentially facilities that needed flooding protection include the electrical substation, compressed air plant, grout plant, bentonite batch plant, and foam batch plant, maintenance shop, supervisor's office, and crew wash house facilities. These facilities would be essential to the continuation of the project under high water flooding levels. A raised work area was designed to house the essential facilities.

We coordinated the construction of the main site pad with the excavation of the cast in place box tunnel. The excavated material was used to build up the main site pad to the 100-year flood elevation of 694 feet. All necessary water and electrical lines were installed prior to the construction of the facilities and substation. Three different ramps were installed to provide access to the pad facilities.

Communication and Data Systems

The entire mining process was setup to utilize automated processes and to take advantage of the Programmable Logic Controllers (PLCs) that are used to control the Tunnel Boring Machine, Compressed Air Plant, Grout Batch Plant, Bentonite Batch Plant, and Foam Batch Plant. The operation was designed to take advantage of these controls to provide information from one plant to another. The supply of the consumable tunneling liquids to the tunnel boring machine was automated and used the level sensors in the tanks to provide real-time data throughout the project. A fiber optic communication cable and various modems were maintained throughout the project to provide a communication path from the tunnel boring machine to the surface plants. Through PLC communication, the batch plants and pumps would automatically fill the TBM tanks when the low preset levels were reached. The pumping of the liquids was continually monitored and logged by utilizing pressure sensors installed in the service lines. Using the collected data, any inconsistencies in the pressures were an easy indication of line blockages, worn pumps, or other causes for concern. The automated batching processes also provided JDMT with labor savings over manually operated batch plants. A data logging system on the surface continuously recorded all relevant data to the systems.

Grout Batching Plant

The grout batching plant used on the project was supplied by TEAM Mixing Technologies, Inc. The grout plant is a PLC controlled plant that is comprised of three 90 metric ton silos, a 600 liter colloidal mixer, 1,200 liter holding tank, an admixture pumping system, and a piston pump to supply the tunnel with batched grout. The batch plant automatically batched the grout and held it in an agitated holding tank until the low tank level signal was received from the tunnel boring machine. When needed, the grout was pumped through a 2½ inch steel pipe to the tunnel heading. The grout batch plant proved to be a highly reliable plant throughout the tunnel drive. The maintenance required was very minimal, thus caused very little downtime to the mining process. The only major maintenance required was replacement of the seals in the pump, which were changed at convenient times when cutterhead maintenance interventions were being performed on the TBM. The system was designed to pump the grout for the first half of the tunnel drive, and then the plant would be moved up to manhole #4. Due to good early performance, the plant was never moved up and eventually the pumping distance proved to be too much for the piston pump and delivery system components. During the last 800 meters of the tunnel drive the grout was carried to the heading in a custom grout transfer car.

Bentonite Slurry Batching Plant

An additional bentonite slurry batch plant was added to the grout batch plant setup in order to provide large amounts of bentonite slurry to the tunnel heading for use in soil conditioning as well as injection around the perimeter of the cutterhead. The pre-batched bentonite slurry was also used for the mixing of the tunnel backfill grout.

Soil Conditioner Batching Plant

The soil conditioner batch plant was manufactured by Traylor Bros., Inc. It is an updated "weather proof" version of a soil conditioner batching plant design that was originally used on the South Bay Ocean Outfall project in San Diego, CA. The Soil Conditioning Batch Plant is a PLC controlled facility that mixes powdered Sodium Carboxy Methyl Cellulose (CMC—Soilax C) and water, along with an anionic surfactant (Soilax S), to produce a pre-batched foam solution. This solution was held in 2 agitation tanks for use in the tunnel. JDMT installed a Moyno pump on the surface to deliver the foam solution to the TBM. The PLC system in the foam batch plant was in constant communication with the PLCs on the Tunnel Boring Machine. When the low level sensors on the foam solution holding tanks were reached, the PLC sent a signal to the foam batch plant. The batch plant then automatically pumped the pre-batched solution to the tunnel heading and filled the holding tank on the TBM via a 3-inch diameter PVC pipe. Aquamine pipe was used for all the plastic pipes to provide a smooth internal surface and high pressure capability.

This foam batching facility was used in lieu of the more common method of using foam solution that is delivered to the tunnel heading via plastic totes or steel drums. The purchase cost of the batching plant was offset by the savings of purchasing the pre-batched solutions. This plant also provided the capabilities to adjust the foam solution components to better match the types of foams needed during the tunnel drive. The varying ground conditions mandated use of different soil conditioning agents. The foam batch plant permitted use of custom foam solutions on an ever changing basis.

Compressed Air Equipment

A fairly large compressed air plant was setup on site in order to provide compressed air in case pressurized face interventions were necessary. The compressed air plant was developed by Traylor Bros., Inc. and used pre-existing controls and equipment. A total of seven compressed air blowers were housed on site in three sea containers to provide large volumes of low pressure air to the tunnel. The compressed air plant and airlock were never installed or operated due to the free air cutterhead interventions that were possible during the project.

FLOODING OF THE SCIOTO RIVER

The Columbus area is essentially a storm water basin for north central Ohio. The Scioto River, the Olentangy River, and the Big Walnut Creek provide transport means for surface water to eventually flow to the Ohio River in Portsmouth, OH in the southern portion of the state. The Scioto River is the main waterway through downtown Columbus, and passes adjacent to the project site. The Scioto River sits roughly a quarter mile from the job site and the entire project site is situated in the Scioto River flood plain. Due to this, the EPA has strict guidelines on the surface development of the area. We were not permitted to install flood protection measures to protect the work site.

Figure 11.   BWARI main site satellite photo

The adjacent Pump Station construction included the construction of a 2.74 meter (108 inch) diameter sewer by cut and cover construction methodology, that terminated on the west side of the Scioto River bank. During the construction of this section of cut and cover tunnel there was a direct path for flood water to enter the area surrounding the BWARI Project. The pre-existing levee which held back any flood waters adjacent to the job site was removed by the pump station contractor during construction (Figure 11).

During 2004, there were two flooding events which brought flood waters into the work area. The first flooding event occurred in January 2004, when the initial site mobilization was proceeding. A second flooding occurred in June 2004, during the excavation for the main shaft and cast-in-place box tunnel section. The flooding covered the job site and the excavation. The flooding occurrence cost almost a two month loss to the dewatering system and drawdown that was underway at the time.

During the period of January 3–14, the Army Corps of Engineers rain gages recorded over eight inches of rainfall in the central Ohio area. This rainfall followed a fairly large snowfall period over the holidays in December 2004. The massive amounts of rainfall, typically what is expected over a two month time frame, combined with the snow melt and saturated soils to provide a dangerous flooding potential. Each river in controlled by dams located on the north side of Columbus. The heavy rains on January 3–5 were too much for the upstream dams to handle and the discharge caused the rivers to flood their banks. At approximately 8 AM on January 4th, the flood waters entered the adjacent work site and began to fill up the excavation for the Pump Station. At that time, all tunneling activities were ceased and the crews worked to secure all equipment and valuable materials to the project. A rear ventilation shaft, located adjacent to a concrete bulkhead inside the cast-in-place box tunnel, was in danger of flooding the BWARI tunnel. The construction method used was a rib and lagging shaft lined with plastic fabric as the excavation was being backfilled. The workers acted swiftly to construct an earthen dam around the shaft top to the 100-year flood elevation. This dam was built due to fears that the rib-and-lagging shaft would not be watertight and would allow the flood waters to fill the BWARI tunnel and severely damage the Tunnel Boring Machine.

Over the 24 hour time frame from January 5–6, the waters continued to rise and eventually covered the BWARI work site with 4 feet of water. The water level in the Scioto River on January 6th was the second highest level on record. Only a previous flood in the early 1900s was more damaging. The photo in Figure 12 shows the main worksite before and after the flooding. Figure 13 is an aerial photograph taken of the jobsite during the flooding.

When the flood waters were receding, more subsequent rains followed from January 9–11, 2004. The high water levels and the saturated soils could not handle any additional rain, and the river levels rose again. The work site remained under water

Figure 12.   Main shaft before and after the flooding

Figure 13.   Aerial photograph of BWARI site and southerly WWTP during flooding

until January 17th, 2004. The river level gradually dropped and allowed the waters to return to its banks. The recovery efforts of the work site were further complicated over the next few days as the receding water froze into thick sheets of ice due to very cold weather. Heavy equipment had to be utilized to strip the site of the ice sheets before tunnel work could continue. In all, the flooding events of January 2005 caused roughly a month delay to the project.

TUNNELING EXPERIENCES AND PRODUCTION RATES

Cutterhead Wear and Maintenance

We experienced relatively low cutter wear over the course of the tunnel drive once the rippers were upgraded over the standard 6.35 cm × 12.7 cm (2.5 inch × 5 inch) rippers approximately halfway through the drive. We worked with Lovat to produce larger rippers that were 7.62 cm × 15.24 cm (3 inch × 6 inch), along with larger adapter boxes to house the rippers. The larger rippers were used on the outer portion of the cutterhead for the second half of the tunnel drive. The ripper wear was greatly reduced and allowed the capability to mine from one manhole to the next, without making cutterhead interventions to change ripper teeth. The second half of the tunnel drive was under much higher water pressures, thus also making cutterhead maintenance much more difficult. We utilized the jet grouted areas around the access and drop shafts to change the rippers in a stabilized zone, with reduced permeability. The cutterhead entries were performed fairly quickly due to access from the surface as well as the tunnel heading. The TBM was advanced into the previously excavated access manhole shaft, with the inside excavated down to springline of the tunnel zone. The excavated inside diameter of the work area was 2 meters. The TBM proceeded to mine into the excavated zone and stopped once the cutterhead was in the desired location, slightly inside the excavated hole. Once the TBM was stopped and the segments were fully backfill grouted to refusal, the flood doors could be opened and the plenum was mostly emptied. The crew was able to redress the cutterhead in free air and have the use of a small crane from the surface to lower materials. This free air access to the upper half of the cutterhead provided the capability to redo the hard facing on the TBM cutterhead, as well as change the face scrapers that are damaged. Two crews were able to perform the maintenance simultaneously using this method. This method was used at all six manhole locations throughout the project. Without access to the front of the cutterhead to maintain the hard facing and scrapers, the cobbles and boulders would have greatly damaged the structure of the cutterhead over the length of the drive. The design of the jet grout blocks to allow a grout seal to be developed behind the trailing shield was a key feature to performing the cutterhead maintenance in a protected environment. During cutter changes performed in areas that were not pre-grouted, the crew used submersible pumps inside the TBM plenum, as well as running the screw conveyor as a pump in order to control the ground water inflow. The water was allowed to run out of the tunnel due to the slight 0.05 percent grade of the tunnel. Due to the lack of surface infrastructure that can be damaged by surface settlement, the compressed air system was never installed in the tunnel during the drive. Cutter changes were only performed in areas that are either pre-stabilized (using jet grouting), or areas that cannot be damaged by surface settlement. Roughly sixty percent of the tunnel alignment was through rural farmland, which would not sustain major damage from a localized surface settlement. Cutterhead maintenance was performed prior to crossing under any critical settlement sensitive areas along the alignment.

Overall Mining Production

The tunneling portion of the BWARI project took roughly 23 months to complete. The tunnel mining took place predominantly on a single shift basis, while the night shift crew was responsible for maintaining the TBM utilities and consumables. Overall, the day shift mined eighty eight percent of the tunnel footage while the night shift accounted for the remaining twelve percent. The average weekly production over the complete tunnel drive was 42 rings or 64 meters (210 feet). The peak mining production was 83 rings (126.4 meters) over the week ending February 19, 2006. The peak daily production was 26 rings on Wednesday, September 6, 2006. On average, the crew would perform a 1.52 meter (5 feet) mining cycle in thirty-five to forty minutes. The mining advance would take roughly ten to fifteen minutes, while erecting the precast concrete segments would take twenty to twenty-five minutes. The remaining five to ten minutes of each cycle was comprised of moving the trailing gear tow chains and tail shield invert cleaning before the ring build.

Tunnel Quality

Throughout the design process of the tunnel boring machine and precast segments, final quality of the finished product was the highest concern. There is a final corrosion protection lining that will be installed now that the tunneling portion is completed. The product chosen by the design team for the project is the Linabond Corrosion Protection System. JDMT recognized during the bidding process that the cost of the system components is greatly magnified by any tunnel lining defects, namely segment joint offsets and spalled concrete. Any defects would have to be filled with more Linabond materials to provide a smooth inner tunnel profile.

The contract specifications allow for up to a 6.35 mm (0.25 inch) offset in the precast concrete segment joints. All the tunneling systems were designed to achieve this joint offset criterion, with goals for the installation of the concrete segments with zero offsets. Considerable time was spent during the tunneling startup to train the crews on the precise installation of the segments. The VMT SLS-T software provided excellent data for the steering of the tapered rings, while the crew quickly learned to install the rings with great precision. The use of the stiff Sofrasar dowels provided a stable connection at the segment circumferential joints. Only one location in the tunnel was found that had measured joint offsets larger than the 6.35 mm criteria. This larger offset was due to slight ring deformation in the first curve while the steering tendencies were fine tuned. Once this deformation was recognized, adjustments were made in the mining process to provide better control of the trailing shield through the curves. Throughout the project, numerous industry representatives provided confirmation that the quality of the segment lining is on par with any comparable project in the world.

SHALLOW SEWER CONSTRUCTION

The project included two shallow sewers that were to be constructed and connected to active sanitary sewer lines after the tunnel construction was finished. This work is under construction at the time of writing this paper. The first shallow sewer (Shook Road Connector) was approximately 10 meters deep to the invert of the sewer. The original contract drawings detailed the sewer to be constructed in tunnel beneath a two lane highway (SR-317), with the balance of the alignment to be constructed using open-cut excavation methods. Due to right-of-way and easement limitations, the entire alignment was constructed in tunnel. The second shallow sewer (Eastport Connector) was approximately 8 meters deep to invert of the sewer and was to be constructed using tunneling methodology. The original contract documents detailed both sewers to be constructed using microtunneling methodology.

Due to the existence of boulders and cobbles along the alignment of both tunnels, it was determined that a microtunneling machine would not be able to mine through the ground without difficulty. Due to vast experience Jay Dee has in tunneling using hand mining methods, this excavation method was chosen for these shallow sewers. Examination of the boring logs in this vicinity, and inspection of the soil excavated during the construction of the drop manholes, the conditions proved the suitability for this construction method.

Both tunnels were excavated as box tunnel cross-section with dimensions of 1.2 meters wide by 1.3 meters high (4 feet by 4.5 feet) and was braced with 1.2 meter (4 feet) long hardwood lagging boards spaced tight throughout the length of tunnel. Alignment control of the tunneling work was established using a laser beam set to grade and located precisely on tunnel centerline down the access shafts. Grade was established from a benchmark transferred from the surface to the bottom of the shaft.

Once tunnel excavation was completed, the tunnel was lined with Hobas fiberglass pipe. The pipes were blocked as per the manufacture recommendation. The pipes were then grouted into place from one end with a structural grout mix. Prior to grouting, a bulkhead was built around the outside of the pipe at both ends of each pipe line. The grout was pumped into the annulus through a grout pipe in one end of the tunnel length. A breather pipe was installed in the top of the bulkhead on the other end of the tunnel. This breather pipe was extended approximately one to two meters above the highest elevation of the tunnel void. When grout flowed out the breather pipe, the annulus was determined to be full and grouting was complete.

ACKNOWLEDGMENTS

The authors would like to acknowledge URS Corporation, prime consultants for the project, for their cooperation and assistance, and the City of Columbus for permission to prepare and present this paper. Finally, the support of the authors' employers is gratefully acknowledged.

REFERENCES

Frank, Glen, and Chapman, David; Geotechnical Investigations for Tunneling in Glacial Soils, 2001 RETC Proceedings, Society for Mining, Metallurgy, and Exploration, Inc, Littleton, Colorado; pages 309–324.

Frank, Glen, and Chapman, David; A New Model for Characterizing the Cobble and Boulder Fraction for Soft Ground Tunneling, 2005 RETC Proceedings, Society for Mining, Metallurgy, and Exploration, Inc, Littleton, Colorado; pages 780–791.

Williamson, G., Traylor, M., and Higuchi, M.; Soil Conditioning for EPB Shield Tunneling on the South Bay Ocean Outfall, 1999 RETC Proceedings, Society for Mining, Metallurgy, and Exploration, Inc, Littleton, Colorado; pages 897–926.

CONSTRUCTION OF THE SACRAMENTO RIVER TUNNELS ON THE LOWER NORTHWEST INTERCEPTOR SEWER, SACRAMENTO, CALIFORNIA

Tolga Togan
Affholder, Inc.

Dan Martz
JF Shea Co., Inc.

Wally Chen
Parsons Brinckerhoff

Matthew Crow
Kellogg Brown & Root

David Young
Hatch Mott MacDonald

Bill Moler
URS Corp.

Steve Norris
Sacramento Regional County Sanitation District

ABSTRACT

Two 610 m (2000 ft) long tunnels were driven through water bearing silts, sands and clays beneath the Sacramento River using a 4.59 m (15.1 ft) diameter Earth Pressure Balance Tunnel Boring Machine (EPBM). Four sheet pile shored shafts were sunk with and without dewatering. The pre-cast concrete segmentally lined and grouted tunnels were driven down at 6% grades beneath the river flood protection levees, buildings, roads, a railroad and adjacent to a Freeway Viaduct. Details of the schedule, cost, shaft construction, TBM, tunnel construction, and installation methodology for the twin 1.676 m (66 in) steel force main pipes and environmental constraints are provided.

INTRODUCTION

The Lower Northwest Interceptor Sewer (LNWI) is part of the Sacramento Regional County Sanitation District's (SRCSD) interceptor system expansion from about 130 km (80 mi) to an estimated 320 km (200 miles) of gravity and force main pipelines to serve the growth projected for its service region. The River Crossings Project is part of the 30.3 km (18.9 mi) LNWI, presented in Figure 1.

The LNWI brings wastewater from the newly developing areas to the north of Sacramento and the City of West Sacramento to the regional wastewater treatment plant south of Sacramento, near Elk Grove. Case histories of the other LNWI underground construction projects are presented elsewhere (Moler et al. 2007; Doig and Paige 2006), and the preliminary design of the LNWI tunnels performed by Montgomery

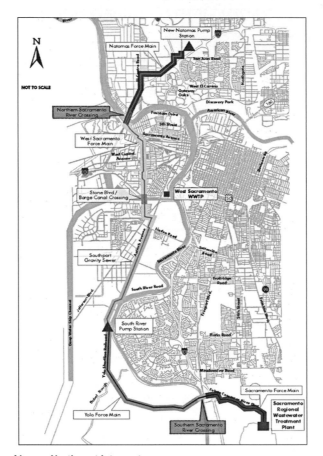

Figure 1. Plan of Lower Northwest Interceptor program

Watson Harza (MWH) and URS was presented by Nagle and Nonnweiler 2003. Program management of LNWI was performed by MWH for the SRCSD.

The paper describes the two tunneled crossings of the Sacramento River, driven with 4.59 m (15.1 ft) diameter Earth Pressure Balance (EPB) Tunnel Boring Machine (TBM). These crossings are currently the largest tunnels constructed under the Sacramento River. Construction was completed by Affholder Inc., Chesterfield, MO, with construction management (CM) for the SRCSD performed by Parsons Brinckerhoff Construction Services Inc. supported by Kellogg Brown and Root, Inc.

PROJECT DESCRIPTION

The pipeline was installed in a tunnel driven under the Sacramento River and levees in two locations, with precautions taken to avoid any ill effects on the river and its flood protection levees. The location of the Northern Sacramento River Crossing (NSRC) and the Southern Sacramento River Crossing (SSRC) are presented in Figure 1. The NSRC is located about 30 m to 120 m (100 ft to 400 ft) east and downstream of the Interstate 80 over-crossing of the Sacramento River. The SSRC is located upstream of the town of Freeport and about 1,200 m (4,000 ft) upstream of the

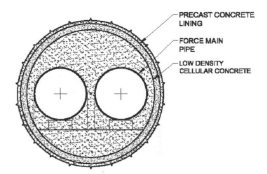

Figure 2. Aerial photograph of NSRC

Figure 3. Schematic of pipes installed in tunnel

Freeport Bridge. Aspects of the detailed design of the River Crossings Project performed by Hatch Mott MacDonald are presented elsewhere (Nonnweiler et al. 2003).

Northern Sacramento River Crossing

The NSRC, Figure 2, is about 752 m (2,466 ft) long overall, and consists of twin
1.524 m (60 in.) internal diameter pipelines.

The twin force main pipelines were installed in a tunnel with a length of 602 m
(1,975 ft) that was lined with pre-cast concrete segments. The remaining 150 m
(492 ft) of pipeline was constructed in temporary tunnel construction shafts and conventional open cut. The space between the pre-cast concrete tunnel lining and the
force main pipes was backfilled with low-density cellular concrete (LDCC). A section
view of the pipes installed in the tunnel is shown in Figure 3.

The tunnel alignment, presented in Figure 2, begins within a launching shaft on the
north side of the river. The vertical profile, Figure 4, shows the tunnel invert level at the
shaft about 8.5 m (28 ft) below grade. The tunnel continues downward at a 6% grade.

The alignment passes through a series of vertical and horizontal curves beneath
the low point in the river before ascending another 6% grade to the receiving shaft. At
its deepest point, the tunnel reaches about elevation −22 m (−72 ft) msl, providing
about 13 m (42 ft) of cover from tunnel crown to river mud line.

The land on the north side of the NSRC is used for a restaurant, single-family
homes, and un-developed city-owned and developer-owned areas. The launching
shaft is located on city-owned land in an open field adjacent to the site of a Native
American village. The levees are constructed of compacted soil with typically 2:1 (horizontal to vertical) slopes and have crest widths of about 8 m (26 ft) at elevations of
about 11.5 m (+38 ft) and 12 m (+39 ft) for the north and south sides of the river,
respectively.

The levee on the north side of the river has a cutoff wall consisting of soil, cement,
and bentonite constructed through the levee and extending 1.5 m to 3 m (5 to 10 ft)
below the natural ground level, which is around elevation +6.7 m (+22 ft). The Garden
Highway is located atop the levee north of the river about 91 m (300 ft) south of the

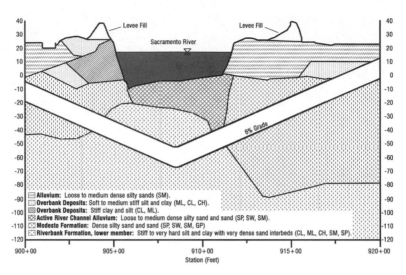

Figure 4. Subsurface profile of NSRC (launching shaft on right)

launching shaft. The south end of the NSRC is located adjacent to the sludge drying beds for the City of West Sacramento's Bryte Bend Water Treatment Plant. Between the treatment plant and the river are a city street, unused open fields and the levee.

Southern Sacramento River Crossing

The SSRC, Figure 5, is about 743 m (2,437 ft) long overall, and comprises twin 1.676 m (66 in) internal diameter pipelines. The construction details are very similar to those of the NSRC with a tunnel length of 635 m (2,082 ft) and the remaining 108 m (355 ft) of pipeline constructed in temporary tunnel construction shafts and conventional open cut segments. The tunnel construction began within a launching shaft on the east side of the river, identified in Figure 5.

The vertical profile, Figure 6, presents the shaft with the tunnel invert at the shaft about 9 m (29 ft) below grade. The tunnel continues downward at a 6% grade. At its deepest point, the excavated tunnel reaches about elevation –23 m (–76 ft), providing 10 m (33 ft) of cover from tunnel crown to river mud line.

The land on both sides of the river at the SSRC is currently being used for agricultural purposes. On the east side, the former Southern Pacific Railroad (SPRR) tracks are located along the levee crest and State Highway 160 (Freeport Boulevard), a two-lane road, running along the landside levee toe. On the west side, South River Road is located along the levee crest.

The levees are constructed of earth and have crest elevations of about +9.5 m (+31 ft). The levee on the east side of the river has a cutoff wall consisting of soil, cement, and bentonite constructed through the levee and extending 1.5 to 3 m (5 to 10 ft) below the natural ground level.

GROUND CONDITIONS

The soils encountered on both river crossings were similar and generally consisted of silts, sands, and clays with infrequent layers containing gravels. Generalized geologic

Figure 5.   Aerial photograph of SSRC

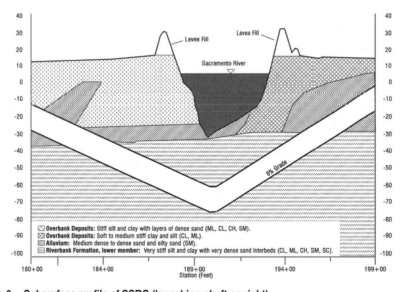

Figure 6.   Subsurface profile of SSRC (launching shaft on right)

profiles developed by the designer's geotechnical consultant, Kleinfelder, for the NSRC and SSRC are shown on Figures 4 and 6 respectively. The maximum size of gravel particles encountered in the geotechnical investigation borings was generally less than 25 mm (1 in.). However, because of reports of larger gravels, cobbles and boulders in other borings drilled in the area, and in other near-by projects, soil layers containing coarse gravels, cobbles and boulders were anticipated. The maximum clast size observed in the tunnel spoil during construction was just over 100 mm (4 in.). A detailed description of the ground conditions, potential for encountering man made obstructions, clay stickiness, groundwater levels and hazardous gasses is provided in Chen 2006.

PERMITS

In order to construct the tunnels, a variety of permits and agreements were obtained by the SRCSD or contractor from the following agencies: CalOSHA; Caltrans; City of Sacramento; City of West Sacramento; County of Sacramento; County of Yolo; Central Valley Regional Water Quality Control Board (RWQCB); California Department of Fish and Game; California Department of Water Resources; State Reclamation Board (SRB); Natomas Mutual Water Company; Reclamation Districts 307, 537 and 1000; Sacramento Regional County Sanitation District; California State Land Commission; U.S. Army Corps of Engineers; and California State Parks Department.

The key permits that allowed tunneling under the Sacramento River were issued by the SRB. The permits as issued, required flood protection collars around the shafts and confined all construction activities, including tunneling and pipelaying, between the Sacramento River levees to the summer months from April 15 to October 30 to avoid flooding.

The permit that had the biggest negative impact during construction was the National Pollutant Discharge Elimination System (NPDES) permit issued by the RWQCB. This permit constrained the Contractor's ability to dewater shaft excavations, and resulted in the NSRC launching shaft and SSRC receiving shaft being constructed in the wet. The provision in this permit that constrained dewatering flows and points of discharge into the Sacramento River included dewatering discharge restrictions at the launching and receiving sites of 1.1 ML/d (0.3 mgd) and 0.4 ML/d (0.1 mgd) respectively. This restriction had a significant impact on removal of the TBM from the SSRC receiving shaft.

ENVIRONMENTAL

An Environmental Impact Report (EIR) was prepared to cover the entire LNWI project, including the crossings of the Sacramento River. The EIR required protection of delineated wetlands and threatened species such as the Valley Elderberry Longhorn Beetle, Swainson's Hawk and Giant Garter Snake at both crossings, as well as potential impacts to cultural resources identified at the NSRC. During construction, the Contractor had to adhere to a series of Mitigation Monitoring and Reporting Program (MMRP) measures enforced by Sacramento County's Department of Environmental Review and Assessment (DERA) including preparation of a Storm Water Pollution Prevention Plan (SWPPP); Frac-Out (release of pressurized tunneling fluids to the surface) Prevention and Response Plan; Levee Protection Plan; Cultural Resources Survey and Monitoring Plan, and coordination of emergency response with local fire departments and police.

COST

Construction Contract Bids

Two bids were received from pre-qualified contractors for the construction contract and a contract for $44,453,700 was awarded to the low bidder, Affholder Inc., Chesterfield, MO. A summary of the bids and a comparison with the estimate by the Engineer is included in Table 1.

Table 1. Summary of the bids and comparison with the engineer's estimate

	Bid Price	%Diff vs Engineer's Estimate
Engineer's Estimate*	$43,478,970	—
Affholder Inc.	$44,453,700	+2%
Kenny Construction Inc.	$58,728,200	+35%

* Without contingency

Table 2. Sample unit rates derived from bid prices

Item	Quantity	Unit Rate
1.524 m (60 in.)/1.676m (66 in.) ID AWWA C200 pipe in open cut	272 m	$884/m*
Connection to adjacent projects	4 ea.	$10,000 ea.*
Shafts excavation and support	14,750 m^3	$200/m^3
4.013 m (158 in.) ID concrete segment lined tunnel	1,239 m	$1,500/m
1.524 m (60 in.)/1.676 m (66 in.) ID AWWA C200 pipe in tunnel[†]	2,477 m	$2050/m[†]

* Price for single pipe
† Price excludes cost of tunneling

Bid Price Metrics

An analysis of the bid prices indicates that the total price for time related overhead (TRO), mobilization, the Disputes Review Board (DRB) and insurance represented 23% of the bid price. Sample unit rates derived from the bid items, excluding TRO, mobilization, DRB and insurance are indicated in Table 2.

Change Orders

During the contract, eight change orders, comprising of 31 items, were issued. The final cost of the contract was $43,439,171, a decrease of $1,014,529 (2.28%) below the original contract price.

The largest additive change order items were, resolution of claims and disputes relating to control of water at the receiving shaft of the SSRC, $228,946, specification of materials testing services to be provided by the Contractor, $180,000 and unforeseen conditions requiring modifications to an existing irrigation pumphouse at the SSRC, $71,875.

Significant deductive change orders resulted from Construction Incentive Change Proposals (CICP) for use of a single TBM instead of the two specified ($1,330,085), and deletion of jet grouting at the launching shaft eye at the SSRC, where ground conditions and dewatering provided stable ground ($165,732).

SCHEDULE

The key construction schedule dates and those actually achieved are summarized in Table 3.

The combination of the schedule and shaft dewatering constraints along with the construction moratorium resulted in requiring the completion of the shafts and tunneling activities within the 2005 construction period. When the construction schedule for

Table 3. Summary of contract schedule milestones and actual dates achieved

Activity	Bid Milestone	Actual
Preliminary Design	—	2000–2001
Detailed Design	—	June 2002–Spring 2004
NTP—Construction Contract	—	August 9, 2004
Start Shaft Construction	—	January 10, 2005
Complete Shaft Construction	July 8, 2005	December 7, 2005
Start Tunneling	—	June 3, 2005
Complete Tunneling	November 16, 2005	December 8, 2005
Start Open Cut	—	February 14, 2006
Substantial Completion	September 1, 2006	July 28, 2006
Field Acceptance	November 30, 2006	August 25, 2006
Final Acceptance	—	September 27, 2006

the NSRC threatened to extend into the winter black-out period, a waiver was obtained from the SRB that allowed completion of the NSRC in 2005.

RELOCATION OF EXISTING FACILITIES

A number of existing facilities required relocation to enable construction of the project including a pump house used for irrigation of agricultural land and domestic water wells.

Agricultural Irrigation Pumphouse

It became evident during the initial mobilization at the SSRC Launching Shaft site that the profile of an agricultural irrigation pumphouse required lowering, as it blocked adequate sight distance to the only entrance to the site. The site team worked together with the pumphouse owner to avoid disruption to work by replacement of the siding on the structure with open mesh fencing to provide visibility while the design, approval, permitting and construction of the new structure was completed.

Domestic Water Wells

Three domestic water wells, located within 30 m (100 ft) of the completed force main required relocation. Well relocation involved drilling new 67 m (220 ft) deep cased wells and installing submersible pumps. In addition pressure tanks, pipework, filters and power supply were installed and the old wells abandoned in accordance with Sacramento County standards. As the wells were located close to the completed pipeline, abandonment required destruction of the existing well casing using drilling methods prior to grouting, rather than the usual use of explosives. There was a significant amount of work in obtaining satisfactory water quality results when connecting the new well to an undefined existing household system that does not fully comply with current codes for new construction. The allowances provided by the contract for the work to be performed under force account proved adequate.

Table 4. Summary of temporary construction shaft features

Shaft Location	Plan Area (m)	Depth (m)	Sheet Length (m)	Base Thickness (m)	Method
NSRC Launching	46 by 11	9	16.8	3	Wet
NSRC Receiving	15 by 8	15	18.3	1.5	Dewatered
SSRC Launching	46 by 11	9	15.2	1.5	Dewatered
SSRC Receiving	15 by 8	9	15.2	3	Wet

Figure 7. NSRC launching shaft (left) and SSRC launching shaft (right)

SHAFT CONSTRUCTION

Shoring

The shoring subcontractor, Blue Iron Inc., drove the AZ-18 steel sheet piles using an ABI T-M20 Sheetpile Rig using vibration after pre-augering holes on the line of the sheets. Typical crew size for pile driving was three. The contractor's shoring design was performed by Brierley Associates, Littleton, CO. The shafts were either excavated in the dry using dewatering prior to casting a reinforced concrete base slab or where dewatering discharges were limited, in the wet, casting a mass concrete base slab underwater as summarized in Table 4.

Excavation of the shafts underwater, Figure 7, required monitoring of the groundwater table and the level of water in the shaft to ensure a positive head on the base of excavation. Prior to placement of the underwater concrete, the subgrade level was checked by an apparatus attached to the excavator arm, surveying from the surface, and underwater observations by divers. Additional trimming was undertaken as necessary. Placement of the underwater concrete used a conventional concrete pump and boom.

In both wet and dry cases a Kobelco 330 excavator was used for initial excavation and a Kobelco 290 excavator with a long reach arm to complete the deeper parts of the excavation. Typical crew size for excavation and support was four.

The 100 year flood level, for the Sacramento River confined between its levees as predicted by the United States Geological Survey can be up to 4 m (13 ft) above the surrounding area. Flood protection shoring was provided around both launching shafts and the SSRC receiving shaft to prevent flooding should inundation of the tunnel occur coinciding with high river levels. The flood protection was designed with a 0.9 m (3 ft) freeboard above the 100 year flood level and comprised 6.1 m (20 ft) long AZ-18 Steel sheet piles supported by earth berms. The completed flood protection shoring and fill can be seen in the right side of Figure 7.

Figure 8.   Jet grouting equipment—batching plant (left) and drilling columns (right)

Backfilling of the excavation utilized similar excavation equipment, with compaction using a combination of an excavator mounted roller and a 10T Ingersoll Rand vibrating sheepsfoot roller. Typical crew size for backfilling was three.

Dewatering

The contract permitted limited dewatering of 1.1 ML/d (0.3 mgd) at the launching shaft sites and 0.4 ML/d (0.1mgd) at the receiving shaft sites for shaft construction, tunnel nuisance water and open cut excavations. Additional dewatering capacity was made available at all sites by utilizing spare capacity and dewatering mains from adjoining contracts on the LNWI Program. Dewatering wells were installed outside of the sheet pile shoring. The recently completed Natomas Force Main (NFM) was used to convey dewatering discharges from the north side of the NSRC to the existing Natomas Pump Station. The Yolo Force Main (YFM) dewatering conveyance line was used to convey dewatering discharges from the west side of the SSRC to a temporary dewatering water treatment facility at the South River Pumping Station, prior to discharge into the Sacramento River. No adverse effects were observed on domestic water wells within 76 m (250 ft) of the dewatering, due to the high soil permeability and close proximity to the river.

Jet Grouting

Jet grouting of the launching and receiving eyes at the shafts was performed by a subcontractor, Layne GeoConstruction, Inc. using a Comacchio MCl200 drill rig and on site batching of materials. Typical crew size was five. Figure 8 presents the typical set-up for the jet grouting operations.

TUNNEL CONSTRUCTION

Tunnel Segment Fabrication

The tunnel lining was formed of gasketed, precast concrete segments that were manufactured by a subsidiary of Affholder at their facility in Sylmar, north of Los Angeles. The reinforced concrete segments were initially cured using electric heating elements attached to the molds prior to stripping and final curing. Two rings complete with packings, dowels and gaskets were shipped on each truck for the 576 km (360 mi) to Sacramento. Figure 9 shows the precast concrete segments and finished tunnel lining.

Figure 9.   Precast concrete lining—straight-sided segments with tapered key

Table 5.   Features of Lovat RME 181 SE

Overall Geometry	Power	Segments
Cut diameter: 4.622 m	Total power available: 800kW	Segment ID: 4.013 m
TBM length: 8.3 m	Power to cutterhead: 200 kW	Thickness: 223 mm
TBM weight:145 tonnes	Variable Frequency Drive	5 piece parallelogram/
Minimum radius: 100 m	Max cutterhead torque:	trapezoidal ring with
Open Face Area: 30%	4.2 MNm (0.97 to 4.65 rpm)	trapezoidal key
	Max. propulsion thrust:	Length: 1.220 m
	1540 tonnes at 340 bar	
Annular Grout	**Cutterhead Tools**	**Screw Conveyor**
Grouting through segments by	Nose cone	Length: 9.5 m (Primary)
two pairs of 7.5 HP/1 HP	30 Ripper teeth	Length: 13 m (Secondary)
Peristaltic pumps for two part	interchangeable with disc	Diameter: 610 mm
accelerated cement fly ash	cutters.	
grout supplied by fixed lines.	64 Scraper teeth	
	Cutterhead carbide plated and	
	tools with carbide inserts.	
Spoil Removal	**Soil Conditioning**	**Plenum Monitoring**
Secondary belt conveyor	Injection ports:	6 ea bulkhead mounted total
discharging onto fixed	Cutterhead face—5	pressure cells
conveyor discharging to	Screw conveyor—4	
surface. Segments brought in		
by diesel powered train.		

Earth Pressure Balance Tunnel Boring Machine

The contract required tunneling with an EPB or Slurry TBM. Affholder chose to use a refurbished Lovat EPB-TBM with the features presented in Table 5 and Figure 10.

Tunnel Excavation

The tunnels were driven with a 4.59 m (181 in.) OD Lovat RME181SE EPB-TBM (Figure 11) operating in closed mode while maintaining a positive face pressure, with removal of excavated soil from the plenum of the TBM using a screw conveyor. Following discharge onto the TBM secondary belt conveyor, a Robbins conveyor fixed to the tunnel crown transported excavated soil from the tunnels down and up the 6% tunnel

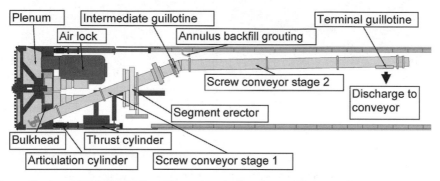

Figure 10. Lovat EPB-TBM (4.59 m OD) used on the Sacramento River crossings

Figure 11. TBM "Mary 1" in Lovat workshops, Toronto, Ontario, Canada

grades, then up the shaft to the surface. Segments, other materials and utilities were brought in by rail using diesel locomotives. Typically the crew size was fifteen on each ten hour shift. The performance of the TBM is reported elsewhere by Chen 2006.

The contractor elected to drive the SSRC followed by the NSRC. First the launching shaft for the SSRC tunneling on the east side of the river was constructed. The tunnel was then driven westwards under the Sacramento River while constructing the receiving shaft on the west side of the river. Mining started on June 3, 2005, reaching the receiving shaft on August 23, 2005, a duration of 12 weeks. The launching shaft for NSRC tunnel on the north side of the river was constructed prior to completion of the SSRC. The NSRC tunnel was driven south while constructing the deeper receiving shaft. Mining started on October 17, 2005, with hole-through at the West Sacramento receiving shaft on December 8, 2005. A maximum weekly production of 152 m (500 ft) a week was achieved, with 24 m (80 ft) being achieved in a single 10 hour shift. Higher overall production rates were not realized due to the short overall length and external logistic problems. Off site soil disposal using regular truck haulage required mixing the excavated soil with dry material or stockpiling on site until dry weather permitted transport.

Removal of TBM from Receiving Shafts

The contract required jet grouting in combination with secondary and tertiary grouting into the tunnel eye to provide groundwater and ground loss control during hole-through. At the SSRC receiving shaft, the jet grouting work was completed along with some probe drilling and supplementary grouting into the tunnel eye. The functional performance of the jet grouting and subsequent grouting into the tunnel eye to control ground water during hole through was difficult to attain, which created a situation that required the TBM to stay in place in the tunnel eye until the groundwater and piping sand could be controlled by dewatering.

Since this occurrence was prior to the construction variance that was later to be granted by the SRB, removal of the TBM from the SSRC receiving shaft was time critical to facilitate a prompt completion of the NSRC. The SRCSD and their program and CM team worked with the contractor in solving the problem. Although the contract limited dewatering discharges to 0.4 ML/d (0.1mgd) from the site to a discharge point near the South River Pumping Station, use was made of excess capacity in the adjacent project, in the YFM dewatering conveyance pipeline. In addition, the SSRC Contractor assisted the YFM Contractor in constructing four temporary earthen lagoons, with a capacity of 185 ML (47 mg), to temporarily store the excess dewatering flows and help mitigate the dewatering impacts to the YFM Contractor.

The groundwater was successfully managed via dewatering with the added capacity. It was inconclusive whether the use of jet grouting with supplementary grouting into the tunnel eye can control groundwater and running sands during hole-through. A clearer means of demonstrating achieving functional performance of the jet grouting and supplementary grouting would have been of benefit, as well as use of a receiving seal as a secondary means of groundwater control.

Tunneling Beneath Sensitive Structures

Surface settlement measurements indicated that settlement was generally within the contract threshold, however higher settlement was observed at the following two locations during tunneling the NSRC.

- Settlement resulted in significant damage to an abandoned house caused by over-excavation during mining and the TBM being stationary beneath the property for a week during tunnel conveyor installation.
- Surface settlement of about 50 mm (2 in.) was observed on the levee beneath Garden Highway, to the north of the Sacramento River. The TBM had been stationary or advancing slowly due to no Sunday work and unavailability of the TBM alignment guidance system. A subsequent study found that there was no indication of significant alteration to the levee structure or stability. As no over-excavation was observed, it appears that the observed settlement was a result of relaxation of the soil around the TBM and consolidation of the loose sand above the TBM. Subsequent tunneling beneath the levees and an occupied house were undertaken without stopping and no significant settlement was observed.

PIPE INSTALLATION

Pipe Installation

Installation of the welded steel pipe manufactured by Northwest Pipe, Inc. in Portland, OR followed tunneling on both crossings. The tunnel construction utilities were

Figure 12.   Pipe installation invert (left), pipe carrier (center), LDCC placement (right)

removed, the tunnel cleaned and the invert prepared as a flat concrete surface with channel rail guides for pipe transportation prior to pipe placement in the tunnel, as shown in Figure 12. Each 12.12 m (40 ft) length of pipe was pushed into the previously laid pipe using a pipe carrier on rails cast into the tunnel invert concrete. Pipelaying commenced at the receiving shaft end utilizing access from the launching shafts.

Pipe Tunnel Transport

After lowering into the shaft with a crane, pipe joints were placed on two dollies in the shaft portal and secured with cable wrapped around both dollies. Pipe was transported down the tunnel on the pipe dollies, Figure 12, propelled by a diesel powered rubber tired telescopic handler. Each dolly was equipped with a hand operated low height 10 T Enerpac RCS series hydraulic cylinder to align the pipe. Once in place, the pipe was blocked using 300 mm (12 in.) timber blocks and wedges from the concrete invert slab to the design grade. Once at grade, the pipe was secured in place with cable tied to the invert concrete slab and pipe joints were tack welded to prevent dislocation during further pipe installation. The final arrangement of the pipes in the tunnel is presented in Figure 12. Installation of the individual pipes was staggered to enable safe access to each pipe end. Typical crew size was eight.

Pipe Joints

Apart from pairs of mechanical couplings to allow for movement at the interface between open cut and tunnel construction, all joints were welded. Field welding procedures, welders and welding operator qualifications were in accordance with AWS D1.1. All welders satisfied project qualification weld tests prior to performing welds on the project, and each weld was tested using magnetic particle and dye penetration methods by independent testing inspectors from the contractor and CM respectively.

Low Density Cellular Concrete Backfill

The carrier pipes were then filled with water after completing welding, weld testing and patching of the joints to prevent floatation during the subsequent low density cellular concrete (LDCC) placement. The LDCC was pumped from the surface to two faces either side of the low point by the subcontractor Pacific International Grout of Bellingham, WA, using on site batched material. Subsequent comprehensive probing of the LDCC indicated that it was free from any cavities.

Pipe Pressure Testing

On completion of patching and cleaning, the pipes were filled with water from dewatering wells. Air was bled from the pipe over a number of days by applying nominal pressure prior to satisfactory pressure testing to static pressures of 10.5 bar (155 psi) and 14 bar (205 psi) at the low points of the NSRC and SSRC respectively.

OPEN CUT CONSTRUCTION

In addition to the tunnel, there were 272 m (892 ft) of carrier pipe to be installed in the ground and in shafts. This open cut pipelaying at the west side of the SSRC was completed first to make use of availability of a dewatering discharge line. Pipelaying for north and south sections of open cut at the NSRC were then completed, followed by the east open cut at the SSRC when weather permitted compaction of backfill. The availability of additional dewatering discharge capacity to that provided by the contract enabled all open cut pipelaying to be completed with dewatering wells. 1:1 side slopes were generally used, however steel sheets were required for the deeper open cut at the south side of the NSRC. The typical crew size was seven, with excavation by a Cat 325 excavator. Controlled density fill (CDF) to provide support around the pipe was in general batched off site, however on site mixing of site won materials was used at the south side of the NSRC using equipment operated by Mountain Cascade, Inc.

PARTNERING

A partnering program was a requirement of SRCSD for all LNWI contracts. The program consisted of a "kick-off" session, monthly written updates, semi-annual review sessions and a final partnering wrap-up meeting. On the River Crossings contract, written monthly reports were supplemented by informal monthly lunches attended by project level management representatives of the Contractor, CM, PM, and Owner. These luncheons encouraged open communication between the partners, and were utilized to review and update "rocks-in-the-road" for the monthly partnering report.

The partnering program called for four levels of participation and issue elevation: (1) field level supervisory personnel involved with day-to-day construction activities; (2) project level management personnel; (3) program and corporate management; and (4) SRCSD, PM and contractor senior management. Issues and potential disputes were resolved at as low a level as possible. If a solution could not be reached the issue was elevated to the next higher level. This procedure worked well and all potential claims were resolved at the 2 or 3 levels, before becoming acrimonious disputes.

PUBLIC OUTREACH

The public outreach effort during construction consisted of notifying appropriate agencies of road closures and after-hours work, as well as informal coordination with project neighbors. These activities were supported by the PM's Public Outreach Manager. An "I Walked Under The Sacramento River" tunnel tour was held for invited public officials, their staff and the media to commemorate the successful completion of the Sacramento River Crossings, with over 200 people walking through the tunnel that day.

CONCLUDING REMARKS

As construction of the project was completed in the Summer of 2006 and the construction contract closed out, the following conclusions may be drawn:

- Tunnels can be driven through alluvial soils beneath the Sacramento River using an EPBM to erect a gasketed precast concrete tunnel lining.
- The sewer tunnel project was successfully completed on schedule and within budget.
- Construction risks can be managed by an owner by the engagement of designers, construction managers and pre-qualified contractor's experienced in underground construction working as a team.

ACKNOWLEDGMENTS

The Authors wish to acknowledge the organization led by Neal Allen the SRCSD and the team directed by Bruce Frost of Affholder, Inc., who constructed the project, without which there would be nothing to report on.

REFERENCES

Moler, W., Lewis, D., Crow, M., Doig, P., Cole, M., and Norris, S. 2007. Case History of Trenchless Construction at the Lower Northwest Interceptor Program, Sacramento, California. Proceedings of the Rapid Excavation and Tunneling Conference, Toronto, Canada. Society for Mining, Metallurgy, and Exploration, Inc. Littleton, CO.

Doig, P.J., and Page, A. 2006. Microtunneling on the Lower Northwest Interceptor, Sacramento, California. Proceedings of the North American Tunneling Conference, Chicago, 2006, L Ozdemir, 437–443. Taylor & Francis, London.

Nagle, G., and Nonnweiler, J. 2003. Preliminary Design of Tunnels for the Lower Northwest Interceptor, Sacramento, California, USA. Proceedings of the Rapid Excavation and Tunneling Conference, New Orleans, 2003, R.A. Robinson and J.M. Marquardt, 1,294–1,300. Society for Mining, Metallurgy, and Exploration, Inc. Littleton, CO.

Nonnweiler, J., Hawley, J., and Young, D. 2003. Design of Lower Northwest Interceptor Tunnels beneath the Sacramento River. Proceedings of the Rapid Excavation and Tunneling Conference, New Orleans, 2003, R.A. Robinson and J.M. Marquardt, 179–190. Society for Mining, Metallurgy, and Exploration, Inc. Littleton, CO.

Chen, W., Crow, M., Young, D., and Norris, S. 2006. Performance of an EPB-TBM beneath the Sacramento River and Levees on the Lower Northwest Interceptor Project, Sacramento, CA, USA. Proceedings of Canadian Tunneling Association Conference, Vancouver, Canada.

FACTORS IN THE VARIABILITY OF MEASURED SURFACE SETTLEMENTS OVER EPB DRIVEN TUNNELS IN SOFT CLAY

J.N. Shirlaw

Golder Associates

S. Boone

Golder Associates

INTRODUCTION

Soft, near normally consolidated clays generally provide ideal conditions for tunneling using Earth Pressure Balance (EPB) shields. It is easy to turn the clay into a low permeability, viscous spoil that can be pressurized to maintain the required face pressure. Despite this, there are records of relatively large, and often quite variable, surface settlements over EPB driven tunnels in soft clay. To understand this variation it is necessary to identify the various factors that contribute to settlement. Tunnels in soft clay are ideal for study these effects in the field, as an unsupported void will close almost immediately, due to the low shear strength of the soil. In the short term, deformation in a soft clay will occur at constant volume, so short term surface settlements can be related to movements at tunnel level.

In this paper, data will be used from EPB driven tunnels in Singapore marine clay. In all of the cases discussed the lining consisted of precast concreted segmental rings erected in the tail of the shield.

The marine clay in Singapore has been studied, among others, by Tanaka et al. (2001). The clay typically has a plasticity index of 40 to 60, and the undrained shear strength (S_U) can be derived from:

$$S_U = 0.22 \, \sigma_{vo}'$$

where σ_{vo}' is the current vertical effective stress.

Figure 1 shows a summary of undrained strength/depth plots used for design on two sections of Singapore's North East Line. The stability of a tunnel in clay can be assessed by comparing the stability number (N) of the tunnel with the stability number at total collapse (N_{TC}). The stability number can be calculated from the equation:

$$N = (\gamma Z + q - P_t)/S_U$$

where:

    γ = the average unit weight of the ground above the springline of the tunnel
    Z = the depth to the springline of the tunnel from the ground surface
    q = the surcharge
    P_t = the support pressure within the tunnel

The stability number at total collapse depends on the size, geometry and depth of the tunnel, and typically ranges between 5 and 9. Charts for assessing the number at total collapse are provided in Kimura and Mair (1981), based on the testing of model tunnels in a geotechnical centrifuge.

For tunnels in the Singapore marine clay, the stability number of an unsupported tunnel would be over nine if the tunnel ois more than 8m deep. As a result, a support pressure

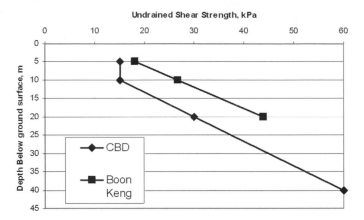

Figure 1. Examples of undrained strenght/depth values used for different sections of the North East Line

has to be provided at all times to avoid collapse of the tunnel. The stability number is so high that the marine clay will squeeze immediately into any unsupported opening.

It is common to convert settlement into 'volume loss.' Volume loss is the volume of the surface settlement trough, over 1m of advance, divided by the volume of 1m length of tunnel. This is typically expressed as a percentage. For soft clay the immediate ground movements take place at constant volume, so the volume of the surface settlement trough can be directly related to the volume of ground 'lost' at tunnel level, through overexcavation or the ground squeezing into any void around the shield or lining.

Before considering the measured settlements over tunnels driven through the marine clay in Singapore, the various possible sources of settlement during EPB tunneling will be considered.

POSSIBLE SOURCES OF SETTLEMENT OVER EPB DRIVEN TUNNELS

Figures 2 and 3 show EPB machines in a shaft, ready for launching. When these machines start tunneling, there will be a number of potential sources for settlement. These are:

1. At the face of the machine.
2. Due to the ground moving in on the skin of the machine. This will occur if the diameter of the size of the hole created by the shield is larger than the diameter of the tailskin.
3. Due to the ground moving into the void around the lining; the lining has to be smaller than the tailskin, so there is a void around the ring as the tail of the machine is shoved clear of the ring. Grout is injected to fill this void, but the effectiveness of the grouting depends on the method and materials used.
4. Due to the ground load causing the lining to compress.
5. Due to consolidation: if the process of tunneling generates excess positive pore pressures or if the lining leaks, there will be long term settlement due to consolidation.

These potential sources of settlement are discussed below. The potential magnitude of each source is illustrated in respect of the machine shown in Figure 2.

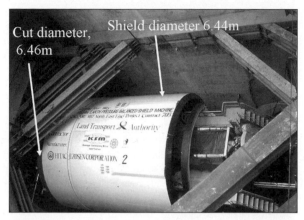

Figure 2.   EPB shield ready for lauching, showing relative size of cut and tailskin

Figure 3.   EPB shield ready for launching, showing relative size of tailskin and lining

At the Face of the Machine

If the pressure applied at the face of the machine is less than the in situ earth pressure, the ground will move in at the face of the machine. The potential magnitude of such movement can be calculated using the results of testing on model tunnels in a geotechnical centrifuge, published by Kimura and Mair (1981). They used the term 'load factor' defined as the ratio of the stability number to the stability number at total collapse. The load factor is, in effect, the inverse of the factor of safety. At a load factor of zero, the face pressure is equal to the total overburden pressure, and there should be no movement of ground at the face of the tunnel. As the load factor increases, the relative volume loss increases exponentially. When the Load Factor reaches one, theoretically there is uncontrollable squeezing of the clay at the face leading to the excavated tunnel being filled with soft clay. This occurred, on several occasions, during tunneling with open face shields in Singapore during the early phases of subway construction in the 1980s. With EPB shields, the screw conveyor housing is of limited size. This, and the ability to use the discharge gate from the screw conveyor to control the rate of discharge, have meant that there has been no case of totally uncontrolled filling of any

EPB machine and tunnel with squeezing clay in Singapore, even at very low pressures. As will be illustrated below, there have, however, been cases of very large volume loss due to tunneling at relatively low face pressures in the soft clay. Even with an EPB machine, the volume loss associated with tunneling at little or no face pressure in the marine clay in Singapore can be hundreds of percent.

The centrifuge data can be used to assess the potential for volume loss at the face of the shield, for a given face pressure. However, during EPB tunneling it is difficult to hold the face pressure constant during each advance. Typically, the face pressure fluctuates during the shield advance, with the degree of fluctuation varying depending on the ground conditions and the skill of the operator. The effect of a fluctuating face pressure is to increase the volume loss at the face, compared with what would occur at a constant pressure equal to the average of the pressures applied. As discussed in Shirlaw (2002), this is because the benefits, in terms of reduced volume loss, of the transient use of higher than average pressure are less than the increased volume loss due to the transient use of lower than the average pressure.

Due to the Ground Moving in on the Skin of the Machine

If the hole that is created by the EPB machine is larger than the shield skin, soft clay will squeeze rapidly to fill the resulting void around the skin. There are four potential reasons for the machine creating a hole of greater diameter than the diameter of the shield skin. These are:

- The diameter of the head of the machine being larger than the diameter of the skin (minimum overcut)
- The use of extendable overcutters, to further increase the excavated diameter
- The machine being driven on a horizontal or vertical curve
- The machine being driven with 'look-up' or 'overhang,' where the axis of the machine is at an inclination to the direction of the tunnel

The Diameter of the Head of the Machine Being Larger than the Diameter of the Skin (Minimum Overcut). To reduce the friction on the sides of the machine, it is common for the head of the machine to be slightly larger than the shield skin. In the case of the shield illustrated in Figure 1, the diameter of the head was 6.46m, while the diameter of the tail skin was 6.44m. The volume loss due to closure of this small overcut can be calculated as:

$$\text{Volume loss due to minimum overcut} = \frac{(3.23^2 \times \pi) - (3.22^2 \times \pi)}{(3.23^2 \times \pi)} = 0.6\%$$

This minimum overcut will close in a soft clay, unless the ground around the shield is supported, such as by the injection of bentonite around the skin.

The Use of Extendable Overcutters, to Further Increase the Excavated Diameter. Many EPB shields are equipped with the facility for increasing the excavated diameter by the use of overcutters. This facility is provided as an aid to steering. For example, if the shield is driven through a mixed face of soft clay overlying hard soil, the machine will tend to ride up over the hard soil, potentially causing the machine to go off the design alignment. One way to counteract this effect is to ream out a larger hole in the hard soil, by the use of the overcutters. The potential effect on the volume loss of the use of overcutters is, however, severe. Typically, overcutters can increase the radius of the hole cut by the machine by 70 to 100mm. In the case of the machine in Figure 2, the effect on the volume loss of overcutting by 70mm would be:

Volume loss due to overcutting by 70mm on radius =

$$\frac{(3.20^2 \times \pi) - (3.23^2 \times \pi)}{(3.23^2 \times \pi)} = 4.6\%$$

This volume loss, if the overcutters are used, is additive to the volume loss due to the minimum overcut.

The Machine Being Driven on a Horizontal or Vertical Curve. Where the machine is driven on a curve, it will carve out an oval hole. The difference between the size of the oval hole and the circular one formed when driving straight can be calculated from the following equation:

Potential volume loss due to curvature =

$$\frac{[(R + r_s)^2 + (L_s/2)^2]^{1/2} - (R + r_s) \times 100}{2r_s}$$

Where R = the radius of the curve,
          r_s = the radius of the machine
          L_s = the length of the machine; for articulated machines it is the length of
             the longest fixed segment of the machine

For the shield shown in Figure 1, for a 200m curve and a fixed length of 8.1m, then:

Potential Volume Loss due to curvature =

$$\frac{[(200 + 3.22)^2 + (8.1/2)^2]^{1/2} - (200 + 3.22)}{2 \times 3.22} \times 100 = 0.62\%$$

Generally, subway tunnels are designed for curves of 300m radius or more, and a 200m curve is extreme for such tunnels. For other types of tunnel, such as sewer tunnels, tighter curves may be used.

This potential volume loss is not additive to that from overcutting. However, not all of the potential volume loss may be realized in practice. Unlike overcutting, the 'ground loss' will not be displaced into the face of the machine. The movement of the machine around a curve will create a void on one side of the shield and displace ground outwards on the other. The disturbance to the ground involved in this process will cause settlement, but the magnitude of this settlement may not be as high as the figure calculated above. The calculated figure can be considered an upper bound for the possible volume loss.

The Machine Being Driven with 'Look-up' or 'Overhang.' This occurs when the axis of the machine is at an inclination to the direction of the tunnel. 'Look-up' is where the machine is tilted upwards, 'overhang' when it is titled downwards. Much of the weight of the machine is in the head, and in a full face of soft clay it is common to drive with 'look-up,' to counteract the tendency of the head of the machine to move downwards, due to its weight. If there is relatively stiffer soil in the invert of a tunnel that is otherwise in soft clay, the machine will tend to ride up over the stiffer material. One way of counteracting this is to drive with overhang, helping to push the machine down into the stiffer material.

Look-up and overhang are measured as the distance of the top and base (respectively) of the cutting head from a plane orthogonal to the direction of tunneling. The potential effect of look-up and overhang on Volume Loss is:

Potential Volume Loss due to overhang or look-up = $\dfrac{O_h \times L_s \times 100}{\pi r_s^2}$

Where O_h is the overhang or look up.

Using the shield in Figure 1 as an example, a 50mm overhang or look-up would give:

Potential Volume Loss due to overhang or look-up = $\dfrac{0.05 \times 8.1 \times 100}{\pi \times 3.23^2}$ = 1.24%

This potential volume loss is not additive to that from overcutting or driving around curves. However, not all of the potential volume loss may be realized in practice. As for driving around curves, the effects of overhang and look-up are not simple void formation, but a combination of ground displacement and void formation. The calculation above can be regarded as an upper bound, rather than an actual volume loss.

Due to the Ground Moving into the Void Around the Lining

At the tail of the shield, the ring is expelled from the tailskin as the machine is advanced. The ring has to be smaller than the tailskin, to allow it to be erected within the tailskin. There will therefore be a void around the ring due to the difference between the external diameters of the shield and the lining ring. Grout should be injected into this void, which will reduce the size of the void that will translate into surface settlement. However, if this is either not done or ineffective:

Potential volume loss due to complete closure of the tail void =

$$\frac{r_s^2 - r_r^2 \times 100}{r_s^2}$$

where r_r is the external radius of the tunnel lining ring.
For the shield shown in Figure 2, this was:

Potential volume loss due to complete closure of the tail void =

$$\frac{3.22^2 - 3.15^2 \times 100}{3.23^2} = 4.27\%$$

This potential volume loss is additive to the potential volume loss from overcutting, curves and overhang/look-up.

The effectiveness of tail void grouting in a soft clay, in terms of reducing this potential volume loss, depends on the method of grouting and the type of grout used. Conventionally, grout has been injected through ports in the rings. In soft, squeezing clay the ground moves down onto the rings before grout can be injected through the rings. If the grout is then injected, the grout will form bulbs or fractures in the ground. The cavity expansion/fracturing will cause heave; in the short term this will help to compensate for the settlement due to the closure of the tail void. However, cavity expansion or fracturing in a soft clay will lead to the development of large, positive excess pore pressures around the tunnel. Dissipation of those positive excess pore pressures will result in settlement. Broms and Shirlaw (1989) found that the dissipation of excess pore pressures generated in this manner, in Singapore, caused surface settlements that were similar in shape to those caused by ground movements at the face of the tunnel. The magnitude of the settlements, over an EPB driven tunnel, were equal to the total volume that would be lost due to closure of the tail void, so the net effect of the grouting on settlement, in the long term, was zero. This finding is consistent with studies by Soga et al., who demonstrated that for normally and near normally consolidated clay, the net effect of compensation grouting, after allowing for consolidation settlement, was close to zero. It is also consistent with the results of a field trial of compensation grouting in the Singapore marine clay, reported by Shirlaw et al. (1998), where the initial effect of the grouting was heave of the ground surface, but the net effect was close to zero.

By placing grout pipes along the tailskin of the machine and through the tailseals, grouting can be carried out simultaneous with the advance (simultaneous tailvoid grouting). Proper use of this method should allow the tail void to be filled as it is created. If the grout can travel around the rings, forming a uniform annulus of grout around each ring, this should not affect the pore pressures in the surrounding soil, provided that the quantity of grout injected is not excessive.

Apart from the method of grouting, the grout mix used is also important. If the grout is subject to bleeding under the injection pressure or after it has set, the part of the benefit of the grout, in terms of reducing surface settlement, may be lost.

Due to the Ground Load Causing the Lining to Compress

The settlement related to ring compression of a circular segmental concrete lining is typically very small.

Due to Consolidation

One type of consolidation settlement, due to the dissipation of positive excess pore pressures generated during tunneling, has already been discussed above. This type of consolidation settlement is just the delayed effect of the void closing at tunnel level. There is another cause for consolidation settlement: a reduction in pore pressures due to seepage into the tunnel. The consolidation settlements resulting from seepage are typically much more widespread than the settlements due to the ground movements directly associated with the tunneling, and are independent of those settlements. The potential settlement due to leakage into a tunnel in a deep deposit of soft clay is, potentially, very large. However, it should not be expressed in terms of volume loss, as the surface settlement trough associated with such consolidation settlement is quite different to that due to that normally found during tunneling. This type of consolidation settlement will therefore not be considered further in this paper.

Summary of Potential Volume Loss for Tunneling in Soft Clay.

The various potential sources for volume loss during EPB tunneling, using the machine shown in Figures 2 and 3 are shown in Table 1.

For the particular shields shown in Figures 2 and 3, the potential volume loss will be that at the face plus between 0.6% and 9.47%. The numbers quoted here are specific to that shield. Generally, the maximum potential volume loss from overcutting and tail void closure will reduce as the size of the shield increases.

A common target for EPB tunneling in urban areas is a maximum of 1% volume loss. In most soils, this is generally achievable. However, for soft clay, with a minimum overcut of 0.6% (in this example), to keep the volume loss down to less than 1% consistently would be extremely difficult.

Table 1.   Potential for minimum and maximum volume loss through various sources, example for EPB machine shown in Figure 2.

Potential source	Minimum volume loss %	Maximum volume loss %	Comment
Movement into the face	0	Infinite	
Minimum overcut	0.6	0.6	
Additional overcut	0	4.6	
Curve	0 (for no curve)	0.62 (for 200m curve)	Not additive to additional overcutting
Overcut/Look-up	0 (for no overcut/ look-up)	1.24 (for 50mm)	Not additive to additional overcutting
Tail Void	0 (for perfect grouting)	4.27 (for ineffective grouting)	

MEASURED VOLUME LOSS DURING EPB TUNNELING THROUGH SINGAPORE MARINE CLAY

EPB tunneling through marine clay in Singapore has been carried out for:

- The Ulu Pandan trunk sewer using a 3.754m diameter machine. About 2,000m of this tunnel was in marine clay or in a mixed face of marine clay and underlying, stiffer strata, and the tunnel was constructed between 1983 and 1985 (Balasubramanian, 1987)

- For the East-West subway line between Lavender and Bugis Stations, two tunnels with a total length of about 1595m. Both tunnels were in either marine clay or in a mixed face of marine clay and underlying, stiffer strata for their whole length. The machines used to excavate these tunnels were 5.930m, with the tunneling carried out in 1986 and 1987 (Elias and Mizuno, 1987)

- For the North East subway line, over a number of sections totaling about 1600m in marine clay or a mixed face of marine clay and stiffer materials. The tunneling was carried out using a number of machines, including that shown in Figure 2, between 1998 and 2000 (Shirlaw et al. 2003)

EPB Tunneling for the Circle Line has also encountered the marine clay. However, this work is still in progress and no data from the tunneling has been published.

For the Ulu Pandan sewer tunnel and the Lavender to Bugis subway tunnels, the face pressure used in the marine clay was higher than the total overburden pressure. A pressure as high as 1.8 times the total overburden pressure was used for the Ulu Pandan sewer tunnel, and up to 1.6 times for the Lavender-Bugis tunnels. Overcutting was used freely during tunneling, as an aid to maintaining alignment. Grout was injected conventionally, through the rings. Although settlements were initially small, large settlements developed over time. The maximum settlement, including up one year of consolidation settlement, over the tunnels was large. Up to 150mm of surface settlement was measured over the centre line of the Ulu Pandan sewer tunnel and up to 157mm over the twin tunnels between Lavender and Bugis. The consolidation settlements in these cases were found to form a narrow settlement trough at the surface, similar to that recorded during active tunneling. The consolidation settlement reflected delayed volume loss at the tunnel, rather than the effects of seepage into the tunnel. In terms of volume loss, the

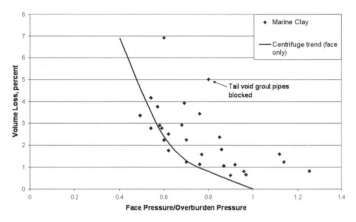

Figure 4.   Volume loss for North East Line tunneling in a full face of marine clay

maximum settlements recorded translate to over 15% for the Ulu Pandan sewer and nearly 10% for the Lavender to Bugis subway tunnels (Shirlaw 1995). These volume losses are equivalent to total closure of the tail void plus the void due to full extension on the overcutters. Broms and Shirlaw (1989) demonstrated that the range of settlements on these tunnels largely fell between that from full closure of the tail void, at the low end, and full closure plus full extension on the overcutters.

The experience from the earlier tunnels in Singapore was used in planning the tunnels for the North East Line. In particular, it was specified that all shields must be equipped for simultaneous tail void grouting. The use of the extendable overcutters was discouraged, as was the use of face pressures in excess of 1.2 times the full over burden pressure. The results of the settlement monitoring over the various sections of the North East Line are summarized in Figure 4 for the tunnel all in marine clay and Figure 5 for the tunnel partly in marine clay and partly in stiffer soils or weathered rock. The volume loss is plotted against the average face pressure applied during the advance for the ring directly below that settlement point, expressed as a proportion of the total overburden pressure at the spring line of the tunnel. Also plotted in Figures 4 and 5 is the anticipated relationship between face pressure and volume loss derived from the model tunnel tests published by Kimura and Mair (1981), for the typical depth, tunnel diameter and undrained strength of the marine clay for the North East Line tunnels. In deriving this line, it has been assumed that the tunnel is fully supported immediately behind the face, i.e., no consideration is given to the losses related to overcut, steering or tail void closure discussed previously.

Where the North East Line tunnels were completely in marine clay, the measured volume loss typically scattered from 0 to 2% of volume loss above the general trend line for face related volume loss. Where full, or close to full, overburden pressure was used the measured Volume Loss was typically in the range 0.5 to 1.5%. In two cases the measured volume loss was over 4% above the general trend line. In one of these cases it is known that all of the simultaneous grouting lines were blocked, and the machine was being advanced with only conventional grouting through the rings.

Where the tunnels were assessed as being partly in marine clay and partly in stiffer soils, many of the points plot in the same general range of 0 to 2% volume loss above the general trend line established for settlement associated with the face. At two locations sinkholes were recorded above the tunnel, in these conditions. This typically occurred as the face was passing from a full face of stiffer soil or weathered rock into

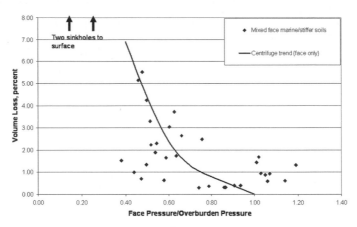

Figure 5.   Volume loss for North East Line tunneling in a mixed face of marine clay and stiffer soils or weathered rock

an incised valley infilled with marine clay. In these cases, it had been possible to advance the tunnel safely with a relatively low face pressure in the stiffer soil or weathered rock, but the face pressure was not adjusted upwards in time to provide the pressure necessary to control the marine clay. A number of points also plot below, sometimes well below, the trend line. This may be partly due to the uncertainty involved in establishing the exact ground conditions when tunneling using an EPB machine. With this type of machine there is rarely the opportunity to inspect the face. The interpretation of the ground conditions was primarily based on borehole data. The precise location of the interface between different strata between boreholes is therefore uncertain. It is probable that some points assessed as being in the mixed face zone were actually in a full face of stiffer soil, and may be why some of the data points fall well below the trend line for the marine clay. It would also be reasonable to expect some reduction in settlement and volume loss as the proportion of marine clay in the face reduces.

Even where the face pressure used was close to the full overburden pressure, and with simultaneous tail void grouting, there was a residual volume loss of up to 2%. Although the actual source of this residual settlement has not been identified from the data, it is likely that the following factors contributed to it:

- Fluctuations of the face pressure around the mean used in the graphs
- Closure of the minimum overcut
- Ground losses associated with steering (curves, overhang, look-up)
- Some residual settlements associated with the tail void

Although the use of simultaneous tail void grouting was clearly effective in reducing the measured volume loss, it is unlikely that this was consistently perfect. One particular concern is the potential for bleeding, and so shrinkage, of the grouts used. The majority of the tunneling for the North East Line was carried out using a cement-silicate, rapid setting grout (Shirlaw et al. 2004). The grout had a similar mix to one reported by Komiya et al. (2001) as having bled significantly under pressure, even after initial setting.

CONCLUSIONS

From the data from the North East Line tunneling in marine clay in Singapore, it can be seen that:

1. The largest potential sources of settlement for EPB shields in marine clay are at the face, due to overcutting and tail void closure; careful control of each these sources is required to keep the Volume Loss to 2% or less

2. The use of simultaneous tail void grouting and minimizing overcutting are important in controlling the settlement over tunnels in soft clay

3. A face pressure of at about 0.9 times the full overburden pressure is needed to keep the Volume Loss to below 2%

4. A residual Volume Loss of up to 2% can be ascribed to closure of the minimum overcut, shield steering and possibly some residual settlement associated with imperfect grouting at the tail void

A volume loss of 2% appears to be achievable for EPB tunneling in the marine clay in Singapore, based on the data. However, it is notable that, in practice, the measured volume loss frequently exceeded this value.

REFERENCES

Balasubramaniam, K. 1987. Construction of effluent outfall pipeline using earth pressure balance shield. *Proceedings of the Fifth International Geotechnical Seminar*, Nanyang Technological Institute, Singapore. 17–40.

Broms, B.B. and Shirlaw, J.N. 1989. Settlements caused by earth pressure balance shields in Singapore. In Tunnels et Micro-tunnels en Terrain Meuble—Du Chantier a la Theorie. Presse de l'Ecole Nationale des Ponts et Chaussees, Paris, 209–219

Elias, V.H. and Mizuno, A. 1987. Tunneling with earth pressure balance shields. Proceedings, Fifth International Geotechnical Seminar, Nanyang Technological Institute, Singapore, 41–58.

Kimura, T. and Mair, R.J. 1981. Centrifugal testing of model tunnels in soft clay. *Proceedings of the 10th International conference on Soil Mechanics and Foundation Engineering,* Stockholm. Publ. Balkema, Rotterdam, Volume 1: 319–322.

Komiya, K. Soga, K., Agaki, H., Jafari M.and Bolton, M.D. 2001. Soil consolidation associated with grouting during shield tunneling in soft clayey ground. *Geotechnique*, 51, number 10, 835–846.

Shirlaw, J.N., 1995. Observed and calculated pore pressures and deformations induced by an Earth Pressure Balance shield: Discussion, Canadian Geotechnical Journal, 32, pp181–189 (1995).

Shirlaw, J.N., Dazhi, W., Ganeshan, V and Hoe, C.S. 2000. A compensation grouting trial in Singapore marine clay. Proc. Geotechnical Aspects of Underground Construction in Soft Ground, Kusakabe, Fujita & Miyazaki (eds) publ Balkema, Rotterdam, 149–154.

Shirlaw, J.N. 2002. Controlling the risk of excessive settlement during EPB tunneling. Keynote lecture, *Proc. Case studies in geotechnical engineering, NTU Singapore.*

Shirlaw, J.N., Ong, J.C.W. Rosser, H.B., Tan, C.G, Osborne, N.H.and Heslop P.J.E. 2003. Local settlements and sinkholes due to EPB tunneling. Proc. ICE, Geotechnical Engineering I 56, October Issue GE4, pp 193–211.

Shirlaw, J.N., Richards, D.P., Raymond, P. and Longchamp, P. 2004. Recent experience in automatic tail void grouting with soft ground tunnel boring machines. Proc. 30th ITA-AITES World Tunnel Congress, Singapore, ed. J.N. Shirlaw, J. Zhao and R. Krishnan. Publ in Tunneling and Underground Space Technology Vol 19, No 4–5, July–September.

Soga, K., Bolton, M.D., Au, S.K.A, Komiya, K., Hamelin, J.P., Van Cotthem, A., Buchet, G. and Michel, J.P. 2000. Development of compensation grouting modelling and control system. Proc. Geotechnical Aspects of Underground Construction in Soft Ground, Kusakabe, Fujita & Miyazaki (eds) publ Balkema, Rotterdam, 425–430

Tanaka, H., Locat, J., Shibuya, S., Tan, T.S. and Shiwakoti, D.R. 2001. Characterization of Singapore, Bangkok and Ariake clays. *Can. Geot. J.* 38:378–400.

THE MADRID RENEWAL INNER RING CALLE 30 WITH THE LARGEST EPB MACHINES—PLANNING AND RESULTS

Enrique Fernandez

Dragados SA

INTRODUCTION

M-30, Madrid's inner ring, covers 42 sq km of the city centre where one million people reside. It was constructed in several phases during the sixties and the eighties. The design criteria applied to those phases were different and were not always homogeneous with urban roads with lighting regulation as well as speedways. These areas had a heavy concentration of access and exits and there had been made many changes on the number of traffic lanes which has caused heavy traffic congestions and high accident rates. The environmental impact and noise pollution in residential areas are also consequences of M-30.

The Madrid Major took a decision at the beginning of the present legislation, based on socio economic and environmental evaluation; rebuild the M-30 with the following sustainable mobility principles:

- Underground transfer of a considerable portion of the main ring and some of its links replacing some parts of the surface with new green areas in order to reduce the environmental impact.

- Increase the traffic capacity, improving the traffic continuity and fluidity and eliminating the lighting regulation areas, which is one of the main congestion problems currently.

- Increase the traffic safety with clear signalling information panels and avoiding difficult maneuvering.

- Redesign the links with the high capacity radial highways and other connecting lanes of the city.

The project is worth 3.7 billion Euros. In 2007, when the construction is completed, it is expected will save 14 million of journey hours that represent 4.5 million Euros saved in fuel consumption per annum, There will be a very significant reduction of 35,000 tons of CO_2 in terms of pollution and 400 less car accidents. Those values will increase even more in the following years.

The areas beside Manzanares River, running parallel to the M-30 on the western side of Madrid, will be highly enhanced. Better water quality and the recovering of the green areas existing in old times will also return for the citizen's enjoyment.

TUNNEL CROSS SECTION OF CALLE 30 SOUTH BY PASS

One of the major challenges of the new Calle 30 is actually not only the related to the main road but also to the link with the A-3, with highway to Valencia. This link is currently the most congested area and has one of the highest traffic densities of Spanish roads, with more than 250,000 vehicles/day. The remodelling of this area is the principle objective of the South by-pass.

The tunnel is the only answer for this by-pass as the road alignment will go among buildings. A twin tube design, one for each direction, with three lanes allowing the circulation of both, light cars and heavy trucks, is the solution that has been

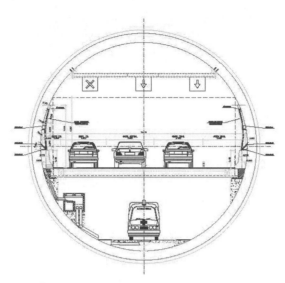

Figure 1.   Tunnel cross section

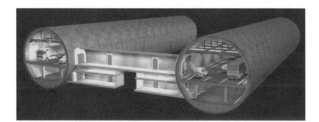

Figure 2.   3D view of tunnels and cross passages

taken. The lanes are 3.5 m wide, 0.5 m in hard shoulders section and 0.8 m in the walkway section.

The gauge height is fixed in 4.5 m and the road is placed 1 m below the tunnel centre. These exigencies require large tunnels, as large as 13.45 m inner diameter. The circular shape of the tunnels allows three traffic lanes and two emergency lanes (5.5 × 4 m) below the road concrete slab as well as ventilation conducts and other facilities. With 3.660 m long, both tubes are connected by 8 cross passages from which 3 of them are vehicular galleries and the rest are pedestrian galleries as shown in the 3D view on Figure 2. A ventilation shaft is also defined on every tube.

The protection against fire is based on polypropylene fibres imbibed on concrete lining are used. Tunnels run along the alignment with a maximum overburden from 30 m up to 65 m as an average, which means only 2 diameters. At tunnel portals, shafts with rectangular shapes, and only one diameter of overburden have been considered in order to reduce their depth. This special requirement has forced ground treatment in the first meters of the tunnels.

EXCAVATION METHOD AND PREVIOUS EXPERIENCES IN METRO MADRID

Excavation Method

As previously indicated, the new design of Calle 30 foresees some of its parts will run underground. The most varied construction methods have been applied in the tunnels, from cut and cover to TBM tunnels as well as the traditional Madrid method.

Cut and cover method is the cheapest according to Madrid's subsoil but it has some big disadvantages in terms of environmental impact during construction. Also, the relocation of facilities in congested areas and higher amount of muck transportation to disposal areas, some 30 km far away from the site, made the tunnel option recommended. This option allows keeping the alignment under buildings as required in some areas.

Traditional Madrid method is a hand mining method used in short tunnels, caverns and adits, based on a multi heading scheme and very well adapted to variable geometry.

In the South By pass, TBM tunnel requires slightly more than 15 m excavation diameter, which obviously can not be digged in open mode. Madrid geology is very well adapted to EPB machines as confirmed on the previous Metro experiences. Thus, two 15 m EPB machines have been assigned on July 2004 and became the largest TBMs in the world.

The Madrid south geology specifically on the project area has the following characteristics:

- Soil deposits from surface to 20 m deep.
- 25 to 30 m of sandy clay called "Peñuela"
- 20 to 25 m of hard clay with gypsum levels (Peñuela yesifera)

In terms of geotechnical parameters of this Peñuela, the cohesion is $\tau = 60$ kPa, the internal friction angle $\phi = 28°$ and the deformation modulus $E = 220$ MPa. The fine material content, sieve 200, goes up to 85–95%. Tunnels are built below groundwater table with a maximum hydrostatic height of 40 m.

The precast concrete ring support is 60 cm thick and 2 m long, with universal ring configuration. The 9 pieces plus key are fully bolted, 3 bolts between segments of the same ring and 4 bolts between segments of adjacent rings. To centre the segments in the right position, two bicones are placed to connect the rings. The longer segments, 4.76 m in outer perimeter, have a weight of 13.1 ton. and 125 ton for the whole ring. Additional invert segment for installation of a double track was also designed.

Metro Madrid Experiences

Madrid Metro started EPB experience in 1995 with Line 4 extension including the cross of the Manzanares River with only 2 m overburden. Since then, more than 65 km have been built and another 78 km will be completed according to the 2003–2007 Extension Plan. On this project, average excavation rates of 607 m/month have been achieved as well as a peak production of 1,230 m/month.

The success of the Metro extension and the experience gained by its contractor and client in the last 10 years with EPB machines has been very helpful to face up to the new challenge. But not only success but mainly the lack of progress and unproductive time consuming has been analyzed in detail in order to improve the efficiency of the machines. It is well known that shield machines do not use more than 50% of the working time in production and more specifically in excavation and ring assembly. Another 50% of the working time is usually spent on maintenance, cutterhead tools replacement, breakdown of TBM and/or facilities, derailment of locomotives, adverse geology, surveying and other minor reasons.

The experience shows that cutterhead tools replacement still remains a matter to be solved by the manufacturers. Despite the latest improvement with cutter tools design enabling a safe replacement from behind, narrow access to the cutter chamber, improvised scaffoldings, tools transfer and difficult bolt removal makes this operation an extremely high time consuming and a waste of production time.

Breakdown of TBM is, in most cases directly related to operation at the top level of the electromechanical exigency. Thus, higher TBM technical characteristics than extrictly required should allow a comfortable operation and increase the production. This solution is also applied to the facilities such as pumping system, ventilation system and backfilling containers capacity.

The third cause of the waste of time is related to derailment of convoys in muck charge transfer from the TBM to the tunnel portal. Actual tendency for muck transport is to be carried out by a continuous conveyor belt system which will be applied. Its efficiency has been proved in previous projects and so convoys will remain only for segment and mortar transportation.

Due to the fact that adverse geology was not expected and the machines had been designed to face the Madrid subsoil working in EPB close mode, the improvement and detailed study of the three main reasons of waste of time captioned above has reverted in a significant increase of productive time.

TECHNICAL SPECIFICATIONS FOR THE LARGE MACHINES

The technical specifications for the new machines were based on those from the Metro extension, where similar ground conditions were faced. These specifications were adapted to bigger diameters and a reserve to guarantee the productivity and reduce the risk of breakdown was added.

Excavation Diameter

The excavation diameter is obliged because of the inner diameter specifications described above. The outer ring became 14.65 m by adding to the 13.45 m required inner diameter the thickness of concrete segments. Consequently the specification was a bigger than 15 m diameter to absorb the shield plate as well as the necessary gap to avoid TBM squeezing due to ground deformation. Also tolerance to allow a 350 m minimum radius on alignment was considered, despite the minimum road radius is 500 m.

The different designs done by the two manufacturers conclude on different excavation diameters. Meanwhile Herrenknecht, the German manufacturer, with sandwich design on tail shield, fixed 15.08 m, the Japanese Mitsubishi, had considered a 15.01 m diameter enough to fulfil the required specifications.

Maximum Thrust

The thrust is provided by a set of cylinders reacting against the segment ring. These cylinders must be able to cope with the following stresses:

- Avoid horizontal ground deformations at the tunnel front
- Compensate the horizontal effective ground pressure
- Compensate the interstitial groundwater pressure
- Eliminate vertical ground settlement ahead the TBM if required

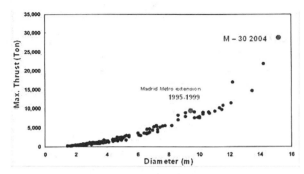

Figure 3. Maximum thrust versus diameter

■ Overcome the shield friction during the advance of the TBM

The TBM used on the Metro extension has a maximum thrust of 100,000 kN. Considering the ratio of cutterhead areas, which is 2.5, 250,000 kN, it has been the minimum required for this project. Based on the fact that sometimes higher thrust was required to face friction troubles, this minimum required was requested to the manufacturers to be increased.

In this way, Mitsubishi shows a maximum thrust of 285,000 kN provided by 57 cylinders of 500 ton each. Similar thrust was provided by Herrenknecht, 276,390 kN, using also 57 cyl. of 485 ton each (Figure 3).

Maximum Torque

The torque was estimated for the Metro Extension Plan according to the formula $T = (\frac{2}{3}) \, \pi\tau \, R^3$ where τ is the ground cohesion (100 kPa or 10 T/m^2). The required torque was 2,175 mT but 2,000 mT was specified.

To determine the minimum requirement for M-30 machines a ratio of excavated volumes was applied. $T = 2,000 \, (15.1/9.4)^3 = 8,240$ mT. This torque allows the following:

■ Excavate the ground with bits and other cutter tools

■ Overcome the friction between ground and cutterhead

■ Overcome radial and longitudinal forces on main bearing

■ Overcome the friction of the sealing system

■ Soil agitation inside the cutter chamber

Figure 4 shows the high nominal torque of the M-30 machines compared with the previous ones.

Other Requirements

Based on the geotechnical conditions of the present project, other requirements were requested to the manufacturers:

■ Maximum working pressure on cutterhead chamber as higher as 6 bar

■ Open ratio on cutterhead over 30%

■ More than 6 earth pressure sensors in chamber to control properly the excavation process

■ Excavation speed of 65 mm/min, means the excavation of one ring of 2 m long will be completed in 30 min.

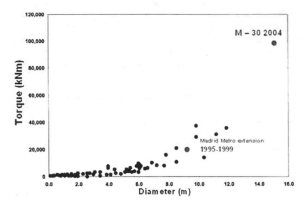

Figure 4. Torque versus diameter

- Maximum boulder size able to pass through the screw conveyor: 700 × 300 × 300 mm
- Foam capacity for ground treatment over 1,000 m³/h
- Mortar pump flow over 50 m³/h to fill the gap behind the pre cast segment rings
- More than 3 rows of brushes on tail shield to avoid water inflows.
- Safety chamber for 20 people and 12 hours survival

DUAL DESIGN

Two EPB machines were awarded, one to Herrenknecht and another to Mitsubishi. In the case of those big TBMs, innovative solutions were applied in order to accomplish the strict specifications provided by the client as well as to allow the contractors to fulfil the tight construction schedule.

The high excavation speed, as a consequence of the high performances in terms of thrust and torque, as well as a revolutionary new cutterhead design and the new devices employed to expedite the ring assembly, are, among others, the solutions that will allow the completion of the project as scheduled.

Both manufacturers submitted these new solutions which, in our opinion, will open the panorama to further development of huge TBMs able to excavate large tunnels.

Herrenknecht Conception

The German design was based on a new idea to deal with the big surface, 181.5 m², of the heading excavated in these tunnels. The main concern was how to handle the muck material from the ground to the conveyor belt avoiding meanwhile the rolling risk of the TBM. For this purpose, cutterhead was divided in two parts; the inner cutterhead rotates in opposite sense than the outer one creating a counter torque.

The inner cutterhead has a 7 m diameter, 38.5 m², has 6 cutting radius and is able to advance independently from the outer one being able, in such capacity to excavate the central portion and facilitating the excavation of the peripheral zone with the outer one.

The outer cutterhead, with a 143 m² surface, has 12 cutting radius and the hole cutter is armoured with bits and disc cutters. In total the cutterhead has 57 double 17" disc cutters, 332 bits, 24 scrappers and one centre cutter. The open ratio designed has been 31.6%.

Figure 5.   Opposite rotation cutterhead

The split cutterhead was applied by Herrenknecht in previous machines but always working in slurry mode. This is the first time that this concept was applied to an EPB machine. The inner cutterhead has a higher rotation speed than the outer one and this reduces the risk of plugging the muck material into the chamber.

The cutterhead is fully hydraulically driven. The inner head has 10 drives meanwhile the outer ring has double bull gear with 24 inner drive units and 32 outer drive units. Taking into account the loosen power on hydraulic scheme, around 32%, the 15.8 MW installed provides 10.7 MW to the cutterhead. This power provides 9,600 mT at 0.81 rpm on the outer ring and 8,450 mT at 1.5 rpm on the inner cutterhead. Exceptional torque is 12,527 and 10,890 mT for outer and inner cutterhead respectively. The shield length is 11.51 m and no articulation was designed. Figure 5 shows a Herrenknecht splitted cutterhead.

There are 3 rows of brushes on tail shield and mucking is organized by three screw conveyors. The first screw conveyor 700 mm diameter, is in charge of mucking the central portion and two other screw conveyors, 1,200 mm diameter, are used to remove the material from the lower part of the cutter chamber. The segment erector has a standard design, centre free and it is a vacuum plate type, it has been proven in several projects before.

Mitsubishi Conception

The Japanese concept is very different from the German. Mitsubishi machine has one compact cutterhead armoured with 44 triple 17" disc cutters, 226 knife edge bits, 420 drag bits, 16 triple trim bits, 16 double trim bits and one centre cutter (Figure 6).

Open ratio designed has been 43%, much higher than Herrenknecht design. Mitsubishi machine is electrically driven by 28 Variable Frequency Drives of 358 kW, means 10 MW. Electrical drive has less loosen power than hydraulic one, then, the nominal torque provided with this system is 8,570 mT at 1.05 rpm and exceptionally 12,700 mT.

Figure 6. EPB Mitsubishi

To prevent the rolling risk, the main weapons were adjustable thrust cylinders and steel profiles along the invert shield, together with an alternate of the cutterhead rotation during excavation.

The rotation speed in big machines is very low, around 1 rpm, thus there is a high plugging risk of muck material into the centre of the chamber where the linear speed of the crumbling blades is not as higher as required. Mitsubishi's solution is the centre agitator, a hydraulically drive agitator 5 m span, installed in the cutting chamber and able to mix and adequate the muck material in order to be removed by the screw conveyor and reducing drastically the risk of plugging. The agitator can rotate at 2 rpm, enough to accomplish the expected requirements.

The shield length is 12.22 m and active articulation was designed. There are 4 rows of brushes on the tail shield. Mucking is organized by one screw conveyors with 1,500 mm diameter.

Other relevant innovation of this machine is the double erector. The segments erector was built with two opposite vacuum plates in such way that the closer plate to the segment feeder will be activated saving time in the cycle time. Figure 7 shows a Mitsubishi TBM longitudinal cross section and Figure 8 shows the machine assembled in the factory.

PROJECT SCHEDULE AND PLANNING

The project has a 33-month schedule, starting from July 2004. Previously, a 6-month design contract was awarded to these manufacturers to confirm the feasibility of the new machines with those exceptional requirements from the client. Two separate contracts have been awarded, one for each tunnel.

The machines were manufactured, within 12 months by Herrenknecht and 17 months by Mitsubishi. During this period, big shafts (40 × 100 m) 30 m deep were

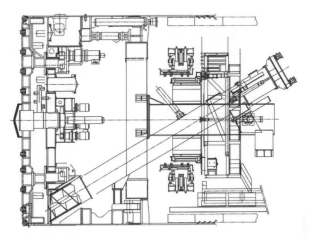

Figure 7. Mitsubishi longitudinal section

Figure 8. Mitsubishi machine at Duro Felguera yard for preassembly and running test

built in both ends of the project. The idea was to separate the activities of both Joint Ventures and avoid interferences that the operation of both TBMs in the same portal could generate.

The expected excavation progress rate was 12 m /day or 360 m/month, which means 10 months to complete the whole tunnel. Additional works such as intermediate slab or cross passages have been made simultaneously to the excavation, some hundreds meters behind the heading, in order to fulfil the tight schedule.

EVENTS AND RESULTS IN SOUTH TUNNEL

Mitsubishi TBM was in charge of the South tunnel excavation (awarded to the Joint Venture of Spanish contractors DRAGADOS and FCC). In order to save costs and reduce the construction schedule, Mitsubishi reached an agreement with a Spanish steel factory, Duro Felguera, to build the machine in Spain and this company took the responsibility of the back up design.

The delay occurred during the manufacturing of the Mitsubishi TBM, and has been due mainly to the limited factories in Europe able to deal with the construction of these enormous pieces like the ones for the main bearing as well as the lack of experience

Figure 9.   Launching shaft for the south tunnel

and adequacy of the Spanish yard for this specific activity of building the largest EPB machines in the world.

Nevertheless, the proper design and the good manufacturing quality allowed the contractor to recover the initial schedule during the excavation of the South tunnel.

After one month transportation from the yard to the site, assembly of TBM at the portal shaft took three months considering the huge size and the site restrictions and access.

Note in Figure 9 the M-30 inner ring road and the subsequent traffic congestion during rush hours, which has considerably affected the normal progress of works.

Excavation of South Tunnel

The excavation started on April 11th 2006 by using temporary conveyor belt mucking system forced by the back up length which does not allow being fully assembly at the launching shaft. After 10 working days and 64 m of tunnel excavation, a one week stop was required to extend the back up installation and permanent facilities.

The progress on the first 240 m of tunnel was conditioned by the learning curve and the conservative approach based on the expectancy of the big machine performances. On May 17th, 15 days stops in excavation were required to complete the shaft installations and permanent facilities. In addition, a failure on the shuttle car system, transferring the precast segments from the segment cars to the Back up feeder forced to modify the design and build a new performing shuttle. On June 1st, the excavation restarted as shown in Figure 10.

After this event, a continuous and regular progress was achieved by a very experienced crew, formerly working in Madrid's Metro projects as well as the very long (28 km) Guadarrama tunnel.

As a summary of the achieved productions during the south tunnel excavation, the figures are the following:

- Average daily production: 18.3 meters/day
- Average monthly production: 557 meters/month
- Maximum daily production: 46 m (October 9th, 2006)
- Best monthly production: 791 m (October 2006)
- Best 30 days production 928 m (From September 22nd to October 21st)

The numbers are self explanatory. After 200 calendar working days, the breakthrough was the celebration day of this successful excavation, started on April 11th, 2006 and completed on October 27th, 2006 (Figure 11).

If the site conditions in terms of space to assemble the full TBM and back up and permanent facilities had not been so restricted and the project had been developed in

Figure 12.   Segment erection

The high performance machines in terms of cutterhead power, torque and thrust has revealed to be very useful to avoid breakdowns and stops when the ground loads increase.

The centre agitator designed to mix the soil in the chamber core where the head rotation is almost indiscernible, proved to be a simple and effective tool to resolve this problem in big machines.

The double erector design has worked more as a spare erector that as formerly conceived and explained on the TBM description above. The time consumed on the inversion of the erector commands to activate the opposite vacuum plate is higher than the savings in the cycle time using the closer plate in each segment erection (Figure 12).

On the other hand, the two vacuum plates availability, allows the crew to complete the ring erection without delays even in the case of one plate failure. Repairing works were made during the daily maintenance period and consequently this has not affected, in general, the excavation progress.

Road Concrete Slab

Originally designed as an in situ build slab, supported by two pillars and leaning against the cantilever structures, a more practical and faster construction system was finally chosen.

A 24 m formwork allows casting in place the reinforced steel concrete cantilevers daily (Figure 13). The reinforced steel is placed in advance by independent scaffolding and after the receiving bars are installed from the back up rear where the segments are drilled and re-bars grouted. The mixer trucks supplied the concrete through the concrete slab where the concrete pump was placed at the extreme, keeping the twin tracks free for the TBM operation. A 200 m of 6" pipe pumped the concrete to the formwork.

Neoprene pads installation was the next step on this activity, followed by the precast concrete slabs being towed from the portal by tractor and placed by forklift and a special device (Figure 14). Precast beams are 2 m wide and the full operation has been made from the upper level without interference with the tunnel excavation.

Progress of the full activity was conditioned by the cast in place of cantilevers, thus 24 m a day have been achieved. The slab is finished by asphalt facing layers.

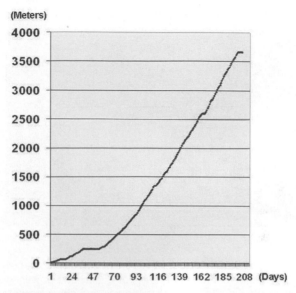

Figure 10.   Cumulative production in South tunnel

Figure 11.   Breakthrough of South tunnel

a more open area, the results obtained in this South tunnel could have been extended to 20.5 m/day, as the 21 days stops indicated could not been required.

Lessons Learned About Excavation

The excavation of the South tunnel of the South by-pass on the Calle 30 project, has confirmed our expectations about big diameter tunnels and the way to deal with them. The consequence to be noted is that the excavation speed is not related at all to the tunnel size, but to the selection of proper equipment in accordance with the ground conditions and the operation by the most skilful crew. As a result of it 60.7% of productive time has been achieved in the overall project excluding the 21 days dedicated to complete the TBM assembly and permanent facilities at the launching shaft.

Figure 13. Formwork for cast in place cantilever

Figure 14. Installation of precast concrete slab

Cross passages have followed the concrete slab installation and executed by conventional excavation, as well as the ventilation shafts.

CONCLUSIONS

The application of the world largest EPB machines on Madrid Calle 30 project is a big step on present tunneling techniques not only for the extremely high requirements in terms of power, thrust and torque but also the facilities and services that must be designed in accordance with the tunnel size.

With this aim, and based on the previous successful experiences in tunneling with TBM, the Madrid Municipality has successfully faced this huge project which will dramatically change the environmental aspect of the city.

Further actions of Calle 30 with similar scheme are the North by-pass of 4.8 km length, the underground transferral of the road currently at grade and regulated by traffic lights and the 2.6 km length connection of this by-pass with A-1, highway to the north of Spain. Additionally, the closure of M-50 of Madrid's outer ring in its northern zone under the "Mount El Pardo" restricted environmental area, by this way, is also under consideration by the Ministry of Civil Works.

Not only road tunnels but multifunctional tunnels in which the whole space of the cross section can contain adequate road lanes, rail tracks, emergency lanes and other facilities like water, electricity or cable services, will allow in the near future further developments of urban tunnels, lowering the infrastructures to the underground and creating new free spaces on the surface which will contribute to the sustainable growth and citizen's life improvement.

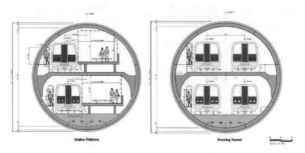

Figure 15.   Tysons Corner tunnel alternative cross sections

Samples like the SMART project in Kuala Lumpur, which combine road and water storm storage in a single tunnel, the Port of Miami tunnel between Dodge and Watson islands or the proposed Tysons Corner tunnel alternative integrated in the Dulles Corridor Metrorail project in Washington D.C. shown in Figure 15, are new solutions in highly congested urban areas.

The latest TBM technology and the innovative designs, provided by engineers following the clients' requirements, have opened a very fruitful future on tunneling, beyond the current expectations.

REFERENCES

Monzon, A., Pardillo, J.M., and others. April 2005 "The M-30 ring road improvement project in terms of a sustainable mobility strategy for Madrid" Revista de Obras Publicas Madrid, Spain.

Bambridge, C., Seng Hoor. L. June 2005 "The Stormwater Management And Road Tunnel (SMART), Kuala Lumpur, Malaysia—A unique use of tunnel space for both traffic and flood water" R.E.T.C. Seattle, U.S.A.

Sauer, G. December 2005 "Tunnel engineering" Transportation and Mobility Conference. Houston, U.S.A.

Melis, M., Arnaiz, M., Fernandez, E. February 2006 "The Madrid renewal inner ring Calle 30 with the largest EPB machines" International Symposium on Underground Excavation and Tunneling ISUET Bangkok, Thailand.

12

Pressurized Face Tunneling II

Chair
S. Yanagisawa
Kiewit-Bilfinger Berger

DESIGN PRINCIPLES FOR SOFT GROUND CUTTERHEADS

Werner Burger

Herrenknecht, AG

ABSTRACT

The last decade has witnessed a rapid development and application of slurry and EPB type tunnel boring machines. Both soft ground types are able to excavate tunnels in unstable ground conditions below the water table with positive face support. The most significant component of the excavation process is the cutterhead and tool configuration were the excavation and face support meets the geological challenge. The TBM downstream processes cannot take place unless the cutterhead and tools effectively do their job. This paper will review the current state of the art design for both slurry and EPB cutterheads and highlight both common as well as geological specific design features.

FACE SUPPORT PRINCIPLES

Two different situations should be considered with respect to cutterheads for soft ground applications:

- Stable tunnel face with no or little water inflow, predominantly above the ground water table
- Unstable tunnel face mostly in combination with significant potential water inflow below the groundwater table

In the first case full face (rotary cutterhead) or partial face (digger/roadheader shield) excavation could be used. This should most likely be done in combination with breast doors for mechanical support during stoppages or even compressed air for ground water control.

The second case requires "positive" face support during excavation or stoppages and chamber access for cutterhead inspection or tool change.

The general principle of the positive face support is that the tunnel face is supported by a pressurized slurry (STBM) or conditioned muck (EPBM) whereas the difference between the two is only the density of the pressurized support medium. The pressurization of the support medium is controlled either by the flow (EPB and STBM without air bubble) or a separate air cushion (STBM with air cushion).

Pressure control in an EPB is based on a low volumetric flow which results from the volume of the excavated ground and injected conditioning materials. The pressure control in a slurry TBM without air bubble is based on a high volumetric flow of the circulating bentonite slurry and the excavated ground. The pressure control in slurry TBMs with air bubble is completely independent from the flow volume and purely based on an air pressure control (Figure 1).

To control the pressure by flow becomes more difficult the higher the flow volume gets. The air pressure control independent from the flow provides the fastest and most accurate reaction. For both principles it is essential that the controlled pressure of the support medium is transferred to the entire tunnel face area in the best possible way. Thus isolating the face from the support medium by closing the cutterhead produces a high risk for non controlled face behavior resulting in settlement or even sinkholes in the worst case.

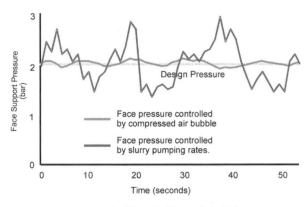

Figure 1. Face support pressure control with and without air bubble

Whatever attempts are made for mechanical face support during excavation, they may not be successful under difficult and unstable ground conditions. In any case there must be a working gap for the tools between the tunnel face and the cutterhead front plate. Even if this gap is kept small it creates sufficient volume for potential ground losses in the range of 5% or more.

The decision to use EPB or STBM technology is strongly influenced by the ground conditions. This topic has been extensively dealt with in numerous publications and is not examined in greater detail here.

EVOLUTION OF CUTTERHEAD DESIGNS

STBMs

In the late 1970s and early 1980s when the first slurry TBMs with remote air bubble were buildt in Europe the idea was to allow the best possible exchange and contact of the supporting slurry fluid to the face. These machines were developed to excavate permeable sands and gravel below the ground water table. For this reason the cutterheads were "star-type" designs carrying simple square tools either in the middle of the arms or a sort of cutting tools at the outer edges of the arms (Figure 2). These type of design worked quite well in sand or gravel and even later when designing the first mixed face cutterheads the cutter discs were mounted onto the arms. Based on the design principle of these cutterheads the implementation of backloading cutters on the arms is difficult for smaller diameters. In the center area it is not possible at all for disc cutters or any type of soft ground tool.

For large diameters like the Grauholz TBM in 1988, arm cross-sections big enough for man access were realized, providing the possibility to install fully backloaded cutters over the entire profile (Figure 3). This principle was used later for the atmospheric cutterchange, keeping the interior area of the arm under atmospheric conditions while changing the cutters from inside. These cutterhead types worked quite well regarding the face support during excavation but had the disadvantage that large face areas were completely open during chamber access under compressed air. And even if the support principle slurry cake–compressed air worked well, there was no additional mechanical protection for the workers which created an "unsafe feeling"

Figure 2.   Early mixshields: HERA Hamburg 1984 (left), Metro Moscow 1985 (right)

Figure 3.   Grauholz 1988 with back-loading cutters

Figure 4.   Essen 1994 with extendable face plates and active center cutter

especially in diameters over 6–7m. For this reason extendable face plates were developed for the use as a mechanical protection during chamber access (Figure 4).

In 1990 the first closed cutterhead on a slurry TBM was used in Mülheim for a tunnel in mostly hard rock conditions and in 1991 in Strasbourg for pure sands and gravel (Figure 5). These projects finally were followed by numerous other applications of closed face cutterheads in all kinds of ground conditions demonstrating that the face support principle with slurry works also with disc type or closed cutterheads.

This development and knowledge opened the door for further improvements especially for mixed ground applications or even hard rock cutterhead configurations with back-loading cutters on slurry machines. Parallel to the development of this wider variation of cutterhead designs improvements on the slurry circuit were achieved. New STBMs allow to inject slurry not only into the air bubble chamber, as on the very first machines, but also into the excavation chamber or through the cutterhead. These modern designs allow a wide variation of flushing configurations and provide sufficient supply of fresh bentonite slurry to the tunnel face (Figure 6).

Figure 5.   Early closed STBM cutterheads: Mülheim 1990 (left), Strasbourg 1991 (right)

Figure 6.   Modern STBM cutterheads: West Side CSO Portland (left), East Side CSO Portland (right)

EPBMs

Developed for the excavation of fine grained soils with mostly low permeability cutterheads for EPBMs not only have to carry the tools for soil excavation but also have to act as a "mixing paddle" to create the soil paste that is needed for face support. As a result of this dual function the majority of the EPBM cutterheads closed disc type designs, creating a primary excavation and mixing area between tunnel face and cutterhead front plate and a mixing chamber between cutterhead back side and shield bulkhead. In very soft and ideal ground conditions that require only little mixing and conditioning even star type designs are occasionally used.

The twin function excavation and mixing creates additional design requirements for EPBM cutterheads. This is related to injection of soil conditioning materials in front and behind the cutterhead as well as all kinds of muck flow considerations to enable the mixing and conditioning process to be optimized. The better this process is put into practice, the easier the face and settlement control can be realized.

As was previously for the slurry TBMs the use of the EPBMs was also extended towards mixed coarser grain sizes or even mixed face ground conditions based on great improvements in conditioning materials and technology (Figure 7). This required the implementation of mixed face tool arrangements with disc cutters and soft ground tools causing additional difficulties to fulfill the mixing and muckflow requirements.

Figure 7.   Early EPBM cutterheads: Taipei Metro 1992 (left), Valencia Metro 1995 with partial mixed face tool arrangement (right)

Figure 8.   EPB cutterheads for variable ground: Porto Metro 2000 (left), Singapore DTSS 2003 (right)

Some of today's EPBM applications even show a full hard rock design for the cutterhead being used on tunnel alignments with variable ground conditions reaching from full face hard rock to mixed face intersections to full face soft ground (Figure 8). These TBMs operate in the hard rock sections in open mode and closed mode in the mixed and soft ground sections. Based on the design requirements for hard rock excavation it is inevitable that the closed mode EPB operation has to accept compromises related to mixing and muck flow considerations finally resulting in lower performance and higher wear compared to a pure EPB design in the soft ground sections or the use of a STBM for this type of variable ground alignments.

SPECIAL DESIGN CONSIDERATIONS

Openings, Opening Ratio

The opening ratio is the percentage of openings of the total face area. This value is of greater importance for EPBMs than for STBMs. The mostly used range for the opening ratio is between 25 and 35 percent. The size and location of the openings on a cutterhead openings are the result of a compromise involving a large number of aspects.

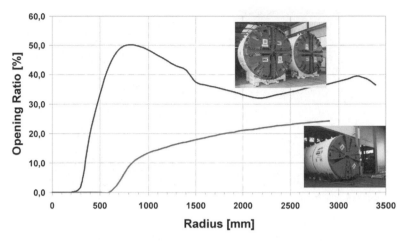

Figure 9.   Opening ratio of different EPBM cutterhead designs with similar specifications—soft ground cutters for full coverage, backloading twin disc cutters in outer face and periphery

The openings have to be large enough to assure the passing of the excavated and in front of the cutterhead already preconditioned soil into the rear mixing chamber without the need of a substantial pressure drop. The measurement of the support pressure is done with pressure cells installed in the rear wall of the mixing chamber (shield bulkhead). Major fluctuations of the earth pressure in front and behind the cutterhead are not acceptable if an accurate face pressure control is required. Pressure cells have occasionally been installed in the cutterhead front plate in the past but the readings are not of such accuracy that they can be used as primary input signals for earth pressure control.

The individual size of the openings has to limit the grain or boulder size that is allowed to enter into the mixing chamber. The basic rule is in most cases that only grain sizes should be able to enter the mixing chamber that could also pass through the follow-up conveying equipment (usually a screw conveyor). This recommendation often results in conflicts in mid- or small-size EPBMs with only limited possibilities for big diameter augers. The use of grizzly bars to limit opening cross sections is common in such cases, but also causes often plugging of the small remaining muck passages.

More important than the overall percentage of the openings is the their location or in other words the set-up of the opening ratio along the radius (Figure 9). Especially the centers of cutterheads are the most critical areas for plugging in slurry and even more in EPB machines. There is very little mixing dynamic close to the center in EPB machines. On slurry machines this can be compensated to some extend by center flushing nozzles reducing the density in the center area.

Certainly, the openings close to the center are of considerably greater importance than the ones at the outer face area. In particular when there is a potential need for disc cutters in the center a compromise has to be found. The cutterhead used for the machines for the East Side Extension in Los Angeles for example kept the possibility to install center housings for disc cutters but started (and finished) using the center openings as muck openings (Figure 10).

Muck Flow

The muck flow in the excavation chamber is of major importance for both STBM and EBPM cutterheads. When mining in cohesive and sticky materials clogging of the

Figure 10. EPBM for the East Side Extension Los Angeles during shop assembly and at the arrival shaft

Figure 11. EPBM for M30 Madrid during shop assembly and in the arrival shaft

cutterhead or excavation chamber is one of the major risks for production. Wherever possible the cutterhead design has to avoid "dead corners" or pockets allowing the muck to stick to and start building bridges and should be as smooth as possible on the front as well as on the rear side. Especially the center area becomes a major issue were there is very little or no tool speed or mixing dynamic. For this particular reason, active center cutters including slurry feed and suction lines have been developed for STBMs to provide higher tool speed and dynamics in the center (Figure 4). An increase of the mining speed of up to 30% has been achieved in cohesive soils. At the same the cutterhead torque has been reduced by about 25%.

The basic idea of an independent center wheel has been realized also for the first time now on the 15.2m EPBM for the M30 project in Madrid, Spain (Figure 11). The inner and outer wheel have completely independent drives and can rotate at different speeds and in different directions. Thus the center wheel with a diameter of 7m provides higher tool speed and mixing dynamic in the inner face area of this giant EPBM. An additional effect of partial torque compensation could be also achieved by rotating the inner and outer wheel in different directions.

Designing the cutterhead for only one working direction would help for optimization of the muck flow but eliminates the possibility of correcting shield roll by changing the direction of rotation. A few unidirectional cutterheads for soft machines have been built but it has to be accepted that in order to counteract the cutterhead torque by

Figure 12.   East Side CSO Portland, tunnel face (left) and boulder entering the excavation chamber (right)

skewing the thrust cylinders the torque transfers into the tunnel lining. Since this places additional demands on the tunnel lining the common solution is the use of bi-directional cutterheads.

Boulders

For both EPBM and STBM machines mining through cobbles and boulders is a demanding but nevertheless quite common requirement. The implementation of cutter discs and soft ground tools is common practice under such conditions (Figure 12). The decision to install cutter discs in the center area is based on a risk evaluation or the predicted boulder frequency. The use of center disc cutters would normally leave very little possibilities for center openings allowing optimized muck flow (Figure 10).

Based on practical experience from numerous job sites it is proven that large boulders can be excavated by the disc cutters as long as they are properly fixed in the surrounding matrix. Depending on the boulder size the excavation process could be regular chipping as known from hard rock excavation or splitting of the boulder in smaller pieces. For the operation of the machine it is essential to understand that the presence of boulders in the face limits the cutterhead penetration to the maximum value allowed by this rock excavation. Even if the boulder (or rock surface) is only a few percent of the face area it is not possible to mine through a granite boulder with a penetration rate of 15 or 25 mm without destroying the cutterdiscs by extreme impact loads. Situations like this frequently happen in some cases even resulting in significant cutterhead damage caused by "grinding" through the boulders (Figure 13).

The disc cutters for the use in pressurized closed mode operation were originally designed similarly to those intended for hard rock TBMs. For higher face pressures improved seal configurations with higher capacity and low friction have been developed that worked successful up to approximately 4 bar. For higher face pressures the use of pressure compensated disc cutters is recommended.

Very good results were achieved with monobloc designs without using exchangeable cutterrings. These cuttertyps are more expensive but less sensitive to secondary wear or damage. Where ground conditions make it feasible twin cutters can probably also be used because of the reduced risk of failing to rotate in soft ground (Figure 14).

Based on extensive experience with STBMs mining through gravel with cobbles and boulders it appears to be likely that in most cases no boulders bigger than 500 mm in diameter or remaining pieces of them would fall out of the face or enter the excavation

Figure 13.   Cutterhead damage as consequence of broken cutter disc

Figure 14.   14" Monobloc cutters as used on East Side CSO Portland

chamber. These findings are based on the fact that entering the chamber to recover boulders has no longer been a frequently needed operation after the introduction of jaw crushers with of this grain size capacity.

Due to the fact that there is still no solution for the installation of a stone crusher mechanism in an EPBM, boulder sizes smaller than 500 mm that must be considered as being critical for these machine types. Whereas there is no real difference between STBMs and EPBMs for the excavation with disc cutters in the face the handling after entering the excavation chamber is critical should the screw conveyor size of the EPBM be incapable of tackling such grain sizes. This problem can be solved in very large machines but becomes more and more critical for diameters below 6m. Here, screw conveyor diameters in excess of 800 mm can scarcely be realized.

CONCLUSION

Hand in hand with the great extension of the application range for STBMs and EPBMs in the last decade significant improvements in the design of cutterheads and all kinds of cutter tools have been achieved. Even though some border lines are being passed like EPBM—diameters beyond 15m or hard rock STBMs for face pressures beyond 10 bar, there is still enough left to improve and develop. Some of the key tasks for the near future may be cobbles, boulders or variable ground conditions for EPBMs, high face pressures for STBMs and most important permanent improvement of personnel safety when entering the chamber under pressurized conditions.

DRIVING THE TWIN BORE RUNNING TUNNELS BY EARTH PRESSURE BALANCE MACHINE ON THE SOUND TRANSIT C-710 BEACON HILL STATION AND TUNNELS PROJECT

Steve Redmond

Obayashi Corp.

Chris Tattersall

Hatch Mott MacDonald

Nestor Garavelli

Obayashi Corp.

Tomyuki Kudo

Obayashi Corp.

Peter Raleigh

Parsons Brinckerhoff

Michael J. Lehnen

Hatch Mott MacDonald

ABSTRACT

Twin bore running tunnels driven approximately 4,500 route feet by a Mitsubishi Heavy Industries (MHI) EPB Tunnel Boring Machine link the West Portal and East Portal on this complex Light Rail Project. One third along the way, the TBM arrives at the Beacon Hill Station which is the deepest underground "binocular" station of its kind in North America and then "walks-across" the station to be re-launched on the other side to complete the drive.

This paper discusses briefly the layout and design considerations but then focuses more deeply on the approach by the Contractor to drive and supply the tunnel. The paper addresses the approach used to launch, re-launch and recover the TBM and touches briefly on muck volume control and monitoring plus settlement monitoring and the performance achieved.

Also of interest is a unique solution to a difficult launch situation involving a secant pile tieback wall designed with several rows of double I-beam steel walers spanning the eye at the West Portal under the I-5 freeway.

TBM RUNNING TUNNEL PLAN AND ORIENTATION

The underground portion of this project connects two surface portions either side of Beacon Hill. It connects a portion which runs south from Union Station with a portion that follows the Rainier Valley south towards Tukwila. Beacon Hill is a glacial deposit rising some 220 ft above the adjacent ground. Commitments to the community required the construction of a station in the vicinity of the commercial center of Beacon Hill (Figure 1).

The vertical alignment selected balanced the geometrical design parameters of the light rail system—maximum allowable grade of 4.5 percent—with the selected

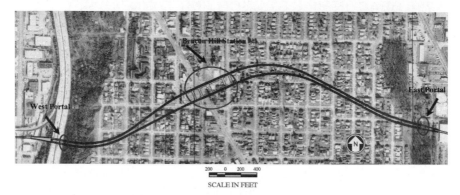

SCALE IN FEET

Figure 1. Beacon Hill Tunnels and station horizontal alignment

elevation for the station platforms, for which geotechnical consideration were paramount. Sequences of hard clays and tills between 150 and 200 ft below grade were projected be the preferred tunneling horizon for the platform tunnels and ventilation adits, all of which were to be constructed using the Sequential Excavation Method (SEM). This meant that the twin running tunnels would need to be driven at the limiting grade for the western portion of the tunnel alignment.

Construction access and community impact considerations steered the Owner towards requiring the tunnel drives to be staged from the west, at a site beneath the I-5 viaduct. The West Portal (WP) located above the valley floor in very soft landslide material made it necessary to construct a mechanically stabilized embankment to provide sufficient space to assemble the TBM prior to walking it under the I-5 overpass to a prepared portal eye for launching.

GEOLOGICAL DESCRIPTION

Beacon Hill is composed of a complex sequence of glacial deposits—primarily clays and tills—with zones of silts, sands and gravels. An extensive boring program concentrated at the station site indicated that the ground was extremely variable over short distances, most likely the result of tectonic activity associated with the site's inferred proximity to the Bremerton Fault. Groundwater investigations revealed a number of disconnected water tables and that dewatering, at least at the station site, would be challenging.

As is common in glacially deposited material, granitic boulders were anticipated based upon excavation records from nearby tunnels. Geotechnical testing also confirmed the recent experience on the Denny Way CSO project that soils would be quite abrasive. However, by-and-large the predominant tunneling medium—hard clays and tills—was considered very-good. Refer to the Geotechnical Baseline Report prepared by Shannon & Wilson for a geotechnical profile along the alignment of the SB Running Tunnel.

DESIGN CONSIDERATIONS

Recent project experience in Toronto and Minneapolis confirmed that single-pass precast gasketed segmental linings were appropriate for this project—providing a high quality finished product with a rapid build time—a fact confirmed by the Owner's peer

Table 1.   Owner-specified TBM features

Feature	Reason
Earth pressure balance TBM	Closed face technology most commonly used in North America, i.e., easier to find experienced EPB operators.
Provision for injection of soil conditioning agents	Ability to prevent EPB pressure loss in sands and reduce cutter head torque in general
Max design operating water pressure of 6 bar	Practical limit for EPB machines without costly special provisions
Provision for manned intervention to forward chamber under pressure: • Airlock • Cutter head doors	Risk reduction measure due to presence of boulders and potentially very high tool wear rates
Annular grouting through TBM tailskin	Improve completeness of annular grouting
Interlocking of annular grouting to TBM advance	Reduce potential for voids behind the lining
TBM-mounted probe drill	Ability to probe without significant interruption to production means better chance of it taking place

review panel. The designers considered a number of excavation approaches including: open face TBMs, digger shields, slurry shields, earth pressure balance shields and even SEM. The selection of closed-face TBM technology was driven primarily by risk considerations related to anticipated zones of running sands, both dry and hydrostatically charged. The following specific issues were considered by the designers in preparing the tunneling specifications:

- Sandy zones
- Relatively high abrasivity
- Static water pressures of up to 60 feet
- Boulders
- North American experience with closed face TBM's

To cover the range of ground conditions, the Owner's designers specified the TBM features listed in Table 1.

LAYOUT, STATUS AND SEQUENCE OF CONSTRUCTION

The WP site sits alongside the noisy I-5 Corridor adjacent to a light industrial area with few residences. For this reason, the contract documents required that the TBM be launched and serviced from the West Portal to avoid disruption to the community at the East Portal.

The contract documents also required the arrival and re-launch of the TBM at the already completed excavation and support of the Beacon Hill Station (BHS) using SEM techniques. Early on in the project Obayashi Corporation promoted the idea of "mining through" the Platform Tunnels with the TBM and later taking down the segments in conjunction with completion of the excavation and support of the Platform Tunnel. This was promoted to de-couple the TBM progress from excavation and support activities at Beacon Hill Station. Obayashi Corporation began to embark on a

Value Engineering Proposal but soon after abandoned the effort since Sound Transit was not receptive to the idea and seemed to discourage it.

The TBM was originally planned to be setup on the Northbound (NB) side for the first drive. After delivery of the TBM in July 2005 to the north side of the MSE wall, it was decided to switch to the Southbound (SB) side for the first drive due to the perceived better ground conditions in the SB side of the SEM workings at the time. A post award geotechnical exploratory program being conducted on site at Beacon Hill at the time revealed different site conditions than had been originally foreseen. The TBM was assembled on a steel cradle positioned on greased steel plates welded which had been bolted down to the topping slab of the MSE wall. The TBM/cradle was jacked across on the steel plates to the SB side and prepared for final assembly and advance to the headwall.

The TBM was launched at the SB side of the WP headwall in January, 2006 and was driven to the limits of the west dished headwall of the SB Platform Tunnel in July 2006 ahead of the completion of the SB Platform Tunnel. The SB Platform Tunnel was excavated up to the TBM cutterhead and the headwall was redesigned with a ring beam collar around the TBM cutterhead. Later the TBM was shoved out into the Platform Tunnel after the excavation and initial ground support was completed.

The TBM was re-launched on the east side of the SB Platform Tunnel on December 18th, 2006 and at the time of this writing the TBM is underway and expected to arrive at the EP by the first of March, 2007.

After what is expected to be a six week turnaround, the TBM should re-launch on the NB side at the WP by mid-April 2007 and is expected to arrive at the completed NB Platform Tunnel at BHS by mid-June. After re-launching in July from the eastern end of the NB Platform Tunnel, the TBM is expected to arrive again at the EP by mid-September and then disassembled for off-site storage.

SITE LAYOUT AND LOGISTICS

The layout of the West Portal is shown in Figure 2. The MSE wall divides the site into two sections; the north portion is associated with muck and grout for the tunnel and the south portion of the site is used for Segment Storage.

The muck from the first 400 ft of mining was discharged to the WP using muck boxes. Once the TBM and all nine trailing gantries were fully within the tunnel, the continuous conveyor was installed and would be used for the remaining portion of the drive. The muck pile has a capacity of approximately of 20 rings (2,000 cy) which gives storage area to support mining three eight hour shifts. As stated previously the area for muck and segments are separated so that the delivery of segments and loading of muck trucks do not interfere with each other. The muck trucks are loaded on the north side of the MSE Wall with a CAT 988. The 200 ton crane is positioned on the south side of the MSE wall allowing the unloading of segments during deliveries and loading of segments onto segment cars for delivery to the TBM. The positioning of the crane also allows for the grout cars to be moved from the tracks on top of the MSE wall to the grout plant for any cleaning that may be needed. An automatic wheel wash was constructed at the south gate to aid wheel wash attendants in the prevention of muck from being tracked out onto the public roadway.

WEST PORTAL LAUNCH INCLUDING HEADWALL CONSIDERATIONS

The West portal head wall was poured in nine lifts creating a permanent head wall for the box structure. The portion of the head wall that coincided with the diameter of

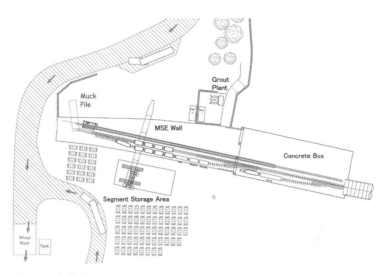

Figure 2.   West Portal site layout

the TBM was modified by cutting the nine whalers and replacing standard rebar with fiberglass rebar. A niche was formed at the center of the headwall to accommodate the nose of the TBM. See later sections of this paper for additional detail.

A steel sealing ring equipped with a rubber seals and supported by "flappers" was presented and bolted to the headwall. The sealing ring closed around the shield and ensured good sealing in case of earth pressure and following burial of the tailshield of the TBM, ensured little grout leakage from the annular space.

A standard steel thrust frame anchored to a reinforced concrete slab was used to launch the machine. The top of the frame was raked back to the slab and the entire frame welded down to steel embedments in the slab. There was great concern about isolating the excavation and support system of the WP Box under the I-5 highway. To accomplish this, a heavily reinforced concrete slab was poured in the invert with reinforced pockets for 2 sets of 12 each cable strand tie-backs deeply seated and anchored into competent soil behind the wall. These tie backs were preloaded to offset the eventual thrust from the TBM and the slab was "kicked" to the headwall to prevent the slab from moving forward prior to preloading of the anchors. Based on the geological conditions at the West Portal and the anticipated load, the foundation of the push frame was designed to resist up to 2,190 kips. This represents approximately 24% of the capacity of the TBM. To accommodate this load, twelve 189 kip six strand tie-downs were installed angled at approximately 30 degrees from horizontal to resist the uplift and the horizontal shear of the frame.

In order to efficiently cut through the headwall and to begin mining, the cutterhead was equipped with seven rolling disc cutters in the last seven gage positions.

"THE DRIVE"

Launching the TBM from the West Portal was challenging. The first 60ft of the Northbound and Southbound (mining west to east) lies in landslide debris and consisted of Class 2 Loose to Stiff Clays and Silt in contact with the full face of the TBM. After the first 60 feet, the TBM was to encounter an overlaying Class 3 Till and Till-Like

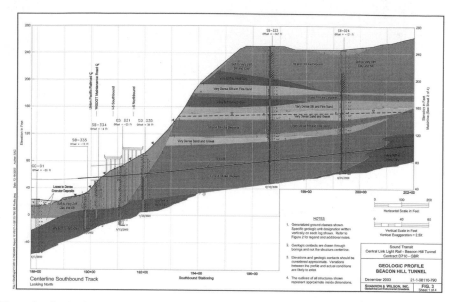

Figure 3.   Geologic profile at West Portal

Deposits in a mixed face and full face Class 3. The launch of the TBM took place under Interstate 5 by first cutting through the 4 foot thick permanent concrete headwall and then the 3 foot diameter secant piles positioned behind the head wall.

After breaking through the concrete wall, the TBM mined through the Class 2 soil. During this period, the pressure and volume of the annular grouting as well as the earth pressure were carefully monitored since Class 2 soil has the potential to shear and cause a slide. After passing the Class 2 soil, the TBM advanced into Class 3 soil. Given the reported strength of the Class 3 Till and given the quickly climbing 120ft–160ft cover over the tunnel permitted a measure of security so no great concern for earth pressure or ground settlement was needed. Nevertheless, the TBM was driven in closed mode with a excellent soil conditioning results use at times water alone, standard anionic surfactant foams, and occasionally a short chain polymer when inrushes of water occurred. See Figure 3.

The first several hundred feet of tunnel was driven using lift-off muck boxes and a 200 MT Kobelco crane. Once the trailing gear was buried, we completed the development of the portal for productive mining including the installation of a Robbins horizontal belt conveyor system. The belt system was selected given Obayashi's recent success using this method in Minneapolis and in consideration of the tunnel grades of up to 4.7%.

The plant and equipment setup for this drive is mostly standard. A 54"-diameter ABC flexible duct carries fresh air to the heading. The rail haulage system comprises 90# Re rail on wood cross ties and three steel ties per panel are installed to control track gauge. The rail packages are prefabricated on a template at the portal and transported into the heading as needed. Lights on 30' centers provide the minimum lighting requirements demanded by WISHA.

There are four 6" diameter thin-wall Victaulic style pipes positioned on the outboard side of the tunnel wall so as not to interfere later with the future walkway concrete operation. The first pipe is dedicated for compressed air supplied by two 650 cfm IR electric compressors positioned up under the I-5 at the portal. The second pipe

Figure 4.   Bring the platform tunnel to the TBM at BH Station

supplies fresh water for TBM cooling, ground conditioning, and wash down. A series of pumps connected to a topping off tank under I-5 provide the water under pressure to the heading. The third pipe is used for hot water return bringing water to a radiator/fan chiller also at the portal. A six inch discharge line is also provided as the fourth pipe.

A 13.2 kV mine feeder cable is supplied in trays laid in "figure 8" patterns and is paid out as the TBM advances. Phone, data communication and mine phone cables are strung out to support those services. A walkway is provided down the center and "step-ups" provided every 250 feet provide a safer haven in the event of a passing locomotive.

A 30ton diesel locomotive transports a complete ring on two low-boy segment cars and the cement/bentonite grout is provided by a weigh-batch plant at the portal and is transported to the TBM grout pumps in a conventional grout car.

Within 250 feet of the west headwall of the "Shifted" Beacon Hill Station, the TBM faced very dense sand and gravel, which did not allow movement even with maximum thrust force (40,000 kN). We injected bentonite around the TBM body for lubrication and tried various methods of soil conditioning, thus giving the TBM ability to move slowly through the sand layer. The TBM arrived to the BH station prior to completion of the Station Platform Tunnel by Sequential Excavation Method (SEM) tunneling. See Figure 4.

"WALK-ACROSS-PLAN" AND RE-LAUNCH AT BEACON HILL STATION

The permanent invert concrete of the SB Platform Tunnel of the Beacon Hill Station was modified so that it could be poured in two lifts to accommodate walking the TBM across the 500 foot long station. Upon reaching the other side of the BHS, the TBM would necessarily be re-launched to continue mining the SB tunnel towards the East Portal.

The Running Tunnel and consequently the TBM was largely offset from centerline of the Platform Tunnel creating tight working conditions where the TBM consequently is up against the sidewall of the Platform Tunnel. Instead of attempting to construct and later remove a time consuming steel shoring and roller system in this space constrained zone, it was decided to put those efforts towards constructing part of the permanent structure. Also, instead of pouring the complete platform tunnel invert concrete necessitating lifting the TBM and trailing gear in the air to match this elevation, the portion of the invert that fall within the envelope of the Running Tunnel was modified to contain a disc shape cradle with the radius of the TBM. Form-Saver rebar couplers in

Figure 5.   TBM cradle in permanent invert concrete at BH Station

the first lift were installed so the remaining portion of the invert could be poured at a later date to meet the requirements of the contract documents. See Figure 5.

Rails were put in place to assist with sliding the TBM across the station. Pockets on five foot centers were carved out of the fresh concrete during the pour to receive a pair of "rail grabbers." A steel table to bridge the tail brushes in the tail of the TBM was used to pass the load from the shoes of the two lower thrust rams of the TBM in the invert to a cross beam catching the rail grabbers. The hydraulic power-pack normally used to power the take-up unit of the horizontal belt was borrowed from the portal to provide a source of oil while indexing the TBM across the slab to the re-launch position.

Upon the premature arrival of the TBM to the Beacon Hill Station and prior to walking across the Station, military-style electrical couplers were installed on most of the electrical lines connecting the TBM to the trailing gear. This was done in anticipation of breaking the TBM from the trailing gear in order to put the push frame behind the shield and to expedite the re-assembly of the TBM on the critical path.

The bridge conveyor between the shield and trailing gear was disconnected and supported on a rail bogie. The shield was moved toward the headwall on the other side of the Station, the push frame was installed behind it, and then the trailing gear was pulled forward to re-unite it with the TBM.

As was the case for the launch of the TBM at the WP a push frame needed to be installed behind the TBM. The frame was modified to accommodate the Station geometry since the Platform Tunnel did not afford us the same working room around the TBM provided in the box structure at the WP. The frame was modified to pass the vertical component of the raked loads on one side of the thrust frame onto the permanent invert. A large hole in the invert was cut and a steel embedment was installed and backfilled with concrete. The thrust frame was welded to this embedment so that the horizontal loads were passed into the shotcrete lining in the invert. As the excavation and support of the Platform Tunnel approached this area, the use of steel fiber reinforced shotcrete was temporarily suspended and replaced with double mats of welded wire fabric for additional strength. A thickened zone was created directly overhead so that a steel column installed between the top of the thrust frame and the shell would take the uplift and not cause the column and base plate to punch through the shotcrete shell. Posting the frame up to the ceiling prevented the need for installing tie-backs to resist the uplift. See Figure 6.

On the outboard side of the Platform Tunnel, there was not enough space to practically install the rakers as described for the inboard side. Consequently, a different

Figure 6.   Thrust frame and TBM at   Figure 7.   East Portal excavation for TBM recovery
east headwall of SB Platform Tunnel

approach was used to pass the loads to the ground. During the excavation and support of the Platform Tunnel, a large niche was constructed in the sidewall in anticipation of accommodating this reaction into the lining. Again, the use of steel fiber reinforced shot-crete was suspended temporarily and switched to the double mats of welded wire fabric for added strength. Dowels were installed during the excavation for the eventual pour back of a concrete (or shotcrete) footing to spread the load back to the lining at both the top and bottom of the thrust frame. Two three foot long beams were attached to the frame in a tangential manner to correspond with the line of forces going toward the lin-ing. Reinforcement wa installed and shotcrete was sprayed to spread the load back into the lining of the Platform Tunnel.

Notches were cut into the skid rails under the TBM prior to its arrival there and 1" diameter steel cables with turnbuckles were pre-hung on the sidewall and crown of the Platform Tunnel on 5' centers. These would later be connected and tensioned around the blind rings ahead of the push frame in a manner similar to that of steel straps on an oak wine barrel. This was done in lieu of trying to "kick" the ring with wooden posts to either the Platform Tunnel sidewalls or to some temporary steel supporting structure.

TBM Recovery at East Portal

When the TBM arrives at the East Portal Headwall and the nose becomes visible the whalers will be cut in accordance with the diameter of the TBM. The design of the headwall at the East Portal is similar to that of the West Portal except that the concrete will not be poured when the TBM arrives. See Figure 7.

Disassembly of the TBM for the SB Running Tunnel (RT) will occur from within the excavation for the Box Structure. The TBM body will be disassembled into five sections (Cutter-Head, Front, Middle and Rear Bodies, and the Erector) and transported to the West Portal on low boys. The nine trailing gears will be transported one at a time through the tunnel by using a flat car and the loci. The disassembly/assembly of the TBM will be done with the help of a 500 ton crane and a 200 ton crane as back up.

Once the TBM disassembly is complete on the south side, the EP Box Structure will be constructed complete with a temporary window in the ceiling at the exit location of the NB Running Tunnel TBM. In this case, the Box Structure will be shored from within to allow the crane and trucks for load-out to work from the top slab of the Box Structure.

Figure 8.   Installation of the belt scale on the Beacon Hill TBM trailing conveyor

MUCK WEIGHT CONTROL

The contract specifications did not explicitly call for the contractor to control the weight of tunnel muck generated by TBM excavations. However, the contractor opted to control muck weight because a continuous conveyor system, rather than a traditional muck train, was being used. The continuous conveyor was favored over a muck train because of the steep grades of the tunnel. At least part of the tunnel alignment would have required loaded trains to start on grades as high as 4.5%.

A back-up mounted static trailing conveyor was used to convey the muck to the continuous conveyor transfer chute. A belt scale system incorporating twin balances was installed on the conveyor, as shown in Figure 8. The excavated weight per foot per ring for the first reach of the southbound tunnel is shown in Figure 9. The weight per linear foot was calculated on a ring-by-ring basis in order to remove the variability of the advance lengths so that rings of differing advance lengths could be compared directly. On Figure 9, the curve indicates the weight measurements obtained. The shaded areas show the anticipated moist unit weight that was expected to be encountered during boring in the range of 107 to 148 lbs/ ft^3. As can be seen, the weight of the excavated material was within anticipated ranges except during a short interval where the dewatered sands were encountered just prior to the breakthrough into the Beacon Hill station.

BACKFILL GROUTING CONTROL

The backfill grout injection was also carefully monitored to ensure that the annular gap created by the difference in diameter between the shield and the extrados of the segmental lining was completely filled. The contract called for immediate filling of the annular gap as the TBM advanced in order to counter settlement and avoid eccentric loading of the lining. Figure 10 shows the grout injection volume per foot per ring along the advance. Once again, the variability of the stroke lengths for each advance had been taken into account, allowing for a ring-by-ring comparison of the grouting volume. As is normally expected, the amount of grouting placed during the first 125 rings varied more than the amount placed in the latter portion of the drive. The previously noted zone of dewatered sand resulted in a TBM steering incident that necessitated the use of the hydraulic over cutter to facilitate recovery of the TBM attitude. The volume of grout increased in this area due to the larger annulus created during the recovery.

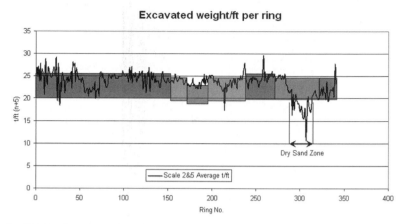

Figure 9.   Weight of muck per foot per ring on the first reach of the southbound running tunnel

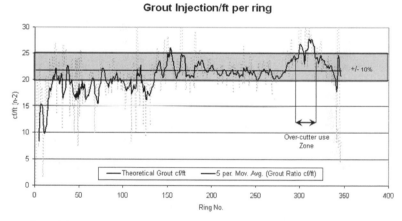

Figure 10.   Grout injection volumes per foot per ring on the first reach of the Beacon Hill Southbound running tunnel

TBM PERFORMANCE

At the end of January, 2006 the TBM was launched in a ceremony marked by great fanfare. The first 350 feet of the drive was completed in six weeks by crews working ten-hour day shifts. During this initial stretch of tunnel, the TBM traversed a short length of landslide debris and then moved into stiff clays. The TBM was halted to permit removal of the thrust frame and initial false rings and to make way for the installation of the continuous conveyor, final track work and switches, and other support equipment needed to streamline tunnel production. The TBM restarted again on April 24, 2006. Crews working the day shift completed the 1,345-foot drive in 12½ weeks, coming to rest in the platform tunnel envelope on July 19, 2006. The platform tunnel was later completed around the cutterhead using the SEM method.

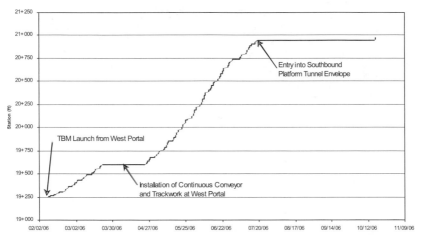

Figure 11. TBM progress for the initial reach of the Southbound tunnel

Figure 11 shows the progress of the initial Southbound reach of the running tunnel from the West Portal to the Beacon Hill station. The first 350 feet was excavated at a rate of 11.7 feet per day, during which muck cars were used to remove tunnel spoil. Following final installation of the conveyor and other tunnel support equipment, the TBM completed the balance of the reach achieving an overall rate of 19.4 feet per day, encountering some difficulties in the dewatered sandy area which were resolved by adapting the ground treatment.

WEST PORTAL BULKHEAD WALL—ADDITIONAL DETAILS

The West Portal (WP) of the Beacon Hill Tunnel Project posed many challenges during the final design phase, due to both the poor ground conditions in the area and its proximity to the I-5 freeway.

One of the most complicated aspects was the bulkhead wall, through which the TBM was launched. The design of this structure is discussed in this section.

Refer to the Geotechnical Baseline Report prepared by Shannon and Wilson for details of the soils conditions in this area. Also see Figure 3 for a geological profile of the West Portal.

Prior arrangements between Sound Transit (ST) and the Washington State Department of Transportation (WSDOT) included the stipulation that the excavation support system at the WP be a prescriptive, owner-designed system. This was due to the fact that the WP structures—both temporary and permanent—would be located below grade and between the piled foundations of three bents of the I-5 freeway, which travels along the base of the west side of Beacon Hill.

Based on the above considerations and engineering analyses, the following decisions were made with regards to the bulkhead wall during the final design phase of the project:

- The bulkhead wall would be a secant pile structure, which would function as both temporary support during construction and as permanent support of the adjacent hillside; designed to ensure the internal stability of the portal structural system during a maximum 100-year design earthquake. Lean concrete piles would be specified in the TBM "eye" to facilitate launch of the machine.

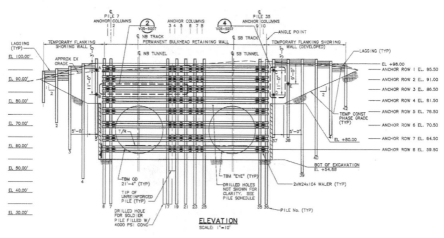

Figure 12. Elevation view of bulkhead wall

- The Till and Till-Like ground behind the bulkhead wall was such that tieback anchors were a feasible method of supporting the wall. Permanent tieback anchor loads ranging from 83 to 128 kips, declination angles and "No-Load" zones would be supplied in the design, and the contractor would be responsible for design of the tiebacks (tendon selection [bar or strand], hole diameter, etc.)

- Steel walers would be designed to span between the tieback anchors in the area of the TBM "eye." The method of removing these walers and supporting the eye prior to the launch of the TBM would be left to the discretion of the contractor. A suggested conceptual approach of constructing a bottom-up temporary concrete structure to be cast against the secant pile wall and supported on its back side by the TBM, with the walers sequentially removed as this structure was built up, was provided in the design.

- The remainder of the excavation support system (along the sides of the excavation) would be comprised of soldier piles and lagging, and would require internal structural bracing since the Landslide Deposits were of such low strength and poor quality that prohibitively long tieback anchor lengths would be required to support the side walls.

- After completion of tunneling operations, a cast-in-place reinforced concrete facade wall would be constructed integral to the secant pile wall, and would be placed over the walers and tieback anchor heads to protect these elements of the structure prior to construction of the box structures, and final backfilling and restoration of the site.

- Long-term drainage behind the bulkhead wall was an important consideration, and therefore drainage piles would be installed behind the wall, filled with free-draining crushed stone type material, and into which perforated pipes would be installed and tied into the subdrainage system under the box structures. The sub-drains would be fitted with cleanouts to facilitate maintenance of the system, and would be tied into storm drainage manholes for conveyance.

The resultant design included in the contract documents is shown in Figure 12, which illustrates a section view of the wall.

A photos of the completed structure is shown in Figure 13.

Figure 13.   Completed bulkhead wall

The Contract Documents provided a "Suggested Plan" by where each wale was to be removed one at a time from the bottom-up and temporary concrete poured up against the cutterhead of the TBM. This was to be repeated until the entire eye was clear of steel walers and the load passed onto the TBM and its thrust frame.

Obayashi was deeply concerned about this method since pouring concrete up against the cutterhead would not only be problematic from a practical point of view but also it would mean having to start the cutterhead up under load. This would interfere with commissioning and Start-Up Testing activities which require free rotation of the cutterhead. Also, the resulting reactive moment would cause the TBM to roll over excessively on the cradle. Obayashi wanted to use a flapper style-sealing ring and this method would not allow us to do so. Also, from a schedule point of view, the removal of the wales with the TBM in place put the TBM Assembly directly on the critical path of the TBM drive.

Obayashi proposed the following plan to facilitate this operation:

- Rather than building a temporary structure, the final cast-in-place facade wall would be sequentially constructed in a bottom-up fashion, with the walers removed one-by-one for the height of the wall necessary to clear the top of the TBM.

- The reinforcing for the wall would be tied into the "stub" pieces of the walers remaining at the tieback anchor head locations so that the wall would span the TBM "eye," and act as a waler.

- In the area of the TBM "eye," Glass Fiber Reinforced Polymer (GFRP) reinforcing bars would be used instead of steel bars. The GFRP bars would be tied to the steel bars outside the area of the "eye" (similar to slurry wall installations), and this would allow the TBM to mine through both the concrete facade wall and the secant pile wall behind. Steel fiber reinforced concrete was first considered but then dropped due difficulty achieving the desired resistance to the specified loads.

A photo of this wall prior to TBM launch is shown in Figure 14.

A truly collaborative effort was put forth by all parties—Obayashi, ST, PB, HMMJ and Shannon & Wilson—during the installation of the secant pile bulkhead wall, and the planning, design and construction of the GFRP concrete facade wall. This effort paid off

Figure 14.   GFRP wall prior to TBM launch

in a partnership that solved problems as they arose, overcame some difficult design and construction issues, and ultimately resulted in a successful launch of the TBM.

CONCLUSION

A lot of hard work provided by the planners, Sound Transit site personnel, the geotechnical investigators, the designers, the members of the Contractor's team, the members of the inspection team, and by the TBM manufacturer has made the execution of the Running Tunnels portion of this complex project a success so far. An interesting aspect of this EPB project is the interaction required at the interface with the conventional excavation of the Beacon Hill Station using the Sequential Excavation Method. As usual the TBM entry and exit conditions provide an interesting challenge and this project is no different. An interesting solution to what seemed at the time an impossible starting situation at the West Portal headwall was provided by the willful collaboration of the Contractor, the geotechnical engineers and the box structure shoring designer.

ACKNOWLEDGMENTS

The authors of this paper would like to make recognition of all those individuals involved deeply in the success of this aspect of the Work but who could not participate in the paper due to practical space considerations. Special recognition is given to Red Robinson and Mike Kucker of Shannon and Wilson who have brought there deep knowledge and understanding of tunneling conditions in the Puget Sound region to the team. Additional recognition is given to Gall Ziedler Consultants, LLC for their assistance to the Contractor.

REFERENCE

C710—Geotechnical Baseline Report; Shannon and Wilson, Dec. 19, 2003.

PLANNING AND CONSTRUCTION OF THE METRO GOLD LINE EASTSIDE EXTENSION TUNNELS, LOS ANGELES, CALIFORNIA

Eli Choueiry
Los Angeles County Metropolitan Transportation Authority

Amanda Elioff
Parsons Brinckerhoff

Jim Richards
Kellogg Brown & Root

Brett Robinson
Traylor Frontier Kemper JV

ABSTRACT

The Los Angeles Metro Gold Line Eastside Extension (MGLEE) extends Metro's light rail system to the densely populated and growing East Los Angeles communities. Much of the 6 mile alignment consists of at-grade track, but the project also features a 1.7 mile underground segment in cut-and-cover and twin-bored tunnel structures. In light of past experiences with open shield construction, The Metropolitan Transportation Authority (Metro) specified, for the first time, use of Pressure Face Tunnel Boring Machines (TBM's) to advance tunnels through soft ground. This paper describes the history of the tunnel design, construction management, and construction methods adopted that resulted in successfully completing the tunneling segment.

INTRODUCTION

The Metro Gold Line Eastside Extension Project (MGLEE) will extend the existing Metro Gold Line Light Rail (LRT), which is presently in operation from Pasadena to Union Station in downtown Los Angeles, to East Los Angeles. Upon completion, the LRT will add 9.6 km (6 mi) to the Metro Rail system, currently comprised of about 118 km (74 mi) of urban rail. Figure 1, Metro Rail System, shows the project location relative to the overall system, and Figure 2, illustrates the project alignment in larger detail. The existing Metro system includes three LRT lines (Blue, Green and Gold) as well as the Red Line Subway (Heavy Rail). The first LRT system, the Blue Line, became operational in 1990, and the first segment of the Red Line system began operations in 1993.

Boyle Heights and East Los Angeles, which the MGLEE will serve, are some of the most densely populated regions in the nation. Furthermore, demographics, transit dependency, low private vehicle ownership ratios, bus service capacity limitations, all were clear indications during the planning phase that a rail transit extension would be the most cost effective system to serve these communities. Closely working with Metro's funding partner, the Federal Transit Administration (FTA), Metro successfully validated the cost effectiveness of the project at $898 million. Conservative ridership projections have been estimated at 23,000 boarding's per day in the year 2020.

The project includes the construction of a dual track system in three cross-sections: aerial (2 bridges), tunnel (2.7 km/1.7 mi) and at-grade (7.1 km/4.4 mi). The

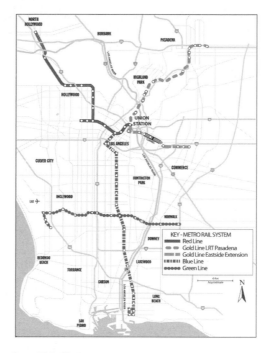

Figure 1. Project location within Metro system

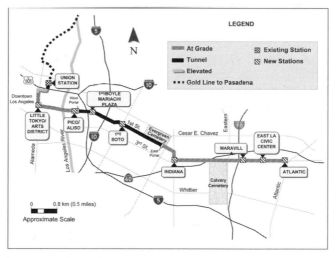

Figure 2. Project plan

at-grade double track sections are constructed for the most part in the center of the city streets. While most of the alignment could be accommodated at grade, parts of 1st Street in the older, narrow streets of Boyle Heights could not be placed at-grade unless parking was eliminated. In this 2.7 km (1.7 mi) section, the LRT system was designed to be underground. Underground structures include two cut and cover subway stations, portals, cut-and-cover sections adjacent to the portals, and twin-bored tunnels (6.5 m/21.4 ft excavated diameter). After over 15 years of planning and design, MGLEE's twin bore tunnels were successfully completed in December 2006.

This paper presents a history of the design, construction management, and construction methods used principally for the tunnels. A "companion" paper, prepared for this conference by Robinson et al. provides more detailed information on the tunnel contractor's means and methods adopted to complete the project.

BACKGROUND

Metro Red Line Subway

Design of tunnels and underground stations in the Los Angeles area has always had to include special considerations for development of seismic design criteria and subsurface gas exclusion. Additional considerations with respect to subsurface gasses began in 1983 after a methane explosion occurred (under an existing building) near one of the proposed subway alignments. This renewed awareness of subsurface gasses prompted, not only re-alignment of the subway system west of downtown, but also special federal legislation to prevent use of federal funds for rail tunnels in Los Angeles' newly created "Methane Zone." Subsequently all Metro subway tunnel designs included "two-pass" lining systems—typically using an initial lining comprising expanded pre-cast concrete segment with final cast-in-place concrete linings of 5.44 m (17 ft 10 in.) I.D. High Density Polyethylene (HDPE) material placed between the linings was added to further protect against gas inflow. Operating systems also included gas detection and automatic ventilation systems. The traditional tunneling method used in Los Angeles' soft ground was open face shields—and these had been used for all soft-ground Metro tunnels. One exception was the tunnel through the Hollywood Hills (4.63 km (2.9 mi) of the 29 km (18 mi) system) that included hard rock tunneling with ribs and mesh initial support.

Red Line Eastside Extension (Suspended Project)

Design and construction of the Red Line system were implemented in discrete manageable sections in terms of design and construction funding. These were termed "Minimal Operating Segments" (MOS). The first operating segment (MOS-1) opened in 1993, followed by completion of MOS-2 and MOS-3 in 1999 and 2000 respectively. The Metro Red Line Eastside Extension was planned to be included in MOS-3 as an extension to the Red Line system to East Los Angeles, with design beginning in 1995. While the extension was ultimately designed using Light Rail Transit (LRT), the previous design as heavy rail is reviewed for background on the development of the design and tunneling specifications. The extension from Union Station to the East was to be accomplished in two phases: the first a 6.4 km (4 mi) tunnel segment having four underground stations. Figure 3 shows the original Red Line alignment against the current MGLEE alignment. The Red Line alignment continued the existing Red Line tunnels under Union Station south from the station, then east under the Los Angeles River, and northeast to end at 1st and Lorena Streets. The second phase—not taken to final design—would bring the underground system to Atlantic Boulevard. This Eastside Extension design progressed

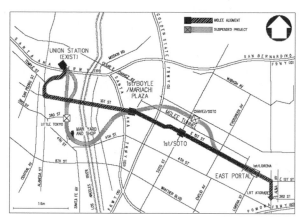

Figure 3. Suspended project alignment versus MGLEE

to 100 percent level, but was "suspended" in 1998 due to funding shortfalls and local legislation preventing use of local transportation funds for subways.

As mentioned, previous Metro soft ground tunnels were constructed using open shields and two-pass lining systems. Final design for the Eastside Extension (Red Line) began in 1995—just after some highly publicized events occurred during MOS-2 and MOS-3 tunnel construction in 1994 and 1995. These included surface settlement, construction cost over runs, and a major sinkhole on Hollywood Blvd. In part due to these events, Metro commissioned a Tunnel Advisory Panel to assess tunneling conditions in Los Angeles with respect to worldwide experience and available tunneling technology. In summary, the panel concluded that by comparison with worldwide experience, geologic conditions in Los Angeles were generally more favorable for tunneling and that tunneling performance was actually "slightly better," based on settlement data. However, use of pressure-face TBMs was recommended for future projects (Eisenstein et al. 1995), in line with the state of the art technology.

Initial design for the Eastside Red Line tunnels included new challenges for Metro given the construction issues on prior tunnels, as well as the discovery of a new site-specific seismic feature and subsurface contamination, in addition to the regional seismic and naturally occurring gas conditions.

Subsurface Conditions. Geologic conditions for most shallow tunnels in the Los Angeles Basin consist of alluvial soils of Holocene and Pleistocene Age. Alluvium generally comprises inter-layered dense sands and medium to hard clays. However, in some locations siltstone bedrock has been uplifted, and occurs at shallow depth—equivalent to tunnel elevations. Furthermore, the area of the Los Angeles River contains boulders and cobbles. Groundwater levels varied along the tunnel such that portions of the tunnel were below the groundwater table. High levels of ground and groundwater contamination were discovered during final exploration within the alignment segment west of the Los Angeles River. This area had been subject to heavy industrial use, including oil storage tanks and a manufactured gas facility. Contamination included hydrogen sulfide (H_2S) levels well above safe limits. H_2S occurred both in dissolved form in groundwater and as gas in unsaturated zones, with maximum measured concentrations of 21,000 ppm accumulating in test wells.

Buried Thrust Fault. During geotechnical explorations for the tunnel in 1996, geologists postulated that an escarpment (approximately 13 m (42 ft) rise in the ground surface over a distance of 200 m (650 ft) in one location) actually represented

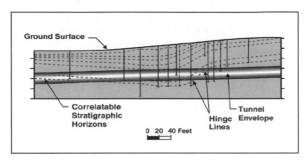

Figure 4.   Stratographic horizons from buried thrust fault movement

a potentially active buried thrust fault, now called the Coyote Pass Escarpment. (A buried thrust fault ruptures at depth and causes surface deformation (bending) but not an
abrupt offset). Extensive geotechnical investigations were carried-out to estimate fault
recurrence intervals and deformations should a seismic event occur along this feature.
Neither the tunnel nor the thrust fault was linear, and the tunnel would cross the fault in
three locations. At two of the three locations, predicted ground distortions would produce strains capable of crushing a concrete lining. At these crossings steel liners were
designed to be installed for additional flexibility. Figure 4 illustrates bending of stratigraphic horizons from thrust fault movement.

Building Protection. The non-linear character of the alignment produced an estimated zone of influence encompassing over 200 buildings of commercial, industrial,
and residential designation. Shallow tunnel depths (9 to 12 m [30 to 40 ft]) adjacent to
the station areas further raised concerns about potential for building settlement.

Pressure Face TBMs. Given the settlement predictions and growing use of
Pressure-Face Machine (PFM) technology worldwide, Metro specified PFMs and a precast concrete, bolted, gasketed lining for the Eastside Extension. Using an estimated
ground loss of 1.0 percent as a design criterion, based on data from other soft ground
tunnel excavations at the time, over 50 structures would require additional protection to
avoid structural and/or architectural damage. A program of compaction grouting (no-
slump grout injected above the tunnel crown from the surface during tunneling) was
designed to reduce settlement to acceptable levels. Similar compaction grouting programs, though not as extensive, had been used for previous Metro tunnel contracts.

For the reach west of the Los Angeles River, slurry machines were to be used exclusively for additional containment of H_2S gas. A program of field exploration, laboratory
testing, and "bench testing" was carried out to identify sources of the gas, estimate quantities, and to develop pre and post H_2S treatment specifications (Jacobs et al., 1999).
Previous investigations for another proposed Metro alignment, the Mid-City Extension of
Segment 3, also produced data for development of tunneling specifications using slurry
TBMs exclusively to contain hazardous (and toxic) gases in the slurry lines until they
could be finally treated at the surface (Elioff et al., 1995). These studies included soil gas
extraction and pre-treatment by injection of hydrogen peroxide and/or air sparging.

Seismic Design. Metro's design criteria for ground shaking uses two design levels: (1) the Operating Design Earthquake (ODE) and (2) the Maximum Design Earthquake (MDE). Under the ODE, structures are to be designed to have little damage and
go into service immediately after an earthquake. Under MDE shaking, structures must
not collapse or cause loss of life. ODE's have an average return interval of 100 years,
while the MDE return period is 2,000 years. A probabilistic Seismic Hazard Analysis is
performed to characterize ground motions in the project location.

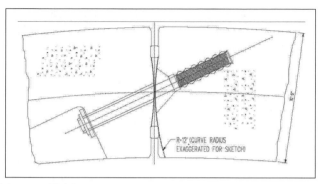

R-12' (CURVE RADIUS
EXAGGERATED FOR SKETCH)

Figure 5. Radial segment joint

Given predicted tunnel distortions (racking) due to ground motion, as well as potential for gas intrusion through segment gaskets during an earthquake, Metro designed a double-gasketed precast segment to add redundancy to the gas exclusion system. The traditional two-pass system with expanded initial liners would not be acceptable with the use of PFMs, since effective use of pressure-face tunneling also requires gasketed segments and grouting immediately after the liners are assembled. No precedent for one-pass liners in gassy ground was found, and Metro undertook further studies to develop a one-pass liner with additional gas sealing properties.

This one pass liner consisted of a double-gasketed segment with convex radial joints—principally to "flex" during earthquakes so that the tunnel remained sealed from gas. Figure 5 illustrates the double gasket concept for a radial segment joint. A six-month, full scale, laboratory testing program was conducted, at the University of Illinois, to evaluate tunnel structural capacity under seismic and ground loads. Tests were performed both with direct axial loading and with a slight rotation of segments to simulate distortion from an earthquake. In both cases, load at failure (due to splitting) was about one-half of the theoretical total load capacity of the segment, but several times the design load. To test segments after cyclic loading to simulate an earthquake, segments were rotated under working loads and no damage to gaskets under cyclic loading was observed. To test leakage subsequent to cyclic loading, water pressure was introduced between gaskets. Initially, under low pressures, no leakage occurred. However, some occurred through bolt pockets at pressure of about 3.4 bar, or over two times the expected field conditions. Similarly, a dilute methane was introduced between the gaskets to examine gas leakage after cyclic loading. Here again, very minor leakage occurred. Figures 6 and 7 are photos of the testing frame used for the gasket gas leakage test and cyclic loading test.

To add further redundancy against potential gas intrusion, the final liner design also included a slightly oversized tunnel—about 15 cm (6 in.) on the radius, to allow for future placement of an HDPE liner with 15 cm (6 in.) additional concrete to secure it in place. As for the existing operating portions of the subway, Eastside Extension design called for continuous gas detection in the operating tunnels and automatic activation of emergency ventilation.

Funding Loss and Project Suspension. Mainly due to some publicized events, including those briefly described, along with an economic recession and the perception of routine construction cost over-runs, local legislation was passed by Los Angeles County voters in 1998 prohibiting use of local funds (the half-cent sales tax) on subway construction. The project was taken to 100 percent design but not advertised for bid.

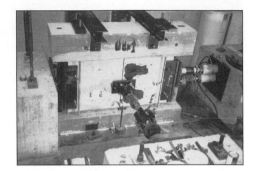

Figure 6.   Testing frame for gasket leakage   Figure 7.   Cyclic loading test

METRO GOLDLINE EASTSIDE EXTENSION

After assessing funding and tunneling safety, Metro conducted a "re-structuring plan" and a "Regional Transit Alternatives Analysis" in 1998 to evaluate available funding and fixed guideway alternatives to heavy rail for three transit corridors, including the Eastside. A new Alternatives Analysis and environmental process was undertaken in 1999–2002, with the preferred alternative for final design being LRT, for the most part at-grade, with 2.7 km (1.4 mi) of tunnel and two underground stations. The LRT project, MGLEE, would extend an existing LRT, the Metro Gold Line, operating from Pasadena to Union Station, to East Los Angeles. Final design began in 2002, and the project was initially issued for bid in 2003.

Design

The MGLEE alignment serves the same communities as the Suspended Project, but extends the trackway an additional 3.2 km (2.0 mi) from Lorena St. to Atlantic Blvd. Major elements of the underground portion included:

- Cut and cover construction of the 1st/Boyle Station (also called Mariachi Plaza Station) and 1st/Soto Station.
- Twin bored tunnels (5.75 m (18 ft-10 in.) I.D.) between Mariachi Plaza and 1st/ Soto Station and between 1st/Soto Station and the East Portal.
- Cut and cover and open cut construction for the West and East Portals and a connecting cut and cover structure between Mariachi Plaza station and the West Portal. This connecting tunnel would be constructed under the existing elevated US 101 freeway.
- Six cross passages and two sump structures (eight cross-connect tunnels total) between the bored tunnels at approximately 200 m (700 ft) spacing.
- A grouting program consisting of: chemical grouting for ground improvement at tunnel breakouts from stations, cross passages, the Interstate 5 freeway crossing, one critical utility crossing and at tunnel break-ins into the 1st/Soto Station and East Portal; and compensation grouting for building protection.

Approximately 1.7 km (6,000 ft) of the MGLEE underground alignment, including the Mariachi Plaza Station, is coincident with the Suspended Project alignment. The MGLEE is in an urban setting and minimal disturbance to community was of primary importance. Project design was by a Joint Venture of Parsons Brinckerhoff, JGM

Engineers, and Barrio Planners, with subconsultants Mactec and Jacobs Associates responsible for geotechnical investigations and building protection studies, respectively.

The new alignment had some advantages in that it was aerial or at-grade west of the Los Angeles River and thus avoided tunneling in highly contaminated areas. In addition the track alignment was mostly in a tangent section as opposed to the numerous curves in the previous project, and it did not cross the Coyote Pass escarpment. Notwithstanding, much of the Suspended Project's tunnel design could be used as a basis for the new project. For example, portions of the geotechnical investigations, pressure-face TBM specifications, double-gasket segment design, etc. were still relevant. The tunnel precast segment diameter (I.D.) was reduced by about 15 cm (6 in.) overall from the Suspended Project's to eliminate the provisional HDPE and cast-in-place liner but added 7.5 cm (3 in.) for the LRT vehicle and overhead power (total of 7.5 cm [3 in.]. on diameter reduction). On the other hand, although the relationship between the tunnel alignment and the Coyote Pass Escarpment changed, additional geotechnical investigations were undertaken to evaluate the 1st/Soto Station design because the station was closely parallel to the feature.

Geologic Conditions. Subsurface conditions along the new underground segment were explored by 71 borings and 14 Cone Penetrometer Tests (CPTs). The borings were drilled to depths ranging from 10 m (33 feet) to 35 m (116 feet), as maximum tunnel invert depth would be about 25 m (80 ft). CPTs were performed to further evaluate a continuous geologic profile for evaluation of soil stabilization alternatives beneath potentially sensitive structures. In addition to the project-specific borings, boring data were available from the prior investigations carried out for the Suspended Project. Nineteen of the Suspended Project borings were used and the boring spacing along the tunnel alignment averaged about 75 m (250 ft). Groundwater and gas monitoring wells were installed in 17 boring locations.

Geologically, the alluvial soil deposits along the MGLEE were broadly differentiated into two units, the Old Alluvium (Pleistocene Age) and Young Alluvium (Holocene Age). Both Old Alluvium and Young Alluvium are heterogeneous and consist of coarse-grained (granular) deposits interbedded with fine-grained deposits. The contact between Old Alluvium and Young Alluvium is generally gradational and is often difficult to distinguish. In very general terms, the Old Alluvium material is more dense or stiffer material than the Young Alluvium. The granular deposits in the Older Alluvium are composed of loose to very dense sand and gravel (with varying amounts of silt and/or clay and cobbles and occasional boulders) that occur in layers ranging from one foot up to about 15 m (50 ft) in thickness. The fine-grained alluvium consists of layers ranging from 15 cm to 20 m (0.5 to 65 ft) thickness of stiff to hard silt, lean clay, sandy clay, and silty clay with sand. It is not uncommon for Standard Penetration Test (SPT) blow counts to exceed 50 in the Old Alluvium. Figure 8, shows the tunnel/underground section profile, geology, and groundwater level.

With the exception of about 1200 ft at either end of the tunnel alignment, the tunnel was below the groundwater level, which reached a high of about 4 m (13 ft) above the tunnel at about mid alignment. The water levels shown on the Contract Drawings represented the highest levels measured over an eight-year period between November 1994 and September 2002, during the investigations for both the suspended and MGLEE projects. Additional readings were taken prior to contract award in 2004 and water levels were found to be stable—even after an El Nino winter.

Subsurface Gas. The project area is located in what was once the Boyle Heights Oil Field and several abandoned oil wells were located in proximity to the alignment. No H_2S gas was detected; however methane was measured with a maximum reading of 1,700 ppm. Cal/OSHA ultimately issued a "Gassy" Tunneling classification.

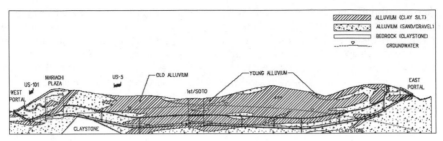

Figure 8. Geologic profile

Tunneling Specifications. As for the Suspended Project, specifications required PFMs for face control in the alluvial ground above and below the groundwater. Selection of the EPB or Slurry TBM was left to the contractor. The primary objectives of the TBM specification were to minimize ground loss and related disturbance to the community and increase tunneling safety. Thus, the specification called for tunneling under positive pressure only (minimum of 0.6 bar (10 psi) above groundwater pressure measured at the invert and maximum of at-rest pressure), as well as immediate backfilling of the tail void by grouting through the tail shield; grouting through segments was not allowed. Other requirements for the TBM tunneling included:

- Supplying two TBMs for accelerated schedule.
- EPB TBMs to have screw conveyor system only.
- Qualifications were given for TBM operators as well as training and certification by the TBM manufacturer.
- New or used machines were acceptable, however if used machines were proposed, they required submittal of basic information for approval prior to bid.
- Compressed air locks to access face.
- Provisions for adding disc cutters, should they be needed in cobbly or bouldery ground.
- Probe hole drilling and grouting through the face.
- Automatic Data Acquisition Systems.
- Ability to "check grout" to evaluate presence of voids above crown.
- "Gassy" tunnel classification—compliance with Tunnel Safety Orders, Title 8, of the State of California for use in Class 1, Division 2, hazardous locations.

Allowable surface settlement based on tunnel depth and first or second tunnel drive was specified. Table 1 shows the action and maximum levels given in the contract. Furthermore, settlement at 1.5 m (5 ft) above the tunnel crown was limited to 3.8 cm and 5 cm (1.5 in. and 2 in.) for action and maximum levels as measured by multipoint borehole extensometers (MPBXs). MPBXs were spaced at about 75 m (250 ft) intervals along the alignment while surface settlement was to be measured at 7 m (20 ft) spacing along the tunnel centerlines, building corners, and points extending to 100 ft offset from each tunnel centerline. Ground surface settlement and extensometer movement were to be surveyed daily or in each shift during tunneling. Metro would monitor the geotechnical instrumentation.

At cross streets, instrumentation arrays, consisting of MPBX's above the tunnel springlines and surface settlement markers (measuring ground surface just below pavement) were placed to determine the overall settlement "trough."

Table 1. Allowable settlement measured at the ground surface above each tunnel centerline

Depth to Tunnel Crown (m/ft.)	First Tunnel Drive		Second Tunnel Drive	
	Action Level (cm/in.)	Maximum Level (cm/in.)	Action Level (cm/in.)	Maximum Level (cm/in.)
3–6.1/10–20	3.0/1.2	3.8/1.5	3.3/1.3	5.0/2
6.4–9.15/21–30	2.5/1.0	3.3/1.3	3.0/1.2	4.3/1.7
9.45–12.2/31–40	2.0/0.8	2.5/1.0	2.5/1.0	3.8/1.5
41/12.5 and over	1.5/0.6	2.0/0.8	2.0/0.8	3.3/1.3

Cross Passages. Six cross passages and two sump structures (within two additional cross passage structures) were to be constructed for emergency exits to the adjacent tunnel, housing equipment for ventilation, communications, fire detection and suppression, electrical distribution and tunnel drainage (sumps). The environmental documents limited active dewatering at the crosspassage locations and chemical grouting was specified to limit groundwater inflows during construction, as well as to stabilize the ground during excavation using sequential excavation techniques. While the precast segmented tunnel lining for the bored tunnel did not require HDPE barriers, given the double gasketed precast segment system, the cross passages required HDPE lining behind a final cast-in-place lining.

Building Protection. Potential effects of tunneling under structures were evaluated through building surveys and analyses. While the MGLEE tunnels were aligned more closely along the public streets than the Suspended Project, the street alignments made some abrupt changes that could not be accommodated within the limits of the TBM tunneling turning radii, and in some cases the tunnel passed directly under or in close proximity to buildings. Furthermore, tunnel depths were shallow (9–10 m [30–35 ft]) at the break-outs from portals and stations. All buildings within 30 m (100 ft) of the tunnels' centerlines were surveyed to identify type of construction and overall condition. Commercial buildings consisted predominantly of old (over 50 years) one and two-story masonry and concrete structures with a few light one-story steel frame structures. The residential structures were primarily single-family wood framed homes, generally less than 140 m² (1,500 sf) in area.

Approximately 220 existing commercial and residential buildings were identified as lying within the zone of influence of tunneling. Given the subsurface conditions and large tunnel diameters, settlement induced by tunneling in some areas could impact a number of buildings, if not adequately controlled. Building construction types along the underground alignment within the potential zone of influence of tunneling are given in Table 2.

Buildings were further analyzed if predicted settlement could produce angular distortions of more than 1/600 or an absolute settlement of over 2.4 cm (1.0 in.), consistent with minor architectural cracking thresholds. This further analysis included a more detailed survey of the building and foundation, evaluation of geologic conditions and tunnel construction sequence. Calculations were made of theoretical ground surface horizontal movement and strain to characterize how each building might behave with ground movement and how damage could manifest. The analysis followed methods described in Boscardin and Cording (1989) and took into account overall magnitude of movements and the building location within the settlement trough. Subsequently, estimated building strains were plotted against distortion to categorize potential damage classification. Where levels exceeded "very slight" (more than fine cracks less than

Table 2. Distribution of building types along the underground alignment

Location along Alignment	Type of Structure			
	Unreinforced Masonry	Concrete/CMU	Steel	Wood
West Portal to 1st/Boyle Station	7	3	0	8
1st/Boyle Station to I-5 Freeway	10	16	3	13
I-5 Freeway to 1st/Soto Station	18	7	3	12
1st/Soto Station to Evergreen St.	3	15	2	50
Evergreen St to Lorena St.	1	4	0	44

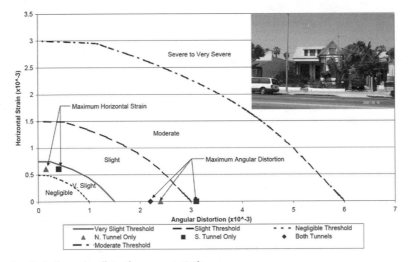

Figure 9. Plot of angular distortion versus strain

1.0 mm), additional building protection was called for. Figure 9 displays a typical plot of strain vs. distortions and damage categories for a single building. The plot shows esti-mated distortions for each tunnel drive and for after both tunnels have been driven.

Considerations for building protection during tunneling included:

- Compaction grouting (post-excavation ground improvement where no-slump grout is injected at high pressure above tunnel to compact ground "loosened" during tunneling).
- Chemical grouting using sodium silicate (pre-injected chemical grout used to add cohesion and strength to granular soils, and reduce water inflow).
- Compensation grouting (pre and post injection of very high slump cementitious grout at high pressures to fracture and lift the ground). Grout is injected about mid distance between the ground surface and tunnel crown elevation.

Compensation grouting was selected as the preferred method for best "real-time" control over settlement of buildings in variable geology. Grout pipes and building moni-toring points could be pre-installed ahead of the tunneling, and the specifications called for grouting to be initiated with the first evidence of structure movement, with action level set at 6.4 mm (0.25 in.) and maximum level 12 mm (0.5 in.). Grouting

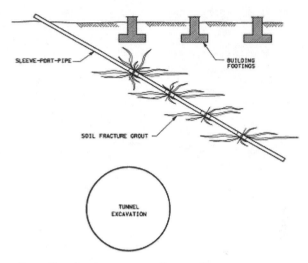

SLEEVE-PORT-PIPE

BUILDING FOOTINGS

SOIL FRACTURE GROUT

TUNNEL EXCAVATION

Figure 10.   Schematic section showing compensation grouting

crews were to be on stand-by at all times when the TBM was within 30 m (100 ft) of a structure to be protected by compensation grouting. Design drawings indicated all grout pipes to be installed from the public right of way (street) to cause minimal disturbance to building occupants. Grout pipe spacing averaged about 2 m (7 ft). Figure 10 shows schematically the grouting method.

Ground stabilization with chemical grout (sodium silicate) was to be used for tunnel breakouts from portals and shafts, protection of the Interstate 5 Freeway overcrossing, a large-diameter water main, and at cross passage locations. Chemical stabilization had been used successfully on previous Metro projects, particularly at freeway crossings, and is now a typical requirement by the state highway department (Caltrans). Figure 11 is a photo of typical cross passage grouting from 1st Street.

With the exception of the north side of Mariachi Plaza Station, existing buildings were not in close proximity of the station excavations. At Mariachi Plaza, one three-story brick building, constructed in 1889, would be approximately 2 m (6 ft) from the station excavation support system. Another structure near the excavation support system was a large limestone kiosk, now the symbol of Mariachi Plaza. This kiosk was a donation the Mexican state of Jalisco, Mexico to honor and encourage mariachi music and Mexican folklore in the City of Los Angeles. Figure 12 depicts the kiosk at Mariachi Plaza. While most of the station initial support systems were expected to be soldier pile and lagging systems, tangent pile walls or closer pile spacing were designed for the area in front of these historically sensitive structures to reduce potential for ground loss. Design loads were also to be increased in these areas.

Seismic Design. Seismic design criteria for ground shaking changed little from the suspended project. For the ODE, ground accelerations were estimated at 0.41g, and MDE was 0.79 g. Predicted "free-field" racking distortions, were 6.4 mm in 14 m (0.25 in. in 45 ft) and 25 mm in 14 m (1 in. in 45 ft) for the ODE and MDE respectively. However, the orientation of the alignment relative to the Coyote Pass Escarpment meant that Soto Station would be built in close proximity to the fault. Figure 13 shows the base of the escarpment with respect to the tunnel and station. A new geotechnical investigation was undertaken to locate the hinge of the fault and estimate distortions in the event of an earthquake. Using the predicted free-field ground motion, designers

Figure 11.   Cross passage grouting

Figure 12.   Mariachi Plaza kiosk

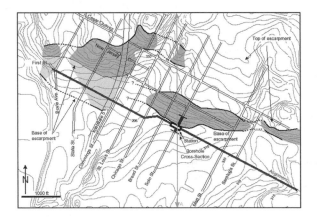

Figure 13.   Plan of escarpment at 1st/Soto Station

used soil-structure interaction methods to develop additional racking criteria for the 1st/Soto Station structure and adjacent tunnel sections near the escarpment (Habimana and Elioff 2006). Results indicated that approximately an additional ten percent distortion would need to be added for ODE and MDE design of Soto Station and 2.5 mm (0.1 in.) for the ODE on top of the predicted distortions from an earthquake on a regional fault. At the tunnel-station interface, designers specified steel liners for additional flexibility in the 6 m (20 ft) beyond the station "box." Beyond the end of the steel liners, the tunnel design prescribed a further 200 ft of "enhanced" pre-cast concrete liners (additional reinforcing), to a location where the fault hinge was estimated to not influence the tunnel.

Station Design. Experience with construction on other Metro Rail projects was that initial excavation support options—generally the contractor's selection would include soldier piles and timber or shotcrete lagging, tangent pile walls, or slurry walls. Soldier piles and lagging systems would require dewatering operations prior to excavation. The Mariachi Plaza station excavations would serve as tunnel access shaft, with mucking operations limited to this location. Twenty-four hour operations were allowed at the mining site.

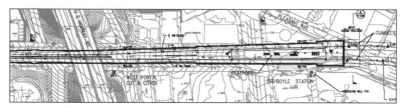

Figure 14.   Mariachi Plaza Station plan

A unique feature of the Mariachi Plaza station was a tapered shape to transition from a cut and cover double track box structure, to a center platform station, to tunnels having a separation of 5.2 m (17 ft). Figure 14 illustrates the Mariachi Plaza Station plan.

Contracting Practice and Contract Packaging

Final design included assessment of construction contract form and packaging, with consideration of bidder qualifications, possible incentives for performance, and use of the Design-Build (DB) contracting methods. Interface issues on previous projects, as well as industry trends led to selection of DB for the at-grade portion of the project and station completion (Contract C0801). Contractor risks in tunneling and underground construction were considered unsuitable for DB and therefore these sections of the work were let as Design Bid Build (DBB) (Contract C0800). A third element of the project, a bridge over the US 101 freeway, was packaged in a separate contract managed by Caltrans (state highway department), who would also be widening/improving the freeway below. The underground contract was advertised first and separately, but was re-bid about 6 months later concurrently with the DB contract, so that a single contract (C0803) could be awarded if this gave lowest overall bid. The award of a single contract would facilitate the management of the interfaces between the DB and the DBB contracts. Initial bid to award duration was over 18 months with the contract NTP in July 2004.

Technical Qualifications. Prior to bidding, contractors were required to pre-qualify, demonstrating that they, or their subcontractors had the experience and capabilities in Pressure Face Tunneling using large diameter TBMs (over 3.7 m/12 ft), tunneling in gassy ground, manufacturing of precast segments, installation of geotechnical instrumentation, handling of contaminated materials and grouting. They were also to submit qualifications of key personnel. Other "pre-qualification" data was for previously used TBMs. Should the contractor consider proposing a used machine for the MGLEE, details of the machines were to be submitted to Metro for acceptance prior to bid, and the contract specifications also stipulated requirements for the machine's refurbishment.

Third Party Coordination and Real Estate

Construction in a congested urban environment, such as Boyle Heights and East Los Angeles typically requires intense interaction and coordination with City and County Agencies. Although this requirement is alleviated to a large extent by constructing underground, there are still significant issues to be addressed relating to the cut and cover tunnel segments and station construction. In addition the tunnel passed under both the US 101 and I-5 freeways.

The relocation of utilities and the obtaining of necessary permits from the City and County agencies, as well as Caltrans with regard to the freeway crossings, can seriously impede the project schedule and result in considerable additional costs due to delays in construction. Metro avoided much of this by embarking during the design stage on close coordination and negotiation with the third party agencies. Some of the components of this process were:

- Metro worked closely with Caltrans to develop, using Caltrans format, Project Reports and Project Study Reports as well as a Cooperative Agreement to provide access onto the State Highway right-of-way and clearly identify and address the impacts of tunnel and cut & cover construction on Caltrans bridges and freeway system.

- Other Third Party Agreements: Metro entered into separate Memoranda of Understanding with the City of Los Angeles, County of Los Angeles and most utility companies to delineate roles and responsibilities for coordination of modifications to existing facilities and cost responsibilities for oversight (engineering, inspection) and self-performed work reimbursement.

- Utility Relocation Advance Contracts: Utility relocations were considered "long lead" items. Metro worked with various agencies and owners to relocate dry utilities (not including wet flow) at the station sites, such that the excavation could begin without undue delay when the Tunnel and Station contract was awarded. These relocations were included under the Memoranda of Understanding.

INTEGRATED PROJECT MANAGEMENT OFFICE

A project of this type is typically multifaceted in terms of the different engineering disciplines, involving specialist types of construction such as tunneling by EPB TBMs as well as standard construction such as utility installation and building construction. Furthermore, the roles and responsibilities of the various engineering professionals differ widely, ranging from design, through project and construction management to commissioning and operations, each requiring a distinct level of skill sets. In addition, construction of an underground project in a congested urban area will have significant community and environmental impacts, which require to be managed and controlled.

The appointment of a team of specialist firms, each with its specific skill sets and areas of responsibility (and liability), can lead to a coordination nightmare and impeded project implementation. For an owner such as the Los Angeles Metro, with considerable expertise in-house, both in the implementation and the operation and maintenance of LRT, as well as established relationships with local communities and third party agencies, the formation of an IPMO is a novel approach with considerable advantages. An IPMO arrangement involves the formation of a separate project office, in which the owner's technical and project management staff allocated to that particular project and those professionals contracted from outside firms work together as an integrated team—all reporting directly to Metro Project Management. Stages in the formation of the team for the MGLEE project were as follows:

1. Metro invited firms to propose for the provision of design services.

2. The design firm was selected and accommodated within Metro's project office. The design and contract documents were completed, incorporating Metro's reviews, contractual practices and operational requirements.

3. On completion of the design, Metro invited Contractors to bid for the construction. At the same time Metro invited firms to provide Construction Management (CM) Staff.

4. Metro selected the CM firm based on qualifications of staff and CM experience to form an integrated team with Metro and design staff in accordance with the skills and staff requirements at each stage of the project.

5. The design firm is retained to provide support services during construction and is accommodated with the CM firm and Metro project staff in a project specific office complex on the construction site. Care is taken to ensure that the organization is seamless with no differentiation based on company or organizational affiliation. The various disciplines incorporated in the team were:
    - Engineering
    - Construction Management
    - Safety
    - Quality Assurance
    - Planning
    - Cost and Schedule
    - Configuration Management
    - Public Relations
    - Environmental
    - Third Party Coordination
    - Procurement

The IPMO for the underground segment of the project primarily comprised the integrated staff of Metro management, design staff (referenced above), and KBR with Carter & Burgess, Inc. for the Construction Management components. Metro staff allocated to the IPMO also performed support functions ranging from safety to third party coordination.

The advantages of such an arrangement are:

- Adoption of common goals.
- Achieve administration efficiencies with little duplication of positions within each company or organization.
- Optimize use of skills and experience within the respective organizations.
- Facilitate interaction between the respective team members: this results in better appreciation of the priorities and faster resolution of issues, taking into account all of the relevant factors (design, schedule and cost), which in turn lead to better Quality Control. This advantage is evidenced in the fast response time on the Contractor's Submittals and Requests for Information (RFI's): as of November 2006, the Contractor made over 2,700 Submittals and 750 RFI's (DB and DBB contracts) and the average response time was about 10 working days on each.
- Facilitated adoption of common criteria, procedures and policies.

The attainment of these objectives also supported the adoption of a partnering approach with the Contractor and third party agencies. Partnering workshops were routinely held and were attended by representatives of the respective organizations and the IPMO, where respective project goals and successes were noted, problems in attaining those goals were identified and working groups appointed to formulate appropriate action plans and find solutions.

CONSTRUCTION

The MGLEE was finally granted $490.7 million from the US Department of Transportation, or slightly over half of the estimated $898 million required for the overall project. Of this, approximately $200 million was estimated for the underground (DBB)

Table 3. Summary of TBM information

Description	Specification
Inside segment diameter	5.74 m (18 ft-10in.)
Segment thickness	267 mm (10-½ in.)
Segment length	1.524 m (5 ft).
Number of segments per ring	5 plus Key
Total length of TBM and backup	105 m (344 ft)
Electrical power installed	3,400 kVA
Shield diameter	6.514 m (256.5 in.)
Total shield length	9.454 m (31 ft)
No. of push cylinders	32 (16 pairs)
Max total thrust force	48,600 kN (2,500 tonf)
Max advance speed	100 mm/min (4 in/min)
Wire tail seal brushes	3 rows
Number of articulation cylinders	12
Cutterhead	Closed Wheel, rotates either direction
No. Disc Cutters	18 double disc—monoblock style
No. Buckets	16
Hydraulic motors	8 motors, Max torque: 112,243 kN/m (3,873,676 lbf/ft)
Erector type	Center free ring with vacuum plate
Grout Injection	4 pairs—with cleanout system—integrated within tail shield.
Screw conveyor	Three screw conveyors in series for total length of 59 m (193 ft). Diameter 900mm (35.4 in)
Foam System	6 ports on cutterhead. 6 independent foam generating systems.
Tunnel Guidance System	ZED Tunnel Guidance System. VMT Ring Building Package.

portion of the project. The construction of both contracts was awarded to a Joint Venture (JV) of Washington Group, Obayashi and Shimmick, with the tunnels subcontracted to Traylor Bros., Inc. and Frontier Kemper (TB/FK) Joint Venture and chemical grouting for building protection and ground improvement subcontracted to Hayward Baker.

Construction began in July 2004 with tunneling beginning in February 2006. As of January 2007, underground stations have been excavated and tunnel construction for the two main bores is 100 percent complete, with cross passage construction now underway. Tunnels were completed using two new Herrenkencht EPB TBMs. Among the unique features of the machines was the long screw conveyor (59 m/193 ft), designed to discharge into the train muck cars a distance away from the face, protecting workers from gas and debris by moving the mucking operations well away from the segment erection and working areas. Summary of other features of the TBM and segment design are given in Table 3. The TBM is shown schematically in Figures 15 and Figure 16 is a photograph of the two TBMs at the Herrenknecht plant located in Schwanau, Germany. Figure 17 shows the double gasketed segments at the TBM heading.

Tunneling. The overall schedule for TBM tunneling was 10 months with advance the following rates for the East Bound and West Bound bores (EB and WB): the best daily production rate for the EB was 28 m (91 ft) with an average of 11 m/day (37 ft/day),

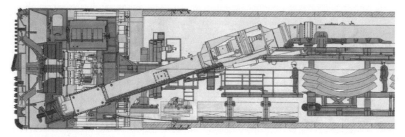

Figure 15. Schematic of TBM

Figure 16. TBMs at Herrenknecht plant

Figure 17. Double-gasketed segments at TBM heading

while production rates for the WB showed a best day of 29 m (96 ft) and an average of 13 m/day (44 ft/day). A principal reason for the lower production rates on the EB was that this was the first tunnel to start and the learning curve was significantly longer. The EB TBM also hit a 700 ft patch of hard/cemented alluvium that hampered production greatly. The WB TBM was able to pass through this same zone of ground with much less difficulty. Initially, each tunnel began mining on a single 8 hr shift, and after about a month boring continued on two 10 hr shifts. Survey of the surface above and in front of the tunnel face was carried out continuously on each shift to carefully monitor for settlement and to assess the need for remedial action, which may include compensation grouting under buildings or adjustment of TBM parameters such as face pressure, foam ratios, backfill grout pressures, etc. The initial surface settlement readings in the 50 ft grouted breakout zone from Mariachi Plaza station showed no measurable settlement. Although this may be generally expected, it is noted that the tunnel started in a 250 m (800 ft) horizontal curve, was shallow at 11 m (35 ft), and traversed under buildings within 60 m (200 ft) from the launching position. Compensation grouting crews were on standby to begin grouting when the TBM was within 30 m (100 ft) of the first building.

After emerging from the grouted zone, again, little settlement was observed, typically less than 0.6 cm (0.25 in.). This was also the case for the extensometer anchors—also showing less than 0.6 cm (0.25 in.) for the bottom anchor which was at 1.5 m (5 ft) above the tunnel crown. The EPB technology had thus performed better than expectations. Figure 18 displays a typical section at an instrumentation array,

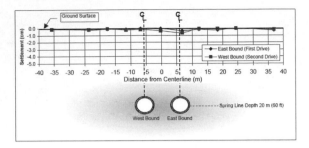

Figure 18.   Typical settlement trough

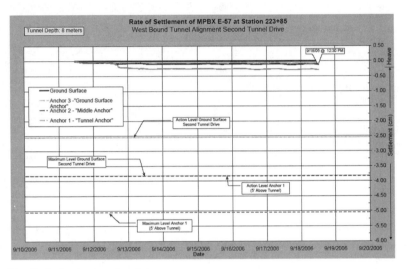

Figure 19.   Typical extensometer plot

while Figure 19 indicates typical extensometer movement above the tunnel centerline. For further details of the construction, refer to Robinson et al.

As-Constructed Tunnels

As with most American tunneling projects, means and methods of construction affect the final design of the respective construction elements and the Contract design specifications. Variations of the design to accommodate the contractor's means and methods are the responsibility of the contractor. For the MGLEE Project, we describe some of the final design variations to achieve the means and methods selected, as well as one of the Value Engineering proposals accepted by Metro.

Precast Segments. As previously described, the contract required precast concrete segments with double gaskets for initial and final lining support. Straight segments were allowed, however tapered segments were to be used at all curves. Final design used all tapered segments with a trapezoidal shaped counterkey and key, and

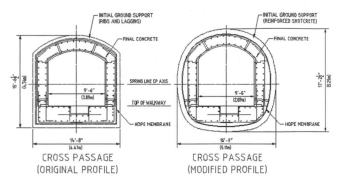

Figure 20.   Cross passage sections

4 parallelogram-shaped segments. Two ring sets were used (up and down) to generally maintain the key segment above the tunnel spring line.

Cross Passages. Cross passage design in the contract included six cross passages and two connecting sump structures. Traditionally, sumps had been constructed separately from the cross passages such that there could be no public access through the sump. A Value Engineering solution using a slightly enlarged cross passage cross-section to house additional equipment and a fire-rated hatch over the sump, as well as provide public access, was accepted by the Los Angeles Fire Department.

For cross passage excavation, Sequential Excavation Methods (SEM) with grouted ground were envisioned. Concerns for excavation with little dewatering led below the groundwater led to Beton und Monierbau of Austria subcontracted by TB/FK to use Finite Element Methods for design of the initial ground support using pipe arches, heading and bench methods and ground support by shotcrete and lattice girders. A modified excavation profile was designed and Figure 20 illustrates the final cross passage excavation.

Tail Shield Grouting. Specifications required backfill grouting through the tail shield in six locations spaced around the shield. Final shield design included eight ports (four pairs). The pairs of ports provided the means to have an active operational port, and a backup port in case of blockages. Trial usage of only the top two ports was proven effective using a two part cementious grout used previously by Traylor Bros., Inc. on the nearby North East Interceptor Sewer Project (Zernich et al.). To examine backfill grouting effectiveness, "check" grout holes were drilled at locations around the tunnel liner and the annulus was found to be fully grouted. Thereafter, only the top two ports (at 10:00 and 2:00 o'clock) were used. However, in case tunnel "squat" or other issues occur, the ability to grout through all ports was maintained. Consequently, a requirement to check grout every other segment was relaxed, and check grouting was used only at a few selected locations.

Compensation Grouting. A major effort for building protection included the compensation grouting program. For the designed program, all grout holes were to be drilled from public right of way—from the streets or alleys fronting the buildings. This called for over 200 grout pipes to be installed for full coverage of all the structures. Instead of the surface installation, Hayward Baker, subcontractor to Traylor Frontier, proposed use of directionally drilled grout pipes installed from either the project worksites or smaller areas within the street. The final number of pipes installed was approximately 35, which significantly reduced impacts to the neighborhood. Figure 21 shows schematically the revised grout pipe installation plan. Monitoring of structures was accomplished using a

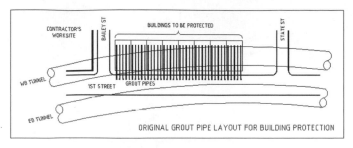

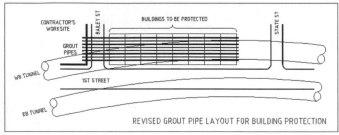

Figure 21.   Compensation grout pipe layouts

total station and proprietary software—which was linked via modem to a data collecting computer in Hayward Baker's Headquarters in Santa Paula, CA. The data could be viewed on site from remote computers using a VPN connection. Ultimately though, no settlement levels were measured that would require grouting to be implemented.

Implications of Success

It has been almost eight years since Metro completed its last tunnel project. Since then, there has been widespread opinion that tunneling for rail transit projects in Los Angeles was over due to unfortunate circumstances related to the past tunneling projects. Because the MGLEE project would not be feasible unless the alignment traversed Boyle Heights via a tunnel section, there was a consensus to proceed with as short a tunnel section as possible in order to obtain political support for the project. It was also recognized that the performance of this tunnel section during construction would either seal the fate of tunneling in Los Angeles or would resurrect the confidence in Metro's ability to successfully construct rail transit tunnels, and perhaps open the opportunity for future extensions and new underground segments. This responsibility became a principal goal and objective for the IPMO team and the Contractor.

Now that the tunnels have been successfully completed, with minimal disruption to the community and no impact to existing infrastructure and buildings, the IPMO project team is extremely elated. The Mayor of Los Angeles and other politicians are already announcing their support for major extension to the rail transit system, including extending the Red Line subway west to Santa Monica, as well as constructing a regional downtown LRT connector.

REFERENCES

Boscardin, M.D., and Cording, E.J. (1989). "Building Response to Excavation-Induced Settlement," Journal of Geotechnical Engineering Division, ASCE, Vol.115, No. 1.

Jacobs et al. "Hydrogen sulfide controls for Slurry Shield Tunneling in Gassy Ground Conditions—A Case History. Proceedings, RETC, 1999.

Eisenstein, Dan, et al., 1995, Los Angeles County Metro Rail Project: Report on Tunneling Feasibility and Performance.

Elioff et al., Geotechnical Investigations and Design Alternatives for Tunneling in the Presence of Hydrogen Sulfide Gas, Proceedings, RETC 1995.

Habimana and Elioff, Numerical Modeling to Estimate the effect of Permanent Distortion Induced by a Blind Thrust Fault Earthquake on a Light Rail Underground Station, Proceedings, North American Tunneling, 2006.

Robinson and Bragard, "Los Angeles Metro Gold Line Eastside Extension—Tunnel Construction Case History," Proceedings, RETC 2007.

Zernich et al., Northeast Interceptor Sewer Case History, Proceedings, RETC 2005.

SLURRY SHIELD TUNNELING IN PORTLAND— WEST SIDE CSO TUNNEL PROJECT

James A. McDonald

Impregilo Healy JV

James J. Kabat

Impregilo Healy JV

ABSTRACT

Impregilo/SA Healy JV recently completed this Project using the first slurry shield TBM's in the United States. The Project used two TBM's to mine 18,000 feet in sands, gravels, and silty alluvial materials under the Willamette River and downtown Portland, Oregon. This paper discusses alignment and geological conditions, equipment selection and plant setup, and tunnel excavation means and methods, parameters, and results. Extensive hyperbaric operations were required and will also be discussed.

PROJECT DESCRIPTION

The West Side CSO Tunnel, Shafts, Pump Station, and Pipeline Project (the "Project") was constructed by the Impregilo/S.A. Healy Co. Joint Venture ("IHJV") for the City of Portland, Oregon, Bureau of Environmental Services (the "Owner"). Design services were provided to the Owner by Parsons Brinckerhoff, and Jacobs Associates was retained as the Owner's Construction Manager.

The unique contract approach for underground work, a Reimbursable Cost plus Fixed Fee type of contract, is the subject of another RETC paper. A preconstruction phase took place in January through August, 2002, and the construction phase covered 48 months from September, 2002, to September, 2006.

The $293 million contract is part of a larger scheme to intercept Combined Sewer Overflows that historically have flowed into the Willamette River during rain events, to capture and store that untreated sewage in an underground tunnel system, and then to pump it to the City's treatment plant when the plant has excess treatment capacity after the rain event. The Project was designed to intercept those overflows from the west side of the Willamette. A similar East Side Tunnel is currently under construction along the east bank of the Willamette.

The Project consisted of four distinct components, each with its own challenges and requirements. Those components were the CSO Tunnel, the drop and access shafts, the Swan Island Pump Station, and the connector/interceptor pipelines constructed by microtunneling and open cut methods. Only the tunnel will be addressed in this paper.

The West Side CSO Tunnel flows from the Clay Street Shaft north to the Nicolai Shaft, and then under the Willamette River to the Swan Island Pump Station (Figure 1). The tunnel has an internal diameter of 14 feet and is supported and lined by a one-pass precast, bolted, gasketed segmental liner. The original plan was to drive the tunnel in two drives from a service shaft on the east side of the river. During the preconstruction phase, the Pump Station was relocated to the site of the service shaft, and the tunnel was shortened by some 4,000 feet to its final alignment with a length of 18,200 feet. The service shaft was eliminated. The Contractor then proposed to excavate the tunnel in two drives from the Nicolai Shaft on the west side of the river.

Figure 1. West Side CSO tunnel alignment

The north drive downstream from the Nicolai Shaft was 4,100 feet in length and offered two special challenges. The crossing under the Willamette had to be accomplished in mixed face conditions with about 40 feet of cover from the tunnel crown to the river bottom. When the TBM reached the Confluent Shaft on the east shore, it had to be turned 45 degrees for the final short (150 feet) drive to the Pump Station. The north drive also had an additional design consideration: at the downstream end of the tunnel under the river, there is a possibility of a condition where the internal pressures could exceed the external pressures. The liner was therefore designed with stainless steel banana bolts at the radial joints.

The south drive was 14,400 feet in length and was driven uphill from the Nicolai Shaft to the Clay Street Shaft, passing through two other shafts along the way, Upshur and Ankeny. Challenges along the south drive would include the numerous curves, the crossings under several bridges and their proximate footings, the crossings of the intermediate shafts, the long drive between Upshur and Ankeny shafts (8,000 feet), boulders at the contacts of different formations, and the Troutdale formation.

PROJECT GEOLOGY

Geotechnical investigation and laboratory testing took place in four phases. The Owner and Contractor jointly developed a Geotechnical Baseline Report during the preconstruction contract period. The following information comes from that report.

The Project area is located near the western edge of the relatively flat Portland Basin of Northwest Oregon. The Portland Basin is underlain by thick sequences of basalt indicating that subsidence was occurring during deposition. As the Portland Basin subsided, Sandy River Mudstone and Troutdale Formation materials were deposited. During Pleistocene glaciation, deep channels were cut into the Troutdale Formation. Catastrophic glacial outburst floods deposited both coarse and fine-grained material. Following the last of the catastrophic floods, significant amounts of alluvium consisting of silty sand and silt were deposited along the Willamette River. During modern times, fill material has been added to most of the areas on each shore of the Willamette.

For the tunnel excavation, the three relevant strata were sand/silt alluvium, gravel alluvium, and Troutdale Formation (Figure 2).

Sand/Silt Alluvium

The alluvial deposits consisted primarily of interbedded sandy silt and silty fine sand. They are typically non-plastic to low plasticity with consistency described as soft to

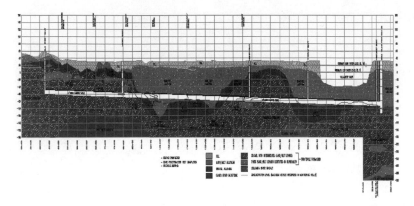

Figure 2. Tunnel geotechnical profile

medium stiff and loose to medium dense. High organic content was possible and boulders were randomly located within the alluvium. Occasional logs were noted in borings.

Gravel Alluvium

The gravel alluvium consisted predominately of poorly graded gravel with cobbles and occasional boulders in a sand matrix. Scattered lenses of silt and sand are interbedded with the coarser material. The gravel, cobbles, and boulders were predominantly basalt. No cementation was observed in the borings, and caving was common in the borings. Boulders of up to 42 inches were expected. Drilling mud loss was common in the upper areas of the gravel alluvium.

Troutdale Formation

The Troutdale Formation consists of very dense poorly graded gravel within a fine grained matrix that ranges from clayey silt to sand. The matrix material provides a weak cementation within the formation. Open graded zones which would have very high groundwater flows were reported. Caving of the borings was also common.

The gravel was very difficult to drill and drive sampling often encountered refusal with less than 6 inches of penetration. Most of the gravel, cobbles, and boulders were strong to very strong Columbia River Basalt, and could be nested within the formation.

Groundwater

Beneath the upper, near surface aquifer lies the Troutdale Formation gravel aquifer. Large groundwater yields were reported from coarse sand and gravel units within the Gravel Alluvium and Troutdale Formation. The aquifer is essentially unconfined and unlimited recharge is provided directly from the Willamette River. The ordinary high and low water levels for the river vary by some 17 feet. The groundwater reflected the elevation of the Willamette. Groundwater elevation could therefore vary by around 30 feet, and was 70–130 feet above the tunnel invert.

Subsurface Gas

Methane was detected in many of the borings, at levels that varied with depth and strata. Due to the presence of methane, the tunnel was classified as "potentially gassy."

EQUIPMENT SELECTION AND PLANT SETUP

Tunnel Boring Machines and Backup

The primary factors in selection of the Tunnel Boring Machine were cohesiveness of the soils, soil permeability, grain size distribution, open zones, face support pressures, presence of subsurface gas, and sensitivity to surface settlement. Analysis of these criteria led IHJV to select a Slurry Shield TBM with a double excavation chamber (Figure 3). In addition to the standard requirements, special characteristics of the TBM's would include:

- Double excavation chamber with compressed air bubble
- Closed cutterhead face with small openings
- Disk cutters to grind boulders combined with ripper teeth to remove fine materials, all back loading
- Jaw crusher in the excavation chamber
- Double manlock for hyperbaric intervention into the face, each equipped for 3 people
- Provisions for annulus grouting through the tail shield
- Automatic guidance and segment setting system
- Designed for maximum hyperbaric pressure of 4.8 bar, including factor of safety
- Slurry circuit of 650 m^3 per hour through 10" pipes

For the shorter north drive, a used Herrenknecht Mixshield TBM was located and procured. This TBM had a slightly larger diameter (5.14 m) than needed and worked with a shorter precast segment (1.1 m) than was desired. It was determined, however, that it would be more cost effective to manufacture precast segments of different dimensions than the south drive rather than do a major rebuild on the TBM. The segments would therefore be 12" thick with length of 1.1 m, and internal diameter of 14 feet.

With the longer south drive, a new TBM and backup was procured from Herrenknecht. This TBM had a diameter of 5.05 m and worked with precast segments of 14 foot I.D., 10" thick and 5 feet long. The backup included a segment quick unloading system and magazine in its standard nine car arrangement.

Separation Plant

Two 800 m^3 per hour Schauenburg separation plants were provided and erected in a warehouse building adjacent to the Nicolai shaft to support the concurrent mining of the two drives. Two centrifuges were provided to help handle the fine materials. A belt conveyor collected the spoils and transported them to a waiting barge in the Willamette River for disposal upriver at a sand and gravel operation. A laboratory was set up in the plant to perform tests on the bentonite slurry and make adjustments as necessary.

Grout Plant

The Specifications originally required the use of a cementitious annular grout and IHJV performed a number of tests with grout mixes and accelerators, and grouting through piping embedded in the TBM tailskin. After numerous trials, an inert mortar composed of sand, flyash, and bentonite was proposed and tested. This concept was ultimately accepted and used throughout the tunnel, although cement was added in the area of break-ins and breakouts to/from the shafts.

Figure 3.   TBM S-231 cutterhead

A batch plant was erected on the surface adjacent to the Nicolai Shaft. A holding tank, with agitating screw feeding a drop pipe, was placed over the edge of the shaft. Custom made grout cars, with a large paddle screw and piston pumps, were fabricated on-site to deliver the grout to the TBM.

Compressed Air Plant

Selection of the compressed air plant was governed by (1) two concurrent TBM drives, (2) need for breathable air, and (3) the need for fail-safe redundancy to maintain the air bubble in the face at all times and to provide additional safety when a hyperbaric intervention was underway. Two completely independent systems were set up around Ingersoll Rand 990 cfm compressors, complete with filters and cooler/dryers, and then a third compressor was added for redundancy. In addition, four receiver tanks were used, with each pair being capable of automatically receiving air from either of the compressor systems. The site industrial air compressor was also plumbed in with the redundant third compressor.

Precast Segment Plant

The precast, bolted, gasketed liners were produced on site in one of the two large warehouses. Because of the two different sized TBM's and the lengths of the two drives, three sets of ring moulds were supplied for the north drive and eight sets of ring moulds were used for the south drive. A universal type of ring layout was used to nego-tiate the numerous curves in each drive. Two castings were made each day in all of the moulds. A steam generator was used in conjunction with retractable tarps fed with the shop gantry cranes to speed the cure of the segments.

TUNNEL EXCAVATION MEANS AND METHODS

TBM Launching

The two Herrenknecht tunnel boring machines (TBMs) were both launched from the Nicolai Shaft. Prior to launching these machines, the Nicolai Shaft base slab was constructed with a cradle to support the TBM during the machine launches. The cradle was later cast into the final drop connect for the shaft after the tunnel was completed. The final shaft lining was also placed to a point one diameter above the crown of the TBM, and two eighteen foot diameter steel liners were installed in the shaft lining for the TBM break-out. These openings extended through the full thickness of the lining to the face of the slurry wall, which was reinforced with fiberglass rebar at this location. Behind the shaft slurry wall at each of the openings, a section of the ground (32 ft. wide × 32 ft. high × 26 ft long) was stabilized with jet grouted columns. A two stage inflatable seal and a rubber double-lip seal were installed in the opening and used to seal between the steel liner and the TBM as it breached the slurry wall. The inflatable seal was then inflated a second time to seal between the shaft lining and the precast segmental tunnel lining.

The north TBM was assembled in the cradle at the base of the shaft. The first two backup cars, containing the controls and the power equipment for the TBM, were lowered into the shaft and temporarily set on top of one another along side of the machine. These cars would remain stationary in the shaft during the initial 265 lft. of excavation. A 290 foot length of hydraulic hoses, and electric and control cables was used to connect the TBM to the power equipment. The cables and hoses were temporarily hung in the shaft on a steel frame attached to a cable hoist. As the TBM advance, the cables and hoses were lowered down the shaft and feed into the tunnel.

A steel thrust frame was Installed behind the TBM and set against the back side of the shaft. The TBM thrust off of this frame as it moved forward into the opening. Temporary pre-cast concrete segments were then installed as the TBM advanced.

When the TBM had moved into the opening with the cutterhead in contact with the shaft slurry wall, steel anti-roll devises were welded on to the central shield of the machine and the first stage of the inflatable seal was inflated with a cement bentonite grout. As the TBM began to excavate through the concrete slurry wall, the bentonite in the working chamber was held at the 50% level to maintain the mucking circuit. After the TBM had mined approximately three feet (partially through the slurry wall), the bentonite level in the working chamber was increased to the full level. Air pressure in the excavation chamber was increased to 0.5 bar. As the TBM completed its excavation through the slurry wall, pressure was gradually increased. When the tail shield of the TBM moved past the inflatable seal, the second stage of the seal was expanded to compress against the precast segments and grouting behind the segments commenced. The TBM continued to operate at the initial slurry parameters until it exited the jet grouted break-out zone. At this point the pressure was again increased and the slurry parameters were adjusted to predetermined values based on the ground conditions described in the geotechnical report. The TBM excavation continued in this manner for a total of 265 feet.

After completing the north drive starter tunnel, the temporary precast segments and the steel thrust frame were removed from the shaft. The trailing gantries carrying the control and power equipment were moved from the shaft into the starter tunnel and connected to the north TBM. The TBM was then left in the "stand by" mode while the excavation of the south drive starter tunnel took place.

The first five backup cars for the south TBM were temporarily placed in the north drive starter tunnel. These cars would remain stationary in the north drive until the first 265 lft. of excavation was completed. The TBM was then assembled in the cradle in the base of the shaft. The launch of the south TBM then proceeded as above.

After excavating 265 feet of the south drive starter tunnel, the temporary precast segments, the steel thrust frame, and the temporary cables and hoses were removed from the shaft. The five backup cars were moved from the north tunnel into the south and permanently connected to the TBM. An additional forty-five feet of tunnel was excavated prior to installing the remaining backup equipment in both the north and south tunnels. At this point, both TBM's were prepared for production mining.

Shaft Crossings

The south TBM intersected two shafts along the tunnel alignment, the Upshur and Ankeny Drop/Access Shafts. Due to the difficult geology at the location of these shafts, the tunnel/shaft connection was designed to allow the TBM to; enter the shaft under pressure, isolate itself from the ground outside, depressurize for TBM maintenance without hyperbaric intervention, and re-pressurize to exit the shaft.

The tunnel/shaft connection structure consisted of three sections: (1) a circular steel TBM entrance chamber with double two stage inflatable seals, (2) a 18.5 ft. wide by 15.5 ft long rectangular room in the center of the shaft, and (3) a circular steel TBM exit chamber with a single two stage inflatable seal. The entrance chamber was longer than the exit chamber to accommodate the double seal and contained a recessed section to protect the seal from damaged by the rotating TBM cutterhead as it entered the shaft. Grout pipes used to inflate the seals were installed through the steel can in the reinforced concrete and extended to the surface. The center room was a reinforced concrete structure that extended to a height of one tunnel diameter above the tunnel crown.

Since the TBM would enter and exit the shaft under pressure, the center room and the entrance and exit chambers would be filled with a material that would resist the operating pressure of the TBM, yet would be easily excavated by the machine. Control Density Fill (CDF), consisting of fly ash, cement, and sand, was placed in the center room to the full height of the room. Compacted sand was carefully placed inside the entrance and exit chambers to protect the inflatable seals and the grout connections. After the sand and CDF was placed in the structures, the shaft was flooded to a minimum elevation of ten feet above the ground water table and left in this condition until the TBM was ready to enter the shaft.

As the TBM started to excavate into each shaft, the water level in the shaft was carefully monitored to insure that water was not leaking through the connection structure into the surrounding ground. If water was leaking out of the shaft, additional grouting would be required to seal these leaks prior to pumping the water out of the shaft. As with the launch of the TBM, fiberglass reinforcement was used in place of steel reinforcing in the slurry wall at the tunnel entrance. As the TBM mined into the entrance chamber and moved forward into the shaft, it removed the sand from the chamber and exposed the inflatable seal. The excavation continued until the cutterhead had penetrated into the CDF in the center of the connection structure. Once the cutterhead was completely through the entrance chamber, the inflatable seals were inflated with bentonite cement grout. As a precaution, a two component hydrophilic grout was injected through grout ports around the perimeter of the TBM near the location of the slurry wall to create a seal between the shaft slurry wall and the TBM.

With the TBM position partially inside the shaft and the seals inflated, the mining pressure was reduced in 0.5 bar increments and the water level in the shaft was lowered fifteen feet every thirty minutes. The slurry level in the excavation chamber of the TBM and the water level in the shaft were closely monitored to detect any possible fluctuation that would indicate infiltration from the ground outside of the shaft and TBM. If the slurry level started to rise or the water level in the shaft increased, the air pressure in the TBM would be increased, water would be pump back into the shaft and additional grouting would be completed.

When the water was completely pumped out of the shaft, and with the mining pressure at zero, the slurry was pumped out of the excavation chamber and an inspection of the cutterhead was conducted. Based on the condition of the cutterhead, it was decided at both shaft locations that work needed to be done that could not take place from inside the tunnel. At the Upshur Shaft, the CDF was excavated to expose the upper portion of the cutter head and repairs were completed in the shaft. At the Ankeny Shaft, the cutterhead was completely exposed, removed from the machine, and shipped to a local fabrication shop for repairs.

After the necessary repairs were completed, sand was placed inside and compacted in front of the cutterhead to a distance of two feet above the machine. A plastic barrier was placed over the sand and CDF was place above the plastic. A concrete cap was placed on top of the CDF near the top of the connection structure. The shaft was then flooded and the TBM began to exit the shaft.

Excavation continued until the TBM and backup had passed the shaft. A separate crew inflated the seal in the exit chamber of the shaft, pumped the water out the shaft, excavated the material from the shaft, and constructed the final drop connection. The tunnel segments at the base of the shaft were left in place and removed after the tunnel was completed.

Hole Throughs

The holing through procedure was similar at each of the three locations, with some slight variations. As with other shafts, a section of the ground (32 ft wide × 32 ft high × 26 ft long) was stabilized with jet grouted columns at the break-in location, and an 18-foot diameter steel liner was cast in place. A two stage inflatable seal was installed in the opening and used to seal between the shaft lining and the TBM as it entered through the slurry wall. The inflatable seal was inflated a second time to seal between the shaft lining and the precast segmental tunnel lining when the tail shield of the TBM moved past the seal.

Around the inside perimeter of the break-in opening, eight grout ports with preventers were equally spaced. The grout ports would be used for the chemical grouting around the TBM when it reached the outside face of the slurry wall. Two grout ports and preventers with indicator pipes were installed near the center of the opening. These indicator pipes would confirm the position of the TBM as it neared the slurry wall. A steel cradle was placed at the bottom of the shaft to receive the TBM as it entered the shaft. As the TBM excavated into the jet grouted break-in zone, the slurry parameters were monitored closely to maintain the quality of the slurry. The cement in the jet grout caused the density and yield value of the slurry to deteriorate rapidly.

The excavation proceeded through the jet grouted zone at an excavation rate starting at 25 mm/min and reduced to 5 mm /min as the TBM approached the slurry wall. When the TBM reached the outside face of the shaft slurry wall, as verified by the indicator pipe, the mining pressure was reduced in 0.1–0.2 bar increments every 10 minutes. The slurry level in the TBM excavation chamber was monitored to insure that the water was not infiltrating in through the face. With the mining pressure at 0.80 bar, the excavation proceeded through the slurry wall. When the point of the cutterhead was within twelve inches of the breaking through the shaft slurry wall, the mining slurry circuit was stopped and the circulation valves were closed. The TBM continued to push through the slurry wall until the entire face was exposed and the cutterhead had moved past the inflatable seal. With the TBM at this location, the 1st stage of the seal was inflated and the excess TBM slurry and muck was removed from the shaft.

At the hole through into the Confluent Shaft, the TBM was pushed forward onto the steel cradle until the back end of the tail shield was at the inflatable seal. The 2nd stage of the seal was then inflated onto the precast segments as the shield was pushed away from the seal. Once the TBM was completely inside the shaft, the TBM and cradle was gradually rotated into alignment for the short drive between the Confluent Shaft and the

Swan Island Pump Station. Steel plates had been placed on the shaft floor prior to the installation of the cradle to provide smooth surface, and hydraulic cylinders were used to reposition the equipment.

For the Swan Island Pump Station and the Clay Street Shaft hole throughs, the TBM was removed in sections as it was pushed into the shaft and the tail shield was left in place to become a part of the final connection of the tunnel to the shaft.

Bridge Undercrossings

The south drive crossed under six of the City of Portland and Multnomah County bridges over the Willamette River. Based on the geologic conditions in the vicinity of each bridge, varying levels of protection was required. Three of the bridges, the Hawthorne, Morrison, and Fremont, were founded in the Troutdale Formation and required no pre-excavation ground improvement. The remaining three bridges, the Burnside, Steel, and Broadway, were founded in the gravel alluvium and the sand/silt alluvium and their deep foundations were either adjacent to or ended above the tunnel. These bridges required three different types of ground improvements prior to the excavation of the tunnel.

The Burnside Bridge required an 8-foot thick, 24-foot wide, and 128 foot long jet grouted slab placed at a depth of five feet above the crown of the tunnel under the foundation of the bridge. Fifteen inclined compensation pipes were also installed under the foundations to be used if settlement was detected during tunneling. For the Steel Bridge, 24 foot wide by 24 foot high jet grouted zones were placed at the tunnel elevation with the zones centered horizontally on the tunnel and two feet above springline vertically. These zones were placed adjacent to the bridge abutment foundations. Four separate jet grouted zones were placed for a total of 260 lineal feet. The Broadway Bridge required a 110 foot long, 71 foot high, and 3 foot thick jet grouted wall to be placed between the bridge piles and the tunnel. This wall extended from depth of thirty five feet to the tunnel invert and was placed twenty one feet from the center of the tunnel.

In addition to these pre-excavation ground improvements, a "Settlement Action Plan" was prepared and put in place prior to excavation of the tunnel. This plan was based on the principle that the best method for limiting potential settlement was to control the source of the settlement. The emphasis was placed on maintaining the correct TBM parameters for the different geological conditions encountered and for the structures situated above the tunnel.

Geotechnical instrumentation was installed along the tunnel alignment to monitor ground movement in the three predominant geological formations. An analysis of the geotechnical data was completed prior to tunneling to determine projected TBM parameters along the tunnel alignment. The following is a list the parameters monitored:

- Mining pressure
- Slurry properties—yield, density, viscosity, and filtrate
- Excavated quantity
- Backfill grout pressure and quantity

The TBM parameters were correlated with the data gathered from the geotechnical instrumentation. The data was analyzed to determine the optimum TBM parameters for each geologic formation. These optimum parameters were then set as the parameters to be used for tunneling under the bridges. Adjustments were made to the parameters as information was gathered from further tunneling.

Along with maintaining the optimum TBM parameters for controlling settlement of the bridges, a plan was prepared for compensation grouting under the foundation of

SOUTH DRIVE - TBM S 231
EXCAVATION PARAMETERS FROM RING 1745 to 1980
PRINT OUT: 9-Mar-05

Ring	Cutter head position	River Water Elevation	Air bubble Pressure	Working Slurry					Fresh Slurry				Max pressure (axis)
				Slurry Density	Yield Value Yv	Vp	Filtrate		Slurry Density	Yield Value Yv	Vp	Filtrate	
		feet	in bar		Pa ≥	cPs ≤	ml ≤		Pa ≥	cPs ≤	ml ≤		in bar
If	REAL WATER LEVEL - ESTIMATED WATER LEVEL	> 3 feet -> Values must be recalcuied by the Technical Dept.											
1745	091-76	11	2.88	1.20	12	8	25		1.02	11	6	12	3.50
1746	091-71	11	2.88	1.20	12	8	25		1.02	11	6	12	3.50
1747	091-66	11	2.88	1.20	12	8	25		1.02	11	6	12	3.50
1748	091-61	11	2.88	1.20	12	8	25		1.02	11	6	12	3.50
1749	091-56	11	2.88	1.20	12	8	25		1.02	11	6	12	3.50
1750	091-51	11	2.88	1.20	12	8	25		1.02	11	6	12	3.50
1751	091-46	11	2.88	1.20	12	8	25		1.02	11	6	12	3.50
1752	091-41	11	2.88	1.20	12	8	25		1.02	11	6	12	3.50

Figure 4. Typical excavation parameters

the bridges, if settlement as measured by the geotechnical instrumentation reached the action level specified in the Contract. A drill and grout crew with the appropriate drill rig for the ground conditions was established and placed on standby when the TBM was within 100 feet of the first bridge foundation. The crew, equipment, and material were available within four hours notice. Plans for potential grout hole locations were prepared for each site, and utility clearance and permits were obtained in advance of the work. Surveyors and instrumentation technicians increased the frequency of their monitoring and reported all measurements directly to the shift superintendent when tunneling under the bridges.

The careful attention to the mining parameters proved to be all that was needed to control the settlement of all of the bridges and structures along the tunnel alignment. The measured settlement was significantly less than the required action level and in most situations was not detected. The drill and grout crew, although placed on standby, was never mobilized.

TUNNEL RESULTS

Machine Operating Parameters

The key to successful operation of the slurry shield TBM is the strict control of the machine operating parameters (Figure 4). These parameters include the slurry face pressure, the grout pressure, and the slurry properties, which include the yield, density, viscosity, and filtrate characteristics of the slurry. Each parameter changes with the change in geology, depth of tunnel, surface conditions, and groundwater fluctuation. An analysis of the geotechnical data, and the surface structures produced a report that provided a range in value of these parameters and for every reach of tunnel. These values were based on the groundwater elevation, and were adjusted if the level fluctuated more that three feet. The following chart is an example of the information provided to the TBM operator on a daily basis.

Productivity

Mining progressed at a steady rate over the course of the project. Table 1 and Figures 5 and 6 show the best production in each drive.

Table 1.

	North Drive	South Drive
Best Day	65 ft/day	130 ft/day
Best Week	383 ft/week	605 ft/week
Best Month	1,375 ft/month	1,805 ft/month

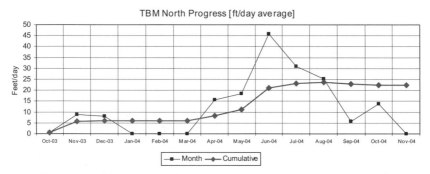

Figure 5. North Drive production summary

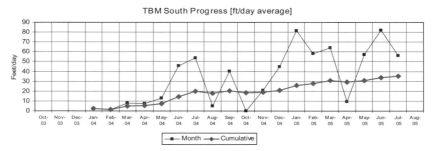

Figure 6. South Drive production summary

HYPERBARIC OPERATIONS

With requirement for continuous face support pressures of 3–3.5 bars and hyperbaric pressures of up to 4 bars, hyperbaric tunneling operations on the Project had to break new ground as far as the American market is concerned. An extensive and comprehensive safety manual was created for this work.

OSHA Variances

First, regulatory concerns had to be considered. OSHA regulations at 1926.803 provide some guidelines for compressed air work, but they are limited to pressures of up to 50 psi (3.5 bar). In general, these regulations consider caisson work or tunnel work where the entire tunnel is pressurized. Use of the slurry shield TBM is not anticipated in the regulations.

Among the regulations that did not fit well with the requirements of the Project were the minimum headroom required in the manlocks, maximum working pressures, and, most importantly, the OSHA decompression tables themselves. Available litera-ture indicated that use of the OSHA tables produced a 33% incidence rate of dysbaric osteonecrosis when used at pressures above 36 psi. And, of course, the maximum pressure considered in the OSHA tables is 50 psi.

Variances from Oregon OSHA were therefore requested and obtained, with the condition of detailed record-keeping and a report to OSHA outlining the effectiveness of the compressed air program. Variances included revisions to the TBM manlock headroom to accommodate the size of the TBM, use of French diving tables with oxy-gen decompression, and manual staged decompression rather than automatic linear decompression.

TBM Design

TBM design, for each machine, included two manlocks equipped for oxygen decompression. Each would allow for decompression of three people, two performing the work duties and one observing and directing the work, and the tandem of man-locks would allow the entry of rescue workers if needed.

Hyperbaric Equipment

In addition to the TBM manlocks, IHJV set up a medical lock on the surface, to be used for training and for medical emergencies or treatment of decompression illness. The medical lock, supplied by HyTech of the Netherlands, was the largest such lock in the Pacific Northwest, capable of handling two stretchers with attendants or up to six people. HyTech also supplied a Diver Attendant Recompression Transportable (DART), a small lock capable of transporting two people, a patient and an attendant, from the TBM to the medical lock. The DART and medical lock were provided with a mating device.

Hyperbaric Organization

For hyperbaric operations, OSHA requires that a medical physician be retained and be available during interventions and to examine all hyperbaric workers. A quali-fied manlock attendant is also required, as is a fully equipped first aid station. IHJV contracted with a local hyperbaric physician to provide his services and those of a Phy-sician's Assistant or EMT as needed. Tenders with diving experience were hired directly and worked on the Project crews in various other capacities.

However, the key person during the intervention is an intervention supervisor who is appointed by the tunnel superintendent. The intervention supervisor may be the tun-nel superintendent, a walker, engineer, or other person who is responsible for the plan-ning and execution of the compressed air work. This person must be in total control of the operation.

On the West Side CSO Project, all hyperbaric operations occurred on day shift. The intervention supervisor, tunnel superintendent, tunnel engineer, and others as necessary, established a detailed plan for the intervention, including tasks, tools, per-sonnel, intervention pressure, durations of interventions, and decompression table to be used. All compressed air workers were given a medical examination. The DART was mounted on a flatcar, along with breathable oxygen tanks for decompression, first aid supplies, and other special equipment. A flatbed truck stood ready at the shaft col-lar to move the DART to the medical lock if required. A tender, along with the physi-cian's assistant and EMT, were in the heading during all interventions. The PA stayed on site for five hours after the last intervention to monitor all of the compressed work-ers. The physician was on call during all operations.

Hyperbaric Results

In total, there were 175 hyperbaric interventions, with 154 interventions on the south drive and 21 on the north drive. 111 of the interventions required pressures in excess of 3.5 bar (50 psi). The maximum intervention pressure was 3.91 bar, and the average intervention time was 57 minutes. The maximum number of interventions in a single day was five.

For the most part, interventions were for inspection or to replace a small number of cutting tools. However, as with all TBM's, some more extensive work was sometimes required. In one case, 35 interventions over eight days were required to replace a large number of cutters. In another case, 28 interventions over eight days were required to replace and weld cutting picks. Repair of the stone crusher required 15 interventions over five days.

In the course of all of these interventions, there was only one case where a worker required a treatment table after completing his decompression. And in that case, the problem was caused by the employee's failure to follow the post-intervention instructions. His decompression illness symptoms arose while playing softball on a hot, sunny day. After returning to the medical lock for treatment, the symptoms disappeared.

CONCLUSION

The slurry shield TBM proved to be appropriate for the non-cohesive, extremely permeable soils in Portland. The tunnel was successfully completed and was invisible to the City of Portland. Due to strict control of the TBM operating parameters and the bentonite slurry, no surface settlement occurred and the tunnel had no impact to any of the structures above or adjacent to it. OSHA regulations related to compressed air work should be reviewed for each specific project, and that work should be addressed very carefully.

REFERENCE

City of Portland, Bureau of Environmental Services: Geotechnical Baseline Report, West Side CSO Tunnel, Shafts, Pump Station, & Pipeline Project, May, 2002.

SOFT GROUND TUNNELING ISSUES HANDLED BY PARTNERING—A WIN-WIN PROJECT IN RANCHO CORDOVA, CALIFORNIA

Tom Martin
URS Corp.

Lysa Voight
Sacramento Regional County Sanitation District

John Forero
Affholder, Inc.

Jack Magtoto
Sacramento Regional County Sanitation District

SUMMARY

Commitment by all parties to partnering goals established early in the contract aided Affholder, Inc. in constructing 2,360m (7,700') of soft ground tunnel in Rancho Cordova, CA. Despite two zones of known contamination, encountering a zone of loose cobbles and boulders in a matrix of flowing sand, some perched water tables and a flooded tunnel after 2 upstream safety plugs failed, the Contractor finished the work 33 days early and only 1.91% over his original bid.

BACKGROUND

The Bradshaw Interceptor Section 8 Project, constructed for Sacramento Regional County Sanitation District (SRCSD) was one segment of the larger Bradshaw/Folsom Interceptor sewer system constructed to convey wastewater from northeastern Sacramento County to the 165 MGD Sacramento Regional Wastewater Treatment Plant located in Elk Grove, CA. The Bradshaw/Folsom link in SRCSD's Master Plan provides increased sewer capacity for existing communities and for planned growth within the region.

The Bradshaw 8 interceptor, designed by Montgomery-Watson-Harza (MWH), required the installation of approximately 2,360m (7,744 lf) of 1.8 m (6' 0") diameter RCP interceptor sewer within the City of Rancho Cordova, California. The RCP was to be installed in a soft ground tunnel constructed using the 2-pass method (first pass using a temporary initial ground support system followed by the second pass consisting of RCP set to line and grade and the annulus grouted), above the regional groundwater table. There were however, zones where the tunnel and some shafts would encounter localized perched water tables increasing the risk of water inflows and/or loss of ground at the work face. Concern about ground stability lead the designer to restrict mining to either an Earth Pressure Balance TBM or a Slurry TBM to mine the three (3) separate tunnel reaches constructed from two (2) work shafts which were also turning structures. Provided the TBM could handling the ground and water conditions without converting to ground support measures, the specifications permitted the contractor to operate in "open mode."

Table 1. Bradshaw 8 project deadlines and completion dates

Dates	Activities
October 13, 2004	Bradshaw 8 Contract awarded to Affholder, Inc. for $19,292,420
November 15, 2004	Notice to Proceed (day #1)
November 16, 2004	Start of Construction
May 28, 2006	End of allowed Contract time (560 days)
April 25, 2006	Substantial Completion achieved (33 days ahead of schedule)

The upstream end of the tunnel connects to the Folsom East Interceptor Section 1B (FE1B) on Citrus Road south of Trade Center Drive while the downstream end of the project joins another Bradshaw Interceptor project (7B) on Kilgore Road just north of International Drive. In addition to constructing a two pass tunnel and installing RCP, Bradshaw 8 included two (2) cast-in-place concrete turning structures and seven (7) access manholes spaced along the alignment. A connecting manhole located at the junction of Bradshaw 7B and Bradshaw 8, included as a bid item in both contracts, was deleted from the tunnel contract and constructed by the open cut contractor who arrived at this connection point last.

CONSTRUCTION MILESTONES

Deadlines and actual completions for Bradshaw 8 are summarized in Table 1.

ROUTE SELECTION AND SETTLEMENT

Because significant development had taken place in Rancho Cordova since 1995 when the SRCSD Bradshaw/Folsom Interceptor System Design Report recommended the initial alignment for Bradshaw 8, it was decided route selection should be revisited. Open cut alternative routes were deemed to have unacceptable impacts to public and traffic with several eliminated because there was insufficient room available for open cut or trenchless construction methods. During a final analysis done by MWH the trench-less construction alternative was accepted as the most favorable with the least impact. As demonstrated in this paper, the decision to tunnel under Rancho Cordova streets was definitely the best solution for SRCSD, Rancho Cordova and the contractor.

Conventional two-pass tunneling was determined to be the preferred construction method however soil sampling identified at two sites along the alignment were likely to contain soil contaminated by hydrocarbons. The contract was modified to allow for stockpiling, handling and transporting offsite any contaminated soils encountered to reduce delays to the work. In addition, the contractor was required to provide 40 hours of HAZWOPER training for any worker that would be exposed to contaminated soils. In the interest of safety, CM staff including tunnel and shaft inspectors were required to complete the same training.

Since SRCSD (owner) had concerns with settlement issues not limited to just the street and nearby utilities a maximum settlement of 19mm (¾") was established for the tunnel excavation. This limit was set low enough that even settlement with that level occurred the affected parties would have limited or no basis to claim damages caused by tunneling. During the tunnel and shaft excavation only a single location settled and that stabilized well below the allowable limit. This settlement occurred in an area of perched groundwater which impacted shaft and tunnel construction and where surface indications such as a depressed street, broken sidewalks, etc. pointed to previous problems before the tunnel was constructed. There were no third party claims for any settlement damage on the Bradshaw 8 project.

PARTNERING

To encourage partnering between the owner, the contractor and the CM, the Specifications (SP-38 Project Partnering Relationship) established that while participation by the contractor was voluntary, a pay item which could fund a facilitator (and needed facilities) up to an amount of $25,000 was included in the bid schedule. To encourage participation actual costs were 100% reimbursable to the contractor provided they were applicable and pre-approved by SRCSD.

Once the work got underway the firm selected to lead the Bradshaw 8 partnering effort was Global Leadership Alliance (GLA) of West Sacramento, CA. and their facilitator, Sam Hassoun. An initial session, lasting one full day, was held on February 18, 2005 at a nearby offsite location. Because the facilitator felt the parties were sincerely committed to partnering on the project, he recommended no other sessions be scheduled unless contractual relationships became strained.

To record the effort and support given to partnering, GLA setup a website for the project and sent an e-mail reminders each month reminding the original attendees to evaluate the relationship. Even without numerical balance between owner, engineer and contractor, monthly ratings remained consistent and definitely tracked how well the partnering was working.

In addition to formal partnering costs, SRCSD was asked to approve funding of onsite BBQ's to promote the partnering concept from SRCSD upper management down to and including the tunnel excavation crew. A total of three BBQ functions, which were always held at noon, are credited with promoting trust and understanding between the parties which continued throughout the duration of the project.

To capture "lessons learned and what might have been done differently," a formal closeout session was scheduled as the project neared completion. Once again, Sam Hassoun from GLA. was selected to guide the session and those in attendance agreed the commitment to partnering played a major role in the success of the project. Because of partnering certain modifications to work hours and sequences were possible and these were unlikely to have been accepted if a strict 'no-give' policy had been in effect.

GLA evaluation form provided a sound plan and a tool for measuring the commitment from each party as well as methods for overcoming 'the bumps in the road' that are generally part of any heavy construction project. Hassoun's role as facilitator was extremely beneficial and provided the initial emphasis necessary to get the concept working. The final cost for partnering, including the three BBQ's was $12,397 or just 49.51% of the bid allowance.

Each Weekly Project Meeting (WPM) included a partnering discussion and the contractor, CM, owner and designer were encouraged to table issues that might be beneficial to the overall success of the project. Issues discussed included waiving requirements to restore shaft detours daily in the early morning hours following excavation at night, allowing two way traffic around the final TBM recovery shaft instead of subjecting the street and local businesses to a complete 14 day closure, installation of a second safety plug by Sacramento County personnel in the upstream interceptor (FE1B) under confined space entry regulations, modification of work hours at some shafts after the contractor demonstrated there would be no negative impact to the public, permitting multiple shaft detours to remain in place instead of requiring a shaft be completed before the next one was started. In each of these situations, cooperation between the contractor and owner, with support and assistance from the design and CM teams, allowed a trial modification to be tested without compromising the finished

Figure 1.   Lovat TBM arrives at main   Figure 2.   Lovat TBM in Turning Structure #1
Bradshaw 8 worksite

product and set the partnering relationship up for success. Not a single complaint was received from the public during Bradshaw 8 construction.

TUNNELING

To mine the Bradshaw 8 tunnel Affholder selected a used EPB TBM manufactured by the Lovat Company of Toronto, Canada. The outside diameter of this TBM was 121" which, even after deducting the allowance for the temporary support system of ribs and boards, gave adequate clearance to set the 1.8m (6') inside diameter RCP that forms the finished interceptor. The TBM had the option for conversion into EPB mode to satisfy contract requirements, however this feature was never necessary even when difficult mining conditions were encountered in reach 3 (Figure 1).

The used TBM was delivered to the main worksite for Bradshaw 8 on March 13, 2005 after an extensive bearing race rebuild in Utah. The following day, the front section was lowered to tunnel level and set on temporary beams in the 31' 6" diameter shaft constructed for Turning Structure #1 (MH #8) (Figure 2). Once setup was complete Affholder began mining on Wednesday March 30, 2005. On that day TBM advance was limited to 3m (10') because the contractor was thrusting against steel cross-beams welded to the temporary horizontal support beams in the shaft invert. Progress was also restricted because just a single muck car could be loaded under the short conveyor and then the car had to be shoved by hand to the opposite side of the shaft before it could be hoisted and dumped.

Once adequate tunnel was mined, the remaining trailing gear was added in increments until a full five (5) car train could be used for muck haulage. Cars continued to be hoisted and dumped by an American 165 ton crawler crane. Affholder adopted a single 9 hour shift including the mandatory half-hour lunch shutdown for underground work in California and stuck with it during tunnel excavation. Muck handling was impacted by lack of a tail tunnel or switch to allow two (2) muck trains underground until the TBM reached and crossed MH #7 located 175m (580') from MH #8. Once all the muck handling facilities were in place, TBM advances of 100' in less than a full 9-hour shift were experienced.

As the TBM crossed MH #7, which had been enlarged to 9.5m (31' 6"), the center drag bits were replaced due to damage from cobbles encountered in the initial tunnel drive. Continuing along reach 1, advances of 30m (100') per day were achieved on 12 occasions and good progress was made by the Lovat TBM operating in "open mode." Perched groundwater was encountered near MH #5 and the Contractor stopped to grout off flows estimated at 1.0 L/s (15 gpm) which were carrying fine material into the tunnel. This delay was minor and soon the TBM continued toward MH #4 (designated emergency egress shaft) where perched groundwater had been encountered just over the tunnel during shaft excavation. Shaft construction halted until the TBM passed on June 10th and the relatively low volume of groundwater had drained into the tunnel where it was pumped outside to a settlement tank followed by discharge into the County system.

The TBM entered the first potentially contaminated zone on 6/24/05 while mining reach 1, however testing determined there was no contamination until near the southern end of the 300m (1,000') zone when hydrocarbons were found in the soil at tunnel level. All material excavated from the contaminated zone was stockpiled in 230 cubic meter (250 c.y.) piles representing approximately 24m (80') of tunnel advance and held there until testing determined if the material was clean or not. Clean muck could be placed in the on-site muck pile. When test results confirmed the presence of hydrocarbons, stockpiled material was re-tested to determine heavy metal content before the designated landfill would accept it for disposal. The delay in removing contaminated material to offsite disposal caused by additional testing was eliminated in the second contaminated zone by modifying the testing protocol.

The TBM reached the southern end of the contract, MH #1, which had been excavated and supported "just in time" at 10:49 a.m. on July 27, 2005. For the next couple of days Affholder concentrated on advancing the TBM into the shaft so the tail shield could be unbolted and the TBM sections could be hoisted and transported. On Saturday, July 30th, using a 125 ton American truck crane, the TBM was hoisted from MH #1 and loaded onto a transport for the trip back to MH #8 where it would be reset for mining reach 2.

Beginning reach 2 on August 10th, the TBM entered the second zone of potentially contaminated material on August 23rd and to speed up disposal for material containing hydrocarbons, testing for heavy metals was done concurrently with hydrocarbon testing to help the contractor maintain progress with his limited storage area. On September 7th the TBM entered MH #10 (Turning Structure #2) and for the next two days the TBM was unbolted and turned to the alignment necessary for mining reach 3.

Reach 3, the shortest section of the Bradshaw 8 tunnel, would prove to be the most difficult to mine due to a short zone of unconsolidated cobbles, boulders and running sand encountered approximately 43m (142') into the drive. Turn-under for reach 3 occurred on September 16th, still within the second contaminated zone and by the 23rd the TBM was clear of the last area containing hydrocarbons. The following day however, Affholder began experiencing problems in advancing the TBM and elected to stop and perform some maintenance while relocating the remaining trailing gear sections from reach 2.

As Affholder resumed mining, the ground was noticeably different from anything encountered to date in Bradshaw 8 consisting of significant amounts of cobbles and small boulders within a matrix of running sand (Figure 3). Over-excavation became a concern with the TBM operator noting all 5 muck cars were being filled (4 cars were adequate previously) but the machine had not advanced enough to install the next 1.5m (5') ring. Voids ahead over the tunnel could be seen through the cutterhead doors which raised the risk of a ground collapse which might reach the surface about 10.5m (35') overhead. Affholder closed the cutterhead doors and attempted to mine but the high percentage of cobbles and boulders in the face material locked the cutterhead and prevented it from rotating in either direction. In an attempt to get the TBM moving, propel pressure was increased but instead of the machine advancing the result was bent ribs and broken lagging as thrust exceeded the capacity of the temporary support system.

At this time, the contractor advised SRCSD and the CM of a "potential Differing Site Condition" (DSC). To stabilize the soil enough to mine through the zone, tentatively identified as unconsolidated dredge tailings left from gold mining operations in the area, Affholder decided to get bids from two firms that could mobilize rapidly and begin a series of pattern grouting from the surface ahead of the TBM. This procedure would fill the voids and consolidate the raveling ground to prevent the fines, cobbles and boulders from running into the tunnel proper.

Affholder selected Coastal Grouting East who mobilized to the Citrus Road site on Saturday October 15th and began drilling grout holes from the surface just ahead of the TBM. Pattern drilling and grouting continued until October 24th when the contractor decided to resume mining. Over-excavation continued to be a concern however and additional grouting immediately ahead of the TBM (and later over the machine) was selected to prevent street settlement. Grouting the voids compensated for the loss of material over and directly in front of the TBM and the grout provided adequate binder to prevent the running sand from flowing to the cutterhead. After stabilizing the cobbles and boulders the Contractor was able to resume his mining operation on October 26, 2005.

Several voids up to 1m (3.3') in height and of undetermined length, located as much as 26–33m (85–100') in front of the TBM were logged during the surface drilling operation and subsequently filled with slurry grout. These voids, located just over the crown of the tunnel, could not have been caused by the TBM, gave confirmation that the area had previously been dredged without subsequent compaction leaving voids several feet below the street. It appears dredge pile surfaces were simply dozed level and subsequent fill placed to build up the area for commercial use. During a series of meetings and negotiations, SRCSD and the CM concluded that Affholder had encountered "Difficult Mining Conditions" in reach 3 which were not predicted in the Geotechnical Baseline Report (GBR) or in descriptions of the soil units to be mined in this section of Bradshaw 8.

Upon receipt of the contractor's claim, the supporting information was reviewed and a settlement was reached in March 2006. No determination was made about whether the conditions encountered constituted a DSC, instead the parties agreed that Affholder had mined through material that delayed their progress, required consolidation grouting ahead of, over, and behind the TBM, which costs their original bid for Bradshaw 8 did not include. All parties accepted the negotiated settlement and because the TBM mining was not on the critical path for project completion, no additional contract time was requested or granted.

Once the TBM resumed mining in reach 3 (October 26, 2005) the advance was 9m (30') which increased to 18m (60') the following day, supporting the theory that "difficult mining conditions" were no longer present in the heading. On November 7th at 2:02 p.m. the Lovat TBM reached the FE1B shaft support system close to design line and grade (Figure 4). The next day shaft supports were removed and the TBM advanced into the shaft where it remained until 11/9/05 when it was hoisted out after the tail shield was unbolted. The TBM was stored in the main yard until relocated to Affholder's storage yard in 2006.

The used TBM preformed very well, suffering only a few minor breakdowns such as a hydraulic drive or conveyor motor. In mining the entire tunnel, ground conditions never required the TBM to be switched from 'open to closed mode.' The "difficult ground conditions" in reach 3 could not have been handled in closed mode because the TBM was unable to advance with the cutterhead doors in the closed or partially open position necessary for operating as Earth Pressure Balance machine.

Overall, the mining of Bradshaw 8 was a success and there was never a need for Affholder to adopt a second shift because production and re-setting the TBM was done in short time frames which eliminated previous concerns over meeting the project schedule.

Figure 3.   Cobbles and small boulders from "difficult mining zone"

Figure 4.   TBM completes reach 3 (FE1B RCP in lower left)

When reach 1 mining finished, the impact to Baseline Schedule Critical Path was nearly a slip of 3 weeks, however once RCP installation got underway it became apparent the contractor could recover all lost time. In fact all Bradshaw 8 work was completed 33 days early despite a complete flooding of the tunnel on December 31, 2005.

BRADSHAW 8 TUNNEL FLOOD

December 2005 was a very wet month in Sacramento and on Saturday, December 31st sometime after 10:00 a.m. the entire Bradshaw 8 tunnel and three (3) connected shafts were flooded with a mixture of sewage and storm water that flowed in from FE1B. By 2:30 p.m. the entire tunnel was inundated after two upstream tunnel plugs, installed as safety measures, failed and allowed overflow from another interceptor located approximately 1.2 km (4,000') from the connection between FE1B and Bradshaw 8 to enter the partially completed tunnel (Figure 5).

The most upstream plug, furnished by Affholder and installed by Sacramento County M&O consisted of segmented steel sections bolted together with a pneumatic tube positioned around the circumference. Inflating this tube to 1.8 bar (27 psi) effectively sealed the 1.8m (6') RCP against flows, however no pins, anchors, struts or bolts were installed to provide a shear reaction in the event the FE1B section flooded. A second plug set in FE1B was made of fiberglass just downstream of the new plug. Two plugs were required to provide redundancy for a system designed as temporary storage of effluent during periods of high flow in the adjacent interceptor. The FE1B system would provide storage until the remaining downstream Bradshaw sections were constructed and the finished system connected to the 165 MGD treatment plant located in Elk Grove.

On the day of the flood, a Saturday, Affholder's Superintendent checked the jobsite early and found no evidence of the plug failures with the tunnel completely dry at MH #8. Later that same day, the Supt. returned to find the three open shafts, MH #7, #8 and #10, flooded to a depth of approximately 3m (10') with a sewage and storm water mixture. Following established job protocol, the appropriate notifications were made and the decision to deal with the flooded tunnel issue when contract work resumed on Tuesday, January 3, 2006 was made by SRCSD in conjunction with the CM and contractor.

Following Affholder's return to work on January 3, 2006, various measures were put in place to ensure risks and exposure to fecal contamination was managed properly and safely. Direction for the contractor was provided by Field Instructions (FI's) issued by the CM after consultation with SRCSD management. Dewatering the flooded tunnel was initially done by a SRCSD M&O crew that used a submergible

Figure 5.   Turning Structure #2 following tunnel flood on 12/31/05

Figure 6.   Tunnel safety plug #3 installed after flooding

pump in nearby manholes pumping the sewage/groundwater mixture to a sanitary sewer while complying with volume restrictions since the entire County sanitary system was running at or near capacity.

Affholder assigned some of their tunnel crew to standby work which included yard cleanup, sorting surplus materials, and bundling supplies on pallets in anticipation of demobilization. Costs for this operation were later split between SRCSD and the contractor since for the most part it was work Affholder would have had to accomplish later in the contract. Keeping the crews employed prevented the contractor from re-training new workers once the ban on underground work was rescinded.

There were several shafts which had not been connected to the tunnel yet and here excavation could proceed while the tunnel was being dewatered and sanitized. Other directed work included surveying the tunnel to determine if incremental settlement occurred in ungrouted joints of RCP when the tunnel filled, purchasing a third tunnel plug and fabricating a shear support system as an additional safety measure before that new plug was installed upstream of the connection between FE1B and Bradshaw 8 (Figure 6).

Dewatering efforts for the flooded shafts, MH #7, #8 and #10 were accomplished along with an excavation to springline at MH #1 so 2 holes could be cored through the precast concrete plug located in the first section of RCP set by Affholder. These holes were later used for ventilation and pumping operations in this section of dead end tunnel.

Added work included washing the tunnel followed by cleanup, disinfecting and removal of sludge and debris for the entire length and this work began on January 9th at MH #8 & #10, continuing until February 9th when the rebar in MH #8 was cleaned and re-tied in place. Safety was a major concern during the cleanup and disinfecting operation so personnel would not be exposed to toxins and other irritants present in the tunnel. All sludge had to be carried back to MH #2 and hoisted from the tunnel before being deposited in a proper landfill. Tunnel ventilation was provided using a 12" hole cored in the concrete plug. By February 28, 2006 all remedial work caused by the tunnel flood of 12/31/05 was corrected and Affholder was able to resume contract work.

SAFETY

The commitment to safety by Affholder paid big dividends on the Bradshaw 8 work. Starting with a full 8 hour training course for every new hire and stressing safe work habits and thinking before acting, the entire tunnel project, which required approximately 200,000 man hours did not have a single Lost Time Accident (LTA). Safety was discussed during a weekly tool box meeting as well as a crew briefing before any new task began, and refresher training, HAZMAT training for personnel exposed to the contaminated soil in conjunction with meaningful awards for working safe keep the workers aware of the risks during the 18 months of work necessary on the project.

CONCLUSION

A heavy construction project can benefit from Partnering, particularly when the parties involved see immediate cooperation, an easing of tension during difficult times, assistance with problems as they arise and accept the premise that on a good project all entities are the winner. This was demonstrated several times during the construction of Bradshaw 8 as the County, CM, Designer and Contractor worked out problems or issues which could be addressed without compromising the finished product, making the work flow better or simply made more sense.

Forcing the Contractor to do something "just because the specifications say so" is onerous and leads to disputes that disrupt the work without adding anything to the finished product. Instead, by merely taking a fresh look at the contract wording or specification requirements and asking "Is this really necessary?,' the Owner will often find things have changed or limitations that seemed sound during the design phase are no longer necessary once the work gets underway. Generally the Contractor has done similar work on other projects and understands what limitations or restrictions are needed as well as those which can be modified without compromising the quality or function of the final product.

An example of this on Bradshaw 8 was the specification requirement for some shaft construction to take place at night and for the Contractor to set out and pick up all related traffic control devices daily. Once Affholder demonstrated they could maintain 2-way traffic around each work site safely, the fact detours remained in place actually made these routes safer for the public who were not faced with constantly changing lane restrictions. This also benefited the Contractor but instead of trying to recover monies saved by leaving barricades, etc. in place, SRCSD considered this an outcome of Partnering and made a challenging construction task easier. This is what elevates a good project to the level of a great project, the ability to look at the overall picture and compromise when it is possible to do so without accepting a final product which is less than what was called for at the time bids were received.

By the account of all the individuals associated with Bradshaw 8, this was a great project, done in complete conformance with the specifications, an effort that dealt with potential disputes quickly and fairly, and one that met the Partnering challenge on a grand scale. Add in a project without a single LTA and one begins to understand why the contract has been referred as "the perfect project." Those that worked on Bradshaw 8 tend to agree, it doesn't get much better than this.

SOIL DEFORMATION ANALYSIS DURING TBM EXCAVATION: COMPARING THREE DIFFERENT TBM TECHNOLOGIES USED DURING THE CONSTRUCTION OF THE TOULOUSE METRO LINE B

Jacques Robert
ARCADIS

Emilie Vanoudheusden
ARCADIS

Fabrice Emeriault
LGCIE, INSA

Richard Kastner
LGCIE, INSA

INTRODUCTION

The subway line B of Toulouse (France) has been realized with three different tunnel boring machines (TBM) techniques: Contracts 2 and 5 are excavated with earth-pressure balanced TBM, Contract 4 with a slurry shield TBM, and Contract 3 with a compressed-air TBM. The first three machines are around 7.7 m in diameter, while the last one is 5.3 m. Moreover, a small part of the line (50 m) has been excavated by a conventional method.

The tunnels run through the Toulouse molasses, composed of hard sandy clay with pockets and lenses of dense sand. Geotechnical investigations have shown that in these formations, K_0 is greater than 1, around 1.8 at a depth of 25 m (Serratrice, 2005) and that geotechnical characteristics are homogenous on the whole layout ($\gamma = 22$ kN/m^3, $S_u = 300$ kPa, c' = 30 kPa, Φ' = 32°). The water table is found between 2 and 4 m below ground level.

As geology, stratigraphy and tunnel cover (between 10 and 20 m) are almost identical, a comparison of movements induced by the excavation by each of the four excavation techniques is possible.

This paper presents the soil mass movement measures in the case of the conventional excavation method and in the case of TBM sections. It shows how tunneling parameters may explain the different surface movements observed during excavation. For the sake of simplicity, only the results of Contract 2 (earth pressure balanced TBM) will be presented.

MONITORING

Monitoring for the Conventional Method Section

For the conventional method section (Vanoudheusden et al. 2005) the mass soil movements are measured with two inclinometers and one multi-point extensometer borehole with automatic data acquisition (Figure 1). The monitoring devices include

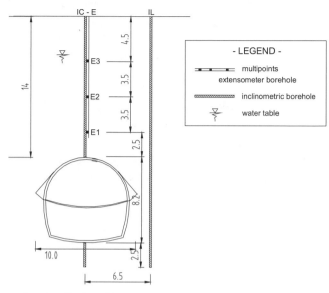

Figure 1. Conventional method section

high precision leveling of 3 profiles on the pavement (Figure 2): the axis profile with points along the axis of the gallery (from C1 to C7), the lateral profile with points along a lateral line of the gallery axis (from L1 to L9) and the transversal profile with points along a transversal line (from T1 to T11). Besides, a building has been equipped to evaluate the variations of horizontal distances (measures with invar thread) and rotations (automatic clinometric gauges) (Figure 2).

Monitoring for TBM Sections

The movements within the soil mass are observed by the instrumentation of several sections, including inclinometers, multi-points borehole extensometers and precise leveling. A typical monitoring section is described in Figure 3 (Emeriault et al., 2005). Additional monitoring consists in strain measurements in the tunnel concrete lining segments and pore pressure variations in the soil.

Monitoring All Along the Line

Besides instrumentation of soil mass movements at different specific sections, vertical surface displacements are recorded all along the line by a high precision leveling of points on the pavement and buildings close to the tunnel. Figure 4 presents location of several leveling points on a part of the tunnel.

The vertical displacement induced by the excavation of the tunnel is calculated with the measurements realized during the operation. This displacement equals the difference between the measurement realized when the front of the TBM is about 50 m before the point and the measurement done when the front is 50 m or 100 m after the point.

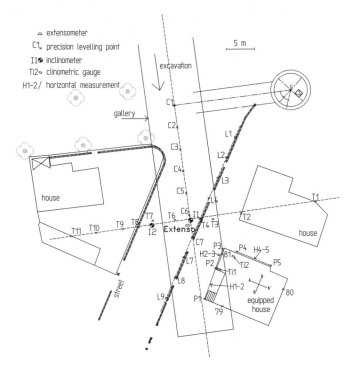

Figure 2.   Location of instrumentation devices for the conventional method section

THE DIFFERENT EXCAVATING METHODS

Conventional Method Characteristics

The part of the tunnel excavated by conventional method is 50 m long, 10 m wide and 8 m in height. It has been excavated sequentially, by two-half-sections, with an excavator.

The upper half section has been excavated at a rate of 1.5 m per day, in 3 shifts; the lower half section has been excavated after the completion of the upper half section of the gallery, at a rate of 2.5 m per day.

The temporary support consisted of HEB 200 steel arches and 20 cm of fiber reinforced shotcrete (Figure 5).

TBM Parameters

The number of TBM parameters is different according to the machine: the Earth Pressure Balanced TBM of Contract 2 records around 150 parameters against more than 300 for the Slurry Shield TBM of Contract 4; in the case of the compressed-air TBM (Contract 3), only about ten parameters are recorded. Different devices measure these parameters at each second.

From the literature, we have chosen the tunneling parameters of interest for this analysis:

- advancement rate,
- chamber confining pressure,

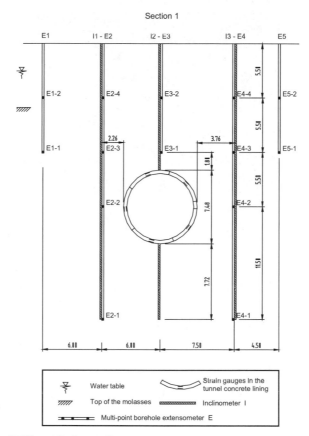

Figure 3. Typical TBM monitoring section

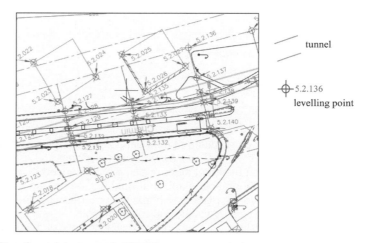

Figure 4. Location of leveling points on a part of tunnel

Figure 5.   The gallery during lower half section excavation

Table 1.   Average values and standard deviations of tunneling parameters of Contract 2

Advancement Rate (mm/min)		Confining Pressure (σ_{v0})		Thrust Pressure (σ_{v0})		Pressure on Torque (σ_{v0})		Number of Pump Shots/ Theoretical Number	
$\bar{x}$	$\sigma/\bar{x}$	$\bar{x}$	$\sigma/\bar{x}$	$\bar{x}$	$\sigma/\bar{x}$	$\bar{x}$	$\sigma/\bar{x}$	$\bar{x}$	$\sigma/\bar{x}$
53.1	0.17	0.44	0.14	31.4	4.6	65.9	10	1.3	0.1

Grouting Pressure (σ_{v0})		Vertical Deviation		Horizontal Deviation		Time for One Ring		Energy for Excavation	
$\bar{x}$	$\sigma/\bar{x}$	$\bar{x}$	$\sigma/\bar{x}$	$\bar{x}$	$\sigma/\bar{x}$	$\bar{x}$	$\sigma/\bar{x}$	$\bar{x}$	$\sigma/\bar{x}$
0.84	0.16	0.034	0.671	0.196	1.836	186.8	183.5	1.01	0.2

- thrust pressure,
- hydraulic pressure responsible for the torque on the cutting wheel,
- volume of mortar injected,
- mortar grouting pressure,
- punctual deviation,
- time to realize one ring (excavation + ring segments installation),
- energy for excavation.

In order to compare values at different points of the excavation line, the parameters of pressure are normalized by vertical initial stress σ_{v0}; the volume of mortar injected is expressed by the number of pump shot required to obtain this volume, and normalized by the theoretical number of pump shot required to fill the annulus void. The punctual deviation (vertical and horizontal) expresses a direction change of excavation line between two consecutive excavated rings.

The average values $\bar{x}$ and standard deviations $\sigma/\bar{x}$ of these normalized parameters are summarized in Table 1 for Contract 2.

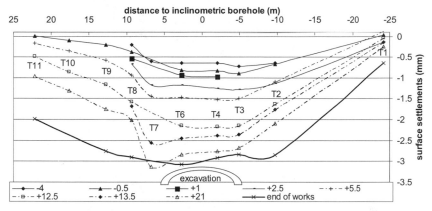

Figure 6.   Transversal profile settlements during upper half excavation and at the end of the excavation of the conventional section

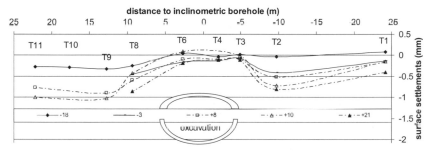

Figure 7.   Transversal profile settlements during lower half excavation of the conventional section

SURFACE DISPLACEMENTS

Conventional Method Section

Figure 6 presents the vertical displacements on the transversal surface profile observed during the upper half section excavation. These displacements are small (maximum 3 mm of settlement).

Until the +1m measure, surface settlements were within 1mm; afterwards, there is a more important increase in the vertical settlement. This increase stops after the +13.5m measure.

The transversal settlement curve is slightly asymmetric; this could be due to the rigidity of the house situated between leveling points T1 and T2.

Figure 7 shows the surface displacements on the transversal profile during the lower half excavation: the 3 points close to the axis of the gallery (T3, T4 and T6) did not move, the others settled after the −3 m measure. The maximum settlement was recorded for points set at around 10 m away from the gallery axis (T2, T8 and T9).

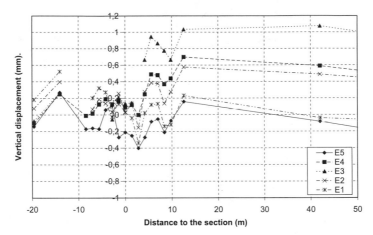

Figure 8. Vertical displacement versus distance to the monitoring TBM section (positive = heave; negative = settlement)

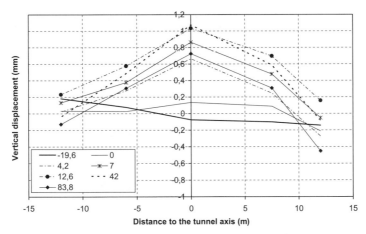

Figure 9. "Settlement" trough for different distances to the monitoring TBM section (positive = heave; negative = settlement)

TBM Sections Observations

On TBM sections, the vertical displacements of the ground surface are essentially measured by precise leveling of the five borehole extensometer heads. In the [–10 m; +12 m] range, one topographic survey is done for each installed tunnel ring (1.0 m or 1.4 m large).

The global soil reaction is the same for all TBM sections, whatever confining technique has been used. Figure 8 and Figure 9 summarize this behavior.

Figure 8 shows that the TBM advance induces upwards movements of the ground surface, with a maximum heave observed directly above the tunnel axis and equal to about 1 mm. After approximately 40 m, the different leveling points start to settle (see Figure 9). The final "settlement" trough (83.8 m) shows a heave of the surface close to the tunnel axis, and a small settlement (0.1 to 0.5 mm) of points with a 12 m offset.

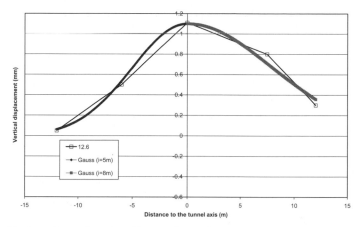

Figure 10. Maximum heave trough and inversed gaussian curve

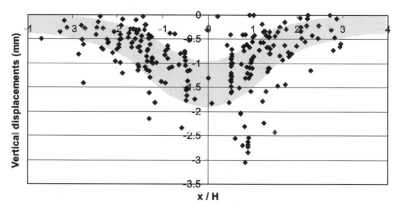

Figure 11. Vertical surface displacements—area "settlement only"

The maximum heave trough (Figure 10) looks like an inversed Gaussian curve, like the description of Peck (1969). The characteristics of the curve are a maximum heave of 1.1 mm and a distance from the tunnel axis to the inflexion point (parameter i) which equals to 5 m on the right side and 8 m on the left side.

Thus, compared to the conventional section, the utilization of TBM reduces the magnitude of displacements, and induces more heave than settlement.

Observations All Along the Line

Analysis of vertical surface displacements has been realized all along the Contract 2 (earth-pressure balanced TBM). Displacements are expressed according to the parameter x/H, where x is the distance to the tunnel axis and H is the thickness of soil above the tunnel.

Results show an atypical displacement of the surface induced by excavation of the tunnel: in parts of the line of this Contract, the maximum settlement is classically observed above the tunnel axis (Figure 11, area "settlement only"), in other parts of the line, the maximum settlement is not above the tunnel axis, but with an offset, and

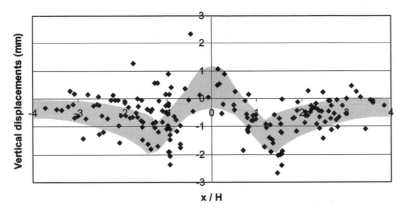

Figure 12.   Vertical surface displacements—area "heave at tunnel axis"

the points close to the tunnel axis show a little heave (Figure 12, area "heave at the tunnel axis").

The shape of the trough drawn with points of the area "heave at the tunnel axis" is consistent with the final trough measured on TBM sections.

These different behaviors are not explained by the layout of the tunnel, and may be explained by the tunneling parameters (See Surface Vertical Displacements section).

HORIZONTAL DISPLACEMENTS IN THE SOIL MASS

Conventional Method Section

Measures by the axial inclinometric borehole show that the soil at the gallery depth moves towards the excavation (Figure 13); the maximum recorded displacement is 3 mm at a depth of 14 m (i.e., at the tunnel crown).

Evolution of horizontal displacement of the lateral inclinometric borehole is presented in Figure 14 for the upper half section and in Figure 15 for the lower half section (the initial reference measure is taken after the end of the upper half section). During the upper half section excavation, the bulb is located at the crown of the gallery and increased slowly; the maximum displacement reaches 3 mm at a depth of 14 m. During the lower half section excavation, we observe the formation of a bulb at the top of the lower half section (17.5 m deep); this bulb increases with the excavation works. It seems to be stable when the excavation works were 14 m after the inclinometric borehole position, and reaches a maximum of 4.6 mm.

TBM Section Observations

The horizontal movements are measured in three inclinometric boreholes located in the tunnel axis (I2, Figure 3) and on both sides of the tunnel (I1 and I3). The bottom of the boreholes is one diameter below the tunnel invert.

The longitudinal horizontal movements are generally small and close to the accuracy of the inclinometric measure (evaluated at 1 mm). This corresponds to a good confining pressure applied on the front face limiting soil displacements toward the excavation.

Figure 16 presents typical transversal displacements measured by lateral inclinometric boreholes. The movements appear when the cutting wheel is 2 m before

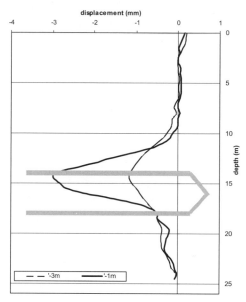

Figure 13.   Horizontal longitudinal displacements of I1—conventional method section

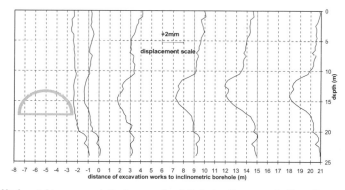

Figure 14.   Horizontal transversal displacements of I2 during the upper half section excavation

reaching the monitoring section and correspond to a convergence towards the tunnel. They increase with the advance of the TBM; around 15 days are necessary to obtain the final stabilized values. The maximum displacement reaches about 6 mm.

The maximum displacement is obtained in the lower part of the tunnel slightly below the axis. Magnitude of horizontal displacements in the case of TBM sections is quite the same than in the case of conventional excavation (5.6 mm at the level of the tunnel axis: 1 mm for the upper half section excavated, and 4.6 mm for the lower half section excavated).

The maximum horizontal movements are twice maximum vertical movements. Numerical analysis has shown the importance of highly overconsolidated character of molasses on this phenomenon.

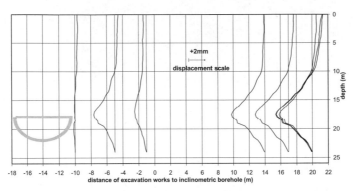

Figure 15.   Horizontal transversal displacements of I2 during the lower half section excavation

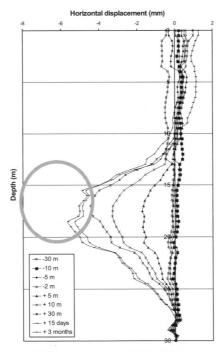

Figure 16.   Horizontal transversal displacements for TBM sections

VERTICAL MOVEMENTS IN THE SOIL MASS

Conventional Method Section

Only one extensometer borehole has been installed in this section.

During the upper half section excavation, measures show that the anchor situated 4.5 m under the ground level has the same settlement as the ground surface point; the soil just above the gallery settles a little bit more.

Displacements of the anchors during the lower half section excavation are not significant.

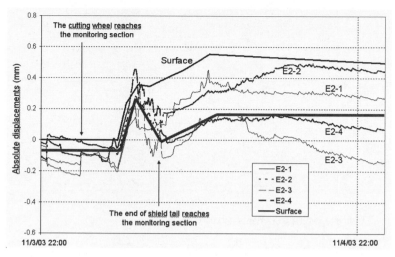

Figure 17.   Absolute vertical displacement of the four anchors of the lateral extensometer E2 versus time

TBM Sections

The vertical displacements at different locations within the soil are measured by five multi-points borehole extensometers E1 to E5 (see Figure 3). The differential vertical displacements between the anchor and the extensometer head are automatically recorded every 5 seconds. The absolute displacements of each anchor are calculated by the addition of the differential displacements to the vertical displacements of the extensometer head measured by precise leveling.

Figure 17 presents the absolute vertical movements of the anchors. The red line shows the global movement: the maximum movement is recorded during the advance of the shield under the section; we observe then a global heave of the anchors followed by a settlement which starts before the end of the shield tail. Mortar groutings stop this settlement, with sometimes a slight heave.

STRAINS IN THE TUNNEL LINING

The variations of strains in the tunnel lining are measured by 5 pairs of vibrating strain gauges installed in the lining segments at the intrados and extrados. An automatic acquisition is installed. Data acquisition starts as soon as the ring is installed; one measurement is taken every 10 minutes.

Reference for the strains is taken before the lining segments are transported in the tunnel entrance shaft and thus before any load is applied. The average strain for each pair of gauges is presented in Figure 18 (negative μ-strains correspond to compression and positive values to extension).

The segment installation process induces positive μ-strains, corresponding to extension. This phenomenon is probably the result of several factors like the temperature of the concrete, the temperature of the strains gauges, the loads applied by the hydraulic jack, but can not be physically explained.

Approximately 24 hours after the installation (which corresponds to a progression of the TBM of 35–40 m), μ-strains start to decrease; thus the lining segments are subjected to compression. Final values are obtained after approximately 2 months.

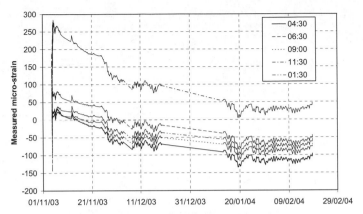

Figure 18. Strain gauges measurements versus time

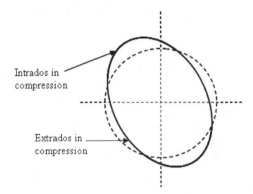

Figure 19. Shape of the tunnel lining deformation

The differences of µ-strains by pair of gauges give qualitative information on the global shape of deformation of the tunnel lining. Figure 19 presents the shape deducted from the measures. This shape is consistent with the convergent horizontal movement and the vertical heave.

PORE PRESSURE VARIATIONS

Two pore pressure gauges have been installed in one cross section. The first (CPI1) was installed 0.75 m above the crown of the tunnel; the second one was located on the tunnel axis and was destroyed by the TBM excavation of the tunnel.

Figure 20 presents the evolution of the pore pressure measured by the two gauges during excavation of the tunnel. The approach of the front of excavation is synonymous with a decrease of pore pressure that is meaning a decompression of the soil; this decrease is maximum for gauge CPI2. When the front is situated 3 m ahead of the gauge an increase of pore pressure is observed, with a maximum when the front is near the sensors. So, there is a compression of the soil just near the confined front of excavation.

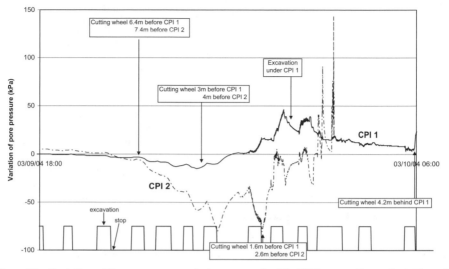

Figure 20. Evolution of the pore pressure during approach of the TBM, excavation and advance of the shield tail

After, during the advance of the shield tail in front of the gauges a quick decrease of the overpressure is observed.

Mortar grouting at the end of the shield tail modify the pore pressure: we observe an overpressure associated to excavation/grouting phase; this overpressure is rapidly dissipated during the stop phase of the TBM.

MAIN PARAMETERS FOR REDUCING SOIL DISPLACEMENTS

Soil Mass Movements

The analysis of three TBM sections has shown the relations between soil mass movements and some tunneling parameters (Emeriault et al., 2005). This analysis is based on surface vertical displacements and soil mass horizontal and vertical displacements. The tunneling parameters taken into account are the advancement rate, the pressure at the front, the mortar grouting pressure and the volume of mortar grouted per ring.

It appears that the convergent horizontal movements can be reduced by use of suited confining pressure and grouting pressure.

Surface Vertical Displacements

The interest of the use of surface vertical displacements is that these data are available all along the line of each Contract. Thus the number of initial data is largely greater than the number of TBM parameters (see TBM Parameters section), and a statistical approach may be used to define the TBM parameters which influence the vertical displacements.

We can find details of this analysis in Vanoudheusden et al. (2006).

This approach used in the case of Contract 2 (earth pressure balanced machine) shows that three out of ten TBM parameters are influent. They are the advancement

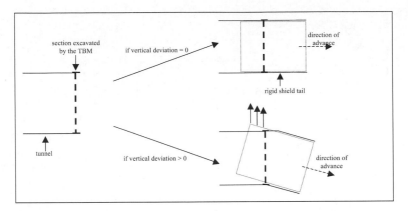

Figure 21. Heaves induced by vertical deviation

rate, the hydraulic pressure responsible for the torque on the cutting wheel and the vertical deviation.

Thus a quick excavation of the tunnel induces a quick filling of void and so reduces the settlements in the soil.

The hydraulic pressure responsible for the torque is dependant of the type of the soil excavated and the filling of the confining chamber by this soil. A low pressure on the torque results in settlements, a high pressure may lead to heaves.

The vertical deviation is usually due to a direction change of the TBM. If this is no deviation, the rigid shield tail advance in straight line in the tube; if there is a deviation the shield tail may create a soil heave close to the crown (Figure 21).

CONCLUSION

The construction of tunnels in Toulouse molasses by a conventional method and three different pressurised closed-face TBM lead to small volume losses and surface settlements, without any risk of damage to existing buildings.

The monitoring of horizontal and vertical ground movements around the tunnel showed an unusual deformation scheme, maximum horizontal movements being twice maximum vertical movements. This is due to the highly overconsolidated character of the molasses, which gave also a very flat settlement or heave (for TBM method) trough in the vicinity of the tunnel axis.

The conventional method gives greater ground settlements than the three TBM methods, this is due to the pressurised close-face effect. The maximum surface settlement is around 3 mm for the conventional method; for the TBM methods, it is usually observed a heave with a maximum about 1 mm in the tunnel axis forward the TBM face and a maximum settlement on the sides of the tunnel of less than 1 mm. The horizontal ground displacement on the sides of the tunnel is quite the same for the four different excavating methods; the maximum is around 6 mm. But there is a great difference for the horizontal ground displacements before the tunnel face in the axis direction; there is no movement with the three TBM methods (pressurised closed-face), and a displacement around 3 mm for the conventional method.

For the earth pressure TBM, three TBM parameters have an influence on ground displacements: the advancement rate, the hydraulic pressure responsible for the torque on the cutting wheel and the vertical deviation. The observed displacements are smaller with a higher advancement rate or hydraulic pressure or vertical deviation.

ACKNOWLEDGMENTS

The authors would like to thank for their financial support the French Ministry of Research, the Réseau Génie Civil & Urbain (RGC&U) and the Société du Métro de l'Agglomération Toulousaine (SMAT). They also would like to acknowledge the help of the different contractors in collecting data.

REFERENCES

Emeriault F., Bonnet-Eymard T., Kastner R., Vanoudheusden E., Petit G., Robert J., Lamballerie J.Y., Reynaud B. 2005. Ground movements induced by earth-pressure balanced, slurry shield and compressed-air tunneling techniques on the Toulouse subway line B. In *Underground Space Use—Proceedings of the 31st ITIA-AITES World Tunnel Congress.* Erden & Solak eds BALKEMA, Istanbul (Turquie), 07/05/2005–12/05/2005. p.841–848.
Peck. 1969. Deep excavations and tunneling in soft ground. In *Proceedings of the 7th International Congress of Soil Mechanics and Foundation Engineering,* Moxico, p.225–285.
Serratrice, J.F. 2005. "Métro de Toulouse—Ligne B. Essais de laboratoire. Avril Août 2005. Rapport du LRPC d'Aix en Provence, CETE Méditerranée, du 12/10/2005." Unpublished report.
Vanoudheusden E., Petit G., Robert J., Kastner R., Lamballerie J.Y. 2005. Impact on the environment of a shallow gallery excavated in Toulouse's molasse by a conventional method. In *Underground Space Use—Proceedings of the 31st ITIA-AITES World Tunnel Congress.* Erden & Solak eds BALKEMA, Istanbul (Turquie), 07/05/2005–12/05/2005. p.1219–1225.
Vanoudheusden E. 2006. Impact de la construction de tunnels urbains sur les mouvements de sol et le bâti existant—Incidence du mode de pressurisation du front. Ph.D. dissertation, INSA de Lyon, France.

13

Rock Tunneling Project Case Histories

Chair
J. MacDonald
Consultant

CONSTRUCTION OF THE POWERHOUSE CAVERN FOR THE PUMPED STORAGE SCHEME KOPS II IN AUSTRIA

Helmut Westermayr

Beton-und Monierbau

Michael Tergl

Beton-und Monierbau

DESCRIPTION OF PROJECT

Geographic Setting

The Kops II plant is a pumped storage scheme of the hydroelectric power producer 'Vorarlberger Illwerke AG' which is currently under construction in the Austrian Province of Vorarlberg. The plant is located at the end of the Montafon valley and will be fed with water coming from the Silvretta mountain range (Figure 1)

Description of Plant

The Kops II plant shall be operated as a pumped storage plant with three highly flexible and quickly controllable power generating units. Each unit shall have a capacity of 150 MW and shall be equipped for both turbine operation and pump operation.

The Kops II plant uses the existing Kops Lake as upper storage reservoir and the existing Rifa balancing reservoir as lower storage reservoir. All major plant components of the Kops II scheme are located below ground. Total investment costs are approx. 360 million Euro.

Construction Lots

The construction works for the Kops II plant, which were subdivided into three lots, were launched in September 2004. It is planned that the first power generating unit shall go on line in the year 2007. The completion of the overall project and the operation start-up of the second and third unit are envisaged for the year 2008.

Lot 1—Pressure Tunnel. The 5.5-km-long pressure tunnel forms the connection between the Kops Lake and the beginning of the pressure shaft. The tunnel was excavated by the use of a double-shield TBM.

Lot 2—Pressure Shaft and Surge Chamber. The approx. 1.1-km-long pressure shaft was excavated by TBM, advancing from bottom to top. Featuring an inclination of 80%, the pressure shaft overcomes a difference in height of approx. 700 m.

Lot 3—Power House Cavern and Tailrace Structure. The power house cavern, which is being built approx. 150 m inside the mountain, forms the core of this plant. The power house is designed to accommodate three power generating units, and the transformer cavern is forseen to accommodate three transformer units. The tailrace structure serves the purpose of conveying the water to the lower storage reservoir during turbine operation and as water intake during pump operation (Figure 2).

Figure 1. Location map within Europe

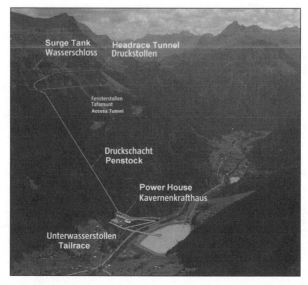

Figure 2. Kops II, overall scheme

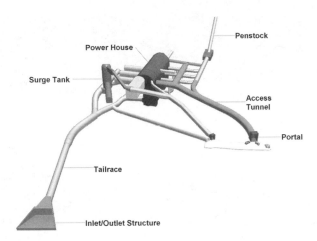

Figure 3. Kops II, Lot 3, underground structures

Geology

All facilities of the Kops II plant are located in the Silvretta crystalline rock series, which mainly consist of solid and hard rocks such as amphibolite, hornblende gneiss and other types of gneiss. In addition to this, less solid mica schist was also encountered.

The geological conditions prevailing during power plant construction were generally found to be quite favourable, although in some areas rather difficult ground conditions were met. In these areas, adequate support measures such as rock bolts, shotcrete, wire mesh and steel arches were used, depending on the local rock mass quality.

Technical Data of Lot 3

Power House Cavern. This cavern is currently one of the largest rock cavities in the world. In order to accommodate the 38-m high power generating units, the cavern had to be 88 m long, 30.5 m wide and 60.5 m high. The excavation volume thus comes to approx. 113,000 m^3.

Transformer Cavern. This cavern is 35 m long, 16 m wide and 19 m high and the excavation volume is approx. 10,000 m^3.

Tailrace Structure. This structure comprises several components and forms a link between the water used in turbine operation and the water conveyed to the Rifa balancing reservoir. These plant components essentially include the inlet and the outlet structure in the Rifa reservoir, the 267-m-long tailrace tunnel, the tailrace surge chamber with a 31-m-high shaft (12 m in diameter) and a 47-m-long chamber, as well as three pressure chambers, each with a length of 45 m, which are linked to the tailrace tunnel via a 77-m-long connection tunnel (Figure 3).

Construction Period

On August 27th, 2004, an award of contract was issued by the client, the 'Vorarlberger Illwerke AG.' The contractually stipulated commencement of the construction works was defined to be September 7th, 2004. The official start of the underground excavation works was October 11th, 2004. As the power plant is scheduled to be taken into operation in the year 2008, this results in an overall construction time of approx. 3.5 years.

PROJECT-SPECIFIC CHARACTERISTICS

Environmental Impact Assessment Procedure

The Kops II plant was the first project for which an environmental impact assessment (EIA) was performed in the province of Vorarlberg.

An EIA serves the objective of identifying, describing and evaluating the impact of a project and of assessing possible mitigation measures. In addition to this, it reflects upon the advantages and disadvantages of the various solutions of such a large-scale project.

In compliance with the Austrian EIA-law, the public was also involved in the decision-making process of this project. For the realisation of the Kops II project, various conditions were imposed. Never before in the history of Vorarlberg, were such stringent ecological requirements to be met, as with the construction of the Kops II plant. The requirements to be satisfied include an environmentally friendly transport of excavation material and a compliance with air, noise and water threshold values in the vicinity of the construction site.

Measurements

The environmental regulations for transport, blasting, construction and muck disposal works were clearly defined and were strictly adhered to. In order to ensure objective measurement results, independent companies were commissioned with these tasks. The expertises encompassing various fields were performed by specialised firms, while the responsibilities to ensure compliance with the existing environmental regulations rested with the relevant authorities.

Measurements and expertises predominantly cover the following fields:

- Dust control on unpaved construction roads
- Pollution control on paved construction roads by means of road sweeping machines
- Provision of particle filters for all diesel-operated machines and facilities above and below ground
- Electric engine pre-heating for diesel-operated machines to prevent emission-intensive cold start procedures
- Minimization of the conveyor systems' discharge heights (for noise and dust reduction reasons)
- Storage of muck material on a storage site—exclusively by the use of wheeled vehicles
- Restriction of noise emissions to ≤45 dB at day time and to ≤35 dB at night time
- Limitation of noise level peaks during blasting to ≤90 dB at day time and to ≤65 dB at night time; as a result of this requirement, blasting works in the vicinity of the portals, were prohibited at night
- Definition of a mean noise level for the ventilation fans located at the portals of ≤98 dB
- Use of biologically degradable formwork oils and release agents
- Containment of conveyor systems including noise insulation
- Continued uswe of a hiking trail which runs aside the construction and the muck disposal site.

The strict compliance with the various requirements stipulated by the relevant authorities was ensured by an independent environmental specialist during the construction works.

Tunnel Systems

Another characteristic of the project were the various tunnels, which in total added up to a length of approx. 1,800 m. Cross-sectional areas ranging between 12 and 52 m^2, as well as declines and inclines up to 37%, pushed conventional tunneling to its technical limits. In addition to this, the excavation works were further complicated by the fact that very few tunnel sections were long enough to ensure high production rates. Numerous bifurcations were another reason why the excavation performance was hardly comparable to that of traditional tunnels.

And last but not least, the cross-sectional shape of the tunnel (horseshoe profile) decisively influenced the muck removal concept as well as the equipment and material logistics, since most tunnels were not wide enough for two vehicles to pass at the same time.

EXCAVATION WORKS

Power House

The excavation of the power house and transformer cavern was characterized by a number of adverse boundary conditions. These boundary conditions were partly the result of tender requirements and partly the result of logistics considerations.

In a first step, the existing pilot tunnel, which is intended to be used as a cable tunnel, had to be enlarged, as the available space was not sufficient for the two ventilation ducts (ø 1,000 mm) and the concrete mixer trucks. In parallel to these measures, the 170-m-long inclinined muck removal tunnel (24%), which has a cross-sectional area of 22 m^2, was constructed.

The transformer cavern and the power house were constructed via a newly built access tunnel which was branching off the pilot tunnel.

Upon completion of the transformer cavern's top heading, the power house cavern's top heading was excavated via two side drifts. Access to the right side drift was gained via a previously built cross-passage. Despite the apparently favourable geological conditions in predominantly compact hornblende gneiss, amphibolite, and quartz gneiss, this excavation was not an easy undertaking. The impact of the pronounced system of tectonic joints, the mica schist layers, the fault layers with clay-mylonites and the cataclasites, each featuring a thickness of 10–20 cm, was not to be underestimated.

In order to prevent sliding of large-size blocks, a strict adherence to the previously defined excavation and support measures was imperative, especially in the 30.5-m-wide, gently-vaulted cavern.

The top heading, with a cross-sectional area of 220 m^2, was divided into three sections. It was only upon installation and pre-stressing of the up to 43-m-long strand anchors with working loads of 1,250 kN in the side drifts that the core of the top heading could be removed (Figures 4 and 5).

Whilst the Swellex anchors and the reinforced concrete were installed as initial support after every round of advance, the up to 16-m-long rock bolts and the second layer of shotcrete followed in the second round of advance.

Due to the limited range of the lifting platforms, the anchor heads of the strand anchors were capped with fire and blast protection covers and the two-layer lining shotcrete was applied before the bench excavation was even started (Figure 6).

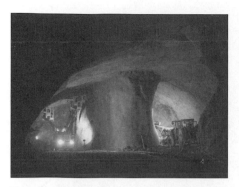

Figure 4. Top heading power house cavern

Figure 5. Excavation of central pillar in top heading of power house cavern

Figure 6. Finished top heading of power house cavern with fire protected anchor heads of prestressed strands

Figure 7. Glory hole used to muck out bench of power house cavern

Upon excavation of the first bench, the concrete beams for the crane were installed and subsequently supported by 68 strand anchors. In order to protect the concrete beams from possible damage caused by blast vibrations during the following bench excavation, the concrete beams were placed on a 1.50-m-high frame structure.

The 12.5 ton construction crane was erected in the middle of June 2005.

The entire muck material of the cavern bench was to be evacuated via a glory hole (ø 2.80 m) and a muck tunnel. To prevent a possible clogging of the shaft, it was constructed in a conical shape (a 4 m diameter at the top and a 6 m diameter and the bottom). The shaft was excavated by drilling blast holes from the top and by then adopting controlled bottom to top blasts. This shaft excavation took approx. 6 days and on March 21st, 2005, muck removal through the shaft commenced (Figure 7).

At the base of the muck tunnel in the junction area to the tailrace tunnel, which branches off at a right angle, a crusher was installed. Based upon the predicted rock characteristics and the limited space, a jaw crusher of the RC 3506S-E type with an 8.5 m³ hopper and a power capacity of 160 kW was chosen. The hopper could directly be fed from the wheel loader.

A discharge conveyor allowed the muck material to be loaded onto an 830-m-long and 800-mm-wide conveyor belt. At the West muck disposal site, a discharge conveyor,

Figure 8. Bench excavation in steps of 5 m height

which could be swivelled by 180° and had a radius of 24 m, distributed the material (2,500 m³ of stockpile capacity). For the installation of the conveyor system, the L188 provincial road had to be underpassed by a pipe culvert and the Ill river had to be crossed by a conveyor bridge. The mucking system was designed for 220 tons/h but effectively handled up to 400 tons/h.

Geotechnical reasons made it necessary that the three large compressed air surge chambers (cross-sectional area: 66 m²) were excavated and supported before the respective bench level was reached and that the two remaining ground pillars (each 11 m in width), located between the chambers, were strengthened by chamber to chamber anchors. The objective of this approach was to reduce stresses and strains, which might have had adverse effects on the large cavern opening, due to a redistribution of loads during the following pressure chamber excavation. A later addi-tional anchoring of the cavern top heading, in case of excessive deformations, would no longer have been possible for geometric reasons.

The pressure chambers were initially envisaged to be built via the muck and tail-race tunnel and the bent connection tunnel (37% ascending). But for construction logistics reasons, they were ultimately advanced via three inclined tunnels, which were running through the power house from the headrace to the tailrace system. The pressure chambers were constructed at the same time as the crane beams (April/May 2005).

The optimum bench excavation height, needed to achieve a reasonable balance between drilling, mucking and support measures, was found to be 5 m. The blast holes for the central bench were drilled vertically; the drilling pattern varied between 2.0/2.0 and 2.50/2.50 m depending on muck grading (40–50 cm maximum size). The blast holes for the side benches were drilled horizontally to minimize overbreak. These benches should at least have been 4 m wide, but ended up being approx. 5–6 m wide, as a result of unfavourable dipping and marked jointing phenomena, as the central bench blasting works often created sloping rock surfaces (Figure 8).

When constructing the cable cellar of the transformer cavern, a relatively low rock mass strength was furthermore observed. The remaining vertical side walls had, despite extremely careful blasting works, still to be excavated and scaled manually, and had in sections still to be supported and anchored.

All ten bench levels were excavated by the end of November 2005, despite the fact that at the pressure chamber level, the normal excavation height of 5 m had to be reduced to half the original height. The remaining excavation works like dewatering pump sumps and local pits were completed in December 2005 and on December 11th, 2005 the so called "target level ceremony" was staged.

The three inclined shafts (78°, depth: 22 m) for the pump mains were excavated top down, allowing the horizontal pump main tunnels to be broken through immediately upon arrival at the respective bench level.

The 31-m-deep pilot shaft for the tailrace surge chamber was accessed using Alimak raise climbers (ø 2.80 m), before the surge chamber (ø 13 m) with the adjacent 47-m-long pressure-balancing chamber was enlarged, removing the muck material via the pilot shaft.

For the tailrace tunnel heading towards the Rifa reservoir, the L188 provincial road, the bed of the Ill river and the dam of the Rifa reservoir had to be underpassed. The approx. 90-m-long soft-rock section, running below the Ill river up to the cut at the Rifa reservoir, which had been completed by then, was excavated under the protection of a pipe arch and two groundwater lowering wells. Breakthrough took place on December 11th, 2005.

The entire excavation works took approximately one month longer than initially anticipated, but it is—in this context to be borne in mind—that some additional services had to be rendered which lay on the critical path. These services included the excavation of an approx. 130-m-long tunnel (cross-sectional area: 19 m²) to accommodate the cooling water and fire-fighting water tank. In addition to this, the concreting works for the cable cellar in the transformer cavern were brought forward and the concrete pavement in the assembly area of the power house was poured on September 3rd, 2005. This also marks the point in time, when the necessary material handling platform in the power house cavern was made available (Figure 9).

LINING AND CONCRETING WORKS

Power House Cavern

The decision, which formwork and concreting system should be adopted in the cavern, was influenced by a series of boundary conditions. In view of the mass concrete and the resulting expected high concrete temperatures, the tender documents predominantly did not permit the use of pumped concrete, but it soon became evident that a consistent use of concrete placed with crane and bucket might put the anticipated construction schedule at risk. It was in light of these facts that an agreement was reached with the client to use pumpable concrete. In addition to this, the start of the lining works in the power house cavern coincided with the start of the installation works of the hydro-mechanical equipment, which furthermore limited the use of the auxiliary construction crane.

The solution adopted finally consisted of two concrete distribution booms, which were positioned in such a way that every corner of the power house cavern could easily be reached. The mix for the pumpable concrete was modified and adjusted several times to keep the concrete temperature low and to consequently minimize the occurrence of shrinkage cracks. And last but not least, an in-situ concrete mixing plant was installed directly on site, to allow a completely independent and flexible response to any conceivable situation.

Figure 9. Power house excavation
completed

Figure 10. Concrete works in power house cavern

With respect to the formwork system, steps were taken to reduce the stock of formwork material to a minimum. Another criterion brought into play, were the substantial dimensions of the structural components and the resulting pressures and loads, which had to be sustained by the formwork. This included slab thicknesses of up to 4.0 m and wall heights of up to 8.5 m, as well as individual structural components with concrete volumes ranging between 400 m^3 and 600 m^3.

For the performance of the lining works, the power house cavern was subdivided into seven levels, with each section comprising up to 45 separate structural components (Figure 10).

Time Schedule and Milestones. December 5th, 2005 marked the official launching of the lining works. It was on this date that the go-ahead signal was given for the installation of the first concrete foundations in the power house cavern.

The construction sequence in the power house cavern had to be arranged in such a way that the assembly/installation and concreting works, could generally be carried out adopting a staggered approach "unit no. 3–unit no. 2–unit no. 1." All other lining works had either to be performed simultaneously, or—depending on the technical feasibility—overlapping with the main lining works in the power house cavern. To better illustrate this concept, works in the transformer cavern were done to compensate a longer interruption or a total standstill of works in the power house cavern. Decisive dates in the power house cavern were influenced by the installation of the tailrace crane support structures and by the operation start-up of unit no. 3. The crane support structures were imperative for the mounting of the heavy-duty crane, and this crane in turn was essential for the installation of the hydro-mechanical equipment. Switching on unit no. 3 in December 2007 at the latest is a crucial precondition for meeting the final completion date.

Figure 11.   Typical concrete slab at unit 3

Difficulties and Particularities. As has already been indicated in the introduction, special challenges, which not only originated in the exceptional dimensions of the cavern, had to be overcome in connection with the lining works. Apart from the impressive dimensions of the structural components, which led to concreting times of 24 hours and more, the demands resulting from mechanical equipment installations were also enormous.

Contrary to previous projects, steel structures, such as bearings, rails, etc., and their pertaining structural components were concreted in one go. With this strategy being pursued, special demands were placed on positioning these steel structures in the exact x, y, z coordinates.

Due to the densely-spaced, three-dimensionally bent reinforcement, this reinforcement installation repeatedly turned out to be more time-consuming than initially anticipated. With numerous structural components being either fully or partially lined with steel, the adequate reinforcement arrangement and the accurate reinforcement installation could often only be achieved by an intensive cooperation and coordination with the respective firms supplying the hydro-mechanical equipment (Figure 11).

Yet another, rather work-intensive obstacle consisted in the number and size of the pipes as well as in the complex pipe arrangement. Challenging formwork ducts and reinforcement arrangements in the vicinity of pipes as well as difficult pipe configurations quite frequently made it necessary to interrupt and resume the formwork and reinforcement works.

The installation and assembly of recesses of various sizes called for the provision of an on-site carpentry work shop with a team focusing on the production of these recesses.

Numerous base plates for structures to be welded on, with individual weights of up to 250 kg as well as the stipulation not to exceed an accuracy tolerance of 5 mm in location and height when concreting these steel plates, required special preparatory works to be made. And the fact that a welding of these plates to the structural reinforcement was prohibited, furthermore complicated the placement and final positioning of these plates (Figure 12).

Auxiliary Structures

All auxiliary structures had—for project schedule reasons—to be completed at the same time as the lining works of the power house cavern. In addition to the concrete

Figure 12.   Pressure-pipe-jet ring for horizontally located Pelton-Turbine

Figure 13.   Tailrace inlet/outlet structure

backfilling of the steel lining and the installation of concrete linings by the use of shutters, several highly complicated transition and narrowing or widening sections had to be constructed. The formwork elements needed for these sections were all produced on site and successfully installed and used. The 40-m-high surge chamber shaft was lined using slipforming.

A special challenge was the above-ground construction of the inlet and outlet structure in the Rifa reservoir. The time window for the construction of this structure was the low water period between January 2006 and May 2006, defined by the client. The extremely cold and long winter with temperatures dropping to –15 °C, pushed the teams to their limits. The frost zone extended to a depth of 1 m below the ground surface. In order to shield the concrete from frost, 3–5 layers of frost protection fleece were used. As the date for the refilling of the Rifa reservoir at the beginning of June 2006 could not be postponed, works partially continued day and night (Figure 13).

CONCLUSION

In the period between November 2004 and December 2005 (13 months) a total volume of 125,000 m^3 was excavated in the power house and the transformer cavern. The excavation works for the tunnels and shafts—all in all 1,782 m—were performed in parallel to the excavation works for the caverns and amounted to a total of 78,000 m^3.

The performance level attained with respect to the lining works is also quite remarkable, as is illustrated by the fact that in the period from December 2005 to May 2007 (1.5 years) 48,000 m^3 of concrete were installed in the cavern. The numerous lining and backfilling works in the tunnels, shafts and other auxiliary structures were carried out simultaneously with the concreting works in the cavern.

This outstanding accomplishment was only made possible by an intense work preparation throughout the project implementation period and by the strong commitment of all specialists involved. At peak times, the number of skilled workers on site was in the range of approx. 130. All works were performed continuously in day and night shifts, only interrupted by the traditional Christmas and Easter holiday.

DESIGN-BUILD OF THE LAKE HODGES TO OLIVENHAIN PIPELINE TUNNEL AND SHAFT

M. Luis Piek
Parsons

Jon Y. Kaneshiro
Parsons

Sean A. Menge
Parsons

Brian Barker
Kiewit Pacific Co.

ABSTRACT

The Lake Hodges to Olivenhain Pipeline project is part of a 40-megawatt pumped storage project linking two reservoirs as part of the San Diego County Water Authority's $939 million Emergency Storage Project. Seeking to expedite project completion, the project was advertised as Design-Build, a first for the owner. Awarded to Kiewit Pacific with Parsons as designer, the 1,750m (5,700ft) long, 4.25-m (14 ft) horseshoe tunnel was excavated with conventional drill-and-shoot technique. The mining method was well suited to deal with the shear zones, the very abrasive rock, and 13.3% and 19.6% grades. At the apex of the alignment, a 60m (200ft) by 3.6m (12ft) diameter raisebore shaft with surge chamber to protect against pump malfunction was constructed. Upon completion of mining, the tunnel was lined with a 3m (10 ft) diameter high strength steel pipe. Scheduled to come on-line by January 1, 2008, the Lake Hodges to Olivenhain pumped storage project is valued at $108 million, delivering a reliable source of water in emergency situations and electricity-generating capability to meet the growing demands of the San Diego region.

INTRODUCTION AND HISTORY

The San Diego region has experienced significant growth in recent decades, and in response, the San Diego County Water Authority (SDCWA) has kept pace by completing several major construction projects throughout the County to secure the drinking water supply. The Olivenhain Dam, completed in 2003 by Kiewit Pacific and the first in a series of contracts in the Escondido area, created a reservoir to increase the total water capacity of the SDCWA. The dam contract also included an intake structure, stub tunnel, gate structure, and operations building in anticipation of a tunnel that would eventually be used to link the lower Lake Hodges reservoir to the upper Olivenhain Dam reservoir.

The second contract, the tunnel linking the two reservoirs, was brought to a 50% design level by MWH and set the horizontal alignment. In order to accelerate the schedule and allow for design optimization, the Lake Hodges to Olivenhain Pipeline (LHOP) was released as a Design-Build project, a first for the SDCWA. The tunnel size required a minimum 3m (10ft) diameter steel liner in the lower reaches and a concrete

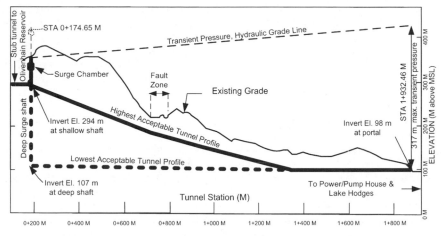

Figure 1. Allowable lowest and highest vertical alignments

or unlined tunnel in the upper reach. The optimum economical vertical alignment was left to the choice of the Design-Builder, and allowed grades between 0.5% and 19.6% (Figure 1). At the apex of the alignment, a 3m (10ft) diameter shaft, varying between 60m (200ft) to 245m (800ft) deep, depending on the vertical alignment chosen, was specified with a surge chamber to protect against pump malfunction.

Aside from the tunnel contract linking the two reservoirs, a third contract for a power generation/pump house was released which will eventually allow the SDCWA to produce electricity during peak times to answer the growing electricity demands of the region. Further, in wet years the Hodges Dam is overtopped by large amounts of runoff entering the lake. By using the Olivenhain Dam Reservoir as a buffer, the water level in Lake Hodges can be maintained at optimal capacity, capturing water that would other-wise be sent further downstream, and eventually lost to sea.

PROPOSAL PHASE—TBM VERSUS DRILL AND BLAST OPTIONS

The design-build approach in the bid phase consisted of review of the existing field conditions, the borehole core samples, contractual requirements, design criteria, milestones and schedules, risk evaluation, various alternatives allowed by the specifi-cations, and constructability. The tender package also provided a Geotechnical Data Report (GDR) and Geotechnical Baseline Report for Bidding (GBR-B). As a part of the preparation for the bid, the tender required that the DB team prepare a GBR-C, with the "C" signifying for "construction" purposes, which was finalized and clarified during the contract negotiations. Also, the contract pricing included a $1 million allowance for differing site conditions to be used as needed for compensation if actual conditions dif-fered from the GBR-C.

During the bid phase, Kiewit Pacific evaluated the several options available to con-struct the tunnel, primarily TBM and conventional drill and blast. The granitic rock was estimated from boring logs as being between very strong to extremely strong, massive to moderately fractured, and highly abrasive. The available grades of the tunnel varied between 0.5% continuously to the shaft at the lowest, or at roughly one-third intervals 0.5%, 13.3% and 19.6% grades at the highest.

Based on surface geology, field inspection, and bore logs of the area, the GBR-B indicated that in the upper reach a shear zone or potential fault was identified underneath an eroded canyon. In addition, groundwater wells along the alignment showed considerable variation in the groundwater table above and below this same section of ground, potentially indicating a groundwater divide caused in part by filling of the Olivenhain Dam Reservoir. The mining method selected would have to handle potentially high groundwater inflows and changing rock characteristics in the shear zones and dikes identified.

The lower alignment would have been ideal for a TBM, though the potentially high groundwater levels and recent experience in southern California under such conditions made this option less appealing. Further, the tolerances for the shaft, maximum deviation from vertical of 0.15% (within 40cm [15in] for a 245m [800ft] deep shaft), would have presented several challenges. While pipeline installation would have been uncomplicated on a 0.5% slope, the resulting elevated internal static pressures and external groundwater head would have resulted in an approximate 20% increase in the steel tonnage versus a higher alignment.

When considering the conventional drill and blast method, several vertical alignments were evaluated based on economy and constructability. The drill and blast method was appealing from the standpoint that if difficult ground conditions were encountered, conventional excavation allows ease of access for a variety of excavation and support methods to allow mining to continue. Mobilization time and costs for this length of tunnel were also found to be competitive against a TBM option. Further, selection of a high alignment, only possible through drill and blast methods, would lessen the impact of groundwater on construction, and when mining through the shear zone would be akin to popping a water balloon from the top rather than the bottom, to slowly drain the water from the ground.

While the upper reaches of the alignment would have allowed a concrete liner or an unlined section, the use of a steel liner over the length of the project was selected to avoid leakage requirements and used essentially as stay in place formwork. Finally, the all steel lined option will work better as a finished system, providing long term operational benefits to the SDCWA.

DESIGN PHASE

The Design-Build team of Kiewit-Parsons was selected as the winning bidder in February of 2005. Working with a very tight construction schedule from Day One, the difficulty was compounded when the steel lining manufacturer, Ameron International (Rancho Cucamonga, CA), found that obtaining firm production dates for the specified ASTM A537 Class 2 high tensile strength steel became all but impossible, as all quench and tempered steel produced domestically had been requisitioned for armor plating by the US Military. Seeking alternate steel products, ASTM A841, a steel type commonly used for offshore oil platforms and ocean pipelines was identified as a potential alternate. After extensive examination by the DB team, the steel was presented for consideration and approval given by the SDCWA to utilize the A841 steel in the project (Piek et al., 2006).

Having never completed a penstock utilizing this steel, and without significant precedent, the Design-Build team and Ameron formed project-specific quality control procedures and quality assurance measures to be used through all stages of fabrication of the steel liner. A major part of this was writing a welding qualification procedure for the longitudinal welds to form the pipe from the steel plate, as well as a welding qualification procedure for the field radial welds and repair welds. A full QA/QC program was formed and executed at the fabrication plant, as well as during field installation of the steel liner, and included the participation of the Contractor, the Designer, the Fabricator, and the field welding subcontractor.

In order to economize the cost of the steel liner, the inherent strength of the rock was utilized through a design process called "load sharing" to reduce the wall thickness of the steel liner (Kaneshiro et al., 2006). Use of load sharing in design resulted in a savings of 12% of the steel tonnage versus traditional pipe design.

CONSTRUCTION PHASE

Portal preparation began in August 2005 with tunnel excavation commencing toward the end of September 2005. Tunnel excavation was performed using, on average, a 46-hole drill pattern set using a Terex MK35 2-Boom drill jumbo and mucking performed with a pair of Caterpillar R1600G LHDs outfitted with 4.7 m^3 (6.2 cy) buckets. When turning under, tunnel excavation immediately tested the mining crew's abilities due to a deeper than anticipated weathering profile for the first 30m (100 ft). After this difficult patch of ground, tunneling progressed quickly thereafter. Groundwater inflows were essentially zero, with the first 550m (1,800ft) of excavation consuming more water for drilling operations than water production out the portal. To decrease mucking times, a mud slab was placed along the length of the tunnel invert, with ditches to control drill water along either side of the slab. Preparation of the tunnel invert prior to placement of the mud slab required extensive blowdown and removal of debris to ensure proper strength to satisfy the load sharing assumptions.

Adapting to performing drill and blast operations on the incline occurred quickly with little loss in production. When nearing the shear zone, a 30m (100ft) long probe hole was advanced beyond the face to forecast potential increases in groundwater production due to the groundwater divide identified during the proposal phase. Monitoring of the surface wells on either side of the shear zone was not possible due to backfilling operations performed by the SDCWA. The shear zone was encountered 38m (125ft) earlier than anticipated, but the rock had sufficient strength to only require patterned split sets and mesh for ground support. More importantly, there were no significant sustained increases in groundwater inflows, indicating the higher alignment was the safer choice. Mining continued on, unimpeded to hole-through, on August 1, 2006.

In order to verify several variables used in design, geophysical testing was performed in competent rock as well as in the shear zone to determine compressive strength of the intact host rock, compressive strength and extent of fissured rock due to drill-and-blast operations, Poisson's ratio of the intact and fissured rock, unit weight, etc.

As previously mentioned, at the junction of the drill and blast tunnel and the previously constructed stub tunnel, a 60m (200ft) deep shaft complete with surge chamber was to be constructed. Sized to accommodate surges and transient fluctuations from daily changes from power generation to pumping, the shaft provided some interesting challenges to construct. The pilot hole driven for the raise bore indicated that the rock at the shaft site was extremely abrasive, and the 35cm (14in) diameter hole tended to drift, and took 3 weeks to complete. A 3.6m (12ft) diameter raised bore head was assembled at the apex of the tunnel and used to excavate the shaft. Although the shaft is less than 100m (300ft) from the shore of the upper reservoir, no leakage was seen through the extremely tight joints of the rock. Further, the rock was very massive with few joints present through the first 55m (175vf) of the shaft, requiring no concrete lining of the shaft. The shaft excavation took four weeks.

Construction of the surge chamber began immediately following shaft excavation. An unusual design, the 8.2m (27ft) diameter surge chamber situated 30m (100ft) down from the top of the shaft proved to be challenging from both a safety and a construction standpoint. Designing and building a custom work deck that would satisfy

Figure 2. Final assembly of the surge chamber work deck

the sequential operational requirements was the first major hurdle to overcome. Working so close to the shaft face on a semi-rigid platform, the miners had to exercise extreme caution when using the necessary mining equipment to drill shot holes and install rock bolts. Finally, shotcrete lining of the surge chamber a depth of 100vf posed in a relatively small diameter shaft challenged the miners as well. With all of the construction obstacles the miners faced, the surge chamber only took slightly more than two weeks to complete. A photo of the assembled work deck prior to installation in the shaft is shown in Figure 2.

In preparation of pipe installation a significant amount of additional effort was required to satisfy the design assumptions for load sharing. This included blowdown of the tunnel invert to sound rock before placement of the concrete slab, geophysical testing of the ground to verify adequate rock strength, backfilling muck bays and electrical nitches with over 350 m^3 (450 cy) of concrete.

Installation of the 12m (40ft) long pipeline sections on the steep inclines was completed by using the same CAT R1600 LHD equipment used during tunnel excavation. Outfitted with a custom built pipe dolly and carrier system, the LHD had sufficient capacity to push the pipe sections, some nearing 27 metric tons (60,000 lb), up the inclines. The dolly and carrier were capable of pipe rotation, side to side movement, and articulation on either end of the pipe. The system proved invaluable to setting the pipe within the specified line and grade tolerances in such tight quarters. Further, the familiarity of the LHDs with the operators, plus the need to mobilize and store less equipment was beneficial. A photo of the pipe carrier and dolly is shown in Figure 3.

Since the tunnel is accessible from only the portal or the shaft, high strength cellular concrete was selected as an annular backfill material to overcome the long pumping distances. Interior stulling and exterior blocking were used to support the pipe during backfill operations. Having a unit weight of 1,040 kg/m^3 (65pcf) and a minimum strength of 4,500 kPa (650psi), the cellular concrete mix design samples were subjected to additional testing prior to acceptance in order to satisfy the assumptions used in design.

Final skin and contact grouting was performed with additional monitoring to fill any voids not filled during annular backfill. Seal welding of the plugs, and epoxy paint touch-up completed the steel liner portion of the project.

Figure 3. Pipe carrier and dolly mounted to LHD

SDCWA AND COMMUNITY OUTREACH

Construction of major tunneling projects within city limits and in close vicinity to communities always present challenges to owners and contractors. The SDCWA embarked on a community outreach plan prior to construction which including community meetings, monthly newsletters, a public relations hotline to register complaints and resolve conflicts, and personal visits to community members. Yet, complaints are almost inevitable and several were registered during the course of the project, most early on. To keep roadway disruptions to minimum, delivery hours were limited to 7AM to 4PM weekdays providing a construction scheduling challenge.

In order to mitigate the impact to the community, blasting operations were limited from 7AM to 7PM until the heading had progressed 245m (800ft) inside the tunnel. A suitable program of blast monitoring was implemented using portable seismographs. Each shot was recorded at the nearest three structures to the blast and analyzed using scaled distance method. Peak particle velocities were successfully controlled by varying the round length and blast pattern. The alignment passed directly beneath several houses and farms with livestock with ground depths around 30m vertically (100vf). As an indication of the level of effort put forth, continuing monitoring was required to determine the impact felt by horses and other livestock located nearby during blasting. No discernable effect was found.

CONCLUSION

Being one of only a handful of large hydroelectric projects commissioned since 1989 in the USA, the LHOP project meets the energy and water storage needs of the San Diego region by connecting the upper Olivenhain Dam Reservoir with Lake Hodges. The SDCWA, recognizing the various techniques that could have been used to construct the project, used this as their first Design Build contract to accelerate the schedule, and select the construction method and final alignment.

Anticipating difficult ground conditions, the Design Build team lowered project risks by choosing to excavate the tunnel using conventional drill and blast methodology, incorporating the highest allowable vertical alignment with grades as steep as 20%. Mining proceeded well through the project and maintained good production rates even on the higher slopes. The raise bore method of shaft construction was well suited for the hard granite, though the surge chamber in the shaft created a variety of safety and construction difficulties.

Faced with unforeseen challenges with obtaining the high strength steel for the project, the DB team identified and obtained approval of for ASTM A841 steel, whose strength is derived from the thermo-mechanical control process. The penstock was constructed to handle pressures in excess of 3,100 kPa (450 psi) with load sharing to shed a portion of the load to the host rock. While further economizing the liner, the design required several additional construction steps to verify the design methodology and assumptions. Steel liner installation was completed using a custom built carrier and dolly system attached to the LHDs originally used for muck removal. Cellular concrete was selected to backfill the annulus of the tunnel due to the long pumping distances.

With the pumphouse scheduled to come on-line by January 1, 2008, the Lake Hodges to Olivenhain Pipeline will deliver a reliable source of water in emergency situations and electricity-generating capability to meet the demands of the growing San Diego region.

ACKNOWLEDGMENTS

The authors would like to show appreciation to several individuals whose efforts went above and beyond what their position required. These people include David Chamberlain of the SDCWA, Mike Shough, Joshua Schurger, Kurt Millsaps, Pedro Moreno, Matt Roberts and Jarrett Carlson of Kiewit Pacific, the Dean's Welding crew, Gregg Korbin, and Kuantsai Lee, Alan Hull, and Sam Thompson of Golder Associates. Your strength and dedication to overcome the management, engineering, and construction challenges faced on a daily basis was essential to the success of this project.

REFERENCES

ASCE [American Society of Civil Engineers] Manuals and Reports on Engineering Practice No. 79 Steel Penstocks (1993).

ASTM [American Society for Testing and Materials] A841/A 841M-03a Standard Specification for Steel Plates for Pressure Vessels, Produced by Thermo-Mechanical Control Process (TMCP).

Cases of ASME [American Society of Mechanical Engineers] Boiler and Pressure Vessel Code—Case 2130-3.

Kaneshiro, J.Y., L. Piek, K. Lee, G. Korbin, J. Carlson, M. Ramsey. (2006) "Emergency Storage Project: Design and construction of the Lake Hodges to Olivenhain Pipeline, San Diego, California," International Tunneling Association Conference, April 2006, Seoul, Korea.

Piek, M.L., J. Carlson, G. Fehr, J.Y. Kaneshiro, (2006). "Lake Hodges to Olivenhain Pipeline Tunnel, Shaft and Site Development. Use of ASTM A841 TMCP Steel" American Society of Civil Engineers Pipeline Conference, August 2006, Chicago, USA.

ELM ROAD GENERATING STATION
WATER INTAKE TUNNEL

Jon Isaacson

Kenny Construction Co.

Brendan Reilly

Bechtel Power Corp.

Paul McDermott

Kenny Construction Co.

ABSTRACT

Kenny/Shea JV is constructing the water intake tunnel for the new Elm Road Generating Station, a supercritical coal-fired power plant in Oak Creek, Wisconsin, being built by Bechtel Power Corporation. The intake tunnel system consists of a temporary mining shaft, pump house and intake channel riser shafts, and four lakebed intake shafts. An 8.33 m diameter, 2,800 m long TBM mined rock tunnel is being constructed to connect the shafts. The intakes are located 2.4 kilometers offshore in Lake Michigan. The shafts and tunnel are being constructed in glacial tills and Silurian Dolomite. The estimated tunnel completion is December 2007.

PROJECT OVERVIEW

We Energies announced "Power the Future" in September 2000 to address power supply and reliability issues for We Energies' customers. The Power the Future program has been investing in additional power generation, improvements to existing power plants, and upgrades to the distribution system. In November 2003 the Public Service Commission of Wisconsin authorized the expansion of the existing Oak Creek Power Plant (OCPP) on the border between Milwaukee and Racine counties along the western shore of Lake Michigan.

In April 2004 Bechtel Power Corporation (BPC) signed a contract with We Power for the largest lump sum turnkey project in Bechtel's 100 year history to engineer, procure, and construct the Elm Road Generating Station (ERGS) facility at a cost of approximately $2 billion. The ERGS facility, shown in Figure 1, will include a 1,230 MW power plant consisting of two pulverized coal fired supercritical reheat boilers with steam turbine generators, coal handling system, flue gas systems, ash handling systems, and all necessary auxiliary and accessory equipment. Unit 1 is scheduled for commercial operation in 2009 and Unit 2 will come on line in 2010.

COOLING WATER INTAKE SYSTEM

The design required that cooling and make-up water for the new facility be taken directly from Lake Michigan through an offshore intake tunnel, shown in Figure 2. The Oak Creek Power Plant (OCPP), an existing facility, currently uses a shoreline surface water intake. An offshore intake location was selected to meet anticipated Environmental Protection Agency (EPA) cooling water intake structure regulations that were

Figure 1.   Artist's illustration of completed ERGS facility and adjacent OCPP along Lake Michigan

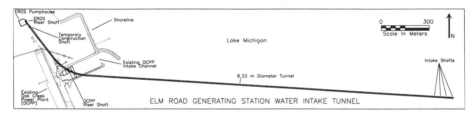

Figure 2.   Plan view of water intake tunnel system proposed to supply ERGS and OCPP

being developed to address the protection of aquatic life. In addition to the aquatic life protection benefits, moving the intake point offshore reduces the potential for silt and vegetation being drawn into the existing plant. By designing the intake system to serve both the new and existing plants, the system would be cleaner, more efficient, and less expensive than traditional cooling tower systems to lower the temperature of the water. BPC developed a conceptual design for drawing water from the Lake utilizing the tunnel connected to four 3 m (12 ft) diameter shafts approximately 2.4 km (1.5 mi) off-shore; each shaft would be connected by buried steel manifold pipes to 2.4 m (8 ft) diameter, 10.4 m (34 ft) long, cylindrical wedge wire screens. The design required a total volume of water inflow through the tunnel of 98.4 m³/s (1,560,000 gpm); this pro-vided 51.7 m³/s (820,000 gpm) to the existing Oak Creek Power Plant and 46.7 m³/s (740,000 gpm) to the new Elm Road Generating Station facility.

In May 2004 BPC issued a Request for Proposal Package from experienced tun-neling contractors for the construction of the Intake Tunnel system, and the other struc-tures associated with the intake system, as a design-build subcontract. After bid review, the subcontract was awarded to Kenny-Shea, a Joint Venture (K/S) in Novem-ber 2004. Work was expected to commence on site in early March 2005.

On November 29, 2004, however, work on the project was delayed when a Dane County Circuit Court, which was reviewing the Certificate of Public Convenience and Necessity (CPCN) for the Oak Creek Power Plant expansion, vacated the Public Ser-vice Commission of Wisconsin's (PSCW) Order authorizing construction and sent the case back to the Commission for additional proceedings. The court determined that the PSCW did not follow proper procedures in approving the power plant expansion project, and should have considered additional sites and generation types. The uncertainty for the go-ahead for the project at this stage resulted in strict financial spending restrictions for all areas of the project. K/S continued the design works and procured equipment and components necessary for the essential marine works scheduled to be performed in 2005. If the marine works did not commence in 2005 there was real potential to delay the Tunnel Flooding date, the major milestone of the subcontract, by up to one year. The Supreme Court of Wisconsin reversed the November ruling on June 28, 2005, and reinstated the PSCW's final decision and order for the construction of the Oak Creek Power Plant expansion. Following this ruling construction works immediately commenced on site.

GEOLOGIC SETTING AND SITE CONDITIONS

The Central Lowlands physiographic province along the western edge of the Michigan Basin is predominantly underlain by Paleozoic shale and dolomite. The Milwaukee region is underlain by Devonian age dolomite and shale, Silurian dolomite, and Ordovician dolomite and shale deposited in shallow inland seas. Although the bedrock surface is typically flat lying, these strata have eastward dips of up to 10°, as shown in Figure 3. Bedrock within the Milwaukee region are overlain by loess, lacustrine clays, glacial tills, and glacial moraine and outwash deposits.

The onshore construction site typically consisted of 24–59 m (80–195 ft) of overburden comprised of loess, soft to medium stiff glacio-lacustine clays, loose saturated beach sands, and stiff to very stiff glacial till with occasional channels/lenses of outwash sands and gravels. The construction site originally was a 37 m (120 ft) tall bluff along the western shoreline of Lake Michigan. The depth of overburden was reduced to 23–26 m (75–85 ft) in the area of the ERGS riser and construction shafts as part of the mass excavation to remove 4.6 million m^3 (6 million yd^3) of the bluff overburden down to a construction grade level for the new ERGS plant of El. 592'.[*] The offshore site was located in 13 m (43 ft) of water where lakebed overburden consisted of a top layer of soft lacustrine clay with stiff to very stiff glacial till beneath. The total depth of overburden at the offshore site was 12–15 m (40–50 ft).

Beneath the overburden onshore, the top of bedrock was between El. 520' and 510'. Offshore the top of bedrock elevations decreased in elevation to between El. 495' and 485'. The top of rock at both sites was comprised of the Silurian Racine formation, with typical compressive strengths between 28 and 34 MPa (4,000 and 5,000 psi), and possessing two distinct facies types; porous "reef" facies, medium to massively bedded, fossiliferous dolomite, and "bank" facies that are more argillaceous dolomite with shale laminations and thin to medium undulating bedding. The Racine reefs are medium to coarse grained and typically exhibit higher hydraulic conductivities (K), pitting and vugs, and solution enlarged, moderately dipping joint sets. The bank facies are finer grained and exhibit lower K values with steeply dipping, tight joint sets. Conductivity values in the Racine formation ranged between 8E–4 and 2E–7 cm/s. Underlying the Racine Formation are the Silurian Waukesha and Brandon Bridge Formations. The Waukesha is a thin to medium bedded, fine grained dolomite with chert nodules and shaley partings. The underlying Brandon Bridge Formation consists of thinly bedded argillaceous, pink to pale green in color, thin shale partings and chert nodules/bands. These two lower formations have compressive strengths typically

---

[*]All elevations referenced within the text are measured in feet referenced to the National Geodetic Vertical Datum (NGVD) 1929.

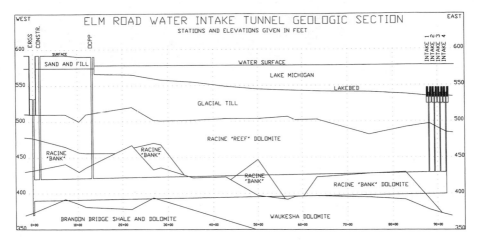

Figure 3.   Geologic cross-section of water intake tunnel system

between 69 and 90 MPa (10,000 and 13,000 psi), possess low hydraulic conductivities (4E–5 to 1E–7 cm/sec), and clay/shale filled steep joints with small apertures.

No known fault features are known to exist along the tunnel alignment. The nearest known faults are two northwest-southeast trending faults 10 km (6 mi) to the west and 26 km (16 mi) southwest of the site. These faults exhibit vertical displacements of the order of 100 feet with the downthrown blocks to the northeast. Geotechnical information obtained along the tunnel alignment indicated two significant joint sets, with steep and moderate dips, plus bedding. These joint sets had spacings of 3–8 m (9–27 ft), and bedding 0.03–3.0 m (0.1–10 ft). Onshore, these sets cross the tunnel at near perpendicular and at 20° from the tunnel axis. Offshore, both joint sets were within 20–30° from the tunnel axis.

In the Milwaukee area solution collapse features have been encountered, which contain breccias with fragments between gravel and boulder size within a clay matrix that can extend for up to 30 m (100 ft) distances. Some brecciated zones and slickensided joints were also found during various investigations along the alignment.

SHAFT DESIGN AND CONSTRUCTION

The intake tunnel system required two permanent onshore shafts, one to supply water to the existing Oak Creek Power Plant and one to supply the new ERGS Plant. A third temporary onshore shaft was required for construction. Hatch Mott MacDonald (HMM) was subcontracted by K/S to perform the hydraulic system design as well as the temporary rock support and final lining designs for the onshore shafts. Ground Engineering Consultants (GEC) was subcontracted by K/S to design the temporary overburden shaft support systems for all onshore shafts. Offshore shafts were designed and constructed by Case Foundation Company (CFC). Overburden excavation and support methods considered for the onshore shafts included soldier pile and lagging, ribs and lagging, sheet piles, and concrete caissons. Conventional drill and blast and raise bore techniques were considered for shaft excavations through rock. Temporary design support in rock by HMM included #8 GR95 rock dowels, 2–3 m (6–10 ft) long (depending on the shaft), installed on 2 m (6 ft) centers and inclined up to 25° from horizontal. These dowels were fully encapsulated in resin

Figure 4. ERGS pumphouse slurry wall excavation and ERGS riser shaft at the base of the excavation

for anchorage and corrosion protection. Convergence monitoring was required to verify the adequacy of the initial support and prior to installation of the final support to verify that no rock loading would be exerted on the final liner.

ERGS Pumphouse and Riser Shaft

The ERGS riser shaft is designed to conduct water from the intake tunnel up to the ERGS Pumphouse. This shaft was the anchor point of the tunnel system, and located at Station 0+00.* Case Foundation Company (CFC) was subcontracted by K/S for the design and construction of the ERGS Pumphouse, shown in Figure 4. The support of excavation consisted of a slurry wall with tieback anchors. The configuration of the Pumphouse went through a series of design developments prior to its final construction. Borings conducted by STS Consultants for CFC demonstrated that bedrock surface at the shaft and Pumphouse location had local undulations of up to 2.7 m (9 ft) within the limits of the slurry wall plan area. The excavation was designed with concrete slurry wall panels extending from the ground surface (El. 592') through loose beach sands and very stiff glacial till to the top of bedrock 23 to 26 m (75 to 85 ft) below. The final configuration was a 44 m by 37 m (143×121 ft) excavation with 45° end walls on the eastern end supported by a 0.9 m (36 in.) slurry wall with four levels of 21 m (70 ft) long cable tendon soil tiebacks, declined 15° from horizontal. The slurry wall panels were excavated using a clamshell to the top of rock and a chisel was used to create a 15 cm (6 in.) key into bedrock. The panel excavations were temporarily supported with bentonite slurry. Reinforcement cages, complete with steel tieback trumpets, were lowered into the excavated panels with steel end stop forms, and concrete was tremmied from the base of the panel to form the wall. Following completion of the twenty panels, a sequenced excavation within the interior of the slurry wall was conducted in order to drill, install, and grout the four levels of tiebacks required through the pre-installed trumpets. The 247 tiebacks installed were tensioned to design loads between 449 and 1179 kN (101 and 265 kips). The excavated grade within the Pumphouse was El. 535.5' to the east and El. 537' to the west, the latter the top of the ERGS riser shaft.

---

*Tunnel stations referenced within the text are given in feet as measured from the ERGS riser shaft centerline.

The founding of the ERGS Pumphouse slurry wall 0.15 m (6 in) into the top of rock formed a positive soil groundwater cutoff for the ERGS riser shaft, which was to be constructed within the Pumphouse structure. Coupled with the very stiff nature of the glacial till below the Pumphouse grade and dewatering wells at the bedrock surface to relieve artesian head, steel ribs and oak lagging were chosen as the temporary overburden support method to drive the 7.3 m (24 ft) diameter shaft to the top of bedrock at El. 513.8'. The bedrock surface varied by 2.6 m (8.5 ft) within the limits of the ERGS riser shaft excavation. Several boulders were encountered in the glacial till during excavation of the overburden, but no significant inflow was encountered at the bedrock interface. A concrete collar was constructed at the base of the rib and timber excavation on the top of rock prior to commencing the drill and blast excavation of the rock portion of the shaft. The 6.1 m (20 ft) diameter rock shaft excavation was then sunk to El. 420' through the Racine and Waukesha Formations to the crown of the tail tunnel below using 2 m (6 ft) drill and blast rounds. Prior to excavation and breakthrough of the ERGS shaft, a sump was excavated below the shaft from within the tail tunnel, at an elevation between El. 390' and El. 370' in the Upper Brandon Bridge Formation. After breakthrough into the tail tunnel, the sump, riser shaft intersection, and riser shaft will be lined with concrete up to El. 543' at a finished diameter of 5.5 m (18 ft). The tunnel/shaft intersection and top 3 m (10 ft) of the liner require steel reinforcement within the concrete liner.

Construction Shaft

A temporary construction shaft was excavated at Station 1+50 to gain initial access to the starter/tail tunnel level between El. 390' and 420'. Without the protection of a slurry wall to provide a hydraulic cut-off in the overburden, and given the loose beach sand aquifer at the site, sinking a concrete caisson was selected for the primary overburden support method at the construction shaft. A 1 m (3 ft) high tapered steel cutting shoe was installed at the leading edge of a reinforced concrete caisson, designed with 0.6 m (2 ft) thick walls and an internal diameter of 9.8 m (32 ft). The caisson was poured in 6.1 m (20 ft) vertical segments. The caisson was initially excavated through the top 6 m (20 ft) of overburden from El. 592' to 572' in the wet using a clam bucket to seat the caisson through the loose beach sand aquifer being recharged by Lake Michigan. Bentonite slurry was injected into the overcut through pre-formed ports to support and lubricate the soil outside the caisson wall. Upon entry into stiff glacial till, a bentonite seal was attained at the cutting edge, the caisson was dewatered, and excavation continued utilizing a backhoe loading muck boxes from within the caisson. The caisson was sunk 22.5 m (74 ft) from the surface to the top of bedrock at El. 518'. The top of rock within the shaft varied by only 0.3–0.6 m (1–2 ft). Upon reaching bedrock, a concrete collar was poured at the cutting shoe and the overcut was grouted to secure the caisson. Drill and blast excavation methods were then used to sink the 9.1 m (30 ft) inside diameter rock shaft thorough the Racine and Waukesha Formations to the tunnel invert level at El. 390'. Temporary rock support dowels 3 m (10 ft) long were installed at the prescribed pattern with mesh. The rock portion of the shaft was lined at a diameter of 8.5 m (28 ft) with reinforced concrete from the top of bedrock to the crown of the starter/tail tunnel at El.420' before excavation continued below that level.

This temporary construction shaft was then used to construct the starter/tail tunnel launch chamber for the tunnel boring machine (TBM) and was the access point for muck conveyance, tunnel crews, supplies, and utilities. Once TBM mining is completed, the onshore tunnel lining will close off the temporary construction shaft and the shaft will be abandoned. The shaft will be backfilled above the lining with rock and a concrete plug was constructed within the lining at the bedrock surface. Soil will then used to backfill the overburden portion of the shaft.

OCPP Riser Shaft

Under the new facility designed by BPC, the Oak Creek Power Plant (OCPP) will be supplied water by the intake tunnel system through the OCPP Riser Shaft at Station 13+33, which was constructed in the existing surface water intake channel of the plant. Water rising through this shaft enters the North Forebay pool, where it is lifted by a BPC constructed booster pump station and discharged into the main intake channel for use by the existing plant without alteration of the existing Pumphouse. The existing near shore water intake of the OCPP is isolated from Lake Michigan by a newly con-structed dike wall structure, which is designed to act as an emergency bypass intake utilizing a series of slide gates. This emergency bypass is required should the offshore screens become blocked and/or the intake tunnel system becomes inoperable.

The intake channel had been dredged prior to OCPP shaft construction by a local marine and pile driving subcontractor, Edward E. Gillen Company (EEG), a subcon-tractor appointed by K/S. The base of the material dredged in the channel was medium stiff to stiff glacial till. Due to the construction of the shaft in an open water channel, a combination of ribs and steel sheeting and a reinforced concrete caisson were selected for the temporary overburden support for the OCPP riser shaft. In order to construct the shaft, a cofferdam first had to be constructed in the inlet channel to gain access to its location. A permanent parallelogram steel sheeted cofferdam was first constructed by EEG. This cofferdam cell later became the aforementioned North Fore-bay. Then a second, 12 m (40 ft) diameter temporary steel sheet piled cofferdam was driven in the center of the first cell. Both sheeting cells were then internally braced, and the space between the inner and outer cells was temporarily backfilled with recycled crushed concrete to form a working surface above the existing lake level. The interior of the sheet piled shaft was then dewatered and excavated to El. 552', 8.2 m (27 ft) below lake level. At that point, the steel cutting shoe of the caisson was placed and the entire 12m (40 ft) length of the 0.6 m (2 ft) thick walled, 9.1m (30 ft) inside diameter reinforced concrete caisson was constructed. The overburden was then excavated using both a clam bucket and backhoe within the caisson down to the top of bedrock. Boulders may have been encountered by the cutting shoe, which may have created voids in the annulus, and may have contributed to inflow into the shaft from water bear-ing layers connected by the caisson overcut during overburden excavation. Upon reaching the top of rock at El. 514', the annulus will be grouted and a collar poured at the cutting shoe to secure the caisson. Conventional excavation using 2 m (6 ft) drill and blast rounds will then used to sink the 8.5 m (28 ft) diameter shaft through the Racine and Waukesha Formations to El. 427'. Primary support in the rock excavation will include 2.7m (9 ft) dowels on a prescribed pattern with steel mesh. A final shaft lin-ing will be poured from El. 431' to 461' at the base of the North forebay. The reinforced concrete final lining will be 7.9 m (26 ft) inside diameter. At the conclusion of the TBM mining, the conveyor and tunnel utilities will be removed below the OCPP shaft within the tunnel (El. 419') and the remaining rock plug will be shot through into the tunnel below.

Offshore Intake Shafts

The new intake tunnel system will be supplied by water from Lake Michigan through a group of four intake shafts approximately 2.4 km (1.5 mi) offshore from the construction site, between tunnel Stations 88+11 and 91+96, connecting the lake with the intake tunnel below. Edward E. Gillen Company (EEG) provided marine support to CFC for intake shaft construction. The intake shafts extend 33.2 m (109 ft) from the lakebed at El. 536' to the intake tunnel crown at El. 427'. Each shaft is 3.7m (12 ft) inside diameter, made of 2.5 cm (1 in.) thick steel coated with 305 microns (12 mils)

Figure 5.   Offshore drilling of Intake Shaft No. 2 through rock; platform for Intake Shaft No. 4 seen in background

dry film thickness of Duraplate 235, a multi-purpose epoxy. The shafts were designed to extend from El. 430' to the lakebed with bulkheads at each end and top flanges for attachment of a tee-section and two lakebed manifold arm sections, each manifold having three screen attachment points. The four shafts are spaced at 39 m (128 ft) centers. The double bulkheaded steel linings were selected to protect the future tunnel excavation beneath the existing shafts in the lakebed. Excavation on the shafts started immediately after the release for construction to commence on site. The shafts were drilled from temporary platforms installed on the lakebed by EEG working for CFC as shown in Figure 5. A 5.3 m (17.5 ft) steel conductor casing was first lowered to the lakebed from the surface to isolate the work from the lake environment. The overburden shaft was then excavated by drilling a pilot hole with a 2.4 m (8 ft) diameter auger in the lacustrine clay and glacial tills down to the top of rock and reaming the hole to the final 4.9 m (16 ft) diameter. A 4.6 m (15 ft) inside diameter soil casing was then seated into the top of rock and grouted into place. The top of rock elevations in the intake field area varied from El 495' to 485'. A Wirth rock drill was then used to cut a 4.5 m (14.7 ft) diameter shaft through the Racine Formation to El. 430', approximately 1 m (3 ft) above the future tunnel crown. The final bulkheaded steel linings were then assembled, lowered into the rock and cased soil shaft, and grouted in place. The temporary construction platform was then removed from the lakebed.

The first two shafts were excavated through overburden and rock during the 2005 marine season but did not receive final linings until the 2006 marine season. The remaining intake shafts were drilled and lined in the second season as well. In 2007, EEG installed buried manifold arms at the top of each of the shafts. A total of 8 arms were installed, each arm having attachment points for 3 screens. The 177×30.5 m (580×100 ft) intake field was then armored with rip-rap to protect the lakebed and buried manifold piping against scour caused by intake velocities.

The intake structures protruding from the lakebed were required to meet certain velocity requirements, remain relatively free of blockage, and protect against the ingestion of debris. The screens also had to be designed with break away flange connections in the event they were struck by a marine anchor or other equipment, while able to resist forces induced by the 100-year storm lakebed velocities. Johnson Screens was selected for the design and fabrication of the wedge wire screen assemblies that were constructed of special alloys that were resistant to zebra muscle attachment. A total of 24 screens were fabricated for the intakes, each with a specific orifice plate

Figure 6.   Intake screen assembly designed and fabricated by Johnson Screens

diameter to ensure uniform water flow into the screens across the intake field. The screen assemblies were lowered to the lakebed and attached by divers to the manifolds. These screen assemblies were installed during the 2007 marine season. A screen assembly is shown in Figure 6.

TUNNEL DESIGN AND CONSTRUCTION

The alignment of the tunnel connecting the intake shafts at the offshore intake field to the OCPP and ERGS riser shafts system was selected to minimize the total length and place the tunnel horizon in the most favorable geology. The resulting intake tunnel is 2,800 m (9,200 ft) long and includes both conventional drill and blast and TBM mined tunnel. The hydraulic analysis by HMM of the intake system velocity and volume demand resulted in a required tunnel diameter of 8.2 m (27 ft). The tunnel grade was designed with an elevation increase of 3 m (10 ft) from the ERGS riser shaft to the offshore intakes to provide drainage during construction. The elevation of the tunnel horizon was selected to place approximately half of the tunnel in the Racine Formation (offshore) and the remaining half in the underlying Waukesha and Brandon Bridge formations in horizons with high RQD values. A Geotechnical Baseline Report (GBR) for the rock horizon was prepared by HMM based on the available geotechnical information and was agreed with BPC prior to subcontract award. This GBR became the basis of the subcontract with regard to the expected ground conditions to be encountered. In this report the Groundwater inflow into the tunnel was baselined at 158 L/s (2,500 gpm).

Based on this GBR the tunnel was assumed to be unlined for 90% of the TBM mined length, exclusive of shaft intersections. Four types of primary support were identified in the design, and are summarized in Table 1.

In areas where Type I support is appropriate, it is considered the permanent support for the tunnel. The tunnel design called for #8 GR95 rock dowels, 3 m (10 ft) long, spaced on a 1.5m (5 ft) center pattern. Type II support supplements Type I and includes supplemental dowel installation and the addition of a C8×11.50 GR50 steel channel every 3 m (10 ft). Type III support includes W6×25 GR 50 segmental steel ribs installed on 2m (6 ft) centers with 10×10 cm (4×4 in.) W6×W6 steel matting to support the rock. Type IV reduces rib spacing to 1 m (3 ft) and uses steel channel lagging instead. Areas where Type III and IV support are erected will receive final concrete linings. K/S installed steel straps and light mesh to supplement safety for the crews and equipment in some areas of the tunnel. The tunnel design life is 100-years. There is no intent to dewater the tunnel following flooding.

Table 1.   Primary support types required by the HMM tunnel design

Support Type	% of Tunnel	Rock Type	Pattern Dowels	Channels and Dowels	Ribs and Mesh	Ribs and Lagging
Type I	93	Competent intact rock	X			
Type II	20	Fractured or thinly bedded rock	X	X		
Type III	5	Shear or brecciated rock zones			X	
Type IV	2	Inflow or ground retention required				X

Probe hole drilling was required every 41.1 m (135 ft) of mined tunnel, requiring a minimum of 4.6 m (15 ft) of probe hole advanced at all times beyond the face. Probes that encountered high inflows (>0.4.7 L/s, 75 gpm) required the drilling of additional grout holes and pre-excavation grouting to reduce tunnel inflows. The probing also identified brecciated or seamy ground in advance of the face.

Equipment Selected

Kenny/Shea JV selected an 8.3m (27 ft 4 in.) diameter Robbins-CTS Model 740-11 Diagonal Thrust TBM from its equipment inventory as the workhorse for the project. The Robbins machine had worked on 4 previous Chicago TARP projects in similar ground types. To supplement the TBM, a continuous tunnel conveyor was used as the muck transport to the surface. This conveyor system utilized a 305 m (1,000 ft) belt storage magazine on the surface adjacent to the construction shaft, a bucketed vertical belt, and an elevated overland belt to deposit spoils in an area designated by BPC near the shaft across major haul and site access roads. The muck was then distributed on-site as fill and road base material.

Starter/Tail Tunnel

Upon completion of the construction shaft, conventional drill and blast excavation of the starter/tail tunnel was undertaken. This section of tunnel was excavated in a horseshoe configuration 9.1 m (30 ft) wide and 9.1 m (30 ft) high between El. 390' to 420 utilizing heading and bench techniques. The tail tunnel was driven down station (northwest) 50 m (165 ft) to the rear of the ERGS riser shaft. The invert of the last 9.1 m (30 ft) of the tail tunnel was then blasted to form the ERGS shaft sump, which became the sump pit for the mining operation and will serve as a debris chamber when the tunnel is in service. The starter tunnel was driven up station (southeast) 70.1 m (230 ft) to allow the installation of the Robbins mining machine, essential trailing gear, tunnel conveyor, vertical conveyor, and muck transfer system at the construction shaft. The starter/tail tunnel was excavated within the Lower Waukesha and Upper Brandon Bridge Formations, leading to a stepped crown that was supported with Type I pattern dowels and supplemented with light mesh for safety. Groundwater inflow into the 120.4 m (395 LF) launch chamber, between Stations 0–15 and 3+80, was low at only 1.9 L/s (30 gpm) mostly from one open, solutioned southeast trending joint. A joint was noted at the base of the ERGS shaft with a vertical displacement of nearly 30 cm (1 ft). An invert mud mat was used as a TBM assembly working surface.

Figure 7.   Offshore TBM mined 8.3 m (27 ft 4 in.) diameter tunnel at Station 19+30 in Waukesha Dolomite

TBM Tunnel

The portion of the tunnel to be mined by TBM extends from Station 3+80 to 92+00, comprising 2,688 m (8,820 LF) of 8.33 m (27.3 ft) diameter tunnel consisting of two straight segments separated by a 49°36', 274 m (900 ft) radius curve. The OCPP riser shaft is located at the midpoint of this curve at Station 13+33, which divides the onshore and offshore sections of the tunnel. After the Robbins TBM had been assembled, mining was commenced. K/S worked two 10 hour mining shifts per day during TBM operations. Advance rates increased as mining moved forward.

Excavation of the onshore TBM tunnel 290.5 m (953 LF) was concluded in 30 calendar days, which included a week where mining was halted to reconfigure the tunnel conveyor and muck transfer system into the TBM portal and complete trailing gear installation. Probe drilling conducted within the onshore section required two pre-excavation grouting efforts near 12+00 in the Waukesha Formation. The 366 m (1,333 LF) of onshore tunnel was mined within the Lower Waukesha and Upper Brandon Bridge Formations with an average advance of 16.2 m/day (53 ft/day) and yielded at total groundwater inflow of only 5.7 L/s (90 gpm).

Excavation of the offshore tunnel which is 2,398 m (7,867 LF) long has encountered increases in groundwater inflow as different strata were encountered. A section of offshore tunnel is shown in Figure 7. The tunnel began mining offshore along the boundary of the Waukesha and Brandon Bridge Formations and transitioned into the Racine Formation starting at Station 44+00. Probing up to that point had yielded little water in these formations and inflow totaled 11.4 L/s (180 gpm). The TBM was mining in a full face Racine reef bank by Station 52+00. Groundwater inflow began to increase, peaking upon entry into mixed and full face Racine reef at Station 53+00, with inflows totaling 22.1 L/s (350 gpm). Inflows doubled to 44.2 L/s (700 gpm) after mining only 244 m (800 LF) of Racine reef, with crown probe holes typically issuing 1.9–3.8 L/s (30–60 gpm). The remaining TBM tunnel will mined through both facies of the Racine Formation to Station 92+00, just past intake shaft 4. Advance rates currently average 29 m/day (95 ft/day) offshore. Areas of straps, light mesh, supplemental dowels and channel installation tend to coincide with areas of the tunnel where the crown elevation is being traversed or within close proximity of a formation boundary or within clusters of intersecting joints. Crown probe inflow is typically higher when the Racine reef formation boundary was just above crown elevation. Gouge filled joints with vertical displacements of up to 20 cm (8 in.) were mapped along the length of the tunnel, most likely related to the geologic faulting near the site.

Intake Shaft Connections

Upon completion of offshore mining, the Robbins machine will be removed from the tunnel through the recently completed OCPP riser shaft. Drill probing will be used to locate the bottom steel bulkhead of the grouted intake liners, 1 m (3 ft) above the tunnel crown. Once located, the rock plugs beneath each of the four shafts will be excavated, exposing the lower hemispherical bulkheads of the steel lined shafts.

Final Lining

Lining will be installed in the starter/tail tunnel area and all shaft intersections. HMM has not yet made a final determination for the actual amount of tunnel to be lined. Once determined, those areas and intersections will be lined with reinforced concrete. Temporary drainage will be provided behind the linings to relieve groundwater pressure until the tunnel is flooded. The reinforced concrete lining will constructed at a 4.62 m (25 ft) inside diameter with tapered edges for the hydraulic transition to unlined sections of tunnel.

Tunnel Flooding

Once lining and operations within the tunnel are completed, the bottom hemispherical bulkheads will be cut out of each of the intake shaft linings, exposing the tunnel to the upper steel bulkhead holding back Lake Michigan at El. 579'. Divers will then open valves in the top bulkhead of the intake shaft, and water will be allowed to enter the OCPP riser shaft, flooding the tunnel. Once flooding is completed, the upper intake shaft bulkheads will be removed utilizing divers and at that point the intake tunnel system will be completed.

ACKNOWLEDGMENTS

The authors wish to thank the Kenny/Shea Joint Venture, Bechtel Power Corporation, and We Energies for their help in the preparation of this paper. Special thanks to Case Foundation Company, Edward E. Gillen Company, Ground Engineering Consultants, Hatch Mott MacDonald, Johnson Screens, Legal Video Services, and STS Consultants for their contributions to the successful completion of this tunnel subcontract.

AN INNOVATIVE APPROACH TO TUNNELING IN THE SWELLING QUEENSTON FORMATION OF SOUTHERN ONTARIO

Michael Hughes

Hatch Energy, Inc.

Paul Bonapace

ILF Beratende Ingenieure ZT GmbH

Stephen Rigbey

Hatch Energy, Inc.

Harry Charalambu

Hatch Ltd.

ABSTRACT

Ontario Power Generation is building a 10.4-km water diversion tunnel in Niagara Falls, Canada. The tunnel is a two-pass, cast-in-place prestressed concrete-lined pressure tunnel with a nominal internal diameter of 12.7 m. It is mostly situated within the Queenston formation which swells in contact with fresh water and has very aggressive groundwater. Tunneling in this material presents challenges in both design and construction. The tunnel design includes for a membrane to eliminate the potential for rock swelling and to protect the concrete liner from the ground conditions. A unique three-stage Geotechnical Baseline Report (GBR) is being used and is expected to provide an equitable method of dealing with any variations in subsurface conditions.

PROJECT DESCRIPTION

The tunnel alignment and vertical profile is shown in Figures 1 and 2. The tunnel is located at a maximum depth of 140 m below ground. Tunneling started at the open outlet canal and is proceeding towards the intake at the International Niagara Control Works (INCW) just upstream from the Niagara Falls. The tunnel will divert an additional 500 m^3/s of flow from above the falls to the existing Sir Adam Beck generating complex to enable an increase in power production. The tunnel is being excavated using an open gripper tunnel boring machine (TBM). With an excavated diameter of 14.4 m, the TBM is currently the largest hard rock machine in the world and the largest TBM project ever undertaken within Ontario. The TBM is supplied by The Robbins Company and the backup is by ROWA Tunneling Logistics with an overall length of about 130 m.

The tunnel will intersect a variety of geological formations, including the Queenston formation in which 80% of the tunnel is situated. This formation is commonly termed as shale, but is mainly comprised of massive mudstones and siltstones that are known to exhibit moderate swelling[1] in contact with fresh water. This potential for swelling poses a significant technical challenge on the project. Furthermore, the regional geology is known to possess relatively high horizontal in situ stresses that will create challenging tunneling conditions. The tunnel is being constructed in two-passes with initial rock support consisting of rock dowels, steel sets, mesh and shotcrete to

Figure 1.   Tunnel alignment

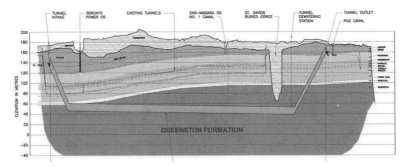

Figure 2.   Tunnel vertical profile

cover the range of expected tunneling conditions. Following the initial lining, an impermeable membrane and a prestressed cast-in-place concrete lining are installed. This
innovative approach to constructing the tunnel is expected to result in a watertight tunnel that protects the rock from swelling and thus eliminates the load buildup in the tunnel lining that would otherwise result from other tunnel construction techniques.

GEOLOGIC SETTING

The Niagara region is underlain by Cambrian, Ordovician and Silurian sedimentary
rocks having a total thickness of approximately 800 to 900 m. The project area has generally well-defined bedding with a southerly dip of about 6 m/km and an east-west strike.
The strata include dolostones, dolomitic limestones, sandstones and shales. The top of
the Queenston formation is at a depth of about 100 to 120 m in the project area.

Although the Queenston formation is mainly a muddy siltstone, a significant percentage of mudstone occurs in the upper sections, and siltstones/sandstones have
been encountered in deep boreholes below the elevation of the tunnel. The formation

is generally massive, with indistinct bedding and few joints. However, major discontinuities occur along primary bedding planes at spacings of about 5 m to somewhat greater than 20 m, and locally affect the rock mass quality. These planes often exhibit features such as gouge or breccia (a few millimetres to 2 to 3 cm) and slickensides that are consistent with lateral structural dislocation.

Laboratory strength testing of intact Queenston samples has shown a wide range of strength values, ranging from 8 to 118 MPa due to variations in lithotypes with average strengths in the range of 38 to 45 MPa. The rock exhibits an anisotropic stiffness, with a fairly consistent vertical modulus in the order of 19 to 25 GPa.

The sedimentary rock strata in the Niagara region are known to possess relatively high horizontal in situ stresses, with a maximum horizontal/vertical stress ratio varying from 3 to 5 in the Queenston formation and 7 to 8 in the overlying rock. Measurements show that maximum horizontal stress in the Queenston formation ranges from 10 to 24 MPa.

Groundwater from the primary bedding planes in the Queenston formation is generally of connate origin and supersaturated with salts, with chloride contents up to 296,000 mg/L and sulphate contents up to 1,860 mg/L having been measured.

Swelling of the shale units in the Niagara region is well documented, and is reported in terms of 'swell potential,' or the strain rate per log cycle of time in days. Based on laboratory measurements of time-dependent deformation by Dr. K.Y. Lo and others, it is found that that the horizontal swelling potential of the Queenston formation is isotropic and that the vertical swelling potential is up to 1.6 times the horizontal swelling potential. The swelling deformation response is stress-dependent, and can be represented by a linear relationship between swelling potential and applied stress in a semi-log plot and that swelling can be completely suppressed under 4 to 5 MPa stress.

The swelling process is associated with ionic diffusion of salts from the connate pore water in the rock, and a corresponding reduction in capillary and surface tension and water uptake within the rocks. Swelling is initiated by the relief of initial stresses to below the swell suppression pressure, accessibility to fresh water and an outward salt concentration gradient from the pore fluid of the rock to the ambient fluid (chloride diffusion). This process is a significant consideration for the tunnel design, as tunnel excavation, followed by the introduction of a significant fresh water source to the Queenston formation would initiate long-term swelling. This would result in significant additional and increasing load acting on the concrete liner over the life of the project. The design adopted would prevent the long-term contact of the Queenston with fresh water around the the tunnel.

SITE INVESTIGATIONS

The geotechnical investigations for the project were carried out in a number of stages. The earliest concept phase investigations were carried out by Ontario Hydro in 1983, and continued through 1993.

The 1992/1993 investigations included the excavation of an adit and a 12-m diameter trial enlargement in the Queenston. Observations indicated that temporary support measures are problematic due to rock mass behavior. Numerous instances of stress-induced and excavation geometry controlled sidewall spalling developed soon after excavation, in the range of 0.1 to 0.5 m deep. Rock bolts at 1.5 m spacing were not successful in controlling the sidewall spalling. In the crown, the rock broke back 0.5 m to an overlying bedding plane within a few hours after excavation. Excavation related stress-controlled slabbing of the rock also occurred in the invert. Degradation of the Queenston developed within days, and eventually resulted in significant spall behind the protective wire mesh.

In those areas where spalling and slabbing did not take place, the zone of active movement was limited to the first 1 to 2 m from the surface of the excavation. Very small 'creep' movements in the order of 5mm or less were measured from 7 to 70 days following excavation.

GEOTECHNICAL BASELINE

Despite extensive investigations, subsurface conditions always remain a significant risk to both design and tunneling on underground projects.. Geotechnical baseline reports that describe the expected conditions and provide baselines against which changed conditions can be measured, have been used to help deal with this situation. However, these reports are traditionally prepared by the owner and do not necessarily address risks related to means and methods of a design/build contract. Therefore, to provide a consistent interpretation of the expected subsurface conditions, a unique three-stage Geotechnical Baseline Report (GBR) was developed for the project which included input from the contractor during the invitation and negotiation stages. The process is outlined as follows:

- An initial GBR (GBR-A), prepared by OPG, was included in the proposal invitation that allowed proponents to revise the GBR as required to suit their proposed means and methods,
- A proposal GBR (GBR-B), updated by the proponent with the proponents proposal and reviewed and evaluated as part of the proposal evaluation with changes made during the negotiation process,
- The final negotiated GBR (the GBR) which forms the basis for the project.

It is believed this collaborative approach will provide an equitable way of dealing with variations in subsurface conditions.

Significant items baselined on the project are:

- Geotechnical parameters—lithology, rock mass and bedding plane properties, permeability, deformability, swelling, etc.,
- Rock classifications (relating to support conditions) along the tunnel as a percentage of total tunnel length,
- Groundwater inflow into the tunnel,
- Overbreak quantities which are expected to occur due to the high horizontal stresses and bedding,
- Grouting and pumping quantities at the open excavations.

SELECTION OF TUNNEL LINING SYSTEM

Lining Type

Due to the rock conditions, the tunnel must be concrete lined to prevent erosion of the rock and enhance flow characteristics. Both one-pass and two-pass concrete lining systems were considered for the tunnel. The one-pass system consists of installing a precast concrete segmental lining immediately behind the TBM cutterhead. To facilitate segment installation, a shielded TBM, jacked off the installed segments, would be utilized. In the double-pass system, an initial lining of ribs, mesh, rock bolts and shotcrete would be installed close behind the TBM cutterhead. The TBM is an open gripper-type machine with associated rock support installation equipment (rock drills, rib erectors, shotcrete, etc.) on the TBM. A final lining of cast-in-place concrete is then placed after initial boring is complete or well advanced.

The one-pass alternative has the advantage that the lining is installed as boring proceeds which could potentially result in a shorter tunnel boring and tunneling schedule. However, significant difficulties were identified with this alternative as follows:

- Due to joints, grout and anchor pockets and possible offset within the precast segments, the surface roughness of precast segments was assessed as inferior to that of a cast-in-place lining.

- The groundwater, which will be in direct contact with the precast lining, is highly aggressive to the lining concrete and reinforcement.

- To mitigate rock swelling, either a very high strength concrete would be required or a compressible annulus grout installed to compensate for swelling deformations. Although grout mortar has been used in the past for deformation resulting from squeezing ground, it would be difficult to also sustain the operational internal water pressure in segmental lining rings.

- A uniform grouting of the annulus around the segmental lining ring cannot be guaranteed, since significant spalling is expected and the rock will fall into the annular gap behind the TBM shield and may cause shadowing and voids when grout is injected.

- A segmental liner cannot sustain hoop stresses and will rely entirely on the rock to sustain internal water pressures. Proper sealing at the segments to prevent leakage is problematic and the option of prestressing the tunnel by borehole grouting into the rock is not practical. The opening of segment joints and the associated loss of water once the tunnel is filled would also have the potential to erode the grout and rock of geological formations which are sensitive to water. High swelling deformations and potential erosion and loss of the ground surrounding the tunnel would be the undesirable result.

The cast-in-place tunnel lining system will take longer to bore and line as it is done in two passes. However, the impacts on schedule have been mitigated by the faster procurement of the simpler TBM required and the possibility for paralleling of construction of other components of the project. Furthermore, the cast-in-place lining solution has the following additional advantages:

- With an adjustable steel formwork, sections of minimum rock support requirements may be utilized to enlarge the tunnel diameter and to provide improved flow capacity.

- The possibility for installation of a waterproofing membrane between the initial lining and the final lining prevents leakage from within the tunnel into the rock to prevent swelling and the aggressive groundwater from contacting the final lining concrete.

- Installation of a comprehensive interface grouting system allows the final lining to be compressed in such a way that the internal water pressure is sustained without the lining concrete being subject to tension and resulting cracking. Hence, the requirement of steel reinforcement in the final lining can be eliminated and potential corrosion of lining reinforcement avoided.

- The tunnel structure is tightly embedded in the surrounding rock mass, thus more efficiently mobilizing the rock to support the liner.

On these bases, a two-pass, cast-in-place tunnel lining system was selected by the design/build contractor.

Membrane

The final lining is separated from the initial lining by a waterproofing membrane system, which prevents the seepage of fresh water from within the tunnel into the rock and protects the final ling from the aggressive groundwater which may damage the concrete. As rock swelling requires both availability of fresh water and a reduction in salinity of the porewater of the rock, the membrane will effectively eliminate the potential for swelling to occur. This eliminates the need for the concrete lining to accept swelling loads or the need for compressible grouts or other materials to reduce lining loads to acceptable limits. Protection from the aggressive groundwater allows the use of normal portland cement concrete.

Two types of membranes are typically used for tunnel membranes—polyvinyl chloride (PVC) and flexible polyolefin (FPO). Although PVC has been the most commonly used membrane material for tunnels, the material was not considered suitable for this application because:

- FPO material has better mechanical properties,
- PVC can breakdown in the high chloride environment found in the groundwater,
- Laboratory tests revealed that PVC cannot fulfil the very strict chloride diffusion criteria required of the membrane material,
- PVC is less fire retardant than FPO and produces noxious gasses when ignited.

Prestressing

An unreinforced concrete lining was assessed to be the most practical lining solution and eliminates potential for damage of the impermeable membrane from the reinforcement during construction. Furthermore, unless very significantly reinforced, the lining will crack under the action of internal pressure with the reinforcement only distributing the cracking and limiting the individual crack widths. Cracking is not considered desirable as it reduces the watertightness of the lining system and can potentially damage the membrane. Unreinforced concrete linings can only prevent cracking under high groundwater pressures or by prestressing the lining.

Tunnel prestressing can be accomplished by post-tensioned cables or by high pressure consolidation grouting of the rock. The high pressure grouting method has been used successfully on a number of projects around the world with the first application of interface grouting reported on the Catskill Water Supply Project[2]. Three grouting systems which provide prestressing effects for the tunnel lining were later developed and applied in various pressure tunnels and shafts for hydropower schemes:

- Grouting through boreholes[3],
- Grouting of a circumferential void provided in the final lining (Kieser System)[4],
- Grouting of the interface between final lining and rock mass (TIWAG System)[5].

While the first system requires various phased injections into a substantial amount of boreholes to achieve the desired performance, the second system involves a rather complicated and expensive construction procedure for the final liner. The third system constitutes a further development of the previous technologies and is simpler in application.

- The interface grouting system was later adapted to be used in combination with synthetic waterproofing membranes[6], which were applied with pressure tunnels traversing poor ground, where significant water losses would otherwise have been encountered. This system was selected for the project by the design/build contractor.

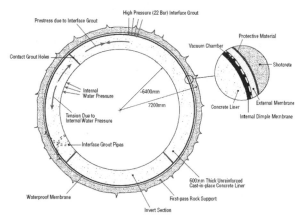

Figure 3.   Details of lining system

DESCRIPTION OF TUNNEL LINING COMPONENTS

The tunnel lining system is shown in Figure 3 with key components described as follows.

Initial Lining

Partial or full-circle ribs, rock dowels and mesh are installed 4 to 7 m behind the cutterhead in the TBM L1 area. A mechanical rib-erector is provided immediately behind the finger shield at cutterhead. In addition, two drill rigs for the installation of rock anchors are positioned in the area in front of the grippers. Auxiliary shotcreting equipment is also available in the cutterhead area for shotcreting of the crown if necessary. Two full-circle shotcrete robots for the application of wet shotcrete are available at the TBM L2 area some 20 to 40 m behind the cutterhead.

Six types of rock support types have been defined to cover the range of rock conditions expected along the length of the tunnel. The rock support types define support measures for the specified rock conditions and range from sealing shotcrete (if required to prevent rock disintegration at shales) to heavier support with full-circle steel ribs, rock dowels and reinforced shotcrete.

Final Lining

The final lining consists of a minimum 600-mm thick, cast-in-place concrete with a design compressive cylinder strength of 38 MPa at 90 days. The lining is prestressed by high pressure interface grouting to counteract the effects of internal water pressures which develop during normal operation. The liner is constructed in two stages using approximately 12.5-m long steel forms. Initially, the invert concrete is placed over a 106° angle. The top concrete is placed subsequent to the invert concrete with some delay. The steel formwork will result in a concrete finish that will meet the stringent smoothness criteria set forth for the flow of water.

Waterproofing Membrane

The waterproofing membrane is required to act as an impermeable layer between the initial and the final linings. The membrane system consists of:

- A layer of regulating shotcrete where necessary to provide the stringent surface requirements necessary for installation of the membrane,
- A membrane-backed geotextile fleece fixed to the shotcrete with concrete nails and Velcro discs,
- A 3-mm single-layer membrane system used in nonswelling rock formations,
- Two mm plus 1.5 mm dimple membrane for the dual-layer membrane system used in swelling formations to facilitate vacuum testing.

The shotcrete smoothes out sharp corners and edges, and prevents damaging the membrane during placement, concreting and tunnel operation. The regulating shotcrete has the same material properties as the initial lining structural shotcrete and is also applied to cover both steel parts protruding from the initial lining surface as well as to smooth out sharp corners in the shotcrete finish. The geotextile fleece protects the waterproofing membrane from damage from contact with the shotcrete. The fleece is backed by a thin plastic membrane which facilitates the flow of interface grout. The waterproofing membrane consists of PFO, which provide resistance to chloride ion diffusion. The surface of the dual-layer membrane system is tested 'full face' with vacuum testing equipment. The inner layer is manufactured with dimples to prevent collapse of space between the layers during vacuum testing.

As with typical membrane installations, each double heat welded joint is tested to ensure adequacy. Using such techniques, as well as simple visual inspection, watertight installations have been produced in Europe. However, due to the need for a fully watertight tunnel to prevent swelling of the rock, additional measures to ensure competency of the membrane are required. For this purpose, a double membrane system is used for installation within the swelling formations. Four-meter long membrane sections are prefabricated outside of the tunnel. The membrane layers are welding longitudinally to form two compartments, one at the invert and another at the crown. Each compartment is 2 m long. Each of these compartments is vacuum tested to ensure that there are no holes in the membrane and repaired or replaced as necessary. With testing of 100% of the surface area of the membrane and repairing or replacing damaged sections, a watertight and damage-free membrane can be provided in its final installation.

Tunnel Prestressing

To fill any voids and imperfections in the final interface lining, the interface between the inside of the membrane and the final lining is first contact grouted over the full circumference using low pressure cement grout. Following this, high pressure interface grouting is carried out through a system of grout hose rings installed at regular intervals between the initial lining and the waterproofing membrane system. Grout hose lines with rubber sleeve valves also at intervals are attached to the surface of the initial lining. Grout blocking rings are provided around the entire circumference of the tunnel to control the flow of grout along the tunnel. These details ensure an even grout flow distribution along the membrane-backed extrados of the geotextile fleece and facilitates the grout filling of joints and cracks in the initial lining and the rock mass with grout.

The interface grouting procedure follows a defined pattern of pressure application with the aid of two or more pumps, creating a consistent flow of grout along the tunnel. Grouting is, in a first phase, carried out at every other interface grouting ring from both ends of the grout hose. The success of grouting is carefully monitored by precise deformation measurements of the final lining around the full circumference of the tunnel. Fixed monitoring sections are used before and after interface grouting at predefined locations. Mobile monitoring sections supported by gantry mounted laser digital meters measure deformations during the interface grouting process. Pumping pressures defined by structural analysis are thus kept within the allowable limits and pumps automatically shut off as soon as threshold values are reached.

After the grout hardens, a locked-in compressive prestress will remain in the lining and between the lining and the rock. This prestress will be reduced to some degree by drying shrinkage of the concrete lining and grout, creep of the concrete and by contraction of the liner due to the drop in temperature resulting from the tunnel being filled with water. These losses are carefully considered in calculations to determine the required initial grouting pressures, which are to be applied to ensure that the desired prestress of the final lining is maintained throughout the envisaged service life. A second phase of grouting through the remaining grout pipe is required if the anticipated initial (short term) grouting pressure is not achieved or if the pressure loss falls below the calculated minimum prestressing pressure determined to sustain the internal water pressure.

TUNNEL LINING DESIGN

Rock Parameters

Rock parameters are determined from site investigations described earlier. Rock mass strengths are determined using the Hoek and Brown formulation[7] using measured laboratory unconfined compressive strength (UCS) and Hoek mi and GSI values determined through testing during the investigations. Effective stresses are used in determining rock mass strength. The computer program PLAXIS is used for finite element analysis which uses a Mohr-Coulomb failure definition. Therefore, equivalent Φ's and c's are calculated from the Hoek-Brown rock mass strengths for input into PLAXIS.

Initial Lining

The rock support is designed using the convergence-confinement method. In this method, the convergence that takes place at and away from the face along the tunnel is determined. The rock support is then designed to stabilize the tunnel at these initial rock convergences. A two-dimensional finite element model along the axis of the tunnel using the computer program PLAXIS is used to determine the convergence of the tunnel at the face, TBM L1 area, and at the TBM L2 area. These convergences are then used as input into two-dimensional finite element analysis of a tunnel cross-section and accounting for the stiffness of the rock support, final convergences and forces in the rock support determined. Additional rock failure mechanisms are considered such as rock blocks and slabbing which are then added to the rock support loading.

Membrane

No membrane is 100% resistant to the movement of molecules and ions. A target diffusion resistance of 10^{-10} cm^2/s for the membrane was determined through analysis to provide a sufficient barrier to chloride ions over the design life of the tunnel. A series of laboratory tests were performed on the membrane material to determine its diffusion resistance. Two types of membranes were tested—a membrane open to diffusion and the prototype membrane which is closed to diffusion. The CEN-EN1931 standard was used to test both membrane types with the open membrane used as a control to ensure that the testing method adequately detected movement of chloride ions through the membrane. The test consists of sealing a membrane over a dish of desiccant that absorbs water vapor while keeping the exposed face of the membrane covered with a NaCl bath. At the end of a test, the desiccant and membrane are both

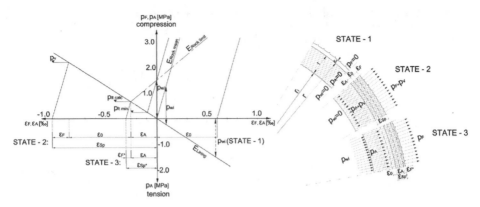

Figure 4.   Effect of internal water pressure on concrete linings

washed and the solution titrated to determine the amount of chloride ions that passed through the membrane. The tests were run for a sufficient length of time to ensure that a detectable quantity of ions would theoretically pass through the membrane at the target diffusion resistance. The actual membrane diffusion resistance was found to be well in excess of that required to form a suitable barrier to chloride diffusion.

Prestressing

The combination of concrete lining, prestressing and rock mass can be visualized in diagrams (Figure 4), proposed by Seeber[8]. The radial pressure (pF, pA) in the rock mass and the concrete lining is indicated in the upper half and the radial tension (pA) in the concrete lining is indicated in the lower half of the diagram. Strains (εF, εA) toward the tunnel center are shown on the left side and strains toward the rock are shown on the right side. The stiffness of the lining ring and of the rock mass is illustrated by the lines drawn through the centre point. The internal water pressure (pWI) at a particular strain is the vertical connection between the two lines representing the stiffness.

Even if contact grouting successfully fills any voids caused by concrete placement, the concrete lining is subject to tension due to internal water pressure contraction by thermal strain, concrete shrinkage and creep (illustrated by 0 in State-1, Figure 4), which induces the concrete lining to crack. The contribution of the rock mass to support the internal water pressure (pWI) when filling the tunnel at State-1 (Figure 4) will not prevent tension development and resultant cracking.

Filling the gap between the rock mass and the final lining with pressurized grout will result in a desirable prestressing effect. This effect can be visualized by moving the line representing the stiffness of the rock mass to the left until the design water pressure can be placed between the two lines without any tension being caused in the concrete lining. This represents the minimum long-term prestress required (State-3, Figure 4). Alternatively, a maximum internal water pressure can be determined from the applied short-term grouting pressure (pV) (State-2, Figure 4) after all losses (ε0) related to thermal contraction, concrete shrinkage and creep have been subtracted.

The interface grouting pressure and resulting tunnel lining prestress is time dependent. The initial short-term grouting pressure is determined to be close to the maximum grouting pressure (pst) that can be structurally applied to the lining. The long-term grouting pressure is determined by subtracting all losses due to cooling, shrinkage and

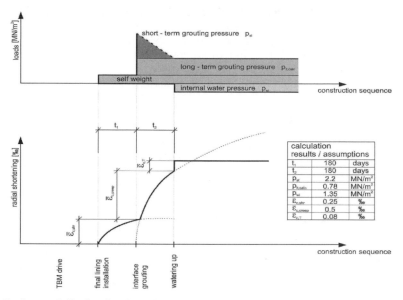

Figure 5.   Losses ($\varepsilon 0$) related to thermal contraction, concrete shrinkage and creep

creep that develop during time t2 as illustrated in Figure 5. Material testing during concrete mix design is used to establish final creep and modulus data. After filling of the tunnel, the concrete will cool and the corresponding temperature strain $\varepsilon C,T$ will tend to induce additional relaxation stresses on filling the tunnel. Concrete creep and shrinkage strains will effectively cease and the concrete lining brought close to a nonstrained condition as internal water pressure and prestress neutralize each other.

CONCLUSIONS

A number of options were available to address the potential of swelling rock on the project. The tunnel lining design selected will eliminate the potential for swelling by isolating the rock off from access to fresh water. This innovative solution to the swelling problem was the result of the designer and contractor developing the solution together within the context of a design/build contract.

ACKNOWLEDGMENT

The authors wish to thank Ontario Power Generation for its kind permission to publish this paper.

REFERENCES

[1]   Rigbey, S. and Hughes, M. Accounting for Time Dependent Deformation in the Niagara Diversion Tunnel Design, 1st Canada-US Rock Mechanics Symposium, May 27 to 31, 2007, Vancouver, BC (in publication).
[2]   Sanborn J.F. and Zipser M.E. : Grouting Operations, Catskill Water Supply, ASCE-Proc. 1920/1.

[3] Hautum F., Der Druckstollen des Pumpspeicherwerkes Reisach-Rabenleite, VDI Zeitschrift, Heft 5, 1957.

[4] Kieser A.: Druckstollenbau, Springer Verlag, 1960.

[5] Lauffer H., Vorspanninjektion für Druckstollen, Der Bauingenieur, 43 Jahrgang, Heft 7, 1968.

[6] Bonapace B., Tests and Measurements for the Pressure Tunnels and Shafts of the Sellrain-Silz Hydroelectric Power Scheme with Extremely High Head, Proceedings 5th ISRM Congress, Melbourne, 1983.

[7] Hoek E, and Brown E.T., Practical Estimates of Rock Mass Strength, Int. J. Rock Mech. Min., Vol 34, No. 8, 1997.

[8] Seeber G., Druckstollen und Druckschächte, ENKE im Georg Thieme Verlag, 1999.

URBAN BLASTING ESSENTIAL TO SCHEDULE-DRIVEN WEST AREA CSO TUNNEL AND PUMPING STATION PROJECT

John MacGregor

Dyno Nobel

James McNally

Obayashi Corp.

Adam Stremcha

Obayashi Corp.

ABSTRACT

With 13.7km of TBM tunneling in Atlanta's hard gneiss well under way, drill and shoot work continues on several sites throughout the city. In order for the City of Atlanta to meet the consent decree imposed by the federal government, an aggressive construction schedule was a necessity. Construction consists of two deep, large diameter tunnels, four construction shafts, a pumping station, one overflow shaft, three large intake structures, and two connecting tunnels. Due to the multitude of blasting in sensitive urban environments, a high standard of seismic and air-overpressure criteria, coupled with time restrictions, was adopted for the project. At any given time, drill and blast operations were under way at four of the seven sites requiring a certain amount of coordination between the TBM excavation, surface work, and the forces of Mother Nature. This paper addresses the challenges faced and the techniques used to safely and successfully complete the rock excavation of the West Area CSO Tunnel and Pumping Station project.

INTRODUCTION

The City of Atlanta chose tunneling as their main technological solution to their current CSO problem. Tunneling was chosen for several reasons, one of which is their successful history with tunneling. The surface area required for an underground storage facility is minimal, which is essential for a major metropolitan area like Atlanta. There will also be little or no impact on the surrounding community. A significant portion of the project is within commercial or industrial neighborhoods where construction is not unusual. Another significant factor behind this solution is that storm water, for the most part, will be captured and treated before being discharged. The tunnel will convey flow from three existing CSO facilities; Clear Creek, Tanyard, and North Avenue, to a new treatment plant at the R.M. Clayton site, eventually being discharged into the Chattahoochee River (Figure 1).

The West Area CSO tunnel will consist of two main sections, the Clear Creek tunnel and the North Avenue tunnel. The Clear Creek tunnel is 8.22m diam. and is approximately 6.4km. It will be excavated using a tunnel boring machine (TBM) and is to have a finished diameter of approximately 7.32m. The lining will consist of both reinforced and non-reinforced concrete. This section of the project also consists of a drill and blast connecting tunnel that is 4.8m square and 160m long, two drop shafts of

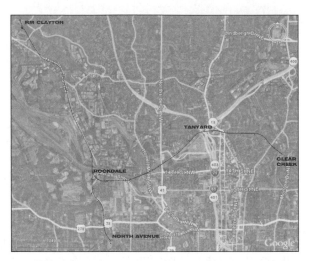

Figure 1. Project overview

3.96m diam., three de-aeration chambers of 9m square and 77m in length, and three construction shafts, one 9m diam. and the others 14m diam.

The North Avenue tunnel will run north to south, connecting the North Avenue CSO facility to the R.M. Clayton facility. The Clear Creek tunnel will intercept the North Avenue tunnel approximately 2.3km south of the R.M. Clayton facility and will be done so using drill and blast techniques. This run of the tunnel is 7.1km in length and will have the same dimensions as mentioned above. This section of the project also consists of a drill and blast tunnel that is 9m square and 250m long, a drop shaft of 3.96m diam., two de-aeration chambers of 9m square and 77m in length, two construction shafts 14m diam., one overflow shaft 9m diam., and a pump station shaft 26.8m diam.

With overburden ranging from 3m to 21m, an estimated 131,000m³ of rock will be excavated using drilling and blasting techniques throughout the project. Excavating such a large amount of material using an aggressive technique in a sensitive, urban area while attempting to conform to the contract specified air overpressure of 130 decibels and peak particle velocity (PPV) of 12.7mm/s, required substantial planning and effective blast design. This paper will discuss the challenges and diversities faced throughout the West Area CSO Tunnel project, with particular attention to sensitive urban blasting and unique underground techniques.

OPERATIONS—SETUP AND STARTUP

The West End CSO Project with its multiple sites spread throughout Atlanta's West downtown area provided a series of blasting challenges. Each site had it own technical challenges with regards to blasting, but the logistical challenge of supply of explosives came first.

While the hours of operation for blasting varied by site, the normal hours of operation were 7AM to 7PM. It was permitted to load holes at any time and detonate them within that window. The mining cycle proceeded 24 hours per day with multiple sites in the drill and shoot development stage at the same time. Before all six shafts were down with associated tunnel work, the blasting of three diversion structures, two with drill and shoot drop shafts, would begin.

The need for an economic explosive supply channel that could react quickly to changing demands, and supply multiple sites around the clock over the length of the job existed. Dyno Nobel's magazine storage was located 80.5km outside Atlanta, not making it feasible for deliveries in such a varying timeframe. With Atlanta's traffic, Atlanta CSO Constructors (ACC) needs, and the associated safety issues in transporting explosives around the clock, ACC, along with Dyno Nobel's assistance, came up with the following solution.

The only site capable of housing a storage magazine with an explosive capacity of 1,800kg was the Rockdale site. This site was the largest piece of property throughout the project and was located in an industrial neighborhood. By barricading the magazines, ACC was able to meet the Bureau of Alcohol, Tobacco and Firearms table of distance requirements to inhabited structures and roadways. A magazine site with security was established to provide a supply of both explosives and detonators. Typical explosives and detonators used throughout the projected included: Dyno Xtra, Dynomax Pro, D-Gel 1000, Dynosplit D, and various NONEL EZ DET, NONEL EZTL, NONEL MS, NONEL LP, and NONEL TD detonators.

Two portable magazines capable of transporting detonators and explosives were purchased to move powder from ACC's on-site Rockdale magazine directly to other locations. A forklift was used to load the portable magazines onto a flat bed truck to transport to other sites. Cranes on-site hoisted magazines off the truck and down the shaft to work area.

To meet the Department of Transportation regulations, ACC had to operate a truck with a current DOT inspection, employ drivers with commercial driver's license and Hazardous Materials Endorsements, and operate as a shipper of hazardous materials and all that encompasses.

The move to be an in house shipper of explosives, along with good coordination with Dyno Nobel, gave the flexibility needed to meet ACC's explosive needs throughout the project.

CONSTRUCTION SHAFTS—CLEAR CREEK

The Clear Creek (CC) site is located at the end of the Clear Creek tunnel run in Midtown Atlanta. On the hill to the north overlooking the site resides a number of commercial buildings including an architectural firm and two recording studios. Piedmont Park is situated west of the site, immediately adjacent to the creek. The existing CSO facility is just south of the site. Geology in the area consists of approximately 6m of fill and alluvium overlying residual soil followed by a weathered transition zone of approximately 2m. The site is also characterized by large undulating foliation fractures. Groundwater inflows are anticipated to be a total 3,785 lpm for all excavations. The site includes a construction shaft for extracting the TBM, a diversion structure, a drop shaft, and a vent shaft. The diversion structure is approximately 119m long and 8.2m wide, with the structure depth varying from 12.2m to 19.8m, upstream to downstream (Figure 2).

Seismic limits were attainable at the Clayton and Rockdale shafts with greater distances to structures; using a combination of NONEL long period (LP) detonators and NONEL 42 millisecond (MS) surface delays. LP detonators, delays #1 through #18, radiated out from a burn cut in the center of the shaft. The 42ms surface delays were used to divide the shaft area into four quadrants. The detonating cord surface times were 0ms for quadrant one, 42ms for quadrant two, 84ms for quadrant three, and 126ms for quadrant four. By using this method, ACC was able to increase the available delay times from 18 to 72. It is important to note, that the surface time should not exceed the first down hole detonator time. In this case the total surface time is 126ms,

Figure 2. Clear Creek site layout

and the #1 LP detonator tied into quadrant one fires at 500ms. All surface detonators are fired before any blast hole detonates.

Air-overpressure and PPV predictions were calculated for the Clear Creek shaft using a formula created from a regression analysis performed on data collected from other sites. ACC's method of initiation with LP detonators was similar, but required six segments to obtain a maximum of 16.4kg per delay. Both noise and vibration limits were exceeded during the first blast. An attempt was made to adjust the drill pattern in subsequent shots with no success. New on-site analysis suggested limiting the maximum pounds per delay to 12.7kg. This criterion limited the blast design to two holes per delay. This was very challenging, as the shot design contained a minimum of 180 holes.

With increasing pressure from the community and project owners, ACC redesigned the shot geometry and the method of initiation. The conventional ring design was changed to a 1.2m by 1.2m square pattern, utilizing a 6.35cm hole diam. with a typical hole depth of 3.2m. This grid was centered on a burn cut with relief holes and merged as best as possible with the 0.75m center-to-center circular perimeter holes (Figure 3).

A short period MS initiation system using NONEL down hole and surface timing was used to obtain a two hole per delay scenario. While this design obtained the desired maximum kg per delay, its timing did not maintain consistent true burdens that the square drill pattern geometry creates. Again air-overpressure and PPV criterion were not met (Figure 4).

A second short period MS design, taking advantage of the true burden created by the square pattern, proved to be the answer. The design shot four holes per delay with a maximum of 25.4kg per delay. Timing between rows was 84ms, and timing between holes was 25ms. The PPV result of this shot was 8.6mm/s, and allowed for acceptable results for the remainder of the shaft.

Once the vibration issue was solved, efforts were focused on reducing the air-overpressure. Detonating over 900kg of powder within 36.6m of a structure site is going to cause a substantial amount of air-overpressure. The short period MS timing, which helped the vibration limit be met, did little to reduce the noise level. However, the business owner did appreciate a shorter shot duration. Millisecond timing reduced the shot duration from 8 seconds to 763 milliseconds.

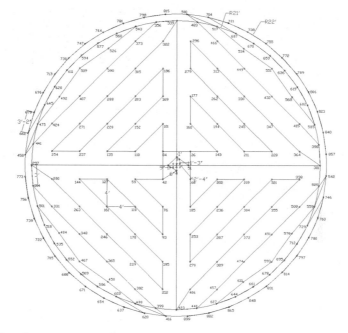

Figure 3. Clear Creek shaft timing (MS) design 1

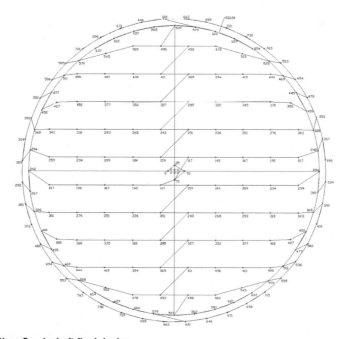

Figure 4. Clear Creek shaft final design

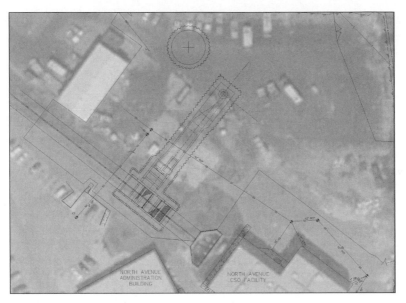

Figure 5.   North Avenue site layout

Modifications of the shaft cover, including a wood lining, a rubber conveyor belt overlay with skirting on the sides, and a layer of hay bails on top, did not dampen the noise enough to meet specifications.

Two methods of noise abatement along with the shaft cover were utilized to reach the air-overpressure goals. The first shot was covered with 1m of manufactured sand and blasting mats. This amount was reduced by trial and error down to approximately 30.5cm. Another method that proved to be successful was using water as an alternative cover to that of sand and mats. Using trial and error, approximately 1m of water proved to obtain the desired results. While both methods added cost and time to the project, they did get the job done in a "comply or shut down" situation.

DIVERSION STRUCTURES—NORTH AVENUE

Located at 1150 North Avenue North West, the North Avenue site is within an existing City of Atlanta facility, with close proximity to surrounding municipal structures. Work within the site included a construction shaft, a dearation chamber, a diversion structure, a drop shaft, and a vent shaft. With an average overburden depth of 4.5 to 9 meters, approximately 17,200m^3 of rock was expected to be removed using drilling and blasting techniques. Roughly 3,670m^3 of this material was to be excavated from the diversion structure itself. With two facilities within 9–18 meters of one end of the diversion structure, it seemed as though excavating this material within the contract specifications would not be feasible. There was also a live 106.7cm RCP sanitary sewer that divided the structure excavation in half. With this sewer running full 100% of the time, it was of the essence that it was not disturbed or broken due to blasting on the site (Figures 5 and 6).

The geology expected to be encountered within the diversion structure excavation varied from gneiss to mylonite, with extensive weathering within the first 1.5m of the transition zone. With this zone sloping from north to south and west to east along the

Figure 6. North Avenue RCP sewer

structure, it was expected that the encountered material would not be of a massive nature and could contain weathered seams throughout the depth of excavation. This posed as a potential issue, not only for drilling, but also for its unpredictable nature. When in situ material is not consistent, designing and executing an effective blast is much more difficult. This type of material typically results in poor fragmentation and unpredictable blast control. Because of these factors, effective shot design, drilling, loading, and cover were essential to effectively and safely drilling and blasting the North Avenue diversion structure.

Due to the site conditions at hand, the final shot designs consisted of a MS design, utilizing DynoMAX Pro. The benefits of this particular product, an extra gelatin nitroglycerin dynamite, is that it has an increased relative weight strength compared to other products used throughout the project. This allowed for smaller diameter holes to be used at an increased spacing, ultimately resulting in less product being used while maintaining desirable fragmentation and control. It is also highly resistant to water and is unlikely to result in hole-to-hole propagation. Due to the contract constraints, the proximity to surrounding structures, and the location of the 106.7cm sanitary sewer, all of these attributes were extremely desirable. The MS design allowed for effective timing and blast control, both essential to success on this particular site. It allowed for two holes per delay on bench rounds and four holes per delay on cut rounds to be shot, while maintaining the contract specifications. These benefits of MS delays allowed for bench shots of varying lengths, compared to that of LP delays. The controllable timing allowed for effective cover to be implemented. Since MS timing is predictable and quick, cover is not thrown from the shot like it would be with LP timing. Using this, controlling of fly rock was not an issue while excavating the diversion structure.

The sequence of shots started in the drop shaft portion of the structure with a cut round. This round consisted of a pattern with an offset burn of six 50.8mm holes. The design allowed for four holes per delay to be used, resulting in 9.5kg per delay. Approximately 0.5–0.75m of sand was then placed over the shot, with it peaking over the burn. Mats were then placed from the exterior edge of the shot and overlapped towards the burn. Placement in this manner allowed for the mats to cover as much material as possible while the shot initiated. After the material had been mucked out creating a free face, a 1.2m by 1.2m square pattern was drilled to fit the structures dimensions, starting

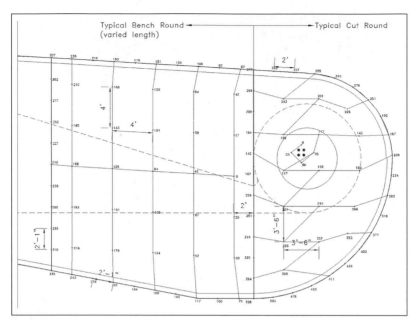

Figure 7.   North Avenue diversion structure—cut and bench rounds

0.6m off of the face. This pattern repeated until holes came within approximately 0.6m of the perimeter, which were also drilled at 0.6m centers (Figure 7).

The MS design of the bench rounds allowed for varied shot lengths. Shooting in this manner allowed for the back section of the shot to be cleared, allowing for mats to be removed, followed by drilling of the next pattern. Thus, the drill, shoot, muck cycle was maximized, getting efficiency out of a one shift operation. During the first round of shots, the pattern was repeated until the back row of holes was parallel to the eastern edge of the sewer. Another cut round was then drilled, centered between the sewer and the existing box culvert, dividing the area into three equal sections. The shot was covered similarly to the first and was then mucked out partially out to allow for the drilling of the two remaining sides. Perimeter holes parallel to the sewer were toed out. After completion of drilling, a free face for each side was then mucked. Both sides were then loaded and timed such that the free faces would shoot towards each other. Additional mats were placed over the free face under the sewer, preventing potential fly rock from being thrown from the toed out perimeter holes.

After the initial round of blasting was completely mucked out and the bottom exposed, a cut round was then drilled in the drop shaft end. Bench rounds continued to the back wall, parallel to the existing box culvert. This pattern was repeated until the appropriate structure elevations were achieved. The largest shot allowed for approximately 600 bank cubic meters to be removed while being well under the contract specifications (Figure 8).

DROP SHAFTS—TANYARD/CLEAR CREEK

The Tanyard (TY) site presented ACC's most challenging blasting thus far. Job specifications for air-overpressure and PPV limitations were identical throughout the job, but the distance to an inhabited structure was significantly reduced at a distance of

Figure 8.   North Avenue diversion structure final excavation stage

23m from center of shaft. The TY site lies about 122m off the Clear Creek tunnel, align-ment approximately half-way through the tunnel drive. Only 8m east of the site runs I-75/I-85, a major artery passing through the heart of Atlanta. Residential houses boarder the west and north side of the site some 8m away. Of these is a house well known for its attached greenhouse and gardens. An existing CSO facility forms the boarders the southern edge. Typical geology in the area consists of 1.5m to 3m of overlaying fill and residual soil followed by an extensively weathered transition zone of approximately 12m. The site is also characterized by a series of closely spaced major lineaments trending north to south. Groundwater inflows are expected to be minimal. Work includes a diversion structure, a drop shaft, and a vent shaft. The diversion struc-ture is approximately 54.9m long and 7m wide. Its deepest end is approximately 23.8m, with the rest of the structure around 14m. The drop shafts at both CC and TY are located in the deep section of each structure and will be the focus of this section of the paper.

Originally the diversion structures were going to be backfilled upon completion and the drop shafts excavated by pulling a vertical raise. In an attempt to excavate the drop shafts prior to the structures completion, conventional drill and blast methods were used. Based on a relatively new method pioneered in Canada, it was feasible that vertical retreat mining (VRM) could be as efficient and cost effective as a conven-tional raisebore. Having successfully excavated the construction shafts, chambers, and diversion structures, the localized geological conditions were well-known at each site and favorable for long-hole drilling. Total depth of rock from the chamber crown to the bottom of the diversion cut was approximately 17m at both sites. The cuts were supported with several tiers of struts and walers. For production purposes a drilling subcontractor was hired to perform the drilling using a Furukawa HCR1500 quarry rig. The drill could only be set in place with its mast standing straight up due to the con-straints imposed by the support of excavation. The approach channels sloped down to the drop shaft on a 27.5 degree incline and a slight over-excavation of these slopes was necessary to level off and provide a large enough footprint for the drill (Figure 9).

A concern of trying to vertically retreat mine the drop shafts was the accuracy of the drilled pattern. The initial thinking was to employ a down-the-hole hammer rig in order to maintain verticality. After a short cost and accessibility analysis, it was decided that a top hammer rig would suffice provided it stayed within the agreed tolerance and

Figure 9. Drill used for drop shafts

fit between the walers and struts. Another concern was the hole diameter. Typical blast hole diameters for the rest of the project had been between 5cm to 6.4cm. The drill selected had a minimum drill diameter of 8.6cm. This was acceptable for two reasons. First, the larger hole diameter meant fewer holes were needed in the pattern to achieve a satisfactory powder factor. The larger hole diameter also meant there was a better chance for maintaining drill alignment.

The pattern consisted of three 12.7cm burn holes drilled in the center and three concentric rings surrounding them. After the initial 12.7cm hole was drilled in the CC drop shaft, a rough survey of the hole was done and the hole had deviated 11.4cm in 17m, well with in the agreed tolerance. Visual checks were done periodically. After blasting of the shaft, it was determined that a majority of the 8.6cm holes had drifted approximately 30.5cm to the south. This discovery led to a more detailed survey of the TY drop shaft. Not only were the holes surveyed on top and bottom, but detailed drill logs were requested from the driller.

Drilling on the CC drop shaft started on a Thursday, mid-morning, and finished late in the evening on Friday. An extra hole was drilled in a low spot just outside the desired pattern to allow the water to drain off and maintain dry holes. Within a 20 hour period, close to 700m was drilled. Drilling of the TY drop shaft took slightly longer even though there were less holes. This was due to greater confinement, both going through the struts and walers and at the bottom of the diversion cut (Figure 10).

Once drilling had finished, each hole was measured prior to stemming to determine the appropriate elevations for terminating the powder column. Each lift would be taken out in 3m increments excluding the final 4.9m round. Borehole plugs normally used to plug 6.4cm holes were shoved within 30cm of the end of the hole, creating a false bottom. On top of this plug 0.6m to 0.9m of stemming was placed. The bottom of each hole was stemmed such that the last stick of powder in every hole was at approximately the same elevation. Each hole was loaded with packaged emulsion, slit and dropped in the hole. The last stick received an initiator pointing down the borehole. On top of the powder column, 1.8m of stemming was then added to lock the explosive in place.

Vibration concerns were minimal based on pounds per delay and distance calculations. After the first three shots at CC, it became apparent that air-overpressure readings were unacceptable. Review of video footage showed much of the stemming being

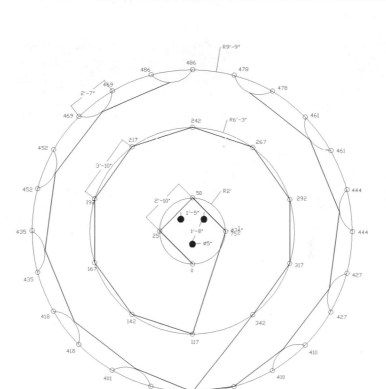

Figure 10. Clear Creek drop shaft layout

blown back out the holes and a significant amount of dust and smoke coming from the 12.7cm dead holes. It appeared that much of the energy was escaping out the burn. On the fourth round, sand bags were used to cover the inner two circles of holes. Then, 4m of sand was used to cover the shock tubes in the center. On top of this and directly over the burn, one steel mat and one rubber mat were placed for a total of approximately 4,000kg. Video footage confirmed the escaping energy was coming from the burn as it put a hole through the steel mat and picked the rubber mat completely off the ground. Air-overpressure readings came back within a permissible range. A similar technique was thus used at TY to control the air blast by using bulk bags filed with sand (Figure 11).

Slight adjustments were made to the final 4.9m rounds. Load-outs had proven successful thus far, but it was not known whether 1.8m of powder would reach up and break another 1.8m of rock. A "satellite" charge of dynamite was placed in the middle of the final 1.8m of stemming with its initiator point up. The dynamite was timed to go off with the rest of the powder column and extra cover placed on top. This small adjustment proved to be very successful. The entire drop shaft was then mucked from the connecting chamber through the construction shaft.

Initial doubts were based on the fact that drill accuracy would not be sufficient to successfully blast the drop shafts. A more diligent survey and analysis of the drilled pattern performed at TY allowed fine tuning of the blast design to compensate for inaccuracies. Three 12.7cm holes had uniformly deviated 1.8m to the left of center, and several 8.9cm holes had wondered almost 1m. Before the drill was pulled out, two

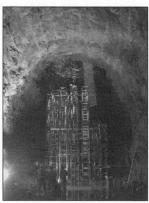
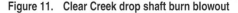

Figure 11.   Clear Creek drop shaft burn blowout

Figure 12.   Clear Creek drop shaft concrete shoring towers

extra 8.9cm holes were drilled around the burn for added relief. It was also shown in the bore logs that only the last 4.5m of several holes had deviated. It was anticipated from the beginning that the presence of tights was very likely; adjustments were made during loading of certain holes, based on their deviation and location within the pattern. After shooting the last round, only one tight area was found (Figure 12).

A brief comparison between the 37m deep NA Drop Shaft, which was excavated by a raise bore subcontractor, and the 17m deep CC Drop Shaft shows a clear advantage to using VRM. The total duration for the subcontract to mob in at NA, drill the pilot hole, pull the raise, and de-mob was approximately 420 hours. This averages out to about 8.8 cm per hour. During the pilot hole drilling and raise boring 2 crews with 4 men each worked around the clock. By comparison the CC Drop Shaft was drilled out in approximately 20 hours with a three-man crew. Subsequent blasting was done in the following four, eight hour work days with an eight man crew for a total of 52 hours. This averages out to close to 33cm per hour. Based on the results at CC and the improvements made at TY, the only real disadvantage of the VRM method seen was the more extensive support of excavation needed due to the over break versus a smooth raised shaft. Although this topic merits a more detailed analysis, the time saving advantage of the VRM method is quite evident.

SPECIAL CASE—TBM BEARING CAVITY

On Tuesday, July 11, 2006, motor number seven tripped five times and metal fragments were discovered on the Herrenknecht 8.2m diam. hard rock TBM nicknamed Rocksand. The following day the decision was made to start preparing to pull the main bearing and replace it. The main bearing and bull gear assembly was 0.9m thick by 4.3m diam. and weighed just over 20,000kg. Having completed 3,247m of the drive, there was no easy access to the front of the machine. One of the first steps would be to excavate a cavity in order to place the new bearing. Once the machine reversed past the cavity, the old bearing could be removed, the new one put on, and the old one placed back in the cavity.

The cavity would be placed just behind the machine to minimize the distance needed to backup the TBM. As luck would have it, the geology in this particular area

Figure 13. Bearing cavity drill-out

was "very good" according to both RMR and Q classifications. Two joint sets were mapped; the predominant one striking sub-parallel to the main tunnel and dipping into the left rib on about a 55 degree angle. Several initial concerns arouse due to the nature of this operation. First was the potential of damaging not only the trailing gear but also the tunnel conveyor. Another concern was a large wedge forming along the previously mentioned joint set and damaging the new bearing. Other concerns included the protection of the 13.2 kV power cable, fan line, rail, and the tunnel utilities.

After some discussion, it was decided the most practical location for the chamber would be the left rib looking up station just above invert level. A 9.1m wide by 2.1m high by 4.9m deep cavern would provide enough room for the main bearing assembly, a manlift, and several other pieces of equipment needed during the replacement. The conveyor structure opposite the planned cavity was removed to prevent damage. Steel pipe was cut in half and bolted over the 13.2 kV cable. All of the shots would be directed down station and away from the trailing deck. The rail and fan line would be replaced if necessary (Figure 13).

Because of the confined working area and the fact that crews would be working in the cavity for several weeks, the top row of holes was pre-split. This allowed for a clean bolting surface and minimized back break. It was also necessary to square off the face in order to facilitate the drilling of subsequent rounds. A somewhat complex pattern of varying hole depths and angles was designed to peel off 9.1m of the lower arch of the tunnel. The rest of the excavation consisted of two heading rounds 2.1m high by 3m wide by 2.4m deep and a final slash round of 6.1m long by 4.9m deep. Ground support was installed prior to excavation and as crews advanced towards the back. A series of 4.6m bolts were installed perpendicular to aforementioned joint set for stability (Figures 14 and 15).

Equipment necessary for the excavation of the bearing cavity consisted of a Plymouth locomotive and a flat car, a knuckle-boomed Tamrock Commando 300, and a Kobelco SR70 excavator.

Drilling began on the 18th of July and the final round was shot on July 25th. Support of the excavation was finished the following day. The successful completion of the bearing cavity allowed crews to safely and efficiently replace the main bearing and bull gear. The only major damage was done to a small section of bagline which was cut out and replaced as the TBM backed up.

Figure 14. Bearing cavity cleaned and   Figure 15. Excavated bearing cavity
supported

CONCLUSION

As blasting continues at various sites, ACC is making significant headway on completing the West Area CSO Tunnel project. Throughout the project, various basting techniques were applied to allow for ease of excavation, while attempting to conform to the restrictive contract specifications. Air-overpressure and PPV specifications were set well below the damaging criteria, making conventional blasting techniques not feasible in many situations. With each blast throughout the project having unique parameters, substantial planning between ACC and Dyno Nobel was required to maximize efficiency while maintaining safety. The blasting discussed throughout this paper accounts for approximately 8% of the rock removed using drilling and blasting techniques, and is only an example of how ingenuity of various techniques was applied.

ACKNOWLEDGMENTS

The authors would like to acknowledge the crews of Atlanta CSO Constructors, whose diligence and constant adaptation to the ever-changing site conditions continues to help in the success of this project; Reliable Drilling, Inc., for allowing ACC to put their drill in locations it was never meant to be; and the construction management team of JDH Joint Venture for their patience.

14

Sequential Excavation Methods

Chair
J. Laubbichler
Dr. Sauer Corp.

CONSTRUCTION OF A MIXED FACE REACH THROUGH GRANITIC ROCKS AND CONGLOMERATE

Michael A. Krulc

Traylor Shea JV

John J. Murray

Jacobs Associates

Michael T. McRae

Jacobs Associates

Kathy L. Schuler

San Diego County Water Authority

ABSTRACT

The San Vicente Pipeline Tunnel Project involves the construction of an 8.5-foot-diameter water pipeline in an 11 mile tunnel in San Diego, California. The tunnel is located in extremely variable geologic conditions of strong granitic and volcanic rocks, weak sedimentary rocks, abrasive conglomerates, and mixed face conditions. One reach of the tunnel is a 5,200-foot mixed face reach that includes Friars Formation Conglomerate overlying granitic rock. Due to the variable ground conditions in this reach, the use of a TBM was deemed too risky and the contract documents required drill-and-blast and hand mining methods. Although it was not required, the contractor elected to construct this reach using NATM methods. This paper describes the geology of the 5,200-foot mixed-face reach, the design evaluation, and the construction challenges faced on this project.

INTRODUCTION

The San Vicente Pipeline (SVP) is an 11-mile, 102-inch-ID raw water transmission pipeline that is part of the San Diego County Water Authority's Emergency Storage Project (ESP). The ESP is intended to ensure that water is available to the San Diego region in the event that an earthquake or severe drought interrupts the delivery of imported water. The ESP involves the construction of a system of reservoirs, pipelines, flow control structures, and pumping stations. As part of this system, the SVP will help supply water in an emergency. It will also improve the operational efficiency of the Water Authority's overall system. The SVP project links the San Vicente Reservoir (SVR), located on the east side of San Diego County, and the Water Authority's Second Aqueduct, located west of Interstate 15 (see Figure 1).

The SVP project includes the construction of: a tunnel that is approximately 57,230 feet long, two vertical riser access structures (one at the west end of the tunnel and the other near the midpoint of the tunnel), an isolation vault and drive-in vault access structure at the east end of the tunnel.

The tunnel is located in extremely variable geologic conditions of strong granitic and volcanic rocks, weak sedimentary rocks, abrasive conglomerates, and mixed face conditions. One reach of the tunnel is a 5,200-foot mixed face reach that includes Friars Formation Conglomerate overlying granitic rock.

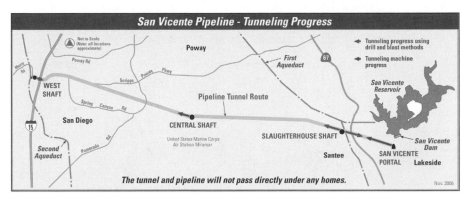

Figure 1. Project location map

This paper discusses some of the important design considerations for the mixed-face reach, designated as Reach R-5, and describes the ground conditions encountered and the construction methods used in the variable conditions. The project is currently under construction, with an estimated completion date of December 2008.

PROJECT AREA GEOLOGY

The project is located in San Diego County in the Peninsular Ranges geomorphic province. Along the western side of the Peninsular Ranges, sedimentary rocks overlie the basement rocks. The sedimentary rocks are generally flat-lying or dip gently to the southwest. They include sandstone, siltstone, claystone, and conglomerate. The basement rocks include Jurassic metamorphic rocks and Cretaceous igneous rocks of the Southern California batholith.

REACH R-5 GEOLOGY

The tunnel alignment within Reach R-5 extends through both the Friars Formation Conglomerate and granitic rocks of the Southern California Batholith. The actual contact of the conglomerate with the granitic rocks in this reach is irregular and undulating (see Figure 2). Therefore, this reach was anticipated to encounter alternating zones of granitic rock and conglomerate in various proportions, including a full face of either rock type or mixed faced conditions at any location within the reach. In this reach, the cover over the tunnel invert ranges from 55 to 240 feet. This reach of the tunnel passes beneath Slaughterhouse Creek, and the entire reach is below the water table. Groundwater levels ranged from about 55 to 90 feet above the tunnel invert.

Friars Formation Conglomerate

The Friars Formation is part of the Eocene La Jolla Group, and conformably underlies the Stadium Conglomerate. It is horizontally bedded to gently dipping (less than about 20 degrees). The Friars Conglomerate consists of medium gray to grayish green, moderately indurated, matrix-supported gravel/cobble conglomerate with occasional boulders. Figure 3 is a photograph of an outcrop that illustrates the high cobble content of this formation. The matrix consists of silty to clayey sand, and the unit contains localized interbeds of sandstone and siltstone. Although it is generally uncemented, it is

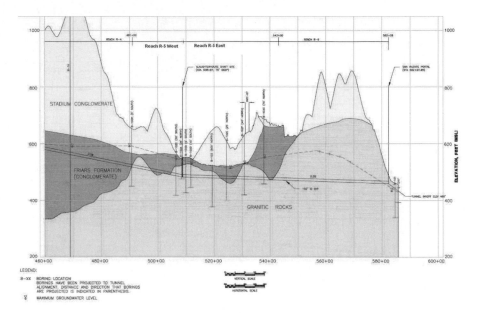

Figure 2.   Reach R-5 geologic profile

Figure 3.   Conglomerate outcrop

weakly to strongly cemented in some places. Induration of this unit is due more to the tightly compacted matrix than to the degree of cementation.

The percentage of clasts (typically gravel size and larger) in the formation varies and the clasts are typically surrounded by the matrix (i.e., matrix-supported). However, there are intervals within the unit in which less matrix is present and the clasts are in contact with each other. The clasts in the formation typically consist of extremely strong (i.e., unconfined compressive strengths up to 60,000 psi), metamorphosed rhyolite, dacite, metasandstone, and quartzite. Figure 4 shows the grain size distribution of the conglomerate and its matrix. The average clast size is typically in the cobble

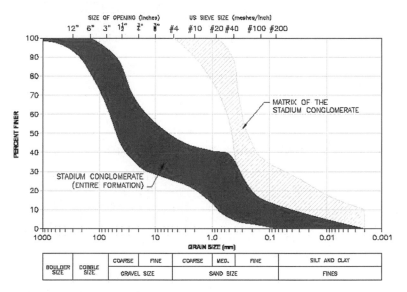

Figure 4. Grain size distribution of the conglomerate and its matrix

range (i.e., 3 to 12 in.). However, boulder-sized clasts of up to 6 feet were encountered in the borings.

The conglomerate exhibits a wide range of strength characteristics depending on the clast compositions, degree of matrix cementation, and grain size distribution. Local quarry operators report that these rocks are some of the most abrasive in the United States.

Cretaceous Granitic Rocks

Cretaceous Granitic Rocks of the Southern California batholith outcrop near the eastern and western ends of the SVP tunnel alignment. They are overlain unconformably by the sedimentary rock units described above. The granitic rocks in the vicinity of Reach R-5 consist of a group of older, variably metamorphosed granitic rocks that possess a metamorphic texture and alteration believed to be related to the emplacement of the large bodies of Cretaceous Granitic Rocks to the east and west. Weak to strong gneissic textures have been observed in these rocks and locally these rocks are schistose. The rock types in this older body include granodiorite, quartz diorite, diorite, tonalite, gneiss, gabbro, schist, and gneissic granodiorite. In addition to being locally altered, these rocks are typically more deeply weathered than the younger granitic rocks to the east and west. As a result, these rocks tend to be weaker.

Unconfined compressive strength (UCS) tests of slightly weathered to fresh granitic rock in this reach exhibited strength values ranging from 5,400 to 28,500 psi with an average value of 14,000 psi. UCS tests of highly to moderately weathered granitic rock in this reach exhibited strength values ranging from about 1,500 to 8,500 psi with an average value of 4,500 psi. Fifty percent of the granitic rock in this reach is anticipated to be fresh to slightly weathered and 50 percent to be highly to moderately weathered.

Discontinuities observed in the granitic rock core within this reach consisted primarily of joints that are smooth to slightly rough and tight to moderately wide. The

Table 1.   Summary of RMR, Q-Value, and RQD for reach R-5 granitic rock

Classification System	Rock Quality Description (percent of cored footage in Reach R-5)				
	Very Good	Good	Fair	Poor	Very Poor
RMR	7%	33%	53%	7%	0%
Q-System	13%	47%	20%	0%	20%
RQD	26%	27%	7%	13%	27%

Table 2 .   RMR, Q-Value, and RQD correlation to Terzaghi's descriptions

Rock Quality Description	RMR Range	Q-Value Range	RQD Range	Approximate Correlation to Terzaghi's Description
Very Good Rock	>80	>100	90–100	Massive or hard and intact
Good Rock	60–80	10–100	75–90	Massive, moderately jointed
Fair Rock	40–60	1–10	50–75	Moderately blocky and seamy
Poor Rock	20–40	0.1–1	25–50	Very blocky and seamy
Very Poor Rock	<20	<0.1	0–25	Crushed

joints typically strike to the northeast and northwest and have steep to vertical dips to the southeast, southwest, and northwest.

DESIGN EVALUATIONS

Design evaluations for Reach R-5 focused on: determining the properties of the granitic rocks and conglomerate formation; determining the location and nature of the contact between the conglomerate and the granite; establishing the proportions of the reach that would encounter conglomerate, granite, or mixed face conditions; establishing the groundwater conditions and potential for groundwater inflows into the tunnel; and identifying appropriate tunnel excavation and support methods. The geotechnical investigation program performed within this reach included nine soil test borings ranging in depth from 70 to 210 feet, one test pit for gradation sampling, and nine seismic refraction lines.

Evaluation of Granitic Rocks

The granitic rocks in this reach were characterized using the Rock Mass Rating (RMR) System developed by Bieniawski (1984); the Q-System developed by Barton, Lien, and Lunde (1974); and the Rock Quality Designation (RQD) developed by Deere et al. (1970). A summary of the results of these analyses is shown in Table 1.

The RMR, Q, and RQD values were divided into several categories as suggested by Bieniawski (1984). Table 2 shows the approximate correlation with rock condition terms developed by Terzaghi (1946) for each category based on correlations developed by Barton (1974) and Deere et al. (1970).

Based on these analyses, it was estimated that: 25 percent of the granitic rock in Reach R-5 would be in rock that is massive, moderately jointed; 45 percent would be in moderately blocky and seamy rock; and 30 percent would be in very blocky and seamy rock.

Evaluation of Friars Formation Conglomerate

The parameters used to characterize the conglomerate were: the density of the formation, the gradation of the clasts, the amount and degree of cementation of the matrix, the strength of the clasts, and the groundwater conditions. The information obtained from the borings in Reach R-5 indicated that the matrix is typically 97 to 100 percent uncemented and 0 to 1 percent strongly cemented. Stand up time tests performed for the nearby San Diego State (SDSU) Tunnel (Boone et al., 2001), excavated in the Stadium Conglomerate were analyzed. The Stadium Conglomerate is a similar formation to the Friars Formation Conglomerate in Reach R-5. The results of the SDSU test pits indicated no signs of collapse for 96 hours in a 2-foot-deep test pit above the groundwater level. A similar test pit in saturated conditions collapsed after about 45 hours. Flooding of the shallow test pits above the groundwater level resulted in collapse after about 8 to 16 hours. This information was supplemented by observations of local quarry exposures that indicated that the Friars Formation Conglomerate is generally a stable formation. Excavated slopes 150 to 200 feet high with an inclination of 0.2:1 (horizontal to vertical) above the groundwater level have been stable for decades with virtually no spalling or raveling of the slopes. These observations generally indicated that stand up time would be good above the groundwater level and less favorable below the groundwater level.

For the section of Reach R-5 encountering conglomerate, firm to slow raveling ground was anticipated above the groundwater table and slow to fast raveling ground was anticipated below the groundwater table. Where the conglomerate matrix is very weak, cobble and boulder clasts are not firmly held in place by the matrix, and it was expected that the clasts would fall from the tunnel roof, sidewalls, and face if not supported.

Geologic Contact

The geologic investigation indicated that the surface of the granitic rock below the conglomerate is very irregular and the tunnel in this reach would encounter granitic rocks and conglomerate in a complex arrangement, including mixed face conditions. Based on the available data, a geologic profile was developed for Reach R-5 as shown in Figure 2.

Near the contact, ground conditions were expected to be highly variable with the granitic rock ranging from slightly to highly weathered. Boulders were expected to be present at the contact. Based on the boulders encountered in the borings, it was estimated that 2.5 percent of the conglomerate (by volume) would consist of boulders and that 50 large boulders (between 2 and 8 feet in maximum dimension) would be encountered in this reach. The large boulders were expected to be encountered mainly along the irregular contact.

The entire reach is below the groundwater table and the high core loss in the borings within this reach indicated the potential for unstable ground conditions and large groundwater inflows, especially close to the contact between the conglomerate and granitic rocks.

Groundwater Inflow Analyses

For the general tunnel alignment, groundwater inflow analyses were performed using the method outlined by Heuer (1995). This method used permeability data from borehole packer tests to develop permeability distributions that were used along with groundwater elevation to estimate steady state groundwater inflows and maximum heading inflows. This method is applicable for distributed sources of water inflow due to rock mass features such as jointed and bedding planes. The method also takes into

Table 3. Summary of results of groundwater inflow analyses

Heuer Method		Probabilistic Method	
Maximum Sustained (Steady-State) Inflow	Instantaneous (Peak) Heading Inflow	Maximum Sustained (Steady-State) Inflow	Instantaneous (Peak) Heading Inflow
104 GPM	43 GPM	300 GPM	225 GPM

account groundwater head above the tunnel and recharge conditions. However, in Reach R-5 it was anticipated that the majority of the inflows would be associated with the contact. Therefore, a probabilistic approach was also used to analyze the potential groundwater inflows in Reach R-5. The probabilistic method accounts for high groundwater inflows that could occur through high permeability features which are more likely to be present at the contact. This approach utilized steady-state and transient equations developed by Goodman (1965). The Goodman equations provide a more conservative estimate of inflow for fracture flow in granitic rock as compared to the Heuer (1995) method. The risk analysis program @Risk (Version 4.5) was used to perform the Monte-Carlo simulation for the probabilistic approach. The predicted groundwater inflows for both the Heuer (1995) and probabilistic methods are shown in Table 3.

Design Stage Recommendations

Hydraulic, operational, and general tunnel alignment and profile constraints (i.e., right of way, shaft locations, etc.) resulted in a relatively narrow alignment/profile window for Reach R-5. Therefore, it was not possible to locate this tunnel reach entirely within either the conglomerate or the granite and some amount of mixed face ground was unavoidable. The risk analyses performed for the project established that using a TBM within this reach would involve a relatively high risk for significant ground related delays. The granitic rocks were judged to be locally too strong and abrasive to be mined efficiently with a roadheader. Further, given that a shaft site was located within this reach and recognizing that it would be possible to commence mining out of this shaft with short lead time equipment, the decision was made to specify that this reach be mined using drill and blast or hand mining methods. This approach was judged to entail significantly less risk due to the adaptability to variable ground conditions. Steel sets were judged to be the most appropriate initial ground support given the variable ground conditions. Presupport measures were anticipated to be required in the mixed face conditions to address low stand-up time and raveling conditions. Presupport was expected to consist of forepoling, crown bars, or combinations thereof. In addition, a probe/drain hole was required to be drilled and maintained ahead of the tunnel face throughout this reach to check for potential groundwater inflows, investigate geologic conditions ahead of the tunnel, and to drain the groundwater ahead of the tunnel excavation. This approach was anticipated to be the most effective and economical way to construct this reach of tunnel.

CONTRACTOR'S EVALUATION OF THE REACH

The contractor, Traylor/Shea JV, recognized that a flexible and adaptive excavation and support approach was needed to cope with the wide variety of ground conditions that were anticipated along Reach R-5. The GBR advised that mixed conglomerate and rock face conditions below the water table could be present anywhere along the reach and that "Conditions will vary widely over short distances and could change rapidly with little warning." The Traylor Shea JV decided that the "toolbox" of ground support measures provided

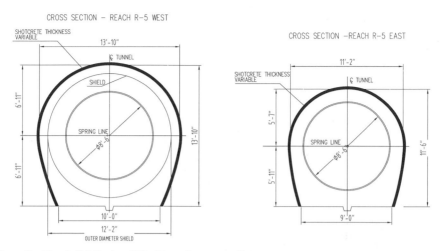

Figure 5.   Reach R-5 East and West tunnel cross sections

by the New Austrian Tunneling Method (NATM) adequately matched the requirements of the reach. The Joint Venture decided to employ the services of Beton-und Monierbau (BeMo) of Innsbruck, Austria to aid in the development of mining methods, selection of equipment, and design of initial tunnel support.

Design

Reach R-5 was divided into two drives; one drive trended east from the Slaughterhouse Shaft and one drive trended west from the Slaughterhouse Shaft (see Figure 2). The east drive, designated as Reach R-5 East, was 3,413 feet in length and the west drive, designated as Reach R-5 West, was 1,787 feet in length. Once completed, the Slaughterhouse Shaft and Reach R-5 West will be used as the primary means of access to launch a digger shield, which will be used to mine another reach of the SVP. Because the 12-foot, 2-inch diameter digger shield will need to be walked through Reach R-5 West to begin mining, this heading was planned to be larger in size than the Reach R-5 East heading, which was sized to accommodate the 8-foot, 6-inch diameter steel pipe and annular space backfill. The Reach R-5 West heading was designed with a nominal height and width dimension of 13 feet, 10 inches. The East heading was designed with a nominal height and width dimension of 11 feet, 2 inches. The tunnel dimensions varied slightly depending on the support classification. A general layout of the tunnel cross sections can be seen in Figure 5.

A ground support design concept was developed for each of the ground conditions that were anticipated during the tunneling, with the main support categories including structural support design for fresh granite, weathered granite, and conglomerate. Each main category was further broken into three sub-categories to address the particularities of the ground behavior within each category. The final design drawings detailed a total of nine support classes, summarized in Table 4.

Support Class 1—Granitic Rocks. Generally, Support Class 1 encompasses the range of ground conditions from fresh granite that is slightly fractured to slightly weathered granite that is moderately fractured. Ground support for this class includes spot installation of rock bolts and a flashcoat of shotcrete or chain link mesh as required as a surface treatment. It was anticipated that material in this class would be excavated using drill-and-blast methods with round lengths of 6 to 10 feet.

Table 4. Support class description

Support Class	Rock Mass Description	Support	Presupport
1.1	Fresh granite, massive to slightly fractured	Spot bolting	None
1.2	Fresh to slightly weathered granite, moderately jointed in the crown	Spot bolting, flashcoat of shotcrete in the crown	None
1.3	Fresh to slightly weathered granite, moderately jointed in the crown and bench	Spot bolting, flashcoat of shotcrete in the crown and bench	None
2.1	Slightly weathered granite, blocky and seamy	Systematic bolting, 2" of shotcrete in the crown and bench	None
2.2	Moderately weathered granite, very blocky and seamy	Systematic bolting, 5" of shotcrete in the crown and bench	None
2.3	Highly weathered granite, very blocky and seamy, disintegrated	Systematic bolting, 3" of shotcrete in the crown and bench, lattice girders	Spiles optional
3.1	Strongly to moderately cemented conglomerate	4" of shotcrete in the crown and bench, lattice girders	Spiles optional
3.2	Moderately to weakly cemented conglomerate	4" of shotcrete in the crown and bench, lattice girders	Pipe umbrella, dewatering
3.3	Uncemented conglomerate	4" of shotcrete in the crown, bench, and invert, lattice girders	Pipe umbrella, dewatering, spiles

Support Class 2—Weathered Granite. Generally, Support Class 2 encompasses the range of ground conditions from slightly weathered granite that is blocky and seamy to highly weathered granite that is disintegrated. Ground support for this class ranges from systematic rock bolting with a flashcoat of shotcrete in the crown and bench to systematic bolting with a structural application of shotcrete in the crown and bench. It was anticipated that material in this class would be excavated using drill-and-blast methods with short round lengths of 4 to 6 ft to limit overbreak.

Support Class 3—Conglomerate. Generally, Support Class 3 encompasses the range of ground conditions from strongly cemented conglomerate to uncemented conglomerate, sand lenses, and mixed granitic and conglomerate face conditions. Ground support for this class includes shotcrete and lattice girder structural support with presupport measures including drain holes, spiling, self-drilling tube spiles, and pipe umbrellas as required. It was anticipated that material in this class would be excavated using an Inter Techmo Commerce (ITC) tunnel excavator.

The NATM approach required the presence of personnel on site who have experience with and knowledge of this method. The Joint Venture employed Albin Reinhart, a NATM engineer from BeMo who, in conjunction with the Joint Venture's engineers and superintendent, decided what support type was appropriate. He also helped to plan mining sequences and worked to increase mining efficiency and reduce cycle times. A technician from BeMo, Manfred Baumgartner, was on site as well. Mr. Baumgartner is experienced in NATM mining methods and equipment operation. He was on site to assist in training the local miners in such things as application of high performance shotcrete, accurate drilling, installation of pipe umbrellas, and operation of the ITC tunnel excavator.

Equipment

Personnel from the Joint Venture took several trips to Central Europe to investigate NATM tunneling equipment and methods. Particular site tours were planned to

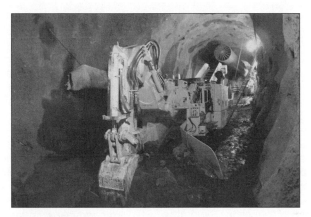

Figure 6. ITC 112 with E3S boom

allow personnel to witness pipe umbrella drilling and operation of an ITC tunnel exca-vator. Neither pipe umbrellas nor ITC tunnel excavators are in common use in the United States at this time, but the Joint Venture thought that these tools were well suited to the potentially unstable conglomerate and mixed face conditions below the water table. Based on the observations made during these trips, the Joint Venture decided to use an ITC excavator to dig the conglomerate sections of Reach R-5. They also prepared to procure the equipment necessary to install pipe umbrellas in case ground conditions warranted their use. An ITC is shown in Figure 6.

A used ITC 112 was purchased and refurbished at the Traylor Bros. Inc. shop in Evansville, Ind. ITC supplied a new E3S boom, which had the capability of tilting side-to-side in order to efficiently dig the horseshoe tunnel shape. Additionally, the travel system was upgraded so the ITC could tram quickly from one heading to another. Other equip-ment purchased to excavate and support the tunnel included: an Atlas Copco Rocket Boomer 282 drill jumbo equipped with Atlas Copco COP1838ME hydraulic drills to drill blast holes as well as drill probe holes, drain holes, and install pipe umbrellas; two Wagner ST3.5 LHDs and one Atlas Copco ST3.5 LHD to load and haul muck from the drill-and-blast sections; and two Dux DT12 8 cy tunnel trucks to haul muck in the highly weathered granite, conglomerate, and mixed face sections where the ITC 112 tunnel excavator was expected to be used. The LHDs loaded excavated material into a heavy-duty muck box in the shaft bottom. The muck box was hoisted by a Liebherr 883 crane. A robotic shotcrete arm was purchased from Shotcrete Technologies and mounted on a New Holland LS185 skid steer loader. The skid steer towed a trailer that carried a Schwing B310 shotcrete pump as well as the accelerator dosing system. A 5 cy Normet Transmixer was used to transport shotcrete to the heading.

CONSTRUCTION CONDITIONS

Shaft Excavation

Shaft sinking operations began on October 10, 2005. Since the shaft was expected to progress through three anticipated ground types (conglomerate, weath-ered granite, and fresh granite) the Joint Venture anticipated that shaft excavation would be a good indication of how the ground would behave during tunneling, and would indicate the nature of the contact between the conglomerate and the granite.

The Slaughterhouse Shaft is 36 feet in diameter and 70 feet deep. Through the alluvium, conglomerate, and weathered granite, shaft support consisted of W6x 20 steel ribs on 5 foot centers with 2 inches of steel-fiber-reinforced shotcrete as lagging. Through the moderately weathered to fresh granite, support consisted of 6-foot-long rockbolts on a 5 foot pattern with 2 inches of plain shotcrete as a surface treatment.

The top 5 feet of the shaft was excavated through alluvium. The conglomerate, which extended to a depth of 35 feet, was excavated quite easily using a CAT 312 excavator. The 5-foot-high lifts stood vertically with no sign of raveling. A lift was typically not left open for longer than four hours. However, on isolated occasions, a lift was left open for roughly 12 hours with no sign of raveling. The groundwater at the time of shaft sinking was 10 feet below the ground surface. Excavation continued through the conglomerate below the water table without incident. Any troublesome groundwater flows were channeled by hoses or panning so as not to affect the shotcrete application.

However, several ground conditions were encountered that could be problematic if encountered during tunneling. Several weakly to moderately cemented sand lenses were encountered that presented a challenge to get shotcrete to adhere when below the water table. This was a minor problem on the shaft walls, but could be a much larger problem if encountered in the crown of the tunnel under much higher water heads. Also, a cluster of large boulders 2 to 8 feet in size were encountered close to the interface with the underlying weathered granite. In the shaft excavation, these boulders could be removed by the excavator, but in the tunnel they would likely need to be blasted. The weathered granite, extending to a depth of 60 feet, could be dug with an excavator, though with difficulty, so a hoe-ram was used until this method was no longer economical. The weathered granite stood with no problem, but it was very blocky, seamy, and friable and exhibited a higher permeability than the conglomerate. The remainder of the shaft through moderately to slightly weathered granite was excavated by drill-and-blast methods. Water inflow through the fresh granite appeared less than through the weathered granite above, and after several weeks most of the water inflow through the fresh granite ceased, while water continued to flow through the weathered granite. The total water inflow upon completion of the shaft excavation was on the order of 50 gpm.

The shaft sinking operations indicated that the transition zones from fresh granite to weathered granite to conglomerate would be marked by increasing water flows and increasingly friable and blocky material. Clusters of 2- to 8-foot boulders could be expected at the interface between the weathered granite and conglomerate, and fast raveling sand lenses could be encountered in the conglomerate. The conglomerate appeared to have good stand up time, though the water head in the shaft was less than would likely be encountered in the tunnel.

Tunnel Excavation

Tunneling began on December 16, 2005 with mining crews working two 10-hour shifts per day, six days per week. It was planned that there would be one mining crew and one fleet of equipment per shift, rather than a crew at each heading. The crew and equipment swung between the east heading and west heading.

Reach R-5 East. Reach R-5 East was on the critical path of the job, and was therefore the focus of the mining effort. Drill-and-blast operations began with an 8-foot round length in the granite. The first 100 feet of the east and west headings were mined using an enlarged section to provide space for parking and passing equipment underground. In the initial 46 feet of excavation, the granite was blocky and seamy and blasting operations led to loosening and separation of blocks in the crown. Because not all of the shotcrete equipment was on site, steel ribs and timber blocking were used

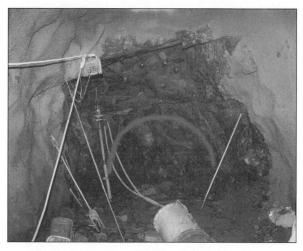

Figure 7. Excessive overbreak near rock-conglomerate interface

to support this section of tunnel. Ground conditions improved to the point where the ribs were no longer necessary and 6 foot Swellex rock dowels were adequate to support the rock. After 100 feet, the tunnel was necked down. It was taken from 16 feet, 6 inches wide by 14 feet tall to 10 feet, 6 inches wide by 11 feet, 2 inches tall. Drilling and blasting continued with 8-foot round lengths and rockbolt support for another 605 feet. The ground varied from slightly weathered granite with some spalling and loosening of blocks in the crown to fresh granite where no spalling was exhibited. When Reach R-5 East had advanced 605 feet, groundwater inflows began to increase, with several small flows from the face and sidewall. Up to this point, groundwater inflow had been limited to dripping in the crown and sidewall that lessened to nothing a week or so after the heading had advanced. A large block that fell out of the sidewall during blasting operations at this location is shown on Figure 7.

As the observed ground conditions became more adverse, the support classification was changed from 1.1 to 2.2 and the round length was shortened to 5 feet. Mining continued without incident for another 44 feet, at which point the first conglomerate was seen in the heading. The conglomerate was exposed in an approximately one foot square patch of the face near the crown. Though it appeared stable, there was a steady flow of water (estimated to be 15 gpm) dripping and streaming from the crown. The granite in the vicinity was weathered, but not disintegrated. Probe holes drilled in the heading did not show a significant weakness in the ground conditions, so it was decided to drill and shoot another round. The support classification was changed to 3.1 and the round length further shortened to 4 feet. Several more rounds were shot, fully exposing the conglomerate in the crown. There were no large boulders present and the ground exhibited good stand up time, with the heading once being left open for six hours with no raveling. The heading advanced 23 feet more with conglomerate in the crown until the face once again returned to rock. It was likely that the transition between rock and conglomerate was not far away, so mining continued with short rounds, though the support class was changed to 2.2. Conglomerate in the heading returned after approximately 46 feet of mining and the support class was again increased to 3.1 and the round length reduced to 4 feet. It was decided to continue blasting operations rather than to try to dig the conglomerate with the ITC tunnel excavator because, at the time, the conglomerate was difficult to dig and a majority of the

Figure 8.   Mixed face condition

face consisted of granite. The proportion of conglomerate in the face continued to increase until on any given round it accounted for 45% to 55% of the face area, with the remaining face area being moderately weathered to slightly weathered granite. Figure 8 shows the mixed face condition, depicting three distinct layers of ground: weathered granite on the bottom, boulder-sized conglomerate in the middle, and cobble-sized conglomerate near the crown.

This roughly half-and-half mixed face condition continued for approximately 725 feet, after which the granite interface dipped down and conglomerate accounted for the full face area. An attempt was made to use the ITC 112 excavator to mine the conglomerate, but it was unable to do much more than scratch the face, and so blasting operations were continued. The inability of the excavator to dig the conglomerate was attributed to the density and cementation of the matrix and the tightly packed nature of the clasts within the matrix. After it became apparent that the conglomerate was relatively stable, the round length was increased to 5 feet with favorable results. Overbreak was limited and the conglomerate rarely raveled. One notable exception was a weakly cemented sand lens encountered in the crown for a length of approximately 20 feet. The sand was slow raveling and continuously dripped water, which resulted in difficulties related to shotcrete adhesion. This problem was remedied by drilling and installing 10-foot-long rebar spiles in advance of the excavation.

To date, excavation has continued, using drill-and-blast methods through the granite, mixed face, and conglomerate areas. The nested boulders encountered along the transition zone between granite and conglomerate during shaft excavation have only been observed in the tunnel in localized areas. The boulders were not as large as those encountered during the shaft excavation and there was enough supporting matrix to hold the boulders in place during blasting.

Groundwater inflow proved to be a good indication of when highly weathered granite and conglomerate was ahead of the face and the transition zones yielded most of the water inflows into this reach. Probe holes were required to be advanced a minimum of 20 feet ahead of the face at all times. These probe holes also acted as drain holes. Typically, groundwater inflows through the probe holes were on the order of 0 to 5 gpm, with probe holes in the transition zones producing a flow on the order of 10 gpm. Generally, the flows diminished to a trickle over the course of several days. Based on piezometer readings along the tunnel alignment, groundwater levels have been drawn down by the tunnel excavation ahead of the tunnel face and this drawdown has tended to improve tunnel stability. Drawdown was typically on the order of 25 to

50 feet and as high as 60 feet in the vicinity of the shaft. At the eastern end of the reach, drawdown was less, possibly as a result of this section of the tunnel being entirely within the conglomerate, which exhibited lower permeability values than the granitic rock.

To date, neither the ITC 112 excavator and tunnel trucks nor the pipe umbrellas have been used. The length of tunnel where full face conglomerate conditions have been present has been relatively short, and in the areas were the conglomerate was full face, it was too strong to be excavated with the ITC 112. Generally, the ground has had stand up time sufficient to allow installation of the initial tunnel support without the need for presupport. Pipe umbrellas have not been installed to date, and drilled spiling was necessary only for a short distance.

Excavation rates have averaged 7.6 feet per day, with the best daily advance being 24 feet. At the time of writing, Reach R-5 East was 92% complete.

Reach R-5 West. Reach R-5 West was not on the critical path of the job, since it was half the length of Reach R-5 East. It was planned that this heading would be mined when equipment and personnel were not being used in the East heading, such as during mucking operations.

The initial 100-foot-long enlarged section was mined with 8-foot rounds and rock bolt support. The tunnel was then necked down to a section 13 feet, 10 inches wide by 13 feet, 10 inches tall. Soon after necking down, the rock became moderately to highly weathered, and blasting operations led to significant loosening and separation of blocks in the crown and sidewall. The round length was shortened to 6 feet and the support classification changed to 2.1, which included 3 inches of shotcrete and 6 foot Swellex rock bolts on a 5 foot pattern. Mining continued in this manner, though it was anticipated that conglomerate could be encountered at any time. However, the buried conglomerate valley that was indicated in the geological profile was not encountered and after approximately 640 feet of mining, the granitic rock became much stronger. The support classification was reduced to 1.3 and the round length increased to 10 feet. Mining has continued to the present time in slightly weathered to fresh granite with little to no groundwater inflow. At the time of this writing, 1,527 feet of the 1,787-foot design length had been mined (86% complete).

Excavation rates have averaged 5.8 feet per day with the best daily advance being 22 feet.

CONCLUSIONS

Although the investigation program performed during the design phase provided a good indication of the nature of the conglomerate and granite that would be encountered in Reach R-5, accurately identifying the exact location and nature of the contact proved difficult. This can be attributed to the distance between the boreholes, the inability to use geophysics along the entire length of the reach, and the variability of the contact. The nature of the contact proved to be more favorable than expected for several reasons. First, the groundwater inflow was less than anticipated with heading inflows in the mixed face condition typically only about 2 to 3 gpm at the face. The standup time of the conglomerate in the mixed face zone was also better than anticipated. This was likely due to the very dense nature of the matrix and the drawdown of the groundwater ahead of the tunnel through the use of probe/drain holes. Finally, the very large boulders (6 to 8 feet in maximum dimension) encountered at the contact within the shaft were not encountered at the contact in the tunnel. Approximately 150 boulders larger than 2 feet were encountered in the reach, but most of these boulders were between 2 and 3 feet in maximum dimension. These boulders were typically located at the contact but did not impact the construction since the reach was constructed entirely by drill-and-blast methods.

The decision to require excavation of Reach R-5 by drill-and-blast or hand mining methods proved to be the correct decision given the highly variable ground conditions and the long section of the reach that encountered mixed face conditions. The decision by the contractor to use a NATM approach also proved to be a good decision given the adaptability of this approach to variable ground conditions. This approach allowed steady progress to be made through the mixed face ground and rapid progress in the granitic rock sections of tunnel. The use of the ITC 112 tunnel excavator was not effective in the conglomerate. The bucket of the excavator could not excavate the conglomerate even in the uncemented to weakly cemented conglomerate because the bucket was too wide to impact only the matrix without encountering the very strong clasts. The clasts in this formation are very tightly packed and require a pick type excavation tool to excavate the matrix around the clasts.

REFERENCES

Barton, N., Lien, R., and Lunde, V., 1974, "*Engineering Classification of Rock Masses for the Design of Tunnel Supports*," Rock Mechanics, V. 6, pp. 189–236.

Bieniawski, Z.T., 1984, *Rock Mechanics Design in Mining and Tunneling*, A.A. Balkema, Boston, 272p.

Boone, S.J., Poschmann, A., Pace, A., and Pound, C., 2001, "*Characterization of San Diego's Stadium Conglomerate for Tunnel Design*." RETC Proceedings, pp. 33–45.

Deere et al., 1970, "Design of Tunnel Support Systems," Highway Research Record, Number 39, National Research Council, Washington, D.C., pp. 26–33.

Deere, D.U., Peck, R.B., Monses, J.E., and Schmidt, B., 1969, "Design of Tunnel Liners and Support Systems," OHSGT Contract 3-0152, PB 183799, NTIS, Springfield, VA, 287 p.

GeoPentech-Golder-Southland, December, 2004 "Geotechnical Data Report," San Vicente Pipeline Project, prepared for the San Diego County Water Authority.

Goodman, R.E., D.G. Moye, A. Van Schalkwyk, and I. Javandel, "Groundwater Inflows Durring Tunnel Driving," Engineering Geology, V. 1, No. 1, pp. 39–56, 1965.

Heuer, R.E., "*Estimating Rock Tunnel Water Inflow*," *Proceedings: Rapid Excavation and Tunneling Conference*, San Francisco, CA: Society for Mining, Metallurgy, and Exploration, Inc. pp. 41–60.

Proctor, R.V., and White, T.L., 1968, Rock Tunneling with Steel Supports, Commercial Shearing, Inc.

Terzaghi, K., 1946, "Rock Defects and Loads on Tunnel Supports," in *Rock Tunneling with Steel Supports*, by R.V. White and T.L. White, Commercial Shearing, Youngstown, Ohio, pp. 17–99.

Jacobs Associates, November 29, 2004 "Geotechnical Baseline Report," San Vicente to Second Aqueduct Pipeline Project, prepared for the San Diego County Water Authority.

CONSTRUCTION OF THE C710 BEACON HILL STATION USING SEM IN SEATTLE— "EVERY CHAPTER IN THE BOOK..."

Satoshi Akai
Obayashi Corp.

Mike Murray
Hatch Mott MacDonald

Steve Redmond
Obayashi Corp.

Richard Sage
Sound Transit

Rohit Shetty
Obayashi Corp.

Gerald Skalla
Dr. Sauer Corp.

Zephaniah Varley
Parsons Brinckerhoff

ABSTRACT

One would be hard pressed to find another project as complex as this one given its configuration, the variety of issues to address and the required construction techniques. Contract C710 is currently under construction as part of Sound Transit's Link Light Rail, connecting downtown Seattle with Sea-Tac Airport. The paper describes the construction-to-date of the deep mined station under Beacon Hill using the Sequential Excavation Method (SEM).

This deep "binocular" station is being mined through some of the most challenging soft ground conditions in the USA. The excavated volume of the station is approximately 60,000 cy (46,000 m^3) and the station comprises a variety of geometries and cross sections ranging from 235 sf (22 m^2) up to 1,670 sf (155 m^2). This underground complex includes platform, concourse, cross-passage and emergency ventilation tunnels together with station egress and ventilation shafts. Various excavation sequences are in use for the different tunnels including the twin-sidewall drifts for the impressive 45 ft (13.7 m) wide by 42 ft (12.8 m) high concourse cross adits.

This station has two train platforms approximately 165 feet (50m) below the surface. This station, constructed in highly variable glacial deposits including water bearing sands and silts approximately 50 ft below multiple perched water tables in an urban setting, creates a very unique and challenging situation. The ground is being managed with various types of ground improvements such as jet grouting and vacuum assisted deep dewatering wells from the surface, and from underground by use of "SEM Toolbox Items" such as barrel vault canopy, grouted pipe and rebar spiling, chemical grout,

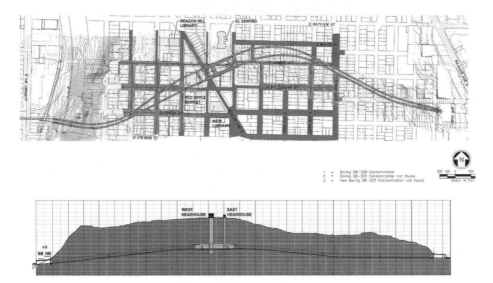

Figure 1.   Preliminary engineering, Beacon Hill Station and tunnel alignment

sheets, vacuum dewatering, and a few others developed on site are used to cope with this ground.

PROJECT BACKGROUND, ORIENTATION, DESIGN CONSIDERATIONS

In 1999 upon completion of Preliminary Engineering, the Sound Transit Board of Directors adopted the Locally Preferred Alternate Alignment for the Central Link Light Rail System from the Sea-Tac Airport (south of Seattle) to Northgate (north of Seattle) which included alignment through Beacon Hill (Figure 1).

As a result of the budget problems associated with the design/build contract from downtown Seattle to the University District in 2000 and other budget problems identified in 2001, the Sound Transit Board of Directors elected to change the focus of the Initial Segment. The Initial Segment construction became Convention Place (downtown Seattle) to SeaTac Airport, thus allowing additional time to develop and consider potential cost saving measures from Convention Place to the University District. As a consequence the Beacon Hill Station was changed from a deferred build-out to a fully operating station (Figure 2).

During the geotechnical exploration, accomplished during 2001 Preliminary Engineering, ground and ground water contaminated with dry cleaning fluids was discovered at the site of the Westerly station entrance shaft. The depth and extent of contamination was sufficient to cause Sound Transit to re-evaluate the location of the station entrances. As a result of the re-evaluation and because the options for shaft locations were restricted, the station was re-configured to include a single large entrance shaft with 4 high speed elevators, emergency egress stairs and emergency ventilation, and a smaller ancillary shaft with only emergency egress and emergency ventilation. The single entrance station necessitated that the elevator shaft connect to the platform tunnels via a pedestrian concourse tunnel near the middle of the platforms. With the large entrance shaft sited at the location of the original eastern entrance shaft and the ancillary shaft

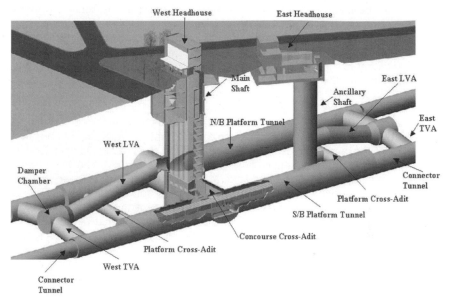

West Headhouse · East Headhouse · Main Shaft · East LVA · Ancillary Shaft · N/B Platform Tunnel · East TVA · West LVA · Damper Chamber · Connector Tunnel · Platform Cross-Adit · S/B Platform Tunnel · Concourse Cross-Adit · Platform Cross-Adit · West TVA · Connector Tunnel

Figure 2.   Isometric of final design station configuration ("as bid")

sited within the street right-of-way a short distance to the east, it was necessary to change the orientation of the station and adjust the alignment of the tunnels.

The data being obtained from the geotechnical exploration program during 2002 Final Design was slightly unusual for the Seattle area, in that there was a complex inter-mixing, rather than layering, of glacially over-ridden clays, silts, tills and sands. Because of this and the size of the shaft (50' diameter), a Sound Transit assembled Peer Review Committee recommended that a test shaft be constructed to check the ground charac-teristics and the ground behavior with respect to the construction methodology, which was to be specified in the station and tunnel construction contract. As a part of the Final Design process, Sound Transit contracted for the construction of an 18' diameter by 150' deep test shaft with two 10' diameter by 15' long tunnel adits. Construction of the test shaft began in January 2003. The test shaft was sited within the foot print of the Station's main shaft in order to provide accurate information on the ground to be excavated as part of the permanent construction and to allow the test shaft to "disappear" during the permanent construction. It was anticipated that the sand in the station area would have adequate "standup" characteristics to allow time for the SEM mining to be utilized. The primary concern was the water laden silts near and in the station tunnels. As excavation progressed the behavior of the sand was not as anticipated. The sand lenses inter-layered within the clays and tills would flow under groundwater heads as little as 1 foot. Because these sand lenses were difficult to dewater, they created significant delays to the excavation progress and significant additional cost to the test shaft. At approximately 100' below ground level a thick flowing sand layer was encountered. The shaft could only be advanced 4 feet in 30 days. Sound Transit made the decision to advance the last 50' of the test shaft with a 6' diameter shaft using a steel casing incrementally vibrated into position and the ground augured out to allow examination of the remaining ground as the conditions would allow. The last 50' of shaft was successfully constructed using this method, thus allowing the silts at Station level to be examined. They were found to behave much better than anticipated.

Based on the information obtained from the test shaft, it was confirmed that the SEM method of excavation was suitable for use in the construction of the Station tunnels. However, due to the flowing sands within the shaft, it was decided that SEM should not be used to construct the shafts or headhouse basements. Sound Transit made the decision to "remove" the ground and groundwater impacts from construction by utilizing slurry walls for the shaft construction and to allow the contractor to use either a "water tight" slurry wall or a "water tight" secant pile wall for the headhouse basement construction.

As a result of encountering some sand layers in the test shaft that had not been encountered in the Final Design geotechnical exploration program, Sound Transit decided to drill two additional exploratory borings: one at the East Damper Chamber (EDC) and one on the West Longitudinal Ventilation Adit (WLVA). The boring on the WLVA indicated that there was a sand layer in the WLVA that had not been previously identified. The boring at the EDC found a full face (30') of sand at tunnel level. This was a complete surprise as all of the previous borings at the Station site and especially two borings located within approximately 30' of the EDC indicated that all of the tunnels should be situated in a full face of clay, silt or glacial till. Because of these last two borings, the conditions encountered in the test shaft and a recently published USGS report on the Seattle-Bremerton Fault, it was concluded that the station was sited in a geologic formation highly unusual for the Seattle region. As a result, there was a high level of uncertainty concerning the ground conditions to be expected in the Station tunnels.

Because the Beacon Hill Station and Tunnel Contract was the critical path of the Initial Segment and there was insufficient time before the scheduled advertisement date of this Contract to perform additional geological exploration, without delaying the opening of the system, Sound Transit made two significant decisions.

1. Jet grouting was added to the Contract to stabilize the sands in the EDC, WLVA and other sand layers within the Station that met a certain criteria.

2. In order to better understand the ground conditions to be encountered in each of the Station tunnels, a supplemental geological exploration program with borings at intervals of approximately 50' along each station tunnel was placed in the Contract. This exploration was to be performed by the Contractor and logged by Sound Transit during the early stages of construction.

This will be addressed in more detail later in the paper.

POST-AWARD GEO-TECHNICAL INVESTIGATION PROGRAM AND GEOLOGICAL DESCRIPTION

The Beacon Hill geology consists of glacial and interglacial soil units with a high degree of variation. This is evident locally to the extent that some units cannot be reliably correlated between adjacent borings. Based on 73 investigation boreholes drilled during the design stages, a simplified geological model was developed and included in the Geological Baseline Report (GBR). The model groups the geologic units into six engineering classes having similar physical and engineering properties as follows:

- Class 1—Loose to Dense Granular Deposits
- Class 2—Soft to Very Stiff Clay and Silt
- Class 3—Till and Till-Like Deposits
- Class 4—Very Dense Sand and Gravel
- Class 5—Very Dense Silt and Fine Sand
- Class 6—Very Stiff to Hard Clay

For reasons described above, the designers specified a post award subsurface drilling program to enable the geotechnical interpretations to be refined. These boreholes, drilled down to platform tunnel level using mud rotary and sonic core recovery methods, were combined with instrumentation including inclinometers, extensometers and piezometers. Over 50 borings were drilled between August 2004 and November 2005 and these were used to refine the extent of jet grouting and dewatering. Core extraction and sampling was conducted under the engineering oversight of Shannon & Wilson (S&W). Periodically S&W updated the geological sections using the latest borehole logs, and presented the sections on an ftp site with access readily available to ST, OC and HMMJ. A major consequence of the updated geological profiles at the east sections of the northbound and southbound platform tunnels resulted in jet grouting of a long thick section of sand present at the tunnel crown level. As one will see later in the paper and in more detail, these updated profiles resulted in the decision to shift the platform tunnels further west away from the predominantly Class 4 soils.

Systematic probing ahead of the face is conducted using air track rigs mounted on Komatsu haulers. These rigs were also used for the installation of the rebar spiles and grouted pipe spiles installed as part of the pre-support ahead of the advancing face. Also, for each tunnel section, a horizontal probe hole is cored to augment the geotechnical interpretations. The probe hole information is reviewed during the daily SEM meetings.

STATION SHIFT, INCLUDING STEPS LEADING UP TO IT

Shortly after award of the Contract in May 2004 the Contractor, Obayashi Corporation, commenced the supplemental exploration program referenced earlier. After completing approximately 50% of the exploration, it became evident that the sand layers within the Station tunnels had a much greater extent than previously anticipated. These sands also had water pressures as high as 50'. Based on this information and the knowledge that the Contractor was told to anticipate predominately clays, tills or silts in the Station tunnels other than the EDC and a short section of the WLVA; that sands similar to these flowed in the test shaft under as little as 1' of water pressure; and that the critical path of the construction was through the Station tunnel excavation; it was essential to stabilize these sands in order to utilize the prescribed method of SEM construction in a cost/time effective manner, and it was important to stabilize these sands off of the construction critical path in order to minimize the construction duration of the Contract.

Sound Transit decided to begin jet grouting of the soils along the North Bound Platform Tunnel before the full exploratory program was complete and began to revise the dewatering program that had been specified in the Contract (Figure 3).

As the supplemental exploration program neared completion, Sound Transit realized that the quantity of sand identified for jet grouting was approximately 300 percent that anticipated in the Contract (Figure 4). Not only would the additional jet grouting cause a significant delay to the construction progress and have a significant cost impact on the Contract, there would also be a significant impact to the local community, because a large portion of the work would have to occur outside of the established worksite and would be within street right-of-way and on private property.

For this and other reasons related to the schedule, Obayashi Corporation prepared a Value Engineering proposal to eliminate the SEM driven Connector Tunnels and to drive three of them with the Earth Pressure Balance (EPB) TBM. This effort was abandoned as the originally perceived benefits were overtaken by events.

Based on these considerations Sound Transit began to review potential changes to the station configuration that would reduce the quantity of jet grouting. Seven

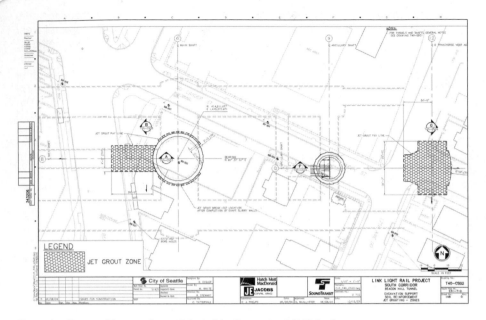

Figure 3. Extent of jet grouting anticipated by the contract (4,994 cy)

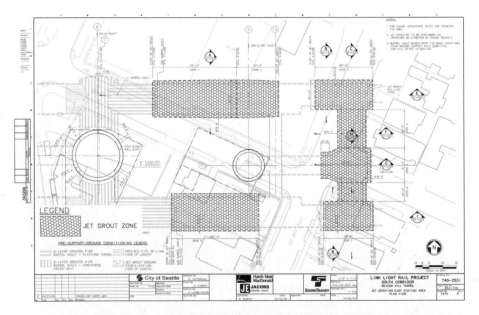

Figure 4. Extent of jet grouting anticipated after the supplemental geological exploration program was complete (14,220 cy)

options were considered that did not shift the location of the two shafts, as the slurry walls had already been constructed and excavation of the main shaft was underway.

Faced with the threat of the ongoing distraction to the Work while jet grouting and given the rising cost per cubic yard of Jet Grouting and the potential savings to the project if jet grouting could me minimized, Obayashi Corporation began sketching out what was to become know later as the "First Shift"; a stepping stone toward what would ultimately become known as the "Ultimate Shift." The first shift was simply reversing the lengths of the Platform Tunnels on either side of the Main Shaft taking advantage of the asymmetry of the Station. In this case, the East Damper Chamber was positioned up tight against the Ancillary Shaft on the East side with no adjoining longitudinal ventilation adit. This reduced the quantity of jet grout and moved the East Damper Chamber out from the street to mostly within the boundaries of the site.

However, following even more exploratory holes at the shifted location of the East Transverse Ventilation Adits and the East Damper Chamber, water filled potentially running sands were discovered, requiring this zone to be jet grouted. A second shift was suggested by Obayashi to eliminate the Damper Chamber and to move the East Transverse Ventilation Adits further west and coincident with the Ancillary Shaft.

Upon consultation with the Contractor and evaluation of the required design effort, the impact to the community, the amount of jet grouting required, and the overall cost of the change, Sound Transit chose to implement what has been characterized as the Ultimate Shift (See Figure 5).

The Ultimate Shift involved:

- sliding the Platform Tunnels and Pedestrian Cross Adits to the west 88 feet,
- eliminating of the Connector Tunnels,
- extending the Platform Tunnels approximately 45 feet at each end to accommodate the ventilation dampers and the ETVA & WTVA connections,
- moving the ETVA to a "sucked-in" position at the east end of Station Platform and connected directly to the Platform Tunnel extensions and the Ancillary Shaft (N&S),
- moving the WTVA to a "sucked-in" position at the west end of Station Platform and connected directly to the Platform Tunnel extensions,
- moving the ventilation dampers from the damper chambers to the draft-relief chambers in the Platform Tunnel extensions,
- eliminating the East Longitudinal Ventilation Adit,
- eliminating the East Damper Chamber,
- reducing the size of the West Damper Chamber, now to be utilized as a connection chamber between the WTVA and the West Longitudinal Ventilation Adit (WLVA),
- increasing the length of the WLVA by 68 feet

By implementing the Ultimate Shift, there was:

- a reduction in the overall quantity of SEM excavation
- minimal impact to the architectural finishes in the Platform Tunnels,
- minimal impact to the ventilation system,
- a reduction in the quantity of dewatering wells required, and
- the greatest reduction in jet grouting.

The final volume of jet grouting required as a result of this Ultimate Shift was only 10% greater than the original bid quantity (Figure 6).

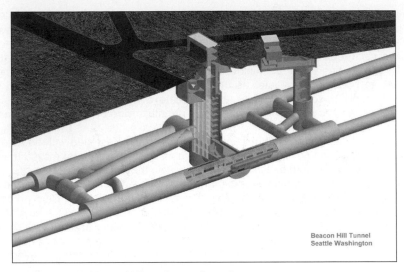

Beacon Hill Tunnel
Seattle Washington

Figure 5.   Isometric of Ultimate Shift station configuration

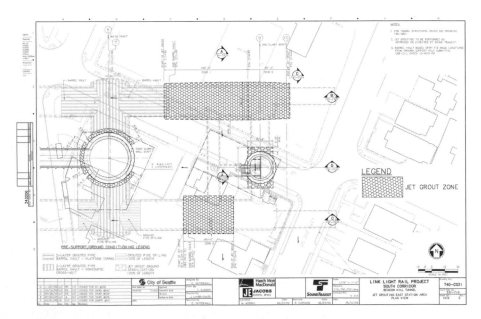

Figure 6.   Extent of jet grouting required after the Ultimate Shift was implemented (5,453 cy)

INITIAL LINING AND SEM TOOLBOX PRE-SUPPORT DESIGN CONSIDERATIONS

The Hatch Mott McDonald/Jacobs (HMMJ) Joint Venture is the lead designer for the Beacon Hill Tunnels and Station. The JV is assisted by the Dr. G. Sauer Corporation (DSC) to provide the SEM design for the Concourse Cross Adit (41 ft wide), the Platform Tunnels (31 ft wide) and the Platform Cross Adits (16 ft wide). Shafts, Ventilation- and Running Tunnels are designed by HMMJ.

General Design Considerations

The above mentioned tunnel sizes required for the Beacon Hill Station set a new record for SEM soft ground tunneling in the United States. Based on the findings of an intensive geological exploratory program SEM tunnel design usually utilizes advanced Finite Element Modeling tools. The appropriate excavation sequences and support measures have to be determined. It also relies heavily on engineering judgment and experienced personnel on site that make sure that key elements are used properly.

Key design elements for excavation of large SEM tunnels in soft ground are:

- Subdivision of the faces into smaller drifts and adjustment of round lengths to be able to control and stabilize the excavation
- Prevent stress concentrations by designing ovoid cross sections with rounded inverts and domed excavation faces and headwalls
- Achieve ring Closure within a minimum of 1.5 times the tunnel diameter to prevent excess settlements
- Timely installation of sealing shotcrete (flashcrete) and the initial support to prevent deterioration and loosening of the soils
- Utilization of the appropriate ground improvement measures, pre-support, and face support tools
- Monitoring of the structure during construction to assure stability and verify design assumptions
- The ability to make adjustments in the field to deal with actual ground conditions encountered
- Experienced Construction Management, Site Supervision and Quality Control to ensure safety and efficiency

The "SEM Toolbox" Approach for Variable Ground Conditions

In the highly variable geology of Beacon Hill different ground types and ground behaviors have been encountered within several feet. Mixed face conditions are present over large portions of the tunnel alignment. That is why no 'prescriptive support' was designed but a baseline scenario was chosen that includes excavation sequence and standard support measures such as, e.g., shotcrete, wire mesh and lattice girders. Additional support measures ("Toolbox Items") are used as needed, to ensure safety of miners and stability of the tunnel.

These Toolbox items include:

- Pre-Support Measures
  - Rebar Spiling
  - Grouted Pipe Spiling
  - Metal Sheets
  - Grouted Barrel Vault/Pipe Arch

- Face Stabilization Measures
  - Face Stabilization Wedge
  - Pocket Excavation
  - Reduction of Round Length
  - Face Bolts
  - Additional Shotcrete and/or additional WWF
- Ground Improvement Measures
  - Gravity and Vacuum Dewatering
  - Permeation Grouting, Fracture Grouting, Jet Grouting
- Annular Support
  - Additional Shotcrete
  - Soil Nails
  - Temporary Invert

The use of toolbox items is in general a joint decision from the contractor's and the owner's representatives but in the end the contractor's experienced SEM personnel decide about quantity and exact location.

JET GROUT AND DEWATERING WELL DESIGN/INSTALLATION

The variable ground conditions with the potential for saturated and unstable sand lenses led to the consideration of large scale ground improvement measures prior to Sequential Excavation below ground. A series of deep dewatering wells and deep vertical jet grouting was specified in the initial contract and subsequently modified to suit ground conditions identified early on through the instrument and probe drilling program described above. Details of the layout, design and installation of these ground stabilization programs are described in another paper in these proceedings; however the following is an overview of the two programs.

The initial Contract Drawings identified a total of 26 deep wells centered principally around the Main Shaft, Concourse Cross Adit and the SBPT, with an additional 12 wells along the NBPT to be installed if ground conditions warranted based on the probing program. Moretrench American Corp. was subcontracted for the vacuum-assisted dewatering well system setup and installation, and Geotech Exploration was subcontracted for well drilling and casing installation.

Work began south of the Main Shaft and along the SBPT with a mud-rotary drill rig using polymer drilling fluid to support 10-inch (25 mm) holes drilled down through the large sand layer into silty clays 130 to 150 ft (40 to 45 m) below ground surface. As well drilling and exploratory probe drilling progressed the locations of the "optional" wells were redefined and several additional wells were added by Change Order. The original "optional" wells along the NBPT west of the CCA were shifted to the eastern portion of the NBPT and additional wells were added along the SBPT, and around the Ancillary Shaft and new ETVA location. In addition the well drilling method was revised from polymer mud-rotary to air-rotary, and several of the well casings were grout-plugged at the bottom instead of sealed with bentonite chips to prevent possible seepage through fissures into and above the SEM excavation crowns. In total 39 of 42 (revised quantity) wells were drilled, including several re-drilled due to utility, tieback and site logistic conflicts.

Several of the wells remain permanently off-line due to dryness, even with the vacuum system, and others cycle quite slowly due to reduced or barely measurable flow volumes. However, typically 28 to 34 wells have been operational from late Fall 2005 to present. The total system discharge has been reduced from a maximum of roughly 50 gpm upon initial system startup (Nov. 2005) to about 10 to 15 gpm in mid-Winter 2007. An approximate "steady-state" rate of about 30 gpm was established after the first month of operation and continued for roughly 12 months before flow rates diminished in late 2006 to the current rate. Drawdown of the groundwater levels in the predominant sand layer above BH Station has been on the order of 20 to 40 feet (6 to 12 m) depending on piezometer location and local fluctuations. This drawdown and the relatively low system flow volume, and the fact that very little groundwater from layers above the tunnel crowns has been seen in SEM tunnel headings is evidence of both a necessary and an effectively installed system that has that has helped station construction to date.

Jet grouting was depicted on the Contract Drawings for the Main Shaft break-out for the CCA's, along the WLVA break-out extending westward from the Main Shaft and for the East Damper Chamber adjacent to the Ancillary Shaft. Jet grouting was intended to "pre-stabilize potentially unstable water-bearing sand layers expected in the crown of..." these various tunnels. [GBR 10.3.3]. In addition to the predefined zones, additional areas of jet grout were to be identified based on the exploratory drilling program. The Joint Venture of Condon—Johnson/Soletanche was subcontracted to perform the jet grouting at BH Station, in addition to the separately bid yet concurrent slurry wall work.

As described above and early on in the exploratory drilling program, a significant sand lens was identified in the NBPT East of the CCA, with a plan location just north of the main work site Noise Wall limits. This sand lens running 150 ft (45m) in length was projected to span the entire width of the Platform Tunnel from just above the crown down to approximately spring-line. Limits for this "new" zone were defined with a starting tunnel stationing and bottom elevation, and extended 4-ft radially from spring-line through and above the crown theoretical excavation section. Since this "new" zone ran along a Platform Tunnel as opposed to the central shaft-connected tunnels, it was deemed more urgent for completion in light of the need to pass the TBM through the Station platform areas. Therefore the initial focus and work sequence for jet grouting was shifted from the predefined zones to the NBPT.

Initial jet grouting work began with a test program designed to demonstrate the effectiveness of both the drilling equipment at depth and soil-column displacement diameters compared to a design parameters. Following a program of 4 shallow (25 ft [8m]) and 2 deep (125 ft [38.5m]) holes, production work began on the NBPT working from west to east. A 4.75-ft (1.45m) triangular spacing pattern was laid out for 6-ft (1.8m) diameter columns to overlap. An air-rotary drill rig was used to pre-drill through the 160 ft (48m) overburden to the top of the grouting zone and open holes were then stabilized with a light bentonite and cement mix. The predrilled holes were then surveyed for alignment (target was 2% or less from vertical) and projected for adjacent column overlap. Once a sufficient number of predrilled holes were completed (typically 15–20 holes in advance of jetting), the remainder of each individual drill hole was finished by the jet grout drill rig.

Upon reaching final drill hole depth, jetting proceeded from the bottom up with typical pull speeds of 4 to 6 inches per minute (10–15 cm/min). Displaced soil cuttings and discharged spoils were pushed out of the top of the hole, and checked for compressive strength. Additionally core-hole samples from completed columns and interstices were extracted for strength testing, and constant head permeability tests on the cored columns were performed to confirm the effective of the finished product. The target compressive strength in-situ was 400 psi (2.8 MPa), however most cores were in

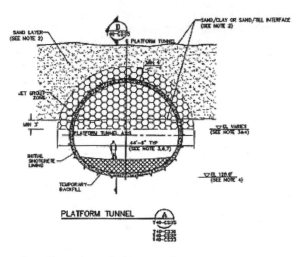

Figure 7. Cross section of jet grout at platform tunnel

the 700 to 1,500 psi (5–10 MPa) range, with a few as high as 4,000 psi (28 MPa). Permeability was consistently below the 3×10^{-6} cm/s limit. Core samples were required for every 200 CY (150 cu-m) of jet grout in place. The total neat line volume of jet grout completed was just over 5,800 CY (4,250 cu-m), distributed over 565 drill holes.

Production jet grouting began in the NBPT progressing from West to East starting in December 2004, and was completed in late July 2005 with a brief 5-week interlude to complete columns around the Main Shaft and WLVA break-outs. Jet grout work then progressed to the SBPT and was completed in early October 2005. After a bit of cleanup and maintenance, the final scope of jet grouting around the Ancillary Shaft for the new ETVA break-outs was started in November and completed in December 2005. In all the jet grouting work was completed in just over 13 months. However there are a few more details to the story than simply the start and finish dates.

As NBPT jet grouting progressed in the winter and spring of 2005, the exploratory boring program began to uncover other significant sand lenses adjacent to the existing zone at the East Damper Chamber and over the SBPT crown east of the CCA. In addition, the WLVA zone was reduced to just the break-out zone at the Main Shaft due to better than expected soil boring results. The new SBPT zone and the deletion of the WLVA zone became topics of Change Order discussions. However the large extent of the potential zone adjacent to and east of the Ancillary Shaft eventually forced the entire station to be shifted to the west. See Station Shift section for more details. The final configuration of the Station and its associated jet grout zones resulted in a total neat line volume increase of 10%. However the increase was not a straight quantity adjustment. The added platform sections were markedly different in profile from the full height WLVA and East Damper Chamber sections, and the timing and identification of the zones as work proceeded led to an adjustment of the contract bid item prices in two separate Change Orders. See Figures 3 and 4 for the original jet grout plan and the final jet grout plan after the Station Shift.

Contractual adjustments aside, the jet grouting program provided a very competent soil stabilization mass, and has been mined with little incident and virtually no water influx to date (Figure 7).

SITE LAYOUT/EQUIPMENT/LOGISTICS

The Contractor has been coping with a large surface plant and equipment spread on a relatively small site. This kind of work requires very comprehensive and dynamic equipment to cover the various activities required in the work cycle. This 5 acre site is bound by several streets in this urban setting creating an interesting challenge when first laying out the site. The physical layout of the shafts and headhouse basements along with the defined North and South entrances at South Lander and McClellan Avenue respectively defined largely the site configuration. First a 30' wide road was installed through the site with asphalt pavement to ease the burden of track-out from trucks. Trucks entering from the south must exit to the north after being loaded or unloaded and after having its wheels washed at a wheel-wash station.

A crane pad along the east side of the shaft spreads the loads out from a 200 MT Kobelco crawler crane and puts the crane within fixed reach of a muck bin for load-out by a Caterpillar 320 excavator. This crane is the principle means of support for the Main Shaft. A 160 MT Kobelco crawler crane positioned on a pad constructed on the south side of the East Headhouse puts the 12 cy lift boxes within reach of the same muck bin. However, to maintain access through the site, a bridge was built over the Southwest corner of the East Headhouse.

A volumetric CemTech shotcrete batching plant on the south side of the Main Shaft provides the heading with its dry steel fiber reinforced flashcrete needs. Meyco GM060 dry pots with pre-dampeners on the surface deliver dry-mix to the heading. A Terex weigh batch plant positioned on the north side of the main shaft provides the heading through a drop pipe with wet steel fiber reinforced production shotcrete. A 7 c.y. Maxon MK-7 at the base of the drop pipe in the Concourse Cross Adit inventories the shotcrete for delivery to one of two shotcrete pumps positioned below it.

Provisions are made on site for a shop, a compressor pad holding three 625 c.f.m. Ingersol-Rand rotary screw compressors with after-coolers and filters, a water treatment area, sand and 3/8 aggregate bins, a containerized system of heated and circulated accelerator tanks, a vacuum pump station for the dewatering system, an on site coring and Q.C. laboratory, a lay down area for lattice girders, and a lay down area for a large array of other materials and supplies.

Other key pieces of equipment required underground on this job include two Leibherr 932 NATM Excavators, a Leibherr 900 NATM Excavator, a Komatsu truck mounted GD 3700 Air Track for probing and spiling, a Cat 953 Trackloader, a Cat 963 Trackloader, and two 4-wheel drive, rubber-tired extend-a-boom style man-lifts.

MAIN SHAFT AND ANCILLARY SHAFT
CONSTRUCTION INCLUDING HEADHOUSE

Observations from the Test Shaft during the final design stage resulted in redesigning both the main shaft and ancillary shaft lining from SEM to using slurry walls.

The main shaft diaphragm is approximately 52 ft in diameter and is 182 ft deep. This work was performed by Soletanche using a hydro-fraise machine mounted on a Leibherr crawler crane cutting a 3 ft-4 inch thick panel. The ancillary shaft diaphragm is approximately 30 ft in diameter and has a depth of 167 ft. This work too was done with the same hydro-fraise used for the main shaft except the cutting wheels were changed to cut a thinner wall at 3 ft-2 inch.

The bentonite slurry transported the cuttings to a separation plant complete with screens, cyclones, and centrifuge for return to the excavation. A Cat 320 was used to muck a pit constructed from the basement of one of the houses demolished to clear the site. Rebar cages were tied on site with block-outs for invert slab niches and with pipe sleeves for instrumentation. The cages with their attachments were lowered into the bentonite and suspended from a structure on the guide walls. Concrete trucks backed up to hoppers setting on tremie pipes to deliver approximately 4,600 cy of concrete to the main shaft and 2,300 cy to the ancillary shaft. Since the upper 60 ft of circular main shaft slurry wall and of the circular ancillary shaft slurry wall was to be demolished while the interior excavation was being done, a lean mix was used in the upper reaches of the slurry wall panels.

The headhouse basement diaphragm wall was 2 ft-8 inch thick and 62 ft deep. This work was performed by Soletanche using a conventional cable grab mounted on a Leibherr crawler crane. The grab deposited the material directly into trucks queued on site. The work was orchestrated so that some of the headhouse wall panels were constructed while some of the main shaft panels were constructed.

The station is mined from two shafts; the main shaft is 52 ft in diameter (46 ft finished diameter) and the ancillary shaft is 30 ft in diameter (26 ft finished diameter). Acting as both entrance/exit and ventilation structures, the main shaft and the ancillary shaft is 182 and 167 ft deep respectively.

The main shaft was excavated using a Hitachi 330 Excavator with breaker and Cat 320 for excavating and loading muck skips. Twelve cubic yard muck skips were lifted to the surface and tipped into a muck bin at the collar using a Kobelco 2000 (200 MT) hydraulic lattice boom crawler crane. The main shaft diaphragm walls extending through the headhouse were demolished as the excavation advanced. This was all carefully choreographed with the installation of several rows of multi-strand tie-backs which extended on average approximately 70 ft into the surrounding ground.

After the headhouse excavation was complete, a cap beam approximately 6 ft tall was formed and poured, tying in dowels protruding from the top of the slurry wall panels. Soon after, the interior excavation of the circular shaft continued down to the bottom using the Cat 320 excavator and the same muck skips described above.

Upon reaching the bottom, the sub-grade was excavated to a "dished" shape, rebar dowels were installed, and "submarine" style concrete invert was poured as a provisional shaft bottom. Upon reaching the design strength, the invert was backfilled with spoils to achieve a working platform for commencement of the break-in to the SEM tunnels.

PRE-SUPPORT BARREL VAULT CANOPY

Two areas were identified on the original contract drawings that required the systemic installation of a lost casing barrel vault (aka pipe arch) canopy. Approximately 7,986 lf of 4" diameter lost casing pipe was called for over the entire 78' length of the both the NB and SB Concourse Cross Adits (CCA's). Also, 6,360 lf of 4" diameter lost casing pipe was called for over the first 40 feet and 60 feet of the NB West and NB East Platform Tunnels respectively.

Based upon results of additional geotechnical exploratory holes described elsewhere in the paper, only a single row of pipes was installed over the SB CCA and pipes over the NB CCA originally specified to be only 30 to 34 feet long were extended the full length of the CCA.

A contract was awarded to NW Cascade of Puyallup, Washington to furnish, drill/install, and grout with micro fine cement the barrel vault over the CCA. Later, a change order was given to do the same over the NB Platform Tunnels.

In the main shaft, the subcontractor drilled with a Klem 806 track drill with duplex head. In lieu of using lost crown bit and impact shoe as expected, "J" teeth were welded into the starter casing and various center steel bits were used both out ahead of the lead casing section and sometimes behind when water was encountered. Pipes where drilled and installed in a primary-secondary pattern to ensure that a pipe was not drilled next to a recently grouted pipe that didn't have at least 24 hours of initial set.

A Interock track drill was used to drill and install the canopy pipes over the NB Platform Tunnels. A smaller drill was required due to the limited space in the crown and the length of the drill slide. Obayashi backfilled the NB Concourse Cross Adit with spoils from the other tunnels to create a work platform for the driller. This bench was taken down in two steps to accommodate the reach of the drill.

After the completion of the hole, the center of the pipe was flushed with water and a Maxibore horizontal inclinometer was used to verify the required drilling tolerance of ½%. The drilling accuracy was excellent with no significant deviations reported.

The lost casings arrived on site with 14mm diameter holes drilled in an alternating fashion every foot of the length of the casing. In it was pressed a simple check valve to ensure that grout would not flow back along the casing and enter into the pipe behind the two-stage packers set up for 5 foot intervals.

A weak "frackable" cement/bentonite grout was injected to fill the annulus between the outside of the casing and the ground to ensure that the stage grouting would catch each length of ground along the pipe and not simply grout at the weakest point. This step was eliminated in the second campaign over the Platform Tunnel. The "J" teeth were welded in with a tighter gage and the ground seemed to squeeze down on the pipe eliminating the annular space.

Microfine cement was injected using a double packer on 5' foot centers starting from the end of the pipe and ending at the collar. A mixture with 0.8 w/c ratio (weight) yielded at least 500 psi cube strengths within 24 hours. A dispersant was also used in the mixture.

This "home grown" variation of the drilling and installation of a lost casing barrel vault canopy was developed jointly between Obayashi and NW Cascade (the driller) and the barrel vault installation has been deemed largely successful on this project.

TEAM ORGANIZATION AND DECISION MAKING INCLUDING DAILY SEM MEETING AND RESS

Minimum experience levels were specified in the Contract for key personnel responsible for the SEM activities. Under the oversight of the Obayashi Tunnel Manager, these key personnel include the SEM Manager, SEM Project Engineer and SEM Superintendents. Obayashi entered into an agreement with Beton and Monierbau USA, Inc. to provide key SEM staff. The SEM Superintendents work shifts to facilitate immediate decision making at the face during the six day 24 hour weekly schedule, and are supported at the headings by Walkers and Shift Engineers. Generally two crews are working three 8 hour shifts. ST recognized the inherent risks involved and agreement was reached during the design stage to have the Designer represented on site during the implementation of the SEM design. Hatch Mott MacDonald/Jacobs Joint Venture (HMMJ) was responsible for the detailed design of all tunnels and portals, shafts and mined station tunnels, including the final lining and waterproofing system. Dr Sauer Corporation (DSC) assisted with the SEM design and waterproofing design for the Station as a sub-consultant to HMMJ. HMMJ and DSC provide experienced SEM engineers and inspectors to support the Construction Management team (Parsons Brinkerhoff) and oversee the SEM activities. S & W is represented on site providing oversight on geotechnical activities.

Each morning joint inspections at the headings are followed by Daily SEM Meetings held at the BH site office. ST, OC and HMMJ/DSC are always represented at these meetings which are conducted in an open constructive partnering manner. At these meetings, the status of construction is discussed followed by the planned activities for the next 24 hours. Any necessary changes to the Construction Work Plans are agreed. Face maps are presented and discussed as well as the results of probing. A review of the latest instrumentation readings is included to confirm stability of the headings. On a weekly basis, shotcrete strength results are presented and discussed. RESS sheets (Required Excavation and Support Sheets) confirm the required support and pre-support for each tunnel section and are used to assist communications.

MID-STREAM CHANGES AND SUBSTITUTIONS

One of the most interesting aspects of this technique of tunneling is the flexibility it gives to the builders. Being that sometimes the best way to tackle a problem is to simply avoid it; much effort was spent reconfiguring the Station elements when possible as witnessed by the wholesale shift of the Station. It is described elsewhere in this paper how the Station was shifted away from a section of ground that otherwise would have required more than 8,800 c.y. of additional jet grout.

Other changes and substitutions include the following:

- As part of the Station Shift, the Contractor requested the Transverse Ventilation Adits be down-sized to match the cross section and lattice girders required of the Longitudinal Ventilation Adits for the sake of simplicity and this change was incorporated in the Station Shift redesign. Also, the West Damper Chamber was reduced in section for economy.

- Obayashi paid for the redesign and for pre-construction performance flexural beam tests in order to substitute steel fiber for double mats of welded wire fabric. This was initially undertaken in the Platform Tunnels and later extended to other tunnels in the Station complex.

- A use of a single row of barrel vault pipes over the NB Platform Tunnel instead of the two rows anticipated and a renegotiation of the unit price was successfully accomplished based upon the reduced bid item quantity.

- Obayashi promoted the excavation and support of the NE Platform Tunnel situated largely in jet grout with a single top heading, bench and invert approach instead of the time consuming dual sidewall drift method as had been specified. It was deemed the benefits did not outweigh the risks and the subject was dropped.

- Obayashi promoted as a substitution a single top heading, bench, and invert approach to the revised "Station Shift" West Junction Chamber from the dual sidewall approach to accommodate the excavators and equipment already existent on site and for the sake of simplicity. This redesign was accepted and executed and the work remains to be done at the time of this writing.

- As part of the "Station Shift," the alignment of the West Passenger Cross Adit (WPCA) was revised to include a curve similar in shape to an "Omega" in order to maintain the original cover between the West Passenger Cross Adit and the West Longitudinal Ventilation Adit (WLVA). Later in the job as it became obvious that the WPCA would be completed before the WLVA. Consequently, Obayashi requested the Designers to reduce cover to nearly zero, allowing a shorter and simpler WPCA alignment. This change has been executed and the WPCA construction is ongoing at the time of this writing.

Table 1. Excavation sequence cross sections

Excavation Method	Twin Sidewall Drift Excavation	Single Sidewall Drift Excavation	Top Heading/ Bench/Invert Excavation
Location & Area	Concourse Cross Adit (1,500sf)	Platform Tunnel (1,000sf)	Ventilation Adit (400sf) Cross Passage (130sf)
Excavation Sequence			
Excavation Support	Shotcrete t=14" Girder @ 3'-3" WWF W12*W12, 6*6	Shotcrete t=14" Girder @ 4' WWF W12*W12, 6*6	Shotcrete t=10" Girder @ 4' WWF W8*W8, 6*6

CONCOURSE CROSS ADIT, PLATFORM TUNNEL, TRANSVERSE ADIT, LONGITUDINAL VENT ADIT, AND PASSENGER CROSS ADIT EXCAVATION SEQUENCES AND CONSTRUCTION

There are various sizes of the tunnels excavated for this Station. The biggest tunnel has a cross sectional area of approximately 1,500 s.f. and the smallest tunnel has a cross sectional area of 130 s.f. The excavation sequence for each tunnel is specified in the contract and the sequence varies with the size of the tunnel. These excavation sequence methods are aimed at minimizing the load distribution by excavating piece by piece. At the same time, it is important to make the ring closure as soon as possible to minimize relaxation of the ground and to minimize the tunnel deformation. Each excavation sequence is described in Table 1.

Twin Sidewall Drift Excavation for Concourse Cross Adit (1,500SF)

A twin Sidewall drift excavation method was applied to the Concourse Cross Adit (CCA). The CCA section is divided into right and left side drift and center drift. In addition, each of the three drifts is further divided into top heading, bench, and invert. In total the CCA section is comprised of nine (9) segments.

First the twin left and right side drifts were excavated followed by the excavation of the center drift. The twin side drift behaves like a pilot tunnel before opening fully the large section. These side drift sections support the load of the center drift top heading the same way as does the anchorage of the classic arch bridge (Figure 8).

One of the first major modifications brought by Obayashi was to change the center drift excavation sequence from Top Heading, Bench, and Invert (as specified) to full top heading all the way to the end, followed by a bench/invert sequence. The reason this was proposed is the top heading in the specified sequence was beyond the practical reach of most excavators. To execute the specified sequence, Obayashi would have to place and then remove for every section a temporary bench and this would not be practical either.

However, this modification has the disadvantage of delaying the ring closure and changes the stress contour for the initial lining. The designer solved this issue by

Figure 8.   Full T/H excavation at center drift   Figure 9.   Multiple face mining platform tunnel

revising the design to include a double layer on WWF on temporary wall and make more rigid the center wall. The result was Obayashi did not need to work with such a high face and this modification improved not only for the constructability and production but also the safety aspect.

Single Sidewall Drift Excavation for Platform Tunnel (1,000SF)

The single sidewall drift excavation method was specified for the Platform Tunnel (PT). The PT section is divided into the 1st drift (left) and the 2nd drift (right) and each drift is also further divided into top heading/bench/invert. In total the PT is comprised of 6 segments per round.

The first drift is excavated as a pilot tunnel and then the second drift follows behind the first drift with a minimum specified offset. This offset rule gives time to make the ring closure and gives the shotcrete time to develop strength in the first drift. Obayashi investigated several ways to try to improve the cycle of the PT in the interest of schedule (it is the longest drive in the Station) and several minor changes were implemented. The single most useful way is of course to increase the round length when possible given the results of probings, face maps, and feedback from the instrumentation in the shell nearby.

In the Design given to Obayashi, the lattice girder serves only as a template and is not taken into account in the structural analysis. Therefore, the span length could be increased if the ground permitted and if this could be done safely. In the end, Obayashi in places adjusted to span length from 4'-0" to 4'-6" or 5'-0".

Given the space constraints and practical limits of equipment resources, the way the excavation crews were organized was analyzed and adjusted from time to time. Since at times there were two heading working both side drifts, there were four available faces (Figure 9). Obayashi found it best in this case to establish excavation crews, girder crews, and shotcrete crews instead of establishing these capabilities in each individual heading team. Obayashi found this to be a more efficient use of the resources and improved the overall progress viewed globally.

Top Heading, Bench, Invert Excavation for Transverse Vent Adit (400SF)

A single drift top heading/bench/invert excavation method was applied to other small tunnels such as Transverse Vent Adit (TVA). The TVA is divided into top heading/bench/invert. This excavation method is the simplest and most productive.

Figure 10.   Double bench/invert excavation

At times the ground conditions afforded Obayashi the ability to excavate the bench and invert together, saving a few steps. Also, at times the ground conditions permitted the excavation of multiple benches instead of just one (Figure 10).

DEWATERING AND INSTRUMENTATION MONITORING AND SETTLEMENT

Geotechnical instruments specified for ground condition monitoring outside the excavated tunnels included extensometers, inclinometers, open standpipe and vibrating wire piezometers, and optical survey points and monuments. Each instrument is read periodically by Shannon & Wilson technicians, with the exception of the survey points checked by CH2M survey crews. Vertical and lateral displacements are tracked from the tunnel invert up through the crown to street level prior to excavation, then just above tunnel crown up to ground surface after SEM mining has been completed for a given tunnel section. Data from the instrument readings are reviewed periodically at the Daily SEM meeting and to-date have shown a maximum of 1.6 inches (4 cm) of vertical settlement over the station tunnels roughly 150 feet (46m) below ground. Street level surface settlement has only reached a maximum of 0.6 inches (1.5 cm), most of which appears to be attributed to recent water line replacement work in Beacon Avenue. In general surface settlement contours have followed the tunnel excavation sequence albeit with very small measurements, and not surprisingly there has been little to no measurable surface settlement over the jet grouted zones. All of this data compares quite favorably to the expected 4 inches (10 cm) anticipated upon completion of tunneling around the Main and Ancillary Shafts (Figure 11).

Similarly dewatering wells have proved quite effective as groundwater levels in the predominant sand layers above the tunnel crowns have dropped between 20 and 40 feet (6 to 12 m). Groundwater levels are measured at 17 individual locations on site, typically with open standpipe bottom casings and vibrating wire piezometers at intermediate elevations in the upper sand layers. The vacuum-assisted dewatering well system has been on-line for over one year and has dropped from an initial 50 gpm, through a steady state of 30 gpm down to 10 to 15 gpm at the time of writing (Figure 12).

CONCLUSION

This project is very complex and has in it just about every chapter in the book.

The SEM approach to tunneling offers the flexibility to steer away from possible trouble and the various and wide ranging techniques now available for ground pretreatment,

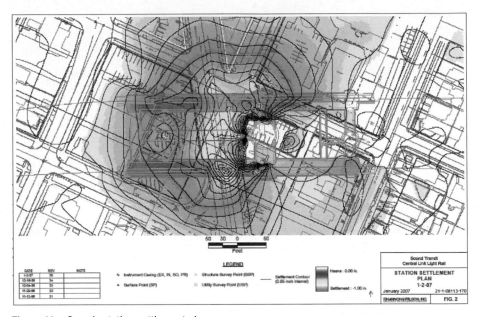

Figure 11.   Sample station settlement plan

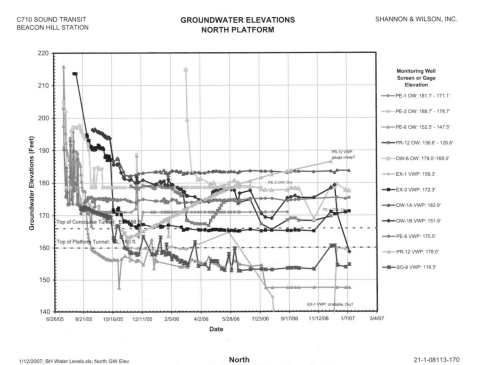

Figure 12.   Sample groundwater elevation for North Platform tunnel

dewatering, pre-support, and ground treatment at the heading allow the team to get through just about anything safely. However, good discipline and having qualified individuals in a partnering environment on site is important for success. The monitoring programs give instant feedback to the team to make necessary adjustments in the field to avoid a problem and likewise allow the team to take advantage of favorable conditions.

This segment of the project brings us a good example of how a group of respectful and dedicated professionals from the Client, the Contractor, the Geotechnical Investigators, the Designers, and the Inspection team can meet the challenges and difficulties sometimes thrown at us as Tunnelers by Mother Nature. The large changes such as the "Station Shift" and others implemented successfully mid-way through the construction of the project are unprecedented in our opinion.

ACKNOWLEDGMENTS

The authors of this paper would like to make recognition of all those individuals involved in the success of a project like this and that are too many to identify individually. Special recognition is given to Red Robinson of Shannon and Wilson who brought to the table his deep knowledge and understanding of tunneling conditions in the Puget Sound region. Also, the Contractor would like to give special thanks to the members of Beton-und-Monierbau USA who aided with their expertise and know-how with special recognition to Christian Neumann who gave the team excellent support. Additional recognition is given to Franz Langer of Dr. Sauer Corporation for his assistance with the preparation of sections of this paper.

EVALUATION OF LARGE TUNNELS IN POOR GROUND— ALTERNATIVE TUNNEL CONCEPTS FOR THE TRANSBAY DOWNTOWN RAIL EXTENSION PROJECT

Steve Klein

Jacobs Associates

David Hopkins

Jacobs Associates

Bradford Townsend

Hatch Mott McDonald

Derek Penrice

Hatch Mott McDonald

Ed Sum

Transbay Joint Powers Authority

INTRODUCTION

The Transbay-Caltrain Downtown Extension (DTX) Project involves the construction of an approximately 1.5-mile rail line that will extend Caltrain commuter service and the future California High Speed Rail system into downtown San Francisco. The underground rail extension will be constructed by mined tunnel and cut-and-cover methods. Tunneling challenges include difficult ground conditions, low rock cover, the presence of historic buildings along the alignment, and the large tunnel span, ranging from approximately 50 to 65 feet. Three mined tunnel construction methods were evaluated: the Stacked Drift Method, the New Austrian Tunneling Method/Sequential Excavation Method (NATM/SEM), and Tunnel Boring Machines (TBM). This paper describes the challenges presented within the mined tunnel section, and summarizes the evaluation of tunneling methods that led to the selection of NATM/SEM.

PROJECT OVERVIEW

The existing Transbay Terminal opened in downtown San Francisco in 1939. This facility is antiquated, and does not meet current building seismic safety standards or the transportation needs of the Bay Area. For these reasons, San Francisco voters approved Proposition H in 1999, which proposed to construct a new or rebuilt regional transit center at the site of the existing terminal and extend Caltrain commuter rail service to the new transit center.

The new transit center is the hub of the Transbay Transit Center (TTC) Program, which will greatly enhance regional transit service by providing connectivity for six Bay Area transit providers at one facility. The program will be the largest inter-modal center west of New York City. The TTC Program is comprised of two major projects, the Transit

Figure 1.   General plan of DTX alignment

Center and associated infrastructure, and the DTX Project (Figure 1). The DTX Project includes:

- A 1.5-mile underground rail extension along Townsend and 2nd Streets to the new TTC,
- A new underground station at the intersection of 4th and Townsend Streets, and
- Improvements to the existing rail yard facility and existing Caltrain terminal at 4th and King Streets.

The DTX Project comprises approximately 3,200 feet of mined tunnel, 4,700 feet of cut-and-cover tunnel, and 2,300 feet of open/retained cut construction. The excavated size of the mined tunnel—a single bore tunnel that could be 65 feet wide and 43 feet high—will be among the largest constructed in the United States.

PLANNING AND DESIGN BACKGROUND

Planning and conceptual design efforts for the project have focused on the development of a project configuration which meets the operational and functional requirements of the rail operators; Caltrain and the California High Speed Rail Authority (CHSRA). The mined tunnel concepts developed thus far include several potential configurations. It is recognized that the selected configuration will need to be refined as the design process progresses.

The configuration and cross-section of the mined tunnel are being determined by on-going parallel studies involving rail operation simulations and rail engineering evaluations. The outcomes of these studies have a significant impact on the tunnel construction methodology and the cost of the project. Some of the goals of these studies are to determine the:

- Rolling stock dynamic envelope and required tunnel clearances,
- Number of tracks required to provide sufficient operational capacity (e.g., two or three tracks), and
- Locations of crossovers and other special track work.

Tunnel Clearances

While the development of tunnel clearances would appear to be a straightforward matter, neither Caltrain nor CHSRA have identified the rolling stock they intend to employ for future operations. In the interim, a composite vehicle envelope has been developed based upon current technology which accommodates a range of vehicles that could potentially be used. Preliminary tunnel clearances have been developed using this vehicle envelope with appropriate provisions to generally satisfy California Public Utilities Commission (CPUC) General Order 26D. However, the CPUC requirements mainly address freight lines and the Transbay Joint Powers Authority (TJPA) Program does not intend to bring freight rail traffic into the new TTC. The current clearance diagrams developed to meet these requirements call for a tunnel 21.5 feet high (above top of the rail) and 40 and 55 feet wide for the two- and three-track options. Passenger dedicated lines (PDLs), like this proposed rail extension, would typically have smaller clearances. In due course, the project will seek a dispensation (variance) from CPUC to reduce clearances to those required for PDLs.

Two- and Three-Track Options

Construction of either a two- or three-track tunnel has obvious cost implications. This choice will also significantly impact the tunnel configuration and selection of the tunneling method. The DTX alignment is largely within the City-owned street right of way. Due to this width constraint a three-track tunnel can only be accomplished using a large single bore. In contrast, the two-track tunnel can be constructed in either a single bore or twin bore configuration, however the practical length of twin bore tunnel that can be constructed is limited to about 2,250 feet due to the necessity for crossovers between adjacent tracks to satisfy rail operations requirements.

GEOLOGIC CONDITIONS

The geologic conditions along the DTX alignment are known to be highly variable and complex. To develop a thorough understanding of the subsurface conditions an extensive geotechnical investigation program has been conducted over the past two years (Arup, 2006). To date, 25 borings have been completed along the mined tunnel section, totaling approximately 2,900 linear feet of borehole. A variety of in situ field and laboratory tests were completed to characterize geologic conditions.

The mined tunnel section of the DTX traverses an area of high bedrock associated with Rincon Hill, a prominent ridge on the southern side of the financial district in downtown San Francisco. Like most of the San Francisco peninsula, bedrock in this area is composed of Jurassic to Cretaceous sedimentary rocks of the Franciscan Formation. These rocks are generally highly fractured and sheared as a result of tectonic movements at the boundary between the Pacific and North American Plates. In the mined tunnel section, the Franciscan bedrock is generally within 5 to 10 feet of the ground surface, and groundwater levels are about 10 to 20 feet below the ground surface.

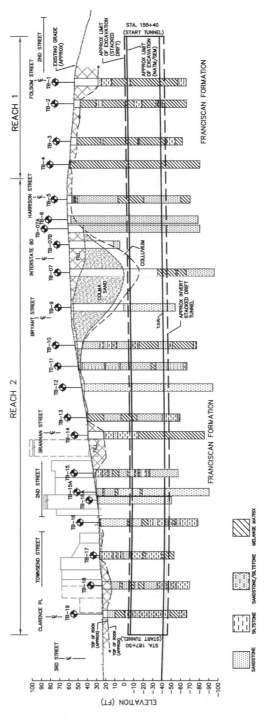

Figure 2. Summary geologic profile for DTX mined tunnel section

Soil deposits along the DTX alignment include artificial fill, Bay Mud and other marine deposits, Colma Formation, and colluvium. These deposits will only be encountered in one area of the tunnel, on 2nd Street between about Bryant Street and Interstate 80 (see Figure 2). In this area a paleovalley of colluvium and Colma Formation extends into the tunnel horizon. Ground improvement (such as jet grouting or similar techniques) may be required to treat these soils prior to tunnel excavation.

Franciscan Formation

The Franciscan Formation is a highly deformed rock mass that includes weak, sheared, fine-grained sediments and stronger rock blocks of various lithologies. The resulting rock mass is extremely variable and it possesses a chaotic structure of disconnected rock blocks surrounded by a weaker matrix. Locally these rocks are referred to as a "mélange." The mélange matrix is composed of stiff clay, sheared shale, and disaggregated rock fragments. Blocks within the mélange are predominantly sandstone and siltstone, although blocks of stronger rocks like chert and greenstone are sometimes present. For the purposes of the conceptual tunnel evaluation, the alignment was split into two distinct reaches, as indicated on Figure 2.

Reach 1. Borings in the northern 800 feet of the mined tunnel, referred to as Reach 1, found that over 85% of the rock mass in this area is composed of mélange matrix. The clay content of some core runs was as high as 90%, and only a small number of discrete blocks of sandstone and siltstone were observed. The mélange matrix in this reach is extremely weak and extensively sheared, and in places is pulverized to crushed rock fragments.

The strength of the clayey mélange matrix was evaluated based on UU triaxial tests. These tests indicate an average compressive strength of about 100 pounds per square inch (psi). The unconfined compressive strength (UCS) of sheared shale samples that could be tested are higher, ranging from about 160 to 400 psi. However, much of the mélange matrix was so pulverized that it could not be tested in the laboratory.

Rock mass quality was assessed from the Rock Quality Designation (RQD) values indicated on the core logs and estimates of the Rock Mass Rating (RMR) and Tunneling Quality Index (Q) determined following the procedures of Bieniawski (1989) and Barton et al. (1974), respectively. For this reach, almost 70% of the core runs had a RQD of zero, and none of the values were above 25, which is the upper bound value for "very poor" quality rock (Deere, 1989). RMR and Q values also indicated low rock quality for this reach with overall ratings of "very poor" for the RMR system and "exceptionally poor" for the Q system.

Reach 2. Borings drilled in the southern 2,400 feet of the mined tunnel section, referred to as Reach 2, found a rock mass that was much different than that encountered in the Reach 1 borings. The rock mass encountered in Reach 2 was mainly strong sandstone and siltstone blocks (about 75%), and contained much less of the weak mélange matrix. The rock in Reach 2 is still highly fractured, but in general, the sandstone and siltstone blocks are much stronger than the mélange matrix.

The sandstone in this reach varies considerably in strength, from weak to strong. High variability in the UCS test results appears to be a result of defects in the rock samples caused by shear failures along preexisting, but healed, fractures, rather than through intact rock. UCS values for the sandstone range from about 1,300 to 19,000 psi, with an average of 7,700 psi, although one test result from a previous investigation program recorded an UCS of 27,200 psi. These results imply that the fresh, unfractured sandstone can be very strong, but most of the test samples contained healed fractures or other defects that tended to reduce their strength. The siltstone is considerably weaker than the sandstone, with strengths ranging from approximately 200 to 4,000 psi.

Figure 3.   Existing buildings on Townsend Street above DTX alignment

RQD values in this reach are typically higher than those in Reach 1. Approximately 45% of the rock core in Reach 2 is classified as "poor" or "fair" (RQD of 25 to 75) and only about 55% as "very poor" (RQD less than 25). Less fracturing and higher rock strength in Reach 2 resulted in higher RMR and Q ratings, but the abundance of discontinuity fillings with sheared shale and clay tended to lower the ratings. Ratings in Reach 2 are generally "fair" to "poor" for the RMR system and "extremely poor" to "very poor" for the Q system.

HISTORIC BUILDINGS

As the DTX alignment transitions from Townsend Street to 2nd Street, the mined tunnel passes beneath 11 existing buildings. Figure 3 is a photograph of some of these buildings, which range in height from one to six stories. The buildings were constructed in the early 1900s. All are part of the Rincon Point/South Beach Historic Warehouse-Industrial District, and may be eligible for the National Historic Register. The mined tunnel will pass beneath these buildings with low rock cover (relative to the tunnel span) that ranges from about 20 to 35 feet. To the north, along 2nd Street, the alignment does not pass directly under any buildings because it is located under the street. However, there are numerous existing buildings on both sides of the street adjacent to the tunnel. Most of them are low rise buildings two to seven stories high, but there are some newer high-rise buildings north of Harrison Street. Limiting ground movements and associated surface settlements to avoid damage to the adjacent and overlying buildings is a critical concern for the project.

MINED TUNNEL ALTERNATIVES EVALUATION

The goal of the mined tunnel alternatives evaluation was to identify the preferred method of constructing the mined tunnel in terms of cost and schedule, while seeking to mitigate the inherent risks posed by the difficult and variable geologic conditions and the physical requirements of the project.

This was accomplished by identifying conservative tunneling methods that are appropriate for the challenging ground conditions and have the ability to control ground movements, reducing the potential for damaging the existing buildings. The combination of challenging ground conditions, existing buildings above the alignment,

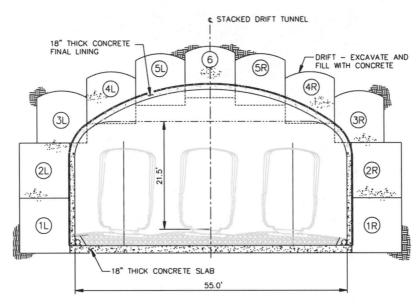

Figure 4.   Typical tunnel section for stacked drift tunnel alternative

low rock cover, and the large tunnel span required an approach that would allow the tunnel to be excavated and supported in stages, thereby limiting the unsupported span and maintaining the stability of the excavation. The Stacked Drift Method and the NATM/SEM satisfy these requirements. Tunnel Boring Machines (TBMs) were also considered, as this method is often a cost effective and an attractive approach for constructing tunnels in urban areas.

Both two- and three-track single-bore tunnels were evaluated for the Stacked Drift Method and NATM/SEM. For the NATM/SEM, a two-track twin-bore tunnel concept was also evaluated. TBMs were considered for a twin-bore tunnel configuration (one track each) only. The inability to develop a three-track tunnel configuration was a limitation of the TBM method. A three-track extension is the current base track configuration for the project.

Stacked Drift Method

The Stacked Drift Method was developed specifically for the construction of large tunnels in poor ground conditions. Basically, this method involves mining a series of interconnected concrete-filled drifts in a certain sequence around the tunnel perimeter. After all of the drifts are backfilled with concrete, a continuous and robust structural arch has been constructed over the core of the tunnel entirely pre-supporting the ground before any excavation of the central core. For the DTX Project this means the central core can be safely excavated without inducing significant ground movements, reducing the risk of surface settlement which could damage historic buildings. Figure 4 shows a typical cross section for the three-track Stacked Drift tunnel alternative. This approach has been successfully used in the U.S. for other large tunnels in difficult, poor-quality ground. One project was the Eisenhower Tunnel on Interstate 70 in Colorado (Hopper et al., 1972). Another was the Rio Piedras Station for the Tren Urbano Project in Puerto Rico, which was a 62-foot-wide by 52-foot-high excavation in soft ground directly beneath several historic structures (Romero et al., 2001).

For the DTX Project, small drifts of approximately 10 feet by 10 feet would be excavated by hand-mining operations using a small roadheader or excavator and a tunnel loader for spoil removal. Some blasting could be required in strong, less fractured sandstone. Given the anticipated difficult ground conditions, 3-foot rounds were assumed, with one steel set being installed in each advance cycle. Shotcrete is provided for ground support between the sets, and timber lagging could be used where it projects inside the final excavation lines and would be removed by subsequent excavation operations. In unstable ground, special precautions could be required, such as the use of spiling installed in advance of the excavation and/or breasting or shotcrete to support the face. Such measures would maintain excavation stability and minimize overbreak that could lead to surface settlement.

The drifts would be mined from the lowest drift to the top drift with drifts at the same elevations being mined simultaneously (see Figure 4). Excavation for each pair of drifts would be completed and the drifts backfilled with concrete before excavation of the next drifts in the sequence began. Constructing each subsequent drift above the previously concreted drifts minimizes potential surface settlement by establishing a stable foundation for the base of the arch. Once all the drifts have been excavated and concreted, the core inside the arch would be removed in several excavation stages and a concrete invert slab cast to serve as a strut at the base of the tunnel. This would limit potential convergence of the sidewalls of the tunnel.

After tunnel clean-up, a cast-in-place concrete final lining would be placed as a finish for the tunnel. Groundwater would be continuously drained to avoid the build-up of hydrostatic pressure on the final tunnel lining. Drainage would be accomplished by installing a drainage geotextile and a PVC waterproofing membrane along the crown, arch, and sidewalls of the tunnel that is connected to an invert drain system. Design of watertight tunnel is not cost-effective because of the flat tunnel invert and high groundwater levels.

Conceptual design analyses included several preliminary structural and geotechnical calculations. For example, soil-structure interaction analyses were conducted to check the stability of the drift arrangement and to evaluate the need for shear reinforcement between adjacent drifts. Settlement estimates were prepared indicating that only minor surface settlement is expected with this method—approximately ½ to ¾ inch.

NATM/SEM

The NATM/SEM is in use worldwide. The method has been used in the United States since the early 1980s, with the first applications being in the Pittsburgh and Washington D.C. areas. Since then, the NATM/SEM has been used on a variety of transit projects in Dallas, Boston, Seattle, and San Juan, Puerto Rico. The basic principle of NATM/SEM design is to allow small ground movements to occur around the tunnel in order to mobilize the strength of the ground. These limited movements significantly reduce the loads on the final lining. Rock bolts, lattice girders, shotcrete, and wire mesh are employed instead of heavy timber or steel supports to develop the strength of the ground without compromising excavation stability.

For large tunnels constructed using a NATM/SEM approach, the tunnel excavation is divided into a number of drifts and the tunnel is gradually enlarged and supported to maintain the stability of the tunnel and control ground movements. Drift sizes and the excavation sequence are based on anticipated ground behavior and construction logistics. The drift advance lengths are primarily controlled by the anticipated stand-up time of the ground and the size of the drifts. For the two-track tunnel, a total excavated width of about 49 feet and an excavated height of about 39 feet would be advanced

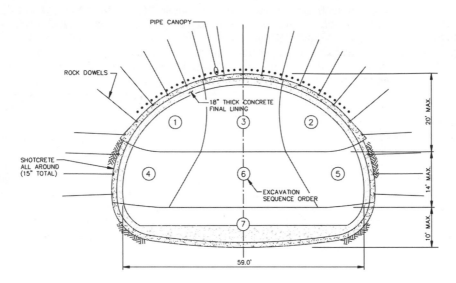

Figure 5. Typical tunnel section for NATM/SEM tunnel alternative

using seven individually mined drifts: three in the top heading, three in a top bench, and one at the invert. The same total number of drifts and sequencing would be used for the three-track tunnel, which has a total excavated width of about 66 feet and an excavated height of about 43 feet. Figure 5 shows a typical cross-section for the three-track tunnel alternative, including the drift excavation sequence. This is a conservative drift configuration which could be necessary when tunneling beneath the existing buildings along Townsend and 2nd Streets. In other areas, a less conservative drift configuration may be feasible allowing more efficient mining rates and reducing the construction schedule. During future design analyses, the drift configuration will be reevaluated and modified based on geologic conditions and potential impacts to surface facilities.

After completion of the tunnel excavation and installation of the shotcrete lining, a waterproof membrane would be installed on the inside of the initial lining before placement of the final concrete lining. The waterproofing system would likely consist of a flexible PVC waterproofing membrane that is heat-welded to create a continuous watertight seal around the tunnel opening. Design of a watertight tunnel is considered feasible structurally for an NATM/SEM tunnel because the invert can be curved to reduce the bending moments induced by the hydrostatic pressure. This approach is beneficial in minimizing long-term impacts on the local groundwater system and reducing tunnel maintenance requirements.

Preliminary numerical analyses using finite difference methods and the computer program FLAC were done to predict the stresses and displacements in the ground and support elements during the various excavation stages. These analyses were used to check the drift size, round length, and excavation and support sequence. Generally, the results indicated that the proposed drift arrangement and sequence were conservative. Surface settlement with NATM was estimated to be slightly greater than the settlement associated with the stacked drift method—approximately ¾ to 1 inch. To mitigate the risk of damage to overlying buildings, it was proposed that additional ground support be provided in the form of a grouted pipe canopy. The pipe canopy

Table 1.   Construction cost and schedule estimates

Alternative Description	Construction Cost	Construction Duration (months)
2-Track Stacked Drift	$297 M	71
3-Track Stacked Drift	$329 M	80
2-Track NATM/SEM	$226 M	48
3-Track NATM/SEM	$253 M	53
2-Track Twin Bore NATM/SEM	$260 M	45

would consist of 3- or 4-inch diameter pipes installed about 12 inches apart in the tunnel arch. The pipes would provide positive presupport/reinforcement for the tunnel arch, reducing loss of ground and minimizing surface settlement. Additional rock bolts, greater shotcrete thickness, and reduced round lengths would be considered for improved control of ground movements within sections of the tunnel where there is a greater risk of damage to adjacent properties and utilities from ground movements.

TBMs

TBMs can be used to construct tunnels in a wide variety of geologic conditions, including the Franciscan Formation. A shielded TBM was used to successfully construct the Richmond Transport Tunnel in San Francisco. It extended through Franciscan mélange and sandstone (Klein et al., 2001). One advantage of TBM methods is that there is minimal ground disturbance during excavation, although some minor vibrations are produced.

TBMs were not evaluated in detail because early evaluations determined that this method was probably not practical for the DTX project. This was primarily because of economic and right-of-way considerations. To accommodate the established tunnel clearances, a 33.5-foot-diameter TBM would be required. Given the short length of the mined tunnel section (only about 3,200 feet) and the cost of a TBM of this size, TBM methods did not appear to be cost-effective. In addition, the public right-of-way beneath 2nd Street is only about 82 feet wide, and twin TBM tunnels located at the edge of the right-of-way on each side would only be about 15 feet at the center. Such a small pillar might be acceptable in competent rock, but is questionable in the weak rock along the DTX alignment, and any settlement resulting from tunnel excavation would affect the existing buildings on 2nd Street unless mitigation measures were implemented. For these reasons TBM methods were not considered further in the conceptual design evaluations and cost and schedule estimates were not developed for this method.

CONSTRUCTION COST AND SCHEDULE ESTIMATES

Conceptual construction cost and schedule estimates were prepared for the two- and three-track configurations using the Stacked Drift Method and NATM/SEM. Table 1 summarizes the construction cost and schedule estimates for each alternative. The cost estimates are production-based and account for all labor, material, and equipment costs. Some basic assumptions are:

- Prevailing labor rates from California Department of Industrial Relations;
- Equipment rates from the U.S. Army Corps of Engineers *Construction Equipment Ownership and Operating Expense Schedule (Region VII)*, published in 2003 for the Western States; and

- Material and subcontract costs based on current market prices. Quotes were obtained for construction equipment and for large-quantity items such as rock bolts, steel sets, and lattice girders.

Cost estimates were prepared in December 2005 dollars and they were not adjusted for inflation. The following were excluded from these estimates:

- Design and construction contingencies;
- Escalation to adjust costs to the time of expenditure;
- Program costs, such as design and construction management fees, right-of-way and land acquisition, permits, etc.; and
- Rail and tunnel operating system costs.

EVALUATION OF ALTERNATIVES

While many constructability factors were considered in the evaluation of the alternatives, ultimately the preferred construction method was determined on the basis of cost and schedule in relation to achieving the owner's requirements. A comparison of the Stacked Drift Method and NATM/SEM approach for the two- and three-track alternatives reveals that the Stacked Drift method is approximately 30 percent more expensive than the NATM/SEM approach and has a construction duration that is about two years longer. Therefore, the NATM/SEM approach has significant cost and schedule benefits over the Stacked Drift Method.

Though it was originally expected that the Stacked Drift Method would result in much less surface settlement than the NATM/SEM approach, preliminary analyses have indicated that by employing a pipe canopy and a conservative excavation and support sequence (including short round lengths), the NATM/SEM approach yields only slightly more settlement (about ¼ inch). In terms of risks the Stacked Drift Method minimizes the area of potential collapse to that of the drift currently being excavated; whereas with NATM/SEM the drifts are larger exposing a greater area that could be susceptible to collapse. This higher risk can be mitigated with NATM/SEM in the same way the potential for settlement is controlled, i.e., by use of a pipe canopy, conservative excavation and support sequence, and shorter round lengths.

Besides the cost and schedule advantages, some other identified advantages of the NATM/SEM over the Stacked Drift Method are:

- Less potential that blasting will be required, as the larger drift sizes will allow the use of larger, high-capacity roadheaders.
- More consistently manageable truck traffic volumes during tunnel excavation (Stacked Drift Method would produce a spike in traffic during excavation of the central core).
- More economical and efficient ground support as ground support measures are tailored to the ground conditions actually encountered.
- Feasible to design a watertight tunnel lining minimizing groundwater impacts and tunnel maintenance costs.

Considering the above comparison, it was concluded that the NATM/SEM approach is more cost effective, faster, and the resulting risks can be effectively mitigated. Therefore, the NATM/SEM approach has been adopted for preliminary design.

CURRENT PROJECT STATUS

The DTX Project is currently in Preliminary Engineering Part 1, which includes various project development studies, alternatives evaluations, conceptual design, and development of associated cost estimates. The objectives of this phase are to define the configuration of the DTX project in terms of the number of rail tracks and structure type and limits, and to refine the project cost and schedule estimates.

The Part 1 studies focus on the Locally Preferred Alternative (LPA) alignment, approved as part of the FEIS/EIR process and shown in Figure 1. Additional funding has also been made available to investigate the feasibility of several value management recommendations conceived to reduce the cost of the DTX project. It is anticipated that these parallel studies will converge in the summer of 2007, resulting in a recommended configuration and overall construction approach for the DTX Project. Preliminary Engineering Part 1 will conclude in late 2007.

While the design studies and conceptual engineering continue, TJPA will be identifying and securing the funding necessary to construct the DTX. Once the funding plan is finalized (which is currently scheduled for 2008), the project will proceed to Preliminary Engineering Part 2 (completion of preliminary engineering) and then detailed design. Construction is currently scheduled to commence in late 2011 or early 2012. If funding is identified earlier, construction could begin sooner.

ACKNOWLEDGMENTS

The authors would like to thank the Transbay Joint Powers Authority for the permission to publish this paper. In addition, the contributions of other key members of the Project Team are gratefully acknowledged: Parsons Transportation Group—prime consultant, rail design, and open cut construction; and Arup—geotechnical investigations, open cut construction, and project management assistance.

REFERENCES

Arup, (2006). *Draft Geotechnical Data Report for the Downtown Extension Project,* Vol. I, II, III, and IV, Prepared for the Transbay Joint Powers Authority, August 15.

Barton, N., R. Lien, and J. Lunde, (1974). "Engineering Classification of Rock Masses for the Design of Tunnel Support," *Rock Mechanics,* Vol. 6(4): 189–239.

Bieniawski, Z.T., (1989). *Engineering Rock Mass Classifications,* Wiley, New York.

Deere, D.U. and D.W. Deere, (1989). *Rock Quality Designation after Twenty Years,* U.S. Army Corps of Engineers, Technical Report GL-89-1.

Hopper, R.C., T.A. Lang, and A.A. Mathews, (1972). "Construction at Straight Creek Tunnel, Colorado," *RETC Proceedings*, Chicago, IL, Vol. 1, Chapter 22, pp. 501–538.

Klein, S., M. Kobler, and J. Strid, (2001). "Overcoming Difficult Ground Condition in San Francisco-The Richmond Transport Tunnel, San Francisco County, California," *Engineering Geology Practice in Northern California*, Association of Engineering Geologists, Special Publication 12, Bulletin 210, pp. 433–442.

Romero, V.S. and P.H. Madsen, (2001). "Design and Construction Performance of a Large Diameter Tunnel Constructed in Soft-Ground by the Stacked Drift Method," *RETC Proceedings*, San Diego, California, Chapter 19, pp. 199–219.

MONITORING SUCCESSFUL NATM IN SINGAPORE

K. Zeidler

Gall Zeidler Consultants, LLC

T. Schwind

Gall Zeidler Consultants, LLC

ABSTRACT

The Fort Canning Tunnel, Singapore, is an approximately 50 feet (15m) wide vehicular tunnel under shallow ground cover that has been constructed according to the principles of the NATM. Its shallow location in soft ground combined with a high groundwater elevation in close vicinity of historically important features called for special design considerations combined with a rigorous monitoring scheme. Surface, subsurface and in-tunnel instrumentation was installed to monitor the performance of the ground and tunnel support during construction. Comparison of actual reading data with the results from the computer modeling carried out during the design phase is provided in this paper.

INTRODUCTION

The Fort Canning Tunnel forms a part of the project "Contract PE101A—Design and Construction of Fort Canning Tunnel and Realignment of Stamford Road," sponsored by the Land Transport Authority (LTA) of Singapore. At Fort Canning Tunnel (FCT) the New Austrian Tunneling Method (NATM) was used for the first time for a large span (14.7m–48ft) tunnel in Singapore. The contract has been tendered in a Design–Build framework and the team with the members Sato Kogyo (S) Ltd. (Contractor), TY Lin Ltd. (Engineer) and Gall Zeidler Consultants (NATM designer) has been awarded the contract. Fort Canning Tunnel is a 180m (590ft) long, three-lane highway tunnel, and has a cross section area of $135m^2$ ($1440ft^2$). It was constructed in residual soils under an overburden between 3m and 9m (10ft and 30ft). The tunnel approaches at the north and the south portal of Fort Canning Tunnel are constructed using cut-and-cover techniques.

LOCAL SETTINGS

Surrounding Structures

Whereas at the northern end of the tunnel no buildings are located in close vicinity to the tunnel structure, a retaining wall for the newly constructed Singapore History Museum is positioned in immediate proximity of the tunnel next to the south portal.

The alignment leads underneath the historic Fort Canning Park with its preserved trees and a historic cemetery. Fort Canning Rise, a public road crosses the tunnel alignment at a vertical clearance of approximately 5 m between the tunnel roof and the road surface.

In vicinity of the south portal, a settlement sensitive Church building is located.

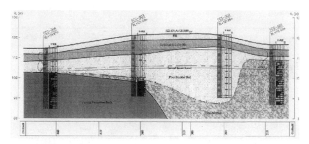

Figure 1.   Original schematic geological profile (from initial site investigation report by Kiso-Jiban Consultants Co. Ltd., 2003)

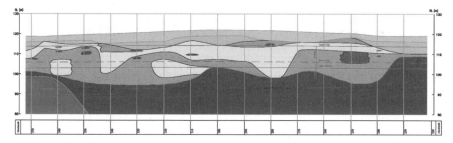

Figure 2.   Updated schematic geological profile (Kiso-Jiban Consultants Co. Ltd., 2004)

Geology

The Fort Canning Tunnel is constructed in the residual soils of the Fort Canning Boulder Beds. The Fort Canning Boulder Bed is a colluvial deposit of Pleistocene age that underlies parts of the central business and commercial district of Singapore. It consists of sandstone boulders in a matrix of hard sandy silt or sandy clay with silt. The matrix is of deep-red, red and white or mottled red, yellow and white color. For classification purposes the residual soil layer was subdivided based on the SPT N-value; RS I (N<15), RS II (15<N<30), RS III (30<N<50) and RS IV (N>50). The residual soils of Fort Canning Boulder Bed are overlain by a man-made fill layer of varying thickness (1m to 5m) and are underlain by the bedrock of the Jurong Formation, a sedimentary rock (Sandstone).

Ground water level at the Fort Canning Hill is established at approximately 1m below surface level, i.e., 2m to 8m above tunnel crown level.

The deposits were found in non-continuous layers and lenses with a varying amount of boulders leading to a inhomogeneous excavation medium with a wide variety of ground properties. Observations in the tunnel face revealed that the ratio between matrix vs. boulders varied between 70% boulders vs. 30% matrix and 30% boulders vs. 70% matrix (Figures 1 and 2).

DESIGN AND CONSTRUCTION

General

The project was tendered based on a Design–Build contract format. The NATM mined construction approach was the Contractor's proposed alternative. The design

Table 1. Soil parameters

Layer		γ (kN/m³)	E (MPa)	φ' (deg)	C' (MPa)	Cu (MPa)	n'
Fill	Original	19/	10/	30/	0.005	0.040/	0.35/
	1st Variation	19/	10/	30/	/0.001/	N/A/	0.35/
	2nd Variation	19	10	30	0.001	N/A	0.2
Residual Soil (RS I)	Original	—/	—/	—/	—/	—/	—/
	1st Variation	20/	15/	30/	0.010/	0.045/	0.35/
	2nd Variation	20	15	30	0.010	0.045	0.2
Residual Soil (RS II)	Original	20/	60/	30/	0.015/	0.180/	0.35/
	1st Variation	20/	30/	30/	0.010/	0.090/	0.35/
	2nd Variation	20	30	30	0.010	0.090	0.2
Residual Soil (RS III)	Original	—/	—/	—/	—/	—/	—/
	1st Variation	20/	60/	30/	0.020/	0.180/	0.35/
	2nd Variation	20	60	30	0.020	0.180	0.2
Residual Soil (RS IV)	Original	20/	100/	30/	0.035/	0.300/	0.35/
	1st Variation	20/	100/	30/	0.035/	0.300/	0.35/
	2nd Variation	20	100	30	0.035	0.300	0.2
Boulder Bed	Original	22/	300/	30/	0.040/	0.500/	0.35/
	1st Variation	—/	—/	—/	—/	—/	—/
	2nd Variation	—	—	—	—	—	—
Jurong Rock Formation	Original	24/	600/	35/	0.040/	1.000	0.3/
	1st Variation	—/	—/	—/	—/	/—/	—/
	2nd Variation	—	—	—	—	—	—

for both the tunnel initial support (temporary works) and the permanent support (permanent works) had to be provided by the Contractor for approval by the LTA and the Building & Construction Authority (BCA).

The tragic incident at the Nicoll Highway in early 2004 triggered a series of additional ground investigations and the request for additional numerical analyses for the design of the Fort Canning Tunnel. These analyses included comprehensive ground parameter studies and sensitivity analyses to ensure that a robust design was developed leading to a save and successful tunnel construction.

Design Parameters

Investigation Program. The soil investigation program included approximately 40 borings along the alignment to retrieve cores for laboratory testing and to carry out borehole tests such as Standard Penetration Tests and groundwater observations.

Variation of Design Parameters

During the design verification phase the geological model along the alignment was modified (Figure 2) and a series of parameter studies was carried out. Furthermore, a weak residual soil layer (RS I) was assumed to be present above the crown of the tunnel and a "worst credible" geological profile was introduced. The corresponding FE model was generated. Table 1 summarizes three sets of parameters used during that phase. These were used for the design analyses at different design stages.

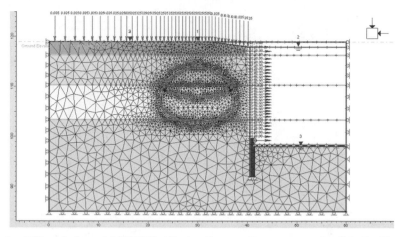

Figure 3.   FEM analysis model (CH 240) at the location near the new support wall for the Singapore Natural History Museum

Design Analyses Results

The finite element program Phase2, V5.04 by Rocscience, Inc. was employed for the analyses of loading conditions, ground response to the tunnel construction and to assess the lining forces. Triangular solid material elements were used to model the ground, and beam elements were used for the representation of the linings. The soil was modeled as an elastic-plastic material using the Mohr-Coulomb failure criterion; the linings were simulated as ideally elastic-plastic materials.

All finite element models for the primary lining design utilize a multi-staged modeling approach. An initial stage describing the in-situ stress state of the soil prior to tunneling formed the start point. The modeling stages were established to asses the individual construction stages. Top heading excavation was modeled by softening the soil within the excavation limits of the heading followed by the excavation of the soil elements within the excavation area. This approach simulates the excavation and resulting relaxation of the soil surrounding the tunnel prior to installation of the shotcrete support.

The increase of compressive strength and stiffness of the shotcrete lining after installation was modeled by gradually increasing the stiffness of the lining elements during the following construction stages. By this method, the interaction between ground deformations of the surrounding soil and the initial shotcrete lining is simulated as the excavation progresses.

The modeling sequence of the excavation and support sequence is concluded with bench and invert excavation and installation of the shotcrete lining in the invert.

The shotcrete lining was checked for its integrity during all the intermediate construction stages. The structural capacity of the shotcrete lining was determined in accordance British Standard BS 8110 and Singapore Standard SS CP65.

In addition to the two-dimensional finite element analyses, a three-dimensional finite element analysis was performed to assess the effects of the AGF pre-support umbrella and to confirm the performance of the two dimensional finite element models. All structural design, however, was based on the results of the two-dimensional finite element analyses. Figures 3 and 4 show typical finite element models.

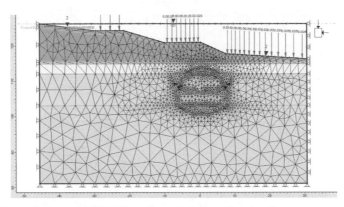

Figure 4.   FEM analysis model (CH 340) underneath the Fort Canning Rise

Table 2.   Summary of FE results

	Surface Settlements	Tunnel Lining Deformation	
CH 240	max. 104 mm	max. 104 mm (roof settlement)	max. 18 mm (settlement @ springline)
CH 340	max. 39 mm	max. 57 mm (roof settlement)	max. 21 mm (settlement @ springline)
Worst credible	max. 114 mm	max. 138 mm (roof settlement)	max. 48 mm (settlement @ springline)

Utilizing the last iteration of geotechnical parameters (2nd Variation in Table 1) in a series of FE analyses, the deformations at ground level and within the tunnel structure are determined in Table 2.

Pre-Support

The Fort Canning Tunnel is constructed under very limited overburden ranging from 3m (9ft) to 9m (19ft); i.e., 20% and 60% of the tunnel width. Due to this shallow overburden an AGF pipe arch was used as a continuous pre-support system over the full length of the tunnel (Figure 5).

The AGF ("All Ground Fastened") pipe arch consisted of a single row of steel pipes drilled at 400mm spacing in the crown of the tunnel top heading. The outer diameter of the steel pipes was 114mm (4.5"), pipe wall thickness 6mm (¼"); overall pipe length is 12.5 m (41ft), installed in four sections (3.5m–3m–3m–3m). The steel pipes were installed at a 7% outwards angle. A 3.5m overlap between succeeding pipe umbrellas was provided. 9m long tunnel sections were excavated followed by the installation of the next AGF umbrella in sequence.

A polyurethane two-component grout was injected through the AGF pipes via grouting ports at 0.25m (¾ ft) spacing along the length of each pipe. The grouting process was both volume and pressure limited for each individual pipe.

Excavation and Support Sequence

The Fort Canning Tunnel was constructed using an excavation and support sequence comprising of top heading excavation (with temporary invert) and combined

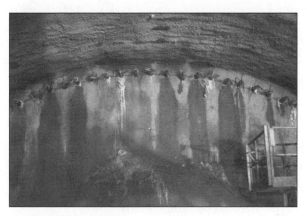

Figure 5. Pre-support umbrella installation

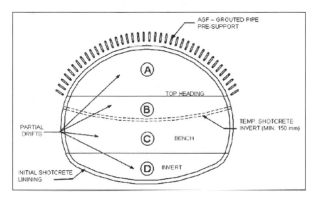

Figure 6. Excavation and support sequence

bench/invert excavation. The top heading extended over the entire width of the tunnel and its height was 6.0m that increased to 6.5m (20ft to 21.5ft) at the AGF installation location. Advance length in the top heading was 1m (3.3 ft). During the trial period, the temporary ring closure of the top heading (temporary invert) was installed in 2m increments at maximum 4m distance behind the excavated tunnel face. The temporary invert installation was increased to 3 m blocks at max. 6 m distance to the face following the trial section. In order to enhance the stability of the tunneling face a face stabilization wedge of 2 m depth at its base is left in place.

The combined bench and invert excavation with two 2m increments in bench followed by the two 2m increments in the invert was carried out in two steps immediately following each other. A minimum distance of 20m between top heading and bench/invert excavation faces was maintained (Figure 6).

Tunnel Support

Initial Support. The primary lining consists of minimum 300mm thick shotcrete with a design compressive strength of 40 N/mm^2. The primary lining is reinforced by one three bar lattice girder installed in every excavation round at 1m spacing and two layers (inside and outside) welded steel mesh of 10mm diameter bars at 200mm

Figure 7.   Initial shotcrete tunnel support in top heading

spacing. The temporary invert had a design shotcrete thickness of minimum 150mm and was reinforced by one layer of welded steel mesh of 10mm diameter bars at 200mm spacing (Figure 7).

Tunnel Final Lining. The Land Transport Authority (LTA) as the Client of Fort Canning Tunnel requested in the project design criteria that no ground water may be drained by the tunnel structure. A waterproofing system that extends around the full tunnel circumference and that consists of a geotextile layer and a welded PVC waterproofing membrane was utilized to achieve this requirement. The waterproofing system is segmented by a series of circumferential and longitudinal water barriers.

OBSERVATIONS DURING CONSTRUCTION

General

Excavation for 180m (590ft) long tunnel started in March 2005 and was completed in March 2006. Excavation was carried out using excavators and breakers.

In general, the ground conditions and ground behavior was found in agreement with the expectations.

Ground Behavior

The deposits were found in non-continuous layers and lenses with a varying amount of boulders leading to inhomogeneous ground conditions with a wide variety of ground properties. Depending on the density of the soil matrix, number and size of boulders in the excavation face and water saturation the ground varied from very hard to very soft (Figure 8).

Most of the sandstone boulders displayed a more-or-less thin weathered mantle. Inside the weathered zone the boulders were very hard generating significant resistance to excavation and breaking. While blasting of the larger boulders was not an option, excavator mounted hoe rams were used to remove the boulders from the excavation face and to break them up. Boulder removal lead sometimes to loosening of the surrounding soil matrix and soil layers due to the vibration energy exerted by the breakers.

While the clay typically yields low permeability, sand and silt admixtures, layers and lenses generated water paths that lead to groundwater discharge through the tunnel face. Despite the limited overburden thickness, the hill foot location in combination

Figure 8.   Boulders in the bench face

with frequent tropical rainfall generated sufficient water recharge to cause wet tunneling conditions. Pump sumps had to be provided to avoid softening of the ground in the bench and invert area.

The inherent stand-up time provided by the ground was considered too limited for a safe tunnel support installation. That led to the decision to employ a pre-support system over the entire length of the tunnel early on in the design phase. The lateral extent of the pre-support arch proved sufficient to safely install the shotcrete tunnel support after each excavation round. The grouting material penetrated the various soil deposits including the very fine silt materials introducing cohesion for improved ground strength.

The slightly domed tunnel face of the top heading was sealed after each excavation round. A flash layer of shotcrete was sufficient to stabilize the face and to prevent desiccation of the soil between the excavation operations. Face instabilities have not been observed. This may have been caused—in parts—by the continuous pre-support but also by the face stabilization wedge. During longer excavation interruptions, such as for the AGF umbrella installation, full shotcrete face support was installed (Figure 9).

Surface Settlements

Surface settlements were monitored using surface monitoring points either anchored into paved surfaces or into the soil. Readings were carried out using precise survey instruments. Generally, settlement readings were carried out on a daily basis. If undue readings were recorded, the reading frequency was increased to twice or several times a day (Figures 10 and 11).

Tunnel Deformation

Tunnel deformation monitoring was carried out with total stations. The daily readings yielded 3-D movement data that were evaluated immediately after taking the readings. In tunnel sections were the full final ring closure had been established and the tunnel lining deflections had ceased, the reading frequency was reduced to once per week (Figures 12 and 13).

Figure 9.   Top heading face shortly after excavation with exposed pre-support pipes

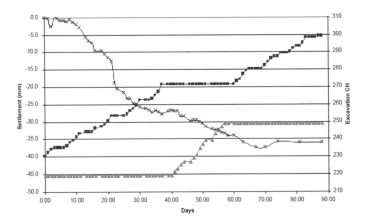

Figure 10.   Surface settlement CH 245 (overburden 5.3m)

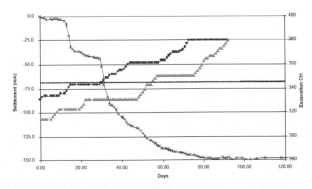

Figure 11.   Surface settlement CH 345 (overburden 5.1m)

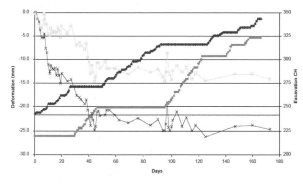

Figure 12.   In-tunnel deformations CH 242 (overburden 4.8m)

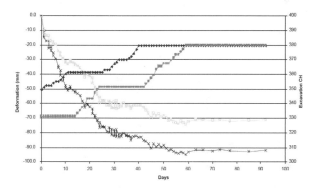

Figure 13.   In-tunnel deformations CH 350 (overburden 4.5m)

COMPARISON OF ACTUAL READING DATA TO DESIGN DATA—
SURFACE SETTLEMENTS AND TUNNEL DEFORMATIONS

Comparison of the analytical predictions with deformation measurements during construction yields partially inconclusive results. Using the ground properties stated above (2nd Variation), surface settlements for CH 240 are considerably over-predicted (104 mm predicted vs. 36 mm measured). Similarly, tunnel lining deformations were predicted too high (104 mm roof settlement and 21 mm settlement at the spring line) compared to the measured data (25 mm roof settlement and 13 mm settlement at the spring line). Back analyses of the FE model at CH 240 show that the introduction of an improved material layer in the tunnel roof modeling the AGF umbrella and an increase of the stiffness of the soils by 50% and the shear strength by a factor of 10 yield results more comparable to the measured deformations (24 mm roof settlement and 15 mm settlement at the spring line). The effect of the AGF pre-support in the model was ignored during the later design stages, to ensure a more conservative design approach (Figures 14 and 15).

For the back analysis, the strength and stiffness parameters of the improved soil have been assessed by multiplying the strength of each component (soil, steel) by the cross sectional area of each component and then dividing the sum of these products by the total area of improved soil under consideration (see Equation 1).

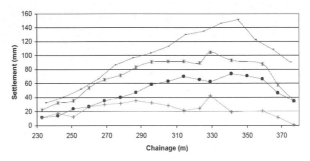

Figure 14.   Surface and tunnel settlement along the alignment

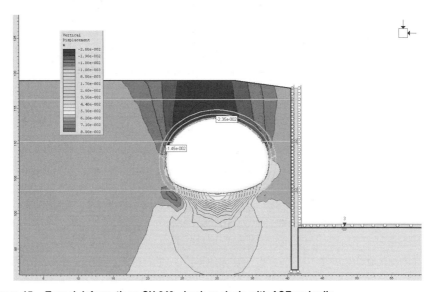

Figure 15.   Tunnel deformations CH 240—back analysis with AGF umbrella

$$P_{AGF} = \frac{(P_P \cdot A_P) + (P_S \cdot A_S)}{(A_P + A_S)} \qquad (1)$$

When using the same soil properties (2nd Variation from Table 1) at CH 350 and ignoring the beneficial effects of the AGF umbrella (as it was done for all final design calculations), the numerical analyses significantly under-predicted the deformations measured in the field. Whereas, following numerical analyses, surface settlements of 39 mm should have been expected for the prevalent soil conditions, surface settlements of up to 148 mm were observed. Tunnel lining deformations at this location were also considerably underestimated. Tunnel settlements of 57 mm in the roof and 21 mm at spring line level are predicted; 93 mm and 71 mm respectively are measured.

A closer match to the actually measured data at this location forms the model used as the worst credible situation at Fort Canning Tunnel. In the latter analysis, which formed the design case for the tunnel lining, surface settlements of 114 mm and vertical tunnel deformations of 138 mm (crown) and 48 mm (spring line) are determined.

Under normally consolidated soil conditions it may be expected that surface settlements are somewhat smaller, or approximately equal for very shallow tunnels than settlements measured in the tunnel crown. The surface settlement values measured at Fort Canning tunnel are locally significantly higher than the tunnel lining deformations (up to 50% more surface settlement). This may be attributed to insufficiently compacted fill above the tunnel alignment that has not been identified during the soil investigation and a short-term consolidation effect triggered by the tunneling operation below. However, measurement data are not yielding completely conclusive indications for this assumption.

CONCLUSION

Categorizing inhomogeneous residual soils—as prevalent at Fort Canning Hill—according to their SPT-N-values, as well as assessing elasticity and Mohr-Coulomb shear parameters in order to predict the ground response to a tunneling operation constitutes a challenging task. Numerical analyses utilizing a wide range of soil parameters are required to reliably predict upper and lower bound of the expected surface and tunnel lining deformations. In the Finite Element analyses performed for Fort Canning Tunnel, the upper bound settlement values were predicted acceptably well while lower bound values were not closely matched by analytical results using the stated elasto-plastic soil parameters. By using SPT N-values to categorize soil strata, an exact prediction of surface and tunnel deformations at locations where detailed geological stratification data were available (i.e., borehole location) however was not achieved.

ACKNOWLEDGMENTS

The authors wish to thank Mr. Osamu Hasegawa, Mr. Hitoshi Suzuki and the engineering staff of Sato Kogyo (S) Ltd. and Mr. Tan See Chee of TY Lin Ltd., as well as the engineering staff of LTA for their continued support during the design and the construction of Fort Canning Tunnel.

REFERENCES

Kiso-Jiban Consultants Co. Ltd., 2003: Fort Canning Tunnel, unpublished Geotechnical Interpretative Report.

Kiso-Jiban Consultants Co. Ltd., 2004: Fort Canning Tunnel, unpublished Geotechnical Interpretative Report on additional soil investigation borings.

Wallis, S., (2006). Singapore tests NATM at Fort Canning, Tunnels & Tunneling International. May 2006.

T. Schwind, K. Zeidler, V. Gall, NATM for Singapore, North American Tunneling 2006, ed. Levent Ozdemir, Taylor & Francis, 2006.

NATM THROUGH CLEAN SANDS—
THE MICHIGAN STREET EXPERIENCE

Paul H. Madsen

Kiewit Construction Co.

Mohamed A. Younis

Black & Veatch

Vojtech Gall

Gall Zeidler Consultants, LLC

Paul J. Headland

Black & Veatch

INTRODUCTION

A 100 ft long, 20 ft high and 18 ft wide NATM tunnel was excavated through grouted clean sands beneath a busy city street, including a variety of existing utility lines. The new tunnel associated with a hospital expansion will serve as a pedestrian tunnel connecting a parking structure with a new hospital facility.

Tunnel construction started on December 15, 2005 when drilling for the grouting program started and was completed on May 26, 2006 when the tunnel final lining was completed. The tunnel was constructed by Kiewit Construction Company as a tunnel subcontractor to The Christman Company who is constructing the hospital facility for Spectrum Health in Grand Rapids, Michigan. URS was the project designer. Gall Zeidler Consultants, LLC were the contractor's NATM consultant.

PROJECT DESCRIPTION

The Michigan Street Pedestrian Tunnel is a component of the Michigan Street Development located along the north side of Michigan Street NE between Coit Avenue NE and North Division Avenue in the downtown area of the Grand Rapids, Michigan. The Michigan Street Development includes a parking deck and four new towers to be constructed over the proposed parking deck. The tower planned furthest east is being developed by Spectrum Health as the Lemmen-Holton Cancer Pavillion. The purpose of the pedestrian tunnel is to connect the proposed Lemmen-Holton Cancer Pavillion with the future Spectrum building to be located on the south side of Michigan Street NE. The southern end of the tunnel was constructed with a bulkhead until the future Spectrum building is constructed at which time interconnection will be required. The tunnel site plan and profile are shown in Figure 1, Site Plan, and Figure 2, Tunnel Profile.

Several challenges were faced during the design and construction of the Michigan Street Pedestrian Tunnel. These challenges included compressed design and construction schedule, maintenance of traffic and protection of old underground utilities. The tunnel was being constructed as part of the Michigan Street Development and coordination with other concurrent construction activities was required. The tunnel was constructed within a specified window of time to allow other associated construction activities to take place without compromising the overall project schedule. The design started in June 2005 and construction was completed by June 2006.

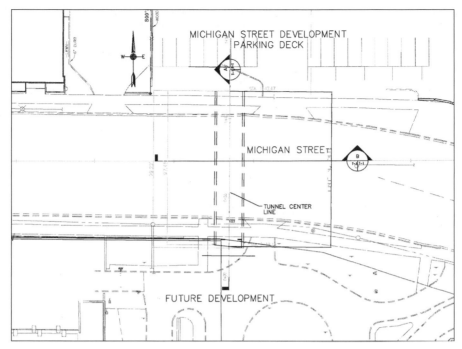

Figure 1.   Site plan

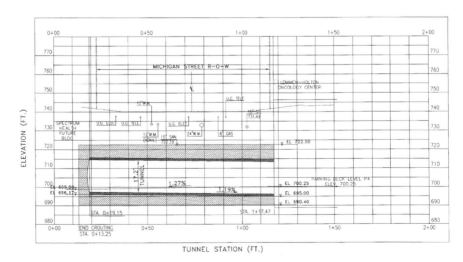

Figure 2.   Tunnel profile

The required internal finished dimensions of the tunnel cross section are 12 feet wide by 12 feet high at a length of approximately 100 feet. The tunnel will be equipped with heating, ventilation, and air conditioning (HVAC), lighting, fire suppression, and drainage systems. The tunnel was constructed as a modified horseshoe shape with an outside height of 19.6 feet and a outside width of 18.5 feet using the New Austrian Tunneling Method (NATM) with ground modification consisting of chemical grouting within the tunnel horizon. Tunnel cover is approximately 22 feet. A NATM design scheme was provided and was successfully implemented during construction with no significant modifications. Details of the support system and the construction sequence are provided below.

DESIGN

Project Requirements/Constraints and Challenges

Surface Settlement. Requirements were that the tunnel construction does not cause any roadway surface settlement. Any roadway surface settlement that occurred as a result of the tunnel construction activities had to be kept within reasonable limits so as not to disrupt traffic, endanger users of Michigan Street NE, or adversely affect underground utilities.

Existing Utilities. Several existing utility lines including several electrical and telephone conduits, a 6-inch gas main, a 24-inch water main, a 12-inch water main, an abandoned 12-inch water main, a 15-inch storm drain pipe, and a 15-inch sanitary sewer pipe are located perpendicular to the tunnel alignment along Michigan Street NE. Of particular concern was the presence of the high-pressure, 24-inch water main and the 15-inch sanitary sewer pipe. Refer to Figure 2, Tunnel Profile, for utility locations.

The 24-inch water main was believed to be a high-pressure cast iron pipe over 100 years old and located approximately 17 feet above the tunnel crown. The 15-inch sanitary sewer pipe was believed to be made of clay and located 7 feet above the proposed tunnel crown. The age of the existing 15-inch sanitary sewer pipe was not known.

Waterproofing. The Owner required that the tunnel be waterproofed to offset any groundwater infiltration (resulting from precipitation and/or utility leakage) into the tunnel. Watertight joints were planned between the tunnel and the parking deck permanent structure and the future Spectrum building.

Subsurface Conditions

Three borings were drilled along the tunnel alignment to investigate soil subsurface conditions. In addition, several borings were drilled in the vicinity of the tunnel alignment. The borings revealed that the subsurface soils consist of a Fill layer, varying in thickness between 0 and 8 feet below ground surface, underlain by Glacial Sand extending to 77 feet below ground surface within the tunnel alignment. The Fill and the Glacial Sand layers consist predominantly of poorly graded sand (SP) and well graded sand (SW) based on visual classification. Standard Penetration Test (SPT) blow counts ranged from 6 to over 100 blows per foot (bpf). The high blow counts may be attributed to the presence of cobbles and boulders. The soil medium within the tunnel horizon was generally classified as medium-dense to dense sand. The Glacial Sands overlie the Glacial Till comprising silty clay to silty sandy clay including layers of sand. Figure 3, Geological Profile, presents geological conditions along the tunnel alignment.

Groundwater was not encountered within the tunnel horizon during the subsurface investigation. It was anticipated that the tunnel would be constructed entirely above the groundwater table.

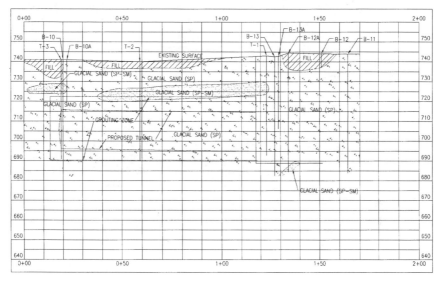

Figure 3. Geological profile

The ground behavior for the Glacial Sand materials within the tunnel envelope, as classified in accordance with the Tunnelman's Ground Classification system (Heuer, 1974), was "slow to fast running."

Tunneling Alternatives Evaluation

Two tunneling construction method alternatives were considered for the tunnel construction: the New Austrian Tunneling Method (NATM) with ground modification and the use of a tunnel excavator shield. The alternative study included tunneling method applicability, pro's, con's, and construction cost estimates for comparison. A recommendation was provided to adopt the New Austrian Tunneling Method (NATM) as the most economically and practically preferred alternative.

Two main issues of concern were identified. These challenges were (1) the presence of boulders or obstructions within the tunnel envelope, and (2) excessive ground surface settlement as a result of tunneling operations.

Boulders and cobbles are commonly encountered within the glacial sand. The difficulty of the removal of these boulders and other obstructions depends on the tunneling method used. Obstructions are more accessible and easier to remove with the NATM. With the excavator shield, the obstruction may need to be broken in place prior to removal due to limited accessibility at the tunnel face. In addition, boulders only partially within the face and extending beyond the limit of shield excavation will cause excess settlement with a shield but not with the NATM. Excessive settlement at utility elevations could lead to undermining utility lines with consequent leakage of water being catastrophic to tunnel stability. To mitigate the effect of tunneling on existing underground utilities, ground modification applications were employed. Based on the ground conditions provided on the boring logs, jet grouting and chemical grouting were considered options to mitigate for settlement near utilities. The use of an excavator shield without any ground modification could lead to excessive ground surface settlement.

NATM Design

Numerical Analysis. Numerical analysis was performed to study and prepare a design for the tunnel structural support system, stand-up time for native soil and grouted soil within the tunnel horizon, and to investigate the estimated surface and underground movement likely to result from tunneling operations. Data obtained from the numerical analysis was utilized to determine the appropriate measures to protect existing underground utilities. In addition, empirical analysis was performed to validate the numerical analysis results.

The preliminary assessment of ground surface settlement resulting from tunneling operations for the proposed tunnel indicates that surface settlement could be controlled to less than 1 inch, providing that proper design is implemented including ground modification such as grouting and good construction workmanship.

Permeation Grouting. The tunnel crown is approximately 22 feet below the existing ground surface and a preliminary estimate of the required mined tunnel height is 19.6 feet (width is 18.5 feet). Therefore, the ratio of cover to tunnel diameter is close to 1:1. The potential for ground settlement during mining in a loose to very dense Glacial Sand without ground treatment was considered significant.

Permeation grouting creates an in-situ cemented soil matrix that increases strength and reduces permeability of the original soils and thereby increases tunnel stability during construction. This method is suitable for granular soils, sand, and gravel with less than 15% of fines that have sufficient void space to be permeated with fluid cement or chemical grout to produce the cemented matrix. Permeation grouting can be performed from vertical or horizontal pipe arrangements. Due to the limited access available on Michigan Street, horizontal grouting pipe arrangement was selected. The suggested design pipe arrangement is presented in Figure 4, Grouting Scheme (Design Phase).

Permeation grouts are injected into the soil void spaces through tube-a-manchette (TAM) pipes. A stabilizer added to the grout controls gel time. The grout bonds with the soil particles, producing a composite soil mass with a higher shear and compressive strength than the un-grouted formation. Permeation grouting also reduces the soil permeability by filling the voids. This grouting technique does not use mechanical means to restructure the soil in the process, and therefore the soil structure remains relatively undisturbed. The operation creates minimal ground disturbances and reduces adverse deformation and damage to the ground formation.

NATM Excavation. The NATM design included excavation sequencing and support system details and installation. The excavation sequence was selected based on the modified ground condition, tunnel cross section size, and settlement requirements. The numerical analysis assisted in selecting the NATM excavation sequence. The excavation sequence consisted of top heading and bench with a 5 ft round length. Details of the excavation sequence are presented in Figure 5. The NATM design was implemented during construction without any significant modifications to round length, top heading dimensions or pre-support philosophy.

Primary and Final Support System. The primary support system consisted of a 7-inch steel fiber reinforced shotcrete initial layer applied directly onto the excavated

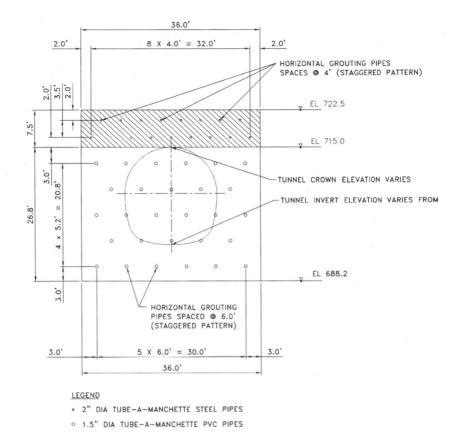

Figure 4. Grouting scheme (design phase)

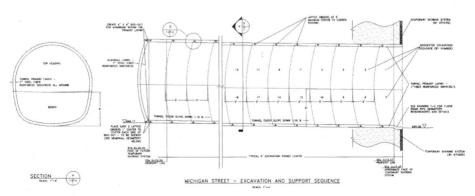

Figure 5. NATM design

grouted soil mass. A PVC waterproofing membrane with welded joints was installed and a final 6-inch steel fiber reinforced shotcrete layer was installed followed by a smoothing 1-inch un-reinforced shotcrete layer. Isolation joints were provided at both ends of the tunnel to structurally isolate the tunnel from the garage parking deck and proposed future Spectrum building. The isolation joints were designed as watertight joints to prevent groundwater infiltration into the tunnel at the joint locations.

Geotechnical Instrumentation. A ground surface movement monitoring program was implemented before, during, and after tunnel construction. The monitoring program included surface settlement monitoring points, multiple point borehole extensometers (MPBXs), and in-tunnel instrumentation including convergence meters and crown settlement markers. Additional instrumentation was needed during construction including inclinometers to monitor the headwall during tunnel breakout and construction. Additionally, an alternative in-tunnel monitoring system consisting of optical targets was implemented during construction.

Contractor Selection

Due to the compressed schedule, three tunneling contractors were invited to bid on the project as the contract documents were being developed. Only two contractors provided bids on the project. The lowest bid was 43% above the engineer's estimate. This was attributed to the constraints to the work area, the schedule as well as the unexpectedly increase in cost for raw materials and in particular the rising cost for permeation grouting. Value engineering proposals were requested from both bidders in an attempt to lower the price. Figure 6, Grouting Scheme shows a modified grouting pipe arrangement proposed by Kiewit that was accepted by the Owner's team. The modified grouting pipe arrangement reduced the lowest bid price by 15%.

PERMEATION GROUTING

Drilling

The grouting sub contractor utilized a crawler rig in combination with a specialized system to advance a 4" casing using down the hole hammer. Weaknesses would occur in the casing after multiple uses causing the casing to shear at the thread to straight casing intersection. This problem was overcome by using a thicker and higher grade casing.

Following the drilling of each hole the TAM pipe was inserted and the annular space between the pipe and the drill hole tremied with a cement bentonite grout, while retracting the casing. Using this system a high degree of drilling accuracy could be achieved since the casing did not rotate. The soil loss is also minimized due to drilling since the spoils are flushed back through the casing using compressed air.

A total of 21 steel TAM grout pipes, 2" diameter, were installed above the crown, and 20 PVC 1.5" TAM pipes were installed throughout the profile as seen on Figure 6. The total drilling length of 4,000' was completed in 18 working days on two 12-hour shifts per day. No obstructions such as cobbles or boulders were encountered during the drilling. However, it was a challenge drilling the pipes between soldier piles, tie back anchors and the MPBXs previously installed from the road surface above the tunnel.

Grouting

The chemical grout mix consisted of 50 percent liquid sodium silicate, five percent reactant and 45 percent water. Prior to grouting laboratory testing was conducted by injecting the mix into medium dense sand samples and curing them to study the

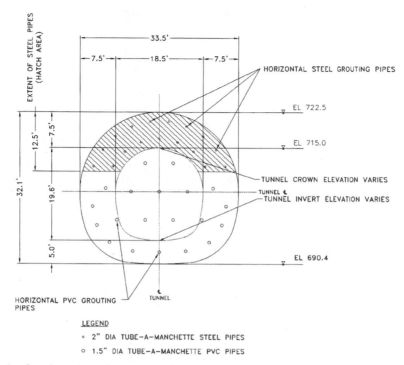

Figure 6. Grouting scheme (proposed and constructed by Kiewit)

unconfined compressive strength performance at seven and 28 days. The 28 day strength requirement was 150 psi. Grouting volume was based on a 30 percent void ratio of the Sands with a void filling factor of nearly 100 percent resulting in a target volume of 188,100 Gallons. Grouting started on February 6, 2006 and was completed in 29 workdays on March 13, 2006. The actual injected volume was 212,000 Gallons. The actual strength of the grouted soil ranged upwards of 500 psi to an estimated 1000 psi causing the tunnel excavation tool described below to almost reach its limit. The maximum distance between the pipes was five to six feet, which was at the limit for obtaining required soil mass permeation.

Verification and Testing

To verify the compressive strength, and the extent of the grouted sand zone, standard penetration tests (SPT) were performed prior to and following grouting. The blow counts typically increased to 50/4" or refusal from pre-grouting values of around 25/6", in the grouted zone. Other tests during construction included gel times of grout sampled every 60 minutes or 1,000 gallons of grout, which ever came first. Gel time requirements ranged from five to fifty minutes.

Hand-written grouting reports were prepared by the technician at the manifold. The report contained information regarding target and actual volumes and pressures for each sleeve, date and time of injection and flow/pressure measurements every 15 minutes.

TUNNEL CONSTRUCTION

Safety

The Michigan Street tunnel was constructed without any Recordable, Reportable or First Aid cases.

This success, for a job with rapidly changing operation that does not allow time for learning curves, can be contributed to a few main factors.

First, Kiewit's safety program, that does not allow an operation to start unless it has an approved Job Hazard Analysis combined with a work plan, reviewed and signed off by the Crew.

Second the entire crew, except for a few, all had experience with working on projects for Kiewit, either locally in Grand Rapids or on other tunnel projects. Third, without jeopardizing the safety of the crew the field supervision was able to direct the work when the ground conditions were challenging.

Excavation Method

The selection criteria for the tunnel excavator were as follows:

- Must be able to efficiently excavate the required profile.
- Have enough power and weight to excavate the chemically grouted sands, while still being able to fit within the tunnel.
- Be readily available.
- Be reliable.
- Be cost effective.

After performing three dimensional simulations to check working range, a new excavator with a modified stick attachment was selected together with a new grinding head attachment to cut a smooth profile in the grouted sands, which behaved like a weak sand stone. The excavation time for a five foot top heading round would be two hours excavating the bulk 80% followed two hours of trimming to the desired profile. Three times during the excavation the excavator would back out of the tunnel to allow for muck to be removed.

The excavation profile together with girder installation was set-out, checked and as-built in real time using a tunnel specific software package onboard a total station. Excavation started on March 18, 2006 and was completed on May 6, 2006. Progress was stopped for a number of days during the period in order to perform additional localized grouting from within the tunnel, and due to finalizing portions of the parking structures work in front of the portal, resulting in 24 actual tunneling days working an average of 10.5 hour shifts.

Shotcrete was supplied by a local readymix supplier and pumped from a lane closure on Michigan Street down to the tunnel where it was applied by a certified nozzleman.

Waterproofing

A tanked, closed perimeter PVC waterproofing system was designed for the tunnel consisting of a layer of fleece followed by a 3mm PVC continuous welded membrane. To facilitate the final lining shotcrete application, applied onto the membrane, BA-Anchors were used to secure a 4×4 W4.0×W4.0 Welded Wire Fabric (WWF) tightly against the membrane for the shotcrete to adhere to.

The WWF was held at a specific distance from the membrane using PVC spacers that also prevent the risk of punctuating the membrane.

The waterproofing system was purchased through a supplier who also supplied required submittals and an experienced superintendent to oversee the installation, which was performed by Kiewit on seven 10 hour shifts, including WWF installation.

INSTRUMENTATION AND MONITORING

System

The instrumentation and monitoring system for the Michigan Street Tunnel was laid out according to typical monitoring requirements and principles for a shallow NATM tunnel driven in soft ground conditions in an urban setting.

The instruments included four arrays of Surface Monitoring Points (SMPs) arranged in lines perpendicular to the tunnel axis with the most distant instrument located about 40 feet from the tunnel centerline, four Multiple Point Borehole Extensometers (MPBXs) located outside the excavation profile, Inclinometers at the soldier pile and lagging wall and four monitoring cross sections within the tunnel with each having five prisms for the optical monitoring of in-tunnel deformations. The instruments were monitored on a three-times-per shift basis during active tunneling and every 24 hours when no mining was on going. Data were evaluated by plotting monitoring values vs. time and vs. tunnel progress to show ground deformation behavior as a function of the tunneling process. Monitoring data were compared to a two-level observation system consisting of threshold and limiting values for each instrument. Monitoring results were graphed by the contractor and shared with the construction manager, the tunnel designer and the contractor's NATM consultant on a daily basis. The contract stipulated the implementation of contingency measures upon reaching of threshold values, which was 12.5 mm for convergence points and 6.25 mm for surface settlement.

Representative Observations

The deformation values for the SMPs in Array #3, located 60 feet from the portal, vs. time for a time frame between late April 2006 and mid May of 2006 illustrate a number of observations made during tunneling. Close to tunnel centerline the increase in settlement was on the order of some 1–3 millimeters as the heading passed the array. Comparing the graphs for the individual SMPs in Array #3 the settlement values decrease clearly with increasing distance from the tunnel centerline. The maxim cumulative surface settlement trough was some 15 millimeters. The cumulative graphs included surface settlement data measured prior to the tunneling operation before mid March 2006 that were caused by movements of the support of excavation wall at the tunnel portal.

An evaluation of the tunnel convergence values showed that the deformation of individual prisms was generally below a 5 millimeter-value although at occasion values close to 10 millimeters were recorded.

Instrumentation Summary

In summary, the instrumentation and monitoring system installed allowed for a detailed evaluation of the deformation regime within the ground affected by the tunneling operation. The deformation values measured are viewed as typical for shallow soft ground tunneling considering size of the tunnel, soil conditions and ground improvement measures implemented.

The MPBXs installed prior to the parking structure excavation reaching full depth turned out to provide valuable information regarding soil movement prior to start of tunnel excavation.

OVERCOMING CHALLENGES IN THE CONSTRUCTION OF THE DUBLIN PORT TUNNEL

Peter Jewell
Kellogg Brown & Root

Tim Brick
Dublin City Council

Stephen Thompson
London Bridge Associates Ltd.

Simon Morgan
London Bridge Associates Ltd.

ABSTRACT

The $980m (€750m) Dublin Port Tunnel, Europe's biggest urban highway tunnel project, was opened in December 2006. The culmination of this immense scheme is set to transform the environment, mobility and road safety of Ireland's capital city. This paper summarizes the key aspects of a project linking both sequential and parallel construction of individual tunnels and describes various approaches to bored, mined and pre-support cut-and-cover tunneling in a wide variety of ground types and depth of cover. Examples of innovative approaches to deal with structural and geotechnical problems while maintaining full operation of both road and rail traffic and the integrity of properties in a dense urban corridor are described.

INTRODUCTION

On 20th December 2006 the Dublin Port Tunnel was officially opened by Mr. Bertie Ahern, the Taoiseach (Head of Government) of the Republic of Ireland. This event represented the culmination of many years of planning, followed by a construction period of nearly six years. At a cost of $980m (€750m), this was the biggest single infrastructure project ever undertaken in the Irish State, besides being Europe's largest urban highway tunnel completed following the introduction of the 2004 European Community's Directive on minimum safety requirements for tunnels in the Trans-European Road Network.

The Port of Dublin is one of Europe's busiest sea ports, and is the Republic of Ireland's principal gateway for global trade. The completion of a tunneled direct motorway link to Dublin Port means that trucks no longer have to negotiate Dublin's city centre, including the quays flanking the River Liffey. The removal of this traffic from congested central streets to a direct link to Dublin's peripheral and Ireland's national motorway system is an important step in making the centre of Dublin less busy, quieter, safer and overall a highly desirable environment in which to live and work.

This paper summarizes the background to and history of the Dublin Port Tunnel Project and describes some of its more interesting and technically challenging features. There is particular emphasis on the bored tunneling and the sequential excavation tunneling undertaken to construct mined lay-bys and vehicle cross passages (VCPs), and constructing a tunnel section with very shallow depth beneath a fully operational main line railway.

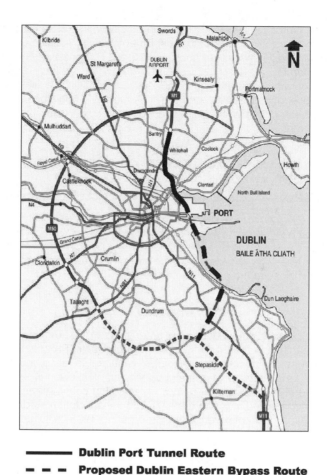

—— Dublin Port Tunnel Route
— — — Proposed Dublin Eastern Bypass Route

Figure 1.   Project location

BACKGROUND

The Dublin Port Tunnel route is shown in Figure 1. It provides a direct connection between the roads which serve the Port of Dublin and Ireland's national M1 and M50 motorways to the north. The dual carriageway tunneled section itself is 4.5 km in length, with a further 1.1 km of main roads at grade, and several miles of associated upgraded local roads. The route itself has been designated as part of the M50 and would eventually form part of the completed peripheral motorway route around Dublin.

The project brings many benefits to the Port of Dublin and Dublin's City Centre. Heavy goods vehicles (HGVs) have been removed from the City Centre and residential areas, facilitating public transport improvements and pedestrian and cycling facilities enhancements in a much improved City Centre environment. Traffic calming measures can now be implemented in more residential areas. Dublin Port's ability to develop as an accessible, thriving 21st century trading facility is greatly improved.

The Dublin Port Tunnel is unusual in that its principal purpose is to carry HGVs, which are permitted to travel toll free. In contrast, in line with a policy of restraining growth of private motor vehicle usage in this large city which offers other efficient public transportation alternatives, cars and most other forms of vehicles other than HGVs are obliged to pay a toll between $4 and $16 (€3 and €12), depending upon the time of day and direction.

Previous experience of tunneling in the Dublin area is limited historic railway tunnels and more recent small sewers and services tunnels. The Dublin Port Tunnel is of much larger diameter than any of these, and had to be constructed through the following geological formations:

- Fill materials of varying composition and thickness
- Glaciomarine and fluvio-glacial deposits—comprising of clays, silts and sands with associated gravel layers
- Glacial deposits—mostly boulder clay consisting of silty sandy clay, containing layers and lenses of sand and gravel deposits
- Bedrock—carboniferous limestone interbedded with mudstone and shale (Dublin Limestone). The bedrock has been faulted and partly folded and uplifted.

The Dublin Limestone is an aquifer. The prevailing groundwater table throughout the tunnel route is close to the ground surface.

HISTORY

Ireland's economic success during the latter years of the 20th century had led to a considerable expansion in the trade handled through Dublin Port, its largest sea port, and the existing road network had become unsuited to accomodating the huge increase in HGV traffic. The relief scheme which was conceived would require many years of planning and development before construction could commence. Feasibility investigations were initiated in the mid-1980s with transportation studies conducted in 1990. Engineering consultants were appointed during 1995. A tolled tunnel scheme was finally approved by Government in 1999 following a Public Inquiry into tolled and non-tolled options.

Tenders for the design and construction of the Dublin Port Tunnel were invited in 2000, and in December of that year a Contract valued at $620m (€448m) was awarded to the international Nishimatsu Mowlem Irishenco (NMI) Consortium, comprising the Nishimatsu Construction Company Limited of Japan, John Mowlem PLC from the UK, and Irishenco from the Republic of Ireland. The Institution of Civil Engineers (ICE) Third Edition Design and Construct Contract form, modified for Irish Conditions, was employed. A notable amendment was the deletion of Clause 12, thus placing on NMI all risk for unforeseen ground conditions. The NMI Consortium's lead designer was Haswell Consulting Engineers from the UK, and they engaged several other design organizations including Carl Bro, Geotechnical Consulting Group, ILF and Geo-Design. The Contract required the appointment of an Independent Design Checker, and NMI appointed Mott MacDonald to this role.

As agent for the National Roads Authority (NRA) of Ireland, Dublin City Council (DCC) acted as Employer for the Contract. DCC appointed Kellogg Brown & Root (KBR) as Construction Supervisor to report on design and construction compliance. The KBR team was supported by Hyder Consulting Limited, London Bridge Associates, the Dr. G. Sauer Company Limited and Jaakoo Pöyry Group (formerly Electrowatt International).

Following further design work and certain on-site utility diversions and other enabling activities, construction work started in earnest in June 2001. Key decisions which needed to be made early in 2001 included finalizing the bored tunnel diameter, so that orders for the two tunnel boring machines (TBMs) supplied by Herrenknecht could be placed. The TBMs were assembled on site in the first part of 2002 and bored tunneling commenced in July of that year. Construction of the bored tunnels was completed in August 2004, with other tunneling, including notably at two mined layby and vehicle cross passage (VCP) locations, continuing between 2003 and 2005. Cut-and-cover tunnel structures and tunnel portals, including the section of tunnels beneath the main Dublin to Belfast and Dublin Area Rapid Transit (DART) railway lines located south of Fairview Park, were constructed between July 2002 and March 2005. Sections of completed tunneling were handed over progressively for the installation of mechanical, electrical and communications equipment, surface finishing and protective coating, invert filling and roadway construction, and other finishing works. Final commissioning of tunnel systems was conducted in November 2006 and the Dublin Port Tunnel was officially opened on December 20, 2006.

TECHNICAL CHALLENGES

The dense urban environment of north Dublin precluded a direct route from Dublin Port to the Irish motorway network from being at grade or elevated. Accordingly a tunneled route was developed. Areas of shallower ground cover and/or unconsolidated subsoils at the northern and southern extremities were to be traversed with cut-and-cover tunnel structures, with the central portion being mined through stiffer glacial clay subsoil and strong limestone rock. The range of geological conditions present on the route gave rise to a number of potential technical problems for the constructors and designers. The resolution of three such problems is described in this paper.

Firstly, the conceptual project scheme was based upon the main road tunnels being mined by conventional sequential excavation tunneling techniques. A major difficulty in this regard was that excavation of the strong to extremely strong limestone rock (with an unconfined compressive strength of up to 250 MPa in places) was originally envisaged to be carried out by drill and blast methods. The tunneled route passed directly beneath more than 400 properties, most of which were residential and privately owned. NMI concluded that excavation by means of two TBMs would be preferable, as the contractual limits on vibration from blasting and working hours would otherwise have severely restricted production rates.

Secondly, the project required the provision of laybys and vehicle cross passages (VCPs) at two locations in the mined tunnel section. The available geotechnical information indicated zones of excessively fractured limestone rock and limited thicknesses of rock cover below the glacial subsoils, particularly at the location known as Cloisters. The decision to advance the main tunnels with TBMs gave the opportunity to initially excavate and provide temporary support to large volumes of the layby areas, thereby simplifying the excavation sequencing and program.

Thirdly, the constrained urban corridor determined that the tunnels would need to pass at very shallow depth beneath the Iarnród Éireann (Irish Rail) main Dublin to Belfast Railway. Iarnród Éireann had specified that the railway lines would need to be kept fully operational during the works and that access at track level was restricted. These constraints, together with the complex ground conditions and groundwater regime at this location, caused NMI to design an elaborate temporary modular steel section supporting structure which was installed sequentially as excavation proceeded downwards beneath a pipe arch roof.

Outside the scope of this paper, the project gave rise to a number of other innovative technical solutions which took account of the site specific geological conditions to provide more cost and/or time effective designs, particularly in respect of temporary works ground support.

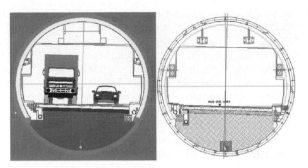

Figure 2.   Bored tunnel design and construction

Figure 3.   Hard rock TBS S-193

BORED TUNNELING

The tunneled portions of the motorway comprise 2.6 km of twin tube in bored tunnel and 1.9 km in a twin tunnel cut-and-cover reinforced concrete box structure. This paper describes the design and construction of the bored tunnel section undertaken by Nishimatsu as part of the NMI Consortium.

Figure 2 shows a typical cross-section of the bored tunnels. The excavated diameter was 11.6 m, making these amongst the largest diameter bored road tunnels in the world. They contained two lane carriageways with 3.65 m wide lanes and 0.25 m margins. There are 0.85 m wide walkways on either side. The vertical clearance is 4.9 m with a vehicular operational height of 4.65 m.

Having decided not to construct the main tunnels by drill and blast techniques, NMI elected to procure two TBMs from Herrenknecht GMBH. The first TBM, known as S-193, illustrated in Figure 3, was a refurbished machine selected for construction through strong rock using cutter disks and picks. This TBM assembly weighed more than 1600 tons, was 56 m long, and was equipped with a series of trailing gantries totaling 100 m in length. The machine was assembled in the working shaft WA2 at Whitehall, and undertook two 2.1km drives. Initially it excavated southwards and was then turned round in a specially constructed chamber at the portal with the south cut-and-cover tunnel section, returning to shaft WA2. At times S-193 excavated more than 450 tons per hour of rock.

Figure 4. Open shield TBM S-194

The second TBM, known as S-194, illustrated in Figure 4, was another refurbished machine, selected for its suitability to excavated glacial clay containing water bearing sand and gravel layers. It carried out two 330 m long drives, the first northwards from shaft WA2, and the second in the reverse direction following turn round. The ground cover was about 20m at shaft WA2, but this reduced to only 9 m at the north reception shaft (only approximately 75% of the excavated diameter).

Because of concerns that there could be abandoned water wells and other anomalies in the ground ahead of the face, investigatory forward probing was stipulated in the contract documents. S-193 was equipped with a probing drill and a Lutz Data Acquisition system. The Tunnel Reflection Tomography system (TRT™) from Skanska was successfully trialled and subsequently adopted for both the main tunnel drives and the other major mined tunnel structures.

Both TBMs were launched from a 56.6 m diameter shaft, 30 m deep, located at site area WA2 at the intersection of Swords Road and Collins Avenue. Launch adits of approximately 12 m length were constructed at the shaft base, using sprayed concrete lining (SCL) sequential excavation techniques, incorporating a canopy of grouted, self-drilling spiles, lattice arch girders, reinforcement mesh and fibre reinforced shotcrete. Following completion of the outward TBM drives, both machines were received through SCL reception adits of approximately 9 m length, constructed as designated turnaround chambers in Fairview Park for S-193 and close to Whitehall Church for S-194. The machines were then turned through 180 degrees, placed within pre-constructed launch adits (12 m long at Whitehall and 56 m long at Fairview) and recommenced tunneling on the return drives.

The bored tunnels were provided with both primary and secondary linings. Figure 5 shows the primary lining, consisting of precast reinforced concrete segments of internal diameter 10.84 m. Each ring was 1.7 m long and 350 mm thick, and in each ring there was a total of six segments plus a key. Individual segments weighed up to 8.5 tons, and more than 21,400 segments were manufactured by Banagher Concrete in County Offally, Ireland, to Nishimatsu's own specification. There were three types of segmental rings (straight, left taper and right taper). A 54 mm taper was provided, to provide for a minimum radius of 720 m.

A number of other primary segmental lining design features were notable. A very high (C60) design strength for concrete was specified. Pulverised fuel ash (PFA) adopted to meet durability requirements.

Radial joints were flat, with 95 mm relief on each edge. A pot/cam system was incorporated on circumferential joints, to profiles previously used successfully by Nishimatsu on the Tokyo Bay project in Japan. The segments contained approximately 94kg/m^3 reinforcement.

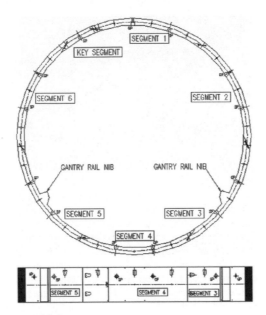

Figure 5. Primary segmental lining

Figure 6. Secondary lining

Nibs were cast into the straight segmental rings only. These were designed to support loads of up to 400 tons from the TBM traveling gantries, and they also formed the base for the secondary lining. The segments needed to be capable of transmitting the 350 tons thrust which each of the 24 ram groups could produce.

The secondary lining, illustrated in Figure 6, was required to provide structural protection for a 100 MW design fire, a watertight structure above the carriageways, and a smooth, coated finish for airflow, light reflection and cleaning. These requirements were met by constructing a 10.29 m diameter insitu unreinforced concrete lining of 275 mm diameter nominal thickness. The design needed to accommodate structural support for mechanical & electrical (M&E) equipment, but the secondary lining was not required as the structural ground support of the tunnel itself; the segmental primary lining was designed to provide this function.

The secondary lining was required to act as both a thermal and a physical barrier to prevent any deterioration to the structural primary lining. A 100 MW design fire is an onerous design criterion, but reflected the fact that the Dublin Port Tunnel is to carry a high proportion of HGVs, and most forms of hazardous cargo are permitted. The contract specified a Modified Hydrocarbon design fire curve to reflect this fire load, with the lining having to sustain a fast temperature rise followed by being subjected to a temperature of 1,200 degrees Centigrade for two hours. Polypropylene fibres of 18 mm microfilament were proposed to be added to the concrete in order to provide fire protection. Physical fire testing was carried out on a series of test cylinders containing varying quantities of fibres. Compliance testing was then undertaken at the TNO laboratory in Delft, The Netherlands, using large (2 m by 2 m) concrete model slabs, with a preload was applied to simulate restraint forces. From the test results, a fibre dosage of 1 kg/m^2 was adopted for the secondary lining, apart from for the layby structures where 2 kg/m^2 was used.

A watertight tunnel structure was achieved by providing four barriers to the passage of groundwater, as follows:

- Grouting the annulus between the excavated face of the ground and the primary lining
- Providing hydrophilic seals in the joints between segments of the primary lining
- Attaching a waterproof membrane and drainage fleece to the primary lining
- Casting the 275 mm secondary lining of dense concrete up to the waterproofing membrane

The requirement for a smooth finish to the tunnel walls was obtained by travelling steel shutters. Durability of the tunnel surface was enhanced by applying a Ceramicoat coating, a magnolia color being stipulated for the walls with the roof of the tunnels being black. Niches were incorporated into the secondary lining for hose reels at approximately every 60 m, fire hydrants (incorporating hose reels) at approximately every 120 m, and emergency call points at approximately every 240 m.

In the bored tunnels, the road construction was directly supported upon selected compacted fill material placed in the tunnel invert. There was no secondary lining below the nibs cast into the primary lining segments. Any water finding its way into the invert fill, whether through the primary lining or from the road deck, would eventually drain into the tunnel invert drains, which ran the length of the bored tunnels. This tunnel seepage water and the separate carriageway drainage water were collected in drainage sumps situated at the low point of the tunnels adjacent to and below the vehicle cross passage (VCP) at Marino, and pumped into the city's main drainage at the south end of the bored tunnel via discharge pipes.

MINED SEQUENTIAL TUNNELING

Mined sequential excavation method (SEM) tunneling on the Dublin Port Tunnel Project was undertaken to provide cross passages, for both pedestrian emergency use, and for vehicles in conjunction with layby enlargements.

Emergency pedestrian cross passages were provided approximately every 250 m in both the cut-and-cover and bored tunnel sections. Between the bored tunnels these cross passages were mined, while in the cut-and-cover sections, they were formed as a part of the cut-and-cover structures. The mined emergency pedestrian cross passages were primary (shotcrete) and secondary (in-situ reinforced concrete) lined forming a domed roof, near vertical sides and flat floor, and as such are relatively conventional structures.

In comparison, the mined laybys and VCPs are major, complex underground excavations with several noteworthy features. In total there are four 40 m long vehicle laybys and cross passages, spaced at approximately 1,100 m intervals. The northern-most (VCP1) was built into the north cut-and-cover tunnel structure itself. The second (VCP2) was formed as a structure built in the base of working shaft WA2 before the shaft was permanently backfilled to the pre-existing ground level. To the south, there are two mined tunnel enlargements for laybys and VCPs located at areas named Cloisters and Marino. These locations were carefully selected to coincide with open spaces on the surface.

In the mined laybys, the TBM excavated tunnel diameter of 11.8 m needed to be increased to an excavated span of 15.7 m, so that the additional layby lane could be accommodated. The VCP design called for 16.5 m long articulated HGVs to be able to pass through. The detailed geometrical constraints, including distance between main tunnel tubes, relative orientations, gradients, main tunnel structure, ground conditions, etc., were different at both mined VCP locations. Each location required its own sepa-rate swept path analysis For the mined VCPs, excavated spans of around 8.1 m were required, enlarging to 10.1 m at the junctions with the laybys. In view of the large spans and uncertain nature of the ground at the two mined layby and VCP locations (VCP3 at Cloisters and VCP4 at Marino), comprehensive geotechnical investigations were con-ducted by Nishimatsu in addition to those undertaken prior to Contract award. Addi-tional information was obtained from logging the TBM tunnel face and using information obtained from tomographic surveys undertaken during the first TBM drive.

Cloisters Layby and VCP3 Complex

This site was characterised by having only a 2 m to 5 m cover of rock, which was low in comparison to most of the length of the main tunnel drives. In view of the vari-ability of ground conditions shown by the additional geotechnical studies, there remained some uncertainty about the nature and variability of ground conditions upon starting the required enlargement work. This was the largest underground cavern undertaken to date in Dublin Limestone. As expected, the rock conditions were vari-able with a number of geological features including faulted and highly fractured ground. Design of the whole complex proceeded using the UDEC analysis program.

The design of the layby tunnels at Cloisters also needed to take account of a design change proposed by NMI Consortium, to relocate equipment rooms for electri-cal switchgear and related equipment. In the concept design these were situated at the ends of the laybys, which would have necessitated increasing the lengths of the sec-tions of bored tunnels to be enlarged. NMI Consortium undertook a value engineering review for these facilities and relocated the equipment rooms into separate adits located between the two layby enlargements. This design change reduced the lengths of bored tunnels to be enlarged and benefited the overall construction schedule.

The layby enlargement of the first (northbound) TBM drive could be commenced only once the TBM and its backup gantries were several hundred metres to the south of the Cloisters site. A detailed program of primary lining segment removal and primary support for the enlarged excavation was implemented. This included support by means of shotcrete, lattice girders, tensioned and untensioned rockbolts working in combina-tion. For the main enlargements, pre-support was provided by rockbolting through the existing segmental lining.

The large size of the junction between the laybys and VCP3 at Cloisters and the poor quality of the rock led to a novel solution of constructing an insitu reinforced con-crete ring beam at the junction eye, as pre-support before the full enlargement of the layby and VCP3.

Figure 7.   Construction of southbound layby at Cloisters

In view of the ground being frequently heavily fractured or containing faulted zones, particular attention was paid to monitoring the response of the ground during excavation. This included deformation monitoring, pressure cells, extensometers and piezometers. Strain gauges were cast into the permanent linings of the electrical service adits to monitor relative strain and stress during the main excavations of VCP3 and the laybys. The planned excavation and primary support sequences were kept under continual review and were modified according to the conditions encountered, and assessment of performance the support structure; in other words the observational method was applied in a fundamental manner. This was based on a range of design support types and additional analysis as the work progressed.

Because there would be a period of time between commencement of the layby enlargement in the northbound main tunnel at Cloisters and the arrival of the TBM on its return journey of excavating the southbound tunnel, the opportunity was taken to excavate the upper portion of the southbound tunnel layby by sequential excavation tunneling from the northbound tunnel in advance of the TBM arrival, which is illustrated in Figure 7. The lower portion of the southbound tunnel layby was subsequently excavated by the TBM.

Marino Layby and VCP4 Complex

At between 10 m and 15 m, the rock cover gave less concern at Marino than at Cloisters, and ground conditions were expected to be better than at Marino. However, the northbound tunnel TBM drive had encountered a faulted zone directly at the planned position of the VCP. Accordingly, the whole complex was relocated 24 m to the south (this distance corresponding to two main tunnel secondary lining pour bays). The Marino site was close to the lowest point of the tunnel bores, so it was decided to incorporate the drainage sumps into the complex.

The design approach was again to undertake extensive modelling using UDEC and to design a range of location specific support. Pre-support was provided from rockbolts drilled through the segmental tunnel lining, and a hybrid solution of reinforcing the junction from the layby to VCP4, as for VCP3, was also initially adopted. As at Cloisters, the ground was again continually observed and monitored as work progressed, and changes made to the degree of primary support as dictated by the conditions revealed. In fact, the rock at Marino was more intact than anticipated and without major faults, having decided to relocate to avoid the known major fault.

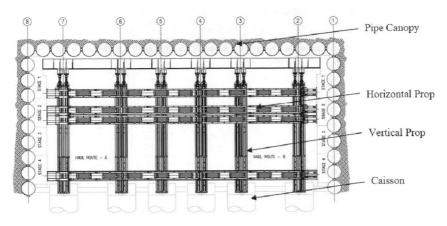

Figure 8. Section through railway crossing underpass main support frame

RAILWAY CROSSING UNDERPASS SEQUENTIAL TUNNELING

The southern cut-and-cover tunnel section passes under the main Dublin to Belfast Railway Line. At this location there are six tracks, two main tracks plus four others to a depot and wash area. In the contract Irish Rail required that no construction work was to take place within the railway land apart from enabling works and monitoring. The distance between the boundaries of railway land resulted in the 60m section of dual two lane tunnel being mined adopting a sequential excavation and support method. The required excavation was approximately 22m wide and up to 15m deep. The minimum distance from the underpass primary support to track level was approximately 2.9m.

The railway embankment consisted of compacted boulder clay constructed originally in a river estuary, with subsequent backfill, including domestic and commercial refuse, placed to each side. The encountered ground conditions were extremely mixed with complex stratigraphy of made ground, marine silt and clay, marine sand and gravel, glaciomarine deposits, glacial sand and gravel and black boulder clay. The groundwater regime was also complex and challenging. The face stability of this mix of material even with ground treatment and dewatering would have been extremely poor with small cover to the operating tracks. Access for extensive ground treatment from the surface area via the tracks was not available.

Nishimatsu adopted a specialised construction methodology that they had successfully used in forming underpasses under road and rail embankments in Japan. This method was adapted to suit the local ground conditions and geology. The principle was to install a large diameter pipe canopy around the excavation, and utilise this canopy as temporary support that spanned on to five heavy frames, with vertical support founded on caisson piles. The frames and support system were installed in advance of the main excavation. Thus this system positively supported the track during the bulk excavation, and controlled the effect of ground movements. Figure 8 depicts the arrangement.

The construction sequence is summarised in Table 1 and illustrated in Figure 9.

To complete the construction of the support system a number of ground improvement methods were required. To ensure a hydraulic cut off was achieved within the excavation, low level dewatering below the level of the permanent tunnel base slab was undertaken, supplemented by external permeation grouting to the side pipes, external horizontal drainage and installation of lengths of jet grouted wall. The internal

Table 1. Railway crossing underpass construction sequence

Stage No.	Construction Activity
i.	Construction of drive and reception shafts at Railway boundary using diaphragm/slurry walling techniques.
ii.	Part excavation of diaphragm wall shafts.
iii.	Installation of 1.22 m diameter pipe roof by pipejacking (auger plus shield).
iv.	Installation of pipe walls by pipejacking (auger plus shield).
v.	Construction of pilot tunnel and header tunnels beneath pipe canopy for installation of the three internal support frames. The shaft diaphragm wall was used for the internal external frames. The maximum pipe span was 17m between frames.
vi.	Caisson construction within header tunnels to install vertical supports and form foundations.
vii.	Installation of support frames, vertical and top members. Support pipe canopy by using jacks.
viii.	Staged benched excavation to base level in four staged levels. In advance of each level local excavations were made to install the horizontal prop for that level and to install preload to reduce ground movements.
ix.	Construction of permanent reinforced concrete tunnel works between frames.
x.	Removal of temporary steelwork and completion of permanent works.
xi.	Backfilling of all voids between permanent works and pipe support structure.

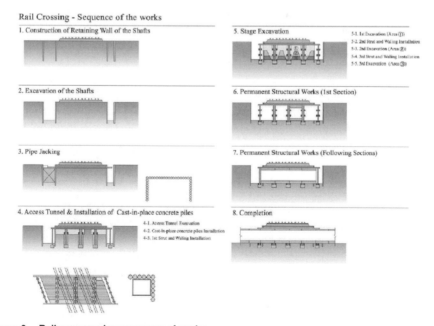

Figure 9.   Railway crossing sequence of works

Pipe canopy consisting of 39, 1.2m dia. pipes, 60m long forming 28m wide opening approx. 14m deep under live railway

Permanent tunnel concrete lining

Temporary shuttering

Drive shaft (cut & cover section)

Figure 10.   Completed railway crossing underpass

excavation was assisted by the use of sand drains, local sump pumping and vacuum dewatering. The predicted movement of the track was assessed for each stage of construction, and local track releveling and ballasting undertaken as required. A fully automated monitoring system was established on the track to ensure that the track remained within the required tolerances. Appropriate alarms levels were established and action plans established.

Figure 10 depicts the works nearing completion. These design and construction works were complex and challenging but were successfully completed by Nishimatsu.

CONCLUSIONS

This paper has highlighted the resolution of some of Dublin Port Tunnel project's design and construction issues, with particular relation to the bored and mined tunneling techniques employed. The successful completion of this, the first major road tunnel in Ireland as well as Ireland's biggest single infrastructure project, is a testament to the foresight, determination and patience of the staff of the NRA and DCC over a very long period, and to the dedication of the design and construction teams of the NMI Consortium.

A large amount of knowledge and experience has been gained in tunnel construction in the ground formations and groundwater regimes in the Dublin area. It is to be hoped that what has been learnt will be of benefit to future underground transportation planned for Dublin. Potential schemes include Dublin Metro lines, with Dublin Metro North planned to have 8 km of tunneling and seven deep stations; the Cross Dublin Railway linking Connelly and Heuston stations; and the M50 Dublin Eastern Bypass continuing south from Dublin Port.

ACKNOWLEDGMENTS

The authors wish to express their appreciation to their respective organizations and NMI Consortium for supporting the preparation of this paper. The authors are also indebted to the NRA as the funding agency and DCC as the Employer for permission for its production and publication.

15

Shafts and Outfalls—
Design and Construction

Chair
P. Rice
Parsons Brinckerhoff

CONSTRUCTION OF NINE SHAFTS FOR THE MANHATTAN SECTION OF THE NEW YORK CITY WATER TUNNEL NO. 3

Shemek Oginski

JF Shea Co., Inc.

Patrick Finn

JF Shea Co., Inc.

Yifan Ding

JF Shea Co., Inc.

ABSTRACT

The Manhattan section of the New York Water Tunnel No. 3 constructed by the Schiavone/Shea/Frontier-Kemper Joint Venture is a $668 million project. The contract includes a tunnel drive of 8.85 km (29,045 feet), tunnel lining of 14.77 km (48,468 feet) and construction of nine shafts ranging in depth from 143 m (469 feet) to 169 m (555 feet). Construction methods for excavation of the shafts are presented including raise bore, ground freezing, soil excavation, blasting, slashing, muck handling, rock support and shaft final concrete lining.

INTRODUCTION

Construction of the New York City Water Tunnel Number 3 began in 1970 and the completion is expected in 2020 at a total cost of $6 billion. Stage 1 is in service since 1998 and runs 13 miles from Hillview reservoir in Yonkers, down Manhattan and into Queens. Stage 2 consists of two sections: the Brooklyn/Queens Section and the Manhattan Section. The 14.8 km (9 miles) of 3 m (10 ft) diameter tunnel in Manhattan was constructed by Schiavone/Shea/Frontier-Kemper Joint Venture (Figure 1.)

The major part of this project are 9 shafts up to 170 m (555 ft) deep, built using unique construction methods and specialized equipment due to unique shaft profiles, geometry, construction requirements and geology. These construction methods included shaft raise boring while freezing the overburden at the same time, special drill and blast operations including blasting the rock plugs and slashing various shaft diameters using specialized shaft jumbo. Shaft construction sequence was imposed by the contract to limit excavation depth and ground exposure, thus requiring switching form excavation phase to concrete placement phase several times during the shaft construction. Finally, five different types of excavation support systems were used for nine distribution chambers due to the varying ground conditions, top of rock elevations, and site specific requirements.

SHAFT SUPPORT AND CONSTRUCTION IN OVERBURDEN

Typical Shaft Overview

Of the nine shafts on this project, eight of them, shafts 24B and 27B thru 33B, had the same basic profiles (Figure 2). The shafts ranged in depth from 150 m to 170 m (465 ft to 555 ft). Starting from the bottom and working up the shaft, the lower portion

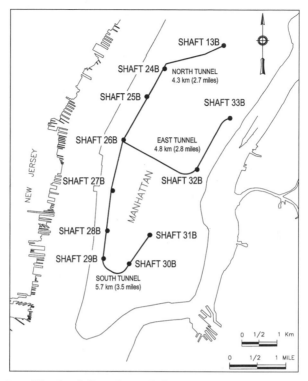

Figure 1.   Water tunnel Number 3, Stage 2 tunnel alignment

of the shaft had a 3.05 m (10 ft) finish concrete liner that connected with the 3.05 m (10 ft) finish tunnel. The shaft then transitioned to a 3.51 m (11ft 6 in) section for 3.05 m (10 ft). Continuing up was a 4.27 m (14 ft) diameter for 29.26 m to 35.36 m (96 ft to 116 ft), then a 6.71 m (22 ft) diameter for 15.24 m (50 ft) and finally a 7.92 m (26 ft) diameter for 36.58 m to 45.72 m (120 ft to 150 ft). Each shaft was topped with a subsurface distribution chamber measuring approximately 10.67 m (35 ft) by 18.29 m (60 ft) and 9.14 m (30 ft) deep.

As previously stated, part of the project consisted of the excavation and lining of nine shafts in Manhattan. Several different methods of overburden support were utilized in this project, ground freezing, secant pile wall, soil mixing, steel sheeting, soldier pile with lagging, and simply sloping back to pour concrete walls.

Ground Freezing and Soil Excavation of Five Southern Shafts

Five of the shafts, 27B thru 31B, were on the south run of the tunnel. The depth of overburden for these five shafts ranged from 18.29 m to 33.53 m (60 ft to 110 ft), and finished shaft lining was 7.92 m (26 ft) diameter. The contract specified that ground freezing be used to support the overburden in these shafts. First, a concrete collar was poured, then the pilot hole for the raise bore was drilled, and the freeze pipes were installed (Figure 3). The freezing operation was performed and monitored by Moretrench American Corporation. They installed 4.5 in OD freeze pipes on 1.07 m (3.5 ft) centers, to circulate the brine. The pipes were drilled on a 6.4 m (21 ft) radius, and installed 4.5 m (15 ft) into the bedrock. The freeze was turned on as the raise bore

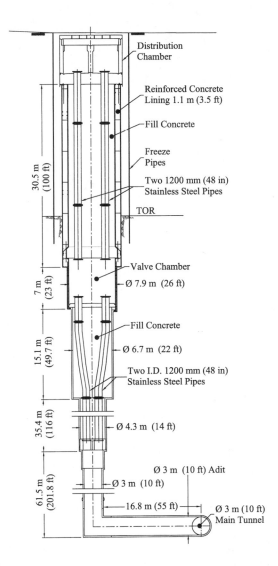

Figure 2.   Shaft 29B profile

operation was mobilized. The time required for getting closure on the freeze wall and completing the raise bore were approximately the same, this allowed the overburden excavation to begin once the raise bore equipment was off site.

A small excavator and a six cubic yard muck boxes were used to excavate frozen soil (Figure 4). At times a rock breaker was installed on the excavator to chip back the frozen ground to the required diameter. Following the mucking cycle, a light gauge wire mesh was pinned to the frozen ground using nails, and then 38 mm (1.5 in) thick two part polyurethane foam was sprayed over the wire mesh. The polyurethane foam had several purposes. First, it was installed to protect workers from any minor unraveling of soil due to surface defrost. Second, it insulated the frozen wall and maintained the

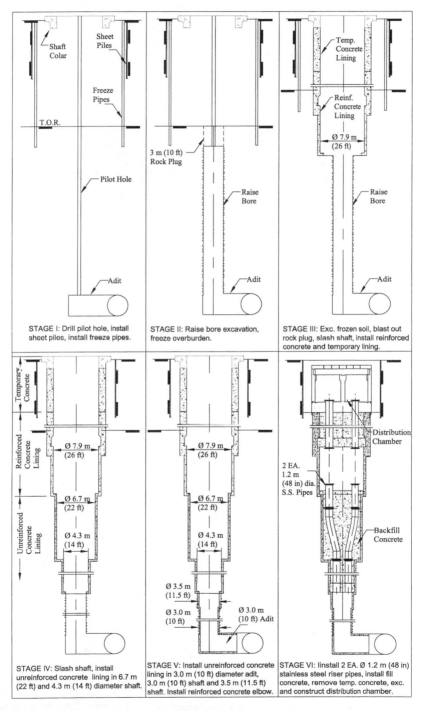

Figure 3. Construction sequence for Shaft 29B

Figure 4.   Excavation of frozen soil

temperature. Finally, it also expanded or contracted with the frozen ground, unlike steel liner plate which may buckle and become a safety issue for workers in the shaft.

Distribution Chambers Excavation

Distribution Chambers for Shafts 27B through 31B. During the excavation of the southern distribution chambers, soil mix and sheet piles, both with three levels of bracing, were used to support the overburden. These support systems were installed prior to any other shaft work being done on the sites so that access around an open shaft was not an issue. Shafts 29B and 28B used 13.7 m (45 ft) long PZC18 steel sheet piles. The sheets were driven 4.6 m (15 ft) below the final excavation elevation. Shafts 27B, 30B and 31B used a soil mixing wall for support. The soil mixing was used to minimize any vibrations that may have caused damage to structures around these three sites. The soil mixing wall was installed using a specialized drill rig with a set of three drill rods. The rods augered thru the soil and injected a bentonite and cement mix into the soil and then a 13.7 m (45 ft) W21×57 steel beam was placed into every other element before the soil mix set up. Once the shafts had been excavated, concrete lined, pipes installed and backfilled, the excavation of the distribution chambers was done. The overburden was excavated to 1 m (3 ft) below the first level of support and then steel whalers were installed. This cycle continued through all three levels of bracing until subgrade elevation was reached. Mini-piles were installed in the bottom of the excavated chamber to support the distribution chamber structure. The chamber is a subsurface cast in place reinforced concrete structure.

Distribution Chambers for Shafts 24B, 25B, 32B and 33B. Shafts located on the northern portion of this project have significantly less overburden, ranging from 1.52 m to 7.62 m (5 ft to 25 ft). For these shafts, 24B, 25B, 32B and 33B, a different approach was taken. The entire distribution chambers was supported and excavated down to a final grade, including rock and overburden, prior to setup of the raise bore operation. Shaft 25B, with the bedrock only 1.52 m to 2.44 m (5 ft to 8 ft) down and no ground water issues, was approached by just sloping the overburden back and pouring a concrete curb wall. As the curb wall was poured, pockets were blocked out to hold beams that spanned the shaft and supported steel decking. The beams and decking were designed to support the weight of two loaded concrete trucks, which were used in the concrete backfill process.

Shafts 24B and 32B utilized soldier pile and lagging, along with two stages of bracing for support of the overburden. A steel casing was installed in the overburden, and a 610 mm (2 ft) rock socket drilled, then W12×65 soldier piles were installed on 1.98 m to 2.44 m (6.5 ft to 8 ft) centers. The rock sockets were backfilled with a 20.67 MPa (3,000 psi) grout and the remainder with a lean mix grout. Prior to excavation a curb wall was formed and poured around the shafts to help support the beams and steel decking used to cover the shafts during the slashing phase. The shafts were excavated to 610 mm (2 ft) below the first bracing level, installing wood lagging between the piles and backpacking with sand and hay as needed. The top bracing level was installed and the chamber excavated to 610 mm (2 ft) below the second bracing level. The second bracing level was installed and excavation resumed to the top of rock. At shaft 24B, rock pins were also installed in any area where the rock was deeper then 6.1 m (20 ft). These grade 150 threaded rock pins were located on each side of a pile, drilled in at a 45 degree angle, 1.8 m (6 ft) embedment into the rock, and used resin epoxy. Once the overburden had been excavated, concrete grout was pumped behind the wood lagging to fill any voids and minimize water inflow. Shotcrete was also used at the interface between the rock and the overburden to stop any running ground and support the weathered rock.

The Shaft 33B distribution chamber was located right next to the historic Queensboro Bridge, approximately 3 m (10 ft) away. To eliminate any settlement, a secant pile wall was specified. In addition, a future bridge footing was incorporated in the design, along a portion of the wall near the bridge. The secant wall was designed with 103 secant pile elements, 53 primary and 50 secondary. Through the overburden, each element was, 762 mm (30 in) in diameter, and spaced at 610 mm (2 ft) center to center. The primary secant elements were cased thru the overburden and drilled 305 mm (12 in) into rock, backfilled with 6.89 MPa (1,000 psi) lean mix, and then the casing was removed. The secondary elements were cased thru the overburden as well, a 610 mm (24 in) diameter hammer drill was then used to extent a rock socket to 915 mm (3 ft) below the bottom of the finished excavation. A W18×76 pile was installed in the secondary element and backfilled with a 6.89 MPa (1,000 psi) lean mix. In the area of the future bridge footing, the W18×76 was replaced with a W18×106 and the 6.89 MPa (1,000 psi) lean mix was replaced with 27.58 MPa (4,000 psi) concrete. In addition 13 HP14×117, were installed as individual caissons and backfilled with 27.58 MPa (4,000 psi) concrete. Once the secant wall was completed, a curb wall was poured around the shaft similar to shafts 24B and 32B. Excavation of the chamber then began; it was excavated down 1.52 m (5 ft) from the existing surface and bracing was installed. The rest of the distribution chamber was then excavated down to rock.

The rock portions of these northern distribution chambers were excavated with drilling and blasting. Typically the entire chamber was predrilled to the subgrade elevation, requiring 550 blast holes ranging in depth from 1.2 m to 5 m (4 ft to 16 ft). Due to the noise and vibration concerns, the New York Fire Department limited the number of holes that could be shot in each blast. The initial blast at each of the chambers started with 20 to 25 blast holes, the amount of powder was also limited resulting in portions of the chambers being shot in multiple lifts. The blast pattern could be modified following each of the blasts depending on the noise and vibration results. The shot rock was loaded with an excavator to a muck box then hoisted to the surface to be trucked away.

Figure 5. Raise bore head

Raise Bore Excavation

The excavation of the hard rock portion of the nine shafts was done in two main stages. The first stage was the raise bore operation and the second stage was a combination of blasting and slashing to widen the raise bore hole to the final excavated dimensions of the shaft. The raise bore created a 3.82 m (12.54 ft) diameter hole that connected the rock surface to the existing tunnel. The blasting and slashing operation followed the path of the raise bore and all the muck generated was pushed into the hole to be hauled out by muck cars in the tunnel.

The first stage of the raise bore operation was to create a 356 mm (14 in) diameter pilot hole through the centerline of the circular shaft. The pilot hole started from the surface and extended down to the existing tunnel. The pilot hole was drilled using a dual rotary drill. Periodical single shot survey and occasional gyroscopic survey were used to ensure that the pilot hole was within maximum tolerances of 0.5% of the depth, thus typically allowing not more than 750 mm (2.5 ft) of deviation at the intersection with the tunnel. Adjustments in the direction of the drilling were done using directional drilling tools. Once the pilot hole was finished, a comprehensive gyroscopic survey was used to map the pilot hole in north-south and east-west directions to verify the required tolerances. The pilot hole was constructed by Ziegenfuss Drilling Inc.

After the pilot hole was completed, the raise bore equipment was assembled on the surface and the raise bore head was transported through the tunnel for assembly right below the pilot hole. The two raise bores machines used were: Ingersoll-Rand RBM211M (300 HP) and RBM 7SP (200 HP). Temporary concrete pads were poured surrounding the pilot hole. The concrete pads were used as a bearing surface to the raise bore equipment. After the raise bore head was assembled, the thrust and torque were gradually increased to production boring rates. Muck fell from the raise bore head and into the tunnel where it was loaded into muck cars and hauled to the dump station at the main shaft. For the five southern shafts with deep overburden, the raise bore operation stopped when the head was within 3.05 m (10 ft) of the rock surface. The head was then lowered down into the tunnel and disassembled. A different approach was taken at the four northern shafts. Since the top of rock was near the surface, the raise bores holed through and the raise bore head was pulled out at the top of the shaft (Figure 5). Frontier-Kemper Constructors, Inc. was directly responsible for the raise bore operation.

Figure 7.   Shaft jumbo

Figure 8.   Shaft headframe

attached to the ends of the beam. Turning sheaves moved along the length of the support beams. The work deck could move along the length of the support beams or perpendicular to them. The headframe was also used to suspend concrete forms and
move them up after each concrete placement.

     Each shaft also had a service crane. The cranes ranged from 100 tons to 250 tons
of capacity. The capacity of the crane was determined not only by the heaviest load
(two level shaft jumbo) but also by the size and orientation of the distribution chamber.
Some of the cranes used were a Manitowoc 777, a Liebherr HS 555 and a Liebherr
HS 895.

Figure 6.   Rock drilling with shaft jumbo

Blasting and Slashing

To widen each shaft to its final dimensions, conventional blasting method was used and the rock was slashed into the raise bore hole. Since each shaft was divided into segments of different diameters, each diameter had its own drill pattern. The typical blast hole drill pattern was a series of concentric circles with the raise bore at its center (Figure 6). The challenge in this stage of the excavation was to meet the strict noise and vibration limits and prevent fly rock. These requirements were most difficult to satisfy when the excavation operation was still near the surface. Heavy-duty steel shaft covers and multiple layers of steel and rubber mats were used when the blasting took place near the surface. The amount of explosive charge per delay was also very limited. The depth of the blast holes was reduced to as little as 1 m (3 ft) to meet the requirements of the contract. As the excavation got deeper, the blast holes were also allowed to be deeper and the excavated shaft diameter also got smaller. As a result, the advance rate downward increased as the excavation operation moved deeper into the shaft. The blasting was stopped when the excavation reached an elevation that was just above the last segment of the shaft whereby the diameter of the raise bore was large enough to satisfy the minimum excavation needed specified in the contract.

Shaft Jumbo. The drilling for both the blast holes and rock bolt holes was done by using a two level shaft jumbo (Figure 7). The maximum excavation diameter of the jumbo was approximately 11 m (36 ft) and the distance between the lower and upper decks was 6.1 m (20 ft). Lateral stabilizers were attached to both the upper and lower levels of the jumbo. The upper deck had removable extension segments that could expand the diameter of the upper deck from 3.66 m (12 ft) to 6.71 m (22 ft). The upper deck was the main staging area of the jumbo. It encompassed the hydraulic power units and a landing for the man cage. Four attachment points for the head frame or the crane were at the same level. The installation of rock bolts and wire mesh took place on the upper deck. The lower deck was only 3.66 m (12 ft) in diameter. Blast hole drilling and rock bolt installation took place in this area. A mucker was attached to the underside of the lower deck. The mucker was used to move the blasted rock into the raise bore hole.

The shaft jumbo was suspended by cables attached to the headframe. The headframe mainly consisted of two paired support beams that moved on rails and a work deck that contained the man cage hoists and the operating panel for the entire headframe (Figure 8). Each pair of support beams had two Timberland winches that were

Muck Handling

With the start of the raise bore, all muck was removed through the tunnel. Muck cars transported the muck to the vertical belt at the main staging area at the bottom of Shaft 26B. There were California switches at key locations in the tunnel that allowed the muck trains to pass each other. The muck trains that supported the raise bore used side dump muck cars and dumped muck directly into a hopper where the muck was taken by an incline 762 mm (30 in) conveyor to the vertical conveyor. The muck trains that supported the blasting and slashing operation used lift off muck cars and the muck went through a crusher to allow for the muck to be hoisted by the vertical conveyor. The vertical belt transported the muck to the surface and dumped it into an enclosed area. CAT 966 loader was used to load the muck into trucks.

Temporary Rock Support

The temporary rock support for all shafts included rock bolts and wire mesh. The rock bolts installed in all shafts, except non circular shaft 25B, were either No. 8 (25 mm) Grade 75 Dywidags fully encapsulated or Swellex friction bolts. Swellex bolts allowed for quick installation and provided consistent performance that met the specified bolt capacity.

For the noncircular Shaft 25B, Williams hollow bar, groutable bolts were used as specified in the contract. Unlike the Dwyidag rock bolts, cement grout was used instead of polyester resin. The cement was pumped through the hollow center of each bolt until the drilled hole was filled. Compared to Swellex rock bolts, the Williams bolts required more installation time because of the added steps related to grouting and tensioning.

CONCRETE LINING

Since the specifications did not allow entire shaft concrete lining placement after all the blasting was done, the concrete lining had to be poured after about 30 m (100 ft) of shaft excavation. As a result, the shafts were constructed in cycles of blasting and placing concrete. With the changing vertical height of the various shaft diameters, but typically not exceeding 30 m (100 ft) excavating and concreting was done one shaft diameter at a time.

For all the shafts, the upper 7.92 m (26 ft) diameter had reinforced concrete lining while all the shaft lining below was unreinforced. Most of the reinforcement was delivered to the job site as prefabricated rebar mats. The height of the rebar mats was either as tall as the shaft form or twice as tall as the form used in the pour with proper splices protruding above each pour. Rubber waterstops were used at construction joint. There were 76 mm (3 in) deep keyways at set intervals along the length of the shaft. These keyways were formed either by styrofoam pieces attached to the outside face of the steel form or the form had steel keyways fabricated as part of the shaft form. All shaft forms were fabricated by Everest Equipment from Quebec, Canada.

The height of the forms used for the shafts were based on the pour sequence of the shafts. The pour sequence was decided based on the need to minimize the number of construction joints and the maximum amount of concrete that could be poured in one day. As a result, each shaft diameter had a different form height that varied from 3.05 m (10 ft) to 9.75 m (32 ft). For the larger diameter portions of the shaft, the form height varied from 3.05 m (10 ft) to 5.18 m (17 ft). For the smaller diameter portions of the shaft, the form height varied from 8.23 m (27 ft) to 9.75 m (32 ft). For some transition areas in the shaft, styrofoam forms were attached to the outside face of the form to create the necessary transition shape.

Installation of Stainless Steel Pipes and Backfill Concrete

Once the shafts were excavated and lined to the tunnel, the piping and backfill started. The lower section of the shafts, 3.05 m (10 ft) and 3.51 m (11.5 ft) diameter sections, remain as a final product with the concrete lining. Starting with the 4.27 m (14 ft) diameter section of the shaft stainless steel pipes were installed along with the concrete fill. There were two stainless steel pipes installed in each shaft, 1,219 mm (48 in) in diameter and up to 15 m (50 ft) in length. Once both lengths of pipe were installed they were backfilled with several lifts of concrete. This piping and backfilling cycle continued throughout the 4.27 m (14 ft), and 6.71 m (22 ft) sections of the shaft to the bottom of the 7.92 m (26 ft) diameter section. The bottom of the 7.92 m (26 ft) section was left open for a valve chamber and then a plug and pipes installed continuing to the distribution chamber. In addition to the stainless steel pipes, an access shaft and an elevator shaft were installed in the backfill concrete thru the 7.92 m (26 ft) diameter section, for access to the valve chamber. This portion of the work was performed by J.P. Picone. They will also install stainless steel piping and fill concrete in the main shaft 26B which was excavated and lined under the previous contract.

CONCLUSIONS

The construction of the New York Water Tunnel Shafts offered Schiavone/Shea/Frontier-Kemper Joint Venture an extremely challenging project. All stages of constructing the nine shafts had to be precisely scheduled as the work in the shafts so heavily depended on the support from the tunnel crews below. As the south tunnel was excavated under the previous contract, the excavation of the south shafts could start immediately with the notice to proceed. All the blasting in the south shafts had to be completed before the concreting operation of the south tunnel could begin. The excavation of the north tunnel and adits was completed next followed by the excavation of shafts 24B and 25B. Once the east tunnel was excavated, excavation of the last two shafts 32B and 33B was completed. The tunnel concrete operation could then move to the north and east tunnel. Job schedule had to be followed very closely as all these operations depended on each other.

Another challenging aspect of construction was the choice of equipment. Cranes, custom made shaft headframes with winches, drill jumbos and concrete forms were purchased or custom built to accelerate the schedule and perform multiple tasks. Our means and methods were also optimized so the specialized equipment could be used multiple times.

DESIGN RISK MITIGATIONS FOR CONTRACT CQ028 OF THE EAST SIDE ACCESS PROJECT

Jeffrey L. Rice
Parsons Brinckerhoff

Avelino Alonso
Parsons Brinckerhoff

Wai-shing Lee
Parsons Brinckerhoff

Samer Sadek
HNTB Corp.

David I. Smith
Parsons Brinckerhoff

INTRODUCTION

East Side Access (ESA) Contract CQ028 is located in the New York City Borough of Queens. The contract involves construction of a 230-meter (760-foot) long open-cut excavation with a permanent deck supporting Amtrak yard tracks over a semi-circular TBM launch wall on the east end, and the 36.5-meter (120-foot) long Northern Blvd. Crossing on the west end. The Northern Boulevard Crossing will be mined under an existing New York City Transit (NYCT) five-track subway tunnel; Northern Boulevard, a heavily traveled five-lane truck route; and an existing NYCT three-track elevated line; connecting to an existing open-cut shaft adjacent to the existing 63rd Street Tunnel bellmouth. When completed, Contract CQ028 will provide access for future Manhattan and Queens tunnel contracts. The footprint of the open-cut was determined by the geometric requirements of the future Plaza Interlocking facility to be constructed within in another future ESA contract. See Figure 1.

This paper discusses design risk mitigations for the key elements of the contract, namely the Northern Blvd. Crossing and the semi-circular TBM Launch Wall.

PROJECT OVERVIEW

The Metropolitan Transportation Authority Capital Construction Company—East Side Access Project for the Long Island Rail Road will provide Long Island commuters with direct access to Manhattan's East Side via a new rail line between the Harold Interlocking in Queens and Grand Central Terminal in Manhattan. The project connects the unused lower level of the existing tunnel under the East River at 63rd Street to new connecting tunnels in Queens and Manhattan.

Contract CQ028 Scope and Construction Schedule

This contract consists of the excavation of an open-cut from the east side of Northern Boulevard, through an existing MTA-owned rail yard and into the western limits of Amtrak's Sunnyside Yard. The contract will connect to a completed open-cut

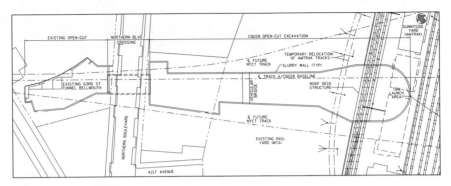

Figure 1. Contract CQ028

through a mined tunnel under Northern Blvd. When complete, the open-cut in the existing rail yard will serve as a muck handling/material delivery/staging area for future Manhattan and Queens tunneling contracts. Upon completion of the major tunneling contracts the open-cut will house the Plaza Interlocking Facility, which includes a two-to-four track interlocking, tunnel ventilation, emergency egress, signal instrumentation control rooms, traction power substations for the Queens tunnels and a future storage yard above, and a bulk power facility substation feeding all of the power requirements for the ancillary spaces throughout the Queens segment.

Contract CQ028 was awarded in April of 2006 and is currently within schedule and budget. The contract is scheduled for an October 2008 completion. The Contractor, Pile Foundations Construction Co., Inc. is currently preparing the final design for the Northern Boulevard Crossing.

SUMMARY OF SITE CONDITIONS

Geotechnical Investigation

More than 80 borings were taken in the project area with most terminating at least 4.5 meters (15 feet) below top of sound rock. Additional field investigations were conducted, including field permeability estimation, geophysical investigation to confirm the elevated structure pile lengths, and suspension P and S velocities. Pump tests were also conducted on site for dewatering work assessment.

Overburden at Project Site

The subsurface strata identified in this area in sequence are Fill (Stratum 1), Organic Deposits (Stratum 8), Mixed Glacial Deposits (Strata 2, 3 and 4), Glacial Till/Reworked Till/Outwash (Stratum 5), Decomposed Rock (Stratum 6) and Bedrock (Stratum 7). The depth to bedrock ranges from approximately 12.2 to 27.4 meters (40 to 90 feet) below the existing ground level. A brief summary of each stratum in general sequence below the ground surface is presented in Table 1. Subsurface profiles at Northern Boulevard Crossing and the TBM Launch Wall are shown in Figures 2 and 3, respectively.

Local Groundwater Conditions

At the project site, groundwater flow is principally southwest towards Newtown Creek and to a lesser degree, west and northwest toward the East River. The depth to

Table 1. Subsurface strata summary

Stratum	Thickness	Description
1	2.7–10 m (9–33 ft)	Brown, gray-brown to yellow-brown or reddish brown, very loose to very dense sands with silts (generally 5–22%), gravel and miscellaneous debris. The Unified Soil Classification Symbol (USCS) group symbols are generally SW and SM or SW-SM. The Standard Penetration Test (SPT) N-values ranged from 2 to >60, but generally were 2 to 50.
8	0–3 m (0–10 ft)	Either black to gray, medium stiff organic silt or a very soft brown peat with organic silts and clay (generally 20–99%). Sands, gravels and fibers were also encountered. The USCS group symbols are generally OL, OH, Pt and ML. The SPT N-values ranged from weight of hammer (WOH) to 34, but generally were WOH to 12.
2	0–10.7 m (0–35 ft)	Brown to gray, loose to very dense coarse to fine micaceous sands with silts (generally 2–28%) and gravel. The USCS group symbols are generally SM or SW-SM. The SPT N-values ranged from 3 to >60, but generally were 3 to 51.
3	0–5.5 m (0–18 ft)	Gray to brown and olive brown, very loose to dense fine micaceous sands and silts (generally 9–62%) with gravels and boulders. The USCS group symbols are generally SM or ML. The SPT N-values ranged from 2 to >60, but generally were 4 to 50.
4	0–12.2 m (0–40 ft)	Brown, gray to olive brown, medium stiff to hard non-plastic to low plasticity silts and clays (generally 50–96%). The stratum was predominantly varved with fine micaceous sands and fine gravel. Gravel, cobbles and boulders are also noted. The USCS group symbols are generally ML to CL. The SPT N-values ranged from 2 to >60, but generally were 6 to 45.
5	0–7.6 m (0–25 ft)	Heterogeneous mixture of brown, gray to reddish gray/brown, medium dense to very dense sands with silts (generally 5–36%), gravel, and boulders. The USCS group symbols are SM, SW-SM and SM/GW. The SPT N-values ranged from 6 to >60, but generally were 12 to >60.
6	0–9.1 m (0–30 ft)	Green to gray, very stiff silts and clays (generally 32–53%) with sands and gravel. The USCS group symbols are generally SM to CL. The SPT N-values ranged from 17 to >60, but generally were 25 to >60.
7		Gray fine- to coarse-grained, unweathered to moderately weathered, strong to very strong Gneiss. RQDs are generally higher than 70%.

groundwater generally varies from 3 feet (in the existing rail yard) to 15 feet (in the Northern Blvd. vicinity) below the ground surface. Cut-and-cover structure construction will occur beneath the water table.

Groundwater Contamination

Sunnyside Yard is listed as a Class II Inactive Hazardous Waste Site by NYSDEC. Several contaminant plumes have been delineated and are being monitored. There are several dissolved contaminant plumes and a floating petroleum product plume in shallow groundwater.

In order to prevent plume migration, groundwater lowering is limited to 2 feet at the excavation support wall. The groundwater cut-off system will be achieved by constructing a relatively watertight slurry wall around the perimeter of the excavation. The cutoff walls will be extended to bedrock.

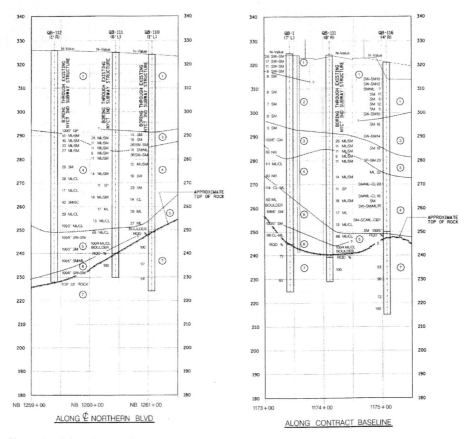

Figure 2.   Subsurface profiles at Northern Boulevard crossing

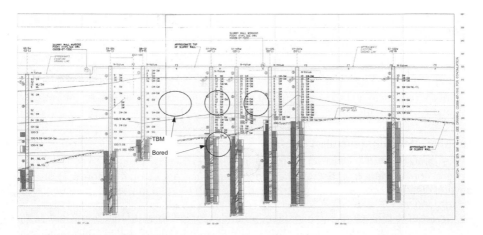

Figure 3.   Surface profile at TBM launch wall

NORTHERN BOULEVARD CROSSING

Perhaps the most challenging portion of CQ028 is the Northern Boulevard Crossing. See Figures 1 and 4. Due to the complexity of this segment, feasibility studies of multiple options were conducted during the design phase.

Challenges in design and construction of this mined tunnel can be categorized into two major categories.

Location Constraints. Construction from within the existing subway structure is prohibited and the CQ026 open-cut is actively used by Manhattan tunneling contracts, thus the only option is construction of a mined tunnel under Northern Boulevard from the CQ028 open-cut. See Figure 4. The elevated structure is supported on steel pile groups. The alignment of the mined tunnel intersects with four groups of piles.

Subsurface Constraints. As discussed in the subsurface condition section, the ground condition at the site is mixed glacial deposits, with high consistency of silt, sand and boulders. The groundwater level is approximately 5 meters (16 feet) below grade and roughly 9 meters (30 feet) above the mined tunnel crown. The hydraulic conductivity varies from 3^{-4} to 3^{-6} m/sec (10^{-4} to 10^{-6} ft/sec). The invert of the tunnel will intersect with the highly undulated rock surface which tends to slope toward the west. Table 2 summarizes the soil parameters used.

Design Schemes

To support the elevated structure, an underpinning scheme was designed using a new piling system to support the elevated structure while mining past and cutting the existing piles. This temporary support system will be removed later and the existing piles supported on the final tunnel structure.

Multiple options were considered for groundwater control and soil strengthening. Jet grouting was excluded due to the difficulty of performing horizontal jet grout columns under high groundwater pressure. The fines contents in the soil layers are high making it difficult to achieve a consistent soil/grout matrix to control groundwater and strengthen the soil by using chemical or cement grouting. An arched pipe option was considered, however the variation in the top of rock and the steel piles and boulders in the soil matrix, ruled this method out.

The study concluded that a mined option utilizing a frozen arch and sequential excavation was the most feasible option capable of the groundwater cutoff and soil strengthening required. In this option, freezing pipes are installed in a semicircular array resulting in a frozen arch for groundwater cut-off. The freezing pipe array provides flexibility to overcome obstacles such as boulders, the steel piles, and drilling misalignment that might occur due to the difficult ground conditions. The frozen arch can also provide a tight rock interface that is flexible in shape.

In order to overcome the heave that may result from the frozen arch, temperature control pipes are required. The contract documents also required compensation grout pipes to be installed prior to all other activity under Northern Blvd. to control potential settlements that might occur during the support system installation and subsequent excavation, and during the thaw of the frozen arch.

After the placement of the frozen arch, mining can be performed using sequential excavation and a temporary shotcrete lining while partially relying on the strength of the frozen arch. During the design phase, multiple options and sequence were considered and optimized. The resulting is scheme indicated on the contract documents as a suggested method of construction. Final support design will be the responsibility of the contractor.

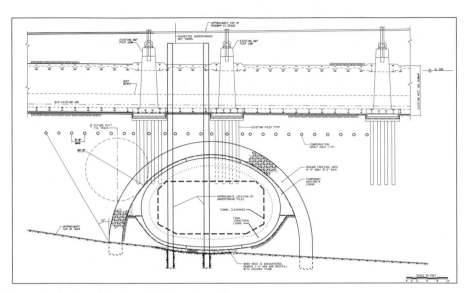

Figure 4.   Section at Northern Boulevard crossing

Table 2.   Soil parameters

Soil Layer	γ (sat) kN/m³ (pcf)	K m/sec (Ft/sec)	υ	E MN/m² (psf)	Su kN/m² (psf)	φ°
Layer 3	19.6 (125)	1×10^{-5} (3×10^{-5})	0.4	13.9 (2.9×10^5)	0	30
Layer 4	19.6 (125)	1×10^{-6} (3×10^{-6})	0.4	27.8 (5.8×10^5)	62 (1,300)	0
Layer 5	21.2 (135)	2×10^{-4} (7×10^{-4})	0.3	69.0 (1.44×10^6)	0	38
Layer 7— Rock	26.7 (170)	—	0.3	6,900 (1.44×10^8)	6,030 (126,000)	—

SEMI-CIRCULAR TBM LAUNCH WALL

The east end of the open-cut excavation will be closed using a 25-meter (82-foot) radius semi-circular slurry wall. Soil and groundwater loads will place the wall in compression, creating a horizontal arching effect that maintains stability without the need for internal bracing and with only a small number of tie-backs to account for the undulating rock surface.

Three 6.55 meter (21.5 foot) diameter bored tunnels will be launched through the semi-circular wall. The TBM for mining the yard-lead track will be assembled in a trench excavated in the rock at the bottom of the open-cut. It will launch into a full face of rock and bore under the semi-circular wall. See Figure 5. The tunnel profiles are arranged to minimize mixed-face tunneling, however although two of the upper tunnels will be launched into a full face of soil, the third will launch into a small amount of rock in the invert. Jet grouting, or other ground treatment methods, will be required to stabilize the soil in mixed-face areas and to prevent soil or groundwater from entering the

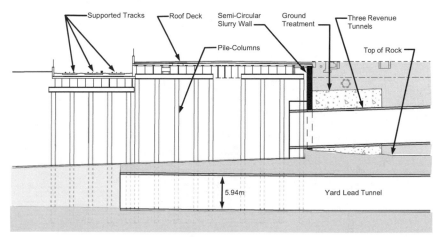

Figure 5. Longitudinal section through TBM launch wall

open-cut. The avoidance of tie-backs and bracing for the arched wall simplifies ground treatment, TBM assembly and launch.

The slurry wall is between 12m and 20m deep, 305mm thick, and uses soldier-pile-tremie-concrete (SPTC) construction with light reinforcing cages that provide additional bending capacity. Where TBMs mine through the wall, fiberglass reinforcement will be used. No capping beam or internal ring beams are incorporated. Panels are 10.2 meter (10.5 feet) wide, constructed in three sub-panels using a conventional grab. A deviation from vertical of 0.5% is permitted and accounted for in the arch design. The wall is keyed a minimum of 150mm (6 inches) into rock to achieve a good seal during wall-rock interface grouting.

A three-dimensional shell-element analysis of the wall was undertaken to determine the effect of the variable rock line, the three holes created by the TBMs, and asymmetrical external loading. This indicated that in general the arch remains stable without internal bracing or external support. However, because the rock line dips down to the north, the arching effect at the bottom of the deepest part of the wall is reduced. The bottom of two panels will be anchored into rock using a single row of tie-backs. Analysis showed that even with imperfect keying into rock, toe deflections and moments in the wall are minimal.

Construction of the TBM launch wall will staged to minimize disruption on two railroad yard tracks which cross the site that are vital for Amtrak's Northeast Corridor and New Jersey Transit services, and demolition of two unused two-storey buildings. Trains will be temporarily diverted onto two shoofly tracks on top of completed sections of slurry wall, prior to bulk excavation. See Figure 6. After all of the slurry walls are finished, a semi-circular roof deck will be constructed across the east end of the open-cut, and the tracks will be returned to their original alignments. The roof deck will be supported on pile-columns located between the future TBM launch locations and in alignment with the future facility to be constructed within the open-cut. Bulk excavation in uniform layers followed by rock blasting will take place below the roof deck. After completion of the open-cut and subsequent bored tunnels, a future contract will construct a permanent internal structure, including track levels, ventilation spaces, emergency exits, a central signal room, and miscellaneous equipment rooms.

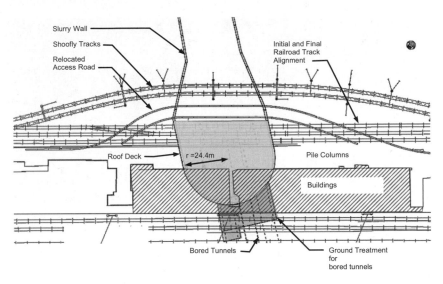

Figure 6. Plan at TBM launch area

Construction Status

The majority of the slurry wall work is complete, including a temporary slurry wall creating an early access shaft for the Northern Boulevard Crossing work. Remediation and demolition of the existing buildings and installation of the temporary shoo-fly tracks are scheduled to begin mid-January, 2007.

The contractor, Pile Foundation Construction Company, Inc. has retained the services of Dr. G. Sauer Corp. for the design of the Northern Blvd. Crossing, including underpinning of the elevated structure columns and modification to the existing slurry wall bracing. Currently, the Northern Blvd. Crossing scheme under development is a hybrid of the ground freeze and grouted pipe arch.

DEVELOPMENT OF SLURRY WALL TECHNIQUE AND EQUIPMENT AND CSM CUTTER SOIL MIXING FOR OPEN-CUT TUNNELS

W.G. Brunner

Bauer Maschinen GmbH

ABSTRACT

The slurry wall technique has undergone an evolution from its invention to its today's status. In the early stages slurry walls were built with cable grab, later on with much stronger hydraulic grab and last but not least with the use of the cutter technique. The cutter allows thick walls with high verticality to extreme depths through hard soil and rock formations. With the rising demand of the construction of watertight diaphragm walls different joint systems have been developed. In addition to the rapid development of excavation systems it was necessary to improve slurry treatment and desanding plants, especially in conjunction with high performance trench cutters.

Deep mixing methods can be used economically for the construction of cut-off or excavation support walls where other systems such as traditional soldier beams and lagging walls would yield unsatisfactory performance, where the installation of vibrated or driven sheet piles could cause vibration induced settlements and where concrete diaphragm walls would be time consuming and expensive. The new CSM method can largely replace the more conventional single or multiple auger methods of soil mixing or jet grouting. The cutter tool is similar to a cutter used in diaphragm walls or cut-off walls and is operated with a Kelly mounted on a drilling rig or rope suspended on a crane jib or mast of a base carrier. With this system much deeper walls can be installed and even hard soil layers can be penetrated.

INTRODUCTION OF SLURRY WALL SYSTEM

The slurry wall or diaphragm wall technique was introduced in Europe in the 1950s for the construction of underground reinforced concrete structural wall systems. The technique initially evolved from the principal idea of stabilising open excavations with thixotropic fluids, a concept invented in the 1940s by Prof. Veder of Austria. During the past 20 years it has undergone a further dramatic evolution not only in terms of analysis and design methodologies, but also in terms of construction technology and construction equipment.

In the early stages of development, slurry walls were constructed by cable grab mounted on tripod rigs on rails. In subsequent years, specialist firms developed alternative excavation equipment such as rope grabs suspended from crawler cranes (Figure 1).

The limitations of rope operated grabs, mainly due to their lack of power and speed when excavating stiff clayey soils and rock formations, were overcome by the development of hydraulically operated grabs capable of generating high closing forces (Figure 2).

In the 1960s, a milestone in the development of slurry wall excavation equipment was reached with the introduction, in Japan, of the reverse circulation technique in the field of slurry wall construction. Whilst one company developed a rig with blade cutters rotating around a vertical axis, another one designed a cutter with two cutting wheels rotating around a horizontal axis. Both systems used either a suction pump located outside the trench or an airlift system to remove the excavated material.

Figure 1.   Crane with rope grab

Figure 2.   Hydraulic grab

Figure 3.   BC 30 at Brombach

Figure 4.   BC cutter wheels

The arrival of the reverse circulation cutter technology in Europe resulted in intense research and development by several European geotechnical and foundation contractors.

DEVELOPMENT OF THE CUTTER

In the early 1980s, the first BC 30 trench cutter was developed and built for the construction of a cut-off wall in fractured sandstone below the main dam of the Brombach storage reservoir in Germany (Figures 3, 4, 5 and 6).

The introduction of the newly developed cutter system on the world market opened up new horizons for the uses and applications of the slurry wall technology:

- **Increase in depth:**
  Whilst the 1950s the average depth of slurry walls was in the range of 20 m in the 1950s, it is now possible to construct walls to depths of 100 m and more.

- **Increase in wall thickness:**
  When the slurry wall technique was first introduced, the equipment capacity was limited to widths of between 500 and 800 mm. Today, projects have been successfully executed with a wall thickness of 2,500 mm.

- **Wall construction in hard soil formations and rock:**
  Trench excavation in rock has been made possible by the introduction of trench cutters capable of producing high torque (hydraulically power is often in excess of 600 kW) and also by the development of cutter wheels with excellent rock crushing capabilities (such as the roller bit).

Figure 5.   BC 30

Figure 6.   BC 30

Figure 7.   CBC/MBC in Singapore

Figure 8.   CBC in Japan

- **Wall construction in limited headroom conditions:**
  Despite the fact that the capacity of trench cutters is still increasing, special cutters have also been developed which do not require more than 6 m headroom. These 'mini' cutters were specifically requested by the Japanese, Korean, Taiwanese, US and recently Singaporean and Italian markets for the construction of diaphragm walls in connection with mass rapid transit systems in the big cities (Figures 7, 8, 9 and 10).

- **Environmentally friendly construction process:**
  Trench cutters have the added qualities of noise and vibrationfree excavation, which is an important environmental factor when working in densely populated areas.

- **High degree of verticality:**
  One of the major advantages of trench cutters over other conventional cutter equipment is their ability to construct trenches to extremely high vertical tolerances. The verticality of cutters is controlled by inclinometers, which are integrated into the cutter system. For stringent verticality requirements a specially developed hydraulically operated cutter steering system is deployed. Adjustments

Figure 9.   MBC in USA   Figure 10.   MBC in Korea

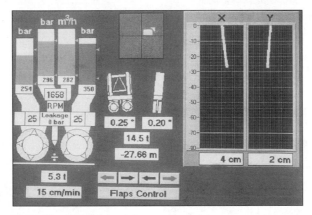

Figure 11.   B-Tronic module

of the cutter position in vertical direction of the trench are made via long hydrauli-cally operated steering plates, whilst adjustments in longitudinal direction are made by pivoting the cutter head through up to ±3 degrees. Throughout the cor-rection process, the rig operator is continuously guided by corrective measures displayed on his on-board monitor.

- **Automatic quality assurance system:**
  The entire slurry trench construction process is monitored and controlled by B-Tronic, a PC-based electronic system. On completion of each diaphragm wall panel a full cutter report can be printed out inside the operator's cab or transmit-ted by remote data transfer direct to the site office, the company's head office or the consultant engineer's office for further analysis and evaluation (Figure 11).

In conclusion, the market required slurry trench or diaphragm wall construction equipment capable of excavating deeper and thicker trenches at greater speed through harder soil formations.

In the early days, the cutter frame as well as the hydraulic hoses and slurry hoses were simply suspended from a rope on a long crane jib and the depth of the trench was limited to twice the length of the jib of the base carrier. As a result of the increas-ing demands on slurry trenches as referred to above, the base machines for cutters also increased until gigantic cranes had to be used. But more and more frequently dia-phragm walls had to be constructed in down-town city centre areas amidst bustling

Figure 12. Slurry recycling plant

traffic and on very restricted sites. This led to the development of smaller, so called 'compact' cutters, with all hoses on hose drums, capable of constructing diaphragm and cut-off walls to greater depths on extremely restricted sites.

In addition to the rapid development of trench cutter excavation systems, it was also necessary to improve slurry treatment, desanding plants as well as decanters, especially in conjunction with high performance trench cutters (Figure 12).

WATER TIGHTNESS AND DIFFERENT JOINT SYSTEMS

Water tightness is particularly important for permanent diaphragm walls and is not always easy to achieve. The degree of water tightness that can be achieved depends primarily on the type of joint between individual panels, the joint details and adequacy of their construction.

Standard Joints Constructed with Stop-End Casings

This is the traditional and conventional system, using rope grabs for the excavation for a slurry wall depth of up to 25 m and for water pressures of 4–6 m. The stop-ends are pulled out with casing extractors. The guide wall has to be designed for the reaction forces resulting from the stop-end casing extractor.

Joints Constructed with Prefabricated Concrete Elements

This system is generally used for a slurry wall depth of up to 25 m or a lifting weight of the pre-cast concrete element of up to 25 tons. For greater depth the pre-cast element can be installed in several sections which will be connected during the installation. The pre-cast element guides the grab during the excavation of the second-ary panel. The water tightness can be improved by placing additional rubber water stops in the concrete panel (Figures 13 and 14).

Joints Predrilled with Prefabricated Concrete Elements

This system is used for a slurry wall depth of up to 25 m or a lifting weight of the precast element of up to 25 tons. For greater depth the precast element can be installed in several sections which will be connected together during the installation.

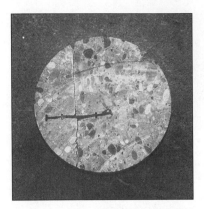

Figure 13. Core of joint with water stop

Figure 14. Installation of pre-cast concrete element

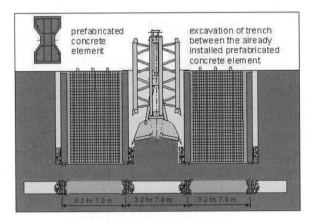

Figure 15. Hydraulic grab excavating a panel

The precast element guides the grab during the excavation of the secondary panel. The water tightness can be improved by placing additional rubber water stops in the concrete panel (Figure 15).

Joints Constructed with Endplate Reinforcement Cage

This system is used for a slurry wall depth of up to 80 m. The endplates guide the grab during the excavation of the secondary panel. The water tightness can be improved by placing rubber water bars at the endplate (Figure 16).

Joints Constructed with Continuous Water Stop

This system developed by Bachy is used for a slurry wall depth of up to 30 m. The water tightness is improved by the rubber water stops of the Continuous Water Stop (CWS) system. 2 CWS forms should be placed in each primary panel. They can only be removed during the excavation of the secondary panels (Figures 17 and 18).

Figure 16.   Endplate reinforcement cage

Figure 17.   CWS forms with two water stops

Figure 18.   Exposed CWS joint with two water stops

Joints Constructed by Overcutting the Primary Panels

This system is used for a slurry wall depth of up to 120 m. The water tightness is improved by cleaning the overcut joint with a brush before installation of the reinforcement cage (Figures 19, 20 and 21).

Due to the cutter technique, the construction of watertight joints is now much easier and much more effective than with the conventional rope grab method. Instead of using stop-end tubes, the cutter encroaches on the adjacent primary panels as it descends during excavation and thereby cuts back a fillet from each end of the previously concreted primary panels. The cut-back or overlap between primary and secondary panels ranges between 200 and 300 mm. Joints produced in this manner are watertight, because they consist of a serrated surface resulting from the formation of grooves cut into the concrete of the primary panels by the cutter wheels. This not only increases the seepage path through the joint, but also provides a clean and roughened surface with which the concrete of the secondary panel is then able to bond extremely well without the risk of the formation of a filter cake.

Limitations of the grab technology are the excavation output reduces with the increasing trench depth as a result of the intermittent nature of the process of

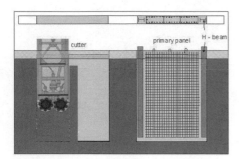

Figure 19.   Overcut joint system with BC

Figure 20.   BC at open-cut tunnel job site in Turin, Italy

Figure 21.   BC at Naples, Italy job site

excavation and in the case of the excavation of hard cemented soil formations. There are difficulties with the construction of joints for very deep slurry walls or cut-off walls.

CONVENTIONAL SOIL MIXING WALLS

In 1962 Norman Liver patented an auger-based soil mixing technique that has become the basis for the world's current technology. The deep mixing method (DMM), as it is known today, is an in situ soil treatment and improvement technology where the soil is blended with binders and/or other materials that are introduced in a slurry or dry form. Typically one or more hollow, rotating mixing shafts tipped with some type of cutting tool are mounted vertically to the mast of a base carrier and are driven into the ground by hydraulic or electrically powered rotary heads. The shafts above the cutting tool may be further equipped with discontinuous auger flights and/or mixing paddles. The binder is injected through the hollow shafts to nozzles located near the cutting tool where it is mixed with the ground to form columns of treated soil. Individual column diameters range from 0.6m to 1.5m and may extend to 40 m in depth. They can be arranged individually or in contiguous or secant patterns to produce a wide range of

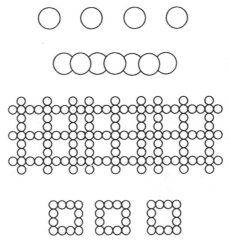

Figure 22.   Types of column arrangement

treated soil structures on both land and marine projects. Typical configurations are: Single elements, rows of overlapping (secant) elements, grids, lattices and blocks (Figure 22). The particular geometry that is chosen is dictated by the purpose of the DMM application and reflects the mechanical capabilities and characteristics of the method used.

The main groups of applications are:

- Hydraulic cut-off walls
- Excavation support walls
- Large volume ground treatment
- Liquefaction mitigation, in situ reinforcement
- Piles, gravity walls
- Environmental remediation.

CUTTER SOIL MIXING (CSM) WALLS

In 2003 Bauer developed the CSM method by exploiting its experience in the manufacture and use of the trench cutter systems to excavate diaphragm walls panels. The CSM method differs from the traditional DMM method in so far as it makes use of two sets of cutting wheels that rotate around a horizontal axis to produce rectangular panels of treated soil rather than one or more vertical rotating shafts that produce circular columns of treated soil.

Two cutter gear boxes are connected to a special mounting that is in turn connected to a robust Kelly bar. The Kelly bar is connected to the mast of a drill rig by two guide sledges that steer and provide crowd and extraction forces and, if necessary, rotation to the cutting tool assembly (Figure 23). As the cutting wheels rotate and penetrate into the ground they break up and loosen the soil. During this phase a fluidifying agent or the binder itself is injected into the area between the two cutting wheels. In the extraction phase the cutting wheels rotate in a mixing mode and blend the binder and soil to form a rectangular panel of treated material.

Figure 23. CSM head

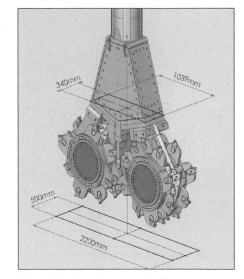

Figure 24. Typical cutter unit with measurements

The parameters of the machines designed and built to-date are (Figure 24):

Length of panel	2,200, 2,400 and 2,800 mm
Width of panel	500 mm to 1,200 mm
Torque output	0 kNm, 50 kNm, 80 kNm
Max wheel speed	40 rpm
Max cutter unit power	150 kW, 200 kW, 300 kW
Weight of cutter unit	3,700 kg, 5,100 kg, 7,400 kg
Maximum depth capability with a single tube Kelly	35 m
Maximum depth capability cable suspended	70 m

On future machines the lengths of the panels could be extended to 3,200 mm and their width capability up to 1,500 mm. The rectangular panel, when compared to a series of contiguous or secant columns, offers a number of distinct advantages: structurally, if we compare the properties of a rectangular shaped panel to those of secant or contiguous columns whose diameter is equivalent to the width of the panel, we find that the former is a much more efficient shape the areas of treated soil in compression and tension are larger and the lever arm of the rectangular section is larger; this implies a higher moment of resistance. When considering a secant column wall, column diameters need to be much bigger than the thickness of the rectangular panel to produce a section of equivalent width. This means that when using the CSM method we need to treat significantly smaller volumes of soil to obtain the same effect. Clearly this implies savings in the total energy expenditure in producing the wall and a saving in the amount of binder that is used (Figure 25). When additional strength or resistance to bending moments is required, the CSM wall can be reinforced efficiently with steel 'H' sections. Given the rectangular shape of the panels, distribution of the steel in the panels can be designed to optimize the quantity of steel (Figure 26).

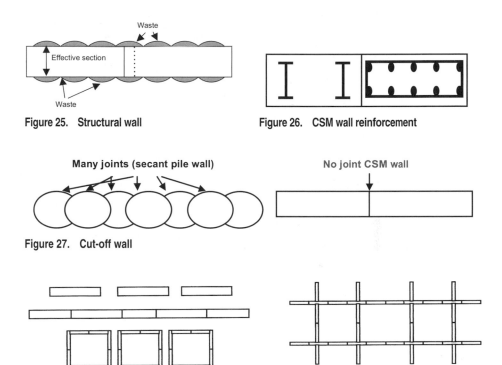

Figure 25.   Structural wall

Figure 26.   CSM wall reinforcement

Figure 27.   Cut-off wall

Figure 28.   Types of panel arrangements

If we design the wall to act as a cut-off wall, the single CSM panel is continuous over a 2,200 mm up to 2,800 mm length whereas an equivalent secant column wall will have at least 3 joints. There is a much lower risk of leakage through the CSM panel (Figure 27). There are a number of other advantages that the CSM method and machinery offers when compared to the traditional rotating augers or paddles, notably: the only moving parts in the CSM method are the cutting wheels, as a result we can mount instruments inside the cutter gearbox support frame that give real time information throughout the treatment depth, information such as verticality, deviations, excess pressure build-up in the surrounding soil, etc.

In addition, by varying the relative speeds of the two cutting wheels the operator can correct any deviation that may occur. Further as the Kelly bar does not rotate there is no energy expenditure like in the traditional DMM methods where a certain amount of energy is lost to overcome friction between the long shafts of the augers and the soil/cement mix. Similarly to when using circular columns, the rectangular panels can be arranged individually, or in contiguous or secant patterns to produce a wide range of treated soil structures. Typical configurations are: single elements, rows of overlapping (secant) elements, grids, lattices and blocks (Figure 28).

The First Test in Aresing, Germany

A round shaped test shaft consisting of 14 panels with a depth of 10 m was installed at our testing area in Aresing (Figures 29 and 30). Additional panels were constructed to be excavated and examined afterwards (Figure 31). The important parameters in the construction of the shaft are summarized in Table 1. Most of the

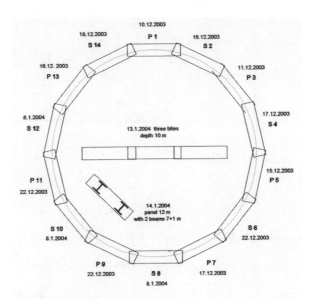

Figure 29. Aresing CSM-shaft with production dates

Figure 30. Shaft showing perimeter and panels   Figure 31. Exhumed panel

research effort on this first test was dedicated to finding a fluidifying agent that would give the best penetration performance during the cutting and mixing phase: we started with a water and stone powder mix and then tried other mixes including bentonite, polymers, the cement binder and finally a mix of bentonite and a fluidifier that gave the best results. Bentonite mud was made up using a water/bentonite ratio of 12 and the grout was made up using a water/cement ratio of 0.5. Average performance rates in executing the panels that made up the shaft perimeter ring were:

Average excavating performance (Down)	18.05 m^2/hr.
Average blending performance (Up)	62.73 m^2/hr
Overall average productivity	13.8 m^2/hr

Table 1. Aresing construction parameters

Element Number	Surface Area (m²)	Excavating Time (min)	Blending Time (min)	Consumption of Bentonite Mud (m³)	Consumption Cement Grout (m³)
P1	44	126	48	10	6.6
P7	41.8	140	57	10	7.2
P9	44	135	42	14.2	7.5
P11	44	133	35	11.8	7.7
S2	44	176	55	14.5	8.2
S4	44	109	46	10	7.0
S6	41.8	153	31	10	6.0
S8	41.8	170	30	9.4	5.3
S10	41.8	130	44	9.5	6.8
S12	41.8	163	36	10.6	6.8
S14	44	185	47	12.2	6.1

Two panels that were excavated using the bentonite plus fluidifier were performed at an average penetration rate of 31 m²/hr; the average blending performance was 67 m²/hr. This implies that an average productivity 21 m²/hr can be achieved. The average consumption of bentonite powder was 44.5 kg/m³ of treated soil and the average consumption of cement was 444 kg/m³ of treated soil. The in-situ soil at the test site is generally sandy gravel with a 2 m to 3 m thick clayey sand layer at approximately 12 m below ground level. The final strength of the soil/cement mix, given by compression strength tests carried out on cores taken in different panels, ranged from 5 MPa to 8 MPa.

Test in Tokyo, Japan

A second test was carried out in conjunction with Messrs. D.K. Com in May 2004 in the Tokyo bay area in Japan where soil conditions are typically 6 m to 9 m of soft clays underlain by fine silty sands (Figure 32). The object of the test, performed strictly according to normal Japanese soil mixing practice for retaining walls, was to create a series of panels of improved soil that were fluid enough to permit the problem-free installation of steel beams. Other requirements were: good homogeneity of the mix, continuity of treatment throughout the panel and good quality joints; strength was of secondary importance (0.49 MPa was sufficient). The tests followed Japanese practice and included:

- using compressed air, injected together with the binder, during both the downward cutting and upward mixing strokes;
- pumping a high water/cement ratio binder and inject cement binder on both the downward and upward strokes.

The main working parameters used for the test were:

- depth of panels—20 m.
- intersection of primary and secondary panels—100 mm.
- cement content: 200–250 kg/m of treated soil.

Figure 32. Top of excavated wall

Table 2. Selected construction parameters

Panel Number	Depth (m)	Total Time (min)	Total Volume of Binder (liters)	Productivity
2/6	20	86	15,200	30.7
2/8b	19.4	48	10,600	53.3
2/7	18.7	44	13,300	56.1
2/9	20	64	12,900	41.3
2/11	20	55	11,000	48.0
2/10	20	71	12,650	37.2
2/13	20	63	13,355	42.0
2/12	20	72	13,300	36.7
2/5	19.4	60	6,990	42.7
2/4	20	57	7,260	46.3

- approx. 70% of theoretical binder volume injected on the down stroke, 30% injected on the up stroke.
- high penetration extraction speed 50–70 cm/min on the downward stroke and 70–120 cm/min upwards using high pump rates of 200–300 L/min).

Observations. Levels of torque during the cutting and mixing phases were low and the speeds of penetration and extraction were sustained throughout. This was due to the nature of the soils, the large volumes of liquid that were injected into the soil/cement mix and the action of compressed air that kept the slurry in the panel turbulent and fluid throughout the process. The degree of blending and the consistency of the mix were very good; a visual check of this quality, done after exposing some of the panels, showed the mix to be very homogeneous, much better than expected. The selected construction parameters are shown in Table 2.

Retaining Wall in Holland

A CSM retaining wall (approx. 700 m^2) was offered as an alternative solution by HOFFMANN GROEP (Helmond) for the construction of a retaining wall to support

Figure 33.   Installation of CSM wall

Figure 34.   Propped reinforced CSM wall

3 sides of an excavation for a 2-storey basement. The original design was for a series of contiguous, soil mix columns reinforced with steel beams (Figures 33 and 34). The site is located in Valkenburg (Netherlands) in a hilly area between Aachen and Maastricht. Some of the main criteria imposed were:

- The work had to be performed from within the site, a confined area; this was particularly restrictive in some of the corners.
- There was minimum clearance to the neighbourhood buildings and structures
- The system had to be vibration-free.
- Strength was of secondary importance. The structural design assumed that horizontal loads would be transferred by the vertical steel beams (270 × 280 mm at 1 m centres) (Figure 35)
- Spoil quantity was to be minimized to reduce the amount of cement used and the removal of spoil.

Soil conditions were generally a fine grained soil (silty fine sand, with clayey lenses) with a stiff clay layer generally between 5 m and 6 m below ground. There was no groundwater. The working parameters were:

- panel size 2.2 m × 0.5 m, 10 m deep
- overlap = 150 mm
- water/cement ratio: Several w/c ratios were tried in order to find an optimum mix. W/C 1.0 proved to be too dry; W/C 1.33 to 1.6—the fluidifying effect was better but there was some concern about final strength; W/C 1.2 was finally adopted as a good compromise between workability and strength.
- Cement content: 250 kg/m^3 of treated soil.
- Use of cement slurry from start (approx. 85% of theoretical volume downwards, 15% upwards)

During the construction of the CSM wall all data were recorded and following observations have been made:

- High torque levels on the mixing wheels were required.

- Generally the wheel mixing and cutting direction was outwards. Sometimes the rotation direction was changed to inwards when extracting the tool.

- Visual check of mixing quality showed very homogeneous spoil with thick, mortar-like consistency. Sometimes small lumps of unmixed soil were found in the spoil.

- Beam Installation: 2 beams (270 × 280 mm) were installed in every panel. Due to the confined size of the site, the beams had to be installed with the auxiliary winch of the BG 15H. It was assumed that the beams would penetrate easily under their self weight. This occurred only at a few locations; mostly they had to be pushed down with the aid of a small backhoe. The stiff clay layer proved to offer the largest resistance to their penetration.

- Performance: The overall production was influenced by many side activities, mainly by the installation of the beams. Nevertheless, average production was 86 m^2/day (8 hours). The average net, cutting and mixing production was 16.4 m^2/hr.

The execution was controlled by the B-Tronic quality control system. The collected working parameters such as depth, rotations per minute, slurry pressure and slurry volume were recorded.

CONCLUSIONS

The cutter technology as well as the cutter soil mixing technique may be used for the construction of retaining walls for open cut or top down excavated tunnels or other deep excavations.

The described advantages of the cutter technique led to the use of cutters in many projects around the world in particular because of high performance, greater depth, greater wall thickness, penetration through and into hard rock, noise and vibration-free excavation and high degree of verticality.

The cutter soil mixing method, CSM, is an innovative method for carrying out Deep Soil Mixing. It constitutes a new item in the Soil Mixing Methods and offers numerous advantages over the more conventional methods of mixing soils using standard rotary tools. The method has been greeted with enthusiasm by the DMM community and promises to develop into a powerful construction tool. The B-Tronic control system as described for the cutter is used for the CSM system as well.

The tests described in this paper were some of the first applications of the CSM method. Bauer is carrying out work using cutting wheels with a different geometry of the cutting and mixing blades in order to improve productivity and to reduce wear. Some of our Japanese clients are applying the method in different soil conditions in Japan in order to optimize the working procedures and parameters and the concentrations of different fluids that can be used during the cutting and blending phases of the work. In the last three years several retaining walls have been carried out using the CSM system in Germany, Netherlands, Belgium, Italy, USA, Canada, Japan and Australia.

DROP SHAFTS FOR NARRAGANSETT BAY COMMISSION CSO ABATEMENT PROGRAM, PROVIDENCE, RHODE ISLAND

Rafael C. Castro
Jacobs Associates

Geoffrey Hughes
Louis Berger Group

Fredrick (Rick) Vincent
Jacobs Associates

Philip H. Albert
Narragansett Bay Commission

ABSTRACT

Seven pairs of drop and ventilation shafts, between two and nine feet in diameter, will convey combined sewer flow to a 230-foot-deep, 26-foot diameter storage tunnel. This paper will present case histories for each site. Discussion will include functional requirements, program and site access constraints, structural design, geology, the variety of excavation and lining construction methods used, and lessons learned. Noteworthy observations include the importance of assuring competent ground support at the soil-bedrock interface and of achieving tight verticality tolerance using both top-down and raise bore excavation methods.

PROJECT OVERVIEW

The Narragansett Bay Commission (NBC) is a regional sewer authority that serves 10 communities in the Providence, Rhode Island metropolitan area. From 1992 to 1999, the NBC developed a comprehensive facilities plan to abate Combined Sewer Overflow (CSO) pollution in the Upper Narragansett Bay. The plan underwent many major evolutions during this period, most notably the addition of a stakeholder's process by which NBC demonstrated a site-specific plan to control the pollutants and meet water quality standards. The favored solution consisted of deep tunnel storage with treatment at existing wastewater facilities. In 1999, a final plan was accepted which included two tunnels, five CSO interceptors, 12 sewer separation projects, a wetland facility, and a wastewater treatment facility upgrade. The program was to be constructed over 20 years in three sequential phases.

Phase 1 of the program included a 16,284-foot-long, 26-foot-diameter, 230-foot-deep storage tunnel; adits; drop shafts; ancillary facilities; and a pump station. Construction of Phase 1 began in 2001 and is expected to operational by 2008. Seven near surface sites to divert and convey combined sewer flow to the deep storage tunnel consist of diversion or interceptor relief structures, consolidation conduits, screening structures, approach and vortex structures, and drop and vent shafts.

In 1992 the NBC retained Louis Berger Group as Program Manager (PM) for the CSO Abatement Program. The PM team included Jacobs Civil Inc. as the primary designer assisted by CH2M Hill Inc. Haley and Aldrich provided geotechnical design and Gilbane/Jacobs Associates provided construction management services.

This paper focuses on design and construction of the drop and vent shafts.

DESIGN

Design Issues

During the design process, it was established that the program should encourage participation by local construction firms and, where possible, the construction contract scopes should be kept within the bidding range of smaller local contractors. To this end the work at each near surface site was to be packaged to bid individually and let out periodically in a series, rather than simultaneously. Appropriate milestones and constraints were thereafter established in the near-surface and tunnel contracts, and the program schedule organized to complete all but one pair of shafts using top down construction methods prior to excavation of the tunnel below.

Functionally, the near surface structures and shafts were designed to transfer up to the peak flows from a 1 year storm. Near surface structure configurations were based on the flow dependent guidelines established by the Iowa Institute of Hydraulic Research (IIHR) and scale model studies. The models dictated the geometric sizes and configurations for the approach channel, tangential inlet vortex generator, drop and vent shafts, deaeration chambers, and connecting adits. One of the shafts was also enlarged to act as an emergency overflow in the event that all gates closed when the tunnel was full (Figure 1).

Based on the hydraulic study and modeling, appropriate nominal dimensions were selected for each shaft and associated collection components. Four 5-foot, one 6-foot and two 9-foot diameter drop shafts were selected. Once the sizes had been established, it was important to consider installation requirements that would maintain the modeling assumptions.

The tangential inlet vortex configuration was favored over the plunge and helical configurations because of its proven record of energy dissipation and control of air entrainment. For the tangential configuration to function properly, it was critical for the verticality of the drop shaft to be maintained within prescribed limits. If the shaft deviated from plumb beyond these limits, the water would detach from the shaft walls and freefall into the chamber below, instead of swirling around the shaft as it descended. The freefall would have the negative effects of entraining additional air into the tunnel that would consume needed capacity, and accelerating long-term scouring of the deaeration chamber below by the erosive action of the falling water. Based on tolerances used for construction on previous shaft projects, the Contract Documents included maximum deviation from vertical of 3 inches in any 100-foot section and 6 inches in the overall depth of the shaft. Specific quality control measures were also established, such as verticality checks at the lesser of every 10 feet of drilling or once per day, and sizing of the excavated opening to meet the final tolerances based on the specific drilling and support methods chosen.

Geology

The region's lower surficial deposits consist of a basal glacial till overlain by discontinuous strata of glaciofluvial and glaciolacustrine cobbles, sands, and silts. Nearer to the surface are estuarine, alluvial, and urban fill deposits, which are generally thicker adjacent to the river and bay. The bedrock consists of a Devonian- to Pennsylvanian-aged sedimentary series that has been variably metamorphosed and deformed. In

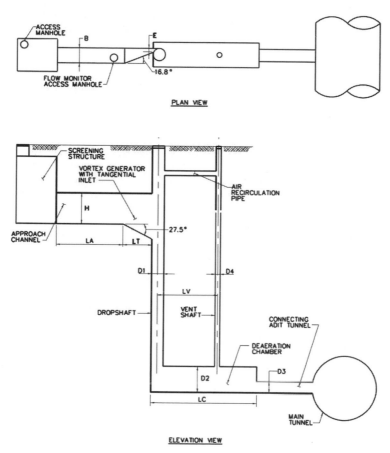

Figure 1.   Components of collection system as described by the IIHR Model with critical dimensions identified (Jain and Kennedy 1983)

addition to relatively competent conglomerates, sandstones, siltstones, and shales with average UCSs of 10 to 13 ksi and a maximum UCS of 25ksi, there lay intermittent deposits of weak graphitic shales with UCSs of less than 1 ksi.

Bedrock topography is extremely variable with soil depths of 26 to 170 feet along the main spine alignment. The tunnel profile was selected so that the crown was two diameters below the lowest apparent top of rock elevation. Apart from regional variability, it was also documented that extreme localized irregularity existed both in the elevation and composition at the soil/rock interface. The thickness of overburden at the drop shaft sites varied between 60 feet and 120 feet, representing one-third to one-half the total depth of the shafts (220 feet).

Groundwater levels were generally within 10 feet of the surface. At a few locations along the alignment, hydrostatic readings in rock were found to be higher than in the soil, with one well at the north end becoming seasonally artesian.

A Geotechnical Baseline Report was generated for each pair of shafts. Special emphasis was placed on the need for sealing off soil at the irregular top of rock, anticipating high hydraulic conductivity in soil and rock, and coping with graphitic shale, weathered rock, and man-made obstructions. Discussions also included appropriate warnings to maintain verticality through the variably resistant rock and soil.

Contract Design

The Contract Documents for the drop and vent shaft linings called for a cast-in-place concrete lining with a minimum compressive strength of 4,000 psi. Drop shafts were to be reinforced with welded wire fabric (WWF) placed at the center of the minimum 6-inch lining. The vent shafts, which were 2 feet in diameter at all sites except one, which was 3.5 feet in diameter, were all unreinforced. The Contract Documents allowed for stay-in-place (SIP) forms, especially for construction of the smaller shafts where manned entry was not possible. Due to the corrosive nature of combined sewage, the SIP forms were allowed to be constructed only of specific non-corrosive materials. To facilitate compatibility of construction methods, the contractor was responsible for the initial soil and rock support design.

In recognition of the variety of installation and design methods that would meet the design intent for the drop and vent shafts, the Contract Documents allowed for alternate liner designs. The alternate design liner criteria included most of the pertinent design assumptions that were used by the original designer, together with limits on material selection for the shaft liners (allowable materials were those that met previously described durability concerns). The contractor was responsible for designing several structural interfaces with near surface connections. The contractor was also required to show calculations that proved that the shaft lining installed would not require bearing support from below in the final condition, as the sequence of the program was such that the shaft would be constructed prior to the tunnel being constructed below. Alternate liner designs were required to be submitted as one package with the shaft excavation submittal to demonstrate that the methods were compatible.

Approved Alternate Lining Designs

At all but one of the seven shaft sites, alternate liners were installed. Two different alternate lining designs were used. One alternate lining concept, developed by GZA GeoEnvironmental of Providence, consisted of a High Density Polyethylene (HDPE) pipe as the inside surface, surrounded by a 4,000 psi grouted annulus. The annulus contained a ⅝-inch thick, ASTM A36 lining steel casing to replace the WWF. The vent shaft at these sites included only the HDPE and annular grout. The second alternate lining concept, developed by RT Group of East Providence, consisted of a prestressed cast cylinder pipe (PCCP) as the final internal lining with a 4,000 psi grouted annulus in both the drop and vent shafts. The drop shaft used an Embedded Cylinder Pipe (ECP), while the vent shaft used a Lined Cylinder Pipe (LCP). The embedded steel cylinder and prestressed steel strands in the PCCP pipes satisfied the reinforcing requirement.

CONSTRUCTION

Construction of the drop and vent shafts was accomplished by various drilling and lining installation methods. In addition to the subsurface conditions, such as top of rock elevation and thickness and density of glacial deposits overlying the rock, drilling methods were selected based on the diameter of shafts and accessibility to the connecting tunnel below. The alternate lining designs significantly reduced installation efforts.

Table 1. Summary results for drilling methods

Method	Drill Head	No. of Shaft Pairs	Finished Shaft Diameters (feet)	Drilled Hole/ Casing Diameters (inches)	Drilled Hole Verticality, feet out of plumb	Time to Complete One Hole (weeks)
Pile Top	Button Bit Rollers	3	2 and 5	48 to 84	2 to 5	4 to 15
Super Top	Cutting Casing/ Clam Shell	2	2, 5, and 6	72 to 98	0.5 to 2	3 to 8.5
Conventional	Augers, core barrels, telescoping casings	1	2 and 9	78 and 162	2 to 4	5 to 6
Raise Bore	Button Bit Rollers	1	3.5 and 9	72 and 144	1.3 to 1.5	8 (freeze) + 3 (pilot & raise)

Drilling Methods

Four different drilling methods were used to excavate the seven pairs of holes in which the drop and vent shaft liners were installed. At the six sites where shafts had to be installed ahead of tunnel construction, three different top-down drilling methods were employed. At the seventh location, the schedule allowed shaft installation after tunnel construction and the contractor chose to freeze the soil and use the raise bore method. Specialty subcontractors were hired by the prime contractors for each of the four methods employed. Table 1 provides a summary of the drilling methods used with production and verticality results for each method.

Drilling Method No. 1—Pile Top. Three pairs of shafts were excavated using "Pile Top" drilling rigs. Two specialty contractors—Treviicos of Boston, MA and New England Foundation Company (NefCo) of Hyde Park, MA—used pile top rigs manufactured by Wirth. These rigs were used to excavate, in the wet, both the overburden and the rock. The pile top rig, as the name suggests, is secured to a steel casing, which also provides the temporary support for the overburden. The steel casings were vibrated through 80 to 110 feet of overburden to the top of rock using large vibratory hammers. The drill head, rotated by a 12-inch-diameter hollow drill stem, is used to excavate the overburden within the installed casing and the rock below the casing. For these projects, the drill heads were relatively flat faced and tooled with button bit rollers. Soil and rock cuttings were removed through the hollow drill stems using an air lift system. The air lift system pushes high pressure air down to the drill head through 2-inch pipes attached to the outside of the drill stem. Immediately above the drill head the air is piped into a special fitting which turns the air 180 degrees back up and inside the drill stem, creating a vacuum affect that sucks the water and cut rock and soil from beneath the drill head up to the surface. This system necessitates relatively higher effort and larger surface space to set up, since it requires large sediment ponds, pumps and piping to remove the rock and soil from the water. Figure 2 shows a photograph of a pile top rig and a drill head as it is lowered into the casing to be connected to the drill on the rig.

Overall, this drilling method worked well for excavating both the overburden and the rock. The first site where this system was used provided valuable information about

Figure 2. Wirth pile top rig

which methods worked and which did not. However, the geology varied enough from site to site that there were more lessons to be learned at the second and third sites. Lessons learned include:

- Trained and well-informed operators are critical to the successful operation of the drill rigs. The first hole drilled with this method achieved twice the production rate of the remaining holes, but the resulting hole was five feet out of plumb, which was too great a deviation to allow installation of the liner within tolerance. The operator had been bearing down with the hydraulic jacks of the drill slide to increase downward pressure on the head in an attempt to achieve higher penetration rates. The correct approach to achieve better tolerance was to allow only the weight of the drill head and stems to apply downward pressure.

- Vibrating the casing to the top of rock proved to be one of the more difficult tasks. A proper seal between the casing and rock was critical to prevent overburden material from running or flowing into the uncased rock excavation. At one site the softer glacial deposit, overlying weathered rock allowed the casing to be installed and seated securely in rock with relative ease. At this location the installation of 100 feet of temporary overburden casing took only one day, with most of the time spent on casing welding. At a different site, denser and thicker glacial deposits overlying a sloping top of rock made it difficult to install the casing and seat it properly in the rock. At this location, repeated cycles of drilling and advancing the casing to properly seat the casing within the rock was required. For each cycle, this involved removing and reinstalling the pile top rig to allow the vibratory hammer to advance. The casing and installation of the casing took several weeks to complete.

- Achieving the desired verticality tolerance with this drill was difficult. However, Contract verticality tolerances could be and were achieved by simply drilling larger holes that would allow the liner to be installed within tolerance. This lesson not only applied to this specific method, but to all subsequent top-down drilled shafts.

Figure 3.   Super top drill rig

The time to excavate a single shaft with this method varied considerably (refer to Table 1). The shortest duration (four weeks) was achieved at two shaft locations where there were no delays and work was performed in two shifts per day. Observations indicate that drilling production rates were *not* dependent on the shaft diameter. At approximately one foot per hour, the penetration rate through rock and soil was consistent for all the shafts except the first shaft, as noted, which averaged double this rate. While the penetration rate in the rock was dependent on the ability of the drill head to break up the rock, the penetration rate for soil excavation was restrained intentionally to avoid plugging the drill stems with spoils. The most significant delays to production were due to either trouble installing steel casing to the top of rock and/or hampered efforts in achieving the verticality tolerance.

Drilling Method No. 2—Super Top. One contractor—Raito of Woburn, MA—used a "super top" drill rig. This drill rig sits on the surface, uses hydraulic grippers to grip the outside of steel casings, and spins and pushes the casings into the ground. The casings are fitted with carbide teeth and spun at approximately one rpm. The soil within the casing is removed using a clam shell while maintaining an appropriate plug of soil and water level within the casing to avoid bottom instability. Rock excavation was also completed with a clam shell after a rock chisel was used to loosen/break up the rock. The welded overburden casing consisted of 40-foot-long 1-inch-think steel sections. The casings used in the rock were slightly smaller in diameter to fit inside the overburden casing and flush-bolted to ease removal prior to installation of the final liner. Spoil handling required less effort than the methods that used air lift systems, since significantly less water was removed along with the soil and rock. Additionally, the excavated rock ranged from small cobble-sized chips to bolder-sized chunks, as opposed to the silty sandy consistency produce by the selected roller bits of the pile top method. Figures 3 is a photograph of this drill rig in operation.

This method also worked well for excavating the overburden and the rock. Holes drilled with this method provided the best verticality results of the four methods (see

Table 1). The primary lesson learned with this method is that it relies heavily on the competency of the ground on which the drill rig is founded. These rigs have limited ability to adjust shaft verticality for an uneven foundation. This was evident where the drill rig experienced some differential settlement during drilling the overburden and caused the drilled hole to be too far out of plumb. The contractor had to backfill the hole and install temporary foundation piles to better support the drill rig. Although it would appear that this method would be well suited for securing a good seal between the soil casing and the top of rock, for one shaft a complete seal was not made, resulting in significant loss of ground. Two conditions that contributed to this are: (1) top of rock was sloped and (2) because of the rig differential settlement problems discussed above, the overburden portion of this particular shaft had to be redrilled through the grout-filled abandoned hole. The combination of these two conditions caused the casing to bind and jam, exceeding the torque limitations of the drill rig and casing left above the top of rock.

Achieving good production rates with this method required good operators trained in the art of using a clam shell and chisel within steel casings. Excavation rates for this method did depend somewhat on the size of the hole that was being excavated. Refer to Table 1 for production rates.

Drilling Method No. 3—Telescoping Casings and Auger. One pair of shafts was constructed using conventional telescoping casings with auger excavation under slurry for the overburden. The rock portion was excavated using an A-frame mounted air lift drill, which was tooled and operated in similar fashion to the pile top rigs. The specialty contractor for this method was Case Foundation of Broomall, PA. The drop shaft for this pair of shafts was a nine-foot-diameter shaft and required a larger drilled hole than the locally-available pile top and super top rigs could drill. For the overburden, slurry was used to excavate below two telescoping casings. The crane-mounted drill used a kelley bar and cross bar to install the casings. The drill tools used were augers, core barrels, and toothed buckets ranging between 20 inches and 10 feet in diameter. This shaft excavation method had the most difficulty in drilling through boulders and obstructions.

The overburden excavation was riddled with difficulties related to excavation of boulders and man-made obstructions, including loss of approximately 90 cubic yards of ground from outside the casing. Penetration rates for the rock portion of the shafts were similar to those on the pile top method—1 foot per hour. Verticality of holes drilled with this method was similar to those achieved with the pile top rigs.

Drilling Method No 4.—Raise Bore with Ground Freezing. At one location the tunneling contractor was able to schedule construction of one pair of drop and vent shafts after tunneling was completed below. The tunneling contractor, Shank/Balfour Beatty (Shank), chose to subcontract with Moretrench of Rockaway, NJ to freeze the 160-foot depth of soil, and with Dynatec of Richmond Hill, Ontario to raise the bore through rock and soil. Freeze holes were drilled from the surface on 4-foot centers along the circumference of both shafts, with additional freeze holes drilled inside of the drop shaft. Raise bore operation consisted of a pilot hole to the tunnel and then reaming the pilot hole up from the tunnel to the surface. Excavated material dropped into the tunnel and was removed with the existing tunnel mucking systems. The verticality achieved at the excavated shafts was largely a function of the verticality achieve by the pilot holes (see Table 1). Figure 4 shows a photograph of the raise bore reamer in the tunnel.

The excavation rate for this method was significantly faster than for the other three. Once pilot holes were established, each shaft took less than one week to excavate. Although ground freeze pipes had been drilled around and within the shaft excavation envelope, one large steel obstruction went undetected and stalled the pilot hole

Figure 4.   Raise bore reamer (144-inch) prior to commencing drop shaft excavation

and reaming operations at the drop shaft. Except for this occurrence, this operation was completed with little delay. Refer to Table 1 for overall production rates.

Difficulties Encountered at Top of Rock

Using Methods 1, 2, and 3, repeated difficulties were experienced related to unstable conditions at the transition between soil and rock. Principally the problems related to insufficient seating of the casing into the irregular, fractured, and weathered horizon at the top of rock. At four of the seven drops shafts this led to significant loss of surrounding soil, backfilling the excavation, and redrilling through the stabilized zone. Only one of these incidents prompted a request for change due to differing site conditions, and that request has been disputed.

Liner Installation

The three types of shaft liners are described in detail above. This section describes liner installation. The precast concrete pipe liner and HDPE-steel composite liner were installed for the shafts that were drilled prior to tunnel construction and used the water that filled the drilled holes to ease the installation process. The cast-in-place shaft liner was installed using slip forming by the tunneling contractor that had access to the tunnel.

Concrete Pipe Liner Installation. Concrete pipe was used to line shafts of 2, 5, 6 and 9 feet in diameter. The pipe string was made of 40-foot-long pieces. The lower piece was fitted with an end cap to allow controlled "sinking" of the pipe string to the bottom of the hole. To accomplish this, water was pumped into the pipe to make it heavier than the water it displaced within the drilled hole. The installation was facilitated with a pair of specially fabricated full circumference clamps, each with four support wings. One clamp would hang pipe in the drill hole while the second clamp was used to lift the subsequent pipe. Before lowering into the drill hole, pipe joints were welded and pressure-tested using the integral external joint test nipples. The interior joints were patched prior to submerging. Once the full length of pipe string was in the drill hole, special care was used to level the top of the pipe by shimming the "hanging"

Figure 5. Welding 5-foot diameter concrete pipe liner installation—note clamps "hanging" pipe vertically

clamp support wings. Figure 5 depicts installation of a 5-foot-diameter concrete pipe. Once the pipe was in place, the annulus between the concrete pipe and the rock or overburden casing was filled with grout to complete the installation.

This liner was the preferred liner by the contractors due to ease of installation. At one location, after the tunnel contractor completed the connection to the base of the completed shaft, a 20-foot-long section of shaft liner failed. The inside mortar and the thin steel cylinder sandwiched between the inside mortar and prestressed strands crushed inward. After some investigation it was theorized that the lower piece of shaft liner pipe experience some damage—although it was not evident at the time—during drill-and-blast tunnel excavation and/or contact grouting after the tunnel liner was installed. It is believed that this damage compromised the pipe integrity. The failure occurred shortly after tunnel liner completion and ground water pressure recovery. More careful, controlled blasting and grouting was done at subsequent connections.

HDPE and Steel Composite Liner. This liner was used for two pairs of shafts with 2-feet and 5-feet diameters and was among the first to be installed at pre-tunnel construction sites. It was installed in similar "sinking" fashion as the concrete pipe liners by including bottom caps on the steel and HDPE casings. First, a welded steel casing was lowered with 3-inch external spacers to ensure proper concrete coverage. Steel wings were welded to the top of this steel casing and shimmed on a level frame to position the liner vertically. An inclinometer tube with large centralizers was then lowered into the casing to confirm that verticality tolerance was achieved. Grout was then tremied to fill the annulus outside the steel casing. The HDPE casing had centralizers that were tight to the inside diameter of the steel casing such that when it was lowered into the steel casing it was deemed to match the confirmed verticality of the steel casing. The annulus between the HDPE and steel casing was tremie-grouted to complete the installation. Refer to Figure 6 for a photograph of this liner installation.

Slip Formed Cast-In-Place Liner. The two shafts drilled using raise bore methods were cast in place using slip forms. Forms were fabricated on site and raised through the excavated holes using hydraulic cable jacks while placing concrete between the forms and the rock or frozen overburden. The average rates of concrete installation at the drop and ventilation shafts were 2.1 and 2.5 feet per hour, and slip forming was completed with little difficulty. The tunneling contractor lined three larger working shafts on this same site using the same method.

Figure 6.   HDPE being installed as the inner member of composite liner

Liner Tolerances. Except for drilling Method 2, all excavation methods struggled to maintain verticality that would result in the achievement of finished lining tolerances. Although attempts were made to excavate larger diameters, many of the shaft liners were installed with greater-than-specified deviations from plumb. With the assistance of Dr. S.C. Jain, who originated the vortex system, the design team scrutinized the hydraulic assumptions and parameters used to define the nominal shaft sizes and concluded that additional deviation could be tolerated in the drop shafts. The allowable deviation ranged from 6 inches to 2 feet, depending on the site-specific flows and the shaft-to-shaft distance at the base.

CONCLUSIONS

The seven pairs of shafts were constructed by four different specialty drilling companies and lined by six different prime contractors. Four distinctly different drilling methods were successfully used to complete the shaft drilling.

The nominal verticality tolerances could not be met by using these methods. However, the use of larger excavations and application of site-specific tolerances enabled the achievement of the functional requirements. The inclusion of performance criteria to encourage alternate lining materials and techniques in lieu of cast-in-place liners significantly aided the top-down construction. Allowing alternate designs for shaft liners proved to be an effective means of meeting both the contractor's individual installation preferences and the owner's ultimate goal of a cost-effective, functional drop and vent shaft system.

REFERENCE

Jain, S.C. and Kennedy, J.F., "Vortex-Flow Drop Structures for the Milwaukee Metropolitan Sewerage District Inline Storage System," IIHR Report No. 264, Iowa Institute of Hydraulic Research, July, 1983.

SHAFT CONSTRUCTION IN TORONTO USING SLURRY WALLS

Vince Luongo

Petrifond Foundation Co., Ltd.

PROJECT DESCRIPTION

The York Durham Sanitary System (YDSS) Interceptor in the Town of Richmond Hill located just north of Toronto, Ontario, Canada will see the construction of a tunnel extending along 19th Avenue for a length of approximately 3.8 Km from Yonge Street, east to Leslie Street and will have a minimum internal diameter of 2.1 m.

The tunnel will connect into four shafts, for which two of the shafts will be constructed using the slurry wall construction method given the high water table and difficult ground conditions encountered at these locations.

The two shafts noted as Access Shaft No. 2 and No. 3 were designed and constructed to meet regulatory conditions, which required a sealed shaft construction method.

GEOGRAPHICAL PROFILE

The ground conditions in the Toronto region consist predominately of glacial till, glaciolacustrine and glaciofluvial sand, silt and clay deposits, and beach sands and gravels.

The shallow geological formation found in the project area consists of four types of deposits: Oak Ridge Morain deposits, Halton Till, Glacial lake deposits and Organic soils.

The Oak Ridge Morain is the dominant deposit encountered along the tunnel alignment. This formation consists of three deposits: The Upper Till, Middle Sand and the Lower Till. Standard penetration tests (SPT) carried out in these deposits measured N-values ranging from 2 to more than 100 blows/300mm which indicate very loose to very dense soil conditions.

The Middle Sand deposit will be the dominant feature within the slurry wall construction at access shafts 2 and 3.

Access Shaft 3

The base of the slurry wall will be 20.5 M below the ground surface. Slurry wall excavation will be within peat and organic soils found at the upper 3 M, followed by 5 M of the Upper Till. The remainder of the excavation will be within the Middle Sand. The piezometric head elevation is found at the ground surface.

Access Shaft 2

The base of the slurry wall will be 22.5 M below the ground surface. Slurry wall excavation will be entirely within the Middle Sand 2.The piezometric head elevation is found at 7 M below the ground surface.

Figure 1. Slurry wall rig equipped with hydraulic clamshell

SHAFT CONSTRUCTION USING SLURRY WALL

Equipment

The choice of equipment is governed by certain site requirements and tasks. Alignment and verticality are the most essential elements for a successful result. The primary slurry wall construction equipment consists of a hydraulic pile rig equipped with specialized clamshell buckets. This rig will excavate through the overburden material to the required depth. The KRC Casagrande hydraulic clamshell bucket is fixed to a rigid Kelly bar enhancing vertical excavation tolerance through the dense soils. Figure 1 shows slurry wall equipment excavating at Shaft #2.

Slurry Wall

Prior to the construction of any slurry wall, guide walls are constructed to facilitate proper vertical and horizontal alignment for slurry wall excavation. The guide walls also prevent soil loss near the surface and act as containment for the introduction of bentonite slurry. At this site the guide walls will be built in circular fashion as per the required shaft footprint, as shown in Figure 2.

Slurry wall panel layout for both Access Shafts #2 and #3 will be constructed using 8 panels and a wall thickness of 600mm, as shown in Figure 3. The primary panels will be excavated followed by the secondary or closing panels. The first step in panel excavation is to position the digging rig in front of the panel. The hydraulic clamshell will excavate the overburden to required depth. Table 1 illustrates the shaft size and depth.

Throughout the excavation, the excavation progress is monitored for verticality, and stability. The panels will be excavated in two passes. As the excavation proceeds panel verticality is monitored at 3 meter intervals.

At the completion of excavation, panel dimensions, verticality and continuity will be checked. Verticality readings of the same joint obtained from two adjacent panels will be compared and the deviation, if any, the panel dimensions will be corrected. Just prior to placement of concrete, the bentonite slurry is cleaned to achieve a sand content below 5%. The reinforcing cage is then placed into the panel trench and suspended off the guidewalls, see Figure 4 for typical reinforcing cage used. Tremie pipes

Figure 2. Guide walls at Access Shaft No. 3

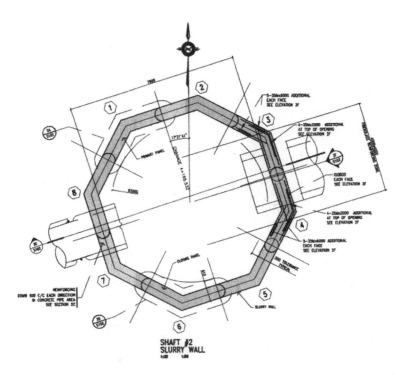

Figure 3. Slurry wall panel layout at Shaft No. 2

Table 1. Shaft sizes and depths

Access Shaft	Outside Diameter (m)	Slurry Wall Depth (m)	Shaft Excavation Depth (m)
No. 2	9.0	22.5	18.0
No. 3	12.0	20.5	15.0

Figure 4. Typical reinforcing cages used for panel construction

will be installed and secured to the top of the guide walls. Each length of tremie pipe installed will be recorded on specific quality control forms. Tremie pipes are 250 mm in diameter, smooth and with watertight joints. The hopper attached at the top of the tremies will be 1.5 M^3 in volume. The bottom tip of the tremie pipe will be positioned 150 mm off the bottom of the panel excavation prior to commencement of the concrete placement.

Concrete will be delivered to the site in transit mixers and discharged directly into hoppers at the top of the tremie pipes. Prior to commencing discharge, the concrete will be checked for slump and air content. Concrete discharge into the hopper at the top of the tremie pipe will be continuous, and controlled. To avoid segregation, a Go-Devil will be introduced into the tremie pipe just prior to concrete discharge. Depth of the concrete will be monitored and sections of the tremie pipe removed regularly, always leaving a minimum 3M and a maximum 10M of pipe embedded in the placed concrete. Concrete volume placed and depth of the placed concrete are recorded and drawn on a chart after the complete truckload discharge of concrete. Concrete is placed continuously until it reaches the cut-off level within the guide walls. Displaced slurry will be pumped off, and stored in ponds for reuse

The slurry wall panel joints will be created using 610 mm diameter joint sectional pipes. These are installed in the open excavation prior to the concreting of primary panels. Once the concrete has set sufficiently they are withdrawn leaving a concave surface into which is poured the next adjacent panel. This provides a watertight joint and continuity given the wider contact between panels.

The general contractor McNally-Aecon JV requested a steel reinforcement free zone in the slurry wall in the areas of TBM penetrations, while assuring the structural integrity of the shaft. Given the TBM cannot mine through a heavily reinforced slurry wall, these areas will be replaced with a fiberglass "soft-eye." The "soft-eye" is designed with the equivalent strength to replace the steel reinforcement.

McNally-Aecon JV will also perform shaft excavation under water. Once reaching the target depth a tremie concrete plug is placed and left to cure. The water within the shaft is then pumped out and the sealed shaft completed.

CONCLUSION

While completing this paper, the slurry wall at access Shaft #2 was completed. The use of a hydraulic clamshell proved to be very effective through the very dense soils. The Kelly mounted hydraulic clamshell also assured good verticality and panel alignment.

Circular self-supporting shafts are most effective providing safe, stable dry unhindered excavation and access for tunnel construction.

16

TBM Case Studies

Chair
P.S. McDermott
Kenny Construction

CONSTRUCTION OF THE WEST AREA CSO TUNNELS AND PUMPING STATION

Ray Hutton

Obayashi Corp.

Darrell Liebno

Obayashi Corp.

Taro Nonaka

Obayashi Corp.

INTRODUCTION

Background

Population growth and increasingly stringent regulatory controls on water quality across North America have resulted in municipal governments investing heavily in infrastructure to control what are known as Combined Sewer Overflows (CSOs). These are events which occur when combined sewers exceed their conveyance capacity during periods of heavy rainfall or snowmelt.

The city of Atlanta, Georgia is one such municipality; with a population of over 450,000 within city limits and a sewer system servicing 1.2 million including adjacent jurisdictions, the City of Atlanta has a Combined Sewer System (CSS) encompassing approximately 15% of its total 3,540 km (2,200 miles) of sewers. Atlanta's CSO discharges to local watercourses began to draw fire from environmental groups in the 1980s. As a result, in the early 1990s, the City constructed a series of CSO control facilities which provided screening and chlorination at seven CSO discharge points across the CSS watershed. Unfortunately, these facilities were not capable of handling all of the overflows generated due to heavy storm events; frequently, untreated water was bypassed around the treatment facilities. The federal Environmental Protection Agency (EPA) deemed that water quality standards were being violated by these discharges, and a suit was brought against the City to force a higher standard of compliance. As a result, the City entered into a federal Consent Decree which would bring CSOs into compliance with federal standards by 2007. A critical component of this plan was the design and construction of the West Area CSO Tunnels and Pumping Station. In February 2003 the City's Department of Watershed Management retained the services of Jordan Jones & Goulding/Hatch Mott MacDonald/Delon Hampton Associates (JDH Joint Venture) to design and provide construction management services for the project, which was designed to capture combined sewage from three of the City's existing CSO facilities on the northern end of the CSS watershed. The design consisted of two connected tunnels which would convey the wastewater originating from the North Avenue, Clear Creek, and Tanyard Creek CSO facilities to a submersible wet well pumping station located at the R.M. Clayton Water Reclamation Center (WRC). Since the majority of the effluent stream consisted of storm water, the City elected to design and construct a separate treatment plant. This plant was designed under a separate contract, and is currently under construction.

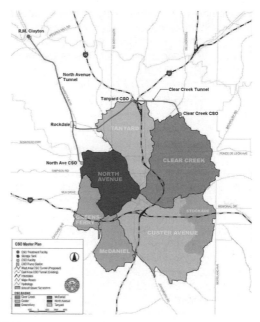

Figure 1.   West Area CSO tunnel alignment

Project Description

The two tunnels have a combined length of 13,350 m (43,800 ft), the majority of which is to be excavated using two hard rock Tunnel Boring Machines (TBMs). The excavated diameter for the TBM tunnels are 8.2 m (27 ft) with a 7.3-m (24-ft) diameter unreinforced concrete lining constructed in areas where required for permanent support or to control long-term groundwater inflows. For bidding purposes, the percentage of lining for each tunnel was established at 50%. The finished grade for all of the tunnels is 0.1%, and all tunnels are to be excavated up-slope.

Access for tunnel construction is provided through 12.2-m (40-ft) finished diameter construction shafts located at each end of the two TBM tunnels. The depths of the four construction shafts vary from 95 m (310 ft) to 50 m (165 ft).

The shorter of the two tunnels, referred to as the Clear Creek Tunnel, runs 6,358 m (20,860 ft) from the Clear Creek CSO facility at the east to the Rockdale Construction Shaft at the west. Approximately midway between construction shafts lies the connection point for the Tanyard Connecting Tunnel which connects the Tanyard Creek CSO facility intake to the Clear Creek tunnel. This connection is concrete-lined to a finished diameter of 3.7 m (12 ft).

Just down-station from the Rockdale Construction Shaft the Clear Creek Tunnel connects to the North Avenue Tunnel. The 6,980-m (22,900-ft) long North Avenue Tunnel runs from south to north, beginning at the North Avenue CSO Facility and ending at the construction shaft located at the R.M. Clayton WRC.

Other work associated with the project includes the construction of the 20.1-m (66-ft) diameter pumping station at the northern terminus of the North Avenue Tunnel; a 7.3-m (24-ft) diameter Overflow Shaft and Outfall Structure; three large open-cut intake structures with tangential vortex drop structures; and several smaller diameter vent shafts.

The project was designed with the flexibility to be let as two separate contracts (i.e., Contract A & Contract B) awarded to different contractors; however, at the tender opening on March 3, 2004, Atlanta CSO Constructors (ACC), a joint venture of Obayashi Corporation and Massana Construction LLC, was low bidder on both contracts with a combined bid price of $210,231,000 USD.

The City issued a contract to ACC, and provided Notice to Proceed on June 1st, 2004; however, due to site access issues, the Contract Start Date was delayed until July 9th, 2004. (See Figure 1.)

GEOLOGY AND GROUNDWATER

Regionally, the Project lies in the Piedmont region south of the Brevard fault zone. The tunnels were designed to be located in the bedrock zone at an average depth of approximately 61.0 m (200 ft) below ground surface. The geology throughout the tunnel area was expected to be a medium-grade metamorphic rock consisting mainly of fine to medium-grained gneiss. The Geotechnical Baseline Report (GBR) prepared prior to tender indicated that the rock was hard and massive, having an average Unconfined Compressive Strength (UCS) of 175 MPa (25,300 psi) with an upper limit expected to be 330 MPa (48,000 psi). The rock was also expected to be abrasive due to an average quartz content of 42%, and Cerchar Abrasivity Index (CAI) values averaging 4.5. These factors, in conjunction with the fact that predominant joint sets run parallel to tunnel alignment, were indications that the ground would yield low instantaneous penetration rates.

The GBR also described the potential for encountering large open-foliation fractures in the rock. These fractures had the capacity to contain large volumes of water which could inundate the TBM with short-term flows up to 190 L/s (3,000 gpm). Once any short-term flows had dissipated, the long-term inflows for the complete length of the unlined tunnels were baselined at 283 L/s (4,500 gpm) for the Clear Creek Tunnel, and 300 L/s (4,800 gpm) for the North Avenue Tunnel. At the time of writing, no major short-term inflows have been encountered during the excavation of either tunnel, and the long-term inflows are moderately lower than the baseline values.

Another issue addressed in the contract documents was the potential for groundwater contamination. The tunnel alignments pass beneath several industrial areas and sites known by regulatory bodies to be contaminated with high levels of Volatile Organic Compounds (VOCs). Groundwater sampling during the design phase also revealed higher than normal levels of VOC contamination. As such, there was a concern that contaminants in the groundwater flowing into the excavated tunnels could volatilize into the tunnel atmosphere and result in a hazardous working environment. In order to deal with this potential, ACC implemented a comprehensive monitoring plan to continuously monitor airborne VOC levels. Additionally, the tunnel ventilation system that was installed provided air flow rates greatly in excess of OSHA requirements in order to dilute and exhaust any harmful gaseous contaminants.

SHAFT CONSTRUCTION

Once site access was provided by the City, ACC proceeded to begin work on the first two of four construction shafts. These shafts were designed to be 12.2-m (40-ft) diameter access shafts to facilitate installation and removal of the two TBMs. The Rockdale and R.M. Clayton Shafts would be the first excavated as these were the mining shafts. Following their excavation, work began on the Clear Creek and North Avenue sites.

Figure 2.   Lowering jumbo—R.M. Clayton construction shaft

Rockdale Construction Shaft

Being the deeper of the two mining shafts at 94.5 m (310 ft), excavation of the Rockdale Construction Shaft was a priority for ACC. Excavation work began in August 2004. Overburden support through the residual soil consisted of steel ribs and lagging. Bedrock elevations were higher than anticipated which necessitated drill-and-blast excavation in some areas prior to installing the first few ribs. Once this initial excavation work was complete, the first two lifts of reinforced concrete lining were placed and rock excavation work proceeded.

The shaft lining was placed concurrent with the excavation. A typical shaft excavation cycle consisted of the following: drilling out a full-face 3.2-m (10.5-ft) round using two Tamrock Commando 300 hydraulic drills; charging the 63.5-mm (2 -in.) diameter blast holes with packaged emulsion explosives; blasting the round; leveling out the muck-pile and dropping the form; pouring a 3.2-m (10.5-ft) lift of concrete; mucking-out the previous round. Operating three shifts per day, and working six days per week, ACC was able to cycle the shaft forms twice per week.

Hoisting support for the shaft excavation operations was performed using a Kobelco 230-t (250-st) crane, and mucking out was performed using 8-m³ (10-yd³) muck-boxes loaded using a LinkBelt 225 hydraulic excavator.

Once the shaft had been excavated to starter tunnel invert elevation in January 2005, top heading excavation began for a 90-m (300-ft) starter tunnel and a 30.5-m (100-ft) tail tunnel. Using a Reedrill MK65 two-boom jumbo with basket, a 6.1-m (20-ft) high by 9.1-m (30-ft) wide top heading rounds were drilled; mucking-out was performed using a Wagner ST-6 LHD. Following top heading excavation, a 3.0-m (10-ft) bench was excavated with vertical holes drilled with Commandos.

In preparation for TBM assembly and launch, an invert mud slab was poured and gripper starter walls were shotcreted-in.

R.M. Clayton Construction Shaft

Prior to any work commencing for the 64.0-m (210-ft) deep R.M. Clayton Construction Shaft, ACC and the City agreed to relocate the shaft by approximately 42.7 m (140 ft) in order to better suit the proposed conveyor layout as well as to reduce the quantity of drill-and-blast tunnel excavation. Once site access was granted, existing buildings demolished, and site grading was under way, preliminary borings at the new location revealed that the residual soil overburden was nearly twice as deep as in the

original location. ACC ordered additional ribs and lagging to cover the total depth of 24.4 m (80 ft). Due to the depth of overburden, ACC elected to excavate the shaft to rock prior to placing any of the 12.2-m (40-ft) diameter concrete lining Excavation began in October 2004, and eight lifts of 3-m (10-ft) high concrete lining were placed bottom-up prior to beginning the rock excavation and lining sequence in a similar manner to that used for the Rockdale Construction Shaft.

Shaft excavation and lining reached the invert elevation in late January 2005, and the subsequent starter tunnel excavation was completed by April 2005.

Equipment for drilling and mucking out was identical to that used at the Rockdale Construction Shaft. (See Figure 2.)

Clear Creek Construction Shaft

Although the excavation of the receiving shafts was not on the Project's critical path, ACC elected to excavate them at both the Clear Creek site as well as the North Avenue site in order to complete excavation of the deaeration chambers and gain access to excavate vent shafts and adits for future connections which were located down-station of these chambers.

The GBR indicated that the Clear Creek Construction Shaft was located adjacent to a flood plain and that overburden soils consisted of alluvial silts and sands. There was a contract requirement for pre-installed support of excavation through the overburden, and ACC chose to install sheet piles to rock. These sheets were installed using a vibro-hammer to refusal which was met at an average depth of 11 m (35 ft).

Excavation and lining began in July 2005 using the same shaft lining forms and excavation equipment as at the R.M. Clayton & Rockdale Construction Shafts.

Once the shaft excavation was complete, excavation began for the top heading of the deaeration chamber which has a finished diameter of 8.3 m (27.3 ft) and an overall length of 112.8 m (370 ft).

Due to easement constraints on the surface, the design called for a tunnel vent shaft to be located approximately 53.3 m (175 ft) beyond the end of the deaeration chamber. In order to gain access to raisebore the shaft prior to holing through with the TBM, ACC excavated a pilot tunnel from the end of the deaeration chamber to the location of the vent shaft. The 4.3-m by 4.3-m (14-ft by 14-ft) horseshoe-shaped pilot tunnel was excavated within the future excavation envelope of the 8.2-m (27-ft) foot diameter TBM, and 1.8-m (6-ft) long fiberglass resin-anchored dowels were used for rock support with the intention that the TBM would be able to excavate through the ground support without danger of damaging continuous conveyor components.

North Avenue Construction Shaft

Excavation of the 64.0-m (210-ft) deep North Avenue Construction Shaft began in October 2005 following the installation of sheet piles. Techniques and equipment used were the same as on the Clear Creek Construction Shaft. Once the shaft invert elevation was reached, excavation for deaeration chamber and adit began. No pilot tunnel was necessary at North Avenue, since the tunnel vent shaft was just beyond the end of the chamber.

Figure 3. TBM shop assembly—Schwanau, Germany

TUNNEL DRIVES

Tunnel Boring Machine

ACC reviewed proposals for supply of the TBMs and selected Herrenknecht AG to provide two identical new 8.23 m (27 ft) diameter TBMs (machine numbers S-288 & S-289). Each of the main-beam hard-rock machines is are equipped with 52 rear-loading 482-mm (19-in.) disc cutters. The cutterhead is powered by nine VFD-controlled 350 kW (476-hp) electric motors equipped with Voith Safeset torque limiters, which rotate the cutterhead at a maximum speed of 7.6 rpm. The maximum cutterhead thrust was 21,600 kN (4.86×10^6 lbf). Overall TBM length with backup is 100 m (328 ft), with a total weight of 1,200 t (1,320 st).

The machines were equipped with roof bolting platforms supplied by ARO, which accommodated two roof drills and a single probe drill. Roof and probe drills were model HD 150s, supplied by Boart Longyear. A ring steel erector was provided for installation of full circle ring steel as required.

Herrenknecht fabricated and initially assembled and tested both TBMs at their factory in Schwanau, Germany (Figure 3). TBMs were disassembled and package for shipment and arrived on schedule in Atlanta, Georgia one year after purchase.

ACC entered into a cutter agreement with Herrenknecht Tunneling Systems USA (HTS) to provide cutter services for the project. HTS established a shop facility near the project to handle cutter rebuilds for both TBMs.

Both TBMs were equipped with guidance systems manufactured by Poltinger Precision Systems, which were supplied and refurbished by Precision Centerline Inc. of Atlanta, Georgia. This equipment was previously used successfully on the Chattahoochee Tunnel and the Nancy Creek Tunnel—projects of similar scope in the Atlanta area.

Conveyor System

For TBM muck removal, ACC opted to use a continuous tunnel conveyor system supplied by DBT American Inc. and vertical conveyors supplied by FKC Lakeshore for both tunnel drives. The horizontal conveyors each had a belt width of 914 mm (36 in), and were designed for a capacity of 725 t/h (800 tph) at a belt speed of 4.0 m/s (780 fpm). The tunnel conveyor for the North Avenue Tunnel drive had a total installed power of 1,120 kW (1,500 hp) consisting of one main drive, four carrying boosters, and one return booster. The Clear Creek Tunnel drive required a total of 970 kW (1,300 hp), with one main drive, four carrying boosters and one return booster. Each system had an advancing tailpiece mounted on the back-up to transfer muck from the

TBM conveyor to the tunnel conveyor. Additionally, a belt storage unit with a storage capacity of 490 m (1,600ft) was supplied for each tunnel drive. Belt storage units were designed and manufactured by DBT. For both tunnel drives, each unit was located underground in the respective starter tunnels, due to shaft depth and site constraints.

The vertical conveyor systems were both S-shaped designs, and were supplied with 1,400-mm (55-in) cross-rigid steel core belting equipped with corrugated sidewalls and cleats. Vertical conveyors were designed for a 725 t/h (800 tph) capacity.

The tunnel muck was discharged from the vertical conveyors to surface conveyors on each site. At the R.M. Clayton Construction Shaft site, ACC designed a static elevated stacker conveyor to transfer tunnel muck over the site access road to the muck storage and load-out area of the site. This configuration allowed the mining operation and muck load-out operations to take place on separate sites. Tunnel muck from the Clear Creek Tunnel was transferred from the vertical conveyor to a series of overland conveyors which transported the material to the disposal site approximately 185 m (600 ft) away. This overland conveyor system included a transfer point equipped with a diverter to allow the material to be transferred to a radial stacker conveyor should contaminated material be encountered.

ACC designed an integrated PLC-based conveyor control system for each tunnel that had the ability to control all the belts including those in the tunnel, in the shaft, and those on the surface. The system was also interlocked with the TBM's PLC start-up logic.

TBM Installation

The first TBM to be assembled was for the Clear Creek Tunnel drive (S-288). Assembly of the TBM started in mid-May 2005 at the Rockdale Construction Shaft site. ACC in conjunction with Herrenknecht and Barnhart Crane & Rigging developed an assembly plan which allowed a large portion of the TBM assembly to be completed on the surface. The depth of the shaft and relatively heavy weights of the individual components (i.e., weighing up to 136 t (150 st)), made a typical assembly sequence unattractive due to the size and availability of cranes large enough to handle the individual pieces at the depth of the shaft and the large number of multiple crane lifts which would have been required. The assembly sequence allowed for assembly of three main sections of the TBM: the complete cutterhead with cutters; the complete cutterhead support with motors and the forward portion of the main beam with drill platform and ring beam erector; and complete assembly of the gripper section of the main beam with gripper carriage, gripper cylinder, gripper shoes and the thrust rams on the surface. The decision to assemble large units on the surface reduced overall assembly time by facilitating work on multiple portions of the TBM simultaneously and by allowing for greater supervision and management of the assembly process.

The heavy lifts used to lower the TBM sections were made by Barnhart. A heavy lift gantry crane and strand jack system was used at the Rockdale shaft and a heavy lift gantry and 450-t (500-st) hoist at the R.M. Clayton site. Barnhart designed and provided the required rigging for the heavy lifts. The heavy lifts ranged from 280 t (310 st) to 180 t (200 st) with a depth of 98 m (320 ft) at the Rockdale Shaft. The trailing deck structures were assembled and loaded with equipment on the surface and lowered as complete units.

The TBMs were advanced into the starter tunnels during assembly utilizing a hydraulic slide beam system supplied by Barnhart in conjunction with a cradle provided by Herrenknecht. The slide system allowed the TBMs to be moved as required during assembly without the energizing the TBMs.

Figure 4. TBM assembly at R.M. Clayton construction shaft

All the components of the first TBM (S-288) were underground by the end of June 2005 and final assembly and testing occurred during July 2005.

At the R.M. Clayton site TBM assembly for the S-289 machine began on the surface in June 2005, Barnhart mobilized to the site in early July and all the major components were lowered into the shaft by early August 2005. TBM assembly and testing was completed for the S-289 TBM in early September 2005 and initial mining began immediately thereafter. (See Figure 4.)

Both TBM drives were started using lift-off muck-boxes to advance the first 150 m (500 ft), at which point the TBMs were stopped to allow for installation of horizontal and vertical conveyor systems.

Progress to Date

Clear Creek Tunnel Drive. As of January 1st 2007, the Clear Creek tunnel drive has completed a total of 4,470 m (14,661 ft). The current overall average daily production is 11.9 m/d (39.1 ft/d), with a best shift of 22.0 m (72.4 ft), best day of 44.5 m (146 ft), and a best week of 187 m (615 ft).

The rock encountered to date has required more support than expected; currently 556 m (1,825 ft) or 12.5% of the tunnel has required full-circle steel rib support. The anticipated or predicted requirement for full circle steel ribs was 3.1% of the tunnel length.

North Avenue Tunnel Drive. As of January 1st 2007, a total of 4,230 m (13,823 ft) of the North Avenue tunnel drive has been completed (Figure 5). The current overall average production is 15.1 m/d (49.5 ft/d), with a best shift of 24.4 m (80.2 ft), best day of 48.2 m (158 ft), and a best week of 204 m (668 ft).

The rock encountered to date has been largely as predicted in the design; currently 204 m (668 ft) of the tunnel has required full-circle steel rib support. Sections of ground requiring steel ribs have general been short in duration and oriented perpendicular to the tunnel, and can be characterized by the dramatic or quick changes in rock conditions.

Progress on the North Avenue drive has been hindered by blocky ground conditions, which were responsible for damage to the vertical conveyor requiring repairs to the vertical belt on two occasions. Additionally, on July 12, 2006 the S-289 TBM suffered a failure of one of the cutterhead drive pinions. The failed pinion in turn damaged the bull gear, which is integral to the main bearing. After review of the damage, the decision was made to replace the main bearing and the integrated bull gear, along with all the remaining pinions, seals and all affected parts. A plan was developed and

Figure 5.   North Avenue TBM tunnel

implemented which involved excavating a storage cavity for the new main bearing behind the TBM, pinning the cutterhead to the face, and backing the TBM up past the cavity to allow access to the new main bearing. A twin beam monorail system was installed in front of the TBM to allow for removal and installation of the main bearing. Herrenknecht assisted with the planning for the main bearing change, made a significant effort to provide replacement parts in a timely and provided field service technicians to assist with the main bearing exchange. The entire operation to replace the main bearing took 52 days.

GROUND SUPPORT

The contract-designed initial rock support for the TBM tunnel was categorized into three types, Type A, Type B, and Type C. Type A defined a four-bolt pattern of 3-m (10-ft) bolts spaced longitudinally at 1.5 m (5 ft). Type B support was a six-bolt pattern of 3-m (10-ft) bolts with a spacing of 1.5 m (5 ft); these bolts were to be installed in conjunction with rolled steel channel (C6×13) and wire mesh (4×4—W6×W×). Type C support required the installation of full-circle ring steel on 1.2-m (4-ft) centers with timber lagging. The selection criteria for ground support installation was based on Rock Mass Rating (RMR) with Type A installed for RMR values greater than 60, Type B installed for RMR's between 41 and 59, and Type C installed for RMR less than 40. The contract documents specified the use of a fully grouted, double corrosion protected CT-Bolt. As the TBM excavated tunnels advanced the relatively horizontal bedding of the rock caused eventual flaking of rock in the tunnel crown, even in areas of Type A-classified ground. This condition was recognized and acknowledged by the City and JDH, and ACC's proposal to add wire mesh to areas of the tunnel meeting these conditions was accepted.

VALUE ENGINEERING

Throughout the course of the Project to date, ACC has proposed to the City various design changes and Value Engineering Change Proposals (VECPs) for modifications to the permanent works. These design changes and VECPs were proposed based on an evaluation of the cost, schedule, constructability, and safety benefits to both the City and ACC. Due to time constraints on the project, the City, the JDH and ACC were required to expedite the preparation, review, negotiation, and approval process for these proposals. In some cases, the City proceeded with the redesign work

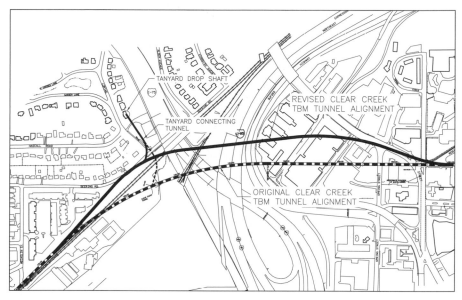

Figure 6. Tanyard tunnel realignment

prior to reaching a final settlement being reached on cost and schedule adjustments. These decisions to proceed were based on a mutual understanding that negotiations would be performed in good faith at a later date. Engineering work related to the redesign of permanent structures and their interaction with existing structures and processes was performed by the JDH Joint Venture working closely with ACC personnel to incorporate their concepts and concerns with regard to constructability, safety, and schedule issues. To date, the VECP process has successfully been implemented for changes associated with tunnel alignment, drop shaft configuration, intake structure sequencing, and grouting programs. The following sections discuss various design changes and VECPs in more detail.

TBM Tunnel Realignment

In addition to the slight change in alignment of the North Avenue tunnel associated with the relocation of the R.M. Clayton Construction Shaft as described above, a more substantial alignment change was proposed for the Clear Creek Tunnel. The impetus for this change was ACC's desire to reduce the length of the drill-and-blast excavation length of the Tanyard Connecting Tunnel. In the original contract design, the 3.7-m (12-ft) finished diameter Connecting Tunnel had an overall length of 145 m (476 ft). Additionally, the contract required ACC to excavate the Connecting Tunnel from within the Clear Creek Tunnel once the machine had mined past the junction. After consideration of several alignment options, ACC proposed one which would reduce the connecting tunnel's overall length by 77 m (253 ft). As a result, the Clear Creek tunnel would be lengthened by 17.7 m (58 ft). ACC presented the proposal in the form of a VECP having an instant savings in excess of $120,000 USD; the City accepted the proposal, recognizing that by reducing the length of connecting tunnel there would also be a reduced impact on the excavation rates of the TBM beyond the junction location. (See Figure 6.)

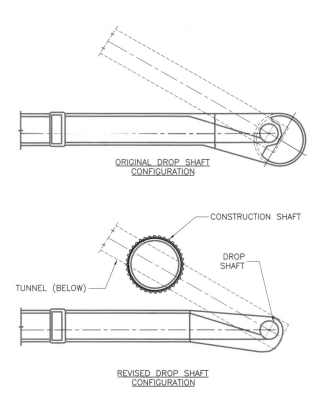

ORIGINAL DROP SHAFT
CONFIGURATION

CONSTRUCTION SHAFT

DROP
SHAFT

TUNNEL (BELOW)

REVISED DROP SHAFT
CONFIGURATION

Figure 7.   Original versus revised drop structure configurations

Separation of the Drop Shafts from the Main Construction Shafts

In the original Contract, the vortex drop structures at both the North Avenue and Clear Creek sites were located within the 12.2-m (40-ft) diameter TBM removal shafts. The finished diameters of these drop structures at the North Avenue and Clear Creek sites were 3.3 m (10.7 ft), and 4.7 m (15.5 ft) in diameter, respectively. Initially, upon commencing work, ACC had proposed the following excavation and lining sequence for the shafts: (1) excavate and place concrete lining in the shafts top-down; (2) excavate deaeration chamber; (3) remove TBM following hole through; (4) form/pour drop structure from bottom-up sequence would allow for a large portion of the permanent lining work to be completed prior to TBM hole-through, as well as ensuring the stability of the excavation. However, the JDH design team insisted that the drop structures be constructed monolithically with the main construction shaft concrete for structural reasons. ACC was concerned that this requirement could cause potential impacts not only to the overall project schedule but also to job safety, constructability, and quality.

Following several discussions with the City and the Engineer, ACC proposed a design change which separated the vortex drop shaft from the main construction shaft completely, and allowed for completion of the drop shafts prior to and independent of the tunnel construction (Figure 7). The following briefly describes the changes proposed:

- Relocation of the construction shafts down-station of the vortex drop structure locations by 20 m (66 ft) at the North Avenue site, and by 30 m (98 ft) at the Clear Creek site from the originally designed locations. The vortex drop structure locations remained unchanged to minimize changes to the intake and diversion structures.

- Construction of the main construction shafts using top-down concrete lining methods based on ACC's planned sequence.
- Excavation of the deaeration chambers using conventional drill and blast techniques and excavation/lining of the vortex drop shaft in advance of the TBM hole through.

Although there was no direct financial savings created by this design change proposal, the proposal was able to provide the following benefits to the City and ACC:

- Reduced the potential for the overall schedule to be impacted resulting from delays in the TBM tunnel excavation by allowing both the construction shaft and the vortex drop shaft to be successfully completed ahead of schedule and prior to completion of tunnel activities.
- Simplified the overall constructability and helped to ensure the safety of the operations and the quality of the final product.
- Provided the City with improved access to the finished tunnel for future inspection and maintenance.

Intake Structure Redesign

Downstream of each of the three existing CSO facilities (North Avenue, Clear Creek, and Tanyard), the combined sewage effluent needs to be captured and conveyed or diverted into the deep tunnels. Generally, these flow diversion structures consist of an intake, a rectangular-shaped approach channel, and a tangential vortex structure leading to the drop shaft connected to the deaeration chamber.

The specific construction details for the intake structures vary from site to site, but the process generally consists of demolishing a section of the existing effluent channel, and constructing a deeper box directly beneath the invert. Contract documents advised that the existing CSO facilities discharged into these effluent channels and that the Contractor should expect some flooding to occur following a rain event, Additionally, they state that the severity of each flooding event depended on the intensity of the rain event. However, the cost of managing and recovering from these flooding events was to be included in the bid price.

ACC spent a large portion of time early in the project studying and analyzing the operation of the existing CSO facilities in an attempt to determine a safe and effective manner in which to construct the intake structures, including meetings with the City's facility operators. The most significant issue revealed in these discussions was the immediate, and almost instantaneous nature of the discharge events, the magnitude of the flows (exceeding 225 m^3/s (7,800 ft^3/s), and the limited or non-existent warning before the occurrence of each event. These factors indicated that the greatest risk during construction was the ability to alert and evacuate personnel from the deep excavations prior to the occurrence of rapid flooding during and after a rain event. After these issues were revealed, concerns for personnel safety, potential schedule impact, and constructability became critical. ACC studied and evaluated options and alternate plans for intake design/construction and met with City and JDH personnel to ascertain re-design constraints. In order to propose alternative designs with the end goal of eliminating or minimizing exposure to flooding events during intake construction, it was determined that the following constraints needed to be met:

- A temporary cofferdam and/or temporary flow diversions could be installed in the existing channel, given that a minimum of 75% of the existing channel flow capacity was maintained.
- Any alternate design was to minimize any additional cost impacts to the City with minimal redesign cost and no schedule impacts.

After evaluating many alternatives with the City's team, the general concept of the redesign was agreed to as follows:

- Modify the shape of the intake structures to better accommodate constructability and accommodate a phased construction sequence.
- Install temporary cofferdams in the existing channels and construct the intake structure in multiple phases while maintaining the required 75% of existing flow capacity.
- Integrate the redesign process with a constructability assessment to ensure that final design could be built in a safe manner.

Following a negotiation process between ACC and the City, it was agreed that all three of the intake structures would be redesigned. The final design for each of the structures took into account all of these factors and created new intake structures which fulfilled the operational goals of the City while at the same time allowing the structures to be constructed in the safest possible manner by minimizing exposure of the workforce to hazards.

Tacoma Drive Surface Grouting Program

Another area where a positive relationship between owner and contractor proved beneficial was through a ground improvement program on a portion of the North Avenue TBM Tunnel where a probe drilling and grouting zone had been specified in the Contract.

In an area defined as the Tacoma Drive Lineament Zone, which falls at approximately the mid point of the North Avenue Tunnel, the GBR noted a stretch of ground that was of high concern for the possibility of affecting the tunneling conditions. According to the GBR, this zone was anticipated to be highly fractured and deeply weathered through the tunnel horizon with the possibility that ground conditions in the tunnel zone could include zones of clay or silt, highly weathered rock requiring use of full circular steel tunnel support, and large volumes of groundwater, having a high probability of being contaminated with hazardous concentrations of VOCs. Based on the GBR, the City designated a 305-m (1,000-ft) long section of tunnel in the area of this lineament zone as a probe-hole drilling and pre-excavation grouting zone under the Contract.

As the tunnel excavation proceeded prior to reaching this area, overall excavation rates were found to be slower than anticipated due to unexpected mechanical downtime and worse than expected ground conditions. These excavation rates, specifically in highly weathered rock conditions having high groundwater inflows, and requiring Type C support were approximately 6.5 m/d (21 ft/d). The low excavation rates coupled with concerns over further slow-downs should a significant amount of grouting ahead of the machine be required through the probe zone, prompted ACC to propose a drilling and pre-grouting program from the surface as an alternative for probing and grouting from the tunnel. The perceived goals of the proposed program were as follows:

- Provide a more precise knowledge of the ground conditions through the zone prior to the TBM's arrival.
- Reduce downtime by reducing or eliminating probing and grouting required from the tunnel.

- Potentially increase the TBM advance rate by reducing the required tunnel support.
- Reduce the risk of hazardous exposure from groundwater contamination and associated downtime.

To minimize cost impacts to the City with respect to the surface grouting program, the City and ACC agreed to use the bid items originally designated for probing and pre-excavation grouting from the tunnel to perform the initial portion of the program and to evaluate the program's success to determine if additional funding was required or warranted.

Upon agreement with the City, ACC entered into a subcontract agreement with Hayward Baker Inc. , (HBI) to perform the work. Through close communication with the Engineer, ACC and HBI established the following general workplan for the operation:

1. Drill probe holes on 15-m (50-ft) centers along the centerline of the tunnel in the area of concern and perform Lugeon tests.

2. Evaluate the probe hole data and Lugeon test results, to determine the areas of greatest concern.

3. In areas of concern, start drilling operation by drilling a pattern of primary holes on a 3.5-m by 3.5-m (11-ft by 11-ft) square pattern. The grout target zone was determined to be a 15-m (50-ft) wide by 18-m (60-ft) high area (1.5 m (5 ft) beyond the tunnel diameter in width by 1.5 m (5 ft) below the tunnel invert to 8 m (26 ft) above the tunnel crown).

4. Grout holes using up-stage grouting with double packers

5. Monitor grout takes, and drill/grout secondary confirmation holes.

Subsequent to the surface probe drilling and water pressure testing of the probe holes, ACC was able to identify that the area of greatest concern was limited to a length of only 150 m (490 ft) as opposed to the 300 m (980 ft) originally designated in the Contract.

Type III cement was utilized for all the surface grouting. Consideration was given to the use of microfine cement as part of the surface grouting program. However, as the program's main focus was to seal off the larger open features in the rock and to expedite the TBM progress while minimizing cost, a joint decision was made not to use microfine cement. The typical grout mixes used during the operation contained a water to cement (W/C) ratio (by weight) ranging from 2:1 to 0.5:1 depending on grout take and pressure. The maximum grout pressure was established as 34 kPa/m (1.5 psi/ft) of vertical depth with an average grout hole depth of 69 m (225 ft). The refusal criterion was set as a flow rate of 0.95 L/min. (0.25 gpm) or less at the maximum grout pressure for 5 minutes.

At the completion of the surface grouting program, the total grout volume injected was 120 m^3 (4,250 ft^3). The maximum grout on a single drill hole was 10.5 m^3 (370 ft^3), or 582 liters (20 ft^3) per vertical meter of the target zone. The overall average was 1,718 liters (61 ft^3) per drill hole, or 95 liters (3.5 ft^3) per vertical meter of the target zone.

As the TBM advanced through the grouted lineament zone, ACC encountered areas of highly weathered rock as anticipated. However, many open joints and features which were sealed with grout material and overall, groundwater inflows were relatively low through the grouted zone. As an indication of the effectiveness of the surface grouting program, the TBM achieved an average advance rate of 16 m (50 ft) per day through the surface grout zone without major delays to production caused by ground conditions. In total only 4 m (12 ft) of Type C support was installed in the grouted zone and no probe drilling or pre-excavation grouting was performed from the TBM.

Figure 8.   Pumping station shaft concrete lining

Although the total cost for the surface grouting program exceeded the originally allocated initial budget, the overrun costs were shared by ACC since the ground improvement benefited the TBM advance rate through the zone.

PUMPING STATION

The Pumping Station consists of a 20 m (66 ft) finished diameter shaft 55 m (180 ft) deep, which will be capable of pumping 3.7 m^3/s (85 GPD). The construction of the pumping station and the associated connecting tunnel and overflow tunnel excavation and lining were subcontracted to W.L. Hailey & Company. Excavation of the 22.5-m (74-ft) diameter shaft was started in early 2005, with overburden excavation to a depth of 15 m (50 ft) supported with ribs and boards, and the final 40 m (130 ft) of drill-and-blast through rock to invert. Excavation of the 8.5 m (28 ft) horseshoe-shaped Overflow and Connecting Tunnel was then driven using drill-blast methods and top heading and bench. Following excavation, the tunnels and junction were lined to a finished diameter of 7.3 m (24 ft) utilizing ACC's tunnel forms procured from Everest Equipment. Concurrent with tunnel lining, the 5.5 m (18 ft) thick base slab for the pump station shaft was placed. Currently, the center dividing walls and the exterior shaft walls are being placed. (See Figure 8.)

ACKNOWLEDGMENTS

The authors wish to thank the JDH Joint Venture and the City of Atlanta for their help in the preparation of this paper as well as for their contributions to the overall success to date of the project.

GUADARRAMA TUNNEL CONSTRUCTION
WITH DOUBLE SHIELD TBMS

F. Mendaña

ADIF Tunneling Advisor

ABSTRACT

The characteristics of the igneous and metamorphic bedrock, extra-hard and abrasive, of the Guadarrama Tunnels, a bi-tube project of length 28.4 km without any intermediate attack, are first commented upon. Next, a detailed discussion is given of the most relevant aspects of the performance of the four double shield type TBMs. The experience of this leads to various recommendations to be borne in mind in future designs of TBM having to work in rock of similar characteristics.

INTRODUCTION

The Guadarrama Tunnels are the most important work of the new Spanish High Speed Rail Network. They form part of the New Rail Access to the North and North-West of Spain, one of 14 projects classified as priority by the European Union, and which forms part of the Atlantic Axis of European Railways.

This work belongs to the Spanish Ministry of Public Constructions (Ministerio de Fomento) which has engaged ADIF ("Administrador de Infraestructuras Ferroviarias"), an institution of mentioned Ministry, responsible for the management of the new Spanish Railway Network, for its design and construction.

The work consists of a base tunnel of length 28.4 km whose route crosses underneath the mountain range of the Sistema Central at a maximum elevation of 1,200 m above sea level, with a maximum overburden of 992 m underneath the massif of Peñalara. The design consists of two main tubes for single track, of excavation diameter 9.51 m, with separation between centrelines of 30 m and connection galleries of 24 m^2 every 250 m, the construction of which is being carried out simultaneously with that of the main tubes for safety reasons (Figure 1). Finally, in the central part of the route the galleries will be excavated every 50 m, adding a parallel tunnel of about 500 m in order to form an emergency area with capacity for 1,200 passengers.

The construction of the civil work on the main tubes, which was completed in March 2005, has permitted a series of problems to be solved. It is the purpose of this Paper to discuss these problems since a knowledge of these solutions will aid the development of the technology of TBMs for extra-hard and abrasive rock.

CHARACTERISTICS OF THE GROUND

Geology of the Route

The Guadarrama massif consists of igneous and metamorphic rock, with a predominance of gneisses followed by granites, and its general geology is familiar.

Having previously chosen a corridor between the town of Soto del Real (Madrid province) and the city of Segovia, the final route of the tunnel was defined by ADIF within the proposals of a Project and Work Tender Competition, which included survey works conducted over the course of a year.

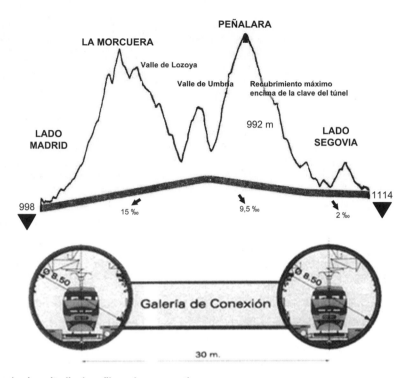

Figure 1. Longitudinal profile and cross section

Those works had to be limited to the drilling of boreholes with geophysical sampling since the environmental conditioning factors were very strict, due to the fact that the Sierra de Guadarrama is a specially protected zone. The geotechnical quality of the massif was assessed as acceptable, though faults were defined that had a very direct effect on the tunnel in the zones with less overburden at the two ends, plus the existence of fractures and interlaying of intrusive dikes along the route.

Five major sectors are distinguished in the Guadarrama massif. From south to north: La Pedriza granites; La Najarra orthogneisses; Peñalara Sector; La Granja granites and the metamorphic complex of Hontoria. The two main faults in the interior of the massif occur in the Sector of La Najarra: the La Angostura fault with an estimated width of 150 m and the La Umbría fault with an estimated width of 600 m. In the initial surveys it was confirmed that the first fault presented a cohesive fault gap deep down, though any appreciable degree of permeability was discarded, and so it was assessed as being a minor difficulty. On the other hand, the surveys of La Umbría gave negative results, with the possibility of presenting a high degree of fracturing, and important flows and a water load of up to 18 Bar.

It was considered advisable to complete the surveys of La Umbría with a vertical shaft and a gallery from which prior consolidation treatment could be carried out. The environmental restrictions did not permit this, and so the final study was delayed until February to September 2004.

The section of study included about 900 m of the route, with the following works being performed: (a) Study of previous records and of the structural geological cartography in a band of width 3 km (2 km to the east and 1 km to the west of the route);

Table 1. Simple compression strength and indirect tension values

Rock	SCS (MPa)		STS (MPa)		$I_{S\,50}$
	Forecast	Real	Forecast	Real	Real
Gneiss	85–110	85–120	8.8	11.5	7.5–9.0
Adamellites	95	80–100	7.2	8–10	6.5–8.0
Porphyry dikes	110	150	12.8	14	9.5–10.0
Leucogranites	81–150	85–100	10.8	10.2	8.0–8.5
Pegmatites	95	—	6.6	9.7	—

(b) Electrical tomography of the surface; (c) Mechanical drilling of five vertical bore-holes, of between 200 and 280 m in depth, and one borehole inclined at 45° and parallel to the route; (d) Geological-geotechnical extraction of cores and taking of samples for tests; (e) Geophysical sampling of the boreholes, and (f) Borehole-borehole and borehole-surface electrical tomography, complemented with seismic tomography. With this it was possible to prepare the longitudinal and transverse profiles and the structural geological cartography at a scale 1/2000, which recommended making a slight variant to the route, with a displacement of about 230 m to the east. With this variant, the following main points were regarded as confirmed:

- The variant ensured the avoidance of low quality ground (Utrillas facies) belonging to the Cretaceous, which the original route would have traversed.

- Although the ground could present a low geotechnical quality along almost 600 m, the length of the section affected by the main fractures was limited to about 250 m. No water load was expected, nor any significant inflows.

- It was planned to construct a "by-pass" gallery, if necessary.

Nevertheless this fault was passed through in the months of January and February 2005, with the aid of means appropriate to the TBMs.

Geotechnical Characterization of the Materials

RMR Index. The results of the tests on borehole samples have been confirmed, with RMR values in general higher than 45. The forecast of an RMR of at least 20 for the sections with faults turned out in general to be pessimistic, and the geomechanical quality of the bedrock had a level of average-good and good.

Simple Compression Strength/Indirect Tension and $I_{s\,50}$ Index. The average forecasts from the initial studies turned out to be much lower than reality, as can be seen in the Table 1.

Abrasivity. The Cerchar Index was in general maintained at values at all times greater than 3, reaching values close to the maximum of 6. In terms of the content of abrasive minerals, the following high values were very frequent (Table 2).

TUNNELERS FOR HARD GUADARRAMA ROCK

The designs of large diameter tunnelers, in other words, those above 8 m, were first developed in 1995 and are intimately connected to tunnel projects of great length, which are basic elements in modern Transport Infrastructure Plans (high speed rail, motorways, metropolitan railways, etc.). However, the technology of the tunnelers themselves has not developed sufficiently fast for being able to make an in-depth study of the new problem presented by the large dimensions of the machines, the industrial

Table 2. Content in abrasive minerals

| | Averages | |
	Forecast	Real
Gneisses	29–33%	48%
Granites	26%	52%
Adamellites	32%	54%

| | Maxima | |
	Forecast	Real
Porphyry dikes	40%	64%
Quartziferous dikes	60%	98%

design of which has involved the extrapolation of the experiences of earlier machines which did not have the success that was expected.

Because of this, although it can be said that the experience of these last few years has meant that certain drawbacks could be regarded as overcome, it nevertheless has to be admitted that this experience has also highlighted the major stumbling blocks that still persist.

As far as hard-rock tunnelers are concerned, at the risk of certain omissions and confining ourselves to newly built machines from the most well-known manufacturers, Tables 3 and 4 contain large-dimension hard-rock tunnelers built in recent years. As can be seen, a large part of these machines correspond to Spanish projects. Standing out among them are the machines of the Guadarrama Tunnels to which the following refers.

Choice of TBM Type

Knowing the characteristics of the bedrock, it was decided that the Guadarrama TBMs would be machines with a cutting wheel for extra-hard and very abrasive rock.

Moreover, compliance with a minimum execution period for the main tubes was set as a target of the utmost interest, though nevertheless avoiding the construction of intermediate attacks in order to respect environmental requirements as much as possible. So, it was decided to carry out the construction with four machines, with two of them attacking each tube from its two mouths.

As a result of all this, hard-rock TBMs of the double shield type were considered to be most suitable, setting the figures of the basic parameters (Thrust and Turning torque) as they appear in Table 3. As complementary equipment, the following were adopted.

- Conveyor belt along each tube for the extraction of rubble from the TBM to the outside. Conveyor belts (three of them hanging and one supported on a light structure) were used of capacity 1,500 t/h (widths of 900 or 1,000 mm) which could deal with lengths of up to 15 km of tunnel.

- Two drilling probes per TBM, with auxiliary equipment for analysis of parameters for surveying the ground from the face and the possible location of micropile umbrellas of fibreglass tube.

The ring type lining, of thickness 320 mm and interior diameter 8.50 m, has a length of 1.60 m and consists of six segments: one base segment (66°) another contiguous one (73.5°), plus four equal segments (49°) and the key piece (245°). Given the magnitude of the radii of the route (R_{min} = 8,400 m) the ring has no conicity.

Table 3. TBMs of large diameter rock

Project and Year of Manufacture	Type of Ground	Tunneler Make (Type)* Units/Ø$_{exc}$/Weight	Head Rotation Power (rpm) Rated Torque	Max Thrust Principal Auxiliary
Pinglin, Taiwan 1996	Sandstones, Slates[†] (>200 MPa)	Wirth (2)	4,000 kW (0–4)	50,500 kN
		2 units/11.75 m /	7,200 kNm	78,700 kN
Tscharner, Switzerland 1999	Limestone, Marl	Herrenknecht (1)	2,100 kW	
		1 unit/9.53 m /		
Guadarrama, Spain 2001	Gneiss, granites[†] (>200 MPa)	Wirth-NFM (2)	4,000 kW (0–5)	10,450 kN
		2 units/9.465 m/ 1,000 t	20,750 kNm (@1.8 rpm)	108,000 kN
Lötschberg, Switzerland 2000/ 2001	Gneiss, granite, granodiorite[†] (> 200 MPa)	Herrenknecht (1)	3,500 kW (0–6)	
		2 units/9.43 m	8,825 kNm (@3.76 rpm)	22,800 kN
Guadarrama, Spain 2001	Gneiss, granites[†] (>200 MPa)	Herrenknecht (2)	4,200 kW (0–5)	82.000 kN
		2 units/9.56 m	20,447 kNm (@1.85 rpm)	101,200 kN
Gotthard (Bodio/ Faido), Switzerland 2002	Gneiss, granite, schists[†] (>200 MPa)	Herrenknecht (1)	4.000 kW	
		2 units/9.33 m		
Abdalajis, Spain 2003	Limestone, Marls	Mitsubishi-Robbins (2)	4,900 kW (0–6)	82,500 kN
		2 units/10.2 m /	18,700 kNm (@2.5 rpm)	117,000 kN
San Pedro, Spain 2004	Gneiss, granite, schists[†] (>150 MPa)	Herrenknecht (1)	~3.500 kW	
		2 units/~9.55 m/	Adopted Lötschberg machines	
Pajares L.2, Spain 2004	Limestone, marls sandstone	Herrenknecht (2)	5,600 kW (0–6 rpm)	67,900 kN
		1 units /10.12 m/ 1,550 t	30,051 kNm	135,800 kN
Perthus, Spain/ France 2004	Granodiorites sand- stone slates[†]	Herrenknecht (2)	4,900 kW (0–5.5 rpm)	64,150 kN
		2 units/9.96 m/ 1,500 t	29,150 kNm	89,000 kN

* Types: (1) Conventional open (2) Double shield.
† Extra-hard and abrasive rocks.

The four TBMs that are constructing the Guadarrama Tunnels consist of two pairs of twins, from the makes Herrenknecht and Wirth respectively. For the excavation of each of the tubes, one TBM from each of those makes is working from the respective mouths (Figures 2 and 3). The most notable aspects of the study and performance of various elements of the machines are now going to be discussed.

Cutting Tools

The cutting tools for extra-hard rock are nowadays exclusively disc cutters. Their basic elements are the hard-metal cutting ring, fastened hot to a steel disc freely rotating on its axis, supported on the housing by means of sets of roller bearings.

The investigation into hard materials led to defining cutting models that accorded with commercial dimensions and capacities from diameters of 15½" (250 kN) to 17" (270 kN) and 19" (300 kN), with other larger prototypes that were successfully tested

Table 4. Large diameter shields for rock

Project and Year of Manufacture	Type of Ground	Tunneler Make (Type)[*] Units/$\varnothing_{exc}$/Weight	Head Rotation Power (rpm) Rated Torque	Max Thrust Principal Auxiliary
Mürgental Switzerland 1997	Marl and sandstone. Molasse	Herrenknecht (3)	3,200 kW	
		1 unit/11.98 m/		
Flüelen 2000	Sandstone, quartz, schists[†]	Herrenknecht (3)	3,200 kW	
		1 unit/11.98 m/		
CRC 320 Hong Kong 2000	Gravels, sands, hard granite[†] (>200 MPa)	NFM / (4)	2,250 kW (0–3)	166,000kN
Metro Porto Portugal 2000/2002	Granites, jabre schists[†] (>200 MPa)	Herrenknecht (4)	2,400 kW (0–4.5)	
		2 units/8.90 m /	12,900 kNm (@2 rpm)	70,600 kN
L-9 Metro Barcelona Spain 2002	Granodiorites, cornubianites, schists[†] (>150 MPa)	Wirth/NFM (4) dual	4,750 kW (0–3.7)	90,000 kN
		1 unit/11.95 m/ 1.472 t	57,000 kNm	110,000 kN
Pajares North L.1 Spain 2004	Limestone, Marl, Sandstone	Herrenknecht (3)	4,900 kW (0–6)	
		1 unit/9.90 m /	30,000 kNm	115,000 kN
Pajares North L.1 Spain 2004	Limestone, Marl, Sandstone	Wirth/NFM (3)	5,600 kW (0–6)	
		1 unit/10,12 m / 1,550 t	23,051 kNm (@2.25 rpm)	135,773 kN
Pajares South L.3 Spain 2004	Limestone, Marl, Sandstone	Wirth/NFM/ (3)	4,900 kW (0–5)	
		1 unit/9.60 m /	25,000 kNm	150,000 kN
Pajares South L.4 Spain 2004	Limestone, Marl, Sandstone	Mitsubishi-Robbins (3)	5,180 kW (0–5)	
		1 unit/9.60 m /	26,000 kNm (@1.9 rpm)	139,500 kN

* Types: (3) Conventional open (4) EPB.
† Extra-hard and abrasive rocks.

in laboratories being abandoned because their excessive weight made their maintenance difficult.

Among the options of Ø 17" and Ø 19", in Guadarrama the former was chosen since, with current technology and for extra-hard and abrasive rocks, it is not clear if there is any advantage in the "ratio" of advance, nor, with wheels of that size, whether the deformation which its structure had to admit would be compatible with the dimensions of the gaps required by housings of cutters of more than 17" in diameter, which creates problems in welding to that structure.

The 17" cutter, of reasonable weight for its replacement (around 160–170 kg), and with a tread separation of 80 mm,[*] has satisfactorily solved the cutting of Guadarrama rock. Other important questions deriving from the experience of this work which must be pointed out are:

- **Control of the contact pressure.** Two of the Guadarrama TBMs had a system for this control, which is of great importance, because, although the pressure is

---

*In a TBM of $\varnothing_{exc}$ ≈ 10 m passing to a tread of 70 mm implies an increase of 12%–14% in the number of cutters, which can be difficult to solve for machines of less than 12 m in diameter, due to excessive weakening of the metallic structure of the wheel.

Figure 2. Double shield Wirth   Figure 3. Double shield Herrenknecht

not measured in each cutter (which would appreciably complicate the designs on the market), it facilitates total control over the real contact thrust since it measures the pressure in the hydraulic cylinders acting as support for the transmission of that thrust to the rotating head.[*]

- **Questions pending solution.** Some difficulties also appeared which are yet to be solved. The main ones, which are due to the dual nature of hardness and abrasivity of the rock, and which have to be a new challenge for the technology of hard metals, are the following: Rapid wear of the broad angle cutting edge and brittle breakages in the narrow angle cutting edges. The effect on the bearing capacity due to the above causes is immediate, since it causes the cutter to because blocked with the consequent increase in pressure on the other cutters, with which their maximum capacity becomes exceeded and there is risk of serious generalized breakdown (Figure 4).

  It is now understood that these phenomena become heightened if the face contains rock of very different characteristics: the thrust of the machines becomes concentrated on a limited number of cutters, exceeding their maximum capacity and accelerating the wear or, more likely, breakage of the cutting edge. In Guadarrama the influence of the orientation of the fracture or fault planes of the massif was also confirmed. With planes "transverse" to the advance, the breakdowns in cutters were less than 40% of those suffered when those planes were orientated in the same direction with respect to the advance.

- **Faults in the cutters**. In Guadarrama it has been verified that a large part of the breakdowns of the cutting tools correspond to faults in the pure mechanical design of them. In particular, the following faults have been repeatedly confirmed in rocks of above 180–200 Mpa: Appreciable increase in temperature in the oil of the bearings, which can reach the point of its decomposition and, in any case, with breakdown of the airtight seal; subsequent loss of the required grease with failure of the support bearings, and improvement of the situation when the cutters are cooled.

  Cooling of the cutters was tested, first of all and in a simple way, with water from the dust collection system. The improvement observed led to the method being perfected, with special foams being injected in the front of the cutting

---

[*]There is also a system of wear control of the hard disc, applied just in some cutters, which can be interesting in rocks of no more than average hardness. The problem of hard and abrasive rocks is, as shall be seen, a different matter.

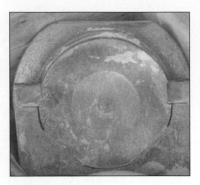 

Figure 4. Wear and breakages of the cutting edges

wheel, using the ducts from the rotary gasket of the TBMs, for which the available section had to be expanded in some cases. The result of cooling with foam has therefore been satisfactory and can be another element to bear in mind in the future design of the wheels and cutters of TBMs.

- **Maximum penetration and consumption of cutters.** The current technology of cutting tools permits a circumferential speed for cutters of $\varnothing$ 17" of around 145–150 m/min as maximum limits. As a consequence, the speed of rotation for large diameter TBMs is defined as:

$$\omega_{max} = \frac{1}{K} \times \frac{145}{\pi\phi_{exc}(\text{in meters})}$$

K is a coefficient of the order of 0.75 to 1.00 for hard and abrasive rocks and $\varnothing_{exc}$ is the TBM diameter.

So, in accordance with the hardness and abrasivity of the rock, both the maximum thrust and the speed of rotation have to be regulated in order to achieve the maximum possible advance under conditions of safety. From the experience of Guadarrama, it can be deduced:

- In very hard and abrasive rocks (above 200 MPa of SCS and 5.5 points in the CAI index[*]) the speed of advance usually has to be limited to figures of between 20 and 30 mm/minute.

- We give this warning because, one way or the other, it has been stated in technical publications on hard and abrasive rocks that the technical and economic limit of mechanical cutting is of the order of 50 mm/minute, which is not true. The experiences of Guadarrama and Lötschberg contradict this technically since, with figures of the order of 25 mm/minute, advances have been maintained of between 18 and 22 m/day.

- From the general point of view, the result of the Guadarrama experiences in similar formations of gneiss and hard granites with some sections of great hardness (SCS > 200 MPa) and abrasivity (CAI > 5.5) is summarised in the following graphs (Figures 5 and 6).

- **Peripheral cutters.** Special mention deserves to be made of the pre-clearance and clearance cutters. It can be said in general that the designs and positioning of the central cutters and of the intermediate strips have been satis-

---

[*]Cerchar Abrasivity Index (maximum approved value: 6).

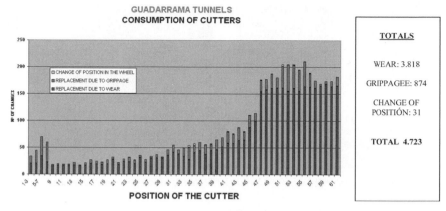

Figure 5.   Consumption of cutters according to position in the cutting wheel

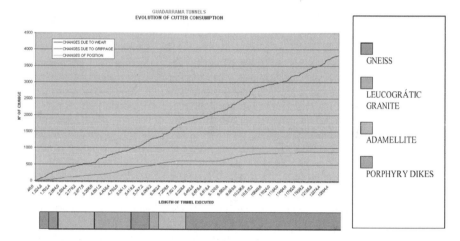

Figure 6.   Evolution of cutter consumption (Batch 4)

factorily solved by the manufacturers in accordance with their earlier experiences. This is not the case with the peripheral cutters, which is where the greatest number of failures have been concentrated, and it can be stated that these failures have been repeated time and time again, sometimes in one or two cutters in particular, and above all in the clearance cutters.

One of the ways open for improving the cutting tools and for the most scientific possible operation of the TBMs is to investigate the process of mechanical cutting of the rock. The most recent studies propose mathematical models which seek to relate geotechnical parameters to the geometry of the cutting disc. Although we believe that the majority of current failures derive from the mechanical design of the cutter, as we have discussed in some detail, nevertheless, other investigation work along these lines could cast some light on the problem of the location and orientation of the peripheral cutters.[*]

*Apart from the empirical formulas of Wang and Ozdemir, or those proposed by Robbins, see also the modern studies of Innaurato et al. 2003 cited in the Bibliography.

Figure 7. Perimetric protection   Figure 8. Protection for the buckets

Monitoring and Control of the Cutting Wheel

As well as the problems of the cutting tools, the head of a TBM for extra-hard rock suffers continual wear which has to be monitored in order prevent a situation being reached in which the structural steel of the wheel becomes affected, which would require a serious general repair, always very difficult and delicate because it has to be done inside the tunnel.[*]

For this, it is worth while highlighting the new designs of some manufacturers which permit the cutting wheel to be advanced (and therefore retracted) by around 500 to 700 mm, which facilitates inspection of the head and its tools with a certain frequency.

It is recommended to conduct a weekly inspection of the cutting wheel in both non-abrasive and homogenous rocks and even in those of medium-high hardness, and in the shields for soft ground. However, in Guadarrama the usual thing was to make inspections in all the advances when excavating extra-hard and abrasive rock. As far other aspects to emphasise are concerned, the following can be pointed out:

- **Protection of the structure.** A modern large-diameter cutting wheel in general has an almost plane profile, which is what experience has been recommending in order to have the least possible effect on the stability of the cut, even in extra-hard and abrasive rocks.

   However, this basic metallic structure has to be able to resist both the stresses of the advance due to cutting of the rock, and the abrasive effect thereof. For this reason, the thick steel sheet (thicknesses of up to 10 cm are frequent), constituting this basic structure, is, depending on the various technologies of the manufacturers, protected with three types of elements:

   - Perimetric protection. This can consist of superimposed reinforcements with hard-metal beads up to screwed pieces by way of "rakes" (Figure 7).
   - Protection of the buckets for loading the excavated material. These consist of plate reinforcements and anti-wear beads, as well as bars for producing the break-up of blocks or lumps of the wrong size (Figure 8).

---

*In Guadarrama, experience with extra-hard and abrasive rocks showed the need to conduct a general check and repair of the head roughly every 4 km of advance, for which sections of competent rock were chosen. Also, on two occasions the wheel had to be removed, securing it to the face of the rock by means of bolts in order to replace elements of the sealing systems for the main bearing. The operation of removal, securing of the wheel and assembly again, which was carried out successfully on both occasions, is extremely delicate.

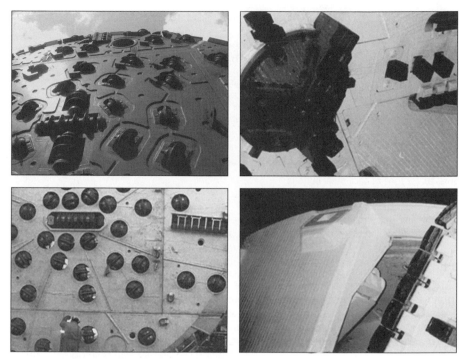

Figure 9. Types of protection of the structural sheet and cutters

      – Protection of the cutters and of the structural sheet in general. Three types of protection are currently used (Figure 9): anti-wear plates welded to the structure and basically protecting the cutters; hard-metal beads and wear sheets for general protection superimposed on top of the basic structure by means of welding.

        In any of these solutions, the row of the disc has to project far enough (around 7 to 10 cm) for permitting the free fall of material to the lower part of the face, where the buckets are to be found, which is how these machines work.

        So, the thickness of the plates, if this is the solution, has to respect that minimum. We have confirmed in two Guadarrama machines that otherwise part of the rubble remains caught against the head, causing a reduction in the penetration due to producing unnecessary wear on account of the abrasivity of the plates until free fall becomes possible. For that same reason, the second type of protection (simple hard-metal beads) has functioned well from the very beginning in the other two Guadarrama machines.

    ■ **Check and repair of the protection.** The experience of these recent years has recommended that, in hard and abrasive rock, and above all in large diameter machines, a check and a general repair as necessary should be made every 3 to 4 km of advance. This is what has been done in the works on the Guadarrama tunnels.

        For the repair of structural protection the following have been used: Replacement of rakes and new hard-metal reinforcement pieces in the peripheries; replacement of the bars for breaking up blocks and other elements, and

Figure 10. Types of hard-steel buttons

finally application of welded hard-steel buttons, of the cylindrical or prismatic types, which can be seen in Figure 10.

Cutting Wheel Drive

In general, in large dimension machines for hard rock all manufacturers have been adopting the same type of drive, based, apart from in details, on the following criteria: FV electric motors, up to eight frequency variators; main bearing with three rows of rollers and reducers in each motor with attack pinions to a crown with internal cogs.

This was what was done when designing the Guadarrama TBMs, whose functioning in general has been satisfactory, with just a few problems to be pointed out:

- **Excessive vibration.** It is not easy to evaluate "a priori" the excessive vibration of a hard-rock machine until the first consequences are suffered, namely: breakdowns in the attack pinion to the wheel and failures in the "seals" of the main bearing.

  The experience of almost two years of functioning of the Guadarrama TBMs enables the causes of those failures in the excavation of extra-hard rock to be identified:

  - Continual working without all the motors running. It is fundamental to have the necessary replacement units (around 80% to 100%) in order to prevent this situation.[*]

  - Maladjustment due to lack of rigidity in the drive support on which the reducer motors are mounted.

  The breakdowns suffered over the course of the first year of work have not been repeated since these causes were fully corrected.

- **Anti-rotation systems.** A hard-rock machine only rotates in the direction in which the buckets collect the excavated material and, therefore, the opposite direction of rotation is done with the wheel empty, or with minimum inertia, solely as an aid in cases of unblocking.

  Hence the particular importance of anti-rotation systems, which are essential in TBMs of the "double shield" type, since at the end of the advance cycle, the machine has to remain perfectly level in order to position the lining rings. This leveling, which in conventional shields is achieved by alternating the direc-

---

[*]The usual practical recommendation that one can work with up to 70% of the motors only, providing that they are distributed symmetrically in the attack to the crown and that the penetration of the TBM is also reduced, is not admissible in extra-hard rock.

tion of rotation along the excavation cycle (since the entrance of material takes place in either direction of rotation of the head), cannot be done like this in a "double shield" other than with an adequate design of the anti-rotation system.

This system can be said to have two possibilities depending on the "mode of operation." Working as a "double shield," the "grippers" constitute the anti-rotation system. However, working as "single shield" the cylinders of the auxiliary thrust equipment have to be able to vary their inclination, in order to counteract the tendency of the TBM to roll.

This system was applied to the Guadarrama machines, which made it possible to carry out small corrections during the advance cycle, getting the "anti-rotation" system to function automatically when a limit previously established in the computer program governing the machine was approached. The results were very acceptable, though it is essential for the operator to have a medium-high level of experience, therefore substantial improvements are to be expected in the future designs of these TBMs.

- **Excavation with extra-cut and curve management.** Modern TBMs for rock have to have the possibility of making an additional cut or extra-cut, whether it be for increasing the excavation diameter of conventional machines, or for excavation in a curve for all kinds of machines, mainly those of the double shield type.

This extra-cut starts with two or more peripheral cutters which can be projected by hydraulic thrust and which slowly carry out an initial enlargement of the excavation. Having achieved this, in order to continue the over-excavation along a section there are two operations that have to be performed: displace the axis of the machine vertically and mechanically attach the cutters that are going to be used to the structure of the head.

Manufacturers solve this in various ways, but what is interesting to state here is that, for the time being, it has not been possible to clearly exceed extra-cuts of the order of 100 mm measured in terms of radius (200 mm in diameter) for excavating a section of tunnel, even if its length is not great. In any case, the yields of the excavation drop appreciably.

Finally, it must also be pointed out that some manufacturers have developed prototypes with rotation of the cutting wheel in order to manage curves. However, although the displacement of the head according to the axis of the machine has been fully achieved (and has been commented upon when dealing with inspections and repairs), rotation towards the sides has not so far been successful, even in rocks of medium hardness.

Works Modes and Overall Yields

As has already been said, for the Guadarrama TBMs is was decided to choose the "double shield" type with the aim of achieving the shortest possible execution period. The purpose of the "double shield" design is to be able to carry out the advance activating just the front shield on the basis of the "grippers," while the rear shield is carrying out the positioning of the lining rings. Putting this another way, it is possible to advance and carry out the lining simultaneously in order to achieve the maximum possible speed of construction of the tube.

All this requires that the greater part of the massif should consist of rock of sufficient competence for supporting the "grippers." But, when passing through faults, it can be indispensable to have to work as a single shield, for which the auxiliary thrust equipment is used located in the rear shield. In this case, the work cycle has the two

Table 5.   Advances according to work "mode"

Attack Faces	Tunneler	Advance Length		
		Double Shield	Single Shield	Total
South Mouth (T.1) West Tube	Herrenknecht	8,228 m	4,487 m	12,715 m
		(64.71%)	(35.29%)	(100%)
South Mouth (T.2) East Tube	Wirth	11,168 m	2,050 m	13,218 m
		(84.49%)	(15.51%)	(100%)
North Mouth (T.3) West Tube	Wirth	11,183 m	2,909 m	14,092 m
		(79.36%)	(20.64%)	(100%)
North Mouth (T.4) East Tube	Herrenknecht	10,151 m	4,142 m	14,293 m
		(71.02%)	(28.98%)	(100%)

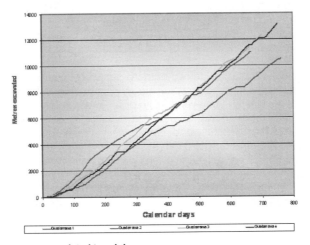

Figure 11.   Advances accumulated to origin

stages typical of a single shield (advance and positioning of the ring) rather than these processes being simultaneous.

Some data relating to the Guadarrama TBMs since the start up to the end of the work is contained in Table 5 and Figure 11.

"Double shield" type TBMs require wide experience from the operators. They have been in use for more than 20 years in diameters of less than 7 m. In 1996 the first large diameter machines were built (more than 9 m) with some major failures. As always, on the basis of faults or errors, whether of design or operation, the technology has improved appreciably and good proof of this is the success which the four machines have had for the Guadarrama Tunnel of the High Speed Line to North and North-West Spain, as shown in Figure 11.

ACKNOWLEDGMENTS

The executives of the ADIF who are managing this work have provided the author with the support and the necessary access to the technical documentation of this work of the Spanish Ministry of Public Constructions (Ministerio de Fomento). The author wishes to thank all of them for their help and collaboration.

The author wishes to mention as well, the help received from the Job Directorate of the ADIF and the companies responsible for the Technical Assistance, as well as from each and all the Spanish Construction companies that are integrated in the temporary joint ventures "Guadarrama Norte" and "Guadarrama Sur" which are performing this unique work.

Specific mention is made to the personal collaboration in the preparation of this document of Engineers Mr. José M. Fernández de Castro, of the organization for the Job Directorate of the GIF, and Mr. Ramón Fernández, of SPICC, S.L.

BIBLIOGRAPHY

Cobreros. J.A., Mendaña, F. and Moreno Cervera, M. 2002. *Design and Construction of the Guadarrama Tunnel. Revista de Obras Públicas No. 3426. October 2002. Madrid.*

Herrenknecht, M. &Rehm U. 2003. *Developments of Mixshields and Hard-Rock TBM. Proceedings ITA April 2003 Amsterdam.*

Innaurato, N., Oggevi C., Oreste P.P. & Virai R. 2003. *Problems Concerning Cutting Tool Performance during TBM Work: Modelisation and Testing of the Rock Under the Action of the Tool. Proceedings ITA April 2003 Amsterdam.*

Mendaña, F. 2003. *Novedades en el Diseño de TBMS: Túneles de Gran Diámetro y Adecuación A Terrenos de Gran Resistencia, Abrasivos o de Baja Capacidad Portante. Tunneling 2003. Symposium October 2003. Madrid.*

Merguerian, C., and Ozdemir, L. 2003. *Rock Mass Properties and Hard Rock TBM Penetration Rate Investigations, Queens Tunnel Complex, NYC Water Tunnel 3, Stage 2. Proceedings RETC June 2003 New Orleans.*

Wolfgang G. and Romualdi, P. 2003. *New Design for a 10 M Universal Double Shield TBM for Long Railway Tunnels in Critical and Varying Rock Conditions. Proceedings RETC June 2003 New Orleans.*

THE RELATIONSHIP BETWEEN TUNNEL CONVERGENCE AND MACHINE OPERATIONAL PARAMETERS AND CHIP SIZE FOR DOUBLE SHIELD TBMS—A CASE HISTORY OF GHOMROUD WATER CONVEYANCE TUNNEL

Ebrahim Farrokh

Sahel Consulting Group

Jamal Rostami

CDM Inc.

ABSTRACT

Selection of tunneling method, specially the use of shielded machines for application in deep rock tunnels can be heavily impacted by presence of ground convergence and squeezing conditions. Tunnel convergence is a function of ground conditions (i.e., rock mass behavior, groundwater, in-situ stresses), size of the opening, and TBM operational parameters. The case study of Ghomroud water conveyance tunnel project in Central Iran, which is under construction using a double shield TBM, is used to examine the effect of rock parameters on tunnel convergence and hence on the need for over excavation and shield lubrication to avoid problems with the shield getting stuck. Although it is impractical to perform a continuous assessment of rock mass conditions within the shield or in the lined tunnel, results of field observations and subsequent studies show that the amount of the tunnel convergence has direct correlation with the percentage of fines and large rock fragments in the muck and TBM thrust/torque/ROP. These relationships have been examined and will be presented in this paper.

INTRODUCTION

Planning of activities and accurate performance/cost estimates for tunneling is heavily influenced by understanding of the ground conditions and their impact on operation. As such, selection of tunneling method and related machinery for application in deep rock tunnels can be affected by the anticipated ground behavior, in particular the possibility of running in to squeezing grounds. This is especially true for complex ground conditions such as rock tunnels in heavily folded and metamorphosed areas, highly variable formations, frequent faults on tunnel alignment, very deep tunnels, and finally in mixed face situations. Often times, when facing such conditions, the choice of the machine is between Open type TBM versus shielded machines, specifically Double Shields (DS).

Successful use of DS-TBMs in many projects clearly indicates the capability of this concept in offering efficient performance in various ground conditions (Grandori 2006). DS-TBMs offer more versatility when it comes to rapid changes in ground conditions and passing through broken zones and faults. This is especially handy in design built projects where the amount of geotechnical information is very limited at the time of design and when the machine is ordered concurrently with the on going explorations. In addition, simultaneous installation of segmental lining as ground support and final lining, or one pass tunneling system is particularly attractive in fast track projects. Yet the main question in selection of shields for any tunneling project remains the possibility of machine seizure in swelling or squeezing grounds. This is not much of a problem in soils and shallow tunnels, but for deep tunnels, which primarily are in rock, the possibility of a machine getting stuck poses a major concern. Such incidents could cause major delays and impose an expensive burden on the tunneling operation. Stoppage of machine in squeezing ground is bad news in many respects. First of all, the passage of time has a negative impact, the ground can only tighten its grip! Also, the process of hand mining for releasing the machine is very labor intensive and thus is very slow and dangerous. Therefore, examining the possibility of machine stoppage due to excessive ground convergence is an important step in tunnel design involving the use of Double Shield machines.

Accurate estimation of ground convergence and squeezing conditions are difficult. There are analytical and numerical methods for estimating ground convergence that are heavily dependent on ground characterization, based on geotechnical information, which are often limited and hard to obtain. Besides, most of the available models for estimation of ground convergence are empirical and hence, site specific. Moreover, using a shielded machine for tunnel construction limits the access to the tunnel walls and face for observation of the ground/rock mass conditions. Despite all these short-comings, some of the models have proven to be useful and effective if correct input data is collected and properly processed. Once the analysis is complete and conclusions were made that the estimated convergence is acceptable, a shield could be selected and introduced to the ground, however, this does not prevent ground from incurring higher convergence.

Naturally, when a shield machine is selected and put to work, monitoring the ground behavior is essential to avoid potential problems associated with long delays in areas of potentially excessive convergence. When ground conditions conducive to convergence are encountered, work schedule should be modified to expedite efficient and uninterrupted tunneling through such grounds and performing more extensive maintenance in stable grounds. The TBM performance is also a function of the machine's operational parameters and interaction between the machine and ground. As such, while tunneling, the observations and measurements are limited to machine parameters, rock cuttings and muck, and some measure of ground convergence to make an attempt to understand and develop a relationship between these parameters. Therefore, determination of the ground conditions based on observations from within the tunnel can provide a powerful tool to the operators. This paper presents a case study of Ghomroud tunnel project in central Iran where study of ground convergence, machine parameters, and rock mass quality was performed to examine their relationship. The results of field measurements and analysis will be discussed and new thoughts and formulas will be presented to facilitate a more accurate estimation of ground convergence from within the tunnel during the excavation.

PROJECT DESCRIPTION

Management of water resources has been the focus of national development plans of Iran in past two decades. This includes construction of many dam and tunnel projects to transfer water from the high elevations in Alborz and Zagross mountains to dry plains of Central Iran for agricultural use as well as support of rapidly growing urban areas. Ghomroud water conveyance tunnel is one of these projects involving a 36 kilometers tunnel from upstream of Dez river around the city of Aligoodarz to Golpayegan reservoir (Figure 1). The tunnel was divided into four construction packages of about 9 km each, put to bid as design build project in 2002. Ghaem Construction Co., a subsidiary of Khatam Corp. won the combined section 3–4 of this project totaling 18 km at the exit end of the tunnel with a portal access. This segment is under construction using a 4.5 m diameter Double Shield TBM. Tunnel will be excavated at a

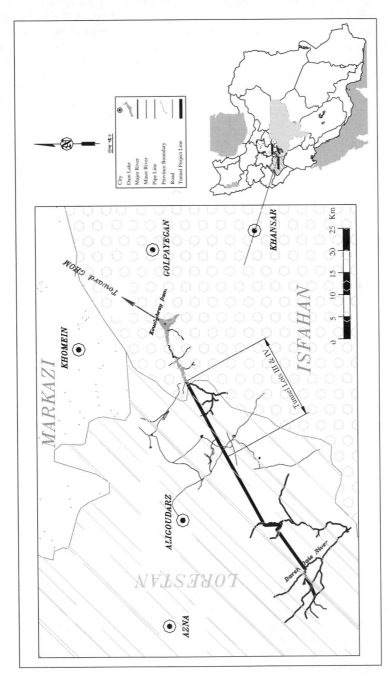

Figure 1.   Location of the Ghomroud Tunnel project site

Table 1.   Technical specifications of Wirth Double-Shield TBM model TB 453 E/TS

Machine diameter with New cutters	4,525 mm
Machine diameter with Worn cutters	4,495 mm
Over cutting (3 cutters)	65 mm
Cutters diameter	17" (432 mm)
Number of disc cutters	36
Average cutter spacing	75 mm
Cutterhead thrust	18,000 kN
Cutterhead power	1,120 kw
Head Rotational Speed	0–12 rpm
TBM Weight (approx.)	255 ton

Figure 2.   Wirth Double Shield TBM model TB 453 E/TS

grade of 0.134% and finished with concrete segmental lining to a diameter of 3.8 m. Machine specifications were developed in the summer of 2002 and a Wirth DS-TBM was selected (Table 1 and Figure 2).

SITE GEOLOGY

The tunnel is located in Sannandaj-Sirjan formation, one of the well-known geological divisions of Iran. This formation consists of series of asymmetric foldings and faults and has gone through mild to high metamorphisms. The lithology of this area consists of a sequence of Jurassic-Cretaceous formations. The Cretaceous formations consist of massive limestone and dolomite layers while the Jurassic formations mainly include slate, schist and metamorphic shale and sandstone units. The majority of the rock masses are weak to fair quality. The intact rock properties of the formations are generally medium to low strength. Anticipated geological conditions along the first six km of tunnel alignment (area of current study) are illustrated in the cross sectional profile of tunnel (Figure 3).

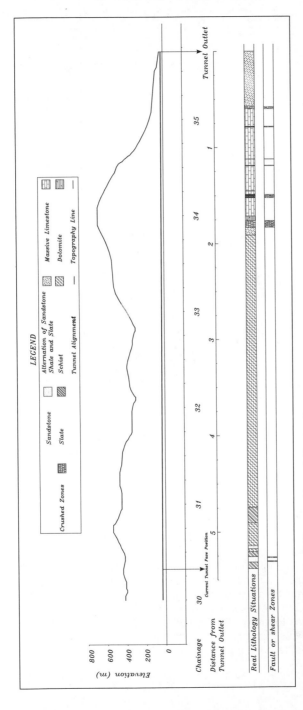

Figure 3.   Geological profile of Ghomroud tunnel alignment

Table 2.   List of the delays during tunneling operation

No.	Station (m)	Stop Date	Compressive Strength (Mpa)	Cutter Head	Shield	Duration (day)
1	22+55	14-Jul-04	15–50	Free	Jammed	13
2	25+36	09-Aug-04	30–60	Free	Jammed	5
3	31+44	09-Sep-04	15–50	Jammed	Free	21
4	32+17	02-Oct-04	20–35	Jammed	Free	53
5	46+72	17-Jan-05	25–30	Free	Jammed	12
6	47+14	30-Jan-05	25–30	Free	Jammed	5

Figure 4.   Portal and segment factory

TUNNELING ACTIVITIES

Site activities, included some earth work to set up tunnel portal, roads, shops, storage areas and segment factory. A segment factory was set up at the portal to fabricate hexagonal shape concrete segmental lining. Figure 4 is picture of the portal and segment factory and Figure 5 is the picture of finished tunnel with final lining. The excavation of tunnel using Drill & Blast method began in fall of 2002. This tunnel was excavated to allow drainage of a water bearing formation and for use as starter tunnel for the TBM. The machine was delivered to the site in fall of 2003. TBM and the trailing gear were assembled at the site and operation commenced in early 2004. Italian contractor, SELI was selected as a JV partner to manage some of the site activities, specially TBM assembly and operation. To date, about 5,400 m of tunnel has been excavated and machine has had several stoppages including major delays due to getting stuck in squeezing grounds as well as face collapses.

The progress of the tunnel has been a bit inconsistent. The average rate of penetration (ROP) has been 4 m/hr and average daily advance rate was about 10 m/day. The best daily advance rate has been 53 m. As for utilization, the overall average utilization from the time that TBM was launched was estimated around 8% and the utilization without the long delays was 18%. Figure 6 presents the monthly advance rate and the

Figure 5. The finished tunnel

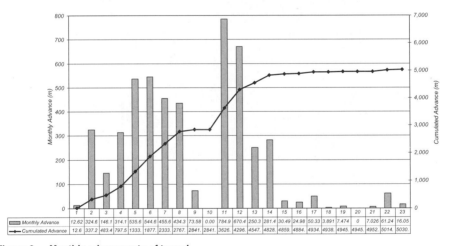

	1	2	3	4	5	6	7	8	9	10	11	12	13	14	15	16	17	18	19	20	21	22	23
Monthly Advance	12.62	324.6	146.1	314.1	535.6	544.6	455.6	434.3	73.58	0.00	784.9	670.4	250.3	281.4	30.49	24.98	50.33	3.891	7.474	0	7.026	61.24	16.05
Cumulated Advance	12.6	337.2	483.4	797.5	1333.	1877.	2333.	2767.	2841.	2841.	3626.	4296.	4547.	4828.	4859.	4884.	4934.	4938.	4945.	4945.	4952.	5014.	5030.

Figure 6. Monthly advance rate of tunnel

cumulative advance of the tunnel. Stoppages caused by the complex geology involving weak and broken grounds, where face collapse and high convergence were experienced. These incidents have caused long delays. Table 2 is a summary of the delays with the station numbers, time, and the duration of the delays.

TUNNEL CONVERGENCE

The excavated radius of the tunnel is a fixed value and can be obtained by the measurement of the tip loss (due to wear) of the gage cutters. This value varies in a small range since cutter inspection and changes are done as a daily routine to assure a minimum tip length on the gage cutters to maintain a safe radius of over cut and clearance for the front shield. The range of permissible tip loss is typically less than 15 mm (0.6 inch). In this tunnel project the maximum permissible tip loss for gage

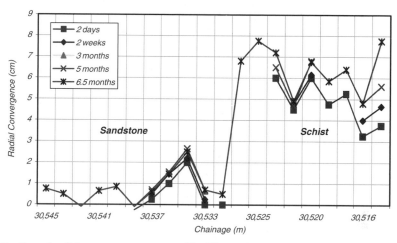

Figure 7. Tunnel radial convergence measured in different times

cutters was decided to be 8 mm. Besides, the double shield machine used in this project has an extendible gage cutter controlled by a hydraulic jack that can extend out and assure clearance for the shield and can be used to make relieve space for changing of the gage cutters. As such, the radius of the excavated tunnel can be measured with a good accuracy. To measure tunnel wall convergence, the subtraction of radius of excavated and converged tunnel was used. This distance is expressed as follows:

$$u_r = R_e - R_c \qquad (1)$$

Where: u_r = Radial tunnel wall convergence
        R_c = Converged tunnel radius
        R_e = TBM excavation radius

Segmental lining configuration included grout holes, through which surrounding walls could be accessed using a probe to make a direct measurement of ground convergence. Therefore, to measure the radial convergence, after erection and extrusion of segments from the tail shield, a scaled instrument was placed in the roof segment for reading the amount of tunnel convergence in the vertical (floor-roof) direction. This procedure was repeated for every segment (1.3 meters segment width) within some stretches of tunnel where measurement was possible. In some sections high tendency for convergence was observed and associated with the rock mass, which had plastic behavior due to existence of graphite on some joint surfaces. Obviously, convergence is a time dependent parameter and this factor had to be taken into consideration. Figure 7 presents the variations in convergence as a function of time for two different rock types, sandstone and graphite schist.

As shown the time dependent portion of convergence varied from 30% in schist, compared to 5–10% in sandstone. In other words, about 70% of convergence in Schist occurs in the first 2 days and remaining 30% in the months to follow. The results for other tunnel sections show the same trends.

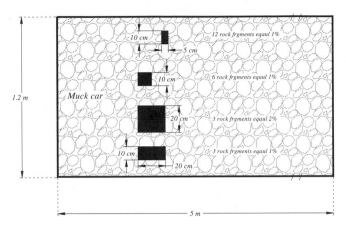

Figure 8. Pattern of rock fragments percentage calculation for rock fragments with length more than 10 cm

MUCK SIZE DISTRIBUTION

Several studies have been performed to provide a good understanding of the size of rock cuttings generated by disc cutters in TBM operation, among them Rostami & Gertsch, 2002, and Bruland 1988. Size of rock chips and in general the size distribution of the muck is a function of rock mass properties where the existence of discontinuities, bedding, joints, or other features impact crack growth and shape and size of rock chips. In jointed rock, the chip size can be larger due to dislocation of blocks created by crossing joints under the static or dynamic loading of the disc cutters. Similarly, it is conceivable that closely spaced joints and bedding, crushed zones, fault breccias, or other geological features can contribute to increased percentage of fines or smaller size muck. Moreover, it is also a fact that cutting is more difficult when the joints are parallel to the tunnel axis, which in turn increases the cutting forces needed for penetrating into the rock surface while increasing the amount of fines and crushed material. Hence, examining the muck generated in the excavation process allows for monitoring the condition of joints in the rock mass. Meanwhile, these parameters are the controlling parameters defining geomechanical behavior of rock masses and directly impact tunnel convergence. Therefore, a study of chip sizes generated by TBM was performed to evaluate the ground conditions, rock mass properties, and their impact on tunnel convergence.

Due to limited access to the face and walls for observation of the rock conditions, also limitations in instrumentation in the lined tunnels to measure convergence, one of the remaining parameters to measure relatively easily, was the muck size distribution. A quantitative system was developed for determination of chip size or muck size distribution. This was done to provide a quantitative measure related to rock mass characteristics for prediction of tunnel convergence and to alert the operator and avoid Shield becoming stuck in the ground. A procedure was developed where size distribution of the muck was estimated by visual observation of the rock fragments in the muck cars. Figure 8 shows the general procedure for this measurement performed in a limited area of the muck car (quarter or ⅛th of the area) and extrapolated to the entire load.

The basis of this approach was that the volume (and hence weight) of fragments of certain size is proportional to the surface area of such fragments in a given cross section. This is a common practice in many applications including evaluation of rock cores, thin section analysis, grain size distribution for minerals and grains in a rock texture, etc. To measure the percentage of certain chip sizes, a picture was taken from the muck and using calibrated ruler, the size of larger particles were measured in 5 cm (2 inch) increments (5, 10, 15 cm etc.). This procedure was typically performed over 2–3 muck cars and the numbers were averaged to represent the percentages for the given stroke. Similarly the percentage of powder or small particles was estimated from the pictures in a separate step. The following measurements were made by analyzing the pictures:

- Percentage of rock fragments in muck cars with length more than 10 cm.
- Percentage of rock fragments in muck cars with length less than 2 cm
- Maximum dimension of rock fragments in muck cars.

The numbers in these criteria (2 and 10 cm) are preliminary and were selected based on the disc cutter tip width (2.5 cm) and average disc cutter spacing on the cutterhead of 7.5 cm, respectively. To define the percentages of each size of rock fragments; the ratio of observed areas for each fraction to the total area was used as follows: (based on the templates shown in Figure 8)

$$R_f = (S/S')*100 \tag{2}$$

Where: R_f = Percent of rock fragments of certain size
   S = Area of rock fragments of certain size
   S' = Area of muck car surface

IMPACT OF GROUND CONDITIONS ON CONVERGENCE

Measured tunnel convergence can be plotted against the percentage of certain size rock fragments. As noted before, the amount of rock fragments of over 10 cm (4 inch) and below 2 cm (0.8 in) were estimated based on visual observations and measurements of the cuttings in the muck cars. Figure 9 presents the final radial convergence versus percentage of given rock fragments. This includes three categories of >10 cm, 2–10 cm and <2 cm. It can be observed that the convergence increases as the amount of rock fragments of over 10 cm and below 2 cm increases. Likewise, the convergence decreases with the increase in portion of rock fragments between 2 and 10 cm.

Figure 10 shows the radial convergence as a function of largest size of rock fragment in the muck pile (some long elongated pieces). As can be seen, despite the dispersed set of data in the chart, a general, albeit, weak correlation can be observed. A simple explanation for this trend is that in jointed rock masses where the existence of discontinuities results in weak and plastic rock behavior, jointing of the face will cause larger size block to get dislocated and makes its way into the muck. Thus, in closely jointed rock masses, the chip formation is incomplete and small fall outs occur more frequently. When such blocks are generated in the face, those that are small enough to go through the grizzlies and bucket screens, will show up in the muck pile. On the other hand, in very weak and broken rock masses and fault zones, very small fragments are dominant and can get excessively crushed when they are being transferred to the muck. In other words, the portion of the rock fragments between 2–10 represents the ground conditions where the face is stable and is actually being excavated, thus rock fragments and chips with the size proportional to disc spacing can dominate the muck.

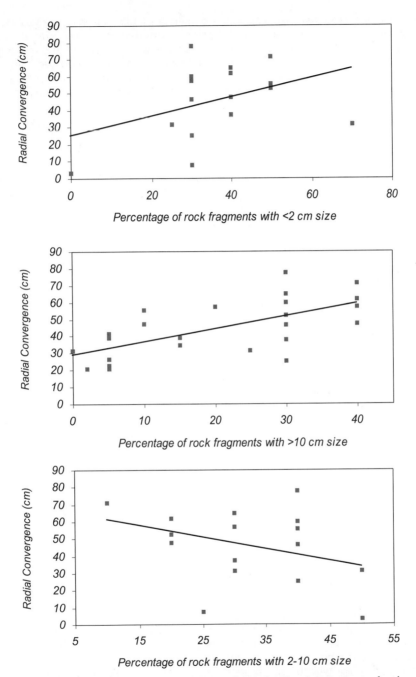

Figure 9. Relationship between radial convergence of tunnel wall and percentages of rock fragments

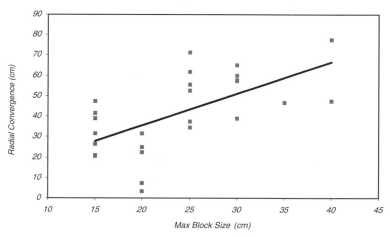

Figure 10. Relationship between maximum size of rock fragments and tunnel walls radial convergence

TBM OPERATIONAL PARAMETERS AND ROCK MASS QUALITY

Rock mass quality is one of the most important factors effecting TBM performance and operational parameters. Thrust force and cutterhead torque are the two most important TBM operational parameters. TBM thrust determines the amount of normal force delivered by the cutters to the face. Torque or rolling force is a function of normal force and depth of penetration. Therefore, Torque is related to applied thrust of the machine through certain established formulas. The ratio of rolling force to normal force is called cutter (or rolling) coefficient. Rostami 1997 (and 2002) offers formulas for estimation of cutter coefficient as a function of the angle of contact area (ϕ) which is determined by penetration. On the other hand, the amount of cutter penetration is related to rock mass and intact rock strength parameters. Several researchers (Kline 1995 and Sapigni et al. 2002) have tried to offer relationships between rock mass strength parameters (such as RMR, Q, etc.), intact rock strength parameters (i.e., compressive strength, tensile strength, etc.) and TBM penetration rate. These studies concluded that the cutter coefficient can be related to rock mass and intact rock strength parameters. In field studies, this relationship is expressed in terms of cutterhead torque and thrust ratios in various rock types and ground conditions.

The existing relationships between cutterhead torque/thrust, and intact rock and rock mass are typically for open type TBMs and need to be adjusted for shielded machines. This refers to interaction between the machine and the ground, frictional forces acting on the shield, as well as mechanical losses within the machine. For example, measurable thrust of the machine is based on the measurement of pressure of the hydraulic cylinders of the propel or thrust system and it could be different than what is delivered to the cutterhead. Likewise, the torque measurements are made either by motor Amps on electric drives or hydraulic pressure of the hydraulic drives. Thus certain adjustments need to be made to derive the torque-thrust curve. Logically, in grounds where convergence is very high and rapid and walls can grip the shields, the amount of thrust delivered to the cutterhead is reduced significantly. Meanwhile, with lower cutterloads, cutter penetration is reduced, leading to lower rolling force and torque (and thus lower torque to thrust ratio). In this respect, very interesting observa-

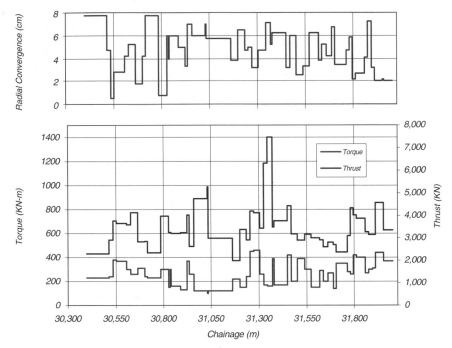

Figure 11. Tunnel radial convergence, average thrust, and torque along the excavated length

tions were made when TBM operational parameters and their relationship with tunnel convergence were examined for Ghomroud tunnel project.

When analyzing field data from the double shield TBM, various plots were made and relationship between tunnel convergence and machine parameters were evaluated. For this purpose, convergence records for many segment were recorded and tunnel areas with the same radial convergence were identified. This was followed by calculating average TBM operational parameters for theses areas and examining the relationship between machine torque to thrust ratio and rock mass quality as represented by tunnel radial convergence. Observation of a generally good trend led to performing additional measurements and analysis. Subsequently, a normalization system was developed and several composite parameters were derived to represent overall operational parameters of the machine. Figure 11 shows the amount of tunnel convergence, average Thrust and Torque for certain stretches of tunnel.

Statistical analysis of the available data indicated that the best combination of TBM operational parameters could be found as described in Equation 3. These combined parameters were named TBM operational index (I_{TBM}).

$$I_{TBM} = TQ^2/Th/Pr \tag{3}$$

Where: TQ = Cutterhead torque (KN-m)
        Th = TBM thrust (KN)
        Pr = Penetration rate (m/hr)

Figure 12 shows the relationship between TBM operational index and tunnel radial convergence. As can be seen, a strong relationship exists between these parameters, indicating a potential for developing reliable predictive capability for tunnel convergence

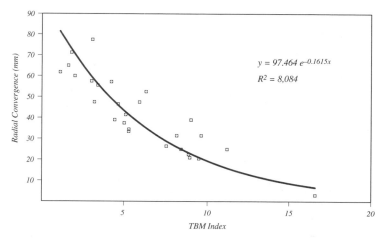

Figure 12.   The relationship between tunnel convergence and I$_{TBM}$ and for Ghomroud tunnel

based on measured machine parameters. This is especially important since most TBMs are equipped with PLC and data acquisition systems that can provide accurate real time reading of machine parameters. Thus, with a simple analysis of the measured parameters, machine's installed monitoring systems can offer some indicators or warnings for rapid tunnel convergence.

DISCUSSIONS

It must be emphasized that the results of the current studies and analysis is preliminary and further investigation on the subject is needed to confirm the hypothesis offered in this paper. The relationship between muck size distribution and convergence is very rough and could be considered too disperse and inconclusive. It could also be misleading due to the limited number of observations. Nevertheless, there seem to be a trend in the data which is logical and further investigations could prove to be worthwhile. This could include performing a number of size distribution measurements by sieve analysis to verify the graphical calculations and also to compare the results with the tunnel convergence.

As for the analysis of tunnel convergence based on TBM operating parameters, this relationship is naturally site specific and therefore should be redefined for other tunneling machines and tunneling projects. A different combination of parameters could also be considered to simplify the parameters used in the formula. The I$_{TBM}$ index in its current form represents actual measurements, but it can be modified to represent its true nature which is a measure of cutting coefficient. A site-specific criteria and relationships can be developed for each project in the initial stages of tunnel excavation and calibrated to provide more accurate criteria to be used as a warning system. Using indications and early warnings could lead the contractor to revisit machine operational parameter, tunneling operation and scheduling, and scheduled stoppages to cope with the ground conditions at and ahead of the tunnel face. The system can also be used to optimize the amount of over cut by machine to prevent machine jamming while cutting the required minimum annulus space to avoid excessive use of grout and minimize related costs.

CONCLUSIONS

The primary focus of this paper was to examine the impact of ground conditions on the operation of Shielded TBMs. Rock fragmentation mechanism by disc cutters in jointed rock masses and its effects on size distribution of muck and operating parameters of TBM can be related to potential convergence of the tunnel. The analysis of data from field measurement of tunnel convergence as well as percent of muck of certain size in Ghomroud tunnel proved to be a useful tool in prediction of rock behavior in tunneling operation using shielded TBMs. It also showed that tunnel convergence had a good agreement with TBM operational parameters such as Thrust /Torque/penetration.

These results are preliminary and require more rigorous follow up studies and analysis. Yet it can provide a useful tool to examine rock quality during tunnel excavation by a shielded TBM and develop an estimate of potential for tunnel convergence while tunneling is in progress. Once these areas are identified, suitable precautionary measures can be taken to avoid long delay and prevent machine from being held by excessive pressures of squeezing grounds.

REFERENCES

Bruland, A, 1998 "Hard Rock Tunnel Boring, The Boring Process," NTNU-Anleggsdrift. *Project Report 1F-98*, 1998.

Grandori, Remo, 2006, "Abdalajis east railway tunnel (Spain)—double shield universal TBM cope with extremely poor and squeezing formations"*Tunneling and Underground Space Technology* Volume 21, Issues 3–4, May–July 2006, pp. 268

Klein, S. Schmoll, M. and Avery, T. "TBM Performance at Four Hard Rock Tunnels in California," *Rapid Excavation and Tunneling Conference (RETC) Proceedings*, chapter 4, pp. 61–75, 1995.

Rostami, J. *"Development of a Force Estimation Model for Rock Fragmentation with Disc Cutters Through Theoretical Modeling and Physical Measurement of Crushed Zone Pressure,"* PhD Thesis, Mining Engineering, Colorado School of Mines, 1997.

Rostami, J., Gertsch, R., Gertsch. L. "Rock Fragmentation by Disc Cutter A Critical Review and an Update," *Proceedings, North American Rock Mechanics Symposium (NARMS)—Tunneling Association of Canada (TAC) meeting*, Toronto, Canada, 9 pages, July 7–10, 2002.

Sapigni, M. Berti, M. Bethaz, E. Busillo, A. and Cardone, G. "TBM performance estimation using rock mass classifications," *International Journal of Rock Mechanics & Mining Sciences 39*, pp. 771–788, 2002.

SUCCESSFUL TUNNELING FOR WATER TUNNEL NO. 3, NEW YORK CITY, NEW YORK

Thomas P. Kwiatkowski

Jenny Engineering Corp.

Joginder S. Bhore

Joginder S. Bhore International, Inc.

Jose Velez

Jenny Engineering Corp.

Christopher D. Orlandi

Jenny Engineering Corp.

ABSTRACT

Three case histories of successful TBM tunneling for New York City Water Tunnel No. 3 are presented. The tunnels, costing about one billion dollars total, are the 5-mile long, 18'-11" OD Brooklyn Tunnel; the 5-mile long, 23'-2" OD Queens Tunnel, and the 9-mile long, 12'-4" OD Manhattan Tunnel. The three tunnels are the first major tunnels constructed in New York City rock utilizing the TBM method of construction. This paper describes and compares the three tunnel projects in terms of geology, construction methods, TBM's, rock and groundwater behavior, rates of advance and other performance data, and challenges and solutions encountered during the projects.

OVERVIEW

The New York City Department of Environmental Protection (NYCDEP) City Tunnel No. 3 is the largest capital construction project in New York City's history. Planning for the tunnel began in the 1960s, and construction was begun in 1970 on Stage 1. The Stage 1 portion of the tunnel begins at the Hillview Reservoir in Yonkers and extends south through the Bronx into Manhattan, and then eastward under the East River and Roosevelt Island into Astoria, Queens. Stage 1 includes a 7.31m (24 ft) diameter concrete lined pressure tunnel and three subsurface valve chambers. Stage 1 was completed in 1984. Stage 2 was started in 1988 with the construction of several shafts. The Stage 2 tunnel will provide water to the Boroughs of Brooklyn, Queens and lower Manhattan. The Brooklyn and Queens sections of the tunnel have been completed, and the Manhattan section of the tunnel is currently under construction. Figure 1 shows the layouts of the Stage 1 and Stage 2 portions of Water Tunnel No. 3. The primary purpose of Water Tunnel No. 3 is to provide a by-pass capability to the City so that the in-service City Tunnels No. 1 and 2 can be inspected.

All of the Stage 1 tunnel excavations were performed utilizing drill-and-blast construction methods. The Stage 2 tunnel excavations were all performed utilizing Tunnel Boring Machines (TBM). The Brooklyn, Queens, and Manhattan Stage 2 sections of City Tunnel No. 3 are the first major tunnels constructed in New York City rock utilizing

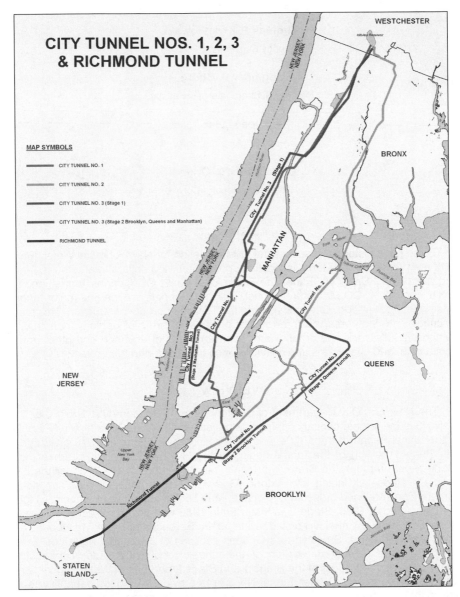

Figure 1.   General view of City Tunnels 1, 2, and 3

the TBM method of construction. The three Stage 2 tunnels have traversed through a wide range of extremely complex geology in Brooklyn, Queens, and Manhattan, and discoveries have been made that might require modification to the current geology maps in use today.

This paper describes and compares the three Stage 2 tunnel projects in terms of geology, construction methods, TBMs, TBM performance data, and challenges and solutions encountered during the projects.

Table 1. Average rock properties

Tunnel	Predominant Rock Type	Average Rock Test Result							
		UCS		PLI		STS		CLI	DRI
		(kPa)	(psi)	(kPa)	(psi)	(kPa)	(psi)		
Brooklyn	Migmatite Gneisses	156,511	22,700	6,895	1,000	NA	NA	9	41
Queens	Migmatite Gneisses	153,064	22,200	8,274	1,200	8,274	1,200	9	51
Manhattan									
South	Mica Schist	31,716	4,600	5,171	750	6,067	880	11	62
	Schistose Gneiss	62,742	9,100						
North	Mica Schist	51,021	7,400	4,964	720	8,963	1,300	11	50
	Granitic Rock	144,100	20,900	7,102	1,030	6,343	920		
Crosstown	Mica Schist	51,021	7,400	4,964	720	8,963	1,300	12	49
	Schistose Gneiss	65,500	9,500	4,137	600	9,653	1,400		

NOTE:
UCS—Unconfined Compressive Strength
PLI—Point Load Index
STS—Splitting Tensile Strength
CLI—Cutter Life Index
DRI—Drill Rate Index

BROOKLYN TUNNEL

The Brooklyn Tunnel is located primarily in the Borough of Brooklyn. It was excavated from Shaft 23B in the Red Hook area of Brooklyn to Shaft 19B, located in Queens. The construction was performed under NYCDEP Contract 543B. The tunnel is approximately 8,784.95m (28,822 ft) long and the depth varies between 167.33m and 203.60m (549 ft and 668 ft) below the ground surface. The excavated tunnel diameter was 5.76m (18.9 ft) and the finished tunnel diameter is 4.88m (16 ft).

The tunnel was driven downhill at a slope of approximately 1.22m (4 ft) per 305m (1,000 ft). The horizontal tunnel alignment contained one major curve approximately 240m (789 ft) in length with a radius of approximately 305m (1,000 ft). The construction contract low bid price submitted by the joint venture of Schiavone Construction Company, Inc. and J.F. Shea Company, Inc. (Schiavone/Shea) was $138,768,000.

Geology

The Brooklyn Tunnel was excavated through predominantly migmatic gneisses bedrock intermixed with lesser amounts of amphibolites and hornblendite (metamorphosed ultramafic rock). Granite, schists, pegmatite, gabbro, serpentinite and diorite were also encountered intermixed to varying degrees. Garnets were encountered in appreciable quantities throughout the various rock types. Foliation trends were northeast, and in general, the rock was hard, abrasive and fractured to varying degrees.

In addition to "thousands" of minor faults with small displacements, three major NNW-trending faults were also located along the tunnel alignment. These major faults were located near the north end of the tunnel and were responsible for significant slow-downs in construction. Average test results for Unconfined Compression Strength (UCS), Point Load Index (PLI), Splitting Tensile Strength (STS), Cutter Life Index (CLI) and Drill Rate Index (DRI) are located in Table 1.

Table 2. TBM data comparison

Tunnel	Diameter		Cutter-head Speed	Cutter Capacity		Thrust		Cutter Size	
	meters	feet	rpm	kN	lb	kN	lb	mm	in
Brooklyn	5.76	18.9	8.9	289	65,000	17,712	4,000,000	432	17
Queens	7.06	23.2	8.3	311	70,000	15,550	3,500,000	483	19
Manhattan	3.84	12.6	12.6	314	70,590	10,404	2,339,000	432	17

Table 3. Tunnel excavation data

Tunnel	Length		Duration of Excavation	Average Rate of Penetration		Number of Cutters Changed
	meters	feet	hours	m/h	ft/h	
Brooklyn	8,784.95	28,822	4,032	2.18	7.14	4,500
Queens	7,658.71	25,127	4,364	1.88	6.16	2,487
Manhattan						
South	5,613.50	18,417	1,327	4.23	13.88	427
North	4,196.18	13,767	1,140	3.68	12.07	333
Crosstown	4,384.85	14,386	955	4.59	15.06	346

TBM Performance Data

The Brooklyn Tunnel was started on July 5, 1994 and holed through on January 31, 1997. The TBM used by the contractor (Schiavone/Shea) was a modified Robbins tunnel boring machine. The machine had a cutterhead speed of 8.90 rpm, a cutter capacity of 289 kN (65,000 lb), a thrust of 17,712 kN (4,000,000 lb) and disk cutter diameter of 432mm (17 in). There were approximately 4,032 total hours on the TBM clock for the 8,784.95m (28,822 ft) mined for an average penetration rate of 2.18 m/h (7.14 ft/h). The average penetration rates in the main rock formations encountered during mining varied from 2.609 m/h to 2.066 m/h (8.56 ft/h to 6.78 ft/h). TBM Data is summarized in Table 2 and Tunnel Excavation Data is summarized in Table 3.

Muck removal was accomplished using six side dump muck cars. The muck cars traveled on a single track from the heading back to Shaft 23B where they were unloaded into a hopper and then to a feeder that dumped onto an inclined conveyor that discharged into a vertical belt conveyor. The vertical conveyor carried the muck upwards a distance of approximately 168m (550 ft) to the ground surface.

QUEENS TUNNEL

The Queens Tunnel is located in the Borough of Queens. It was excavated from Shaft 19B, located near Newtown Creek, to Shaft 16B which is the terminus of the Stage 1 Tunnel. The construction was performed under NYCDEP Contract 542. The tunnel is 7,658.71m (25,127 ft) long and the depth varies between 203.61m and 238.35m (668 ft and 782 ft) below the ground surface. The excavated tunnel diameter is 7.06m (23.17 ft) and the finished tunnel diameter is 6.1m (20 ft).

The tunnel was driven downhill at a slope of approximately 1.31m (4.3 ft) per 305m (1,000 ft). The horizontal tunnel alignment contained one major curve approximately 621m (2,037 ft) in length with a radius of approximately 305m (1,000 ft) and two minor curves approximately 73m and 141m (239 ft and 462 ft) each with a radius of 305m (1,000 ft). The construction contract low bid price submitted by the joint venture of Grow Tunneling Corporation, Perini Corporation, and Skanska USA (Grow, Perini and Skanska, JV) was $172,497,000.

Geology

The Queens Tunnel was constructed through rock types similar to those of the Brooklyn Tunnel; migmatite gneisses bedrock intermixed with lesser amounts of amphibolites, hornblendite and mafic dikes (metamorphosed ultramafic rock). Granite, schists, pegmatite, gabbro and diorite were also encountered intermixed to varying degrees. Garnets were encountered in appreciable quantities throughout the various rock types. Five igneous dikes of a red dactite rock (hornblende dactite porphyry) were also encountered during tunneling operations.

In general, the rock was hard, abrasive, and fractured, as well as seamy and blocky to varying degrees along the tunnel alignment. Multiple NNE-trending faults were also located along the tunnel alignment. Similar to the Brooklyn tunnel, "thousands" of minor faults with small displacements were encountered. Average test results for Unconfined Compression Strength (UCS), Point Load Index (PLI), Splitting Tensile Strength (STS), Cutter Life Index (CLI) and Drill Rate Index (DRI) are located in Table 1.

TBM Performance Data

The Queens Tunnel was started on October 31, 1996 and holed through on October 21, 1999. The TBM used by the contractor (Grow, Perini and Skanska, JV) was a Robbins HP 235-282. The TBM was equipped with a computer TBM data logger (PLC system), which automatically maintained an operations record of various machine components and movements. The machine had a cutterhead speed of 8.3 rpm, a cutter capacity of 311 kN (70,000 lb), a thrust of 15,550 kN (3,500,000 lb), and disk cutter diameter of 482.6mm (19 in). Mining took approximately 4,364 hours for 7,658.71m (25,127 ft) for an average penetration rate of 1.88 m/h (6.16 ft/h) based on the reduced results of the TBM data logger. TBM Data is summarized in Table 2 and Tunnel Excavation Data is summarized in Table 3.

Muck removal was performed using a 0.609m (2 ft) wide horizontal belt conveyor system which ran within the excavated tunnel from the TBM trailing gear to Shaft 19B. The horizontal conveyor discharged to a vertical belt conveyor that carried the muck upwards to the ground surface, a distance of approximately 204m (668 ft).

MANHATTAN TUNNEL

The Manhattan Tunnel is located primarily in lower Manhattan south of Central Park. It consists of three tunnel sections: South, North and Crosstown (east). All tunnel excavations began at Shaft 26B. The South Tunnel was excavated under Contract 538B, and will be lined under the subsequent Contract 538C. The North and Crosstown Tunnels were both excavated and will be lined under Contract 538C. The 538C Contract also includes the construction of nine shafts and the completion of one shaft. The South Tunnel alignment generally travels to the south, then curves to the northeast and terminates at Shaft 31B. The North Tunnel proceeds in a generally northerly direction from Shaft 26B, terminating at the existing Shaft 13B stub tunnel within Central Park. The

Crosstown tunnel proceeds in a generally easterly direction from Shaft 26B to the vicinity of Second Avenue, and then curves to the north terminating at Shaft 33B.

The South Tunnel is approximately 5,613.50m (18,417 ft) long and its depth varies between 156.36m and 175.87m (513 ft and 577 ft) below the ground surface. The North Tunnel is approximately 4,196.18m (13,767 ft) long and its depth varies between 140.82m and 163.68m (462 ft and 537 ft) below the ground surface. The Crosstown Tunnel is approximately 4,384.85m (14,386 ft) long and its depth varies between 142.34m and 175.87m (467 ft and 577 ft) below the ground surface. The excavated diameters of the three TBM tunnels are 3.84m (12.6 ft) and the finished tunnel diameter will be 3.05m (10 ft).

Shaft 26B is a low point for the system, and all three tunnels were driven uphill. The horizontal alignment of each of the three Manhattan tunnel sections contained at least one major curve. The South Tunnel has two major curves approximately 353m and 455m (1,157 ft and 1,494 ft) in length with a 155m (510 ft) tangent between the curves. The North Tunnel has one major curve approximately 253m (829 ft) in length, and the Crosstown Tunnel (has one major curve approximately 47 m (1,571 ft) in length. All the curves have a radius of 305m (1,000 ft).

The construction contract low bid prices was $85,620,000 for Contract 538B and $668,533,000 for Contract 538C. The joint venture of Schiavone Construction Company, Inc., J.F. Shea Company, Inc., and Frontier-Kemper Constructors, Inc. (Schiavone/Shea/Frontier-Kemper) was the successful bidder for both contracts.

Geology

The bedrock encountered in the three tunnel excavations in Manhattan was found to be much softer and less abrasive than the rock types encountered in the Brooklyn and Queens tunnels. The South and Crosstown Tunnels were excavated through predominantly mica schist and schistose gneiss rock with lesser amounts of pegmatite and serpentinite. The North Tunnel was excavated through predominantly mica schist and granitic rock with lesser amounts of schistose gneiss, pegmatite and serpentinite. "Thousands" of minor faults with small displacements, similar to those found in the Brooklyn and Queens tunnels were encountered by all three Manhattan tunnels, with only a few major faults having an impact on the mining.

Average test results for Unconfined Compression Strength (UCS), Point Load Index (PLI), Splitting Tensile Strength (STS), Cutter Life Index (CLI) and Drill Rate Index (DRI) are located in Table 1.

TBM Performance Data

The TBM used for excavating both contracts for the Manhattan Tunnel was a Robbins Model 1215-257 tunnel boring machine. The machine has a cutterhead speed of 12.6 rpm, a cutter capacity of 314 kN (70,590 lb), a maximum available cutterhead thrust of 10,404 kN (2,339,000 lb), and disk cutter diameter of 431.8 mm (17 in). The South Tunnel was started on September 11, 2003 and TBM excavation was completed on July 27, 2004. The average penetration rate for the mining was 4.23 m/h (13.88 ft/h). After completion of the South Tunnel mining, the TBM was refurbished. The North Tunnel was started on February 11, 2005 and was completed on October 30, 2005. The average penetration rate was 3.68 m/h (12.07 ft/h). The Crosstown (east) Tunnel was started on January 4, 2006 and was completed on August 4, 2006. The average penetration rate was 4.59 m/h (15.06 ft/h). TBM Data is summarized in Table 2 and Tunnel Excavation Data is summarized in Table 3.

Muck removal in all three tunnel sections was accomplished using side dump muck cars. The muck cars traveled on a single track from the heading back to Shaft

26B where they were unloaded into a hopper and then to a feeder that dumped onto an inclined conveyor that discharged into a vertical belt conveyor. The vertical conveyor carried the muck upwards to the ground surface, a distance of approximately 176m (577 ft).

COMPARISONS

The Brooklyn and Queens Tunnel were generally constructed in Migmatite Gneiss with unconfined strengths in the 155,000 kPa (22,000 psi) range. Cutter life indexes (CLI) were 9 and drilling rate indexes (DRI) were 41 and 51 for Brooklyn and Queens, respectively. The Manhattan Tunnel was constructed generally in Mica Schist and Schistose Gneisses with some granitic rocks. Unconfined compression strengths ranged from 30,000 to 144,000 kPa (4,500 psi to 21,000 psi) with CLIs of 11 and 12 and DRIs of 49 to 62.

These values correlate well to penetration rates as well as cutter changes. The Brooklyn and Queens Tunnels averaged 2.18 and 1.88 m/h (7.14 ft/h and 6.16 ft/h) respectively. The Manhattan Tunnel had average penetration rates between 3.68 and 4.59 m/h (12.07 ft/h and 15.06 ft/h). The higher the rock strength, the lower the DRI and lower the penetration rates. Cutter usage on the Brooklyn and Queens Tunnels far exceeded that of the Manhattan Tunnel. This was predicted by the higher CLI in the Manhattan Tunnel and the lower CLI in the Brooklyn and Queens Tunnels.

Tables 1, 2, and 3 summarize the data for the three TBM projects.

GEOLOGIC IMPACTS TO MINING

The geology of the three tunnels did vary significantly. However, many of the rock types and geologic characteristics were similar. For example, all three tunnels encountered geologic formations which included "thousands" of small displacement faults in a geologic sense. However, in the tunneling sense, these small displacements in the rock formation were generally healed and provided no adverse effects to mining operations. That is not to say that Manhattan or the surrounding boroughs do not contain faults of significance to tunneling.

The Brooklyn Tunnel encountered a significant fault along its alignment. This fault zone contained brecciated rock material as well as clay materials. Tunnel excavation was difficult and a tunnel interliner was required to withstand the significant internal pressure in this zone. The Queens Tunnel and the Manhattan Tunnel also encountered some faults that did result in tunneling difficulties. However, the majority of both tunnels were excavated through bedrock that contained thousands of low displacement geologic faults with no adverse effects to the tunneling.

For the Manhattan Tunnel, an extensive geotechnical investigation and testing program was conducted. The program consisted of over 100 borings, some over 183m (600 ft) deep, and laboratory testing for conventional rock properties as well as specialized rock characteristic testing (CLI and DRI) specifically for TBM construction conducted by the Norwegian Institute. A Geotechnical Interpretive Report (GIR) was produced and included with the contract documents. This report defined geology and predicted anticipated conditions.

As presented earlier, mining penetration rates for the Manhattan Tunnel continued to increase over the duration of the project. This can be attributed to many factors, some of which include familiarity and acceptance of mining equipment and methods by mining personnel, excellent mining equipment and maintenance procedures and the

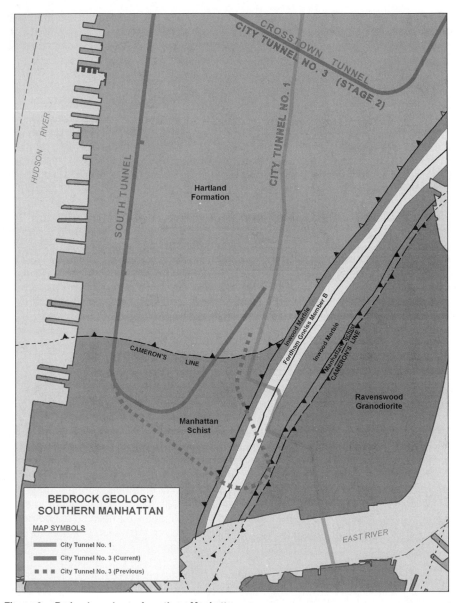

Figure 2. Bedrock geology of southern Manhattan

favorable geology. Even though the Manhattan Tunnel encountered thousands of geologic faults, only a few created minimal fall out or blocky conditions and only in one area did a combination of 3 faults produce a fault zone that did stop mining for several days. The geology of Manhattan encountered by Water Tunnel No. 3 is a very good tunneling medium.

The NYCDEP engineers did recognize the effect of geology on tunneling when planning the Manhattan tunnels. This is indicated by the extensive geotechnical investigation and laboratory testing program undertaken prior to bidding. In addition, the early alignments for the South Tunnel took it further east for water distribution reasons, as shown in Figure 2. But evaluation of the geologic maps and the results of a horizontal boring through the significant synclinal feature prompted the NYCDEP to modify the final South Tunnel alignment to miss this feature.

The synclinal feature included marbles and gneisses with the contact zones between the features exhibiting poor rock quality and high groundwater permeabilities. This is evident in the alignment of Water Tunnel No. 1, which is also shown on Figure 2. The stepped alignment of Water Tunnel No. 1 enabled the tunnel to cross the poor quality rock area perpendicular to the feature's orientation, thereby minimizing the length of difficult tunneling conditions. The NYCDEP's decision to avoid these poor rock conditions with Water Tunnel No. 3 resulted in significant cost savings as well as a shorter construction schedule.

SUMMARY

Three successful TBM bored tunnels have been constructed in New York City over the past 12 years. Although the complex geology of New York City can be challenging for this type of construction, the construction of these three tunnels shows that success can be achieved.

TBM EXCAVABILITY: PREDICTION AND MACHINE–ROCK INTERACTION

Z.T. Bieniawski
Universidad Politécnica de Madrid

Benjamín Celada
Universidad Politécnica de Madrid

José Miguel Galera
Geocontrol S.A.

ABSTRACT

TBM case histories from over 400 tunnel sections are analyzed presenting a comprehensive application of the Rock Mass Excavability (RME) index (Bieniawski et al. 2006). The paper introduces two new adjustment factors, called the Factor of Tunnel Length and the Factor of Crew Effectiveness, which were established to represent more realistically the interaction of the rock mass characteristics, the TBM parameters and the performance of the crew operating the machine. A new correlation is provided for predicting the average rate of advance of double-shield TBMs working specifically in the double-shield mode.

PREVIOUS INVESTIGATIONS

Excavability is defined as the ease of excavation and was investigated as early as Kirsten 1982. TBM excavability or performance prediction models were studied by Barton 2000, Alber 2001, Bieniawski 2004, Blindheim 2005, and others. In this research, all potential methods in existence which might assess rock mass excavability for TBM tunneling were studied.

In essence, five approaches were reported in the technical literature, each a rock mass classification index modified for machine excavability: RQD_{TBM} (Deere), N (Kirsten), RMR_{TBM} (Bieniawski), RMi_{TBM} (Palmström) and Q_{TBM} (Barton). The most publicized of these—and claimed as the only successful approach—was Q_{TBM}. We examined Q_{TBM} as an option but could not make it work for our case histories because of a problem with the definition of rock mass strength σ_{MASS} which is based on "inversion of σ_c to a rock mass strength, with correction for density"—rendering it unacceptable, not to mention it being too complex featuring 21 parameters! Subsequently, the key objection to Q_{TBM} was provided by a major study from Norway (where the Q-system was invented) published by Palmström and Broch, 2006. They concluded: "Q_{TBM} is complex and even misleading and shows low sensitivity to penetration rate; the correlation coefficient with recorded data is even worse than conventional Q or RMR or with other basic parameters like the uniaxial compressive strength of the intact rock. It is recommended that the Q_{TBM} should not to be used." This finding is clearly supported by Figure 1.

After such overwhelming evidence, we concluded that modifying an existing rock mass quality classification, be it the RMR or Q, for determining rock mass excavability was not appropriate as an effective approach for modern engineering practice. It is

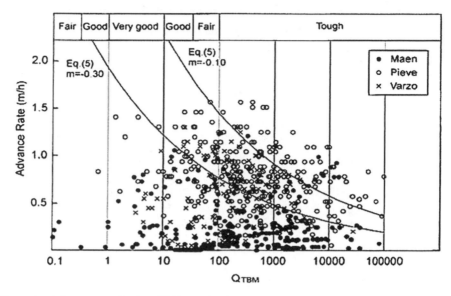

Figure 1.   Advance rates for three TBM tunnels plotted against Q_{TBM} (Sapigni et al. 2002)

doubtful that one formula can include all the factors pertinent to rock mass quality as well as those influencing TBM choice and performance.

Accordingly, research devoted to rock mass excavability was initiated in 2004 with the objective of establishing an index, similar to the RMR, but which was specifically directed to prediction of rock mass excavability, rather than rock mass quality. This work was aimed at the selection of the appropriate method of tunnel excavation, having considered rock mass-machine interaction, using tunnel boring machines or conventional mechanized excavation. The RME concept proposed first by Bieniawski, Celada, Galera and Álvarez (2006) was based on analyses of 387 sections of Spanish tunnels comprising 22.9 km in length. They involved three tunnels: Guadarrama (ADIF 2005), Abdalajís (Grandori 2004, Weber et al. 2005) and Line 9 of the Barcelona metro (De la Valle 2002). In each case, the tunnels studied included detailed data on rock mass characteristics and TBM parameters, as depicted in the RME Input Data Form in Figure 2. For record purposes, the original ratings of the five parameters inherent in the RME are given in Table 1. Note that the stand up time parameter is derived from the RMR system (see Bieniawski et al. 2006). Based on these criteria, the originally derived relationship between the RME and the average rate of advance (ARA), expressed in meters/day, for single-shield and double-shield TBMs, is shown in Figure 3. Other correlations have also been obtained for such TBM parameters as the specific energy of excavation, the torque and thrust applied to the TBM cutterhead (see Appendix 2).

NEW LINES OF INVESTIGATION

While the original RME concept offered promising possibilities, from the point of view of practical applications, it had limitations in some aspects, due to the fact that the excellent correlation between the RME and the ARA was based on data from TBMs of both the double-shield and single shield types; Figure 3 does not differentiate between

---

INPUT DATA FORM for Rock Mass Excavability

Name of Tunnel ....

Initial chainage of section:..............................Final chainage of section..

Length of section:.............................m (should be > 40 m)

Duration of excavation (days):.. (number + 1 decimal)

Average Rate of Advance ARA = ...m/day

Lithology:...Average depth:...................m

ROCK MASS PARAMETERS

Uniaxial compressive strength of intact rock (σ_c):..MPa

Drilling Rate Index DRI:................ Type of homogeneity at excavation face:............................

N° of joints per meter:...............Rock Mass Rating *RMR*: range.........................average...............

Orientation of discontinuities with respect to tunnel axis

(perpendicular, parallel or oblique):..

Stand up time:...................hours   Groundwater inflow at tunnel face:.............liters/sec

Rock Mass Excavability RME range......................................average...................

TBM PARAMETERS

Average speed of cutterhead rotation: ....rpm   Applied Thrust:...................m . kN

Specific Penetration:.....................................mm /rev

Rate of Penetration:..................................... mm /min

N° cutters changed:..   Rate of TBM utilization: ....%

Figure 2.   Input data form for determination of the Rock Mass Excavability (RMR) index

the rates of advance by a distinct type of TBM. Yet this kind of information is helpful for selection of the appropriate type of machine for the anticipated tunneling conditions.

Therefore, during the past year, specific investigations were carried out to ensure more practical benefits, namely:

1. To incorporate the effect of the construction process, differentiating the rate of advance obtainable, in the same rock conditions, as a function of the tunnel length excavated, and thus improving interaction of the TBM, its crew and the terrain.

2. To differentiate between the specific types of TBMs employed, such as double-shields, (working in the modes of either double-shield or single-shield), single-shield machines, and open TBMs.

3. To adjust the ratings of the RME, increasing the number of tunnel sections studied with the aim of improving the broad representation of the RME.

The results obtained from the above studies are summarized in the following sections.

Table 1.   Original ratings of the parameters comprising the Rock Mass Excavability (RME_{06}) index

Uniaxial compressive strength of intact rock [0–15 points]								
σ_{ci} (MPa)	<5		5–30		30–90	90–180		>180
Rating	0 (*1)		10		15	5		0
Drillability [0–15 points]								
DRI	>80		80–65		65–50	50–40		<40
Rating	15		10		7	3		0
Discontinuities in front of the tunnel face [0–40 points]								
Homogeneity			Number of joints per meter			Orientation with respect to tunnel axis		
Homogeneous	Mixed	0–4	4–8	8–15	15–30	>30	Perpendicular	Oblique	Parallel
Rating	10	0	5	10	20	15	0	10	5	0
Stand up time [0–25 points]								
Hours	<5		5–24		24–96	96–192		>192
Rating	0		2		10	15		25
Groundwater inflow [0–5 points]								
Liters/sec	>100		70–100		30–70	10–30		<10
Rating	0		1		2	4 (**0)		5

* For double-shield and single-shield.
** For argillacious rocks.

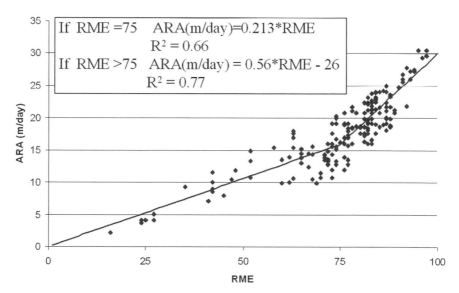

If RME =75   ARA(m/day)=0.213*RME
                        $R^2 = 0.66$
If RME >75   ARA(m/day) = 0.56*RME - 26
                        $R^2 = 0.77$

Figure 3.   Correlation between the RME_{06} index and the average rate of advance (m/day) for single-shield and double-shield TBMs

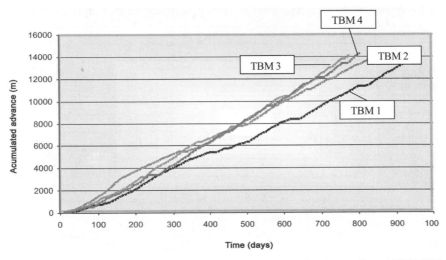

Figure 4.   Rates of advance obtained by the four TBMs excavating the Guadarrama Tunnel (ADIF, 2005)

Table 2.   Performance data from the four TBMs at the Guadarrama Tunnel

TBM	Manufacturer	Total Length Excavated (km)	Length Excavated in 700 Days (km)	Average Advance for Length, km (m/day)
1 (S)	Herrenknecht	13.3	9.4	14.5
2 (S)	Wirth	13.6	11.5	16.6
3 (N)	Wirth	14.1	12.3	18.2
4 (N)	Herrenknecht	14.3	12.0	17.7

The Effect of the Construction Process

The construction of the Guadarrama tunnels involved two tubes, each 9.5 m in diameter and 28 km in length, using four double-shield TBMs; two manufactured by Herrenknecht and the other two by Wirth. Figure 4 presents the rates of advance achieved by each of the four TBMs. In addition to this Figure, Table 2 summarizes the pertinent information about the performance of each TBM.

In considering these results, it should be noted that if one analyzes the length of tunnel excavated by each machine after 700 days of construction, one notices appreciable differences in performance. For example, TBM#3 excavated 12.3 km in that time, while TBM#1 managed only 9.4 km. Thus, in those 700 days, TBM#3 achieved an advance 31% greater than that by TBM#1. On the other hand, if one examines the average rate of advance over the whole tunnel length of the four tubes (excavated by April 5, 2005), the differences between the TBMs are also significant. The performance of TBM#3 over the whole length of tunnel was 18.2 m/day, compared to 14.5 m/day from TBM#1. According to that criterion, TBM#3 achieved a rate of advance 25% better than TBM#1. One might think that these differences were due to the TBM being made by different manufacturers. However, TBM#3, made by Wirth, advanced in 700 days only 2.5% more than TBM#4 by Herrenknecht, and its performance in m/day was better by only 2.8% than that of TBM#4.

Table 3. Factor of tunnel length excavated (FL)

Tunel Length Excavated (km)	Adjustment Factor (F_L)
0.5	0.50
1.0	0.86
2.0	0.97
4.0	1.00
6.0	1.07
8.0	1.12
10.0	1.15
12.0	1.20

The results in Table 2 indicate that the performance of the TBMs in the North portal did not depend on who manufactured the machines. In addition, at the Guadarrama tunnel, after each TBM had excavated between 13 km and 14 km, the characteristics of the rock mass conditions were essentially similar, with RMR being 60 to 80 (with the exception of the 500 meters of the fault La Umbría). In view of this, it seems evident that the differences in the performance between the TBMs working at Guadarrama was due to the skills and competence of the crews operating these machines.

The second interesting aspect, evident at Guadarrama, was the finding that, as the distance tunneled increased, the performance of the machines also increased gradually.

In fact, an analysis of the TBM performance obtained in the successive sections of the tunnel, from the portal, showed that the performance increased of the order of 15% when the excavation reached some 10 km. This "learning curve" effect was significant, so given that the RME reflects the excavability of the terrain, it is reasonable to introduce an adjustment to the predicted ARA obtained from a given RME, incorporating an effect of the length of the tunnel excavated and the influence of the crew skills, when dealing with the TBM and the terrain. Thus, one can represent this situation as:

$$ARA_T = \frac{ARA_R}{F_L \times F_C} \qquad (1)$$

where:

ARA_T = predicted 'true' value of ARA from the correlation with RME;

ARA_R = recorded average rate of advance, m/day, actually achieved in a tunnel section;

F_L = factor of experience as a function of tunnel length excavated; and

F_C = factor of effectiveness by the crew handling the TBM and the terrain.

Based on the results obtained during the construction of the tunnels of Guadarrama and Abdalajís, Tables 3 and 4 show the values appropriate for the coefficients F_L and F_C.

The Effect of TBM Type

For the most beneficial application of the RME index, one should establish a separate correlation between the RME and the average rate of advance, ARA, for each type of TBMs in use: double-shield, single-shield and open TBM. Due to space limitation, we are reporting here only the results obtained to date specifically for double-shield TBMs. The correlation between the RME and the ARA for single-shield TBMs and open TBMs is pursued currently.

Table 4. Factor of crew effectiveness (F_C)

Effectiveness of the Crew Handling TBM and Terrain	Adjustment Factor (F_C)
Less than efficient	0.88
Efficient	1.00
Very efficient	1.15

REVISED RME INPUT RATINGS

Revising the RME ratings as more case histories become available is a logical procedure for "fine tuning" of the obtained correlations. As more case histories were collected, a revision became essential also bearing in mind that double-shield TBMs work in two modes. One is when they actually work as double-shield machines; in this form the placement of the lining segments does not stop the TBM advance—the utilization of the machine is at its maximum. In the second or the single-shield mode, the pressure by the grippers may induce rock falls on the telescopic shield which interfere with machine advance and cause delays. The decision to use such a TBM in either mode is often taken without clear guidelines which can lead to errors and uncertainties. For this reason, to ensure the most trustworthy data, we decided to analyze first the case histories of double-shield TBMs working only in the double-shield mode.

RME_{07}

The data base for double-shield TBMs included 175 tunnel sections, of which 157 had RME>70 which corresponds to tunneling conditions most appropriate for double-shields. Parametric sensitivity analyses were performed varying the ratings of the five input parameters to maximize the coefficient of correlation between RME and ARA_{07} until it exceeded the coefficient for RME_{06}. In addition, for each input parameter, the continuous graphical form, as presented in Figure 5, was used. In total, 22 such correlation cycles were analyzed, resulting in the ratings presented in Table 5. This produced the RME_{07} correlation depicted in Figure 6.

It should be noted that the shown correlation was derived for tunnels with diameters close to 10 meters. In order to take into account the influence of different tunnel diameters, D, a coefficient k_D was proposed previously (Bieniawski et al. 2006) as follows:

$$k_D = -0.007D^3 + 0.1637D^2 - 1.2859D + 4.5158$$

Thus, for D = 10 m, k_D = 1, while for D = 8 m, k_D = 1.12 but for D = 12 m, k_D = 0.5, that is, one-half of the coefficient for D = 10 m.

By comparison with RME_{06}, the new RME_{07} features the following significant modifications:

1. The maximum rating for the strength of intact rock σ_c is increased from 15 to 25 points;

2. The maximum rating for the effect of discontinuities at the tunnel face is decreased from 40 to 30 points as the stand up time parameter is also influenced by the presence of discontinuities.

The modifications to RME_{07} input ratings were the result of not only the analyses performed but also due to a survey of world TBM experts sharing their observations on

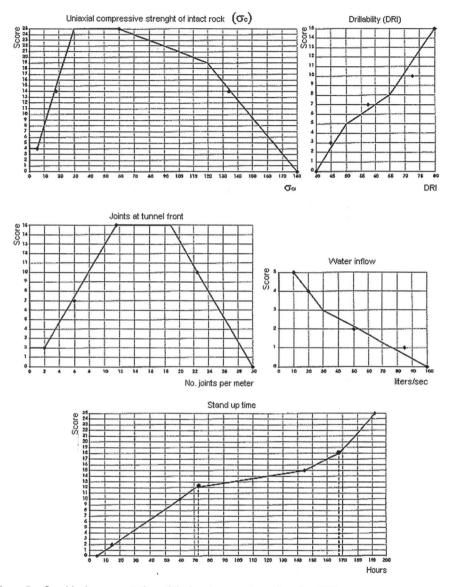

Figure 5. Graphical representation of the input parameter ratings for RME$_{07}$

RME based on their extensive tunneling experience. In particular, Dott.Ing. Remo Grandori of SELI made valuable suggestions.

Examination of Figure 6 reveals that most double-shield TBMs are particularly effective in the range RME$_{07}$ > 70 while none can be found with RME$_{07}$ < 50. Of course, it is realized that the actual relationship in Figure 5 would follow a curve rather than a simplification of a straight line which is used at present until more case histories are analyzed.

Table 5. Modified input ratings for RME$_{07}$

Uniaxial Compressive Strength of Intact Rock [0–25 points]				
σ$_c$ (MPa)	<5	5–30	30–90	90–180	>180
Average rating	4	14	25	14	0

Drillability [0–15 points]				
DRI	>80	80–65	65–50	50–40	<40
Average rating	15	10	7	3	0

Discontinuities at Tunnel Face [0–30 points]									
Homogeneity			Number of joints per meter				Orientation with respect to tunnel axis		
Homogeneous	Mixed	0–4	4–8	8–15	15–30	>30	Perpendicular	Oblique	Paralell
Average rating 10	0	2	7	15	10	0	5	3	0

Stand-Up Time [0–25 points]				
Hours	<5	5–24	24–96	96–192	>192
Average rating	0	2	10	15	25

Groundwater Inflow [0–5 points]				
Liters/sec	>100	70–100	30–70	10–30	<10
Average rating	0	1	2	4(0*)	5

* For argillacious rocks.

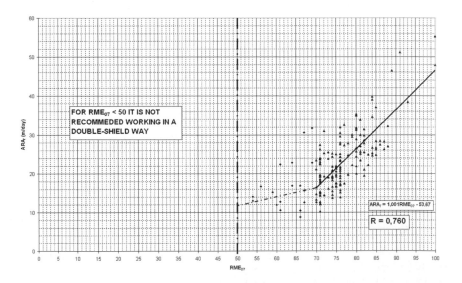

Figure 6. Correlation between the RME$_{07}$ and the average rate of advance for double-shield TBMs

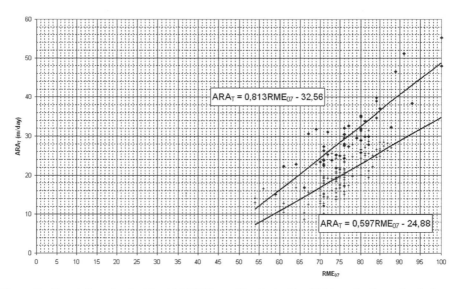

Figure 7. Correlations for double-shield TBMs working in double-shield mode. Upper line is for σ_c < 45 MPA, the lower line is for σ_c > 45 MPA.

SPECIFIC CONSIDERATIONS FOR DOUBLE-SHIELD TBMS

The achieved coefficient of correlation, shown in Figure 6 as R = 0.76 in the region RME_{07} > 70, is very promising but an appreciable scatter of data suggests further room for improvement. For this reason, it can be noted from Figure 5 that for a given value of RME_{07} one case may have σ_c < 60 MPa and the other σ_c > 60 MPa. Although in both cases RME_{07} could be the same, experience shows that for σ_c < 60 MPa, a higher rate of advance may be expected than for the case of σ_c > 60 MPa. In fact, this has been clearly demonstrated at the Guadarrama Tunnel where for rock masses with σ_c < 30 MPa the best rates of advance were achieved, namely, between 40 to 50 meters/day.

In accordance with these facts, we arrived at the conclusion that for a double-shield TBM working in the double-shield mode, the correlation between RME_{07} and ARA_T could be defined more realistically by considering two ranges of the rock material strength: one for σ_c < 45 MPa and the one for σ_c > 45 MPa.

This is shown on Figure 7 which introduces the two correlations:

For σ_c < 45 MPa: $ARA_T = 0.813\ RME_{07} - 32.56$ R = 0.865

For σ_c > 45 MPA: $ARA_T = 0.597\ RME_{07} - 24.88$ R = 0.744

The above correlations are very useful for predicting the rates of advance for double-shield TBMs working in the double-shield mode. This improves and clarifies the chart presented in Figure 6.

Along this line of reasoning, other investigations are currently in progress specifically directed to application of the RME_{07} to single-shield TBMs and double-shield TBMs working in the single-shield mode, as well as open-type TBMs. Also being studied are applications of RME_{07} to tunnels constructed by conventional drill-and-blast and other mechanized methods.

CONCLUSIONS

Case histories data base has been increased for tunnels constructed by TBMs working in double-shield mode to present a more comprehensive application of the Rock Mass Excavability (RME) index. Two new adjustment factors are introduced, called the Factor of Tunnel Length Excavated (F_L) and the Factor of Crew effectiveness (F_C), which were established to represent more realistically the interaction of the rock mass characteristics, the TBM parameters and the performance of the crew operating the machine. A new correlation is provided for predicting the average rate of advance of double-shield TBMs working in the double-shield mode.

ACKNOWLEDGMENTS

This paper was made possible by dedicated work of the professional staff at Geocontrol S.A. in Madrid, Spain, including the first two recipients of the Bieniawski Scholarship for tunneling research at the Superior School of Mines, Universidad Politécnica de Madrid: Doña María Álvarez Hernández and currently Don José Carballo Rodríguez.

These Bieniawski Scholarships were generously funded by Geocontrol S.A.

REFERENCES

ADIF 2005. *"El Túnel de Guadarrama."* Ed. Entorno Gráfico, Madrid.

Alber, M. 2000. "Advance rates for hard rock TBMs and their effects on project economics." *Tunneling & Underground Space Technology,* Vol. 15 (1), pp. 55–60.

Barton, N. 2000. *"TBM tunneling in jointed and faulted rock."* A.A. Balkema Publishers, Rotterdam.

Bieniawski, Z.T. 2004. "Aspectos clave en la elección del método constructivo de túneles." Proc. Jornada Técnica, Madrid, *Ingeopress,* No. 126, pp. 50–68.

Bieniawski, Z.T., Celada, B., Galera, J.M. and M. Álvares 2006. "Rock Mass Excavability (RME) index: a New Way to Selecting the Optimum Tunnel Construction Method," *Proc. ITA World Tunneling Congress,* Seoul.

Blindheim, O.T. 2005. "A critique of Q_{TBM}." *Tunnels & Tunneling Int.,* June 2005, pp. 32–35.

De la Valle, N. 2002. "Barcelona's new backbone runs deep." *Tunnels & Tunneling Int.,* March 2002.

Grandori, R. 2004. "The Abdalajís Tunnel (Málaga, Spain): The new double-shield universal TBM challenge." *Proc. Int. Congess on Mechanized Tunneling,* Torino, pp. 35–42.

Kirsten, H.A.D. 1982. "A classification for excavation in natural materials" *Civil Engr in S. Africa,* July.

Palmström, A. and Broch, E. 2006. "Problems with Q_{TBM}." *Tunneling & Underground Space Technology.*

Palmström, A. 1995. "RM_i—a rock mass classification system for rock engineering purposes." Ph.D. thesis, University of Oslo.

Sapigni, M., Berti, M., Bethaz, E., Bustillo, A., and Cardone, G. 2002. "TBM performance estimation using rock mass classifications." *Int. J. Rock Mech. & Mining Sciences,* 39, pp. 771–788.

Weber, W., Daoud, H. and Fernández, E. 2005. "Challenging TBM Tunneling at Abdalajís." *Tunnels & Tunneling Int.* May 2005.

APPENDIX 1: AN EXAMPLE OF RME CALCULATIONS

Figure 8 presents an example of actual procedure for calculating RME_{07} for one of the case histories plotted in Figures 6 and 7.

RME CALCULATION

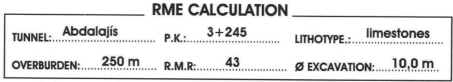

TUNNEL: Abdalajís   P.K.: 3+245   LITHOTYPE.: limestones

OVERBURDEN: 250 m   R.M.R: 43   Ø EXCAVATION: 10,0 m

1.- Stand up time estimation.

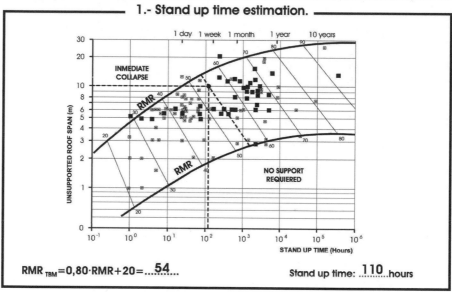

$RMR_{TBM} = 0,80 \cdot RMR + 20 = $ **54**   Stand up time: **110** hours

2.- RME Parameters evaluation.

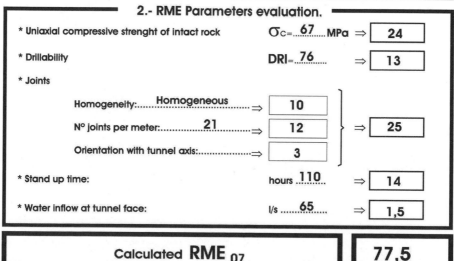

* Uniaxial compressive strenght of intact rock $\sigma_c =$ **67** MPa $\Rightarrow$ **24**

* Drillability $DRI =$ **76** $\Rightarrow$ **13**

* Joints

    Homogeneity: Homogeneous $\Rightarrow$ **10**

    Nº joints per meter: **21** $\Rightarrow$ **12** $\Rightarrow$ **25**

    Orientation with tunnel axis: $\Rightarrow$ **3**

* Stand up time: hours **110** $\Rightarrow$ **14**

* Water inflow at tunnel face: l/s **65** $\Rightarrow$ **1,5**

Calculated RME_{07} **77,5**

Figure 8.   Calculations for RME_{07}

APPENDIX 2: OTHER CORRELATIONS WITH THE RME INDEX FOR SINGLE-SHIELD AND DOUBLE-SHIELD TBMS

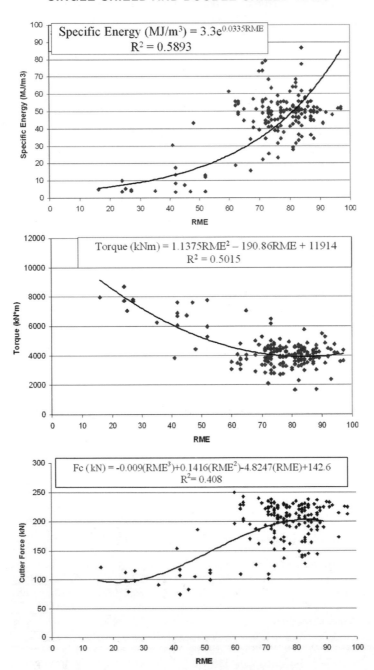

USE OF TUNNEL BORING MACHINES AT DEPTH: EXTENDING THE LIMITS

Bruce Downing

Golder Associates

Trevor Carter

Golder Associates

Richard Beddoes

Golder Associates

Allan Moss

Rio Tinto

Peter Dowden

Mechanical Tunneling Consulting Services

ABSTRACT

Traditionally (with a few exceptions) tunnel boring machines (TBMs) have been utilized on tunnel projects where the risks associated with deep tunnels (rock burst, squeeze, etc) are low (or expected to be low). This has generally been due to the extremely adverse consequences of entrapping or damaging a TBM under such conditions; but also due to the relative difficulty of dealing with these ground conditions in the confined working area of a TBM, in comparison to a drill and blast heading. This paper review the constraints of operating a TBM in a deep tunnel and discusses related project experience, particularly recent and current experience with TBMs in the deep Alpine Base Tunnels. Potential future applications of TBMs in the development of large, deep underground mines are also discussed.

INTRODUCTION

This paper reviews current and historic limitations on the application of tunnel boring machines (TBMs) for deep rock tunnel excavations. Such limitations have resulted in a number of long deep rock tunnels being driven by drill and blast, which, if located at shallower depth, would have been driven with TBMs. Recent advances in the design and manufacture of TBMs and recent project experience has expanded the potential for future applications of TBMs in the excavation of deep rock tunnels for civil engineering and mining projects. An example of a potential mining application will be discussed in this paper.

For the purposes of this paper, deep rock tunnels are defined as those where a particular set of unique challenges arise which include high stress (and overstress) of the rock, high ambient temperatures, high groundwater pressures and potential difficulty of access. For the purposes of this paper, a distinction is made between deep rock tunnels and applications at shallower depths which may encounter related difficulties. An example could include a shallow tunnel in weak rock or soil, which may experience overstress and squeezing conditions. For the purposes of this paper, tunnels greater than approximately 1,000 m in depth will be considered to be deep tunnels.

CHALLENGES RELATED TO DEEP ROCK TUNNELS

Unique challenges that are typically only encountered in the excavation of deep rock tunnels and which must be addressed when considering potential TBM applications are as follows.

- **High rock stress.** Depending on the strength and quality of the rock, high rock stress may lead to one of two possible difficulties. Where rock is relatively strong and massive (unfractured), rock bursting or spalling may occur. Where the rock material is weak, or the rock mass is relatively highly fractured, large strains may occur. Predicting whether or not any given tunneling situation will result in conditions that will entrap a machine requires that careful evaluation be made of the likely alignment geology, geotechnical properties and stress regime through which the tunnel will be driven.

- **High temperature.** Temperature at the tunnel level will be dependent on the geothermal gradient at the project site. For deep tunnels, as defined in this paper, temperatures of the rock and groundwater would generally exceed 40°C.

- **High pressure groundwater.** Generally deep tunnels are well below the groundwater table and thus can be at very high pressure. However, due to the depth and high in-situ stress, most rock fractures are tight, generally resulting in relatively low hydraulic conductivity of the rock mass. As a result, groundwater inflows to a tunnel heading may not be high, or if high initially, may not be sustained.

- **Access and logistics.** Particularly for situations where a tunnel is launched from the base of a deep shaft, access for equipment and continuous maintenance of material supply will be significant issues. Where use of a TBM is being contemplated, component sizing (for hoisting in the shaft) and machine assembly must be considered.

GEOMECHANICS ASSESSMENT NEEDS FOR DEEP TUNNELING

Characteristically, two forms of rock behavior problems can occur due to high stress conditions (Carter, 2005; Caret et al., 2007):

- For competent, brittle rock masses, typically characterized by rocks with high HoekBrown m_i values (typically >15 [Hoek, 1988, 2002; Diederich et al., 2007]), relatively high intact strengths (>150 MPa typically), and with only limited jointing (s > 0.1, Q > 10 approximately [Barton, 1976; Grimstad and Barton, 1995]); under high stress levels, such rock can exhibit spalling or slabbing, and in extreme situations, bursting.

- For less competent rock masses (with Q and intact strength values well below those indicated above) significant tunnel deformations may occur if such ground is subjected to very high stress. The magnitude of the deformation for low Q values and rock mass strengths at great depth, may be sufficient to create uncontrollable squeezing" conditions.

For the first case of competent rock masses (bursting, spalling etc), rock masses can be characterized at a preliminary engineering level by estimating the rock mass strength from m_i, UCS, GSI and estimating SRF (Carter et al., 2007; Diederichs et al., 2007) using the Kirsten (1988) modifications to Barton's SRF factor, calculated as follows:

$$SRF = 0.244k^{0.346}(H/\sigma_C)^{1.322} + 0.176 (\sigma_C/H)^{1.413}$$

Table 1.   Evaluation of the potential for slabbing or rock burst in competent rock masses

Stress Levels Competent Rock	σ_{cmass}/σ_v	$\sigma_\theta/\sigma_{cmass}$	SRF (1977)	SRF (1993)
Low Stress	>200	<0.01	2.5	2.5
Medium Stress	200–10	0.01–0.3	1	1
High Stress	10–5	0.3–0.4	0.5–2	0.5–2
Moderate Slabbing (after >1 hour in massive rock	5–3	0.5–0.65	5–9	5–50
Slabbing and Rock Bursts after a few minutes in massive rock	3–2	0.65–1.0	9–15	50–200
Heavy Rock Bursts (strain bursts) and immediate dynamic deformation	<2	>1.0	15–19	200–400

Source: Compiled from various sources, principally Barton, 1977; Grimstad and Barton, 1993; Alber, 1988; Carter, 2005.

where
  k = principal field stress ratio, H the cover depth (m)
  σ_C = uniaxial compressive strength of the intact rock (MPa)

The potential for rock bursting and slabbing for competent rock masses may then be assessed by entering the SRF value, or ratio of rock mass strength to vertical stress (or radial induced stress) on Table 1.

One of the earliest attempts at defining limits of problematic TBM workability was made by Alber, 1988 by comparing rock mass strength at the tunnel periphery to the induced tangential stress (defined as $\sigma_{CM}/\sigma_{\theta Max}$ as estimated from the Kirsch solution for a circular tunnel). Other methods have been mostly empirical (Singh et al., 1995, Hoek, 1999, 2000). Most recent assessments for weak ground have concentrated on peripheral strain (Hoek and Marinos, 2000). This latter method, which is based on comparing induced closure for different ratios of rock mass strength to insitu stress (σ_{cm}/p_o) is plotted on Figure 1. Rock mass strength may be calculated by a number of different methods, the following formula requires the input parameters m_i (Hoek Brown m value), GSI and σ_c (intact rock strength).

$$\sigma_{cm} = (0.0034 m_i^{0.8}) \sigma_c (1.029 + 0.025 e^{(-0.1mi)})^{GSI}$$

GROUNDWATER

Forecasting the magnitude of convergence and closure is just one aspect of importance to assessing the viability of using a certain type of TBM for driving a given alignment. Of the other geotechnical issues of importance to the design of a machine (e.g., rock hardness and abrasivity; structural fabric, and the necessity of designing rock reinforcement installation and backup systems that will efficiently carry out rock support and muck handling) dealing with water constitutes one of the main worries for deep tunnels.

If water bearing faults are encountered during tunneling under high cover, then not only is the magnitude of inflow a problem, but grouting and sealing issues are also problematic as grouting is notoriously difficult against high inflow pressures. A fully sealed "can" such as in a closed face shield machine cannot provide full enough

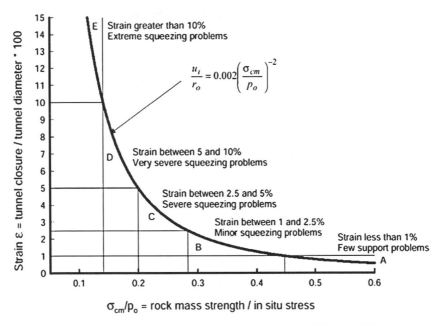

$$\frac{u_i}{r_o} = 0.002\left(\frac{\sigma_{cm}}{p_o}\right)^{-2}$$

σ_{cm}/p_o = rock mass strength / in situ stress

Figure 1.　Strain-based ground behavior classification (after Hoek and Marinos, 2000)

control to prevent ingress to the machine, as pressures will far exceed the maximum possible with current brush and tail seal technology.

The onus is therefore on grouting to reduce hydraulic conductivity, and on forward probing to detect adverse water conditions ahead of the machine, so that drainage or other remedial measures can be undertaken. For the Lötschberg Base Tunnel, the karstified and fractured limestones and marls of the Doldenhorn Nappe, which were encountered in the tunnel over a length of 3.2 km, were anticipated to be subject to high groundwater inflows (pressures up to 60 bar) (Pesendorfer, 2006). As a result, this tunnel section was intensively explored during tunnel construction with long (300 m) horizontal cored boreholes, drilled at the tunnel face. A program of tests (open hole tests and packer tests) was developed that minimized time required at the face, usually being completed within four to five hours. Single and cross-hole tests provided data on local and far field hydraulic transmissivity, storativity and formation pressure distributions. Analyses were carried out at site to immediately identify zones of high risk for the tunnel. This testing program resulted in the identification of one highly transmissive zone (approximate inflows of 50 L/s under a pressure of 35 bar). This zone was grouted with 50 tonnes of cement through a pattern of more than 50 grouting holes. A second identified zone of high inflow (approx. 30 L/s under a pressure of 52 bars) was determined to be treatable using drainage, and excavation was performed without grouting but at a reduced rate of approximately 2 m/day. Based on this hydrogeologic exploration program, tunneling and grouting costs could be kept to a minimum as grouting was only undertaken when necessary. This high risk zone was completed with six months ahead of the construction program. Although this tunnel section was not excavated by TBM, due to the anticipated high risk of inundation, the success of this method of exploration, in such difficult circumstances, holds promise for similar applications in TBM tunnels where the risk of high groundwater inflows is anticipated.

CASE HISTORIES OF TBM USE AT DEPTH/SQUEEZING CONDITIONS

While there are a number of case histories of tunnel construction in overstressed rock using conventional means (drill and blast), there are relatively few case histories of TBM applications in such conditions. The Gotthard Base Tunnel stands out as the most recent and best example of TBM excavation of a deep tunnel in overstressed rock. Most of this tunnel is being excavated by TBMs (50 km of the 57 km total tunnel length is being excavated by TBM). The tunnel is subdivided into five sections—Erstfield, Amsteg, Sedrun, Faido and Bodio. All sections except Sedrun are excavated by unshielded hard rock TBMs. The Sedrun section is being excavated by conventional means because of anticipated poor rock conditions, the potential for high groundwater inflows and the possibility of adverse ground deformations. At the end of 2006, the Amsteg and Bodio sections had been completely excavated, while the Sedrun and Faido sections were partially complete and Erstfield had not yet been started.

The sections which are beneath the greatest rock cover are the Amsteg and Faido sections, which are beneath cover of 2,200 m and 2,500 m respectively. Through these sections the TBM tunnels are driven at a slightly larger diameter (9.4 m diameter for Faido and Amsteg, compared with 8.8 m diameter for the adjacent Bodio section) as a means to allow for the expected ground movement so that closure can be controlled without impacting the structural envelope required for train operation. The Amsteg section was driven through a geological formation known as the Aar Massif, which consists of gneiss and granite, and features the Intschi Zone, a zone of particularly poor tunneling conditions. The remainder of the Faido section that has not yet been excavated will be driven through the Gotthard Massif, consisting of gneiss and schist. Rock support consists of systematic rock bolts, steel mesh and shotcrete installed behind the cutterhead. Yielding support with sliding steel sets are being installed in areas as required. Challenging conditions encountered to date in Faido include a rock bursting zone (Eberhardt, 2001; Schmalz, 2006; Diederichs, 2007), with one event at a measured magnitude of 2.4 (Richter Scale), thought to have been caused by stress re-distribution associated with the adjacent tunnel drive. This burst was reported to have resulted in only minor damage to the tunnel lining. In June 2005, the west tunnel TBM on the Amsteg Section encountered a 50 m zone of "hydrothermally decomposed granite" at a depth of 2,200 m. This zone stalled TBM progress for a period of five months. The tunnel was restarted following completion of an adit from the east tunnel, from which grouting was carried out. From this adit, a pre-excavation heading was developed through this zone of adverse rock, through which the TBM was traversed. Despite this setback, the Amsteg section was completed six months ahead of schedule in October 2006. Alptransit reports a total project cost of SF8.14 billion ($US 6.5 billion), of which 5.4% (SF 423 million [$US 337 million]) is allocated for unforeseen geological occurrences (Alptransit website).

In contrast to the Gotthard Base Tunnels, where difficulties have mostly been associated with competent rock, the Yacambu Irrigation project in Venezuela probably represents the most adverse weak rock conditions where a TBM has been applied. This project consists of a 4.8 m diameter, 24 km long tunnel, which was to be driven using a variety of methods, including TBMs. In 1975, two shielded TBMs (short single shields) were launched, with ground support consisting of steel ribs and shotcrete. In 1979, one of the TBMs was engulfed by squeezing ground during a holiday break. Ground movement was particularly severe at the invert. Portions of the TBM were cut into pieces and removed. The second TBM was abandoned in a side drift shortly afterwards. Two further TBMs were brought to site, specially designed to deal with the Bocono Fault (a major Andean plate structure). Due to funding constraints, these TBMs were not started and all remaining construction was carried out by drill and blast or with roadheaders. The most severe squeezing conditions occurred in sheared, graphitic phyllite with UCS values as low as 1 MPa with cover of 600 m. In these areas, strain values in excess of 30% were observed. A prototype TBM design to complete the tunneling was proposed in the early 1990s. The proposed machine was a single shield, incorporating a tapered shield, which was strengthened to accommodate loads up to 18 MPa and convergence within the shield of up to 10%. Due to funding constraints the machine was never adopted and used on the project.

In the Evinos-Mornos project in Greece, constructed between 1992 and 1995, four TBMs were used to excavate a 30 km long, 4 m diameter water supply tunnel under cover of up to 1,300 m. It was reported that more than 30% of the tunnel was in very poor quality rock (RMR of Class V) and 650 m of this tunnel was in squeezing ground (Grandori, 1995). The project is of interest since both open and shielded TBMs were used. For ground support, the open TBMs used steel ribs and lagging in the poor ground. In squeezing ground, shotcrete was applied immediately behind the cutterhead. Some steel sets deformed under the applied loads, thus requiring replacement. A final cast in place concrete liner was placed in sections of the tunnel excavated with the open TBMs. Two double-shield TBMs were also used on the project and utilized a concrete segmental lining erected immediately behind the shield. It was reported that all machines coped with the ground reasonably well, with the open TBMs performing better in the good ground and poorer in the poor ground. Average advance rates of 15 and 21 m/day were reported for the two open TBMs and 17 and 26 m/day for the two double-shield TBMs. Squeezing ground movements of up to 15 cm were reported 1 m from the face.

At the San Manuel Mine, located 40 km northeast of Tucson, Arizona, between 1993 and 1996, a 4.6 m diameter Robbins hard rock open TBM was used to drive almost 10,000 m of development drifts at a depth of over 1,000 m. The drifts were excavated through curves as tight at 107 m radius and at grades up to 5.7%. The rock consisted of generally poor quality quartz monzonite/dacite porphyry. It was reported that RMR values typically were between 40 and 50, with occasional areas with RMR values between 60 and 70. Rock quality was worst at the contacts between the monzonite and the dacite porphyry. Support was initially expected to consist of rock bolts, however it was discovered that most of the drift required support with ring beams with heavy welded wire mesh for lagging. It was also found that the initial TBM design was not well suited to the ground conditions, therefore after completion of approximately 1.5 km of tunneling, major modifications to the TBM were carried out underground. Modifications were made to the drive motors (to increase torque), to the trailing fingers (to reduce ground falls), to the cutterhead roof support (to reduce problems with re-starting the cutterhead) and modifications to the cutterhead buckets. Average daily advance rates were 6.5m before the modifications and were improved to 22.6 m/day after the modifications. TBM assembly was carried out underground, with all loads being delivered underground via a shaft. The TBM was designed with the components sized to fit the capacity of the shaft hoist (maximum size of 21.8 tonne, 2 m × 4 m × 12 m). Assembly was substantially completed in 114 hours, which was believed to be a record for a deep TBM application at the time. In general, the mine operator reported the experience with the TBM to be favorable.

Other notable applications of TBMs in deep, overstressed rock tunnels include the Stillwater Tunnel in Utah, the Misicuni Tunnel in Bolivia, the Chinox tunnel in Guatemala, Pinglin in Taiwan and applications in South African Gold Mines (Pickering, 1999), the discussion of which is beyond the scope of this paper.

DESIGN CONSIDERATIONS FOR TBMS FOR USE
IN OVERSTRESSED ROCK

TBM Selection Issues

Although modern TBM's can cope with a wide variety of conditions, many compromises are made in machine design in order to balance cuttability and thrust requirements against production rate and ground support needs. This is especially the case for projects involving deep tunnels. Open machines provide the optimum design arrangement for achieving rapid progress rates, but only function at their very best in good ground conditions. While fully shielded machines have a number of advantages in continuously unstable conditions, especially when a full tunnel lining is necessary to support the tunnel and keep it dry, shields are generally unsuitable for squeezing situations, as they can rapidly become trapped. Segmented shields, which can contract to a limited extent, and may be able to move ahead fast enough to escape entrapment have been proposed as a solution to this problem, but none are known to have actually been built for this specific intent. Similar machines have been shown to be difficult to steer. They are also unsuited for wet conditions.

The roof and side support shoes on an open TBM can be retracted by some distance to avoid entrapment. The shoes are relatively short (a couple of meters or so), and close to the face, where the squeeze is not fully developed. If squeeze develops while the machine is stationary, the cutterhead should be kept revolving to remove incoming material. Another advantage of the open TBM is that it is easy to retract the head from the face by using the gripper/propel system, should face deformation impede startup.

The potential for rock burst may mandate the use of rear-loading cutters for worker protection. It might be advisable to block the temporary opening left when the cutter is removed to avoid rock shooting inside the head. With a shielded TBM, spalling can compromise the operation of the machine and in severe cases can even trap it.

Proper cutterhead design is very important where the tunnel crown tends to collapse as it is cut. There should be a minimum longitudinal gap between the outermost gauge cutters and the front of the roof support on an open machine or the leading edge of the shield on a shield machine. If there is an exposed section of head rotating behind the gage, any rock fall will tend to jam its rotation, and result in pulling down excessive material from the crown. This was a particular problem for the TBM used at the San Manuel Mine, where the cutterhead became jammed due to rockfall on a number of occasions. In addition, the face structure of the TBM should be as smooth as possible. These conditions virtually mandate a rear loading cutter design, which automatically results in a smooth face with very little cutter housing protrusion. The muck buckets should also be designed with a smooth profile to muck efficiently and avoid tearing the ground.

Muck buckets can be designed to incorporate closable entry doors, which can control the inflow of flowing material, which can overwhelm the conveyor system, and can also precipitate tunnel crown collapse.

Variable speed cutterhead drives are essential in loose ground conditions, both in limiting the disturbance of fractured rock and in controlling muck flow into the head.

Rock Support Considerations for TBM Tunnels

From the viewpoint of tunnel support, a TBM driven tunnel has a number of advantages over a drill and blast tunnel. Rock mass damage is lessened for a machine driven tunnel, leading to reduced requirements for ground support. Also, where high stresses may be an issue, a circular tunnel is generally more stable than a horseshoe shaped tunnel with a flat floor.

For an open TBM, conventional tunnel support (rock bolts, wire mesh, shotcrete or ring beams) is a flexible system generally allowing adjustments to be made to suit most geological conditions. With an open TBM, support can be installed relatively close to the face, which is particularly important in poor or squeezing ground. Open TBMs can be designed to allow installation of rock bolts, ring beams and shotcrete immediately behind the cutterhead, and ahead of the grippers, as adopted for the Gotthard tunnels. Conventional support can be designed to be ductile and allow tunnel deformation without collapse or loss of support integrity. In contrast, shielded TBMs do not allow for early installation of support, and typically do not allow for adjustment of rock support to suit varying conditions.

In the case of an open TBM, rock spalling and bursting can be dangerous to workers, as it occurs violently without warning. Ring beams and mesh in combination or liner plates, should contain these potential projectiles. The flexible steel fingers usually attached to the back of the roof support can be extended around the side shoes to provide maximum protection during the installation of the rings and mesh or plate. Some permanent protective shielding (not contacting the wall) could be installed around critical areas of an open machine.

Bore deformation in the invert area could lead to movement of the rail track or difficulty in laying it, affecting the alignment of the back-up system. Yielding ring beam systems that can accommodate some ground movement but can keep the bore substantially circular would be effective. If squeeze is anticipated, running clearances between the back-up and the tunnel wall, or ring beams, should be sized accordingly.

Machine design to cope with potentially squeezing ground concentrates on being able to provide some degree of overcut on the tunnel so that modest convergence can be tolerated before impinging on the shield or roof and side supports. This is accomplished by mounting a number of gage cutters on hydraulically adjustable swinging arms. In operation the cutter is mechanically locked in position to assure a rigid load path. The allowable movement can be as much as 100 mm, achieved in increments of 25 mm when extending from the nominal bore.

Other TBM Design Issues

High ground and groundwater temperatures are commonly encountered in deep tunnels. Uncontrolled, these high temperatures can adversely impact the ability of personnel to work safely and effectively, but normally do not significantly affect the operation of equipment (depending on the actual temperature) as much of the equipment already operates at a high temperature even under normal circumstances, in particular the cutters and drive systems. As a consequence, the drive and hydraulic systems are routinely water cooled.

Experience gained with operating TBMs in high temperatures in South Africa (Pickering 1999) and in the Gotthard and Loetschberg Base Tunnels, where TBMs have been operated at temperatures in excess of 45°C, has allowed development of conditioned cabs and crew rest areas and mitigation of ambient heat build-up with the use of ventilation and chilled ventilation systems.

TBMs that require transport into the tunnel via a shaft will impose restrictions on the maximum size of any one component. This is a common type of restriction faced in TBM projects, whether the restriction is due to difficult road access to site or transport via a restricted underground opening. The TBM at the San Manuel Mine was designed for this purpose, and, as mentioned previously, was assembled underground in a remarkably short period of time. Generally, the largest component pieces in the TBM (and therefore likely to dictate access) are the main bearing, the cutterhead support, and the forward section of the main beam for an open TBM. The main bearing is usually about 50% of the bore diameter. The cutterhead is commonly shipped in pieces (consisting of the cutterhead center section, which mounts onto the main bearing, and four, or more, outer segments) and then assembled at site. Obviously the more components that can be pre-assembled before delivery the quicker will be the assembly.

POTENTIAL USES OF TBMS IN BLOCK CAVE MINES

The mining industry finds itself in the position of having decreasing numbers of large orebodies that are amenable to open pit mining and is under pressure to respond to societal concern that impacts of mining should be minimized. One of the results of this trend is a greater interest in large, deep low grade orebodies that can be exploited using bulk mining methods, and in particular, block caving. Block caving requires development of an extensive extraction level below the ore body for the block of ore to be mined. Mining is achieved by creating an undercut of sufficient extent that the rock mass caves without any artificial rock breakage, such as blasting, being required.

Block caving is an extremely capital intensive method of mining, but with the significant advantage that the production costs are comparatively low once the infrastructure has been constructed, the undercut has been fully developed and caving has been initiated. The majority of capital must be sunk before any ore is produced. This places an extremely high emphasis on efficient development of the mine infrastructure since the return on investment is very dependent on the time required between breaking ground and the first production of ore.

Large block cave mines require an extensive tunnel network to provide access to the undercut and underlying production level, ventilation reticulation and ore transport. Most of these tunnels must be completed before any production is possible, unlike other mining methods in which progressive development during the production phase is possible. Traditional mine development uses tunnels that are 4 m to 6 m in span and generally of square or rectangular profile. Common mining equipment can advance tunnels of this size as a single face at typical rates of 5 m to 7 m per day, slow by the standards of TBMs. Since very little typical mine development equipment (drill jumbos, bolters etc) is used once production begins there is an opportunity to look at more efficient methods of tunneling during the construction phase. TBMs become competitive under these conditions.

Figure 2 shows a typical block cave mine layout. All of the ring drives surrounding the orebody footprint offer opportunities for TBMs provided horizontal curves are designed appropriately.

Large block caving mines are likely to require numerous headings of the traditional 4 m to 6 m in size to provide sufficient cross sectional area for fresh and return ventilation, large conveyor belts and equipment movement. Smaller numbers of large diameter tunnels (~10 m diameter) may offer capital and operating cost advantages since they reduce friction losses in ventilation (saving chilling costs in deep, high temperature environments) and allow a greater proportion of their cross section to be used for services such as water, power and compressed air. This presents a further opportunity for TBMs, as the comparative advantage over drill and blast in terms of advance rate will widen with larger diameter tunnels. Lastly, TBMs are conducive to installation of life-of-mine ground support systems under the very high stress conditions encountered at the depths of 1,500 to 2,500 m likely to become common for mining in the future.

To date TBMs have seen very little application in the mining industry. Only one example of a TBM being used for block caving mine development is described in the literature (the San Manuel Mine). However, the number of block caving mines is likely to

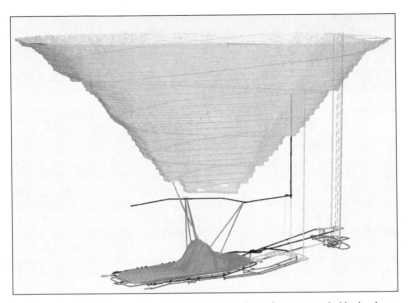

Figure 2.   Example of a block cave layout indicating cave footprint, surrounded by haulage, services and ventilation drives which could be driven by TBMs. The main access shafts are shown offset from the footprint area and the draw zone of the cave.

increase in the future and the mining industry is beginning to recognize that the "rock factory" philosophy to construction has merits for large, long-life mines. In this philosophy, design and construction standards are based upon the requirement to move large quantities of ore at the maximum speed and minimum cost possible. This requires maintenance-free roadways with surfaces that minimize equipment wear, high quality ventilation, and services that can achieve levels of availability common in industrial plants. TBMs, which have achieved these goals for transportation and public infrastructure projects, have a potentially bright future in this arena.

CONCLUSIONS

In recent years increasing use has been made of TBMs for the excavation of deep rock tunnels. In particular, the ongoing use of TBMs in the Alpine Base Tunnels has demonstrated the suitability of TBMs in such applications. In these tunnels, TBMs have been applied in all but the most severe of squeezing conditions. Although there have been some notable failures of TBMs in deep rock tunnels, the great majority of applications have been successful. The increasing body of experience is providing the necessary information on ground behavior and machine design required to extend the use of TBMs to other deep tunnel applications.

Case histories have demonstrated the importance of developing an adequate understanding of the ground conditions prior to construction. The full range of expected ground conditions should be understood, with TBM designs to accommodate the expected range of ground behavior. The potential for squeezing, spalling or bursting must be investigated and understood. An incomplete understanding of possible ground conditions and likely ground behavior during tunneling can be detrimental to the project, and may lead to costly and time consuming machine modifications underground, slow progress and in extreme cases, loss of the TBM.

While there are many potential future applications of TBMs in the construction of tunnels for civil engineering purposes, one important developing application is in the development of large scale, deep underground mines. This application is in its infancy, but may become widespread once the concept has been successfully proven.

REFERENCES

Alber, M. (1988). *Design of High Speed TBM Drives.* Proc. Canadian Tunneling; pp.181–187. (From selected papers from the 15th Can. Tunneling Conference, Vancouver).

Alptransit website. www.alptransit.ch

Barton, N. (1976). *Recent Experiences with the Q-system of Tunnel Support Design. Proc. Symp. on Exploration for Rock Engineering.* Johannesburg, pp. 107–117.

Barton, N. (1993). *Updating the Q-System for NMT. Proc. Int. Symp. On Sprayed Concrete,* Fragernes.

Carter, A.R. 1996. *Magma Puts the Pedal to the Metal.* Engineering and Mining Journal, January 1996.

Carter, T.G., 1990. *Use of Forward Probe Coring Methods and Remote Sensing for Investigation of Geological Conditions Ahead of Drill and Blast and TBM Tunnel Drivages,* Proc. ISRM International Symposium on Static and Dynamic Considerations in Rock Engineering, Mbabane, Swaziland, pp. 91–101.

Carter, T.G., Carvalho, J.L, & Diederichs, M.S. (2007) *A unified procedure for prediction of strength and post yield behaviour for rockmasses at the extreme ends of the integrated GSI and UCS rock competence scale.* Proc. ISRM 11th International Congress on Rock Mechanics, Lisbon, Portugal. 4pp.

Carter, T.G., Steels, D., Dhillon H.S., & Brophy D. (2005). *Difficulties of Tunneling under High Cover in Mountainous Regions,* 2005. Proc. Int. AFTES Congress, Tunneling for a Sustainable Europe, Chambery, pp. 349–358.

Carvalho, J.L., Carter, T.G. & Diederichs, M.S, 2007. *An approach for prediction of strength and post yield behaviour for rock masses of low intact strength.* Proc. 1st Can-US Rock Symposium. June, Vancouver. 8pp.

Diederichs, M.S. 2007. *Mechanistic Validation and Practical Application of Damage and Spalling Prediction Criteria for Deep Tunneling.* The 2003 Canadian Geotechnical Colloquium. In Press: Canadian Geotechnical Journal.

Diederichs, M.S, Carvalho, J.L., Carter, T.G. 2007. *A modified approach for prediction of strength and post yield behaviour for high GSI rockmasses in strong, brittle ground.* Proc. 1st Can-US Rock Symposium. June, Vancouver. 8pp.

Eberhardt, E. (2001). *Numerical Modelling of Three-Dimension Stress Rotation ahead of an Advancing Tunnel Face.* Int. J. Rock mech. Min. Sci. Vol. 38, pp. 499–518.

Einstein, H.H. and Bobet, A. 1997. *Mechanized Tunneling in Squeezing Rock-From Basic Thoughts to Continuous Tunneling.* Tunnels for People, Golser, Hinkel & Schubert (eds), Balkema, Rotterdam.

Fasching A. and Bofer, M. (1997). *A Tunnel Under the Andes.* Tunnels and Tunneling.

Grandori, R. et al., 1995. "Construction of a 30 km Long Hydraulic Tunnel in Less than Three Years Under the Most Adverse Geological Conditions." *Rapid Excavation and Tunneling Conference.*

Schmalz, G., DeDomenicis, S.L. "Gotthard Base Tunnel—the worlds longest railway tunnel," in the *Tunneling Association of Canada 19th National Conference,* September 2006.

Grimstad, E. and Barton, N. (1995). *Rock Mass Classification and the Use of NMT in India.* Proc. Int. Conf. On Design and Construction of Underground Structures, New Delhi, India.

Hoek, E. (1999). *Putting numbers to geology—an engineer's viewpoint.* Q. Jnl Eng. Geol. & Hydrogeol. Vol. 32, no. 1, pp. 1–19 (19)

Hoek, E. (2000). Big Tunnels in Bad Rock, The Terzaghi lecture—presented at the ASCE Civil Engineering Conf., Seattle, October 18–21, 2000.

Hoek, E. and Brown, E.T. (1988). *The Hoek-Brown failure criterion—a 1988 update.* Proc.15th Canadian Rock Mech. Symp. (ed. J.H. Curran), Toronto: Civil Engineering Dept., University of Toronto, pp. 31–38.

Hoek, E. and Marinos, P. (2000). *Predicting Tunnel Squeezing.* Tunnels and Tunneling International, Part 1—November 2000, Part 2—December.

Kirsten, H.A.D., (1988). *Discussion Contribution relating to the Norwegian Geotechnical Institute Q System.* In Rock Classification Systems for Engineering Purposes. ASTM STP 984, ed. Kirkaldie, L., Philadelphia, pp. 86–88.

Moergeli, A. (2005a), "Risk management in action—controlling difficult ground by innovation," Rapid Excavation and Tunneling Conference.

Moergeli, A. (2005b), Presentation of "Risk management in action—controlling difficult ground by innovation," Rapid Excavation and Tunneling Conference.

Neil, D.M., Harami, K.Y., Hanson, D.R., Morino, A., Kalamaras, G.S., & Grasso P. (1999). *State-of-the-art 3D tomography techniques for geotechnical evaluation of site and alignment conditions during tunneling World Tunnel Congress ITA.* Oslo, 29 May–3 June 1999, pp. 121–128.

Pesendorfer, M. (2006); *Hydrogeologic Exploration and Tunneling in a Karstified and Fractured Limestone Aquifer (Lötschberg Base Tunnel, Switzerland).* PhD Thesis, Swiss Federal Institute of Technology (ETH) Zürich. Diss. ETH Nr. 16724

Pickering, R.G.B., et al. (1999). "Practical Feasibility of using TBMs in Deep Level Gold Mines," *Rapid Excavation and Tunneling International.* April Edition pp. 34–37 Conference.

Singh, B., Jethwa, L.L and Dube, A.K. (1995) *A Classification System for Support Pressure in Tunnels and Caverns.* J. Rock Mech and Tunneling Technology, Vol. 1, #1, pp. 13–24.

VanDerPas, Ed. and Alum, R. (1995). "TBM Technology in Deep Underground Copper Mine." Rapid Excavation and Tunneling Conference.

Vigl, L., Jager, M. (1997). "Double shield TBM and Open TBM in squeezing rock—a comparison." In International Conference on Tunnels for People, Golser, Hinkel & Schubert (eds), Balkema, Rotterdam.

17

Tunnel and Shaft Rehabilitation

Chair
J. Rostami
CDM

EMERGENCY REPAIRS TO BEACH INTERCEPTOR TUNNEL

David Jurich

Hatch Mott MacDonald

Joseph N. McDivitt

South Coast Water District

INTRODUCTION

South Coast Water District (SCWD) operates the Beach Interceptor Sewer Tunnel (BIST) in South Laguna, Orange County, California between Monarch Beach and Aliso Beach (Figure 1). The 10,200-foot long BIST houses a 24-inch gravity Techite sewer pipeline critical to the SCWD Sanitary Sewer Collection System. The original tunnel and sewer pipeline were constructed in 1954 and the sewer line was replaced in 1974. Periodic inspections by sewer pipeline maintenance crews documented deterioration of specific intervals of the tunnel. Rotting timber supports and rock falls in the tunnel as a result of the deterioration were endangering the sewer pipeline and a 731-foot section of tunnel located at Thousand Steps Beach was determined to be in immediate need of repairs. SCWD and consultants Hatch Mott MacDonald (HMM) and Tetra Tech prepared a Work Plan for emergency tunnel stabilization and pipe protection measures that was implemented in January 2007. Normal access to the beach and the section of tunnel requiring repairs was limited to public stairs referred to as Thousand Steps and, therefore, all materials and equipment needed for the work were transported to the beach by ocean-going landing craft. Additionally, the emergency repairs had to be completed without taking the pipeline out of service. Numerous environmental, logistical and operational requirements had to be overcome to complete the challenging emergency repairs.

GEOLOGICAL SETTING

The Laguna Beach area is located in the Peninsular Ranges of Southern California. The project area is located on the southwest flank of Niguel Hill along marine wave-cut cliffs and slopes. The tunnel was excavated in the San Onofre Formation, a marine deposit, characterized by abrupt lateral and vertical changes in lithologies that vary from boulder to cobble and gravel size breccias to medium coarse sandstones with gravel, sandy claystone and siltstone. The interval of the tunnel that required repairs is located within the San Onofre breccia, a moderately-to-well cemented, poorly bedded to massive unit with angular to subangular boulder and cobble sized clasts of schist. The sandy silt matrix of the breccia is composed of quartz, feldspar and schist grains (Figure 2).

GENERAL TUNNEL DESCRIPTION

The approximately 6-feet high by 6-feet wide 10,200-foot-long BIST was hand mined in 1954 at the base of the coastal cliffs. Primary access is through twenty-one access adits from public and private beaches. The tunnel is located in an easement that extends beneath private property with residential structures situated 50 to 80 feet above and horizontally in close proximity to the tunnel alignment. Most of the tunnel is horseshoe shaped and unsupported, but approximately 3,000 feet is supported by timber

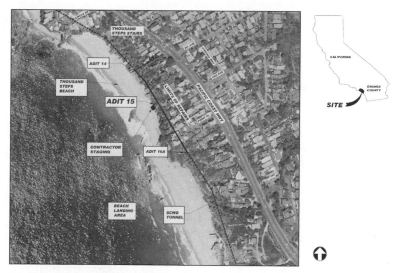

Figure 1. Location of beach interceptor tunnel repairs

Figure 2. Exposure of San Onofre Formation

struts and wooden lagging (AKM Consulting Engineers, 1998). Timber supports are positioned at a spacing that ranges from one to three feet. Each timber support consists of 8-inch square side posts and an 8-inch square horizontal crown beam.

The tunnel was constructed to house a 20-inch diameter vitrified clay pipe (VCP). In 1974 the VCP sewer pipeline was demolished in place and replaced with the existing 24-inch diameter Techite pipe. Over the entire length of the tunnel, several timber

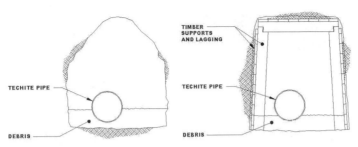

Figure 3. Typical unsupported and timber-supported beach interceptor tunnel cross sections

supports have been replaced and fifteen short intervals of the tunnel totaling approximately 1,000 feet have been reinforced with shotcrete. No design details are available
for the shotcrete. Lastly, one 350-foot interval of the tunnel has collapsed and remains
impassable. Typical cross sections of the unsupported and timber supported tunnel
are presented in Figure 3.

Rock mass conditions along most of the tunnel are good. The San Onofre breccia
is classified as firm to slow raveling in the absence of water (Terzaghi's Tunnelman's
Ground Classification System as modified by Heuer). Bedding structures and discontinuities within the breccia are poorly developed to absent. Groundwater seepage is
present to varying degrees and the tunnel is generally damp with some isolated dripping to wet intervals.

TUNNEL CONDITION AT THOUSAND STEPS BEACH

The SCWD began an annual pipeline inspection program in 2003 and maintenance crews are making pipeline and lateral connection repairs to correct deficiencies
observed during the inspections. Maintenance crews entering the tunnel at Thousand
Steps Beach have reported ongoing rock falls and deterioration of timber supports and
lagging in the tunnel. SCWD hired consulting engineer HMM in 2005 to inspect the
tunnel at Thousand Steps Beach, complete an assessment of ground and support elements, and prepare a design for tunnel repairs to 750 feet of the tunnel between adit
14 and adit 16A at Thousand Steps Beach.

HMM completed detailed geological mapping, three exploratory coreholes in the
tunnel crown, and a condition assessment of the timber supports. Detailed mapping
revealed that between adit 14 and adit 16A the 750 feet of tunnel has 5 to 265-foot
alternating intervals of unsupported ground and ground supported with timbers or
shotcrete with the following totals: 229 feet of the tunnel is unsupported, 463 feet is
supported by timbers on one to three-foot centers, and 39 feet is shotcrete lined.
Unsupported intervals exhibit rock weathering two to six inches deep with loosening of
the matrix and detachment of cobble to boulder size clasts. Timber supports exhibit
deterioration and wooden lagging is generally rotten and has failed at several locations
in the side walls and crown. Failures of the wooden lagging have resulted in small rock
falls, one to two cubic feet in volume, composed of sand to gravel sized weathered rock
fragments with occasional cobble to boulder sized clasts (Figure 4). Additionally, timber supported intervals of the tunnel were found to have overhead accumulations of
broken rock, as much as 6 feet high, loading the timber supports. Most of the tunnel
was damp to wet with groundwater seepage saturating the several inches of fill material and crushed VCP that covers the invert.

Figure 4.   Rock fall at failure of wooden lagging

Based on observations made during the inspections, the deteriorating condition of the tunnel was considered to be due to a combination of age, presence of groundwater and ongoing weathering of the rock mass. The rock mass quality and localized extent of the tunnel deterioration were not considered to constitute a risk to the stability of the residential structures located on top of the cliff 50 to 80 feet above and horizontally in close proximity to the tunnel alignment. HMM recommended repairs be made to the BIST at Thousand Steps Beach and developed bid documents for SCWD. Repairs were delayed due to access and permitting issues and HMM inspected the tunnel again in 2006. HMM observed continued deterioration with numerous rock falls that occurred since the 2005 inspection and recommended emergency repairs to specific intervals of the tunnel.

PROJECT SETTING AND CONSTRAINTS

Access to Thousand Steps Beach is limited to: (1) a set of public stairs, approximately 165 feet in height, from the Pacific Coast Highway (PHC), (2) a steep narrow road that switchbacks from the PCH through an easement on private property, and (3) the ocean. Permits for the tunnel repair issued by the Coastal Development Commission and Orange County allow the use of Thousand Steps Stairs, but requires continuous public access be maintained during construction activities. Therefore, Thousand Steps Stairs can be used only for worker access and not for movement of heavy construction equipment or any materials that cannot be transported by hand.

Historically, SCWD used the road easement through the private property for small work crews and limited construction equipment to perform maintenance and make repairs to the pipeline. However, due to developing access issues that were expected to require a lengthy legal effort to resolve and the emergency nature of the repairs, SCWD elected to not pursue use of the private road for substantial construction traffic required to make the tunnel repairs and to reserve its use for emergency situations.

Construction permits for the tunnel repairs limit the size of the contractor staging and work area on Thousand Steps Beach and the hours of operation to 8:00 AM to 5:00 PM. There is not an adequate supply of construction power or water, and there is no telephone service at the beach. A public bathroom is located at the base of Thousand Steps Stairs and available to workers. No hazardous materials are permitted on

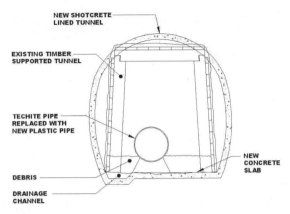

Figure 5.   2005 tunnel repair concept

the beach and double-walled containment is required for all potentially toxic materials. The permits also prohibit work being performed during the tourist season when beach usage and surfing is at a peak. Construction activities cannot start until after Labor Day weekend and are to be completed before Memorial Day weekend.

The tunnel is located within a 10-foot wide easement; all repairs are to be made within the existing easement. Two short adits, (14 and 15) and one portal (referred to as adit 16A) located in easements through private property provide access to the tunnel at Thousand Steps Beach. Repair activities are to be conducted in such a way to minimize impacts to private property.

The pipeline in the tunnel conveys by gravity flow on average 1,100,000 gallons of sewage daily from an upstream pump station, a hospital, and several private residences by lateral connections directly to the interceptor pipeline. There is no alternative pipeline route and tunnel repairs cannot interrupt service or must include a temporary bypass if service is required to be interrupted. Emergency Techite pipeline repair tools and materials must be on site during construction.

DESIGN CRITERIA AND TUNNEL REPAIR CONCEPTS

The 2005 repair concepts developed by HMM included:

- Enlargement of adit 15 and the tunnel between adit 14 and adit 16a to a shotcrete lined horseshoe shape
- Construction of a concrete floor slab with a formed drain for seepage control
- Replacement of the Techite pipe with a new plastic pipe

This repair concept (Figure 5) was a significant exercise and required multiple staging areas on Thousand Steps Beach and installation of a temporary bypass for the sewer pipe.

SCWD received only one bid for the repairs that was rejected and the work was postponed. SCWD and HMM then developed the following revised design criteria for the required emergency repairs to the tunnel as a result of the identified access and impacts issues and the risks associated with the sewer temporary bypass:

- Repair those intervals in need of immediate attention and reinforce other intervals as required
- The Techite pipe should be protected from further rock falls

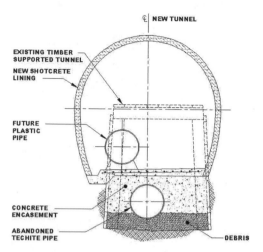

Figure 6. 2006 tunnel repair concept

- The repaired tunnel must be sufficient in height for workers to walk upright and with adequate width to permit passage of a pushcart
- The design life of repairs shall be greater than 50 years
- The repairs shall be completed without taking the Techite pipe out of service
- The repairs shall allow for future rehabilitation of the entire tunnel currently under study by SCWD
- The repairs shall allow for the future replacement of the Techite pipe with a new plastic pipe

The 2006 repair concept included:

- Immediate repair at the intersection of the tunnel and adit 15 consisting of encasing the Techite pipe in concrete and constructing an enlarged shotcrete lined horseshoe shaped tunnel above the concrete encasement (Figures 6 and 7). Exact length of repair to be determined in the field as ground is exposed behind the timber supports and wooden lagging.
- Install permanent shotcrete lining for adit 15
- Limited replacement of select timber supports or reinforcement with additional timber supports between adit 14 and adit 16a.
- Encase Techite pipe in concrete between adit 14 and adit 16a (approximately 750 feet)

Concepts for the future rehabilitation of the tunnel have not been developed, but it is anticipated that mechanized excavation would be preferred as the safer, more efficient and cost effective construction method. Temporary bypasses of the Techite pipe would be necessary in order to implement mechanized construction methods. Temporary bypasses and pump stations, while an accepted method for handling sewage, have an added risk of leakage or system failure and are not preferred.

The 2006 concept requires a larger quantity of rock excavation and concrete materials, but provides an improved work area and construction efficiencies for the immediate repairs and future tunnel rehabilitation. Encasement of the Techite pipe in concrete requires reduced worker activity and material handling around the unprotected pipeline

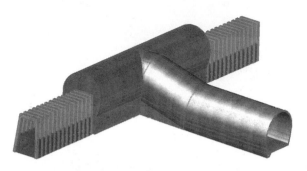

Figure 7. Isometric of tunnel and adit 15 repairs

Figure 8.   Workboat landing at Thousand Steps Beach

during the repairs and is considered a more robust system of protecting the pipeline once constructed. The concrete encasement also provides an improved platform for the future tunnel enlargement by use of mechanized equipment.

Repairs are to be supported entirely by an ocean workboat capable of landing on the beach (Figure 8). The work will be completed from a single staging area in front of adit 15, field located above high tide, and enclosed with screened fence to minimize visual impact. The repairs include improvement of the adit 15 portal structure.

CONSTRUCTION CONTRACT AWARD AND TUNNEL REPAIRS

SCWD selected Sancon Engineering (Sancon) as the construction contractor to make the emergency tunnel repairs. The work was initiated on a fast track basis with SCWD obtaining the necessary permits, Tetra Tech provided environmental services and prepared a Work Plan for distribution to SCWD staff, HMM on site representation, and Sancon staff. The Work Plan provided a single source of relevant information for all parties and included:

- Description of the Site, Access and Environmental Requirements
- Description of the Work
- Environmental Protection Measures

- Project Permits
  - CEQA Notice of Exemption
  - Coastal Development Permit—City of Laguna Beach
  - Encroachment Permit—County of Orange
  - Mooring Permit—U.S. Coast Guard
  - Tunnel Classification—Division of Occupational Safety
- Contact Information for SCWD Staff and Regulatory Agencies

SCWD implemented an extensive public outreach program months prior to construction that included mailings and door hangers for local residents, press releases, signage at Thousand Steps Beach, a project information pamphlet, and a page at SCWD's website dedicated to the project (www.scwd.org). Weekly planning, coordinating, and progress meetings held at SCWD's office in the months preceding Sancon mobilization to the site were critical to successful execution of the work.

Sancon completed shotcrete test panels in vertical and overhead positions using a prepackaged shotcrete mix and certified two nozzlemen on December 16, 2006 at a SCWD maintenance facility. Shotcrete core samples tested in the laboratory exceeded the required unconfined compressive strength of 6,000 psi. Sancon also conducted tests and selected appropriate construction equipment for work at the adit and for transporting equipment and materials from the workboat to the staging area. A successful test landing of the workboat was conducted January 3, 2007 to confirm the landing site, equipment function, and procedures. SCWD issued Sancon Notice to Proceed on January 8, 2007 and construction is scheduled for completion by May 25, 2007.

REFERENCES

AKM Consulting Engineers, August 1998, Draft Preliminary Engineering Report, Beach Interceptor Sewer Tunnel Investigation, prepared for South Coast Water District.

Hatch Mott MacDonald, 2005, Tunnel Inspection Report, Sanitary Sewer Interceptor Tunnel, Laguna Beach, California, prepared for South Coast Water District.

Hatch Mott MacDonald, 2006, Tunnel Inspection Report, Sanitary Sewer Interceptor Tunnel, Laguna Beach, California, prepared for South Coast Water District.

South Coast Water District, 2006, Beach Interceptor Tunnel Emergency Stabilization and Sewer Pipe Protection, South Laguna Beach, California, Design Drawings and Specifications, prepared by Hatch Mott MacDonald.

South Coast Water District, 2007, Work Plan—Beach Interceptor Tunnel Emergency Stabilization and Sewer Pipeline Protection, South Laguna Beach, California, prepared by Tetra Tech and Hatch Mott MacDonald.

HEARTLAND CORRIDOR TUNNELS

Frank P. Frandina
Hatch Mott MacDonald

Michael J. Loehr
Hatch Mott MacDonald

Paul E. Gabryszak
Hatch Mott MacDonald

ABSTRACT

The Heartland Corridor Clearance Improvement Project is part of a regional Public-Private initiative to allow double stack container trains a direct route from the Port of Norfolk through Virginia, West Virginia, Eastern Kentucky and into Central Ohio. In the Preliminary Engineering stage, 30 existing tunnels were assessed and 28 were found to need clearance improvements to allow the use of double stack trains. Alternative tunnel enlargement methods were developed to achieve the desired clearance improvements, while maintaining structural integrity and minimizing disruption to train traffic. This paper summarizes the geometric and condition investigations of the existing tunnels, as well as describes the preferred methodologies to achieve the desired clearance. The clearance improvements will be implemented beginning in the summer of 2007.

INTRODUCTION

Over the past 20 years, containerized cargo has become the standard of international trade. Daily, huge ships deliver containers to US ports where they are placed onto trains to intermodal terminals where they are then transferred to trucks for delivery. However, most of WV and Eastern KY do not have close access to intermodal terminals, creating a disadvantage to an already economically depressed region.

Today, the most direct rail route from the Port of Norfolk to Ohio and Chicago is Norfolk Southern's Heartland Corridor. However, the existing vertical clearances on the Heartland Corridor do not permit the use of the more efficient double-stacked container trains through numerous tunnels and other obstructions. Double stack trains must travel over a route that is 375 km (233 miles) longer than the Heartland Corridor, adding a day to shipping times. A 2003 study by the Nick J Rahall Jr. Appalachian Transportation Institute concluded that a new intermodal facility in Prichard WV, served by double stack trains would reduce shipping costs in Western WV, Eastern KY and Southern OH by $450 to $650 per container.

Hatch Mott MacDonald (HMM) performed a Preliminary Engineering study for the clearance improvement project in 2005. Following that study, a public private partnership (PPP) agreement was developed between the Norfolk Southern Corporation, the Federal Government and the States of West Virginia, Virginia and Ohio to increase the clearances on the Heartland Corridor. HMM was then selected to begin final design of the clearance improvement project.

Norfolk Southern has earned unprecedented recognition for safety winning the E.H. Harriman Memorial Gold Medal Award for employee safety in the railroad industry for 17 years in a row. The Gold Medal recognizes NS as the railroad with the lowest employee personal injury ratio among the nation's major railroads and a project priority is to extend this enviable safety record through the tunnel reconstruction period.

This paper focuses on the tunnel reconstruction necessary to safely achieve the clearance improvements with minimal disruption to railroad operations. The project also includes clearance modifications to six thru truss bridges and additional work at undercrossings, slide fences, utility lines and other facilities, not addressed in the paper.

PRELIMINARY INVESTIGATIONS

An initial series of field investigations was conducted over a six-week period in March and April 2005. The primary objectives of these investigations were to determine the extent of the clearance infringements, determine geology, establish preliminary geotechnical information, and assess existing structural conditions including materials, construction and condition.

The investigations focused on 30 tunnels at 28 sites that range in length from 53 m (174 lf) to 1.0 km (3,302 lf), with a total length of 10.1 km (31,112 lf). Ten of the tunnels were constructed for a single track; the other 20 were constructed with a width for two tracks. However, three of the double width tunnels presently have a single track. Most tunnels are concrete lined, but one is unlined and four are masonry lined over some portion of their length. Twenty-three of the 28 sites are in West Virginia, four are in Virginia, and one is in Kentucky.

In the course of the study, investigations were also conducted at three abandoned tunnels for possible rehabilitation and reactivation. Three new tunnels were considered in the single-track areas to minimize railroad disruptions during reconstruction. Five additional tunnels on a parallel route were also investigated for an alternative alignment.

To assess the clearance deficiencies, NSR developed a composite clearance template for the railroad cars that would be used on the Heartland Corridor. The template consists of the maximum dimensions of a Hi-Cube "Superbox" Car, a Conventional Multi-Level Car, a "Q" Car Multi-Level, an AAR Plate "F" Car, and a Double Stack Car. The composite template includes an allowance for dynamic movement of the cars and an air gap.

An on-track laser clearance measurement car was used to measure the tunnel dimensions referenced to the centerline of the track and the plane of the top of the rail. The laser car data collected was processed to yield a composite cross section of the tunnel liner at 15 m (50 foot) increments for each location. This information was analyzed to develop a "best fit" mathematical centerline for each tunnel. The width of the composite clearance template was then adjusted to reflect the increased lateral offsets produced by the track curvature due to the long cars chording around the curves. The appropriate composite clearance envelope was then placed on the tracks on each of the tunnel cross sections to check for encroachments with the tunnel liner. Overall, 28 of the tunnels have vertical clearance deficiencies. Two of those tunnels have relatively minor deficiencies with less than a 150 mm (6 inch) encroachment. Seventeen tunnels have significant clearance deficiencies over 600 mm (2 feet) for at least a portion of their length resulting in the need for liner replacement.

The existing track through the tunnels, as well as the approaches, were located both horizontally and vertically by conventional instrument surveys. Additional topographical survey was performed at locations where daylighting the existing tunnel or constructing a new tunnel was considered.

A Geoprobe investigation was performed in the tunnels. The Geoprobe investigation consists of advancing a sampler into the roadbed and underlying soils for purposes of observation and description. The probes were located at approximately 30 m (100 foot) spacing throughout the tunnel. The probe was advanced to a depth of 1.5 m (5 feet) or refusal. Refusal was determined in the field based upon the Geoprobe behavior and the advance rate. Typically refusal was encountered when the sampler did not advance at all for a period of 10 to 15 seconds.

The geotechnical investigations include additional Geoprobes, rock strength testing, ballast testing, rock mass quality rating, and water pH testing. The Geoprobes were taken at 30 m (100 foot) intervals on the approaches to the tunnel at tunnels where track lowering is being considered. The rock mass quality rating and the rock compressive strength were determined from rock samples taken from locations where tunnel liner removal or new excavations are contemplated. Samples of ballast were taken for lab testing to assist in the assessment of premature fouling in some tunnels. The pH was tested in tunnels that had standing water to see if the water was unusually corrosive.

A second series of field investigations was conducted in June, July, and August 2005 to provide additional data needed for the preliminary engineering phase and to better quantify the construction cost estimates for clearance improvements at each tunnel. Structural investigations included drilling through the lining to determine its thickness, quality, strength, and the presence of voids behind the lining. Cores were taken to determine concrete strength and in the case of the masonry tunnels, to assess the depth of decomposition. A video camera was inserted through selected holes in the liner to view conditions behind the lining. The thickness of the wall footings was measured and the elevation of the bottom of the wall footings was located to determine the necessity for underpinning. The findings were summarized in the individual "Condition Assessment Reports" that were prepared for each tunnel.

In the final design phase, additional cores through the lining were taken with video observation, primarily to better define the nature of the material above the liners. More comprehensive geological surveys of the rock cuts approaching each portal and of adjacent road cuts were also conducted to develop geological profiles and better plan the additional cores.

DESIGN CONDITIONS

Geologic Conditions

Tunnels located between Radford, Virginia and the Virginia-West Virginia state line are located in the Valley and Ridge Physiographic Province. This province is characterized by long parallel ridges underlain by strong sandstone and separated by valleys underlain by shale or karst solutioned limestone, or dolomite. The tunnels were excavated through low ridges and knobs supported by dolomite of the Ordovician and Cambrian age Beekmantown, Copper Ridge, and Elbrook Formations.

The tunnels between Bluefield and Huntington West Virginia are located in the Appalachian Plateaus Physiographic Province which is characterized by deeply incised plateaus of flat-lying sedimentary rock. The tunnels are excavated in the Pocahontas, New River, Kanawha, and Allegheny Formations which are predominately fine- to medium-grained sandstone with interbeds of siltstone, shale, and coal.

Structural Conditions

Based on the few record drawings available for a limited number of tunnels, the typical double width tunnel linings were designed to be lightly reinforced concrete 865 mm (2'-10") thick at the base of the wall tapering to 660 mm (2'-2") in the arch, and the single width tunnel linings were intended to be 610 mm (2'-0") thick at the base tapering to 450 m (1'-6") in the arch. Four tunnels were brick lined with similar dimensions. Lining thicknesses were measured at 123 locations in the Preliminary Engineering phase. The results show that in general, the design minimum thicknesses were

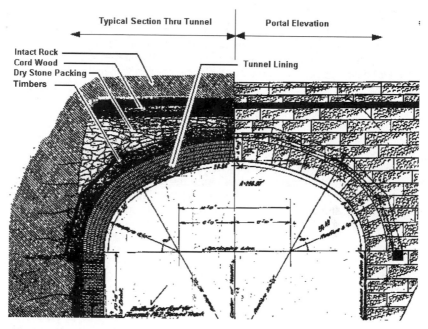

Figure 1.

found but with considerable variation within each tunnel with many thicknesses being considerably thicker and a few thinner than anticipated.

The drawings at some sites also showed that concrete packing was specified to fill any over excavation. Many cores went through concrete many feet thick, confirming that over excavated areas were filled with concrete. Another drawing shows shale excavated a few feet above the extrados of the lining to a sandstone bed with the void between the extrados and sandstone filled with "dry packed stone" and "cordwood" (see Figure 1). Visual inspections at those portals confirm a shale bed for a few feet above the crown overlain with a more competent sandstone bed.

Many of the cores identified voids behind the lining, sometimes with wood and/or loosened rock. Apparently, timbers and lagging were used where necessary to stabilize the excavations until the concrete (or brick) liner was installed. Since record drawings were found for only 8 tunnels, it is not known whether full packing was specified at each site. In addition to determining the concrete thickness, the cores and subsequent video logging through the core holes also gave great insight into the presence of any voids and the nature of material that may have loosened into a cavity.

As a result of the investigation program, it was apparent that the design would have to accommodate the possibility of voids of variable thicknesses and also the possibility of loosened material ranging from a few cobbles, many feet of loosened shale or even dry packed stone overlaid with cordwood.

OBSERVATIONAL METHOD

There are three basic approaches that can be used to design and construct these clearance improvements while recognizing the observed variation in existing lining thickness.

Minimum Investigation Approach

A minimum investigation approach would be safe, but very conservative and costly. The approach would assume the worst possible conditions that could be expected at each tunnel location based on the limited data collected in the Preliminary Engineering and from that data, specify a uniform improvement method. This first approach will result in a safe facility with minimal railroad disruption in the design phase, but with unnecessary cost and disruption to the railroad during the construction phase.

Extensive Testing Program Approach

The second approach would be to conduct an exhaustive lining thickness and rock mass evaluation prior to finalizing the design. This approach would result in a series of differing improvement methods along the length of the tunnel based on the results of the detailed investigations. The second approach has the potential to reduce the construction cost but at the expense of exhaustive design phase investigations and resultant disruptions to the railroad.

Observational Approach

The third approach is to adopt an "observational" approach where a 'cafeteria menu' of improvement methods is provided in the contract documents with the specific criteria as to the conditions where each method can be implemented. Based on the field investigations at each tunnel location, the contract documents would provide the anticipated limits of each of the improvement methods. Contractors would bid based on the anticipated limits and expected conditions. But the specified construction process incorporates confirmations of the expected conditions (lining thickness, joint planes, etc.) with payment mechanisms that allow adjustments to the quantity for each of the improvement methods resulting in appropriate compensation to the contractor when conditions differ from those anticipated in the bid.

The observational approach:

- Substantially reduces the cost of design phase investigations within the tunnels and resultant railroad disruptions by limiting the amount of supplemental data required.
- Contractors' bids will be lower since the bid documents incorporate equitable payment provisions to address the inevitable changing conditions in the tunnels.
- Norfolk Southern gets the clearance improvements at the lowest total cost since the final lining modification is matched foot by foot to the actual requirements along each foot of tunnel.
- Requires Construction Phase monitoring by engineering and inspection staff familiar with tunneling.

Norfolk Southern has agreed to adopt the "observational" approach where a 'cafeteria menu' of improvement methods is provided in the contract documents with the specific criteria as to the conditions where each method can be implemented and payment mechanisms that provide appropriate compensation to the contractor when conditions differ from those anticipated in the bid.

CLEARANCE IMPROVEMENT METHODS

The extent of the clearance deficiency at any point is determined by superimposing the composite clearance template and the final track geometry on the existing cross section as determined from the laser survey data. The extent of encroachment varied by location as well as through each tunnel. As such, a number of improvement methods were developed, each tailored to address a specific range of clearance deficiencies as efficiently as possible. Three main families of clearance improvements methods were identified:

- Track modifications
- Notching or replacing the lining
- Elimination or replacement of the tunnel

Track Modifications

These alternatives are generally the lowest in cost and provide the least disruption to railroad operations and therefore are the preferred methods for improving clearances. However, the amount of clearance gained using these methods is limited and therefore their potential usage is restricted to areas that only have minor interferences.

Track Realignment. Three of the tunnels were originally constructed to a double track width, but presently have a single track. A lateral shift of the track can gain additional clearance at minimal cost.

Track Lowering. In this alternative, the track would be removed and conventional excavation equipment would be used to remove the ballast and sub-ballast and reconstruct the track at a lower elevation. Alternatively, where subsurface conditions are favorable, a track under-cutter can be used to lower the track while the track remains in place. A new drainage system would be constructed at the lower elevation to minimize future track maintenance. Depending on the depth of excavation relative to the wall footings, stabilization or underpinning of the wall footing may be required.

Notching and Liner Replacement Alternatives

Depending on the depth of the tunnel lining encroachment, the methods developed to gain the required clearances are generally notching of the tunnel lining or full replacement of the liner crown. The decision to specify notching or liner replacement is primarily based on the thickness of concrete liner remaining after a notch is made at the required depth, with consideration of the presence and size of voids or loosened material behind the lining, structural condition of the lining, and geotechnical parameters of the rock above the lining.

Prior to notching the lining, probe holes will be taken to confirm the concrete thickness and identify any voids that would then be grouted in advance of the notching operation. Where the depth of the encroachment is so great that the lining must be removed, the replacement method must be tailored to suit the nature of the material above the lining. Where minor voids and occasional cobbles are expected above the lining, no special provisions are anticipated. Where loose material of substantial depth is probable, pre-support such as grouting, spiles or other measures will be required.

Minor Notching. The initial assessments identified clearance deficiencies in many tunnels at the 11 o'clock and 1 o'clock positions in the tunnels. Where the clearance deficiencies are a maximum of a few inches, the existing concrete liner can be notched to accommodate the desired clearance with the lining retaining enough strength to support the tunnel without any additional reinforcement. Initial calculations indicate that about 4 to 6 inches can be locally removed before shear stresses in the concrete at the notch exceed acceptable values and additional support measures are required. In the final design phase, the calculations will be refined on a site specific basis wherever notching is proposed.

Figure 2 shows a typical double-width tunnel with minor notching. The single-width tunnels would be similar. The notch could be created by a roadheader. Developed for the

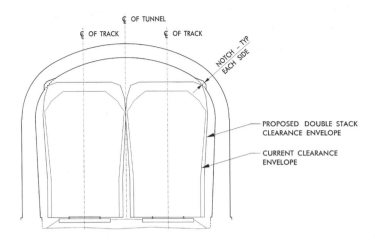

Figure 2.

mining and tunneling industries, roadheaders can be used to grind concrete, although reinforcing steel will hamper progress. A rubber tired or rail-mounted roadheader would work during available track windows, clearing the track as required for trains.

In recent years, hydro-demolition has been used for controlled demolition of concrete bridge decks and parking decks. Reinforcing is exposed by the high-pressure water jets as the concrete is eroded away. The reinforcing can then be cut with torches or saws. Hydro-demolition has many advantages, but large quantities of water are required for the process and must ultimately be collected and disposed of. Hydro-demolition may also prove advantageous in gaining lateral clearance where horizontal encroachment is a concern, provided the large volumes of water can be adequately dealt with.

During construction, holes will be drilled into the lining in the area of the notch to verify concrete thickness and determine if voids are present behind the lining. Should the lining thickness be less than the design calculations were based on, then additional support measures may need to be installed. If voids are present, they will be grouted solid prior to notching.

Deep Notching. Deep notching is defined herein as similar to the minor notch, except that the strength of the arch is compromised due to the depth of the notch, requiring additional reinforcement to support the tunnel.

Figure 3 shows a deep notching concept where untensioned rock dowels are installed to develop arching action within the rock surrounding the tunnel, thereby allowing a stress transfer of the current tunnel lining loads from the concrete liner to the reinforced rock arch as the lining is notched. The concept requires that the concrete arch be capable of supporting at least its own weight and should not be hung from the rock dowels. As with all alternatives, site-specific calculations will confirm the applicability of this concept in the final design phase. Use of tensioned rock bolts and corrosion resistant materials will also be assessed.

During construction, holes for dowels will be drilled into the lining and will also verify concrete thickness and determine if voids are present behind the lining. Should the lining thickness be less than the design calculations were based on, then additional support measures may need to be installed. If voids are present, they will be grouted solid prior to notching.

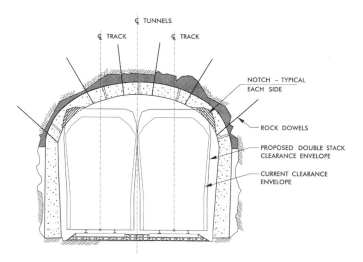

Figure 3.

Liner Replacement. In some tunnels, the clearance deficiency is so great that the entire concrete arch must be removed plus some rock above the existing lining. The challenge with the liner replacement will be to develop a scheme where the work can be done incrementally during the available track windows, while providing time within the windows to support the work in progress, inspect stability and vacate the work area before the next train passes. In the double track tunnels, the design must also permit the incremental construction over a single track at one time.

Figure 4 shows a liner replacement concept. Concrete saws (or rock saws), line drilling, or other methods, would be used to cut or perforate the lining to create a weakened plane at predetermined spacing depending on the rock's ability to span between the old and new lining. Next, a section of lining would be removed along with any rock above the lining scheduled for removal. Rock bolts and mesh would be installed immediately as initial support and in advance of the next train. That operation would be followed in later track windows by additional support measures including multiple layers of shotcrete reinforcement, lattice girders, etc., depending on the level of permanent support required for the actual structural and ground conditions encountered.

Tunnel Elimination or Replacement Alternatives

Daylighting. One approach to accommodate higher clearance freight through the tunnels is to eliminate the tunnels. Essentially the tunnel would be converted to an open cut. Starting from the surfaces above the tunnel, ground would be removed independent of the railroad operations. Once the excavation exposed the roof of the tunnel, the tunnel lining would be removed during scheduled track availability windows. Prior to the arch removal, rock bolts would stabilize the tunnel walls. Alternatively, the rock cut could be widened slightly and the walls removed with the roof.

Due to potential significant impacts on the surface, additional ROW necessary for the sloped cuts, and possible permitting delays, daylighting was not competitive at most locations. However, it is being investigated further to eliminate a short tunnel at one site.

New Tunnel on New Alignment. A new tunnel on a new alignment is self explanatory. While the cost initially appears to be uncompetitive with other alternatives, a new

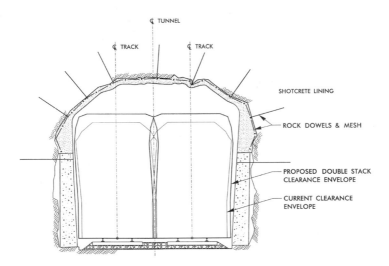

Figure 4.

tunnel might be competitive at the single track tunnels where track time will be at a pre-mium and liner replacement would be required, due to the tunnel contractor's ability to work continuously, independent of railroad operations. Due to additional ROW neces-sary and possible permitting delays, this was not competitive at any locations.

New Cut on New Alignment. This alternative is similar to the new tunnel alterna-tive where the high cost is somewhat offset by the significantly improved productivity by working independent of railroad operations. Due to potential significant impacts on the surface, additional ROW necessary for the sloped cuts, and possible permitting delays, this was not competitive at any locations.

PROJECT IMPLEMENTATION

The design of the modification methods were developed to maximize production for the contractor while minimizing impacts to the railroad. The designs allow for accommodating unexpected conditions with contingent designs that can address the full range of probable conditions. Particular attention will be made to scheduling of the daily construction activities so that work is completed within the daily work windows and trains are not delayed unexpectedly.

The tunnel modifications costs were developed for several scenarios; 8-hour track outage, 10-hour track outage, and 12-hour outage of only one of the two tracks. Trains will be operating on the adjacent track, although at reduced speeds. The incremental changes in the length of the outage affected both the costs per day and the overall costs by varying the production rate. An 8-hour outage has a maximum probable work-ing production window length of 6 hours to allow for mobilization and demobilization of the equipment into the tunnel. A 12-hour outage produces a maximum probable work-ing production window length of 10 hours. For a location with a liner replacement, the existing liner must be removed, the parent rock removed and stabilized, and the initial support installed within in the 6 to 10 hour maximum probable working production.

NSR will model the operations on the railroad in response to the different outage scenarios to determine their effect on railroad operations and the increased costs both on the daily level, and on the overall project duration. The project schedule spreads the construction over 36 months with the construction work generally limited to two areas of the railroad at any one time to allow current rail operations to continue. Once construction begins, extremely close cooperation between the railroads's Transportation Department and the construction crews will be required to minimize unscheduled delays.

Resource Availability

Liner Replacements will require overhead concrete demolition equipment combined with standard tunneling/mining equipment for operations such as rock bolting and shotcreting. Some contractors may elect to use readily available roadheaders for the controlled demolition of the concrete lining and rock above the concrete lining. While specialized equipment such as sophisticated high-speed two-boom drill jumbos and robotic shotcrete applicators will undoubtedly be used when available, less sophisticated equipment is readily available.

Notching requires the use of a specialized hydro-demolition machine, roadheader or other equipment to notch the concrete. A regional contractor has two suitable hydro demolition units and could lease a third unit with a few months notice to meet program requirements. The notching operation will also require a drill jumbo for roof bolting and they should be readily available.

Daylighting requires standard mass excavation equipment and concrete demolition equipment that should be readily available.

Most tunnels will receive new drainage systems. Readily available rock saws are anticipated to be utilized to cut the trenches along with bolting and shotcreting equipment where underpinning of the walls is required.

Skilled operators are needed to safely operate any equipment, but the local labor pool includes many miners who either already possess the necessary skills or can be trained in short order. Tunnelers traditionally will relocate as they follow work around the country so an expanded skilled labor pool is available from regional and national labor pools.

SUMMARY

The design uses an observational approach to safely accommodate the entire range of conditions that may be encountered while minimizing the potential for daily construction activities overrunning the track availability windows and resultant unexpected train delays.

The Heartland Corridor Clearance Improvement Project will stimulate the economies of West Virginia, Eastern Kentucky, and Southern Ohio by allowing the more efficient transportation of container freight via double stack trains. In addition, container traffic between Norfolk and Chicago will have a shorter route shaving a day off the shipping time.

ACKNOWLEDGMENTS

The Authors would like to acknowledge the Norfolk Southern Railway's Office of Bridges and Structures for their cooperation and contributions to design of the Heartland Corridor Clearance Improvement Project and for the permission to publish this paper.

INSPECTION OF A BRICK-LINED AQUEDUCT

Julie Frietas

Parsons Brinckerhoff

Fidel L. Mallonga

Parsons Brinckerhoff

Arne Fareth

New York City Department of Environmental Protection

Francis Lo

New York City Department of Environmental Protection

Paul S. Fisk

NDT Corp.

INTRODUCTION

This report describes the inspection of parts of the New Croton Aqueduct (NCA) undertaken for the New York City Department of Environmental Protection (NYCDEP) by Parsons Brinckerhoff Quade & Douglas, Inc., (PB) in association with Montgomery Watson Engineers of New York, P.C. (MWH). Support for the inspection was provided by AFC Enterprises, Inc., under Construction Contract CRO-333, "WM-11 New Croton Aqueduct Rehabilitation and Inspection Program."

The scope of the inspection work under CRO-333 included assessing the condition of the 10.3 km (6.4 mile) long, pressurized section of NCA and inspection of shafts, head houses and blow-off structures in both the pressurized and gravity sections. The CRO-333 inspection continued investigation work begun under Contract CRO-196, which ran from 1993 to 1997. The CRO-333 inspections were conducted from November 2004 to September 2005.

Background

The New Croton Aqueduct was put into service in 1891 to deliver water from the Croton River Watershed in Westchester County to consumers in New York City. Figure 1 shows the location plan of Croton Water Supply System.

The design flow into Jerome Park Reservoir was 12.7 m³/s (290 mgd) and into the 135th Street Gate House, further south, was 10.9 m³/s (250 mgd). The NCA has been operated as a gravity conduit from Croton Lake to Gate House No. 1, just north of Jerome Park Reservoir, since its original construction. The length from Gate House No. 1 to 135th Street, which includes a major siphon under the Harlem River, has been operated under pressure. Croton Lake gate house, the north end of the alignment lies at Station 4+75; Shaft 33, the south end of the alignment lies at Station 1628+05.

The diameter of the tunnel is 3.2 m (10.5 ft) in the siphon and 3.7 m (12.3 ft) in the Bronx and Manhattan sections. The pressurized sections, constructed in rock, consist of a brick lining surrounded by a mortared and grouted rubble backing. Figure 2 depicts a typical tunnel cross section.

The objective of the shaft and tunnel inspections was to locate and document defects, such as leaks, deterioration of lining material, cracks, spalls, delaminations,

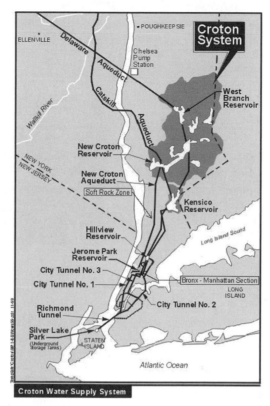

Figure 1. Croton water supply system

heaves and voids. Information gathered during the visual inspection was used to establish focal areas for both future repair contracts and further field investigation, which included coring, probe drilling, fiber optic scope inspection, and test grouting.

INSPECTION OF THE TUNNEL AND SHAFTS

Visual

Tunnel inspection was performed between Stations 1266+54 and 1479+30 in the Bronx and Stations 1503+17 and 1627+85 in Manhattan. Manned inspections were conducted at nine shafts. Figure 3 shows a tunnel inspection team.

Brick-lined portions of the tunnel and shafts were sounded and the entire tunnel alignment was visually inspected. Potential voided areas were recorded in the field log and marked on the wall for possible future probe hole drilling. Where leaks were observed, the location was identified and inflow volumes were estimated and recorded. Observations were recorded on a field log with reference to stationing and position on the tunnel circumference. The invert was typically covered with water 0.1 to 0.2 m (4 to 8 in) deep, so the invert inspection was less concentrated and it was not typically sounded, but defects were noted when observed.

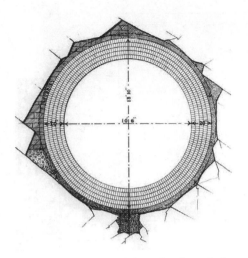

Figure 2.   Cross section of the Harlem River Siphon

Figure 3.   Inspection team within Manhattan pressurized section

Remote video inspections were performed at sixteen shafts. Real time visual inspection was performed from the surface collar to the tunnel invert using a remotely operated camera, averaging at a rate of 0.03 m/s (5 ft/min).

Remote Operated Vehicle (ROV) Inspection of the Harlem River Siphon

The inspection of the Harlem River Siphon and appurtenant shafts and structures was conducted by an underwater remotely operated vehicle (ROV) which was equipped with high resolution video and dual imaging sonar instruments. The objective was to verify shaft and siphon lining materials and construction features, to assess sediment levels and debris accumulations, to locate defects, and to obtain video and sonic records of the structures.

Nondestructive Geophysical Testing

Nondestructive geophysical testing was performed within Gould's Swamp Siphon (GSS), a siphon under a marsh in the gravity section of the NCA, a 'soft rock zone' from Station 777+00 to Station 780+00, sections of the Bronx and Manhattan pressurized Aqueduct and locations within the Aqueduct of known filled shafts. Nondestructive testing methods included sonic/ultrasonic direct and refracted stress wave, seismic refraction, electromagnetic conductivity (EM), electric resistivity and ground penetrating radar (GPR). Surveys were concentrated on portions of the tunnel crown.

The purpose of the nondestructive geophysical testing was to assess the relative strength and integrity of the tunnel liner and of the materials behind the liner, typically a mortared and grouted rubble. Analysis of the data focused on anomalies, indicative of deteriorated materials, poorly grouted sections, metallic objects, and air or water filled voids. Nondestructive geophysical testing was also used to locate and estimate the sizes of the abandoned shafts.

Figure 4.   Test grouting operations in the Bronx pressurized section

Probe Hole and Fiber Optic Testing

A fiber optic testing program was conducted at locations that appeared from the visual and geophysical investigations to merit further investigation. Probe holes were drilled and inspected with fiber optics to evaluate the extent of voids behind the brick liner, quantify the thicknesses of the brick liner and of the mortared and grouted rubble zone, to investigate geophysical anomalies and visual observations, and to verify historical information. Probe holes located within a test grouting zone were incorporated into the grouting program.

Diamond Core Drilling and Water Pressure Testing

Continuous diamond core drilling was performed in selected areas of the Aqueduct to obtain samples of the tunnel lining, mortared and grouted rubble zone, and foundation bedrock for material identification and laboratory testing to determine engineering properties. The data obtained were used to calibrate and help in the interpretation of the nondestructive geophysical test data.

Single packer water pressure tests were conducted in selected core holes, to obtain data on the permeability of the brick liner, the mortared and grouted rubble behind the liner, and the bedrock foundation. The testing results were used to establish the fluid pressures required for grout injection during the test grouting program and after grouting to evaluate the efficiency of the grouting operations.

Test Grouting Program

The test grouting program was performed in both the Bronx and Manhattan pressurized tunnels and at two shaft locations. Major objectives of grouting are to fill voids and fractures and to reduce water infiltration into the tunnel (Figure 4).

This program was performed to identify optimal grouting materials, methods and procedures, and to provide a basis for estimating quantities and setups for the rehabilitation contract.

Contact grouting was used to fill voids between the tunnel liner and the surrounding mortared and grouted rubble zone or voids between the rubble zone and the foundation rock mass. Consolidation grouting was used to fill geologic features, such as open cracks and joints in the surrounding foundation rock mass. Upon completion of the test grouting, a series of check holes were drilled within the test grouted areas to determine the effectiveness of the work.

Table 1. Summary of defects by section

Section	Length (ft)	Area of Hollow Sound SF	Leakage (gpm)	Cracks (ft)	Length of Mortar Loss (ft)	Number of Grout Holes
Manhattan	12,485	933	108	26	252	100
Bronx	21,260	5,470	118	240	282	9,319
Total	33,745	6,403	226	266	534	9,419

Investigation of a Soft Rock Zone

A series of investigations were conducted over a length of tunnel described during construction as a "soft rock zone." Papers published in the mid 1890s by engineers J.P. Carson and Edward Wegman describe the difficulties encountered tunneling through this zone and the use of timber cribbing to support the tunnel during construction. Investigations included an in-tunnel structural inspection, a subsurface investigation, a nondestructive geophysical survey and field reconnaissance work.

SUMMARY OF FINDINGS

Overview

The comprehensive field investigation program was used to assess the current condition of the New Croton Aqueduct and to design a rehabilitation program. Overall, the tunnel and shafts are in very good condition, with reparable leaks or defects at some locations. The visual inspection identified tunnel defects. The geophysical inspection determined the condition of the brick liner and the grouted and mortared rubble backing. Visual and geophysical inspection findings frequently complement the laboratory test results. The test grouting program successfully stopped or decreased inflows in the test areas. The locations of abandoned shafts were verified using geo-physical methods. Investigations of the soft rock zone determined that the area is geo-logically stable and that the tunnel is structurally sound.

Visual Inspection

Tunnel. A total length of 10.3 km (6.4 miles) of tunnel and 25 shafts were inspected. Overall, the tunnel was found to be in very good condition. However, some areas exhibited significant defects ranging from areas of hollow sound, leakages, cracks and mortar loss. Table 1 summarizes the defects found in each section.

The data acquired from the visual inspection was used not only to evaluate the condition of the tunnel and shafts, but also to designate zones for test grouting, provide a means of correlation with the geophysical data, and to locate core holes. The visual and geophysical data were compiled in an Excel database, processed through CADD and presented on location plans, as seen in Figure 5.

Two zones of major defects are described in more detail below:

- Station 1312+01 to 1312+35: A delaminated, hollow sounding area of 11.9 m² (119 SF) exists on the west sidewall (Figure 6). A detailed structural inspection and investigative core drilling was performed, revealing approximately 10.7 m (35 feet) of cracking and three localized areas of loose and disintegrated brick. This defect is near the NW striking Mosholu Fault, which crosses the Aqueduct

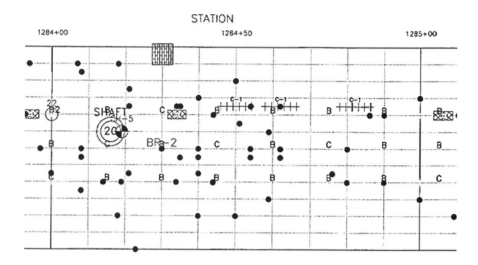

Figure 5.   CADD plot of compiled geophysical and visual inspection data

Figure 6.   Delaminated, hollow sounding area

near station 1310+00, and it was suspected that the delamination could have been caused by movement along the fault. However, the defect appears to be unrelated to the nearby fault, because petrographic examination of one core in the area indicated ASR gel (Alumino-Silica Reaction) in the mortar.

- Station 1594+57 to 1596+91: This section is lined with steel and did not exhibit any major defects other than surface rust and minor tuberculation. Sonic and ultrasonic test data suggested zones 609.6 m (2000 ft) long border each side of this section and consist of loose rubble or weathered, fractured bedrock. Historical reports documented difficult tunneling through this area caused by bad ground conditions.

Shafts. There are a total of 52 shafts along the Aqueduct's alignment, ranging from large chambers in cut and cover sections to small diameter shafts up to 112.8 m (370 ft) deep. Sixteen of these shafts were filled at the end of construction and are difficult to detect from the ground surface or from within the tunnel. Eight shafts were lined with a combination of brick and cast iron or stainless steel. Most are rimmed with granite collars, typically 0.4 m (14 in) thick, at the surface. Collars at the tunnel intersection are up to 0.6 m (2 ft) thick. Most shafts had steel ladders from the ground surface to the tunnel crown which often exhibited oxidation, tuberculation and/or section loss.

The shafts were in very good to excellent condition, overall. Most shaft surfaces were very dry and clean of debris or biofilm above the normal Aqueduct water level. Areas immersed during normal operating conditions usually exhibited a layer of black biofilm, similar to that found in the tunnel. Shaft defects were similar to those encountered in the tunnel. The most prevalent, but minor, defect documented was calcite leachate or efflorescence. Point inflows ranged from less than 3.2×10^{-5} to 2.8×10^{-4} m^3/s (0.5 to 4.5 gpm).

ROV Inspection of the Harlem River Siphon Complex

The Harlem River Siphon and associated structures were successfully inspected by a ROV and found to be aligned according to historical drawings and without evidence of major displacement or defects. The brick lined siphon is covered with silt and deposits that appear to be biofilm. Cast iron lined areas, such as the upper portion of Shaft No. 24, were coated with rust and lightly bonded tubercles. A DVD record and sonar plots were produced from the investigation.

Geophysical Investigation

The sonic/ultrasonic velocity measurements were made to determine the dynamic moduli (deformation) values and based on an empirical relationship, the equivalent strength of the brick liner, the filler materials behind the liner and the foundation bedrock. The measurements indicate the condition of the tunnel liner and of materials behind the liner, whether rock or mortared and grouted rubble. GPR measurements were made to detect open or water filled voids behind the brick liner, adverse geologic conditions and the presence of steel or wood. EM measurements were made to detect changes in the bedrock conductivity that might be associated with weathered zones, fault/shear zones, intruded dikes or lithological changes.

The sonic/ultrasonic testing data shows liner compressional wave velocities are in the range of 2,438 to 3,657 m/s (8,000 to 12,000 ft/sec) with corresponding shear wave velocity values in the range of 1,371 to 1,676 m/s (4,500 to 5,500 ft/sec). The liner and behind the liner velocity data were grouped into and presented in five categories based on their compressional wave velocity. Comparatively, the sections in the Bronx and Manhattan differed slightly in regards to liner compressional wave velocity data.

Bronx. The velocity measurements indicate approximately 85% of liner velocity values are in the range of 2,438 to 3,657 m/s (8,000 to 12,000 ft/sec), a range indicating good condition. Behind-the-liner velocities range from 2,438 to 3,657 m/s (8,000 to 12,000 ft/sec), over 66% of the length, indicating that materials behind the liner contain little voiding or cracking and that the bedrock is relatively competent.

Localized, high amplitude reflectors of GPR, generally behind the liner and less than 3.0 m (10 ft) in length, are believed to be entrapped water. GPR also showed areas behind the liner with abrupt signal changes from high to low amplitude, indicative of well drained rubble or unfractured bedrock. GPR indicated an area 18.3 m (60 ft) in length, possibly shored with timber. The timbers appear to be spaced approximately 2.4 to 3.0 m (8 to 10 ft) apart, 0.9 to 1.5 m (3 to 5 ft) behind the brick liner, mainly in the crown. A metallic anomaly was also located from the EM data in the crown along a 9.1 meter (30 ft) tunnel section.

Manhattan. The velocity measurements indicate approximately 93% of liner velocity values are in the range of 3,048 to 3,657 m/s (10,000 to 12,000 ft/sec) and that the liner is in good condition. Behind-the-liner velocities in the range of 2,438 to

3,657 m/s (8,000 to 12,000 ft/sec) over 77% of the tunnel length indicate that the material behind the liner contains little voiding or cracking and is surrounded by relatively competent bedrock. A 609.6 m (2,000 foot) zone at each end of the steel lined section had lower behind the liner velocities, indicative of loose rubble or weathered, fractured bedrock. A 914.4 m (3,000 ft) weak zone, including an area of leakage, was also identified. A 71.3 m (234 ft) section of the tunnel is steel lined. EM data spikes at four locations possibly indicate water filled cracks or joints in the bedrock that are beyond the depth of investigation afforded by the other geophysical methods.

Probe and Core Hole Drilling and Fiber Optic Testing

A total of 124 probe and test grout holes were drilled in Manhattan, 96 in the Bronx, six at Shaft No. 20 and 12 at Shaft No. 21. Probe holes were drilled to depths ranging from 0.25 to 3.5 m (10 in to 11.5 ft). A total of 14 NX and BX-sized core holes were also drilled. Probe and core holes drilled during this inspection provided data on behind the liner conditions, such as dimensions of voids identified during the visual inspection soundings, thickness of liner materials and bedrock type. In addition, they supplied data for the calibration of geophysical anomalies, access to the test grout zones, water pressure tests, and core samples for lab testing.

The majority of probe holes were drilled between the spring line and crown and only occasionally below the spring line. Fiber optic investigations revealed the tunnel brick liner varied from 0.2 to 0.6 m (7 in to 24 in) thick, the surrounding mortared and grouted rubble varied from 0.03 to 1.1 m (1 inch to 45 in) thick, and the bedrock foundation consisted of gneiss, schist, or granitic pegmatite. Fiber optic scope investigations and core recovery revealed small voids in about one-half of the holes. Voids ranged from 0.03 to 0.25 m (1 to 10 in) thick and occurred mostly in the mortared and grouted rubble zone. The three check holes, drilled after test grouting, showed 'new' grout within the mortared and grouted rubble zone.

Water Pressure Testing

Calculations to correlate the water flow test data to material permeability information showed that the permeability of the zone encompassing the brick-rubble-bedrock was usually higher than that of the bedrock zone by two orders of magnitude prior to grouting and one order of magnitude after grouting. This indicates that it is possible to grout the rubble without grouting the bedrock as well.

Results of Laboratory Testing of Core Samples

A laboratory testing program was carried out to determine basic properties of the brick and mortar liner, of the mortared and grouted rubble and of the foundation bedrock materials. Testing included measurements of direct shear strength, compressive strength, Young's Modulus of Elasticity, Poisson's Ratio, splitting tensile strength and density. Dynamic tests included P-wave velocity, S-wave velocity and elastic modulus tests.

Laboratory and field geophysical data sets for the liner and mortared and grouted rubble fill are largely complementary, though differed occasionally by an order of magnitude. Elastic and engineering property measurements were typical for gneiss and schist. Data was calculated assuming a Poisson's ratio of 0.3.

A total of 15 rubble and rock samples were analyzed petrographically using plane polarized light (PPL) and crossed polarized light (CPL). Samples of the mortared and grouted rubble were composed of a sanded hydraulic cement and rubble concrete with no lime or pozzolanic additions and exhibited no visible or microscopic signs of physical distress. Bedrock types identified were typical for the area. Bedrock from the three Bronx cores were fine to medium grained, moderately to highly indurate rock, and ranged from biotite-garnet schistose gneiss to biotite-amphibolite gneissose schist. One core from the Bronx was determined to be a moderately fractured, coarse-grained granite pegmatite with calcite-healed microcracks, and one core from Manhattan yielded medium-grained, biotite-muscovite-garnet schistose gneiss.

Reactions between the brick and cementitious paste were identified in core samples at three locations. A recrystallized, carbonate gel deposit filled the cracks between adjacent brick units. No reactive aggregates were detected, and the reaction appears to be occurring between the cement paste and the adjacent brick. The substance is similar to ASR gel. It was further noted that very fine layers along the crack surfaces near the mortar contact are composed of the reaction gel which gradually diminishes with depth in the brick.

Test Grouting

Eight tunnel segments and two shaft locations in the Bronx and Manhattan pressurized sections were included in the test grouting program. The following brief summary lists the results of the test grouting:

- A total of 832 cement sacks were injected into six test grouting areas in the tunnel over 170.9 m (561 ft) of section in Manhattan. A total of 691 cement sacks were injected into two test grouting areas in the tunnel over 88.7 m (291 ft) of section in the Bronx. A total of 188 cement sacks were injected into Shaft Nos. 20 and 21.

- The water to cement ratio for grout mixtures was varied between 1:1 and 0.8:1 by volume. Locations with groundwater inflows of 6.3×10^{-5} m^3/s (1 gpm) or greater took the most grout. Grouting successfully reduced or eliminated groundwater inflow.

- Check holes were drilled and the core recovery was inspected to verify grouting extents and effectiveness within a test grout zone.

INVESTIGATION OF THE SOFT ROCK ZONE

Structural

The tunnel has hairline diagonal cracks at the beginning and end of the soft rock zone, consistent with theoretical shear distributions. It does not indicate a significant finding since the tunnel is continuously supported, but less stiffly than within the adjoining hard rock sections.

Subsurface Investigation

Two surface boreholes drilled in the zone showed in situ soils to be mainly fine sand and some 'limestone' sand and gravel, in accordance with historical records. Boring logs also indicate similar soil compositions, mainly sand–gravel mixtures with layers of silts and clays. The in-situ density of the soils within the elevation of the tunnel envelope ranged from medium dense to very dense. Water level measurements indicate that the water table is shallow and matches the piezometric head within the silty sand strata found at depth.

Geophysical Testing

The geophysical data obtained in the area confirms that the initial excavation was supported by timber cribbing. Data from the GPR survey indicate a soil-like rock or timbering behind the liner spanning from station 41.1 m (135 ft) along the east spring line, 48.8 m (160 ft) along the west spring line and 57.9 m (190 ft) along the east invert.

Sonic/ultrasonic and seismic data indicate tunnel liner velocities are in the same range as those outside the soft rock zone. The data indicates an 'intact' liner that has not been substantially weakened over time except perhaps near the boundaries of the soft rock zone where a loss of shear wave velocity is indicative of some cracking or deterioration within the liner. This condition at the boundaries is further confirmed with the structural inspection. It also appears that the tunnel invert produces several resonant frequency values that correspond to timber lengths of about 3.7 to 4.9 m (12 to 16 ft), in reasonably good condition.

Field Reconnaissance

It was not possible to determine the limits of a less stable geologic zone from a reconnaissance of the surface features. However, based on historical documents and the field investigation, the soft rock zone is thought to be either a solution enlarged joint in a limestone formation trending N90E degrees, filled with soils of glacial origin or a combination of weathered and host rock.

CONCLUSIONS

This investigation found that the brick lined tunnel and shafts are in good condition and performing as designed 120 years ago. The brick liner is 0.2 to 0.6 m (7 to 24 in) thick, has Young's Modulus values in the range of 11×10^9 to 20×10^9 Pa (1.6×10^6 to 2.9×10^6 psi) and has isolated areas of leakage, cracking and reaction between the brick and cementitous paste. The mortared and grouted rubble fill between the brick liner and bedrock ranges from 0.03 to 1.1 m (1 to 45 in) thick, has voids ranging from 0.03 to 0.25 m (1 to 10 in) thick, is approximately 2 orders of magnitude more permeable than the bedrock, and can be grouted. The good condition and continued use of the New Croton Aqueduct is a testimony to the skills of the engineers and craftsman of 120 years ago.

ACKNOWLEDGMENTS

We would like to acknowledge the efforts of our colleagues throughout the work and for their help with this manuscript: Walter Herrick, Taehong Kim, Masrour Kizilbash, Kevin Liu, James Marold, Jose Morales, Amitabha Mukherjee, Kyle Ott, Carl Panutti, Donna Roberts, Roger Schiller, Nikolas Sokol, Mark Stephani and Carroll Stewart and Roy Taliaferro.

BIBLIOGRAPHY

Carson, J.P. 1890. Notes on the Excavation of the New Croton Aqueduct. *Transactions of the American Institute of Mining Engineers.* Vol XIX: 705–716, 732–752.
City of New York Aqueduct Commission Reports. 1887. *City of New York Aqueduct Commission Reports on the New Croton Aqueduct 1883–1887.* New York: Douglas Taylor Printer and Publisher. (pp. 72–77).
Wegmann, Edward. 1896. *The Water Supply of the City of New York 1658–1895.* New York: John Wiley & Sons.

MIDDLE ROUGE PARKWAY INTERCEPTOR EXTENSION— A CASE HISTORY

Harry R. Price

NTH Consultants, Ltd.

Charles J. Roarty, Jr.

NTH Consultants, Ltd.

Behrooz Tahmasbi

Wayne County Department of the Environment

ABSTRACT

The 2,408 meter (7,900-foot) long Middle Rouge Parkway Interceptor Extension varies in size from 1.52-meter (5-foot) to 2.36 meter (7-foot 9-inch) inside diameter. Most of the sewer was constructed in tunnel through fine sands and silts. Piping of these soils through the concrete liner eventually resulted in a loss of soil support, settlement and/or severe distress of the sewer, and the formation of several sinkholes. The repair and restoration was generally performed in the sewer and included cast-in-place concrete, shotcrete, cured in place pipe, soil stabilization, cement grouting, dewatering, and chemical grouting. As a part of the work, temporary pump stations and flow diversions were required to perform the work while minimizing discharges to the adjacent Rouge River.

INTRODUCTION AND BACKGROUND

In the late 1920s and 1930s, the first sewage disposal project(s) were completed in Detroit, Michigan and along the Rouge River Valley, located in the northwestern portion of Wayne County, Michigan. The original Middle Rouge Parkway Interceptor (MRPI) was constructed to provide combined sewer service to Northville, Plymouth, Northville Township and Plymouth Township, Michigan with an overall service area of approximately 256 square kilometers (100 square miles). The flow from the MRPI drained to Wayne County's Oakwood Interceptor where it was transported to the Detroit Waste Water Treatment Plant.

The Middle Rouge Parkway Interceptor Extension (MRPIE) was constructed in 1954 and 1955 to provide wet-weather relief to the MRPI, provide an additional outlet to the Detroit system, and provide service to an additional 41 square kilometers (16 square miles) of service area. The wet weather flow from the MRPI discharges from the Middle Rouge Junction Chamber through a siphon under the Rouge River and into the MRPIE at Manhole No. 11. Under an agreement with the Detroit Water and Sewerage Department (DWSD), the MRPIE was also designed to receive dry and wet weather flow from DWSD's Northwest Interceptor (NWI), which connects multiple combined sewer outfalls along a branch of the Rouge River. The Wayne County Department of Public Works (WCDPW) increased the inside diameter of the sewer from 1.52-meter (5-foot) to 2.36-meter (7-foot 9-inch) between Evergreen Road and the Southfield Freeway, thereby allowing the City of Detroit to abandon a pumping station at the intersection of Warren and Pierson Roads which previously forced all flow to the east to DWSD's Southfield Sewer.

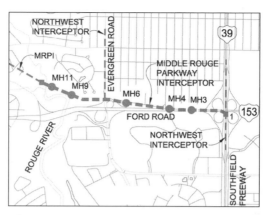

Figure 1. Project area plan

Immediately prior to the construction of the MRPIE, DWSD redirected the flow from the northern reach of their Southfield Sewer to the east in preparation for the construction of the Southfield Freeway. A portion of the former Southfield Sewer was abandoned and the reach downstream of the MRPIE connection (south of Ford Road) became a part of DWSD's Northwest Interceptor. With the construction of a gravity connection from the former Warren—Pierson pump station through the MRPIE, DWSD's Northwest Interceptor system was completed.

The MRPIE surcharges during wet weather events when flow volumes in excess of capacity were stored in the system downstream. In the event the MRPIE could not receive sufficient flow volumes, both the MRPI and the NWI were relieved through a series of combined sewer overflows that discharged directly to the Rouge River. In an effort to reduce discharges to the Rouge River from the MRPI, Wayne County augmented the MRPIE in the early 1990s with the construction of the parallel Middle Rouge Relief Interceptor. This new interceptor connects to the MRPIE upstream of the NWI connection at a new Junction Chamber 18A and discharges further downstream to DWSD's system. In cooperation with the Michigan Department of Environmental Quality (MDEQ) in the late 1990s, several outfalls from the MRPI to the Rouge River were closed and a new pump station was constructed upstream of the MRPIE to maintain the contract volume of sewage discharge from MRPI to Detroit's system even during periods of wet weather. The general alignment of the MRPIE, NWI, major roads and freeways are shown in Figure 1, Plan View of Project Area. Figure 2, prepared by Applied Science, shows a schematic of the entire sewer network, the Middle Rouge Relief Interceptor, Pump Station and Junction Chamber 18A were constructed subsequent to the 1988 emergency Repairs.

The MRPIE was originally constructed along and under minor branches of the Rouge River, through undeveloped property, and within existing roadway right-of-ways. As the area developed, existing roads were replaced with limited access parkways or below grade freeways to handle increased traffic volumes. Employees of the Wayne County Road Commission (WCRC) first noted isolated areas of settled pavement in the vicinity of the MRPIE in the late 1970s to early 1980s. The WCRC filled the depressions and repaired the associated pavement until June 1988 when it advised the WCDPW that the noted settlement might be associated with the MRPIE sewer.

This paper chronicles the investigations and repairs to the MRPIE. The phasing of the work was affected by wet and dry weather flow control issues, the desire to minimize discharges of raw sewage to the Rouge River, the construction of MRRI and

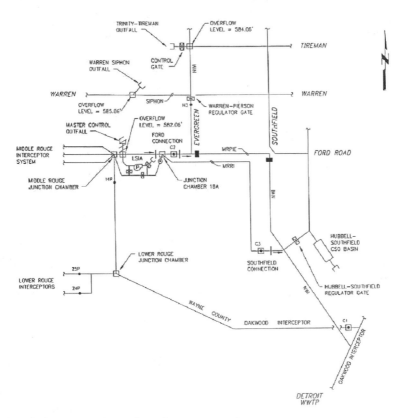

Figure 2. Outlet sewer system schematic

sewage pumping station, impacts on local traffic patterns and coordination between various jurisdictional authorities.

GEOLOGIC SETTING

The upstream end of the Middle Rouge Parkway Interceptor and MRPIE lies within the southern flank of the Inner Defiance Moraine in Northville/Northville Township. The sewers were generally constructed through lacustrine and glacio-fluvial deposits from the Wisconsin glaciation consisting of an overburden of clay overlying deposits of fine sands and silts. The MRPIE was generally constructed through or slightly below the interface between the upper clays and the underlying fine sands and silts. The groundwater level was slightly above the crown of the MRPIE.

INSPECTION AND REPAIR CHRONOLOGY

1988 Emergency Repair

Shortly after notification by the WCRC that surficial settlements may be associated with the MRPIE, the WCDPW retained NTH Consultants, Ltd. (NTH) to perform an engineering evaluation of the MRPIE. In September 1988, NTH commenced a geotechnical investigation and physical exploration of the MRPIE in the vicinity of the noted settlement areas. To facilitate the internal inspection of the MRPIE, DWSD's Warren-Pierson Gate was closed to provide a work period of approximately four hours under dry weather conditions before the stored flow in the Northwest Interceptor would reach the elevation of the Warren-Pierson Overflow to the Rouge River.

To the west of Manhole No. 7 the sewer liner consisted of reinforced concrete pipe. The sewer liner for the NWI connection and the MRPIE east of Manhole No. 7 was a nominal 35.6-cm (14-inch) thick cast-in-place unreinforced concrete. Within the unreinforced concrete liner, it did not appear that waterstop was used between the construction joints. Between Manhole 6 and Manhole 7 there were two areas of significant distress, one beneath the Evergreen off-ramp and one at the Alter Road cul-de-sac. The first area was 41 meters (133 feet) long and the second was 25 meters (81 feet) long. Both of these areas were characterized by closely spaced diagonal cracks that ran from the upper quarter point to the lower quarter point on both sides of the sewer. NTH concluded that fine sands and silts had piped through these cracks. In response to the loss of lateral support the horizontal axis of the sewer increased in length while the vertical axis of the sewer decreased in length. Our opinion was based, in part, on the U.S. Army Corps of Engineers Technical Report GL-81-2 from a similar sewer failure northeast of Detroit that concluded soil fines could pass through cracks as small as 0.01-inch in width under water heads as low as 2-feet. The initial piping was likely through the construction joints in the unreinforced concrete liner due to normal groundwater infiltration or drainage of surrounding soils following surcharge events. As the sewer deformed in response to the loss of lateral support, cracking occurred and ultimately widened thereby providing avenues for additional ground loss.

The WCDPW directed NTH to prepare contract documents for the repair of the MRPIE following our November 1988 study report. In addition to the two repair areas described above, an additional subsidence area located adjacent to Manhole No. 4 appeared to be the result of a densification of backfill material as a result of increased traffic loadings resulting from a widening of Ford Road over a former greenbelt area.

On the day before Thanksgiving 1988, a sinkhole occurred along the off ramp from Westbound Ford Road to Evergreen Road and its rapid expansion engulfed the off ramp. DWSD's Warren-Pierson Gate was closed for four hours on Thanksgiving Day and the physical exploration revealed that there had been a marked increase in the distress to the concrete liner of the sewer. Existing diagonal cracks had widened, the volume of groundwater inflow had increased, and signs of concrete crushing were evident at the springline of the sewer. A photograph of the diagonal cracking taken after the installation of the structural steel ribs is shown in Figure 3.

A local tunnel contractor mobilized on Thanksgiving Day under an emergency contract and repairs were started on the day after Thanksgiving. The MDEQ required that repairs proceed with no overflows into the Rouge River and the installation of temporary shoring and stabilization grouting began on alternating 4-hour shifts. It soon became apparent that stabilization efforts were insufficient to arrest the incremental structural distress. The MDEQ was advised that if the gate were not maintained in a closed position to allow round-the-clock work, a catastrophic failure would occur. The MDEQ acknowledged the necessity for the continued closure of the Warren-Pierson gate; however, they requested the WCDPW undertake all measures possible to mitigate the discharge to the Rouge River.

The only viable alternative to mitigate these flows was to reactivate the abandoned Warren Pierson pump station and force the flow east to the parallel Southfield Sewer. Applied Science, under subcontract to NTH, developed a pump station design that used four 10,000 gpm rated pumps. Three of the pumps were positioned in the wet well of the abandoned pump station, while the fourth was positioned in a manhole

Figure 3.   Diagonal cracking in 1988 repair

immediately upstream of the pump station. The pumps and necessary piping were installed in less than 7 days by pipe fitters working 24-hours per day. Power was provided using temporary generators. Given the constraints of the pumps and the size of the wet well, the pump station was designed to handle approximately 75-percent of the maximum dry weather flow of the Northwest Interceptor. As a result, the discharges to the Rouge River were limited to roughly twice a day at the peak flow periods. In addition to the temporary pump station to isolate the work area from NWI flows, upstream gates were closed and downstream segmental bulkheads were installed to isolate the work area from wet weather flows from the Middle Rouge Parkway Interceptor sewer and back flow from the wastewater treatment plant, respectively. The segmental steel bulkhead erected downstream of Manhole 6 was designed for a limiting surcharge condition equal to the ground surface.

The emergency round-the-clock stabilization effort in the MRPIE consisted of the installation of temporary steel ribs and skeleton lagging. Concurrent with the installation of the ribs, chemical grout was injected through the concrete liner to both seal the leaks and to provide some stabilization of the adjacent soils. A two-components Acrylimide grout was selected because of its ability to penetrate and stabilize the silty sands, no premature reaction of the materials within the wet environment of the sewer, the ability to control set time, and the still viscous grout return at adjacent joints or cracks confirmed the penetration of the grout. Once the area between Manhole Nos. 6 and 7 was stabilized with chemical grout and the temporary support ribs, cement grouting was employed to fill any voids present. Once the voids were filled, a 2.05-meter (6-feet 9-inch) inside diameter reinforced concrete repair liner was installed. Concurrent with these repairs, the third subsidence area was also repaired. The repairs to the subsidence area at Manhole 4 consisted of cement grouting to densify the backfill material. A total of 5,700 liters (1,500 gallons) of chemical grout (Acrylimide) and 73 cubic meters (2,565 cubic feet) of cement grout were placed during the emergency repair. The 15-cm thick (6-inch) reinforced concrete repair liner extended over 107 meters (350 lineal feet) of the sewer starting at a point approximately 33 meters (110 feet) upstream of Manhole 6.

In conjunction with the emergency repair operations, an inspection of the MRPIE was performed from the Northwest Interceptor (Structure 1) under the Southfield Freeway to the siphon under the Rouge River (Manhole 11). Between Manholes 7 and 10, in the reinforced concrete pipe section, a number of leaks were encountered which were also chemically grouted and limited section of the sewer had settled. In the unreinforced concrete lined section between Structure No. 1 and Manhole No. 6, the sewer was in generally good structural condition with a limited number of isolated circumferential cracks. After the emergency repairs were completed, our April 1989 condition survey report recommend development of a schedule to repair a depressed area upstream of Manhole 9 and to grout active leaks at joints and cracks between Structure No. 1 and Manhole No. 6.

1990 Scheduled Repair

Our subsequent investigation of the depressed area located upstream of Manhole No. 9 indicated the cause of settlement was consolidation of the soils supporting the sewer. The sewer in the vicinity of the depressed area was constructed through an abandoned channel of the Middle Rouge River. In the early 1970s when the alignment of Ford Road was straightened, the excavated material was utilized to back fill the abandoned river channel. Once the channel was filled, additional material was placed to facilitate a residential development located along the top of the river valley. As a result of the surcharge from the recently placed fill material, the soils at the bottom of the channel consolidated.

NTH considered several alternatives for the restoration of this depressed area, including in place replacement, pipe jacking and the relining with a cured-in-place pipe (CIPP) liner. Based on both cost and feasibility of construction CIPP was selected. NTH prepared plans and specifications for this work that was performed by Insituform of Michigan. Twenty-five meters (83 feet) of Insituform liner was installed in September of 1990 with only the diversion of the wet weather flow from the MRPI at the Middle Rouge Junction Chamber required.

The recommended leaking joint repair work between Structure No. 1 and Manhole No. 6 was deferred until after the Middle Rouge Relief Interceptor construction because flow control measures associated with the construction imposed further limitations on the available four-hour windows to work in this reach of the sewer.

1993 Emergency Repair

In March 1993, a new sinkhole developed over the MRPIE between Manhole Nos. 3 and 2. The sinkhole day-lighted at the base of a utility pole in the median of Ford Road dropping a pair of traffic signals to within a foot of the pavement. Probing of the four-lane roadway to determine the limits of the void commenced immediately. After waiting four days for dry weather flow conditions, a man-entry inspection was performed working east from Manhole No. 4 with gates closed at the Middle Rouge Junction Chamber and the abandoned Warren-Pierson pump station. The inspection revealed a boil at a construction joint in the invert of the unreinforced concrete sewer at the approximate location of the sinkhole. This boil was free flowing, however, the external void fortunately was limited to the greenbelt side of the sewer.

The flow control required to facilitate the 1993 emergency repair was complicated by DWSD who was in the process of investigating and performing repairs to their Southfield Sewer. In order to gain access for inspection and repair of the Southfield Sewer, DWSD had constructed temporary bulkheads to redirect flow from the Southfield Sewer west to the NWI and then through the MRPIE. As a direct result, the available storage time above the Warren Pierson Gate was reduced to approximately 2-hours. The MDEQ would not permit the performance of the emergency work with more than the previously allowed twice-daily dry weather discharges of sewage into the Rouge River. In conjunction with other work in the DWSD system, the Greater Detroit Regional Sewer System (GDRSS) model had been developed. Applied Science used

Figure 4.   Segmental bulkhead at Manhole 6

the GDRSS model to develop the necessary flow control measures for the repair to eliminate dry weather overflow events and minimize wet weather overflow events and/or residential basement flooding.

The WCDPW diverted flow from MRPI to the County's Oakwood Interceptor at the Middle Rouge Junction Chamber, a temporary bulkhead was installed in the MRPIE at Manhole 6 to force the NWI flow upstream in the MRPIE to the new Junction Chamber 18A (constructed with the MRRI), and the flap gates at the JC-18A were chained open to allow flow to be diverted to the MRRI. In order to install the upstream bulkhead, the Warren Pierson gate was closed for a period of 24 hours. A photograph of a segmental bulkhead is shown in Figure 4. A downstream bulkhead was also constructed in the 3-meter (10-foot) diameter NWI beneath the Southfield Freeway to prevent backflow from the plant into the work area during wet weather events.

Once flow was diverted to the MRRI, the emergency repair in the vicinity of the sinkhole and limited chemical and cement grouting of actively leaking joints between Manhole No. 3 and Structure No. 1 was completed (Phase 1). The flow diversion measures were removed and the leaking joint repair work was scheduled for June 1993 after the spring rainy season had ended. At the request of the MDEQ, the second phase of the work was postponed until September 1993 to allow the MDEQ to complete flow monitoring within the sewers contributory to MRPIE. During Phase 1 emergency repairs, the section of sewer that extended from Manhole 3 to a point 79 meters (260 feet) east was cement grouted. The total grout take in this reach was 5 cubic meters (179 cubic feet), with the greatest take being centered on the sinkhole.

1993 Scheduled Repair

The scheduled (Phase 2) MRPIE repairs were started on September 8, 1993. All remaining joints between Structure 1 and Manhole No. 3 were either chemically or cement grouted. The grout type was selected based on the pattern of cracking and on data developed during the 1988 repairs. If there was a series of leaks in a given reach, there was a high potential there was a void outside of the liner capable of being cement grouted. Using these criteria, areas were selected for cement and/or chemical grout. An evaluation of grout takes at the completion of grouting generally supported the historic data trend. Once the work to the east of Manhole No. 3 was completed, the grouting operation shifted to the west with the initial plan of reaching Manhole No. 6. Because of budget constraints, all of the joints between Manholes 3 and 4 were grouted, only actively leaking cracks were grouted between Manhole Nos. 4 and 5, and no joint grouting was accomplished between Manhole Nos. 5 and 6.

A total of 26,166 liters (6,906 gallons) of chemical grout was placed between Structure No. 1 and Manhole No. 5. The average grout take was 132 liters (35 gallons) per joint. A total of 3.8 cubic meters (135 cubic feet) of additional cement grout was placed adjacent to the Phase 1 cement grouting.

A section of thin concrete liner at the crown of the MRPIE sewer was discovered during grouting work immediately adjacent to Structure 1, the MRPIE/NWI connection. NTH recommended that this area be structurally relined and that all remaining ungrouted joints between Manhole Nos. 4 and 6 be grouted. Subsequent to the completion of the grouting in 1993, DWSD made additional modifications to their system and also completed the repairs and relocation of a portion of their Southfield Sewer. In addition, the Michigan Department of Transportation planned to replace the existing bridges at Ford Road and the Southfield Freeway. As a result the completion of the grouting program and the installation of the repair liner at the MRPIE/NWI was postponed until DWSD completed their work on the Southfield Sewer and MDOT completed their Bridge replacement at Ford Road.

1996 Inspection

At the request of the WCDPW, NTH performed a supplemental inspection of the MRPIE between the NWI connection and the bulkhead ring upstream of Manhole No. 6. Based on this report dated November 22, 1996, a scope of repairs was developed and used to prepare a set of conceptual repair drawings. This work was placed on hold pending completion of MDOT's repair of the Ford Road Bridges over the Southfield Freeway.

2001 Scheduled Repair

As the work on the MDOT Bridge replacement was nearing a close in 1999, NTH started to prepare the necessary contract documents for the completion of the grouting operations and the construction of the relining of a limited section of sewer at the MRPIE/NWI contact. An initial set of documents was prepared that was submitted to the MDEQ for review. With the modifications performed by DWSD to their outfalls and the work that the WCDPW had completed, MDEQ indicated that they would not allow any dry weather discharges into the Rouge River even for the installation of the upstream isolation bulkhead.

Once again Applied Science, using the GDRSS, evaluated the potential flow control measures. A schematic of the sewer network used in the analysis is shown on Figure 2. Based on their analyses, if a partial height bulkhead was constructed in the Warren Siphon Outfall the dry weather flows could be contained. In addition DWSD would have to set their control gates at the Trinity-Tireman Outfall and the West Chicago Outfalls at the 8/10th points of the NWI at these outfalls.

Prior to finalizing the design of the necessary components, an agreement had to be reached with the MDEQ and DWSD for the flow control measures. In addition an agreement with MDOT was required because lane closures were required to access manholes to enter the MRPIE on both Ford Road and the Southfield Freeway are under their jurisdiction. The agreements were in place by September 2000 and the repair project was bid in November 2000. Contract work started at the end of January

2001 and was completed by the end of March 2001. The work was completed within the timeframe agreed upon by WCDPW, DWSD, MDOT, and MDEQ.

During the 2001 repairs, a total of 24,095 liters (6,361 gallons) of chemical grout was placed. A total of 110 square meters (1,179 square feet) of the MRPIE tunnel were relined with 12.7 cm (5-inch thick) wire mesh reinforced shotcrete and 9.3 square meters (100 square feet) of the Northwest Interceptor sidewall opposite the eye of the MRPIE sewer was relined with 5 cm (2-inches) of steel fiber reinforced shotcrete. A total of 22 cubic meters (790 cubic feet) of cement grout was placed adjacent to Manhole No. 6.

CONCLUSIONS

The geologic setting in which MRPIE was constructed, together with the state of the tunneling art at the time of construction resulted in conditions favorable to piping of soils. We recommend that future sewers constructed in similar conditions should be provided with suitable waterstops. In addition, the concrete liners should be reinforced to mitigate any cracking should the liner deflect.

For existing sewer systems constructed in tunnel of unreinforced concrete without waterstop through wet sands and silts, changing usage patterns will affect the behavior of the sewer system. As new development occurs and overflows to surface waters are restricted, older systems are surcharged more often. If cracks exist in the tunnel liner, a process of exfiltration and infiltration can occur with each surcharge event and potentially accelerate migration of fines into the tunnel.

To minimize the impact on the public, all work was done in tunnel and all materials, including segmental bulkhead components, were designed to pass through a standard 1.2-meter (4-foot) diameter manhole. Because access manholes for work between Structure No. 1 and Manhole No. 3 were located within active traffic lanes, non-emergency work was performed primarily at night. The repair of the Middle Rouge Parkway Interceptor Extension was complicated by the needs of the various jurisdictional authorities. Further, in particular the period between 1990 and 1993, the repair work was delayed while a parallel relief sewer was constructed. While in many cases the relationship between the various jurisdictional authorities was cooperative, several meetings with all parties present to discuss the overall impact of their combined decisions would have been helpful.

The total construction and engineering costs for the 13-year long MRPIE repair program were $3.2M and $750K, respectively.

BIBLIOGRAPHY

Corps of Engineers 1981, Technical Report GL-81-2, 15 Mile Road/Edison Corridor Sewer Tunnel Failure Study, Detroit, Michigan

Mazola, A.J. 1969, Geology for land and Ground-Water Development in Wayne County, Michigan, State of Michigan Department of natural Resources

Sherzer, W.H. 1916, Geologic Atlas of the United States—Detroit Folio No. 205, U.S. Geologic Survey, Washington, D.C.

Wayne County Board of Public Works 1964, Report to the Board of Supervisors, Rouge Valley and Down-River Sewage Disposal Projects, Wayne County, Michigan

REHABILITATION OF A BRICK-LINED AQUEDUCT

Nikolas Sokol

Parsons Brinckerhoff

Taehong Kim

Parsons Brinckerhoff

Carl Pannuti

New York City Department of Environmental Protection

Walter Herrick

Parsons Brinckerhoff

BACKGROUND

This paper discusses work performed as part the rehabilitation of the gravity flow section of the New Croton Aqueduct (NCA) during an approximately two year period between 2004 and 2006. The goal of the project was to restore the gravity flow section of the NCA to optimal operating condition, extending the lifespan of this significant water supply for the New York City metropolitan area. Rehabilitation of the NCA was purposely scheduled before the completion of the new NCA Water Treatment Plant now under construction (Figure 1).

The NCA is a 3.73 to 4.34 m (12.25 to 14.25 ft) diameter brick-lined circular and horseshoe shaped water tunnel constructed between 1885 and 1891. The NCA conveys water for approximately 50 km (31 miles) by gravity and pressurized flow from the New Croton Reservoir in Westchester County, NY to distribution and reservoir points along its alignment, terminating at a distribution station at 135th Street and Convent Avenue in Manhattan, NY. Prominent features of the NCA include a large intake facility at New Croton Reservoir, 38.6 km (24 miles) of gravity flow horseshoe shaped aqueduct, 11.3 km (7 miles) of pressurized flow circular aqueduct, 2 siphons, and 36 open shafts and headhouse chambers which are used as sampling, access and blow-off locations. Until a partial inspection in 1982, the NCA had flowed continuously since its completion nearly 100 years prior. Flow tests have confirmed that the NCA can convey up to 12.7 m^3/s (290 mgd) (Figure 2).

Cover depths of NCA vary considerably, ranging from a few to hundreds of meters, typically of hard rock. Primarily mined using means conventional for the late 1800s, construction of the NCA also employed cut and cover methods in several low-lying sections of the alignment, totaling approximately 1.6 km (1 mile). The NCA crosses through numerous lithologic changes, fault/shear zones and under several significant water bodies including the Pocantico, Saw Mill and Harlem Rivers and the Tarrytown Reservoir. The NCA passes under the Harlem River as a siphon at a depth of approximately 122 m (400 ft) below grade (Figure 3).

Soon after construction the Aqueduct Commission Reports on the New Croton Aqueduct documented prominent defects such as large voids behind the lining. More recently, large-scale inspections of the gravity flow section performed during the 1990s and a 2004 inspection of portions of the pressurized sections of the NCA have revealed additional defects such as open and/or deteriorated masonry joints/cracks, leaks ranging from trickles to several liters per minute into and out of the NCA, missing bricks, formed openings, and in one place, a rupture through the liner with discernable

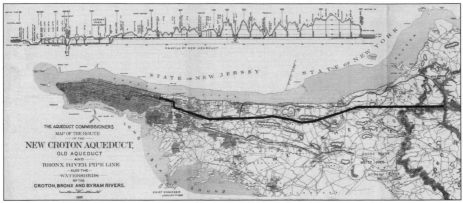

Source: The City of New York Aqueduct Commission, 1887.

Figure 1.   Location map showing the alignment of the New Croton Aqueduct as it passes from the New Croton Reservoir on the far right (northeast) of the map, through Westchester County, the Bronx, and into Manhattan. North points to the upper right corner of the map.

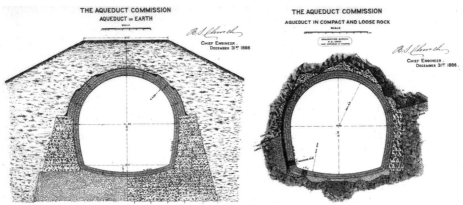

Source: The City of New York Aqueduct Commission, 1895.

Figure 2.   Diagram showing the horseshoe-shaped brick liner of the New Croton Aqueduct in the gravity flow portions of the alignment. The profile on the left shows a typical cut-and-cover section and the profile on the right is typical for the mined sections.

offset (Freitas et al. 2007). Despite these defects, the generally good condition of the NCA is remarkable. It is not uncommon to traverse several kilometers through the Aqueduct without noting any significant defect (Figure 4).

PROJECT OVERVIEW

The inspection program performed during the 1990s served as the primary basis for the rehabilitation work described here. Visual inspection, geophysical surveys and borings within and outside the NCA were used to determine the quantity and extent of defects and to verify ground conditions. The results of the inspection were compiled

Source: The City of New York Aqueduct
Commission, 1895.

Figure 3. Historic photograph showing men
lounging within a void between the NCA liner
and the mined rock tunnel

Figure 4. Heavy groundwater inflow
encountered after drilling into a water-filled void.
Inflow rates subsided over the following 48 hours
and were significantly reduced after grouting.

and used to develop contract documents for the rehabilitation of about 38.6 km
(24 miles) of the gravity flow section of the NCA. The work entailed different kinds of
environmentally sensitive and confined space activities, including: cleaning and washing approximately 39,929 m (131,000 ft) of horseshoe shaped tunnel; dewatering,
inspecting and rehabilitating the 346 m (1135 ft) long Gould Swamp Siphon; filling
voids and controlling water through contact grouting; making major tunnel liner repairs;
masonry repointing; tunnel liner crack repair; removing accumulated sediment and
debris; waterproofing a shaft chamber; plugging a connecting tunnel to an older abandoned aqueduct; disinfecting the entire NCA and restoring it to service. The contract
drawings identified 74 underground work zones in the contract. When existing conditions did not match the drawings, the Field Engineers inspecting the work were
allowed to extend or reduce the work as necessary based on actual conditions found.

Work support activities included site preparation, tunnel ventilation, pumping of
water inflow, maintenance of the Resident Engineer's office, and regular documentation of the rehabilitation by professional photographers. Related surface work included
replacing shaft covers, rehabilitation of a shaft portal and developing access roads and
staging areas. Out of scope work included repairing significant portions of the floor at
the Old Croton Lake Gatehouse and rebuilding a buttress wall inside of the waste
chamber of one of the major access shafts/headhouses.

The entire project was executed with heightened environmental constraints. All materials used in the rehabilitation were NSF approved for potable water supply and all wastes
were to be transported out of the Aqueduct on a regular basis. Additionally, for the entirety
of the project the Contractor had to be prepared to demobilize all in-tunnel activities and
disinfect the Aqueduct in the event of an emergency reactivation of the NCA. An emergency reactivation did occur, halting the underground activities for several months.

CONTACT GROUTING

A large-scaled contact grouting program was performed in the horseshoe shaped
gravity flow sections of the NCA with the objective of filling identified voids behind the
liner, filling fractures in broken mortared and grouted rubble and foundation bedrock
materials, and reducing water infiltration into the tunnel. A description of grouting procedures follows.

Figure 5. Hand-operated electric rotary-impact drilling was the typical method for advancing grout and vent holes through the brick liner

Drilling

The locations of grout and vent holes were selected by the Field Engineer and drilled in areas with groundwater inflow higher than approximately 6.3×10^{-5} m^3/s (1 gpm), with geophysical anomalies, and with defects interpreted from visual observations, soundings and/or remote instrumentation inspections. A typical pattern of grout and vent holes was roughly 1.5 to 2.4 m (5 to 8 ft) horizontally and vertically between holes. Grout and vent holes were drilled to provide grout pathways into suspected void areas behind the tunnel liner. Holes were drilled to depths ranging from 0.5 to 1.8 m (1.5 to 6 ft), with an average hole depth of approximately 0.9 m (3 ft). Holes were drilled through the brick and mortar liner and approximately 15 to 30 cm (6 to 12 inch) into the mortared and grouted rubble behind the liner or into bedrock. A hole would be abandoned if wood was encountered for more than 10 cm (4 in) or unconsolidated material was encountered for more than 5 cm (2 in). All holes were drilled using an electric rotary-impact drill to produce a 1.5-inch diameter hole. After drilling, each hole was washed until the return flow was clear and then fitted with a packer. Packers were the rubber, manually expanded type, equipped with shut-off valves and positive hose connectors. If groundwater issued from a hole after drilling, a rough estimate of the inflow rate was recorded. A total of 616.3 m (2,022 ft) of grout and vent holes were drilled for the grouting program (Figure 5).

Grouting

The water to cement ratio of the grout mix was directed by the field engineer and typically varied between 1:1 (i.e., a mixture of 113.6 liters (30 gallons) of water and 4 bags of regular Type II cement) and 0.8:1 (i.e., a mixture of 113.6 (30 gallons) of water and 5 bags of regular Type II cement) by volume. After each grout mixing, the density of grout was measured with a mud balance and ranged between 1.6 and 1.8 g/cm^3 in most cases. Injected volumes varied considerably, and voids often required more than one tank of grout. The grout mix (e.g., from a ratio of 1:1 to a ratio of 0.8:1) was thickened after injecting a pre-specified number of sacks of cement into an individual grout hole. The cumulative grout volume injected was tracked for each work area by recording the number of tanks and batch mixes used. A total of 7,445 sacks of portland cement grout were used during the grouting program.

Grout was delivered using a dual-tank, hydraulic grout plant. The dual, independent tank system allowed mixing of approximately 0.23 m^3 (8 ft^3) of Portland cement grout in the secondary tank while the grout was pumped from the primary tank into the void. A manually controlled manifold on the grout plant allowed circulation of grout from the primary tank through the hopper and back to the primary tank. Grout was pumped through 15.2 to 30.5 m (50 or 100 ft) lengths of 1.5-inch I.D. hose to packers installed in the grout holes.

The grout delivery pressure was closely monitored at all times using pressure gauges located both at the grout plant and at the grout hose and packer connection. Grouting pressures typically ranged from 69 to 207 kPa (10 to 30 psi), which was generally adequate to enable the grout to flow into the void, but not enough to build up excessive pressure behind the liner, which is easily achievable when grout injection pressure is added to existing groundwater pressure. Total pressure was restricted to a maximum 345 kPa (50 psi) to limit pressure build up behind the liner which could cause cracking or bulging of the liner. When a void was encountered and grout flowed easily, the delivery pressures tended to be on the order of 69 to 138 kPa (10 to 20 psi), the pressure required to pump grout through the length of hose and raise the grout to the elevation of the packer and into the grout hole. As a void became filled, the injection resistance, and therefore the pumping pressure, would increase and begin to approach the 345 kPa (50 psi) maximum. When a few minutes had passed after the grouting pressure had neared or achieved the 345 kPa (50 psi) maximum, the hole was considered grouted and was closed off.

Grouting generally proceeded from the lowest to the highest elevation of grout hole. Vent holes facilitated the monitoring of the upward and outward travel of grout behind the liner as grouting progressed from lower to upper elevation holes. When grout of a density equivalent to that of the batched grout was observed to flow from a vent hole, communication between the active grout hole and vent hole was considered to have been achieved, and the valve on the vent hole packer was closed. When grout hole to grout hole communication occurred the packer valve was temporarily closed, later to be connected to in the appropriate sequence. Observable communication between the active grout hole with adjacent grout holes, vent holes, and minor cracks or holes in the face of the brick liner provided a valuable indicator of the horizontal and vertical progression of grout within the void behind the liner (Figure 6).

JOINT/CRACK MASONRY REPAIR

Prominent joints/cracks had been observed in previous inspections and specified for repair. The lengths of the joints/cracks varied from several centimeters to several meters and varied in thickness from hairline to about 2.5 cm (1 inch). All joints/cracks exceeding 0.6 cm (0.25 inch) in width were scheduled for repair. A total of 2,374.7 m (7,791 ft) of joints/crack repair work was completed in 34 work zones throughout the gravity flow section of the NCA (Figure 7).

All joints/cracks were prepared by saw-cutting through the joint/crack a groove perpendicular to the surface to a depth approximately 2 times the width of the joint/crack. In the case where there was a particularly long joint/crack, the groove length was limited to 6.1 m (20 ft) at any time to maintain lining stability. The groove was then thoroughly cleaned of all dust and debris with clean water. The cleaning of the groove was timed such that at the time of tuck-pointing the groove was damp but free of standing water. The groove was tuck-pointed in approximately 1.3 cm (0.5 inch) lifts, and each lift was well tamped to ensure a good bond and to minimize overall mortar shrinkage (Figure 8).

Some work zones required both joint/crack repair and contact grouting. In certain cases, an ad-hoc repair method was implemented and resulted in successfully filling the open joint/crack from behind concurrent with the contact grouting. While grout was

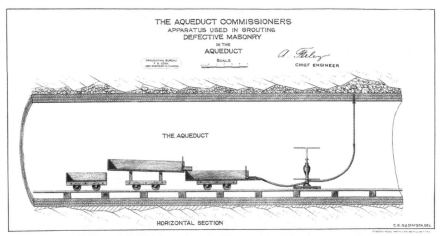

Source: The City of New York Aqueduct Commission, 1895.

Figure 6.   (Top) Historic diagram from the Aqueduct Commission Reports showing typical apparatus for grouting voids behind the liner of the New Croton Aqueduct in the late 1800s; (Bottom) Typical grouting set-up used in the recent rehabilitation of the New Croton Aqueduct

Figure 7.   Stair-step joints/cracks of variable widths through the brick masonry of the New Croton Aqueduct liner

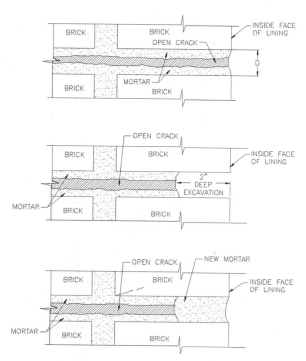

Figure 8. Typical joint/crack repair sequence

being pumped, one or two laborers would pay close attention to any leaks of grout through joints/cracks in the liner. If a grout leak sprung, the laborer would immediately plug the leak with oakum. If a particularly long joint/crack started leaking, the grouting pressure would be reduced and the laborer would persistently follow along the joint/crack, plugging it with oakum as the grout began to issue from it. After allowing time for the grout to set, the oakum was then removed and if needed the joint/crack repair was completed with tuck pointing.

MAJOR LINER REPAIR

Major liner repair work consisted of cast-in-place reinforced concrete for filling of existing formed openings located at or near the crown of the tunnel and areas of brick heave located within the tunnel invert. Areas requiring brick replacement were minimal, totaling only 14.8 m² (159 ft²), the majority of which occurred at one location in a headhouse chamber.

For heaved invert repair work, temporary cofferdams with sandbags and a pumping system were utilized to provide dry working conditions. A total of 3.6 m³ (127 ft³) of concrete was placed into formed openings and heaved liner repairs. A description of lining repair procedures follows.

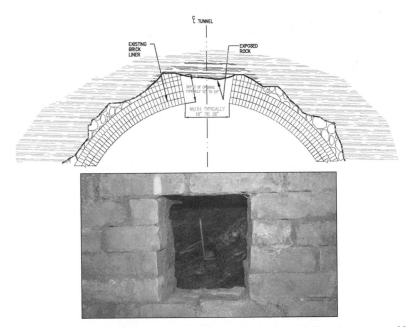

Figure 9. (Top) A diagram from the contract drawings showing the anticipated geometry of formed opening in the crown of the New Croton Aqueduct. (Bottom) Close-up photograph of an actual formed opening encountered; very similar. The gneissic banding of the bedrock beyond the liner can be seen through formed opening.

Formed Openings

Existing formed openings in the tunnel crown were rectangular in shape and sized typically 0.4 to 0.9 m (16 to 36 inches) in each direction and were 0.3 to 0.6 (12 to 24 inches) deep (Figure 9).

Holes of a 2-inch diameter and a minimum embedment of 1.5 m (5 ft) were drilled into the bedrock radially to match the liner arch curvature (see Figure 10). To anchor the repair concrete in place, hot-double-dipped fully galvanized #8 dowels were installed in the holes in a manner such that the bars were completely encapsulated with epoxy without air pockets. A form made of ¾-inch thick plywood was placed over the formed opening and supported by falsework bearing on the tunnel invert, to avoid gravitational loading on the adjacent brick liner. Concrete was then pumped through a grout pipe in the plywood form and through into the formed opening. A ventilation (or breathing) pipe allowed the air to escape and for a minor amount of post-grouting, if needed. To ensure the newly placed concrete would cure under dry conditions, groundwater was sealed off from entering the area using oakum and thick mortar.

Heaved Invert

Most of heaved lining areas were located in the tunnel invert, and the invert water had to be diverted to provide dry working conditions for the repair work. Temporary cofferdams and submersible pumps were used to ensure that the entire working zone stayed dry. In the brick heave areas, the area of demolition and excavation was extended a minimum of 0.3 m (1 ft) beyond the farthest point of heaving. All unsound concrete, deteriorated bricks and substrata were removed by using

Figure 10.   (a) Photograph showing a heaved invert zone prior to rehabilitation. (b) The same zone during rehabilitation work.

pneumatic chipping hammer until sound bedrock was exposed (approximately a 0.9 m [3 ft] depth below the usual invert level). Hot-double-dipped fully galvanized #8 bars were installed to anchor the repair concrete in place as per the procedure described in the formed opening repair.

After installing the galvanized #8 bars, drainage stone and geotextile were placed on the bedrock surface. The thickness of drainage layer, approximately 0.6 m (2 ft), was such that the placement of concrete was at a thickness equal to or greater than the design thickness of the tunnel liner, or about 0.3 m (1 ft). Hot-double-dipped galvanized #6 reinforcement bars were assembled in a lattice with 20.3 cm (8 inch) spacing. Prepackaged high performance concrete was then placed and completely consolidated by an internal vibrator. After two days of curing in a dry condition, the temporary cofferdam and submersible water pumps were removed from the work area.

WATERPROOFING

The NCA rehabilitation contract included the reconstruction and waterproofing of the chamber walls and invert at the Shaft 9 headhouse. The work included control of water, the cleaning of all surfaces in the chamber, removal and disposal of sediment and debris from the main and waste chambers, removal of loose mortar from and the repointing of stone and brick masonry joints, repointing using waterproof mortar, and the application of two coats of waterproofing material.

The work commenced with the power-washing of the entire chamber down to the standing water level, approximately 0.6 m (2 ft) above the chamber invert. All masonry joints were checked for missing, loose or deteriorated mortar and repointed with a waterproof mortar in accordance with the procedures described in the joint/crack masonry repair section.

To address the lower 0.6 m (2 ft) of the chamber, it was necessary to dam and divert the approximately 7.6×10^{-3} m^3/s (100 gpm) flow of water passing through the chamber, as was typical when the NCA was off line. A water-filled plastic dam was placed across the invert approximately 3 m (10 ft) north of the chamber. The base of the dam was sealed using dry portland cement. A section of plastic conduit approximately 15.2 m (50 ft) long was laid across the invert and propped up by cement blocks. The conduit carried most of the water through the chamber, although minor trickles continued to pass under the dam and also entered the chamber through very small holes in the brick wall. This water was diverted along the chamber invert by a temporary channel made from dry Portland cement and then pumped out of the lowest point of the chamber.

After all of the masonry joints in need of repointing had been repaired, two coats of a polymer modified Portland cement waterproofing slurry compound were applied to all surfaces in the chamber below the elevation of the crown of the NCA. The waterproofing was applied during the winter, and because of the temperature-sensitive nature of the waterproofing material, the Contractor locally heated the active work areas of the chamber. The application of the waterproofing material was allowed only if the surrounding air temperature was expected to remain above 4.4°C (40°F) for a minimum of 48 hours. The Contractor had several options for applying the waterproofing and chose to use rollers. This method was successful for the bulk application of the compound, but it was necessary to use brushes locally to ensure complete coverage. The thickness of each coat followed the recommendation by the manufacturer of the waterproofing compound.

SUMMARY

The successful rehabilitation of the 24-mile-long gravity flow section of the New Croton Aqueduct resulted from good management and positive working relationships between the Engineers and the Contractor. The rehabilitation, including notable out of scope work, was completed on schedule in approximately 17 months with crews working a typical 8-hour work shift five days a week.

ACKNOWLEDGMENTS

The authors wish to thank Donna Roberts, Deputy Project Manager for the NCA rehabilitation, for her invaluable assistance in the preparation of this paper. They would also like to thank the New York City Department of Environmental Protection for allowing the publishing of project information. The success of the NCA rehabilitation is due in no small part to the efforts of John Beasley, Julie Freitas, Lou Gonzalez, Masrour Kizilbash, Phil Mallonga, Jim Marold, Jamie Monte, Jose Morales, Amitabha Mukherjee, Mark Stephani and Chris Wang.

REFERENCES

City of New York Aqueduct Commission. *Reports on the New Croton Aqueduct—1883 to 1887.* New York City, NY.

City of New York Aqueduct Commission. *Reports on the New Croton Aqueduct—1887 to 1895.* New York City, NY.

Freitas, J., A. Fareth, P. Fisk, F. Lo, F. Mallonga. 2007. Inspection of a Brick Lined Aqueduct. *2007 RETC*, June 2007. Toronto, Canada.

18

Tunnel Lining Technology

Chair
C. Hebert
Traylor Shea Ghazi Precast JV

DESIGN, TESTING, AND PRODUCTION OF STEEL-FIBER REINFORCED CONCRETE SEGMENTAL TUNNEL LINING FOR THE EAST SIDE COMBINED SEWER OVERFLOW TUNNEL

R.F. Cook

Parsons Brinckerhoff Quade & Douglas, Inc.

K. Wongaew

Parsons Brinckerhoff

J.E. Carlson

Kiewit-Bilfinger Berger

C.R. Smith

FICE

T. Cleys

City of Portland

ABSTRACT

Steel fiber reinforced concrete segmental rings are the final tunnel lining for approximately 85% of the alignment for a 6.706-meter (22-foot) internal diameter tunnel being built as part of the East Side Combined Sewer Overflow Tunnel Project. This paper describes the approach adopted for the design of the steel fiber reinforced concrete segments, the development of the performance criteria for the segments, and the testing program used to confirm the performance of the materials and segments in the laboratory and in the field.

INTRODUCTION

This paper describes the design, testing, and production of the pre-cast steel fiber reinforced concrete (SFRC) segments that are the final lining for a 6.7-meter (22-foot) internal diameter tunnel for the East Side Combined Sewer Overflow (ESCSO) Tunnel Project. The 9.7-kilometer (6-mile) long tunnel is being built along the east bank of the Willamette River in Portland Oregon for the City of Portland, Bureau of Environmental Services. The tunnel is from 18.3 (60) to 48.8 meters (160 feet) below ground surface and in water bearing alluvium deposits. The SFRC segmental rings are the tunnel lining for approximately 85% of the alignment where the tunnel is located within highly competent Troutdale Formation.

Steel fiber reinforced concrete is a composite material with a performance affected by the interaction of the concrete mix, material selection and the characteristics of the fiber including fiber type, aspect ratio, fiber quantity and fiber strength (King, 2005). The steel fibers improve the concrete's physical properties, especially its flexural strength and its resistance to cracking, abrasion, and erosion. The strengthening mechanism of the fibers results from the transfer of stresses between the concrete and the fiber by interfacial shear or by mechanical interlock. The fiber and the concrete

1192

matrix share tensile stresses until the concrete cracks, after which the stresses progressively transfer to the fibers. Ultimate failure occurs when the fibers pull out or when the fibers break.

Currently, there is no applicable U.S. code for the design of SFRC. The American Concrete Institute provides recommendations for the application of SFRC in ACI 544 but these are not yet codified. The recommendations state, "For structural applications, steel fibers should be used in a role supplementary to reinforcing bars" however if conventional reinforcement bars are not used "the reliability of the members should be demonstrated by full-scale load tests, and the fabrication should employ rigid quality control." For the tunnel sections located within the Troutdale Formation, the structural analyses indicate that no reinforcement is required in the segments to carry the permanent loads including the impacts from seismic events. Thus, the design approach used ACI 318 Chapter 22, Structural Plain Concrete and further analysis or testing for compliance with ACI 544.4 was not required. However, because of the limited experience in the U.S. with SFRC tunnel lining segments and none for their use as a final lining, the project has carried out a test program demonstrating the adequacy of the SFRC segments as well as testing to develop a suitable mix design meeting project requirements.

The concrete mix was developed to meet the structural and durability performance criteria established during design and using available materials in the Portland area. For the steel fibers, a study was performed to determine their performance requirements. Based on the study, it was established that the fibers should be Type I (cold drawn wire) with end deformation and conform to ASTM A 820, Standard Specification for Steel Fibers for Fiber Reinforced Concrete. The length of steel fibers should be in the range of 1.9 (0.75) to 6.4 centimeters (2.5 in) and have an aspect ratio (fiber length divided by its diameter) in the range of 50 to 100. Preliminary assessments indicated that fiber content should be in the range of 17.8 kg/m^3 (30 lb/yd^3) and 53.4 kg/m^3 (90 lb/yd^3) and a test program was required to select the fiber type and optimum content.

TUNNEL LINING DESIGN

The ring configuration consists of seven trapezoidal segments with a smaller trapezoidal key. The segments are 35.6 centimeters (14-inches) thick with a skewed longitudinal joint of five degrees. The ring length is 1.5 meters (5 feet). Each segment has an EPDM gasket on its extrados to seal the lining against water ingress. The segments form a tapered ring that can negotiate the 304.8 meter (1,000-foot) radius curves along the designed tunnel alignment with the capability to negotiate tighter radii for alignment corrections, if necessary. A vacuum erector installs each ring within the tail shield of the tunnel-boring machine. Once installed, each ring is backfill grouted through the tail shield to avoid voids behind the lining.

Dowels provide connections between segments on the circumferential joints. Each dowel comprises a steel shank within a plastic sleeve with specified pullout strength. The dowels facilitate segment alignment during erection and maintain gasket compression during ring building. They can also help maintain ring position following seismic events.

Pullout bolts connect across the skewed longitudinal joints of segments to facilitate ring building.

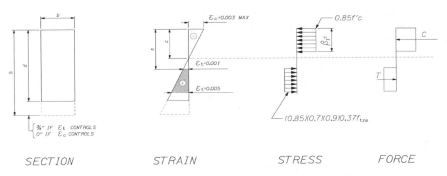

Figure 1. Stress-strain relationship for the design of the concrete cross section

STRUCTURAL PERFORMANCE

In the absence of an applicable U.S. design code for SFRC, the design used an approach developed from ACI 318-02, Chapter 22—Structural Plain Concrete and existing "guidelines" for SFRC (ACI 544, 1996; RILEM, 2000; German Concrete Association, 1992). For design, the concrete flexural tensile stresses are limited to less than the SFRC post-crack residual strength (King, 2005) in order that the section remains uncracked and behaves elastically. Figure 1 shows the stress-strain relationship used for the design of the concrete cross-section. Factors applied to the stress block (German Concrete Association, 1992) account for:

- Long-term stability of SFRC composite material
- Geometric difference between the structure and the test specimens
- Variability between the mix design tests and the production QC tests

PERMANENT LOADING ANALYSIS

The structural analysis for the tunnel lining during operating (long-term) conditions accounted for:

- External ground and groundwater loads—estimated from the tunnel profile with the tunnel between 18.3 (60) and 48.8 meters (160 feet) deep and below the groundwater table.
- Ring distortion resulting from its erection and based on an allowable ring distortion of 1.7-centimeter (0.66-inch) diametric ovaling and 0.9-centimeter (0.375-inch) segment lipping.
- Seismicity based on an earthquake having a 2% probability of excedance in 50 years.
- Transient internal water pressures based on the transient hydraulic gradient estimated from the 25-year design storm event with a water elevation 1.6 meters (5.2 feet) above the overflow elevation to the river.

The ground-tunnel lining interaction was analyzed using various approaches:

- Closed Form Solutions (Muir Wood, 1975; Wu and Penzien, 1997)
- FLAC numerical analyses
- Beam-ground spring analysis using SAP90 software.

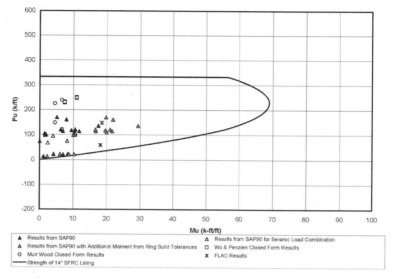

Figure 2.   Moment-thrust interaction diagram

The design load, consisting of the load combinations with the load factors defined in ACI 318-02, is compared with the structural capacity of the segment section. The structural capacity of the section was based on the key principles identified in ACI-318, Chapter 10, for members subject to combined flexure and thrust with the use of the stress-strain relationship in Figure 1 and a resistance factor of $\varphi = 0.55$.

The results are summarized in the moment-thrust interaction diagram, Figure 2.

As a check, the approach outlined by Hashash et al., 2001 was used to confirm that ovaling deformations of the tunnel induced by vertically propagating shear waves, and flexural and axial deformations induced by horizontally propagating waves oriented 45 degrees to the longitudinal axis resulting from seismic ground shaking remain within the maximum allowable strain for concrete (0.3%). The results indicate free-field combined strains of about 0.1% and with the effect of ground-structure interaction taking into account, the strain "felt" by the tunnel reduces to about 0.02%.

TEMPORARY LOADING ANALYSIS

The structural analysis for the temporary loading conditions applied to the tunnel lining accounted for:

- Removal of segments from molds
- Handling and stacking in the storage yard
- Transportation
- Lining erection
- Tunnel-Boring Machine Thrust Loads
- Backfill grouting

As a result of segment removal from molds, handling and stacking requirements, segments were checked for flexure and shear. For installation underground, the segments were also checked for bursting under the action of the loads from the thrust rams of the tunnel-boring machine. To ensure that the segments remain elastic and uncracked under temporary construction loading conditions, the stresses are required to be maintained below the average residual flexural strength as well as below the compressive and tensile strengths of the SFRC.

Based on the structural analyses, the minimum performance requirements for SFRC segments were established as:

- Compressive Strength, f'_c, of 6,000 psi at 28 days
- Modulus of Elasticity, E_c', of 4,415,200 psi at 28 days
- Splitting tensile strength, f'_t, of 690 psi at 28 days
- First peak strength, f_1, 660 psi at 28 days
- Average Residual-Strength, $f'_{cr.3.0}$, of 290 psi at 28 days

DURABILITY PERFORMANCE CRITERIA

As well as storing CSOs, the tunnel will transport dry weather (sanitary) flows. Corrosion associated with the transport of domestic wastewater result primarily from acid attack of moistened cementitious and metallic surfaces exposed to the atmosphere where sulfate present in biological systems has been reduced to hydrogen sulfide under anaerobic conditions. Analysis of Portland wastewater has indicated that sulfide build-up can occur at very low and very high flows. Since low flows will exist in the tunnel from dry weather flow, then sulfate resistant materials are required.

To meet the sulfate resistant needs, two types of cement were considered: Type II (moderate sulfate resistant) cement with added fly ash and Type V (sulfate resistant) cement. Type II cement with fly ash added was specified for the cement because of its workability characteristics, its reduced heat of generation, its reduced thermal volume change, its reduced permeability, as well as its protection against sulfate attack characteristics.

To further protect against corrosion, the segment thickness was increased by 1.9 centimeters (0.75 inches) to provide a sacrificial zone on the tension side of the section. Steel fibers within this zone can corrode but the discontinuous characteristics of the fibers limit the potential for deep seated corrosion. As further protection measures, the performance requirements of SFRC were specified to minimize permeability, shrinkage cracking, and abrasion-erosion loss.

PRODUCTION

The ESCSO project was awarded in March of 2006 to Kiewit Bilfinger Berger AJV (KBB), a joint venture between Kiewit Construction Co. and Bilfinger Berger. Prior to finalization of the contract KBB had previously been working on the project under a provisional service agreement with the City of Portland. CRS Engineering Consultants, who had been a member of the City of Portland's Technical Review Committee for the SFRC lining, were subsequently employed by the Joint Venture to design the segment plant, develop the mix design and associated testing, as well as implementation of the production system and the initial training of production management and personnel.

MIX DESIGN

The contract required two mixes, one for the normally reinforced segments and one for the steel fiber reinforced segments. The specification requirements for both mixes were similar but the SFRC mix had a number of additional requirements noted

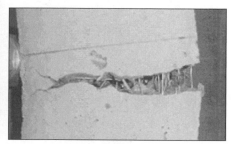

Figure 3.   UBC's closed loop machine   Figure 4.   Test beam at failure

above. Recently, ASTM introduced a new active standard, ASTM C1609, for Flexural Performance of Fiber Reinforced Concrete (Using Beam with Third Point Loading). This new test is improved in that it required the use of a closed loop servo controlled testing system. The flexural beam performance was to ASTM 1609 with the First-peak strength, and the Residual Flexural Strength, at 3.0mm deflection was specified.

To obtain a mix design for the normally reinforced segments was not a difficult issue so the main thrust of the development program was to refine the SFRC mix. This program was aimed at obtaining a balanced, efficient fiber mix using locally available materials that met the specified requirements and was suitable for the envisaged production methodology; this took a number of months.

In addition to designing a mix to meet the strength parameters it was necessary to determine the type and supplier of the fiber to be used. The fibers were required to be to ASTM A820 Type 1 and have a minimum tensile strength of 145,000 psi. Three suppliers were pre-selected for testing following a review of the available products.

- Dramix 80/60 BN
- Propex FE 085 50
- Sika SH 80/60

CONCRETE TESTING

An on site laboratory was set up to carry out the comprehensive program of testing, numerous basic mix designs were tested to obtain the necessary performance and production data a well as a range of different fiber dosages. A large number of cylinders and flexural beams were made and tested. These beams were tested to ASTM C1609, using a "closed loop" system which is significantly more accurate than the traditional "open" system (Figures 3 and 4). The nearest available machine was at the University of British Columbia in Vancouver, Canada and we are indebted to Professor Nemy Banthia for his assistance in these tests. A sampling of typical results is given in Figure 5 and Table 1.

After a detailed review it was determined that the structural performance of all three fibers tested was satisfactory at the same dosage and in the same material matrix. The choice of supplier was therefore simply one of price. It is important to understand that the flexural performance of individual types of fiber and the fiber concrete as a whole is significantly affected by the cement and aggregate matrix, it is therefore critical to tune the materials as well as the quantity of fibers to obtain an efficient mix

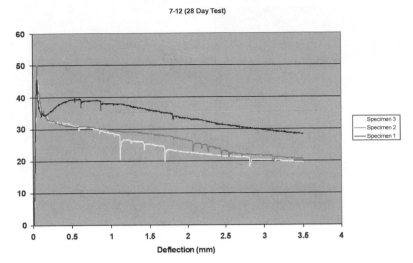

Figure 5. Typical development beam test printout

Table 1. Typical development beam test data

	Series 7–12			
	Specimen 1	Specimen 2	Specimen 3	Average
No. of Fibers	140	110	108	119.3
P_1 (kN)	45.6	50.1	48.1	47.9
f_1 (MPa)	6.18	6.69	6.42	6.39
δ_1 ($_{or}\delta_p$) (mm)	0.053	0.051	0.056	0.053
$P_{150,0.75}$ (kN)	39.0	30.6	30.4	33.3
$f_{150,0.75}$ (MPa)	5.20	4.08	4.06	4.45
$P_{150,3.0}$ (kN)	29.8	21.5	20.4	23.9
$f_{150,3.0}$ (MPa)	3.97	2.87	2.72	3.19
Toughness $_{150,3.0}$ (J)	104.68	83.53	78.03	88.75

design. After the testing program determined that 32.6 kg/m³ 55 lbs per cubic yard was necessary to achieve the desired flexural performance (Figure 6).

JOINT TESTING

A custom fabricated form was manufactured by the supplier of the production segment molds; it was made of the same materials and to the same tolerances as the production molds and was equipped with an identical vibrator. This test mold produced castings equal to the design segment thickness of 35.6 centimeters (14 inches); one face of the mold had the gasket groove and stress relief pattern machined to the same dimensions as the typical segment. Also included in the mold, were interchangeable fittings to represent either a bolted, doweled, or flush connection.

The joint testing program was divided up into two phases; during the first phase, six specimens were cast without inserts, using the proposed SFRC mix design, for three

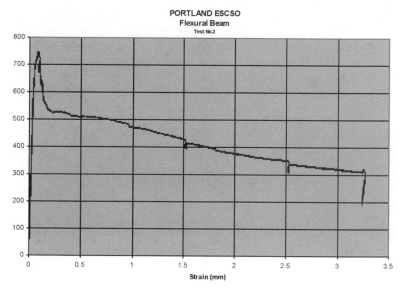

Figure 6.   Typical beam test result

tests, in the flush condition, offset and offset-rotated one degree. These initial tests were to be completed prior to commissioning of the precision planetary batch plant. These early tests would be representative of the chemistry and curing of the production SFRC segments. The second phase would use the actual SFRC mix design produced by the precision batch plant with bolt pockets or dowels as configured in the final design. By completing some of the testing early in the initial phase, the Project gained confidence in the overall SFRC approach and the designer obtained practical test results confirming the design assumptions per ACI 318.

Each physical representation, flush, bolted or doweled connection, is to be tested in each of three geometries, perfectly aligned, offset, and offset-rotated one degree. To measure the response of the SFRC, each block incorporated an array of strain gauges cast into the joint face area as well as a matrix of button strain gauges on the surface.

In short, nine pairs of specimens were to be tested in the three physical representations and each for three different geometries. During each of the nine tests, the loads would be cycled three times, generating a small encyclopedia of response data. The University of Illinois at Urbana-Champaign was selected to perform these tests based on past experience with other tunnel segment testing.

The University of Illinois' has a large-stroke, 1,360,777.1-kilogram (3,000,000-lb) test machine. Cycling each test three times checked for repeatability and confirmed that the design stresses were within the elastic range. The design criterion for the test was 365,776.9 kilograms (806,400 lbs); this was achieved and the results are illlustrated in Figures 7, 8, and 9. At the end of the three cycles of the first test, one pair of specimens was taken to ultimate failure, which occurred at 815,105.5 kilograms (1,797,000 lbs).

FIXING TESTS

In addition to testing the concrete and the actual joints for compression at full scale, it was a requirement of the contract to test a sample of the bolts and dowel fix-

Figure 7.   Joint test mould

Figure 8.   3,000,000 lb test machine

Figure 9.   Failure

ings used in the segments. Bolt tensile strength was a simple test and the specified load was achieved.

To test the dowels for tensile and shear capacity in the embedded condition, requires a more sophisticated test arrangement. The selected dowel caps were cast into concrete block with the same dimensions as the segment thickness, a dowel was pressed fully home and then a tensile or shear load was applied.

Initial results from the these tests did not meet expectations and required further tests and a re-evaluation of the design loads as well as a review of the buildability requirements.

The preliminary results for the tensile tests are shown in Figure 10.

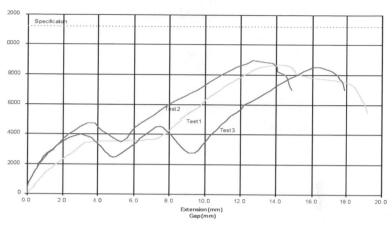

Figure 10.   Tensile test results

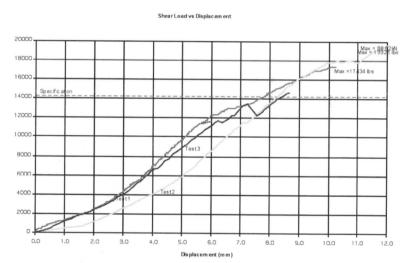

Figure 11.   Shear test results

SHEAR TEST

The shear test was carried out in double shear with the dowels driven fully home and restrained. It was felt that this configuration represented as closely as possible the actual applied loading condition in the tunnel from the erection procedure and from possible seismic loading.

The results from the three shear tests were close to the expected figures, as shown in Figure 11. The shear test is illustrated in Figure 12.

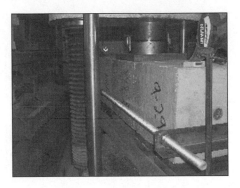

Figure 12. Shear test

PRODUCTION

A total of approximately 6,000 rings are required to be produced over 180 weeks to support the tunneling operation. A number of different scenarios and rates of production were considered but the most economical production was achieved with a segment plant designed for 8 rings per day. The option of working two shifts with fewer molds was considered but it was less economical and single production with less molds barely covered average tunnel production. More importantly, the choice to operate on a single shift basis yielded the advantage of allowing for natural curing of the segments, thus minimizing any temperature induced stresses.

The designated plant location is the Terminal 1 warehouse and surrounding dockside area in North West Portland. The warehouse is a classic timber framed structure 121.9-meters (400-foot) long with two 36.6-meters (120-foot) wide bays. The one major limiting factor is the clear height to the bottom of the timber roof trusses which is only 6.7 meters (22 feet) above finished floor. The segment plant was designed to fit inside one 121.9-meters (120-foot) wide bay of the existing building to produce 64 segments per day, working single shift. The other bay held the reinforcement fabrication plant, steel fiber storage, fitting shops and general storage as well as the materials testing laboratory. The batch plant and segment storage are located outside the warehouse.

To complete 64 segments in less than one shift, and given a batch time between three to four minutes, a batch plant capable of mixing just over one segment at a time is required. With the largest segment at just over 1.5 m^3 (2 yd^3), the batch plant selected is a Wiggert 1.7 m^3 (2.25 yd^3) planetary style high precision mixer system. It was supplied from Germany through the American agent to Wiggert, Advanced Concrete Technologies, of Portsmouth, New Hampshire. The batch plant is erected so that the discharge point from the batch hopper is just above the 6.7-meters (22-foot) level. It discharges into a transfer bucket running from outside of the building to above the concrete delivery crane which can be positioned directly below the transfer bucket.

The production bay in the warehouse is split into two bays, casting and curing. Each bay contains a bridge crane equipped with vacuum lifters for the handling of segments. The casting bay is also equipped with the concrete delivery crane; which consists of a second bridge crane with an integrated concrete bucket, operating at a lower elevation. The three bridge cranes all operate on a steel substructure of columns and beams comprising the runway structure, with the two bays utilizing common center columns. The two segment handling bridge cranes and supporting steel runway structure has been supplied by North American Industries, of Woburn, Massachusetts. The concrete delivery crane has been supplied by Kubat GmbH, of Altshausen, Germany.

Figure 13. Production bay

Of special note is the size of the concrete delivery crane. KBB specified that the concrete capacity of the delivery crane should be two batches, or two segments. By making the delivery crane capacity twice as large as a segment volume, the number of cycles the delivery crane is required to make down the casting bay is cut from 64 to 32 each shift. Each travel of a fully loaded crane can thus fill two segments. In addition the transfer bucket under the batch plant has a capacity of 3.4 m^3 (4.5 yd^3) and acts as a holding hopper to transfer the two batches of 1.7 m^3 (2.25 yd^3) such that the concrete delivery crane does not have to wait for the batch plant to complete the second batch.

Segment molds are supplied by Euroform S.r.L, of Gessate, Italia. Through Euroform, the Joint Venture also obtained both intrados and extrados vacuum lifting devices, manufactured by ATA—Produzione Ventose S.r.L, of Bientina, Italia. Euroform also provided the custom built segment flipping table, which was positioned between the casting and curing bays at the opposite end of the bays from the batch plant. This means that segment lifting is always moving away from the concrete delivery to prepared molds allowing parallel production operations (Figure 13).

The cure bay is parallel and adjacent to the casting bay. The purpose of the cure bay was to allow for strength gain of the segments in a controlled environment for an additional day before dispatch to the storage yard. During this extra day on the production floor, gaskets will be attached and any minor repairs performed.

Once the segments reach two day maturity and the gaskets are applied, the bridge crane returns to lift and load them unto custom built trailers. Pairs of trailers are pulled via tractor to the outdoor storage yard. The tractor and trailer equipment and their turning radii are inspired by airport baggage handling tugs and trailers and have proved efficient on previous operations. It was an important part of the concept that no forklifts were to be used to handle the segments at any time in order to minimize damage.

Once in the storage yard, a 0.9 ton (metric) [40 ton (short, US)] Mi-Jack rubber tired gantry crane is used to lift segments off the trailer and onto precast concrete cradles arranged on the ground. The cradles are designed to hold a complete ring in a single stack. The device used for lifting segments on or off trailers is a custom built four legged mechanical clamp. This device was sized to give the tightest ring to ring stacking arrangement possible and has the capacity to lift half a complete ring at a single lift onto the delivery trucks. Hebetechnik Porath GmbH of Kevelaer, Germany, provided the design and fabrication of the lifting clamps. Once 28 day age is obtained, the same gantry crane equipped with the same mechanical clamp lifts four segments at a time to load street legal tandems for delivery to the operations shaft.

QUALITY ASSURANCE

In order to control the quality of the production a comprehensive QA system was introduced covering all aspects of the manufacturing process. Obviously, with the production of SFRC segments, a critical component to a successful quality management program is the validation of flexural strength. Therefore the Joint Venture decided upon equipping an on site laboratory with both compression and flexural testing equipment. Of specific note is the specially designed Instron Corporation of Norwood, Massachusetts, flexural beam test machine with a bespoke software control system. This extensive laboratory enables a high degree of production control to be achieved with frequency of testing well in excess of that called for in the contract.

SUMMARY

At the time of writing the segment fabrication facility erection is complete and batch plant commissioning is nearing completion. Full scale joint testing program has completed the first phase, and casting has begun for the second phase, including the segment hardware. The hardware tests have been completed.

REFERENCES

American Concrete Institute, 1996, State-of-the-Art Report on Fiber Reinforced Concrete, Reported by ACI Committee 544, ACI 544.1R-96.

American Concrete Institute, 2002, Building Code Requirements for Structural Concrete, Reported by ACI Committee 318, ACI 318-02.

German Concrete Association, 1992, Design Principles of Steel Fibre Reinforced Concrete for Tunneling Works, DBV-Recommendation.

Hashash Y.M., J.J. Hook, B. Schmidt and J. I-Chiang Yao, 2001, Seismic Design and Analysis Underground Structures, Tunneling and Underground Space Technology, 16.

King M.R., 2005, The Design and Use of Steel Fiber Reinforced Concrete Segments, Proceedings of the Rapid Excavation and Tunneling Conference, Seattle, Washington.

Muir Wood A.M., 1975, The Circular Tunnel in Elastic Ground, Geotechnique, 25.

RILEM TDF-162, 2000, Test and Design Methods for Steel Fiber Reinforced Concrete—Method, International Union of Laboratories and Experts in Construction Materials, Systems and Structures.

Wu, C.L. and Penzien, 1997, Stress Analysis and Design of Tunnel Linings, Proceedings of the 1997 Rapid Excavation and Tunneling Conference.

DEVELOPMENT OF STEEL-FIBER REINFORCED HIGH FLUIDITY CONCRETE SEGMENT AND APPLICATION TO CONSTRUCTION

Hiroshi Dobashi

Metropolitan Expressway Co. Ltd.

Mitsuro Matsuda

Metropolitan Expressway Co. Ltd.

Yoshinari Kondo

Obayashi Corp.

Aki Fujii

Obayashi Corp.

INTRODUCTION

The Metropolitan Expressway Central Circular Route with a total length of approximately 47 km is located outside the Inner Circular Route of Tokyo (at a radius of approximately 8 km from the city center). Sections with a length of approximately 26 km, slightly less than 60% of the total length, on the east and north sides are in service. Completing the Central Circular Route (Figure 1) is expected to enable through traffic concentrated at the city center to get around, and to form a well balanced transport network, which will alleviate traffic congestion considerably. The Central Circular Shinjuku Route, a section with a length of approximately 11 km on the west side, is now being constructed at a fast pace. The route is being built mainly by the shield tunneling method to carry out work smoothly under a trunk highway carrying heavy traffic and to reduce adverse effects of tunneling work on residents in the neighborhood.

Shield tunnels have recently been becoming longer and deeper, and no secondary linings have been used in an increasing number of cases because of the overcrowding of cities or the reduction of budgets for public works. Economical, durable and high-quality segments have been required for linings of shield tunnels. Segments are made of concrete, steel, cast iron or combination thereof. Concrete segments are generally used in shield tunnels of medium to large diameter. Concrete segments are highly durable and resistant to compression, so they are highly resistant to earth pressure, water pressure and jack thrust. They are, however, heavy and have low tensile strength, so they are brittle and their corners are likely to spall. Concrete segments should therefore be handled carefully.[1]

The authors have developed a steel fiber reinforced highly flowable concrete segment (referred as SFRC segment) that overcomes the weaknesses of concrete segments[2] to construct economical, durable and high-quality linings. SFRC segments have been applied to outbound sections SJ51 through SJ53 of the Central Circular Shinjuku Route. Unlike conventional concrete segments (referred as conventional RC segments), SFRC segments are made of steel fiber reinforced self-compacting highly flowable concrete. SFRC segments require less reinforcing bars than conventional RC segments because steel fibers are expected to work as reinforcements.

Steel fiber reinforced concrete is a composite material that overcomes the brittleness of plain concrete, and has higher resistance to bending, tension and shear than

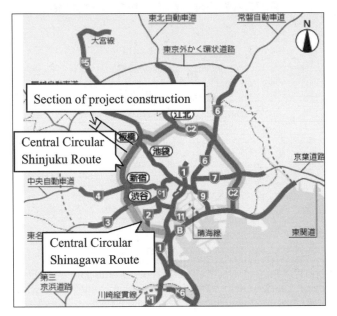

Figure 1.   Project outline

conventional concrete. As uniformly distributed steel fibers provide great deformability and prevent concrete spalling, steel fiber reinforced concrete has been used for shotcrete, secondary linings in mountain tunnels[3] and cast-in-place concrete for the extruded concrete lining (ECL) method. As the steel fibers exposed on the surface of the concrete protect embedded steel fibers and reinforcement from corrosion,[4] SFRC is expected to be a durable and high-quality material for tunnel linings. As highly flowable concrete requires no compaction and contributes to the saving of labor for segment manufacturing, technologies have recently been developed to use highly flowable concrete for segments.[5] Using the characteristics described above, we applied SFRC segments to primary linings in a shield tunnel for the first time as economical and durable concrete segments.

In sections SJ51 through SJ53, a tunnel is constructed over a length of approximately 2,020 m on the Central Circular Shinjuku Route using one of the largest mud shields in the world. No secondary lining is applied in the tunnel. Specifications for the tunnel are listed in Table 1. SFRC segments are used in a 90-m-long section (in 60 rings).

This paper first describes the characteristics and structure of SFRC segment as a lining material in a shield tunnel. Explained next are the structural performance and fire resistance performance of SFRC segments identified in full-scale performance tests. Finally, the first application of SFRC segments to a shield tunnel with an outer diameter of 11.8 m is discussed.

Table 1.   Tunnel specifications

Project	Tunnel construction in work sections SJ51 through SJ53 (clockwise route)
Owner	Metropolitan Expressway Company Limited
Outer Diameter of Shield	ϕ 12.02m
Tunnel Length	2,018.8m
Primary Lining	RC segments: 811 rings (SFRC segments: 60 rings) Outer diameter; 11,800mm Girder height: 450 mm. Width: 1,500 mm Ductile segments: 302 rings Outer diameter: 11,800mm Girder height: 400 mm. Width: 1,200 mm Steel segments; 287 rings Outer diameter: 11,800mm Girder height; 400 mm. Width: 1,500 mm
Soil Type	Musashino gravel layer, Tokyo layer, Tokyo gravel layer, Kazusa group
Depth of Ground Cover	7.3 to 23.4 m

Photograph 1.   Steel fibers (bundles with hooks at both ends)

Photograph 2.   SFRC segment

OUTLINE OF SFRC SEGMENT

Characteristics of SFRC Segment

SFRC segments are made of short steel fibers (Photograph 1) and self-compacting highly flowable concrete (Photograph 2). SFRC segments make it possible to reduce main reinforcement and eliminate transverse reinforcement and hoop reinforcement. Figure 2 shows the structural outlines of SFRC segment and conventional RC segment. SFRC segments have the following characteristics.

- Higher durability and water cutoff
  - Uniform distribution of steel fibers reduces the crack width because cracks are dispersed more uniformly.
  - Concrete delamination or spalling on the edge of a segment is prevented because steel fibers are distributed evenly in corners or on the edge.

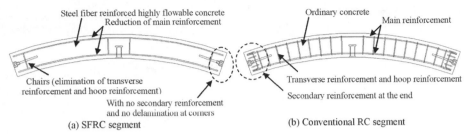

Figure 2. Comparison of segments

- Higher structural performance
  - With steel fibers mixed, tensile strength of concrete can be counted on in structural analysis. Then, the amount of main reinforcement can be reduced.
  - Steel fibers increase shear strength of concrete and ensure stress distribution along the width. As a result, no transverse reinforcement is required.
  - With steel fibers mixed, reinforcement is expected to be constrained by steel fibers. No hoop reinforcement is therefore required.
- Resistance to fire
  - Depth of explosive fracture is smaller in SFRC segments than in conventional concrete or highly flowable concrete segments. SFRC segments have greater residual strength after they are exposed to fire.
- Simpler process of segment manufacturing
  - Reduction of main reinforcement and elimination of transverse reinforcement and hoop reinforcement contribute to a simpler process of reinforcement assembly.
  - Using highly flowable concrete eliminates the need of vibrating tables. Then, the weight of formwork can be reduced. Segment production cost can be reduced further in the future by adopting on-site manufacturing of segments.

Materials Used

Concrete materials and mix proportions of SFRC segments and conventional RC segments used in the project are shown in Table 2. The design strength of concrete was f'ck = 48 N/mm^2 for both types of segments.[6]

Steel fibers with a diameter of 0.6 mm and a length of 30 mm were equipped with hooks at both ends and bound together (Photograph 1) for uniform distribution of fibers, self-filling performance and easy handling. A volumetric percentage of steel fibers was set to be 0.8% based on the results of compressive strength, flexural strength and shear strength tests conducted at varying mix proportions.[7]

OUTLINE OF PERFORMANCE VERIFICATION TESTS

When applying steel fiber reinforced concrete to segments in shield tunnels, tests are required to verify whether the designated performance can be achieved or not while fully considering the load bearing mechanism and structural characteristics of segments.

For developing the SFRC segment, we conducted the tests listed in Table 3. This paper describes in detail the tests shown in the shaded boxes in the table.

Table 2. Concrete mix proportions

Type of Concrete	Maximum Diameter of Coarse Aggregate (mm)	Slump (cm)	All-Content (%)	W/P (%)	s/a (%)	Unit Volume (kg/m^3)					Admixture	Steel Fiber Content	
						W	C	F	S	G		Volumetric Percentage (%)	Unit Volume (kg/m^3)
SFRC segment	20	67.5±5.0	3.0±1.5	29.0	67.4	180	310	310	1,037	501	SP	0.8	63
Conventional RC segment	20	3.0±1.5	2.0±1.0	29.0	41.0	125	215	215	765	1,116	AE	—	—

Table 3. List of verification tests

Objective of Development of SFRC Segment	Test	Results of Evaluation
Higher durability and water cutoff	Unit test to verify flexural strength	Cracks were found to be distributed in specimens.
	Verification of filling with and distribution of steel fibers	Steel fibers were found to be distributed uniformly based on the X-ray photographs.
	Thrust test	No cracks occurred due to shield jack thrust during excavation.
Increased structural performance	Bending or compression rupture test	Segments were found to have higher deformability than conventional type even where main reinforcement was reduced.
		Steel fibers were found to be effective for constraining reinforcement. (Eliminating hoop reinforcement is possible.)
	Unit test to verify flexural strength	Segments were found to have the same strength as conventional type even where main reinforcement was reduced.
	Main reinforcement stress distribution test	Main reinforcement was found to be distributed more uniformly than in conventional type even where no transverse reinforcement was applied. (Eliminating transverse reinforcement is possible.)
	Ring joint shear test	Joints were found to have increased shear strength and be effective for transmitting shear.
	Joint pull-out test	Effectiveness for distributing stress around joints owing to the mixing of steel fibers was verified.
	Joint bending test	In the case only with main reinforcement and steel fibers, joints maintained strength until the crushing of concrete on the convex side (in compression) after the yielding of anchor bars as designed.
Ensuring of fire resistance	Fire resistance test	Thermal history of internal concrete was found to be equivalent to that in the conventional type where the specimen was coated with fireproof materials.
		Depth of explosive fracture was found to be smaller in SFRC segments without fireproof coating than in conventional type. SFRC segments had greater residual compressive and flexural strengths when heated.

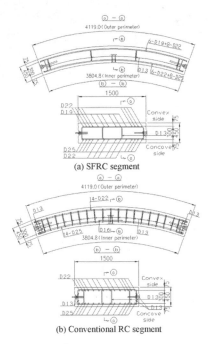

(a) SFRC segment

(b) Conventional RC segment

Figure 3.   Reinforcement arrangements in segments

FULL-SCALE PERFORMANCE TEST OF SFRC SEGMENTS

Bending Test[8]

Outline and Objective of Test. The objective of the bending test is to verify whether or not SFRC segments with reduced main reinforcement relying on the reinforcing effect of steel fibers have flexural strength equivalent to or higher than that of conventional RC segments. A Type-A segmental ring evenly divided into nine sections with an outer diameter of 11,800 mm, a girder height of 450 mm and a width of 1,500 mm, of the same shape as the primary lining in the tunnel in work sections SJ51 through SJ53 (outbound route) is used in the test for applying of the segmental ring in actual construction. The area of main reinforcement in SFRC segmental ring is reduced by 10.6% from the conventional RC segmental ring while relying on steel fibers' effectiveness for reinforcing. For main reinforcement assembly, four hoop-shaped reinforcing bars are arranged inside the main reinforcement on the outside and inside of the segmental ring. Reinforcement arrangements in SFRC and conventional RC segmental rings are shown in Figure 3. Areas of reinforcement are shown in Table 4.

For determining the area of main reinforcement in the SFRC segment, tensile stress that steel fiber reinforced concrete is counted on to resist is calculated by equation (1) and flexural strength calculated by equation (2). The area of main reinforcement is set so that the flexural strength can be equivalent to that of conventional RC segments. For the tensile stress that steel fiber reinforced concrete is counted on to resist, a mean tensile stress during a period until the crack width at the extreme tension fiber reaches the marginal crack width wk along the tension softening curve[9] (Figure 4) is assumed.

Table 4. Area of reinforcement in segments

	Item	RC Segment	SFRC Segment
Area of reinforcement	Main reinforcement on the outside	14-D22	8-D22 + 6-D19
	Main reinforcement on the inside	14-D25	8-D25 + 6-D22
	Area ratio of main reinforcement	100%	89.4%
	Relative weight of reinforcement (steel fibers are considered)	100%	66.0% (96.7%)

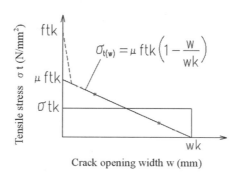

$$\sigma_{t(w)} = \mu\, ftk\left(1 - \frac{w}{wk}\right)$$

Crack opening width w (mm)

Figure 4.   Tension softening curve

$$\sigma_{td} = \sigma_{tk}/\gamma_s = 0.80 \ (\text{N/mm}^2)$$

$$\sigma_{tk} = \tfrac{1}{2} \cdot \mu \cdot f_{tk} = 0.84 \ (\text{N/mm}^2) \tag{1}$$

where
   σ_{td} = Tensile stress used when calculating flexural strength
   μ = Residual strength (= 0.55)[9]
   γ_s = Material factor (= 1.05)
   σ_{tk} = Tensile stress of concrete
   f_{tk} = Characteristic value for tensile strength of concrete ($0.23f'ck^{2/3} = 3.04$ N/mm^2)

For calculating flexural strength, concrete compressive stress of concrete is assumed to be distributed in the shape of a rectangle,[10] and tensile stress of concrete is also assumed to be distributed in the shape of a rectangle (Figure 5).[11]

$$Mu = C\left(\frac{h}{2} - 0.4x\right) + T_{s'}\left(\frac{h}{2} - d'\right) - \left(\frac{C_{t1}(h/2 - x)}{2} + \frac{C_{t2}(h/2)}{2}\right) + Ts\left(d - \frac{h}{2}\right) \tag{2}$$

where
   Mu = Flexural strength
   $T_{s'}$ = Resultant force of compressive stress of reinforcement
   C_{t1} = Resultant force of tensile stress of concrete on the compression side for the centroid axis
   h = Height of cross section

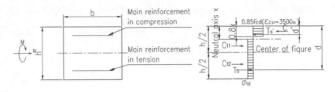

Figure 5. Flexural strength calculation model

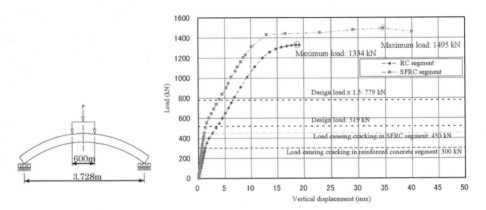

Figure 6. Test outline   **Figure 7. Load displacement**

d, d' = Effective heights
    C = Resultant force of compressive stress of concrete
    T_s = Resultant force of tensile stress of reinforcement
    C_{t2} = Resultant force of tensile stress of concrete on the tension side for the
          centroid axis
    x = Position of neutral axis

    A test outline is given in Figure 6. Point loads were applied at two points, and two
supports are allowed to move so that simple bending forces act on the specimen. The
spacing between supports is 3,728 mm and the distance between the loading points
was 600 mm.[12] Loads are applied at an interval of 25 kN until cracking occurs. Subsequently, loads are applied generally at an interval of 50 kN until failure occurs.
    Test Results and Discussions. The relationship between the load and the dis-
placement at midpoint of the specimen is shown in Figure 7. Results for the conventional RC segments used in the project are also shown in the figure for comparison.
The figure additionally provides the design load (allowable load) on conventional RC
segment and a load 50% greater than the design load.
    Cracking occurred in the SFRC segment under a load of 450 kN, and in the con-
ventional RC segment under a load of 300 kN. Then, the gradient of the curve
decreased subsequently due to the reduction in stiffness as a result of cracking. The
gradient was nearly the same for both curves until the load reached 1,000 kN. For the
SFRC segment, displacement started increasing at a load of 1,400 kN. A maximum
load of 1,495 kN was reached at a displacement of 34 mm. For the conventional RC

 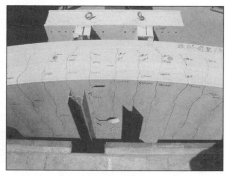

(a) SFRC segment   (b) Conventional RC segment

Photograph 3. Cracks

segment, displacement started increasing at a load of 1,200 kN. A maximum load of 1,334 kN was reached at a displacement of 19 mm. The safety factor against failure (maximum load/design load) was 2.88 for the SFRC segment and 2.57 for the conventional RC segment. In the SFRC segment, reinforcement may have yielded at a load of 1,400 kN. Displacement increased considerably after yielding. Thus, it was confirmed that mixing steel fibers enhanced the ductility of the concrete segment. Cracks after loader occurred in the specimen at the same intervals as those of transverse reinforcing bars in the conventional RC segment. In the SFRC segment on the other hand, cracks were well dispersed over the width (Photograph 3).

The test results have confirmed that SFRC segments have bending capacity equivalent to or higher than that of conventional RC segments, and are more durable.

Test to Verify Stress Distribution of Main Reinforcement Along the Width of Segment[13], [14]

Outline and Objective of Test. In the test, in consideration of the effectiveness of a staggered arrangement of segmental rings, stress distribution along the width of the segment is verified while the segment is treated as a beam member that resists shear force, which is transmitted from radial joints. A specimen with a thickness (h) of 300 mm and a width (B) of 1,000 mm is used to reflect the width/thickness ratio B/h of the segment actually adopted in the project. The arrangements of reinforcing bars in SFRC segment and conventional RC segment are shown in Figures 8 and 9, respectively for comparison.

A test outline is given in Figure 10. Simple bending loads are applied at jigs that are connected to four inserts placed at the interface between rings using M24 (10.9) bolts. The supports at both ends are allowed to move. The spacing between supports is 2,648 mm and the spacing between the loading points is 745.2 mm.

Loads are applied at an interval of 10 kN until cracking occurs. Subsequently, loads are applied generally at an interval of 20 kN until failure occurs.

Stress distribution is measured using four strain gauges attached to each inner main reinforcing bar (Figure 11).

Test Results and Discussions. Stress distributions in main inner reinforcing bars at typical three load levels are shown in Figure 12 for SFRC segment and conventional RC segment. No difference was observed between SFRC and conventional RC segments at a low load level of 100 kN. At higher load levels, however, main reinforcement bore less

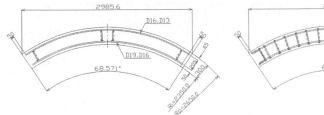

Figure 8. Arrangement of reinforcing bars in SRFC segment

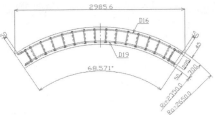

Figure 9. Arrangement of reinforcing bars in conventional RC segment

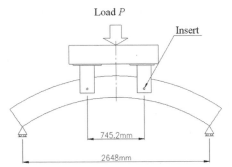

Figure 10. Test outline

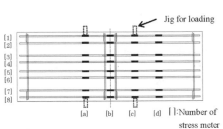

Figure 11. Positions of stress meters for reinforcement on the concave side

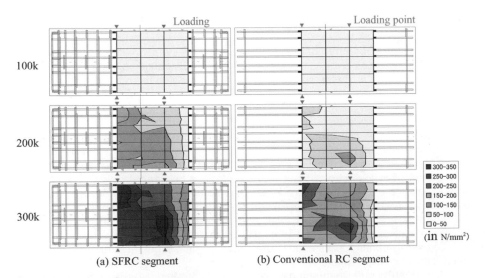

(a) SFRC segment   (b) Conventional RC segment

Figure 12. Stress distribution contours

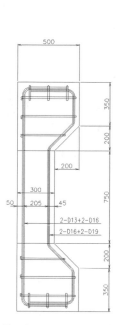

Figure 13.   Arrangement of reinforcing bars in the specimen

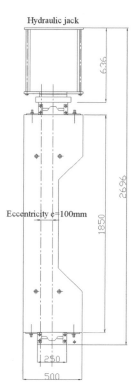

Figure 14.   Test outline

stress in the SFRC segment than in the conventional RC segment. Moderate stress concentration was found around the loading points in both types of segments. Stress was, however, transmitted to main reinforcement at the middle of the segment along the width. It is evident that uniformly and randomly mixed steel fibers provided functions of both main and transverse reinforcements. The test results show that steel fibers can provide the effectiveness conventionally provided by transverse reinforcement and that transverse reinforcement can be eliminated in SFRC segments.

Flexural Compression Failure Test[13], [14]

Outline and Objective of Test. Axial forces are generally predominant in shield tunnels. The objective of flexural compression failure test is therefore to verify the behavior at the time of compressive failure. The specimen is of a plate type with a thickness of 300 mm and a width of 500 mm. It has a thinner concrete cover for constraining compressive reinforcement than the segment used in the project (Figure 13). No hoop reinforcement is arranged at the middle. A test outline is shown in Figure 14. Bending moment and axial compression are applied by applying eccentric axial compression at the middle of the planar specimen.[15] Eccentricity is 100 mm. Loads are applied at an interval of 50 kN until cracking occurs. Then, loads are applied generally at an interval of 100 kN until failure occurs.

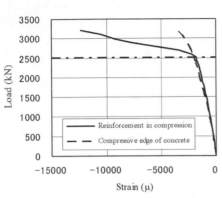

Figure 15. Load-displacement diagram

Photograph 4. Crushing of concrete

Test Results and Discussions. The relationship between the load and compressive strain of both reinforced concrete is shown in Figure 15. Compressive reinforcement yielded nearly at a load P of 2,500 kN, and concrete crushed at a maximum load P of 3,208 kN (Photograph 4). It was confirmed that compressive reinforcement did not occur buckling even after yielding because it was constrained by steel fibers and that compressive capacity was determined by the strength of concrete.

The test results have confirmed that steel fibers can be counted on to function as conventional hoop reinforcement and that hoop reinforcement can be eliminated in SFRC segments.

Verification of Placement with and Distribution of Steel Fibers[7]

Outline and Objective of Test. To maximize the strength of steel fiber reinforced highly flowable concrete, adequately distributing steel fibers is necessary. The distribution of steel fibers is verified by manufacturing a full-scale model of segment with the mix proportion described above and examining X-ray photographs of a piece cut out of the model.

Test Results and Discussions. Concrete was placed at the center of the form on the convex side (Figure 16). X-ray photographs were taken of a piece that was cut out of the segment (Photograph 5). The position of the piece is shown in Figure 16. The X-ray photographs show that steel fibers were distributed uniformly around the reinforcing bars and in the corners. No fibers formed any balls. It was thus confirmed that steel fibers were well distributed. Uniform distribution of fibers was also confirmed in pieces cut out of other areas.

Fire Resistance Test[16]

Outline and Objective of Test

1. **Case without fireproofing.** In order to verify fire resistance of steel fiber reinforced highly flowable concrete and ordinary concrete used in conventional RC segments, the depth of explosive fracture when the fire curve of RABT (Figure 17) is applied is measured using an unreinforced specimen (1,400 × 1,000 × 300) without any fireproofing. The residual strength of heated pieces was verified as compared with unheated pieces.

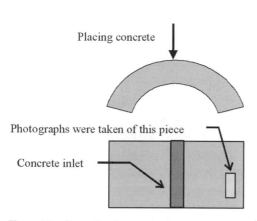

Figure 16.   Concrete placement

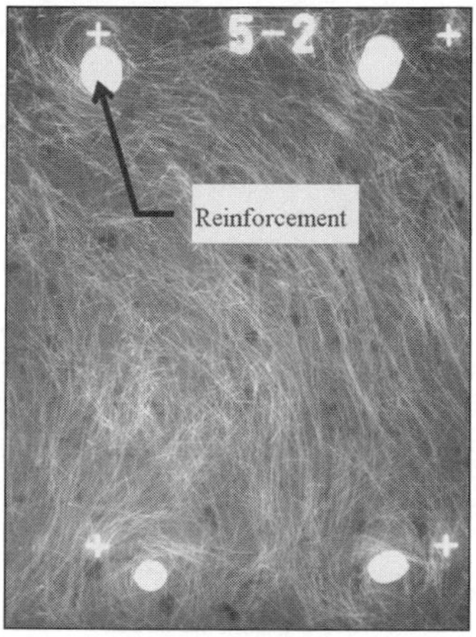

Photograph 5.   Distribution of steel fibers

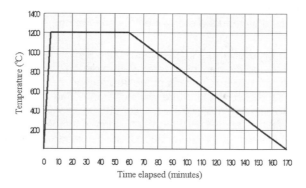

Figure 17.   RABT fire curve

2. **Case with fireproofing.** As in case (1), the steel fiber reinforced highly flow-able concrete and ordinary concrete specimens with fireproofing are sub-jected to RABT fire curve, and temperature histories of concrete surface and internal reinforcement are measured. Two types of fireproofing materials are used: type A (fireproof panel with a thickness of 27 mm) and type B (shot-crete coating with a thickness of 30 mm).

Test Results and Discussions

1. **Case without fireproofing.** The results of the heating test are shown in Figure 18. Explosive fracture occurred immediately following heating in both cases. The depth of explosive fracture was 35 mm in the steel fiber reinforced highly flowable concrete, half of 69 mm in the conventional RC. The former proved to be highly resistant to explosive fracture. This is probably because mixing steel fibers enhanced resistance to concrete delamination and because expansion pressure due to vapor pressure, a cause of explosive fracture, was dissipated out of the specimen through the steel fiber interface. Ratios of residual compressive strength and flexural strength were higher in

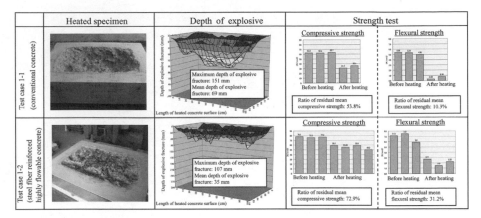

Figure 18.   Results of fire resistance test (without fireproof coating)

the steel fiber reinforced highly flowable concrete than in after heating conventional RC. Then, it has been found that the strength of the SFRC segment is not likely to be reduced even when the specimen is subjected to thermal hysteresis.

2. **Case with fireproofing.** Figure 19 and Table 5 show respectively changes in temperature and the maximum temperature during the heating test on the surface of concrete on the back of the fireproofing, in internal reinforcement, on the surface of concrete at the joints of fireproofing and in ceramic inserts. In the case with fireproofing, the temperature of structural members was lower than the allowable level (concrete surface temperature: 350°C, reinforcement surface temperature: 300°C) at all measurement points. It has been confirmed that applying fireproofing considering the construction conditions could ensure fire resistance.

PRACTICAL APPLICATION OF SFRC SEGMENTS

As previously described, the structural performance of SFRC segments was verified in full-scale tests. Then, 60 rings of SFRC segments were used in the field in work sections SJ51 through SJ53 (outbound route). Photograph 6 shows assembled SFRC segmental rings in the tunnel.

Shown in Table 6 are structural drawings and materials used for SFRC segments and reinforced concrete segments adopted in the project. In the SFRC segment with reduced main reinforcement and no transverse reinforcement, the total quantity of reinforcement was 34% smaller than in the conventional RC segment. The rate of reduction was approximately 11% for main reinforcement and 87% for transverse reinforcement. Since only 60 rings of SFRC segments were made in the project, the formwork for conventional RC segments was used without reducing the weight. Visual inspection of the surface of SFRC segment after the assembly of a segmental ring found that concrete was well placed throughout the segment and neither the loss of concrete in corners nor cracking occurred. Steel fibers were distributed to the edge of SFRC segment, so the concrete in the SFRC segment had greater resistance to tension than that in the conventional RC segment. Concrete was therefore unlikely to spall in corners.

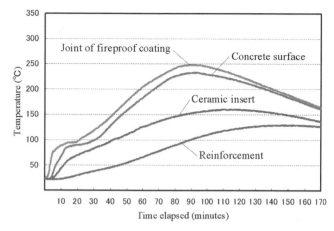

Figure 19.   Results of measurement of member temperature

Table 5.   Maximum temperature at each measurement point

Type of Fireproof Coating		Test Case	Concrete Surface	Reinforcement	Joint of Fireproof Coating	Ceramic Insert
Type A (thickness: 27.5 mm)	2-1	Conventional concrete	233	130	249	161
	2-3	Steel fiber reinforced highly flowable concrete	238	115	239	156
Type B (thickness: 30 mm)	2-2	Conventional concrete	180	111	183	137
	2-4	Steel fiber reinforced highly flowable concrete	169	93	172	134

 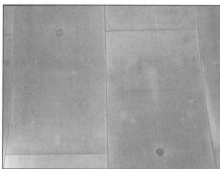

Photograph 6.   Assembled SFRC segmental rings

Stress in the main reinforcement was measured in the field to verify the stress distribution in SFRC segments assembled. Stress in main reinforcement, earth pressure and water pressure were measured in a ring of reinforced concrete segments next to an SFRC segmental ring to which measuring instruments were attached to compare stress in SFRC and reinforced concrete segmental rings and to identify external loads

Table 6. Comparison of structures

Structure		RC Segment						SFRC Segment					
Structural Drawing													
Concrete	Design strength	48N/mm²						48N/mm²					
	Steel fiber content	0%						0.8%					
	Slump	3.0 ±1.5cm						67.5 ±5.0cm					
	Air content	2.0 ±1.5%						3.0 ±1.5%					
	Maximum size of coarse aggregate	20mm						20mm					
	W/(C + Sg)	29.1%						22.2kg (D13)					
	S/a	41.0%						67.4%					
	Quantity per unit value (kg/m³)	SF	W	C	Sg	S	G	Ad	SF	W	C	Sg	S	G	Ad
		—	125	215	215	765	1,116	3.44	63	180	310	310	1,037	501	5.39
Reinforcement	Quantity of reinforcement on the concave side	209.5kg (14-D25), As = 7,093.8mm²						188.3kg (8-D25 + 6-D22), As = 6,376.2mm²					
	Quantity of reinforcement on the convex side	169.0kg (14-D22), As5,419.4mm²						150.2kg (8-D22 + 6-D19), As4,815.8mm²					
	Quantity of transverse reinforcement and secondary reinforcement	167.8kg (D16, D13)						22.2kg (D13)					
	Total quantity of reinforcement	546.3kg						360.7kg					

NOTE: SF = Steel fiber; W = Water; C = Cement; Sg = Blast furnace slag; S = Fine aggregate; G = Coarse aggregate; Ad = Admixture.

charged to the segments. Ground settlement was also measured at the same location using multi-layer settlement gauges. Measurement points and the positions of stress gauges for reinforcement in the SFRC segment are shown in Figures 20 and 21, respectively.

Changes in earth and water pressure gauge readings after the assembly of the ring are shown in Figures 22 and 23. Earth pressure increased to a maximum value of 0.6 MPa due to the load applied by the tail brush and the loads during construction work such as backfilling with grout. Earth pressure reading reached approximately

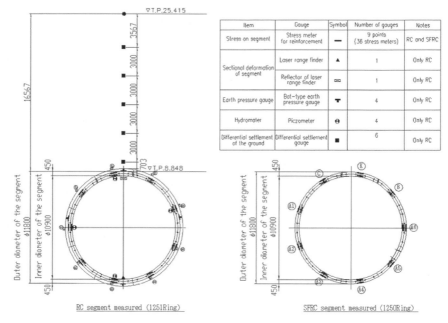

Item	Gauge	Symbol	Number of gauges	Notes
Stress on segment	Stress meter for reinforcement	—	9 points (36 stress meters)	RC and SFRC
Sectional deformation of segment	Laser range finder	▲	1	Only RC
	Reflector of laser range finder	▭	1	Only RC
Earth pressure gauge	Bat-type earth pressure gauge	▼	4	Only RC
Hydrometer	Piezometer	⊖	4	Only RC
Differential settlement of the ground	Differential settlement gauge	■	6	Only RC

RC segment measured (1251Ring)

SFRC segment measured (1250Ring)

Figure 20. Cross section for measurement

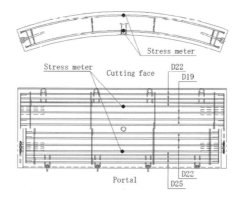

Figure 21. Stress meters for reinforcement installed in SFRC segment

0.17 MPa at the tunnel crown and became stable beyond the 15th day after ring assembly. Water pressure remained nearly constant. Water pressure was 0.11 MPa at the tunnel crown. Vertical earth pressure was therefore assumed to be approximately 0.06 MPa. Design vertical load at the tunnel crown was 0.24 MPa due to full overburden depth and the water pressure was 0.08 MPa. It was found that actual earth pressure was lower than the design value, and that water pressure had predominant effects on the behavior of the ring.

Stress in reinforcement was measured using the stress gauges installed on the top (K), on sides (A6 and A1) and at the bottom (A3) of the ring. Changes in stress gauge reading with time are shown in Figure 24. Figure 25 shows distributions of

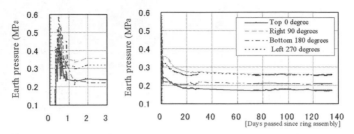

Figure 22.   Earth pressure

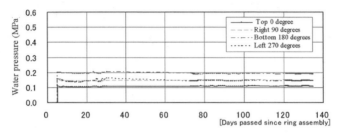

Figure 23.   Water pressure

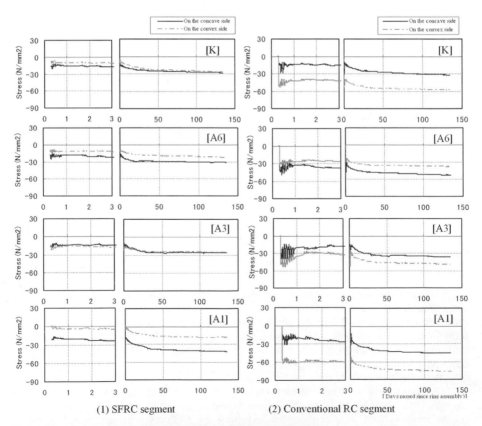

(1) SFRC segment   (2) Conventional RC segment

Figure 24.   Changes in stress meter reading with time reinforcement

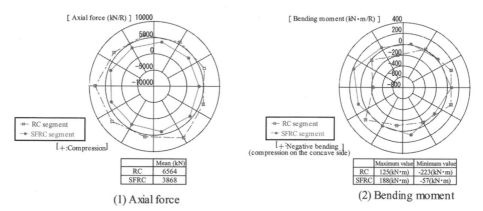

(1) Axial force

(2) Bending moment

Figure 25.   Stress resultant

stress obtained from the stress gauges installed in the ring when measurements were stabilized (on the 135th day after ring assembly).

Figure 24 shows that all the stresses in reinforcement were compressive. It is evident that stresses changed after ring assembly due to the effects of shield thrust, pressure of the tail seal and the loads during construction work such as backfilling with grout. Then, compressive stress gradually increased and became constant about 50 days after ring assembly. Since the tendency was the same either in the SFRC or conventional RC segment, stress distribution may have been similar. Positive flexural stress (tension on the inside surface at the crown and the bottom) and negative flexural stress (compression on the inside surface) at the spring line occurred in both types of segments in a nearly same mode. Changes in stress in reinforcement, the difference in stress between inner and outer rebars and the stress resultant were smaller in the SFRC segment than in the conventional RC segment.

Then, it was confirmed that SFRC segments can provide performance equivalent to or higher than that of conventional RC segments in field applications.

CLOSING REMARK

It has been confirmed that the SFRC segment, in which the quantity of main rebars is reduced and transverse and hoop rebars are eliminated in consideration of the reinforcement effects of steel fibers, is equivalent to or better than the conventional RC segment in terms of load bearing capacity, restraint on compressive rebars and effect of transmitting stress along the segment width. SFRC segments have been applied to a project. Uniform distribution of steel fibers has been verified based on the observation of neither spalling nor cracking at corners, and the performance of SFRC segments has been found to be equivalent to or better than that of conventional RC segments based on the stress gauge measurements of rebars. SFRC segments are expected to be used widely as shield tunnel linings because of their unique performance. Studies will be made to reduce the cost of manufacturing SFRC segments by manufacturing them at the project site.

REFERENCES

[1] Japan Society of Civil Engineers: Standard Specifications for Tunnel—Shield tunneling method, July 1996.

[2] Aki Fujii et al.: Outline of development of steel fiber reinforced highly flowable concrete segments, Proceedings of the 59th Annual Conference of Japan Society of Civil Engineers, 6-001, September 2004.

[3] Toshio Yasuda et al.: A consideration on the characteristics of steel fiber reinforced highly flowable concrete applied to secondary linings, Proceedings of the 58th Annual Conference of Japan Society of Civil Engineers, Section 6, pp. 7–8, September 2003.

[4] Kazusuke Kobayashi et al.: Effectiveness of steel fiber reinforced concrete for preventing the corrosion of reinforcement in oceans, Proceeding of Japan Society of Civil Engineers, No. 414, V-12, pp. 195–203, February 1990.

[5] Kazunori Hanami et al.: Development of highly flowable concrete segments (1)—Manufacturing system, Proceedings of the 53rd Annual Conference of Japan Society of Civil Engineers, VI-25, pp. 50–51, October 1998.

[6] Takumi Uehara et al.: A study on mix proportion design for highly flowable steel fiber reinforced concrete, Proceedings of the 57th Annual Conference of Japan Society of Civil Engineers, Section 5, pp. 991–992, September 2002.

[7] Tomoko Urano et al.: Strength of steel fiber reinforced highly flowable concrete and applicability to reinforced concrete segments, Proceedings of the 59th Annual Conference of Japan Society of Civil Engineers, 6-002, September 2004.

[8] Aki Fujii et al.: Bending test for steel fiber reinforced highly flowable concrete segmental ring with an outer diameter of 11.8 m, Proceedings of the 60th Annual Conference of Japan Society of Civil Engineers, 6-118, September 2005.

[9] Japan Society of Civil Engineers: Concrete library 97 Design guidelines for steel fiber reinforced highly flowable concrete column members (draft), November 1999.

[10] Japan Society of Civil Engineers: Standard Specifications for Design and Construction of Concrete Structures—Structural performance check, March 2002.

[11] Japan Iron and Steel Federation: Design and construction manual for steel fiber reinforced concrete Tunnels, second version, Gihodo Shuppan publishing, November 2002.

[12] Japan Sewage Works Association: Standard segments for shield tunneling, July 2001.

[13] Masahiro Yoshida et al.: Full-scale performance test of steel fiber reinforced highly flowable concrete segments, Proceedings of the 59th Annual Conference of Japan Society of Civil Engineers, 6-003, September 2004.

[14] Masahiro Yoshida et al.: Full-scale performance test of steel fiber reinforced concrete segments, Proceedings of Tunnel Engineering, Japan Society of Civil Engineers vol. 14, pp. 151–158, November 2004.

[15] Japan Tunneling Association: Report on full-scale performance verification tests of segments used for the Tokyo Wan Aqua-Line 1990, May 1991.

[16] Masataka Hayashi et al.: Fire resistance test for steel fiber reinforced highly flowable concrete segments, Proceedings of the 59th Annual Conference of Japan Society of Civil Engineers, 6-323, October 2004.

SEISMIC RESPONSE OF PRECAST TUNNEL LININGS

Gary J.E. Kramer
Hatch Mott MacDonald

H. Sedarat
SC Solutions

A. Kozak
SC Solutions

A. Liu
Hatch Mott MacDonald

J. Chai
Santa Clara Valley Transportation Authority

ABSTRACT

Precast concrete segmental lining systems have been gaining widespread use in seismically active areas and a single-pass version of such a system has been developed for use on the Silicon Valley Rapid Transit (SVRT) Project in San Jose, California. The response of tunnel linings to seismic shaking is often predicted using engineering judgment, elastic closed-form solutions and numerical modeling. Such analyses generally assume a continuous structural system and hence do not explicitly model the behavior of the jointed segmental lining.

A number of analyses were performed to predict radial and circumferential joint behavior during seismic ovaling and wave propagation. These analyses included a complex, three-dimensional finite element based ground-structure interaction model that incorporates inelastic constitutive soil behavior, cracked concrete properties and no-tension, frictional segment joint surfaces. The paper discusses the results of the analyses and provides guidelines for the design of precast systems in zones of high seismicity.

BACKGROUND

SVRT Project

The Santa Clara Valley Transportation Authority (VTA) intends to construct the Silicon Valley Rapid Transit (SVRT) Project in San Jose, California. This will be 26.2 km (16.3 mile) extension of the Bay Area Rapid Transit (BART) heavy rail rapid transit system from its planned terminus at the end of the Warms Springs Extension in Fremont, to San Jose. The SVRTP will be constructed and owned by VTA but operated and maintained by BART. To facilitate this, the SVRT system will be designed and constructed in accordance with the requirements of the BART Facilities Standards (BFS).

This alignment includes an 8.2 km (5.1 miles) long Tunnel Segment consisting of twin bored tunnels and cut-and-cover structures through downtown San Jose (see Figure 1). The remaining 6,700 m (22,800 feet) of the alignment will be twin tunnels constructed by a closed face tunnel boring machine (TBM) to interconnect the stations

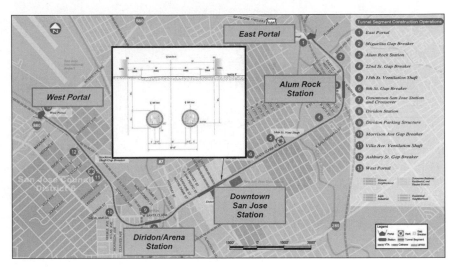

Figure 1.   Plan view of SVRT tunnel segment alignment

and portals. A joint venture consisting of Hatch Mott MacDonald of Pleasanton, California and Bechtel Infrastructure of San Francisco, California has been retained by the
VTA to prepare the design and manage the construction of the SVRT Tunnel Segment.
The Preliminary Engineering Phase has been completed and the project is now moving into final design.

The VTA has decided to proceed with the advance purchase of two 6.2 m (20 foot
4 inch) diameter TBMs and the tunnel lining system and provide those items to the
main tunnel contractor as part of their contracting strategy for the project (Biggart
et al., 2005). The TBMs will be used to install a single pass lining system consisting of
gasketed, pre-cast concrete segmental rings in lightly over-consolidated alluvial soils.
Typical ground cover above the tunnel crown will be 12 m (40 feet) but may increase to
as much as 27 m (90 feet) to avoid specific obstructions such as bridge and retaining
wall foundations.

Regional Geology and Ground Conditions

The Tunnel Segment of the SVRT Project is located in the Santa Clara Valley,
which is bounded by San Francisco Bay to the north, the Diablo Range to the northeast and the Santa Cruz Mountains to the southwest. The valley is covered by alluvial
fan, levee and active stream channel deposits with marine estuary deposits along the
Bay margins. The Tunnel Segment is located on alluvial deposits that are underlain, at
depths greater than 300 m (1,000 feet) by Tertiary-age bedrock.

The alluvial deposits within the depth of the bored tunnel are variable across the
alignment but generally consist of lightly over-consolidated stiff silty clay with some thin
sand layers interbedded within. This clay layer is commonly separated into an upper
and lower clay by the highly permeable upper aquifer which can be described as a
granular layer with either dense silty sand or silty gravel. Below the upper aquifer is a
highly variable interlayering of the clay and granular sand and gravel strata that corresponds to the depths of the cutoff wall toe. The clay layers within these interlayered
strata serve as the major aquitard to the lower aquifer which is well below the bottom of

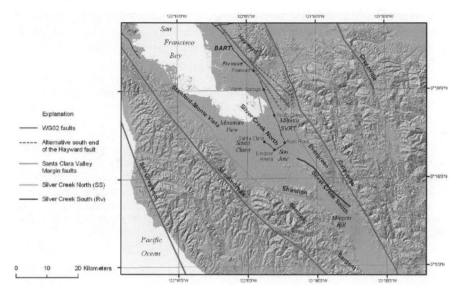

Figure 2.   Detailed map of faults in the Santa Clara Valley

the cutoff wall. The ground water table varies along the alignment but is typically 3 to 5 m (10 to 15 feet) below ground surface.

Regional Seismicity and Faulting

San Jose is located in the San Francisco Bay area, which is a highly active seismic region, bounded by the San Andreas Fault to the west and the Hayward and Calaveras Faults to the east (see Figure 2). Each of these faults has produced damaging earthquakes in the past and thus dominate the ground motion shaking hazard. Their maximum magnitude events are defined as follows: San Andreas (Magnitude 8.0), Hayward (Magnitude 7.25), and Calaveras (Magnitude 7.0). The only known occurrence of a fault intersection with the tunnel alignment occurs at the northern reach of the Silver Creek Fault. A study concluded that the potential for offset of this fault is negligible.

SVRT Segmental Lining Design

Two geometric configurations were developed during the preliminary engineering effort—rectangular and trapezoidal. A detailed comparison of the two systems was performed including an evaluation of their predicted seismic behaviors. Aspects of each configuration include:

- 5.436 m (17 foot, 10 inch) internal diameter, 1.510 m (5 foot) long, 254 mm (10 inch) thick,
- Double taper of 25mm (1 inch) to accommodate minimum 300 m (1,000 foot) alignment radius curve,
- Partial penetration grout ports to encourage tail shield grouting,
- Packers on all radial and circumferential joints, rebar reinforcement, vacuum lifting method,
- Bolts and guide rods at radial joints, connecting dowels at circumferential joints,

- Single EPDM gasket and foam strips located near the extrados of the ring and
- Bolted radial joints staggered to avoid erection distortion and cruciform joints at gaskets.

Rectangular Segment Configuration. Rectangular segments and consist of five 60-degree segments, one 40-degree segment, and a 20-degree key. There are 24 dowels per ring at the circumferential joints centered at 15-degee spacing.

Trapezoidal Segment Configuration. Trapezoidal/parallelogram segments consist of four 67.5-degree segments and two 45-degree segments. There are 16 dowels per ring at the circumferential joints centered at 22.5-degee spacing.

The trapezoidal arrangement is a universal ring and rectangular configurations included a left/right ring and a universal configuration. Fiber reinforced segments were considered but were found to be uneconomic for seismic conditions requiring too many segments to make up a ring. The project team is currently evaluating the use of a hybrid reinforcement (rebar plus fiber) segment. Reinforcement in the segments consists of six No. 4 bars (on each face) with minimum yield strength of 515 MPa (75 ksi). Specified concrete 28 day compressive strength is 41 MPa (6,000 psi).

PRECAST TUNNEL LINING (PCTL) SYSTEMS IN SEISMIC ZONES

Existing BART tunnels were mined in soft ground in the 1960s and 1970s using open-faced shields to install a single-pass lining system consisting of full circle steel segments. These structures performed well during the 1989 Loma Prieta earthquake and accordingly, the MUNI Turnback constructed in the early 1990s utilized a similar system for the relatively short length of mined tunnels. It is only in the past two decades that precast tunnel lining (PCTL) systems have gained widespread use throughout the United States including seismically active areas such as Seattle, Portland, San Diego and Los Angeles for sewage, water transport, and highway and transit applications. However, all Bay Area transit tunnels were built before the development and widespread use of PCTL in the United States.

While permitting the use of PCTL systems, the BART Facilities Standards (BART, 2005) requirements for segmental linings were more appropriate to tunnels lined with steel segments. For example, they require the use of A325 bolts along all segment joints, bolts located at 7.5 degree intervals along the circumferential joints, full moment resisting connections at radial joints and similar details. Since the SVRTP will be adopting a PCTL that utilizes connection devices differently than those used for fabricated steel construction, the SVRTP design team requested that BART permit adoption of connection details that are more compatible with the state of the practice for PCTL.

BART staff were certainly receptive to the proposed changes to permit the use of PCTL systems. However, they had little experience with the system in general and inquired about their previous use in seismic zones. Concern was expressed regarding the ability of the joints and connecting devices to maintain a stable system during earthquakes. Correspondingly, HMM/Bechtel embarked on two studies:

1. **Precast Concrete Tunnel Lining (PCTL) Seismic Performance Study.** This documented the use of PCTL in literally hundreds of applications in seismically active areas in the United States and internationally. In addition, the study summarized the observed and documented performance of PCTL systems (particularly transit tunnels) subjected to significant seismic shaking. In all cases, PCTL systems successfully withstood such events with little to no damage. For more details of this study, reference is made to Dean et al. (2006) and Young and Dean (2006).

2. **Seismic Compliance of PCTL Systems for the SVRTP.** This summarized a number of studies and calculations relating to the compliance of the lining system to meet project seismic design criteria. This included analyses of the jointed tunnel lining system under seismic ovaling and longitudinal wave propagation effects. The results of that work are summarized herein.

SEISMIC ANALYSIS OF TUNNEL LININGS

Hashash et al. (2001) presented an excellent state-of-the-art review of the design and analysis of tunnels subject to earthquake shaking. They indicated that in the absence of ground failure effects (fault rupture, liquefaction and slope instability), that the emphasis of seismic design of underground structures involves consideration of the transient ground deformation induced by seismic wave passage (shaking). Under such wave passage, underground structures undergo three primary modes of deformation during seismic shaking: compression-extension, longitudinal bending and ovaling (racking).

The design of tunnels to withstand seismic shaking is very different from the seismic design of surface structures when the emphasis is on the inertial effects of the structure itself. For most underground structures, the inertia of the surrounding soil is large relative to the inertia of the structure and hence the tunnel excavation will have negligible effect on the seismic wave induced ground displacements. Fundamental to underground seismic design is consideration of the free field deformations of the ground mass around the tunnel and its interaction with the tunnel structure. The primary principle of seismic design for tunnels is to ensure that the structure is able accommodate the movements of the ground mass associated with wave passage without excessive permanent displacement or damage to the tunnel itself.

A number of authors in the literature (Peck et al. (1972), Owen and Scholl (1981) Merritt et al. (1985), St. John and Zahrah (1987), Monsees and Merritt (1988), Wang (1993), Penzien and Wu (1998), Konagai (1998), Penzien (2000)) have presented closed form or numerically modeled solutions that predict the response of tunnels to seismic effects (Figure 3). Generally, these solutions considered the problem using similar behavioral assumptions:

- Linear elasticity of the ground mass,
- Linear elasticity of the lining,
- Idealized conditions at the ground/lining interface represented by two extremes: full-slip and no-slip,
- Separate analyses of static and seismic effects—meaning that seismically displacements and loads were separately assessed and then superimposed on the existing static state of stress in the lining and ground mass and
- Continuity of the tunnel lining (no joints).

All of the authors correctly indicate that such assumptions provide conservative estimates of the predicted internal lining loads for tunnels subjected to seismic shaking. However, over-estimation of the stiffness and the effects will tend to underestimate the strains and displacements of the system. The stability of a segmental PCTL system depends upon maintaining continuous contact and alignment along the joints between segments and between rings. It is essential that the analysis method predict both forces and displacements as accurately as possible. Consequently, a more rigorous approach than presented in the literature was felt necessary.

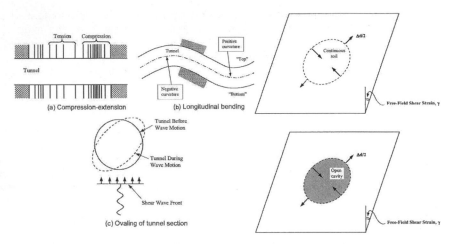

Figure 3. Seismic shaking modes and free field shear distortion in perforated and non-perforated ground ovaling (after Owen and Sholl, 1981, and Wang, 1993)

SVRT NUMERICAL MODELING

Despite documented experience of successful PCTL behavior during major seismic events, it is intuitive that overloading and damage to tunnel lining systems will occur if the magnitude of seismic straining is high enough. When considering only seismic ovaling effects, linear elastic modeling of continuous tunnel liners do predict tensile dynamic stresses. Owen and Scholl (1981) indicated that these stresses subtracted from the compressive static stresses reduce the lining's capacity and resulting stresses may be tensile.

Development of tensile strains would be problematic in a jointed tunnel liner that has relatively weak connecting devices. Therefore, to establish the stability of the SVRT lining design under seismic shaking required development of a model that could explicitly analyze and predict joint behavior failure effects. These effects include the creation of a kinematically admissible mechanism, joint opening, segment slip or fallout and the development of high shear loads on relatively weak circumferential connecting devices. Such a model would be used to analyze the PCTL system under the range of credible seismic shear strains associated with the project.

Model Development

HMM/Bechtel retained the services of SC Solutions of Sunnyvale, CA to perform seismic modeling of the SVRT tunnel lining design. Specific goals for this study included:

- Development of soil-structure interaction finite element models to predict the response of the PCTL system during seismic ovaling conditions using two soil shear strains: 0.2% and 0.5%,

- Use the models to compare the performance of Trapezoidal and Rectangular lining geometric configurations,

- Determine if any general instability (excessive "birds-mouthing," slipping, separation and rotation of the joints) will occur within the range of expected seismic shear strains,

- Establish the need and structural demand for any inter-segment connecting devices and
- Provide internal loads for the structural design of the segments.

Project Design Time Histories and Development of Maximum Shear Strains

The design acceleration response spectra for the project site were developed at depth by taking the envelope of a 475-year probabilistic ground motion hazard and the deterministic median ground motions for the three dominating faults described in the previous section. In general, the probabilistic hazard governed across the entire frequency range. From these design spectra, three sets of three-direction (fault normal, fault parallel and vertical) spectrum-compatible time histories were developed. The three sets of time histories were generated with initial time histories from actual recorded earthquakes that are representative of the controlling earthquake events as defined by the probabilistic seismic hazard analysis.

The time histories generated were then used as input in the site response program SHAKE (SHAKE 2000), along with the representative soil profiles and associated dynamic soil properties (including strain based deformation moduli), to generate a profile of maximum shear strength versus depth for a range of soil conditions and input ground motions. Preliminary analyses, to be confirmed during final design by more detailed site-specific analyses, show that the expected maximum shear strains at tunnel depth for project seismic events could range from 0.2% to 0.4%.

Software Used

The finite element computer program ADINA ("Automatic Dynamic Incremental Nonlinear Analysis") was used for this study. ADINA is a general purpose finite element system for advanced engineering linear and nonlinear analysis of structural, geotechnical, heat transfer, and fluid flow problems and is developed and maintained by ADINA R&D, Inc. (www.adina.com).

Modeled Effects

The superimposition of results obtained from separate static and seismic analyses is valid under the linear elasticity conditions for the ground mass and tunnel liner. However, tunnel excavation quite often over-stresses the ground mass in a zone around the tunnel resulting in inelastic behavior of the ground mass. Similarly, the behavior of a jointed lining system will be dependent upon of the current level of compressive loading and strains in the lining. Therefore, the pre-existing ground stress state (stress path history) and lining loads must be known to properly model the soil structure interaction response to seismic ovaling. Consequently, it is important that the modeling should also simulate the tunnel excavation and construction sequence (commonly referred to as static loads). Further, effects such as localized interface slip, strain-dependent soil moduli, frictional properties of the concrete segments and the presence of inter-segment connecting devices may affect the response of the PCTL system to shaking.

Correspondingly, SC Solutions developed a series of models to simulate the response of the proposed tunnel lining and ground mass system under the following effects:

1. The loading imposed during construction—initial in-situ stresses, response due to excavation by earth pressure balance TBM, annulus grouting and TBM thrust loads,

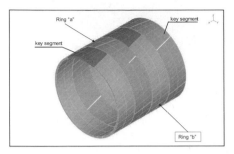 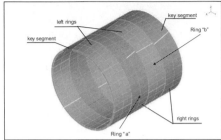

Figure 4. FEM lining models for trapezoidal and rectangular geometric configurations

2. Nonlinear soil-structure constitutive behavior with representative soil strength and stiffness,

3. The soil-structure interaction between lining segments and the ground mass (partial slip versus full-slip/no-slip conditions) and

4. Explicit modeling of no-tension, frictional behavior along radial and circumferential segment joints by using contact surfaces.

Mesh Development

A three-dimensional finite element mesh was developed following the geometry of the tunnel lining inside of a soil block 24m × 24m × 6m deep (80 × 80 × 20 feet). The water table is assumed to be 1.6m (8 feet) below the ground surface and initial stresses were based on a tunnel depth of 15m (50 feet) below the ground surface. The soil block surrounds the concrete lining with a gap between them. Both the concrete lining and the soil around it are modeled with the ADINA three-dimensional solid 27-node isoparametric elements with Gauss numerical integration of the order 3 × 3 × 3.

Figure 4 shows the finite element models of the lining for rectangular and trapezoidal configurations with staggered radial joints between adjacent rings. The two exterior rings are boundary rings that minimize the effect of the boundary conditions on the two middle rings "a" and "b." Only the responses of these two middle rings are presented in this report.

To establish the validity of this mesh, a number of larger geometry, two-dimensional models were developed to provide a comparison and calibration basis. Results of comparison analyses between the two meshes were used to justify the accuracy of use of the reduced width but more detailed three-dimensional geometry. In addition, a number of validations were performed on the model to verify that it could predict birdsmouthing and full separation of the segments (should they happen).

Boundary Conditions

To properly model the interaction, a number of different boundary conditions were imposed on the mesh.

Base Boundaries. The base of the soil block is fixed in all directions for the entire analysis, during construction sequence and to the end of ovaling.

Top and Side Boundaries During Construction. The side boundaries are fixed in the horizontal direction during the construction sequence. This boundary is controlled by specifying a horizontal displacement time-function, which has a value of zero

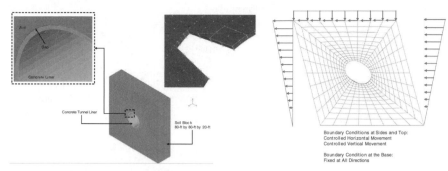

Figure 5.   Details of mesh development and boundary conditions

during the construction sequence. The sides are free in the vertical direction. At the top surface, the equivalent overburden pressure is applied.

Top and Side Boundaries During Ovaling (After Construction). To simulate ovaling, horizontal displacements are applied at the sides and top surface of the mesh (see Figure 5). As the soil block racks it should also move vertically down proportionally to the cosine of the push angle. Stiff vertical truss elements are added to the model (after construction) using the ADINA birth option so that these vertical displacements can be applied.

Lining Ring End Boundaries. The four-ring model of the tunnel is only a portion of the entire SVRT tunnel system. The boundary rings sole purpose is to provide proper boundaries to the middle rings. Periodic boundary conditions which require that the displacement at one end be equal to those at the other end of the model were used to simulation a tunnel that is infinitely long.

Modeling of Joints and Connecting Devices Between Pre-cast Segments

Joint planes or contact between segments were modeled as no-tension, frictional surfaces that would allow slip along and separation (gapping) between these contact surfaces. The packers at the radial joint reduce the stress concentration between the segments. Therefore, they were considered as secondary elements in this study and were not modeled.

The full width concrete section was used for the analysis meaning that the beneficial effects of stress relief grooves were ignored. This modeling assumption overestimates the stiffness of the tunnel lining at these localized regions. Since the deformation of the lining is controlled by the soil movement the overestimation of stiffness in the lining will not affect the deformations of the tunnel lining. However, it results in localized overestimation of lining forces at these regions.

Frictional Surfaces. Radial and circumferential joints were modeled with ADINA contact surfaces using a coefficient of friction of 0.65 was used for both radial and circumferential joint surfaces.

Circumferential Dowels. To establish structural demand on circumferential joint dowels, dowels were simulated by the ADINA nonlinear springs using material properties from manufacturer's test results and the performance idealized for simulation in this study. Radial Joint Bolts: Radial joint bolts were intentionally not modeled to not limit the model predictions of any joint opening.

Table 1.   Soil behavior parameters used in the study

Soil Behavior Parameter	MC Sand	MC Clay-1	MC Clay-2
Total unit weight, γ	19.7 kN/m^3 (125 pcf)	19.7 kN/m^3 (125 pcf)	19.7 kN/m^3 (125 pcf)
In-situ Horizontal Pressure Coefficient (K_0)	0.65	0.65	0.60
Effective Friction Angle, (Φ')	37.5 degrees	Not used	Not used
Effective Cohesion, (c')	0	Not used	Not Used
Undrained Shear Strength (S_u)	0	105 kPa (2.2 ksf) (at tunnel depth)	105 kPa (2.2 ksf)
Effective tensile strength, T'	0.5 kPa (10 psf)	Not used	Not used
Poisson's ratio, ν	0.35	0.49	0.49
Elastic modulus, E_S	335 MPa (7,000 ksf)	275 MPa (5,700 ksf)	40 MPa (830 ksf)

Material Behaviors and Parameters

The work described herein was performed concurrently with geotechnical investigations associated with the Preliminary Engineering Phase of the SVRT Project. Consequently, only a limited number of "representative" soil conditions (Table 1) were selected as a basis for modeling. Soil behavior was considered to be elasto-plastic following a Mohr-Coulomb failure criteria formulation. Three types of Mohr-Coulomb soil properties (denoted "MC Sand," "MC Clay-1," and "MC Clay-2") were used in this study. Only undrained properties of the clay were used for the analyses and pore pressure dissipation was not modeled. The "Clay-2" model is considered to be the most representative soil type for this site.

Concrete Lining. Each segment of the concrete lining was initially modeled with linear elastic material (uncracked) properties. Cracked behavior was considered in the study and is discussed in the Analysis Results section of this paper. The modulus of elasticity of the concrete was 38,600 MPa (5600 ksi) and Poisson's ratio was 0.15 and unit weight of 24.4 kN/m^3 (155 pcf). For the computation of strains and cracked bending moments using the deformation-based approach, the expected strength factor for concrete was equal to 1.3 and that for steel was equal to 1.1.

Modeling Sequence

The analysis was performed in two distinct steps: Construction Excavation Sequence and Seismic Ovaling. The Construction Excavation Sequence was modeled in the following manner:

- Gravity, hydrostatic pressure and initial stresses are applied throughout the soil mesh elements.

- The hydrostatic pressure effect is defined in terms of a pressure load applied to the lining and a corresponding reduction in the density of the soil.

- The phenomenon of the soil relaxing onto the tunnel shield before grouting is characterized using a force-displacement relationship. The gap size corresponds to the volume loss of the opening of 0.25% of the notional excavated volume.

- At the lining interface, two sets of pre-construction reactions were placed at the excavation perimeter and against the lining. These reactions represent the initial in-situ stress state and lining self-weight.

- Gravity and initial stresses are maintained while pre-construction reactions are gradually reduced to simulate soil response due to excavation. Interface springs allow the soil to move not more than the specified gap value corresponding to the convergence from the excavation perimeter to the TBM shield "a" (see Figure 5). The annulus between the converged ground and lining extrados is filled with pressurized grout that prevents the soil from further radial convergence. During this step the grout can only resist normal forces (no tangential forces).

- The interface normal springs are changed to nonlinear compression-only springs without any gap. At this step tangential springs are added. The tangential springs represent the friction between soil and lining. This is to simulate the stage where grout has cured. Note that the gravity and hydrostatic pressure are kept constant for the rest of the analysis.

- A uniform, locked-in TBM thrust of 11,100 kN (2500 kips) was longitudinally applied to the liner. This was felt to be a representative value allowing for the effects of joint packers and concrete creep.

The Seismic Ovaling Analysis was performed after the construction sequence is completed. Ovaling is the application of a pseudo-static displacement-controlled push to the boundaries that corresponds to the range of seismic shear strains (see Figure 5). This type of displacement-controlled loading in an elastic, homogenous media will result in pure shear stress in the soil that, in turn, causes the largest distortion of the lining (from an initial circular section to an "oval" section). This analysis is highly dependent on the interaction of soil and lining and their material properties. The soil shear strain was increased incrementally and statically to a maximum value of 0.5%. Two strain levels were defined for presentation of results: a 0.2% soil shear strain value which is considered representative of an Operating Design Earthquake (ODE) and a 0.5% soil shear strain value which was considered to exceed the upper bound of all conceivable events including the Maximum Design Earthquake (MDE).

NUMERICAL ANALYSIS RESULTS

Lining Displacements and Diametric Strain

The maximum displacements of the lining system predicted by the ADINA analyses of construction and seismic ovaling conditions have been summarized in Table 2. Displacements from the analyses have been quantified in terms of diametric strain which is the ratio of change in diameter to the original diameter $\Delta D/D$. Note that the tabulated diametric strains during ovaling were obtained by subtracting the diametric strain at the end of construction sequence from the total value during ovaling. Therefore, the strain values at 0.2% and 0.5% in this table are only due to ovaling.

As suggested by Wang (1993), the free-field diametric strain assuming the presence of a pre-existing (unlined) perforated ground due to tunnel excavation, then the diametric strain in lining is predicted to be:

$$\frac{\Delta D}{D} = \pm 2\gamma_{max}(1 - \nu_m)$$

where ν_m is the soil Poisson's Ratio.

Table 2. Diametric strain of the lining ($\Delta D/D$)

Soil	Ring	Segment	Ovaling Strain = 0.2%			Ovaling Strain = 0.5%		
			Non-linear FEM	Free Field Perforated Ground	Elastic, Lining Full Slip	Non-linear FEM	Free Field Perforated Ground	Elastic, Lining Full Slip
MC-Sand	a	Rectangular	0.237%	0.26%	0.25%	0.587%	0.69%	0.66%
		Trapezoidal	0.232%			0.574%		
	b	Rectangular	0.234%			0.580%		
		Trapezoidal	0.232%			0.574%		
MC-Clay 1	a	Rectangular	0.198%	0.20%	0.19%	0.410%	0.51%	0.52%
		Trapezoidal	0.195%			0.399%		
	b	Rectangular	0.196%			0.409%		
		Trapezoidal	0.195%			0.399%		
MC-Clay 2	a	Rectangular	0.140%	0.20%	0.15%	0.37%	0.51%	0.41%
		Trapezoidal	0.137%			0.35%		
	b	Rectangular	0.139%			0.37%		
		Trapezoidal	0.137%			0.35%		

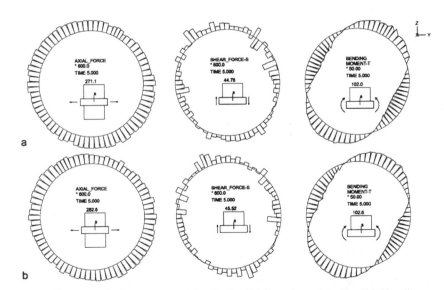

Figure 6. Axial force, shear and moment distribution (0.2% strain, rectangular, MC Clay-2)

Table 2 also includes diametric strains for the unlined tunnel cavity determined on this basis and a tunnel with elastic liner in elastic ground and full interface slip calculated per Wang (1993). By observation, at 0.2% ovaling strain levels, it can be seen that the jointed tunnel lining deformations using non-linear behavior are comparable to those for an elastic, continuous lining (full interface slip). However, at 0.5% strain levels, the jointed tunnel lining deformations are less than predicted by the free-field perforated ground or the elastic continuous lining. This felt to be due to the development of relatively uniform internal compressive thrusts in the lining (preventing tensile strains) as shown in Figure 6.

Table 3. Summary of maximum thrust forces, uncracked and cracked section moment

Soil	Segment Configuration	Max. Axial Force kN (kip) / ring	Max. Uncracked Moment kN-m (kip-ft) / ring	Maximum Cracked Moment kN-m (kip-ft) / ring	Cracked to Uncracked Moment Ratio
End of Construction					
MC Sand	Rectangular	752 (169)	18 (13)	16 (12)	1.0
	Trapezoidal	818 (184)	26 (19)	24 (18)	1.0
MC Clay-1	Rectangular	712 (160)	16 (12)	16 (12)	1.0
	Trapezoidal	765 (172)	23 (17)	22 (16)	1.0
MC Clay-2	Rectangular	1,085 (244)	77 (57)	75 (55)	1.0
	Trapezoidal	1,125 (253)	76 (56)	73 (54)	1.0
0.2% Soil Shear Strain					
MC Sand	Rectangular	2,366 (532)	254 (187)	188 (139)	0.74
	Trapezoidal	2,397 (539)	245 (181)	190 (140)	0.77
MC Clay-1	Rectangular	1,450 (326)	182 (134)	130 (96)	0.72
	Trapezoidal	1,503 (338)	180 (133)	133 (98)	0.73
MC Clay-2	Rectangular	1,259 (283)	140 (103)	118 (87)	0.85
	Trapezoidal	1,290 (290)	146 (108)	119 (88)	0.82
0.5% Soil Shear Strain					
MC Sand	Rectangular	4,768 (1,072)	594 (438)	354 (261)	0.60
	Trapezoidal	4,884 (1,098)	570 (420)	366 (270)	0.64
MC Clay-1	Rectangular	1,899 (427)	359 (265)	174 (128)	0.48
	Trapezoidal	1,966 (442)	349 (257)	178 (131)	0.51
MC Clay-2	Rectangular	1,579 (355)	324 (239)	149 (110)	0.46
	Trapezoidal	1,610 (362)	339 (250)	151 (111)	0.45

Axial Forces and Moments

Internal axial forces (thrusts) and bending moments obtained in the analysis are found in Table 3 and Figure 6 shows a typical distribution of internal thrusts, shears and bending moments in the lining during seismic ovaling. Interestingly, the analyses predict very little difference in internal forces and joint openings between the trapezoidal and rectangular geometric configurations. Prior to the study, it was felt that the trapezoidal rings would show significantly higher internal forces due to torque and other effects at the skewed radial joints. However, it is now apparent that the impact of this is smaller than originally expected. The reasons for this are felt to be due to inter-ring connectivity (connecting devices and friction) and axial force loads across joints that provide a degree of axial continuity and stiffness across the joints.

All of the ADINA analyses in this study were performed assuming an initially uncracked concrete section for the pre-cast concrete segments. However, using uncracked sectional properties over-estimates the stiffness of the lining system and

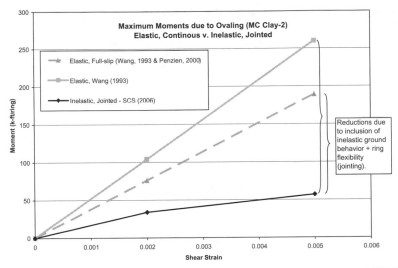

Figure 7. Bending moment comparison—elastic, continuous versus inelastic, jointed (MC Clay-2)

the lining bending moments. For the given stiffness of the tunnel lining and surrounding soil, the soil shear strain predominantly controls the tunnel deformation whether the concrete tunnel lining is uncracked or cracked. Also, joint rotations are the dominant flexural relief within the ring—flexural strains concentrate at these locations. Knowing this, the cracked-behavior bending moments within the lining were estimated using a deformation-control formulation. As expected, the cracked and uncracked bending moments are similar for small soil shear strain, but start deviating as the soil shear strain increases.

The reinforcing design for the segments was initially selected for construction handling and TBM thrust loads. This condition governed the design as thrusts and bending moments produced under seismic loading did not require any adjustment to reinforcing levels or detailing.

Comparisons were made of lining bending moments due to ovaling that were predicted using inelastic ground with jointed lining behavior and those predicted by elastic ground with a continuous lining. For the comparison, elastic, continuous closed form solutions found in Penzien (2000) and Wang (1993) were utilized with the same ground stiffness properties described in Table 1. A cracked moment of inertia was used in the calculations, equal to half the gross moment of inertia. A typical example is shown in Figure 7, from which it can be seen that inclusion of inelastic and jointed behavior can lead to significant reductions in the predicted design bending moment.

Relative Rotation and Gapping Between Segments

The separation of the contact surfaces were quantified by measuring the "birds-mouthing" (relative rotation between the joint surfaces or angle α), joint separation (loss of contact across the joint or gap δ) as shown in Figure 8 and joint contact forces as shown in Figure 9. Maximum predicted joint rotation angles and gap widths during ovaling for a are given in Figure 10 which show that at 0.2% and 0.5% shear strains, "Birds-mouthing" (rotation of joints) will occur (as expected) but that no joint separation (total loss of contact between the segments) was observed.

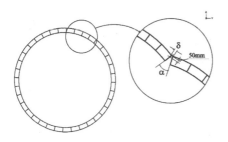

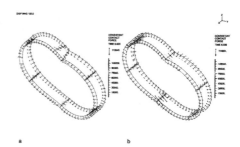

Figure 8. Relative rotation, α, and separation, δ, of radial joint sides

Figure 9. Joint contact forces, rectangular—MC Clay-2, 0.5% solid shear strain

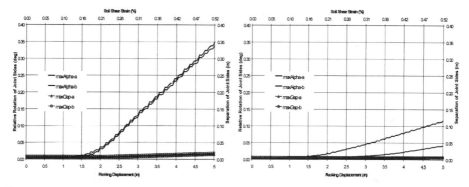

Figure 10. Radial joint contact—rectangular and trapezoidal rings—MC Clay-2

This study found that there is no total loss of contact for any of the configurations when Mohr-Coulomb soil is used. When using the Mohr-Coulomb soil model, thrust in the liner is always compressive even under 0.5% soil shear strain. Correspondingly, no radial joint contact was broken in the non-linear soil model analyses. During validation runs at the ovaling strain levels, a linear elastic soil model with no soil or interface failure criteria does predict tensile thrusts in the lining suggesting that joint contact could be broken. However, comparison of the normal forces relative to tangential shears along the segment interface shows that localized interface slip should occur. When a using a Mohr-Coulomb failure criterion, this slip is predicted and no tensile thrusts or gapping occurs.

Friction and Dowel Forces

During the study, predicted friction forces and dowel forces at the circumferential surface between rings "a" and "b" were determined. At this surface, the friction force caused by the locked in TBM thrust and the resistance from dowels connect the two rings. When the ratio of the tangent friction force to normal contact force is less than the friction coefficient (0.65) there would be no sliding. Figure 11 shows typical friction force to normal force ratios along the contact surface against the 0.65 threshold value.

It is important to note that these ratios are for each finite element within a precast segment. Therefore, even if this value is close to 0.65 it does not mean that the entire

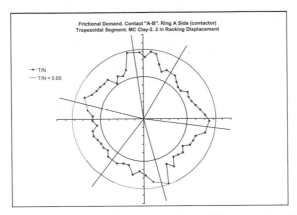

Figure 11. Frictional demand/capacity between rings "a" and "b" trapezoidal—MC Clay-2, 0.2% soil shear strain

segment slides. In general, the trapezoidal segment configuration shows slightly less sliding than the rectangular segment configuration.

Maximum shear forces in the dowels are shown in Table 4 which, as expected, are larger when sliding is greatest. Throughout the analyses, the dowels do not fully engage because the friction at the circumferential surface is not overcome. The maximum force in the dowels for all the cases was 6.0 kN (1.34 kips) which compared to the yield value of about 52 kN (11.7 kips). This indicates that the dowels do not significantly contribute to the sliding resistance along the joint and that there would be little demand placed on them during ovaling.

The ability of joints to collect strain without generating load and hence accommodate seismic deformations is one of the inherent advantages of precast segmental tunnel liners. This study has modeled the connecting devices as flexible elements while establishing that the segment ring geometry remains stable, joints remain in-contact and ovaling does not impose significant force or displacement demand on the connecting devices. However, care should be exercised to select devices that are compatible with such assumptions. Use of axially stiff connection devices (e.g., steel bolts) along the circumferential joints or moment resisting bolt arrangements at radial joints could impose high, localized forces on the lining under seismic straining resulting in cracking or spalling.

The SVRT Project will specify appropriate force-displacement characteristics for circumferential connecting devices that are compatible with anticipated seismic strains produced under the combined effects of compression-extension and longitudinal bending effects, yet still have sufficient tensile and shear capacity for construction loads and to maintain full gasket closure during ring builds.

CONCLUSIONS

The following conclusions can be drawing from this study:

1. Detailed non-linear three-dimensional finite element analyses of the rectangular and trapezoidal tunnel lining configurations were successfully performed to evaluate the state of stresses and strains (forces and deformations) in the lining during construction and ovaling.

Table 4. Maximum circumferential joint dowel forces

Soil	Segment Configuration	End of Construction kN (lb)	0.2% Soil Shear Strain kN (lb)	0.5% Soil Shear Strain kN (lb)
MC Sand	Rectangular	0.40 (90)	1.90 (428)	4.29 (965)
	Trapezoidal	0.25 (56)	2.47 (555)	5.97 (1,343)
MC Clay-1	Rectangular	0.40 (91)	1.38 (311)	3.25 (731)
	Trapezoidal	0.25 (56)	2.53 (568)	5.39 (1,212)
MC Clay-2	Rectangular	1.16 (260)	1.62 (365)	2.52 (567)
	Trapezoidal	0.77 (173)	2.04 (459)	4.73 (1,065)

2. Due to friction contact along joint surfaces, construction loads and the inclusion of non-linear behavior, no gapping between segments occurs for anticipated ovaling strain levels. Consequently, the stability of the PCTL jointed tunnel lining surrounded by a nonlinear soil was maintained for the range of anticipated ovaling shear strains (0.2% to a maximum credible shear strain of 0.5%) for both rectangular and trapezoidal segment configurations.

3. Predicted internal thrusts and bending moments were not significantly different between rectangular and trapezoidal segment geometric configurations and hence, seismic loads are not considered to be a differentiator in selection of either option for the SVRT Project. Comparisons of lining bending moments using inelastic ground with jointed lining behavior and those predicted by elastic ground with a continuous lining indicate that there are benefits to the inclusion of the effects versus using more simplified methods.

4. Inter-segment connecting devices such as dowels, bolts, and guide-rods (largely used for construction situations only) maintain segment alignment during installation and provide a level of redundancy with respect to stability of the segment positions. However, seismic ovaling places negligible demand on connecting devices located in the circumferential or radial joints. Therefore, their inclusion is not required for the seismic stability of jointed lining system. However, such devices should be detailed to have appropriate force-displacement characteristics compatible with anticipated seismic strains.

5. This study concentrates on jointed behavior during ovaling, however, the principles and conclusions regarding the flexibility and stability of the jointed PCTL system are applicable to other forms of shaking such as compression-extension and longitudinal bending effects

6. The reinforcing design for the segments was initially selected for construction handling and TBM thrust loads. The thrusts and bending moments produced under seismic loading did not require any adjustment to reinforcing levels or detailing.

ACKNOWLEDGMENTS

The authors would like to thank their many colleagues at VTA, Hatch Mott MacDonald, Bechtel and SC Solutions who assisted in the development of this work. In particular, we would like to acknowledge Arch Walters of VTA; Aaron Miller, Tomas Gregor, John Hawley and Andrew Hindmarch of HMM; and Phoebe Cheng, Alex Krimotat, Vince Jacob and Tom Ballard of SC Solutions for their valued contributions to this work.

REFERENCES

ADINA R&D, "Automatic Dynamic Nonlinear Incremental Analysis," Watertown, MA 02172.

Bay Area Rapid Transit (BART) Facilities Standard (BFS), Release 1.2 (2005).

Dean, A., Young, D.J. and Kramer, G.J.E., (2006). "The Use and Performance of Precast Concrete Tunnel Linings in Seismic Areas," International Association of Engineering Geologists, 2006 Proceedings, paper number 679.

Hashash, Y.M.A., Hook, J.J., Schmidt, B., and Yao, J.I.-C., (2001). "Seismic Design and Analysis of Underground Structures," Tunneling and Underground Space Technology, 16, 247–293.

Konagai, K., (1998). "Diagonal Expansion and Contraction of a Circular Tunnel During Earthquakes," J. Structural Eng./Earthquake Eng., JSCE, Vol. 15, No. 1, pp. 91–95.

Merritt, J.L., Monsees, J.E., Hendron, A.J., Jr., (1985). "Seismic Design of Underground Structures." Proceedings of the 1985 Rapid Excavation Tunneling Conference, Vol. 1, pp. 104–131.

Monsees, J.E., Merritt, J.L., (1988). "Seismic Modelling and Design of Underground Structures." Numerical Methods in Geomechanics. Innsbruck, pp. 1833–1842.

Owen, G.N. and Scholl, R.E., (1981). "Earthquake Engineering of Large Underground Structures." Report No. FHWA_RD-80_195. Federal Highway Administration and National Science Foundation.

Peck, R.B., Hendron, A.J., Mohraz, B., (1972). "State of the Art in Soft Ground Tunneling." Proceedings of the Rapid Excavation and Tunneling Conference. American Institute of Mining, Metallurgical and Petroleum Engineers, New York, pp. 259–286.

Penzien, J. and Wu, C., (1998). "Stresses in Linings of Bored Tunnels." Int. J. Earthquake Eng. Struct. Dyn., Vol. 27, pp. 283–300.

Penzien, J., (2000). "Seismically-Induced Racking of Tunnel Linings." Int. J. Earthquake Eng. Struct. Dyn., Vol. 29, pp. 683–691.

Shake2000, Modified by Deng, N. & Ostadan, F., (2000). Geotechnical and Hydraulic Engineering Services, Bechtel National Inc., San Francisco, California.

St. John, C.M. and Zahrah, T.F., (1987). "Aseismic Design of Underground Structures." Tunneling Underground Space Technol. Vol. 2 No. 2., pp. 165–197.

Wang, J.N., (1993). Seismic Design of Tunnels: A State-of-the-Art Approach, Monograph, Monograph 7. Parsons, Brinckerhoff, Quade and Douglas Inc, New York.

Young, D.J. and Dean, A. (2006), "Seismic Performance of Precast Concrete Tunnel Linings," Tunneling Association of Canada, 19th National Conference Proceedings, September, 2006.

STEEL-FIBER REINFORCED SELF-COMPACTING CONCRETE ON THE SAN VICENTE AQUEDUCT TUNNEL

M.R. King

Halcrow Group Ltd.

C.D. Hebert

Traylor Shea Ghazi Precast JV

ABSTRACT

The 56,000' (17 km) long San Vicente Aqueduct Pipeline tunnel is being constructed using precast concrete rings to act as preliminary ground support through two sections of the bored tunnel with lengths of 9,450' (2.9 km) and 33,700' (10.27 km). The six segment bolted trapezoidal ring was manufactured using a combination of steel fiber reinforced concrete and self compacting concrete in vertical moulds. This paper briefly describes the design and function of the segments, and reviews the development stages of the concrete mix, and the advantages and disadvantages of the adopted manufacturing method.

INTRODUCTION

The segment manufacturer for the San Vicente to Second Aqueduct Pipeline Project (SVP) decided to make use of simplified precasting methods for the production of over 66,000 segments to be used for the primary lining of the tunnels. The methods examined and adopted included the use of steel fibers to replace traditional reinforcement bars, vertical molds to minimize surface finishing, and self compacting concrete to minimize vibration and compaction effort. Although the concrete mix design was a success, there were difficulties with developing a fully effective self compacting concrete along with all of the other constraints imposed on a typical segmental lining casting regime. The final mix exhibited imperfect compaction without some external vibration effort, but was fully successful in limiting hand working for surface finish and exceeded all of its strength performance requirements.

OUTLINE OF THE SCHEME

The SVP consists of a water transmission pipeline running from the San Vicente Pump Station to the Rancho Penasquitos Pressure Control and Hydroelectric Facility, and is part of the system required to enable water supply to the San Diego region in the event of supply interruptions. The SVP is being constructed predominantly in tunnel with a total length of approximately 10.8 miles (17.4 km), constructed mainly through Sedimentary rocks. Different tunneling construction methods are being adopted along the route to suite local geological conditions, with a bored tunnel and segmental lining adopted as the primary support for approximately 9,450 feet (2.9 km) of bored tunnel through Reach 2, and 33,700 feet (10.27 km) through Reach 4. The ground traversed within these sections is predominantly weaker Sedimentary formations and Conglomerates between higher strength Granitic rocks either side of the two Reaches. A raw water transmission pipe will be placed through the tunnel, which will form the long term support for the ground loads and internal water pressure. The precast concrete lining therefore

Table 1. Concrete performance requirements

Parameter	Strength
Characteristic compressive cylinder strength	6,000 psi (41.3 N/mm^2)
Characteristic flexural strength at first crack	610 psi (4.2 N/mm^2)
Characteristic residual (post-crack) flexural strength	440 psi (3.0 N/mm^2)
Characteristic tensile strength	420 psi (2.8 N/mm^2)
Early age (stacking) compressive strength	1,400 psi (9.6 N/mm^2)

has a functional requirement of maintaining suitable support to the ground until the permanent support can be installed.

The tunnel is being built with a 6 segment trapezoidal segmental lining with an
internal diameter of 10' 6" (3,200 mm) and a thickness of 7" (178 mm). The ring has a
length of 4 feet (1,220 mm) and a taper of ¾" (20mm) across the diameter. Two spear
bolts are used across each longitudinal joint, and the ring has a total of 12 dowels
across the circumferential joint. Blind end grout sockets are cast into the centre of each
segment. More details of the scheme may be found in Reference 1 and Reference 2.

The design of the segmental lining is based upon the use of steel fiber reinforced
concrete (SFRC), with the performance requirements listed in Table 1.

The segments are being erected behind open face shields manufactured by CTS.

The Project owner is San Diego County Water Authority and the Project Engineer
is Jacobs Associates. The Contractor is Traylor Bros., Inc. and JF Shea Company
(TSJV), the segments were manufactured by Traylor Shea Ghazi Precast, and the segment designer was Halcrow Group Ltd.

BACKGROUND TO SELF-COMPACTING CONCRETE

Self-compacting concrete (SCC) can be described as a concrete that will flow and
consolidate under its own weight to completely fill the form void, while maintaining
homogeneity, without the need for an external vibrating energy source. The advantages in terms of noise reduction, health and safety and cost saving on equipment purchase, operation and maintenance are obvious. However, there can also be quality
advantages in some cases as poker vibrated concrete may visually appear well compacted, but the effort may be uneven through the section, and poor compaction may
occur around congested reinforcement, or within complex form details for example, or
there could be some reinforcement displacement due to excessive effort.

There have been specialist applications of self compacting concrete for some
time, examples including the high cement content concrete for placing underwater, or
for diaphragm walls and piles, and the development of water reducing admixtures has
aided the workability of concretes while allowing the cement contents to be reduced to
a level that would not result in heat of hydration problems. One of the first superplasti-
cisers was developed in the 1960s, but development of modern self compacted concrete started in the 1980s, and the first project use of a modern SCC design was in the
1990s in Japan. The material also gained popularity, particularly with precast manufacturers and floor slab contractors, around North America and Europe during the 1990s,
and the precast industry is now the biggest user of SCC.

Figure 1.   Vertical mold arrangement

STEEL-FIBER REINFORCED SELF-COMPACTING CONCRETE

Steel-fiber reinforced self-compacting concrete has many similarities with normally compacted SFRC with the requirements for strength and toughness, for example, being unchanged for design purposes, and assuming suitable mix design, effectively being equal in terms of being able to achieve the required performance. Some of the issues that do, however, need to be considered, include:

- Fiber orientation—possible difference between vibrated concrete and more free flowing concrete. Preferential orientation of fibers is always a concern but in segments the additional energy during vibration may enhance the strength by orienting the fibers in the most efficient direction. This is less likely to occur with SCC and is complicated on this project with the use of vertical moulds (see Figure 1).

- Workability is generally affected by the addition of fibers—the concrete requiring modified mix designs. This is made worse in SCC compared to "plain" concrete due to the enhanced flow requirements with no segregation.

- Testing—standard L box tests used in SCC with bar grills are used for concrete to be placed around reinforcement, but are not entirely appropriate or relevant with non bar reinforced elements.

MIX DESIGN

The development of the mix design for the segments was complicated by three of the primary items noted above (vertical segment molds, the use of SCC, and the use of steel fibers). While these materials and processes are not new, none had been employed at the segment casting facility in Littlerock, CA, and combining all three is a first for the segment industry.

In addition to meeting the required design parameters, the concrete had to provide the high-early strengths dictated by the production schedule and casting process, and had to be robust enough to deal with the variability inherent in the local materials and in any high production casting facility.

One of the initial steps in developing the mix design was to contact the admixture suppliers with experience in SCC. Although the casting yard had poured tens of thousands of yards of segment concrete on previous jobs, the mixes for those projects were typically a "dry cast" concrete, and the universal advice was to not to adopt these mixes for SCC, but rather to make a thorough analysis of the requirements, process, and materials, and based on this analysis, design a new mix. The design strength parameters developed by Halcrow that had to be met are in Table 1.

The materials incorporated into the mix were fairly standard. The decisions that had to be made related to the steel fibers, the high-range water reducing admixture (HWRA), and the mix proportioning. The choice of steel fibers was largely based on the experience of other projects with fiber reinforced segments, such as the CTRL project in the UK, and also on the recommendations of the technical departments of the fiber suppliers. During the trial mix testing, the steel fiber concrete was tested to ensure that the concrete met the flexural and tensile requirements set forth by the design. A dosage of 50 lbs/yrd^3 (30 kg/m^3) proved to be optimal, and the fiber employed was the Dramix RC 80/60 BN as supplied by Bekaert. This fiber is 60mm long, 0.75mm in diameter, and has a tensile strength of 150 KSI (1,050 N/mm^2). The fiber has hooked ends, and comes glued in bundles. Each pound of fiber contains approximately 2,087 fibers, so each yard of concrete has over 100,000 fibers (130,000 fibers/m^3).

The HWRAs tested were of the polycarboxylate family, and were designed to be used specifically for SCC concrete. Testing was undertaken with Glenium 3400 NV, provided by Degusssa/Master Builders (now BASF), and Viscocrete 2100, provided by Sika. Dosages tested ranged from 5 to 10 ounces per hundred weight (oz/cwt) of cementitious material, but the testing narrowed the ideal range for this application down to 5.5 to 8.5 oz/cwt. The performance of these two admixtures was very similar, but it was subjectively felt that the Viscocrete 2100 provided a more consistent workability. With either admixture it was apparent that small changes in the dosing could have big impacts on the rheology of the concrete, and during the field trials changes in the dosing were limited to 0.5 oz/cwt per test. The sensitivity of the mix was likely related to the local materials. The use of a viscosity modifying admixture (VMA) was suggested, and tested, but the benefits observed in the testing did not justify the additional cost.

Field trials were started 7 months prior to segment production. Over this period 50 batches of concrete were produced, and of these 33 were sampled and tested. It was originally thought that 2 months of trial tests would be adequate, but getting a mix suitable for the project requirements proved to be challenging. Prior to starting the field trials a series of tests was performed on the local aggregates. This testing included specific gravities, gradations, and aggregate void contents. This data, along with the details of the local cement, fly ash, and design requirements were given to the admixture suppliers for their analysis. The suppliers employed a variety of tools, including in-house computer programs specific to SCC, to make recommendations for the mix designs for the field trials. Key parameters that were considered included the aggregate properties, the ratio of fine aggregate to coarse aggregate, the paste volume, and of course the water/cement ratio and admixture dosing. Early trials utilized 1" (25 mm) aggregate, ⅜" (6 mm) aggregate, and sand. These mixes tended to segregate and bleed as shown in the slump flow test pictured in Figure 2 (left). The pile of rock in the middle, along with the bleeding at the edge of the sample, is obviously undesirable. A better example of SCC is shown in Figure 2 (right), but some piling of the 1" rock is still evident.

Based on the early testing, it was decided to focus on a mix with only ⅜" (6 mm) aggregate and sand, and a series of tests was made with sand and rock in varying ratios. A sand to aggregate ratio of 45/55 provided the most consistent workability and consistency. The best indicator of the quality of the SCC mix rheology was the slump flow test combined with the Visual Stability Index (VSI) test. This is performed almost exactly as a normal concrete slump test, but instead of measuring the slump, the diameter of the resulting circle of concrete is measured. After the diameter of the concrete is measured, the stability of the mixture is rated using established guidelines (Reference 3). For example, the best rating, a VSI rating of 0, indicates no evidence of segregation in the slump flow patty, in the mixer drum, or in the sampling wheelbarrow. The worst rating, a VSI rating of 3, indicates obvious segregation as evidenced by a large

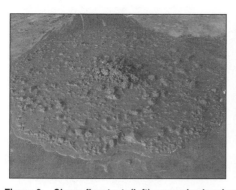 

Figure 2.   Slump flow test: (left) poor mix showing segregation; (right) improved mix

mortar halo, and/or a large aggregate pile in the center of the slump flow patty, and a thick layer of paste on the surface of the resting SCC in the mixer drum or the sampling wheelbarrow. Figure 2a above would be rated a 3, and Figure 2b would be rated a 1.5. The importance of having a cohesive and stable mix was amplified by the fact that the segments were cast in vertical molds. This could be considered as pouring a 7" (178 mm) thick by 4' (1.2 m) high wall, and the danger was that an unstable mix would easily segregate, and the top edge (the circle joint) would be only mortar, and would not contain any rock or steel fiber. This was in fact seen in a number of test segments where the top 5 inches (125mm) was only mortar. This obviously was not acceptable, but was a valuable lesson in what to avoid. The final accepted range for the slump flow was 17" to 26", measured within 10 minutes of batching. Slump flows below 17" required additional vibration, and those above 26" were more susceptible to segregation and bleeding.

Once the stability and workability of the mix were satisfactory, focus changed to the consolidation and finish of the concrete. Both of these are directly tied to the slump flow and workability of the concrete, but any changes made to the mix, especially relating to achieving a suitable finish, could not have a negative impact on the stability of the mix. It was felt that vibration of the segments could be eliminated with a proper SCC mix, but in this case it did not seem to be possible to totally eliminate the vibration. Vibration was found to be necessary both to speed up the consolidation of the concrete, and to help reduce the surface air voids. Without vibration the air voids could not be eliminated, and one of the contributing factors to this was the steel fibers. It was believed that the fibers tended to impede the flow of air bubbles up and out of the mold. The use of shorter fibers between 1" and 1 ¼" (25 mm to 30 mm) was recommended, but the use of such fibers would have decreased the structural strength of the segments so was not an option. For the purposes of this project the aesthetics of the segments were not critical, so due to time constraints segment production was started, and continued, with the minor shallow air voids present. For reference, relative to a typical dry cast concrete segment, the vibration duration and intensity were reduced by approximately 50%. Figure 3 is a picture of the typical segment finish, along with a close up of an air void.

The final mix design used in production is listed in Table 2. The water cement ratio shown is the maximum. During production both the water cement ratio and admixture dosage were varied as necessary to maintain the desired concrete characteristics. The typical w/c ratio was 0.31, with an HWRA dosage of 6.5 oz/cwt.

 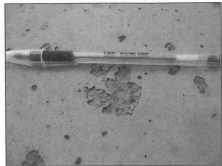

Figure 3. Segment finish: (left) typical segment; (right) air void

Table 2. Mix design quantities

Component	Quantity (lb/yd³)	Quantity (kg/m³)
Cement	700	420
Fly Ash—Class F	235	140
Steel Fibers	50	30
Wash Concrete Sand	1,196	715
⅜" Aggregate	1,446	860
Water (W/C=0.34)	333	200
HWRA	6.00 fl.oz/cwt cementitious	4.7 cc/kg

Table 3. Test results

Test	Design Requirement	Actual Average Result
7-Day compressive strength	None specified	5,930 psi (40.8 N/mm²)
28-Day compressive strength	6,000 psi (41.3 N/mm²)	7,360 psi (50.7 N/mm²)
56-Day compressive strength	None specified	9,430 psi (64.9 N/mm²)*
Flexural strength at 1st crack	610 psi (4.2 N/mm²)	1,162 psi (8.0 N/mm²)
Residual flexural strength	440 psi (3.0 N/mm²)	536 psi (3.7 N/mm²)
Tensile strength	420 psi (2.8 N/mm²)	780 psi (5.4 N/mm²)
Fiber Content	50 lbs/yrd³ (30 kg/m³)	53.3 lbs/yrd³ (31.7 kg/m³)

* This result is from a limited sample of 48 cylinders taken over the first 2 months of production.

TESTING

The testing regimen included the standard tests for concrete, and some additional testing dictated by the use of the steel fibers. All sampling, and the manufacture of beams and cylinders was undertaken by Traylor Shea Ghazi QC staff, and the testing was performed by an independent testing laboratory. Table 3 lists the test, the required result, and the average result through the first eight months of casting.

The fiber content was tested weekly by taking a random sample from the production concrete, filling a ¼" cubic foot bucket, and then washing out all the non-fiber

Figure 4. Steel fiber distribution in production segment

materials. This was done on an inclined table with a magnetic plate at the bottom to collect the steel fibers. The fibers were then dried, weighed, and compared to the required amount for the sample size.

Figure 4 shows a production segment broken to show the distribution of the fibers through the section.

CASTING

Segment production was undertaken at a twin carrousel plant located near Los Angeles, California. One of the carrousels was dedicated to the San Vicente Project, and the other serviced other projects. The molds and carrousel were provided by CBE of France, and the concrete batch plant was supplied by ACT/Wiggert. The facility was originally constructed in 2001 for the NOS-ECIS project, and has since been updated and modified for various projects. The segment casting process was as follows. The segments were cast vertically, with two molds per trolley (see Figure 1). There were a total of 144 molds on 72 trolleys. The molds were filled with concrete from an overhead bucket that traveled from the on-site batch plant into the casting building. While being filled, the mold trolley was lowered onto a vibrating table. Once the molds were filled, the casting line advanced, and the top surface of the segment (in the case of edge casting, the circle joint), was hand finished and the molds cleaned. After finishing, the mold trolley would enter a steam curing chamber, and would exit the chamber approximately 6 to 7 hours later. Once the segments exited the steam curing chamber the molds were opened up (the intrados of the mold face moved outward), and the segments removed from the mold with a vacuum lifter (see Figure 5). The segments were stacked on an exit line, and date stamped and inspected. The segments were then moved down the exit line to the outside stacking crane, and then moved into the storage yard with a 15 ton forklift (see Figure 6). The segments remained in the storage yard until the 28-day design strength was achieved, and were then shipped to the tunnel site on flatbed trailers. The segments were produced on a two shift per day basis, five days per week. The average daily production was 48 rings, and there were approximately 11,000 rings required for the project. Casting was finished in April 2007.

CONCLUSIONS

The use of steel fiber reinforced self compacting concrete in vertical segment molds provided some unique challenges. The application of these technologies,

Table 4. Advantages and disadvantages of vertical molds

Advantages	Disadvantages
Less finishing of the segment	The top edge (circle joint) is hand finished, and after curing is not a smooth surface compared to the formed surfaces. Depending on the project this could have a negative effect on the sealing performance of a gasketed segment, and could cause point loading along this edge.
The option of having two molds per trolley	Difficulty in incorporating a gasket groove, due to the top edge being open. A flip down groove former could be used, but will likely invite some problems of its own
Fewer moving mold parts so less maintenance and decreased chance of mold tolerance problems	Molds are not adjustable. As all the bearing surfaces are fixed this should not be a problem, but places additional emphasis on the mold manufacturer to produce accurate and sturdy molds.
Decreased cost of molds	Requires special concrete placing equipment, and extra care must be taken to properly consolidate the concrete without causing segregation
The ability to cast two molds per cycle, doubling output	

Table 5. Advantages and disadvantages of SCC

Advantages	Disadvantages
No vibration (not true in this case, but certainly less than with typical concrete)	Can require more time to develop the desired mix vs. a traditional mix
Quicker concrete placement	The batching and placing process and equipment may have to be modified to be compatible with SCC
Less wear and tear on the molds and batch plant	Personnel must be educated and trained. Treat this as a new material, not just a new mix
Less noise	Can be less forgiving than traditional concrete, depending on the local materials and the use of other admixtures (i.e., Viscosity modifying admixtures)
Better consolidation around mold details	
Easier finishing	

Table 6. Advantages and disadvantages of SFRC

Advantages	Disadvantages
Cost savings vs. traditional reinforcement	Design codes not current with industry practice, leading to difficulty of acceptance in some cases
Automatic dosing and record keeping (with proper equipment)	Requires new equipment for accurate and efficient dosing
Reinforcement provided in all areas of the segment, no need for special reinforcement around bolt pockets, grout sockets, etc	Requires an additional set of performance testing, which local laboratories may be unable to perform
Improved durability and corrosion resistance	Requires careful analysis of all segment handling, from demolding to segment erection
Improved toughness of concrete	Produces a segment "weaker" than a fully reinforced section.
Readily available, no shortages	

Figure 5. Vacuum lifting out of mold   Figure 6. Handling and stacking

Figure 7. Completed section of tunnel

whether individually, or combined as in this case, must be considered carefully against the design requirements of the project, the concrete production capabilities, and the casting process. Tables 4 to 6 list some of the advantages and disadvantages of each of these three technologies based on the experience from this project.

For each of these technologies there are many more advantages and disadvantages than listed here; these are merely the major ones that were considered, or that came to light during the project. While there were some hurdles to overcome, combining these three technologies was successful and enabled the project to meet the structural, schedule, and commercial goals, and produced a finished product suited to its intended final use (Figure 7).

REFERENCES

1. Burke, J., The Saviour at San Vicente, Tunneling and Trenchless Construction, December 2006.
2. Klein, S., Hopkins, D., McRae, M., Ahinga, Z., Design Evaluations for the San Vicente Pipeline Tunnel, RETC 2005 Proceedings, ed. Hutton, D., Rogstad, W.D.
3. Interim Guidelines for the Use of Self-Consolidating Concrete in Precast/Prestressed Concrete Instituted Member Plants, Precast/Prestressed Concrete Institute, TR-6-03.

WATERPROOFING OF A SUBSEA TUNNEL
WITH A UNIQUE SPRAYABLE MEMBRANE—
THE NORDÖY ROAD TUNNEL, FAROE ISLANDS

Sigurd L. Lamhauge

Nordöytunnilin pf

Karl Gunnar Holter

BASF UGC Europe

Svein E. Kristiansen

Föröyakonsortiet

ABSTRACT

An innovative sprayable waterproofing membrane in combination with the traditional Nordic method of waterproofing with polyethylene foam sheets was used for the waterproofing of the tunnel. The sprayable membrane was integrated in the final sprayed concrete lining in a sandwich structure, hence constituting a composite waterproof lining based on sprayed concrete. The sprayable membrane was applied in areas with small to moderate water ingresses, making up approximately 50% of the total area requiring waterproofing in the tunnel. State-of-the-art application technology with a computerized spraying robot was used. A consistent application of the required thickness at a high speed was thereby achieved.

BACKGROUND

The Project

The Nordöy road tunnel passes below the fiord between the two islands of Eysturöy and Bordöy in the Faroe Islands. The second largest town Klaksvik in the Faroe Islands, located on Bordöy with 5,000 inhabitants is thereby linked to the rest of the archipelago with a fixed road connection. The total population of the Faroe Islands is approximately 50,000. Figure 1 shows the location of the project.

The design of the tunnel was done according to the standards of the Norwegian Road Authorities. The tunnel is entirely located in rock and was excavated by traditional drill-and-blast methods. Rock support was based on the single shell lining method with sprayed concrete and rock bolts. For control of the water seepage into the tunnel, a scheme with percussive probe drilling ahead of the tunnel face was carried out throughout the excavation of the tunnel.

Action criteria for pre-injections in order to reduce the water seepage were based on data from the probe drillings ahead of the tunnel face. Pre-injections were carried out successfully with cementitious grouts reducing the seepage into the tunnel to a cost-effective level.

The total length of the tunnel is 6,155 m with a cross section of 64 m^2 (two lanes). The maximum depth under the sea is 150 m with a minimum rock cover of approximately 40 m.

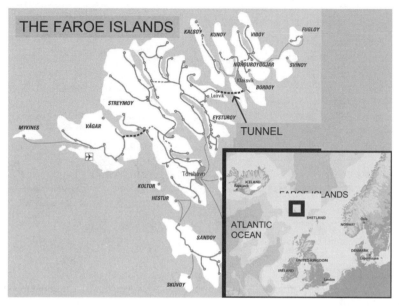

Figure 1.   The Faroe Islands are located in the North Atlantic between Iceland and Norway

Geological Conditions

The Faroe Islands consist of volcanic rocks which are typical for the North Atlantic basaltic area. The rock types in the area are extrusive basalts approximately 50 million years old.

The encountered rock mass consisted of a stratigraphy of basaltic rocks in which distinct layers could be identified with thicknesses that varied from a few meters up to approximately 50 m. Each basalt layer represented an event of a lava flow. A geological longitudinal section is shown in Figure 2. One basaltic layer which was encountered in about 700 m of the tunnel length had a relatively soft and porous character and required extensive need for pre-injections and continuous rock support with sprayed concrete and rock bolts. Otherwise the rock mass encountered in the tunnel consisted of very competent and massive basalts with compressive strengths in the range of 80–90 MPa.

REQUIREMENTS FOR THE WATERPROOFING OF THE TUNNEL

The tunnel was constructed as a drained structure. This means a certain amount of water was allowed to seep into the tunnel, hence a global build-up of water pressure in the immediate rockmass around the tunnel was avoided.

The waterproofing of the tunnel in this context addressed two main issues:

- Water ingress reduction in the entire tunnel in order to reduce the need for pumping (by means of pre-injections)
- Waterproofing of the tunnel contour to avoid the dripping of water onto the actual road surface

The latter issue is the topic of this paper. The basic requirement was to achieve a dry road surface. Dry in this context meant no dripping points which directly could hit

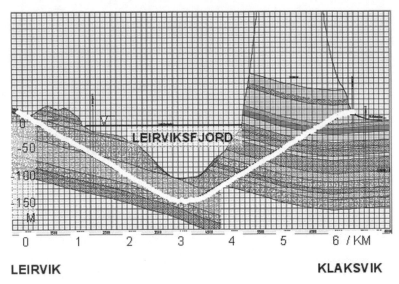

LEIRVIK   KLAKSVIK

Figure 2.   Geological longitudinal section of the Nordöy subsea road tunnel. The stratigraphy of the basaltic rocks is schematically indicated.

the road surface. This requirement was later adjusted so that single dripping points with a maximum of 6 drips per minute could be accepted. Drips or small single water ingress points in the walls were acceptable, since the water would enter into the drainage system in the invert without causing a problem in the actual road surface.

Waterproofing of the tunnel surface was to be installed selectively in the areas where the seepage exceeded the limit of 6 drips per minute. The requirement for the installed waterproofing was a completely dry surface where the waterproofing actually had been installed.

The original design and the tender documents required a waterproofing system based on Norwegian standards which basically consisted of a drainage shield with polyethylene (PE) foam sheets covered with sprayed concrete. The total cost of the waterproofing of the tunnel made up approximately 10% of the total project cost.

The experiences from the first subsea tunnel built in the Faroe Islands, the Vaga Tunnel, completed in 2002, were that seepages tend to diminish significantly over time. The first four years after opening of this tunnel, the seepage into the tunnel had been reduced by 20%. This was caused by minerals in the groundwater which deposit on the joints, and hence reduced the conductivity of the rock mass.

The Faroe Islands has a relatively mild climate. Significant freezing occurs only under adverse weather conditions. The waterproofing was therefore designed with frost insulation in only 200 m of the tunnel from both portals.

In the last few years there has been a trend to avoid flammable materials in underground construction. The flammable PE foam sheets were therefore not in accordance with a modern design philosophy with non-flammable materials. For this reason the owner was interested in alternative technical solutions for the waterproofing which could fit in with the single shell tunnel lining design. The goal was to reduce the amount of the PE foam waterproofing with more than 50%.

Several alternative technical solutions were considered. A technical solution with a sprayable waterproofing membrane in a composite liner based on sprayed concrete

Figure 3.   Traditional waterproofing system. The drainage shield which consists of polyethylene foam (PE foam) sheets has been installed. The PE foam sheets were subsequently covered with 60 mm fiber reinforced sprayed concrete.

was chosen. Favourable experiences with sprayable waterproofing membrane had recently been made in Switzerland (Meier et al. 2005) and in the United States (Blair et al. 2005). This system could be made an integral part of a single shell tunnel lining based on sprayed concrete and rock bolts. Hence, this technical solution offered significant financial benefits over a technical solution with PVC sheet membrane and cast-in-place concrete lining.

THE FINAL DESIGN OF THE WATERPROOFING OF THE TUNNEL

The final solution for the waterproofing of the tunnel comprized both the traditional system with PE foam sheets and the innovative system with a composite liner with sprayable membrane and sprayed concrete.

The decisions as to which system to apply in the different parts of the tunnel was a process which involved both the main contractor and the owner. Criteria were established for when the two systems were to be applied. The choice of system was based on where it had significant advantages over the other of the two systems. The composite liner with sprayable waterproofing membrane offered the following advantages:

- possibility to locally waterproof smaller areas without the need to bring the waterproofing all the way down to the invert
- no inflammable materials
- reduced total lining thickness

The traditional waterproofing system with PE foam sheets was especially well suited in very wet areas with dense and concentrated drips, as well as the portal areas where there was a need for frost insulation. Figure 3 shows a photo of the traditional waterproofing system partially installed.

In addition, the cost-effectiveness based on the required amount of additional sprayed concrete was to be taken into account. Both the contractor and the owner had a financial benefit of utilizing the sprayable waterproofing membrane instead of the traditional PE foam waterproofing.

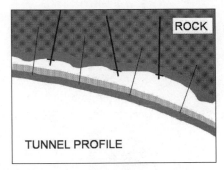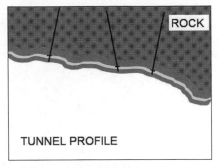

Figure 4.   Basic layout of the two utilized waterproofing systems, cross section of detail close to rock surface. Rock support with sprayed concrete and rock bolts is shown. Left: Traditional waterproofing system with a drainage shield with 50mm thick polyethylene (PE) foam sheets covered with 60 mm thick sprayed concrete leaving a gap for drainage behind the shield. Right: Innovative composite liner with sprayable waterproofing membrane in a sandwich structure between the rock support sprayed concrete and a protective layer of sprayed concrete.

Figure 5.   System layout of composite liner with sprayable waterproofing membrane and sprayed concrete

Properties of the Innovative Waterproofing System

The alternative waterproofing system is basically a composite liner which consists of a spray applied membrane in a sandwich structure between two sprayed concrete layers. The membrane has a normal thickness of 3 mm and a minimum thickness of 2 mm in order to be waterproof.

The main property of this membrane is that it bonds to both the substrate and the concrete which is applied onto it. This again leads to the important technical feature that the two interfaces between membrane and concrete on either side are impermeable. Hence, no water can migrate along these interfaces. Figure 4 shows a principal comparison of the two waterproofing systems. Figure 5 shows a sample cut through the composite liner with spray applied waterproofing membrane and sprayed concrete.

The membrane-concrete interface has a tensile bonding strength of 1–1.2 MPa when completely cured. The curing of the membrane starts immediately after the spray application. The curing process consists of the hydration of the cement part of the product and evaporation of the excessive water. Through this process a continuous waterproof membrane is formed.

The curing process is irreversible. Hence, a subsequent feeding of water onto the membrane after curing, will not lead to a change of the waterproofing properties.

It is important that the membrane thickness does not exceed approximately 5 mm (other than local spots). Excessive membrane thickness might lead to very slow curing and if exposed to water from the substrate, the bonding and waterproofing properties might be compromised.

Waterproofing with a sprayable membrane embedded in a composite liner creates an undrained surface impermeabilization. This might lead to a local build-up of groundwater pressure immediately behind the membrane. When designing the waterproofing of the tunnel one therefore had to bear in mind that a part of the tunnel contour had to remain drained, or a drainage measure had to be considered based on the actual situation.

In most cases, cracks and joints in the rock behind the membrane/concrete sandwich provided sufficient drainage to the invert ("umbrella" solution) without any significant pressure build-up.

APPLICATION OF THE SPRAYABLE WATERPROOFING SYSTEM

Sequence of Operations

In order to provide suitable conditions for the spray application it was necessary to establish a proper substrate. The two main issues of the substrate preparation were substrate roughness improvement and removal of wet spots. After the substrate had met the specified requirements, the application of the waterproofing membrane could commence.

Improvement of the Substrate Roughness

Achieving a continuous membrane with the required thickness, depends on the quality of the substrate. An acceptable roughness in the cm order is achieved by utilizing a sprayed concrete with maximum grain size of 4 mm. A good substrate can also be achieved with 8 mm if the aggregate is very well graded and the actual application of sprayed concrete is particularly aimed at achieving a sufficiently good surface finish.

The surface characteristics regarding edges, holes and other irregularities were important to improve in a way that a thin continuous membrane was feasible. This involved the filling and smoothening out of holes and "trenches" as well as edges, sharp points and major irregularities in the substrate surface. Additional sprayed concrete had to be applied locally to the substrate for this purpose.

Handling of Wet Spots

The wet areas which occurred where the sprayable membrane was intended had to be treated before the application of the membrane. Basically wet spots had to be drained or injected.

The methods which were utilized for the removal of surface water ingress points were installation of drainage half pipes, polyurethane injections, drainage holes with

Figure 6. Mechanized application of sprayable waterproofing membrane with a computerized spraying robot

drainage hoses and priming of the surface immediately before the application with a special rapid curing membrane.

Mechanized Application of the Waterproofing Membrane

The mechanized application of the sprayable waterproofing membrane utilized a state-of-the-art robotic (fully automatic) spraying machine. This machine was originally designed for the robotic application of sprayed concrete with a very precise layer thickness. The features of this machine, in particular the scanning of the substrate and the possibility to achieve a constant speed of the nozzle over the surface, made it suited for the robotic application of the waterproofing membrane.

The actual application of the membrane took place with the dry spraying method. That means the product was delivered using a rotor machine and transported pneumatically in a powdered state through a 32 mm diameter hose to the nozzle. The nozzle for membrane spraying is a standard nozzle for dry mortar spraying. Water is added at the nozzle through the precise adjustment possible with a needle valve. The equipment used for the application is very similar to that of dry sprayed shotcrete, but with minor modifications in order to achieve a low output which could be also be varied.

The mechanized spraying of the membrane took place in the area which was scanned immediately in advance. In the given tunnel dimensions with a cross section of about 64 m^2, it was possible to spray sections of 3 m tunnel length covering the entire tunnel contour. The robotic part of the spraying process was the automatic movement of the boom and the nozzle. Feeding of air, water and the output of the product had to be adjusted manually. Figure 6 shows mechanized application.

The constant velocity and smoothness of the spray manipulator movement, as well as the ability to follow the surface topography made it possible for all given points in the scanned area to receive the exact same amount of membrane. These machine parameters had to be defined in advance. This was done in an initial exercise in order to optimize the mechanized part of the application.

When, during the application some irregularities on the surface were encountered, the necessary adjustments of the machine parameters could be carried out online with subsequent immediate response of the machine and consequential best results.

When starting spraying on each shift, one only had to fine-tune the air, amount of water (in order to achieve to correct mixture ratio) and the output of the dry spraying machine. When the crew had gained sufficient experience with the equipment, capacities of approximately 800 m^2 per twelve-hour shift were realized.

Application of the Covering Layer of Sprayed Concrete

The completion of the composite liner consisted of the application of a final inner layer of sprayed concrete. The minimum thickness of the inner layer sprayed concrete was 4 cm. Steel fiber reinforced sprayed concrete was used for this purpose. This layer was applied after the membrane had reached a sufficient level of curing.

The curing state of the membrane was measured indirectly by measuring the hardness of the membrane surface (Shore A hardness). Criteria based on measured Shore A hardness values were specified for the application of sprayed concrete onto the membrane. Under the given humidity, temperature and ventilation conditions in the tunnel sufficient curing was reached 10–14 days after the spraying of the membrane.

Quality Control Issues

In order to assure that the proper preparations were undertaken and that the actual application of the membrane was done correctly a detailed procedure for the practical works was made. The procedure required the proper preparation works like substrate roughness improvement with sprayed concrete, different surface treatment of seepage points in order to assure feasible conditions for the membrane application.

Furthermore, the details in the application of the membrane were described, in particular how to achieve the correct sprayed thickness on the wall. This included thickness control measurements and the monitoring of the membrane powder consumption. The procedure also specified remedial measures to deal with seepages which occurred through the membrane after application.

EXPERIENCES

The waterproofing of this tunnel with a combination of the two systems with the traditional PE foam system and the composite sprayable liner system was successful. The main experience gained with the sprayable membrane system was that the importance of preparing the substrate properly cannot be underestimated. It was essential to have a smooth substrate with a limited roughness in order to achieve a continuous membrane. In this way a continuous regular composite liner was achieved. A good example of this can be seen in Figure 7.

Also in areas where the waterproofing system was applied only locally in the tunnel contour or in smaller areas, the result was good. No effects of migrating water seepages to adjacent areas causing a possible need for waterproofing in a previously dry area were observed.

The mechanized application of the membrane proved to be cost-effective regarding overall application speed as well as reaching optimum material consumption. With the mechanized application a consistent membrane thickness within the intended range was achieved. This again depended much on the substrate quality.

In areas with large substrate irregularities the robotic application could not be successfully completed so an additional application of the membrane with the same robotic equipment in manual-mode together with computerized support had to be done.

The areas which were sufficiently treated with drainage measures showed no or only very little need for remedial works.

Figure 7. Completed composite liner

Some difficulties with the system during the application were experienced. These were mainly seepages penetrating through the lining in areas where a continuous membrane was not achieved during the spray application. This partially resulted in need for remedial local injection works. One important experience is that seepages through the membrane occur within the first day after application. These cases were targeted directly in a simple manner with local injections since the water seeping points in the membrane could be easily found.

CONCLUDING REMARKS

The cost-effectiveness of the waterproofing system with a composite liner with sprayable membrane depends mainly on two main issues; creating a smooth enough substrate and the extent of wet areas which impose a need for wet spot treatment prior to the application of the membrane.

The experienced advantages with robotic spraying definitely favour this application method. Particularly the consistent membrane thickness which can be achieved with the lowest possible material consumption is a significant advantage. High application capacity with low demand on amount of manpower was realized.

REFERENCES

Blair, A.J. & Lacerda, L.L. 2005. Construction of the Wolf Creek Upper Narrows tunnel waterproof membrane—a comparison of tunnel projects utilizing the spray-on waterproof membrane. *Rapid excavation & tunneling conference 2005, Seattle.* Littleton, USA. SME.
Meier, W. & Holter, K.G. & Häusermann, S. 2005. Waterproofing of an emergency escape tunnel by employing an innovative sprayable membrane: the Giswil highway tunnel project, Switzerland. *Proceeding of the 31st ITA-AITES world tunnel congress 2005, Istanbul Turkey* pp 571–575. Leiden. Balkema.

19

Underground Project
Case Studies

Chair
P. Madsen
Kiewit Construction Co.

BREAKTHROUGH OUTAGE AT THE SECOND MANAPOURI TAILRACE TUNNEL PROJECT, NEW ZEALAND

Ken Smales

Meridian Energy Ltd.

Ron Fleming

URS New Zealand Ltd.

Charlie Watts

Meridian Energy Ltd.

ABSTRACT

The Manapouri Power Station is an underground hydro-electric station located in the South Island of New Zealand, and was constructed in the 1960s. Between 1997 and 2002, a second 10m diameter tailrace tunnel was constructed to increase the output and to improve the overall hydraulic efficiency and output of the station.

A critical economic and project approval consideration for the second tunnel was the outage time required for the breakthrough into the power station. The power station is in part dedicated to providing power to an aluminum smelter with the balance to the National Grid, and apart from the physical problems associated with providing an alternative power supply to the smelter, the loss of generation revenue and supply risk were significant issues.

Innovative planning reduced the planned outage time from several months to 21 days, and the breakthrough was eventually achieved in 11 days as a result of additional value engineering measures implemented during construction.

This paper describes the engineering optimization that took place during the planning stages of the project and later during construction, as well as describing the construction methodology used.

INTRODUCTION

Manapouri Power Station is an underground hydro electric facility initially commissioned in 1969 and now owned and operated by Meridian Energy Ltd. The facility is located in the Fiordland National Park in the south west of the South Island of New Zealand and was developed primarily for the purpose of providing power to an aluminum smelter at Bluff, on the south coast (Figure 1).

The underground powerhouse, as originally designed and constructed, contains 7×100 MW Francis turbines, fed via seven vertical penstocks from the West Arm of Lake Manapouri through a head of 178 metres (584 feet). The draft tubes from the seven machines discharge into a draft tube manifold chamber, thence through a single 10 km long × 9.4 m diameter concrete lined tailrace tunnel (Tunnel No. 1) to the outfall to the sea at Deep Cove.

A 2 km long spiral tunnel provides road access into the power house from the intake area at the West Arm of Lake Manapouri. An important feature of the complex is the sloping tailrace surge chamber tunnel which connects the powerhouse access tunnel (and hence indirectly the powerhouse) to the draft tube manifold.

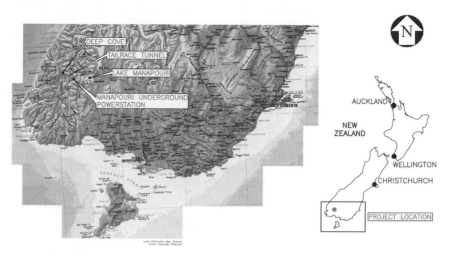

Figure 1.   Project location

Following the commissioning of all seven machines in the early 1970s, it became apparent that the hydraulic head loss in the tailrace tunnel was higher than expected during the original design, resulting in higher operating water levels in the draft tube manifold, and potentially higher surge levels than calculated during design. With the tailrace surge chamber having a direct link to the powerhouse via the powerhouse access tunnel, the peak power output had to be limited by operational procedures to prevent flooding of the powerhouse.

The original design also allowed for the raising of Lake Manapouri to provide additional head. The lake raising was subsequently stopped on environmental grounds, which also contributed to the reduction in power output.

As a consequence of these two factors, the maximum generation capacity was limited to 590 MW compared with the 700 MW installed capacity.

A number of studies were subsequently carried out to find a solution to the tailrace friction loss problem. All solutions required extensive station outage time for construction and it was not until the early 1990s that an economic methodology was devised to minimize this outage time.

The following describes the methodology adopted in resolving the problem, from evolution during the feasibility study through design and construction.

INITIAL FEASIBILITY PLANNING

Desktop studies carried out at various stages to consider an economic method of modifying the tailrace tunnel to increase its hydraulic capacity and reduce the head loss identified several options. These included: reaming out the existing Tunnel No. 1, deepening Tunnel No. 1 and constructing a duplicate tunnel.

These options all suffered from one major unacceptable drawback; the power station outage period required to undertake the works or to breakthrough into the draft tube manifold. Depending on the option, the outage durations varied from about 5 months to several years. The power station is important to providing power to the aluminum smelter and apart from the practical problems of providing an alternative supply from elsewhere in the New Zealand electricity supply system, the loss of generation revenue and supply risk to the aluminum smelter were significant issues which needed to be resolved before any tailrace enhancement project could be implemented.

In 1992, a feasibility study was commissioned by the Owner (then Electricity Corporation of New Zealand-ECNZ), utilizing a small team of experienced tunnel engineers. This team considered and then discarded the options involving modifications to the existing tailrace tunnel due to the extended power station outage implications, before concentrating on a duplicate tunnel option. This study was able to demonstrate the technical feasibility and economic viability of a duplicate tunnel option, with the following changes from the previous studies being key defining factors:

- **TBM Technology.** The principal rock types along the tunnel alignment are Paleozoic gneiss with intrusions of granitic rocks, gabbro and diorite. Over 50% of the tunnel was expected to be in very hard rock, with UCS in the range of 150 MPa to 240 MPa, and joints widely spaced. Previous studies had mainly assumed drill and blast methods, due to the lack of track record and hence lack of confidence that a 10 m TBM could excavate rock of this hardness. During this study it was able to be demonstrated that the rock was within the capabilities of an open hard rock TBM of this diameter and could be excavated at significantly reduced cost and construction duration when compared with drill and blast.

- **Tunnel Stabilization and Lining.** Tunnel No. 1 was excavated by drill and blast techniques and fully concrete lined. It was realized that for the duplicate tunnel to be economic, an alternative to concrete lining had to be developed. The geological logs of Tunnel 1 were analysed in detail and it was determined that some 56% of the tunnel would be in excellent rock conditions, (designated as Type T1, little or no stabilization required) and only about 9% would be Type T3 conditions (requiring steel set support). Apart from five known fault zones, the rock was generally hard and massive, with relatively widely spaced jointing. This led to the conclusion that concrete lining would not be required through most of the tunnel. A state-of-the-art rock stabilization and lining system, primarily based on rockbolts and shotcrete, was therefore developed to a concept design level, and provided the basis for the later detailed design.

- **Power Station Breakthrough Outage Time.** The previous duplicate tunnel options required an extensive power station outage period for the Tunnel No. 2 breakthrough, to firstly remove the TBM at the draft tube manifold and secondly to construct an upstream stoplog structure which was required to facilitate Tunnel No. 2 de-watering for future maintenance inspection. The study team placed considerable emphasis on resolving this outage time issue and succeeded in reducing the time from several months to a planned 21 day period.

The resolution of these issues was instrumental in proving the project's economic viability and resulted in the decision to advance the project to the detailed design and tendering stage.

PROJECT LAYOUT

The overall power station layout is shown in Figure 2. The elements relevant to this paper are the Draft Tube Manifold, the Tailrace Surge Chamber, the Powerhouse Access Tunnel and the existing Tailrace Tunnel No. 1.

The duplicate tailrace tunnel concept comprised a second 10 km long tunnel located approximately 50 metres to the south of Tunnel No. 1, linking into the downstream section of the draft tube manifold through a bifurcation. Refer to Figure 3. The hydraulic analysis indicated that the tunnel should be an excavated diameter of 10 metres, after allowing for the various stabilization and lining categories.

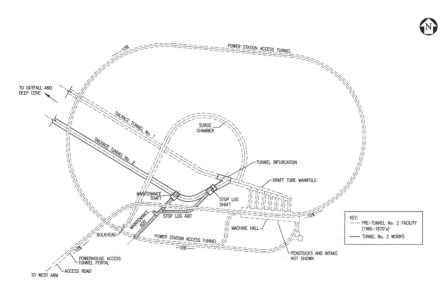

Figure 2.   Plan of Manapouri Power Station

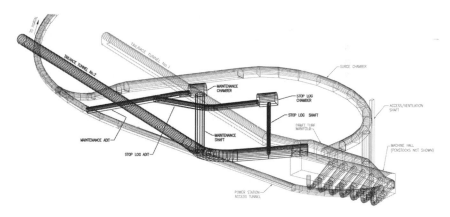

Figure 3.   Isometric layout of draft tube manifold and tailrace tunnels

Tunnels No. 1 and No. 2 are inverted siphons with a decline at each end and a 1 in 1,000 decline from the upstream through the main length of each tunnel to a low point near the outfall. The tunnels each have a submerged outfall to the sea, with the Tunnel No. 2 low point being at EL −44 masl. Apart from a steeper decline for Tunnel No. 2 at the downstream end, the vertical alignment of both tunnels is identical to simplify hydraulic surge characteristics between the two tunnels. At the upstream end, the invert of the Draft Tube Manifold is EL −3 masl.

Tunnel No. 1 has no stoplog or gate provisions for dewatering, being open to the sea at the outfall. Provision had been incorporated in the Draft Tube Manifold to install a 3+ metre high bulkhead across the invert, which permitted dewatering of the mani-

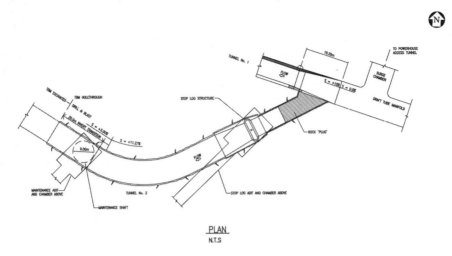

Figure 4.   Headworks plan

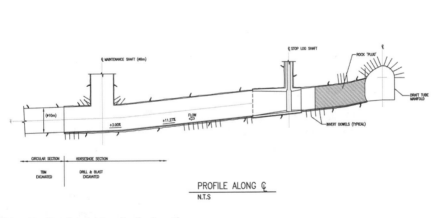

Figure 5.   Headworks longitudinal section

fold for maintenance access to the draft tubes, but not to dewater Tunnel No. 1. The inability to dewater and inspect Tunnel No. 1 was a considerable point of frustration during the various studies carried out into the tailrace tunnel head loss, because it was not possible to clearly determine whether the head loss was due to lining roughness or a major lining failure and blockage. Whilst video inspections using ROVs had been carried out over the first one kilometer from each end, it was not until part way through the construction of Tunnel No. 2 that the ROV technology was available to inspect the full length of tunnel. The video subsequently indicated that no blockages existed and the lining was generally in good structural condition. It was concluded that excessive lining roughness was the main cause of the higher than expected head loss in Tunnel No. 1.

This inability to inspect Tunnel No. 1 led to the decision that Tunnel No. 2 would incorporate a stoplog structure at each end to permit dewatering of Tunnel No. 2 for periodic inspection and maintenance work. The decision to adopt an essentially "unlined" rock tunnel design also determined that maintenance inspections and the ability to carry out repairs would be essential through the life of the tunnel.

Figures 4 and 5 shows the layout of the headworks comprising the Draft Tube Manifold, Tunnel Nos. 1 and 2 and the Tailrace Surge Chamber. The Draft Tube Manifold is a 12 metre wide chamber supported with rockbolts and Pneumatically Applied Mortar (PAM). The height of the chamber varies, and the Tunnel No. 1/No. 2 bifurcation is located in the downstream transition section between the draft tube and the Tunnel No. 1 entry.

TUNNEL NO. 2 BREAKTHROUGH CONCEPT

Excavation of Tunnel No. 2 would be achieved with a 10 metre diameter hard rock open TBM, which would permit flexibility during the excavation in the choice and extent of the tunnel stabilization and lining required. TBM excavation needed to commence from the downstream end at Deep Cove due to the impracticality of providing the TBM construction access from the power station complex at the upstream end.

Earlier studies had assumed that the TBM would break through into the Draft Tube Manifold, before being dismantled and removed via the Tailrace Surge Chamber. This would require Tunnel No. 1 to be partially dewatered, requiring a cofferdam to be constructed at the outfall, then a period of several weeks to dismantle and remove the TBM, before completing the tunnel disestablishment. With the decision being taken to install upstream and downstream stoplog structures in Tunnel No. 2, an additional period of about six months was required to construct the upstream stoplog structure after the TBM had been removed. The total power station outage period assuming this approach was estimated to be approximately 8 months.

Purely on an economic analysis basis, the cost of this extensive outage period in terms of loss of power generation revenue was estimated to be in excess of NZ$100 M at that time, adding 50% to the estimated cost of the project. In addition, extensive transmission system modifications would be required to guarantee an alternative supply of power from the national electricity grid to the aluminum smelter during this period. The project could not be justified with these cost loadings.

Considerable value engineering effort was therefore spent during the feasibility stage of the project in resolving this breakthrough outage issue, to reduce the power station outage period to an economically acceptable duration.

The major change in approach comprised re-scheduling the work so that all of the headworks construction except for the final "plug," could be completed independently of the TBM excavation, off the project critical path, and without the need to cease power generation through Tunnel No. 1. This would be achieved through the following design changes:

- **Headworks Access and Maintenance Shaft.** Providing an additional 6.5 metre diameter access tunnel and 9 metre diameter × 50 metre deep shaft from the powerhouse access tunnel to gain access to the upstream section of Tunnel No. 2, to allow this tunnel construction to proceed ahead of and independently of the TBM operation. This access would later be used to remove the TBM as well as providing long term maintenance access to Tunnel No. 2.

- **Drill and Blast Excavation.** Constructing the upstream 120 metres of Tunnel No. 2 excavation by drill and blast methodology rather than TBM excavation, leaving a 15 metre long rock "plug" for the final breakthrough into the Draft Tube Manifold.

- **Stoplog Shaft and Structure.** Constructing the stoplog access tunnel, stoplog shaft and stoplog structure during this period, installing and commissioning the stoplogs prior to the breakthrough. It was expected that the tunnel plug would be removed prior to TBM breakthrough into this upstream section, therefore the stoplogs needed to be fully tested and commissioned to allow Tunnel No. 1 (and power station) operation to resume after completion of the plug removal works. Whilst the stoplog commissioning could not be achieved under full operational conditions, the stoplog would be pressure tested against the rock plug to prove its water tightness.

- **Tunnel Plug.** Following completion of these headworks, removal of the tunnel plug and construction of the bifurcation. This work would comprise installing an earthfill cofferdam at the Tunnel No. 1 outfall, lowering the water level in the Draft Tube Manifold, excavation of the 15 metre long rock plug (Figure 6), installation of rockbolts and shotcrete lining, lowering of the stoplogs on completion and re-watering the Draft Tube Manifold. Extensive construction planning from first principles was undertaken on the work required to remove this plug to determine the construction duration required to a high level of certainty. It was estimated that this activity could be achieved within a 21 day period, which included a six day contingency allowance.

- **TBM Holethrough.** The TBM would holethrough into the headworks area independently of the Tunnel No. 1 operation, protected in the dry by the upstream stoplog. It would then be dismantled and removed via the Maintenance Shaft.

The success of this concept centred around implementing all the headworks construction except the final breakthrough off the project critical path and without the need for an extended power station outage. Separating it from the TBM critical path provided timing flexibility for the plug removal, allowing a period to be chosen in advance by the Owner when the power outage would have least impact, both physically and commercially, on the power supply system.

This change in approach would potentially reduce the power station outage period from about 8 months to 21 days, as the only work required during a station outage was the work associated with the draft tube manifold dewatering and removal of the 15 metre long rock plug.

CONTRACT PROVISIONS

Site investigations, preliminary design and detailed design was undertaken by URS (formerly Woodward-Clyde) during 1995–1996. URS developed the tender documents with input from Meridian Energy Ltd. and proposed an Interactive Tender Process for the project, as one of several risk management provisions.

The form of contract used for the construction of the Second Manapouri Tailrace Tunnel was a lump sum contract with a schedule of unit prices. The General Conditions of Contract were based on NZS 3910:1987 *Conditions of Contract for Building and Civil Engineering Construction,* modified to suit project specific provisions.

The contract recognized the high risk nature of underground works, the isolation and environmental sensitivity of the site and included risk sharing provisions such as: pay items for different categories of tunnel stabilization and lining, a Geotechnical Baseline Report, Geotechnical Data Report, Escrow Bid Documents and a Disputes Review Board.

The Interactive Tender Process provided for input from selected tenderers commencing at the 90% detailed design stage prior to production of the final documents.

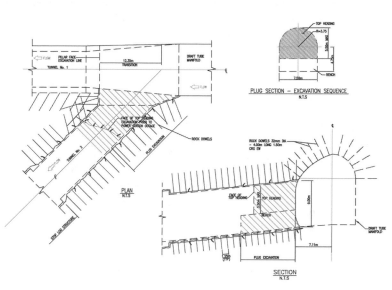

Figure 6.   Plug geometry

This provided valuable constructability feedback from the tenderers into the final tender documents including verification of the practicality of the breakthrough outage concept.

With due regard for the importance and potential impact on the Owner of the breakthrough outage works, a separate "21 Day Outage" specification was developed and this work was identified in the contract as a Separable Portion of the works. Liquidated Damages of NZ$250,000 per day were attached to this separable portion, but an early completion bonus of NZ$125,000 per day was also offered by the Owner for this Separable Portion.

The specification identified some key requirements and constraints for the 21 day outage works as follows:

- It placed considerable emphasis on the need for extensive planning of the outage works by the contractor, requiring a detailed submittal to the Engineer 6 months prior to the event. The submittal was required to cover the detailed methodology, sequencing, equipment and personnel, sufficient to demonstrate that the work could be completed within the 21 day period.

- The contractor was required to develop a separate outage schedule to a timescale of shifts, which had to be reviewed one month prior to the work and then updated daily during the outage works.

- The contractor was required to develop a contingency plan to allow for suspension and demobilization of the works at short notice to permit the power station to return to operation.

- The outage had to take place prior to TBM breakthrough and only in specified low power use months (summer).

- Strict limits were placed on blasting vibrations to protect the completed stoplog structure and to avoid damage to the powerhouse machines some 50 m distant.

- Restrictions were placed on the face area able to be blasted, the round length, the loading and the shooting sequence to avoid damage to the bifurcation pillar, which was expected to be intersected by some major rock joints and a crush zone.
- Instrumentation and monitoring of rock divergence as excavation progressed.

CONSTRUCTION PHASE

General Background

The construction contract for the project was awarded to a Joint Venture comprising Fletcher Construction from New Zealand, Dillingham from the USA and Ilbau from Austria (FDI) in February 1997 for a lump sum price of approximately NZ$190 million.

Following site establishment work, the first round for the headworks was fired on 9 June 1997 in the Powerhouse Access Tunnel with the commencement of the Maintenance Adit. Excavation of the headworks adits, Maintenance Shaft and the drill and blast section of Tunnel No. 2 was completed in April 1998, leaving a 15 m rock plug, as designed, between Tunnel No. 2 and the Draft Tube Manifold.

The rock in the vicinity of the plug and the bifurcation was known to contain a sheared zone, crossing the proposed Tunnel No. 2 obliquely, with the potential for a direct leakage path to Tunnel No. 1. The sheared zone was also close to the bifurcation "nose," raising stability concerns for this area. The designers required the zone to be formation grouted with an umbrella array of holes through and around the plug, prior to completion of this section of Tunnel No. 2 excavation and ahead of the plug excavation. This work was satisfactorily completed, with formation grouting taking some 16 tonnes of cement grout.

Construction of the Headworks Stoplog Structure followed with the majority of the work being completed by February 1999. FDI then chose to delay completion of the headworks in favor of concentrating resources in the TBM tunnel at Deep Cove. The timing of the plug removal was critical in that it could not be scheduled for winter months due to the high power demand on the power station. As a result, the start date for the 21 Day Outage was delayed until 2 December 2000.

Pre-Outage Works

Draft Tube Manifold Dewatering. The plug removal required the lowering of the water level in the Draft Tube Manifold to gain access to the upstream side of the plug. The original design concept was to construct an earth cofferdam across the Tunnel No. 1 outfall channel at the commencement of the outage to permit lowering of the water level in Tunnel No. 1 and the Draft Tube Manifold by pumping at the outfall.

Subsequent experience during the excavation and dewatering of the adjacent Tunnel No. 2 portal at Deep Cove through the alluvial material raised concerns over the potential for excessive leakage by-passing this cofferdam and preventing the required dewatering in the short time available. Further investigations were therefore requested by FDI and carried out by Meridian. These confirmed that a groundwater seal around the cofferdam may be difficult to achieve and hence the pumping capacity required and the dewatering time required would be difficult to pre-determine. This led Meridian to consider other dewatering options to reduce this risk.

Given the likely difficulties in achieving a seal at the downstream end, the further studies concentrated on a mechanism for sealing the upstream end of Tunnel No. 1, at a location immediately downstream of the proposed Tunnel No. 1/No. 2 bifurcation. At that location Tunnel No. 1 is a horseshoe profile, fully concrete lined. The static water depth in the Draft Tube Manifold is 3 metres when the power station is not operating and any temporary bulkhead in Tunnel No. 1 would be totally submerged, with the tunnel invert level at that location being approximately 8 metres below static water level. Construction of the bulkhead would therefore need to be undertaken under water.

Meridian commissioned a separate design for an underwater bulkhead, with the bulkhead being designed as a sectionalized steel bulkhead to be floated in and assembled by divers. Apart from the innovative design required for the bulkhead, there were considerable challenges to overcome in terms of fixing the bulkhead to the tunnel lining, achieving a watertight seal and ensuring that installation and removal of the bulkhead could be achieved within a few hours.

Several short power station outages were required ahead of the main outage, to determine the exact tunnel dimensions, to extract cores from the lining to confirm the competency of the concrete and later to install shear brackets for the bulkhead. In addition, a mock-up of Tunnel No. 1 was fabricated and then submerged in Lake Manapouri to an equivalent depth, which allowed the divers to become familiar with the handling and installation of the floating bulkhead sections ahead of the actual event. These trials led to minor modifications to the design but also identified a few installation constraints that needed to be resolved. This trial was invaluable in sorting out minor design and installation issues and led to the rapid installation of the bulkhead during the actual event, with few difficulties.

At the commencement of the actual power station outage, a team of 18 divers successfully installed the Tunnel No. 1 bulkhead within 15 hours and dewatering was completed 3 hours later. Removal of the bulkhead was also successfully achieved.

Geotechnical Conditions. The investigations and design phase had anticipated that the geological conditions of the plug would consist of gneiss and granitic intrusive rocks which contained a 60m wide sheared zone anticipated to intercept the bifurcation pillar of the plug. However, during excavation of Tunnel 2 good quality rock was intercepted adjacent to the plug. To further evaluate the geological condition of the plug and the anticipated sheared zone, two cored holes (using blow out protection) were drilled from Tunnel 2 towards the Draft Tube Manifold. These cored holes intercepted generally good quality rock and a 1m wide sheared zone. Based on the drill hole information the plug thickness was carefully reduced from 15m to 10m prior to commencement of the outage.

As shown in Figure 7, a sheared zone was intercepted in the bifurcation "nose," but at this location was only approximately 1m wide and did not significantly affect the excavation of the plug. Five 7.5m long rock bolts were drilled through this sheared zone to support the pillar. Due to the better than anticipated geological conditions 3 multiple position borehole extensometers that were planned to be located within the bifurcation pillar were not installed, but were replaced by convergence monitoring. This change in instrumentation significantly reduced the plug excavation period.

Outage Works

The outage work commenced late on 1 December 2000 with the shutdown of the Manapouri Power Station. The works comprised the following elements:

- Installation of the Tunnel No. 1 bulkhead and dewatering of the Draft Tube Manifold as described above.
- Excavation and stabilization with rockbolts of the top heading of the plug from Tunnel No. 2, which commenced as soon as the Power Station shutdown occurred. Holethrough was achieved in the afternoon of Day 2.
- Rockbolt installation around the bifurcation perimeter from the Draft Tube Manifold, which commenced as soon as the manifold had been dewatered.

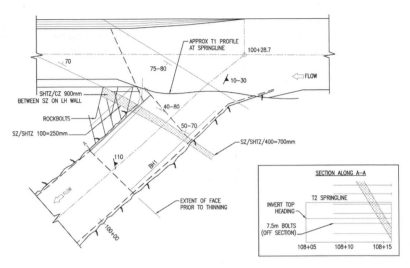

Figure 7.   Geology plug

Figure 8.   View of completed bifurcation from upstream

- Removal of the bottom bench of the plug
- Shotcrete lining of the plug and bifurcation
- Installation of invert anchor dowels and concrete invert
- Lowering and checking the Headworks Stop logs for damage
- Removal of the Tunnel No. 1 bulkhead and re-watering of the Draft Tube Manifold

On 10 December 2000, FDI completed all their work, but a problem with machine maintenance work being undertaken concurrently by Meridian on the draft tube gates that seal off the turbines for maintenance, resulted in a delay to the programme of 20 hours. Once this was completed, the Tunnel No. 1 bulkhead was removed and the Power Station was returned to generation on 12 December 2000.

Putting aside the delay caused by the power station maintenance work, the actual duration required to complete the plug removal work was 10.2 calendar days, resulting in a substantial bonus for FDI (Figure 8).

CONCLUSION

The methodology for achieving the breakthrough from the new Tunnel No. 2 into the existing Draft Tube Manifold was a critical element for the Second Manapouri Tailrace Tunnel project. With of loss of revenue calculated at about $0.5 M per day and the risk problems associated with supplying power to the aluminum smelter from an alternative source, the time required for the breakthrough had to be reduced to a minimum to ensure that the project was financially viable.

An innovative approach to the breakthrough works was initiated during the feasibility stage of the project, with the result that the overall project was able to be proven economically and hence gain financial approval to advance into detailed design and tendering phase.

Additional innovations during construction planning post contract award, provided greater certainty in the timing of the breakthrough works and resolved the remaining concerns which could have resulted in time overruns. This was an invaluable value engineering or fine tuning process.

Detailed planning of the works was required from the contractor commencing six months prior to the planned outage. Additional investigations and a trial bulkhead assembly were undertaken, both of which proved extremely worthwhile. This planning by the contractor FDI was extremely thorough and involved a wide team from Meridian, designers and diving subcontractors with the result that the execution of the work was outstanding. The breakthrough outage works were achieved in 11 days, some 10 days ahead of programme.

In terms of the overall Second Manapouri Tailrace Tunnel Project, 2MTT was successfully commissioned on 5 May 2002. Two hours into commissioning, generation levels had increased to 714 MW, at last achieving the original design generation of 700 MW, some 30 years after the original commissioning.

Measurement of the head in the Draft Tube Manifold at the highest generation level indicated a head of 10.7 m, compared with the design prediction of 12.5 m, indicating that the actual friction loss through Tunnel No. 2 was less than expected. In comparison the head when only Tunnel No. 1 was operating was 30 m, confirming the success of the new tunnel.

In conclusion, the Second Manapouri Tailrace Tunnel project has been an outstanding success in terms of both increased generation output and the cost effectiveness of that additional capacity. The project has received several engineering excellence awards as a result.

REFERENCES

Standards Association of New Zealand, NZS 3910:1987 *Conditions of Contract for Building and Civil Engineering Construction.*

The Implementation and Results of an Interactive Tender Process (T.F Martin, B.P. Heer, M. France, M.J. Bell, 1997)

The Second Manapouri Tailrace Tunnel: Undertaking a Large Diameter TBM Project (B.P. Heer, E Giles, Odd Askilsrud, 1998)

Manapouri Power Station, Second Tailrace Tunnel—Engineering Geological Construction Report, (URS April 2003)

CASE HISTORY OF TRENCHLESS CONSTRUCTION AT THE LOWER NORTHWEST INTERCEPTOR PROGRAM, SACRAMENTO, CALIFORNIA

William Moler

URS Corp.

Debi Lewis

MWH

Matthew Crow

Kellogg Brown & Root

Patrick Doig

Hatch Mott MacDonald

Michael Cole

EPC Consultants, Inc.

Craig Pyle

Yolo Force Main

Steven Norris

Sacramento Regional County Sanitation District

ABSTRACT

The recently completed Lower Northwest Interceptor Program (LNWI) in Sacramento, California comprises, ten pairs of microtunnels totaling 4,369 m (14,417 ft), three horizontal directionally drilled bores totaling 956 m (3,154 ft), and two large diameter tunnels under the Sacramento River totaling 1,198 m (3,953 ft). This paper briefly summarizes surface and subsurface conditions, and planning and design considerations. It then goes on to describe problems encounterd with each trenchless crossing, as well as lessons learned with respect to shaft construction, ground control at shafts during tunnel break-in and break-out, frac-out and heave, settlement, obstructions, coordination with permitting agencies, and application of geotechnical baseline reports.

INTRODUCTION

LNWI Program Description

The Lower Northwest Interceptor (LNWI) is a 30 Km (19 mi) long pipeline constructed for the Sacramento Regional County Sanitation District (SRCSD) that will convey sewage from the growing northern Sacramento County area and the City of West Sacramento to the Sacramento Regional Wastewater Treatment Plant in Elk Grove as shown in Figure 1.

The LNWI is comprised of ten projects including two new pump stations, five pipeline projects and 14 tunneled crossings. The projects were grouped into seven

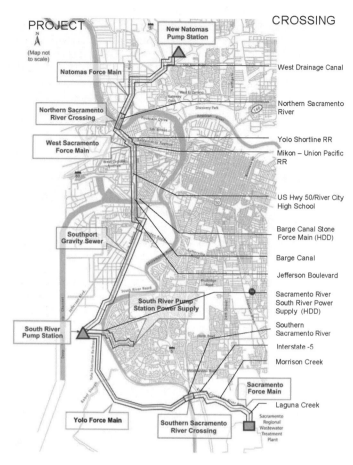

Figure 1. Lower Northwest Interceptor

design and construction contracts, and five construction management contracts, all of which were coordinated by Montgomery Watson Harza (MWH) as Program Manager (Program).

Surface and Subsurface Conditions

The topography of the LNWI alignment is relatively flat except for canals and river levees, and railroad and highway embankments. Land use along the alignment ranges from new housing developments adjacent to a freeway interchange at the northern end, to rural farm land crossed by drainage and irrigation canals, to an urban environment where the alignment passes through the middle of the City of West Sacramento, to wetlands.

Geological conditions along the alignment are characterized by Holocene age (less than 10,000 years old) fluvial and overbank deposits in a generally complex sequence of discontinuous lenses of inter-bedded clays, silts, sands and fine gravel. The subsurface profile is divided into shallow deposits comprised of active Sacramento River sediments (Qsc) and recent alluvium (Qa), underlain by Basin Deposits

(Qb) generated by meandering and seasonal flooding of regional streams and rivers, and the older River Bank Formation (Qrl) formed by alluvial fan material emanating from the Sierra-Nevada Mountains to the east. (Nagel and Nonnweiler, 2003)

Tunneling conditions ranged from very stiff to hard clays and silts, sometimes exhibiting squeezing behavior, to loose to very dense silty sand, and medium dense to very dense gravel occasionally exhibiting flowing characteristics. Groundwater along the alignment is uniformly high, ranging from a depth of 3 m (10 ft) during the dry summer months to ground surface during the wet winter season.

Geotechnical Baseline Reports (GBR) were prepared for all LNWI contracts including open cut excavations as well as trenchless crossings.

PLANNING AND DESIGN CONSIDERATIONS

During preliminary planning and design the location and length of the trenchless crossings were determined considering a number of physical and environmental constraints that would otherwise preclude open cut trenching methods. These constraints included highways, railroads, and waterways. The feasibility of various alternative construction methods for the proposed trenchless segments were evaluated in order to develop preliminary cost estimates and to identify key design issues. The preliminary design is described in Nagel and Nonnweiler, 2003, and Nonnweiler et al., 2003.

During final design some crossings originally considered to be trenchless were converted to open trench. The longest and most complicated tunnel crossings, the Northern and Southern Sacramento River Crossings, the Barge Canal, and the U.S. 50/River City High School Tunnel, were the subject of a number of alternative designs and risk assessment workshops.

CONTRACTING AND BIDDING

All LNWI construction contracts required that tunneling contractors/subcontractors be pre-qualified. Specifications for tunnel boring machines (TBM), microtunneling, horizontal directional drilling (HDD) equipment, and shaft excavation and shoring varied from contract to contract. All contracts contained requirements for contractors and tunneling subcontractors to participate in a formal partnering process. The contracts also required establishment of a Disputes Review Board (DRB).

Table 1 summarizes, for each LNWI crossing, project name, design consultant, construction manager, prime contractor and tunneling subcontractor, length of crossing, tunnel diameter, tunneling method and equipment, casing type and diameter where applicable, carrier pipe diameter and type, total cost of crossing excluding mobilization and incidental costs, and unit cost per linear foot of tunnel.

CONSTRUCTION

Following is a description of the problems experienced during construction of each trenchless crossing. Production data including: start and completion dates, pipe or segment installation time, penetration rate, and tunneling efficiency are summarized in Table 2.

West Drainage Canal

No significant problems were encountered while driving the twin West Drainage Canal microtunnels and no settlement was recorded. However, arcs had to be cut in

Table 1. LNWI tunnel summary

INTERCEPTOR SEGMENT	CROSSING	DESIGN CONSULTANT	PRIME CONTRACTOR/ TUNNELING SUBCONTRACTOR	TUNNEL LENGTH (LF)	TUNNEL DIA (IN)	TUNNEL METHOD	TUNNEL EQUIPMENT	CASING OD (IN)	CASING TYPE	CARRIER PIPE ID (IN)	CARRIER PIPE TYPE	COST TOTAL ($)	COST UNIT ($/LF)*
Natomas Force Main	West Drainage Canal	Black & Veatch	Mountain Cascade/ Vadnais	380	82	Micro Tunnel/Pipe Jack	Soltau RVS 600A-S		Permalok Steel	Twin 60	Ameron 60-375 RCCP	1,857,600	2,494
Northern Sacramento River Crossing	Sacramento River	Hatch Mott MacDonald	AFHolder	1,975	198	EPBM	Lovat TBM	181	Pre-Cast Concrete Segments	Twin 60	Northwest Pipe C200 WSP	11,658,391	5,894
West Sacramento Force Main	Yolo Shortline RR	CH2MHill	Mountain Cascade/Michaels	228	84	Micro Tunnel/Pipe Jack	Ackerman	78	Permalok Steel	Twin 60	Ameron C300 RCCP	1,118,600	2,442
	Milton – Union Pacific RR			394				78	Permalok Steel			2,435,449	3,440
	Milton – Union Pacific RR Inclined Risers							N/A					
	U.S. Hwy 50/High School			1,402	84			78	Permalok Steel			6,838,535	2,045
	Barge Canal			1,492					N/A			6,330,666	2,178
	Barge Canal Inclined Risers			313	36	Horizontal Directional Drill				Twin 24	HDPE	656,000	400
	Barge Canal Stone Force Main		Mountain Cascade/Cherrington	827	84	Micro Tunnel/Pipe Jack	Ackerman			Twin 60	Ameron C300 RCCP	2,424,545	2,582
	Jefferson Boulevard		Mountain Cascade/Michaels	313									
South River Pump Station Power Supply	Sacramento River	HDR	The HDD Company	1,500	36	Horizontal Directional Drill				24	PVC	738,000	542
Yolo Force Main	Babel Slough	CDM	Las Vegas Paving/Vadnais	490	75	Micro Tunnel/Pipe Jack	Soltau RVS 600A-S	82.5	Permalok Steel	Twin 66	Ameron C300 RCCP	2,684,475	5,535
Southern Sacramento River Crossing	Sacramento River	Hatch Mott MacDonald	AFHolder	2,083	198	EPBM	Lovat TBM	181	Pre-Cast Concrete Segments	Twin 66	Northwest Pipe C200 WSP	11,915,795	5,712
Sacramento Force Main	Interstate-5	URS	Rados/Fowler	640	83	Micro Tunnel/Pipe Jack	Soltau RVS 800	82.5	Permalok Steel	Twin 66	Ameron C200 WSP	3,345,997	2,614
	Interstate-5 Inclined Risers			236				N/A			Ameron C200 WSP	1,203,600	2,550
	Morrison Creek			880							Ameron C300 RCCP	8,077,088	2,729
	Laguna Creek			600									

* Cost includes: 1) shaft construction including ground treatment; 2) tunneling; and 3) lining or casing where applicable. Cost does not include: 1) mobilization/demobilization or other general costs; and carrier pipe.

Table 2. LNWI tunnel production summary

CROSSING	START DATE	COMPLETION DATE	NO. TUNNELS	TOTAL LENGTH (ft)	CASING/PIPE/ SEG OD (IN)	NO. PIPES/SEGMENTS	TOTAL TIME (h)	AVG PIPE/SEGMENT CYCLE TIME (Min)	BEST CYCLE TIME (Min)	TOTAL EXCAVATION TIME (h)	AVG EXCAVATION TIME (Min)	BEST EXCAVATION TIME (Min)	AVG PENETRATION RATE (Ft/Min)	EFFICIENCY (TIME EXCAVATED/TOTAL TIME)	EXCAVATION RATE (Cy/Hr)
West Drainage Canal	7/31/2005	8/10/2005	2	760	78	38	119	188	56	47	74	30	0.27	39%	20
Sacramento River	10/17/2005	12/8/2005	1	1,975	182	494	891	108	20	265	32	10	0.12	30%	50
Yolo Shortline RR	3/5/2005	5/30/2005	2	456	68	23	140	30	70	42	40	25	0.50	30%	28
Mokon - Union Pacific RR	6/6/2005	6/29/2005	2	788	68	39	190	30	70	57	45	25	0.45	30%	25
Mokon - Union Pacific RR Inclined Risers	7/11/2005	8/4/2005	2	400	68	20	70	30	70	21	45	25	0.44	30%	25
U.S. Hwy - 50 River City HS	10/3/2005	12/12/2005	2	2,852	68	139	384	24	60	212	30	18	0.68	55%	38
Barge Canal	4/25/2006	6/18/2006	2	2,320	68	116	372	24	60	186	35	18	0.57	50%	32
Barge Canal Inclined Risers	1/27/2006	3/22/2006	2	600	68	30	80	24	60	40	35	18	0.57	50%	32
Barge Canal Stone Force Main (HDD)	2/12/2005	2/24/2005	2	2,916	36	N/A	230	N/A	N/A	N/A	N/A	N/A	N/A	N/A	N/A
Jefferson Boulevard	11/22/2004	2/19/2005	3	936	68	47	364	120	90	89	50	30	0.40	24%	22
Sacramento River South River Pump Station Power Supply (HDD)	7/22/2005	10/19/2005	1	1,500	36	N/A		N/A	N/A	N/A	N/A	N/A	N/A	N/A	N/A
Babel Slough	7/13/2005	8/25/2005	2	980	75	48	402	502	118	212	265	35	0.08	53%	6
Sacramento River	6/3/2005	8/23/2005	1	2,083	182	521	1389	160	25	349	40	15	0.10	25%	40
Interstate-5	10/27/2005	2/4/2006	2	1,280	82.5	56	500	536	170	328	351	45	0.07	66%	5
Interstate-5 Inclined Risers			2	472	75		No Data Available								
Morrison Creek	6/6/2005	8/15/2005	2	1,760	75		No Data Available								
Laguna Creek	9/17/2005	10/5/2005	2	1,200	75		No Data Available								

the waler flanges at the break-out and break-in locations in the launching and receiving shafts to accommodate the tunnel seals. In addition a diagonal strut interfered with placing casing pipe directly into the launching shaft and the receiving shaft was too small to accommodate the entire machine, and therefore the MTBM had to be disassembled before removal. (Doig et al., 2006)

Northern Sacramento River

Issues during construction of the Northern Sacramento River Crossing (NSRC) tunnel (Togan et al., 2007) included:

- A permit from the State Reclamation Board (SRB) prohibited any construction activity between the levees of the Sacramento River between 1 November and 15 April. To avoid schedule risk related to this blackout, the contract documents required the contractor to mobilize two TBMs. Following award, the contractor successfully proposed a Construction Incentive Change Proposal (CICP) that would allow him to mobilize a single TBM. This resulted in a net saving of $1,330,085 to the project. The contractor's revised scheduled called for mining the crossings sequentially during the Summer of 2005 starting with the Southern Sacramento River Crossing (SSRC) and completing the NSRC prior to the blackout period. However, delays experienced with the SSRC related to removing the TBM from the receiving shaft, caused a delay in mobilization for the NSRC. The revised schedule showed that tunneling activity would extend into the blackout period. Following several meetings and review of long range weather forecasts, the SRB agreed to allow the contractor to complete the tunnel and install carrier pipe into the blackout period.

- The TBM was stopped for a period of 5 d with the cutter head directly beneath an abandoned house belonging to the City of Sacramento at a depth of approximately 7m (23 ft) from tunnel crown to ground surface, in order to install the TBM's trailing gear prior to tunneling beneath the levee, an occupied house, and the river. This stoppage resulted in loss of face pressure and settlement of 50 mm (2 in.) adjacent to the house which caused damage to the wooden structure. The contractor is responsible for either repairing or demolishing the house depending on the result of negotiations with the City of Sacramento.

- Immediately after the earlier settlement and prior to tunneling beneath an occupied house, the TBM was stopped over a weekend with the cutter head directly under the north levee of the Sacramento River at a depth of 17 m (56 ft) below the crest. The resulting dissipation of face pressure, and slow intermittent progress resulting from unavailability of computerized alignment information, caused observed surface settlement of up to 50 mm (2 in.).

Yolo Short Line Railroad

The Yolo Short Line Railroad (YSRR) twin, single pass microtunnels were driven without significant difficulties.

Mikon (Union Pacific Railroad)

The Mikon Crossing twin, double pass microtunnels and single pass inclined risers were driven without problems.

U.S. Highway 50/River City High School

Significant issues that arose during construction of the twin U.S. Highway 50/River City High School (Hwy 50/RCHS) microtunnels included:

- The Hwy 50 crossing and RCHS were bid as two separate tunnels sharing the same launching shaft south of Hwy 50. The Contractor offered a CICP to micro-tunnel the entire stretch in a straight 424 m (1,400 ft) alignment. These are among the longest microtunnels constructed in the United States to date.

- Because of the length of the bores, the tunneling subcontractor installed two intermediate jacking stations each separated by 180 m (600 ft), but they were not utilized.

- The microtunnels were launched from a shaft dewatered and shored with sheet piles just north of Hwy 50. Problems were encountered with flowing sand at the break-out points because of inadequate dewatering in mixed faced (clay over sand) conditions. This resulted in inflow of water and sand into the shaft, and settlement just outside the sheets between the shaft and the freeway. The shaft had to be filled with water on several occasions to equalize hydrostatic pressure in order to allow contact grouting from the surface to be performed.

- A significant technical problem to be overcome was that the portion of the tun-nel under Hwy 50 was the first part of the crossing to be mined but also the part that had to have a steel casing. The contractor chose to microtunnel/pipe jack 60 in. RCCP pipe under the highway and the high school to a point 68 m (225 ft) short of the receiving shaft. At that point the RCCP was fitted with a specially made flange that allowed connection to 72 in. ID Permalok steel casing. The microtunnels were then holed-through to the receiving shaft. The steel carrier pipe was then installed inside the casing pipe, welded and grouted into place.

- Ground heave and "frac-out" of slurry as well as bentonite grout occurred during mining of both drives under the high school track, basketball courts and tennis courts. These frac-outs occurred about 15 m (50 ft) behind the cutter head and were thought to be caused by a combination of latent face pressure from the MTBM and grout pressure trapped under about 4.5 m (15 ft) of clay cover. All fluid was cleaned up and heave repaired by the contractor.

Barge Canal WSFM

Problems encountered while constructing the twin single pass Barge Canal micro-tunnels included:

- Excavation of the launching shaft on the south side of the Barge Canal took about 1 yr longer than originally anticipated. Although the shaft was designed to be constructed in the wet with a mass concrete tremie seal in the invert, the Con-tractor chose to dewater the shaft and excavate in the dry. Problems were encountered with split sheets while dewatering and excavating the shaft for the first time even though the ground at the break-out points on either side of the shaft had been jet grouted. This led to a series of partially effective measures to control the inflow of water and soil over the next 12 months, and repeated flood-ing and unwatering of the shaft. These measures included additional jet grouting, construction of a concrete diaphragm wall adjacent to the sheet piles, contact grouting from the surface and from ports drilled into the sheet pile walls. After reaching the invert elevation a structural concrete slab was poured with relief holes, while the shaft continued to be dewatered. Because of concerns about the condition of the mixed face ground (sand and gravel over clay) at the break-out points in the shaft, the Contractor chose to install concrete collar seals poured inside steel forms with hinged steel doors at each tunnel eye, flood the shaft and cut out the sheets underwater with divers to guard against flow of cohesionless material into the shaft. Finally, because of the thin concrete invert slab, the designer became concerned with rebound of soil under the bottom of the shaft, and that subsequent re-consolidation during shaft backfilling would cause the carrier pipes to deflect beyond design tolerances. Therefore, the shaft was back-filled with light weight controlled density fill, produced with light weight aggregate, to reduce load on the foundation. According to the original schedule the Barge Canal Crossing was to be the second crossing to be microtunneled but ended up being the last. The contractor bore responsibility for most of the delays and cost involved, however change orders resulting from shaft excavation totaled $450,000 and 35 d of additional contact time. (Mueller et al., 2006)

- Lower than anticipated rates of advance were achieved in both bores because of sticky clay conditions near the receiving shaft.

- A steel obstruction of unknown nature was encountered in the east bore approximately 130 ft short of the receiving shaft. The MTBM pushed through and over the obstruction with only minor damage to the outside of the forward shell. The obstruction caused a 1 d delay.

- Frac-out of slurry occurred approximately 45 m (150 ft) short of the receiving shaft when there was about 6 m (20 ft) of cover over the top of the MTBM.

Barge Canal Stone Force Main (HDD)

While constructing the twin HDD drives for the City of West Sacramento's twin Stone Force Main crossing of the Barge Canaldrilling fluid migrated horizontally a distance of about 9 m (30 ft) to the Barge Canal launching shaft. This occurred at a point near the southern end of the crossing when the drill string was at a depth of about 8 m (26 ft). A claim for a Differing Site Condition (DSC) for alleged gravel was settled for $10,000.

Jefferson Boulevard

Problems that occurred during construction of the twin, single pass Jefferson Boulevard microtunnels for the WSFM and the single microtunnel for the Stone Force Main included:

- The first microtunnel for the West Sacramento Force Main (WSFM) was the west bore under Jefferson Boulevard. Approximately 150 mm (6 in.) of roadway heave was experienced probably due to excessive face pressure in conjunction with relative shallow clay cover.

- During the second day of tunneling for the first bore, an unmarked low pressure, 4 in. diameter natural gas line owned by Pacific Gas and Electric (PG&E) was punctured by the cutter head. This resulted in closure of Jefferson Boulevard over a weekend while the pipeline was re-routed and the abandoned pipeline removed. SRCSD has submitted a claim to PG&E to cover additional contractor costs and the City of West Sacramento emergency response due to striking the unmarked gas-line. PG&E has submitted a counter claim against the contractor.

- Frac-out of slurry was observed at the south edge of Jefferson Boulevard near the receiving shaft for both drives. The frac-out probably occurred in the center of the roadway and was associated with the heave but slurry was trapped by the asphalt and only appeared at the roadway shoulder.

Sacramento River, South River Pump Station Power Supply (HDD)

A multi-chamber HDPE conduit pipe was pulled inside asteel casing installed under the Sacramento River byHDD as part of the Sacramento Municipal Utility District's (SMUD) power supply to the LNWI South River Pump Station. During grouting of the annular space inside the steel casing, one of the four conduit chambers partially filled with grout. After repeated attempts to ream out the blockage with no success, the contractor chose to install an additional conduit at his own expense.

Babel Slough

The first (west) microtunnel drive of the Babel Slough crossing encountered a nest of buried railroad ties that had been discarded in a previously unidentified dump over a distance of approximately 160 ft starting at a point about 12 m (40 ft) from the launching shaft. The MTBM cut through the wood and completed the remainder of the tunnel without incident. The second bore did not encounter any obstruction. Approximately one day was lost due to slow advance rate while miningthrough the ties. A claim for DSC was settled for $10,500.

Southern Sacramento River

Problems during construction of the SSRC(Togan et al., 2007) tunnel involved:

- Bubbles were observed in the Sacramento River as the tunnel face passed underneath. As there was was no evidence of frac-out of ground conditioner or grout from segment grouting, these bubbles appear to have been associated with EPB face pressure.

- On the west side of the river, as the TBM approached the receiving shaft at a depth of about 15 m (50 ft), the tunnel passed from relatively cohesive clay to sand. This resulted in a maximum of 100 mm (4 in.) of surface settlement before the TBM operator was able to make the necessary adjustment to face pressure. The settlement did not result in any damage as it occurred in an agricultural field, and further tunneling demonstrated classic ground control including minor heave associated with low cover.

- As the TBM approached the receiving shaft on the west side of the Sacramento River probe holes were drilled into the 9 m (30 ft) long jet grouted block of soil at the tunnel eye behind the sheet pile shaft wall. The contract mandated contact grouting of the jet grout block to control water inflow to specified requirements, was not fully complied with, nor was a formal seal, similar to the launching shaft seal, installed at the hole-through point as initially proposed by the contractor. A port with a guillotine gate was installed through the sheets to observe ground and water conditions. The shaft was initially dry, however as the TBM advanced toward the shaft water inflow increased and was periodically arrested by contact grouting. Finally the bottom of the TBM punctured the sheet pile wall accompanied by an inflow of water and sand into the shaft. The sheets were cut out and the TBM pushed into the shaft. However, water and sand continued to flow into the shaft at the base of the TBM. The ground loss resulted in settlement of the ground surface outside the sheetpile shaft wall. In order to control ground loss, the shaft was flooded to equalize hydrostatic pressure. Disposal of dewatering flow from the shaft into the adjacent river was constrained by the permit from the Regional Water Quality Control Board. Therefore, an alternative solution was developed whereby ground water was pumped into temporary holding ponds and then into the dewatering conveyance pipe from the adjacent Yolo Force Main Project and conveyed to a dewatering treatment facility 8 km (5 mi) away prior to discharge into the river. Because of these problems removal of the TBM, and consequently mobilization to the NSRC was delayed by three weeks. A settlement of $240,000 was reached with the contractor.

- The TBM's trailing gear became stressed as it navigated the 300 m (1,000 ft) radius vertical curve between opposing 6% grades under the river. Bolts had to be replaced and welds strengthened in order to proceed.

Interstate 5 (I-5)

Issues during construction of the twin Interstate-5 microtunnels and inclined risers included (Doig et al., 2006):

- Hot-tap observation holes were drilled through the sheet pile shaft walls prior to cutting out to launch the MTBM. Water and soil flowed into the shaft through a small gap < 25 mm (<1-in.) between the sheets and the jet grouted block outside the shaft. Multiple stages of grouting with micro-fine cement were required to fill the gap and seal off flow. It was not clear whether the gap was the result of jet grout not effectively reaching the sheets or the sheets having deflected away from the jet grouted ground during shaft construction.

- There was a problem with the Permalok casing "egging" at some of the joints during mining. Efforts to identify the cause of the problem were unsuccessful, despite intervention by the pipe manufacturer. Ultimately, the egging was accepted as the pipe had sufficient clearance to allow installation of the carrier pipe.

- On the first run under I-5, the MTBM got stuck approximately 6 m (20 ft) short of the receiving shaft. Efforts to free the machine were unsuccessful. The tunnel rose sharply along its length so that the receiving shaft was only 7 m (24 ft) deep. As the tunnel was clear of the freeway by this time and under a field, the machine was dug out and the pipe finished in open-cut. The machine head was seen to be heavily plugged, which could indicate that the slurry cleaning system had not been operating effectively.

Morrison Creek

Problems recorded during construction of the twin Morrison Creek microtunnels included:

- During the first (west) drive, the MTBM became stuck about 0.5 m (1.5 ft) short of the receiving shaft. The reason for this was that the crew took Sunday off and when they returned to work they were unable to budge the machine. As with the I-5 drive, the head was plugged. After several days of trying to resume tunneling, the machine was excavated and removed. This unanticipated excavation occurred very near a protected wetland but was completed without incident.

- The MTBM rate of advance was significantly impacted by to a controversy between the prime contractor and the tunneling subcontractor regarding removal of spoil from the slurry settlement tank. Because solids were not removed in a timely manner slurry denser than desired was re-circulated to the face which resulted in reduced penetration rates.

Laguna Creek

Mining production during construction of the twin, single pass Laguna Creek crossing was impacted because three drive motors had to be switched out over the course of the two microtunnel drives.

LESSONS LEARNED

Following is a summary of the lessons learned from the case histories of the LNWI trenchless crossings previously described.

- Tunneling contractors/subcontractors and other specialty sub-contractors should be pre-qualified including experience of the contractor, qualifications of operators, and history of equipment.
- Ground improvement by jet grouting at tunnel eyes does not in itself guarantee success in controlling inflow of water and ground movement. Specifications should call for a careful, step-by-step, systematic approach of probing and contact grouting of the jet grouted ground behind shaft shoring accompanied by other methods if necessary, prior to launching the TBM.
- Launching and receiving shafts should be dewatered to minimize the possibility of flowing ground during tunnel break-out and break-in.
- Formal seals or double sheets should be specified at all shaft break-in and break-out points to minimize the potential for inflow of water and soil.
- No problems with frac-out into waterways was experienced on any of the HDD crossings, contrary to fears of a number of the regulatory agencies.
- Frac-out can be experienced on microtunnels with shallow cover and face, and grouting pressures must be carefully controlled. Implementation of contingency plans proved crucial.
- Shaft design and construction by the prime contractor should be carefully coordinated with the tunneling subcontractor. Contractors should ensure that subcontractor designed shafts take into account the dimensions of the tunneling, pipe jacking equipment, and seals, as well as casing and carrier pipe.
- GBRs proved to be a useful tool in defining subsurface conditions for comparison with conditions encountered but were not utilized to resolve any formal disputes on any of the trenchless crossings. However, the inclusion of GBRs in all LNWI contract documents may have avoided disputes, as the definition of expected conditions was clear.
- When in sticky clay, mining should continue.
- Coordination with permitting agencies should be transparent over the course of design and construction to facilitate modifications to permits if required by unforeseen conditions.
- TBM stoppage under sensitive features such as levees, roads, and buildings, should be minimized by specifying continuous operation.
- Contractors should not be allowed to construct shafts without an approved submittal.
- Use of a conveyor mucking system alleviated safety concerns while mining on 6% grades.
- Even though the specifications called for settlement monitoring and contractor corrective action depending on the amount, settlement usually occurred faster

than it could be measured, and stopping the TBM to assess the problem usually resulted in exacerbating the problem.

- Public outreach issues due to vibratory sheet pile driving are significant. Noise mitigation for 24 h tunneling should be considered, including the use of new, quiet generators.

- Construction risks can be managed by an owner by engaging designers, construction managers, and pre-qualified contractors experienced in underground construction, working as partners.

ACKNOWLEDGMENTS

The authors would like to thank the SRCSD for permission to publish this paper. Thanks also to members of the LNWI Program, construction managers, and contractors for their support and cooperation in its preparation.

REFERENCES

Nagel, G. and J. Nonnweiler. 2003. Preliminary Design of Tunnels for the Lower Northwest Interceptor, Sacramento, California, USA, Proceedings of the Rapid Excavation and Tunneling Conference, New Orleans, 2003, Society for Mining, Metallurgy, and Exploration, Inc. Littleton, CO

Nonnveiler, J., J. Hawley, and D. Young. 2003. Design of Lower Northwest Interceptor Tunnel Beneath the Sacramento River, Proceedings of the Rapid Excavation and Tunneling Conference, New Orleans, 2003, Society for Mining, Metallurgy, and Exploration, Inc. Littleton, CO

Mueller, C., and G. Klein. 2006. Managing Design Changes and Risk During Construction. Proceedings of the North American Tunneling 2006 Conference, Chicago, 2006,

Doig, P., and A. Page. 2006. Microtunneling on the Lower Northwest Interceptor, Sacramento, California, Proceedings of the North American Tunneling 2006 Conference, Chicago, 2006,

Togan, T., Martz, D., Chen, W., Crow, M., Young, D., Moler, B., Norris, S. 2007. Construction of the Sacramento River Tunnels on the Lower Northwest Interceptor Sewer, Sacramento, CA, USA. Proceedings of the Rapid Excavation and Tunneling Conference, Toronto, Society for Mining, Metallurgy, and Exploration, Inc. Littleton, CO

DESIGN AND CONSTRUCTION OF
DULLES AIRPORT TUNNELS

David P. Field
Hatch Mott MacDonald

Allan Sylvester
Clark Construction Group

Diane R. Hirsch
Metro Washington Airports

Paul E. Gabryszak
Hatch Mott MacDonald

ABSTRACT

With construction of the first phase of the Automated People Mover (APM) Tunnels nearing completion, this paper will address how knowledge from recent Tunnel Boring Machine (TBM) and New Austrian Tunneling Method (NATM) tunnel construction at the airport was used to refine the methodology for the next phase of tunnels currently in design. The benefits include an enhanced understanding of the ground conditions; allowing optimization of excavation sequences and support requirements; a more effective instrumentation and monitoring program; and improved construction details and related contract provisions.

PROJECT DESCRIPTION

Between 1995 and 1998, passenger numbers at Washington Dulles International Airport (IAD) increased by 28 percent, from 12.5 million to nearly 16 million, surpassing the national average growth rate. In 2000, the Airport surpassed the 20 million passenger mark. In 2003, even with the reduction in air travel following September 11, 2001, Dulles served 17 million passengers.

To accommodate the projected growth, the Metropolitan Washington Airports Authority (MWAA) established the Dulles Development (d2) Program to upgrade their facilities including modernization of the Main Terminal, new midfield concourses, additional runways, two new parking garages, a new airport traffic control tower and an Automated People Mover System (APM) to connect travelers to the current and expanded facilities. Under full implementation, this development will increase IAD's capacity to serve up to 55 million passengers per year.

The d2 Program initially commenced in 2001 and included the design of approximately 52,500 feet of APM and associated service tunnels. The network of tunnels included separate Domestic and International APM systems, pedestrian walkways, airport vehicle service, baggage transfer and utility tunnels. Due to funding constraints associated with reduced traffic post 9/11, MWAA was forced to re-evaluate the planned program of tunnels and certain components were deferred at a 60% level of design.

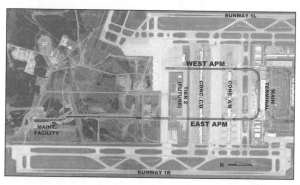

Figure 1. APM system plan

Components of the tunnel program that were carried forward into final design and construction include:

- West Utility Building Tunnel—construction completed in 2006
- West Domestic APM (WAPM)—turned over for APM fitout in January 2007
- East Domestic APM (EAPM)—turned over for APM fitout commencement in July 2006

The Domestic APM tunnels were initially a single construction package, but due to the size of construction and the addition of advanced excavation and station construction packages, it was split into East and West Corridors (Figure 1).

This paper will focus on the design considerations and construction of the East and West Domestic APM Tunnels. Three tunnel construction methodologies were adopted: NATM, TBM, and Cut and Cover. The primary focus of the paper will be to address the design basis of NATM and TBM driven tunnels, adopted adjustments to the proposed construction methods and refinements being carried forward into the phase of tunnel work currently under design.

DESIGN BASIS

The WAPM tunnel package is comprised of 2,100 feet of twin running tunnels inclusive of a 600 feet long advance excavation and support of excavation package for the Tier 1 West Station. The majority of the running tunnels were completed using NATM with 680 feet drives north of the Tier 1 station towards the Main Terminal Station and 250 feet long tail track tunnels south of the Tier 1 Station.

The EAPM tunnel package comprises a total tunnel corridor length of 8,745 feet, inclusive Tier 1 East Station, Tier 3 Station Shell and Tier 4 Egress Structure. The Tier 2 East Station is currently under construction by others. The twin TBM drives were launched from the Tier 2 East Station excavation driving 1,500 feet, prior to being skidded through the Tier 1 East Station excavation to be re-launched for the remaining 650 feet of north drive under Taxilane A. Two, 400 feet NATM drives complete the tunnels from the TBM reception shaft to the Main Terminal Station under the MU2 Baggage Handling Facility. Both the WAPM and EAPM included significant cut and cover structures which are not discussed herein.

Table 1. Derivation of design parameters for adopted support classes

Support Class	GSI Range		RMR (89) Range		Q' Range		UCS	
	Low	High	Low	High	Low	High	psi	MPa
I	40	50	45	55	0.64	1.95	7,000	43
II	25	40	30	45	0.12	0.64	5,000	30
III	15	25	20	30	0.04	0.12	3,000	18

Geology/Ground Conditions/Ground Cover

Dulles Airport is located within the Piedmont Physiographic Province along the eastern edge of a geologic structural basin known as the Culpeper Basin. The basin is a north to northeast trending fault-bounded trough that lies within the upper Precambrian and lower Paleozoic crystalline rocks of the Piedmont. The basin was formed as a result of continental rifting during the Triassic and Jurassic Periods of the Mesozoic Era. The rock formation that directly underlies the Airport site is the Balls Bluff Siltstone. The rocks of the Balls Bluff are generally comprised of grayish and reddish calcareous siltstone with thin layers or limestone, dolomite, mudstone and shale.

Typically, the bedrock is overlain by a transitional sequence of weathered material which transitions from soil to rock of varying degrees of weathering. This transition is gradual in nature with little distinction between layers. The rock lithology generally consists of interbedded mudstone and siltstone with lesser amounts of claystone and sandstone. Most of the rock is characterized as blocky or massive sedimentary rock, having the texture and composition of shale, but lacking its fine lamination or fissility.

For design purposes, the rock has been categorized into two layers; an upper weathered fractured zone and the bedrock. The upper weathered fracture zone is the transitional layer between the saprolite and bedrock. Additionally, the bedrock itself varies in quality. Therefore, three rock mass classifications were assigned to the bedrock for the purposes of design. Classifications were established based on assessments of the Rock Quality Designation (RQD), the degree of weathering, and the discontinuity spacing. These assessments were used in developing ratings of rock mass quality, such as the Rock Mass Rating (RMR), the Tunneling Quality Index (Q) and the Geological Strength Index (GSI). Tables 1 and 2 summarize the derivation of design parameters and rock mass classifications used for design.

In addition, the geology also consisted of a number of joint sets, faults and shear zones located on the property. Although the number of anticipated occurrences of these discontinuities was expected to be infrequent, contingency measures were designed to manage these conditions during construction.

Groundwater levels across the project site varied from 1' to 30' below ground surface but were typically located 10' below ground surface. Groundwater levels fluctuated with seasonal variation, rainfall, temperature, leaking utilities, construction dewatering and pumping from nearby wells. The hydraulic conductivity of the rock is generally low and is typically limited to joint flow. However, in the areas of faults and shear zones, high rate, short term inflows, or flush flows were anticipated to be encountered. The steady-state infiltration rates were anticipated to be in the order of 10 to 50 gpm per 1,000 linear feet of tunnel. The flush flow rates were anticipated to be on the order of 50 to 200 gpm emanating from a very localized zone, however they were expected to be transient in nature and decrease rapidly with time.

Table 2. Rock mass classification

Rock Class	General Description	Key Parameters			
		Major Structural Features	GSI	Groundwater Inflows	Typical Ground Behavior
I	Blocky, well interlocked, slightly weathered, moderately rough, wide to moderately spaced discontinuity surfaces	Local specific shear surfaces-may be sub-horizontal, sub-vertical or inclined	>40	Low	Elastic, minor individual block failures
II	Blocky-seamy, slightly to moderately weathered, stress relieved rock, with moderate to closely spaced discontinuity surfaces	Local shear surfaces and stress relief due to unloading / weathering adjacent to rockhead	25–40	Low-moderate	Mostly elastic, some yielding, loosening in crown and face
III	Blocky-crushed, sheared/faulted, slicken-sided, mineralized and/or clay coated, weathered surfaces or weak saprolite/residual soil/fill	Major shear/fault zones or residual material/fill	<25	Moderate	Elasto-plastic, loosening and progressive deformation

Loading

Determination of tunnel loadings took into consideration a number of factors, including but not limited to: construction methodologies, tunnel size, tunnel depth, rock cover above tunnel crown and rock quality.

All tunnel structures, irrespective of construction method, were designed to withstand the following loads where they exist, and as influenced by the considerations previously identified.

- Live and dead loads
- Impact and derailment loads
- Seismic effects
- Earth pressure to future grade
- Hydrostatic pressure and buoyancy
- Surcharge loading (Vertical and Lateral) due to aircraft, vehicles, and construction equipment
- Loads due to adjacent building foundations, tunnels, and excavations
- Temperature loads, shrinkage and creep
- Differential settlement

In addition, the initial or temporary support for the NATM tunnels were designed to withstand the following loads:

- Self-weight of initial support elements
- Earth pressure, with appropriate ground-structure interaction
- Effects from changes in pore pressure due to draining of groundwater

Table 3.   Table of excavation and support classes

Domestic APM Tunnels—Excavation & Support Classes				
Rock Mass Class	Initial Support		Advance Length	Excavation Sequence
	Shotcrete	Other Support Elements		
I	6" SFR	Lattice girders & fast acting pattern rock dowels	5'-0" max	Top heading & invert
II	8" SFR	Lattice girders & grouted rebar spiles	4'-0" max	Top heading & invert
III	8" SFR	Lattice girders & canopy tubes	3'-0" max	Top heading & invert

- Surcharge loading due to aircraft and airside vehicles
- Loads due to adjacent building foundations
- Loads and effects of adjacent tunnels and excavations
- Construction loads

Similarly, the NATM and TBM permanent tunnel linings and cut-and-cover structures were designed to withstand the additional loads:

- Thrust ram loads (TBM only)
- Hydrostatic pressure and buoyancy

CONTRACT PROVISIONS

NATM Excavation and Support Classes and Sequences

Three excavation and support classes were developed to suit the range of ground conditions and permanent loading anticipated to be encountered along the tunnel alignments. These three support classes generally correspond to the three rock mass classifications previously identified. Initial lining shotcrete thickness, pre-support elements, and advance lengths were varied to address the anticipated ground conditions. Table 3 provides a representative comparison between support classes for a given tunnel.

The pre-support elements consisting of rock dowels, grouted rebar spiles and grouted canopy tubes were selected based on anticipated ground conditions. Rock dowels were proposed for Class I as the ground in the crown of the tunnel was expected to be blocky and well interlocked. Grouted rebar spiles were chosen for Class II where the spacing of discontinuities were expected to be close enough that rock dowels alone would be ineffective. Lastly, grouted canopy tubes were selected for Class III support where the anticipated condition of the material in or immediately above the crown of the tunnel would exhibit soil-like behavior rather than rock.

The excavation sequences for all NATM tunnels at Dulles involved subdividing the cross section of the tunnel into manageable drifts. The size and advance length of the drift was limited to maintain stability of the excavation and to control deformations. As shown in the above table, due to the relatively small size of the Domestic APM Tunnels with an excavated diameter of approximately 22', two drifts, top heading and invert, were all that was required to maintain stability, regardless of the rock mass classification.

Deformations were controlled by adjusting the advance lengths per round based on the stiffness of the surrounding rock mass. For some of the larger tunnels, such as the Pedestrian Tunnel with an excavated width in excess of 40' and relatively shallow cover, additional drifts were introduced, including splitting of the top heading into two drifts, establishment of a bench drift in addition to the invert drift.

All drifts were designed to be driven in the same direction and construction of the top heading was to be finalized prior to subsequent bench and/or invert excavations. However, an allowance was included in the specifications to permit the contractor to begin excavation of subsequent drifts prior to completion of the top heading in an alternating sequence, provided a minimum specified distance was maintained between progressing faces.

Contingency Measures

Although an extensive site-wide geotechnical investigation program was carried out for the project, the exact ground conditions to be encountered on a foot-by-foot basis during tunneling would not be known until excavation commenced. Therefore, a number of contingency measures were developed to augment, modify or replace the typical excavation and support requirements for the three support classes identified above should actual ground conditions deviate radically from the anticipated conditions.

The primary contingency measure adopted was the requirement for a "toolbox" containing local support measures from which the contractor could draw from to suit a wide-range of conditions, such as localized shear zones, unusually poor ground conditions, mixed face conditions, etc. Local support measures are supplemental items above and beyond the prescribed support measures for a given support class, including: rock bolts, rebar spiles, canopy tubes, face bolts, probe holes, core holes, shotcrete, and lattice girders.

In accordance with good practice and quality control, as well as to prevent dictating the contractor's means and methods, specifications were developed that required the contractor to prepare and submit a contingency plan prior to start of tunnel excavation. The requirements of the contingency plan were based on a detailed assessment of ground and site conditions that may require additional measures beyond the regular measures described herein. Conditions that needed to be assessed included groundwater, soil and lining conditions as well as equipment, traffic, grouting, seepage control and maintenance of existing utilities and facilities. Risks associated with these conditions, their likelihood and counter measures to either avoid or mitigate these situations were to be described in detail. As a minimum, the Contingency Plan had to include the following:

- Name and qualification of key personnel
- Description of conditions considered to require contingency measures
- Method and equipment of implementation of contingency measures and their immediate availability
- Plans outlining measures to be undertaken in the event of Contingency Procedure implementation including: Over-break, Face instability, Unexpected inflow of groundwater, Surface/building deformation beyond specified limits, Tunnel deformation beyond specified limits
- Detailed description of response procedure (chain of command) to monitoring values exceeding the Limiting Values specified in Instrumentation and Monitoring requirements
- Detailed description of coordination with and response procedure to monitoring carried out by other disciplines

- Methods of verification of the successful implementation of contingency measures
- Procedures detailing surveillance during stoppages
- Procedures detailing measures to be applied for the resumption of tunneling operation after stoppages

Geotechnical Baseline Report

As required by the MWAA Design Manual and in keeping with current industry practice, a Geotechnical Baseline Report was prepared as part of the Contract Documents for each tunneling contract package. Ground conditions and anticipated ground behavior for the three excavation classes and sequences for NATM construction and for TBM performance. The Contract Document identified the anticipated distribution of excavation and support classes for the purposes of establishing a clear baseline for contractor's proposals and the ability to fairly assess revisions to the anticipated quantities for payment purposes.

Instrumentation and Monitoring

A comprehensive monitoring program was developed to verify the adequacy of support and identify potential impacts to structures, pavements, utilities, and ground surface during tunnel construction. The specified instrumentation and monitoring program consisted of the following key components:

- Surface and shallow subsurface monitoring points
- Building monitoring points
- Extensometers
- In-tunnel convergence monitoring points
- Piezometers
- Inclinometers
- Vertical and Horizontal multipoint extensometers

The purpose of geotechnical instrumentation program was developed to:

- Establish pre-construction baseline data for comparison with construction and post-construction data.
- Provide early information on the interaction of the construction process with, and its effect on, ground and structures.
- Provide a forewarning of unforeseen conditions that may require remedial or precautionary measures.
- Permit timely implementation of proper procedures, such as change in excavation class and additional local support measures, as and when required to prevent damage to structures, equipment and utilities.
- Document ground movement and structure movement, if any that may occur as a result of construction operations.

Surface and shallow subsurface settlement monitoring were employed to identify potential impacts to structures, pavement, and utilities and to identify adverse trends in results and the implementation of remedial measures.

NATM in-tunnel convergence monitoring was undertaken throughout the life of the contract to ensure the structural adequacy of the initial and final linings of the tunnels was maintained and deformations were controlled. Daily summaries of geotechnical records and monitoring measurements, together with the evaluation of results, comparisons with design expectations, schedules of measurements taken, and summaries of the progress of construction during the measurement period was compiled and distributed among the personnel responsible for the works. This information assisted with the determination of appropriate excavation and support classes and/or contingency methods to be used in a timely manner. In the event that serious discrepancies arose between the measured and calculated results, a back analysis could have been initiated to confirm the discrepancies such that consideration could be given to the re-analysis of the tunnel structure. All instrumentation readings were automated and reported in real-time to a web site for remote accessibility by the contractor's NATM engineer and the Contracting Officers Technical Representative.

No specific in-tunnel instrumentation and monitoring requirements were anticipated for the TBM-driven tunnels, although a rigorous quality control of ring erection and contact grouting was specified to limit ground deformation and hence surface settlement.

CONSTRUCTION—PHASE 1

The tunnel construction contracts for the APM System were awarded in two packages; the West APM Tunnels and the East APM Tunnels. The West APM contract was awarded in 2004 to Clark/Shea JV for $78.6 million, to be completed in 36 months. The East APM contract was awarded 6 months later to Atkinson/Clark/Shea JV for $231 million, with a construction duration of 42 months. Having the same contractor perform tunneling on both projects through different JV teams facilitated the learning process as it relates to tunneling methods, equipment, and personnel.

During construction of the NATM tunnels for both tunnel contracts, the contractor elected to make adjustments to the excavation sequences and support measures specified in the contract documents. The adopted changes are outlined in the following paragraphs.

Construction Sequences

In-lieu of excavating both top heading and bench concurrently with a lag of 12 feet between top heading and bench, the top heading was excavated for the full length of the drive followed by the excavation of the invert working back towards the beginning of the drive. This allowed optimization of available construction equipment and a reduction of temporary construction materials. This construction sequence also minimized the potential for damaging the shotcrete invert from construction traffic. The top heading was excavated using an AM 75 roadheader and the invert was excavated using the smaller AM50, allowing the AM75 to move onto adjacent tunnel drives (Figure 2).

To accommodate the tail-end of the APM crossovers south of the Tier 1 East and West Stations, an enlarged section of NATM was designed. The anticipated sequence on the East Station, where the TBM tunnel interfaces with the Station, was to enlarge this section once the TBM had holed through prior to being skidded through the station excavation for re-launch at the north end of the station. The contractor elected to mine the NATM enlargement prior to the passage of the TBM. This had the benefit of continuity of available NATM equipment and labor from other NATM segments of the project rather than having to remobilize a second NATM operation.

Adjacent to the Main Terminal, a staggered portal for the NATM tunnels was included in the Contract Documents due to overlying utilities. Concerns during construction regarding the nature of the rock pillar between the portals led the contractor to propose squaring-off the tunnel portals. Elimination of advanced utility work above

Figure 2. Top heading and bench excavation

the tunnels provided the necessary schedule relief to allow this option to be developed. The conditions encountered in the field at the portal excavation with an unfavorable orientation of bedding joints and discontinuities validated the proposed revision.

Pre-Support Elements Installation

The primary refinement of pre-support installation was made in Class III. To minimize potential overbreak associated with excavation beneath the canopy tubes, the design envisaged installing the canopy tubes from within the excavation perimeter prior to installation of the lattice girder. This required leaving a short stub of each canopy tube projecting into the initial lining envelope, which would have to be cut off prior to lattice girder installation. While it was recognized that this might reduce the potential for overbreak, the installation sequence was refined to install the canopy tubes outside the normal excavation profile in a limited saw-tooth to facilitate installation and eliminate the need to cut the tubes following installation.

PHASE 2—DESIGN

Project Description

Design of an extension of the APM System is ongoing on the west side of the Airport. This is a 2,500 feet extension of the system from Tier 1 to Tier 2 and comprises twin NATM tunnels with a similar configuration to Package 3 East and 3 West. Wherever possible, lessons learned during construction are being adopted into the design process. These refinements consist of both the adoption of refinements made to the construction sequences and modifications to the design as a result of review of the practicalities of installations and construction processes. The design assumptions made during the design of the Phase 1 tunnels has been borne out during construction such that the provisions for each excavation and support class were appropriate for the ground conditions. The following paragraphs highlight some of the proposed design refinements being carried forward into the Phase 2 designs.

Adopted Refinements

Top Heading Advance Full Drive Followed by Invert Excavation Retreating Included as a Contractor Option. Similar to the Phase 1 design, the sequence of excavation being developed for Phase 2 consists of a top heading drive and an invert drive. During Phase 1, a design was developed that would not preclude the contractor from initiating excavation of the invert prior to completion of the top heading. To accomplish this, the design assumed that the invert excavation to be driven in the same direction as the top heading, and could be done concurrently, provided adequate separation was maintained between the two advancing headings. The primary advantage envisioned behind this excavation sequence was in terms of schedule, as there would be minimal lag time between completion of the heading drive and invert drive.

However, during construction, the contractor proposed completing the drive of the top heading prior to starting invert excavation and initiating the invert drive from the end of the top heading drive. The primary advantage realized using this approach was in terms of invert protection. Given the amount of traffic generated by muck removal operations, it would have been difficult to maintain the integrity of the bottom of the round tunnel after the removal of the invert material without the application of an expensive and potentially time consuming invert protection layer. By retreating the invert excavation from the end of the top heading drive, the contractor was able to limit the amount of invert protection required while maintaining the schedule.

Inclusion of Grouted Pipe Spiles In-lieu of Canopy Tubes for Class III Support. As stated previously, Class III support was designed where the anticipated condition of the material in or immediately above the crown of the tunnel would exhibit soil-like behavior rather than rock. This assumption was proven to be overly conservative during construction. The conditions encountered did not exhibit a level of rock decomposition that would warrant the structural stiffness provided by the grouted canopy tubes. However, conditions did exist that necessitated grouting. As difficulties were encountered in installing the fairly large diameter canopy tubes in between the relatively close spaced lattice girders, the current design utilizes hollow-core, self drilling spiles. This small diameter product can be installed quickly and easily through the lattice girders without restricting grouting operations.

Refined Waterproofing Provisions with Respect to Accessibility of Remedial and Contact Grout Pipes. As groundwater levels fluctuate significantly with seasonal variations in precipitation and dewatering operations at adjacent construction operations, the complete recharge of the water table may not occur until long after construction of the tunnels is complete. As a result, should a leak in the waterproofing system exist, it may not present itself until after the tunnels are fitout. While the waterproofing system designed for the Phase 1 tunnels does not preclude remedial grouting operations intended to stop leaks that may be present, it has become apparent that accessibility to the grout ports will be hampered by the fitout of MEP systems.

The current design will be coordinated with MEP systems such that access to remedial and contract grout pipes will be maintained, even after fitout. Where necessary, grout pipes will be extended around or beyond system components that would otherwise make access difficult.

Inclusion of Flexible Provisions to Allow for Addition of COTR/Contractor Agreed Additional Instrumentation and Monitoring. While a comprehensive and prescriptive instrumentation and monitoring program was developed during Phase 1, the design did not include provisions for identification, implementation, or payment for additional measures, should conditions arise that would necessitate such measures. Similar to the contingency plan for additional local support measures described above, provisions are being introduced into the current contract documents that will establish criteria for identifying when additional instrumentation and monitoring measures should be implemented and how such additional measures will be reimbursed.

Revised Location of Invert/Arch Construction Joint. In response to the contractor's proposed means and methods for forming the cast in place final lining, the construction joint between that arch and invert was moved one foot above the invert to provide a kicker for the arch form. This has been adopted in the phase two design, maintaining the joint radial to the arch to allow for effective load transfer between arch and invert.

Inclusion of Channel Inserts to Simplify MEP Installation and Minimize the Need for Drilling. Experience has shown that installation MEP system components within a tunnel can be difficult, if not detrimental, if provisions are not allotted. The primary means of fastening system components to the tunnel lining involves drilling and anchoring into the existing lining. This method can be time consuming and interferences with reinforcing within the lining are often problematic. Even more concerning, drilling operations into a tanked structure has the potential to puncture the waterproofing system, which can lead to costly and disruptive repairs.

To simplify the installation of MEP system components and reduce potential for compromising the waterproofing system, channel inserts will be cast into the final lining. The channel inserts, spaced every 5 feet throughout the tunnel and encompassing the tunnel circumference from invert to invert, will allow for the utmost flexibility and direct fixation of system components.

ACKNOWLEDGMENTS

The Authors would like to thank the Metropolitan Washington Airports Authority for the permission to publish this paper and also acknowledges the contributions of Stephanie Crawley and Frank P. Frandina, PE, of Hatch Mott MacDonald.

PORTLAND, OREGON'S ALTERNATIVE CONTRACT APPROACH—A FINAL SUMMARY

Paul Gribbon

City of Portland, Bureau of Environmental Services

Greg Colzani

Jacobs Associates

Julius Strid

EPC Consultants, Inc.

Jim McDonald

Impregilo Healy JV

ABSTRACT

Portland, Oregon's $370 million West Side Combined Sewer Overflow Tunnel Project has completed the final phase of construction and commissioning. The construction contract, which began in September 2002, was a reimbursable cost plus fixed fee contract that followed a qualifications-based contractor procurement process and a pre-construction planning phase. This paper addresses the final results of the contracting method and provides a lessons-learned case history. Specific items addressed include the benefits realized from the contracting method, areas of improvement, final costs compared to the original contract estimates, design changes and schedule impacts, and subcontract management issues.

PROGRAM BACKGROUND

In August of 1991, the City's Bureau of Environmental Services (BES) entered into a Stipulation and Final Order (SFO) for combined sewer overflow (CSO) abatement with the Oregon Department of Environmental Quality (DEQ). The agreement required the City to control 55 combined sewer outfalls by December 1, 2011, with intervening major deadlines to complete specific parts of the work. The West Side CSO Project was required to meet the regulatory milestone of December 1, 2006 for control of outfalls along the west side of the Willamette River. The project consisted of a combination of near surface pipelines, a 5.5 km (18,180 feet) long 4.3 m (14 feet) diameter soft ground tunnel, and a pump station that transports 832 million liters per day (220 MGD) of CSO flow from combined sewer areas in west Portland to the City's existing wastewater treatment plant as shown in Figure 1.

CONTRACT DEVELOPMENT AND CONTRACTOR PROCUREMENT PROCESS

BES utilized a Cost Reimbursable Fixed Fee Contract (CRFF), loosely modeled on a construction manager/general contractor (CM/GC) approach. The contractor selection process was modeled after the City's selection process for Professional, Technical and Expert Services that requires interested parties to respond to a Request for Qualifications followed by a Request For Proposal (RFP) and subsequent interview.

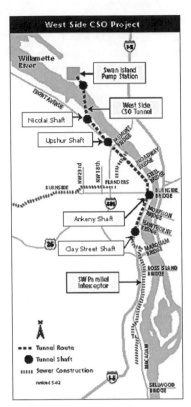

Figure 1. WSCSO project map

Under this approach, the prime contractor was selected using a qualifications-based process that considered project personnel, project approach, subcontracting approach, partnership, safety, Minority/Women/and Emerging Small Businesses (M/W/ESB) practices and a fixed fee. The fixed fee (established by proposing contractors during the procurement phase) was applied over the life of the contract and included all contractor off-site and on-site overhead including management staff, general conditions costs, equipment insurance and profit. Labor, equipment, and materials were paid on a cost-reimbursable basis. Subcontracts were procured using a competitive process and reimbursed on lump sum or unit price basis.

As is explained below, the contract was executed in two parts, a pre-construction planning phase and a construction contract. Notice-to-Proceed for the pre-construction planning phase was issued in January 2002 and ended in August 2002. Notice-to-Proceed for the construction contract was issued on September 12, 2002 with an original completion date of August 12, 2006 (47 month contract period). This date was extended to October 1, 2006 and Substantial Completion was achieved on September 14, 2006, well within the stipulated SFO milestone date of December 1, 2006.

The original construction contract amount was $293,060,874 in 2002 dollars, which was divided into two main parts: a fixed fee of $58,702,000 and an estimated reimbursable cost (ERC) of $234,358,874. In addition to the contract amount, a contingency of $17 million was carried equivalent to 6% of the contract amount based on a joint risk analysis (discussed below) performed during the pre-construction phase. The construction budget in 2002 dollars with contingency was $310 million.

PRE-CONSTRUCTION SERVICES PHASE

Following contractor selection, and prior to executing a construction contract, a five-month pre-construction services consultant contract was executed with Impregilo/ SA Healy Joint Venture (IHJV), the selected contractor. IHJV physically moved into the existing project office with BES, the lead designer Parsons Brinckerhoff (PB) and the construction management staff of Jacobs Associates (JA). During this phase IHJV's primary objectives were to provide a review of the design (which was overall about 75% complete) focusing on constructability, perform value engineering and a joint project risk assessment, develop the ERC for construction, a cost-loaded schedule, and assist BES in producing a final set of general contract provisions for the construction contract. This effort resulted in a large advantage to BES in that a realistic construction budget was set and several features of the design were changed to make the project more practical from a constructability standpoint. The most significant change made during this period was the re-design of the lower pump station outer walls. The initial design concept for the walls separated the final structure from the concrete slurry walls. This resulted in concrete wall thicknesses of up to 2.7 meters (nine feet) in the lower portion of the pump station at a depth of 49 meters (160 feet). Based on IHJV's concern with the pre-cooling and post-cooling required to keep the heat of hydration of the concrete to a manageable level, the lower walls were redesigned as a composite wall with the pump station slurry walls. This resulted in a 0.9 meter (three foot) reduction in the wall thickness with a corresponding direct cost savings as well as significantly reducing the potential for a large defective specification claim.

Based on the previous example and a number of other suggested design changes, it became apparent that the original five-month duration of the pre-construction phase was insufficient. Activities that were thought to be concurrent actually were sequential. Cost-saving ideas by IHJV had to be reviewed by PB and BES. When acceptable, design drawings were changed, new cost estimates for that item of work were developed, and the new estimate then became a revision to the ERC. As a result, three additional months had to be added to the pre-construction phase of the work. For a project of this size, a nine to ten month pre-construction phase (begun no later than the 60% design stage) would have been preferable.

CONSTRUCTION CONTRACT

The $293,060,874 CRFF construction contract utilized by BES shared and distributed project risk in several non-traditional ways. Several of the key risk-sharing provisions are discussed below:

- Contractor Fixed Fee—Guarantees contractor's profit provided project is completed within allotted time.
- Differing Site Conditions Clause—Type I differing site conditions (DSC) are not recognized for the prime contractor's work, since cost reimbursement is already occurring. However, if a Type II DSC is encountered, and it affects the schedule's critical path by a period of at least seven days, additional fee and time is allowed. Since subcontractors are compensated through lump sum or unit price subcontracts, a standard DSC clause recognizing both Type I and II DSC's is included in the contract for subcontractors.

- Compensation—The contractor is reimbursed twice a month for the amount of documented reimbursable costs, plus a fee payment that is based on a percent of the fixed fee equal to the percent of work complete.
- Subcontracting—Subcontracts were competitively bid as needed during the course of the contract by IHJV, rather than having subcontractors selected in advance. The pluses and minuses of this are discussed further on in this paper

At the time the construction contract Notice-to-Proceed (NTP) was issued, the tunnel design was 100% complete and the pump station design was approximately 80% complete. As previously stated, the NTP was issued in September 2002. However, the final pump station drawings were eventually issued in February 2003, five months after the original construction NTP. Consequently, from an Owner perspective, the contracting approach saved approximately ten months of schedule (September 2002 to July 2003) prior to the start of construction.

RISK ASSESSMENT AND OUTCOME

A formal risk assessment workshop was held during the pre-construction phase. A professional facilitator was engaged, and team members from the Owner, Design Engineer, Contractor, Construction Manager, and the City's advisory board were included in the workshop. During the workshop, a total of 251 risks were identified for the categories of work:

Access/Permit Risks	35
Tunnel Construction Risks	47
Ground Improvement Construction Risks	58
Shaft Construction Risks	25
Pump Station Construction Risks	20
Microtunneling Construction Risks	33
Completion/Startup/Operation/Maintenance Risks	10
Financial/Other Risks	23
Total Identified Risks	251

These risks were each assigned a value of potential cost impact and probability of occurrence, each being scored over a range of 1 to 5, with 5 indicating the highest level of cost impact and the highest probability of occurrence. The two factors were then multiplied and a risk matrix was developed with resultant scores of 1 to 25.

Upon rating the various identified risks, a series of risk mitigation proposals were developed to reduce risk to the project during the pre-construction phase. Methods of mitigation included revisions to the design and the contract documents, additional geotechnical investigation or instrumentation, and development of a subcontracting plan that addressed packaging of subcontracts and management of subcontractor claims and disputes. In addition, the Access/Permit and Financial/Other risks were managed by including the potential costs in the Estimated Reimbursable Cost.

For those risks that could not be entirely mitigated or priced, a contingency allowance was developed and approved by the City Council along with the construction contract. This was done by evaluating each category of risk and estimating a total amount expected to be expended if the risk event were to occur (approximately $34 million). The

Table 1. Contingency versus actual expenditures

#	Item	Contingency Allowance	Actual Cost Difference from Estimated
1	Access/Permit Risks	—	—
2	Tunnel Construction Risks	$18,420,000	$(380,000)
3	Ground Improvement Construction Risks	1,447,880	5,720,000
4	Shaft Construction Risks	7,305,000	2,325,000
5	Pump Station Construction Risks	2,000,000	3,280,000
6	Microtunneling Construction Risks	2,400,000	10,700,000
7	Completion/Start Up/Operation/Maintenance Risks	2,500,000	
	Reimburseable indirect costs		(6,100,000)
	Accumulative total risk allocation	$68,145,761	
	Evaluated Risk Contingency (50%)	$17 million	
	Total Escalation	$14.5 million	$15.3 million
	Total Contingency and Actual Cost Difference	$31.5 million	$15.5 million
	Amount project is under budget (escalated)		($16 million)

total of these estimated items were then evaluated using two statistical models, a 'Monte Carlo' model and an evaluation of the range of risk with probability of it occurring. Both of these statistical methods arrived at the same result, the amount of contingency to reserve would be 50% of the total estimated risks. The contingency amount, about $17 million, represented just under 6% of the construction contract value.

We compared the estimated contingency developed from the risk register with the actual unforeseen events during the project. This is shown in Table 1.

The result is that the method of contingency calculation nearly matched the actual additional costs incurred, $17 million estimated compared to the actual of $15.5 million. We have also shown, in Table 1, the estimated escalation prepared at the beginning of the project, which was $14.5 million. The actual escalation experienced was $15.3 million. Combining both the escalation and the contingency, the project came in <$16 million> under the original budget amount.

SUBCONTRACTING ISSUES AND PRACTICE

Subcontractor costs were reimbursable to the prime contractor at face value. The general contract conditions required that all subcontracts be procured on a competitive basis, that they be approved by the Owner, and that they be paid as lump sum or unit price. Because the Project design and the Project means and methods were not completely defined at the time of Notice-to–Proceed (NTP), and due to the length and size of the Project, subcontractors were procured as they were needed, rather than all subcontracts being procured upfront. There were advantages and disadvantages to this. First, additional opportunities for minority, women, and emerging small businesses continued to be made available during the course of the contract. Subcontractors did not have to price their work out several years in advance, allowing for subcontracting pricing with less risk contingency. All potential subcontractors got a shot at the work. Competitive pricing was obtained, providing cost benefits to the Owner. Disadvantages: (1) the prime contractor was not able to have any subcontractor identified at the time of NTP thus slowing initial start-up work; (2) unlike a traditional design-bid-build contract, the project budget established for many areas of work performed by subcontractors was based upon estimates instead of subcontractor bids or supplier quotes, with potential growth in the overall project cost as subcontracts were identified; (3) it was difficult to procure subcontractors for work that was not clearly defined, due to the requirement for lump sum or unit priced contracts; and (4) the prime contractor had to create and employ a sometimes lengthy and cumbersome process to ensure competition, public acceptance, and the Owner's approval. As a consequence of this latter point, the prime contractor did not have the latitude to select subcontractors he was familiar with based on his past experiences, and often there was only one or two quotations.

Subcontract costs in excess of the subcontract amount were only reimbursable if the increase was due to a differing site condition, the subcontractor performed extra work as directed by the Owner, or there was an increase in quantities for unit price pay items. The general conditions were later revised to allow increases in subcontract due to adding work, with the Owner's approval, that had been intended to be self-performed or performed by another subcontractor. All increases in subcontract amount required approval of the Owner.

As for the prime contractor, his Fixed Fee does not change as a result of changes to a subcontract, unless all three of these conditions apply: the subcontractor encounters a differing site condition, the contractor paid the subcontractor, and the additional work resulted in an increase in time of the overall project schedule. The prime contractor receives no other markup on subcontractor changes.

In most cases, everything works according to plan. Changes occurred, change orders were proposed and approved and issued to the subcontractor, and the subcontractors were paid. On a project of this size, very few subcontractors affect the critical path, so the Fixed Fee was not adjusted.

In one case, however, the contract team was severely tested, and the success of the Project was at risk. The pump station, 41 meters (135 feet) in diameter and 49 meters (160 feet) deep, was to be constructed in a shaft supported by 61 meters (200 feet) deep diaphragm walls. A groundwater cutoff was to be constructed with a jet grout curtain from the bottom of the diaphragm walls to bedrock at depth of 101 meters (330 feet). After the primary curtain was grouted, measured inflows exceeded the allowable amount. A secondary curtain was installed, followed by two rounds of remedial grouting. Still, the pump tests were showing inflows in excess of the allowable amount. And, by now, the schedule had slipped by eight months. A differing site condition was claimed by the subcontractor. It was not certain that water could be cut off enough to build the Pump Station.

Additional grouting could be required. Large differing site condition costs could be incurred. The schedule could slip further. Perhaps the pump station would have to be redesigned or even not built. Because of the contract approach and the partnership of the team, alternatives and solutions were openly discussed between Owner and prime. It was decided to install four large dewatering wells and begin excavation. A partial redesign of the pump station would move the Operations Building from the top of the shaft to an adjacent building, allowing concurrent construction. Schedule analysis showed a savings of at least four months from this redesign. Excavation took two months less than scheduled. The initial concrete work completed the recovery of the lost schedule time. The differing site condition was not accepted, but the subcontractor's issues were addressed and an amicable agreement was reached.

The outcome could have been very different. The Owner could have been inflexible on the specified water inflows, or refused to redesign the pump station. Unknown, but large amounts of, expensive grout could have been ordered, adding a minimum of three months of further delay. At these depths, the delays could even be much more. The subcontractor could have refused to perform further without a huge change order. The prime could have pushed the differing site condition and earned a large additional fee, but damaged the partnership and ensured a late Project. In the end, each member of the team assumed its risk and the Project came first. The contractor assumed its risk of time and got back on schedule. The Owner assumed its risk of cost and rightly decided that it would cost less to proceed with some uncertainty than to eliminate all uncertainty up front. And the subcontractor was paid for his efforts and relieved of excess risk.

CONTRACT COST

The cost tracking system utilized by both IHJV and BES tracked the actual costs for each item of work against the ERC budget for that item as the cost was incurred. Semi-annual estimate-at-completion cost projections were performed based upon learned history. The project cost control program included a series of BES checks and balances such as review and approval of all subcontracts and subcontract modifications, periodic field audits of contractor activity, review and approval of all purchases over $50,000, and biweekly reviews of cost reimbursement requests (payment applications) and construction schedules prior to releasing payments.

The breakdown of the original ERC budget in 2002 dollars versus actual payments as of November 2006 is shown below:

Description	Estimated Reimbursable Cost (2002)	Actual Reimbursed Cost (2006)
Main Tunnel	$56,426,113	$56,038,197
Nicolai Shaft	12,491,294	14,021,055
Clay Street Shaft	8,947,244	9,437,201
Ankeny Shaft	7,434,986	7,619,932
Upshur Shaft	8,143,753	6,558,632
Confluent Shaft	9,065,632	10,591,933
Swan Island Pump Station	59,901,514	71,557,760
Pipelines	25,826,435	35,838,060
Indirects	15,289,078	8,577,790
Plant and Equipment	30,829,825	28,157,015
Totals	**$234,358,874**	**$248,397,575**

The system was tested through an external independent audit in November 2004. The findings of the audit included the discovery of a minor number of double billings which were offset by some items which were found to be unbilled by IHJV. Overall, the audit was highly complementary of the IHJV cost tracking system and the overall project management of the contract by both parties.

CHANGE ORDERS, CONSTRUCTION CHANGE DIRECTIVES AND VALUE ENGINEERING

The CRFF delivery method is unique in the area of change management and value engineering. Change orders under this contract are issued only for changes to the General Conditions, an extension of contract time or an increase of contractor

Fixed Fee. There were nine change orders to the prime contract over the four-year contract term. Only one increased the fee due to extra work requested by BES, adding a 115 kV substation originally planned for installation by Portland General Electric. This high-voltage substation was incorporated within the original budget and schedule parameters of the project. Two other change orders added non-compensable time to the contract. The remaining change orders were issued to improve or clarify clauses in the General Conditions with no impact to contract cost or time.

The primary mechanism for change in the contract was the Construction Change Directive (CCD). CCD's were issued to direct changes in the design, changes based upon RFI clarifications, changes resulting from value engineering, and changes due to field conditions. The contract allowed for the rapid implementation of changed work that was self-performed by IHJV because there was no negotiation of cost, overhead or profit since the contractor is simply reimbursed for all direct costs associated with the change. However, for changed work performed by a subcontractor, the process was the same as under a lump sum contract, other than there being no prime contractor markup.

There were a total of 164 CCD's issued on the project, most involving changes to subcontracts, especially the pump station mechanical and electrical work, which was all subcontracted work. In addition, there were 326 subcontract amendments issued involving a total of 200 subcontracts. This was an area of some management difficulty, since although the Owner was required to approve all subcontract amendments in advance, in practice, some of the subcontractor extra work was authorized verbally with the actual subcontract amendment issued long after the work had been performed.

RECOMMENDED CHANGES TO CONTRACT LANGUAGE

Like most delivery methods, refinements in the contract provisions can be made. Some suggestions from both IHJV and BES perspective are provided below.

IHJV Recommendations

The contract language relative to subcontracting can be confusing. The prime contractor is working under a Reimbursable Cost plus Fixed Fee arrangement and the general conditions are geared toward that. Subcontractors, on the other hand, work under conventional fixed price contracts. Primes generally include flowdown clauses in subcontracts, tying subcontractors to the prime contract. This clause is used because the contractor does not want to miss any requirements that the Owner wants to specifically impose on subcontractors, such as prevailing wages, insurance requirements, etc, as well as a host of job-specific requirements. When a subcontractor reads these general conditions, substituting himself for "Contractor," his responsibilities and duties may not be clear. For example, if a subcontractor advances some theory by substituting himself for the cost reimbursable prime, a contractual difficulty may result. If a subcontractor has lower tier sub-subcontracts, the situation may become more clouded. A solution may be to combine the specific flowdown requirements in a distinct general condition paragraph, with references to the applicable clauses. Standard subcontract documents should also be reviewed and revised as necessary to clarify the two different delivery methods that are combined in one Project.

Another area of confusion is in the difference between extra work and changed work. Extra Work is defined as something outside of the scope of work but desired by the Owner. Changes do not necessarily fall within this definition. One area where this is significant is that, as noted above, increases in subcontract amounts are only reimbursable if the cost is due to a differing site condition, Extra Work directed by the Owner, or variations in unit price items. The cost of changed work is not specifically reimbursable. Fortunately, reasonable parties did not allow this to become an issue. There are other areas in the contract where this distinction could be problematic.

Bonding of the project is another area that could be reviewed. In a cost reimbursable contract, the risk to the Owner is changed significantly. The real risk is the cost of the fee, and the work included therein since, in the case of default, the contract terms would remain the same for the bonding company or the successor contractor.

BES Recommendations

The roles of both Contractor and Owner regarding the administration of subcontracts should be studied to better define the expectations of both parties, specifically, the Owner's involvement in subcontract decisions regarding extra work, reassignment of work between subcontracts, and the settlement of subcontractor claims. Additionally, the contract attempts to differentiate how change clauses of the contract apply to the prime contractor and subcontractors. However, in practice, the management of this portion of the contract has been confusing.

The article stating that the costs due to the fault or negligence of the Contractor and subcontractors including the costs for the correction of damaged, defective or nonconforming work are non-reimbursable under the terms of the contract is standard design bid build (DBB) boilerplate language and is difficult to administer fairly in a CRFF environment. In a DBB contract, the contractor is afforded the opportunity to include risk or contingency in his bid and Owners often prefer to stay away from any involvement in Contractor means and methods. However, the CRFF language generally prohibits the contractor from including such risks in the fixed fee and in practice the Owner actually shares responsibility for the means and methods of construction. Interpreted literally, the contractor could be liable for any inefficiency experienced in his normal operations, which the Owner could then refuse to reimburse. A practical and reasonable application of this clause, including agreement on its application between Owner management and field staff, is therefore essential.

Suggestions for Contractors in Considering this CRFF Approach

It is important to realize that this Cost Reimbursable—Fixed Fee type of arrangement is not without risk. While it is often referred to as a cost-plus contract, it is clearly not.

The Fixed Fee can only be increased in limited circumstances—Extra Work (work not provided for in the Contract Documents) by the Contractor or subcontractor, differing site conditions encountered by a subcontractor, Type II differing site conditions encountered by the Contractor, excusable delay due to fault or negligence of Owner—and only if an extension to Contract Time results from these circumstances. Type I differing site conditions or other excusable delays do not trigger an increase in fee, nor do changes in the work, although the Owner does pay the reimbursable costs. Contractors should not expect to be able to increase profits through increases in the fee.

The primary risk of the contractor is in control of the work schedule. If the work is completed early, the contractor still receives his full fee. The profit that is included in the Fixed Fee is therefore relatively secure.

If the work is completed late, the same fee applies, although overhead costs have been extended. Of course, the Owner is committed to completing the work on time. To that end, the Owner pays the reimbursable costs during a period of acceleration, if approved, so the contractor has some leeway in his ability to recover schedule.

Things to consider in pursuing this type of contract would therefore include:

- Adequate contract time is provided for the scope of work
- Reputation of the Owner for fair play

- Willingness to trust the Owner
- Willingness to commit resources for a modest profit
- Willingness to enter into a mutually beneficial partnership to complete the work

To a large extent, success under the contract depends on the good faith of the parties. There are a number of areas in the contract where a contractor could take advantage of the Owner, or the Owner could take advantage of the contractor. But there are checks and balances, and an unreasonable approach by either party will surely lead to disaster.

Suggestions for Owners in Considering this CRFF Approach

Based on our experience with this contract and as previously stated, the success is dependent on the relationship developed between the Owner and its agents and the Contractor. This type of contract is not suitable for everyone or every project. The Owner is taking the financial risk for the Contractor's means and methods, which may be considered unacceptable by many Owners, while the Contractor absorbs the risk of completing within the contract time. However, we have found through this contract that there has been an incentive on the part of both parties to resolve issues swiftly and little to no benefit to either party for position letters to be written at the outset of a project challenge, event, or difficulty. But if the Owner mindset is present that the Contractor will always operate purely on self-interest, it would be difficult for an Owner to put in sufficient contractual wording to protect itself from any and all situations.

In considering the use of a CRFF contract, an Owner has a number of issues to consider. The first is the legislative environment regarding exceptions to low bid public works contracts and the contractual limitations of the exceptions, if any.

The second issue for consideration is the local political will. In our case, most members of the Portland City Council had prior experience with alternate construction contracts in the bureaus assigned to them and were therefore familiar with the process, advantages, and pitfalls, although none had experience on a major underground project. The perception that this type of contract is a "blank check" without protection for the Owner's budget may be insurmountable in some political climates.

Another consideration is development of a selection process that makes very clear the criteria and scoring mechanism used for selection, use of a knowledgeable and impartial selection committee, and payment of a stipend to short-listed but ultimately unsuccessful firms to offset some of the costs of participation in the procurement process.

Ultimately, an Owner must be willing to move forward without the typical risk allocation protection clauses available in a traditional lump sum or unit price contract. Acceptance of risk for a contractor's means and methods may be the biggest hurdle to overcome. In summary, prospective Owners should ensure that:

- Adequate contract time and budget is available for the scope of work
- Impartial technical resources are available to select a contractor based upon qualifications
- There is a willingness to trust the selected Contractor with the Owner's money
- There is acceptance of the contracting method at all levels of the Owner's project staff.

SUMMARY

As of the writing of this paper:

- The project was substantially completed and fully operational 2 months prior to the mandated deadline.

- The original contract time was extended by a total of 50 days. Contract cost, taking inflation into account over the 47-month contract time, was within 1% of the original contract amount.

- The original aspirational goal for minority, women, and emerging small business contracting of $13 million was exceeded by $6+ million.

- Only one subcontract remains open pending a financial settlement. There were no prime contractor claims during the life of the contract.

- A Dispute Review Board, although noted as an expectation in the Request for Proposals, was determined to be unnecessary and was never utilized.

- No litigation attorneys received any fees as a result of this contract

REFERENCES

Gribbon, P., Colzani, C., McDonald, J. 2005. Portland Oregon's Alternative Contract Approach—A Work in Progress. 2005 RETC, June 2005, Seattle, WA.

SINGAPORE'S DEEP TUNNEL SEWERAGE SYSTEM— EXPERIENCES AND CHALLENGES

Robert H. Marshall

CH2M HILL/Parsons Brinckerhoff JV

Richard F. Flanagan

CH2M HILL/Parsons Brinckerhoff JV

ABSTRACT

The Republic of Singapore's Public Utilities Board is implementing the Deep Tunnel Sewerage System Project. The project's deep tunnel component, 48 km of deep, large diameter sewer tunnels has been successfully completed (the works are in the defects liability period). Six International Design/Build tunnel contracts utilized eight Earth Pressure Balance TBMs ranging up to 7.23 m in diameter to deal with varied ground types ranging from deep soft clays to fresh granite. Numerous access/work shafts were constructed and using various support systems.

This paper provides an assessment/overview of the tunnel contracts, tunnel excavation methods and equipment, and the shaft systems. Suggested enhancements to similar future projects are provided.

INTRODUCTION

Singapore's Deep Tunnel Sewerage System (DTSS) is a replacement conveyance, treatment, and disposal system. It will ultimately convey wastewater by gravity to two new Water Reclamation Plants (WRPs) and their individual outfalls by implementing two connected tunnel systems. When fully completed, the DTSS will replace over 130 existing pumping stations and six existing Water Reclamation/Waste Water Treatment Plants.

A series of new link sewers up to 3 m in internal diameter will intercept existing sewers/pumping stations and transfer flows into vortex generators and down drop shafts (with energy dissipation) into deep sewer tunnels up to 6 m in finished diameter. The deep tunnels will then convey the wastewater to one of two new compact state-of-the-art WRPs, each at the extremities of Singapore Island (Figure 1) in reclaimed land away from population centers. The tunnel systems will work entirely by gravity, eliminating the need for the existing pumping stations. The two tunnel systems (North and South Tunnel Systems) will be linked by a 3.3 m finished diameter Spur Tunnel to permit balancing flows between the two new water reclamation plants enabling high capacity utilization, efficient future expansions in both plants and maintenance facilitation. Treated effluent will be reclaimed and reused or be discharged through deep-sea outfalls into the Straits of Singapore thereby eliminating the current discharge into the shallow Straits of Johor.

The DTSS Phase I, comprising the North Tunnel and the Spur Tunnel (Figure 2), has recently been completed (the works are in the defects liability period). The North Tunnel was constructed under five contracts (Contracts T-01 to T-05) and is approximately 38.5 km long. The tunnel alignment follows the Seletar, Central, and Pan Island

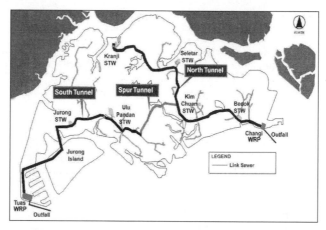

Figure 1. Singapore deep tunnel sewerage system overall implementation

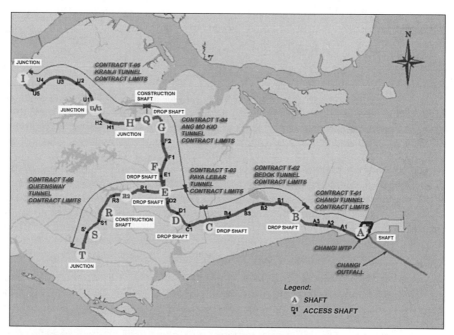

Figure 2. Singapore deep tunnel sewerage system Phase I—North and Spur Tunnel layout

Expressways from Kranji to the under-construction Changi WRP. The maximum excavated tunnel diameter is 7.2 m while finished diameters range from 3.6 m to 6.0 m. Tunnel depths vary from 18 m to 50m. The Spur Tunnel (Contract T-06) is 9.6 km long and is aligned along Braddell, Lornie, Adam, and Farrer Roads to the intersection of Queensway/Commonwealth Avenue. Depths of this 3.3 m finished diameter (4.46 m excavated) tunnel vary from 22 to 41 m.

Table 1.   Singapore deep tunnel sewerage system tunnel contract summary

Contract	Excavated Diameter (m)	Finish Diameter (m)	Length (km)	Contractor	Contract Award (1999/2000)
T-01 Changi Tunnel	7.23	6.00	5.8	Woh Hup–Shanghai Tunnel Engineering JV	$(S) 80.6 M
T-02 Bedok Tunnel	7.16	6.00	7.7	Penta Ocean Construction Co. Ltd.	$(S) 95.6 M
T-03 Paya Lebar Tunnel	7.20	6.00	5.1	Kumagai–SembCorp JV	$(S) 68.8 M
T-04 Ang Mo Kio Tunnel	5.47	4.30	7.3	Samsung Corporation	$(S) 74.2 M
T-05 Kranji Tunnel	4.93	3.60	12.6	Philipp Holzmann–SembCorp JV	$(S) 139.5 M
T-06 Queensway Tunnel	4.46	3.30	9.6	Ed. Zublin AG	$(S) 91.8 M
Total Length			48.1	Total Contract Sum	$(S) 550.5 M

DTSS PHASE I MAIN TUNNEL SYSTEM DETAILS

Project Implementation

CH2M HILL/Parsons Brinkerhoff Joint Venture was awarded the Consultancy Services Contract by the Public Utilities Board (PUB) of the Republic of Singapore in 1997 for the feasibility study for the whole DTSS scheme, preliminary design for the deep tunnels, Changi WRP, outfall, link sewers and the program management of six design and construct tunnel contracts. The first tender for the design and construct tunnel contracts was invited in March 1999 and an award was made in December 1999. The sixth and final contract was awarded in March 2000. The successful contractors represent companies from Singapore, China, Japan, Korea and Germany. Tunnel designers from Singapore, the United Kingdom, the United States, and Austria were appointed by the contractors. A summary of the six design and construct contracts are provided in Table 1.

Ground Conditions in the North and Spur Tunnels

The bedrock formations are the Bukit Timah Granite and the Jurong Formation (sedimentary). The overlying more recent formations are the Old Alluvium and Kallang Formations. The Jurong Formation was laid down in a trough within the rising granitic hills. A downwarp occurred later and the trough filled with the Old Alluvium. River valleys were cut into the Old Alluvium/older rocks due to sea level changes resulting from glacial advance and retreat. The Kallang Formation filled the valleys.

The Bukit Timah Granite is found in the western one-third of the North Tunnel, between Seletar and Kranji, and the majority of the Spur Tunnel. The granite encountered has various weathering grades from I to VI. Weathering Grade I is fresh or faintly weathered rock; Grade VI is completely disintegrated into residual soil. The weathering profile can be deep and commonly with sudden changes from Grades V to II/I, i.e., highly weathered rock/residual soil is often found directly overlying nearly fresh rock.

The Jurong Formation is found in a limited section (south end) of the Spur Tunnel. It comprises sedimentary rocks, mudstone, siltstone, sandstone and conglomerate,

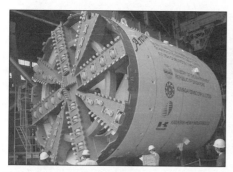

Figure 3. Typical high cutterhead opening ratio earth pressure balance TBM

Figure 4. Typical low cutterhead opening ratio hard rock/earth pressure balance TBM

interbedded in random sequences. The Jurong Formation rocks are highly variable and mostly fractured, with generally poor rock quality.

The Old Alluvium is found in the eastern two-thirds of the North Tunnel and the Northern end of the Spur Tunnel. It consists of indistinctly bedded silty sand, sandy clay or clayey sand with a very stiff to very dense consistency. The materials are usually cohesive and sometimes cemented, but pockets of cohesionless sand are found. It is sometimes considered to be a weak sandstone. The Old Alluvium material is very abrasive to excavation tools because the sand grains are angular with significant amounts of quartz.

The Kallang Formation is encountered in most of the shafts in the eastern two-thirds of the North Tunnel. The tunnels enter into the Kallang Formation soft clays or loose sands at several locations in the middle and eastern DTSS sections. Very extensive land reclamation has been done over the years. The DTSS tunnels are below any reclamation and fill areas, but some shafts penetrate through reclaimed land.

Construction Methods

Tunnels. A total of eight Earth Pressure Balance Tunnel Boring Machines (EPB TBMs) were used to drive the 48km of tunnels, one per contract for Contracts T-01 to T-04 and two TBMs per contract for Contracts T-05 and T-06 due to the much longer lengths of tunnels. Contracts T-01 to T-04 tunneled mostly through Old Alluvium, a generally competent tunneling material. However, the tunnel is located up to 45 m below the groundwater table and required constant control of the tunnel face to prevent face instability/groundwater drawdown, hence the need for EPB TBMs. The contractors, for these "soft ground" tunnel contracts, typically used TBMs with cutterheads with high opening ratios and configured with picks and scrapers (Figure 3).

Contracts T-05 and T-06 tunneled through granite with the rock quality ranging from excellent (up to 300 MPa strength) to very poor (completely decomposed). Portions of the tunnel passed through appreciable lengths of soil-like materials and/or mixed face conditions (hard rock/soil). The very south section of the Spur Tunnel also passed through fractured sedimentary rock. On both these Contracts, EPB TBMs were needed to deal with the soft ground and mixed face conditions but they were generally configured as hard rock machines. They utilized cutterheads with low opening ratios and predominantly fitted with disc cutters (Figure 4). TBM details and performance are provided in Table 2.

Table 2. Singapore deep tunnel sewerage system—tunnel excavation performance

Contract	Maximum Cover (m)	TBM	Progress/Week Plan/Actual (m)	Best Rate Day/Wk/Mon (m)	Ground Conditions
T-01	39	Herrenknecht	76/52	27/132/483	Old Alluvium
T-02	50	NKK	72/51	28/141/602	Old Alluvium, minor Granite
T-03	33	Kawasaki	66/59	32/150/536	Old Alluvium, Kallang
T-04	43	Mitsubishi	70/49	30/137/479	Old Alluvium, minor Kallang
T-05 (N)	52	Herrenknecht	123/43	42/210/631	Fresh to Completely Weathered Granite
T-05 (S)	43	Herrenknecht	110/39	24/131/434	Fresh to Completely Weathered Granite, Old Alluvium
T-06 (N)	49	Herrenknecht	106/58	31/144/625	Fresh to Completely Weathered Granite, Old Alluvium
T-06 (N)	36	Herrenknecht	58/54	40/160/480	Fresh to Completely Weathered Granite, Fractured to Weathered Sedimentary series (Jurong)

As can be seen from Table 2, none of the tunnels achieved their planned rates of progress. The primary reasons for this underachievement were as follows:

Contract T-01: Delays occurred due to compatibility problems between the segments and the TBM thrust rams. Modifications to the design of the segments were required (additional reinforcement);

Initial ring building problems—rings become "iron-bound" on bends;

Logistics problems were experienced due to a single track supply into TBM and locomotive reliability problems.

Contract T-02: An unexpected granite intrusion within the tunnel horizon was discovered during the post-contract award site investigation work. To deal with the granite, a minor realignment of the tunnel, major modifications to the TBM and a major ground treatment operation conducted from the surface were required (marine clays were overlying the granite). Although the TBM modifications (some of which were undertaken immediately before entering the granite and immediately on leaving the granite) and the slow progress of the TBM in the granite delayed tunneling by approximately 9 months, the measures taken to deal with the granite were appropriate and very successful. There was no impact on the nearby expressway.

Contract T-03: Ground loss occurred due to encountering flowing Kallang Formation material when breaking out of the soft eye of the TBM launch chamber—the material flowed around the outside of the TBM body;

Excessive wear to the TBM tools and cutter-arms in the Old Alluvium required repairs to be undertaken from a chamber excavated in front of the TBM cutterhead.

Contract T-04: Excessive wear to the TBM tools and cutter-arms in the Old Alluvium required repairs to be undertaken from a chamber excavated in front of the TBM cutterhead;

Excavated material became "baked" solid within the cutterhead chamber (plenum) requiring breaking out using hand tools;

A ground loss incident due to over-excavation in Kallang Formation materials led to minor surface settlement of the inner lane of an expressway (a partial lane closure was put in place as a precaution but was removed within 12 hours after carriageway repairs had been completed);

Partial failure of the TBM bulkhead resulted in (for a short time) uncontrolled ground loss into the TBM. Tunneling with full earth pressure in highly abrasive materials caused excessive wear on the 40mm thick steel plate of the bulkhead. Freezing of the ground around the TBM cutterhead using nitrogen was conducted from the surface in order to stabilize the ground and permit cutterhead entry for repairs to the bulkhead.

Contract T-05: Over-optimistic planned rates of progress. Planned average rates for the T-05 drives at over 110m per week were significantly higher than the other DTSS contractors had predicted for their drives (refer to Table 2);

Difficulties in excavating TBM launch chambers in mixed face (fresh granite/completely decomposed granite) conditions required the use of freezing using nitrogen fed by gravity into horizontal freeze holes drilled around the tunnel profile;

Excessive wear on TBM discs and cutterhead. Cutterhead interventions for tool changes were required much more frequently than expected by the contractor. Major stoppages were required for repairs to the cutterhead and screw conveyor. The delays due to more frequent interventions were exacerbated due to the difficulties in accessing the cutterhead in up to 5 bars of water pressure. Minor ground loss was experienced on a couple of occasions when the chamber was emptied under free air. Compressed air could not be used alone since man entry at pressures above 3.2 bars was not permitted. Initially, freezing around the cutterhead was used. Although successful, this method proved too costly and time consuming for frequent interventions. Eventually, the contractor used a combination of localized dewatering and compressed air at approximately 1.5 bars to facilitate safe entry into the cutterhead. Dewatering was possible since there were no compressible materials such as marine clays present on Contract T-05 alignment.

Contract T-06: Problems with excessive wear on TBM discs, cutterhead and screw conveyor (requiring frequent cutterhead interventions and stoppages for major repairs) were very similar to the T-05 experience (similar ground conditions and same TBM manufacturer) but without the major problems with cutterhead entry, due to the reduced ground cover on the spur tunnel alignment;

Excessive wear of the brush seals on the rear of the TBM tailskin resulted in a sudden inrush of water and some ground loss into the TBM under a critical structure. Delays occurred while the brushes were repaired and the structure was assessed for potential damage.

Although the planned rates of progress were not achieved, the performance of the TBM tunneling work was considered to be very successful by both Singapore and world standards, considering the extremely difficult and variable ground conditions and high ground water pressures encountered. No serious ground loss incidents occurred in over 48km of large diameter tunneling and there was minimal negative publicity. Startup testing of the tunnel flows was met on schedule.

Main Construction Shafts. Each contract has a main construction shaft (except Contract T-05 which has two) with diameters ranging from 12 m to 16 m and depths varying between 26.6 m and 44.5 m. Shafts were successfully excavated in soft ground, hard rock and through transitions between soft ground and rock using a variety of techniques, as detailed in Table 3.

No major problems were experienced during shaft sinking due to the conservative design of temporary works (use of diaphragm walls; use of braced sheet pile walls in conjunction with jet grouting and a recharge system, etc.), comprehensive instrumentation and monitoring, and effective site supervision.

Structural Lining/Corrosion Protection Lining. A structural lining comprising bolted, precast concrete segments was built within the TBM tailskins. The annular void behind the segmental rings was fully grouted as the rings exited the tailskin (all grouting was undertaken through grout pipes in the TBM tailskins, not through grout holes in the segments). Compression gaskets of EPDM tested up to 10 bars pressure were used to ensure a watertight lining. Segments were 1.5 m long in the 6 m finished diameter (7.2 m excavated diameter) tunnels down to 1.3 m long in the smaller tunnels. Contractors used high quality, high strength (60 N/mm^2) segments reinforced by steel cages. Segment manufacturers were based in Singapore, Malaysia and Indonesia.

Hydrogen sulphide gas which corrodes concrete tunnel lining is produced more rapidly in sewage in tropical areas such as Singapore. To ensure that the 100 year durability requirement of the DTSS is achieved, the precast concrete structural lining is protected by an inner corrosion protection lining (CPL) comprising a high density polyethylene (HDPE) primary liner 2.5 mm thick that is exposed to the sewage and is anchored to a secondary lining of cast insitu concrete, typically 225 mm thick.

The CPL system is placed inside the precast concrete segments on completion of tunnel excavation, thus the CPL installation fell onto the critical path for completion of

Table 3.   Singapore deep tunnel sewerage system tunnel—main construction shafts

Contract	Shaft Diameter (m)	Shaft Depth (m)	Support Type in Soft Ground	Support Type in Hard Ground
T-01	15.5	44.4	Sheet piles/continuous insitu concrete rings	Continuous insitu concrete rings
T-02	16.0	38.4	Sheet piles/continuous insitu concrete rings	Continuous insitu concrete rings
T-03	12.2	33.0	Sheet piles, two rows of jet grout columns, 7 levels of polygonal steel walers	Continuous shotcrete rings
T-04	16.0	36.5	Diaphragm wall	Diaphragm wall
T-05 (N) T-06 (S)	12.0	45.5	Soil mix piles with continuous insitu concrete rings	Pattern rock bolting, fiber reinforced shotcrete
T-06 (N&S)	13.0	26.6	Diaphragm wall	Diaphragm wall

the tunnels. All contractors used steel forms of varying lengths and types. For example, Contract T-01 and T-02 both have a finished diameter of 6m., but T-01 used 2 No. 10m long forms for the crown and side walls and poured the invert separately. The 10m forms could achieve a pour per day thus 120m per week could be achieved as a maximum in the T-01 tunnel. Contract T-02 poured the full circumference monolithically using two 37.5m long forms. These forms could achieve three pours per week, thus the maximum weekly total in T-02 tunnel was 225m.

LESSONS LEARNED ON DTSS PHASE 1

Collaborative Approach

At the start of the DTSS Project consultancy services, the PUB and consultant CH2M/PB JV attended a series of workshops to get to know each other, share perceptions of the project, to align goals, agree strategies, to ensure open communication between team members and to define responsibilities. A Team Charter was signed to signify each individual's "buy-in" to the project goals and objectives. This Charter helped both Client and consultant to remain focused on achieving common goals and objectives throughout all project phases and undoubtedly contributed to the success of the project.

Six Design/Build Contracts

Design/build contracts were preferred in order to provide a single point of contact for both the contractor and the contractor's designer (responsible for the detailed design). The schedule was reduced compared with traditional design/bid/build contracts in that the procurement of the TBMs and the detailed design of the works were undertaken concurrently. The use of a thorough and effective pre-qualification and tendering process, resulting in the selection of suitably experienced contractors and their designers, contributed to the success of the Design and Build approach on the DTSS.

Level of Preliminary Design Detail

The degree to which the works were designed and specified in the tender documents for the six design/build contracts proved to be appropriate. The documents provided adequate control over the detailed design and construction methodology by the contractors, particularly for the critical main tunnel drives. For example, the contractors were required to submit detailed method statements and risk assessments for review and acceptance prior to commencing any new or critical operation. The documents were also sufficiently flexible to give the contractors the opportunity for design development/innovation in the permanent works design, the choice of suppliers of TBMs and other equipment, the choice of detailed segmental lining design and suppliers, and the choice of temporary works designs and construction methodology for temporary and permanent shafts and other items such as TBM launch chambers. It is recommended that in future similar applications that minimum thickness for the precast segments and corrosion protection linings be specified.

Location of Permanent Maintenance Shafts

Although maintenance access into the tunnel is expected, using today's technology with remotely operated equipment, with hindsight, the spacing between shafts could have been increased (from the 1.5 km specified in the DTSS documents) thereby reducing the number of shafts required (there were 38 shaft sites in total on the DTSS). Also, by locating the tunnel under road verges or road drainage reserves, shafts can be located directly over the main tunnel thus eliminating the cost and risk associated with the construction of the connecting adits.

Tunneling Under Major Roadways

The DTSS tunnels were initially located beneath the traffic lanes of the main expressways to avoid sterilizing land for future development. Early in the construction phase it was recognized that there was a need to mitigate the risks from tunnel level ground loss causing surface settlement or collapse of the roadways, also to provide contingency surface access for measures such as ground treatment. Where possible, the tunnel alignment was adjusted within the road reserve to locate it under verges, the drainage reserve or the shoulder lane rather than directly under the traffic lanes. Crossings of traffic lanes were minimized or reduced in length for the same reasons. The DTSS contract documents specified stringent requirements for the installation of ground monitoring instrumentation along all roads in the tunnel "zone of influence." These precautions contributed to the success of tunneling operations.

Numbers of TBMs

Eight TBMs were specified to excavate 48km of large diameter tunnel in ground that at times was difficult. Assuming that land for TBM drive sites is readily available (which was not the case in crowded Singapore, hence the long drives), a more risk averse (but more expensive) approach would have employed more TBMs with shorter excavation lengths per TBM, thus resulting in more efficient tunneling operations (e.g., shorter material haul lengths, less concern with TBM wear in abrasive ground, etc.).

TBM Requirements and Type

The use of closed face TBMs was specified in the contract documents. All six contractors chose to use EPB TBMs rather than Slurry TBMs. Although difficult ground conditions were encountered, the EPB technology proved to be successful without causing any major surface incidents. EPB TBMs would be recommended for similar future projects based on the DTSS Phase I experience. It is also recommended that for future similar applications, minimum TBM power (for specific diameters) and infinite variable speed cutterhead drives be specified. Lessons learned from the performance of the TBMs on the DTSS were that more effective wear protection was needed on the TBMs in the very abrasive ground conditions together with better provisions for undertaking TBM cutterhead tool changes under high ground water pressures.

Corrosion Protection Lining

All six DTSS contractors chose HDPE for the primary corrosion protection lining rather than PVC (the documents provided a choice of using either PVC or HDPE). HDPE has much less onerous fire/fume problems and is more convenient for handling. Proper design of the concrete form and the HDPE installation and placement on the form was critical, especially considering issues with tensioning, wrinkling in the crown and concrete properties (e.g., flowability). Because of the success of installing the HDPE on the DTSS tunnels (this material had no known case history with applications in tunnels up to 6m diameter) and the fire/fume problems associated with PVC, the use of HDPE would be recommended for future works.

Site Investigation

Contractors bidding for the DTSS were required to interpret the provided factual information (in form of a Geotechnical Data Report) and produce a "Preliminary Geotechnical Interpretative Report" (PGIR). This PGIR became one of the key documents (but not the only document) after contract award in assessing the encountered ground conditions and the design of the tunnel support and lining. The PGIR was very useful during bid evaluation in assessing whether the bidders really understood the ground conditions and the effect that they would have on their proposed construction methods and program. As a further risk mitigation measure, the documents also required the contractors to undertake further site investigation post contract and submit a Final Geotechnical Interpretative Report" (FGIR). It was during this further investigation that the granite intrusion was discovered in Contract T-02. Had the granite not been discovered in advance of TBM tunneling commencing then the delays and additional cost would have been far greater than those actually experienced on Contract T-02.

Risk Sharing Approach

PUB's risk sharing approach led to the inclusion of an additional clause in the Particular Conditions of Contract permitting additional costs and time for dealing with Unforeseen Physical Conditions (UPC). The onus was placed on the contractor to demonstrate that the encountered ground conditions were truly unforeseeable. The inclusion of this clause should result in lower tender prices. The Contract assigned to the contractor all risks within the contractor's control, such as selecting the appropriate construction equipment, operating the equipment in the known ground conditions (including the frequency of TBM tool changes, ground treatment where necessary to effect entry into the cutterhead for tool changes, the amount and type of spoil conditioner required, etc.), temporary and permanent works design, methods of construction of shafts and adits, dealing with authorities and third parties, supply of labor, etc.

Insurance

PUB had the foresight to arrange for a Project-wide combined Contractor's All Risks and Third Party Insurance and a Workmens' Compensation Insurance. By procuring insurances itself, the Client ensures that the proper insurances are in place, are controlled, and are taken out of the contractor's tender price. On DTSS, a very cost effective program for insurance was obtained, better than the contractors could have achieved by themselves.

Supervision of Workmanship

Most Design/Build contracts require the contractor's designer to provide independent site supervision of workmanship (i.e., raising and closing NCRs, verifying materials compliance, etc.—effectively providing an independent Resident Engineers (RE) team). Theoretically, the consultant's (on DTSS, CH2M/PB JV) RE teams should not be required to undertake detailed inspection of all construction activities, but just random checks on workmanship and record keeping to ensure that the contractor's designers RE teams are doing their job properly. A problem arises with this arrangement because the designer's site supervision team is not truly independent. The failure of "self certification" by the contractor's team has been observed on many Design/Build contracts. On DTSS, the consultant's RE teams on site had to take on a more substantial hands-on supervision role to ensure that the finished quality of the works met the Client's requirements.

Supervision of Safety

Safety of the general public and of the contractors' staff and operatives was a prime concern throughout the DTSS construction phase. The DTSS contractors were responsible for the overall safety of the construction operations but the PUB and their consultant were deeply involved in all safety aspects including regular safety meetings and inspections, tool box talks, emergency planning, mock rescues, review of risk assessments, attendance at incidents, incident and safety statistics reporting, etc. The proactive approach to safety by all parties resulted in an excellent safety record on the DTSS.

CONCLUSION

By proactively dealing with the challenges encountered and lessons learned during all phases of the DTSS Phase 1 Tunnels, the following positive results were achieved:

- All tunnel contracts were completed on time for the commissioning of Changi WRP. The quality of construction is very good—the 100-year durability of the system will be achieved.
- The tunneling works have not caused any danger to the public and there was very little damage to third party infrastructure. There were fewer than 10 incidents on the whole 48 km of tunnel that required a temporary road closure to resurface minor road settlements. There was no danger to road users at any time and temporary traffic lane closures did not exceed 12 hours in duration. In addition, there was very little environmental impact during construction with very few complaints being received from the public.
- The health and safety performance was very good. At the peak of construction, over 2000 personnel were working on the project and over 18 million man

hours were expended on the project by the six contractors with only 103 lost time accidents.

Much of the success of the project can be attributed to getting it right in the feasibility and preliminary design phases such that when the contracts were awarded, most project risks had been identified and dealt with, and the proper controls were built into the contract documents. Another key factor was the selection of experienced and competent contractors. Proactive involvement and effective communications throughout the DTSS construction phase by the PUB and its consultant also contributed to the positive outcome.

INDEX

A

Al Naboodah Engineering Services, 208–214
Alaskan Way Viaduct and Seawall Replacement, 496–506
Alliancing, 422–423, 424–427
Amsterdam (Netherlands) Central Station, 369–380
Anacostia River CSO tunnels (Washington, DC), 638–646
ARCADIS, 160–167
Arrowhead Tunnels Project (California), 229–231, 233–243, 252–254
  automated real-time probe drill monitoring system for TBM tunneling, 430–441
  difficult geology and ground conditions, 231–233
  hybrid TBM design, 243–252
ASDAM project (Antwerp, Belgium), 176–184
Association of British Insurers, 12–13
A3 Hindhead tunnel (UK), 673–684
Atlanta, Georgia, West Area CSO Tunnel and Pumping Station, 913–926, 1064–1078
Atlanta CSO Constructors, 913–926
Australia, EastLink freeway tunnels (Melbourne), 534–550
Austria, Kops II pumped storage plant, 870–881

B

BART. *See* Bay Area Rapid Transit
Bauer Maschinen GmbH, 1039
Bay Area Rapid Transit (California)
  design of underground structures for Silicon Valley segment, 2–11
  seismic testing for precast concrete segmental linings, 1225–1242
Bay Area Water Supply and Conservation Agency (California), 91
Bay Tunnel (San Francisco Bay Area), 685–693
Beach Interceptor Sewer Tunnel (Orange County, CA), 1144–1151
Beacon Hill Station and Tunnels (Seattle, WA), 346–347
  dewatering wells, 346, 348–352, 952–954
  EPBM-driven twin bore running tunnels, 793–807
  jet grouting, 346, 352–358, 952–954
  SEM excavation, 346–347, 358–359, 943–963
Bechtel Civil, 2–11
Bechtel Infrastructure Corp., 564
Beck, Adam, 185–186
Belgium, Project ASDAM (Antwerp), 176–184
Bieniawski classification, 267, 269, 279
Big Walnut Augmentation/Rickenbacker Interceptor (Columbus, OH), 704–711
  EPBM designed for work in ground with high boulder concentrations, 215–228
  tunnel construction, 712–740
Bologna, Italy, 381–395
Boston (Massachusetts) Central Artery/Tunnel 9A4 ground freezing, 361–362
Bradshaw Interceptor Section 8 Project (Rancho Cordova, CA), 843–851

British Columbia, Galore Creek Mine access road tunnel, 332–344
British Tunneling Society, 13
BWARI. *See* Big Walnut Augmentation/ Rickenbacker Interceptor (Columbus, OH)

C

Cady Marsh Flood Relief Tunnel (East Chicago, Indiana), 80–90
Caldecott Tunnels fourth bore (Oakland, California), 616–626
California
  automated real-time probe drill monitoring system for TBM tunneling (Arrowhead Tunnels Project), 430–441
  Bradshaw Interceptor Section 8 Project (Rancho Cordova), 843–851
  Caldecott Tunnels fourth bore (Oakland), 616–626
  Claremont Tunnel Seismic Upgrade Project, 148–159
  design challenges for New Crystal Springs water tunnel, 91–103
  design of underground structures for BART segment, 2–11
  emergency repairs to Beach Interceptor Sewer Tunnel (Orange County), 1144–1151
  hybrid TBM design for unique ground conditions in Arrowhead Tunnels Project, 229–254
  JWPCP Tunnel and Ocean Outfall Project (Los Angeles County), 594–604
  Lake Hodges to Olivenhain Pipeline (San Diego County), 882–888
  Lower Northwest Interceptor Sewer (Sacramento), 741–756, 1274–1285
  Metro Gold Line Eastside Extension construction challenges and innovations (Los Angeles), 472–494
  Metro Gold Line Eastside Extension EPBM tunneling (Los Angeles), 808–829
  NATM in excavation of linear accelerator tunnels, 21–31
  San Vicente Pipeline Tunnel Project (San Diego), 928–942, 1243–1251
  seismic testing of precast concrete segmental linings for Silicon Valley BART segment, 1225–1242
Chicago (Illinois) Transit Authority
  cut and cover alternative for Tunnel Connection project, 64–68
  mining alternative for Tunnel Connection project, 59–64
  tunneling alternatives for subway Tunnel Connection project, 56–57, 59, 68
  tunneling history, 57–58
China, Liaoning Dahuofang Reservoir Water Transfer Project, 551–561
Claremont Tunnel Seismic Upgrade Project (California), 148–159

Columbus, Ohio
  Big Walnut Augmentation/Rickenbacker
    Interceptor tunnel construction, 712–740
  Big Walnut Outfall Augmentation Sewer,
    704–711
  EPBM designed for work in ground with high
    boulder concentrations, 215–228
Communication technology
  Ethernet-based, 466–471
  old systems, 466
Construction manager at risk. *See* General
    contractor/construction manager contracts
Construction manager/general contractor contracts.
    See General contractor/construction manager
    contracts
Contracting methods, 418, 428–429
  alliancing, 422–423, 424–427
  cost reimbursable fixed fee, 1297–1307
  design-bid-build (DBB), 418, 420–421
  design-build (DB), 421–422
  design-build for Lake Hodges to Olivenhain
    Pipeline tunnel and shaft, 882–888
  design-build for Niagara Tunnel Project, 288–299
  early contractor involvement (ECI), 424
  general contractor/construction manager
    (GCCM), 422–424
  incorporating alliancing and GCCM methods in
    DB or DBB, 427
  innovations, 418–419
  and SEP-14 and SEP-15, 419
  See also Design and planning
Cost reimbursable fixed fee contracts, 1297–1307
Costa Rica, La Joya hydropower plant, 266–279
Croydon Cable Tunnel Project, 117–136
CSM method. *See* Cutter soil mixing
CTA. *See* Chicago (Illinois) Transit Authority
Cut and cover tunnels
  as alternative for subway project in soft ground,
    56–68
  in construction of Dublin Port Tunnel, 1008–1010
  in construction of Transbay-Caltrain Downtown
    Extension Project, 964
  designing to protect adjacent structures during
    tunneling for light rail, 70–79
  in Leipzig City-Tunnel project, 197–200
  proposed for Alaskan Way Viaduct and Seawall
    Replacement (Seattle), 496–506
  support of excavation (SOE) systems, 77
Cutter soil mixing, 1039–1046

D
DB. *See* Design-build contracts
DBB. *See* Design-bid-build contracts
Deep mixing method, 1038–1039
Deep Soil Mix, 5
Deep Underground Science and Engineering
    Laboratory
  candidate sites, 695–696
  proposed large-span cavern construction at
    depth, 694–702
Design and planning
  alignment challenges and complex ground
    conditions for new Bay Area water tunnel,
    91–103
  for Anacostia River CSO tunnels, 638–646
  A3 Hindhead tunnel (UK), 673–684
  for cavern support in subway project, 32–43

Chicago subway tunneling alternatives, 56–68
Croydon Cable Tunnel Project, 117–136
engineering optimization to reduce outage time
  in construction of power station tunnel,
  1262–1273
groundwater control for TBM tunneling in soft
  ground, 80–90
for Kensico-City deep rock water tunnel, 520–533
Lake Mead Intake No. 3 Project (Nevada),
  647–662
for large-span cavern construction at depth
  (DUSEL), 694–702
for NATM excavation of linear accelerator
  tunnels, 21–31
New Crystal Springs Bypass Tunnel (Bay Area),
  91–103
for planned Bay Tunnel (San Francisco Bay
  Area), 685–693
to protect adjacent structures during light rail
  tunneling, 70–79
soil abrasiveness test methodology for TBM
  tunneling, 104–116
for TBM tunneling through difficult ground
  (Gotthard Base Tunnel), 44–55
tunnel interaction assessment, 137–146
underground structures for Silicon Valley Rapid
  Transit Project, 2–11
for Upper Rouge Tunnel, 627–636
See also Contracting methods; Risk
  management
Design-bid-build contracts, 418, 420–421
Design-build contracts, 421–422
  for Deep Tunnel Sewerage System Project
    (Singapore), 1308–1319
  Lake Hodges to Olivenhain Pipeline tunnel and
    shaft, 882–888
  Niagara Tunnel Project, 288–299
Detroit, Michigan
  rehabilitation of Middle Rouge Parkway
    Interceptor Extension, 1172–1180
  Upper Rouge Tunnel project, 627–636
Dewatering
  Beacon Hill Station project, 346–359, 952–954,
    961
  Lower Northwest Interceptor Sewer, 750
Difficult ground
  and abandonment of TBM project due to flooding
    with gas-laden groundwater, 160–167
  encountered in construction of water bypass
    tunnel through fault zone, 148–159
  EPBM designed for work in ground with high
    boulder concentrations, 215–228
  ground freezing in Leipzig City-Tunnel project,
    200–206
  hybrid TBMs in ground with variable geologic
    conditions, high water pressures, and
    significant faults, 229–254
  methane gas mitigation measures at stormwater
    and sewage tunnel, 168–175
  microtunneling in sand, gravel, and sandstone
    for island service tunnels, 208–214
  shield tunneling through sandy soil under
    existing urban structures (Antwerp, Belgium),
    176–184
  TBM experience at Gotthard Base Tunnel, 44–55

tunneling through faults in volcanic rock with low cohesion and high water inflows, 266–279
See also Faults; Ground conditions; Time-dependent ground behavior
District of Columbia Water and Sewer Authority CSO tunnels, 638–646
DMJM+HARRIS•Arup Joint Venture, 32–43, 663–672
DMM. *See* Deep mixing method
Double-shield TBMs
in extra-hard, abrasive bedrock at Guadarrama Tunnels, 1079–1081
relationship between tunnel convergence and machine operational parameters, 1094–1108
and Rock Mass Excavability index, 1127
in rock tunneling for Kops II project, 870–881
in volcanic rock with low cohesion and high water inflow, 266–279
Drill and blast
in construction of Lake Hodges to Olivenhain Pipeline tunnel and shaft, 882–888
in construction of Sir Adam Beck II intake tunnels, 185–196
in Liaoning Dahuofang water tunnel project, 551, 561–562
in urban setting for West Area CSO Tunnel, 913–926
Dublin Port Tunnel (Ireland), 998–1010
Dulles Airport Automated People Mover Tunnels (Washington, D.C.), 1286–1296
Dulles Corridor Metrorail Project (Washington, D.C.), 564–578
Dulles Transit Partners, 577
Durban Harbour Tunnel Project, 579–593
DUSEL. *See* Deep Underground Science and Engineering Laboratory

E

Early contractor involvement, 424
for A3 Hindhead tunnel (UK), 673–684
Earth pressure balance TBMs
considered for JWPCP Tunnel and Ocean Outfall Project, 600, 602–604
in construction of Big Walnut Augmentation/Rickenbacker Interceptor, 726–733
in construction of Big Walnut Outfall Augmentation Sewer, 704–711
in construction of Deep Tunnel Sewerage System Project, 1308–1319
in construction of Lower Northwest Interceptor Sewer, 741–756
in construction of Silicon Valley Rapid Transit Project underground structures, 5
in construction of twin bore running tunnels (Beacon Hill Project), 793–807
design principles for soft ground cutterheads, 784–792
designed for work in ground with high boulder concentrations, 215–228
ground settlement monitoring system used during tunnel construction, 381–395
Madrid Calle 30 project, featuring world's largest TBMs, 769–782
mining through cobbles and boulders, 791–792
for planned Bay Tunnel (San Francisco Bay Area), 685–693

in soft ground tunneling for Metro Gold Line Eastside Extension (Los Angeles), 808–829
in soft ground tunneling for sewer tunnel, 843–851
soil deformation analysis, 852–867
study of surface settlements over EPB driven tunnels in soft clay, 757–768
with unique features for Metro Gold Line Eastside Extension Project, 472, 481–486
East Bay (California) Municipal Utility District, Claremont Tunnel Seismic Upgrade Project, 148–159
East Side Access Project (New York)
geotechnical monitoring during construction, 407–416
open-cut shaft construction at Northern Boulevard, 1023–1030
East Side CSO Tunnel Project (Portland, OR), 1192–1204
EastLink freeway tunnels (Melbourne, Australia), 534–536
excavation and support procedures, 536–539
groundwater contol, 544–547
hydrogeological assessment, 539–544
precast invert lining, 547–550
Eastside LRT Constructors, 472–494
ECI. *See* Early contractor involvement
ECO Grouting Specialists Ltd., 398
Elm Road Generating Station (Oak Creek, WI) water intake tunnel, 889–900
Enwave Project (Toronto), 329–330, 331
EPBMs. *See* Earth pressure balance TBMs
Ethernet communication systems, 466–471

F

Faroe Islands, membrane lining for Nordöy road tunnel, 1252–1260
Faults
and construction of water bypass tunnel, 148–159
defined, 280
hybrid TBMs for ground with significant faults, 229–254
tunneling through faults in volcanic rock, 266–279
See also Mountain faults
Fort Canning Tunnel (Singapore), 976–987
France, Metro Line B (Toulouse), 852–867

G

Galore Creek Mine (B.C.) access road tunnel, 332–333, 343–344
design and constructability issues, 340–343
geotechnical conditions, 336–340
under-glacier alignment, 332, 333–335
Gassy conditions
abandonment of TBM project due to flooding with gas-laden groundwater, 160–167
mitigation measures at stormwater and sewage tunnel (Mill Creek Tunnel), 168–175
Upper Rouge Tunnel, 627–636
GCCM. *See* General contractor/construction manager contracts
General contractor/construction manager contracts, 422–424
Georgia, West Area CSO Tunnel and Pumping Station (Atlanta), 913–926, 1064–1078
Germany, Leipzig City-Tunnel project, 197–206

Ghella SpA, 266–267
Ghomroud water conveyance tunnel (Iran), 1094–1108
Golder Associates, 381–395
Gotthard Base Tunnel (Switzerland), 44–55
Grand Rapids, Michigan, Michigan Street Pedestrian Tunnel, 988–997
Ground conditions
  automated real-time probe drill monitoring system for TBM tunneling, 430–441
  comparative soil deformation analysis for three TBM methods and convention tunneling, 852–867
  complex ground and seismic conditions for JWPCP Tunnel and Ocean Outfall Project, 594–604
  complexities and design challenges for new Bay Area water tunnel, 91–103
  EPBM tunneling in alluvial soils under Sacramento River, 741–756
  EPBM tunneling in soft ground for sewer tunnel, 843–851
  extra-hard, abrasive igneous and metamorphic bedrock at Guadarrama Tunnels, 1079–1081
  ground settlement monitoring system used during tunnel construction, 381–395
  groundwater control for TBM tunneling in soft ground, 80–90
  monitoring with TBM Excavation Control System, 442–456
  NATM and TBM tunneling in bedrock for Dulles Airport Automated People Mover Tunnels, 1286–1296
  NATM construction in soft ground (Fort Canning), 976–987
  NATM in soft ground for Dulles Corridor Metrorail Project, 564, 572–576
  NATM tunneling in clean sands (Michigan Street Pedestrian Tunnel), 988–997
  NATM tunneling through granitic rock and conglomerate, 928–942
  relationship between tunnel convergence and double-shield TBM operational parameters, 1094–1108
  sandstone and sand in A3 Hindhead tunnel (UK), 673–684
  slurry TBM tunneling in non-cohesive, permeable soils (West Side CSO Tunnel Project), 830–842
  soft ground in Harbor Siphon project, 605, 606–610
  soft ground tunneling for Anacostia River CSO tunnels, 638–646
  study of surface settlements over EPB driven tunnels in soft clay, 757–768
  TBM tunneling in extreme topographic and climatic conditions, 507–519
  TBMs in rock tunneling for Dulles Corridor Metrorail Project, 564, 572–576
  Transbay-Caltrain Downtown Extension Project, 964, 966–969
  tunneling alternatives for subway project in soft ground, 56–68
  tunneling through bedrock, soft soil, and seismically active conditions (Bay Tunnel), 685–693
  weak sedimentary rock and seismic criteria (SLAC tunnels), 21–31
  widely varied conditions for Deep Tunnel Sewerage System Project (Singapore), 1308–1319
  See also Difficult ground; Faults; Rock tunneling; Time-dependent ground behavior
Ground freezing
  in Leipzig City-Tunnel project, 200–206
  and problem of moving groundwater, 360–368
Ground improvement
  dewatering wells (Beacon Hill Station), 346, 348–352, 952–954
  hot bitumen grouting through water-bearing zone under high pressure and extreme flow conditions, 396–406
  jet grouting (Beacon Hill Station), 346, 352–358, 952–954
  "sandwich" retaining wall with jet grout, 369–380
Groundwater control
  dewatering in Beacon Hill Station project, 346–359, 952–954, 961
  EastLink freeway tunnels, 544–547
  and environmental issues in design and construction of high voltage tunnel, 117–136
  failure due to difficult ground conditions, 160–167
  in La Joya hydropower plant tunneling, 266–279
  Lower Northwest Interceptor Sewer, 749–750
  moving groundwater and effect of ground freezing, 360–368
  and TBM tunneling in soft ground, 80–90
  See also Mountain faults: hydrogeology of
Grouting
  in Big Walnut Outfall Augmentation Sewer, 704–711
  hot bitumen grouting through water-bearing zone under high pressure and extreme flow conditions, 396–406
  jet and backfill grouting for Big Walnut Augmentation/Rickenbacker Interceptor, 718–720, 730–731, 734
  jet grout in sandwich retaining wall under existing station, 369–380
  jet grouting in Beacon Hill Station project, 346–359, 952–954
  Metro Gold Line Eastside Extension, 472, 474–475, 484–486
  in New Croton (brick-lined) Aqueduct Rehabilitation and Inspection Program, 1183–1185
  permeation grouting in control of moving groundwater, 361
  permeation grouting of clean sands (Michigan Street Pedestrian Tunnel), 988, 992, 994–995
  for planned Lake Mead Intake No. 3 Project, 657–658
Guadarrama Tunnels (Spain), 1079–1093

H
Harbor Siphon water tunnel project (New York City), 605–615
Hatch Energy, 292, 298
Hatch Mott MacDonald, 292
  design of underground structures for Silicon Valley Rapid Transit Project, 2–11
H.D.D. *See* Horizontal directional drilling

Heartland Corridor Clearance Improvement Project, 1152–1161
Herrenknecht TBMs
  AVND 2500 Mixshield TBM, 208–215
  in construction of Dublin Port Tunnel, 1002–1005
  in construction of West Area CSO Tunnel and Pumping Station, 924–925, 1069
  double-shield TBMs in extra-hard, abrasive bedrock at Guadarrama Tunnels, 1079–1081
  EPBMs for Los Angeles Metro Gold Line tunnels, 481–486
  hybrid design for unique ground conditions in Arrowhead Tunnels Project, 243–253
  mix-shield slurry TBM for Durban Harbour project, 579–593
  one of world's two largest TBMs for Madrid Calle 30 project, 774–782
Horizontal directional drilling, 208–214
Hot bitumen grouting, 396–406
Hydro One tunnel (Toronto), 330–331
Hydrogeological assessment (EastLink freeway tunnels), 539–544
Hydroshield TBMs
  in ASDAM project (Antwerp, Belgium), 178, 181–183
  in Leipzig City-Tunnel project, 197

I
Iceland, Kárahnjúkar Hydroelectric Project, 507–519
Illinois, Chicago subway tunneling alternatives, 56–68
Indiana, East Chicago groundwater control, 80–90
Insurance. See Risk management
International Tunneling Association, 12
International Tunneling Insurance Group, risk management code, 12–20, 288, 295
Iran, Ghomroud water conveyance tunnel, 1094–1108
Ireland, Dublin Port Tunnel, 998–1010
ITA. See International Tunneling Association
Italy, Bologna tunnel construction, 381–395
ITIG. See International Tunneling Insurance Group

J
Japan, Tokyo Metropolitan Expressway Central Circular Route, 1205–1206
Jines Construction, 398
Joint Code of Practice for Risk Management of Underground Works in the UK, 12–20
JWPCP Tunnel and Ocean Outfall Project (Los Angeles County)
  ground and seismic conditions, 594–604
  TBMs considered, 600, 602–604

K
Kansas City Trans Missouri Tunnel ground freezing, 363
Kárahnjúkar Hydroelectric Project (Iceland), 507–519
Kensico-City Tunnel (New York), 520–533
Kentucky, and Heartland Corridor Clearance Improvement Project, 1152–1161
Kops II pumped storage plant (Austria), 870–881

L
La Joya (Costa Rica) hydropower plant, 266–279
Lake Hodges to Olivenhain Pipeline (San Diego County, CA), 882–888

Lake Mead Intake No. 3 Project (Nevada), 647–662
Landsvirkjun, 507
LCLS. See LINAC Coherent Light Source tunnels
LCPC abrasimeter test, 108–109, 115
Leaky Feeder communications technology, 466
Leipzig (Germany) City-Tunnel project
  cut and cover tunnels, 197–200
  ground freezing, 197, 200–206
  and hydroshield TBMs, 197
Liaoning Dahuofang Reservoir Water Transfer Project (China)
  drill and shoot tunneling, 551, 561–562
  TBM tunneling, 551–561
LINAC Coherent Light Source tunnels, 21–31
Linings
  concrete liner for Metro Gold Line Eastside Extension, 472, 493–494
  concrete liner permeability subject to internal water pressures, 140–141
  Durban Harbour project, 583, 591–592
  impermeable membrane and prestressed cast-in-place lining as solution to swelling problem (Niagara Tunnel Project), 901–912
  one-pass (Washington MATA), 570–571
  precast concrete segmental (Big Walnut Augmentation/Rickenbacker Interceptor), 722–726
  precast concrete segments (Lower Northwest Interceptor Sewer), 741, 750
  precast invert lining (EastLink freeway tunnels), 547–550
  precast segmental (Harbor Siphon project), 605, 613–614, 615
  pumpable concrete for Kops II project, 877–881
  seismic testing of precast concrete segmental linings for Silicon Valley BART segment, 1225–1242
  for significant seismic demand (Oakland, California), 616–626
  sprayable waterproofing membrane, 1252–1260
  steel fiber reinforced concrete segmental rings (East Side CSO, Portland), 1192–1204
  steel fiber reinforced self-compacting segmental concrete rings, 1243–1251
  testing and application of steel fiber reinforced highly flowable concrete segments, 1205–1224
Liver, Norman, 1038
Long Island (New York) Rail Road
  geotechnical monitoring during East Side Access construction, 407–416
  open-cut shaft construction in East Side Access Project, 1023–1030
Los Angeles, California
  Metro Gold Line Eastside Extension construction challenges and innovations, 472–494
  Metro Gold Line Eastside Extension EPBM tunneling, 808–829
Los Angeles County, California, JWPCP Tunnel and Ocean Outfall Project, 594–604
Lovat Company, 846
Lower Northwest Interceptor Sewer (Sacramento, CA)
  EPBM tunneling in alluvial soils under Sacramento River, 741–756
  trenchless crossings, 1274–1285

M

Madrid (Spain) Calle 30 highway tunnels, 769–782
Manapouri Power Station (New Zealand),
    1262–1273
Massachusetts, Boston Central Artery/Tunnel 9A4,
    361–362
Melbourne, Australia, EastLink freeway tunnels,
    534–550
Metro Gold Line Eastside Extension Project (Los
    Angeles, CA), 472–494
Metropolitan Washington Airports Authority, 564,
    1286
Metropolitan Water District of Southern California.
    See Arrowhead Tunnels Project (California)
Michigan
    Michigan Street Pedestrian Tunnel (Grand
        Rapids), 988–997
    rehabilitation of Middle Rouge Parkway
        Interceptor Extension, 1172–1180
    Upper Rouge Tunnel project (Detroit), 627–636
Microtunneling, 208–214
Middle Rouge Parkway Interceptor Extension
    (Michigan), 1172–1180
Mill Creek Tunnel (Ohio), 168–175
Mined tunnels, 56–68
Missouri, Kansas City Trans Missouri Tunnel, 363
Mitsubishi
    EPBM in construction of twin bore running
        tunnels (Beacon Hill Project), 793–807
    one of world's two largest TBMs for Madrid Calle
        30 project, 774–782
Monitoring
    automated real-time probe drill monitoring
        system for TBM tunneling, 430–441
    gas mitigation (Mill Creek Tunnel), 171–175
    geotechnical (East Side Access), 407–416
    ground settlement during tunnel construction
        (Bologna), 381–395
    of NATM construction in soft ground with high
        groundwater elevation and proximity to
        historically important features, 976–987
    of NATM tunneling in clean sands (Michigan
        Street Pedestrian Tunnel), 988–997
    TBM Excavation Control System (seismic
        reflector tracing with TBM data), 442–456
Mott MacDonald, 509
Mountain faults, 280, 286
    blocky rock zone, 281
    and construction methodologies, 285–286
    crushed rock and sand zone, 282
    detection of, 284–285
    engineering geology of, 280–283
    hydrogeology of, 280, 282–283
    squeezing clay zone, 281
MSHA. See US Mine Safety and Health
    Administration

N

Narragansett Bay Commission CSO Abatement
    Program (Providence, RI), 1047–1057
National Grid Company (UK), 117–136
NATM. See New Austrian Tunneling Method
Netherlands, Amsterdam Central Station, 369–380
Nevada, Lake Mead Intake No. 3 Project, 647–662
New Austrian Tunneling Method
    for Caldecott Tunnels fourth bore, 616–626

    in construction of Dulles Airport Automated
        People Mover Tunnels, 1286–1296
    in construction of Michigan Street Pedestrian
        Tunnel, 988–997
    in construction of Transbay-Caltrain Downtown
        Extension Project, 964–975
    for cross passage excavation in soft ground and
        below water table (Metro Gold Line Eastside
        Extension), 472, 490–493
    in excavation of linear accelerator tunnels, 21–31
    ground settlement monitoring system used
        during tunnel construction, 381–395
    in San Vicente Pipeline Tunnel Project, 928–942
    in soft ground for Dulles Corridor Metrorail
        Project, 564, 572–576
    monitoring of construction in soft ground with
        high groundwater elevation and proximity to
        historically important features, 976–987
    See also Sequential Excavation Method
New Croton Aqueduct Rehabilitation and Inspection
    Program (New York)
    inspection phase, 1162–1171
    rehabilitation phase, 1181–1190
New Crystal Springs Bypass Tunnel (California),
    91–103
New York City
    design and planning of Kensico-City water
        tunnel, 520–533
    Harbor Siphon water tunnel project, 605–615
    New Croton Aqueduct Rehabilitation and
        Inspection Program, 1162–1171, 1181–1190
New York City Transit Authority
    cavern support design (Second Avenue Subway
        project), 32–43
    Second Avenue Subway progress report, 663–
        672
New York City Water Tunnel No. 3, 1109–1110
    Brooklyn Tunnel, 1109, 1111–1112
    ground freezing, 363–365
    Manhattan Tunnel, 1109, 1113–1114
    Queens Tunnel, 1109, 1112–1113
    shaft construction methods, 1012–1022
New York Metropolitan Transportation Authority
    geotechnical monitoring during East Side
        Access construction, 407–416
    open-cut shaft construction in East Side Access
        Project, 1023–1030
New Zealand, Second Manapouri Tailrace Tunnel
    project, 1262–1273
Niagara Tunnel Project (Ontario), 312–314, 320–321
    design-build contract, 288–299
    ground and rock support, 317–320
    High Performance Main Beam TBM, 314–317,
        901
    innovative lining solution to swelling problem,
        901–912
    See also Sir Adam Beck II Hydroelectric Project
        (Ontario)
Nordöy road tunnel (Faroe Islands), 1252–1260
North Shore Connector Project (Pittsburgh, PA),
    70–79
Northeast Ohio Regional Sewer District, 168–175
NovaGold Resources, 332
NTNU abrasion test, 108, 109–110, 113–114
NTNU soil abrasion test, 111–116

O

Oak Creek, Wisconsin, Elm Road Generating Station, 889–900
Oakland, California, Caldecott Tunnels, 616–626
Ohio
Big Walnut Augmentation/Rickenbacker Interceptor (Columbus) tunnel construction, 712–740
Big Walnut Outfall Augmentation Sewer (Columbus), 704–711
EPBM designed for work in ground with high boulder concentrations (BWARI Project), 215–228
gas mitigation measures for Cleveland area stormwater and sewage tunnel, 168–175
and Heartland Corridor Clearance Improvement Project, 1152–1161
Ontario
construction of Richmond Hill shafts for York Durham Sewage System Interceptor, 1058–1062
construction of Sir Adam Beck II intake tunnels, 185–196
design-build contract for Niagara Tunnel Project, 288–299
Enwave Project (Toronto), 329–330, 331
Hydro One tunnel (Toronto), 330–331
innovative lining solution to swelling problem (Niagara Tunnel Project), 901–912
Niagara Tunnel Project description, 312–321
Spadina Subway Extension Environmental Assessment Study (Toronto), 300–311
tunnel permitting and environmental issues, 322–331
York Durham Sewage System Interceptor Trunk Sewer, 322, 323–328
Ontario Power Generation, 288–299, 312–321, 901–912
Orange County, California, Beach Interceptor Sewer Tunnel, 1144–1151
Oregon
East Side CSO Tunnel Project (Portland), 1192–1204
West Side CSO Tunnel Project (Portland), 830–842, 1297–1307

P

Palm Jumeirah Island service tunnels (UAE), 208–214
Parsons Brinckerhoff, 496
Partnering. *See* Public-private partnerships
Pennsylvania, Pittsburgh light rail, 70–79
Pipe jacking. *See* Microtunneling
Pittsburgh, Pennsylvania. *See* Port Authority of Allegheny County (Pennsylvania)
Port Authority of Allegheny County (Pennsylvania), Pittsburgh light rail, 70–79
Portland, Oregon
cost reimbursable fixed fee contract for West Side CSO Tunnel Project, 1297–1307
reinforced concrete segmental lining (East Side CSO), 1192–1204
slurry TBMs in West Side CSO Tunnel Project, 830–842
Poyry Energy Ltd., 509
Providence (Rhode Island) CSO Tunnel ground freezing, 365–367

Public-private partnerships
Bradshaw Interceptor Section 8 Project, 843–851
Dulles Corridor Metrorail Project, 564, 576–578
for Heartland Corridor Clearance Improvement Project, 1152–1161

R

Rancho Cordova, California, Bradshaw Interceptor Section 8 Project, 843–851
Rehabilitation
emergency repairs to Beach Interceptor Sewer Tunnel, 1144–1151
of Middle Rouge Parkway Interceptor Extension (Michigan), 1172–1180
New Croton (brick-lined) Aqueduct Rehabilitation and Inspection Program (inspection phase), 1162–1171
New Croton (brick-lined) Aqueduct Rehabilitation and Inspection Program (rehabilitation phase), 1181–1190
rail tunnel clearance expansion (Heartland Corridor project), 1152–1161
Relationship contracting. *See* Alliancing
Rhode Island
construction of drop and vent shafts for Narragansett Bay Commission CSO Abatement Program, 1047–1057
Providence CSO Tunnel ground freezing, 365–367
Rings
Big Walnut Outfall Augmentation Sewer, 707–708
design for Big Walnut Augmentation/Rickenbacker Interceptor, 723–725
Risk management
codes (UK and ITIG), 12–20
crossing difficult ground safely, 44–55
and geotechnical monitoring during East Side Access construction (New York), 407–416
for open-cut shaft construction in East Side Access Project, 1023–1030
Washington Suburban tunnel case history, 17–18
RME index. *See* Rock Mass Excavability index
Roadheader guidance system, 457–465
The Robbins Company
CTS Model 740-11 Diagonal Thrust TBM, 898
double-shield 194-272 TBM, 267–274
High Performance Main Beam TBM for Niagara Tunnel Project, 312–321, 901
model 236 main-beam, gripper-type TBM, 507–519
TBMs for Liaoning Dahuofang water tunnel project, 556–561
TBMs for New York City Water Tunnel No. 3, 1112, 1113, 1114
Rock Mass Excavability index, 1118–1130
Rock tunneling
in construction of Deep Tunnel Sewerage System Project (Singapore), 1308–1319
double-shield TBMs in extra-hard, abrasive bedrock at Guadarrama Tunnels, 1079–1081
drill and blast in urban setting for West Area CSO Tunnel, 913–926
Dulles Airport Automated People Mover Tunnels, 1286–1296
Elm Road Generating Station water intake tunnel, 889–900

hydraulic, mechanical, and hydromechanical assessment of tunnel interaction, 137–146

Kops II pumped storage plant (Austria), 870–881

Lake Hodges to Olivenhain Pipeline tunnel and shaft, 882–888

limitations and potential future applications of TBMs for deep rock tunneling, 1131–1142

NATM in construction of San Vicente Pipeline Tunnel Project, 928–942

New York City Water Tunnel No. 3, 1109–1117

Niagara Tunnel Project, 312–321, 901–912

outage time reduction in Second Manapouri Tailrace Tunnel project, 1262–1273

Rowa Tunnel Logistics AG, 314, 901

S

Sacramento, California

EPBM tunneling for Lower Northwest Interceptor Sewer, 741–756

trenchless crossings for Lower Northwest Interceptor Sewer, 1274–1285

Sacramento (California) Regional County Sanitation District, Bradshaw Interceptor Section 8 Project, 843–851

San Diego County (California) Water Authority

Lake Hodges to Olivenhain Pipeline tunnel and shaft, 882–888

NATM rock tunneling for San Vicente Pipeline Tunnel Project, 928–942

segmental concrete lining for San Vicente Pipeline Tunnel Project, 1243–1251

San Francisco, California

comparison of mined tunnel alternatives (Transbay-Caltrain project), 969–975

Transbay-Caltrain Downtown Extension Project, 964–975

San Francisco (California) Public Utilities Commission

design challenges for New Crystal Springs water tunnel, 91–103

planned Bay Tunnel (water tunnel), 685–693

San Jose, California

design of underground structures for Silicon Valley BART segment, 2–11

seismic testing of precast concrete segmental linings for Silicon Valley BART segment, 1225–1242

San Vicente Aqueduct Pipeline Tunnel Project (San Diego, CA)

NATM rock tunneling, 928–942

steel fiber reinforced self-compacting segmental concrete linings, 1243–1251

Santa Clara Valley (California) Transportation Authority

design of underground structures, 2–11

seismic testing for precast concrete segmental linings, 1225–1242

SAS. See Second Avenue Subway project (New York)

Schonian, Erich, 401

Seattle, Washington

Alaskan Way Viaduct and Seawall upgrade, 496–506

Beacon Hill Station dewatering and jet grouting, 346–359, 952–954, 961

EPBM-driven twin bore running tunnels (Beacon Hill Project), 793–807

SEM excavation of Beacon Hill Station, 346–347, 358–359, 943–963

Second Avenue Subway project (New York), 663–664, 672

cavern support design, 32–43

station construction and refurbishment, 667–672

TBM contract, 664–667

Second Manapouri Tailrace Tunnel project (New Zealand), 1262–1273

Seismic design

Alaskan Way tunnel (Seattle), 496–506

Caldecott Tunnels fourth bore, 617–618, 624–625

Claremont Tunnel Seismic Upgrade Project (California), 148–159

JWPCP Tunnel and Ocean Outfall Project, 597–598

for planned Bay Tunnel (San Francisco Bay Area), 685–693

Stanford Linear Accelerator tunnels (California), 21–31

and testing for precast concrete segmental linings, 1225–1242

Seismic reflector tracing, 443–448. See also TBM Excavation Control System

SELI SpA, 266–267

SEM. See Sequential Excavation Method

SEP-14 and SEP-15. See Special Experimental Projects 14 and 15 (Transportation Research Board)

Sequential Excavation Method

Beacon Hill Station project, 346–359, 943–963

in construction of Dublin Port Tunnel, 1005–1007

in planned A3 Hindhead tunnel (UK), 673–684

TBM guidance system, 457–465

See also New Austrian Tunneling Method

Sequential Support Method. See Sequential Excavation Method

Shafts

construction of drop and vent shafts for Narragansett Bay Commission CSO Abatement Program, 1047–1057

for construction of West Area CSO Tunnel and Pumping Station (Atlanta), 1064–1068

methods used in construction of nine shafts for New York City Water Tunnel No. 3, 1012–1022

New Croton Aqueduct Rehabilitation and Inspection Program, 1162–1171, 1181–1190

open-cut shaft construction in East Side Access Project, 1023–1030

slurry wall construction of shafts for York Durham Sewage System Interceptor, 1058–1062

Shield tunneling

mix-shield slurry TBM for Durban Harbour project, 579–593

in sandy soil under existing urban structures (Antwerp, Belgium), 176–184

and steel fiber reinforced highly flowable concrete segments, 1205–1208

Silicon Valley Rapid Transit Project (California)

design of underground structures, 2–11

seismic testing for precast concrete segmental linings, 1225–1242

Singapore

Deep Tunnel Sewerage System Project, 1308–1319

monitoring of NATM construction in challenging conditions (Fort Canning Tunnel), 976–987

North East Line and study of surface settlements over EPB driven tunnels in soft clay, 757–768

Sir Adam Beck II Hydroelectric Project (Ontario)
and Adam Beck, 185–186
concrete operations, 194–196
drill and blast construction of intake tunnels, 191–193
major components, 187–189
and Niagara River, 186–187
shafts, 189–191
See also Niagara Tunnel Project (Ontario)

SLAC. *See* Stanford Linear Accelerator Center

SLS-TM Roadheader Guidance System, 457–465

Slurry TBMs
considered for JWPCP Tunnel and Ocean Outfall Project, 600, 602–604
in construction of Metro Line B (Toulouse), 852–867
design principles for soft ground cutterheads, 784–792
mining through cobbles and boulders, 791–792
mix-shield slurry TBM for Durban Harbour project, 579–593
in non-cohesive, permeable soils for West Side CSO Tunnel Project (Portland), 830–842
sealed and pressurized-face slurry TBMs, 165–167

Slurry walls, 1031–1032
in construction of shafts for York Durham Sewage System Interceptor, 1058–1062
cutter soil mixing, 1039–1046
soil mixing, 1038–1039
and trench cutters, 1032–1035
water tightness and joint systems, 1035–1038

South Africa, Durban Harbour Tunnel Project, 579–593

South Coast Water District (Orange County, CA), Beach Interceptor Sewer Tunnel, 1144–1151

Southern Nevada Water Authority, Lake Mead Intake No. 3 Project, 647–662

Spadina Subway Extension (Toronto, Ont.), 300–311

Spain
Guadarrama Tunnels, 1079–1093
Madrid Calle 30 highway tunnels, 769–782

Special Experimental Projects 14 and 15 (Transportation Research Board), 419

S.Ruffillo, 381

STAAD, 41

Stanford Linear Accelerator Center, 21–31

STBMs. *See* Slurry TBMs

Strabag AG, 298, 312, 314, 315–316, 320–321

Switzerland, Gotthard Base Tunnel, 44–55

T

Taiwan-East Railway Bureau, Yung-Chung Tunnel, 396–406

TBM Excavation Control System, 442–456

TBMs. *See* Tunnel boring machines

Telephones, 466

Time-dependent ground behavior, 255
and advance rate, 256–257
and chemical processes, 255, 256, 263–264
and consolidation, 255–256, 263–264
and creep, 255, 256, 263–264

and shallow soft-ground tunnels, 258–261
and squeezing rock, 261–263
and stand-up time, 257–258

Tokyo, Japan
Metropolitan Expressway Central Circular Route, 1205–1206
steel fiber reinforced highly flowable concrete linings for Central Circular Route, 1205–1224

Toronto, Ontario
Enwave Project, 329–330, 331
Hydro One tunnel, 330–331
slurry wall construction of Richmond Hill shafts for York Durham Sewage System Interceptor, 1058–1062
Spadina Subway Extension Environmental Assessment Study, 300–311

Toronto Transit Commission, 300–311

Toulouse, France, Metro Line B, 852–867

Transbay-Caltrain Downtown Extension Project (San Francisco, CA)
comparison of mined tunnel alternatives, 969–975
NATM and cut and cover tunneling in construction of, 964–975

Transportation Research Board (U.S.), 419

Traylor Frontier Kemper Joint Venture, 472–494

Tunnel boring machines
Anacostia River CSO tunnels, 643
analysis and solution for project abandoned because of difficult ground and flooding, 160–167
and automated real-time probe drill monitoring system, 430–441
comparative soil deformation analysis for three TBM methods and convention tunneling, 852–867
considerations for Upper Rouge Tunnel, 632, 634–635
in construction of Big Walnut Outfall Augmentation Sewer, 704–711
in construction of Dublin Port Tunnel, 1002–1005
in construction of Dulles Airport Automated People Mover Tunnels, 1286–1296
in construction of Kops II tunnels and powerhouse cavern, 870–881
in construction of Second Avenue Subway (New York), 664–667
in construction of West Area CSO Tunnel and Pumping Station, 924–925, 1064–1078
design and construction of high voltage tunnel in Groundwater Source Protection Zone, 117–136
designing to protect adjacent structures during tunneling (Pittsburgh light rail), 70–79
double-shield TBM in volcanic rock with low cohesion and high water inflow, 266–279
EPBMs and slurry TBMs considered for JWPCP Tunnel and Ocean Outfall Project, 600, 602–604
in extreme topographic and climatic conditions for Icelandic hydroelectric project, 507–519
groundwater control for TBM tunneling in soft ground, 80–90
guidance system for partial face machines, 457–465

hybrid design for ground with variable geologic conditions, high water pressures, and significant faults, 229–254

in Liaoning Dahuofang water tunnel project, 551–561

limitations and potential future applications for deep rock tunneling, 1131–1142

monitoring ground conditions with TBM Excavation Control System, 442–456

and need for gas mitigation measures (Mill Creek Tunnel), 168–175

outage time reduction in Second Manapouri Tailrace Tunnel project, 1262–1273

for planned Bay Tunnel (San Francisco Bay Area), 685–693

planned for Kensico-City deep rock water tunnel, 520–533

for planned Lake Mead Intake No. 3 Project, 651–656

pressurized face TBM for soft ground tunneling, 605, 612–613, 615

primary and secondary wear, 104–107

Robbins High Performance Main Beam TBM, 312–321, 901

Robbins model 236 main-beam, gripper-type TBM, 507–519

in rock for Dulles Corridor Metrorail Project, 564, 572–576

in rock for three tunnels of New York City Water Tunnel No. 3, 1109–1117

and Rock Mass Excavability index, 1118–1130

rock tunneling for Elm Road Generating Station water intake tunnel, 889–900

soil abrasiveness test methodology, 104–116

in tunneling through difficult ground (Gotthard Base Tunnel), 44–55

See also Double-shield TBMs; Earth pressure balance TBMs; Hydroshield TBMs; Shield tunneling; Slurry TBMs

Tunnel interaction, 137–140

evaluation of mechanical and hydraulic properties, 140–142

hydraulic, 144, 145–146

hydrojacking potential, 145–146

mechanical, 142–144, 145–146

Tunneling

and underground communication systems, 466–471

See also Microtunneling; Rock tunneling; Shield tunneling; Tunnel boring machines; Tunnel interaction

U

UDEC. See Universal Distinct Element Code

Ultra-low frequency communications technology, 466

Underground communication infrastructures

Ethernet-based, 466–471

old technologies, 466

Union Fenosa, 266–267

United Arab Emirates, Palm Jumeirah service tunnels, 208–214

United Kingdom

Croydon Cable Tunnel Project, 117–136

planned A3 Hindhead tunnel, 673–684

Universal Distinct Element Code, 40

UNWEDGE, 40–41

Upper Rouge Tunnel (Detroit, Michigan), 627–636

US Army Corps of Engineers, and Cady Marsh Flood Relief Tunnel, 80–90

US Mine Safety and Health Administration, 466

V

Virginia, and Heartland Corridor Clearance Improvement Project, 1152–1161

VMT GmbH, 457–465

W

Washington, D.C.

NATM and TBM tunneling for Dulles Airport Automated People Mover Tunnels, 1286–1296

NATM and TBM tunneling for Dulles Metrorail project, 564–578

past tunneling experience, 566–571

See also District of Columbia Water and Sewer Authority

Washington (D.C.) Metropolitan Area Transit Authority, 564

Washington (D.C.) Suburban Sanitary Commission, 17–18

Washington (state)

Alaskan Way Viaduct and Seawall upgrade (Seattle), 496–506

Beacon Hill Station dewatering and jet grouting (Seattle), 346–359, 952–954, 961

EPBM-driven twin bore running tunnels (Beacon Hill Project, Seattle), 793–807

SEM excavation of Beacon Hill Station (Seattle), 346–347, 358–359, 943–963

Washington Group International, 564

West Area CSO Tunnel and Pumping Station (Atlanta, GA), 913–926, 1064–1078

West Side CSO Tunnel Project (Portland, OR)

cost reimbursable fixed fee contract, 1297–1307

slurry TBMs in non-cohesive, permeable soils, 830–842

West Virginia, and Heartland Corridor Clearance Improvement Project, 1152–1161

The Wirth Company

double-shield TBMs in extra-hard, abrasive bedrock at Guadarrama Tunnels, 1079–1081

TBM for Liaoning Dahuofang water tunnel project, 556–561

Wisconsin, Elm Road Generating Station (Oak Creek), 889–900

Y

York Durham (Ontario) Sewage System Interceptor Trunk Sewer, 322, 323–328

Yung-Chung Tunnel (Taiwan), 396–406